Springer Monographs in Mathematics



For further volumes published in this series,
www.springer.com/series/3733

R.M. Dudley • R. Norvaiša

# Concrete Functional Calculus

Springer

R.M. Dudley
Department of Mathematics
Massachusetts Institute of Technology
Cambridge, MA 02139-4307
USA
rmd@math.mit.edu

R. Norvaiša
Institute of Mathematics and Informatics
LT-08663 Vilnius
Lithuania
norvaisa@ktl.mii.lt

ISSN 1439-7382
ISBN 978-1-4614-2740-7 ISBN 978-1-4419-6950-7 (eBook)
DOI 10.1007/978-1-4419-6950-7
Springer New York Dordrecht Heidelberg London

Mathematics Subject Classification (2010): 28Bxx, 45Gxx, 46Txx, 60Gxx



Springer is part of Springer Science+Business Media (www.springer.com)

# Preface

The reader of this book will need some background in real analysis, as in the first half of the first author's book *Real Analysis and Probability.* The book should be accessible to graduate students with that background as well as researchers. An interest in probability will also help with motivation.

Although on some topics we may have presented more or less final results, there are others that leave openings for research. The impetus for much of the work came from others' work in mathematical statistics, but we do not include any applications to statistics.

The book is mainly about some aspects of nonlinear analysis, some not much studied and some others previously studied but not in the same ways, and their applications to probability, as in the final Chapter 12. More specifically (to explain the book's title) we consider existence and smoothness questions for some concrete nonlinear operators acting on some concrete Banach spaces of functions. The book has relatively small overlaps, of the order of one or two chapters, with any previous book except for two lecture note volumes by the authors.

Here is a first example of what is done and distinctive in this book. If $F$ and $G$ are two functions such that $F$ is defined on the range of $G$, one can form the composition $H(x) \equiv F(G(x)) \equiv (F \circ G)(x)$. When one mentions differentiability and composition, mathematicians tend to think of the chain rule, which is indeed an important fact, but we consider differentiability of the *two-function composition operator* we call $TC$ which takes the pair of functions $(F, G)$ into the function $H$. To take the derivative of this operator, we will assume $F$ and $G$ take values in Banach spaces. The domain of $G$ need not not have a linear or topological structure (it may be a measure space). What the differentiation will mean at some $F, G$ is to take functions $f$ and $g$ approaching 0 in corresponding spaces, and to represent the increment $(F + f) \circ (G + g) - F \circ G$ as $A(f) + B(g)$ plus a remainder, where $A$ and $B$ are linear operators and the remainder becomes small in norm relative to $f, g \to 0$. The operator $TC$ is linear in $F$ for fixed $G$, but the remainder contains a term $f \circ (G + g) - f \circ G$ which still depends on both $f$ and $g$.

It seems to us that $TC$ is a very natural operator deserving attention. It is treated in Chapter 8. Relatively more attention, e.g. Appell and Zabrejko [3], has been given to the operator $G \mapsto F \circ G$ for fixed (nonlinear) $F$, and its extension to the case where $F$ is a function of two variables, say $x$ and $y$, and one forms a new function $H(x) = F(x, G(x))$. Such operators are treated in Chapters 6 and 7.

Among the most familiar of all Banach spaces are the $L^p$ spaces, of equivalence classes of functions $f$ such that $|f|^p$ is integrable for a given measure $\mu$, where $1 \le p < \infty$. A question is: for given $p$ and $s$ in $[1, \infty)$, for what space(s) $\mathbb{F}$ of functions from $\mathbb{R}$ into $\mathbb{R}$ do we have Fréchet differentiability (defined in Chapter 5) of $TC$ from $\mathbb{F} \times L^s$ into $L^p$, say where the measure space is $[0, 1]$ with Lebesgue measure? It turns out that for $1 \le s < p$, we get degeneracy: even for fixed $F$, it must be constant (Corollary 7.36). For $1 \le p < s < \infty$, consider the following. Let $f$ be a real-valued function on $\mathbb{R}$ and $1 \le p < \infty$. Let $v_p(f)$ be the supremum of $\sum_{j=1}^{n} |f(x_j) - f(x_{j-1})|^p$ over all partitions $x_0 = a < x_1 < \cdots < x_n = b$ and all $n$, called the $p$-variation of $f$. Finiteness of this is ordinary bounded variation when $p = 1$. Let $\mathbb{F} = \mathcal{W}_p(\mathbb{R})$ be the space of all $f$ such that $v_p(f) < \infty$, with the norm $\|f\|_{[p]} = v_p(f)^{1/p} + \|f\|_{\sup}$ where $\|f\|_{\sup} := \sup_x |f(x)|$. Theorem 8.9 shows that $TC$ is Fréchet differentiable from $\mathcal{W}_p(\mathbb{R}) \times L^s$ into $L^p$ of a finite measure space $(\Omega, \mathcal{S}, \mu)$ at suitable $F$ and $G$. Namely, $F$ is differentiable, and its derivative $F'$ satisfies a further condition. The image measure $\mu \circ G^{-1}$ has a bounded density with respect to Lebesgue measure. The theorem gives a bound on the remainder in the differentiation of a given order in terms of $s$ and $p$. The two paragraphs preceding Theorem 8.9 indicate how some of the conditions assumed are necessary or best possible and in particular, optimality of the $\mathcal{W}_p$ norm on $f$ for bounding the remainder term $f \circ (G + g) - f \circ G$ (Proposition 7.28).

Countably additive signed measures are familiar objects in real analysis. For purposes of this book we need to consider Stieltjes-type integrals $\int f dg$ where neither the integrator $g$ nor the integrand $f$ is of bounded variation. We found it useful to consider as integrators interval functions defined as follows. An *interval function* $\mu$ is a function such that $\mu(A)$ is defined for all intervals $A$, which may be restricted to subintervals of some given interval. Then $\mu$ is called *additive* if $\mu(A \cup B) = \mu(A) + \mu(B)$ for any two disjoint intervals $A, B$ such that $A \cup B$ is an interval, and $\mu$ is called *upper continuous* if $\mu(A_n) \to \mu(A)$ whenever intervals $A_n$ decrease down to $A$. These properties would follow from, but do not imply, existence of an extension of $\mu$ to a countably additive signed measure. For example, if $F$ is a right-continuous function with left limits, not necessarily of bounded variation, one can define $\mu_F((c, d]) := F(d) - F(c)$ and define $\mu_F$ for other intervals by taking limits, giving an additive, upper continuous interval function $\mu_F$. For $\mu$ upper continuous we define $v_p(\mu)$, for $\mu$ which may be Banach-valued, as the supremum of $\sum_i \|\mu(A_i)\|^p$ over all finite disjoint collections of intervals $A_i$. For not necessarily additive interval functions, the $p$-variation needs to be defined differently.

Beside composition, another natural operator is the product integral, which takes the coefficient functions (such as $C(t)$ in (1.10) below) of a system of linear ordinary differential equations into a solution, under suitable conditions. Despite linearity of the system, the operator is nonlinear in the coefficients. Some entire books have been written on this operator. The product integral will take values in Banach algebra, namely, a Banach space in which a multiplication is defined satisfying usual algebraic conditions and $\|xy\| \leq \|x\|\|y\|$. A precise definition and some of the theory of Banach algebras are given in Chapter 4. An interval function $\mu$ with values in a Banach algebra will be called *multiplicative* if $\mu(A \cup B) = \mu(B)\mu(A)$ holds for any disjoint intervals $A, B$ such that $A \cup B$ is an interval, with $s < t$ for each $s \in A$ and $t \in B$. For $1 \leq p < 2$, a product integral operator $\mu \mapsto \prod(\mathbb{1} + \mathrm{d}\mu)$ is defined from additive to multiplicative interval functions and is an entire analytic function (as defined in Chapter 5) with respect to $p$-variation norms (Theorem 9.51) and serves to solve differential and integral equations. Whereas, for $p > 2$, finite $p$-variation of an additive, upper continuous $\mu$ does not imply that the product integral even exists (Theorem 9.11).

Since $p$-variation gives sharp results about two natural operators, let's return to point functions $f$ and consider the space $\mathcal{W}_p(\mathbb{R})$ for $1 \leq p < \infty$. If $G$ is any homeomorphism of $\mathbb{R}$, in other words, a continuous, strictly monotone (increasing or decreasing) function from $\mathbb{R}$ onto itself, the map $f \to f \circ G$ preserves $\mathcal{W}_p$ and its norm. Invariance under this very large group holds for the spaces of all bounded continuous functions or all bounded functions with the norm $\|f\|_{\mathrm{sup}}$. Other commonly considered spaces such as Sobolev spaces are of course highly useful, but they are invariant under much more restricted transformations. Whereas, the supremum norm gives no control of oscillations of a function and the $\mathcal{W}_p$ norms do. We suppose the good properties of $\|\cdot\|_{[p]}$ will give other uses than those we have found.

We also treat integrals, although with relatively little attention to the Lebesgue integral. Rather, let at first $f$ and $g$ be real functions of a real variable and consider Stieltjes-type integrals $(f, g) \mapsto \int_a^b f \mathrm{d}g$ where neither $f$ nor $g$ is necessarily of bounded variation. The given bilinear functional can be defined on various domains $f \in \mathbb{F}$, $g \in \mathbb{G}$ as will be seen. If $f$ and $g$ have suitable infinite-dimensional ranges, the integral can also be extended.

If $1 \leq p < \infty$, $1 \leq q < \infty$, $v_p(f) < \infty$, $v_q(g) < \infty$, and $p^{-1} + q^{-1} > 1$, then a Stieltjes-type integral $\int f \mathrm{d}g$ can be defined, as had been shown by E. R. Love and L. C. Young in the late 1930's, with a corresponding inequality we call the *Love–Young* inequality (Corollary 3.91) and use often. Because of bilinearity, the differentiability is then immediate and simple.

Chapter 12 on probability and $p$-variation includes results from several research papers. Among others, Theorem 12.27 gives bounded $p$-variation of the sample paths of Markov processes (with values in metric spaces) for $2 < p < \infty$ under a mild condition on expected lengths of increments, shown to be sharp. Corollary 12.43 extends the celebrated Komlós–Major–Tusnády theorem on convergence of the classical empirical process to a Brownian bridge from the supremum norm to $p$-variation norms for $2 < p < \infty$ with slower, but sharp, rates of convergence. Theorem 12.40 gives a sharp bound on the growth of the $p$-variation of the classical empirical process for $1 \le p < 2$. These facts were published in three papers in *Annals of Probability* (one co-authored by one of us). Proposition 12.54 gives an example due to Terry Lyons, showing that for some processes $X_t$ and $Y_t$ which are each Brownian bridges, but which have an unusual joint distribution, an integral $\int_0^1 X_t dY_t$ cannot be defined by any of the usual methods. The full work of Lyons and co-workers on "rough paths" is in progress and is beyond the scope of this book. In some cases where compact or "Hadamard" differentiability had been proved in the statistics literature, it might well be replaced by Fréchet differentiability with respect to $p$-variation norms, with a gain of remainder bounds. This and other problems we here leave as opportunities for readers.

Some parts of the book appeared, in different forms, in our earlier lecture note volumes Dudley and Norvaiša 1998 [55] and 1999 [54]. Improvements on or corrections to earlier results of ours or others are incorporated. The book also includes some new results published here for the first time as far as we know, as will be mentioned in the text or Notes for each.

*Guide to the reader*: Starred sections, specifically Sections *2.7 and *3.4, are not referred to in later chapters. Chapters of the book depend on earlier chapters as follows: Chapters 1 through 3 are basic in that Chapter 1 is a rather short introduction, and all later chapters refer to Chapter 3 many times each and directly or indirectly to Chapter 2. In the further sequence of chapters 4 through 8, each chapter has many references to the preceding one and directly or indirectly to intervening chapters. The Appendix relates only to Chapter 7. Chapter 12 on stochastic processes refers to, beyond Chapter 3, only two propositions in Chapter 9. Chapter 11, on Fourier series, refers only to the basic Chapters 1-3. Chapter 10, on nonlinear differential and integral equations, refers twice to Chapter 9, once to Chapter 7, once to Chapter 5, and many times to Chapter 3. Chapter 9, on multiplicative interval functions, the product integral, and linear differential and integral equations, refers to Chapters 1 through 6.

*Acknowledgments.* We thank B. M. Garay, Richard Gill, and Timothy Nguyen for advice and help with some points in the book.

MIT, Cambridge, Massachusetts — Richard M. Dudley[1]
Institute of Mathematics and Informatics, Vilnius — Rimas Norvaiša[2]
February 2010

[1] R. M. Dudley was partially supported by U. S. National Science Foundation Grants.

[2] R. Norvaiša was partially supported by U. S. National Science Foundation Grants, by Poduska Family Foundation grants and by Lithuanian State Science and Studies Foundation grants Nos. T-21/07, T-16/08, and T-68/09.

# Contents

# 1

# Introduction and Overview

In this book we are mainly concerned with derivatives of certain specific nonlinear operators on functions. One operator is composition, $TC : (f, g) \mapsto f \circ g$, where $(f \circ g)(x) \equiv f(g(x))$. This operator is linear in $f$ for fixed $g$, but not in $g$ even for fixed nonlinear $f$. We call this operator the *two-function composition operator*, or $TC$ for short, to distinguish it from the more-studied operator $g \mapsto f \circ g$ for fixed $f$, which we call the (autonomous) *Nemytskii operator* $N_f$. The chain rule, on differentiating $x \mapsto f(g(x))$, where $x$, $f$, and $g$ may all have values in Banach spaces, is a very important fact, but it is not directly about either $TC$ or $N_f$. To differentiate $TC$ will mean to approximate $(f + h) \circ (g + k) - f \circ g$, asymptotically as $h$ and $k$ approach 0 in suitable senses, by a sum of linear operators of the functions $h$ and $k$. The operator $TC$ will be treated in Chapter 8 and the Nemytskii operator in Chapters 6 and 7.

We will also consider solutions of certain ordinary differential equations and integral equations, for functions possibly having values in Banach spaces, and representing such solutions by way of nonlinear operators. A first introduction is given in Section 1.2. A basic nonlinear operator giving solutions of linear equations, the product integral, is developed in Chapter 9 and applied to solving equations in Sections 9.11 and 9.12. Chapter 10 treats nonlinear integral equations for possibly discontinuous functions.

A basic operator on functions on an interval $[a, b]$ is the bilinear Riemann–Stieltjes integral operator $(f, g) \mapsto \int_a^b f\, dg$, to be treated in the following section and then more fully in Chapters 2 and 3. Some general facts about Banach algebras are reviewed in Chapter 4. Chapter 5 treats differentiability in general Banach spaces.

The word "concrete" in the title is meant to convey that we consider not only specific operators as mentioned but also specific function spaces, most notably $p$-variation spaces, as will be mentioned in Theorem 1.4 and (1.20), and frequently throughout most of the book.

R.M. Dudley and R. Norvaiša, *Concrete Functional Calculus*, Springer Monographs in Mathematics, DOI 10.1007/978-1-4419-6950-7_1,

## 1.1 How to Define $\int_a^b f\,dg$?

Two classical ways of defining integrals $\int_a^b f\,\mathrm{d}g$ for real-valued functions $f$ and $g$ on an interval $[a,b]$ with $-\infty < a < b < +\infty$ are the Riemann–Stieltjes and Lebesgue–Stieltjes integrals. But, as will be seen later in this section and in Chapter 2, there are multiple ways of defining such integrals. Relations between them are given in Section 2.9. Different forms of integral can be useful for different purposes.

The Riemann–Stieltjes integral is defined as follows. A finite sequence $\kappa = \{t_i\}_{i=0}^n$ for a positive integer $n$ is called a *partition* of $[a,b]$ if $a = t_0 < t_1 < \cdots < t_n = b$. If $\{t_i\}_{i=0}^n$ is a partition of $[a,b]$ then an ordered pair $(\{t_i\}_{i=0}^n, \{s_i\}_{i=1}^n)$ is called a *tagged partition* of $[a,b]$ if $s_i \in [t_{i-1}, t_i]$ for $i = 1,\dots,n$. For a tagged partition $\tau = (\{t_i\}_{i=0}^n, \{s_i\}_{i=1}^n)$, $|\tau| := \max_{1\le i\le n}(t_i - t_{i-1})$ is called the *mesh* of $\tau$. Let $f$ and $g$ be real-valued functions on $[a,b]$. For a tagged partition $\tau = (\{t_i\}_{i=0}^n, \{s_i\}_{i=1}^n)$ of $[a,b]$, the sum

$$S_{RS}(f,\mathrm{d}g;\tau) := \sum_{i=1}^n f(s_i)\big[g(t_i) - g(t_{i-1})\big]$$

is called the Riemann–Stieltjes sum based on $\tau$. The *Riemann–Stieltjes* integral of $f$ with respect to $g$ is defined and equals $C \in \mathbb{R}$ if for each $\epsilon > 0$ there exists a $\delta > 0$ such that $|C - S_{RS}(f,\mathrm{d}g;\tau)| < \epsilon$ for each tagged partition $\tau$ of $[a,b]$ with mesh $|\tau| < \delta$. Then we let

$$(RS)\int_a^b f\,\mathrm{d}g := C = \lim_{|\tau|\to 0} S_{RS}(f,\mathrm{d}g;\tau).$$

Closely related to the Riemann–Stieltjes integral is the integral obtained by replacing the limit as the mesh approaches 0 by the limit in the sense of refinements of partitions, as follows. A partition $\kappa$ is a *refinement* of a partition $\lambda$ if $\lambda \subset \kappa$ as sets. Similarly, a tagged partition $\tau = (\kappa, \xi)$ is a *tagged refinement* of a partition $\lambda$ if $\kappa$ is a refinement of $\lambda$. Now for $f$ and $g$ as before, the *refinement Riemann–Stieltjes* integral of $f$ with respect to $g$ is defined and equals $C \in \mathbb{R}$ if for each $\epsilon > 0$ there exists a partition $\lambda$ of $[a,b]$ such that $|C - S_{RS}(f,\mathrm{d}g;\tau)| < \epsilon$ for each tagged partition $\tau$ which is a tagged refinement of $\lambda$. Then we let

$$(RRS)\int_a^b f\,\mathrm{d}g := \lim_{\tau} S_{RS}(f,\mathrm{d}g;\tau) := C.$$

If the Riemann–Stieltjes integral exists then so does the refinement Riemann–Stieltjes integral with the same value, but not conversely (see Proposition 2.13 below and the example following it).

The following concept will be used to formulate sufficient conditions for existence of integrals. Let $f$ be any real-valued function on an interval $[a,b]$

with $-\infty < a \le b < +\infty$ and let $0 < p < \infty$. If $a < b$, for a partition $\kappa = \{t_i\}_{i=0}^n$ of $[a,b]$, the *p-variation sum* for $f$ over $\kappa$ will be defined by

$$s_p(f;\kappa) := \sum_{i=1}^n \left|f(t_i) - f(t_{i-1})\right|^p. \tag{1.1}$$

The *p-variation* of $f$ on $[a,b]$ is defined as $v_p(f;[a,b]) := 0$ if $a = b$ and

$$v_p(f) := v_p(f;[a,b]) := \sup_\kappa s_p(f;\kappa) \tag{1.2}$$

if $a < b$, where the supremum is over all partitions $\kappa$ of $[a,b]$. Then $f$ is said to be of *bounded p-variation* on $[a,b]$, or $f \in \mathcal{W}_p[a,b]$, if and only if $v_p(f) < +\infty$. Let $V_p(f) := V_p(f;[a,b]) := v_p(f;[a,b])^{1/p}$. A function of bounded $p$-variation is clearly bounded. For any bounded real-valued function $f$ on a set $S$ let $\|f\|_{\sup} := \|f\|_{S,\sup} := \sup_{x\in S}|f(x)|$. For $1 \le p < \infty$ let

$$\|f\|_{(p)} := \|f\|_{[a,b],(p)} := V_p(f), \quad \|f\|_{[p]} := \|f\|_{(p)} + \|f\|_{\sup}. \tag{1.3}$$

It is easily seen that $\|\cdot\|_{(p)}$ is a seminorm, and so $\|\cdot\|_{[p]}$ is a norm on $\mathcal{W}_p[a,b]$, using the Hölder and Minkowski inequalities. Recall the Hölder inequality for finite sums: if $p^{-1} + q^{-1} \ge 1$ and $1 \le p,q < \infty$, then for any nonnegative numbers $\{a_i, b_i\}_{i=1}^n$,

$$\sum_{i=1}^n a_i b_i \le \Big(\sum_{i=1}^n a_i^p\Big)^{1/p}\Big(\sum_{i=1}^n b_i^q\Big)^{1/q}. \tag{1.4}$$

This fact is well known for $p^{-1} + q^{-1} = 1$ and follows for $p^{-1} + q^{-1} \ge 1$ since $(\sum_{i=1}^n a_i^p)^{1/p}$ is a nonincreasing function of $p$. Recall also the Minkowski inequality for finite sums: for any $1 \le r < \infty$ and any nonnegative numbers $\{a_i, b_i\}_{i=1}^n$,

$$\Big(\sum_{i=1}^n (a_i + b_i)^r\Big)^{1/r} \le \Big(\sum_{i=1}^n a_i^r\Big)^{1/r} + \Big(\sum_{i=1}^n b_i^r\Big)^{1/r}. \tag{1.5}$$

The best known case of $p$-variation is for $p = 1$. A function is said to be of *bounded variation* iff it is of bounded 1-variation, and its *total variation* is its 1-variation.

Let $f \in \mathcal{W}_1[a,b]$ and let $F(t) := v_1(f;[a,t])$ for $a \le t \le b$. Then as is well known and easily checked, $F$ and $F-f$ are nondecreasing with $f \equiv F-(F-f)$. Conversely, if $G$ is any nondecreasing function on $[a,b]$, then $v_1(G;[a,b]) = G(b) - G(a)$. For any two functions $g$ and $h$, $v_1(g+h) \le v_1(g) + v_1(h)$. Thus, $f \in \mathcal{W}_1[a,b]$ if and only if $f = G - H$ for two nondecreasing real-valued functions $G$ and $H$ on $[a,b]$.

From basic measure and integration theory, e.g. [53, Theorem 3.2.6], recall that there is a 1–1 correspondence between nondecreasing functions $g$ on $(a,b]$, right-continuous on $(a,b)$, and finite countably additive measures $\mu$ on the Borel sets of $(a,b]$, given by $\mu((c,d]) = g(d) - g(c)$, $a \le c \le d \le b$. The same equations give a 1–1 correspondence between finite countably additive signed measures $\mu$ on the Borel sets of $(a,b]$ and functions $g$ of bounded variation on $(a,b]$, right-continuous on $(a,b)$. Let $\mu := \mu_g$ for the $\mu$ corresponding to such a $g$. Then the *Lebesgue–Stieltjes* integral is defined for $a \le b$ by

$$(LS)\int_a^b f\,\mathrm{d}g := \int_{(a,b]} f\,\mathrm{d}\mu_g \tag{1.6}$$

whenever the right side is defined as a Lebesgue integral, for example, for any bounded, Borel measurable function $f$. For $a = b$, the integral is evidently 0. (Some other authors give definitions which differ as to the point $a$.)

Let $h(t) := 1_{[1,2]}(t)$ for $0 \le t \le 2$. Then the Riemann–Stieltjes integrals $(RS)\int_0^2 h\,\mathrm{d}h$ and $(RRS)\int_0^2 h\,\mathrm{d}h$ are not defined. Clearly $(LS)\int_0^2 h\,\mathrm{d}h$ is defined and equals 1. In this case the Riemann–Stieltjes integrals are remarkably weak.

On the other hand, recall that a real-valued function $f$ on an interval $[a,b]$ is said to satisfy a *Hölder condition* of order $\alpha$, where $0 < \alpha \le 1$, if for some $K < \infty$, $|f(t) - f(s)| \le K|t-s|^\alpha$ for any $s,t \in [a,b]$. It is easily seen that a function $f$ on $[a,b]$, Hölder of order $\alpha$, is of bounded $1/\alpha$-variation. Indeed, this is a result of the bound

$$s_{1/\alpha}(f;\kappa) \le K^{1/\alpha}\sum_{i=1}^n \Big(|t_i - t_{i-1}|^\alpha\Big)^{1/\alpha} = K^{1/\alpha}(b-a), \tag{1.7}$$

valid for any partition $\kappa = \{t_i\}_{i=0}^n$ of $[a,b]$. We have the following:

**Proposition 1.1.** *If $f$ and $g\colon [a,b] \to \mathbb{R}$ are Hölder of orders $\alpha$ and $\beta$ respectively with $\alpha + \beta > 1$, then the Riemann–Stieltjes integral $(RS)\int_a^b f\,\mathrm{d}g$ exists.*

Proposition 1.1 is a special case of the following:

**Proposition 1.2.** *If $f$ and $g\colon [a,b] \to \mathbb{R}$ are continuous, $f \in \mathcal{W}_p[a,b]$ and $g \in \mathcal{W}_q[a,b]$ with $1 \le p < \infty$, $1 \le q < \infty$, and $p^{-1} + q^{-1} > 1$, then the Riemann–Stieltjes integral $(RS)\int_a^b f\,\mathrm{d}g$ exists.*

Proposition 1.2, in turn, is a special case of Corollary 3.91 in light of Definition 2.41 below.

If $g$ is not of bounded 1-variation then $\int_a^b f\,\mathrm{d}g$ is not a Lebesgue–Stieltjes integral. If $f$ is also not of bounded 1-variation, one cannot use integration by parts to obtain a Lebesgue–Stieltjes integral. For any $\alpha$ with $0 < \alpha < 1$ there exist functions $g$, Hölder of order $\alpha$, which are not of bounded 1-variation. Examples can be given by way of lacunary Fourier series, e.g. in the proof of Theorem 3.75. So in defining integrals $\int_a^b f\,\mathrm{d}g$, neither the (refinement) Riemann–Stieltjes nor the Lebesgue–Stieltjes integral is adequate in general. If $f$ and $g$ are regulated functions, then there is an integral (originating in the paper [254] of W. H. Young) which works in the cases we have mentioned. A function $g$ on $[a,b]$ is called *regulated* if for each $t \in [a,b)$, the limit $g(t+) := \lim_{u\downarrow t} g(u)$ exists, and for each $s \in (a,b]$, the limit $g(s-) := \lim_{u\uparrow t} g(u)$ exists. If $g$ has bounded $p$-variation for some $0 < p < \infty$ then $g$ is regulated (as will be shown in Proposition 3.33). For any regulated function $g$ on $[a,b]$, let $\Delta^+ g(t) := g(t+) - g(t)$ if $a \le t < b$, and let $\Delta^- g(s) := g(s) - g(s-)$ if $a < s \le b$. A tagged partition $(\{t_i\}_{i=0}^n, \{s_i\}_{i=1}^n)$ is called a *Young tagged partition* if $t_{i-1} < s_i < t_i$ for each $i = 1, \ldots, n$. Let $f$ be any function on $[a,b]$, and let $g$ be a regulated function on $[a,b]$. Given a Young tagged partition $\tau = (\{t_i\}_{i=0}^n, \{s_i\}_{i=1}^n)$, the sum

$$S_{YS}(f, dg; \tau)$$
$$:= \sum_{i=1}^n \Big\{ f(t_{i-1})\Delta^+ g(t_{i-1}) + f(s_i)[g(t_i-) - g(t_{i-1}+)] + f(t_i)\Delta^- g(t_i) \Big\}$$

is called the *Young–Stieltjes sum* based on $\tau$. The *refinement Young–Stieltjes* integral of $f$ with respect to $g$ is defined and equals $C \in \mathbb{R}$ if for each $\epsilon > 0$ there exists a partition $\lambda$ of $[a,b]$ such that $|C - S_{YS}(f, dg; \tau)| < \epsilon$ for each Young tagged partition $\tau$ which is a refinement of $\lambda$. Then we let

$$(RYS) \int_a^b f\,dg := \lim_{\tau} S_{YS}(f, dg; \tau) := C.$$

If the refinement Riemann–Stieltjes integral exists then so does the refinement Young–Stieltjes integral with the same value, but not conversely (see Proposition 2.18 below and the example following it). Let $g$ be a nondecreasing function on $[a,b]$, right-continuous on $[a,b)$, and let $\mu_g$ be the Lebesgue–Stieltjes measure on $[a,b]$. Then for any $\mu_g$-measurable function $f$ on $[a,b]$, $(RYS) \int_a^b f\,dg = (LS) \int_a^b f\,dg$ whenever both integrals exist (see Propositions 2.27 and 2.28 below). The next fact follows from Corollary 3.91:

**Proposition 1.3.** *If $f \in \mathcal{W}_p[a,b]$ and $g \in \mathcal{W}_q[a,b]$ with $1 \le p < \infty$, $1 \le q < \infty$, and $p^{-1} + q^{-1} > 1$, then the refinement Young–Stieltjes integral $(RYS) \int_a^b f\,dg$ exists, and there is a constant $K_{p,q}$, depending only on $p$ and $q$, such that*

$$\left| (RYS) \int_a^b f\,dg \right| \le K_{p,q} \|f\|_{[p]} \|g\|_{(q)}.$$

The integrals $(A) \int_a^b f\,dg$ have been defined for $A = RS$, $RRS$, and $RYS$ so far only when $a < b$. If $a = b$ then we let $\int_a^a f\,dg := 0$ for each of the three integrals and call any function on a singleton regulated.

*Interval functions*

The Lebesgue–Stieltjes integral is essentially the Lebesgue integral with respect to a countably additive (signed) measure. Similarly, we can consider extended Riemann–Stieltjes integrals with respect to a function $\mu$ defined on sets, namely subintervals of a nonempty interval $J$, and thus to be called an *interval function on* $J$. An interval function $\mu$ on $J$ with values in a vector space will be called *additive* if $\mu(A \cup B) = \mu(A) + \mu(B)$ whenever $A$, $B$ are disjoint subintervals of $J$ and their union $A \cup B$ also is an interval.

The following is a concrete variant of a general integral introduced by A. N. Kolmogorov in [120]. Let $J$ be a nonempty interval, open or closed at either end. An ordered collection $\{A_i\}_{i=1}^n$ of disjoint nonempty subintervals $A_i$ of $J$ is called an *interval partition* of $J$ if their union is $J$ and $s < t$ for all $s \in A_i$ and $t \in A_j$ whenever $i < j$. An interval partition which consists only of open intervals and singletons is called a *Young interval partition.* If $\{A_i\}_{i=1}^n$ is an interval partition of $J$ then an ordered pair $(\{A_i\}_{i=1}^n, \{s_i\}_{i=1}^n)$ is called a *tagged interval partition* of $J$ and $\{s_i\}_{i=1}^n$ a set of *tags* for $\{A_i\}_{i=1}^n$ if $s_i \in A_i$ for $i = 1, \ldots, n$. An interval partition $\mathcal{A}$ is a *refinement* of an interval partition $\mathcal{B}$, written $\mathcal{A} \sqsupset \mathcal{B}$, if each interval in $\mathcal{A}$ is a subinterval of an interval in $\mathcal{B}$.

Similarly, a tagged interval partition $\mathcal{T} = (\mathcal{A}, \xi)$ is a *tagged refinement* of an interval partition $\mathcal{B}$ if $\mathcal{A}$ is a refinement of $\mathcal{B}$. Let $f$ be a function on $J$ and let $\mu$ be an additive interval function on $J$, both real-valued. For a tagged interval partition $\mathcal{T} = (\{A_i\}_{i=1}^n, \{s_i\}_{i=1}^n)$ of $J$, the sum

$$S_K(f, \mathrm{d}\mu; \mathcal{T}) = \sum_{i=1}^n f(s_i)\mu(A_i)$$

is called the *Kolmogorov sum* for $f$ based on $\mathcal{T}$. The *Kolmogorov integral* of $f$ with respect to $\mu$ is defined and equals $C \in \mathbb{R}$ if for each $\epsilon > 0$ there exists an interval partition $\mathcal{A}$ of $J$ such that $|C - S_K(f, \mu; \mathcal{T})| < \epsilon$ for each tagged interval partition $\mathcal{T}$ of $J$ which is a tagged refinement of $\mathcal{A}$. Then we let

$$\oint_J f \, \mathrm{d}\mu := \lim_{\mathcal{T}} S_K(f, \mathrm{d}\mu; \mathcal{T}) := C.$$

The Kolmogorov integral can be related to the refinement Young–Stieltjes integral as follows. An interval function $\mu$ on $[a, b]$ will be called *upper continuous* if $\mu(A_n) \to \mu(A)$ for any sequence of subintervals $A_1, A_2, \ldots$ of $[a, b]$ such that $A_n \downarrow A$. For any regulated function $g$ on $[a, b]$, there exists an additive upper continuous interval function $\mu_g$ on $[a, b]$ such that $\mu_g((s, t)) = g(t-) - g(s+)$ and $\mu_g(\{t\}) = g(t+) - g(t-)$ for $s < t$ in $[a, b]$, setting $g(a-) := g(a)$ and $g(b+) := g(b)$ in this case. For $a = b$ let $\mu_g(\{a\}) := 0$. Let $g$ be a regulated function on $[a, b]$, and let $\mu_g$ be the corresponding additive upper continuous interval function on $[a, b]$. There is a 1–1 correspondence between tagged Young partitions $\tau$ of $[a, b]$ and tagged Young interval partitions $\mathcal{T}$ of $[a, b]$, with $S_{YS}(f, \mathrm{d}g; \tau) \equiv S_K(f, \mathrm{d}\mu_g; \mathcal{T})$. The existence of the Kolmogorov integral with respect to an upper continuous additive interval function depends only on Kolmogorov sums which are based on tagged Young interval partitions, as shown in Proposition 2.25 below. Therefore

$$(RYS)\int_a^b f\,\mathrm{d}g \;=\; \oint_{[a,b]} f\,\mathrm{d}\mu_g$$

if either side is defined. Note that in an integral $(RYS)\int_a^b f\,\mathrm{d}g$, if for some $t \in (a,b)$, $g(t)$ has a value different from $g(t-)$ or $g(t+)$, the value $g(t)$ has no influence on the value of the integral.

In this book we mainly use interval functions and the Kolmogorov integral. The inequality of Proposition 1.3 extends to the Kolmogorov integral using the $p$-variation for interval functions defined as follows. Let $\mu$ be an additive interval function on a nonempty interval $J$ and let $0 < p < \infty$. For an interval partition $\mathcal{A} = \{A_i\}_{i=1}^n$ of $J$, let $s_p(\mu;\mathcal{A}) := \sum_{i=1}^n |\mu(A_i)|^p$. The $p$-variation of $\mu$ on $J$ is an interval function $v_p(\mu) = v_p(\mu;\cdot)$ on $J$ defined by

$$v_p(\mu;A) := \sup_{\mathcal{A}} s_p(\mu;\mathcal{A})$$

if $A$ is a nonempty subinterval of $J$, where the supremum is over all interval partitions $\mathcal{A}$ of $A$, or as 0 if $A = \emptyset$. We say that $\mu$ has bounded $p$-variation if $v_p(\mu;J) < \infty$. For a subinterval $A \subset J$, let $V_p(\mu;A) := v_p(\mu;A)^{1/p}$. The class of all additive and upper continuous interval functions on $J$ with bounded $p$-variation is denoted by $\mathcal{AI}_p(J)$. The following analogue of Proposition 1.3 for the Kolmogorov integral is a special case of Corollary 3.95.

**Theorem 1.4.** *If $\mu \in \mathcal{AI}_p[a,b]$ and $f \in \mathcal{W}_q[a,b]$ with $1 \le p < \infty$, $1 \le q < \infty$, and $p^{-1} + q^{-1} > 1$, then the Kolmogorov integral $\oint_{[a,b]} f\,\mathrm{d}\mu$ exists, and there is a constant $K_{p,q}$ depending only on $p$ and $q$ such that*

$$\Big|\oint_{[a,b]} f\,\mathrm{d}\mu\Big| \le K_{p,q}\|f\|_{[a,b],[q]}V_p(\mu;[a,b]).$$

In Chapter 2, integrals will be defined where integrands $f$ (and for bilinear integrals also $g$) and interval functions $\mu$ can all have values in Banach spaces.

## 1.2 Some Integral and Differential Equations

Consider a linear integral equation

$$f(t) = 1 + (RYS)\int_0^t f\,\mathrm{d}h, \qquad 0 \le t \le 2,$$

with respect to a function $h\colon [0,2] \to \mathbb{R}$, and/or a linear Kolmogorov integral equation

$$f(t) = 1 + \oint_{[0,t]} f\,\mathrm{d}\mu, \qquad 0 \le t \le 2, \tag{1.8}$$

where $\mu(A) := \delta_1(A) := 1_A(1)$ for any interval $A \subset [0,2]$ and $h(t) := 1_{[1,\infty)}(t)$. Either equation gives $f(1) = 1 + f(1)$, a contradiction. If instead we take the integral equation

$$f(t) = 1 + \fint_{[0,t)} f\,\mathrm{d}\mu, \qquad 0 \le t \le 2,$$

then it has the solution $f(t) = 1$ for $0 \le t \le 1$, $f(t) = 2$ for $1 < t \le 2$. Equations of the given form can be solved rather generally; see Section 9.11.

*Product integration*

Consider a linear $k$th order ordinary differential equation

$$\frac{\mathrm{d}^k x(t)}{\mathrm{d}t^k} + u_{k-1}(t)\frac{\mathrm{d}^{k-1}x(t)}{\mathrm{d}t^{k-1}} + \cdots + u_0(t)x(t) = v(t), \qquad a \le t \le b, \tag{1.9}$$

where for the present, $x$ and the coefficients $u_j$ and $v$ are real-valued. As is often done in differential equations, one can write an equivalent first-order linear vector and matrix differential equation

$$\mathrm{d}f(t)/\mathrm{d}t = C(t) \cdot f(t), \tag{1.10}$$

where $f(\cdot)$ is the $(k+1) \times 1$ column vector and $C(\cdot)$ is the $(k+1) \times (k+1)$ matrix-valued function defined respectively by

$$f(t) = \begin{pmatrix} 1 \\ x(t) \\ x'(t) \\ \cdots \\ x^{(k-1)}(t) \end{pmatrix} \quad \text{and } C(t) = \begin{pmatrix} 0 & 0 & 0 & 0 & \ldots & 0 \\ 0 & 0 & 1 & 0 & \ldots & 0 \\ 0 & 0 & 0 & 1 & \ldots & 0 \\ \cdots & \cdots\cdots & \cdots\cdots & \cdots\cdots & \cdots\cdots & \cdots\cdots \\ 0 & 0 & 0 & 0 & \ldots & 1 \\ v & -u_0 & -u_1 & -u_2 & \ldots & -u_{k-1} \end{pmatrix}$$

with $x^{(j)}(t) := \mathrm{d}^j x(t)/\mathrm{d}t^j$, $v := v(t)$, and $u_j := u_j(t)$ for $j = 0, 1, \ldots, k-1$. It is easy to check that for two $3 \times 3$ matrices $A$, $B$ of the form of $C$, so that $k = 2$, we have $AB = BA$ if and only if $A = B$. The same is true for any $k \ge 2$: consider the next-to-last row of the products. Thus commuting matrices will be obtained only for differential equations with constant coefficients if $k \ge 2$.

If (1.10) holds at a point $t$, then

$$f(t+s) - f(t) = sC(t) \cdot f(t) + o(s) \qquad \text{as } s \downarrow 0, \qquad \text{or}$$

$$f(t+s) = \big(I + sC(t)\big) \cdot f(t) + o(s) \qquad \text{as } s \downarrow 0, \tag{1.11}$$

where $I$ is the $(k+1) \times (k+1)$ identity matrix. Suppose that $C(\cdot)$ is continuous on an interval $[a,b]$ and let $h(t) := \int_a^t C(s)\,\mathrm{d}s$, so that $h$ is a $C^1$ function with $h(a) = 0$. Let $a = t_0 < t_1 < \cdots < t_n = t$ be a partition of $[a,t]$ where $a < t \le b$. Then (1.11) implies that approximately

$$f(t) \doteq (I + h(t_n) - h(t_{n-1})) \cdots (I + h(t_1) - h(t_0)) f(a) \tag{1.12}$$

for a fine enough partition. Taking a limit of such products (without the $f(a)$ factor) as the mesh of the partition goes to 0, we get a matrix called $\prod_a^t (I + \mathrm{d}h)$, the product integral of $h$ from $a$ to $t$. The product integral will give us the value of the solution of the differential equation (1.9) in terms of the values of $x(\cdot)$ and its derivatives through order $k-1$ at $t=a$, via

$$f(t) = \Big(\prod_a^t (I + \mathrm{d}h)\Big) f(a).$$

If the increments of $h$ commute then the product integral with respect to $h$ over an interval $[a,t]$ is the exponential $\exp\{h(t)-h(a)\}$, as follows from Theorem 9.40 below and continuity of $h$. We will show in Section 9.12 that $f$ may have values in a Banach space $X$, and so $h$, as well as the product integral with respect to $h$, will have values in the Banach algebra $L(X,X)$ of bounded operators from $X$ into itself.

So far, $h$ has been a $C^1$ point function. In Chapter 9, we will see how the product integral can be defined for interval functions with values in any Banach algebra, of bounded $p$-variation for $1 \le p < 2$. The product integral will give, in Sections 9.11 and 9.12, solutions of integral equations, where the integrals in the equations are Kolmogorov integrals, defined briefly in Section 1.1 and treated more fully in Section 2.3.

## 1.3 Basic Assumptions

Let $\mathbb{K}$ be either the field $\mathbb{R}$ of real numbers or the field $\mathbb{C}$ of complex numbers. Let $X, Y$ and $Z$ be Banach spaces over $\mathbb{K}$. The norm on each will be denoted by $\|\cdot\|$. For intuition, have in mind the case $X = Y = Z = \mathbb{K}$ with $\|x\| = |x|$ for all $x$. Let $B(\cdot,\cdot)$ be a bounded bilinear operator from $X \times Y$ into $Z$, where "bounded" means that for some $M < \infty$,

$$\|B(x,y)\| \le M\|x\|\|y\| \tag{1.13}$$

for all $x \in X$ and $y \in Y$. In the case $X = Y = Z = \mathbb{K}$ we will take $B(x,y) \equiv xy$. If (1.13) holds for a given $M$ we will say that $B$ is $M$-bounded. If the three spaces $X, Y$, and $Z$ are all different then by changing the norm to an equivalent one on any one of the spaces by a constant multiple, we can assume $M=1$. For a fixed $B$ we will write $x{\cdot}y := B(x,y)$. For later reference we summarize some assumptions:

$$\begin{cases} X, Y, Z \text{ are Banach spaces over } \mathbb{K}, \\ X \times Y \ni (x,y) \mapsto x{\cdot}y \in Z \quad \text{is 1-bounded and bilinear.} \end{cases} \tag{1.14}$$

For example, let $\mathbb{B}$ be a Banach algebra over $\mathbb{K}$ with a norm $\|\cdot\|$, as treated in Chapter 4. Then we can take $X = Y = Z = \mathbb{B}$ and $\cdot$ as the multiplication in $\mathbb{B}$. By an equivalent renorming of $\mathbb{B}$ one can take the multiplication to be 1-bounded, as will be seen in Theorem 4.8.

*Reversing the order of integrand and integrator*

Under the assumption (1.14), we will be giving various definitions of integrals

$$\int_a^b f\cdot \mathrm{d}g \equiv \int_a^b B(f, \mathrm{d}g) \quad \text{and} \quad ⨍_J f\cdot \mathrm{d}\mu \equiv ⨍_J B(f, \mathrm{d}\mu),$$

where $-\infty < a \le b < +\infty$, $f\colon [a,b] \to X$, $g\colon [a,b] \to Y$, $J$ is a subinterval of $[a,b]$ and $\mu$ is an interval function on $[a,b]$ with values in $Y$. Then, for each definition of integral $\int_a^b f\cdot \mathrm{d}g$, we will have the corresponding definition

$$\int_a^b \mathrm{d}f\cdot g \equiv \int_a^b B(\mathrm{d}f, g) := \int_a^b \tilde{B}(g, \mathrm{d}f), \tag{1.15}$$

where $\tilde{B}(y,x) := B(x,y)$ is a bounded bilinear operator: $Y\times X \to Z$ and the integrals on the left are defined if and only if the integral on the right is. The integrals $\int_a^b f\cdot \mathrm{d}g$ with $a < b$ will be defined as limits of certain sums, but the sums for $f$ and $g$ will in general not be symmetric in $f$ and $g$, even if $X = Y$ and $B(y,x) \equiv B(x,y)$. Likewise, given a definition of $⨍_{[a,b]} f\cdot \mathrm{d}\mu$ and an $X$-valued interval function $\nu$ on $[a,b]$, we will write

$$⨍_J \mathrm{d}\nu\cdot g \equiv ⨍_J B(\mathrm{d}\nu, g) := ⨍_J \tilde{B}(g, \mathrm{d}\nu). \tag{1.16}$$

## 1.4 Notation and Elementary Notions

*Spaces of operators*

Let $X$ and $Y$ be normed linear spaces. A linear function $T$ from $X$ into $Y$ is called a *bounded linear operator* iff

$$\|T\| := \sup\{\|Tx\| : x \in X, \|x\| \le 1\} < \infty. \tag{1.17}$$

Then $\|T\|$ is called the *operator norm* of $T$. The set of all bounded linear operators from $X$ into $Y$ will be called $L(X,Y)$. It is easily seen to be a normed linear space with the operator norm. If $Y$ is a Banach space, then so is $L(X,Y)$.

*Spaces of bounded functions*

For a function $f$ from a nonempty set $S$ into a normed space $X$, let

$$\|f\|_{\sup} := \|f\|_{S,\sup} := \sup\{\|f(x)\| : x \in S\}.$$

Restricted to functions for which it is finite, i.e., bounded functions, $\|\cdot\|_{\sup}$ is called the *sup norm*. The normed space of all bounded $X$-valued functions on $X$ is denoted by $\ell^\infty(S;X)$. Also, the *oscillation* of $f$ on $S$ is defined by

$$\mathrm{Osc}_S(f) := \mathrm{Osc}\,(f;S) := \sup\big\{\|f(s) - f(t)\|\colon s,t \in S\big\}.$$

*Hölder classes*

Let $X$, $Y$ be normed spaces, let $U$ be a subset of $X$ with more than one element, and let $0 < \alpha \le 1$. A function $f\colon U \mapsto Y$ is said to be *Hölder of order* $\alpha$, or simply $\alpha$*-Hölder*, if

$$\begin{aligned}\|f\|_{(\mathcal{H}_\alpha)} &:= \|f\|_{U,(\mathcal{H}_\alpha)}\\ &:= \sup\bigl\{\|f(x)-f(y)\|/\|x-y\|^\alpha\colon\ x,y\in U,\ x\neq y\bigr\} < \infty. \qquad (1.18)\end{aligned}$$

An $\alpha$-Hölder function on $U$ is clearly continuous on $U$ and will sometimes be called $\alpha$*-Hölder continuous.* The class of all $\alpha$-Hölder functions from $U$ into $Y$ is denoted by $\mathcal{H}_\alpha(U;Y)$. In the case $U = X = Y = \mathbb{R}$ we write $\mathcal{H}_\alpha = \mathcal{H}_\alpha(\mathbb{R};\mathbb{R})$.

*Banach spaces of functions*

Let $S$ be a nonempty set, and let $X$ be a Banach space over $\mathbb{K}$. A set $\mathbb{F}$ of $X$-valued functions on $S$ is a vector space over $\mathbb{K}$ if it is a vector space with respect to pointwise operations on $S$, that is, for $f, g \in \mathbb{F}$, a scalar $r \in \mathbb{K}$, and any $s \in S$,

$$(rf)(s) = rf(s) \qquad \text{and} \qquad (f+g)(s) = f(s) + g(s).$$

For example, the set $X^S$ of all $X$-valued functions on $S$ is a vector space. If $\mathbb{F} \subset X^S$ and $\|\cdot\|$ is a norm on $\mathbb{F}$, then $(\mathbb{F}, \|\cdot\|)$ will be called a *Banach space of* $X$*-valued functions* on $S$ iff $\mathbb{F}$ is a vector space and $(\mathbb{F}, \|\cdot\|)$ is a Banach space. If also $X = \mathbb{K}$ then $(\mathbb{F}, \|\cdot\|)$ will be called a *Banach space of functions.*

*Intervals*

An interval in $\mathbb{R}$ is a set of any of the following four forms: for $-\infty \le u \le v \le +\infty$,

$$\begin{aligned}(u,v) &:= \{t\in\mathbb{R}\colon\ u<t<v\},\\ [u,v) &:= \{t\in\mathbb{R}\colon\ u\le t<v\}, \quad \text{with } -\infty<u,\\ (u,v] &:= \{t\in\mathbb{R}\colon\ u<t\le v\}, \quad \text{with } v<+\infty, \text{ and}\\ [u,v] &:= \{t\in\mathbb{R}\colon\ u\le t\le v\}, \quad \text{with } -\infty<u\le v<+\infty.\end{aligned}$$

For each of the four cases, if the interval is nonempty, $u$ is called its *left endpoint* and $v$ its *right endpoint.* For any $u \in \mathbb{R}$, $[u,u] = \{u\}$ is a singleton and $(u,u] = [u,u) = (u,u) = \emptyset$. An interval is called *bounded* if it is empty or its left and right endpoints are finite. An interval will be called *nondegenerate* if it contains more than one point. Let $J$ be an interval in $\mathbb{R}$. The class of all subintervals of $J$ will be denoted by $\mathfrak{I}(J)$. The subclass of $\mathfrak{I}(J)$ consisting of nonempty open intervals and singletons will be denoted by $\mathfrak{I}_{os}(J)$. If $J = [a,b]$ then we write $\mathfrak{I}[a,b]$ and $\mathfrak{I}_{os}[a,b]$. For example, if $a < b$, $\mathfrak{I}_{os}[a,b] = \{(u,v), \{u\}, \{v\}\colon\ a \le u < v \le b\}$.

An interval $A$ will be called *right-open at* $v$ or in symbols $A = [\![\cdot, v)$ if $A = [u, v)$ or $(u, v)$ for some $u < v$. Similarly, $A$ will be called *right-closed at* $v$, or $A = [\![\cdot, v]\!]$, if $J = [u, v]$ for some $u \leq v$ or $A = (u, v]$ for some $u < v$. An interval $A$ will be called *left-open at* $u$ or $A = (u, \cdot]\!]$ if $A = (u, v)$ or $(u, v]$ for some $u < v$, and $A$ will be called *left-closed at* $u$, or $A = [u, \cdot]\!]$, if $A = [u, v]$ for some $u \leq v$ or $A = [u, v)$ for some $u < v$. For $a \leq u < v \leq b$, any of the four intervals $[u, v]$, $(u, v]$, $[u, v)$ or $(u, v)$ will be denoted by $[\![u, v]\!]$. Here $[\![-\infty, v]\!]$ or $[\![u, +\infty]\!]$ will always mean $(-\infty, v]\!]$ or $[\![u, +\infty)$, respectively. For two disjoint nonempty intervals $A$ and $B$, $A \prec B$ will mean that $s < t$ for all $s \in A$ and $t \in B$.

*Point partitions*

Let $(S, <)$ be any linearly ordered set containing more than one point. Then a *point partition* $\kappa = \{t_i\}_{i=0}^n$ of $S$ is a finite sequence of elements of $S$ such that $(i)$ $t_0 < t_1 < \cdots < t_n$, $(ii)$ if $S$ has a smallest element $a$, then $t_0 = a$, and $(iii)$ if $S$ has a largest element $b$, then $t_n = b$. Let PP$\,(S)$ denote the set of all point partitions of $S$.

Thus for a nondegenerate interval $J \subset \mathbb{R}$, $\kappa = \{t_i\}_{i=0}^n \subset J$ is a point partition of $J$ if $t_0 < t_1 < \cdots < t_n$, and if $J = [a, b]$, a closed bounded interval, then $t_0 = a$ and $t_n = b$. At the beginning of Section 1.1 point partitions of the closed interval $J = [a, b]$ were defined and called partitions. Most of the further terminology related to point partitions was already given near the beginning of Section 1.1 and is repeated here for the reader's convenience. For $\kappa := \{t_i\}_{i=0}^n \in$ PP$\,[a, b]$ with $-\infty < a < b < +\infty$, the *mesh* of $\kappa$ is $|\kappa| := \max_{1 \leq i \leq n}(t_i - t_{i-1})$. A point partition $\kappa$ is a *refinement* of a point partition $\lambda$ if $\lambda \subset \kappa$ as a set. Let $\kappa = \{t_i\}_{i=0}^n$ be a partition of $[a, b]$, and let $s_i \in [t_{i-1}, t_i]$ for $i = 1, \ldots, n$. Then $\tau = (\{t_i\}_{i=0}^n, \{s_i\}_{i=1}^n)$ is called a *tagged partition* of $[a, b]$, and $\tau$ is a *tagged refinement* of a point partition $\lambda$ if $\kappa$ is a refinement of $\lambda$. We will also say that the tagged partition $\tau$ consists of the *tagged intervals* $([t_{i-1}, t_i], s_i)$, $i = 1, \ldots, n$. If a tagged partition $\tau = (\{t_i\}_{i=0}^n, \{s_i\}_{i=1}^n)$ is such that $s_i \in (t_{i-1}, t_i)$ for $i = 1, \ldots, n$, then $\tau$ is called a *Young tagged point partition.* The *mesh* $|\tau|$ is defined as $|\kappa|$.

*Interval partitions*

Let $J$ be a nonempty interval in $\mathbb{R}$. Recall the terminology defined in the subsection "Interval functions" of Section 1.1. Let IP$\,(J)$ be the set of all interval partitions of $J$. If $J$ is a bounded, nondegenerate interval $[t_0, t_k]$ and $\mathcal{A} = \{A_i\}_{i=1}^n$ is a Young interval partition $\{\{t_0\}, (t_0, t_1), \{t_1\}, \ldots, (t_{k-1}, t_k), \{t_k\}\}$ of $J$, then $\{(t_{i-1}, t_i)\}_{i=1}^k$ sometimes will be written instead of $\mathcal{A}$ and we denote by $(\{(t_{i-1}, t_i)\}_{i=1}^k, \{u_i\}_{i=1}^k)$ a corresponding tagged Young interval partition. (The singletons $\{t_i\}$ with their uniquely determined tags $t_i$ are omitted from the notation. Here $n = 2k + 1$.) Similarly, if $J$ is left-open and/or right-open the same notation will be used where now $\{t_0\} \notin \mathcal{A}$ and/or $\{t_k\} \notin \mathcal{A}$.

*Unconditional convergence*

A sum $\sum_j z_j$ of elements $z_j$ of a normed space $Z$ with norm $\|\cdot\|$ is said to converge *unconditionally* in $Z$ to $S$ if and only if for every $\epsilon > 0$ there is a finite set $E$ of values of $j$ such that for every finite set $A \supset E$, $\|S - \sum_{j\in A} z_j\| < \epsilon$. If $Z$ is complete, then $\sum_j z_j$ converges unconditionally to some limit if and only if for every $\epsilon > 0$ there is a finite set $E$ such that for every finite set $B$ of positive integers disjoint from $E$, $\|\sum_{j\in B} z_j\| < \epsilon$.

*p-variation*

Let $(S, <)$ be a linearly ordered set containing more than one point, $(X, \|\cdot\|)$ a normed space over the field $\mathbb{K} = \mathbb{R}$ or $\mathbb{C}$, and $0 < p < \infty$. If $f\colon S \to X$ and $\kappa = \{t_i\}_{i=0}^n$ is a point partition of $S$, let $s_p(f;\kappa) := \sum_{i=1}^n \|f(t_i) - f(t_{i-1})\|^p$ be the *p-variation sum* for $f$ over $\kappa$. Then the *p-variation* of $f$ is defined by

$$v_p(f) := v_p(f;S) := \sup\{s_p(f;\kappa)\colon \kappa \in \mathrm{PP}(S)\}. \tag{1.19}$$

If $S$ is a singleton then let the $p$-variation of $f$ be $v_p(f;S) := 0$.

Let $V_p(f) := V_p(f;S) := v_p(f;S)^{1/p}$, let $\mathcal{W}_p(S;X)$ be the set of all functions $f\colon S \to X$ such that $v_p(f) < \infty$, and let $\mathcal{W}_p(S) := \mathcal{W}_p(S;\mathbb{R})$. In the case $S = [a,b]$ and $X = \mathbb{R}$, we recover the notion already defined by (1.2). If $1 \le p < \infty$ we define

$$\|f\|_{(p)} := \|f\|_{S,(p)} := V_p(f;S), \quad \|\cdot\|_{[p]} := \|\cdot\|_{(p)} + \|\cdot\|_{\sup}. \tag{1.20}$$

Just as for real-valued functions on a closed interval, $\|\cdot\|_{(p)}$ is a seminorm and $\|\cdot\|_{[p]}$ is a norm on $\mathcal{W}_p(S;X)$.

**Remark 1.5.** It is easily seen that the supremum in the definition of $v_p(f)$ is unchanged by restrictions *(ii)* and *(iii)* in the definition of a point partition of $S$.

Let $J$ be a nonempty interval. An interval function on $J$ with values in $X$ is any function from subintervals of $J$ to $X$. Let $\mu$ be an interval function on $J$ with values in $X$ and let $0 < p < \infty$. For an interval partition $\mathcal{A} = \{A_i\}_{i=1}^n \in \mathrm{IP}(J)$, let

$$s_p(\mu;\mathcal{A}) := \sum_{i=1}^n \Big\|\mu\Big(\bigcup_{j=1}^i A_j\Big) - \mu\Big(\bigcup_{j=1}^{i-1} A_j\Big)\Big\|^p,$$

where a union over the empty set of indices is defined as the empty set. The $p$-variation of $\mu$ on $J$ is an extended real-valued interval function $v_p(\mu) = v_p(\mu;\cdot)$ on $J$ defined by

$$v_p(\mu;A) := \sup\{s_p(\mu;\mathcal{A})\colon \mathcal{A} \in \mathrm{IP}(A)\} \le +\infty$$

if $A$ is a nonempty subinterval of $J$, or as 0 if $A = \emptyset$. The class of all interval functions on $J$ with values in $X$ such that $\sup\{v_p(\mu;A)\colon A \in \mathfrak{I}(J)\} < \infty$ is denoted by $\mathcal{I}_p(J;X)$ and $\mathcal{I}_p(J) := \mathcal{I}_p(J;\mathbb{R})$.

Just before Theorem 1.4, definitions of $s_p$ and $v_p$ for an interval function $\mu$ were given in case $\mu$ is additive. In that case, the definitions are easily seen to agree. In general, however, they do not. The definitions just given will be the ones used for general $\mu$, e.g. multiplicative $\mu$ in Chapter 9.

If $1 \le p < \infty$ let

$$\|\mu\|_{J,(p)} := \sup\{v_p(\mu; A)^{1/p} \colon A \in \mathfrak{I}(J)\}. \tag{1.21}$$

By Hölder's and Minkowski's inequalities ((1.4) and (1.5), respectively), it is easily seen that on $\mathcal{I}_p(J;\mathbb{R})$, $\|\cdot\|_{J,(p)}$ is a seminorm.

*Classes of measurable functions*

Let $(\Omega, \mathcal{S}, \mu)$ be a measure space. A subset $B$ of $\Omega$ is a *$\mu$-null set* if there is an $A \in \mathcal{S}$ such that $B \subset A$ and $\mu(A) = 0$. We say that a statement holds *$\mu$-almost everywhere*, or *a.e.* $(\mu)$, if it holds outside of a $\mu$-null set. The *completion* of $\mu$ is the extension of $\mu$ to a measure on the smallest $\sigma$-algebra including $\mathcal{S}$ and containing all $\mu$-null sets, e.g. [53, §3.3]. Let $(X, \|\cdot\|_X)$ be a Banach space. A function $f\colon \Omega \to X$ is *$\mu$-measurable* if for every Borel set $B \subset X$, $f^{-1}(B)$ is measurable for the completion of $\mu$. For $1 \le p < \infty$, $\mathcal{L}^p(\Omega; X) := \mathcal{L}^p(\Omega, \mathcal{S}, \mu; X)$ denotes the set of all $\mu$-measurable functions $f$ from $\Omega$ into $X$ such that

$$\|f\|_p := \|\|f\|_X\|_p := \Big(\int_\Omega \|f(\omega)\|_X^p \mathrm{d}\mu(\omega)\Big)^{1/p} < \infty. \tag{1.22}$$

As usual, $L^p(\Omega; X) := L^p(\Omega, \mathcal{S}, \mu; X)$ is the set of all $\mu$-equivalence classes of functions in $\mathcal{L}^p(\Omega; X)$. A function $f\colon \Omega \to X$ is called *$\mu$-essentially bounded* if

$$\|f\|_\infty := \inf_N \sup\big\{\|f(\omega)\|_X \colon \omega \in \Omega \setminus N\big\} < \infty, \tag{1.23}$$

where the infimum is taken over all $\mu$-null sets $N$. The set of all $\mu$-essentially bounded functions from $\Omega$ into $X$ is denoted by $\mathcal{L}^\infty(\Omega; X) := \mathcal{L}^\infty(\Omega, \mathcal{S}, \mu; X)$, and $L^\infty(\Omega; X) := L^\infty(\Omega, \mathcal{S}, \mu; X)$ is the set of all $\mu$-equivalence classes of functions in $\mathcal{L}^\infty(\Omega; X)$.

It is easy to check that since $\|\cdot\|_p$ is a norm on $L^p(\Omega, \mathcal{S}, \mu; \mathbb{R})$ (Minkowski's inequality for integrals, e.g. [53, Theorem 5.1.5]), it is also a norm on $L^p(\Omega, \mathcal{S}, \mu; X)$ for $1 \le p < \infty$. Completeness also follows as for real-valued $L^p$ e.g. [53, Theorem 5.2.1], so $(L^p(\Omega, \mathcal{S}, \mu; X), \|\cdot\|_p)$ is a Banach space.

If $\Omega$ is the set of positive integers, $\mu$ is the counting measure, and $X = \mathbb{R}$, then $\mathcal{L}^p$ can be identified with $L^p$ and is called $\ell^p$. That is, $\ell^p$ for $1 \le p \le \infty$ is the Banach space of sequences $x = \{x_j\}_{j \ge 1}$ of real numbers with the norm

$$\|x\|_p = \Big(\sum_{j\ge 1} |x_j|^p\Big)^{1/p} \quad \text{if } p < \infty \text{ and } \quad \|x\|_\infty = \sup_{j \ge 1} |x_j|.$$

A useful generalization of the spaces $\mathcal{L}^p(\Omega; X)$ is the class of Orlicz spaces defined as follows. Recall that a set $C$ in a vector space is called *convex* iff whenever $u, v \in C$ and $0 \le \alpha \le 1$, we have $\alpha u + (1-\alpha)v \in C$. Then a real-valued function $f$ on $C$ is called *convex* if $f(\alpha u+(1-\alpha)v) \le \alpha f(u)+(1-\alpha)f(v)$ for all $u, v \in C$ and $0 \le \alpha \le 1$, and $g\colon C \to \mathbb{R}$ is called *concave* iff $-g$ is convex. The class of all functions $\Phi\colon [0,\infty) \to [0,\infty)$ that are strictly increasing, continuous, convex, unbounded and 0 at 0 will be denoted by $\mathcal{CV}$. Note that $\Phi(u) = u^p$, $u \ge 0$, is in $\mathcal{CV}$ if and only if $p \ge 1$.

For $\Phi \in \mathcal{CV}$, the space $\mathcal{L}^\Phi(\Omega, \mathcal{S}, \mu; X) := \mathcal{L}^\Phi(\Omega; X)$ of all $\mu$-measurable functions $f\colon \Omega \to X$ such that for some $c > 0$,

$$\int_\Omega \Phi(\|f\|/c)\,\mathrm{d}\mu < \infty,$$

is called an *Orlicz space*. Then $L^\Phi(\Omega, \mathcal{S}, \mu; X) := L^\Phi(\Omega; X)$ is the set of all $\mu$-equivalence classes of functions in $\mathcal{L}^\Phi(\Omega; X)$. $L^\Phi(\Omega; X)$ is a Banach space with the Luxemburg norm

$$\|f\|_\Phi := \inf\Big\{c > 0\colon \int_\Omega \Phi(\|f\|/c)\,\mathrm{d}\mu \le 1\Big\}, \tag{1.24}$$

e.g. [52, Theorem H.5]. If $\Omega = \mathbb{N}$, $\mathcal{S}$ is the class of all subsets of $\Omega$, and $\mu$ is equal to 1 on each singleton, then the Orlicz space $\mathcal{L}^\Phi(\Omega, \mathcal{S}, \mu; \mathbb{R})$ is called an Orlicz sequence space and is denoted by $\ell^\Phi$.

Let the field $\mathbb{K}$ be $\mathbb{R}$ or $\mathbb{C}$. If $(\Omega, \mathcal{S}, \mu)$ is a finite measure space, then $L^0(\Omega; \mathbb{K}) := L^0(\Omega, \mathcal{S}, \mu; \mathbb{K})$ denotes the linear space of all $\mu$-equivalence classes of $\mu$-measurable functions from $\Omega$ to $\mathbb{K}$, metrized by the metric

$$d_0(f, g) \;:=\; \int_\Omega |f-g|/(1+|f-g|)\,\mathrm{d}\mu.$$

It is well known that $L^0$ under $d_0$ is a topological vector space, and that for $\mu$ nonatomic it is not locally convex. A sequence $\{f_n\}$ of $\mu$-measurable functions converges to a $\mu$-measurable function $f$ in *$\mu$-measure* if for each $\epsilon > 0$,

$$\lim_{n\to\infty} \mu\big(\{\omega \in \Omega\colon |f_n(\omega) - f(\omega)| > \epsilon\}\big) = 0.$$

Convergence in $L^0(\Omega; \mathbb{K})$ with respect to the metric $d_0$ is equivalent to convergence in $\mu$-measure.

# 2

# Definitions and Basic Properties of Extended Riemann–Stieltjes Integrals

## 2.1 Regulated and Interval Functions

*Regulated functions*

Let $X$ be a Banach space, and let $J$ be a nonempty interval in $\mathbb{R}$, which may be bounded or unbounded, and open or closed at either end. Recall that an interval is called nondegenerate if it has nonempty interior or equivalently contains more than one point. Let $\bar{J}$ be the closure of $J$ in the extended real line $[-\infty, \infty]$. A function $f$ on $J$ with values in $X$ is called *regulated on* $J$, just as for real-valued functions in Chapter 1, if the right limit $f(t+) := \lim_{s\downarrow t} f(s)$ exists in $X$ for $t \in \bar{J}$ not equal to the right endpoint of $J$, and if the left limit $f(t-) := \lim_{s\uparrow t} f(s)$ exists in $X$ for $t \in \bar{J}$ not equal to the left endpoint of $J$. If $J = \{a\} = [a, a]$ is a singleton these conditions hold vacuously and we say that $f$ is regulated. The class of all regulated functions on $J$ with values in $X$ will be denoted by $\mathcal{R}(J; X)$.

Let $a < b$ throughout this paragraph (defining quantities $f_-$, $f_+$, and $\Delta$). For a regulated function $f$ on $J = [\![a, b]\!]$, define a function $f_-^{(a)}(t) := f(t-)$ for $t \in (a, b]\!]$ or $f(a)$ if $t = a \in J$. Similarly define $f_+^{(b)}(t) := f(t+)$ for $t \in [\![a, b)$ or $f(b)$ if $t = b \in J$. Define $\Delta^+ f$ on $J$ by $(\Delta^+ f)(t) := f(t+) - f(t)$, called the right jump of $f$ at $t$, for all $t \in J$ except the right endpoint. Similarly, define a function $\Delta^- f$ on $J$ by $(\Delta^- f)(t) := f(t) - f(t-)$, called the left jump of $f$ at $t$, for all $t \in J$ except the left endpoint. Also, let $\Delta^{\pm} f := \Delta^+ f + \Delta^- f$, called the two-sided jump, on the interior of $J$. For a regulated function $f$ on $J = [\![a, b]\!]$, let $\Delta_J^+ f(t) := \Delta^+ f(t)$ if $t \in [\![a, b)$, or 0 if $t = a \notin J$ or $t = b$; $\Delta_J^- f(t) := \Delta^- f(t)$ if $t \in (a, b]\!]$, or 0 if $t = b \notin J$ or $t = a$; and

$$\begin{aligned} \Delta_J^{\pm} f(t) &:= \Delta_J^- f(t) + \Delta_J^+ f(t) \\ &= \begin{cases} \Delta^+ f(a) \ \text{ if } t = a \in J, \text{ or } 0 \text{ if } t = a \notin J; \\ \Delta^{\pm} f(t) \ \text{ if } t \in (a, b); \\ \Delta^- f(b) \ \text{ if } t = b \in J, \text{ or } 0 \text{ if } t = b \notin J. \end{cases} \end{aligned} \tag{2.1}$$

R.M. Dudley and R. Norvaiša, *Concrete Functional Calculus*, Springer Monographs in Mathematics, DOI 10.1007/978-1-4419-6950-7_2,

The definition of $\Delta_J^{\pm}$ is made so that some later formulas will not need to have special endpoint terms.

A *step function* from an interval $J$ into a Banach space $X$ is a finite sum $\sum_{j=1}^{m} 1_{A(j)}x_j$, where $x_j \in X$ and $A(j)$ are intervals, some of which may be singletons. Clearly a step function is regulated. The next fact shows that a function is regulated if and only if it can be approximated uniformly by step functions.

**Theorem 2.1.** *Let $X$ be a Banach space and $-\infty < a < b < +\infty$. The following properties are equivalent for a function $f\colon\ J := [\![a,b]\!] \to X$:*

*(a) $f \in \mathcal{R}([\![a,b]\!];X)$;*
*(b) for each $\epsilon > 0$, there exists a Young interval partition $\{(t_{i-1},t_i)\}_{i=1}^n$ of $[\![a,b]\!]$ such that $\mathrm{Osc}\,(f;(t_{i-1},t_i)) < \epsilon$ for each $i \in \{1,\dots,n\}$;*
*(c) $f$ is a uniform limit of step functions.*

*Proof.* $(a) \Rightarrow (b)$. Let $\epsilon > 0$. By definition of $\mathcal{R}(J;X)$, if $J = (a,b]\!]$, $f(a+)$ exists, and if $J = [\![a,b)$, $f(b-)$ exists. Thus there exist $a_1 > a$ and $b_1 < b$ such that $\mathrm{Osc}(f;(a,a_1)) < \epsilon$ and $\mathrm{Osc}(f;(b_1,b)) < \epsilon$. So it suffices to consider the case that $[\![a,b]\!]$ is a closed, bounded interval $[a,b]$ (specifically, $[a_1,b_1]$). For each $s \in (a,b)$, choose $\delta_s > 0$ such that $A_s := (s-\delta_s, s+\delta_s) \subset [a,b]$ and the oscillation of $f$ over the open intervals $(s-\delta_s,s)$ and $(s,s+\delta_s)$ is less than $\epsilon$. For the endpoints, choose $\delta_a$ and $\delta_b \in (0,b-a)$ such that $\mathrm{Osc}\,(f;(a,a+\delta_a)) < \epsilon$ and $\mathrm{Osc}\,(f;(b-\delta_b,b)) < \epsilon$. Letting $A_a := [a,a+\delta_a)$ and $A_b := (b-\delta_b,b]$, the sets $\{A_s\colon\ s \in [a,b]\}$ form a cover of the compact interval $[a,b]$ by relatively open sets. Therefore there is a finite subcover $A_{s_0}, A_{s_1}, \dots, A_{s_m}$ of $[a,b]$ with $a = s_0 < s_1 < \dots < s_m = b$. Take $t_0 := s_0$, $t_1 \in A_{s_0} \cap A_{s_1} \cap (s_0,s_1)$, $t_2 := s_1, \dots, t_{2m-2} := s_{m-1}$, $t_{2m-1} \in A_{s_{m-1}} \cap A_{s_m} \cap (s_{m-1},s_m)$, and $t_{2m} := s_m$. Thus $(b)$ holds with $n = 2m+1$.

$(b) \Rightarrow (c)$. Given $\epsilon > 0$, choose a Young interval partition $\{(t_{i-1},t_i)\}_{i=1}^n$ of $J$ as in $(b)$. Define a step function $f_\epsilon$ on $J$ by $f_\epsilon(t) := f(s_i)$ with $s_i \in (t_{i-1},t_i)$ if $t \in (t_{i-1},t_i)$ for some $i \in \{1,\dots,n\}$ and $f_\epsilon(t_i) := f(t_i)$ for $i \in \{0,\dots,n\}$ if $t_i \in J$. Then $\|f_\epsilon(t) - f(t)\| < \epsilon$ for each $t \in J$. Since $\epsilon$ is arbitrary, $(c)$ follows.

$(c) \Rightarrow (a)$. Given $\epsilon > 0$, choose a step function $f_\epsilon$ such that $\|f_\epsilon(t) - f(t)\| < \epsilon$ for each $t \in J$. Then for any $s,t \in J$, $\|f(t) - f(s)\| < 2\epsilon + \|f_\epsilon(t) - f_\epsilon(s)\|$. Since $f_\epsilon$ is regulated, the right side can be made arbitrarily small for all $s,t$ close enough from the left or right to any given point of $J$, proving $(a)$. The proof of Theorem 2.1 is complete. □

The following is an easy consequence of the preceding theorem.

**Corollary 2.2.** *If $f$ is regulated on $[\![a,b]\!]$ then $f$ is bounded, Borel measurable, and for each $\epsilon > 0$,*

$$\mathrm{card}\Big\{u \in [\![a,b]\!]\colon\ \text{either } \Big\|\Delta^+_{[\![a,b]\!]} f(u)\Big\| > \epsilon \text{ or } \Big\|\Delta^-_{[\![a,b]\!]} f(u)\Big\| > \epsilon\Big\} < \infty.$$

*Proof.* By implication $(a) \Rightarrow (b)$ of Theorem 2.1, there is a Young interval partition $\{(t_{i-1}, t_i)\}_{i=1}^n$ of $[\![a, b]\!]$ such that $\mathrm{Osc}(f; (t_{i-1}, t_i)) \leq 1$ for each $i = 1, \ldots, n$. Let $t \in [\![a, b]\!]$ and let $i$ be such that $t = t_i$ or $t \in (t_{i-1}, t_i)$. Then $\|f(t)\| \leq 1 + \max_{1 \leq i \leq n} \|f(t_{i-1}+)\| + \max_{0 \leq i \leq n} \|f(t_i)\|$, proving the first part of the conclusion. The third part also follows from Theorem 2.1$(b)$ since each jump can be approximated arbitrarily closely by increments of $f$. It then follows that $f$ is continuous except on a countable set, and so it is Borel measurable, completing the proof. □

### *Interval functions*

Let $X$ be a Banach space with norm $\|\cdot\|$, let $J$ be a nonempty interval in $\mathbb{R}$, possibly unbounded, and let $\mathfrak{I}(J)$ be the class of all subintervals of $J$. Any function $\mu\colon \mathfrak{I}(J) \to X$ will be called an *interval function on* $J$. The class of all $X$-valued interval functions on $J$ is denoted by $\mathcal{I}(J; X)$, and it is denoted by $\mathcal{I}(J)$ if $X = \mathbb{R}$. An interval function $\mu$ on $J$ will be called *additive* if $\mu(A \cup B) = \mu(A) + \mu(B)$ whenever $A, B \in \mathfrak{I}(J)$ are disjoint and $A \cup B \in \mathfrak{I}(J)$. If $\mu$ is an additive interval function then clearly $\mu(\emptyset) = 0$. An additive interval function $\mu$ on $J$ is uniquely determined by its restriction to the class $\mathfrak{I}_{os}(J)$, i.e. the class of all open subintervals and singletons of $J$.

For intervals as for other sets, $A_n \uparrow A$ will mean $A_1 \subset A_2 \subset \cdots$ and $\cup_{n=1}^\infty A_n = A$, while $A_n \downarrow A$ will mean $A_1 \supset A_2 \supset \cdots$ and $\cap_{n=1}^\infty A_n = A$, and $A_n \to A$ will mean $1_{A_n}(t) \to 1_A(t)$ as $n \to \infty$ for all $t \in \mathbb{R}$.

**Definition 2.3.** 1. An interval function $\mu$ on $J$ will be called *upper continuous* if $\mu(A_n) \to \mu(A)$ for any $A, A_1, A_2, \ldots \in \mathfrak{I}(J)$ such that $A_n \downarrow A$.

2. An interval function $\mu$ on $J$ will be called *upper continuous at* $\emptyset$ if $\mu(A_n) \to \mu(\emptyset)$ for any $A_1, A_2, \ldots \in \mathfrak{I}(J)$ such that $A_n \downarrow \emptyset$.

If $J$ is a singleton then the conditions of Definition 2.3 hold vacuously and each interval function on $J$ is upper continuous as well as upper continuous at $\emptyset$.

In this section we establish relations between the class of additive upper continuous interval functions on $J$ and classes of regulated functions on $J$. But first, here is an example of an interval function which is not upper continuous.

**Example 2.4.** Let $J$ be a nondegenerate interval. The Banach space of all real-valued and bounded functions on $J$ with the supremum norm is denoted by $\ell^\infty(J)$. Recall that for Lebesgue measure $\lambda$ on $J$, $L^\infty(J, \lambda)$ is the Banach space of $\lambda$-equivalence classes of $\lambda$-essentially bounded functions with the essential supremum norm. For an interval $A \subset J$, let $\mu(A) := 1_A$. Then for $1_A$ as a member of $\ell^\infty(J)$ or of $L^\infty(J, \lambda)$, $\mu$ is an additive interval function on $J$, but not upper continuous at $\emptyset$.

For a regulated function $h$ on $[\![a, b]\!]$ with values in $X$ there is a corresponding additive interval function $\mu_h := \mu_{h,[\![a,b]\!]}$ on $[\![a, b]\!]$ defined by

$$\begin{aligned} \mu_{h,[\![a,b]\!]}((u,v)) &:= h(v-)-h(u+) \text{ for } a\le u<v\le b, \\ \mu_{h,[\![a,b]\!]}(\{u\}) &:= \Delta^{\pm}_{[\![a,b]\!]}h(u) \text{ for } u\in[\![a,b]\!], \end{aligned} \tag{2.2}$$

if $a<b$, and $\mu_{h,[a,a]}(\{a\}) := \mu_{h,[a,a]}(\emptyset) := 0$ if $a=b$. For $a<b$ we have $\mu_{h,(a,b]}(\{b\}) = \Delta^{\pm}_{(a,b]}h(b) = \Delta^{-}h(b)$. It follows that if $h$ on $(a,b]$ is right-continuous on $(a,b)$ then for $a\le c<d\le b$ we have $\mu_h((c,d]) = h(d)-h(c)$. Thus the definition (2.2) of $\mu_h$ just given extends the one given earlier for the definition (1.6) of the Lebesgue–Stieltjes integral.

Notice that $\mu_h$ does not depend on the values of $h$ at its jump points (points where it has a non-zero left or right jump) in $(a,b)$. In the converse direction, for any interval function $\mu$ on $[\![a,b]\!]$, define two functions $R_{\mu,a}$ and $L_{\mu,a}$ on $[\![a,b]\!]$ by

$$\begin{aligned} R_{\mu,a}(t) &:= \begin{cases} \mu(\emptyset) & \text{if } t=a\in[\![a,b]\!] \\ \mu([\![a,t]) & \text{if } t\in(a,b]\!], \end{cases} \\ L_{\mu,a}(t) &:= \begin{cases} \mu([\![a,t)) & \text{if } t\in[\![a,b) \\ \mu([\![a,b]) & \text{if } t=b\in[\![a,b]\!]. \end{cases} \end{aligned} \tag{2.3}$$

(Recall that if $\mu$ is additive, $\mu(\emptyset)=0$.) If $\mu$ is upper continuous then $R_{\mu,a}$ and $L_{\mu,a}$ are both regulated point functions on $[\![a,b]\!]$, as will be shown in Proposition 2.6 when $a<b$. The converse is not true as the following shows:

**Example 2.5.** Let $a<b$, $\mu([u,v)) := \mu((u,v]) := 0$ and $\mu([u,v]) := 1$ for $a\le u\le v\le b$, and $\mu((u,v)) := -1$ if also $u<v$. Then $\mu$ is an additive interval function on $[a,b]$, $R_{\mu,a} = 1_{(a,b]}$ and $L_{\mu,a} = 1_{\{b\}}$ are regulated, but $\mu$ is not upper continuous at $\emptyset$. For $h=R_{\mu,a}$ or $L_{\mu,a}$, $\mu\ne\mu_h$.

For $x\in J$, we will say that the singleton $\{x\}$ is an *atom* of $\mu$ if $\mu(\{x\})\ne 0$. The following gives a characterization of additive upper continuous interval functions on $J$. In particular, such interval functions cannot have more than countably many atoms.

**Proposition 2.6.** *Let $J$ be a nondegenerate interval and let $\mu\in\mathcal{I}(J;X)$ be additive. The following five statements are equivalent:*

*(a) $\mu$ is upper continuous;*
*(b) $\mu$ is upper continuous at $\emptyset$;*
*(c) $\mu(A_n)\to 0$ whenever open intervals $A_n\downarrow\emptyset$, and*

$$\operatorname{card}\{u\in J\colon \|\mu(\{u\})\|>\epsilon\}<\infty \qquad \textit{for all } \epsilon>0; \tag{2.4}$$

*(d) $\mu(A_n)\to\mu(A)$ whenever intervals $A_n\uparrow A$;*
*(e) $\mu(A_n)\to\mu(A)$ whenever intervals $A_n\to A\ne\emptyset$.*

*If $J=[\![a,b]\!]$ then the above statements are equivalent to each of the following two statements:*

*(f) $R_{\mu,a}$ is regulated, $R_{\mu,a}(x-) = \mu([\![a,x))$ for $x\in(a,b]\!]$, and $R_{\mu,a}(x+) = \mu([\![a,x])$ for $x\in[\![a,b)$;*

(g) $L_{\mu,a}$ *is regulated,* $L_{\mu,a}(x-) = \mu([\![a,x)\!))$ *for* $x \in (a,b]\!]$*, and* $L_{\mu,a}(x+) = \mu([\![a,x])$ *for* $x \in [\![a,b)$.

*Proof.* $(a) \Leftrightarrow (b)$. Clearly, $(a)$ implies $(b)$. For $(b) \Rightarrow (a)$, let intervals $A_n \downarrow A$. Then $A_n = B_n \cup A \cup C_n$ for intervals $B_n \prec A \prec C_n$, with $B_n \downarrow \emptyset$ and $C_n \downarrow \emptyset$. So $\mu(A_n) \to \mu(A)$ by additivity.

$(b) \Rightarrow (c)$. The first part of $(c)$ is clear. For the second part, suppose there exist $\epsilon > 0$ and an infinite sequence $\{u_j\}$ of different points of $J$ such that $\|\mu(\{u_j\})\| > \epsilon$ for all $j$. Then there are a $u \in \bar{J}$ and a subsequence $\{u_{j'}\}$ such that either $u_{j'} \downarrow u$ or $u_{j'} \uparrow u$ as $j' \to \infty$. In the first case, by additivity, $\mu(\{u_{j'}\}) = \mu((u, u_{j'}]) - \mu((u, u_{j'})) \to 0$ as $j' \to \infty$, because both $(u, u_{j'}] \downarrow \emptyset$ and $(u, u_{j'}) \downarrow \emptyset$. This contradiction proves (2.4) in the case $u_{j'} \downarrow u$. The proof in the case $u_{j'} \uparrow u$ is symmetric.

For $(c) \Rightarrow (b)$, let intervals $A_n \downarrow \emptyset$. Then for some $u \in \bar{J}$, for all sufficiently large $n$, either $A_n$ is left-open at $u$, or $A_n$ is right-open at $u$. Using additivity, in each of the two cases $\mu(A_n) \to 0$ follows by $(c)$.

For $(b) \Rightarrow (e)$, let intervals $A_n \to A \neq \emptyset$. If $A = (u,v)$ then $\{u\} \prec A_n \prec \{v\}$ for $n$ large enough. For such $n$, there are intervals $C_n$ and $D_n$ with $\{u\} \prec C_n \prec A_n \prec D_n \prec \{v\}$ and $C_n \cup A_n \cup D_n = A$. Also for such $n$, we have either $C_n = \emptyset$ or $C_n = (u, \cdot]\!]$, and either $D_n = \emptyset$ or $D_n = [\![\cdot, v)$. Clearly $C_n \to \emptyset$. If $N := \{n\colon\ C_n \neq \emptyset\}$ is infinite, there is a function $j \mapsto n(j)$ onto $N$ such that $C_{n(j)} \downarrow \emptyset$. Thus $\mu(C_n) \to 0$. Similarly $\mu(D_n) \to 0$. Therefore $\mu(A_n) = \mu(A) - \mu(C_n) - \mu(D_n) \to \mu(A)$. Similar arguments apply to other cases $A = (u,v]$, $[u,v)$, or $[u,v]$.

Clearly, $(e)$ implies $(d)$. For $(d) \Rightarrow (b)$, let intervals $A_n \downarrow \emptyset$. Then $A_1 = B_n \cup A_n \cup C_n$ for some intervals $B_n, C_n$ with $B_n \prec A_n \prec C_n$, and $B_n \uparrow B$, $C_n \uparrow C$ for some intervals $B, C$ with $A_1 = B \cup C$. Since $\mu$ is additive, $\mu(A_n) \to 0$.

$(b) \Leftrightarrow (f) \Leftrightarrow (g)$. The implications $(b) \Rightarrow (f)$ and $(b) \Rightarrow (g)$ are clear. We prove $(f) \Rightarrow (b)$ only, because the proof of $(g) \Rightarrow (b)$ is similar. Let intervals $A_n \downarrow \emptyset$. Then for some $u$ and all sufficiently large $n$, either $A_n$ is left-open at $u \in [\![a,b)$, or $A_n$ is right-open at $u \in (a,b]\!]$. By assumption, in the first case we have $\lim_n \mu(A_n) = R_{\mu,a}(u+) - R_{\mu,a}(u) = 0$ and in the second case we have $\lim_n \mu(A_n) = R_{\mu,a}(u-) - R_{\mu,a}(u-) = 0$. The proof of Proposition 2.6 is complete. □

The class of all additive and upper continuous functions in $\mathcal{I}(J;X)$ will be denoted by $\mathcal{AI}(J;X)$. For the next two theorems, recall the definition (2.2) of the interval function $\mu_h$ corresponding to a regulated function $h$.

**Theorem 2.7.** *For any regulated function* $h$ *on a nonempty interval* $[\![a,b]\!]$*, the interval function* $\mu_h$ *is in* $\mathcal{AI}([\![a,b]\!];X)$*, and the map* $h \mapsto \mu_h$ *is linear.*

*Proof.* We can assume that $a < b$. Additivity is immediate from the definition of $\mu_h$, as is linearity of $h \mapsto \mu_h$. For upper continuity it is enough to prove statement $(c)$ of Proposition 2.6. Let open intervals $A_n \downarrow \emptyset$. Thus there exist

$\{u, v_n \colon n \geq 1\} \subset [\![a,b]\!]$ such that for all large enough $n$, either $A_n = (v_n, u)$ with $v_n \uparrow u$ or $A_n = (u, v_n)$ with $v_n \downarrow u$. Then $\mu_h(A_n) = h(u-) - h(v_n+) \to 0$ as $n \to \infty$ in the first case and $\mu_h(A_n) = h(v_n-) - h(u+) \to 0$ as $n \to \infty$ in the second case. Since $\mu(\{t\}) = \Delta^{\pm}_{[\![a,b]\!]} h(t)$ for $t \in [\![a,b]\!]$, $\mathrm{card}\{t \in [\![a,b]\!] \colon \|\mu(\{t\})\| > \epsilon\} < \infty$ for all $\epsilon > 0$ by Corollary 2.2. Thus $\mu$ is upper continuous by Proposition 2.6, and Theorem 2.7 is proved. □

**Theorem 2.8.** *For an additive $\mu \in \mathcal{I}([\![a,b]\!]; X)$ with $a < b$, the following statements are equivalent:*

*(a) $\mu$ is upper continuous;*
*(b) $\mu = \mu_h$ on $\mathfrak{I}[\![a,b]\!]$ for $h = R_{\mu,a}$;*
*(b′) $\mu = \mu_h$ on $\mathfrak{I}[\![a,b]\!]$ for $h = L_{\mu,a}$;*
*(c) $\mu = \mu_h$ on $\mathfrak{I}[\![a,b]\!]$ for some $h \in \mathcal{R}([\![a,b]\!]; X)$ with $h(a) = 0$ if $a \in [\![a,b]\!]$ and $h(a+) = 0$ otherwise.*

*Proof.* (a) ⇒ (b): suppose that an additive interval function $\mu$ on $[\![a,b]\!]$ is upper continuous. Let $h := R_{\mu,a}$. Then by (a) ⇔ (f) of Proposition 2.6, and by the definition of $R_{\mu,a}$, $h$ is regulated and right-continuous on $(a,b)$, and $\mu_h = \mu$ on all intervals $[\![a,t]$ and $[\![a,t)$, $t \in [\![a,b]\!]$. Since $\mu$ and $\mu_h$ are additive, $\mu = \mu_h$ on $\mathfrak{I}[\![a,b]\!]$. A proof that (a) ⇒ (b′) is similar.

The implications (b) ⇒ (c) and (b′) ⇒ (c) are immediate. The implication (c) ⇒ (a) follows from Theorem 2.7, proving the theorem. □

By the preceding theorem, for each $\mu \in \mathcal{AI}([\![a,b]\!]; X)$ with $a < b$ there is a regulated function $h$ on $[\![a,b]\!]$ such that $\mu = \mu_h$. The following will be used to clarify differences between all such $h$.

**Definition 2.9.** For any intervals $A \subset J$ and Banach space $(X, \|\cdot\|)$, $c_0(A) := c_0(A, J) := c_0(A, J; X)$ is the set of functions $f \colon J \to X$ such that for some sequence $\{t_n\}_{n=1}^{\infty} \subset A$, $f(t_n) \to 0$ as $n \to \infty$ and $f(t) = 0$ if $t \neq t_n$ for all $n$.

Notice that if $a < b$ and $f \in c_0((a,b), [a,b]; X))$ then $f$ is regulated on $[a,b]$ with $f_-^{(a)} \equiv f_+^{(b)} \equiv 0$ on $[a,b]$.

**Proposition 2.10.** *Let $g, h \in \mathcal{R}([a,b]; X)$. Then $\mu_g = \mu_h$ on $\mathfrak{I}[a,b]$ if and only if for some constant $c$ and $\psi \in c_0((a,b)) = c_0((a,b), [a,b]; X)$, $g - h = c + \psi$ on $[a,b]$. Moreover, if $\mu_g = \mu_h$ on $\mathfrak{I}[a,b]$ and $g(a) = h(a)$ then $c = 0$.*

*Proof.* We can assume that $a < b$. Since $\psi(s+) = \psi(t-) = 0$ for $a \leq s < t \leq b$, the "if" part holds. For the converse implication suppose that $\mu_g = \mu_h$. Let $D_g$ be the set of all points $t \in (a,b)$ such that either $\Delta^- g(t) \neq 0$ or $\Delta^+ g(t) \neq 0$. Define $D_h$ similarly. Let $D := D_g \cup D_h$ and let $c := g(a) - h(a)$. Define $\psi$ on $[a,b]$ by $\psi(t) := g(t) - h(t) - c$ for $t \in D$ and 0 elsewhere. Since $\mu_g = \mu_h$ on $\mathfrak{I}[a,b]$, $g(u-) = c + h(u-)$ and $g(v+) = c + g(v+)$ for $a \leq v < u \leq b$. Thus

$g-h=c$ on $[a,b]\setminus D$ and $g-h=c+\psi$ on $D$. By Corollary 2.2, $\psi\in c_0((a,b))$, completing the proof of the converse implication. □

The regulated function $h$ in Theorem 2.8(*c*) can be chosen uniquely if it satisfies additional properties. For $a<b$, let $\mathcal{D}([\![a,b]\!];X)$ be the set of all $h\in\mathcal{R}([\![a,b]\!];X)$ such that $h$ is right-continuous on $(a,b)$ and either $h(a)=0$ if $a\in[\![a,b]\!]$ or $h(a+)=0$ if $a\notin[\![a,b]\!]$. Thus $h\in\mathcal{D}([a,b];X)$ need not be right-continuous at $a$, just as $R_{\mu,a}$ is not if $\{a\}$ is an atom for $\mu\in\mathcal{AI}([a,b];X)$.

**Corollary 2.11.** *Let $a<b$ and let $X$ be a Banach space. The mappings*

$$\mathcal{D}([\![a,b]\!];X)\ni h\mapsto\mu:=\mu_h\in\mathcal{AI}([\![a,b]\!];X) \tag{2.5}$$

*and*

$$\mathcal{AI}([\![a,b]\!];X)\ni\mu\mapsto h:=R_{\mu,a}\in\mathcal{D}([\![a,b]\!];X) \tag{2.6}$$

*are one-to-one linear operators between the vector spaces $\mathcal{AI}([\![a,b]\!];X)$ and $\mathcal{D}([\![a,b]\!];X)$. Moreover, the two mappings are inverses of each other.*

*Proof.* If $h\in\mathcal{D}([\![a,b]\!];X)$ then $\mu_h\in\mathcal{AI}([\![a,b]\!];X)$ by Theorem 2.8($c$) $\Rightarrow$ ($a$). If $\mu_g=\mu_h$ on $\mathfrak{I}[\![a,b]\!]$ for some $g,h\in\mathcal{D}([\![a,b]\!];X)$ then $h(x)-g(x)=\mu_h([\![a,x])-\mu_g([\![a,x])$ for each $x\in(a,b]\!]$, and so $g\equiv h$ on $[\![a,b]\!]$. Thus the mapping (2.5) is one-to-one. It clearly is linear.

If $\mu\in\mathcal{AI}([\![a,b]\!];X)$ then $R_{\mu,a}\in\mathcal{D}([\![a,b]\!];X)$ by Proposition 2.6($f$). By the same proposition, if $R_{\mu,a}=R_{\nu,a}$ on $[\![a,b]\!]$ for some $\mu,\nu\in\mathcal{AI}([\![a,b]\!];X)$ then $\mu=\nu$ on $\mathfrak{I}[\![a,b]\!]$. Thus the mapping (2.6) also is one-to-one and linear, proving the first part of the conclusion.

To prove the second part of the conclusion, first let $h\in\mathcal{D}([\![a,b]\!];X)$. By (2.3), (2.2), and Proposition 2.6(f), if $t\in(a,b]\!]$ then we have $R_{\mu_h,a}(t)=h(t)$. Also, if $a\in[\![a,b]\!]$ then $R_{\mu_h,a}(a)=0=h(a)$. Thus the composition of (2.5) with (2.6) maps $h\in\mathcal{D}([\![a,b]\!];X)$ into itself. Since both maps are one-to-one and (2.6) is onto, the proof of the corollary is complete. □

Next is a property of an upper continuous additive interval function similar to the property in Theorem 2.1($b$) for regulated functions. For an interval function $\mu$ on $[\![a,b]\!]$ and an interval $J\subset[\![a,b]\!]$, let

$$\mathrm{Osc}(\mu;J):=\sup\bigl\{\|\mu(A)\|\colon\ A\in\mathfrak{I}[\![a,b]\!],\ A\subset J\bigr\}. \tag{2.7}$$

In the case $J=[a,b]$, we also write $\|\mu\|_{\sup}:=\mathrm{Osc}(\mu;[a,b])$.

**Corollary 2.12.** *Let $X$ be a Banach space, $\mu\in\mathcal{AI}([\![a,b]\!];X)$ with $a<b$ and $\epsilon>0$. There exists a Young interval partition $\{(t_{i-1},t_i)\}_{i=1}^n$ of $[\![a,b]\!]$ such that $\mathrm{Osc}(\mu;(t_{i-1},t_i))\le\epsilon$ for each $i=1,\dots,n$. In particular, $\mu$ is bounded.*

*Proof.* Let $h := R_{\mu,a}$ be the function from $[\![a,b]\!]$ to $X$ defined by (2.3). By implication $(a) \Rightarrow (f)$ of Proposition 2.6, $h$ is a regulated function on $[\![a,b]\!]$, and so by implication $(a) \Rightarrow (b)$ of Theorem 2.1, there exists a Young interval partition $\{(t_{i-1},t_i)\}_{i=1}^n$ of $[\![a,b]\!]$ such that $\mathrm{Osc}(h;(t_{i-1},t_i)) < \epsilon$ for each $i = 1,\ldots,n$. By implication $(a) \Rightarrow (b)$ of Theorem 2.8, for an interval $A \subset (t_{i-1},t_i)$, $\mu(A) = \mu_h(A)$, and so $\|\mu(A)\| \le \epsilon$, proving the first part of the conclusion. The second part follows similarly because $h$ is bounded by Corollary 2.2, completing the proof of the corollary.

## 2.2 Riemann–Stieltjes Integrals

Suppose that the basic assumption (1.14) holds. Let $f, h$ be functions defined on an interval $[a,b]$ with $-\infty < a \le b < +\infty$ and having values in $X, Y$, respectively. If $a < b$, given a tagged partition $\tau = (\{t_i\}_{i=0}^n, \{s_i\}_{i=1}^n)$ of $[a,b]$, define the *Riemann–Stieltjes sum* $S_{RS}(\tau) = S_{RS}(f, \mathrm{d}h; \tau)$ based on $\tau$ by

$$S_{RS}(f,\mathrm{d}h;\tau) := \sum_{i=1}^n f(s_i)\cdot\big[h(t_i) - h(t_{i-1})\big].$$

The *Riemann–Stieltjes* or *RS* integral $(RS)\int_a^b f\cdot\mathrm{d}h$ is defined to be 0 if $a = b$ and otherwise is defined if the limit exists in $(Z, \|\cdot\|)$ as

$$(RS)\int_a^b f\cdot\mathrm{d}h := \lim_{|\tau|\downarrow 0} S_{RS}(f,\mathrm{d}h;\tau). \tag{2.8}$$

The *refinement Riemann–Stieltjes* or *RRS* integral $(RRS)\int_a^b f\cdot\mathrm{d}h$ is defined to be 0 if $a = b$ and otherwise is defined as

$$(RRS)\int_a^b f\cdot\mathrm{d}h := \lim_{\tau} S_{RS}(f,\mathrm{d}h;\tau) \tag{2.9}$$

provided the limit exists in the refinement sense, that is, $(RRS)\int_a^b f\cdot\mathrm{d}h := A$ if for every $\epsilon > 0$ there is a point partition $\lambda$ of $[a,b]$ such that for every tagged partition $\tau = (\kappa, \xi)$ such that $\kappa$ is a refinement of $\lambda$, $\|S_{RS}(f,\mathrm{d}h;\tau) - A\| < \epsilon$.

Integrals $\int_a^b \mathrm{d}h\cdot f$ for both the *RS* and *RRS* integrals are defined symmetrically via (1.15). If $a < b$, the Riemann–Stieltjes sum based on a tagged partition $\tau = (\{t_i\}_{i=0}^n, \{s_i\}_{i=1}^n)$ of $[a,b]$, with the integrand and integrator interchanged, will be denoted by

$$S_{RS}(\mathrm{d}h, f;\tau) := \sum_{i=1}^n \big[h(t_i) - h(t_{i-1})\big]\cdot f(s_i).$$

Then the integral $\int_a^b \mathrm{d}h\cdot f$ is defined in the sense of *RS* or *RRS* if and only if the limit (2.8) or (2.9), respectively, exists with $S_{RS}(f,\mathrm{d}h;\tau)$ replaced by $S_{RS}(\mathrm{d}h, f;\tau)$.

**Proposition 2.13.** *The refinement Riemann–Stieltjes integral extends the Riemann–Stieltjes integral.*

*Proof.* For $a = b$, both integrals are 0. For $a < b$, and a partition $\lambda$, a refinement $\kappa$ of $\lambda$ has mesh $|\kappa| \le |\lambda|$, so the conclusion holds. □

**Example 2.14.** Let $f$ be a real-valued function on $[a, b]$ and let $\ell_c := 1_{[c,b]}$ be the indicator function of $[c, b]$ for some $a < c < b$. For any tagged partition $\tau = (\{t_i\}_{i=0}^n, \{s_i\}_{i=1}^n)$ of $[a, b]$, we have $S_{RS}(f, \mathrm{d}\ell_c; \tau) = f(s_i)$ for $t_{i-1} < c \le t_i$ and $s_i \in [t_{i-1}, t_i]$. The integral $(RS) \int_a^b f \,\mathrm{d}\ell_c$ exists and equals $f(c)$ if and only if $f$ is continuous at $c$. Taking $c$ to be a partition point, it follows that the integral $(RRS) \int_a^b f \,\mathrm{d}\ell_c$ exists and equals $f(c)$ if and only if $f$ is left-continuous at $c$. Therefore $(RRS) \int_a^b 1_{[a,c]} \,\mathrm{d}\ell_c$ exists and equals 1, while the same integral in the Riemann–Stieltjes sense does not exist. On the other hand, the integral $(RRS) \int_a^b \ell_c \,\mathrm{d}\ell_c$ also does not exist.

Two functions $f$, $h$ on an interval $[a, b]$ will be said not to have *common discontinuities* if for each $t \in [a, b]$, at least one of $f$ and $h$ is continuous at $t$. Also, $f$ and $h$ will be said to have no *common one-sided discontinuities* on $[a, b]$ if at least one of $f$ and $h$ is left-continuous at each $t \in (a, b]$ and at least one is right-continuous at each $t \in [a, b)$.

**Proposition 2.15.** *Assuming* (1.14), *for any functions $f$ and $h$ from $[a, b]$ into $X$ and $Y$, respectively, we have:*

(*a*) *If $(RS) \int_a^b f \cdot \mathrm{d}h$ exists then $f$ and $h$ have no common discontinuities;*
(*b*) *If $(RRS) \int_a^b f \cdot \mathrm{d}h$ exists then $f$ and $h$ have no common one-sided discontinuities.*

*Proof.* For (*a*) suppose that the integral $(RS) \int_a^b f \cdot \mathrm{d}h$ exists. For each $t \in [a, b]$ and a partition $\kappa = \{x_i\}_{i=1}^n$ of $[a, b]$, there is an index $i$ such that $t \in [x_{i-1}, x_i]$, where we can take $i$ to be unique for partitions with arbitrarily small mesh. Subtracting two Riemann–Stieltjes sums based on the partition $\kappa$ with small enough mesh $|\kappa|$ and with all terms equal except for the $i$th term, one can conclude that $\|[f(y_i') - f(y_i'')] \cdot [h(x_i) - h(x_{i-1})]\|$ is arbitrarily small for any $y_i', y_i'' \in [x_{i-1}, x_i]$. Thus either $f$ or $h$ must be continuous at $t$.

The proof of (*b*) is similar except that one needs to restrict to partitions containing a point $t \in [a, b]$. □

When $f$ or $h$ is continuous, the $RS$ and $RRS$ integrals coincide if either exists, as we will show in Theorem 2.42. Some classical sufficient conditions for existence of the two integrals are expressed using the total variation or the semivariation of the integrator. Let $X$, $Y$, and $Z$ be Banach spaces related by means of a bounded bilinear operator $B$ from $X \times Y$ into $Z$ as in assumption

(1.14) and let $h\colon [a,b] \to Y$ with $-\infty < a < b < +\infty$. The *semivariation* of $h$ on $[a,b]$ is defined by

$$\begin{aligned} w(h;[a,b]) :=& w_B(h;[a,b]) := \sup\Big\{\Big\|\sum_{i=1}^n x_i{\cdot}[h(t_i)-h(t_{i-1})]\Big\|: \\ & \{t_i\}_{i=0}^n \in \mathrm{PP}\,[a,b],\ \{x_i\}_{i=1}^n \subset X,\ \max_i \|x_i\| \le 1\Big\}, \end{aligned} \tag{2.10}$$

where $\mathrm{PP}\,[a,b]$ is the set of all point partitions of $[a,b]$. Then $h$ is said to be of *bounded semivariation* on $[a,b]$ if $w_B(h;[a,b]) < +\infty$. As the notation $w_B(h;[a,b])$ indicates, the semivariation depends on the Banach spaces $X$, $Y$, and $Z$ and the bilinear operator $B\colon (x,y) \mapsto x{\cdot}y$. Recall that the total variation of $h$ on $[a,b]$ is the 1-variation of $h$ on $[a,b]$, that is,

$$\begin{aligned} v(h;[a,b]) &:= \sup\{s_1(h;\kappa)\colon \kappa \in \mathrm{PP}\,[a,b]\} \\ &= \sup\Big\{\sum_{i=1}^n \|h(t_i)-h(t_{i-1})\|\colon \{t_i\}_{i=0}^n \in \mathrm{PP}\,[a,b]\Big\}. \end{aligned}$$

The total variation $v(h;[a,b])$ does not depend on $X$, $Z$ or $B$. It is clear that $w_B(h;[a,b]) \le v(h;[a,b]) \le +\infty$. Also, if $X = Y'$, $Z = \mathbb{R}$ and $B(x,y) = x(y)$, then $w_B(h;[a,b]) = v(h;[a,b])$. However in general, it is possible that $w_B(h;[a,b]) < \infty$ while $v(h;[a,b]) = +\infty$, as the following example shows:

**Example 2.16.** For each $t \in [0,1]$, let $h(t) := 1_{[0,t]}$. Then $h$ is a function from $[0,1]$ into $Y = Z = \ell^\infty[0,1]$, the Banach space of real-valued and bounded functions on $[0,1]$ with the supremum norm. For a point partition $\{t_i\}_{i=0}^n$ of $[0,1]$ and for any finite sequence $\{x_i\}_{i=1}^n \subset X = \mathbb{R}$ of real numbers in $[-1,1]$, we have

$$\Big\|\sum_{i=1}^n x_i[h(t_i)-h(t_{i-1})]\Big\|_{\sup} = \Big\|\sum_{i=1}^n x_i 1_{(t_{i-1},t_i]}\Big\|_{\sup} = \max_i |x_i| \le 1.$$

Thus $h$ has bounded semivariation on $[0,1]$, with $w(h;[0,1]) = 1$. Also, $h$ has unbounded $p$-variation for any $p < \infty$. Indeed, if $0 < p < \infty$ and $\kappa = \{t_i\}_{i=0}^n$ is a point partition of $[0,1]$, then

$$s_p(h;\kappa) = \sum_{i=1}^n \|1_{(t_{i-1},t_i]}\|_{\sup}^p = n.$$

Since $n$ is arbitrary, $v_p(h;[0,1]) = +\infty$. The function $h$ is not regulated on $[0,1]$ since it clearly does not satisfy condition $(b)$ of Theorem 2.1. Moreover, for a real-valued function $f$ on $[0,1]$, the integral $(RS)\int_0^1 f\,dh$ exists if and only if $f$ is continuous. Indeed, if $f$ is continuous, then given $\epsilon > 0$, there is a $\delta > 0$ such that $|f(t)-f(s)| < \epsilon$ whenever $s,t \in [0,1]$ and $|s-t| < \delta$. Let $\tau = (\{t_i\}_{i=0}^n, \{s_i\}_{i=1}^n)$ be a tagged partition of $[0,1]$ with mesh $|\tau| < \delta$.

For each $t \in [0,1]$ there is an $i \in \{1,\ldots,n\}$ such that either $t \in (t_{i-1}, t_i]$, or $t \in [0, t_1]$. In either case $|S_{RS}(f, \mathrm{d}h; \tau)(t) - f(t)| = |f(s_i) - f(t)| < \epsilon$, and so $(RS)\int_0^1 f\,\mathrm{d}h$ exists and equals $f$. Now, if $f$ is not continuous then for some $t \in [0,1]$, the difference $|S_{RS}(f, \mathrm{d}h; \tau)(t) - f(t)|$ cannot be arbitrarily small for all tagged partitions with small enough mesh, proving the claim.

Now we are prepared to prove the following:

**Theorem 2.17.** *Assuming* (1.14), *if $f$ or $h$ is of bounded semivariation and the other is regulated, and they have no common one-sided discontinuities, then* $(RRS)\int_a^b f\cdot\mathrm{d}h$ *exists.*

*Proof.* First suppose that $h$ has bounded semivariation on $[a,b]$ and $f$ is regulated. Given $\epsilon > 0$, by statement (*b*) of Theorem 2.1, there exists a partition $\lambda = \{z_l\}_{l=0}^k$ of $[a,b]$ such that

$$\mathrm{Osc}(f; (z_{l-1}, z_l)) < \epsilon \qquad \text{for } l = 1, \ldots, k. \tag{2.11}$$

Since $f$ and $h$ have no common one-sided discontinuities, there exists a sequence $\mu = \{u_{l-1}, v_l \colon\ l = 1, \ldots, k\} \subset (a,b)$ such that for each $l = 1, \ldots, k$, $z_{l-1} < u_{l-1} < v_l < z_l$,

$$\min\bigl\{\mathrm{Osc}(f; [z_{l-1}, u_{l-1}]), \mathrm{Osc}(h; [z_{l-1}, u_{l-1}])\bigr\} < \epsilon/k \tag{2.12}$$

and

$$\min\bigl\{\mathrm{Osc}(f; [v_l, z_l]), \mathrm{Osc}(h; [v_l, z_l])\bigr\} < \epsilon/k. \tag{2.13}$$

Let $\tau = (\{t_j\}_{j=0}^m, \{s_j\}_{j=1}^m)$ be a tagged refinement of $\lambda \cup \mu$. For each $j = 1, \ldots, m$, let $\tau_j = (\{t_{ji}\}_{i=0}^{n_j}, \{s_{ji}\}_{i=1}^{n_j})$ be any tagged partition of $[t_{j-1}, t_j]$. Then $\cup_j \tau_j$ is an arbitrary tagged refinement of $\tau$. If $t_{j-1} \in \{z_0, \ldots, z_{k-1}\}$ then

$$\begin{aligned} &f(s_j)\cdot[h(t_j) - h(t_{j-1})] - S_{RS}(\tau_j) = \\ &\quad \bigl[f(s_j) - f(t_j)\bigr]\cdot\bigl[h(t_j) - h(t_{j-1})\bigr] \\ &\quad + \sum_{i=1}^{n_j} \bigl[f(t_j) - f(s_{ji})\bigr]\cdot\bigl[h(t_{ji}) - h(t_{j,i-1})\bigr]. \end{aligned}$$

The first product on the right side, say $U_j$, and the $i = 1$ term, say $V_j$, by (2.12), have the bound

$$\|U_j\| + \|V_j\| \le 4(\epsilon/k)\max\{\|f\|_{\sup}, \|h\|_{\sup}\}. \tag{2.14}$$

At most one of $t_{j-1}$ and $t_j$ is a $z_i$. If $t_j \in \{z_1, \ldots, z_k\}$ then

$$\begin{aligned} &f(s_j)\cdot[h(t_j) - h(t_{j-1})] - S_{RS}(\tau_j) = \\ &\quad \bigl[f(s_j) - f(t_{j-1})\bigr]\cdot\bigl[h(t_j) - h(t_{j-1})\bigr] \\ &\quad + \sum_{i=1}^{n_j} \bigl[f(t_{j-1}) - f(s_{ji})\bigr]\cdot\bigl[h(t_{ji}) - h(t_{j,i-1})\bigr]. \end{aligned}$$

The first product on the right side and the $i = n_j$ term have the same bound as in (2.14) by (2.13). We apply the preceding $2k$ representations with the bounds (2.14). We bound the sum of the remaining terms for all $j$ by $\max_{l=1,\dots,k} \mathrm{Osc}(f;(z_{l-1},z_l))w_B(h;[a,b])$, and using (2.11) we get the bound

$$\big\|S_{RS}(\tau) - S_{RS}(\cup_j \tau_j)\big\| < \epsilon w_B(h;[a,b]) + 8\epsilon \max\{\|f\|_{\sup}, \|h\|_{\sup}\}.$$

Since $\epsilon > 0$ is arbitrary, the integral $(RRS)\int_a^b f{\cdot}dh$ exists.

Now suppose that $f$ has bounded semivariation and $h$ is regulated. Given $\epsilon > 0$, choose a partition $\lambda = \{z_l\}_{l=0}^k$ of $[a,b]$ such that (2.11) with $h$ instead of $f$ holds. Then choose a sequence $\mu = \{u_{l-1}, v_l\colon\ l = 1,\dots,k\}$ such that (2.12) and (2.13) hold. Again let $\tau = (\{t_j\}_{j=0}^m, \{s_j\}_{j=1}^m)$ be a tagged refinement of $\lambda \cup \mu$, and for each $j = 1,\dots,m$, let $\tau_j = (\{t_{ji}\}_{i=0}^{n_j}, \{s_{ji}\}_{i=1}^{n_j})$ be a tagged partition of $[t_{j-1}, t_j]$. If $t_j \notin \{z_1,\dots,z_k\}$, then summing by parts we have

$$\begin{aligned} f(s_j){\cdot}[h(t_j) - h(t_{j-1})] - S_{RS}(\tau_j) = \\ \big[f(s_j) - f(s_{j1})\big]{\cdot}[h(t_j) - h(t_{j-1})\big] \\ + \sum_{i=1}^{n_j-1} \big[f(s_{j,i+1}) - f(s_{ji})\big]{\cdot}\big[h(t_{ji}) - h(t_j)\big]. \end{aligned}$$

If also $t_{j-1} \in \{z_0,\dots,z_{k-1}\}$, then for the first product on the right side, say $T_j$, by (2.12), we have the bound

$$\|T_j\| \le 2(\epsilon/k)\max\{\|f\|_{\sup}, \|h\|_{\sup}\}. \tag{2.15}$$

The norm of the sum of the terms $T_j$ for which neither $t_{j-1}$ nor $t_j$ is a $z_i$ has the bound $\epsilon w_B(f;[a,b])$ by (2.11) with $h$ instead of $f$. If $t_j \in \{z_1,\dots,z_k\}$, then summing by parts we have

$$\begin{aligned} f(s_j){\cdot}[h(t_j) - h(t_{j-1})] - S_{RS}(\tau_j) = \\ \big[f(s_j) - f(s_{j,n_j})\big]{\cdot}[h(t_j) - Ah(t_{j-1})\big] \\ + \sum_{i=1}^{n_j-1} \big[f(s_{j,i+1}) - f(s_{ji})\big]{\cdot}\big[h(t_{ji}) - h(t_{j-1})\big]. \end{aligned}$$

The first product on the right side has the same bound as in (2.15) by (2.13). Applying the preceding representations for each $j = 1,\dots,m$, using the bounds (2.15) and (2.11) with $h$ instead of $f$, we get the bound

$$\big\|S_{RS}(\tau) - S_{RS}(\cup_j \tau_j)\big\| \le 4\epsilon \max\{\|f\|_{\sup}, \|h\|_{\sup}\} + 2\epsilon w_B(f;[a,b]).$$

As in the first part of the proof the integral $(RRS)\int_a^b f{\cdot}dh$ exists, proving the theorem. $\square$

## 2.3 The Refinement Young–Stieltjes and Kolmogorov Integrals

Let $f\colon [a,b] \to X$ and $h \in \mathcal{R}([a,b];Y)$. Recall from Section 1.4 that for a point partition $\{t_i\}_{i=0}^n$ of $[a,b]$ with $a < b$, a tagged partition $(\{t_i\}_{i=0}^n, \{s_i\}_{i=1}^n)$ is called a *Young tagged point partition* if $t_{i-1} < s_i < t_i$ for each $i = 1, \ldots, n$. Given a Young tagged point partition $\tau = (\{t_i\}_{i=0}^n, \{s_i\}_{i=1}^n)$, define the *Young–Stieltjes sum* $S_{YS}(f, \mathrm{d}h; \tau)$ based on $\tau$ by, recalling the definition (2.1),

$$\begin{aligned} &S_{YS}(f, \mathrm{d}h; \tau) \\ &:= \sum_{i=0}^{n} \big(f{\cdot}\Delta^{\pm}_{[a,b]}h\big)(t_i) + \sum_{i=1}^{n} f(s_i){\cdot}\big[h(t_i-) - h(t_{i-1}+)\big] \\ &= \sum_{i=1}^{n} \Big\{[f{\cdot}\Delta^{+}h](t_{i-1}) + f(s_i){\cdot}\big[h(t_i-) - h(t_{i-1}+)\big] + [f{\cdot}\Delta^{-}h](t_i)\Big\}. \end{aligned} \tag{2.16}$$

The *refinement Young–Stieltjes* or *RYS* integral $(RYS)\int_a^b f{\cdot}\mathrm{d}h$ is defined as 0 if $a = b$ or as

$$(RYS)\int_a^b f{\cdot}\mathrm{d}h \ := \ \lim_{\tau} S_{YS}(f, \mathrm{d}h; \tau)$$

if $a < b$, provided the limit exists in the refinement sense. The integral $(RYS)\int_a^b \mathrm{d}f{\cdot}h$ is defined, if it exists, via (1.15).

**Proposition 2.18.** *Let $f\colon [a,b] \to X$ and $h \in \mathcal{R}([a,b];Y)$. If $a < b$, given a Young tagged point partition $\tau = (\kappa, \xi)$ of $[a,b]$, the Young–Stieltjes sum $S_{YS}(f, \mathrm{d}h; \tau)$ can be approximated arbitrarily closely by Riemann–Stieltjes sums $S_{RS}(f, \mathrm{d}h; \tilde{\tau})$ based on tagged refinements $\tilde{\tau}$ of $\kappa$ such that all tags $\xi$ of $\tau$ are tags of $\tilde{\tau}$. Thus, if $(RRS)\int_a^b f{\cdot}\mathrm{d}h$ exists then so does $(RYS)\int_a^b f{\cdot}\mathrm{d}h$, and the two are equal.*

*Proof.* The last conclusion holds if $a = b$ with both integrals defined as 0, so let $a < b$. For a Young tagged point partition $\tau = (\kappa, \xi) = (\{x_i\}_{i=0}^n, \{y_i\}_{i=1}^n)$ of $[a,b]$, take a set $\mu = \{u_{i-1}, v_i\}_{i=1}^n \subset (a,b)$ such that $x_0 < u_0 < y_1 < v_1 < x_1 < \cdots < x_{n-1} < u_{n-1} < y_n < v_n < x_n$. Let $\tilde{\tau} := (\kappa \cup \mu, \kappa \cup \xi)$, where each $x_i$, $i = 1, \ldots, n$, is the tag for both $[v_i, x_i]$ and $[x_i, u_i]$. For $i = 1, \ldots, n$, letting $u_{i-1} \downarrow x_{i-1}$ and $v_i \uparrow x_i$, it follows that the Riemann–Stieltjes sums $S_{RS}(f, \mathrm{d}h; \tilde{\tau})$ converge to the Young–Stieltjes sum $S_{YS}(f, \mathrm{d}h; \tau)$ and the conclusions follow. □

**Example 2.19.** As in Example 2.14, for $a < c < b$, let $\ell_c$ be the indicator function of $[c,b]$ and let $f$ be a real-valued function, both on $[a,b]$. For each Young tagged point partition $\tau = (\kappa, \xi)$ such that $c \in \kappa$, we have $S_{YS}(f, \mathrm{d}\ell_c; \tau) = f(c)$. Therefore the integral $(RYS)\int_a^b f\,\mathrm{d}\ell_c$ exists and has value $f(c)$. In particular, $(RYS)\int_a^b \ell_c\, d\ell_c$ exists and has value 1, while the $RRS$ integral does not exist, as shown in Example 2.14. More generally, for any real number $r$, let $\ell_c^r(x) := \ell_c(x)$ for $x \neq c$ and $\ell_c^r(c) := r$, so that $\ell_c^1 \equiv \ell_c$. Then $(RYS)\int_a^b f\, d\ell_c^r$ exists and has value $f(c)$ for any $r$.

In fact, the $RYS$ integral $\int_a^b f\, dh$ does not depend on the values of the integrator $h$ at its jump points in $(a, b)$. This is because the value of the Young–Stieltjes sum (2.16) does not depend on the values of $h$ at jump points in the open interval $(a, b)$. The same is not true for the $RRS$ integral. Indeed, by changing the value of the integrator $h$ at a jump point $t$ of both $f$ and $h$ one can make $h$ discontinuous from the left or right at $t$ and so destroy the necessary condition of Proposition 2.15. The $RS$ integral, like the $RYS$ integral, does not depend on values of the integrator at its discontinuity points. This holds because the $RS$ integral is defined only under restrictive conditions. Specifically, by Proposition 2.15, if $(RS)\int_a^b f\, dh$ exists and $h$ has a jump at $t \in (a, b)$ then $f$ must be continuous at $t$. For $u \in \mathbb{R}$, let $h_u := h$ on $[a, t) \cup (t, b]$ and $h_u(t) := u$. Then for two Riemann–Stieltjes sums we have

$$S_{RS}(f, dh; (\kappa, \xi)) - S_{RS}(f, dh_u; (\kappa, \xi)) = \begin{cases} [f(s') - f(s'')]\cdot[h(t) - u] & \text{if } t \in \kappa, \\ 0 & \text{if } t \notin \kappa, \end{cases}$$

where $s' \in [t - \delta, t]$ and $s'' \in [t, t + \delta]$ for $\delta := |\kappa|$. Due to continuity of $f$ at $t$ the preceding difference can be made arbitrarily small if the mesh $|\kappa|$ is small enough.

**Theorem 2.20.** *Assuming* (1.14), *if $f$ or $h$ is of bounded semivariation and the other is regulated, then* $(RYS)\int_a^b f\cdot dh$ *exists.*

*Proof.* We can assume $a < b$. First suppose that $h$ has bounded semivariation on $[a, b]$. Given $\epsilon > 0$, by statement $(b)$ of Theorem 2.1, there exists a partition $\lambda = \{z_l\}_{l=0}^k$ of $[a, b]$ such that

$$\operatorname{Osc}(f; (z_{l-1}, z_l)) < \epsilon \qquad \text{for } l = 1, \ldots, k. \tag{2.17}$$

Let $\tau = (\{t_j\}_{j=0}^m, \{s_j\}_{j=1}^m)$ be a Young tagged point partition which is a refinement of $\lambda$. For each $j = 1, \ldots, m$, let $\tau_j = (\{t_{ji}\}_{i=0}^{n_j}, \{s_{ji}\}_{i=1}^{n_j})$ be a Young tagged point partition of $[t_{j-1}, t_j]$. Then $\cup_j \tau_j$ is a Young tagged refinement of $\tau$ and

$$\begin{aligned} &S_{YS}(f, dh; \tau) - S_{YS}(f, dh; \cup_j \tau_j) \\ &= \textstyle\sum_{j=1}^m \Big\{ \displaystyle\sum_{i=1}^{n_j} [f(s_j) - f(s_{ji})] \cdot [h(t_{ji}-) - h(t_{j,i-1}+)] \\ &\qquad + \sum_{i=1}^{n_j - 1} [f(s_j) - f(t_{ji})] \cdot [h(t_{ji}+) - h(t_{ji}-)] \Big\}. \end{aligned}$$

Let $t_{j,i-1} < v_{j,2i-1} < v_{j,2i} < t_{ji}$ for $i = 1, \ldots, n_j$ and $j = 1, \ldots, m$. Consider sums

$$\sum_{r=2}^{2n_j} x_{j,r} \cdot [h(v_r) - h(v_{r-1})],$$

where $x_{j,2i} = f(s_j) - f(s_{ji})$ and $x_{j,2i+1} = f(s_j) - f(t_{ji})$ for $i = 1, \dots, n_j$ except that for $i = n_j$, $x_{j,2i+1} = 0$. Letting $v_{j,2i-1} \downarrow t_{j,i-1}$ and $v_{j,2i} \uparrow t_{ji}$, by (2.17), we get the bound

$$\left\| S_{YS}(f, \mathrm{d}h; \tau) - S_{YS}(f, \mathrm{d}h; \cup_j \tau_j) \right\| < 2\epsilon w_B(h; [a, b]).$$

Given any two Young tagged point partitions $\tau_1$ and $\tau_2$ of $[a, b]$, there exists a Young tagged refinement $\tau_3$ of both and

$$\|S_{YS}(\tau_1) - S_{YS}(\tau_2)\| \le \|S_{YS}(\tau_1) - S_{YS}(\tau_3)\| + \|S_{YS}(\tau_3) - S_{YS}(\tau_2)\|.$$

Thus by the Cauchy test under refinement, $(RYS) \int_a^b f \cdot dh$ exists.

Now suppose that $f$ has bounded semivariation and $h$ is regulated. With the same notation for Young tagged point partitions $\tau$ and $\tau_j$, $j = 1, \dots, m$, we have

$$\begin{aligned} &S_{YS}(f, \mathrm{d}h; \tau) - S_{YS}(f, \mathrm{d}h; \cup_j \tau_j) \\ &= \sum_{j=1}^{m} [f(s_j) - f(s_{j1})] \cdot [h(t_j-) - h(t_{j-1}+)] + \sum_{j=1}^{m} d_j, \end{aligned}$$

where for each $j = 1, \dots, m$,

$$\begin{aligned} d_j &:= (f \cdot \Delta^+ h)(t_{j-1}) + f(s_{j1}) \cdot [h(t_j-) - h(t_{j-1}+)] + (f \cdot \Delta^- h)(t_j) \\ &\quad - S_{YS}(f, \mathrm{d}h; \tau_j) \\ &= f(s_{j1}) \cdot [h(t_j-) - h(t_{j-1}+)] - f(s_{j,n_j}) \cdot h(t_j-) + f(s_{j1}) \cdot h(t_{j-1}+) \\ &\quad + \sum_{i=1}^{n_j-1} \Big\{ [f(t_{ji}) - f(s_{ji})] \cdot h(t_{ji}-) + [f(s_{j,i+1}) - f(t_{ji})] \cdot h(t_{ji}+) \Big\} \\ &= \sum_{i=1}^{n_j-1} \Big\{ [f(t_{ji}) - f(s_{ji})] \cdot [h(t_{ji}-) - h(t_j-)] \\ &\qquad + [f(s_{j,i+1}) - f(t_{ji})] \cdot [h(t_{ji}+) - h(t_j-)] \Big\}. \end{aligned}$$

By Theorem 2.1(*b*), given $\epsilon > 0$ one can choose a partition $\lambda = \{z_l\}_{l=1}^{k}$ of $[a, b]$ such that (2.17) with $h$ instead of $f$ holds. If the Young tagged point partition $\tau$ is a refinement of $\lambda$, then by approximations as in the first half of the proof, we get the bound

$$\left\| S_{YS}(f, \mathrm{d}h; \tau) - S_{YS}(f, \mathrm{d}h; \cup_j \tau_j) \right\| < 3\epsilon w_B(f; [a, b]).$$

As in the first part of the proof the integral $(RYS) \int_a^b f \cdot \mathrm{d}h$ exists by the Cauchy test under refinement, proving the theorem. □

*The Kolmogorov integral*

Let $f$ be an $X$-valued function on a nonempty interval $J$, which may be a singleton, and let $\mu$ be a $Y$-valued interval function on $J$. For a tagged interval partition $\mathcal{T} = (\{A_i\}_{i=1}^n, \{s_i\}_{i=1}^n)$ of $J$, define the *Kolmogorov sum* $S_K(J,\mathcal{T}) = S_K(f, \mathrm{d}\mu; J, \mathcal{T})$ based on $\mathcal{T}$ by

$$S_K(f, \mathrm{d}\mu; J, \mathcal{T}) \ := \ \sum_{i=1}^{n} f(s_i)\cdot\mu(A_i). \tag{2.18}$$

For an additive interval function $\mu$ on $J$ and an interval $A \subset J$, the *Kolmogorov integral* $⨎_A f\cdot\mathrm{d}\mu$ is defined as 0 if $A = \emptyset$, or if $A$ is nonempty, as the limit

$$⨎_A f\cdot\mathrm{d}\mu \ := \ \lim_{\mathcal{T}} S_K(f, \mathrm{d}\mu; A, \mathcal{T})$$

if it exists in the refinement sense, that is, $⨎_A f\cdot\mathrm{d}\mu = z \in Z$ if for every $\epsilon > 0$ there is an interval partition $\mathcal{B}$ of $A$ such that for every refinement $\mathcal{A}$ of $\mathcal{B}$ and every tagged interval partition $\mathcal{T} = (\mathcal{A}, \xi)$, $\|S_K(f, \mathrm{d}\mu; A, \mathcal{T}) - z\| < \epsilon$. Integrals of the form $⨎_A \mathrm{d}\nu\cdot h$ are defined, if they exist, via (1.16).

If $A = \{a\} = [a,a]$ is a singleton then $(\{a\},\{a\})$ is the only tagged interval partition of $A$, and so the Kolmogorov integral $⨎_A f\cdot\mathrm{d}\mu$ always exists and equals $f(a)\cdot\mu(\{a\})$. Note that the Kolmogorov integral over a singleton need not be 0, whereas for the integrals with respect to point functions considered in this chapter, such integrals are 0.

**Theorem 2.21.** *Let $f$ be an $X$-valued function on a nonempty interval $J$, and let $\mu$ be a $Y$-valued additive interval function on $J$. For $A, A_1, A_2 \in \mathfrak{I}(J)$ such that $A = A_1 \cup A_2$ and $A_1 \cap A_2 = \emptyset$, $⨎_A f\cdot\mathrm{d}\mu$ exists if and only if both $⨎_{A_1} f\cdot\mathrm{d}\mu$ and $⨎_{A_2} f\cdot\mathrm{d}\mu$ exist, and then*

$$⨎_A f\cdot\mathrm{d}\mu \ = \ ⨎_{A_1} f\cdot\mathrm{d}\mu + ⨎_{A_2} f\cdot\mathrm{d}\mu. \tag{2.19}$$

*In particular, if the integral $⨎_J f\cdot\mathrm{d}\mu$ is defined, then $\mathfrak{I}(J) \ni A \mapsto ⨎_A f\cdot\mathrm{d}\mu$ is a $Z$-valued additive interval function on $J$.*

*Proof.* We can assume that $A_1$ and $A_2$ are nonempty. Let $⨎_A f\cdot\mathrm{d}\mu$ be defined for a given $A \in \mathfrak{I}(J)$ and let $B$ be a nonempty subinterval of $A$. To show that $⨎_B f\cdot\mathrm{d}\mu$ is defined we use the Cauchy test. Given any two tagged interval partitions $\mathcal{T}_1$, $\mathcal{T}_2$ of $B$, let $\mathcal{T}_1'$, $\mathcal{T}_2'$ be extensions of $\mathcal{T}_1$, $\mathcal{T}_2$, respectively, to tagged interval partitions of $A$ having the same subintervals of $A \setminus B$ and tags for them. Then

$$S_K(B, \mathcal{T}_1) - S_K(B, \mathcal{T}_2) \ = \ S_K(A, \mathcal{T}_1') - S_K(A, \mathcal{T}_2').$$

By assumption, the norm of the right side is small if $\mathcal{T}_1'$ and $\mathcal{T}_2'$ are both refinements of a suitable partition $\mathcal{B}$ of $A$. Taking the trace $\mathcal{B}_B$ of $\mathcal{B}$ on $B$, it follows that the norm of the left side is small for tagged interval partitions $\mathcal{T}_1$, $\mathcal{T}_2$ which are refinements of $\mathcal{B}_B$. Thus the integral $\ddagger_B f{\cdot}\mathrm{d}\mu$ exists by the Cauchy test.

Let $A, A_1, A_2 \in \mathfrak{I}(J)$ be such that $A = A_1 \cup A_2$ and $A_1 \cap A_2 = \emptyset$. Let $\mathcal{T}_1$ and $\mathcal{T}_2$ be tagged interval partitions of $A_1$ and $A_2$, respectively. Then $\mathcal{T} := \mathcal{T}_1 \cup \mathcal{T}_2$ is a tagged interval partition of $A$ and

$$S_K(A,\mathcal{T}) \;=\; S_K(A_1,\mathcal{T}_1) + S_K(A_2,\mathcal{T}_2).$$

Thus if the two integrals on the right side of (2.19) exist then the integral on the left side exists, and the equality holds. The converse follows from the first part of the proof, proving the theorem. □

The following shows that the interval function $\ddagger_A f{\cdot}\mathrm{d}\mu$, $A \in \mathfrak{I}(J)$, is upper continuous if $f$ is bounded and $\mu$ is upper continuous in addition to the assumptions of the preceding theorem.

**Proposition 2.22.** *Let $f$ be a bounded $X$-valued function on a nonempty interval $J$, and let $\mu$ be a $Y$-valued upper continuous additive interval function on $J$. If the integral $\ddagger_J f{\cdot}\mathrm{d}\mu$ is defined, then $\mathfrak{I}(J) \ni A \mapsto \ddagger_A f{\cdot}\mathrm{d}\mu$ is a $Z$-valued upper continuous additive interval function on $J$.*

*Proof.* The interval function $\ddagger_A f{\cdot}\mathrm{d}\mu$, $A \in \mathfrak{I}(J)$, is additive by Theorem 2.21. Let $A_1, A_2, \ldots \in \mathfrak{I}(J)$ be such that $A_n \downarrow \emptyset$, and let $s_n \in A_n$ for each $n$. Since $f$ is bounded and $\mu$ is upper continuous, it is enough to prove that

$$D_n := \ddagger_{A_n} f{\cdot}\mathrm{d}\mu - f(s_n){\cdot}\mu(A_n) \to 0 \quad \text{as } n \to \infty. \tag{2.20}$$

Let $J = [\![a,b]\!]$. For some $u \in J$ and all sufficiently large $n$, either $A_n$ is left-open at $u \in [\![a,b)$ or $A_n$ is right-open at $u \in (a,b]\!]$. By symmetry consider only the first case. Let $\epsilon > 0$. There exists a tagged interval partition $\mathcal{T}_0$ of $(u,b]\!]$ such that any two Kolmogorov sums based on tagged refinements of $\mathcal{T}_0$ differ by at most $\epsilon$. Let $n_0$ be such that $A_{n_0}$ is a subset of the first interval in $\mathcal{T}_0$, and let $n \geq n_0$. Choose a Kolmogorov sum based on a tagged interval partition $\mathcal{T}_n'$ of $A_n$ within $\epsilon$ of $\ddagger_{A_n} f{\cdot}\mathrm{d}\mu$. Let $\mathcal{T}_1$ and $\mathcal{T}_2$ be two tagged refinements of $\mathcal{T}_0$ which coincide with $\mathcal{T}_n'$ and $(A_n, \{s_n\})$, respectively, when restricted to $A_n$, and are equal outside of $A_n$. Then the norm of $D_n$ does not exceed

$$\Big\| \ddagger_{A_n} f{\cdot}\,\mathrm{d}\mu - S_K(A_n,\mathcal{T}_n')\Big\| + \big\|S_K((u,b]\!],\mathcal{T}_1) - S_K((u,b]\!],\mathcal{T}_2)\big\| \leq 2\epsilon.$$

Since this is true for each $n \geq n_0$, (2.20) holds, proving the proposition. □

The following is a consequence of Proposition 2.6 and the preceding proposition.

**Corollary 2.23.** *Let $f$ be a bounded $X$-valued function on an interval $J := [\![a,b]\!]$ with $a < b$, and let $\mu$ be a $Y$-valued upper continuous additive interval function on $J$. If the integral $⨎_J f{\cdot}\mathrm{d}\mu$ is defined, then for each $u, v \in J$ such that $u < v$,*

$$\lim_{t\uparrow v} ⨎_{[\![a,t]\!]} f{\cdot}\mathrm{d}\mu = ⨎_{[\![a,v)} f{\cdot}\mathrm{d}\mu, \qquad \lim_{t\downarrow u} ⨎_{[\![a,t]\!]} f{\cdot}\mathrm{d}\mu = ⨎_{[\![a,u]} f{\cdot}\mathrm{d}\mu,$$

$$\lim_{t\downarrow u} ⨎_{[\![t,b]\!]} f{\cdot}\mathrm{d}\mu = ⨎_{(u,b]\!]} f{\cdot}\mathrm{d}\mu, \qquad \lim_{t\uparrow v} ⨎_{[\![t,b]\!]} f{\cdot}\mathrm{d}\mu = ⨎_{[v,b]\!]} f{\cdot}\mathrm{d}\mu.$$

Next we show that in defining the Kolmogorov integral $⨎_J f{\cdot}\mathrm{d}\mu$ with respect to an *upper continuous* additive interval function $\mu$ it is enough to take partitions consisting of intervals in $\mathfrak{I}_{os}(J)$ (open intervals and singletons). We can assume that $J$ is nondegenerate. First consider a closed interval $J = [a,b]$ with $a < b$. For a point partition $\{t_i\}_{i=0}^n$ of $[a,b]$, the collection

$$\Big\{\{t_0\}, (t_0,t_1), \{t_1\}, (t_1,t_2), \dots, (t_{n-1},t_n), \{t_n\}\Big\} \tag{2.21}$$

of subintervals of $[a,b]$ is called the *Young interval partition of* $[a,b]$ *associated to* $\{t_i\}_{i=0}^n$. A *tagged Young interval partition of* $[a,b]$ is a set (2.21) together with the tags $t_0$, $s_1$, $t_1$, $s_2$, $\dots$, $s_n$, $t_n$, where $t_{i-1} < s_i < t_i$ for $i = 1, \dots, n$. For notational simplicity the tagged Young interval partition will be denoted by $(\{(t_{i-1},t_i)\}_{i=1}^n, \{s_i\}_{i=1}^n)$.

Now a *Young interval partition* of any nonempty interval $J$ is any interval partition consisting of singletons and open intervals. For a closed interval this coincides with the previous definition. A *tagged Young interval partition* of any interval $J$ will be a Young interval partition together with a member (tag) of each interval in the partition. A tagged Young interval partition of any nondegenerate interval is given by specifying the open intervals in it and their tags, just as when the interval is closed. Let $\{t_i\}_{i=0}^n$ be a partition of $[u,v]$ with $u < v$, so that $u = t_0 < t_1 < \cdots < t_n = v$. Then the *associated Young interval partition of* $[\![u,v]\!]$ is the set (2.21) if $[\![u,v]\!] = [u,v]$, or is the set (2.21) except for the left and/or right singletons $\{u\}$ or $\{v\}$ respectively if $[\![u,v]\!]$ is a left and/or right open interval. Given a Young interval partition (2.21) of $[a,b]$ and any subinterval $[\![u,v]\!]$ of $[a,b]$, the *trace Young interval partition of* $[\![u,v]\!]$ is the one formed by nonempty intervals obtained by intersecting each subinterval in (2.21) with $(u,v)$ and adjoining the endpoints $\{u\}$ and/or $\{v\}$ if $[\![u,v]\!]$ is a left and/or right closed interval.

Let $J = [\![a,b]\!]$ with $a < b$, let $f$ be an $X$-valued function on $J$, and let $\mu$ be a $Y$-valued additive interval function on $J$. For a tagged Young interval partition $\mathcal{T} = (\{(t_{i-1},t_i)\}_{i=1}^n, \{s_i\}_{i=1}^n)$ of $J$, let

$$\begin{aligned} S_{YS}(f,\mathrm{d}\mu;J,\mathcal{T}) &:= S_K(f,\mathrm{d}\mu;J,\mathcal{T}) \\ &= \sum_{i=1}^n f(s_i){\cdot}\mu((t_{i-1},t_i)) + \sum_{i=0}^n f(t_i){\cdot}\mu(\{t_i\}\cap J). \end{aligned} \tag{2.22}$$

Assuming that $\mu$ is upper continuous, this sum can be approximated by Riemann–Stieltjes sums as follows. Let $h := R_{\mu,a}$ be the function defined by (2.3) and let $\{u_{i-1}, v_i\}_{i=1}^n \subset (a,b)$ be such that $t_0 < u_0 < s_1 < v_1 < t_1 < \cdots < t_{n-1} < u_{n-1} < s_n < v_n < t_n$. Letting $\kappa := J \cap \{t_i\}_{i=0}^n$, $\tau := (\kappa \cup \{u_{i-1}, v_i\}_{i=1}^n, \kappa \cup \{s_i\}_{i=1}^n)$ is a tagged partition of $[c,d] \subset J$, where $c = a$ if $a \in J$, or $c = u_0$ otherwise, and $d = b$ if $b \in J$, or $d = v_n$ otherwise, and each $t_i$ for $i = 1, \ldots, n-1$ is the tag both for $[v_i, t_i]$ and for $[t_i, u_i]$. Next, as in the proof of Proposition 2.18, letting $u_{i-1} \downarrow t_{i-1}$ and $v_i \uparrow t_i$ for each $i = 1, \ldots, n$, the Riemann–Stieltjes sum $S_{RS}(f, \mathrm{d}h; \tau)$ converges to $S_{YS}(f, \mathrm{d}\mu; J, \mathcal{T})$. Thus we have proved the following:

**Lemma 2.24.** *For $f$, $\mu$, $h$, $J$ and $\mathcal{T}$ as above, the sum $S_{YS}(f, \mathrm{d}\mu; J, \mathcal{T})$ can be approximated arbitrary closely by Riemann–Stieltjes sums $S_{RS}(f, \mathrm{d}h; \tau)$ based on tagged partitions $\tau$ of subintervals of $J$ such that all tags of $\mathcal{T}$ are tags of $\tau$.*

Recall that the class of all $Y$-valued additive and upper continuous interval functions on a nonempty interval $J$ is denoted by $\mathcal{AI}(J;Y)$. Now we can show that in defining the Kolmogorov integral with respect to an additive upper continuous interval function it is enough to take Young interval partitions.

**Proposition 2.25.** *Let $J$ be a nondegenerate interval, let $\mu \in \mathcal{AI}(J;Y)$, and let $f\colon J \to X$. The integral $⨍_J f \cdot \mathrm{d}\mu$ is defined if and only if the limit $\lim_{\mathcal{T}} S_{YS}(f, \mathrm{d}\mu; J, \mathcal{T})$ exists in the refinement sense, and then the integral equals the limit.*

*Proof.* It is enough to prove the "if" part. Let $J = [a,b]$ and let $\epsilon > 0$. Then there are a $z \in Z$ and a Young interval partition $\mathcal{B}$ of $[a,b]$ such that for any Young interval partition $\mathcal{Y}$ which is a refinement of $\mathcal{B}$ and any tagged Young interval partition $\mathcal{T} = (\mathcal{Y}, \xi)$, $\|z - S_{YS}(f, \mathrm{d}\mu; J, \mathcal{T})\| < \epsilon$. Let $\mathcal{A}$ be a refinement of $\mathcal{B}$ consisting of arbitrary subintervals of $[a,b]$. Let $\{u_i\}_{i=0}^n$ and $\{v_j\}_{j=0}^m$ be the sets of endpoints of intervals in $\mathcal{A}$ and $\mathcal{B}$, respectively. For each $i \in \{0, \ldots, n\}$ define a sequence $\{t_{ik}\}_{k \geq 1}$ as follows. If $\{u_i\}$ is a singleton in $\mathcal{A}$ then let $t_{ik} := u_i$ for all $k$. If an interval $[\![\cdot, u_i]$ is in $\mathcal{A}$ let $t_{ik} \downarrow u_i$, or if some $[u_i, \cdot]\!] \in \mathcal{A}$ let $t_{ik} \uparrow u_i$, where in either case $t_{ik}$ are not atoms of $\mu$. This can be done by statement (c) of Proposition 2.6. For $k$ large enough, $a = t_{0k} < t_{1k} < \cdots < t_{nk} = b$. Then there is a unique Young interval partition $\mathcal{Y}_k$ of $[a,b]$ associated to $\{t_{ik}\}_{i=0}^n$. Since each singleton $\{v_j\}$ equals $\{u_i\} \in \mathcal{A}$ for some $i$, $\mathcal{Y}_k$ is a refinement of $\mathcal{B}$. The intervals in $\mathcal{Y}_k$ converge to the intervals in $\mathcal{A}$ as $k \to \infty$, except for singletons in $\mathcal{Y}_k$ which are not atoms of $\mu$, and hence do not contribute to sums. Let $\eta$ be a set of tags for $\mathcal{A}$. Then each tag $s$ of $\eta$ is eventually in the interval of $\mathcal{Y}_k$ corresponding to the interval of $\mathcal{A}$ containing $s$. For such $k$, to form a set $\eta_k$ of tags for $\mathcal{Y}_k$, take $\eta$ and adjoin to it, for each $i$ (if any exist) with $\{u_i\} \notin \mathcal{A}$, a $t_{ik}$ with $\mu(\{t_{ik}\}) = 0$.

Thus there are Young–Stieltjes sums $S_{YS}(f, \mathrm{d}\mu; J, (\mathcal{Y}_k, \eta_k))$ converging to $S_K(f, \mathrm{d}\mu; J, (\mathcal{A}, \eta))$, proving the theorem in the case $J = [a,b]$. If $J = (a,b]\!]$

and/or $J = [\![a,b)$ then the proof is the same except that in the first case, $(u_0, c]\!] \in \mathcal{A}$ for some $c$ and we define $t_{0k} := u_0 = a$ for all $k$, while in the second case, $[\![d, u_n) \in \mathcal{A}$ for some $d$ and we define $t_{nk} := u_n = b$ for all $k$. The proof of Proposition 2.25 is complete. □

For the next statement recall that $R_{\mu,a}$ is defined by (2.3).

**Corollary 2.26.** *Let $J = [\![a,b]\!]$ with $a < b$, let $\mu \in \mathcal{AI}(J;Y)$, let $h := R_{\mu,a}$, and let $f\colon J \to X$. Also for each $t \in [a,b]$, let*

$$\tilde{h}(t) := \begin{cases} h(t) & \text{if } t \in J, \\ h(a+) & \text{if } t = a \notin J, \\ h(b-) & \text{if } t = b \notin J, \end{cases} \quad \text{and} \quad \tilde{f}(t) := \begin{cases} f(t) & \text{if } t \in J, \\ 0 & \text{if } t = a \notin J, \\ 0 & \text{if } t = b \notin J. \end{cases} \tag{2.23}$$

*Then $⨎_J f{\cdot}\mathrm{d}\mu = (RYS)\int_a^b \tilde{f}{\cdot}\mathrm{d}\tilde{h}$ if either side is defined.*

*Proof.* By Proposition 2.6($f$), $h$ is regulated, and so $\tilde{h}$ is well defined. By Theorem 2.8($b$), $\mu = \mu_h$ defined by (2.2). Since there is a one-to-one correspondence between tagged Young partitions $\tau$ of $[a,b]$ and tagged Young interval partitions $\mathcal{T}$ of $J$ with $S_{YS}(\tilde{f}, \mathrm{d}\tilde{h}; \tau) = S_{YS}(f, \mathrm{d}\mu_h; J, \mathcal{T})$, the conclusion follows by Proposition 2.25. □

Given a regulated function $h$ on $[a,b]$, by Theorem 2.7 there is an additive, upper continuous interval function $\mu_h := \mu_{h,[a,b]}$ corresponding to $h$ defined by (2.2) if $a < b$ or as 0 if $a = b$. Let

$$⨎_a^b f{\cdot}\mathrm{d}h := ⨎_{[a,b]} f{\cdot}\mathrm{d}\mu_{h,[a,b]}, \tag{2.24}$$

provided the Kolmogorov integral is defined.

**Proposition 2.27.** *Let $h$ be a $Y$-valued regulated function on $[a,b]$, and let $f\colon [a,b] \to X$. Then $⨎_a^b f{\cdot}\mathrm{d}h = (RYS)\int_a^b f{\cdot}\mathrm{d}h$ if either side is defined.*

*Proof.* We can assume that $a < b$. Since there is a one-to-one correspondence between tagged Young partitions $\tau$ and tagged Young interval partitions $\mathcal{T}$ with $S_{YS}(f, \mathrm{d}h; \tau) = S_{YS}(f, \mathrm{d}\mu_h; [a,b], \mathcal{T})$, the conclusion again follows by Proposition 2.25. □

In the special case that the interval function $\mu$ is a Borel measure on $[a,b]$, we will show that the Kolmogorov integral $⨎_{[a,b]} f\,\mathrm{d}\mu$ agrees with the Lebesgue integral $\int_{[a,b]} f\,\mathrm{d}\mu$ whenever both are defined.

**Proposition 2.28.** *Let $\mu$ be a finite positive measure on the Borel sets of $[a,b]$, and let $f$ be a real-valued $\mu$-measurable function on $[a,b]$. Then $⨍_{[a,b]} f\,\mathrm{d}\mu = \int_{[a,b]} f\,\mathrm{d}\mu$ whenever both integrals exist.*

*Proof.* We can assume that $a < b$. Suppose that the two integrals exist and let $\epsilon > 0$. Since $\mu$ is an additive and upper continuous interval function on $[a,b]$, by Proposition 2.25, there is a Young interval partition $\lambda$ of $[a,b]$ such that if $(\{(t_{i-1},t_i)\}_{i=1}^n, \{s_i\}_{i=1}^n)$ is a tagged Young partition which is a refinement of $\lambda$, then

$$\Big|\sum_{i=0}^n f(t_i)\mu(T_i) + \sum_{i=1}^n f(s_i)\mu(T_{n+i}) - ⨍_{[a,b]} f\,\mathrm{d}\mu\Big| < \epsilon, \tag{2.25}$$

where $T_i := \{t_i\}$ for $i = 0,\dots,n$ and $T_{n+i} := (t_{i-1},t_i)$ for $i = 1,\dots,n$. Then $\{T_0,\dots,T_{2n}\}$ is a decomposition of $[a,b]$ into disjoint measurable sets. For $i = 1,\dots,n$, let $m_i := \inf\{f(s)\colon\ s \in T_{n+i}\}$ and $M_i := \sup\{f(s)\colon\ s \in T_{n+i}\}$. Letting $f(s_i) \downarrow m_i$ and $f(s_i) \uparrow M_i$ in (2.25) for each $i = 1,\dots,n$, we get that (2.25) holds with each $f(s_i)$ replaced by $m_i$, or each $f(s_i)$ replaced by $M_i$. Therefore, it follows that

$$\begin{aligned}
-\epsilon &\le \sum_{i=0}^n f(t_i)\mu(T_i) + \sum_{i=1}^n m_i\mu(T_{n+i}) - ⨍_{[a,b]} f\,\mathrm{d}\mu \\
&\le \int_{[a,b]} f\,\mathrm{d}\mu - ⨍_{[a,b]} f\,\mathrm{d}\mu \\
&\le \sum_{i=0}^n f(t_i)\mu(T_i) + \sum_{i=1}^n M_i\mu(T_{n+i}) - ⨍_{[a,b]} f\,\mathrm{d}\mu \le \epsilon.
\end{aligned}$$

Since $\epsilon$ is arbitrary, $\int_a^b f\,\mathrm{d}\mu = ⨍_{[a,b]} f\,\mathrm{d}\mu$. The proof of Proposition 2.28 is complete. □

**Corollary 2.29.** *For a regulated real-valued function $f$ on $[a,b]$ and a real-valued function $h$ on $[a,b]$, right-continuous on $[a,b)$, if $(LS)\int_a^b f\,\mathrm{d}h$ exists then so does $⨍_a^b f\,\mathrm{d}h$ and the two integrals are equal.*

*Proof.* We can assume that $a < b$. Since $(LS)\int_a^b f\,\mathrm{d}h$ exists, $h$ is of bounded variation, and so $h = h^+ - h^-$, where $h^+$ and $h^-$ are nondecreasing. By Theorem 2.20, $(RYS)\int_a^b f\,\mathrm{d}h^+$ and $(RYS)\int_a^b f\,\mathrm{d}h^-$ exist. Thus by Proposition 2.27, $⨍_a^b f\,\mathrm{d}h^+$ and $⨍_a^b f\,\mathrm{d}h^-$ exist. Since $h$ is right-continuous at $a$, $\mu_{h,[a,b]}(\{a\}) = 0$. The conclusion then follows from (2.24), the obvious linearity of $h \mapsto \mu_h$ (Theorem 2.7), and Proposition 2.28. □

The following is a change of variables theorem for Kolmogorov integrals.

**Proposition 2.30.** *For an interval $J$, let $\theta$ be a strictly increasing or strictly decreasing homeomorphism from an interval $J$ onto $\theta(J)$. Let $\mu \in \mathcal{I}(\theta(J);Y)$ be additive and upper continuous, $\mu^{\theta}(A) := \mu(\theta(A))$ for $A \in \mathfrak{J}(J)$, and let $f\colon \theta(J) \to X$. Then*

$$\oint_{\theta(J)} f{\cdot}\mathrm{d}\mu = \oint_{J} f{\circ}\theta{\cdot}\mathrm{d}\mu^{\theta}$$

*if either side is defined.*

*Proof.* The statement follows from the one-to-one correspondence between tagged interval partitions of $J$ and $\theta(J)$, and equality of corresponding Kolmogorov sums. □

*The Bochner integral*

Next we define the Bochner integral of a Banach-space-valued function with respect to a measure and compare it to the Kolmogorov integral. Let $S$ be a nonempty set, $\mathcal{A}$ an algebra of subsets of $S$, and $(X, \|\cdot\|)$ a Banach space. An $\mathcal{A}$-*simple* function $f\colon S \to X$ will be a function $f := \sum_{i=1}^{n} x_i 1_{A_i}$ for some $x_i \in X$, $A_i \in \mathcal{A}$, and finite $n$. If $\mu$ is a subadditive function from $\mathcal{A}$ into $[0,\infty]$, $f$ will be called $\mu$-*simple* if $\mu(A_i) < \infty$ for each $i$.

To define the Bochner integral, for a $\mu$-simple function $f = \sum_{i=1}^{n} x_i 1_{A_i}$ and $A \in \mathcal{S}$, let $(Bo)\int_A f\,\mathrm{d}\mu := \sum_{i=1}^{n} x_i \mu(A_i \cap A)$. It can be shown that $(Bo)\int f\,\mathrm{d}\mu$ is well defined for $\mu$-simple functions, just as for real-valued functions. If $(X,d)$ is any metric space, then the function $(x,y) \mapsto d(x,y)$ is jointly continuous from $X \times X$ into $[0,\infty)$. If $(X,d)$ is separable, then $d(\cdot,\cdot)$ is also jointly measurable for the Borel $\sigma$-algebra on each copy of $X$ (e.g. [53, Propositions 2.1.4, 4.1.7]). If $X$ is not separable, the joint measurability may fail (e.g. [53, §4.1, problem 11]). If $(S,\mathcal{S},\mu)$ is a measure space and $(X,\|\cdot\|)$ is a Banach space, then a $\mu$-measurable function $f$ from $S$ into $X$ will be called $\mu$-*almost separably valued* if there is a closed separable subspace $Y$ of $X$ such that $f^{-1}(X \setminus Y)$ is a $\mu$-null set.

**Definition 2.31.** Let $(S,\mathcal{S},\mu)$ be a measure space and let $X = (X,\|\cdot\|)$ be a Banach space. A function $f\colon S \to X$ will be called *Bochner $\mu$-integrable* if there exist a sequence of $\mu$-simple functions $\{f_k\}_{k\ge1}$ and a sequence of $\mu$-measurable real-valued functions $g_k \ge \|f - f_k\|$ such that $\lim_{k\to\infty}\int_S g_k\,\mathrm{d}\mu = 0$.

**Theorem 2.32.** *Let $f\colon S \to X$ be Bochner $\mu$-integrable. Then*

*(a) $f$ is $\mu$-almost separably valued and measurable for the completion of $\mu$;*
*(b) For each $A \in \mathcal{S}$ the limit*

$$\lim_{k\to\infty}\int_A f_k\,\mathrm{d}\mu \tag{2.26}$$

*exists and does not depend on the choice of $\{f_k\}$ satisfying the definition.*

*Proof.* (*a*): Let $\{f_k\}_{k\ge1}$ and $\{g_k\}_{k\ge1}$ be as in Definition 2.31. Taking a subsequence, we can assume that $\int g_k\,d\mu < 4^{-k}$ for all $k$. Then there is a set $B \in \mathcal{S}$ with $\mu(B) = 0$ such that for $s \notin B$, $g_k(s) \to 0$, and so $f_k(s) \to f(s)$. The set of finite rational linear combinations of elements in the union of ranges of all $f_k$ is countable, so its closure is a separable subspace of $X$, in which $f(s)$ takes values for $s \notin B$. Thus $f$ is $\mu$-almost separably valued and measurable for the completion of $\mu$, since each $f_k$ is $\mu$-measurable, e.g. [53, Theorem 4.2.2]. Also,

$$\lim_{k\to\infty}\int_S \|f - f_k\|\,d\mu = 0, \tag{2.27}$$

where the Lebesgue integrals are well defined.

(*b*): For $g$ and $h$ $\mu$-simple, and any measurable $A \subset S$, $\|(Bo)\int_A[g-h]\,d\mu\| \le \int_S \|g-h\|\,d\mu$. It follows that that if the Bochner integral of $f$ exists then the limit $(Bo)\int_A f_k\,d\mu$ exists and does not depend on the sequence of $\mu$-simple functions $\{f_k\}$ satisfying (2.27). So the theorem is proved. □

If $f$ is Bochner $\mu$-integrable, the limit

$$(Bo)\int_A f\,d\mu := \lim_{k\to\infty}\int_A f_k\,d\mu \tag{2.28}$$

is called the *Bochner integral* with respect to $\mu$ of $f$ over the set $A \in \mathcal{S}$.

**Proposition 2.33.** *Let $(S, \mathcal{S}, \mu)$ be a measure space and $(X, \|\cdot\|)$ a separable Banach space. Let $f$ be a $\mu$-measurable function from $S$ into $X$. Then $f$ is Bochner $\mu$-integrable if and only if $\int_S \|f\|\,d\mu < \infty$. Moreover, if $f$ is Bochner $\mu$-integrable, then*

$$\Big\|(Bo)\int_S f\,d\mu\Big\| \le \int_S \|f\|\,d\mu. \tag{2.29}$$

*Proof.* Suppose that $f$ is Bochner $\mu$-integrable, and let $\{f_k\}$ be a sequence of $\mu$-simple functions satisfying Definition 2.31. By (2.27), for some $k$ large enough, we have $\int \|f - f_k\|\,d\mu < 1$, and then $\int \|f\|\,d\mu < 1 + \int \|f_k\|\,d\mu < \infty$. Moreover, for each $k$, we have

$$\Big\|(Bo)\int_S f d\mu\Big\| \le \Big\|(Bo)\int_S [f - f_k]\,d\mu\Big\| + \int_S \|f_k\|\,d\mu.$$

Letting $k \to \infty$ on the right side, the first term tends to zero by (2.28) and the second term tends to the right side of (2.29) by (2.27), proving (2.29).

For the converse implication, we can assume that $\int \|f\|\,d\mu \le 1$. Let $\{x_j\}_{j=1}^\infty$ be a dense sequence in $X$. For $k = 1, 2, \ldots$, let $A_k := \{s \in S\colon \|f(s)\| > 1/k\}$. Then $\mu(A_k) \le k < \infty$. If $s \notin A_k$, set $g_k(s) := 0$. If $s \in A_k$, let $g_k(s) := x_j$ for the least $j$ such that $\|x_j - f(s)\| \le 1/k$. Then $g_k$ is measurable and $\|g_k(s)\| \le 2\|f(s)\|$ for all $s \in S$, so $\int \|g_k\|\,d\mu \le 2$. Also, $\|(g_k - f)(s)\| \le 1/k$ for all $s$, so by dominated convergence, with

$$\|(g_k - f)(s)\| \le (1/k)1_{A_k}(s) + 1_{A_k^c}(s)\|f(s)\| \le \|f(s)\|$$

for all $s$, $\int \|g_k - f\| \, d\mu \to 0$ as $k \to \infty$. Let $h_{kr}(s) := g_k(s)$ if $g_k(s) = 0$ or $x_j$ for some $j \le r$; otherwise let $h_{kr}(s) := 0$. Then $h_{kr}$ is a $\mu$-simple function and $\|h_{kr}(s)\| \le \|g_k(s)\|$ for all $s$, so by dominated convergence, $\int \|h_{kr} - g_k\| \, d\mu \to 0$ as $r \to \infty$. Let $f_k = h_{kr}$ for $r$ large enough so that $\int \|h_{kr} - g_k\| \, d\mu < 1/k$. Then $\int \|f_k - f\| \, d\mu \to 0$ as $k \to \infty$, so $f$ is Bochner $\mu$-integrable. □

We show next that for a regulated function the Bochner and Kolmogorov integrals both exist and are equal.

**Proposition 2.34.** *Let $X$ be a Banach space, $f$ an $X$-valued regulated function on $[a,b]$, and $\mu$ a finite positive measure on the Borel sets of $[a,b]$. Then $(Bo)\int_{[a,b]} f \, d\mu$ and $\displaystyle⨍_{[a,b]} f \, d\mu$ both exist and are equal.*

*Proof.* We can assume $a < b$. The range of $f$ is separable, so we can assume that $X$ is separable. Since $f$ is bounded by Corollary 2.2, the Bochner integral $(Bo)\int_a^b f \, d\mu$ exists by Proposition 2.33. The Kolmogorov integral $⨍_{[a,b]} f \, d\mu$ exists by Theorem 2.20, Corollaries 2.11 and 2.26, and Proposition 2.27 with $Y = Z = \mathbb{R}$ and $B(x,y) \equiv yx$, $x \in X$, $y \in \mathbb{R}$. To show that the two integrals are equal, let $\epsilon_k \downarrow 0$. By Theorem 2.1 and the definition of $⨍$, there exists a nested sequence $\{\lambda_k\}_{k \ge 1}$ of Young interval partitions $\lambda_k = \{(t_{i-1}^k, t_i^k)\}_{i=1}^{n(k)}$ of $[a,b]$ such that for each $k \ge 1$, $\max_i \mathrm{Osc}(f; (t_{i-1}^k, t_i^k)) < \epsilon_k$ and for any tagged Young partition $\tau_k = (\lambda_k, \{s_i\}_{i=1}^{n(k)})$,

$$\Big\| ⨍_{[a,b]} f \, d\mu - S_{YS}(f, d\mu; \tau_k) \Big\| < \epsilon_k.$$

For each $k \ge 1$, fix a tagged partition $\tau_k$ and let

$$f_k := \sum_{i=1}^{n(k)} f(s_i^k) 1_{(t_{i-1}^k, t_i^k)} + \sum_{i=0}^{n(k)} f(t_i^k) 1_{\{t_i^k\}}. \tag{2.30}$$

Then for each $k \ge 1$, $f_k$ is a $\mu$-simple function, $\int_{[a,b]} f_k \, d\mu = S_{YS}(f, d\mu; \tau_k)$ and $\|f - f_k\|_{\sup} < \epsilon_k$. Thus the two integrals are equal, proving the proposition. □

We will need the following extension of Lebesgue's differentiation theorem to the indefinite Bochner integral.

**Theorem 2.35.** *Let $\lambda$ be Lebesgue measure on $[a,b]$ and let $f\colon [a,b] \to X$ be Bochner $\lambda$-integrable. Then for $\lambda$-almost all $t \in (a,b)$,*

$$\lim_{s \to t} \frac{1}{t-s} \Big[ (Bo)\int_{[a,t]} f \, d\lambda - (Bo)\int_{[a,s]} f \, d\lambda \Big] = f(t).$$

*Proof.* By (2.29), it is enough to prove that for $\lambda$-almost all $t \in (a,b)$,

$$\lim_{u\downarrow 0} \frac{1}{u} \int_{A_t} \|f - f(t)\|\, d\lambda = 0, \tag{2.31}$$

where $A_t = [t-u, t]$ or $A_t = [t, t+u]$. By Theorem 2.32(*a*), $f$ is $\lambda$-almost separably valued. Let $\{x_n\}$ be a countable dense set in $f([a,b] \setminus N)$ for some $\lambda$-null set $N \subset [a,b]$. By the Lebesgue differentiation theorem (e.g. Theorem 7.2.1 in Dudley [53]), we have

$$\lim_{u\downarrow 0} \frac{1}{u} \int_{A_t} \|f - x_n\|\, d\lambda = \|f(t) - x_n\|$$

for almost all $t \in (a,b)$ and for all $n$. For any $t \notin N$ such that this holds for all $n$, it follows that

$$\begin{aligned} \limsup_{u\downarrow 0} \frac{1}{u} \int_{A_t} \|f - f(t)\|\, d\lambda &\le \limsup_{u\downarrow 0} \frac{1}{u} \int_{A_t} \|f - x_n\|\, d\lambda + \|x_n - f(t)\| \\ &= 2\|x_n - f(t)\| \end{aligned}$$

for all $n$. Given $\epsilon > 0$, one can choose $n$ such that $\|x_n - f(t)\| < \epsilon/2$, proving (2.31) for $\lambda$-almost all $t \in (a,b)$. □

### **The Bartle integral*

Next we define the Bartle integral of a vector-valued function with respect to an additive vector measure and compare it to the Kolmogorov integral. As in the basic assumption (1.14), let $X$, $Y$, and $Z$ be three Banach spaces and $B(\cdot,\cdot)\colon X \times Y \to Z$, $x{\cdot}y = B(x,y)$, a bounded bilinear operator with norm $\le 1$. Let $S$ be a nonempty set, let $\mathcal{A}$ be an algebra of subsets of $S$, and let $\mu$ be a finitely additive function from $\mathcal{A}$ into $Y$. The *semivariation* of $\mu$ over a set $A \in \mathcal{A}$ is defined by

$$w(\mu; A) := w_B(\mu; A) := \sup\Big\{\Big\|\sum_i x_i{\cdot}\mu(A_i)\Big\|\Big\},$$

where the supremum is taken over all partitions $\{A_i\}$ of $A$ into finitely many disjoint members of $\mathcal{A}$ and all finite sequences $\{x_i\} \subset X$ satisfying $\|x_i\| \le 1$. As the notation $w_B(\mu; A)$ indicates, the semivariation depends on the Banach spaces $X$ and $Z$ and the bilinear operator $B\colon (x,y) \mapsto x{\cdot}y$. We say that $\mu$ has *bounded semivariation* if $w_B(\mu) := w_B(\mu; S) < \infty$. The *variation* of $\mu$ over a set $A \in \mathcal{A}$ is defined by

$$v(\mu; A) := \sup\Big\{\sum_i \|\mu(A_i)\|\Big\},$$

where the supremum is taken over all partitions $\{A_i\}$ of $A$ into finitely many disjoint members of $\mathcal{A}$. (The variation $v(\mu; A)$ does not depend on $X$, $Z$, or $B$.) We say that $\mu$ has *bounded variation* if $v(\mu) := v(\mu; S) < \infty$. The semivariation $w_B(\mu; \cdot)$ is a monotone, subadditive function on $\mathcal{A}$, while the variation $v(\mu; \cdot)$ is a monotone, additive function on $\mathcal{A}$. It is clear that if $A \in \mathcal{A}$ then $0 \le w_B(\mu; A) \le v(\mu; A) \le +\infty$. Also, if $X = Y'$, $Z = \mathbb{R}$, and $B(x, y) = x(y)$, then $w_B(\mu) = v(\mu)$. However, in general, it is possible that $w_B(\mu; \cdot)$ is not additive and that $w_B(\mu; A) < \infty$ while $v(\mu; A) = +\infty$, as the following examples show:

**Example 2.36.** $(a)$ Let $\mathcal{A}$ be the algebra generated by subintervals of $[0, 1]$, and let $\lambda$ be Lebesgue measure. Let $\mu$ be the extension to $\mathcal{A}$ of the additive interval function of Example 2.4 having values in $L^\infty([0, 1], \lambda)$, that is, for $A \in \mathcal{A}$, $\mu(A)$ is the equivalence class containing $1_A$. Then $\mu$ is an additive function from $\mathcal{A}$ to $Z = Y = L^\infty$. For a set $A \in \mathcal{A}$, a finite partition $\{A_i\}$ of $A$ into finitely many disjoint sets in $\mathcal{A}$ and for any finite sequence $\{x_i\} \subset X = \mathbb{R}$ of real numbers in $[-1, 1]$, we have

$$\Big\| \sum_i x_i \mu(A_i) \Big\|_\infty = \Big\| \sum_i x_i 1_{A_i} \Big\|_\infty = \max_i |x_i| \le 1.$$

Thus $\mu$ has bounded semivariation, with $w(\mu; A) = 1$ for $\lambda(A) > 0$ and $w(\mu; A) = 0$ otherwise. Thus $w(\mu; \cdot)$ is not additive. Also, $\mu$ has unbounded variation. Indeed, if $A \in \mathcal{A}$ and $\lambda(A) > 0$, then one can take a countable partition $\{B_i\}$ of $A$ into disjoint sets each with $\lambda(B_i) > 0$. Then for $n > 1$, the sequence $\{A_i\}_{i=1}^n$ defined by $A_1 := B_1, \ldots, A_{n-1} := B_{n-1}$ and $A_n := \cup_{k \ge n} B_k$ is a finite partition of $A$ and

$$\sum_{i=1}^n \|\mu(A_i)\|_\infty = \sum_{i=1}^{n-1} \|1_{B_i}\|_\infty + \|1_{\cup_{k\ge n} B_k}\|_\infty = n.$$

Since $n$ is arbitrary, $v(\mu; A) = +\infty$. It is easy to see that the extension to $\mathcal{A}$ of the interval function $\mu$ of Example 2.4 with values in $\ell^\infty[0, 1]$ has the same properties.

$(b)$ Let $H$ be a real Hilbert space, $X = \mathbb{R}$, and $Y = Z = L(H, H)$, the Banach space of bounded linear operators from $H$ into itself, with $x \cdot y := xy$. Let $S$ be a set and $\mathcal{A}$ an algebra of subsets of $S$. Then a finitely additive *projection-valued measure* will be a function from $\mathcal{A}$ into $Y$ whose values are orthogonal projections onto closed linear subspaces (which may be $\{0\}$ or $H$) such that if $A$ and $B$ in $\mathcal{A}$ are disjoint, then $\mu(A \cup B) = \mu(A) + \mu(B)$, where $\mu(A)$ and $\mu(B)$ are projections onto orthogonal subspaces. It is easily seen that $w(\mu; A) = 1$ if $\mu(A)$ is a projection onto a non-zero subspace, since otherwise $w(\mu; A) = 0$. If $\|\mu(A_i)\| = 1$ for infinitely many disjoint subsets $A_i$ of a given set $A$, then clearly $v(\mu; A) = +\infty$, so $\mu$ has unbounded variation.

Let $\mu$ be an additive $Y$-valued function on an algebra $\mathcal{A}$ of subsets of $S$. A function will be called *$\mu$-simple* if it is $w(\mu)$-simple. It is convenient to define an "outer" form of $w(\mu;\cdot)$ on all subsets of $S$ as follows. If $B\subset S$, then $w^\star(\mu;B)$ is defined to be the infimum of $w(\mu;A)$ over all $A\in\mathcal{A}$ such that $B\subset A$. Then $w^\star(\mu;\cdot)$ agrees with $w(\mu;\cdot)$ on $\mathcal{A}$ and $w^\star(\mu;\cdot)$ is a monotone and subadditive function on the class of all subsets of $S$. We say that a subset $B$ of $S$ is a *$\mu$-null* set if $w^\star(\mu;B)=0$. Notice that this definition agrees with the one given in the case when $\mu$ is a measure on a $\sigma$-algebra. The definition of a $\mu$-essentially bounded function (see Section 1.4) extends to the present case using the new meaning of $\mu$-null sets. A sequence of functions $\{f_k\}_{k\geq 1}$ from $S$ to $X$ will be said to *converge in outer $\mu$-semivariation* to a function $f\colon S\to X$ if for each $\epsilon>0$, $w^\star(\mu;B_k)\to 0$ as $k\to\infty$, where $B_k:=\{x\in S\colon \|f_k(x)-f(x)\|>\epsilon\}$. For a $\mu$-simple function $f=\sum_{i=1}^n x_i1_{A_i}$ for some $n<\infty$, $x_i\in X$, and $A_i\in\mathcal{S}$, the *indefinite integral* $I(f,\mathrm{d}\mu)$ is defined by

$$I(f,\mathrm{d}\mu)(A):=\int_A f\cdot\mathrm{d}\mu:=\sum_{i=1}^n x_i\cdot\mu(A\cap A_i)$$

for each $A\in\mathcal{S}$. The indefinite integral of a $\mu$-simple function is independent of the representation in its definition.

The following definition is due to Bartle [11, Definition 1].

**Definition 2.37.** Assuming (1.14), let $\mathcal{A}$ be an algebra of subsets of $S$ and let $\mu\colon\mathcal{A}\to Y$ be an additive function. A function $f\colon S\to X$ is called *Bartle $\mu$-integrable* over $S$ if there is a sequence $\{f_k\}_{k\geq 1}$ of $\mu$-simple functions from $S$ to $X$ satisfying the following conditions:

(*a*) the sequence $\{f_k\}$ converges in outer $\mu$-semivariation to $f$;
(*b*) the sequence $\{I(f_k,\mathrm{d}\mu)\}$ of indefinite integrals has the property that for any $\epsilon>0$ there is a $\delta>0$ such that if $A\in\mathcal{A}$ and $w(\mu;A)<\delta$, then $\|I(f_k,\mathrm{d}\mu)(A)\|<\epsilon$ for all $k\geq 1$;
(*c*) given $\epsilon>0$ there is a set $A_\epsilon\in\mathcal{A}$ with $w(\mu;A_\epsilon)<\infty$ and such that if $B\in\mathcal{A}$ and $B\subset S\setminus A_\epsilon$ then $\|I(f_k,\mathrm{d}\mu)(B)\|<\epsilon$ for all $k\geq 1$.

We show next that uniformly for $A\in\mathcal{A}$, the sequence $\{I(f_k,\mathrm{d}\mu)(A)\}_{k\geq 1}$ converges to a well-defined limit

$$(Ba)\int_A f\cdot\mathrm{d}\mu:=\lim_{k\to\infty}I(f_k,\mathrm{d}\mu)(A),$$

which will be called the *Bartle integral* of $f$ over the set $A$.

**Proposition 2.38.** *If $f$ is Bartle $\mu$-integrable then for each $A\in\mathcal{A}$ the sequence of indefinite integrals $\{I(f_k,\mathrm{d}\mu)(A)\}_{k\geq 1}$ in condition (b) converges in the norm of $Z$ uniformly for $A\in\mathcal{A}$ and the limit does not depend on the sequence of $\mu$-simple functions $\{f_k\}_{k\geq 1}$ satisfying the conditions of Definition 2.37.*

*Proof.* If $w(\mu,\cdot)\equiv 0$, then $I(f_k,\mathrm{d}\mu)\equiv 0$ and the conclusion holds. Assume that $w(\mu,\cdot)\not\equiv 0$. Let $\{f_k\}$ be a sequence of $\mu$-simple functions satisfying the conditions of Definition 2.37 and let $\epsilon > 0$. Take $\delta$ as in condition $(b)$ and $A_\epsilon \in \mathcal{A}$ with $w(\mu; A_\epsilon) \neq 0$ as in $(c)$. By condition $(a)$ there is a positive integer $K_\epsilon$ such that $w(\mu; U_{n,m}) < \delta$ for each $n, m \geq K_\epsilon$, where $U_{n,m} \in \mathcal{A}$ and $U_{n,m} \supset \{x \in S\colon \|f_m(x) - f_n(x)\| > (\epsilon/w(\mu; A_\epsilon))\}$. Let $A \in \mathcal{A}$ and let $n, m \geq K_\epsilon$. Then we have

$$\begin{aligned}
&\Big\| \textstyle\int_A f_n \cdot \mathrm{d}\mu - \displaystyle\int_A f_m \cdot \mathrm{d}\mu \Big\| \\
&= \|I(f_n - f_m, \mathrm{d}\mu)(A)\| \\
&\leq \|I(f_n - f_m, \mathrm{d}\mu)(A \cap U_{n,m})\| + \|I(f_n - f_m, \mathrm{d}\mu)(A \cap U_{n,m}^c \cap A_\epsilon^c)\| \\
&\quad + \|I(f_n - f_m, \mathrm{d}\mu)(A \cap U_{n,m}^c \cap A_\epsilon)\| \\
&\leq 2\epsilon + 2\epsilon + (\epsilon/w(\mu; A_\epsilon)) w(\mu; A \cap U_{n,m}^c \cap A_\epsilon) \leq 5\epsilon.
\end{aligned}$$

This proves the existence and the uniformity of the limit.

To prove the second part of the conclusion, let $\{f_k\}$, $\{g_k\}$ be two sequences of $\mu$-simple functions satisfying the conditions of Definition 2.37, and let $\epsilon > 0$. We can and do assume that for each $k \geq 1$, $f_k$ and $g_k$ have constant values on the same sets, that is for some partition $\{A_i^k\}$ of $S$, $f_k = \sum_i x_i^k 1_{A_i^k}$ and $g_k = \sum_i y_i^k 1_{A_i^k}$. For the two sequences $\{f_k\}$, $\{g_k\}$, take the minimum of two values of $\delta$ as in condition $(b)$ and the union of two sets $A_\epsilon$ with $w(\mu; A_\epsilon) \neq 0$ as in $(c)$. By condition $(a)$, there is an integer $K_\epsilon > 0$ such that $w(\mu; U_k) < \delta$ for each $k \geq K_\epsilon$, where $U_k \in \mathcal{A}$ and $U_k \supset \{x \in S\colon \|f_k(x) - g_k(x)\| > \epsilon/w(\mu; A_\epsilon)\}$. Let $A \in \mathcal{A}$. Then for each $k \geq K_\epsilon$,

$$\begin{aligned}
&\Big\| \textstyle\int_A f_k \cdot \mathrm{d}\mu - \displaystyle\int_A g_k \cdot \mathrm{d}\mu \Big\| \\
&= \|I(f_k - g_k, \mathrm{d}\mu)(A)\| \\
&\leq \|I(f_k - g_k, \mathrm{d}\mu)(A \cap U_k)\| + \|I(f_k - g_k, \mathrm{d}\mu)(A \cap U_k^c \cap A_\epsilon^c)\| \\
&\quad + \|I(f_k - g_k, \mathrm{d}\mu)(A \cap U_k^c \cap A_\epsilon)\| \\
&\leq 2\epsilon + 2\epsilon + (\epsilon/w(\mu; A_\epsilon)) w(\mu; A \cap U_k^c \cap A_\epsilon) \leq 5\epsilon.
\end{aligned}$$

This proves the second part of the conclusion of Proposition 2.38. □

For the vector measure $\mu$ of Example 2.36$(a)$ with values in $L^\infty([0,1], \lambda)$, a function $f\colon [0,1] \to \mathbb{R}$ is Bartle $\mu$-integrable if and only if there exists a sequence $\{f_k\}_{k \geq 1}$ of real-valued and $\mu$-simple functions on $[0,1]$ such that $\|f_k - f\|_\infty \to 0$ as $k \to \infty$. Indeed, if $f_k$ are $\mu$-simple functions satisfying condition $(a)$ of Definition 2.37, then for $A \in \mathcal{A}$, $\int_A f_k \, \mathrm{d}\mu = f_k 1_A$. Thus if $f$ is Bartle $\mu$-integrable then by the preceding proposition, $(Ba) \int_A f \, \mathrm{d}\mu = f 1_A$ and $\|f_k - f\|_\infty \to 0$ as $k \to \infty$, since $w(\mu; A) = 0$ implies $\lambda(A) = 0$. The converse is clear by checking the conditions of Definition 2.37.

If $\mu$ is the vector measure $A \mapsto 1_A$ as in Example 2.36$(a)$ but with values in $\ell^\infty[0,1]$, then a function $f\colon [0,1] \to \mathbb{R}$ is Bartle $\mu$-integrable if and only if

$f$ is regulated on $[0,1]$ as follows. The argument of the preceding paragraph gives that $f\colon [0,1] \to \mathbb{R}$ is Bartle $\mu$-integrable if and only if there exists a sequence $\{f_k\}_{k\geq 1}$ of real-valued and $\mu$-simple functions on $[0,1]$ such that $\|f_k - f\|_{\sup} \to 0$ as $k \to \infty$, since in the present case $w(\mu; A) = 0$ implies that $A$ is empty. Moreover, each $\mu$-simple function is a step function, and so $f$ is regulated on $[0,1]$ by Theorem 2.1.

**Proposition 2.39.** *Let $\mathcal{A}$ be an algebra of subsets of a set $S$, and let $\mu$ be a $Y$-valued additive function on $\mathcal{A}$ with bounded semivariation. If $f\colon S \to X$ is $\mu$-essentially bounded and there is a sequence of $\mu$-simple functions converging to $f$ in outer $\mu$-measure, then $f$ is Bartle $\mu$-integrable and*

$$\Big\|(Ba)\int_S f\cdot\mathrm{d}\mu\Big\| \leq \|f\|_\infty w(\mu; S). \tag{2.32}$$

*Proof.* Let $\{f_k\}_{k\geq 1}$ be a sequence of $\mu$-simple functions which converge to $f$ in outer $\mu$-measure. Let $\epsilon > 0$, let $M := \|f\|_\infty + 2\epsilon$, and let $N$ be a $\mu$-null set such that $\{x \in S\colon \|f(x)\| > \|f\|_\infty + \epsilon\} \subset N$. Then for each $k \geq 1$,

$$\{x \in S\colon \|f_k(x)\| > M\} \subset \{x \in S\colon \|f_k(x) - f(x)\| > \epsilon\} \cup N. \tag{2.33}$$

For each $k \geq 1$, let $f_k^M := f_k 1_{A_k}$, where $A_k := \{x \in S\colon \|f_k(x)\| \leq M\}$. Then

$$\begin{aligned}&\{x \in S\colon \|f_k^M(x) - f(x)\| > \epsilon\}\\ &\quad\subset \{x \in S\colon \|f_k(x) - f(x)\| > \epsilon\} \cup \{x \in S\colon \|f_k(x)\| > M\}.\end{aligned}$$

Since $w^\star(\mu; \cdot)$ is monotone and subadditive, by (2.33) it then follows that condition $(a)$ of Definition 2.37 holds for the sequence $\{f_k^M\}_{k\geq 1}$ of $\mu$-simple functions. Since the norm of each value of each $\mu$-simple function $f_k^M$ is bounded by $M$, it follows that for each $k \geq 1$ and any $A \in \mathcal{A}$,

$$\|I(f_k, \mathrm{d}\mu)(A)\| \leq M w(\mu; A), \tag{2.34}$$

proving condition $(b)$. Condition $(c)$ holds trivially because $w(\mu; S) < \infty$. Therefore the function $f$ is Bartle $\mu$-integrable. The bound (2.32) follows from (2.34) and Proposition 2.38 because $\epsilon$ is arbitrary. The proof of the proposition is complete. □

Each additive interval function $\mu\colon \mathfrak{I}[a,b] \to Y$ extends to an additive function $\tilde{\mu}$ on the algebra $\mathcal{A}[a,b]$ generated by subintervals of $[a,b]$. We show next that if $\tilde{\mu}$ has bounded semivariation, then any regulated function on $[a,b]$ is Bartle $\mu$-integrable and the Bartle integral agrees with the Kolmogorov integral in this case.

**Proposition 2.40.** *Assuming* (1.14), *let $\mu$ be a $Y$-valued additive interval function on $[a,b]$ such that $\tilde{\mu}$ is of bounded semivariation, and let $f$ be an $X$-valued regulated function on $[a,b]$. Then $(Ba)\int_{[a,b]} f\cdot\mathrm{d}\tilde{\mu}$ and $⨎_{[a,b]} f\cdot\mathrm{d}\mu$ both exist and are equal.*

*Proof.* By Theorem 2.1, $f$ is bounded and is a uniform limit of step functions, and so it is Bartle $\mu$-integrable by Proposition 2.39. Let $\epsilon_k \downarrow 0$. By Theorem 2.1 once again, there exists a nested sequence $\{\lambda_k\}_{k\geq 1}$ of Young partitions $\lambda_k = \{A_{ki} := (t_{i-1}^k, t_i^k)\}_{i=1}^{n(k)}$ of $[a,b]$ such that for each $k \geq 1$, $\max_i \mathrm{Osc}(f; A_{ki}) < \epsilon_k$. As in the proof of Proposition 2.34, for each $k \geq 1$, fix a tagged partition $\tau_k = (\{A_{ki}\}_{i=1}^{n(k)}, \{s_i^k\}_{i=1}^{n(k)})$ and let $f_k$ be defined by (2.30). Then for each $k \geq 1$, $f_k$ is a step function, $\int_{[a,b]} f_k \, d\tilde{\mu} = S_{YS}(f, d\mu; \tau_k)$, and $\|f - f_k\|_{\sup} < \epsilon_k$. By Proposition 2.38, we have

$$(Ba)\int_{[a,b]} f{\cdot}d\tilde{\mu} = \lim_{k\to\infty}\int_{[a,b]} f_k{\cdot}d\tilde{\mu} = \lim_{k\to\infty} S_{YS}(f, d\mu; \tau_k). \tag{2.35}$$

For any $k \geq 1$ and any tagged refinement $\tau = (\{A_j\}, \{x_j\})$ of $\tau_k$, we have

$$\begin{aligned}&\big\|S_{YS}(f, d\mu; \tau) - S_{YS}(f, d\mu; \tau_k)\big\| \\ &= \Big\|\sum_{i=1}^{n(k)} \sum_{x_j\in A_{ki}} \big[f(x_j) - f(s_i^k)\big]{\cdot}\mu(A_j)\Big\| \leq \epsilon_k w(\tilde{\mu}; [a,b]).\end{aligned}$$

Thus the Kolmogorov integral $⨏_{[a,b]} f\cdot d\mu$ exists and equals the Bartle integral (2.35), completing the proof. □

## 2.4 Relations between *RS*, *RRS*, and *RYS* Integrals

By Propositions 2.13 and 2.18, we have

$$(RS)\int_a^b f{\cdot}dh \longrightarrow (RRS)\int_a^b f{\cdot}dh \overset{h\in\mathcal{R}}{\longrightarrow} (RYS)\int_a^b f{\cdot}dh. \tag{2.36}$$

Here $\longrightarrow$ means that existence of the integral to the left of it implies that of the integral to the right of it, with the same value, and $\overset{h\in\mathcal{R}}{\longrightarrow}$ means that the implication holds for $h$ regulated. In this section we consider the question under what conditions on $f$ and $h$ the two arrows in (2.36) can be inverted.

By Proposition 2.15, if the Riemann–Stieltjes integral exists then the integrand and integrator cannot have common discontinuities. This condition is also sufficient for the first arrow in (2.36) to be inverted (Theorem 2.42). A necessary condition for the existence of the refinement Riemann–Stieltjes integral is that the integrand and integrator cannot have common one-sided discontinuities. This condition is also sufficient if the integrand and integrator satisfy suitable $p$-variation conditions, as we will show in Corollary 3.91. Thus it is tempting to guess that the second arrow in (2.36) is invertible provided $f$ and $h$ have no common one-sided discontinuities. However, Example 2.44

below shows that this is not true in general. We end this section with a result giving a general sufficient condition for the second arrow in (2.36) to be invertible (Proposition 2.46).

The following definition of full Stieltjes integral defines it as the (RYS) integral provided condition (b) holds.

**Definition 2.41.** Let $f$ and $h$ be regulated functions on $[a, b]$ with values in $X$ and $Y$, respectively. We say that the *full Stieltjes integral* $(S)\int_a^b f{\cdot}\mathrm{d}h$ exists, or $f$ is full Stieltjes integrable with respect to $h$ over $[a, b]$, if $(a)$ and $(b)$ hold, where

$(a)$ $(RYS)\int_a^b f{\cdot}\mathrm{d}h$ exists,
$(b)$ if $f$ and $h$ have no common one-sided discontinuities, then $(RRS)\int_a^b f{\cdot}\mathrm{d}h$ exists.

Then let $(S)\int_a^b f{\cdot}\mathrm{d}h := (RYS)\int_a^b f{\cdot}\mathrm{d}h$.

By (2.36), whenever the two integrals in $(a)$ and $(b)$ exist, they have the same value. Next we prove a relation stated above between $RS$ and $RRS$ integrals.

**Theorem 2.42.** *Let $f$ and $h$ be bounded functions from $[a, b]$ into $X$ and $Y$ respectively. The integral $(RS)\int_a^b f{\cdot}\mathrm{d}h$ exists if and only if both $(RRS)\int_a^b f{\cdot}\mathrm{d}h$ exists and $f$, $h$ have no common discontinuities. Moreover, if the two integrals exist then they are equal.*

*Proof.* Propositions 2.15 and 2.13 imply the "only if" part. For the converse implication suppose that the integral $(RRS)\int_a^b f\,dh$ exists and the two functions $f, h$ have no common discontinuities. Given $\epsilon > 0$, there exists a partition $\lambda = \{z_j\}_{j=0}^m$ of $[a, b]$ such that

$$\left\| S_{RS}(f, \mathrm{d}h; (\kappa, \xi)) - (RRS)\int_a^b f\,\mathrm{d}h \right\| < \epsilon/2 \tag{2.37}$$

for each tagged partition $(\kappa, \xi)$ such that $\kappa$ is a refinement of $\lambda$. Then choose a $\delta > 0$ such that for each interval $\Delta_j := [z_j - \delta, z_j + \delta] \cap [a, b]$, $j = 1, \ldots, m-1$,

$$\mathrm{Osc}\,(f; \Delta_j)\,\mathrm{Osc}(h; \Delta_j) < \epsilon/(2m).$$

Let $d := \min_{1 \le j \le m}(z_j - z_{j-1})$ and let $\tau = (\{x_i\}_{i=0}^n, \{y_i\}_{i=1}^n)$ be a tagged partition with mesh $|\tau| < d \wedge \delta$. Then each interval $[x_{i-1}, x_i]$ contains at most one point of $\lambda$. For $j = 1, \ldots, m-1$, let $i(j)$ be the index in $\{1, \ldots, n-1\}$ such that $z_j \in (x_{i(j)-1}, x_{i(j)}]$. Let $(\kappa, \xi)$ be the tagged partition of $[a, b]$ obtained from $\tau$ by replacing each tagged interval $([x_{i(j)-1}, x_{i(j)}], y_{i(j)})$, $j = 1, \ldots, m-1$, by the pair of tagged intervals $([x_{i(j)-1}, z_j], z_j)$, $([z_j, x_{i(j)}], z_j)$ if $z_j < x_{i(j)}$, or by the tagged interval $([x_{i(j)-1}, x_{i(j)}], z_j)$ otherwise. Then $\kappa$ is a refinement of $\lambda$ and

$$\begin{aligned}&\big\|S_{RS}(f,\mathrm{d}h;\tau)-S_{RS}(f,\mathrm{d}h;(\kappa,\xi))\big\|\\&\le\sum_{j=1}^{m-1}\big\|f(y_{i(j)})-f(z_j)\big\|\,\big\|h(x_{i(j)})-h(x_{i(j)-1})\big\|<\epsilon/2.\end{aligned}$$

This in conjunction with (2.37) implies that $(RS)\int_a^b f{\cdot}\mathrm{d}h$ exists and equals $(RRS)\int_a^b f{\cdot}\mathrm{d}h$. The proof of Theorem 2.42 is complete. □

In view of the preceding theorem the notion of the full Stieltjes integral can be extended as follows:

**Corollary 2.43.** *Let $f$ and $h$ be regulated functions on $[a,b]$ with values in $X$ and $Y$, respectively. The full Stieltjes integral $(S)\int_a^b f{\cdot}\mathrm{d}h$ exists if and only if (a), (b) of Definition 2.41 and (c) hold, where*

*(c) if $f$ and $h$ have no common discontinuities, then $(RS)\int_a^b f{\cdot}\mathrm{d}h$ exists.*

*Moreover, whenever two or more of the integrals in (a), (b), and (c) exist, they have the same value.*

The following shows that there are regulated functions $f,h\colon [0,1]\to\mathbb{R}$ having no common discontinuities such that the integral $\int_0^1 f\,\mathrm{d}h$ exists and equals 0 as a refinement Young–Stieltjes integral but it does not exist as a refinement Riemann–Stieltjes integral.

**Example 2.44.** The functions $f$ and $h$ will be defined on $[0,1]$ and equal to 0 everywhere outside countable disjoint subsets $A$ and $B$, respectively. Let $B$ be the union of the sets $\{i/p\colon i=1,\dots,p-1\}$ over the prime numbers $p\ge 5$, $p\in\{5,7,11,\dots\}$. Define the set $A$ recursively along prime numbers $p\ge 5$ by choosing irrational points $a_{p,i}$ in each interval $((i-1)/p,i/p)$, $i=1,\dots,p$, different from all such points previously defined. For each prime $p\ge 5$, let $f(a_{p,2j}):=h(2j/p):=0$ and $f(a_{p,2j-1}):=h((2j-1)/p):=1/\sqrt{p}$ for $j=1,2,\dots$ and $2j<p$. The functions $f$ and $h$ each converge to 0 along any 1–1 enumeration of the sets $A$, $B$, respectively. Thus $f$ and $h$ are regulated and have no common discontinuities. The integral $(RYS)\int_0^1 f\,\mathrm{d}h$ exists and equals 0; in fact, all its approximating Young–Stieltjes sums equal 0. For any partition of $[0,1]$ there are Riemann–Stieltjes sums which are 0. For the partition $\kappa_p:=\{i/p\}_{i=0}^p$ of $[0,1]$, for each prime $p\ge 5$, consider the Riemann–Stieltjes sum

$$\begin{aligned}S_p:=\sum_{j=1}^{[p/2]-1}&\Big\{f(a_{p,2j-1})\Big[h\Big(\frac{2j-1}{p}\Big)-h\Big(\frac{2j-2}{p}\Big)\Big]\\&+f(a_{p,2j})\Big[h\Big(\frac{2j}{p}\Big)-h\Big(\frac{2j-1}{p}\Big)\Big]\Big\}=\frac{[p/2]-1}{p}.\end{aligned}$$

Since $S_p\to 1/2$ as $p\to\infty$, the Riemann–Stieltjes integral does not exist. The refinement Riemann–Stieltjes integral also does not exist by Theorem 2.42.

**Lemma 2.45.** *For regulated functions $f$ and $h$ on $[a,b]$ the following two statements hold:*

(1) *If at each point of $[a,b)$ at least one of $f$ and $h$ is right-continuous, then the integral $(RRS)\int_a^b f\cdot dh$ exists whenever in its definition the Riemann–Stieltjes sums converge for tagged partitions $(\{x_i\}_{i=0}^n, \{y_i\}_{i=1}^n)$ with tags $y_i \in (x_{i-1}, x_i]$, $i = 1, \dots, n$.*

(2) *If at each point of $(a,b]$, at least one of $f$ and $h$ is left-continuous, then the integral $(RRS)\int_a^b f\cdot dh$ exists whenever in its definition the Riemann–Stieltjes sums converge for tagged partitions $(\{x_i\}_{i=0}^n, \{y_i\}_{i=1}^n)$ with tags $y_i \in [x_{i-1}, x_i)$, $i = 1, \dots, n$.*

*Proof.* We prove only (1) because a proof of (2) is symmetric. Thus we have that there is an $A \in Z$ with the following property: given $\epsilon > 0$ there exists a partition $\lambda = \{z_j\}_{j=0}^m$ of $[a,b]$ such that

$$\big\|S_{RS}(f, dh; (\kappa, \xi')) - A\big\| < \epsilon \tag{2.38}$$

for each tagged partition $(\kappa, \xi') = (\{x_i\}_{i=0}^n, \{y_i\}_{i=1}^n)$ such that $\kappa$ is a refinement of $\lambda$ and $y_i \in (x_{i-1}, x_i]$ for $i = 1, \dots, n$. Choose a set $\zeta = \{u_j\}_{j=1}^m \subset (a,b)$ such that $z_{j-1} < u_j < z_j$ for $j = 1, \dots, m$ and

$$\max_{1\le j\le m} \operatorname{Osc}(f; [z_{j-1}, u_j]) \operatorname{Osc}(h; [z_{j-1}, u_j]) < \epsilon/m. \tag{2.39}$$

Suppose that $(\kappa, \xi) = (\{x_i\}_{i=0}^n, \{y_i\}_{i=1}^n)$ is a tagged partition of $[a,b]$ such that $\kappa$ is a refinement of $\lambda \cup \zeta$. Define $\xi_e := \{t_i\}_{i=1}^n$ by $t_i := y_i$ for $i$ odd, $t_i := x_i$ for $i$ even. Define $\xi_o := \{s_i\}_{i=1}^n$ by $s_i := y_i$ for $i$ even, $s_i := x_i$ for $i$ odd. Also, let $\xi_r := \{x_i\}_{i=1}^n$. Then letting $S(\xi) := S_{RS}(f, dh; (\kappa, \xi))$, we have $S(\xi) + S(\xi_r) = S(\xi_e) + S(\xi_o)$. Thus

$$\big\|S(\xi) - A\big\| \le \big\|S(\xi_r) - A\big\| + \big\|S(\xi_e) - A\big\| + \big\|S(\xi_o) - A\big\|. \tag{2.40}$$

By (2.38) with $\xi' = \xi_r$, the first term on the right does not exceed $\epsilon$. If each $t_i$ in $\xi_e$ is not $x_{i-1}$ then the same bound holds for the second term on the right. Otherwise, let $I_e$ be the set of indices $i \in \{1, \dots, n\}$ such that $x_{i-1} = y_i = t_i \in \xi_e$. Thus for each $i \in I_e$, $i$ must be odd, so $t_{i-1} = x_{i-1} = t_i$. Let $J \subset \{0, \dots, n\}$ be the set of indices $i$ such that $x_i \in \lambda$. For $i \in I_e$, if $i-1 \notin J$ then we can and do replace the two tagged intervals $([x_{i-2}, x_{i-1}], t_{i-1})$ and $([x_{i-1}, x_i], t_i)$ by the tagged interval $([x_{i-2}, x_i], x_{i-1})$ without changing the Riemann–Stieltjes sum and where now the tag is in the interior of the interval and the partition is still a refinement of $\kappa$. For $i \in I_e$, if $i-1 \in J$ then if we replace $t_i$ $(= x_{i-1})$ by $x_i$, we change the sum by at most $\epsilon/m$ for each $i$ by (2.39), and since there are at most $m$ such values of $i$, we change the total sum by at most $\epsilon$. Therefore it follows by (2.38) that $\|S(\xi_e) - A\| \le 2\epsilon$. Similarly one can show that the last term in (2.40) does not exceed $2\epsilon$, so $\|S(\xi) - A\| < 5\epsilon$. Since $\epsilon$ is arbitrary, the integral $(RRS)\int_a^b f\cdot dh$ exists and equals $A$, proving the lemma. □

**Proposition 2.46.** *Let regulated functions $f$, $h$ on $[a,b]$ be such that $f$ is left-continuous on $(a,b]$ and $h$ is right-continuous on $[a,b)$, or vice versa. If (RYS) $\int_a^b f{\cdot}\mathrm{d}h$ exists then so does (RRS) $\int_a^b f{\cdot}\mathrm{d}h$, and the two integrals have the same value.*

*Proof.* We can assume that neither $f$ nor $h$ is identically 0, since if one is, the conclusion holds. We prove the proposition when $f$ is left-continuous on $(a,b]$ and $h$ is right-continuous on $[a,b)$. A proof of the other case is symmetric. In that case by (1) of Lemma 2.45 and since $f$ is left-continuous on $(a,b]$, it is enough to prove the convergence of Riemann–Stieltjes sums based on Young tagged partitions of $[a,b]$. To this aim we show that differences between Riemann–Stieltjes and Young–Stieltjes sums based on the same Young tagged partition can be made arbitrarily small by refinement of partitions. Let $\epsilon > 0$. By definition of the refinement Young–Stieltjes integral there exists a partition $\lambda = \{z_j\}_{j=0}^m$ of $[a,b]$ such that

$$\left\| S_{YS}(f,\mathrm{d}h;\tau) - (RYS)\int_a^b f{\cdot}\mathrm{d}h \right\| < \epsilon \tag{2.41}$$

for any Young tagged partition $\tau = (\kappa,\xi)$ such that $\kappa$ is a refinement of $\lambda$. Choose a set $\zeta = \{v_{j-1}, u_j\colon j = 1,\dots,m\} \subset (a,b)$ such that for each $j = 1,\dots,m$, $z_{j-1} < v_{j-1} < u_j < z_j$,

$$\max(\|h\|_{\sup}\mathrm{Osc}\,(f;[u_j,z_j]), \|f\|_{\sup}\mathrm{Osc}\,(h;[z_{j-1},v_{j-1}])) < \epsilon/(2m). \tag{2.42}$$

Let $(\kappa,\xi) = (\{x_i\}_{i=0}^n, \{y_i\}_{i=1}^n)$ be a Young tagged partition of $[a,b]$ such that $\kappa$ is a refinement of $\lambda\cup\zeta$. Then $S_{RS}(f,\mathrm{d}h;(\kappa,\xi)) - S_{YS}(f,\mathrm{d}h;(\kappa,\xi)) = S_1 + S_2$, where

$$\begin{aligned} S_1 &:= \sum_{i=1}^n \big[f(y_i) - f(x_i)\big]{\cdot}\big[h(x_i) - h(x_{i-1})\big], \\ S_2 &:= \sum_{i=1}^n \big[f(x_i) - f(y_i)\big]{\cdot}\big[h(x_i-) - h(x_{i-1})\big]. \end{aligned} \tag{2.43}$$

Let $\xi' = \{y_i'\}_{i=1}^n$ be another set of tags for $\kappa$ such that $x_{i-1} < y_i' < x_i$ for $i = 1,\dots,n$. By (2.41) with $\tau = (\kappa,\xi)$ and $\tau = (\kappa,\xi')$, we get the bound

$$\left\| \sum_{i=1}^n \big[f(y_i') - f(y_i)\big]{\cdot}\big[h(x_i-) - h(x_{i-1})\big] \right\| < 2\epsilon.$$

Letting $y_i' \uparrow x_i$ for $i = 1,\dots,n$, it follows that in (2.43), $\|S_2\| \le 2\epsilon$. Let $J \subset \{0,\dots,n\}$ be the set of indices $i$ such that $x_i \in \lambda$, and let $I := \{0,\dots,n\} \setminus J$. To bound $\|S_1\|$ define a partition $\kappa' = \{x_i'\}_{i=0}^n$ of $[a,b]$ by $x_i' := x_i$ if $i \in J$ and take any $x_i' \in (x_i, y_{i+1})$ if $i \in I$. Then $\kappa'$ is a refinement of $\lambda$ and $(\kappa',\xi)$ is a Young tagged partition. Let $\xi' = \{y_i'\}_{i=1}^n$ be another set of tags for $\kappa'$

defined by $y_i' := y_i$ if $i \in J$ and $y_i' := x_i$ if $i \in I$. By (2.41) with $\tau = (\kappa', \xi)$ and $\tau = (\kappa', \xi')$, we get the bound

$$\Big\|\sum_{i\in I}\big[f(y_i) - f(x_i)\big]\cdot\big[h(x_i'-) - h(x_{i-1}')\big]\Big\| < 2\epsilon.$$

Letting $x_i' \downarrow x_i$ for $i \in I$, it follows that the sum of all terms in $S_1$ with $i \in I$ does not exceed $2\epsilon$. By (2.42), the norm of each term of $S_1$ with $i \in J$ does not exceed $\epsilon/m$, and the norm of the total sum of such terms does not exceed $\epsilon$. Therefore in (2.43), $\|S_1\| \le 3\epsilon$. The proof of Proposition 2.46 is complete. □

## 2.5 The Central Young Integral

L. C. Young [244] defined, up to endpoint terms given by Young [247], an integral which we will call the central Young integral, or the *CY* integral. The idea of the *CY* integral is to use the *RRS* integral, avoiding its lack of definition when $f$ and $h$ have common one-sided discontinuities by taking a right-continuous version of $f$ and left-continuous version of $h$, or vice versa, and adding sums of jump terms to restore the desired value of $\int f\cdot dh$. For regulated functions the *CY* integral extends the *RYS* integral, as will be shown in Theorem 2.51, and thus in turn the *RRS* integral by Proposition 2.46.

Define functions $f_+^{(b)}$ and $f_-^{(a)}$ on $[a,b]$ with $a < b$ by

$$f_+^{(b)}(x) := \begin{cases} f_+(x) := f(x+) := \lim_{z\downarrow x} f(z) & \text{if } a \le x < b,\\ f(b) & \text{if } x = b,\end{cases}$$

and

$$f_-^{(a)}(x) := \begin{cases} f_-(x) := f(x-) := \lim_{z\uparrow x} f(z) & \text{if } a < x \le b,\\ f(a) & \text{if } x = a.\end{cases}$$

Recalling the definitions of $\Delta^+$, $\Delta^-$, and $\Delta^\pm$ in and before (2.1), the *CY* integral has two equivalent forms, defined next:

**Definition 2.47.** Assuming (1.14), let $f$ and $h$ be regulated functions on $[a,b]$ with values in the Banach spaces $X, Y$ respectively. If $a < b$ define the $Y_1$ integral

$$\begin{aligned}(Y_1)\int_a^b f\cdot dh &:= (RRS)\int_a^b f_+^{(b)}\cdot dh_-^{(a)} - \sum_{[a,b]} \Delta_{[a,b]}^+ f\cdot\Delta_{[a,b]}^\pm h\\ &= (RRS)\int_a^b f_+^{(b)}\cdot dh_-^{(a)} - [\Delta^+ f\cdot\Delta^+ h](a)\\ &\quad + [f\cdot\Delta^- h](b) - \sum_{(a,b)} \Delta^+ f\cdot\Delta^\pm h\end{aligned}\tag{2.44}$$

if the $RRS$ integral exists and the sum converges unconditionally in $Z$ as defined in Section 1.4. If $a = b$ define the $Y_1$ integral as 0. Similarly, if $a < b$ define the $Y_2$ integral

$$\begin{aligned}(Y_2)\int_a^b f{\cdot}\mathrm{d}h &:= (RRS)\int_a^b f_-^{(a)}{\cdot}\mathrm{d}h_+^{(b)} + \sum_{[a,b]} \Delta_{[a,b]}^- f{\cdot}\Delta_{[a,b]}^{\pm} h \\ &= (RRS)\int_a^b f_-^{(a)}{\cdot}\mathrm{d}h_+^{(b)} + [f{\cdot}\Delta^+ h](a) \\ &\quad + [\Delta^- f{\cdot}\Delta^- h](b) + \sum_{(a,b)} \Delta^- f{\cdot}\Delta^{\pm} h \end{aligned} \tag{2.45}$$

if the $RRS$ integral exists and the sum converges unconditionally in $Z$. If $a = b$ define the $Y_2$ integral as 0.

Integrals $(Y_j)\int_a^b \mathrm{d}f{\cdot}h$, $j = 1, 2$, are defined symmetrically, if they exist, as in (1.15).

It will be shown that the $Y_1$ integral exists if and only if the $Y_2$ integral does, and if so, the two integrals have the same value. This will be done by giving an alternative representation of the two integrals in terms of the refinement Young–Stieltjes integral. To this aim, for $a < b$ we define the functions $f_-^{(a,b)}$ and $f_+^{(a,b)}$ on $[a, b]$ respectively by

$$f_-^{(a,b)}(x) := \begin{cases} f(a) & \text{if } x = a, \\ f(x-) & \text{if } x \in (a,b), \\ f(b) & \text{if } x = b, \end{cases} \quad \text{and} \quad f_+^{(a,b)}(x) := \begin{cases} f(a) & \text{if } x = a, \\ f(x+) & \text{if } x \in (a,b), \\ f(b) & \text{if } x = b. \end{cases} \tag{2.46}$$

**Proposition 2.48.** *Assuming* (1.14) *and* $a < b$, *for regulated functions* $f$ *and* $h$ *on* $[a, b]$ *with values in* $X$ *and* $Y$, *respectively,*

$$(Y_1)\int_a^b f{\cdot}\mathrm{d}h = (RYS)\int_a^b f_+^{(a,b)}{\cdot}\mathrm{d}h - \sum_{(a,b)} \Delta^+ f{\cdot}\Delta^{\pm} h \tag{2.47}$$

*and*

$$(Y_2)\int_a^b f{\cdot}\mathrm{d}h = (RYS)\int_a^b f_-^{(a,b)}{\cdot}\mathrm{d}h + \sum_{(a,b)} \Delta^- f{\cdot}\Delta^{\pm} h \tag{2.48}$$

*if either side is defined, where the right side is defined if the refinement Young–Stieltjes integral exists and the sum converges unconditionally in* $Z$.

*Proof.* We prove only (2.47) because a proof of (2.48) is symmetric. By Propositions 2.18 and 2.46, the refinement Riemann–Stieltjes integral in (2.44) exists if and only if it exists as a refinement Young–Stieltjes integral, and then both have the same value. Let $\tau = (\{x_i\}_{i=0}^n, \{y_i\}_{i=1}^n)$ be a Young tagged partition of $[a, b]$. Then

$$S_{YS}(f_+^{(b)}, \mathrm{d}h_-^{(a)}; \tau) = \sum_{i=1}^{n} f(y_i+)\cdot\big[h(x_i-) - h(x_{i-1}+)\big] + \big[f_+\cdot\Delta^+ h\big](a)$$
$$+ \sum_{i=1}^{n-1} \big[f_+\cdot\Delta^{\pm} h\big](x_i)$$
$$= S_{YS}(f_+^{(a,b)}, \mathrm{d}h; \tau) + \big[\Delta^+ f\cdot\Delta^+ h\big](a) - \big[f\cdot\Delta^- h\big](b).$$

Therefore $(RYS)\int_a^b f_+^{(b)}\cdot\mathrm{d}h_-^{(a)}$ exists if and only if $(RYS)\int_a^b f_+^{(a,b)}\cdot\mathrm{d}h$ does. Also, (2.47) holds, proving the proposition. □

The following statement relates the two refinement Young–Stieltjes integrals in (2.47) and (2.48) with the integral $(RYS)\int_a^b f\cdot\mathrm{d}h$.

**Lemma 2.49.** *For regulated functions $f$ and $h$ on $[a,b]$ with $a<b$, the following three statements are equivalent:*

*(a) $(RYS)\int_a^b f\cdot\mathrm{d}h$ exists;*
*(b) $(RYS)\int_a^b f_+^{(a,b)}\cdot\mathrm{d}h$ and $(RYS)\int_a^b[f_+^{(a,b)}-f]\cdot\mathrm{d}h$ both exist;*
*(c) $(RYS)\int_a^b f_-^{(a,b)}\cdot\mathrm{d}h$ and $(RYS)\int_a^b[f-f_-^{(a,b)}]\cdot\mathrm{d}h$ both exist.*

*If any one of the three statements holds then*

$$\begin{aligned}(RYS)\int_a^b f\cdot\mathrm{d}h &= (RYS)\int_a^b f_+^{(a,b)}\cdot\mathrm{d}h - \sum_{(a,b)}\Delta^+ f\cdot\Delta^{\pm} h\\ &= (RYS)\int_a^b f_-^{(a,b)}\cdot\mathrm{d}h + \sum_{(a,b)}\Delta^- f\cdot\Delta^{\pm} h,\end{aligned} \tag{2.49}$$

*where the two sums converge unconditionally.*

*Proof.* We prove only the implication $(a)\Rightarrow(b)$ and (2.49) because the converse implication follows by linearity of the $RYS$ integral and a proof of the rest is symmetric. Thus suppose that $(a)$ holds. It is enough to show that $(RYS)\int_a^b[f_+^{(a,b)}-f]\cdot\mathrm{d}h$ exists and equals the sum in (2.49). First we show that the sum in (2.49) converges unconditionally. Let $\epsilon>0$. By definition of the refinement Young–Stieltjes integral there exists a partition $\lambda$ of $[a,b]$ such that

$$\Big\|S_{YS}(f, \mathrm{d}h; \tau) - (RYS)\int_a^b f\cdot\mathrm{d}h\Big\| < \epsilon \tag{2.50}$$

for any Young tagged partition $\tau=(\kappa,\xi)$ of $[a,b]$ such that $\kappa$ is a refinement of $\lambda$. Let $\zeta=\{u_j\}_{j=1}^m\subset(a,b)$ be a set disjoint from $\lambda$. Choose a refinement $\kappa=\{t_i\}_{i=0}^n$ of $\lambda$ such that each $u_j$ is in $(t_{i(j)-1}, t_{i(j)})$, $j=1,\dots,m$, for some $t_{i(j)-1}, t_{i(j)}\in\kappa\setminus\lambda$, with $i(j)<i(j+1)-1$ for $j=1,\dots,m-1$. Define

$\xi' = \{s'_i\}_{i=1}^n$ and $\xi'' = \{s''_i\}_{i=1}^n$ by $u_j = s'_{i(j)} < s''_{i(j)} < t_{i(j)}$ for $j = 1, \ldots, m$ and take any $s'_i = s''_i \in (t_{i-1}, t_i)$ for the other indices $i$. By (2.50) with $\tau = (\kappa, \xi')$ and $\tau = (\kappa, \xi'')$, we get that

$$\begin{aligned}
&\Big\| \sum_{j=1}^{m} \big[f(u_j) - f(s''_{i(j)})\big] \cdot \big[h(t_{i(j)}-) - h(t_{i(j)-1}+)\big] \Big\| \\
&\quad = \big\| S_{YS}(f, \mathrm{d}h; (\kappa, \xi')) - S_{YS}(f, \mathrm{d}h, (\kappa, \xi''))\big\| < 2\epsilon.
\end{aligned}$$

Letting $s''_{i(j)} \downarrow u_j$, $t_{i(j)} \downarrow u_j$ and $t_{i(j)-1} \uparrow u_j$ for $j = 1, \ldots, m$, as is possible, it follows that $\| \sum_\zeta \Delta^+ f \cdot \Delta^{\pm} h\| \le 2\epsilon$ for any finite set $\zeta \subset (a, b)$ disjoint from $\lambda$. Thus the sum in (2.49) converges unconditionally in $Z$.

Now to prove the existence of $(RYS) \int_a^b [f_+^{(a,b)} - f] \cdot \mathrm{d}h$ notice that for any Young tagged partition $\tau = (\{t_i\}_{i=0}^n, \{s_i\}_{i=1}^n)$ of $[a, b]$,

$$\begin{aligned}
&\Big\|S_{YS}(f_+^{(a,b)} - f, \mathrm{d}h; \tau) - \sum_{(a,b)} \Delta^+ f \cdot \Delta^{\pm} h\Big\| \\
&\le \Big\| \sum_{i=1}^{n} \Delta^+ f(s_i) \cdot \big[h(t_i-) - h(t_{i-1}+)\big] \Big\| \\
&\quad + \Big\| \sum_{(a,b)} \Delta^+ f \cdot \Delta^{\pm} h - \sum_{i=1}^{n-1} [\Delta^+ f \cdot \Delta^{\pm} h](t_i) \Big\|.
\end{aligned}$$

Using (2.50) and the unconditional convergence just proved, one can make the right side arbitrarily small by way of refinements of partitions, proving the lemma. □

Now we are ready to prove the equality of $Y_1$ and $Y_2$ integrals.

**Theorem 2.50.** *Let $f \in \mathcal{R}([a, b]; X)$ and $h \in \mathcal{R}([a, b]; Y)$. Then $(Y_1) \int_a^b f \cdot \mathrm{d}h = (Y_2) \int_a^b f \cdot \mathrm{d}h$ if either side is defined.*

*Proof.* We can assume that $a < b$. Suppose that the $Y_2$ integral exists. Hence by Proposition 2.48, the refinement Young–Stieltjes integral $(RYS) \int_a^b f_-^{(a,b)} \cdot \mathrm{d}h$ exists and (2.48) holds. Then by Lemma 2.49 applied to $f = f_-^{(a,b)}$, the integral $(RYS) \int_a^b f_+^{(a,b)} \cdot \mathrm{d}h$ exists and

$$(RYS) \int_a^b f_-^{(a,b)} \cdot \mathrm{d}h = (RYS) \int_a^b f_+^{(a,b)} \cdot \mathrm{d}h - \sum_{(a,b)} \Delta^{\pm} f \cdot \Delta^{\pm} h,$$

where the sum converges unconditionally. Also, by linearity of unconditional convergence, the sum

$$\sum_{(a,b)} \Delta^+ f \cdot \Delta^{\pm} h = \sum_{(a,b)} \Delta^{\pm} f \cdot \Delta^{\pm} h - \sum_{(a,b)} \Delta^- f \cdot \Delta^{\pm} h$$

converges unconditionally. Therefore the right side of (2.47) is defined, proving the existence of the $Y_1$ integral and the relation, by (2.48),

$$(Y_2)\int_a^b f{\cdot}\mathrm{d}h = (RYS)\int_a^b f_-^{(a,b)}{\cdot}\mathrm{d}h + \sum_{(a,b)} \Delta^- f{\cdot}\Delta^{\pm}h$$

$$= (RYS)\int_a^b f_+^{(a,b)}{\cdot}\mathrm{d}h - \sum_{(a,b)} \Delta^+ f{\cdot}\Delta^{\pm}h = (Y_1)\int_a^b f{\cdot}\mathrm{d}h.$$

A proof of the converse implication is symmetric and we omit it. □

Since the $Y_1$ and $Y_2$ integrals have been shown to coincide for regulated functions $f$ and $h$, if either is defined, we now define the *central Young integral*, or $CY$ integral, by

$$(CY)\int_a^b f{\cdot}\mathrm{d}h := (Y_1)\int_a^b f{\cdot}\mathrm{d}h = (Y_2)\int_a^b f{\cdot}\mathrm{d}h. \tag{2.51}$$

The $CY$ integral extends the $RYS$ integral, as the following shows.

**Theorem 2.51.** *Assuming* (1.14), *for* $f \in \mathcal{R}([a,b];X)$ *and* $h \in \mathcal{R}([a,b];Y)$ *the following hold:*

(*a*) *if* $(RYS)\int_a^b f{\cdot}\mathrm{d}h$ *exists then so does* $(CY)\int_a^b f{\cdot}\mathrm{d}h$, *and the two are equal;*
(*b*) *letting* $\hat{f} := [f_-^{(a,b)} + f_+^{(a,b)}]/2$ (*cf.* (2.46)), *if* $a < b$ *and* $(CY)\int_a^b f{\cdot}\mathrm{d}h$ *exists then so does* $(RYS)\int_a^b \hat{f}{\cdot}\mathrm{d}h$, *the sum* $\sum_{(a,b)}[f - \hat{f}]{\cdot}\Delta^{\pm}h$ *converges unconditionally in* $Z$, *and*

$$(CY)\int_a^b f{\cdot}\mathrm{d}h = (RYS)\int_a^b \hat{f}{\cdot}\mathrm{d}h + \sum_{(a,b)} [f - \hat{f}]{\cdot}\Delta^{\pm}h. \tag{2.52}$$

*Proof.* For (*a*), suppose that $(RYS)\int_a^b f{\cdot}\mathrm{d}h$ exists. By the implication (*a*) ⇒ (*b*) and relation (2.49) of Lemma 2.49, the right side of (2.47) is defined. Hence the $CY$ integral exists and equals the $RYS$ integral, proving (*a*).

For (*b*), suppose that $(CY)\int_a^b f{\cdot}\mathrm{d}h$ exists. By Proposition 2.48, the right sides of (2.47) and (2.48) are defined. By linearity, $(RYS)\int_a^b \hat{f}{\cdot}\mathrm{d}h$ exists and the sum $\sum_{(a,b)}[f-\hat{f}]{\cdot}\Delta^{\pm}h$ converges unconditionally in $Z$. Then (2.52) follows by adding (2.47) and (2.48). □

If a function $f$ has values at its jump points in $(a,b)$ which are normalized, that is, $f = \hat{f} = [f_-^{(a,b)} + f_+^{(a,b)}]/2$, then by the preceding theorem $(CY)\int_a^b f{\cdot}\mathrm{d}h$ and $(RYS)\int_a^b f{\cdot}\mathrm{d}h$ both exist or both do not. In general the equivalence is not true, as the following shows:

**Proposition 2.52.** *There exist functions $f$ and $h$ on $[0,1]$, where $f \in c_0((0,1))$ (cf. Definition 2.9) and $h$ is continuous, for which $(Y_2)\int_0^1 f\,dh$ exists, while $(RYS)\int_0^1 [f - f_-^{(0,1)}]\,dh$ and hence $(RYS)\int_0^1 f\,dh$ do not exist.*

*Proof.* Let $f(t) := k^{-1/2}$ if $t = 1/(3k)$ for $k = 1, 2, \ldots$, and $f(t) := 0$ otherwise. Let $h(1/(3k+1)) := 0$ and $h(1/(3k-1)) := k^{-1/2}$ for $k = 1, 2, \ldots$. Let $h(0) = h(1) := 0$ and let $h$ be "linear in between," i.e., linear on each closed interval where $h$ is so far defined only at the endpoints. Since $h$ is continuous and $f_-^{(0)} \equiv 0$, $(Y_2)\int_0^1 f\,dh$ exists and is 0. Since $f_-^{(0,1)} \equiv 0$, it is enough to show the nonexistence of the integral $(RYS)\int_0^1 f\,dh$. Let $\lambda = \{u_j\}_{j=0}^m$ be any partition of $[0,1]$. Take the smallest $m$ such that $t_m := 1/(3m+1) \le u_1$. If $\kappa_m := \lambda \cup \{t_m\}$ then the contribution to any Young–Stieltjes sum based on $\kappa_m$ coming from $[0, t_m]$ is 0. For $n > m$, consider partitions

$$\kappa_n := \kappa_m \cup \{1/(3k+1), 1/(3k-1) \colon k = m+1, \ldots, n\}.$$

We form a Young–Stieltjes sum based on $\kappa_n$ by letting it be the same as one for $\kappa_m$ on $[t_m, 1]$, and by evaluating $f$ at $1/(3k)$ for $k = m+1, \ldots, n$. Then the part of our Young–Stieltjes sum for $\kappa_n$ coming from $[0, t_m]$ is

$$\begin{aligned}\sum_{k=m+1}^{n} f\left(\frac{1}{3k}\right)&\left[h\left(\frac{1}{3k-1}\right) - h\left(\frac{1}{3k+1}\right)\right]\\ &+0\cdot\left[h\left(\frac{1}{3k-2}\right) - h\left(\frac{1}{3k+1}\right)\right]\\ &= \sum_{k=m+1}^{n} k^{-1/2}k^{-1/2} \to \infty\end{aligned}$$

as $n \to \infty$. Thus two Young–Stieltjes sums for $(RYS)\int_0^1 f\,dh$, both based on refinements of $\lambda$, differ by an arbitrarily large amount. So $(RYS)\int_0^1 f\,dh$ does not exist. □

## 2.6 The Henstock–Kurzweil Integral

A *gauge function* on $[a,b]$ is any function defined on $[a,b]$ with strictly positive values. Given a gauge function $\delta(\cdot)$ on $[a,b]$ with $a < b$, a tagged partition $(\{x_i\}_{i=0}^n, \{y_i\}_{i=1}^n)$ of $[a,b]$ is *$\delta$-fine* if $y_i - \delta(y_i) \le x_{i-1} \le y_i \le x_i \le y_i + \delta(y_i)$ for $i = 1, \ldots, n$. Let $TP(\delta, [a,b])$ be the set of all $\delta$-fine tagged partitions of $[a,b]$.

**Lemma 2.53.** *Let $\delta(\cdot)$ be a gauge function defined on an interval $[a,b]$ with $a < b$. Then:*

(*a*) *$TP(\delta, [a, b])$ is nonempty, i.e. there exists a tagged partition*

$$\zeta := (\{x_i\}_{i=0}^n, \{y_i\}_{i=1}^n) \in TP(\delta, [a, b]).$$

(*b*) *For any finite set $F \subset [a, b]$, $\zeta$ can be chosen so that $F \subset \{x_i\}_{i=0}^n$.*
(*c*) *In (b), $\zeta$ can instead be chosen so that $F \subset \{y_i\}_{i=1}^n$.*

*Proof.* (*a*): The system of open intervals $\{(y - \delta(y), y + \delta(y)) \colon y \in [a, b]\}$ is an open cover of $[a, b]$. Since $[a, b]$ is compact, there is a finite subcover $\{J_i := (y_i - \delta(y_i), y_i + \delta(y_i)) \colon i = 1, \dots, n\}$ of $[a, b]$ such that any $n - 1$ of the $J_i$ do not cover $[a, b]$. Since $y_i$ for $i = 1, \dots, n$ are distinct, we can assume that $y_1 < \cdots < y_n$. Also, $J_i \cap J_{i+1} \neq \emptyset$ for $i = 1, \dots, n - 1$, $J_i \cap J_j = \emptyset$ for $|i - j| > 1$, $i, j = 1, \dots, n$, $a \in J_1$ and $b \in J_n$. Hence one can find numbers $a = x_0 < x_1 < \cdots < x_n = b$ such that $x_i \in J_i \cap J_{i+1}$ for $i = 1, \dots, n - 1$ and $y_i \in [x_{i-1}, x_i]$ for $i = 1, \dots, n$. Thus $(\{x_i\}_{i=0}^n, \{y_i\}_{i=1}^n) \in TP(\delta, [a, b])$, proving (*a*).

(*b*): We can assume that $F \subset (a, b)$. Let $F = \{\xi_j\}_{j=1}^k$ where $a < \xi_1 < \cdots < \xi_k < b$. Let $\xi_0 := a$ and $\xi_{k+1} := b$. By part (*a*), each interval $[\xi_j, \xi_{j+1}]$, $j = 0, 1, \dots, k$, has a $\delta$-fine tagged partition. Combining these, we get one for $[a, b]$, proving (*b*).

(*c*): Given a finite set $F \subset [a, b]$, $[a, b] \setminus \bigcup_{y \in F} (y - \delta(y)/2, y + \delta(y)/2)$ is a finite union of closed subintervals $[a_j, b_j]$ of $[a, b]$. Each of these subintervals has a $\delta$-fine tagged partition. Also, $[a, b] \setminus \bigcup_j [a_j, b_j]$ is a finite union of disjoint intervals (which may be open or closed at either end). Decomposing such intervals into smaller ones, we obtain a set of such intervals each containing exactly one $y \in F$ and included in $(y - \delta(y), y + \delta(y))$. Putting together the tagged partitions of each $[a_j, b_j]$ and the tagged intervals indexed by $y \in F$, we get a $\delta$-fine tagged partition of $[a, b]$ in which each $y \in F$ is a tag, proving (*c*). □

**Definition 2.54.** Assuming (1.14), let $f \colon [a, b] \to X$ and $h \colon [a, b] \to Y$. If $a < b$, the *Henstock–Kurzweil integral*, or $HK$ integral, $(HK) \int_a^b f \cdot dh$ is defined as an $I \in Z$, if it exists, with the following property: given $\epsilon > 0$ there is a gauge function $\delta(\cdot)$ on $[a, b]$ such that $\|S_{RS}(f, dh; \tau) - I\| < \epsilon$ for each $\delta$-fine tagged partition $\tau$ of $[a, b]$. If $a = b$, let $(HK) \int_a^a f \cdot dh := 0$. The integral $(HK) \int_a^b df \cdot h$ is defined, if it exists, via (1.15).

**Proposition 2.55** (Cauchy test). *Let $a < b$, $f \colon [a, b] \to X$, and $h \colon [a, b] \to Y$. The Henstock–Kurzweil integral $(HK) \int_a^b f \cdot dh$ is defined if and only if given $\epsilon > 0$ there is a gauge function $\delta(\cdot)$ on $[a, b]$ such that*

$$\|S_{RS}(f, dh; \tau_1) - S_{RS}(f, dh; \tau_2)\| < \epsilon \tag{2.53}$$

*for each $\tau_1, \tau_2 \in TP(\delta, [a, b])$.*

*Proof.* To prove the "if" part, for each $n \geq 1$, choose a gauge function $\delta_n(\cdot)$ on $[a,b]$ such that (2.53) holds with $\epsilon = 1/n$ for each $\tau_1, \tau_2 \in TP(\delta_n, [a,b])$. Replacing $\delta_n$ by $\min\{\delta_1, \ldots, \delta_n\}$, we can assume that $\delta_1 \geq \delta_2 \geq \cdots$. For $n = 1, 2, \ldots$, let $S_n := S_{RS}(f, \mathrm{d}h; \tau_n)$ for some $\tau_n \in TP(\delta_n, [a,b])$. Then $\{S_n \colon n \geq 1\}$ is a Cauchy sequence in $Z$, and hence convergent to a limit $I$. Given $\epsilon > 0$ choose an $n$ such that $n > 2/\epsilon$ and $\|I - S_n\| < \epsilon/2$. Let $\delta := \delta_n$. Then for each $\tau \in TP(\delta, [a,b])$,

$$\big\|I - S_{RS}(f, \mathrm{d}h; \tau)\big\| \;\leq\; \|I - S_n\| + \|S_n - S_{RS}(f, \mathrm{d}h; \tau)\| < \epsilon/2 + 1/n < \epsilon.$$

Thus $(HK)\int_a^b f{\cdot}\mathrm{d}h$ is defined. Since the converse implication is clear, the proof is complete. □

**Proposition 2.56.** *Let $f\colon [a,b] \to X$ and $h\colon [a,b] \to Y$. Let $a < c < b$. Then $(HK)\int_a^b f{\cdot}\mathrm{d}h$ exists if and only if $(HK)\int_a^c f{\cdot}\mathrm{d}h$ and $(HK)\int_c^b f{\cdot}\mathrm{d}h$ both exist, and then we have*

$$(HK)\int_a^b f{\cdot}\mathrm{d}h \;=\; (HK)\int_a^c f{\cdot}\mathrm{d}h + (HK)\int_c^b f{\cdot}\mathrm{d}h. \tag{2.54}$$

The proof of this property is easy once we observe the following:

**Lemma 2.57.** *For any gauge function $\delta(\cdot)$ on $[a,b]$ and $c \in [a,b]$, there is a gauge function $\delta'(\cdot)$ on $[a,b]$ such that $\delta' \leq \delta$ and each $\delta'$-fine tagged partition has $c$ as a tag, in the interior of its interval in the relative topology of $[a,b]$.*

*Proof.* Suppose that $a \leq c \leq b$ and $\delta(\cdot)$ is a gauge function on $[a,b]$. Let $\delta'(c) := \delta(c)$, and let $\delta'(x) := \min\{\delta(x), |c-x|/2\}$ for $x \in [a,b] \setminus \{c\}$. For a $\delta'$-fine tagged partition $(\{x_i\}_{i=0}^n, \{y_i\}_{i=1}^n)$ of $[a,b]$, $c = y_i \in (x_{i-1}, x_i)$ if $c \in [x_{i-1}, x_i] \cap (a,b)$, $c = y_1 < x_1$ if $c = a$, or $c = y_n > x_{n-1}$ if $c = b$. □

**Proof of Proposition 2.56.** Suppose that $I_1 := (HK)\int_a^c f{\cdot}\mathrm{d}h$ and $I_2 := (HK)\int_c^b f{\cdot}\mathrm{d}h$ both exist. Given $\epsilon > 0$, take gauge functions $\delta_1(\cdot)$ on $[a,c]$ for $I_1$ and $\epsilon/2$, and $\delta_2(\cdot)$ on $[c,b]$ for $I_2$ and $\epsilon/2$. By the preceding lemma, there is a gauge function $\delta(\cdot)$ on $[a,b]$ such that $\delta \leq \delta_1$ on $[a,c)$, $\delta \leq \delta_2$ on $(c,b]$, $\delta(c) \leq \min\{\delta_1(c), \delta_2(c)\}$, and $c$ is a tag for each $\delta$-fine tagged partition of $[a,b]$. Let $\tau = (\{x_i\}_{i=0}^n, \{y_i\}_{i=1}^n)$ be a $\delta$-fine tagged partition of $[a,b]$ and let $c = y_i$ for some $i \in \{1, \ldots, n\}$. Writing $f(y_i){\cdot}[h(x_i) - h(x_{i-1})] = f(y_i){\cdot}[h(y_i) - h(x_{i-1})] + f(y_i){\cdot}[h(x_i) - h(y_i)]$, we have $S_{RS}(f, \mathrm{d}h; \tau) = S_{RS}(f, \mathrm{d}h; \tau_1) + S_{RS}(f, \mathrm{d}h; \tau_2)$ for suitable $\tau_1 \in TP(\delta_1, [a,c])$ and $\tau_2 \in TP(\delta_2, [c,b])$. Thus $\|I_1 + I_2 - S_{RS}(f, \mathrm{d}h; \tau)\| < \epsilon$, proving the existence of $(HK)\int_a^b f{\cdot}\mathrm{d}h$ and relation (2.54).

Now suppose that $I := (HK)\int_a^b f{\cdot}\mathrm{d}h$ exists. Let $\delta(\cdot)$ be a gauge function on $[a,b]$ for $I$ and $\epsilon/2$. By the preceding lemma we can and do assume that

$c$ is a tag for each $\delta$-fine tagged partition of $[a,b]$. Let $\delta_1(\cdot)$ and $\delta_2(\cdot)$ be the restrictions of $\delta(\cdot)$ to $[a,c]$ and $[c,b]$, respectively. Let $\tau_1', \tau_1'' \in TP(\delta_1,[a,c])$ and let $\tau_2 \in TP(\delta_2,[c,b])$. Then $\tau' := \tau_1' \cup \tau_2$ and $\tau'' := \tau_2'' \cup \tau_2$ are two $\delta$-fine tagged partitions of $[a,b]$, and

$$\|S_{RS}(f,\mathrm{d}h;\tau_1') - S_{RS}(f,\mathrm{d}h;\tau_1'')\| = \|S_{RS}(f,\mathrm{d}h;\tau') - S_{RS}(f,\mathrm{d}h;\tau'')\| < \epsilon.$$

Thus $(HK)\int_a^c f\cdot\mathrm{d}h$ exists by the Cauchy test. Similarly it follows that $(HK)\int_c^b f\cdot\mathrm{d}h$ exists. Since $\tau_1 \cup \tau_2 \in TP(\delta,[a,b])$ for each $\tau_1 \in TP(\delta_1,[a,c])$ and $\tau_2 \in TP(\delta_2,[c,b])$, (2.54) holds. □

The Henstock–Kurzweil integral shares some elementary properties with other Stieltjes-type integrals considered above, as will be shown in Theorems 2.72 and 2.73. Essentially the next statement says that the $HK$ integral exists if and only if corresponding improper integrals exist. Proposition 2.76 will show that this property is not shared by the $RRS$, $RYS$, or $CY$ integral.

**Proposition 2.58.** *Assuming* (1.14), *let* $f\colon [a,b] \to X$ *and* $h\colon [a,b] \to Y$ *with* $a < b$. *The integral* $(HK)\int_a^b f\cdot \mathrm{d}h$ *is defined if and only if the integral* $(HK)\int_x^b f\cdot\mathrm{d}h$ *is defined for each* $x \in (a,b)$ *and the limit*

$$\lim_{x\downarrow a}\left\{(HK)\int_x^b f\cdot\mathrm{d}h + f(a)\cdot[h(x)-h(a)]\right\} \tag{2.55}$$

*exists. Also, the integral* $(HK)\int_a^b f\cdot\mathrm{d}h$ *is defined if and only if* $(HK)\int_a^x f\cdot\mathrm{d}h$ *is defined for each* $x \in (a,b)$ *and the limit*

$$\lim_{x\uparrow b}\left\{(HK)\int_a^x f\cdot\mathrm{d}h + f(b)\cdot[h(b)-h(x)]\right\}$$

*exists. In either case, the limit and the integral over* $[a,b]$ *are equal.*

*Proof.* We prove only the first part of Proposition 2.58 because a proof of the second part is symmetric. Suppose that $I := (HK)\int_a^b f\cdot\mathrm{d}h$ is defined. Then $I(x) := (HK)\int_x^b f\cdot\mathrm{d}h$ is defined for each $x \in (a,b]$ by Proposition 2.56. Given $\epsilon > 0$, let $\delta(\cdot)$ be a gauge function on $[a,b]$ such that any two Riemann–Stieltjes sums based on $\delta$-fine tagged partitions differ by at most $\epsilon$. For $x \in (a, a+\delta(a)]$, by Proposition 2.56 and Lemma 2.53($a$), there is a $\delta$-fine tagged partition $\tau_x$ of $[a,x]$ such that the Riemann–Stieltjes sum based on $\tau_x$ is within $\epsilon$ of $(HK)\int_a^x f\cdot\mathrm{d}h$. Also, let $\tau^x$ be any $\delta$-fine tagged partition of $[x,b]$ and $\sigma_x := (\{a,x\},a)$. Let $\tau_1$ and $\tau_2$ be two tagged partitions of $[a,b]$ which coincide with $\tau_x$ and $\sigma_x$, respectively, when restricted to $[a,x]$, and which both equal $\tau^x$ when restricted to $[x,b]$. Then $\tau_1$ and $\tau_2$ are $\delta$-fine and

$$\begin{aligned}\|I - I(x) - f(a)\cdot[h(x)-h(a)]\| &\le \Big\|(HK)\int_a^x f\cdot\mathrm{d}h - S_{RS}(f,\mathrm{d}h;\tau_x)\Big\| \\ &\quad + \|S_{RS}(f,\mathrm{d}h;\tau_1) - S_{RS}(f,\mathrm{d}h;\tau_2)\| \\ &< 2\epsilon.\end{aligned}$$

Since $x \in (a, a+\delta(a)]$ and $\epsilon > 0$ are arbitrary, the limit (2.55) exists and equals $I$. Thus the "only if" part of the proposition holds.

To prove the converse implication we need the following auxiliary statement, sometimes called Henstock's lemma.

**Lemma 2.59.** *Suppose $a < b$ and $I := (HK)\int_a^b f\cdot dh$ is defined. Given $\epsilon > 0$, let $\delta(\cdot)$ be a gauge function on $[a,b]$ for $I$ and $\epsilon$, and let $(\{x_i\}_{i=0}^n, \{y_i\}_{i=1}^n)$ be a $\delta$-fine tagged partition of $[a,b]$. Then for any subset $J \subset \{1,\ldots,n\}$,*

$$\Big\|\sum_{i\in J}\Big\{f(y_i)\cdot[h(x_i)-h(x_{i-1})]-(HK)\int_{x_{i-1}}^{x_i} f\cdot dh\Big\}\Big\| \le \epsilon. \tag{2.56}$$

*Proof.* Let $k :=\operatorname{card} J'$ for $J' := \{1,\ldots,n\}\setminus J$, and let $\epsilon'$ be an arbitrary positive number. By Proposition 2.56, for each $i \in J'$, there is a $\delta$-fine tagged partition $\tau_i$ of $[x_{i-1}, x_i]$ such that

$$\left\|S_{RS}(f, dh; \tau_i) - (HK)\int_{x_{i-1}}^{x_i} f\cdot dh\right\| < \epsilon'/k.$$

Let $\tau := \bigcup_{i\in J'}\tau_i \cup \bigcup_{i\in J}(\{x_{i-1}, x_i\}, y_i)$. Then $\tau$ is a $\delta$-fine tagged partition of $[a,b]$, and the left side of (2.56) is equal to

$$\Big\|S_{RS}(f, dh; \tau) - (HK)\int_a^b f\cdot dh - \sum_{i\in J'}\Big\{S_{RS}(f, dh; \tau_i) - (HK)\int_{x_{i-1}}^{x_i} f\cdot dh\Big\}\Big\|,$$

which is less than $\epsilon + \epsilon'$. Since $\epsilon'$ is arbitrary, (2.56) holds. □

**Continuation of the proof of Proposition 2.58.** Suppose that the limit (2.55) exists. Let $I$ be its value. Then $I(x) := (HK)\int_x^b f\cdot dh$ is defined for each $x \in (a,b)$.

Given $\epsilon > 0$, we will construct a gauge function $\delta(\cdot)$ on $(a,b]$ such that for each $x \in (a,b)$, $a+\delta(x) < x$ and $\|I(x) - S_{RS}(f, dh; \tau^x)\| \le \epsilon$ for any $\delta$-fine tagged partition $\tau^x$ of $[x,b]$. Suppose we have such a gauge function $\delta(\cdot)$ on $(a,b]$. Define $\delta(a)$ so that $\|I - I(x) - f(a)\cdot[h(x)-h(a)]\| < \epsilon$ for any $x \in (a, a+\delta(a)]$. Let $\tau = (\{x_i\}_{i=0}^n, \{y_i\}_{i=1}^n)$ be a $\delta$-fine tagged partition of $[a,b]$. Then by construction of $\delta(\cdot)$, we have that $y_1 = a$, $\tau^{x_1} := (\{x_i\}_{i=1}^n, \{y_i\}_{i=2}^n)$ is a $\delta$-fine tagged partition of $[x_1, b]$, and

$$\begin{aligned}\big\|I - S_{RS}(f, dh; \tau)\big\| &\le \big\|I(x_1) - S_{RS}(f, dh; \tau^{x_1})\big\| \\ &\quad + \big\|I - I(x_1) - f(a)\cdot[h(x_1)-h(a)]\big\| \\ &< 2\epsilon.\end{aligned}$$

Thus the integral $(HK)\int_a^b f\cdot dh$ is defined and equals $I$.

To construct the desired gauge function $\delta(\cdot)$ on $(a, b]$, let $\{u_n: n \geq 0\}$ be a decreasing sequence of numbers such that $u_0 = b$ and $\lim_{n\to\infty} u_n = a$. Given $\epsilon > 0$, let $\delta_1(\cdot)$ be a gauge function on $[u_1, u_0]$ for $(HK) \int_{u_1}^{u_0} f \cdot dh$ and $\epsilon/2$. For $n \geq 2$, let $\delta_n(\cdot)$ be a gauge function on $[u_n, u_{n-2}]$ for $(HK) \int_{u_n}^{u_{n-2}} f \cdot dh$ and $\epsilon/2^n$. For $z \in (u_1, u_0]$ define $\delta(z)$ so that $0 < \delta(z) \leq \delta_1(z)$ and $u_1 < z - \delta(z)$. For $z \in (u_n, u_{n-1}]$ with $n > 1$, define $\delta(z)$ so that $0 < \delta(z) \leq \delta_n(z)$ and $u_n < z - \delta(z) < z + \delta(z) < u_{n-2}$. Thus $\delta(\cdot)$ is defined on $(a, b]$. Let $x \in (a, b)$ and let $\tau^x = (\{x_i\}_{i=0}^m, \{y_i\}_{i=1}^m)$ be a $\delta$-fine tagged partition of $[x, b]$. For each $n \geq 1$, let $I_n := \{i: y_i \in (u_n, u_{n-1}]\}$, and let $J_n := \cup_{i\in I_n}[x_{i-1}, x_i]$. By construction, $\tau^x$ restricted to $J_n$ is a $\delta_n$-fine tagged partition of $J_n$. Since $J_1 \subset (u_1, u_0]$ and $J_n \subset (u_n, u_{n-2})$ for each $n \geq 2$, by Lemma 2.59, for each $n \geq 1$,

$$\Big\| \sum_{i\in I_n} \Big\{ f(y_i) \cdot [h(x_i) - h(x_{i-1})] - (HK) \int_{x_{i-1}}^{x_i} f \cdot dh \Big\} \Big\| \leq \frac{\epsilon}{2^n}.$$

Thus

$$\Big\| (HK) \int_x^b f \cdot dh - S_{RS}(f, dh; \tau^x) \Big\| \leq \sum_{n=1}^{\infty} \epsilon/2^n = \epsilon.$$

As noted earlier, this completes the proof of Proposition 2.58. □

**Proposition 2.60.** *Let $h$ be a real-valued function on $[a, b]$. If $(HK) \int_a^b f\, dh$ exists whenever $f = 1_J$ for an interval $J$, then $h$ is regulated on $[a, b]$.*

*Proof.* Suppose not. Then we can assume that for some $u_k \downarrow c \in [a, b)$, the $h(u_k)$ do not converge. If $h(u_k)$ are unbounded let $\epsilon = 1$. If they are bounded let $\epsilon := 3^{-1}[\limsup_{k\to\infty} h(u_k) - \liminf_{k\to\infty} h(u_k)]$. Let $f_c := 1_{(c,b]}$. Take a gauge function $\delta(\cdot)$ on $[a, b]$ for $\epsilon$ and $(HK) \int_a^b f_c\, dh$. By Lemma 2.57, there exist a gauge function $\delta' \leq \delta$ and a $\delta'$-fine tagged partition $\tau$ of $[a, b]$ with $c$ as the tag for an interval $[x_{i-1}, x_i]$ and $c < x_i$. For all $k$ large enough we have $u_k < x_i$. Then take a $\delta'$-fine tagged refinement $\tau_k$ of $\tau$ where $c$ is the tag for $[x_{i-1}, u_k]$, by Lemma 2.53(a) for $[u_k, x_i]$. The corresponding Riemann–Stieltjes sums

$$S_{RS}(f_c, dh; \tau_k) = h(b) - h(u_k)$$

vary by more than $2\epsilon$ as $k \to \infty$, a contradiction. □

Suppose a continuous function $f: [a, b] \mapsto \mathbb{R}$ is differentiable everywhere on $(a, b)$. It may be that $f'$ is not Lebesgue integrable, e.g. if $[a, b] = [0, 1]$, $f(x) = x^2 \sin(\pi/x^2)$, $0 < x < 1$, $f(x) = 0$ elsewhere. Between 1910 and 1920, Denjoy and Perron (see the Notes) defined integrals such that for every such $f'$, $\int_a^x f'(t)\, dt = f(x) - f(a)$, $a \leq x \leq b$. The next fact shows that the later-invented Henstock–Kurzweil integral has this property:

**Theorem 2.61.** *Let $f$ be a continuous function from $[a,b)$ to $\mathbb{R}$ having a derivative $f'(x)$ for $a < x < b$ except for at most countably many $x$. Then $(HK)\int_a^x f'(t)\,\mathrm{d}t$ exists and equals $f(x) - f(a)$ for $a \le x < b$.*

*Proof.* The statement follows from Theorems 7.2 and 11.1 of Gordon [84]. □

For another example, $f(x) := x^2 \sin(e^{1/x})$, $x \neq 0$, $f(0) := 0$, satisfies the conditions of Theorem 2.61 although $f'$ has very wild oscillations.

## *2.7 Ward–Perron–Stieltjes and Henstock–Kurzweil Integrals

Before defining the Ward–Perron–Stieltjes integral we need to define what are called major and minor functions. Given two real-valued functions $f$, $h$ on $[a,b]$, we call $M$ a *major function* of $f$ with respect to $h$ if $M(a) = 0$, $M$ has finite values on $[a,b]$, and for each $x \in [a,b]$ there exists $\delta(x) > 0$ such that

$$\begin{array}{lll} M(z) \ge M(x) + f(x)[h(z) - h(x)] & \text{if} & x \le z \le \min\{b, x + \delta(x)\}, \\ M(z) \le M(x) + f(x)[h(z) - h(x)] & \text{if} & \max\{a, x - \delta(x)\} \le z \le x. \end{array} \tag{2.57}$$

Thus for a major function $M$ there exists a positive function $\delta_M(\cdot) = \delta(\cdot)$ on $[a,b]$ which satisfies (2.57). Let $\mathcal{U}(f,h)$ be the class of all major functions of $f$ with respect to $h$, and let

$$U(f,h) := \begin{cases} \inf\{M(b) \colon M \in \mathcal{U}(f,h)\} & \text{if } \mathcal{U}(f,h) \neq \emptyset, \\ +\infty & \text{if } \mathcal{U}(f,h) = \emptyset. \end{cases}$$

A function $m$ is a *minor function* of $f$ with respect to $h$ if $-m \in \mathcal{U}(-f,h)$. Let $\mathcal{L}(f,h)$ be the class of all minor functions of $f$ with respect to $h$. Thus $m \in \mathcal{L}(f,h)$ provided $m(a) = 0$, $m$ has finite values on $[a,b]$, and there exists a positive function $\delta(\cdot) = \delta_m(\cdot)$ on $[a,b]$ such that

$$\begin{array}{lll} m(z) \le m(x) + f(x)[h(z) - h(x)] & \text{if} & x \le z \le \min\{b, x + \delta(x)\}, \\ m(z) \ge m(x) + f(x)[h(z) - h(x)] & \text{if} & \max\{a, x - \delta(x)\} \le z \le x. \end{array} \tag{2.58}$$

Let

$$L(f,h) := \begin{cases} \sup\{m(b) \colon m \in \mathcal{L}(f,h)\} & \text{if } \mathcal{L}(f,h) \neq \emptyset, \\ -\infty & \text{if } \mathcal{L}(f,h) = \emptyset. \end{cases}$$

Then the following is true:

**Lemma 2.62.** $L(f,h) \le U(f,h)$.

*Proof.* The statement is true if either of the sets $\mathcal{L}(f,h)$ or $\mathcal{U}(f,h)$ is empty. Suppose that $\mathcal{L}(f,h)$ and $\mathcal{U}(f,h)$ are nonempty. Let $m \in \mathcal{L}(f,h)$ and $M \in \mathcal{U}(f,h)$. Also, let $\delta := \delta_m \wedge \delta_M$ and $w(x) := M(x) - m(x)$ for $x \in [a,b]$. By (2.57) and (2.58), for each $x \in (a,b]$, it follows that

$$\begin{array}{lll} w(z) \geq w(x) & \text{if} & x \leq z \leq \min\{b, x+\delta(x)\}, \\ w(z) \leq w(x) & \text{if} & \max\{a, x-\delta(x)\} \leq z \leq x. \end{array}$$

Therefore $\inf\{M(x)\colon M \in \mathcal{U}(f,h)\} - \sup\{m(x)\colon m \in \mathcal{L}(f,h)\}$ is a nondecreasing function of $x$. Then the statement of the lemma holds because $m(a) = M(a) = 0$. □

Now we are ready to define the Ward–Perron–Stieltjes integral.

**Definition 2.63.** If $U(f,h) = L(f,h)$ is finite then the common value will be denoted by $(WPS)\int_a^b f\,dh$ and called the *Ward–Perron–Stieltjes integral*, or the $WPS$ integral.

Note that if $a = b$ then $U(f,h) = L(f,h) = 0$, and so the integral $(WPS)\int_a^a f\,dh$ equals 0. First we show that the Ward–Perron–Stieltjes integral extends the refinement Riemann–Stieltjes integral.

**Theorem 2.64.** *If* $(RRS)\int_a^b f\,dh$ *exists then so does* $(WPS)\int_a^b f\,dh$*, and the two are equal.*

*Proof.* We can assume that $a < b$. For $a \leq u < v \leq b$, let $M(u,v)$ be the supremum of all Riemann–Stieltjes sums $S_{RS}(f, dh; \tau)$ based on tagged partitions $\tau$ of $[u,v]$, with $M(u,u) \equiv 0$. Then

$$\begin{array}{lll} M(u,y) \geq M(u,x) + f(x)[h(y) - h(x)] & \text{if} & u \leq x \leq y, \\ M(u,y) \leq M(u,x) + f(x)[h(y) - h(x)] & \text{if} & u \leq y \leq x. \end{array} \tag{2.59}$$

Suppose $(RRS)\int_a^b f\,dh$ exists and $\epsilon > 0$. Then there exists $\lambda = \{z_j\colon j = 0,\dots,m\} \in \mathrm{PP}\,[a,b]$ such that

$$(RRS)\int_a^b f\,dh - \epsilon < S_{RS}(f, dh; \tau) < (RRS)\int_a^b f\,dh + \epsilon \tag{2.60}$$

for each tagged refinement $\tau$ of $\lambda$. For each $x \in (a,b]$, let $j(x)$ be the largest integer such that $z_{j(x)} \leq x$. Let $M(a) := 0$, and for each $x \in (a,b]$ let

$$M(x) := \sum_{j=1}^{j(x)} M(z_{j-1}, z_j) + M(z_{j(x)}, x),$$

where the sum over the empty set is 0. Define a positive function $\delta(\cdot)$ on $[a,b]$ by $\delta(x) := \min\{x - z_{j(x)}, z_{j(x)+1} - x\}$ if $x \in (z_{j(x)}, z_{j(x)+1})$, $\delta(a) := z_1 - a$,

$\delta(b) := b - z_{m-1}$, and $\delta(z_j) := \min\{z_j - z_{j-1}, z_{j+1} - z_j\}$ for $j = 1, \dots, m-1$. By (2.59), the functions $M$ and $\delta$ so defined satisfy (2.57). Hence $M$ is a major function of $f$ with respect to $h$. By (2.60), it follows that $M(b) \le (RRS) \int_a^b f\, dh + \epsilon$. Similarly one can define a minor function $m$ such that $m(b) \ge (RRS) \int_a^b f\, dh - \epsilon$. Since $\epsilon$ is arbitrarily small, this proves the theorem. □

Next is the main result of this section, which says that the Henstock–Kurzweil and Ward–Perron–Stieltjes integrals coincide.

**Theorem 2.65.** *The integral* $(WPS)\int_a^b f\, dh$ *exists if and only if* $(HK)\int_a^b f\, dh$ *exists, and then their values are equal.*

*Proof.* We can assume that $a < b$. Suppose $(WPS) \int_a^b f\, dh$ is defined. Then given $\epsilon > 0$, there exist a major function $M \in \mathcal{U}(f, h)$ and a minor function $m \in \mathcal{L}(f, h)$ such that

$$M(b) - \epsilon < (WPS) \int_a^b f\, dh < m(b) + \epsilon. \tag{2.61}$$

Let $\delta := \delta_M \wedge \delta_m$ and let $\tau = (\{t_i\}_{i=0}^n, \{s_i\}_{i=1}^n)$ be any $\delta$-fine tagged partition of $[a, b]$, which exists by Lemma 2.53(a). By (2.57), we have

$$\begin{aligned} M(t_i) - M(s_i) &\ge f(s_i)[h(t_i) - h(s_i)], \\ M(s_i) - M(t_{i-1}) &\ge f(s_i)[h(s_i) - h(t_{i-1})] \end{aligned} \tag{2.62}$$

for each $i = 1, \dots, n$. Adding up the $2n$ inequalities (2.62), we get the upper bound $M(b) \ge S_\delta(\tau)$ for the Riemann–Stieltjes sum $S_\delta(\tau) := S_{RS}(f, dh; \tau)$. Similarly, using (2.58), we get the lower bound $S_\delta(\tau) \ge m(b)$. By (2.61), it then follows that

$$(WPS) \int_a^b f\, dh - \epsilon < S_\delta(\tau) < (WPS) \int_a^b f\, dh + \epsilon.$$

Thus $(HK) \int_a^b f\, dh$ exists and has the same value as the $(WPS)$ integral.

Now suppose $(HK) \int_a^b f\, dh$ is defined. Given $\epsilon > 0$ there exists a gauge function $\delta(\cdot)$ such that for each $\delta$-fine tagged partition $\tau$, the Riemann–Stieltjes sum $S_\delta(\tau) := S_{RS}(f, dh; \tau)$ has bounds

$$(HK) \int_a^b f\, dh - \epsilon < S_\delta(\tau) < (HK) \int_a^b f\, dh + \epsilon. \tag{2.63}$$

For each $x \in (a, b]$, let $\delta_x(\cdot)$ be the gauge function $\delta(\cdot)$ restricted to the interval $[a, x]$, and $m(x) := \inf\{S_{\delta_x}(\tau) \colon \tau \in TP(\delta_x, [a, x])\}$, $M(x) := \sup\{S_{\delta_x}(\tau) \colon \tau \in TP(\delta_x, [a, x])\}$. Let $m(a) = M(a) := 0$. Then $m(x)$ and $M(x)$ have finite values for each $x \in (a, b]$. Let $x \in (a, b)$ and $z \in (x, x + \delta(x) \wedge$

$b]$. Then for each $\tau = (\{a = t_0, \ldots, t_m = x\}, \{s_i\}_{i=0}^m) \in TP(\delta_x, [a,x])$, $(\{t_0, \ldots, t_m, z\}, \{s_1, \ldots, s_m, x\}) \in TP(\delta_z, [a,z])$. Thus

$$S_{\delta_x}(\tau) + f(x)[h(z) - h(x)] \leq M(z)$$

for any $\tau \in TP(\delta_x, [a,x])$. Thus the first inequality in (2.57) holds. Similarly one can show that the second one and (2.58) also hold. Therefore $m$ and $M$ are minor and major functions respectively. By Lemma 2.62, the definitions of $L(f,h)$ and $U(f,h)$, and (2.63), it then follows that

$$(HK)\int_a^b f\,\mathrm{d}h - \epsilon \leq m(b) \leq L(f,h) \leq U(f,h) \leq M(b) \leq (HK)\int_a^b f\,\mathrm{d}h + \epsilon.$$

Since $\epsilon > 0$ is arbitrary, $L(f,h) = U(f,h)$ and equals the $HK$ integral. The proof of Theorem 2.65 is complete. □

Next we show that for integrators in the class $c_0((a,b)) = c_0((a,b),[a,b];\mathbb{R})$ with $a < b$ (cf. Definition 2.9), the Henstock–Kurzweil and Ward–Perron–Stieltjes integrals differ from the refinement Young–Stieltjes, central Young, and Kolmogorov integrals.

**Proposition 2.66.** *For $f \in \mathcal{R}[a,b]$ and any $h \in c_0((a,b))$ if $a < b$ or $h \in \mathcal{R}[a,a]$, i.e. $h$ is any function $\{a\} \to \mathbb{R}$ if $a = b$,*

$$(RYS)\int_a^b f\,\mathrm{d}h = (CY)\int_a^b f\,\mathrm{d}h = \oint_{[a,b]} f\,\mathrm{d}\mu_{h,[a,b]} = 0.$$

*Proof.* This follows from the definitions of the integrals because $h_-^{(a)} \equiv h_+^{(b)} \equiv \Delta_{[a,b]}^{\pm}h \equiv \mu_{h,[a,b]} \equiv 0$ if $a < b$ and $\mu_{h,[a,a]} \equiv 0$. □

We will show in Proposition 2.68 that the analogous statement for the Henstock–Kurzweil integral does not hold.

**Lemma 2.67.** *For each $h \in c_0((a,b))$ with $a < b$ there exists a right-continuous $f \in \mathcal{R}[a,b]$ such that $\Delta^- f \equiv h$ on $(a,b)$.*

*Proof.* Let $h = \sum_i c_i 1_{\{\xi_i\}}$. For each $i \geq 1$, let $f_i(x) := 0$ if $x \in [a,\xi_i) \cup [\xi_i + \delta_i, b]$ for some $\delta_i > 0$, let $f_i(\xi_i) := c_i$, and let $f_i$ be linear on $[\xi_i, \xi_i + \delta_i]$. If $\delta_i \to 0$ fast enough as $i \to \infty$ so that each $(\xi_i, \xi_i + \delta_i)$ contains no $\xi_j$ with $|c_j| > |c_i|/2$, then $f := \sum_i f_i$ converges uniformly on $[a,b]$ and $f$ has the stated properties. □

**Proposition 2.68.** *There exist $h \in c_0((0,1))$ and $f \in \mathcal{R}[0,1]$ such that $(HK)\int_0^1 f\,\mathrm{d}h$ is undefined.*

*Proof.* For $k$ odd, $k = 1, 3, \ldots, 2^n - 1$, $n = 1, 2, \ldots$, let $h(k/2^n) := 1/2^{n/3}$ and $h = 0$ elsewhere. Then $h \in c_0((0,1))$. By Lemma 2.67, take a right-continuous $f \in \mathcal{R}[0,1]$ such that $\Delta^- f(k/2^n) = 1/2^{n/3}$ for $k = 1, 3, \ldots, 2^n - 1$ and $n = 1, 2, \ldots$.

Let $\delta(\cdot)$ be any strictly positive function on $[0,1]$. By the category theorem (e.g. Theorem 2.5.2 in Dudley [53]), there is an interval $J = [c,d] \subset [0,1]$ of length $\epsilon$ for some $\epsilon > 0$ such that $\delta(y) > \epsilon$ for a dense set $S$ of $y$ in $J$, and the endpoints of $J$ are not dyadic rationals. So, all tagged partitions of $J$ with tags in $S$ are $\delta$-fine.

For $r = 1, 2, \ldots$, let $M(r)$ be the set of all binary rationals $k/2^n$ in $J$ for integers $n$, $1 \le n \le r$, and odd integers $k \ge 1$. As $n \to +\infty$, the number of such $k/2^n$ for a fixed $n$ is asymptotic to $2^{n-1}\epsilon$. For a given $r$ write $M(r) = \{x_{2j-1}\}_{j=1}^m$ where $x_1 < x_3 < \cdots < x_{2m-1}$ for some $m = m(r)$. For each $j \in \{1, \ldots, m-1\}$, choose $x_{2j} \in (x_{2j-1}, x_{2j+1})$ which is not a dyadic rational, and let $x_0 := c$, $x_{2m} := d$, recalling that $J = [c,d]$. For each $j \in \{1, \ldots, m\}$, choose $y_{2j-1} \in S \cap (x_{2j-2}, x_{2j-1})$ and $y_{2j} \in S \cap (x_{2j-1}, x_{2j})$ close enough to $x_{2j-1}$ so that $|[f(y_{2j}) - f(y_{2j-1}) - \Delta^- f(x_{2j-1})]h(x_{2j-1})| < 1/|M(r)|$, where $|M(r)| = \mathrm{card}\,(M(r))$. Then $\zeta_r := (\{x_j\}_{j=0}^{2m}, \{y_j\}_{j=1}^{2m})$ is a $\delta$-fine tagged partition of $J$ and

$$\Big| - S_{RS}(f, \mathrm{d}h; \zeta_r) - \sum_{j=1}^{m} \Delta^- f(x_{2j-1}) h(x_{2j-1}) \Big| < 1.$$

Since the latter sums are unbounded as $r$ increases, so are the former Riemann–Stieltjes sums. Then combining $\zeta_r$ with a $\delta$-fine tagged partition of $[0,c]$ (if $0 < c$) and a $\delta$-fine tagged partition of $[d,1]$ (if $d < 1$), which exist by Lemma 2.53(a), we get unbounded Riemann–Stieltjes sums for $\delta$-fine tagged partitions of $[0,1]$. Since the gauge function $\delta(\cdot)$ on $[0,1]$ is arbitrary, $(HK)\int_0^1 f\,\mathrm{d}h$ does not exist. □

Under some restrictions on the integrator, the Henstock–Kurzweil integral will be shown to extend the refinement Young–Stieltjes integral. Note that in the following, $h$ may be a function $R_{\mu,a}$ as in Proposition 2.6(f), right-continuous on $(a,b)$ but not necessarily at $a$, so the *RYS* integral can be replaced by the Kolmogorov integral according to Corollary 2.26.

**Theorem 2.69.** *Suppose $f$ and $h$ are regulated on $[a,b]$, and if $a < b$, $h$ is right-continuous or left-continuous on $(a,b)$. If $(RYS)\int_a^b f\,\mathrm{d}h$ exists then $(HK)\int_a^b f\,\mathrm{d}h$ exists and has the same value.*

*Proof.* We can assume that $a < b$ and $h$ is right-continuous on $(a,b)$ by symmetry. Given $\epsilon > 0$, let $\lambda = \{z_j\}_{j=0}^m$ be a partition of $[a,b]$ such that for any Young tagged refinement $\tau$ of $\lambda$,

$$\Big| S_{YS}(f, \mathrm{d}h; \tau) - (RYS)\int_a^b f\,\mathrm{d}h \Big| < \epsilon. \tag{2.64}$$

Recall $\Delta^+_{(a,b)}f$ defined before (2.1) and $f_+^{(a,b)}$ defined in (2.46). Then $\Delta^+_{(a,b)}f \equiv f_+^{(a,b)} - f$ on $[a,b]$. By Lemma 2.49 we can assume that (2.64) holds also for $f$ replaced by $f_+^{(a,b)}$ or by $\Delta^+_{(a,b)}f$, taking $\lambda$ as a common refinement $\lambda = \lambda' \cup \lambda''$ of partitions $\lambda'$ for $f_+^{(a,b)}$ and $\epsilon/2$, and $\lambda''$ for $\Delta^+_{(a,b)}f$ and $\epsilon/2$. So it suffices to prove the theorem for $f_+^{(a,b)}$ and for $\Delta^+_{(a,b)}f$.

Define a gauge function $\delta(\cdot)$ on $[a,b]$ as follows: if $t \notin \lambda$, let $\delta(t) := \min_j |t-z_j|/2$. Then any $\delta(\cdot)$-fine tagged partition must contain each $z_j$ as a tag. For $j = 0, \dots, m$, define $\delta(z_j)$ such that $\delta(z_j) \le \min_{i \ne j} |z_i - z_j|/3$ and

$$\|f\|_\infty \sum_{j=1}^{m} \Big\{ \mathrm{Osc}\big(h; [z_j - \delta(z_j), z_j)\big) + \mathrm{Osc}\big(h; (z_{j-1}, z_{j-1} + 2\delta(z_{j-1})]\big) \Big\} < \epsilon. \tag{2.65}$$

Let $\tau = \big(\{t_i\}_{i=0}^n, \{s_i\}_{i=1}^n\big)$ be any $\delta(\cdot)$-fine tagged partition of $[a,b]$ and let

$$S_\delta(\tau) := S_{RS}(f, \mathrm{d}h; \tau) = \sum_{i=1}^{n} f(s_i)\big[h(t_i) - h(t_{i-1})\big] \tag{2.66}$$

be the Riemann–Stieltjes sum based on $\tau$. We can assume that $s_i = t_i = s_{i+1}$ never occurs, since if it did, the $i$th and $(i+1)$st term in $S_\delta(\tau)$ could be replaced by $f(s_i)[h(t_{i+1}) - h(t_{i-1})]$, and we would still have a Riemann–Stieltjes sum based on a $\delta(\cdot)$-fine tagged partition equal to $S_\delta(\tau)$. Since the tags $\{s_i\}_{i=1}^n$ must contain the points of $\lambda$, for each index $j \in \{0, \dots, m\}$ there is an index $i(j) \in \{1, \dots, n\}$ such that $z_j = s_{i(j)}$. Let $\mu := \{i(j) \colon j = 0, \dots, m\} \subset \{1, \dots, n\}$. For $j = 1, \dots, m-1$, by definition of $\delta(\cdot)$, we must have

$$t_{i(j)-1} < s_{i(j)} < t_{i(j)}. \tag{2.67}$$

We also have $i(0) = 1$, $s_1 = z_0 = t_0 = a$ and $i(m) = n$, $s_n = z_m = t_n = b$.

Now consider the case that $f$ is right-continuous on $(a,b)$, which holds when $f = f_+^{(a,b)}$. We will show that one can find a Young–Stieltjes sum, based on a refinement of $\lambda$, which is arbitrarily close to the Riemann–Stieltjes sum $S_\delta(\tau)$. To this aim we will replace the values $h(t_i)$ in $S_\delta(\tau)$, $i = 1, \dots, n-1$, by values $h(x_i)$ at continuity points $x_i$ of $h$, close to $t_i$. Then the Young–Stieltjes sum terms $f(x_i)\Delta^\pm h(x_i)$ will be 0. Define a Riemann–Stieltjes sum $T$ by, for each $i = 1, \dots, n-1$ such that $t_i < s_{i+1}$, replacing $t_i$ in $S_\delta(\tau)$ by a slightly larger $x_i > t_i$ which is a continuity point of $h$ with $x_i < s_{i+1}$. Then $|T - S_\delta(\tau)| < \epsilon$. Define another Riemann–Stieltjes sum $U$ equal to $T$ except that for each $i$ such that $s_{i+1} = t_i$ and $1 \le i \le n-1$, we replace $s_{i+1}$ by $y_{i+1}$ and $t_i$ by $x_i$ with $t_i < x_i < y_{i+1} < t_{i+1}$, where $x_i$ is a continuity point of $h$ and $y_{i+1} - t_i$ is as small as desired. Since $f$ is right-continuous on $(a,b)$, we can make $|U - T| < \epsilon$. In either case we can assume that $s_i \le t_i < x_i \le s_i + 2\delta(s_i)$ for $i = 1, \dots, n-1$. Let $x_0 := a$ and $x_n := b$. Let $y_i := s_i$ for each $i = 2, \dots, n-1$ for which $y_i$ was not previously defined. Thus $y_{i(j)} = s_{i(j)}$ for

$j = 1, \ldots, m-1$ by (2.67). Choose any $y_1 \in (a, t_1)$ and $y_n \in (x_{n-1}, b)$. Let $V := S_{YS}(f, \mathrm{d}h; \zeta)$ be the Young–Stieltjes sum based on the Young tagged point partition $\zeta := (\xi, \eta) := (\{x_i\}_{i=0}^n, \{y_i\}_{i=1}^n)$. Then

$$\begin{aligned}
&|V - U| \\
&= |f(a)\Delta^+ h(a) + f(y_1)[h(x_1) - h(a+)] - f(a)[h(x_1) - h(a)] \\
&\quad + f(b)\Delta^- h(b) + f(y_n)[h(b-) - h(x_{n-1})] - f(b)[h(b) - h(x_{n-1})]| \\
&= |[f(y_1) - f(a)][h(x_1) - h(a+)] + [f(y_n) - f(b)][h(b-) - h(x_{n-1})]| < 2\epsilon
\end{aligned}$$

by (2.65). Now let $\kappa := \lambda \cup \xi = \{u_l\}_{l=0}^k$, a partition of $[a, b]$ with $k = m+n-1$. Each interval $(x_{i-1}, x_i)$ for $i = 2, \ldots, n-1$ contains $z_j$ if $i = i(j) \in \mu$, and otherwise contains no $z_j$ and becomes an interval $(u_{l-1}, u_l)$ for some $l$. In the latter case let $v_l := y_i$. In the former case we have for some $l$, $u_{l-2} = x_{i-1} < z_j = u_{l-1} < x_i = u_l$. Choose any $v_{l-1} := w_j$ with $u_{l-2} < v_{l-1} < u_{l-1}$ and $v_l := r_j$ with $u_{l-1} < v_l < u_l$. Let $W := S_{YS}(f, \mathrm{d}h; (\kappa, \sigma))$ be the Young–Stieltjes sum based on the Young tagged point partition $(\kappa, \sigma)$, where $\sigma := \{v_l\}_{l=1}^k$. Then

$$\begin{aligned}
|W - V| &= \Big| \sum_{j=1}^{m-1} \Big\{ f(w_j)[h(z_j-) - h(x_{i(j)-1})] + f(z_j)\Delta^- h(z_j) \\
&\qquad + f(r_j)[h(x_{i(j)}-) - h(z_j)] - f(z_j)[h(x_{i(j)}-) - h(x_{i(j)-1})] \Big\} \Big| \\
&\le \sum_{j=1}^{m-1} |[f(w_j) - f(z_j)][h(z_j-) - h(x_{i(j)-1})]| \\
&\qquad + \sum_{j=1}^{m-1} |[f(r_j) - f(z_j)][h(x_{i(j)}-) - h(z_j)]| < 2\epsilon
\end{aligned}$$

by (2.65). Here $W$ is a Young–Stieltjes sum based on the Young tagged refinement $(\kappa, \sigma)$ of $\lambda$, so $|W - (RYS)\int_a^b f\,\mathrm{d}h| < \epsilon$. Thus $|S_\delta(\tau) - (RYS)\int_a^b f\,\mathrm{d}h| < 7\epsilon$. The conclusion follows when $f$ is right-continuous on $(a, b)$.

Next consider the case that $f \in c_0((a, b))$ (cf. Definition 2.9), which holds when $f = \Delta^+_{(a,b)} f$. There is a set of tags $\{w_j\}_{j=1}^m$ for the partition $\lambda = \{z_j\}_{j=0}^m$ such that $(\lambda, \{w_j\}_{j=1}^m)$ is a Young tagged point partition of $[a, b]$ and $f(w_j) = 0$ for all $j$. Thus by (2.64),

$$\Big| \sum_{j=1}^{m-1} f(z_j)\Delta^- h(z_j) - (RYS)\int_a^b f\,\mathrm{d}h \Big| < \epsilon. \tag{2.68}$$

By (2.67), $t_i \in \lambda$ only for $i = 0$ and $n$. For any set $\nu \subset \{1, \ldots, n-1\}$, consider partitions $\kappa = \{u_l\}_{l=0}^k$, $k = m+n-1$, of $[a, b]$ consisting of $\lambda$, $t_i$ for $i \in \nu$, and for each $i \in \{1, \ldots, n-1\} \setminus \nu$, a continuity point of $h$ close to $t_i$. Let $l(1) < \cdots < l(n-1)$ be such that $\{u_{l(i)}\}_{i=1}^{n-1}$ are the $u_l$ not in $\lambda$. Let

$\sigma = \{v_l\}_{l=1}^k$ be a set of tags for $\kappa$ such that $(\kappa, \sigma)$ is a Young point partition of $[a, b]$ and $f(v_l) = 0$ for all $l$. Then from (2.68) and (2.64) we obtain

$$\Big|\sum_{i\in\nu} f(t_i)\Delta^- h(t_i)\Big| < 2\epsilon. \tag{2.69}$$

Consider also partitions $(\kappa, \sigma)$ defined in the same way except that for $\xi \subset \{1, \dots, n-1\}$, if $i \in \xi \setminus \mu$ and $i$ is even, $i \notin \mu$ implies $l(i-1) = l(i) - 1$. We take $u_l = u_{l(i)}$ to be a continuity point of $h$ a little larger than $t_i$, while $u_{l(i)-1}$ is a continuity point of $h$ a little smaller than $t_{i-1}$ and $v_{l(i)} = t_{i-1}$. For each $i = 1, \dots, n-1$ such that $u_{l(i)}$ is not yet defined, let it be a continuity point of $h$ near $t_i$, which is then true for all $i = 1, \dots, n-1$. Let $f(v_l) = 0$ for other $l$ as before. For even $i \in \xi \setminus \mu$, letting $u_{l(i)} \downarrow t_i$ and $u_{l(i)-1} \uparrow t_{i-1}$, it follows that

$$\Big|\sum \{f(t_{i-1})[h(t_i) - h(t_{i-1}-)]\colon\ i \in \xi \setminus \mu,\ i \text{ even}\}\Big| \le 2\epsilon.$$

The same holds likewise for $i$ odd. Thus

$$\Big|\sum_{i\in\xi\setminus\mu} f(t_{i-1})[h(t_i) - h(t_{i-1}-)]\Big| \le 4\epsilon.$$

With $\nu = \{i-1\colon\ i \in \xi \setminus \mu\}$ in (2.69) this gives

$$\Big|\sum_{i\in\xi\setminus\mu} f(t_{i-1})[h(t_i) - h(t_{i-1})]\Big| \le 6\epsilon.$$

If $s_i = t_{i-1}$ then $i \notin \mu$ by (2.67). Thus

$$\Big|\sum \{f(t_{i-1})[h(t_i) - h(t_{i-1})]\colon\ t_{i-1} = s_i,\ \ i = 2, \dots, n-1\}\Big| \le 6\epsilon.$$

Here $f(t_{i-1})$ could be replaced by $f(t_{i-1}) - f(y_i)$, where $t_{i-1} < y_i < t_i$ and $f(y_i) = 0$. So the Riemann–Stieltjes sum (2.66) differs by at most $6\epsilon$ from the Riemann–Stieltjes sum

$$S' := \sum_{i=1}^n f(w_i)[h(t_i) - h(t_{i-1})],$$

where $w_i := y_i$ if $t_{i-1} = s_i$ and $i = 2, \dots, n-1$, $w_i := s_i$ otherwise. Then $t_{i-1} < w_i \le t_i$ for $i = 2, \dots, n-1$. The rest of the proof follows as in the case when $f$ is right-continuous and $t_i < s_{i+1}$ for all $i = 1, \dots, n-1$, so that $U = T$ and $y_i = s_i$ for all $i = 2, \dots, n-1$. The proof of Theorem 2.69 is complete. □

The following gives a partial converse to the preceding theorem. We say that the Young–Stieltjes sums for $f$ and $h$ are *unbounded on any subinterval of* $[a, b]$ with $a < b$ if for all $a \le c < d \le b$, $\sup\{|S_{YS}(f, \mathrm{d}h; \tau)|\} = +\infty$, where the supremum is taken over all Young tagged partitions $\tau$ of $[c, d]$. Then the sums based on partitions which are refinements of any given partition of $[c, d]$ are also unbounded.

**Proposition 2.70.** *Let $a < b$. Given $f$ left- or right-continuous on $(a, b)$ and $h \in \mathcal{R}[a, b]$, suppose that the Young–Stieltjes sums for $f$ and $h$ are unbounded on any subinterval of $[a, b]$. Then $(HK)\int_a^b f\,\mathrm{d}h$ does not exist.*

*Proof.* Suppose that $(HK)\int_a^b f\,\mathrm{d}h$ exists. Let $\delta(\cdot)$ be a gauge function and $I$ a number such that for any $\delta$-fine tagged partition $(\{x_i\}_{i=0}^n, \{y_i\}_{i=1}^n)$ of $[a, b]$,

$$\Big|\sum_{i=1}^{n} f(y_i)[h(x_i) - h(x_{i-1})] - I\Big| < 1. \tag{2.70}$$

By the category theorem (e.g. Theorem 2.5.2 in Dudley [53]), there are an interval $[c, d]$ and a positive integer $m$ such that $\delta(x) > 1/m$ for all $x$ in a dense set $S$ in $[c, d]$. Replacing $[c, d]$ by a subinterval of itself, we can assume that $0 < d - c < 1/m$ and that $c, d \in S$. Thus every tagged partition $(\{\xi_i\}_{i=0}^k, \{\eta_j\}_{j=1}^k)$ of $[c, d]$ with tags $\eta_j \in S$ for $j = 1, \ldots, k$ is $\delta$-fine. Take a $\delta$-fine tagged partition as in (2.70), where we can take some $y_i = c < y_\ell = d$ by Lemma 2.53(c). We can also take $x_i = c$ and $x_\ell = d$, as follows. If initially $y_i = c < x_i$, then let $x'_r := x_r$ for $r = 0, 1, \ldots, i-1$ and $y'_r := y_r$ for $r = 1, \ldots, i-1$. Let $x'_i := y'_i := c$ and for $r = i, \ldots, n$, let $y'_{r+1} := y_r$ and $x'_{r+1} := x_r$. Then $(\{x'_i\}_{i=0}^{n+1}, \{y'_i\}_{i=1}^{n+1})$ is a $\delta$-fine tagged partition of $[a, b]$ giving the same sum as in (2.70). We can take $d$ as some $x_\ell$ likewise. Let $\kappa$ be the resulting $\delta$-fine tagged partition of $[a, b]$. We show next that the part of $\kappa$ corresponding to $[c, d]$ can be replaced by a tagged partition of $[c, d]$ such that the corresponding Riemann–Stieltjes sum is arbitrarily large and the resulting tagged partition of $[a, b]$ is $\delta$-fine.

For any $M < \infty$, by assumption, there is a Young tagged partition $\tau = (\{t_j\}_{j=0}^k, \{s_j\}_{j=1}^k)$ of $[c, d]$ such that $|S_{YS}(f, \mathrm{d}h; \tau)| > M$. Since $h$ is regulated, there exist $v_j < t_j$, $j = 1, \ldots, k$, and $u_j > t_j$, $j = 0, \ldots, k-1$, such that $t_{j-1} < u_{j-1} < s_j < v_j < t_j$ for $j = 1, \ldots, k$ and

$$\begin{aligned}|f(c)[h(u_0) - h(c)] &+ \sum_{j=1}^{k-1} f(t_j)[h(u_j) - h(v_j)]\\ &+ \sum_{j=1}^{k} f(s_j)[h(v_j) - h(u_{j-1})] + f(d)[h(d) - h(v_k)]| > M.\end{aligned}$$

Now since $f$ is right- or left-continuous, and $S$ is dense, we can replace $s_j$ by some $s'_j \in S$ as close to $s_j$ as desired, making a small change in the sum. Likewise we can replace $t_j$ by some $t'_j \in S$, where $u_{j-1} < s'_j < v_j < t'_j < u_j$ for each $j = 1, \ldots, k-1$. Recall that $c, d \in S$; the endpoints $t_0, t_k$ are not replaced. We thus obtain a Riemann–Stieltjes sum $\Sigma$ for a $\delta$-fine tagged partition $\zeta$ of $[c, d]$ with $|\Sigma| > M$. Then joining $\zeta$ with the $\delta$-fine tagged partitions of $[a, c]$ (if $a < c$) and $[d, b]$ (if $d < b$) given by the fixed tagged partition $\kappa$, we get unbounded Riemann–Stieltjes sums for $\delta$-fine tagged partitions of $[a, b]$, contradicting (2.70). □

Ward [238] stated and Saks [201, Theorem VI.8.1] gave a proof of the fact that the integral $(WPS) \int_a^b f\, dh$ is defined provided the corresponding Lebesgue–Stieltjes integral $(LS) \int_a^b f\, dh$ is defined, and then they are equal. By Theorem 2.65, the Henstock–Kurzweil integral is in the same relation with the $LS$ integral. The following gives conditions under which the converse holds.

**Theorem 2.71.** *Let $f$ be nonnegative on $[a, b]$ and let $h$ be nondecreasing on $[a, b]$ and right-continuous on $[a, b)$. Then $(HK) \int_a^b f\, \mathrm{d}h$ exists if and only if $(LS) \int_a^b f\, \mathrm{d}h$ does, and then the two are equal.*

*Proof.* If $a = b$ both integrals are 0, so we can assume $a < b$. The McShane integral is defined as the $HK$ integral except that in a tagged partition $(\{t_i\}_{i=0}^n, \{s_i\}_{i=1}^n)$ the tags $s_i$ need not be in the corresponding intervals $[t_{i-1}, t_i]$. By Corollary 6.3.5 of Pfeffer [185, p. 113], under the present hypotheses $(HK) \int_a^b f\, \mathrm{d}h$ exists if and only if the corresponding McShane integral exists, and then the two are equal. The equivalence between the McShane and Lebesgue–Stieltjes integrals when $f \geq 0$ and $h$ is nondecreasing was proved by McShane [167, pp. 552, 553]. Also, it follows from Theorem 4.4.7, Proposition 3.6.14, and Theorem 2.3.4 of Pfeffer [185]. The proof of the theorem is complete. □

## 2.8 Properties of Integrals

In this section, by "$\not\!\!\int$ integral" we will mean an integral $\not\!\!\int f \cdot \mathrm{d}h$ as opposed to an integral $\not\!\!\int f \cdot \mathrm{d}\mu$ for an interval function $\mu$. Suppose that the basic assumption (1.14) holds. For functions $f \colon [a, b] \to X$ and $h \colon [a, b] \to Y$, consider the following properties of an integral $\int_a^b f \cdot \mathrm{d}h$:

I. For $u_1, u_2 \in \mathbb{K}$ and $f_1, f_2 \colon [a, b] \to X$,

$$\int_a^b (u_1 f_1 + u_2 f_2) \cdot \mathrm{d}h = u_1 \int_a^b f_1 \cdot \mathrm{d}h + u_2 \int_a^b f_2 \cdot \mathrm{d}h,$$

where the left side exists provided the right side does.

II. For $v_1, v_2 \in \mathbb{K}$ and $h_1, h_2\colon [a,b] \to Y$,

$$\int_a^b f{\cdot}\mathrm{d}(v_1h_1 + v_2h_2) = v_1\int_a^b f{\cdot}\mathrm{d}h_1 + v_2\int_a^b f{\cdot}\mathrm{d}h_2,$$

where again the left side exists provided the right side does.

III. For $a < c < b$, $\int_a^b f{\cdot}\mathrm{d}h$ exists if and only if both $\int_a^c f{\cdot}\mathrm{d}h$ and $\int_c^b f{\cdot}\mathrm{d}h$ exist, and then

$$\int_a^b f{\cdot}\mathrm{d}h = \int_a^c f{\cdot}\mathrm{d}h + \int_c^b f{\cdot}\mathrm{d}h. \tag{2.71}$$

IV. If $a < b$ and $\int_a^b f{\cdot}\mathrm{d}h$ exists then for each $t \in [a,b]$, taking a limit for $s \in [a,b]$ if $t = a$ or $b$, $I(t) = I(f,\mathrm{d}h)(t) := \int_a^t f{\cdot}\mathrm{d}h$ exists and

$$\lim_{s\to t}\Big\{I(t) - I(s) - f(t){\cdot}[h(t) - h(s)]\Big\} = 0. \tag{2.72}$$

Properties I and II imply that the operator $V \times W \ni (f,h) \mapsto \int_a^b f{\cdot}\mathrm{d}h \in Z$ is bilinear for any function spaces $V, W$ on which it is defined. If property IV holds, the indefinite integral $I(f,\mathrm{d}h)(t)$, $t \in [a,b]$, is a regulated function whenever the integrator $h$ is regulated, and then

$$\Delta^- I(t) = f(t){\cdot}\Delta^- h(t) \quad\text{and}\quad \Delta^+ I(s) = f(s){\cdot}\Delta^+ h(s) \quad\text{for}\quad a \le s < t \le b.$$

**Theorem 2.72.** *The RS, RRS, RYS, S, CY, HK and* $\oint$ *integrals satisfy properties* I *and* II.

*Proof.* We prove only property I because the proof of property II is symmetric. We can assume that $a < b$. Let $u_1, u_2 \in \mathbb{K}$, $f_1, f_2\colon [a,b] \to X$, and $h\colon [a,b] \to Y$. For the *RS*, *RRS*, and *HK* integrals, property I follows from bilinearity of the mapping $(f,h) \mapsto f{\cdot}h$ from $X \times Y$ to $Z$, and from the equality

$$S_{RS}(u_1f_1 + u_2f_2, \mathrm{d}h; \tau) = u_1S_{RS}(f_1, \mathrm{d}h; \tau) + u_2S_{RS}(f_2, \mathrm{d}h; \tau),$$

valid for any tagged partition $\tau$ of $[a,b]$. The same argument with Riemann–Stieltjes sums replaced by Young–Stieltjes sums (2.16) implies property I for the *YS* integral, and so it holds for the full Stieltjes *S* integral. Also, property I holds for the $\oint$ integral by Proposition 2.27.

For the *CY* integral given by the $Y_1$ integral (2.44), let $f_1$, $f_2$ be regulated functions. On the interval $[a,b)$, we have $(u_1f_1 + u_2f_2)_+ = u_1(f_1)_+ + u_2(f_2)_+$ and $\Delta^+(u_1f_1 + u_2f_2) = u_1\Delta^+ f_1 + u_2\Delta^+ f_2$. Property I for the *CY* integral follows, since it holds for the *RRS* integral and by the linearity of unconditionally convergent sums. The proof of Theorem 2.72 is complete. □

**Theorem 2.73.** *The RRS, RYS, S, CY, HK, and* $\oint$ *integrals satisfy property* III.

*Proof.* Let $a < c < b$ and let $f\colon [a,b] \to X$ and $h\colon [a,b] \to Y$. Suppose that $(RRS)\int_a^c f{\cdot}dh$ and $(RRS)\int_c^b f{\cdot}dh$ exist. For a tagged partition $\tau_1 = (\{x_i\}_{i=0}^n, \{t_i\}_{i=1}^n)$ of $[a,c]$ and a tagged partition $\tau_2 = (\{y_j\}_{j=0}^m, \{s_i\}_{j=1}^m)$ of $[c,b]$, let $\tau := (\{x_i\}_{i=0}^n \cup \{y_j\}_{j=1}^m, \{t_i\}_{i=1}^n \cup \{s_j\}_{j=1}^m)$, a tagged partition of $[a,b]$. Then the equality

$$S_{RS}(f,\mathrm{d}h;\tau) = S_{RS}(f,\mathrm{d}h;\tau_1) + S_{RS}(f,\mathrm{d}h;\tau_2) \tag{2.73}$$

yields that $(RRS)\int_a^b f{\cdot}dh$ exists and (2.71) holds for the $RRS$ integral. For the converse suppose that $(RRS)\int_a^b f{\cdot}dh$ exists. Applying the Cauchy test for pairs of tagged partitions $\tau'$ and $\tau''$ of $[a,b]$ which are refinements of $\{a,c,b\}$ and induce the same tagged partition of $[c,b]$, it follows that $(RRS)\int_a^c f{\cdot}dh$ exists. Likewise $(RRS)\int_c^b f{\cdot}dh$ exists. The proof of property III for the $RRS$ integral is complete.

The proof of property III for the $RYS$ integral is the same except that the equality

$$S_{YS}(f,\mathrm{d}h;\tau) = S_{YS}(f,\mathrm{d}h;\tau_1) + S_{YS}(f,\mathrm{d}h;\tau_2)$$

for Young tagged partitions as in (2.16) is used instead of (2.73). Thus property III also holds for the $\oint$ integral by Proposition 2.27. Since property III holds for the $RRS$ and $RYS$ integrals it also holds for the full Stieltjes integral $(S)\int_a^b f{\cdot}dh$ by definition 2.41 of the $(S)$ integral and since $f$ and $h$ have a common one-sided discontinuity on $[a,b]$ if and only if they also have one on at least one of $[a,c]$ or $[c,b]$.

Since property III for the $HK$ integral is proved by Proposition 2.56, it is left to prove property III for the $CY$ integral. To this aim suppose that functions $f$, $h$ are regulated and that the $Y_1$ integral $(Y_1)\int_a^b f{\cdot}dh$ exists. By the same property for the $RRS$ integral already proved, it follows that the following three integrals exist and

$$(RRS)\int_a^b f_+^{(b)}{\cdot}\mathrm{d}h_-^{(a)} = (RRS)\int_a^c f_+{\cdot}\mathrm{d}h_-^{(a)} + (RRS)\int_c^b f_+^{(b)}{\cdot}\mathrm{d}h_-.$$

On the right side the integrand $f_+$ over $[a,c]$ depends on $f(c+)$ and the integrator $h_-$ over $[c,b]$ depends on $h(c-)$. For the two integrals $(RRS)\int_a^c f_+{\cdot}\mathrm{d}h_-^{(a)}$ and $(RRS)\int_a^c f_+^{(c)}{\cdot}\mathrm{d}h_-^{(a)}$, the difference between Riemann–Stieltjes sums for each based on the same tagged partition $\tau = (\{t_i\}_{i=0}^n, \{s_i\}_{i=1}^n)$ of $[a,c]$ is

$$\begin{aligned} f(s_n+){\cdot}[h(t_n-) - h(t_{n-1}-)] - f_+^{(c)}(s_n){\cdot}[h(t_n-) - h(t_{n-1}-)] \\ = \Delta^+ f(t_n){\cdot}[h(t_n-) - h(t_{n-1}-)] \end{aligned}$$

if $s_n = t_n$ or is 0 if $s_n \in [t_{n-1}, t_n)$. Taking a limit under refinement of tagged partitions of $[a,c]$, we can get $t_{n-1} \uparrow t_n = c$, and so $h(t_n-) - h(t_{n-1}-) \to 0$, while $f$ is bounded. Therefore the following two integrals both exist and are equal:

$$(RRS)\int_a^c f_+\cdot \mathrm{d}h_-^{(a)} = (RRS)\int_a^c f_+^{(c)}\cdot \mathrm{d}h_-^{(a)}.$$

Next, for the two integrals $(RRS)\int_c^b f_+^{(b)}\cdot \mathrm{d}h_-$ and $(RRS)\int_c^b f_+^{(b)}\cdot \mathrm{d}h_-^{(c)}$, the difference between Riemann–Stieltjes sums for each based on the same tagged partition $\tau = (\{t_i\}_{i=0}^n, \{s_i\}_{i=1}^n)$ of $[c,b]$ is

$$\begin{aligned} f(s_1+)\cdot[h(t_1-) - h(c-)] - f(s_1+)\cdot[h(t_1-) - h(c)] \\ = f(s_1+)\cdot\Delta^- h(c) \to f(c+)\Delta^- h(c) \end{aligned}$$

as $t_1 \downarrow t_0 = c$, which one can obtain by taking a limit under refinements of tagged partitions of $[c,b]$. Therefore the following two integrals both exist and the equation

$$(RRS)\int_c^b f_+^{(b)}\cdot \mathrm{d}h_- = (RRS)\int_c^b f_+^{(b)}\cdot \mathrm{d}h_-^{(c)} + f(c+)\Delta^- h(c)$$

holds. By the definition of the $Y_1$ integral (2.44), it then follows that the $Y_1$ integral exists over the subintervals $[a,c]$ and $[c,b]$, and we have

$$\begin{aligned} (Y_1)\int_a^b f\cdot \mathrm{d}h &= (RRS)\int_a^c f_+\cdot \mathrm{d}h_-^{(a)} + (RRS)\int_c^b f_+^{(b)}\cdot \mathrm{d}h_- \\ &\quad -[\Delta^+ f\cdot\Delta^+ h](a) - \sum_{(a,c)} \Delta^+ f\cdot\Delta^\pm h - [\Delta^+ f\cdot\Delta^\pm h](c) \\ &\quad - \sum_{(c,b)} \Delta^+ f\cdot\Delta^\pm h + [f\cdot\Delta^- h](b) \\ &= (RRS)\int_a^c f_+^{(c)}\cdot \mathrm{d}h_-^{(a)} - [\Delta^+ f\cdot\Delta^+ h](a) - \sum_{(a,c)} \Delta^+ f\cdot\Delta^\pm h + [f\Delta^- h](c) \\ &\quad +(RRS)\int_c^b f_+^{(b)}\cdot \mathrm{d}h_-^{(c)} - [\Delta^+ f\cdot\Delta^+ h](c) - \sum_{(c,b)} \Delta^+ f\cdot\Delta^\pm h + [f\Delta^- h](b) \\ &= (Y_1)\int_a^c f\cdot \mathrm{d}h + (Y_1)\int_c^b f\cdot \mathrm{d}h. \end{aligned}$$

The converse implication follows by applying the same arguments, and so property III holds for the $CY$ integral. The proof of Theorem 2.73 is complete. □

The Riemann–Stieltjes integral does not satisfy property III. Indeed, for the indicator functions $f := 1_{[1,2]}$ and $h := 1_{(1,2]}$ defined on $[0,2]$, the $RS$ integral exists over the intervals $[0,1]$ and $[1,2]$, but not over the interval $[0,2]$. Notice that $f$ and $h$ for this example have a common discontinuity at $1 \in [0,2]$. This feature of the Riemann–Stieltjes integral was observed by Pollard [187], who also showed that the refinement Riemann–Stieltjes integral does satisfy property III. A weaker form of this property can be stated as follows:

III′. For $a < c < b$, if $\int_a^b f \cdot \mathrm{d}h$ exists then both $\int_a^c f \cdot \mathrm{d}h$ and $\int_c^b f \cdot \mathrm{d}h$ exist, and (2.71) holds.

Then we have the following result:

**Proposition 2.74.** *The RS integral satisfies property* III′.

*Proof.* Suppose that $(RS) \int_a^b f \cdot \mathrm{d}h$ exists and let $a < c < b$. Adjoining equal tagged subintervals of $[c, b]$ to any two given tagged partitions $\tau_1'$, $\tau_2'$ of $[a, c]$, one can form two tagged partitions $\tau_1$, $\tau_2$ of $[a, b]$, without increasing $|\tau_1'| \vee |\tau_2'|$, and such that $S_{RS}(\tau_1') - S_{RS}(\tau_2') = S_{RS}(\tau_1) - S_{RS}(\tau_2)$. Thus one can apply the Cauchy test to prove the existence of $(RS) \int_a^c f \cdot \mathrm{d}h$. Similarly one can show that $(RS) \int_c^b f \cdot \mathrm{d}h$ also exists. The additivity relation (2.71) then follows from the equality $S_{RS}((\kappa, \xi)) = S_{RS}((\kappa', \xi')) + S_{RS}((\kappa'', \xi''))$, valid for any tagged partitions such that $\kappa = \kappa' \cup \kappa''$, $\xi = \xi' \cup \xi''$, and $\kappa', \kappa''$ are partitions of $[a, c]$ and $[c, b]$ respectively. □

**Theorem 2.75.** *The RS, RRS, RYS, and HK integrals satisfy property* IV.

*Proof.* Suppose that $(RS) \int_a^b f \cdot \mathrm{d}h$ exists. The indefinite integral $I_{RS}(u) := (RS) \int_a^u f \cdot \mathrm{d}h$ is defined for $u \in (a, b]$ by Proposition 2.74 and is 0 when $u = a$ by the definition of the $RS$ integral. Also by Proposition 2.74, we have $I_{RS}(u) - I_{RS}(v) = \operatorname{sgn}(u - v)(RS) \int_{u \wedge v}^{u \vee v} f \cdot \mathrm{d}h$. In light of symmetry of proofs, we prove only that

$$\lim_{v \uparrow u} \Big\{ (RS) \int_v^u f \cdot \mathrm{d}h - S_{RS}(f, \mathrm{d}h; \sigma_v) \Big\} = 0 \tag{2.74}$$

for any $a < u \le b$, where for $a < v < u$, $\sigma_v$ is the tagged partition $(\{v, u\}, \{u\})$ of $[v, u]$. Let $u \in (a, b]$. Given $\epsilon > 0$ there is a $\delta \in (0, u - a]$ such that any two Riemann–Stieltjes sums based on tagged partitions of $[a, u]$ with mesh less than $\delta$ differ by at most $\epsilon$. For $u - \delta \le v < u$, choose a Riemann–Stieltjes sum $S_{RS}(\tau_v) = S_{RS}(f, \mathrm{d}h; \tau_v)$, based on a tagged partition $\tau_v$ of $[v, u]$, within $\epsilon$ of $(RS) \int_v^u f \cdot \mathrm{d}h$. Let $\tau'$ and $\tau''$ be two tagged partitions of $[a, u]$ with mesh less than $\delta$ which are equal when restricted to $[a, v]$ and which coincide with $\tau_v$ and $\sigma_v$, respectively, when restricted to $[v, u]$. Then

$$\begin{aligned} &\Big\| (RS) \int_v^u f \cdot \mathrm{d}h - S_{RS}(f, \mathrm{d}h; \sigma_v) \Big\| \\ &\le \Big\| (RS) \int_v^u f \cdot \mathrm{d}h - S_{RS}(\tau_v) \Big\| + \| S_{RS}(\tau') - S_{RS}(\tau'') \| \\ &< 2\epsilon \end{aligned} \tag{2.75}$$

for each $v$ with $u - \delta \le v < u$. Since $\epsilon > 0$ is arbitrary, (2.74) holds, proving property IV for the $RS$ integral.

If $(RRS)\int_a^b f{\cdot}dh$ exists, we argue similarly. To prove (2.74) with $RRS$ instead of $RS$, let $\epsilon > 0$. Then there is a partition $\lambda = \{t_j\}_{j=0}^m$ of $[a,u]$ such that any two Riemann–Stieltjes sums based on tagged refinements of $\lambda$ differ by at most $\epsilon$. For $v \in [t_{m-1}, u)$, choose a Riemann–Stieltjes sum, based on a tagged partition $\tau_v$ of $[v,u]$, within $\epsilon$ of $(RRS)\int_v^u f{\cdot}dh$. Let $\tau'$ and $\tau''$ be two tagged refinements of $\lambda$ which coincide with $\tau_v$ and $\sigma_v$, respectively, when restricted to $[v,u]$. Then (2.75) with $RRS$ instead of $RS$ holds for each $v \in [t_{m-1}, u)$, proving property IV for the $RRS$ integral.

Suppose now that $(RYS)\int_a^b f{\cdot}dh$ exists. Let $u \in (a,b]$, and for $a < v < u$, let $\sigma_v$ be a Young tagged partition of the form $(\{v,u\},\{s\})$ of $[v,u]$. Since $h$ is regulated by definition of the $RYS$ integral, we have

$$\lim_{v\uparrow u}\Big\{S_{YS}(f,dh;\sigma_v) - f(u){\cdot}[h(u)-h(v)]\Big\}$$

$$= \lim_{v\uparrow u}\Big\{[f(s)-f(u)]{\cdot}[h(u-)-h(v+)] + [f(v)-f(u)]{\cdot}\Delta^+h(v)\Big\} = 0.$$

Therefore and by property III again, we have to show that

$$\lim_{v\uparrow u}\Big\{(RYS)\int_v^u f{\cdot}dh - S_{YS}(f,dh;\sigma_v)\Big\} = 0.$$

The proof is the same as for the $RRS$ integral except that Riemann–Stieltjes sums are replaced by Young–Stieltjes sums. Since property IV for the $HK$ integral follows from Proposition 2.58, the proof of Theorem 2.75 is complete. □

Property IV for the $HK$ integral agrees with the "only if" part of Proposition 2.58, which says that existence of the $HK$ integral implies existence of improper versions. The following shows that the "if" part of Proposition 2.58 does not hold for the $RRS$ integral, for the $RYS$ integral by Propositions 2.18 and 2.46, nor for the $CY$ integral by its definition.

**Proposition 2.76.** *There are continuous real-valued functions $f$, $h$ on $[0,1]$ such that $\lim_{t\uparrow 1}(RRS)\int_0^t f\,dh$ exists, but $(RRS)\int_0^1 f\,dh$ does not.*

*Proof.* Let $s_m := 1 - 1/m$ for $m = 1, 2, \ldots$. For each $m$, let $f(s_{4m-2}) = f(s_{4m}) := 0$, $f(s_{4m-1}) := m^{-1/2}$, $f(0) = f(1) := 0$, and let $f$ be linear and continuous on intervals between adjacent points where it has been defined. Let $h(s_{4m-3}) := 0$, so $h(0) = 0$, $h(s_{4m-2}) = h(s_{4m}) := m^{-1/2}$, $h(1) := 0$, and let $h$ also be linear between adjacent points where it has been defined. Then $f$ and $h$ are both continuous. Since $h$ is constant on each interval where $f$ is non-zero, we have for $0 < t < 1$ that $(RRS)\int_0^t f\,dh = 0$, taking a partition containing $t$ and all points $s_m < t$.

But for any partition $\kappa = \{t_i\}_{i=0}^n$ of $[0,1]$, there exist arbitrarily large Riemann–Stieltjes sums for $f$, $h$ based on refinements of $\kappa$, as follows. Let $m_0$

be the smallest $m$ such that $s_{4m-3} > t_{n-1}$. Form a partition $\kappa_N$ by adjoining to $\kappa$ all points $s_{4m-3}, s_{4m}$ for $m = m_0, m_0+1, \ldots, m_0+N$, and form Riemann–Stieltjes sums containing terms $f(s_{4m-1})[h(s_{4m}) - h(s_{4m-3})] = 1/m$ for all such $m$, and 0 terms $f(s_{4m})[h(s_{4m+1}) - h(s_{4m})]$. As $N \to \infty$ these sums become arbitrarily large, as claimed. □

The following fact is a change of variables formula for the full Stieltjes integral.

**Proposition 2.77.** *Let $\phi\colon [a,b] \to \mathbb{R}$ be a strictly monotone continuous function with range $[c,d] := \operatorname{ran}(\phi)$. Assuming (1.14), let $f\colon [c,d] \to X$ and $h\colon [c,d] \to Y$. Then $(S)\int_a^b (f\circ\phi)\cdot\mathrm{d}(h\circ\phi)$ exists if and only if $(S)\int_c^d f\cdot\mathrm{d}h$ does, and*

$$(S)\int_a^b (f\circ\phi)\cdot\mathrm{d}(h\circ\phi) = \begin{cases} (S)\int_c^d f\cdot\mathrm{d}h & \text{if } \phi \text{ is increasing,} \\ -(S)\int_c^d f\cdot\mathrm{d}h & \text{if } \phi \text{ is decreasing.} \end{cases}$$

*Proof.* We can assume that $a < b$. First suppose that $\phi$ is increasing. The inverse function $\phi^{-1}$ of $\phi$ is also continuous and increasing from $[c,d]$ onto $[a,b]$. Therefore for the two integrals the limits under refinement of partitions both exist or not simultaneously, and if they do exist they are equal, proving the first part of the proposition.

Now suppose that $\phi$ is decreasing on $[a,b]$. Then letting $\theta(x) := -x$ for $x \in [-b,-a]$, it follows that $\widetilde{\phi} := \phi\circ\theta$ is an increasing continuous function on $[-b,-a]$. Let $\tau = (\{t_i\}_{i=0}^n, \{s_i\}_{i=1}^n)$ be a tagged partition of $[a,b]$. Then $\widetilde{\tau} := (\{-t_{n-i}\}_{i=0}^n, \{-s_{n-i}\}_{i=0}^{n-1})$ is a tagged partition of $[-b,-a]$ and

$$S_{RS}(f\circ\phi, \mathrm{d}h\circ\phi; \tau) = -S_{RS}(f\circ\widetilde{\phi}, \mathrm{d}h\circ\widetilde{\phi}; \widetilde{\tau}).$$

Also, if $\sigma$ is a tagged refinement of $\tau$ then $\widetilde{\sigma}$ is a tagged refinement of $\widetilde{\tau}$. Thus

$$(RRS)\int_a^b f\circ\phi\cdot\mathrm{d}h\circ\phi = -(RRS)\int_{-b}^{-a} f\circ\widetilde{\phi}\cdot\mathrm{d}h\circ\widetilde{\phi},$$

where the two integrals exist or not simultaneously. Now applying the first part of the proposition to the integral on the right side, it follows that

$$(RRS)\int_{-b}^{-a} f\circ\widetilde{\phi}\cdot\mathrm{d}h\circ\widetilde{\phi} = (RRS)\int_{\phi(b)}^{\phi(a)} f\cdot\mathrm{d}h,$$

provided either integral is defined. The same is true for refinement Young–Stieltjes integrals, proving the proposition. □

Next is a fact which allows interchange of integration with a linear map for a scalar-valued integrator.

**Proposition 2.78.** *Let $\mathbb{K}$ be either the field $\mathbb{R}$ or the field $\mathbb{C}$, and let $L$ be a bounded linear mapping from a Banach space $X$ to another Banach space $Z$. For $f\colon [a,b] \to X$ and $h\colon [a,b] \to \mathbb{K}$, if $(RS)\int_a^b f{\cdot}\mathrm{d}h$ exists then so does $(RS)\int_a^b (L\circ f){\cdot}\mathrm{d}h$ and*

$$L\Big((RS)\int_a^b f{\cdot}\mathrm{d}h\Big) = (RS)\int_a^b (L\circ f){\cdot}\mathrm{d}h. \tag{2.76}$$

*Proof.* We can assume that $a < b$. For any tagged partition $\tau$ of $[a,b]$, we have

$$L\big(S_{RS}(f,\mathrm{d}h;\tau)\big) = S_{RS}(L\circ f,\mathrm{d}h;\tau).$$

Since $L$ is continuous, if $(RS)\int_a^b f{\cdot}\mathrm{d}h$ exists then the right side has a limit as the mesh of $\tau$ tends to zero and (2.76) holds, proving the proposition. □

Next it will be seen that the previous proposition does not extend to cases where both the integrand and integrator have multidimensional values, for any integral we consider.

**Example 2.79.** Let $X = Y = \mathbb{R}^2$, let $Z = \mathbb{R}$, and let the bilinear form from $X \times Y$ into $Z$ be the usual inner product. Let $g$ and $h$ be any real-valued functions on $[0,1]$ and $\nu$ any real-valued interval function on $[0,1]$. Let $f(t) := (g(t),0) \in \mathbb{R}^2$ and $H(t) := (0,h(t)) \in \mathbb{R}^2$ for any $t \in [0,1]$. For any interval $J \subset [0,1]$ let $\mu(J) := (0,\nu(J)) \in \mathbb{R}^2$. Let $L(x,y) := (y,x)$ for each $(x,y) \in \mathbb{R}^2$. Then for each form of integration we have defined, the integrals $\int f{\cdot}\mathrm{d}H$ and $\int f{\cdot}\mathrm{d}\mu$ exist and are 0, where $\int = \int_0^1$ or $\int_{[0,1]}$, but the integrals $\int (L\circ f){\cdot}\mathrm{d}H$ and $\int (L\circ f){\cdot}\mathrm{d}\mu$ may not exist, or may exist with any value.

For all the integrals $\int_a^b f{\cdot}\mathrm{d}g$ defined so far and properties proved for them, we have the symmetrical properties of integrals $\int_a^b \mathrm{d}f{\cdot}g$ by (1.15).

*Integration by parts*

We start with the classical integration by parts formula for the Riemann–Stieltjes and refinement Riemann–Stieltjes integrals.

**Theorem 2.80.** *Assuming* (1.14), *let $f\colon [a,b] \to X$ and $g\colon [a,b] \to Y$. For $\# = RS$ or $\# = RRS$, if $(\#)\int_a^b f{\cdot}\mathrm{d}g$ exists then so does $(\#)\int_a^b \mathrm{d}f{\cdot}g$, and*

$$(\#)\int_a^b f{\cdot}\mathrm{d}g + (\#)\int_a^b \mathrm{d}f{\cdot}g = f(b){\cdot}g(b) - f(a){\cdot}g(a) =: f{\cdot}g\Big|_a^b. \tag{2.77}$$

*Proof.* We can assume that $a < b$. Let $\tau = (\{t_i\}_{i=0}^n, \{s_i\}_{i=1}^n)$ be a tagged partition of $[a,b]$, and let $s_0 := t_0 = a$, $s_{n+1} := t_n = b$. Summing by parts, we have

$$\sum_{i=1}^{n} \big[f(t_i) - f(t_{i-1})\big]\cdot g(s_i) = f\cdot g\big|_{t_0}^{t_n} - \sum_{i=0}^{n} f(t_i)\cdot\big[g(s_{i+1}) - g(s_i)\big]. \tag{2.78}$$

For $\# = RS$, notice that if $\max_i(t_i - t_{i-1}) < \delta$ for some $\delta > 0$, then $\max_i(s_{i+1} - s_i) < 2\delta$. Therefore if $(RS)\int_a^b f\cdot\mathrm{d}g$ exists then by (2.78) and (1.15), $(RS)\int_a^b \mathrm{d}f\cdot g$ exists and (2.77) holds for $\# = RS$.

For $\# = RRS$, notice that the sum on the right side of (2.78) can be written as the sum

$$\sum_{i=0}^{n} \Big\{f(t_i)\cdot\big[g(s_{i+1}) - g(t_i)\big] + f(t_i)\cdot\big[g(t_i) - g(s_i)\big]\Big\} =: S.$$

Thus if $\{t_i\}_{i=0}^n$ is a refinement of a partition $\lambda$ of $[a,b]$ then $S$ is a Riemann–Stieltjes sum based on a tagged refinement of $\lambda$. Hence if $(RRS)\int_a^b f\cdot\mathrm{d}g$ exists then by (2.78) and (1.15), $(RRS)\int_a^b \mathrm{d}f\cdot g$ exists and (2.77) holds for $\# = RRS$. The proof is complete. □

Next is an integration by parts formula for the central Young integral as defined in Section 2.5.

**Theorem 2.81.** *Assuming* (1.14), *let* $f \in \mathcal{R}([a,b];X)$ *and* $g \in \mathcal{R}([a,b];Y)$. *If* $a < b$, *suppose that at least one of the conditions* (a) *or* (b) *holds:*

(a) *the sums* $\sum_{(a,b)} \Delta^+ f\cdot\Delta^+ g$ *and* $\sum_{(a,b)} \Delta^- f\cdot\Delta^- g$ *converge unconditionally in* $Z$;

(b) *on* $(a,b)$, $f = [f_- + f_+]/2$ *and* $g = [g_- + g_+]/2$.

*If* $(CY)\int_a^b f\cdot\mathrm{d}g$ *exists then so does* $(CY)\int_a^b \mathrm{d}f\cdot g$, *and*

$$(CY)\int_a^b f\cdot\mathrm{d}g + (CY)\int_a^b \mathrm{d}f\cdot g = f\cdot g\big|_a^b + A, \tag{2.79}$$

*where* $A = 0$ *if* $a = b$ *and otherwise* $A = -\big[\Delta^+ f\cdot\Delta^+ g\big](a) + \big[\Delta^- f\cdot\Delta^- g\big](b) + B$ *with*

$$B := \begin{cases} -\sum_{(a,b)} \Delta^+ f\cdot\Delta^+ g + \sum_{(a,b)} \Delta^- f\cdot\Delta^- g & \text{if (a) holds} \\ 0 & \text{if (b) holds.} \end{cases}$$

*Proof.* We can assume that $a < b$. By the definitions of the $CY$ integral (2.51) and the $Y_1$ integral (2.44),

$$(CY)\int_a^b f\cdot\mathrm{d}g = (RRS)\int_a^b f_+^{(b)}\cdot\mathrm{d}g_-^{(a)}$$
$$-\big[\Delta^+ f\cdot\Delta^+ g\big](a) + \big[f\cdot\Delta^- g\big](b) - \sum_{(a,b)} \Delta^+ f\cdot\Delta^\pm g,$$

where the $RRS$ integral exists and the sum converges unconditionally in $Z$. If $(a)$ holds, by linearity of unconditional convergence, the sum $\sum_{(a,b)} \Delta^{\pm}f{\cdot}\Delta^{-}g$ converges unconditionally in $Z$, and

$$\sum_{(a,b)} \Delta^{\pm}f{\cdot}\Delta^{-}g - \sum_{(a,b)} \Delta^{+}f{\cdot}\Delta^{\pm}g = \sum_{(a,b)} \Delta^{-}f{\cdot}\Delta^{-}g - \sum_{(a,b)} \Delta^{+}f{\cdot}\Delta^{+}g. \quad (2.80)$$

If $(b)$ holds, since "$\cdot$" is bilinear,

$$\Delta^{+}f{\cdot}\Delta^{\pm}g = \Delta^{+}f{\cdot}(2\Delta^{-}g) = (2\Delta^{+}f){\cdot}\Delta^{-}g = \Delta^{\pm}f{\cdot}\Delta^{-}g.$$

Thus again the sum $\sum_{(a,b)} \Delta^{\pm}f{\cdot}\Delta^{-}g$ converges unconditionally in $Z$, and

$$\sum_{(a,b)} \Delta^{\pm}f{\cdot}\Delta^{-}g = \sum_{(a,b)} \Delta^{+}f{\cdot}\Delta^{\pm}g. \quad (2.81)$$

Also, in case $(a)$ or $(b)$ the integration by parts theorem for the $RRS$ integral (Theorem 2.80 with $\# = RRS$) yields that $(RRS)\int_a^b \mathrm{d}f_{+}^{(b)}{\cdot}g_{-}^{(a)}$ exists and

$$\begin{aligned}(RRS)\int_a^b f_{+}^{(b)}{\cdot}\mathrm{d}g_{-}^{(a)} + (RRS)\int_a^b \mathrm{d}f_{+}^{(b)}{\cdot}g_{-}^{(a)} &= f_{+}^{(b)}{\cdot}g_{-}^{(a)}\Big|_a^b \\ &= f(b){\cdot}g(b-) - f(a+){\cdot}g(a).\end{aligned} \quad (2.82)$$

Hence by definition of the $Y_2$ integral (2.45), $(CY)\int_a^b \mathrm{d}f{\cdot}g$ exists and

$$\begin{aligned}(CY)\int_a^b \mathrm{d}f{\cdot}g = (RRS)&\int_a^b \mathrm{d}f_{+}^{(b)}{\cdot}g_{-}^{(a)} \\ &+\big[\Delta^{+}f{\cdot}g\big](a) + \big[\Delta^{-}f{\cdot}\Delta^{-}g\big](b) + \sum_{(a,b)} \Delta^{\pm}f{\cdot}\Delta^{-}g.\end{aligned}$$

Then by (2.80) in case $(a)$ or (2.81) in case $(b)$, it follows that

$$\begin{aligned}(CY)\int_a^b f{\cdot}\mathrm{d}g + (CY)\int_a^b \mathrm{d}f{\cdot}g = (RRS)&\int_a^b f_{+}^{(b)}{\cdot}\mathrm{d}g_{-}^{(a)} + (RRS)\int_a^b \mathrm{d}f_{+}^{(b)}{\cdot}g_{-}^{(a)} \\ &+f{\cdot}g\Big|_a^b - f(b){\cdot}g(b-) + f(a+){\cdot}g(a) \\ &-\big[\Delta^{+}f{\cdot}\Delta^{+}g\big](a) + \big[\Delta^{-}f{\cdot}\Delta^{-}g\big](b) + A.\end{aligned}$$

This together with (2.82) implies formula (2.79), proving the theorem. □

For the refinement Young–Stieltjes integral, an integration by parts theorem analogous to the preceding one does not hold. Recall that by Proposition 2.52 there are real-valued functions $f$ and $h$ on $[0,1]$, where $f$ is in $c_0((0,1))$ and $h$ is continuous, for which $(CY)\int_0^1 f\,\mathrm{d}h$ exists, while $(RYS)\int_0^1 f\,\mathrm{d}h$ does not exist. For this pair of functions, $(RYS)\int_0^1 h\,\mathrm{d}f$ exists and equals 0 because $f_{-}^{(0)} \equiv f_{+}^{(1)} \equiv 0$, and $\sum \Delta^{+}f\Delta^{+}h = \sum \Delta^{-}f\Delta^{-}h = 0$ since $h$ is continuous. However, by Theorem 2.51$(a)$ and Theorem 2.81, we have the following corollary:

**Corollary 2.82.** *Under the conditions of Theorem* 2.81, *if* $(RYS)\int_a^b f{\cdot}\mathrm{d}g$ *and* $(RYS)\int_a^b \mathrm{d}f{\cdot}g$ *both exist then* (2.79) *with* $CY$ *replaced by* $RYS$ *holds.*

For a regulated function $f$ on $[a,b]$ with $a<b$ and $\hat{f} := [f_-^{(a,b)} + f_+^{(a,b)}]/2$, we have

$$\sum_{(a,b)} [f-\hat{f}]{\cdot}\Delta^{\pm}g + \sum_{(a,b)} \Delta^{\pm}f{\cdot}[g-\hat{g}] = -\sum_{(a,b)} \Delta^{+}f{\cdot}\Delta^{+}g + \sum_{(a,b)} \Delta^{-}f{\cdot}\Delta^{-}g,$$

provided that the sums converge unconditionally. Therefore by Theorem 2.51($b$) and Theorem 2.81, the following form of integration by parts for the refinement Young–Stieltjes integral holds:

**Corollary 2.83.** *Assuming* (1.14) *and* $a<b$, *let* $f\in\mathcal{R}([a,b];X)$ *and* $g\in\mathcal{R}([a,b];Y)$ *be such that at least one of the two conditions* ($a$) *and* ($b$) *of Theorem* 2.81 *holds. If either* $(CY)\int_a^b f{\cdot}\mathrm{d}g$ *or* $(CY)\int_a^b \mathrm{d}f{\cdot}g$ *exists then* $(RYS)\int_a^b \hat{f}{\cdot}\mathrm{d}g$ *and* $(RYS)\int_a^b \mathrm{d}f{\cdot}\hat{g}$ *both exist, and*

$$(RYS)\int_a^b \hat{f}{\cdot}\mathrm{d}g + (RYS)\int_a^b \mathrm{d}f{\cdot}\hat{g} = f{\cdot}g\big|_a^b - \big[\Delta^{+}f{\cdot}\Delta^{+}g\big](a) + \big[\Delta^{-}f{\cdot}\Delta^{-}g\big](b), \tag{2.83}$$

*where* $\hat{f}=f$ *and* $\hat{g}=g$ *if the condition* ($b$) *of Theorem* 2.81 *holds.*

*Bounds for integrals*

Here we give bounds for integrals assuming that either the integrand or integrator has bounded variation.

**Theorem 2.84.** *Assuming* (1.14), $f\colon [a,b]\to X$, *and* $h\colon [a,b]\to Y$, *we have*

($a$) *if* $f$ *is regulated and* $h$ *is of bounded variation then the full Stieltjes integral* $(S)\int_a^b f{\cdot}\mathrm{d}h$ *is defined and for each* $t\in[a,b]$,

$$\Big\|(S)\int_a^b f{\cdot}\mathrm{d}h - f(t){\cdot}[h(b)-h(a)]\Big\| \le \mathrm{Osc}(f;[a,b])v_1(h;[a,b]); \tag{2.84}$$

($b$) *if* $h$ *is regulated and* $f$ *is of bounded variation then the full Stieltjes integral* $(S)\int_a^b f{\cdot}\mathrm{d}h$ *is defined and for each* $t\in[a,b]$,

$$\Big\|(S)\int_a^b f{\cdot}\mathrm{d}h - f(t){\cdot}[h(b)-h(a)]\Big\| \le \mathrm{Osc}(h;[a,b])v_1(f;[a,b]). \tag{2.85}$$

*Proof.* We can assume that $a<b$. In both cases if $f$ and $h$ have no common discontinuities then the integral $\int_a^b f{\cdot}\mathrm{d}h$ exists in the Riemann–Stieltjes sense by Theorems 2.42 and 2.17. If $f$ and $h$ have no common one-sided discontinuities then the integral $\int_a^b f{\cdot}\mathrm{d}h$ exists in the refinement Riemann–Stieltjes sense by Theorem 2.17. Under the assumptions of the present theorem the integral

$\int_a^b f{\cdot}\mathrm{d}h$ always exists in the refinement Young–Stieltjes sense by Theorem 2.20, and so the full Stieltjes integral $(S)\int_a^b f{\cdot}\mathrm{d}h$ is defined.

To prove $(a)$ it then suffices to prove it for the refinement Young–Stieltjes integral. Letting $\tau = (\{t_i\}_{i=0}^n, \{s_i\}_{i=1}^n)$ be a Young tagged partition of $[a,b]$ and $t \in [a,b]$, we have the bound

$$\begin{aligned}&\big\|S_{YS}(f,\mathrm{d}h;\tau) - f(t){\cdot}[h(b)-h(a)]\big\| \\ &\le \Big\|\sum_{i=0}^n [f(t_i)-f(t)]{\cdot}(\Delta^{\pm}_{[a,b]}h)(t_i)\Big\| + \Big\|\sum_{i=1}^n [f(s_i)-f(t)]{\cdot}[h(t_i-)-h(t_{i-1}+)]\Big\|.\end{aligned}$$

For each $i=1,\dots,n$, letting $u_{i-1}\downarrow t_{i-1}$ and $v_i \uparrow t_i$, one can approximate $\Delta^+h(a)$ by $h(u_0)-h(a)$, $\Delta^-h(b)$ by $h(b)-h(v_n)$, $\Delta^{\pm}h(t_i)$ by $h(u_i)-h(v_i)$ for $i=1,\dots,n-1$ and $h(t_i-)-h(t_{i-1}+)$ by $h(v_i)-h(u_{i-1})$ for $i=1,\dots,n$. For an arbitrary $\epsilon>0$, this gives the further bound

$$\begin{aligned}&\le \epsilon + \max_{0\le i\le n}\|f(t_i)-f(t)\|\Big(\|h(u_0)-h(a)\|+\|h(b)-h(v_n)\|+\sum_{i=1}^{n-1}\|h(u_i)-h(v_i)\|\Big) \\ &\quad + \max_{1\le i\le n}\|f(s_i)-f(t)\|\sum_{i=1}^n\|h(v_i)-h(u_{i-1})\| \le \epsilon + \mathrm{Osc}(f;[a,b])v_1(h;[a,b]).\end{aligned}$$

Since $\epsilon>0$ is arbitrary, the limit under refinements of Young tagged partitions gives the bound (2.84) for the refinement Young–Stieltjes integral, proving $(a)$.

As in $(a)$, it will suffice to prove $(b)$ only for the refinement Young–Stieltjes integral. Let $\tau = (\{t_i\}_{i=0}^n, \{s_i\}_{i=1}^n)$ be a Young tagged partition of $[a,b]$. For $j=1,\dots,2n$, let $\Delta_j f := f(u_j)-f(u_{j-1})$ with $\{u_j\}_{j=0}^{2n} = \{t_0,s_1,t_1,\dots,s_n,t_n\}$ and let $\{v_j\}_{j=1}^{2n} := \{t_0+,t_1-,t_1+,\dots,t_n-\}$. Summation by parts gives

$$S_{YS}(f,\mathrm{d}h;\tau) = f(b){\cdot}h(b) - f(a){\cdot}h(a) - \sum_{j=1}^{2n}\Delta_j f{\cdot}h(v_{j-1}).$$

Adding to and subtracting from the right side first $f(b){\cdot}h(a)$ and then $f(a){\cdot}h(b)$, we get two representations

$$\begin{aligned}S_{YS}(f,\mathrm{d}h;\tau) &= f(b){\cdot}[h(b)-h(a)] - \sum_{j=1}^{2n}\Delta_j f{\cdot}[h(v_{j-1})-h(a)] \\ &= f(a){\cdot}[h(b)-h(a)] - \sum_{j=1}^{2n}\Delta_j f{\cdot}[h(v_{j-1})-h(b)].\end{aligned} \tag{2.86}$$

Thus for $t\in\{a,b\}$, we have the bound

$$\big\|S_{YS}(f,\mathrm{d}h;\tau) - f(t){\cdot}[h(b)-h(a)]\big\| \le \mathrm{Osc}(h;[a,b])v_1(f;[a,b]). \tag{2.87}$$

Since $\tau$ is an arbitrary Young tagged partition of $[a,b]$, (2.85) holds if $t \in \{a,b\}$. Suppose that $t \in (a,b)$ and let $t = t_\nu$ for some $\nu \in \{1,\ldots,n-1\}$. Then $\tau_1 := (\{t_i\}_{i=0}^{\nu}, \{s_i\}_{i=1}^{\nu})$ and $\tau_2 := (\{t_i\}_{\nu}^{n}, \{s_i\}_{\nu+1}^{n})$ are Young tagged partitions of $[a,t]$ and $[t,b]$ respectively. Applying the first representation in (2.86) to $\tau_1$ and the second representation to $\tau_2$, it follows that

$$
\begin{aligned}
S_{YS}(f,\mathrm{d}h;\tau) &= S_{YS}(f,\mathrm{d}h;\tau_1) + S_{YS}(f,\mathrm{d}h;\tau_2) \\
&= f(t)\cdot[h(b)-h(a)] - \sum_{i=1}^{\nu} \Delta_j f\cdot[h(v_{j-1})-h(a)] \\
&\quad - \sum_{i=\nu+1}^{n} \Delta_j f\cdot[h(v_{j-1})-h(b)],
\end{aligned}
$$

and so (2.87) holds for any $t \in (a,b)$ and any $\tau$ containing $t$ as a partition point. Thus (2.85) holds for any $t \in [a,b]$. The proof of Theorem 2.84 is complete. □

**Corollary 2.85.** *Under the assumptions of the preceding theorem, the indefinite full Stieltjes integral $I_S(f,\mathrm{d}h)(t) := (S)\int_a^t f\cdot\mathrm{d}h$, $t \in [a,b]$, is defined and has bounded variation.*

*Proof.* We can assume $a < b$. The indefinite Riemann–Stieltjes integral exists by the preceding theorem and property III′ (see Proposition 2.74). The indefinite refinement Riemann–Stieltjes and Young–Stieltjes integrals exist by the preceding theorem and property III (see Theorem 2.73). Let $\kappa = \{t_i\}_{i=0}^{n}$ be a partition of $[a,b]$. Then using additivity of the three integrals proved as the already mentioned properties III′ and III, we get the bound

$$
\begin{aligned}
s_1(I_S(f,\mathrm{d}h);\kappa) &\le \sum_{i=1}^{n} \Big\| (S)\int_{t_{i-1}}^{t_i} f\cdot\mathrm{d}h - f(t_{i-1})\cdot[h(t_i)-h(t_{i-1})] \Big\| \\
&\quad + \sum_{i=1}^{n} \|f(t_{i-1})\|\,\|h(t_i)-h(t_{i-1})\| \\
&\le \Big(\mathrm{Osc}(f;[a,b]) + \|f\|_{\sup}\Big) v_1(h;[a,b]),
\end{aligned}
$$

completing the proof of the corollary. □

*A substitution rule*

Here is a substitution rule for the Riemann–Stieltjes integrals with integrators having bounded variation.

**Proposition 2.86.** *For a Banach space $X$, let $h \in \mathcal{W}_1[a,b]$, $g \in \mathcal{R}([a,b];X)$, and $f \in \mathcal{R}[a,b]$ be such that the two pairs $(h,g)$ and $(h,f)$ have no common discontinuities, and $gf\colon [a,b] \to X$ is the function defined by pointwise multiplication. Then the following Riemann–Stieltjes integrals (including those in integrands) are defined and*

$$(RS)\int_a^b \mathrm{d}I_{RS}(g,\mathrm{d}h)\cdot f = (RS)\int_a^b gf\cdot\mathrm{d}h = (RS)\int_a^b g\cdot\mathrm{d}I_{RS}(\mathrm{d}h,f), \quad (2.88)$$

*where $I_{RS}(g,\mathrm{d}h)(t) := (RS)\int_a^t g\cdot\mathrm{d}h \in X$ and $I_{RS}(\mathrm{d}h,f)(t) := (RS)\int_a^t f\,\mathrm{d}h$ for $t \in [a,b]$, and $\cdot$ denotes the natural bilinear mapping $X \times \mathbb{R} \to X$.*

*Proof.* We can assume that $a < b$. Since the pairs $(g,h)$ and $(f,h)$ have no common discontinuities, the indefinite integrals $I_{RS}(g,\mathrm{d}h)$ and $I_{RS}(\mathrm{d}h,f)$ exist and are in $\mathcal{W}_1([a,b];X)$ and $\mathcal{W}_1[a,b]$, respectively, by Corollary 2.85. Since the discontinuities of $I_{RS}(g,\mathrm{d}h)$ and $I_{RS}(\mathrm{d}h,f)$ are subsets of those of $h$, and so the pairs $(I_{RS}(g,\mathrm{d}h),f)$ and $(g,I_{RS}(\mathrm{d}h,f)$ have no common discontinuities, the leftmost and the rightmost integrals in (2.88) exist by the same corollary. Since $gf$ is a regulated function on $[a,b]$ and the pair $(gf,h)$ have no common discontinuities, the middle integral in (2.88) exists by the same corollary. We prove only the first equality in (2.88), since the proof of the second one is symmetric. Let $\tau = (\{t_i\}_{i=0}^n, \{s_i\}_{i=1}^n)$ be a tagged partition of $[a,b]$. Then

$$\|S_{RS}(dI_{RS}(g,\mathrm{d}h),f;\tau) - S_{RS}(gf,\mathrm{d}h;\tau)\| \le \|f\|_{\sup}R(\tau),$$

where

$$R(\tau) := \sum_{i=1}^n \Big\|(RS)\int_{t_{i-1}}^{t_i} g\cdot\mathrm{d}h - g(s_i)\cdot\big[h(t_i) - h(t_{i-1})\big]\Big\|.$$

It is enough to prove that

$$\lim_{|\tau|\downarrow 0} R(\tau) = 0. \quad (2.89)$$

Let $\epsilon > 0$. By Theorem 2.1, there exists a point partition $\{z_j\}_{j=0}^m$ of $[a,b]$ such that

$$\mathrm{Osc}(g;(z_{j-1},z_j)) < \epsilon \qquad \text{for } j \in \{1,\dots,m\}. \quad (2.90)$$

Let $a \le u \le z_j \le v \le b$ for some $j$. Splitting the interval $[u,v]$ into the parts $[u,z_j]$ and $[z_j,v]$ if $z_j$ is not an endpoint of $[u,v]$, and then applying properties III′ and IV for the Riemann–Stieltjes integral (see respectively Proposition 2.74 and Theorem 2.75), we have that

$$\Big\|(RS)\int_u^v g\cdot\mathrm{d}h - g(z_j)\cdot[h(v) - h(u)]\Big\|$$

is small provided $v - u$ is small enough. Therefore and since $g$ and $h$ have no common discontinuities, there exists a $\delta > 0$ such that if $z_j \in [t_{i-1},t_i]$ for some $j$ and mesh $|\tau| < \delta$, then

$$\Big\|(RS)\int_{t_{i-1}}^{t_i} g{\cdot}\mathrm{d}h - g(s_i){\cdot}[h(t_i)-h(t_{i-1})]\Big\|$$
$$\le \Big\|(RS)\int_{t_{i-1}}^{t_i} g{\cdot}\mathrm{d}h - g(z_j){\cdot}[h(t_i)-h(t_{i-1})]\Big\|$$
$$+\big\|[g(z_j)-g(s_i)]{\cdot}[h(t_i)-h(t_{i-1})]\big\| < \epsilon/(m+1).$$

This together with (2.90) and the bound of Theorem 2.84 yields

$$R(\tau) \le \epsilon v_1(h;[a,b]) + (m+1)\epsilon/(m+1) = \epsilon\big[v_1(h;[a,b])+1\big]$$

if the mesh of the tagged partition $\tau$ is less than $\delta$. The proof of (2.89), and hence of Theorem 2.86, is complete. □

*A chain rule formula*

Here we give a representation of the full Stieltjes integral as defined in Definition 2.41, with respect to a composition $F\circ f$, where $f$ has bounded variation and $F$ is smooth.

**Theorem 2.87.** *Let $f\colon [a,b]\to\mathbb{R}^d$ be of bounded variation, let $F$ be a real-valued $C^1$ function on $\mathbb{R}^d$ and let $h\in\mathcal{R}[a,b]$. Then the composition $F\circ f$ is of bounded variation and*

$$(S)\int_a^b h\,\mathrm{d}(F\circ f) = (S)\int_a^b h(\nabla F\circ f){\cdot}\mathrm{d}f + \sum_{(a,b]} h\Big\{\Delta^-(F\circ f) - (\nabla F\circ f){\cdot}\Delta^- f\Big\}$$
$$+\sum_{[a,b)} h\Big\{\Delta^+(F\circ f) - (\nabla F\circ f){\cdot}\Delta^+ f\Big\}, \tag{2.91}$$

*where the two sums converge absolutely if $a<b$ and equal 0 if $a=b$.*

*Proof.* We can assume that $a<b$. Since $F$ is a Lipschitz function on any bounded set (the range of $f$), $F\circ f$ is of bounded variation. Thus the two full Stieltjes integrals in (2.91) exist by Theorems 2.20, 2.17, and 2.42. Again since $F$ is Lipschitz and $\nabla F$ is bounded on the range of $f$, the two sums in (2.91) converge absolutely. By Propositions 2.13 and 2.18, it is enough to prove the equality (2.91) for the $RYS$ integrals. Let $\tau = (\{t_i\}_{i=0}^n, \{s_i\}_{i=1}^n)$ be a Young tagged partition of $[a,b]$. Also let

$$S_-(\tau) := \sum_{\{t_1,\dots,t_n\}} h\Big\{\Delta^-(F\circ f) - (\nabla F\circ f){\cdot}\Delta^- f\Big\}$$

and

$$S_+(\tau) := \sum_{\{t_0,\dots,t_{n-1}\}} h\Big\{\Delta^+(F\circ f) - (\nabla F\circ f){\cdot}\Delta^+ f\Big\}.$$

Then we have the identity

$$S_{YS}(h, \mathrm{d}F\circ f;\tau) = S_{YS}(h(\nabla F\circ f), \mathrm{d}f;\tau) + S_-(\tau) + S_+(\tau) + R(\tau), \quad (2.92)$$

where $R(\tau)$ is the sum

$$\sum_{i=1}^{n} h(s_i)\Big\{\big[F(f(t_i-)) - F(f(t_{i-1}+))\big] - (\nabla F(f(s_i)))\cdot\big[f(t_i-) - f(t_{i-1}+)\big]\Big\}.$$

Let $\epsilon > 0$ and let $B$ be a ball in $\mathbb{R}^d$ containing the range of $f$. Since $\nabla F$ is uniformly continuous on $B$, there is a $\delta > 0$ such that $\|\nabla F(u) - \nabla F(v)\| < \epsilon$ whenever $\|u - v\| < \delta$ and $u, v \in B$. Moreover, since $f$ is regulated, by Theorem 2.1($b$), there exists a partition $\lambda = \{z_j\}_{j=0}^{m}$ of $[a,b]$ such that $\mathrm{Osc}(f;(z_{j-1}, z_j)) < \delta$ for $j = 1,\dots,m$. Let $\tau$ be a Young tagged refinement of $\lambda$. Then for each $i = 1,\dots,n$, by the mean value theorem for $\lambda \mapsto F(\lambda f(t_i-) + (1-\lambda)f(t_{i-1}+))$, $0 \le \lambda \le 1$, and by the chain rule of differentiation there is a $\lambda_i \in [0,1]$ such that

$$F(f(t_i-)) - F(f(t_{i-1}+)) = \nabla F(\theta_i)\cdot\big[f(t_i-) - f(t_{i-1}+)\big],$$

where $\theta_i := \lambda_i f(t_i-) + (1-\lambda_i)f(t_{i-1}+)$. Thus

$$\begin{aligned}\big|R(\tau)\big| &\le \sum_{i=1}^{n} |h(s_i)|\big\|\nabla F(\theta_i) - \nabla F(f(s_i))\big\|\,\big\|f(t_i-) - f(t_{i-1}+)\big\| \\ &\le \epsilon\|h\|_{\sup}\sum_{i=1}^{n}\big\|f(t_i-) - f(t_{i-1}+)\big\| \le \epsilon\|h\|_{\sup} v_1(f;[a,b]).\end{aligned}$$

Since each of the four other sums in (2.92) converges under refinements of partitions to the respective four terms in (2.91), the stated equality (2.91) holds, proving the theorem. □

Taking the $C^1$ function $F$ defined by $F(u,v) := uv$, $u, v \in \mathbb{R}$, we obtain from the preceding theorem:

**Corollary 2.88.** *Let $f, g$ be two real-valued functions on $[a,b]$ having bounded variation, and let $h \in \mathcal{R}[a,b]$. Then the product function $fg$ has bounded variation, and*

$$\begin{aligned}(S)\int_a^b h\,\mathrm{d}(fg) = (S)\int_a^b hf\,\mathrm{d}g + (S)\int_a^b hg\,\mathrm{d}f \\ - \sum_{(a,b]} h\Delta^- f\Delta^- g + \sum_{[a,b)} h\Delta^+ f\Delta^+ g,\end{aligned} \quad (2.93)$$

*where the two sums converge absolutely if $a < b$ and equal $0$ if $a = b$.*

*Proof.* We can assume that $a < b$. The two full Stieltjes integrals on the right side of (2.93) exist by Theorems 2.20, 2.17, and 2.42. By linearity, their sum gives the integral $(S) \int_a^b h(\nabla F \circ (g, f)) \cdot \mathrm{d}(g, f)$ with $F(u, v) := uv$ for $u, v \in \mathbb{R}$. Thus (2.93) follows from (2.91), proving the corollary. □

Taking $h \equiv 1$ in the preceding corollary, we recover (for $f, g$ of bounded variation) the integration by parts formula for the *RYS* integral obtained in Corollary 2.82.

## 2.9 Relations between Integrals

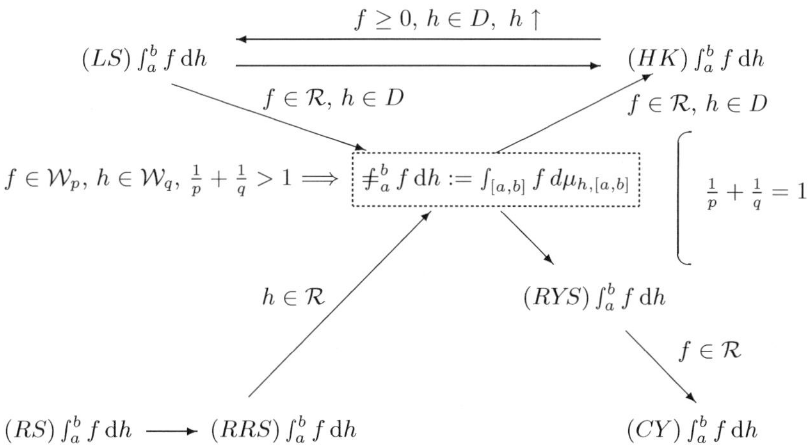


**Fig. 2.1.** Implications for integrals

In this section we explore whether, under some hypotheses, if one integral exists, then so does another, with the same value. Figure 2.1 summarizes some implications and Table 2.1 gives references for proofs. In Figure 2.1, $\longrightarrow$ means that existence of the integral to the left of it implies that of the integral to the right of it, with the same value. For "$\longleftarrow$" left and right are interchanged. The marking "$f \in \mathcal{R}$" or "$h \in \mathcal{R}$" means that the implication holds for regulated $f$ or $h$, respectively. The marking "$h \in D$" means that the implication holds for a regulated and right-continuous $h$. Finally, "$\Longrightarrow$" means that the condition to the left of it implies existence of the integral to the right of it. The condition $\frac{1}{p} + \frac{1}{q} = 1$ on the right has no arrows from (or to) it,

**Table 2.1.** References to proofs of the implications shown in Figure 2.1:

| | |
|---|---|
| $(RS)\int_a^b f\,\mathrm{d}h \longrightarrow (RRS)\int_a^b f\,\mathrm{d}h$ | Proposition 2.13 |
| $f\in\mathcal{W}_q,\ h\in\mathcal{W}_p,\ \frac{1}{p}+\frac{1}{q}>1 \Longrightarrow \fint_a^b f\,\mathrm{d}h$ | Corollary 3.95 |
| $(RRS)\int_a^b f\,\mathrm{d}h \longrightarrow \fint_a^b f\,\mathrm{d}h,\quad h\in\mathcal{R}$ | Propositions 2.18 and 2.27 |
| $(LS)\int_a^b f\,\mathrm{d}h \longrightarrow \fint_a^b f\,\mathrm{d}h,\quad f\in\mathcal{R},\ h\in D$ | Corollary 2.29 |
| $(LS)\int_a^b f\,\mathrm{d}h \longleftrightarrow (HK)\int_a^b f\,\mathrm{d}h,\ f\ge 0,\ h\in D,\ h\uparrow$ | Theorem 2.71 |
| $\fint_a^b f\,\mathrm{d}h \longrightarrow (HK)\int_a^b f\,\mathrm{d}h,\ \ f\in\mathcal{R},\ h\in D$ | Theorem 2.69 |
| $\fint_a^b f\,\mathrm{d}h \longrightarrow (RYS)\int_a^b f\,\mathrm{d}h$ | Proposition 2.27 |
| $(RYS)\int_a^b f\,\mathrm{d}h \longrightarrow (CY)\int_a^b f\,\mathrm{d}h$ | Theorem 2.51($a$) |
| $f\in\mathcal{W}_p,\ g\in\mathcal{W}_q,\ \frac{1}{p}+\frac{1}{q}=1,\ 1<p<\infty$ does not imply that any of the above integrals exists | Proposition 3.104 |

signifying that there exist $f\in\mathcal{W}_p$ and $h\in\mathcal{W}_q$ with $\frac{1}{p}+\frac{1}{q}=1$, $1<p,q<\infty$, such that $\int_a^b f\,\mathrm{d}h$ does not exist for any of the definitions given. The figure is for real-valued functions and for $a<b$, although the implications not about $LS$ or $HK$ integrals hold for $f$ and $h$ $X$- and $Y$-valued respectively.

**Proposition 2.89.** *Let $\mu$ be an upper continuous additive $X$-valued interval function on $[a,b]$, and let $\nu$ be a $Y$-valued interval function on $[a,b]$ such that $\nu(\emptyset)=0$. Let $h$ be the point function on $[a,b]$ defined by $h(t):=\mu([a,t])$ for $a\le t\le b$, and let $g$ be a $Y$-valued regulated function on $[a,b]$ such that $\nu([a,t))=g(t-)$ for $a<t\le b$. The Kolmogorov integral $\fint_{[a,b]}\nu([a,\cdot))\cdot\mathrm{d}\mu$ exists if and only if $(RRS)\int_a^b g_-^{(a)}\cdot\mathrm{d}h$ does, and if they exist then for each $a\le t\le b$,*

$$\fint_{[a,t]}\nu([a,\cdot))\cdot\mathrm{d}\mu=(RRS)\int_a^t g_-^{(a)}\cdot\mathrm{d}h.$$

*Also for $a\le c<b$, $\fint_{(c,b]}\nu([a,\cdot))\cdot\mathrm{d}\mu$ exists if and only if $(RRS)\int_c^b g_-^{(c)}\cdot\mathrm{d}h$ does, and if they exist then for each $t\in(c,b]$,*

$$\fint_{(c,t]}\nu([a,\cdot))\cdot\mathrm{d}\mu=(RRS)\int_c^t g_-^{(c)}\cdot\mathrm{d}h.$$

*Proof.* We can assume that $a<b$. By Proposition 2.6($f$), since $\mu$ is upper continuous and additive, $h$ is regulated and right-continuous on $[a,b]$. By Propositions 2.18 and 2.46, since for any $a\le c\le t\le b$, $g_-^{(c)}$ is left-continuous on $(c,t]$, the integral $(RRS)\int_c^t g_-^{(c)}\cdot\mathrm{d}h$ exists if and only if $(RYS)\int_c^t g_-^{(c)}\cdot\mathrm{d}h$ does, and the two integrals have the same value if they exist. Since $\nu(\emptyset)=0$ and $\Delta^+h(a)=0$, we have that

$$S_{YS}(\nu([a,\cdot)),\mathrm{d}\mu;[a,t],\tau)=S_{YS}(g_-^{(a)},\mathrm{d}h;[a,t],\tau)$$

for any tagged Young interval partition $\tau$ of $[a,t]$. Therefore by Proposition 2.25 and the definition of the (RYS) integral, the Kolmogorov integral

$\fint_{[a,t]} \nu([a,\cdot))\cdot d\mu$ exists if and only if $(RRS) \int_a^t g_-^{(a)}\cdot dh$ does, which follows from existence of either integral for $t = b$, and the two integrals have the same value for each $t$ if they exist, proving the first part of the lemma. To prove the second part, again let $a \le c < t \le b$. Since $\Delta^+ h(c) = 0$, we have that

$$S_{YS}(\nu([a,\cdot)), d\mu; (c,t], \tau) = S_{YS}(g_-^{(c)}, dh; [c,t], \tau)$$

for any tagged Young interval partition $\tau$ of $[\![c,t]\!]$. Therefore by Proposition 2.25 again, the Kolmogorov integral $\fint_{(c,t]} \nu([a,\cdot))\cdot d\mu$ exists if and only if $(RYS) \int_c^t g_-^{(c)}\cdot dh$ does, which follows from existence of either integral for $t = b$, and the two integrals have the same values for each $t$. The proof of the proposition is complete. □

The following relation between integrals of particular forms is used in Chapter 9. It follows from the preceding proposition and from definitions (1.15), (1.16) of the integrals with the reversed order.

**Proposition 2.90.** *Let $\mu$ be an upper continuous additive $X$-valued interval function on $[a,b]$, and let $\nu$ be a $Y$-valued interval function on $[a,b]$ such that $\nu(\emptyset) = 0$. Let $h$ be the point function on $[a,b]$ defined by $h(t) := \mu([a,t])$ for $a \le t \le b$, and let $g$ be a $Y$-valued regulated function on $[a,b]$ such that $\nu([a,t)) = g(t-)$ for $a < t \le b$. The Kolmogorov integral $\fint_{[a,b]} d\mu\cdot\nu([a,\cdot))$ exists if and only if $(RRS) \int_a^b dh\cdot g_-^{(a)}$ does, and if they exist then for each $a \le t \le b$,*

$$\fint_{[a,t]} d\mu\cdot\nu([a,\cdot)) = (RRS)\int_a^t dh\cdot g_-^{(a)}.$$

*Also for $a \le c < b$, $\fint_{(c,b]} d\mu\cdot\nu([a,\cdot))$ exists if and only if $(RRS) \int_c^b dh\cdot g_-^{(c)}$ does, and if they exist then for each $t \in (c,b]$,*

$$\fint_{(c,t]} d\mu\cdot\nu([a,\cdot)) = (RRS)\int_c^t dh\cdot g_-^{(c)}.$$

The additivity property of the Kolmogorov integral in conjunction with the preceding proposition implies the following additivity property.

**Corollary 2.91.** *Let $h \in \mathcal{R}([a,b];X)$ be right-continuous and $g \in \mathcal{R}([a,b];Y)$. For $a < c < b$, the integral $(RRS) \int_a^b dh\cdot g_-^{(a)}$ exists if and only if both $(RRS) \int_a^c dh\cdot g_-^{(a)}$ and $(RRS) \int_c^b dh\cdot g_-^{(c)}$ exist, and then*

$$(RRS)\int_a^b dh\cdot g_-^{(a)} = (RRS)\int_a^c dh\cdot g_-^{(a)} + (RRS)\int_c^b dh\cdot g_-^{(c)}.$$

*Proof.* Let $\mu := \mu_{h,[a,b]}$ and $\nu := \mu_{g,[a,b]}$ be the interval functions defined by (2.2), and let $\tilde{h}(t) := \mu([a,t])$ for $a \le t \le b$. Using Proposition 2.6 and the definition (2.3) of $R_{\mu,a}$, it then follows that $\tilde{h} = R_{\mu,a} = h - h(a)$, $\mu$ is an upper continuous additive interval function, $\nu(\emptyset) = 0$, and $g(t-) = R_{\nu,a}(t-) = \nu([a,t))$ for $a < t \le b$. Therefore by additivity of the Kolmogorov integral (Theorem 2.21 extended to the integral (1.16) with the reversed order) and by Proposition 2.90, it follows that $(RRS)\int_a^b \mathrm{d}h{\cdot}g_-^{(a)}$ exists if and only if both $(RRS)\int_a^c \mathrm{d}h{\cdot}g_-^{(a)}$ and $(RRS)\int_c^b \mathrm{d}h{\cdot}g_-^{(c)}$ exist, and then

$$\begin{aligned}(RRS)\int_a^b \mathrm{d}h{\cdot}g_-^{(a)} &= (RRS)\int_a^b \mathrm{d}\tilde{h}{\cdot}g_-^{(a)} = \oint_{[a,b]} \mathrm{d}\mu{\cdot}\nu([a,\cdot))\\ &= \oint_{[a,c]} \mathrm{d}\mu{\cdot}\nu([a,\cdot)) + \oint_{(c,b]} \mathrm{d}\mu{\cdot}\nu([a,\cdot))\\ &= (RRS)\int_a^c \mathrm{d}h{\cdot}g_-^{(a)} + (RRS)\int_c^b \mathrm{d}h{\cdot}g_-^{(c)}.\end{aligned}$$

The proof is complete. □

## 2.10 Banach-Valued Contour Integrals and Cauchy Formulas

Throughout this section we assume that $a < b$. For suitable curves $\zeta(\cdot)$ in the complex plane $\mathbb{C}$ and functions $f$, which may be Banach-valued, integrals $\int_{\zeta(\cdot)} f(\zeta)\,\mathrm{d}\zeta$ will be defined and treated. The curves will be just as in the classical theory of holomorphic functions into $\mathbb{C}$, as follows.

**Definition 2.92.** A *curve* is a continuous function $\zeta(\cdot)$ from a nondegenerate interval $[a,b]$ into $\mathbb{C}$. A $C^1$ *arc* is a curve $\zeta(\cdot)$ such that for $\zeta \equiv \xi + \mathrm{i}\eta$ with $\xi$ and $\eta$ real-valued, the derivative $\zeta'(t) = \xi'(t) + \mathrm{i}\eta'(t)$ exists and is continuous and non-zero for $a < t < b$, and has limits $\zeta'(a+) = \lim_{t\downarrow a}\zeta'(t) \ne 0 \ne \zeta'(b-) = \lim_{t\uparrow b}\zeta'(t)$. A *piecewise* $C^1$ curve is a curve $\zeta$ on $[a,b]$ such that for some partition $a = t_0 < t_1 < \cdots < t_n = b$, $\zeta(\cdot)$ is a $C^1$ arc on each $[t_{i-1}, t_i]$, and such that for each $i = 1,\dots,n-1$, the ratio $\zeta'(t_i+)/\zeta'(t_i-)$ is not real and negative, nor is $\zeta'(a+)/\zeta'(b-)$. A *closed curve* is a curve $\zeta(\cdot)$ from $[a,b]$ into $\mathbb{C}$ such that $\zeta(a) = \zeta(b)$. The closed curve $\zeta(\cdot)$ is *simple* if $\zeta(s) \ne \zeta(t)$ for $a \le s < t < b$.

If $z_0$ is a point not in the range of a closed curve $\zeta(\cdot)$, that is, $z_0 \notin \mathrm{ran}\,(\zeta)$, then we can write $\zeta(t) - z_0 = r(t)\mathrm{e}^{\mathrm{i}\theta(t)}$ for some real $r(t) > 0$ and $\theta(t)$ which are continuous functions of $t$, with $\theta(b) - \theta(a) = 2n\pi$ for some integer $n$, called the *winding number* $w(\zeta(\cdot), z_0)$ of $\zeta$ around $z_0$.

For example, if $[a,b] = [0,1]$ and $\zeta(t) = \mathrm{e}^{2\pi\mathrm{i}t}$, then $w(\zeta(\cdot), z) = 1$ if $|z| < 1$ and 0 if $|z| > 1$.

**Proposition 2.93.** *For a piecewise $C^1$ curve $\zeta(\cdot)$ on $[a,b]$ (not necessarily simple) and a point $z \notin \operatorname{ran}(\zeta)$, we have*

$$w(\zeta(\cdot), z) = \frac{1}{2\pi\mathrm{i}}(RS)\int_a^b \frac{\mathrm{d}\zeta(t)}{\zeta(t)-z}.$$

*Proof.* In the representation $\zeta(t) = z + r(t)\mathrm{e}^{\mathrm{i}\theta(t)}$, $r(\cdot) > 0$ and $\theta(\cdot)$ are piecewise $C^1$ functions on $[a,b]$. Using elementary calculations, we get

$$\begin{aligned}\frac{1}{2\pi\mathrm{i}}\int_a^b \frac{\mathrm{d}\zeta(t)}{\zeta(t)-z} &= \frac{1}{2\pi\mathrm{i}}\Big[\int_a^b \frac{\mathrm{d}r(t)}{r(t)} + \int_a^b \frac{\mathrm{d}e^{i\theta(t)}}{\mathrm{e}^{\mathrm{i}\theta(t)}}\Big]\\ &= \frac{1}{2\pi\mathrm{i}}\Big[\log\frac{r(b)}{r(a)} + \mathrm{i}(\theta(b)-\theta(a))\Big] = w(\zeta(\cdot), z)\end{aligned}$$

since $r(b) = r(a)$, proving the proposition. □

A topological space $(S, \mathcal{T})$ is *connected* if it is not the union of two disjoint nonempty open sets. A subset $C \subset S$ is *connected* iff it is connected in its relative topology. Holomorphic functions with values in a Banach space are defined as follows:

**Definition 2.94.** Let $X$ be a complex Banach space and let $U$ be an open connected set in $\mathbb{C}$. A function $f\colon U \to X$ has a *Taylor expansion* around a point $u \in U$ if there are an $r > 0$ and a sequence $\{h_k\}_{k\geq 0} \subset X$ such that for all $z$ with $|z-u| < r$, we have $z \in U$ and

$$f(z) = \sum_{k=0}^{\infty}(z-u)^k h_k, \tag{2.94}$$

where the series converges in $X$ and is called the *Taylor series* of $f$ around $u$. A function $f\colon U \to X$ is *holomorphic* on $U$ if it has a Taylor expansion around each point of $U$.

We assume that holomorphic functions are defined on connected open sets, even if this assumption is not used until we deal with analytic continuation, as in Theorems 5.33 and 5.34, for the following reasons.

Let $U$ and $V$ be disjoint nonempty open subsets of $\mathbb{C}$. Let $g$ and $h$ be any two holomorphic functions from $U$ and $V$ respectively into $\mathbb{C}$. Let $f(z) := g(z)$ for $z \in U$ and $f(z) := h(z)$ for $z \in V$. Then $f$ would be "holomorphic" on $U \cup V$. But suppose $g$ and $h$ can each be extended to be entire functions, holomorphic on all of $\mathbb{C}$, and $g \neq h$ on $\mathbb{C}$. When a function $g$ has an entire extension, such an extension is unique and it seems unnatural to define a holomorphic function in a way that could contradict an entire extension.

More specifically, consider sums $f(z) = \sum_{k\geq 1} a_k/(z-z_k)$ with $\sum_k |a_k| < \infty$, called Borel series. Suppose that $|z_k| \geq 1$ for all $k$ and that the set of all limits of subsequences $z_{k_j}$ with $k_j \to \infty$ as $j \to \infty$ is the unit circle $T^1 := \{z : |z| = 1\}$. Then the series converges on $U \cup V$ where $U := \{z : |z| < 1\}$ and $V := \{z : |z| > 1\} \setminus \{z_j\}_{j \geq 1}$. The sum $f(z)$ is holomorphic on $U$ and on $V$. Each of $U$ and $V$ is open and connected, and they are disjoint. It can happen that $f \equiv 0$ on $U$ but $f \not\equiv 0$ on $V$, e.g. [196, Theorem 4.2.5]. So again, analytic continuation from $U$ would conflict with the values on $V$.

Other holomorphic functions on open sets $U \subset \mathbb{C}$ have no entire extension and non-unique maximal holomorphic extensions in $\mathbb{C}$. Let $\mathbb{C} = \{z = x + iy \colon x, y \in \mathbb{R}\}$ and $U := \{z \in \mathbb{C}\colon x > 0\}$. Consider the function $g(z) = \sqrt{z}$ for $z \in U$ (which has a branch point at 0). It has a holomorphic extension to the complement $\{z \in \mathbb{C}\colon y = 0, x \leq 0\}^c$ and another to $\{z \in \mathbb{C}\colon x = 0, y \leq 0\}^c$. Neither extension can be extended holomorphically to any further point. To obtain natural domains for holomorphic functions with branch points one takes Riemann surfaces (cf. [1, 2nd ed., 1966 §3.4.3]).

We show next that a Taylor expansion if it exists is unique and the series converges absolutely and uniformly.

**Lemma 2.95.** *Let $X$ be a Banach space over $\mathbb{K}$ and let $\{h_k\}_{k \geq 0} \subset X$. If the power series $\sum_{k \geq 0} t^k h_k$ converges absolutely and its sum is equal to zero for each $t \in \mathbb{K}$ such that $|t| \leq \delta$ for some $\delta > 0$, then $h_k \equiv 0$ for $k = 0, 1, \ldots$.*

*Proof.* Taking $t = 0$ we get $h_0 \equiv 0$. Suppose there is a least integer $k_0 \geq 1$ such that $h_{k_0} \neq 0$. For each $t$ with $0 < |t| < \delta$, we have $\sum_{k \geq k_0} t^{k-k_0} h_k = 0$, and so

$$\|h_{k_0}\| \leq \sum_{k > k_0} \|h_k\| |t|^{k-k_0} = \frac{1}{\delta^{k_0}} \sum_{k > k_0} \|h_k\| \delta^k \left(\frac{|t|}{\delta}\right)^{k-k_0} \leq \frac{C}{\delta^{k_0}} \frac{|t|}{(\delta - |t|)},$$

where $C := \sup\{\|k_k\| \delta^k \colon k \geq 0\} < \infty$. Letting $|t| \to 0$ it follows that $h_{k_0} = 0$, a contradiction, proving that $h_k = 0$ for all $k = 1, 2, \ldots$.

**Proposition 2.96.** *Let $X$ be a complex Banach space, and let $U$ be an open connected set in $\mathbb{C}$. If $f\colon U \to X$ has a Taylor expansion* (2.94) *around $u \in U$ and the Taylor series converges on $B(u, r)$ for some $r > 0$, then it converges absolutely on $B(u, r)$ and uniformly on $\bar{B}(u, s)$ for any $s < r$. In particular, $f$ is continuous on $B(u, r)$ and its Taylor expansion around $u$ is unique.*

*Proof.* If $0 < s < r$ then $s^k \|h_k\| \to 0$ as $k \to \infty$, and so $\limsup_{k \to \infty} \|h_k\|^{1/k} \leq 1/s$. Since $s \in (0, r)$ is arbitrary, $\limsup_{k \to \infty} \|h_k\|^{1/k} \leq 1/r$. Thus the Taylor series (2.94) converges absolutely by the root test for series of positive numbers. By Lemma 2.95, the Taylor expansion (2.94) is unique. It also follows that the Taylor series converges uniformly on $\bar{B}(u, s)$ for any $s < r$, and so $f$ is continuous on $B(u, r)$, proving the proposition. □

Let $\zeta(\cdot)$ be a piecewise $C^1$ curve from $[a, b]$ into $\mathbb{C}$. Let $X$ be a Banach space over $\mathbb{C}$ and let $f$ be a continuous function into $X$ from a set in $\mathbb{C}$ including the range of $\zeta(\cdot)$. Then the contour integral over $\zeta(\cdot)$ is defined by

$$\oint_{\zeta(\cdot)} f(\zeta)\,\mathrm{d}\zeta := (RS)\int_a^b f(\zeta(t))\cdot\mathrm{d}\zeta(t),$$

where $\cdot$ denotes the natural bilinear mapping $X \times \mathbb{C} \to X$.

The Cauchy integral formula will first be extended to Banach-valued functions for curves which are circles.

**Proposition 2.97.** *Let $f$ be a holomorphic function from an open disk $U := \{z\colon |z-w| < r\} \subset \mathbb{C}$ into a complex Banach space $X$. Let $0 < t < r$ and $\zeta(\theta) := w + te^{\mathrm{i}\theta}$ for $0 \le \theta \le 2\pi$. Then for any $z \in \mathbb{C}$ with $|z-w| < t$,*

$$f(z) = \frac{1}{2\pi\mathrm{i}}\oint_{\zeta(\cdot)} \frac{f(\zeta)\,\mathrm{d}\zeta}{\zeta - z}. \tag{2.95}$$

*Proof.* Let $L$ be any continuous $\mathbb{C}$-valued linear functional on $X$ ($L \in X'$), so that for some $K < \infty$, $|L(x)| \le K\|x\|$ for all $x \in X$. For each convergent series $f(v) = \sum_{k=0}^{\infty}(v-w)^k h_k$, we have a convergent series $L(f(v)) = \sum_{k=0}^{\infty} L(h_k)(v-w)^k$, and so $L\circ f$ is holomorphic from $U$ into $\mathbb{C}$. Thus (2.95) holds for $L\circ f$ in place of $f$ by the classical Cauchy integral formula (e.g. Ahlfors [1, 2nd ed., 1966 §4.2.2, Theorem 6]). Now $L$ can be interchanged with the integral by Proposition 2.78. If $x, y \in X$ and $L(x) = L(y)$ for all $L \in X'$, then $x = y$ by the Hahn–Banach theorem. The conclusion follows. □

For a function $f$ from an open set $U \subset \mathbb{C}$ into a complex Banach space $X$ and $z \in U$, the *derivative* $f'(z)$ is defined, if it exists, as

$$\lim_{w\to z} \frac{f(w) - f(z)}{w - z} \in X.$$

Let $f^{(1)} := f'$ and $n \ge 1$. If the $n$th derivative $f^{(n)}$ is defined in a neighborhood of $z$, then $f^{(n+1)}(z)$ is defined as $(f^{(n)})'(z)$, if it exists.

The classical Cauchy integral formula for derivatives extends directly to Banach spaces:

**Theorem 2.98.** *Under the hypotheses of Proposition 2.97, $f^{(n)}(z)$ exists for all $n = 1, 2, \ldots$ and*

$$f^{(n)}(z) = \frac{n!}{2\pi\mathrm{i}}\oint_{\zeta(\cdot)} \frac{f(\zeta)\,\mathrm{d}\zeta}{(\zeta - z)^{n+1}}.$$

The conclusion follows from Proposition 2.97 and the next lemma since a holomorphic function is bounded on $C$ by Proposition 2.96.

**Lemma 2.99.** *Suppose that $g$ is continuous on the range $C$ of a piecewise $C^1$ curve $\zeta(\cdot)$. Then for each $n \ge 1$, the function*

$$I_n(z) = I_n(g; z) := \oint_{\zeta(\cdot)} \frac{g(\zeta)\,\mathrm{d}\zeta}{(\zeta - z)^n}$$

*has a derivative on the complement $C^c$ of $C$, and $I_n'(z) = nI_{n+1}(z)$.*

*Proof.* We prove first that $I_1$ is continuous on $C^c$. Let $z_0 \in C^c$ and let $\delta > 0$ be such that the open ball $|z - z_0| < \delta$ does not intersect $C$. By restricting $z$ to the smaller ball $|z - z_0| < \delta/2$, we have that $|\zeta(t) - z| > \delta/2$ for each $t \in [a, b]$. By linearity of the $RS$ integral, we have

$$I_1(z) - I_1(z_0) = (z - z_0) \oint_{\zeta(\cdot)} \frac{g(\zeta)\,\mathrm{d}\zeta}{(\zeta - z)(\zeta - z_0)}.$$

Using the bounds (2.84), it follows that

$$|I_1(z) - I_1(z_0)| \le |z - z_0|(12/\delta^2)\|g\circ\zeta\|_{[a,b],\sup} v_1(\zeta; [a, b]),$$

and so $I_1 = I_1(g)$ is continuous at $z_0$, and this is true for all continuous $g$ on $C$.

Applying the first part of the proof to the function $h(\zeta) := g(\zeta)/(\zeta - z_0)$, we conclude that

$$\frac{I_1(z) - I_1(z_0)}{z - z_0} = \oint_{\zeta(\cdot)} \frac{g(\zeta)\,\mathrm{d}\zeta}{(\zeta - z)(\zeta - z_0)} = I_1(h; z) \to I_1(h; z_0) = I_2(g; z_0)$$

as $z \to z_0$. This proves the conclusion in the case $n = 1$.

The conclusion for an arbitrary $n$ is proved by induction. Suppose that for some $n \ge 2$, $I_{n-1}'(g; z) = (n - 1)I_n(g; z)$ for $z \in C^c$, for all continuous $g$ on $C$. Again let $z_0 \in C^c$. Using the identity

$$\begin{aligned} &I_n(z) - I_n(z_0) \\ &= \Big[\oint_{\zeta(\cdot)} \frac{g(\zeta)\,\mathrm{d}\zeta}{(\zeta - z)^{n-1}(\zeta - z_0)} - \oint_{\zeta(\cdot)} \frac{g(\zeta)\,\mathrm{d}\zeta}{(\zeta - z_0)^n}\Big] \\ &\quad + (z - z_0) \oint_{\zeta(\cdot)} \frac{g(\zeta)\,\mathrm{d}\zeta}{(\zeta - z)^n(\zeta - z_0)}, \end{aligned}$$

one can conclude that $I_n$ is continuous at $z_0$. Indeed, defining $h(\zeta) := g(\zeta)/(\zeta - z_0)$, the first term is equal to $I_{n-1}(h; z) - I_{n-1}(h; z_0)$ and tends to zero as $z \to z_0$ by the induction hypothesis, while in the second term the integral is bounded for $z$ in a neighborhood of $z_0$ as shown in the first part of the proof. Now dividing the identity by $z - z_0$ and letting $z \to z_0$, the quotient in the first term tends to a derivative $I_{n-1}'(h; z_0)$, which by the induction hypothesis equals $(n - 1)I_n(h; z_0) = (n - 1)I_{n+1}(g; z_0)$. The remaining factor $I_n(h; z)$ is continuous as before, and so has the limit $I_n(h; z_0) = I_{n+1}(g; z_0)$. Thus $I_n'(z_0)$ exists and equals $nI_{n+1}(z_0)$. The proof of the lemma is complete. □

**Proposition 2.100.** *Let $X$ be a complex Banach space, and let $f$ be a holomorphic function on a disk $B(u,r) \subset \mathbb{C}$ with values in $X$ for some $u \in \mathbb{C}$ and $r > 0$. Then for each $z \in B(u,r)$,*

$$f(z) = f(u) + \sum_{k \geq 1} \frac{(z-u)^k}{k!} f^{(k)}(u)$$

*is the Taylor expansion of $f$ around $u$.*

*Proof.* Let $0 < s < t < r$, let $\zeta(\theta) := u + te^{\mathrm{i}\theta}$ for $0 \leq \theta \leq 2\pi$, and let $z \in B(u,t)$. Then by the Cauchy integral formula (Proposition 2.97),

$$f(z) = \frac{1}{2\pi \mathrm{i}} \oint_{\zeta(\cdot)} \frac{f(\zeta)\, \mathrm{d}\zeta}{\zeta - z}.$$

For each $\zeta \in \operatorname{ran}(\zeta)$, the series

$$\frac{1}{\zeta - z} = (\zeta - u)^{-1} \sum_{k=0}^{\infty} \left( \frac{z-u}{\zeta - u} \right)^k$$

converges absolutely, and uniformly if $|z-u| \leq s$. Since $f$ is continuous on the range $\operatorname{ran}(\zeta)$ by Proposition 2.96, it is bounded. Then term-by-term integration yields

$$f(z) = \sum_{k=0}^{\infty} \frac{(z-u)^k}{2\pi \mathrm{i}} \oint_{\zeta(\cdot)} \frac{f(\zeta)\, \mathrm{d}\zeta}{(\zeta - u)^{k+1}}.$$

Using the bound (2.84) for the *RS* integral and the root test, it follows that the series converges absolutely. Now the conclusion follows by the Cauchy integral formula for derivatives (Theorem 2.98) and by the uniqueness of a Taylor expansion (Proposition 2.96). □

**Proposition 2.101.** *If $f$ is holomorphic from an open connected set $U \subset \mathbb{C}$ into a Banach space $X$ and $w \in U$, then for*

$$g(z) := \begin{cases} \frac{f(z)-f(w)}{z-w}, & \text{if } z \neq w, \\ f'(w), & \text{if } z = w, \end{cases}$$

*$g$ is also holomorphic on $U$.*

*Proof.* It is easily seen that $g$ is holomorphic on $U \setminus \{w\}$, since $z \mapsto 1/(z-w)$ is, so it suffices to find a power series expansion for $g$ around $w$. Let $f(z) = \sum_{k=0}^{\infty} (z-w)^k x_k$ for $|z-w| < r$, where $r > 0$ and $x_k := f^{(k)}(w)/k! \in X$ by Proposition 2.100. Then $g(z) = \sum_{k=0}^{\infty} (z-w)^k x_{k+1}$, also for $|z-w| < r$, with $g(w) = x_1 = f'(w)$, finishing the proof. □

A topological space $(S, \mathcal{T})$ is *locally connected* iff $\mathcal{T}$ has a base consisting of connected open sets. Clearly, any normed vector space with its usual topology is connected and locally connected. In particular, $\mathbb{R}$ and $\mathbb{C}$ are locally connected.

**Theorem 2.102.** *For any topological space $(S, \mathcal{T})$ and any nonempty connected set $A \subset S$ there is a unique connected set $B \supset A$ such that $B$ is a maximal connected set for inclusion. Such a set $B$ is always closed.*

*Proof.* The collection of connected sets including $A$ is partially ordered by inclusion. The union of any inclusion-chain of connected sets is easily seen to be connected. Thus by Zorn's lemma there is a connected set $B_1 \supset A$ which is a maximal connected set for inclusion. Suppose $B_2$ is another such set. The union of two non-disjoint connected sets is easily seen to be connected. This gives a contradiction unless $B_1 = B_2$, so $B$ is unique. Moreover, the closure of any connected set is easily shown to be closed. So by maximality, $B$ is closed and the theorem is proved. □

A maximal connected subset $B$ of $S$ is called a *component*. If $A$ is a singleton $\{x\}$, it is always connected, and the $B$ from the preceding theorem is called the *component of* $x$.

**Proposition 2.103.** *Let $(S, \mathcal{T})$ be a locally connected topological space. Then any component $F$ of $S$ is open as well as closed. Thus $S$ has a unique decomposition as a union of disjoint open and closed components.*

*Proof.* Let $F$ be a component of $S$ and $x \in F$. Let $V$ be a connected open neighborhood of $x$ by local connectedness. Then $F \cup V$ is connected since $F$ and $V$ are and $F \cap V \supset \{x\} \neq \emptyset$. So $V \subset F$ and $F$ is open. The rest follows. □

Clearly, an open set in a locally connected space is locally connected, so it is in a unique way a union of disjoint nonempty connected open sets.

An open set $U \subset \mathbb{C}$ will here, following Ahlfors [1, 2nd ed., §4.4.2], be called *simply connected* if and only if $U^c$ has no bounded component. A more general definition is based on the following.

**Definition 2.104.** A closed curve $\zeta(\cdot)$ from $[a, b]$ into a set $U$ is *null-homotopic in* $U$ if there exists a jointly continuous function $H$ from $[a, b] \times [0, 1]$ into $U$ such that $H(\cdot, 0) \equiv \zeta(\cdot)$, for some $w \in U$, $H(\cdot, 1) \equiv w$, and for $0 \leq u \leq 1$, $H(a, u) = H(b, u)$, so that each $H(\cdot, u)$ is a closed curve.

**Theorem 2.105.** *Let $U \subset \mathbb{C}$ be a connected nonempty open set. Then the following are equivalent:*

(a) *any closed curve $\zeta(\cdot)$ with range in $U$ is null-homotopic in $U$;*

(*b*) *for every closed curve* $\zeta(\cdot)$ *into* $U$ *and* $z \in U^c$, $w(\zeta(\cdot), z) = 0$;
(*c*) $U^c$ *has no bounded component*;
(*d*) $U$ *is homeomorphic to the open unit disk* $\{z \in \mathbb{C}\colon |z| < 1\}$.

*Proof.* The implications $(d) \to (a) \to (b)$ and $(c) \to (b)$ are easy. The other implications are not as easy and are given in Newman [178, §§6.6 and 7.9]. Complex analysis texts also give proofs, of the stronger fact that in (*d*), unless $U = \mathbb{C}$, the homeomorphisms can be taken to be holomorphic with holomorphic inverses (Riemann mapping theorem, see e.g. Beardon [12, Theorem 11.1.1 and §11.2]). □

In Theorem 2.105, (*a*) is the definition of simply connected for general topological spaces. It is not equivalent to (*c*) in $\mathbb{R}^3$, for example.

Here is a Cauchy integral theorem for Banach-valued functions.

**Theorem 2.106.** *Let $f$ be holomorphic from an open connected set $U \subset \mathbb{C}$ into a complex Banach space $X$ and let $\zeta(\cdot)$ be a piecewise $C^1$ closed curve whose range is included in $U$. If (i) the winding number $w(\zeta(\cdot); z) = 0$ for each $z \notin U$, e.g. if (ii) $U$ is simply connected, then*

$$\oint_{\zeta(\cdot)} f(\zeta)\,\mathrm{d}\zeta = 0. \tag{2.96}$$

*Proof.* Condition (*ii*) implies (*i*) by Theorem 2.105, $(c) \Rightarrow (b)$, or $(a) \Rightarrow (b)$ for the general definition of simply connected. So we can assume (*i*). By Proposition 2.96, $f$ is continuous on $U$, and so the integral (2.96) is defined. For any continuous linear functional $L \in X'$, $z \mapsto L(f(z))$ is holomorphic from $U$ into $\mathbb{C}$. Thus by Proposition 2.78, we have

$$L\Big(\oint_{\zeta(\cdot)} f(\zeta)\,\mathrm{d}\zeta\Big) = \oint_{\zeta(\cdot)} L(f(\zeta))\,\mathrm{d}\zeta.$$

The integral on the right is zero by the Cauchy theorem for multiply connected sets (e.g., Ahlfors [1, 2nd ed., §4.4.4 Theorem 18]). Thus by the Hahn–Banach theorem (e.g. [53, 6.1.5 Corollary]), it follows that (2.96) holds, proving the theorem. □

Here are Cauchy integral formulas with winding numbers.

**Theorem 2.107.** *Let $f$ be holomorphic from a connected and simply connected open set $U \subset \mathbb{C}$ into a complex Banach space $X$. Let $\zeta(\cdot)$ be a piecewise $C^1$ closed curve whose range $C$ is included in $U$ and let $z \in U \setminus C$. Then*

$$w(\zeta(\cdot), z) f(z) = \frac{1}{2\pi \mathrm{i}} \oint_{\zeta(\cdot)} \frac{f(\zeta)\,\mathrm{d}\zeta}{\zeta - z}, \tag{2.97}$$

*and for* $n = 1, 2, 3, \ldots$,

$$w(\zeta(\cdot), z) f^{(n)}(z) = \frac{n!}{2\pi \mathrm{i}} \oint_{\zeta(\cdot)} \frac{f(\zeta)\,\mathrm{d}\zeta}{(\zeta - z)^{n+1}}. \tag{2.98}$$

*Proof.* On the right side of (2.97) write $f(\zeta) = [f(\zeta) - f(z)] + f(z)$. For the $f(\zeta) - f(z)$ term, the integral is 0 by Proposition 2.101 and Theorem 2.106. For the $f(z)$ term, we get the left side of (2.97) by Proposition 2.93, proving (2.97). We thus have (2.98) for $n = 0$. For a given $\zeta(\cdot)$ and $z \notin C$, $w(\zeta(\cdot), u)$ is constant for $u$ in a neighborhood of $z$, so the multiplication by $w(\zeta(\cdot), u)$ can be interchanged with $\mathrm{d}^n/\mathrm{d}u^n|_{u=z}$. On the right, we can use induction on $n$ and Lemma 2.99, so (2.98) is proved. □

If $w(\zeta(\cdot), z) = 1$, we get the familiar form of the formula, now for rather general closed curves $\zeta(\cdot)$.

## 2.11 Notes

*Notes on Section 2.1.* More properties of a real-valued function equivalent to being regulated are given by Theorem 2.1 in [54, Part III].

The correspondence between right-continuous regulated point functions and additive upper continuous interval functions given by Corollary 2.11 is similar to the well-known correspondence between right-continuous point functions of bounded variation and $\sigma$-additive set functions on the Borel $\sigma$-algebra.

*Notes on Section 2.2.* In T. J. Stieltjes's treatment of his integral $\int_a^b f\,\mathrm{d}h$ in [227], the integrand $f$ is continuous, while the integrator $h$ is a monotone increasing function. The Riemann–Stieltjes integral $(RS)\int_a^b f\,\mathrm{d}h$, for any $f$, $h$ such that it exists, appeared in G. König [122], who apparently had been using the $RS$ integral in his lectures for some time, as stated in [228, p. 247] and in [168, p. 314]. Szénássy [228] has a chapter on König, where on p. 247 of the English translation, König's general definition of the $RS$ integral is quoted. Interest in Stieltjes-type integrals flourished after F. Riesz [193] showed that the Stieltjes integral provides a representation of an arbitrary bounded linear functional on the space $C[0,1]$ of all continuous functions on $[0,1]$. Pollard [187] introduced the refinement Riemann–Stieltjes integral, showed that it extends the Riemann–Stieltjes integral, and proved that for $h$ nondecreasing and $f$ bounded, the integral $(RRS)\int_a^b f\,\mathrm{d}h$ exists if and only if the Darboux–Stieltjes integral does. The latter integral is defined as $\sup_\kappa L(f,h;\kappa) = \inf_\kappa U(f,h;\kappa)$ if the equality holds, where for a partition $\kappa = \{t_i\}_{i=0}^n$ of $[a,b]$, $m(f;A) := \inf\{f(t)\colon t \in A\}$ and $M(f;A) := \sup\{f(t)\colon t \in A\}$,

$$\begin{cases} L(f,h;\kappa) := \sum_{i=1}^n m(f;[t_{i-1},t_i])[h(t_i) - h(t_{i-1})], \\ U(f,h;\kappa) := \sum_{i=1}^n M(f;[t_{i-1},t_i])[h(t_i) - h(t_{i-1})]. \end{cases} \tag{2.99}$$

F. A. Medvedev [168] gives an extensive account of the early history of Stieltjes-type integrals, and in particular gives some details about König's work on the *RS* integral.

S. Pollard [187] gave a detailed treatment of the refinement Riemann–Stieltjes integral. Earlier, a theory of limits based on refinements was treated by E. H. Moore [175]. Thus the *RRS* integral sometimes is called the Pollard–Moore–Stieltjes integral (see e.g. [94, p. 269]). The books of Gochman [82] and Hildebrandt [97, §§9–18] also contain detailed expositions of the *RRS* integral.

For real-valued functions, Propositions 2.13 and 2.15 were proved in Pollard [187, p. 90] and Smith [223], respectively.

*Notes on Section 2.3.* The refinement Young–Stieltjes integral developed in stages over a long period. This integral was rediscovered by several authors, and hence it is known by different names. According to T. H. Hildebrandt [94], the *RYS* integral originated in the work of W. H. Young [254]. In this work W. H. Young extended his approach to defining the Lebesgue integral, and the *RYS* integral appeared there as an auxiliary tool. At that time there was considerable interest in enlarging the class of functions integrable with respect to a monotone function $h$. Lebesgue [134], for example, showed that such a class could be the class of all Lebesgue summable functions with respect to $\mathrm{d}h$ when a Stieltjes-type integral is defined by means of the Lebesgue integral using a change of variables formula. Lebesgue suggested that it would be difficult to extend the Stieltjes integral to such general integrands by any other means. Recall that Radon's work which led to the Lebesgue–Stieltjes integral appeared a little later [190]. However, W. H. Young [254] showed that his method of monotonic sequences used in connection with the Lebesgue integral extends to an integration with respect to any function $h$ of bounded variation almost without changes. The main change concerned the definition of the integral $\int f\,\mathrm{d}h$ for a step function $f$. In this case, W. H. Young set

$$\int_a^b f\,\mathrm{d}h := \sum_i \big\{[f\Delta^+h](x_{i-1}) + c_i[h(x_i-) - h(x_{i-1}+)] + [f\Delta^-h](x_i)\big\}, \tag{2.100}$$

provided $f(x) = c_i$ for $x \in (x_{i-1}, x_i)$. The method of monotonic sequences of W. H. Young was later extended by Daniell [37] to integrals of functions defined on abstract sets. A concise theory based on an integral of Stieltjes type was provided by W. H. Young's son L. C. Young in the form of a textbook [243], first published in 1927 (see also the presentation in Hildebrandt [97]). The integral, defined as a limit of Young–Stieltjes sums if it exists when the mesh of tagged partitions tends to zero, was mentioned by R. C. Young [251]. Full use of the refinement Young–Stieltjes integral, in the context of Fourier series, is due to L. C. Young [247].

Motivated by the weakness of the Riemann–Stieltjes integral, Ross [195] suggested an extension of the Darboux–Stieltjes integral, replacing (2.99) by sums reminiscent of Young–Stieltjes sums. Namely, let $f$ be a real-valued function on $[a,b]$, and let $h$ be a nondecreasing function on $[a,b]$. For a partition $\kappa = \{t_i\}_{i=0}^n$ of $[a,b]$, let

$$\begin{cases} L(f,\mathrm{d}h;\kappa) := J(f,\mathrm{d}h;\kappa) + \sum_{i=1}^n m(f;(t_{i-1},t_i))[h(t_i-) - h(t_{i-1}+)] \\ U(f,\mathrm{d}h;\kappa) := J(f,\mathrm{d}h;\kappa) + \sum_{i=1}^n M(f;(t_{i-1},t_i))[h(t_i-) - h(t_{i-1}+)], \end{cases}$$

where $J(f,\mathrm{d}h;\kappa) := \sum_{i=1}^n f(t_i)\Delta_{[a,b]}^{\pm}h(t_i)$, and $m(f,A)$, $M(f,A)$ are defined as in (2.99). We say that $f$ is Ross–Darboux–Stieltjes, or *RDS*, integrable on $[a,b]$ with respect to $h$, if $U(f,\mathrm{d}h) := \inf_{\kappa} U(f,\mathrm{d}h;\kappa) = \sup_{\kappa} L(f,\mathrm{d}h;\kappa) =: L(f,\mathrm{d}h)$, and then let $(RDS)\int_a^b f\,\mathrm{d}h := U(f,\mathrm{d}h) = L(f,\mathrm{d}h)$. By Theorem 35.25 in [195], a bounded function $f$ on $[a,b]$ is *RDS* integrable with respect to $h$ if and only if there is $C \in \mathbb{R}$ such that given $\epsilon > 0$ there exists $\delta > 0$ such that $|S_{YS}(f,\mathrm{d}h;\tau) - C| < \epsilon$ for each tagged Young interval partition $\tau = (\{t_i\}_{i=0}^n, \{s_i\}_{i=1}^n)$ of $[a,b]$ such that $\max\{h(t_i-) - h(t_{i-1}+) \colon i = 1,\dots,n\} < \delta$. Then using Theorem 2.1($b$), it is easy to see that if $(RDS)\int_a^b f\,\mathrm{d}h$ exists then so does $(RYS)\int_a^b f\,\mathrm{d}h$ and the two are equal.

The Bartle integral given by Definition 2.37 is a general bilinear integral of a vector function with respect to an additive vector measure. According to Diestel and Uhl [41, p. 58], "Bartle [11] launched a theory of integration that includes most of the known integration procedures that have any claim to quality. His integral specializes to include the Bochner integral but does not include the general Pettis integral. Possibly workers in the theory of vector measures would be better off if they attempted to use the Bartle integral rather than inventing their own." A survey of the history of vector integration can be found in Hildebrandt [95] and Bartle [11].

There are several different constructions of an integral of a Banach-space-valued function with respect to a nonnegative finite scalar-valued measure $\mu$. The Bochner integral given by Definition 2.31 is one of the best known among such integrals. S. Bochner [21] defined his integral. The extension of Lebesgue's differentiation theorem to Banach-valued functions, Theorem 2.35, is given e.g. by Diestel and Uhl [41, Theorem II.2.9, p. 49].

A function $f$ taking values in a Banach space $X$ is integrable with respect to $\mu$ in the sense of Dunford [56] if there exists a sequence of $\mu$-simple functions $\{f_k\}_{k\geq 1}$ such that it converges to $f$ $\mu$-almost everywhere, and the sequence $\{\int_A f_k\,\mathrm{d}\mu\}_{k\geq 1}$ converges in $X$ for each $A \in \mathcal{S}$. The integral $(D)\int_A f\,\mathrm{d}\mu$ is then defined to be the limit $\lim_{k\to\infty}\int_A f_k\,\mathrm{d}\mu$, and is called the *second Dunford integral*. The integral does not depend on $\{f_k\}$ since it is a special case of the Bartle integral by Theorem 9 in [11], which is well defined by Proposition 2.38. The second Dunford integral clearly extends the Bochner integral. Hildebrandt [95, p. 123] showed that for $\mu$-measurable functions $f$, the second Dunford integral coincides with several other integrals. Example 7 of Birkhoff [16, p. 377] gives a $\mu$-measurable function $f$ such that $\|f(\cdot)\|$ is not summable but is integrable in the sense of Dunford. For a scalar-valued measurable function $f$ and a vector-valued measure $\mu$, an integral analogous to the second Dunford integral is treated in Dunford and Schwartz [57, Definition IV.10.7].

*Notes on Section 2.4.* The "if" part of Theorem 2.42 was proved in Smith [223, p. 495].

*Notes on Section 2.5.* L. C. Young [244, p. 263] defined an extension of the Riemann–Stieltjes integral for regulated complex-valued functions $f$ and $h$ by

$$(Y_0)\int_a^b f\,dh := (RRS)\int_a^b f_+\,dh_- + \sum_{a\le t\le b}\big[f(t)-f(t+)\big]\big[h(t+)-h(t-)\big]$$

if the $RRS$ integral exists and the sum converges absolutely. However, the value of the $Y_0$ integral depends on $h(a-)$, $f(b+)$, and $h(b+)$, which need not be defined. If in the $Y_0$ integral we replace $f(t+)$ by $f_+^{(b)}(t)$, $h(t+)$ by $h_+^{(b)}(t)$, and $h(t-)$ by $h_-^{(a)}(t)$, then we get the $Y_1$ integral defined by (2.44). L. C. Young in 1938 gave such a convention for values at the endpoints in footnote 6 on p. 583 of [247], so one may suppose that he had these definitions in mind already in his 1936 paper.

The $Y_1$ and $Y_2$ integrals have been defined and proved equal when both exist in Dudley [49]. An extension of the $Y_1$ and $Y_2$ integrals to integrals of the form $\int f{\cdot}\mathrm{d}h{\cdot}g$ with two integrands $f$ and $g$ was given in [54, Part II]. In Section 9.8 below such an extension is given for the $RYS$ and Kolmogorov integrals.

Theorem 2.50 for functions with values in a Banach algebra is a special case of Theorem 3.7 in [54, Part II].

*Notes on Sections 2.6 and *2.7.* Saks [201] defines integrals due to Perron and Denjoy and gives references to their work, published beginning in 1912–1915. The Perron integral and one form of the Denjoy integral turned out to be equivalent to the Henstock–Kurzweil integral for integrals $\int_a^b g(x)\,\mathrm{d}x$: Gordon [84, Chapter 11].

Ward [238] in 1936 defined an integral called a Perron–Stieltjes integral which includes both the Lebesgue–Stieltjes and refinement Riemann–Stieltjes integrals. Theorem 2.64 is Theorem 5 of Ward [238]. Kurzweil [129, Section 1.2] in 1957 defined the Henstock–Kurzweil integral and proved Theorem 2.65 on the equivalence of the $WPS$ and $HK$ integrals [129, Theorem 1.2.1]. Kurzweil ([129], [130], [131]), Henstock ([90], [91], [92]), and McShane [166] consider extended integrals for functions $U(\cdot,\cdot)$ of two variables, the second and third authors also on general spaces $X$, where $(HK)\int_a^b f\,\mathrm{d}h$ is the special case $U(x,y)\equiv f(y)h(x)$ and $X=[a,b]$.

A discussion of Proposition 2.58 and its proof when $h(x)=x$ can be found in McLeod [165, Sections 1.5, 2.8, and 7.3].

While there has been relatively little literature about the refinement Young–Stieltjes and central Young integrals, there has been much more about Henstock–Kurzweil (or gauge) integrals, e.g. Lee Peng-Yee [184]. A 1991 book by Henstock [92] has a reference list of more than 1200 papers and books, mainly on the theme of "non-absolute integration," if not necessarily about the $HK$ integral itself. In fact, Henstock [92] treats integration over general spaces ("division spaces").

Schwabik [211, Theorem 3.1] proved Theorem 2.69 when $h$ is of bounded variation and $f$ is not necessarily regulated but either (a) $f$ is bounded, or (b) $f$ is arbitrary and for any $t$ if $h(t-) = h(t+)$ then $h(t) = h(t-)$.

*Notes on Section 2.8.* According to Hildebrandt [94, p. 276], the integration by parts formula (2.79) for the refinement Young–Stieltjes integral was proved by de Finetti and Jacob [66] assuming that $f$ and $g$ are of bounded variation. Hewitt [93] proved an integration by parts formula similar to (2.83) for the Lebesgue–Stieltjes integral. Corollary 2.83 contains the result of Love [145], who calls the refinement Young–Stieltjes integral the refinement Ross-Riemann–Stieltjes integral, or the $R^3S$-integral.

*Notes on Section 2.10.* Ahlfors [1, 2nd ed., §4.4.4] wrote: "It was E. Artin who discovered that the characterization of homology by vanishing winding numbers ties in precisely with what is needed for Cauchy's theorem. This idea has led to a remarkable simplification of earlier proofs."

Proposition 2.93 is given e.g. in Ahlfors [1, 2nd ed., §4.2.1].

# 3

# $\Phi$-variation and $p$-variation; Inequalities for Integrals

## 3.1 $\Phi$-variation

Let $\mathcal{V}$ be the class of all functions $\Phi$: $[0,\infty) \to [0,\infty)$ which are strictly increasing, continuous, unbounded, and 0 at 0. Let $\mathcal{CV}$ be the subclass of convex functions in $\mathcal{V}$. Let $X$ be a Banach space with norm $\|\cdot\|$, let $J$ be a nonempty interval in $\mathbb{R}$, let $f$ be a function defined on $J$ with values in $X$, and let $\Phi \in \mathcal{V}$. Recall that an interval is called *nondegenerate* if it contains more than one point. If $J$ is nondegenerate, for each $\kappa = \{t_i\}_{i=0}^n \in \mathrm{PP}\,(J)$, the class of all point partitions of $J$ as defined in Section 1.4, let $s_\Phi(f;\kappa) := \sum_{i=1}^n \Phi(\|f(t_i) - f(t_{i-1})\|)$ be the $\Phi$-variation sum.

**Definition 3.1.** Let $X$ be a Banach space, let $J$ be a nonempty interval, and let $f\colon J \to X$. The *$\Phi$-variation* of $f$ on $J$ is defined by

$$v_\Phi(f) \;:=\; v_\Phi(f;J) \;:=\; \sup\bigl\{s_\Phi(f;\kappa)\colon\ \kappa \in \mathrm{PP}\,(J)\bigr\} \tag{3.1}$$

if $J$ is nondegenerate, or as 0 if $J$ is a singleton. We say that $f$ has bounded $\Phi$-variation if $v_\Phi(f) < \infty$. The class of all functions $f\colon J \to X$ with bounded $\Phi$-variation will be denoted by $\mathcal{W}_\Phi = \mathcal{W}_\Phi(J;X)$. Let $\mathcal{W}_\Phi(J) := \mathcal{W}_\Phi(J;\mathbb{R})$.

If $\Phi(u) \equiv u^p$, $u \ge 0$, for some $0 < p < \infty$, we write $v_p := v_\Phi$, $\mathcal{W}_p := \mathcal{W}_\Phi$, and $V_p(f) = v_p(f)^{1/p}$. So $v_p(f)$ is the $p$-variation of $f$ as defined by (1.2) for $X = \mathbb{R}$ and in Section 1.4 for general $X$. Thus the $p$-variation of $f$ on a nonempty interval $J$ is defined as $v_p(f;J) := 0$ if $J$ is a singleton, and if $J$ is nondegenerate as

$$v_p(f) = v_p(f;J) = \sup\bigl\{s_p(f;\kappa)\colon\ \kappa \in \mathrm{PP}\,(J)\bigr\}. \tag{3.2}$$

In the latter case, for $\kappa = \{t_i\}_{i=0}^n \in \mathrm{PP}\,(J)$, $s_p(f;\kappa) = \sum_{i=1}^n \|f(t_i) - f(t_{i-1})\|^p$ is the $p$-variation sum as defined by (1.1) for $X = \mathbb{R}$ and in Section 1.4 in general.

A function $\Phi \in \mathcal{V}$ is said to satisfy the *$\Delta_2$ condition* if for some $1 \le D < \infty$,

R.M. Dudley and R. Norvaiša, *Concrete Functional Calculus*, Springer Monographs in Mathematics, DOI 10.1007/978-1-4419-6950-7_3,

$$\Phi(2u) \le D\Phi(u) \qquad \text{for all } u \ge 0. \tag{3.3}$$

If $\Phi$ is convex, so that $\Phi \in \mathcal{CV}$, then $D \ge 2$. If $\Phi(u) \equiv u^p$, $u \ge 0$, for some $0 < p < \infty$ (convex if and only if $p \ge 1$) then the $\Delta_2$ condition holds with $D = 2^p$. We have the following:

**Proposition 3.2.** *If $\Phi \in \mathcal{V}$ satisfies the $\Delta_2$ condition then for any nonempty interval $J$ and Banach space $X$, $\mathcal{W}_\Phi(J;X)$ is a vector space.*

*Proof.* We can assume that $J$ is nondegenerate. For $s, t \in J$ and $h\colon J \to X$ let $\Delta h := h(t) - h(s)$. Then for any $f, g \in \mathcal{W}_\Phi(J;X)$, by (3.3) and since $\Phi$ is increasing, we have

$$\begin{aligned}\Phi(\|\Delta(f+g)\|) &\le \Phi(2\max\{\|\Delta f\|, \|\Delta g\|\}) \le D\Phi(\max\{\|\Delta f\|, \|\Delta g\|\})\\ &\le D(\Phi(\|\Delta f\|) + \Phi(\|\Delta g\|)).\end{aligned}$$

Summing over nonoverlapping subintervals of $J$, it follows that $v_\Phi(f+g) \le D(v_\Phi(f) + v_\Phi(g))$, so $f + g \in \mathcal{W}_\Phi(J;X)$. Thus, or directly, $2f, 4f, 8f$, etc. are in $\mathcal{W}_\Phi(J;X)$, and since $\Phi$ is increasing it follows that $cf \in \mathcal{W}_\Phi(J;X)$ for any constant $c$. The proof is complete. □

Any convex function $\Phi$ on $[0,\infty)$ with $\Phi(0) = 0$ has the superadditivity property: for any $u, v \ge 0$,

$$\Phi(u+v) \ \ge\ \Phi(u) + \Phi(v), \tag{3.4}$$

which is immediate if $u = v = 0$, and for $u + v > 0$ follows from $\Phi(u) \le u\Phi(u+v)/(u+v)$ and $\Phi(v) \le v\Phi(u+v)/(u+v)$.

For a monotone function $f$, it is easy to evaluate $v_\Phi(f)$ explicitly:

**Example 3.3.** Let $\Phi \in \mathcal{CV}$. By (3.4), it follows that if $f$ is real-valued and monotone on an interval $J = [a,b]$ with $-\infty < a \le b < \infty$, then

$$v_\Phi(f;[a,b]) = \Phi(|f(b) - f(a)|). \tag{3.5}$$

Now let $f$ be piecewise monotone on $[a,b]$ with $a < b$, so that for a partition $\{t_i\}_{i=0}^n$ of $[a,b]$, $f$ is nondecreasing on each interval $[t_{2j}, t_{2j+1}]$ and nonincreasing on each interval $[t_{2j-1}, t_{2j}]$. Assume also that there is a constant $c$ such that $f(t_i) < c$ and $f(t_k) > c$ for each even $i \le n$ and odd $k \le n$. Then

$$v_\Phi(f;[a,b]) = \sum_{j=1}^n \Phi(|f(t_j) - f(t_{j-1})|). \tag{3.6}$$

To see this, note that for any point partition $\tau$ of $[a,b]$, the refinement $\kappa$ obtained by adjoining all $t_i$ has $s_\Phi(f;\kappa) \ge s_\Phi(f;\tau)$. Then, by (3.5), $s_\Phi$ is not decreased by deleting all points other than the $t_j$.

Finally, for a continuous, increasing function $g$ on $(0,1]$ such that $g(t) \to 0$ as $t \downarrow 0$, let $f(t) := g(t)\cos(\pi/t)$ for $0 < t \le 1$ and $f(0) := 0$. Then

$$v_\Phi(f;[0,1]) = \sum_{i=1}^{\infty} \Phi(g(1/i) + g(1/(i+1))). \tag{3.7}$$

This can be verified as (3.6) was proved, starting with any partition $\{x_i\}_{i=0}^n$ of $[0,1]$, adjoining points $1/j$ for $j = 2,\ldots,m$ with $1/m < x_1$, then deleting any $x_i \ne 1/j$ for $j = 1,\ldots,m$ and letting $m \to \infty$.

In general, the class $\mathcal{W}_\Phi$ need not be a vector space, as the following shows.

**Example 3.4.** Here are examples where $\mathcal{W}_\Phi$ is not a vector space and $\Phi$ does not satisfy the $\Delta_2$ condition. Let $\Phi(u) := \mathrm{e}^{-1/u}$ for $0 < u \le 1/4$ and $\Phi(0) := 0$. Let $d$ be the left derivative of $\Phi$ at $1/4$. Then letting $\Phi(u) := \Phi(1/4)+d(u-1/4)$ for $u > 1/4$ we get that $\Phi$ is convex and in $\mathcal{CV}$. Let $\Psi(u) := \Phi(u)+((u-1/4)_+)^2$, where $v_+ := \max(v,0)$. Then $\Psi$ is strictly convex and also in $\mathcal{CV}$.

Let $g(t) := 1/[c\log(1/t)]$ for $0 < t \le 1/4$ and $g(t) := 0$ elsewhere on $[0,\infty)$, where $c$ is a constant for which we will consider different values. Let $f(t) := g(t)\cos(\pi/t)$ for $0 < t \le 1$ and $f(0) := 0$. Then, as in (3.7),

$$v_\Phi(f;[0,1]) = \Phi(g(1/4)) + \sum_{j=4}^{\infty} \Phi(g(1/j) + g(1/(j+1))). \tag{3.8}$$

Now $g(1/j) = 1/(c\log j)$, so $g(1/j)+g(1/(j+1))$ is asymptotic to $2/(c\log j)$ as $j \to \infty$. It follows that the series in (3.8) converges if $c > 2$ and diverges if $c < 2$. Thus taking $c = 3$, $f \in \mathcal{W}_\Phi[0,1]$ but $3f \notin \mathcal{W}_\Phi[0,1]$. The same statements hold for $\Psi$ in place of $\Phi$.

We will show that the set defined next is a vector space including $\mathcal{W}_\Phi$.

**Definition 3.5.** Let $J$ be a nonempty interval and let $X$ be a Banach space. For $\Phi \in \mathcal{V}$, let $\widetilde{\mathcal{W}}_\Phi = \widetilde{\mathcal{W}}_\Phi(J;X)$ be the set of all functions $f$ from $J$ to $X$ such that $cf \in \mathcal{W}_\Phi(J;X)$ for some $c > 0$. For $\Phi \in \mathcal{CV}$ and $f \in \widetilde{\mathcal{W}}_\Phi(J;X)$, let

$$\begin{aligned} \|f\|_{(\Phi)} &:= \|f\|_{J,(\Phi)} := \inf\{c>0:\ v_\Phi(f/c) \le 1\} \qquad \text{and} \\ \|f\|_{[\Phi]} &:= \|f\|_{J,[\Phi]} := \|f\|_{J,(\Phi)} + \|f\|_{\sup}, \end{aligned}$$

where $\|f\|_{\sup} = \sup_{t\in J}\|f(t)\|$ is the sup norm of $f$. In the special case where $\Phi(u) \equiv u^p$, $u \ge 0$, for some $1 \le p < \infty$, let $\mathcal{W}_p(J;X) := \widetilde{\mathcal{W}}_\Phi(J;X)$, $\|f\|_{(p)} := \|f\|_{J,(p)} := \|f\|_{J,(\Phi)}$, and $\|f\|_{[p]} := \|f\|_{J,[p]} := \|f\|_{\sup} + \|f\|_{J,(p)}$. Also for $X = \mathbb{R}$, let $\widetilde{\mathcal{W}}_\Phi(J) := \widetilde{\mathcal{W}}_\Phi(J;\mathbb{R})$ and $\mathcal{W}_p(J) := \mathcal{W}_p(J;\mathbb{R})$.

We have $\mathcal{W}_\Phi \subset \widetilde{\mathcal{W}}_\Phi$, and the inclusion can be strict as in Example 3.4.

**Remark 3.6.** It is easy to verify that for $1 \le p < \infty$, the definition of $\|f\|_{J,(p)}$ agrees with the one given in (1.3) and (1.20). Thus, as noted there, by the Hölder and Minkowski inequalities, $\|\cdot\|_{J,(p)}$ is a seminorm and $\|\cdot\|_{J,[p]}$ is a norm on $\mathcal{W}_p(J)$. The same also holds on $\mathcal{W}_p(J;X)$.

We show next that $\|\cdot\|_{[\Phi]}$ is a norm, for general $\Phi \in \mathcal{CV}$, and that it makes $\widetilde{\mathcal{W}}_\Phi$ a Banach space. Throughout this chapter, $\Phi^{-1}$, or sometimes $\phi$, denotes the inverse function of $\Phi$ (not $1/\Phi$).

**Theorem 3.7.** *Let $J$ be a nondegenerate interval, $X$ a Banach space, and $\Phi \in \mathcal{CV}$.*

*(a) For $f \in \widetilde{\mathcal{W}}_\Phi(J;X)$, $\|f\|_{(\Phi)} = 0$ if and only if $f$ is a constant.*
*(b) For any non-constant $f \in \widetilde{\mathcal{W}}_\Phi(J;X)$,*

$$v_\Phi(f/\|f\|_{(\Phi)}) \le 1. \tag{3.9}$$

*In particular, for any $c, d \in J$, $\|f(d) - f(c)\| \le \Phi^{-1}(1)\|f\|_{(\Phi)}$.*
*(c) $\|\cdot\|_{(\Phi)}$ is a seminorm and $\|\cdot\|_{[\Phi]}$ a norm on $\widetilde{\mathcal{W}}_\Phi(J;X)$.*
*(d) $(\widetilde{\mathcal{W}}_\Phi(J;X), \|\cdot\|_{[\Phi]})$ is a Banach space.*

*Proof.* $(a)$ Clearly if $f \equiv x_0$ for some $x_0 \in X$ then $f \in \mathcal{W}_\Phi(J;X)$ and $\|f\|_{(\Phi)} = 0$. If $f$ is not constant, then for some $s \ne t$ in $J$, $f(s) \ne f(t)$, so if $u_n \downarrow 0$, then $\Phi(\|f(s) - f(t)\|/u_n) \to \infty$, and $\|f\|_{(\Phi)} > 0$. So $(a)$ holds.

$(b)$ Let $u_0 := \|f\|_{(\Phi)} > 0$. Then for some $u_n \downarrow u_0$, $v_\Phi(f/u_n) \le 1$. Suppose $v_\Phi(f/u_0) > 1$. Then for some partition $\lambda$ of $J$, $s_\Phi(f/u_0;\lambda) > 1$. Since $\Phi$ is continuous, for some $n$ large enough, $s_\Phi(f/u_n;\lambda) > 1$, a contradiction. So the first part of $(b)$ holds. The second part also holds since $\Phi(\|f(d) - f(c)\|/\|f\|_{(\Phi)}) \le v_\Phi(f/\|f\|_{(\Phi)})$.

$(c)$ For any $f \in \widetilde{\mathcal{W}}_\Phi(J;X)$ and $c \in \mathbb{R}$, it is easy to see that $\|cf\|_{(\Phi)} = |c|\|f\|_{(\Phi)}$. If also $g \in \widetilde{\mathcal{W}}_\Phi(J;X)$, we want to show that

$$\|f+g\|_{(\Phi)} \le \|f\|_{(\Phi)} + \|g\|_{(\Phi)}. \tag{3.10}$$

If $f$ or $g$ is constant then $\|f+g\|_{(\Phi)} = \|f\|_{(\Phi)} + \|g\|_{(\Phi)}$, so assume that neither is constant. For any $u, v > 0$ and $0 < \alpha < 1$, since $\Phi$ is convex,

$$v_\Phi(\alpha(f/u) + (1-\alpha)(g/v)) \le \alpha v_\Phi(f/u) + (1-\alpha)v_\Phi(g/v). \tag{3.11}$$

Letting $u = \|f\|_{(\Phi)}$ and $v = \|g\|_{(\Phi)}$ and applying (3.9) gives $v_\Phi(\alpha(f/u) + (1-\alpha)(g/v)) \le 1$. Then for $\alpha := u/(u+v)$ we get $v_\Phi((f+g)/(u+v)) \le 1$, so (3.10) is proved and $\|\cdot\|_{(\Phi)}$ is a seminorm. Since $\|\cdot\|_{\sup}$ is a norm it follows that $\|\cdot\|_{[\Phi]}$ is a norm, proving $(c)$.

$(d)$ Let $\{f_n\}_{n\ge 1}$ be a Cauchy sequence for $\|\cdot\|_{[\Phi]}$. Since the space of bounded functions $\ell^\infty(J;X)$ is a Banach space, the $f_n$ converge uniformly on $J$ to a function $f$. Given $\epsilon > 0$, take $n_0(\epsilon)$ large enough so that for $m, n \ge n_0(\epsilon)$, $\|f_m - f_n\|_{[\Phi]} < \epsilon$. Then $\|f_n - f\|_{\sup} \le \epsilon$ and

$$v_\Phi((f_m - f_n)/\epsilon) \le 1. \tag{3.12}$$

For $n \ge n_0(\epsilon)$ we have $v_\Phi((f_n - f)/\epsilon) \le 1$, since otherwise there is a partition $\lambda$ of $J$ with $s_\Phi((f_n - f)/\epsilon; \lambda) > 1$ and then $s_\Phi((f_n - f_m)/\epsilon; \lambda) > 1$ for $m$ large enough, contradicting (3.12). Thus $\|f_n - f\|_{(\Phi)} \le \epsilon$ and $\|f_n - f\|_{[\Phi]} \le 2\epsilon$. So $f_n \to f$ for $\|\cdot\|_{[\Phi]}$ and $(d)$ is proved. □

Let $X_1, \dots, X_k$, $Z$ be vector spaces. A map $L\colon X_1 \times \cdots \times X_k \to Z$ is called *$k$-linear* if for each $j = 1, \dots, k$, $(x_1, \dots, x_k) \mapsto L(x_1, \dots, x_k)$ is linear in $x_j$ when $x_i$ for $i \ne j$ are fixed. If $X_1, \dots, X_k$, $Z$ are all Banach spaces and $0 \le M < \infty$, $L$ will be called *$M$-bounded* if

$$\|L(x_1, \dots, x_k)\| \le M\|x_1\| \cdots \|x_k\| \tag{3.13}$$

for all $x_j \in X_j$, $j = 1, \dots, k$.

**Theorem 3.8.** *Let $k = 1, 2, \dots,$ let $X_1, \dots, X_k$, $Z$ be Banach spaces, let $0 \le M < \infty$, and let $L$ be $k$-linear and $M$-bounded from $X_1 \times \cdots \times X_k$ into $Z$. Let $\Phi \in \mathcal{CV}$, let $J$ be a nondegenerate interval, and let $f_j \in \widetilde{\mathcal{W}}_\Phi(J; X_j)$ for $j = 1, \dots, k$. Let $h(s) := L(f_1(s), \dots, f_k(s)) \in Z$ for each $s \in J$. Then $h \in \widetilde{\mathcal{W}}_\Phi(J; Z)$ and*

$$\|h\|_{[\Phi]} \le M\|f_1\|_{[\Phi]}\|f_2\|_{[\Phi]} \cdots \|f_k\|_{[\Phi]}. \tag{3.14}$$

*Proof.* For $k = 1$, since $L$ is linear and $\Phi$ is increasing, $\|h\|_{(\Phi)} \le \|Mf_1\|_{(\Phi)} = M\|f_1\|_{(\Phi)}$. Thus $h \in \widetilde{\mathcal{W}}_\Phi(J; Z)$ and (3.14) holds with $k = 1$.

Suppose $k \ge 2$. We can and do assume $M = 1$. For a proof by induction on $k$, let $k = 2$, $X = X_1$, $Y = X_2$, and $x{\cdot}y = L(x, y)$ as in (1.14). Let $f := f_1$ and $g := f_2$, so that $h \equiv f{\cdot}g$. If $f$ or $g$ is constant then (3.14) holds since $\|f{\cdot}g\|_{(\Phi)} \le \|f\|_{\sup}\|g\|_{(\Phi)}$ or $\le \|f\|_{(\Phi)}\|g\|_{\sup}$, respectively, and so we can assume that $f$ and $g$ are both non-constant. For $s < t$ in $J$, we have

$$\begin{aligned}\|h(t) - h(s)\| &\le \|f(t){\cdot}(g(t) - g(s))\| + \|(f(t) - f(s)){\cdot}g(s)\| \\ &\le \|f\|_{\sup}\|g(t) - g(s)\| + \|g\|_{\sup}\|f(t) - f(s)\|.\end{aligned}$$

Let $\kappa = \{t_i\}_{i=0}^n$ be a partition of $J$. For $i = 1, \dots, n$, let $\Delta_i f := f(t_i) - f(t_{i-1})$ and $\Delta_i g := g(t_i) - g(t_{i-1})$. Also, let $\alpha := \|g\|_{\sup}\|f\|_{(\Phi)}$ and $\beta := \|f\|_{\sup}\|g\|_{(\Phi)}$. Since $\Phi$ is convex, we then have

$$\begin{aligned}s_\Phi((f{\cdot}g)/(\alpha+\beta); \kappa) &\le \sum_{i=1}^n \Phi\Big(\Big\{\alpha\|\Delta_i f\|/\|f\|_{(\Phi)} + \beta\|\Delta_i g\|/\|g\|_{(\Phi)}\Big\}/(\alpha+\beta)\Big) \\ &\le \Big\{\alpha v_\Phi(f/\|f\|_{(\Phi)}) + \beta v_\Phi(g/\|g\|_{(\Phi)})\Big\}/(\alpha+\beta) \le 1,\end{aligned}$$

where the last inequality holds by (3.9). Thus

$$\|f{\cdot}g\|_{(\Phi)} \leq \alpha + \beta = \|g\|_{\sup}\|f\|_{(\Phi)} + \|f\|_{\sup}\|g\|_{(\Phi)}.$$

Clearly $\|f{\cdot}g\|_{\sup} \leq \|f\|_{\sup}\|g\|_{\sup}$, and so (3.14) holds for $k = 2$.

Now let $k \geq 3$, and assume that the theorem holds for $k-1$ in place of $k$. The linear space of all $k$-linear and $M$-bounded maps from $X_1 \times \cdots \times X_k$ into $Z$ for $0 \leq M < \infty$ is a normed space, with norm the infimum of all $M$ for which (3.13) holds, and is evidently a Banach space. Call this space and its norm $(\mathcal{L}(X_1, \ldots, X_k; Z), \|\cdot\|)$. For any $x_j \in X_j$, $j = 1, \ldots, k$, let $H(x_2, \ldots, x_k)(x_1) := L(x_1, x_2, \ldots, x_k)$. Then $H$ is evidently a $(k-1)$-linear map from $X_2 \times \cdots \times X_k$ into the space $L(X_1; Z)$ of bounded linear operators from $X_1$ into $Z$. Since for each $x_j \in X_j$, $j = 2, \ldots, k$,

$$\|H(x_2, \ldots, x_k)\| = \sup_{\|x_1\| \leq 1} \|L(x_1, \ldots, x_k)\| \leq \|x_2\| \cdots \|x_k\|,$$

$H$ is 1-bounded and so $H \in \mathcal{L}(X_2, \ldots, X_k; L(X_1; Z))$ with $\|H\| \leq \|L\|$. Let $g(s) := H(f_2(s), \ldots, f_k(s))$ for each $s \in J$. Then by the induction assumption, $g \in \widetilde{\mathcal{W}}_\Phi(J; L(X_1, Z))$ and

$$\|g\|_{[\Phi]} \leq \|f_2\|_{[\Phi]} \cdots \|f_k\|_{[\Phi]}. \tag{3.15}$$

Now $(A, x_1) \mapsto Ax_1 =: A{\cdot}x_1$ is a bilinear, 1-bounded mapping from $L(X_1; Z) \times X_1$ into $Z$, and $h(s) = g(s)(f_1(s)) = g(s){\cdot}f_1(s)$ for each $s \in J$. By the case $k = 2$, we get

$$\|h\|_{[\Phi]} \leq \|g\|_{[\Phi]}\|f_1\|_{[\Phi]}.$$

This together with (3.15) proves (3.14) and the theorem by induction. □

If $X$ is the set of real numbers, then functions with values in $X$ can be multiplied pointwise. So, we have the following special case of Theorem 3.8:

**Corollary 3.9.** *Let $J$ be a nondegenerate interval and $\Phi \in \mathcal{CV}$. If $f, g \in \widetilde{\mathcal{W}}_\Phi(J)$ then $fg \in \widetilde{\mathcal{W}}_\Phi(J)$ and*

$$\|fg\|_{[\Phi]} \leq \|f\|_{[\Phi]}\|g\|_{[\Phi]}. \tag{3.16}$$

*Proof.* For $x, y \in \mathbb{R}$, $(x, y) \mapsto xy$ is a bilinear and 1-bounded mapping from $\mathbb{R} \times \mathbb{R}$ into $\mathbb{R}$. Thus by Theorem 3.8, the pointwise product $fg$ of two functions $f, g \in \widetilde{\mathcal{W}}_\Phi(J)$ also is in $\widetilde{\mathcal{W}}_\Phi(J)$ and (3.16) holds, proving the corollary. □

As soon as Banach algebras are defined, in Section 4.1, Corollary 3.9 will be extended to the space $\widetilde{\mathcal{W}}_\Phi(J; \mathbb{B})$ of functions with values in a Banach algebra $\mathbb{B}$, in Proposition 4.13.

Let $\chi := \chi(\cdot)$ be the identity function on $[0, 1]$.

**Proposition 3.10.** *For $\Phi \in \mathcal{V}$, the following are equivalent:*

(*a*) $\Phi(u) = O(u)$ *as* $u \downarrow 0$;
(*b*) $\chi \in \mathcal{W}_\Phi[0,1]$;
(*c*) $\chi \in \widetilde{\mathcal{W}}_\Phi[0,1]$;
(*d*) $f \in \mathcal{W}_\Phi[0,1]$ *for some non-constant continuous* $f$.

*Proof.* Assuming (*a*), take $\delta > 0$ and $K < \infty$ such that $\Phi(u) \le Ku$ for $0 \le u \le \delta$. Then it is straightforward to check that $v_\Phi(\chi/(1/\delta)) \le K\delta$, so (*c*) holds. The implication (*c*) $\Rightarrow$ (*d*) is immediate (as is (*b*) $\Rightarrow$ (*c*)). For (*d*) $\Rightarrow$ (*c*) suppose $f$ is continuous and $f(0) < f(1)$. We can assume that $f(0) = 0$. By the intermediate value theorem, find $t_q = t(q)$ such that $f(t_q) = qf(1)$ for each rational $q \in [0,1]$, with $t_q < t_r$ for $q < r$. Then for any partition $\kappa = \{s_i\}_{i=0}^n$ of $[0,1]$ with all $s_i$ rational, $s_\Phi(f(1)\chi;\kappa) = s_\Phi(f;\lambda)$, where $\lambda = \{t(s_i)\}_{i=0}^n \cup \{1\}$ is a partition of $[0,1]$. Since $\Phi$ is continuous, it follows that $f(1)\chi \in \widetilde{\mathcal{W}}_\Phi$.

For (*c*) $\Rightarrow$ (*b*) let $c\chi \in \mathcal{W}_\Phi$ for some $c > 0$. Take an integer $m > 0$ such that $m^{-1} < c$. Then for any partition $\kappa = \{t_i\}_{i=0}^n$ of $[0,1]$, we have $s_\Phi(\chi;\kappa) \le S_1 + S_2$, where

$$S_1 := \sum_{j=1}^m \sum \Big\{\Phi(t_i - t_{i-1})\colon\ (j-1)/m \le t_{i-1} < t_i \le j/m\Big\} \le m v_\Phi(c\chi)$$

and

$$S_2 := \sum_{j=1}^m \Big\{\Phi(t_i - t_{i-1})\colon\ t_{i-1} < j/m < t_i\Big\} \le m\Phi(1),$$

where summands are replaced by 0 if terms satisfying their conditions do not exist, so $v_\Phi(\chi) \le m[v_\Phi(c\chi) + \Phi(1)] < \infty$.

For (*b*) $\Rightarrow$ (*a*) suppose that $\Phi(u_n)/u_n \to +\infty$ for some $u_n \downarrow 0$. Then letting $\lfloor x \rfloor$ be the largest integer $\le x$,

$$\sum_{i=1}^{\lfloor u_n^{-1} \rfloor} \Phi(iu_n - (i-1)u_n) = \lfloor u_n^{-1} \rfloor \Phi(u_n) \ge (u_n^{-1} - 1)\Phi(u_n) \to +\infty$$

as $n \to \infty$, a contradiction. The proof is complete. □

*Complementary functions*

For a convex function $\Phi \in \mathcal{V}$, let

$$\Phi^*(v) := \sup\big\{uv - \Phi(u)\colon\ u \ge 0\big\}, \qquad v \ge 0. \tag{3.17}$$

The function $\Phi^*$ so defined may be zero and may have infinite values on $(0,\infty)$. Both occur, for example, if $\Phi$ is the identity function, for $v \le 1$ or $v > 1$ respectively. In any case, $\Phi^*$ is convex as a supremum of affine functions. Clearly $\Phi^*(0) = 0$ and $\Phi^*$ is nondecreasing. Let $v_0 := \inf_{u>0} \Phi(u)/u$ and

$v_1 := \sup_{u>0} \Phi(u)/u$. Since $\Phi$ is convex, $u \mapsto \Phi(u)/u$ is nondecreasing, with $\Phi(u)/u \uparrow v_1 \le +\infty$ as $u \uparrow +\infty$. We have $\Phi^*(v) < +\infty$ for all $v \in [0, v_1)$, a nondegenerate interval or half-line, and $\Phi^*(v) = +\infty$ for all $v > v_1$. At all points $v$ of the interval $[0, v_1)$, $\Phi^*$ has a finite right derivative $D^+\Phi^*(v)$ and for $v \in (0, v_1)$ also a finite left derivative $D^-\Phi^*(v)$ (e.g. [53, Corollary 6.3.3]). In particular, $\Phi^*$ is continuous on the interval $[0, v_1)$. Also, we have $\Phi^*(v) = 0$ for all $v \in [0, v_0]$. It will be shown that $\Phi^* \in \mathcal{V}$ if we assume in addition that $v_0 = 0$ and $v_1 = +\infty$, that is,

$$u^{-1}\Phi(u) \downarrow 0 \qquad \text{as } u \downarrow 0 \qquad \text{and} \qquad u^{-1}\Phi(u) \uparrow +\infty \qquad \text{as } u \uparrow \infty. \tag{3.18}$$

The class of all functions $\Phi \in \mathcal{CV}$ satisfying (3.18) will be denoted by $\mathcal{CV}^*$. For example, the function $\Psi$ in Example 3.4 is in $\mathcal{CV}^*$.

For $\Phi \in \mathcal{CV}^*$, the function $\Phi^*$ defined by (3.17) will be called the *complementary function to* the function $\Phi$. Just by the definition of the complementary function we get the W. H. Young inequality:

$$uv \le \Phi(u) + \Phi^*(v) \qquad \text{for } u, v \ge 0. \tag{3.19}$$

For example, let $\Phi(u) := u^{1+\alpha}/(1+\alpha)$ for an $\alpha$ with $0 < \alpha < \infty$. Then it is easy to check that $\Phi^*(v) = \alpha v^{(1+\alpha)/\alpha}/(1+\alpha)$ for all $v \ge 0$, so (3.19) gives for all $u, v \ge 0$,

$$uv \le \frac{u^{1+\alpha}}{1+\alpha} + \frac{\alpha v^{(1+\alpha)/\alpha}}{1+\alpha}. \tag{3.20}$$

This last inequality appears as a lemma in one proof of Hölder's inequality, e.g. [53, Lemma 5.1.3]. In another example, (3.19) holds for the functions given by $\Phi(u) = \mathrm{e}^u - u - 1$, $u \ge 0$ and $\Phi^*(v) = (1+v)\log(1+v) - v$, $v \ge 0$.

Note that in (3.20), in the notation of (3.19), $\Phi \in \mathcal{CV}^*$, $\Phi^* \in \mathcal{CV}^*$, and $(\Phi^*)^* = \Phi$. The latter two facts hold generally if $\Phi \in \mathcal{CV}^*$, as the following shows:

**Proposition 3.11.** *For any $\Phi \in \mathcal{CV}^*$, its complementary function $\Phi^*$ is in $\mathcal{CV}^*$, and $\Phi^{**} := (\Phi^*)^* = \Phi$.*

*Proof.* As was noted after (3.17), it is clear that $\Phi^*(0) = 0$ and $\Phi^*$ is nondecreasing. If $\Phi^*(v) = 0$ then $v \le \Phi(u)/u$ for each $u > 0$, and hence $v = 0$ by the first relation in (3.18). For each $v \ge 0$, let $u_v := \sup\{u > 0\colon\ v - \Phi(u)/u \ge 0\}$. By the second relation in (3.18), $u_v$ is finite and $\Phi^*(v) = \sup\{uv - \Phi(u)\colon\ 0 \le u \le u_v\} < \infty$ for each $v \ge 0$. As also noted after (3.17), $\Phi^*$ is convex, and it is continuous and strictly increasing, that is, $\Phi^* \in \mathcal{CV}$. To prove (3.18) with $\Phi^*$ in place of $\Phi$, if $v \to 0$ then $u_v \to 0$, and then $\Phi^*(v)/v \to 0$ since $\Phi^*(v) \le u_v v$. If $v \to \infty$ then by (3.19) for any $u$, $\Phi^*(v)/v \ge u - \Phi(u)/v$, and so $\Phi^*(v)/v \to \infty$, proving the first part of the proposition.

For the second part, for any $V > 0$, let $V \in [D^-\Phi(u), D^+\Phi(u)]$ for some $u > 0$. Due to convexity of $\Phi$,

$$\frac{\Phi(x) - \Phi(u)}{x - u} \ge V \qquad \text{or} \qquad \frac{\Phi(x) - \Phi(u)}{x - u} \le V$$

according as $x > u$ or $x < u$. In either case $\Phi(x) - \Phi(u) \ge V(x - u)$ for each $x \ge 0$, and so $uV - \Phi(u) \ge xV - \Phi(x)$ for each $x \ge 0$. Thus by (3.17), $uV - \Phi(u) = \Phi^*(V)$, giving equality in (3.19) for $V$ in place of $v$, and the inequality $\Phi^{**}(u) \ge uV - \Phi^*(V) = \Phi(u)$. Since $u > 0$ is arbitrary, we have that $\Phi^{**} \ge \Phi$. The reverse inequality follows from the W. H. Young inequality (3.19), proving that $\Phi^{**} = \Phi$. The proof of the proposition is now complete. □

Alternatively, the complementary function to $\Phi$ can be defined using the right derivative $P := D^+\Phi$ of $\Phi$, which exists since $\Phi$ is convex. Thus $P$ is a nondecreasing and right-continuous function on $[0, \infty)$. The conditions (3.18) are equivalent to $P(0) = 0$ and $P(t) \uparrow +\infty$ as $t \uparrow \infty$. Consider the right inverse $Q$ of $P$ defined by $Q(t) := \sup\{s \ge 0\colon P(s) \le t\}$ for $t \ge 0$. Then, like $P$, $Q$ has the properties that $Q\colon [0, \infty) \to [0, \infty)$ is right-continuous and nondecreasing, $Q(0) = 0$, and $Q(t) \uparrow +\infty$ as $t \uparrow \infty$. Let $\Psi(v) := \int_0^v Q(t)\,dt$ for $v \ge 0$. Let $A := \{(s,t) : 0 \le s \le u,\ 0 \le t \le P(s)\}$ and $B := \{(s,t) : 0 \le t \le v,\ 0 \le s \le Q(t)\}$. If $P(s) \le t$, then $s \le Q(t)$ by definition of $Q(t)$. Thus $A \cup B \supset R := [0,u] \times [0,v]$ and by comparison of areas,

$$uv \le \int_0^u P(s)\,ds + \int_0^v Q(s)\,ds = \Phi(u) + \Psi(v) \qquad \text{for } u, v \ge 0. \tag{3.21}$$

If $Q(t) > s$, then $P(s) \le t$ since $P$ is nondecreasing. Thus $A \cap B \subset \{(s,t) : s = Q(t) \text{ or } t = P(s)\}$, and the area of $A \cap B$ is always zero.

Suppose $v = P(u)$ for a given $u$. Then if $t < v$, we have $Q(t) \le u$, so $A \cup B \subset R \cup \{(s,t) :\ t = v\}$. Or, suppose $u = Q(v)$ for a given $v$. Then if $s < u$, we have $P(s) \le v$, so $A \cup B \subset R \cup \{(s,t) :\ s = u\}$. So if $v = P(u)$ or $u = Q(v)$, then $(A \cup B) \setminus R$ also has area 0, and (3.21) becomes an equality, that is,

$$\begin{aligned} uP(u) &= \Phi(u) + \Psi(P(u)) \quad \forall\, u \ge 0, \quad \text{and} \\ vQ(v) &= \Phi(Q(v)) + \Psi(v) \quad \forall\, v \ge 0. \end{aligned} \tag{3.22}$$

For the complementary function $\Phi^*$ defined by (3.17), (3.21) implies that $\Psi \ge \Phi^*$. Conversely, the second equation in (3.22), setting $u := Q(v)$, implies that $\Psi \le \Phi^*$. Thus $\Psi \equiv \Phi^*$ and so inequality (3.21) agrees with inequality (3.19). As shown in the proof of Proposition 3.11, equality holds in (3.19) if $P(u-) \le v \le P(u)$ for a given $u$. Symmetrically, since $Q(v) \equiv D^+\Psi(v)$, the same holds when $Q(v-) \le u \le Q(v)$ for a given $v$.

Let $\Phi \in \mathcal{CV}^*$ and let $\Phi^*$ be its complementary function. For an $X$-valued function $f$ defined on a nondegenerate interval $J$, let

$$\|\!|f|\!\|_{J,(\Phi)} := \sup\Big\{\sum_{i=1}^{n} \|f(t_i) - f(t_{i-1})\|\beta_i : \\ \{t_i\}_{i=0}^{n} \in \mathrm{PP}\,(J),\ \sum_{i=1}^{n} \Phi^*(\beta_i) \le 1,\ \beta_i \ge 0\Big\}, \tag{3.23}$$

where PP $(J)$ is the class of all partitions of $J$. It is easy to check that $\|\!|\cdot|\!\|_{(\Phi)} := \|\!|\cdot|\!\|_{J,(\Phi)}$ is 0 only on constants. It is a seminorm equivalent to $\|\cdot\|_{(\Phi)}$ on $\widetilde{\mathcal{W}}_\Phi(J)$ by the next proposition.

**Proposition 3.12.** *For $\Phi$ in $\mathcal{CV}^*$ and $f \in \widetilde{\mathcal{W}}_\Phi(J)$, $f \mapsto \|\!|f|\!\|_{(\Phi)}$ is a seminorm and we have*

$$\|f\|_{(\Phi)} \le \|\!|f|\!\|_{(\Phi)} \le 2\|f\|_{(\Phi)}. \tag{3.24}$$

*Proof.* We can assume that $f$ is non-constant, so $0 < \|f\|_{(\Phi)} < +\infty$ and $\|\!|f|\!\|_{(\Phi)} > 0$. To prove the second inequality in (3.24), let $\|f\|_{(\Phi)} = 1$. For $\{t_i\}_{i=0}^{n} \in \mathrm{PP}\,(J)$, let $\Delta_i f := f(t_i) - f(t_{i-1})$, $i = 1, \ldots, n$. Then by the W. H. Young inequality (3.19) and (3.9), we have

$$\sum_{i=1}^{n} \|\Delta_i f\|\beta_i \le v_\Phi(f) + \sum_{i=1}^{n} \Phi^*(\beta_i) \le 2$$

for any $\{\beta_i\}_{i=1}^{n}$ such that $\beta_i \ge 0$ and $\sum_{i=1}^{n} \Phi^*(\beta_i) \le 1$. Since the partition $\{t_i\}_{i=0}^{n}$ is arbitrary, the second inequality in (3.24) holds. Thus $\|\!|f|\!\|_{(\Phi)} < \infty$ and since the other properties of a seminorm clearly hold by the definition (3.23), $\|\!|\cdot|\!\|_{(\Phi)}$ is a seminorm on $\widetilde{\mathcal{W}}_\Phi(J)$.

To prove the first inequality in (3.24), let $\|\!|f|\!\|_{(\Phi)} = 1$. Let $\{t_i\}_{i=0}^{n} \in \mathrm{PP}\,(J)$, and let $\beta_i := P(\|\Delta_i f\|)$ for $i = 1, \ldots, n$, where $P = D^+\Phi$ is the right derivative of $\Phi$. Since $f$ is non-constant, not all $\Delta_i f = 0$. By the first equality in (3.22), for each $i = 1, \ldots, n$,

$$\|\Delta_i f\|\beta_i = \Phi(\|\Delta_i f\|) + \Phi^*(\beta_i). \tag{3.25}$$

Also, $s_n := \sum_{i=1}^{n} \Phi^*(\beta_i) \le 1$. Suppose not. By convexity of $\Phi^*$, and since $s_n > 1$, we have $\sum_{i=1}^{n} \Phi^*(\beta_i/s_n) \le \sum_{i=1}^{n} \Phi^*(\beta_i)/s_n \le 1$. Thus $1 = \|\!|f|\!\|_{(\Phi)} \ge \sum_{i=1}^{n} \|\Delta_i f\|(\beta_i/s_n)$, and so by (3.25),

$$s_n \ge \sum_{i=1}^{n} \|\Delta_i f\|\beta_i = \sum_{i=1}^{n} \Phi(\|\Delta_i f\|) + s_n,$$

a contradiction since $\|\Delta_i f\| > 0$ for some $i$, proving the claim that $s_n \le 1$. Then by (3.25) again, we have

$$1 = \|\!|f|\!\|_{(\Phi)} \ge \sum_{i=1}^{n} \|\Delta_i f\|\beta_i \ge \sum_{i=1}^{n} \Phi(\|\Delta_i f\|).$$

Since $\{t_i\}_{i=0}^n$ is an arbitrary partition of $J$, it follows that $v_\Phi(f;J) \le 1$, and so $\|f\|_{(\Phi)} \le 1 = \|\!|f|\!\|_{(\Phi)}$, proving the proposition. □

As mentioned between (3.19) and (3.20), if $\Phi(u) \equiv u^p/p$, $1 < p < \infty$, where $p$ here equals $\alpha+1$ there, $\Phi \in \mathcal{CV}^*$ and its complementary function is $\Phi^*(v) \equiv v^q/q$, where $p^{-1} + q^{-1} = 1$. For such $\Phi$, we have $\|f\|_{(\Phi)} = (1/p)^{1/p}\|f\|_{(p)}$. Moreover, considering $\beta_i = q^{1/q}\|f(t_i) - f(t_{i-1})\|^{p-1}/\|f\|_{(p)}^{p-1}$, $i = 1, \dots, n$, for $\{t_i\}_{i=0}^n \in \mathrm{PP}(J)$, it is easy to check that $\sum_{i=1}^n \Phi^*(\beta_i) \le 1$. It follows that $\|\!|f|\!\|_{(\Phi)} \ge q^{1/q}\|f\|_{(p)}$. Conversely, for any $\beta_i \ge 0$ such that $\sum_{i=1}^n \Phi^*(\beta_i) \le 1$, we have by the Hölder inequality

$$\begin{aligned}\sum_{i=1}^n \|f(t_i) - f(t_{i-1})\|\beta_i &\le \Big(\sum_{i=1}^n \|f(t_i) - f(t_{i-1})\|^p\Big)^{1/p}\Big(\sum_{i=1}^n \beta_i^q\Big)^{1/q} \\ &= q^{1/q}\Big(\sum_{i=1}^n \|f(t_i) - f(t_{i-1})\|^p\Big)^{1/p}\Big(\sum_{i=1}^n \Phi^*(\beta_i)\Big)^{1/q} \\ &\le q^{1/q}\|f\|_{(p)}.\end{aligned}$$

Thus $\|\!|f|\!\|_{(\Phi)} \le q^{1/q}\|f\|_{(p)}$ and $\|\!|f|\!\|_{(\Phi)} = q^{1/q}\|f\|_{(p)}$.

If $p = q = 2$, then $\|\!|f|\!\|_{(\Phi)} = \sqrt{2}\|f\|_{(2)} = 2\|f\|_{(\Phi)}$. Furthermore, if $p \downarrow 1$ then both $(1/p)^{1/p} \to 1$ and $q^{1/q} \to 1$. These examples show that the inequalities (3.24) are sharp.

The seminorm $\|\cdot\|_{(\Phi)}$ resembles the Luxemburg norm $\|\cdot\|_\Phi$ on an Orlicz space defined by (1.24). The seminorm $\|\!|\cdot|\!\|_{(\Phi)}$ also has an analogue for Orlicz spaces, defined as follows. Let $(\Omega, \mathcal{S}, \mu)$ be a measure space. For $\Psi \in \mathcal{V}$, let $B^\Psi(\Omega, \mathcal{S}, \mu; \mathbb{R})$ be the set of $\mu$-measurable functions $g\colon \Omega \to \mathbb{R}$ such that $\int_\Omega \Psi(|g|)\,d\mu \le 1$. Let $\Phi \in \mathcal{CV}^*$ and let $\Phi^*$ be its complementary function. For a $\mu$-measurable function $f$ on $\Omega$ with values in $X$, let

$$\|\!|f|\!\|_\Phi := \sup\Big\{\int_\Omega \|f\|g\,d\mu\colon\ g \in B^{\Phi^*}(\Omega, \mathcal{S}, \mu; \mathbb{R}),\ g \ge 0\Big\} \le +\infty. \tag{3.26}$$

It will be shown that this gives a norm equivalent to $\|\cdot\|_\Phi$ on the Orlicz space $L^\Phi(\Omega, \mathcal{S}, \mu; X)$.

**Proposition 3.13.** *Let $(\Omega, \mathcal{S}, \mu)$ be a measure space and let $\Phi$ be in $\mathcal{CV}^*$. For $f \in L^\Phi(\Omega; X)$, we have*

$$\|f\|_\Phi \le \|\!|f|\!\|_\Phi \le 2\|f\|_\Phi, \tag{3.27}$$

*and $\|\!|\cdot|\!\|_\Phi$ is a norm on $L^\Phi(\Omega, \mathcal{S}, \mu; X)$.*

*Proof.* We will first prove the second inequality in (3.27). For some $c$ with $0 < c < \infty$ we have $\int \Phi(\|f\|/c)d\mu \le 1$. Then by the W. H. Young inequality (3.19), for any nonnegative $g \in B^{\Phi^*}(\Omega, \mathcal{S}, \mu; \mathbb{R})$, we have

$$\int_\Omega c^{-1}\|f\|g\,\mathrm{d}\mu \le \int_\Omega \Phi(c^{-1}\|f\|)\,\mathrm{d}\mu + \int_\Omega \Phi^*(g)\,\mathrm{d}\mu \le 2,$$

so $\int \|f\|g\,\mathrm{d}\mu \le 2c$. Letting $c \downarrow \|f\|_\Phi$, the second inequality in (3.27) holds. Thus $\|\!|f|\!\|_\Phi < +\infty$. It follows directly from the definition (3.26) that $\|\!|\cdot|\!\|_\Phi$ is a seminorm on $L^\Phi(\Omega, \mathcal{S}, \mu; X)$. Since it vanishes only for functions $f = 0$ a.e. $(\mu)$, it is a norm.

It remains to prove the first inequality in (3.27). Let $\|\!|f|\!\|_\Phi = 1$. If $\int \Phi(\|f\|)\,\mathrm{d}\mu = 0$ then $\|f\|_\Phi \le 1$. Thus we can and do assume $\int \Phi(\|f\|)\,\mathrm{d}\mu > 0$. Let $g := P(\|f\|)$, where $P = D^+\Phi$ is the right derivative of $\Phi$. By the first equality in (3.22),

$$\|f\|g \equiv \Phi(\|f\|) + \Phi^*(g). \tag{3.28}$$

We claim that $\rho := \int_\Omega \Phi^*(g)\,\mathrm{d}\mu \le 1$. Suppose not. By convexity of $\Phi^*$, and since $\rho > 1$, $\int_\Omega \Phi^*(g/\rho)\,\mathrm{d}\mu \le \int_\Omega \Phi^*(g)\,\mathrm{d}\mu/\rho = 1$. Thus $\int_\Omega \|f\|(g/\rho)\,\mathrm{d}\mu \le \|\!|f|\!\|_\Phi = 1$, and so by (3.28),

$$\rho \ge \int_\Omega \|f\|g\,\mathrm{d}\mu = \int_\Omega \Phi(\|f\|)\,\mathrm{d}\mu + \rho.$$

This is a contradiction since $\int_\Omega \Phi(\|f\|)\,\mathrm{d}\mu > 0$, proving the claim $g \in B^{\Phi^*}(\Omega, \mathcal{S}, \mu; \mathbb{R})$. Then by (3.28) again, we have

$$1 = \|\!|f|\!\|_\Phi \ge \int_\Omega \|f\|g\,\mathrm{d}\mu \ge \int_\Omega \Phi(\|f\|)\,\mathrm{d}\mu,$$

and so $\|f\|_\Phi \le 1$, proving the proposition. □

The inequalities (3.27) are sharp, as will be shown by the same examples used for the $p$-variation norms with $\Phi(u) \equiv u^p/p$, $1 < p < \infty$, $\Phi^*(v) \equiv v^q/q$, and $p^{-1} + q^{-1} = 1$. In the present case, for such $\Phi$, we have $\|f\|_\Phi = (1/p)^{1/p}\|f\|_p$. Taking $g := q^{1/q}\|f\|^{p-1}/\|f\|_p^{p-1}$, it follows that $\|g\|_{\Phi^*} = 1$, and hence $\|\!|f|\!\|_\Phi \ge q^{1/q}\|f\|_p$. The converse inequality holds by the Hölder inequality as before, and so $\|\!|f|\!\|_\Phi = q^{1/q}\|f\|_p$. Now taking $p = q = 2$ or $p \downarrow 1$ shows that inequalities (3.27) are sharp.

For a real-valued function $f$, we have

$$\|\!|f|\!\|_\Phi = \sup\left\{\left|\int_\Omega fg\,\mathrm{d}\mu\right| : g \in B^{\Phi^*}(\Omega, \mathcal{S}, \mu; \mathbb{R})\right\}. \tag{3.29}$$

Indeed, for real-valued and $\mu$-measurable $f$ and $g$,

$$\left|\int_\Omega fg\,\mathrm{d}\mu\right| \le \int_\Omega |f||g|\,\mathrm{d}\mu = \left|\int_\Omega f\tilde{g}\,\mathrm{d}\mu\right|,$$

where $\tilde{g} := |g|\,\mathrm{sgn} f$. Then $\int \Phi^*(|g|)\,\mathrm{d}\mu = \int \Phi^*(|\tilde{g}|)\,\mathrm{d}\mu$ and so (3.29) follows. The right side of (3.29) is called an Orlicz norm, e.g. [128, Section II.9].

*Moduli of continuity*

A *modulus of continuity* will be a continuous, nondecreasing function $\phi$ from $[0, \infty)$ onto itself which is subadditive, so that for all $u, v \geq 0$,

$$\phi(u+v) \;\leq\; \phi(u) + \phi(v). \tag{3.30}$$

If $\phi$ is a modulus of continuity then $\phi$ is zero only at zero. If not then $\phi(u) = 0$ for some $u > 0$. Then by (3.30), $\phi(nu) = 0$ for each positive integer $n$, and so $\phi(u) = 0$ for each $u > 0$, a contradiction. Next, for $0 < u < v$, we have

$$\frac{\phi(v)}{v} \;\leq\; 2\frac{\phi(u)}{u}. \tag{3.31}$$

To prove this, let $n$ be the least integer such that $nu \geq v$. Then $nu \leq 2v$, and by (3.30), it follows that

$$\phi(v)/v \;\leq\; \phi(nu)/v \;\leq\; n\phi(u)/v \;\leq\; 2\phi(u)/u,$$

as stated. Moreover, if $\phi$ is a modulus of continuity then

$$\lim_{u\downarrow 0} \frac{\phi(u)}{u} = \sup_{0<u<u^*} \frac{\phi(u)}{u} \qquad \text{for any } 0 < u^* < \infty. \tag{3.32}$$

To prove this, let $0 < u^* < \infty$ and let $0 < v < u^*$. For any $0 < u \leq v$, let $n$ be the least integer such that $v = nu + \delta(u)$ with $0 \leq \delta(u) < u$. Then by (3.30),

$$\frac{\phi(v)}{v} \leq n\frac{u}{v}\frac{\phi(u)}{u} + \frac{\phi(\delta(u))}{v}.$$

Letting $u \downarrow 0$ it follows that $nu/v \to 1$ and $\phi(\delta(u))/v \to 0$. Thus $\phi(v)/v \leq \liminf_{u\downarrow 0} \phi(u)/u$. Since $v \in (0, u^*)$ is arbitrary, we have

$$S := \sup_{0<v<u^*} \frac{\phi(v)}{v} \leq \liminf_{u\downarrow 0} \frac{\phi(u)}{u} \leq \limsup_{u\downarrow 0} \frac{\phi(u)}{u} \leq S,$$

proving (3.32).

Property (3.31) can be strengthened for concave functions as follows:

**Proposition 3.14.** *Let $\phi$ be a continuous, nondecreasing, concave function from $[0, \infty)$ onto itself. Then $\phi$ is a modulus of continuity and $\phi(u)/u$ is nonincreasing in $u > 0$.*

*Proof.* Let $0 \leq \alpha \leq 1$. By concavity, $\phi(\lambda x) \geq \lambda\phi(x)$ for $\lambda = \alpha$ or $1 - \alpha$ and each $x \geq 0$, so $\phi(x) \leq \phi(\alpha x) + \phi((1-\alpha)x)$. Taking $x := u + v$ and $\alpha := u/(u+v)$, it follows that $\phi$ is subadditive and so a modulus of continuity. Also, by concavity, we have $\phi(\lambda u)/(\lambda u) \geq \phi(u)/u$ for any $0 < \lambda \leq 1$ and $u > 0$, proving the proposition. □

If $\Phi \in \mathcal{V}$ is convex, then its inverse $\phi = \Phi^{-1}$ will be a continuous, nondecreasing and concave function from $[0, \infty)$ onto itself, and so a modulus of continuity by Proposition 3.14. The class of all functions $\Phi \in \mathcal{V}$ such that $\Phi^{-1}$ is a modulus of continuity will be denoted by $\mathcal{VM}$.

The following shows that moduli of continuity can be taken to be concave up to a factor of 2.

**Proposition 3.15.** *Let $\phi$ be any modulus of continuity. Then there is a real-valued least concave function $\overline{\phi} \geq \phi$ with $\overline{\phi} \leq 2\phi$. Here $\overline{\phi}$ is a modulus of continuity and is strictly increasing.*

*Proof.* For $t \geq 0$ let

$$\overline{\phi}(t) \ := \ \inf\bigl\{a + bt : \ a + bu \geq \phi(u) \ \text{ for all } \ u \geq 0\bigr\},$$

or $+\infty$ if the given set is empty (it will be shown to be always nonempty). Then clearly $\phi(t) \leq \overline{\phi}(t)$ for all $t \geq 0$. Since $\phi$ is subadditive we have for each $t > 0$ and positive integer $m$ that $\phi(mt) \leq m\phi(t)$. Thus $\phi(u) \leq m\phi(t)$ for $(m-1)t < u \leq mt$, or equivalently $m - 1 < u/t \leq m$. Thus $\phi(u) \leq (ut^{-1} + 1)\phi(t)$ for all $u \geq 0$. It follows that $\overline{\phi}$ has finite values with $\overline{\phi}(u) \leq (ut^{-1} + 1)\phi(t)$ for all $u \geq 0$. Setting $u = t$ we get $\overline{\phi}(t) \leq 2\phi(t)$ as stated.

Since $\overline{\phi}$ is the infimum of a nonempty set of concave functions and is nonnegative, it is a real-valued concave function. Let $\tilde{\phi}$ be the infimum of all concave functions $\geq \phi$, which is then the least concave function $\geq \phi$. So $\phi \leq \tilde{\phi} \leq \overline{\phi}$. Suppose $\tilde{\phi}(t) < \overline{\phi}(t)$ for some $t$. Since $\tilde{\phi}$ is concave, its subgraph $\{(u, v) : \ u \geq 0, \ v \leq \tilde{\phi}(u)\}$ is a convex set. It has a support line $v = a + bu$ passing through the point $(u, v) = (t, \tilde{\phi}(t))$ [53, Theorem 6.2.7]. Then $a + bu \geq \tilde{\phi}(u) \geq \phi(u)$ for all $u \geq 0$, so $\overline{\phi}(t) \leq \tilde{\phi}(t)$, a contradiction. So $\tilde{\phi} \equiv \overline{\phi}$ and $\overline{\phi}$ is the least concave function $\geq \phi$.

Any concave function on an interval is continuous, in fact has finite one-sided derivatives, on the interior of the interval. The function $\overline{\phi}$ is continuous at 0 and goes to $+\infty$ at $+\infty$ by the inequalities $\phi \leq \overline{\phi} \leq 2\phi$. It follows from Proposition 3.14 that $\overline{\phi}$ is a modulus of continuity. Then since it is concave, it must be strictly increasing. This finishes the proof. □

## 3.2 Interval Functions, $\Phi$-variation, and $p$-variation

In Section 2.1, for interval functions, we considered the properties of additivity and upper continuity, and proved several facts. In this section, $\Phi$-variation and $p$-variation are considered for such functions, which are not necessarily additive. In Chapter 9 we will deal with multiplicative interval functions.

Let $J$ be a nonempty interval in $\mathbb{R}$, possibly unbounded, and let $X$ be a Banach space. Recall that the class of all interval functions defined on the set $\mathfrak{I}(J)$ of subintervals of $J$ with values in $X$ is denoted by $\mathcal{I}(J;X)$ (as just after Corollary 2.2). Recall (Section 1.4) that an *interval partition* of $J$ is a finite sequence $\{A_i\}_{i=1}^n$ of disjoint nonempty intervals with $J = \cup_{i=1}^n A_i$ and $A_i \prec A_j$ whenever $i < j$, and IP $(J)$ denotes the set of all interval partitions of $J$. The $\Phi$-variation for interval functions is defined similarly to the $\Phi$-variation for point functions, as follows.

Recall that $\mathcal{V}$ is the class of all functions $\Phi\colon [0,\infty) \to [0,\infty)$ which are strictly increasing, continuous, unbounded, and 0 at 0. Let $\mu \in \mathcal{I}(J;X)$ and $\Phi \in \mathcal{V}$. For $\mathcal{A} = \{A_i\}_{i=1}^n \in \mathrm{IP}\,(J)$, let the $\Phi$-variation sum be

$$s_\Phi(\mu;\mathcal{A}) := \sum_{i=1}^n \Phi\Big(\Big\|\mu\Big(\bigcup_{j=1}^i A_j\Big) - \mu\Big(\bigcup_{j=1}^{i-1} A_j\Big)\Big\|\Big), \tag{3.33}$$

where a union over the empty set of indices is defined as the empty set. For an additive interval function $\mu$ and $\mathcal{A} = \{A_i\}_{i=1}^n \in \mathrm{IP}\,(J)$, clearly

$$s_\Phi(\mu;\mathcal{A}) = \sum_{i=1}^n \Phi\big(\|\mu(A_i)\|\big). \tag{3.34}$$

Recall that in Chapter 2 we defined interval functions taking values in a Banach space, such as $\mathbb{R}$. An *extended real-valued interval function* is defined as a function from $\mathfrak{I}(J)$ for some $J$ into $[-\infty,\infty]$, for example into $[0,\infty]$ as in the following definition.

**Definition 3.16.** Let $\Phi \in \mathcal{V}$ and let $\mu$ be an $X$-valued interval function on a nonempty interval $J$. The *$\Phi$-variation* of $\mu$ on $J$ will be an extended real-valued interval function $v_\Phi(\mu) := v_\Phi(\mu;\cdot)$ on $J$ defined by

$$v_\Phi(\mu;A) := \sup\big\{s_\Phi(\mu;\mathcal{A})\colon\ \mathcal{A} \in \mathrm{IP}\,(A)\big\} \le +\infty \tag{3.35}$$

if $A \in \mathfrak{I}(J)$ is nonempty, or as 0 if $A = \emptyset$. We say that $\mu$ has *bounded $\Phi$-variation* on $J$ if $\sup\{v_\Phi(\mu;A)\colon A \in \mathfrak{I}(J)\} < \infty$. The class of all interval functions $\mu \in \mathcal{I}(J;X)$ with bounded $\Phi$-variation on $J$ will be denoted by $\mathcal{I}_\Phi(J;X)$, and $\mathcal{I}_\Phi(J)$ will denote $\mathcal{I}_\Phi(J;\mathbb{R})$. In the special case where $\Phi(u) \equiv u^p$, $u \ge 0$, for some $0 < p < \infty$, we write $s_p := s_\Phi$, $v_p := v_\Phi$, and $\mathcal{I}_p := \mathcal{I}_\Phi$. We call $v_p(\mu)$ the *$p$-variation* of $\mu$. Thus the $p$-variation of $\mu$ on $J$ is an extended real-valued interval function $v_p(\mu) = v_p(\mu;\cdot)$ on $J$ defined by

$$v_p(\mu;A) := v_p(\mu;A,X) := \sup\big\{s_p(\mu;\mathcal{A})\colon\ \mathcal{A} \in \mathrm{IP}\,(A)\big\} \tag{3.36}$$

if $A \in \mathfrak{I}(J)$ is nonempty or as 0 if $A = \emptyset$.

Recall that, as defined just before Theorem 2.7, $\mathcal{AI}(J;X)$ is the vector space of additive and upper continuous interval functions on $J$ with values in $X$. Let $\mathcal{AI}_\Phi(J;X) := \mathcal{I}_\Phi(J;X) \cap \mathcal{AI}(J;X)$ and $\mathcal{AI}_p(J;X) := \mathcal{I}_p(J;X) \cap \mathcal{AI}(J;X)$. Also, let $\mathcal{AI}_\Phi(J) := \mathcal{AI}_\Phi(J;\mathbb{R})$ and $\mathcal{AI}_p(J) := \mathcal{AI}_p(J;\mathbb{R})$.

**Remark 3.17.** If $\mu$ is additive then by (3.34), $s_p(\mu;\mathcal{A}) = \sum_{i=1}^n \|\mu(A_i)\|^p$ for $\mathcal{A} = \{A_i\}_{i=1}^n \in \mathrm{IP}\,(J)$. In this case, $v_\Phi(\mu;J) = \sup\{v_\Phi(\mu;A)\colon A \in \mathfrak{I}(J)\}$. Let $\mu$ be an interval function on $\mathfrak{I}(J)$ where $J = [\![a,b]\!]$ with $a < b$. Then all the (unions of) intervals on which $\mu$ is evaluated in the definition (3.33) are either empty or of the form $[\![a,t_i]\!]$ and moreover are left-open if and only if $J$ is.

Suppose that $v_\Phi(\mu;J) < +\infty$. Let $\nu$ be another interval function on $J$ such that $\nu([\![a,t]\!]) = \mu([\![a,t]\!])$ for $a \le t \le b$ but $\nu$ is defined arbitrarily on other intervals. Then $s_\Phi(\nu;\mathcal{A}) = s_\Phi(\mu;\mathcal{A})$ for every interval partition $\mathcal{A}$ of $J$ and $v_\Phi(\nu;J) = v_\Phi(\mu;J) < +\infty$. But even if $v_\Phi(\mu;A) < +\infty$ for all $A \in \mathfrak{I}(J)$, for example if $\mu$ is additive, the same need not hold for $\nu$ if $J$ is nondegenerate.

Thus we see that the condition $v_\Phi(\mu;J) < +\infty$, by itself, is not restrictive except for values on intervals having the same left endpoint (and open- or closedness there) as $J$. Thus, for general (non-additive) interval functions, we take suprema over $A \in \mathfrak{I}(J)$ in defining "bounded $\Phi$-variation," $\mathcal{I}_\Phi(J;X)$, $\|\mu\|_{J,(p)}$, and some related notions later. The definitions with suprema over all subintervals of $J$ will be useful especially for multiplicative and related interval functions in Chapter 9.

If $v_\Phi(\mu;[a,b]) < \infty$, the point functions $L_{\mu,a}$ and $R_{\mu,a}$ defined by (2.3) are of bounded $\Phi$-variation, and hence regulated. If $\mu$ is upper continuous and either $L_{\mu,a}$ or $R_{\mu,a}$ has bounded $\Phi$-variation, then $v_\Phi(\mu;J) < \infty$. Without upper continuity, even for $\mu$ additive, this implication can fail, as Example 2.14 shows. Note also that the mapping $h \mapsto \mu_h$ from point functions to interval functions defined by (2.2) always gives an additive interval function. Corollary 2.11 gives a mutual relation between suitable point functions and additive interval functions. For a non-additive interval function $\mu$ on $J = [a,b]$ with $a < b$, when we take $\sup_{A\in\mathfrak{I}(J)} v_\Phi(\mu;A)$, we would need to consider not only functions $L_{\mu,a}$ or $R_{\mu,a}$, but also $L_{\mu,c}$ or $R_{\mu,c}$ for $a < c < b$.

### *$\Phi$-variation seminorms and norms*

The interval function $\mu_f$ defined by (2.2) with $f$ as defined in Example 3.4 shows that, in general, $\mathcal{I}_\Phi$ need not be a vector space. As in Definition 3.5, we will extend $\mathcal{I}_\Phi$ to a vector space. Recall that $\mathcal{CV}$ is the class of convex functions $\Phi \in \mathcal{V}$.

**Definition 3.18.** Let $J$ be a nonempty interval and let $X$ be a Banach space. For $\Phi \in \mathcal{V}$, let $\widetilde{\mathcal{I}}_\Phi(J;X)$ be the set of all interval functions $\mu \in \mathcal{I}(J;X)$ such that $c\mu \in \mathcal{I}_\Phi(J;X)$ for some $c > 0$. For $\Phi \in \mathcal{CV}$ and $\mu \in \widetilde{\mathcal{I}}_\Phi(J;X)$, let

$$\|\mu\|_{(\Phi)} := \|\mu\|_{J,(\Phi)} := \sup\{\inf\{c>0\colon v_\Phi(\mu/c;A) \le 1\}\colon A \in \mathfrak{I}(J)\} \qquad \text{and}$$
$$\|\mu\|_{[\Phi]} := \|\mu\|_{J,[\Phi]} := \|\mu\|_{J,\sup} + \|\mu\|_{J,(\Phi)},$$

where $\|\mu\|_{J,\sup} = \sup\{\|\mu(A)\|\colon A \in \mathfrak{I}(J)\}$ is the sup norm of $\mu$. In the special case where $\Phi(u) \equiv u^p$, $u \ge 0$, for some $1 \le p < \infty$, let $\|\mu\|_{(p)} := \|\mu\|_{J,(p)} := \|\mu\|_{J,(\Phi)}$ and $\|\mu\|_{[p]} := \|\mu\|_{J,[p]} := \|\mu\|_{J,\sup} + \|\mu\|_{J,(p)}$.

**Remark 3.19.** As for point functions in Remark 3.6, one can verify that for $1 \le p < \infty$ the definition of $\|\mu\|_{J,(p)}$ just given agrees with (1.21), so that

$$\|\mu\|_{J,(p)} = \sup\{v_p(\mu; A)^{1/p} \colon A \in \mathfrak{I}(J)\}.$$

As noted there, by the Hölder and Minkowski inequalities, $\|\cdot\|_{J,(p)}$ is a seminorm and $\|\cdot\|_{J,[p]}$ a norm on $\mathcal{I}_p(J; X)$.

By Theorem 2.8, an additive upper continuous interval function $\mu$ has the representation $\mu_h$ defined by (2.2) for some point function $h$, and the sup norm of $\mu$ is the oscillation of $h$. We will show in Proposition 3.22 that $\|\cdot\|_{(\Phi)}$ is a seminorm and $\|\cdot\|_{[\Phi]}$ is a norm, both on $\widetilde{\mathcal{I}}_\Phi(J; X)$, for general $\Phi \in \mathcal{CV}$.

The next fact shows that Theorem 3.7(*a*) for point functions extends to general interval functions, whereas a constant additive interval function must be 0.

**Proposition 3.20.** *Let $J$ be a nonempty interval and let $\Phi \in \mathcal{CV}$. For any $\mu \in \widetilde{\mathcal{I}}_\Phi(J; X)$, $\|\mu\|_{(\Phi)} = 0$ if and only if $\mu$ is constant. For $\mu \in \widetilde{\mathcal{I}}_\Phi(J; X)$ additive, $\|\mu\|_{(\Phi)} = 0$ if and only if $\mu \equiv 0$.*

*Proof.* If $\mu$ is constant then clearly $\|\mu\|_{(\Phi)} = 0$. If $\|\mu\|_{(\Phi)} = 0$ and $\mu$ is not constant, take $A \in \mathfrak{I}(J)$ with $\mu(A) \neq \mu(\emptyset)$. Thus if $c_n \downarrow 0$ then

$$v_\Phi(\mu/c_n; A) \ge s_\Phi(\mu/c_n; \{A\}) = \Phi(\|\mu(A) - \mu(\emptyset)\|/c_n) \to \infty,$$

a contradiction, completing the proof since if $\mu$ is additive then $\mu(\emptyset) = 0$. □

The next fact extends statement (*b*) of Theorem 3.7 to interval functions and makes further statements in the additive case. Let $\widetilde{\mathcal{AI}}_\Phi(J; X) := \widetilde{\mathcal{I}}_\Phi(J; X) \cap \mathcal{AI}(J; X)$.

**Proposition 3.21.** *Let $J$ be a nonempty interval and let $\Phi \in \mathcal{CV}$ with inverse $\phi$. For any $\mu \in \widetilde{\mathcal{I}}_\Phi(J; X)$ with $\|\mu\|_{(\Phi)} > 0$,*

$$v_\Phi(\mu/\|\mu\|_{(\Phi)}; A) \le 1 \tag{3.37}$$

*for each $A \in \mathfrak{I}(J)$. In particular, if $\mu \in \widetilde{\mathcal{AI}}_\Phi(J; X)$ then $\|\mu\|_{\sup} \le \phi(1)\|\mu\|_{(\Phi)}$ and so*

$$\|\mu\|_{(\Phi)} \le \|\mu\|_{[\Phi]} \le (1 + \phi(1))\|\mu\|_{(\Phi)}.$$

*In particular, for $\mu \in \mathcal{AI}_p(J; X)$ we have $\|\mu\|_{\sup} \le \|\mu\|_{(p)}$ and*

$$\|\mu\|_{(p)} \le \|\mu\|_{[p]} \le 2\|\mu\|_{(p)}.$$

*Proof.* Let $\|\mu\|_{(\Phi)} > 0$. Suppose that $v_\Phi(\mu/\|\mu\|_{(\Phi)}; A) > 1$ for some $A \in \mathfrak{I}(J)$. Then for some interval partition $\mathcal{A}$ of $A$, $s_\Phi(\mu/\|\mu\|_{(\Phi)}; \mathcal{A}) > 1$. Since $\Phi$ is continuous, $s_\Phi(\mu/u; \mathcal{A}) > 1$ for some $u > \|\mu\|_{(\Phi)}$. Since $\Phi$ is increasing,

$v_\Phi(\mu/v; A) > 1$ for $0 < v \le u$, a contradiction, proving the first part of the proposition. For the second part let $\mu$ be additive. If $\|\mu\|_{(\Phi)} = 0$ then $\|\mu\|_{\sup} = 0$ by Proposition 3.20. If $\|\mu\|_{(\Phi)} > 0$ then for any $A \in \mathfrak{I}(J)$, $\Phi(\|\mu(A)\|/\|\mu\|_{(\Phi)}) \le v_\Phi(\mu/\|\mu\|_{(\Phi)}; A) \le 1$. The remaining statements follow, so the proof of the proposition is complete. □

Now we are ready to prove that the $\Phi$-variation for interval functions defines a seminorm and a norm.

**Proposition 3.22.** *Let $J$ be a nonempty interval and let $\Phi \in \mathcal{CV}$. Then on $\widetilde{\mathcal{I}}_\Phi(J; X)$, $\|\cdot\|_{(\Phi)}$ is a seminorm and $\|\cdot\|_{[\Phi]}$ is a norm. On $\widetilde{\mathcal{AI}}_\Phi(J; X)$, $\|\cdot\|_{(\Phi)}$ is a norm. In particular, $\|\cdot\|_{(p)}$ is a norm on $\mathcal{AI}_p(J; X)$.*

*Proof.* For any $\mu \in \widetilde{\mathcal{I}}_\Phi(J; X)$ and $c \in \mathbb{R}$ it is easy to see that $\|c\mu\|_{(\Phi)} = |c|\,\|\mu\|_{(\Phi)}$. If also $\nu \in \widetilde{\mathcal{I}}_\Phi(J; X)$, we want to show that

$$\|\mu + \nu\|_{(\Phi)} \le \|\mu\|_{(\Phi)} + \|\nu\|_{(\Phi)}. \tag{3.38}$$

If $\|\mu\|_{(\Phi)} = 0$ then $\mu$ is constant by Proposition 3.20, and $\|\mu+\nu\|_{(\Phi)} = \|\nu\|_{(\Phi)}$ by definition (3.33) of $\Phi$-variation sums. Thus (3.38) holds when one of the two terms on the right side is 0, so assume that neither is 0. Let $u := \|\mu\|_{(\Phi)}$ and $v := \|\nu\|_{(\Phi)}$. Since $\Phi$ is convex, by (3.37), for each $A \in \mathfrak{I}(J)$, we have

$$v_\Phi((\mu+\nu)/(u+v); A) \le \frac{u}{u+v} v_\Phi(\mu/u; A) + \frac{v}{u+v} v_\Phi(\nu/v; A) \le 1.$$

Thus (3.38) is proved and $\|\cdot\|_{(\Phi)}$ is a seminorm. Since $\|\cdot\|_{\sup}$ is a norm it follows that $\|\cdot\|_{[\Phi]}$ is a norm. The statements about the additive case follow from those in Propositions 3.20 or 3.21, finishing the proof. □

Again as for the $\Phi$-variation for point functions, we have Banach spaces of interval functions with bounded $\Phi$-variation.

**Proposition 3.23.** *Let $J$ be a nonempty interval, let $X$ be a Banach space, and let $\Phi \in \mathcal{CV}$. Then $(\widetilde{\mathcal{I}}_\Phi(J; X), \|\cdot\|_{[\Phi]})$ is a Banach space.*

*Proof.* A $\|\cdot\|_{[\Phi]}$-Cauchy sequence $\{\mu_k\}_{k\ge1}$, being Cauchy for $\|\cdot\|_{\sup}$, converges uniformly to some interval function $\mu$ on $J$. Thus for any $A \in \mathfrak{I}(J)$, any interval partition $\mathcal{A}$ of $A$ and $c > 0$, $s_\Phi(\mu_j/c; \mathcal{A}) \to s_\Phi(\mu/c; \mathcal{A})$ and $s_\Phi((\mu_k - \mu_j)/c; \mathcal{A}) \to s_\Phi((\mu_k - \mu)/c; \mathcal{A})$ for each $k$ as $j \to \infty$, so $\mu \in \widetilde{\mathcal{I}}_\Phi(J; X)$ and $\|\mu_k - \mu\|_{(\Phi)} \to 0$ as $k \to \infty$. □

As in the proof of Proposition 3.2, it follows that $\widetilde{\mathcal{I}}_\Phi = \mathcal{I}_\Phi$ if $\Phi \in \mathcal{V}$ satisfies the $\Delta_2$ condition. In particular, $(\mathcal{I}_p(J; X), \|\cdot\|_{[p]})$ is a Banach space for $1 \le p < \infty$.

**Corollary 3.24.** *Let $\Phi \in \mathcal{CV}$ and let $\mu \in \widetilde{\mathcal{I}}_\Phi(J;X)$ be additive. Then for any nonempty intervals $A \prec B$ such that $A \cup B \in \mathfrak{I}(J)$,*

$$\|\mu\|_{A\cup B,(\Phi)} \le \|\mu\|_{A,(\Phi)} + \|\mu\|_{B,(\Phi)}. \tag{3.39}$$

*Proof.* Writing $\mu_C(D) := \mu(C \cap D)$ for intervals $C$ and $D$, (3.39) follows from subadditivity of $\|\cdot\|_{A\cup B,(\Phi)}$ since $\mu_{A\cup B} = \mu_A + \mu_B$, $\|\mu_A\|_{A\cup B,(\Phi)} = \|\mu\|_{A,(\Phi)}$, and $\|\mu_B\|_{A\cup B,(\Phi)} = \|\mu\|_{B,(\Phi)}$. □

*The $p^*$-variation*

Let $J$ be a nonempty interval and let $\mu$ be an additive and upper continuous interval function on $J$ with values in a Banach space $X$, that is, $\mu \in \mathcal{AI}(J;X)$. Recall that for $x \in J$, the singleton $\{x\}$ is an atom of $\mu$ if $\mu(\{x\}) \neq 0$. By Proposition 2.6(c), the set of all atoms of $\mu$ is at most countable. Hence for $0 < p < \infty$, and any set $A \subset J$, the sum

$$\sigma_p(\mu;A) := \sum_A \|\mu\|^p := \sum_{t\in A} \|\mu(\{t\})\|^p \tag{3.40}$$

equals 0 if $A$ is empty, is well defined, and satisfies $\sigma_p(\mu;J) \le v_p(\mu)$. Also, let

$$v_p^*(\mu;A) := \inf_{\mathcal{B}\in \mathrm{IP}\,(A)} \sup_{\mathcal{A}\sqsupset\mathcal{B}} s_p(\mu;\mathcal{A}) \tag{3.41}$$

if $A \in \mathfrak{I}(J)$ is nonempty, or 0 if $A = \emptyset$, and

$$\sigma_p^*(\mu;A) := \sup_{\mathcal{B}\in \mathrm{IP}\,(A)} \inf_{\mathcal{B}\sqsubset\mathcal{A}} s_p(\mu;\mathcal{A}) \tag{3.42}$$

if $A \in \mathfrak{I}(J)$ is nonempty, or 0 if $A = \emptyset$, where $\mathcal{A} \sqsupset \mathcal{B}$ or equivalently $\mathcal{B} \sqsubset \mathcal{A}$ means that an interval partition $\mathcal{A}$ is a refinement of an interval partition $\mathcal{B}$. Then $\sigma_p(\mu) = \sigma_p(\mu;\cdot)$ restricted to $\mathfrak{I}(J)$, $v_p^*(\mu) = v_p^*(\mu;\cdot)$, and $\sigma_p^*(\mu) = \sigma_p^*(\mu;\cdot)$ are interval functions on $J$. In general we have the following relations:

**Lemma 3.25.** *Let $J$ be a nonempty interval, let $\mu \in \mathcal{AI}(J;X)$, and let $0 < p < \infty$. Then for each $A \in \mathfrak{I}(J)$,*

$$\sigma_p(\mu;A) \le \sigma_p^*(\mu;A) \le v_p^*(\mu;A) \le v_p(\mu;A), \tag{3.43}$$

$$v_p^*(\mu;A) = \inf_{\mathcal{B}\in \mathrm{YP}\,(A)} \sup_{\mathcal{A}\sqsupset\mathcal{B}} s_p(\mu;\mathcal{A}), \tag{3.44}$$

*and*

$$\sigma_p^*(\mu;A) = \sup_{\mathcal{B}\in \mathrm{YP}\,(A)} \inf_{\mathcal{B}\sqsubset\mathcal{A}} s_p(\mu;\mathcal{A}), \tag{3.45}$$

*where* YP $(A)$ *is the class of all Young interval partitions of $A$.*

*Proof.* We can assume that the interval $A$ is nondegenerate. For (3.43), if $\sigma_p(\mu; A) = +\infty$ then for each positive integer $N$ there is a point partition $B = \{z_j\}_{j=0}^m$ of $A$ such that $\sum_B \|\mu\|^p > N$. Then for any interval partition $\mathcal{A}$ which is a refinement of the Young interval partition $\{(z_{j-1}, z_j)\}_{j=1}^m$ of $A$, $s_p(\mu; \mathcal{A}) \geq \sum_B \|\mu\|^p$, and so $\sigma_p^*(\mu; A) = +\infty$. Suppose that $\sigma_p(\mu; A) < +\infty$. Given $\epsilon > 0$ there exists a point partition $B = \{z_j\}_{j=0}^m$ of $A$ such that $\sum_B \|\mu\|^p \geq \sigma_p(\mu; A) - \epsilon$. Then for any refinement $\mathcal{A}$ of $\{(z_{j-1}, z_j)\}_{j=1}^m$, we have $s_p(\mu; \mathcal{A}) \geq \sum_B \|\mu\|^p$, and so the first inequality in (3.43) holds. To prove the second inequality one can assume that $v_p^*(\mu; A) < +\infty$. Suppose that $v_p^*(\mu; A) < \sigma_p^*(\mu; A)$. Then there exist interval partitions $\mathcal{B}_1$ and $\mathcal{B}_2$ of $A$ such that $\sup\{s_p(\mu; \mathcal{A})\colon \mathcal{A} \sqsupset \mathcal{B}_1\} < \inf\{s_p(\mu; \mathcal{A})\colon \mathcal{A} \sqsupset \mathcal{B}_2\}$. Let $\mathcal{B}_3$ be a refinement of $\mathcal{B}_1$ and $\mathcal{B}_2$. Then

$$\sup_{\mathcal{A} \sqsupset \mathcal{B}_3} s_p(\mu; \mathcal{A}) \leq \sup_{\mathcal{A} \sqsupset \mathcal{B}_1} s_p(\mu; \mathcal{A}) < \inf_{\mathcal{A} \sqsupset \mathcal{B}_2} s_p(\mu; \mathcal{A}) \leq \inf_{\mathcal{A} \sqsupset \mathcal{B}_3} s_p(\mu; \mathcal{A}),$$

a contradiction, proving the second inequality in (3.43). Since the third inequality is obvious, (3.43) holds.

For (3.44) and (3.45), let $\mathcal{B} \in \mathrm{IP}(J)$ and $\mathcal{B} \sqsubset \mathcal{C} \in \mathrm{YP}(J)$. Then $\sup\{s_p(\mu; \mathcal{D})\colon \mathcal{D} \sqsupset \mathcal{C}\} \leq \sup\{s_p(\mu; \mathcal{D})\colon \mathcal{D} \sqsupset \mathcal{B}\}$ and $\inf\{s_p(\mu; \mathcal{D})\colon \mathcal{D} \sqsupset \mathcal{C}\} \geq \inf\{s_p(\mu; \mathcal{D})\colon \mathcal{D} \sqsupset \mathcal{B}\}$. Thus "$\geq$" in (3.44) and "$\leq$" in (3.45) hold instead of "$=$". Since the reverse inequalities are obvious, the proof is complete. □

The following fact shows that for $1 < p < \infty$ and $\mu$ real-valued the first inequality in Lemma 3.25 becomes an equation:

**Proposition 3.26.** *Let $\mu \in \mathcal{AI}(J; \mathbb{R})$ and let $1 < p < \infty$. Then $\sigma_p^*(\mu; A) = \sigma_p(\mu; A)$ for each $A \in \mathfrak{I}(J)$.*

*Proof.* By Lemma 3.25, it is enough to prove that $\sigma_p^*(\mu; A) \leq \sigma_p(\mu; A)$. We can assume that the interval $A$ is nondegenerate and $\sigma_p(\mu; A) < +\infty$. Let $\epsilon > 0$. We will prove that for each $\mathcal{B} \in \mathrm{IP}(A)$ there exists a refinement $\mathcal{A}$ of $\mathcal{B}$ such that $s_p(\mu; \mathcal{A}) \leq \sigma_p(\mu; A) + \epsilon$. Choose a Young interval partition $\{(z_{j-1}, z_j)\}_{j=1}^m$ of $A$ which is a refinement of $\mathcal{B}$ such that

$$\sum_{j=1}^m \sigma_p(\mu; (z_{j-1}, z_j)) < \epsilon/8.$$

This is possible because $\sigma_p(\mu)$ is an additive interval function on $J$, and $\sigma_p(\mu; A)$ can be approximated arbitrarily closely by finite sums $\sum_B |\mu|^p$. Let $\delta_j := \sigma_p(\mu; (z_{j-1}, z_j))$. Given $j$, it is enough find a partition $\mathcal{A}_j$ of $(z_{j-1}, z_j)$ such that

$$s_p(\mu; \mathcal{A}_j) \leq \epsilon_j := 4\delta_j + \epsilon/(2m).$$

If $\mu_j := \mu((z_{j-1}, z_j)) = 0$ we can take $\mathcal{A}_j = \{(z_{j-1}, z_j)\}$. Otherwise by symmetry we can assume $\mu_j > 0$. Choose $M \geq 1$ large enough so that

$2\max\{M^{-p}, \mu_j/M^{p-1}\} \le \delta_j + \epsilon/(2m)$. If $\mu_j \le 1/M$, taking $\mathcal{A}_j := \{(z_{j-1}, z_j)\}$ gives $s_p(\mu; \mathcal{A}_j) = \mu_j^p \le M^{-p} \le \epsilon_j$ as desired, so we can assume $\mu_j > 1/M$. Let $t_0 := z_{j-1}$. Given $t_0, \ldots, t_k$ such that $\mu((z_{j-1}, t_k]) + 1/M < \mu_j$, let $t_{k+1} := \inf\{t\colon \mu((t_k, t]) > 1/M\}$. Then by upper continuity of $\mu$ and Proposition 2.6(e), $\mu((t_k, t_{k+1})) \le 1/M \le \mu((t_k, t_{k+1}])$. Thus $\mu((z_{j-1}, t_k]) \ge k/M$ for each $k$, so $K$, the largest $k$ for which $t_{k+1}$ is defined by the recursion, satisfies $(K+1)/M < \mu_j$. Then $t_{K+1} < z_j$ since there is some $t < z_j$ with $\mu((t_k, t]) > 1/M$ by Proposition 2.6(e). Set $t_{K+2} := z_j$, and let $\mathcal{A}_j$ be the Young interval partition $\{(t_{i-1}, t_i)\}_{i=1}^{K+2}$ of $(z_{j-1}, z_j)$. For each $i = 1, \ldots, K+1$, $1/M - \mu(\{t_i\}) \le \mu((t_{i-1}, t_i)) \le 1/M$, so $|\mu((t_{i-1}, t_i))| \le \max\{1/M, \mu(\{t_i\})\}$. We have $\mu((z_{j-1}, t_{K+1}]) \le \mu((z_{j-1}, t_K]) + 1/M + \mu(\{t_{K+1}\}) < \mu_j + \mu(\{t_{K+1}\})$ and $\mu((z_{j-1}, t_{K+1}]) \ge \mu_j - 1/M$. Now $\mu((z_{j-1}, t_{K+1})) < \mu_j - (1/M) + 1/M = \mu_j$, so $\mu([t_{K+1}, t_{K+2})) > 0$ and $-\mu(\{t_{K+1}\}) < \mu((t_{K+1}, t_{K+2})) \le 1/M$ and $|\mu((t_{K+1}, t_{K+2}))| \le \max\{1/M, \mu(\{t_{K+1}\})\}$. Thus

$$\sum_{k=1}^{K+2} |\mu((t_{k-1}, t_k))|^p \le (K+2)/M^p + 2\delta_j$$

and

$$\begin{aligned} s_p(\mu; \mathcal{A}_j) &\le (K+2)/M^p + 3\delta_j < \mu_j/M^{p-1} + (7/2)\delta_j + \epsilon/(4m) \\ &< 4\delta_j + \epsilon/(2m) = \epsilon_j. \end{aligned}$$

The proof of Proposition 3.26 is complete. □

**Lemma 3.27.** *Let $J$ be a nonempty interval, $\mu \in \mathcal{AI}(J; X)$, and $0 < p < \infty$. For $A \in \mathfrak{I}(J)$,*

$$\sigma_p^*(\mu; A) = v_p^*(\mu; A) < +\infty \tag{3.46}$$

*if and only if the limit*

$$\lim_{\mathcal{A}} s_p(\mu; \mathcal{A}) \tag{3.47}$$

*exists under refinements of partitions $\mathcal{A}$ of $A$ (and is finite). Moreover, for $A, B \in \mathfrak{I}(J)$, if $B \subset A$ and* (3.46) *holds then* (3.46) *holds with $B$ in place of $A$.*

*Proof.* We can assume that the interval $A \in \mathfrak{I}(J)$ is nondegenerate. Let (3.46) hold and $\epsilon > 0$. By definitions (3.41) and (3.42), there exist interval partitions $\mathcal{B}_1$ and $\mathcal{B}_2$ of $A$ such that

$$\sup_{\mathcal{A} \sqsupset \mathcal{B}_1} s_p(\mu; \mathcal{A}) < v_p^*(\mu; A) + \epsilon \quad \text{and} \quad \inf_{\mathcal{A} \sqsupset \mathcal{B}_2} s_p(\mu; \mathcal{A}) > \sigma_p^*(\mu; A) - \epsilon.$$

Taking a refinement $\mathcal{B}_3$ of $\mathcal{B}_1$ and $\mathcal{B}_2$, we have for each refinement $\mathcal{A}$ of $\mathcal{B}_3$,

$$\sigma_p^*(\mu; A) - \epsilon < s_p(\mu; \mathcal{A}) < v_p^*(\mu; A) + \epsilon.$$

Since $\sigma_p^*(\mu; A) = v_p^*(\mu; A)$, it then follows that the limit (3.47) exists. Conversely, suppose that the limit (3.47) exists but $\sigma_p^*(\mu; A) \neq v_p^*(\mu; A)$, and so by Lemma 3.25,

$$\sigma_p^*(\mu; A) < v_p^*(\mu; A). \tag{3.48}$$

The existence of a finite limit implies that $v_p^*(\mu; A) < +\infty$. Let $\mathcal{B}$ be an interval partition of $A$. Recursively, there are interval partitions $\mathcal{B} \sqsubset \mathcal{A}_1 \sqsubset \mathcal{A}_2 \sqsubset \cdots$ such that $s_p(\mu; \mathcal{A}_k) > v_p^*(\mu) - 1/k$ for $k$ odd, and $s_p(\mu; \mathcal{A}_k) < \sigma_p^*(\mu) + 1/k$ for $k$ even. Thus by (3.48), the limit (3.47) cannot exist. This contradiction completes the proof of the first part.

To prove the second part, let $A, B \in \mathfrak{I}(J)$, $B \subset A$ and let (3.46) hold. By the first part, (3.47) holds and given $\epsilon > 0$, there is an interval partition $\mathcal{A}_1 \in \mathrm{IP}\,(A)$ such that $\|s_p(\mu; \mathcal{A}) - s_p(\mu; \mathcal{A}_1)\| < \epsilon$ for each refinement $\mathcal{A}$ of $\mathcal{A}_1$. We can assume that there is a subfamily $\mathcal{B}_1$ of $\mathcal{A}_1$ which is an interval partition of $B$. Let $\mathcal{B}$ be a refinement of $\mathcal{B}_1$, and let $\mathcal{A} := \mathcal{B} \cup (\mathcal{A}_1 \setminus \mathcal{B}_1)$. Then $\mathcal{A}$ is a refinement of $\mathcal{A}_1$ and

$$\|s_p(\mu; \mathcal{B}) - s_p(\mu; \mathcal{B}_1)\| = \|s_p(\mu; \mathcal{A}) - s_p(\mu; \mathcal{A}_1)\| < \epsilon.$$

Therefore the limit $\lim_{\mathcal{B}} s_p(\mu; \mathcal{B})$ exists under refinements of partitions $\mathcal{B}$ of $B$ and (3.46) holds with $B$ in place of $A$ by the first part, proving the lemma. □

Proposition 3.26 and Lemma 3.27 show that for $1 < p < \infty$ and $X = \mathbb{R}$ the property defined next is equivalent to existence of the limit (3.47) of $p$-variation sums.

**Definition 3.28.** Let $J$ be a nonempty interval, let $\mu \in \mathcal{AI}(J; X)$, and let $0 < p < \infty$. We will say that $\mu$ has *$p^*$-variation*, and write $\mu \in \mathcal{AI}_p^*(J; X)$, if $\sigma_p(\mu; J) = v_p^*(\mu; J) < +\infty$. If $X = \mathbb{R}$ then we write $\mu \in \mathcal{AI}_p^*(J)$.

The following gives a useful characterization of $p^*$-variation.

**Lemma 3.29.** *Let $J$ be a nondegenerate interval, let $\mu \in \mathcal{AI}(J; X)$, and let $1 < p < \infty$. Then $\mu$ has $p^*$-variation if and only if for every $\epsilon > 0$, there exists a Young interval partition $\{(z_{j-1}, z_j)\}_{j=1}^m$ of $J$ such that*

$$\sum_{j=1}^m v_p(\mu; (z_{j-1}, z_j)) < \epsilon. \tag{3.49}$$

*Proof.* It is easy to see that for any interval partition $\mathcal{A} = \{A_i\}_{i=1}^n$ of $J$,

$$\sum_{i=1}^n v_p(\mu; A_i) = \sup\big\{s_p(\mu; \mathcal{B})\colon\ \mathcal{B} \in \mathrm{IP}\,(J),\ \mathcal{B} \sqsupset \mathcal{A}\big\}.$$

In (3.41), by (3.44), we can equivalently take $\mathcal{B} \in \mathrm{YP}\,(J)$. It then follows that

$$v_p^*(\mu; J) = \inf\Big\{ \sum_{i=1}^n v_p(\mu; A_i) \colon \ \{A_i\}_{i=1}^n \in \mathrm{YP}\,(J) \Big\}. \tag{3.50}$$

Let $\mu$ have $p^*$-variation. Then given $\epsilon > 0$, there exists a Young interval partition $\{(z_{j-1}, z_j)\}_{j=1}^m$ of $J$ such that for $B := \{z_j\}_{j=0}^m \cap J$,

$$\sum_{j=1}^m v_p(\mu; (z_{j-1}, z_j)) + \sum_B \|\mu\|^p < v_p^*(\mu; J) + \epsilon/2$$

and $\sum_B \|\mu\|^p > \sigma_p(\mu) - \epsilon/2$. Since $v_p^*(\mu) = \sigma_p(\mu)$, (3.49) follows.

Conversely, suppose that for every $\epsilon > 0$ there is a Young interval partition $\{(z_{j-1}, z_j)\}_{j=1}^m$ of $J$ such that (3.49) holds. Then $v_p^*(\mu; J) < +\infty$. Suppose that $\sigma_p(\mu; J) \neq v_p^*(\mu; J)$. Then by Lemma 3.25, $v_p^*(\mu; J) - \sigma_p(\mu; J) \geq C$ for some positive constant $C$. Let $\{(z_{j-1}, z_j)\}_{j=1}^m$ be any Young interval partition of $J$, and let $B := \{z_j\}_{j=0}^m \cap J$. Then by (3.50),

$$\sum_{j=1}^m v_p(\mu; (z_{j-1}, z_j)) \geq v_p^*(\mu; J) - \sum_B \|\mu\|^p \geq v_p^*(\mu; J) - \sigma_p(\mu; J) \geq C > 0,$$

a contradiction for $\epsilon < C$, proving that $\mu$ has $p^*$-variation. □

*Comparing the $\Phi$-variation for interval and point functions*

Recall the notations $[\![a, b)$ and $[a, b]\!]$ and the like from Section 1.4. In Proposition 2.6 and Theorem 2.8 we characterized $\mathcal{AI}([\![a, b]\!]; X)$, the class of all additive upper continuous interval functions defined on a given interval $J = [\![a, b]\!]$ with $a < b$, and having values in a Banach space $X$. With each such interval function $\mu$ one can associate a class of regulated functions $h$ on $[a, b]\!]$ such that $h(a) = 0$ and $\mu = \mu_h$ on $\mathfrak{I}(J)$, where $\mu_h$ is defined by (2.2). In general for a regulated function $h$, boundedness of the $\Phi$-variation of the corresponding interval function $\mu_h$ does not imply that the $\Phi$-variation of $h$ is also bounded. For example, let a function $h$ equal 0 everywhere except on a sequence $\{x_n\}$ of points of $(a, b)$ with $h(x_n) \to 0$ slowly as $n \to \infty$ (see Proposition 2.10). Then the interval function $\mu_h$ is identically 0. However, $v_\Phi(\mu_h; J) = v_\Phi(h; J)$ provided $h(a) = 0$ and $h$ is either right-continuous or left-continuous at each point. This is a consequence of the following fact about interval functions which are not necessarily additive.

**Proposition 3.30.** *Let $\Phi \in \mathcal{V}$, let $\mu$ be an $X$-valued interval function on $J = [\![a, b]\!]$ with $a < b$, and let $f$ be a regulated function on $J$ such that $f(a) = \mu(\emptyset)$ if $J = [a, b]\!]$, $f(b) = \mu(J)$ if $J = [\![a, b]$, $f(t-) = \mu([\![a, t))$ for $t \in (a, b]\!]$, and $f(t+) = \mu([\![a, t])$ for $t \in [\![a, b)$. Then $v_\Phi(\mu; J) \leq v_\Phi(f; J)$. If in addition for each $t \in (a, b)$, $f$ is either right-continuous or left-continuous at $t$, then $v_\Phi(\mu; J) = v_\Phi(f; J)$.*

*Proof.* Suppose that $v_\Phi(f;J)<\infty$. Let $\mathcal{A}=\{A_i\}_{i=1}^n$ be an interval partition of $J$. For $i\in\{1,\dots,n-1\}$, let $\{t_{i,m}\}_{m\geq 1}\subset J$ be such that $t_{i,m}\downarrow t$ for the right endpoint $t$ of $A_i$ if $A_i$ is right-closed, or $t_{i,m}\uparrow t$ otherwise, where $t$ is then the left endpoint of the left-closed interval $A_{i+1}$. For $i\in\{0,n\}$, let $\{t_{i,m}\}_{m\geq 1}\subset J$ be such that $t_{0,m}\downarrow a$ if $J=(a,b\rrbracket$, or $t_{0,m}=a$ for all $m$ otherwise, and let $t_{n,m}\uparrow b$ if $J=\llbracket a,b)$, or $t_{n,m}=b$ for all $m$ otherwise. For $m$ large enough, $t_{0,m}<t_{1,m}<\cdots<t_{n,m}$, and for such $m$ let $\kappa_m:=\{t_{i,m}\}_{i=0}^n\in\mathrm{PP}\,(J)$. Then $s_\Phi(f;\kappa_m)\to s_\Phi(\mu;\mathcal{A})$ as $m\to\infty$. Since $\mathcal{A}\in\mathrm{IP}\,(J)$ is arbitrary, $v_\Phi(\mu;J)\leq v_\Phi(f;J)$.

To prove the reverse inequality if for each $t\in(a,b)$, $f$ is either left- or right-continuous at $t$, we can assume that $v_\Phi(\mu;J)<\infty$. Let $\kappa=\{t_i\}_{i=0}^n\in\mathrm{PP}\,(J)$. For $i=1,\dots,n$, let $A_i$ be an interval with endpoints $t_{i-1}$ and $t_i$. More specifically, for $i=1,\dots,n-1$, take $A_i$ to be right-closed and $A_{i+1}$ to be left-open if $f$ is right-continuous at $t_i$, or otherwise take $A_i$ to be right-open and $A_{i+1}$ to be left-closed. Take the left endpoint of $A_1$ to be $a$ if $J=[a,b\rrbracket$. Otherwise take $A_0=(a,t_0)$ and $A_1$ left-closed if $f$ is left-continuous at $t_0>a$, or $A_0=(a,t_0]$ and $A_1$ left-open if $f$ is right-continuous at $t_0$. Finally, take the right endpoint of $A_n$ to be $b$ if $J=\llbracket a,b]$. Otherwise take $A_{n+1}=(t_n,b)$ and $A_n$ right-closed if $f$ is right-continuous at $t_n<b$, or $A_{n+1}=[t_n,b)$ and $A_n$ right-open if $f$ is left-continuous at $t_n$. Then $\mathcal{A}:=\{A_i\}$ is an interval partition of $J$ and $s_\Phi(f;\kappa)\leq s_\Phi(\mu;\mathcal{A})\leq v_\Phi(\mu;J)$. Since $\kappa$ is arbitrary the conclusion follows, and the proof is complete. □

If in the preceding proposition the interval function $\mu$ on $J=\llbracket a,b\rrbracket$ is additive and $f=R_{\mu,a}$ or $f=L_{\mu,a}$, where $R_{\mu,a}$ and $L_{\mu,a}$ are both defined by (2.3), then $\mu$ must be upper continuous by Proposition 2.6. In this case the following provides a relation between the $\Phi$-variations of interval functions $\mu$ and the point functions $R_{\mu,a}$ and $L_{\mu,a}$ over subintervals of $J$.

**Proposition 3.31.** *Let $\Phi\in\mathcal{V}$ and $\mu\in\mathcal{AI}(\llbracket a,b\rrbracket;X)$. Let $B\in\mathfrak{I}\llbracket a,b\rrbracket$ be nonempty. Then the following hold:*

(*a*) *$v_\Phi(R_{\mu,a};B)\leq v_\Phi(\mu;B)$, and $v_\Phi(R_{\mu,a};B)=v_\Phi(\mu;B)$ except when $B$ is an atom of $\mu$ or when $B$ is left-closed and its left endpoint is an atom of $\mu$ in $(a,b\rrbracket$;*

(*b*) *$v_\Phi(L_{\mu,a};B)\leq v_\Phi(\mu;B)$, and $v_\Phi(L_{\mu,a};B)=v_\Phi(\mu;B)$ except when $B$ is an atom of $\mu$ or when $B$ is right-closed and its right endpoint is an atom of $\mu$ in $\llbracket a,b)$.*

*Proof.* If $B$ is a singleton $\{x\}\subset\llbracket a,b\rrbracket$ then $v_\Phi(\mu;\{x\})=\Phi(\|\mu(\{x\})\|)$ and $v_\Phi(R_{\mu,a};\{x\})=v_\Phi(L_{\mu,a};\{x\})=0$. Thus the equalities in (*a*) and (*b*) hold in this case if and only if $B=\{x\}$ is not an atom of $\mu$. Now let $B=\llbracket u,v\rrbracket$ for some $u,v\in\llbracket a,b\rrbracket$, $u<v$, and let $\mathcal{A}=\{A_i\}_{i=1}^n$ be an interval partition of $B$. Let $\{u_i\}_{i=1}^n\subset[u,v]$ be such that $\cup_{1\leq j\leq i}A_j=\llbracket u,u_i\rrbracket$, $i=1,\dots,n$. Here $u_1=u$ if $B$ is left-closed and $A_1=\{u\}$, and $u_1>u$ otherwise. For each

$i = 2, \ldots, n$ and for $i = 1$ if $u_1 > u$, let $\{t_{i,m}\}_{m \ge 1}$ be a sequence of points in $[u_{i-1}, u_i]$ such that $u_{i-1} < t_{i,m} \uparrow u_i$ if $A_i$ is right-open, and $t_{i,m} = u_i$ for each $m$ otherwise. Also if $u_1 > u$ let $\{t_{0,m}\}_{m \ge 1}$ be a sequence of points in $[u, t_{1,1})$ such that $t_{0,m} \downarrow u$ if $B$ is left-open and $t_{0,m} = u$ for each $m \ge 1$ if $B$ is left-closed. If $u_1 = u$ then take a sequence $\{t_{1,m}\}_{m \ge 1}$ of points in $(u, t_{2,1})$ such that $t_{1,m} \downarrow u$, and let $t_{0,m} = u$ for each $m \ge 1$. By Proposition 2.6($f$), first we have $\lim_{m \to \infty}[R_{\mu,a}(t_{i,m}) - R_{\mu,a}(t_{i-1,m})] = \mu(A_i)$ for each $i = 2, \ldots, n$.

Let $T := \lim_{m \to \infty} \big[R_{\mu,a}(t_{1,m}) - R_{\mu,a}(t_{0,m})\big]$. Second, if $u_1 > u$ then we have

$$T = \begin{cases} \mu(A_1 \setminus \{u\}) & \text{if } u > a \text{ and } B = [u, v\rrbracket, \\ \mu(A_1) & \text{if } u > a \text{ and } B = (u, v\rrbracket, \text{ or } u = a \in \llbracket a, b\rrbracket. \end{cases}$$

And third, if $u_1 = u$, or equivalently $A_1 = \{u\}$, then we have

$$T = \begin{cases} 0 & \text{if } u > a, \\ \mu(\{a\}) & \text{if } u = a \in \llbracket a, b\rrbracket. \end{cases}$$

Thus in any case ($u_1 > u$ or $u_1 = u$), $T = \mu(A_1)$ if $B = (u, v\rrbracket$ with $u > a$ or if $B = \llbracket a, v\rrbracket$. If $B = [u, v\rrbracket$ with $u > a$ then $T = \mu(A_1)$ if and only if $\mu(\{u\}) = 0$. Since $\Phi$ is continuous, whenever $T = \mu(A_1)$ it follows that

$$s_{\Phi}(\mu; \mathcal{A}) = \lim_{m \to \infty} \sum_{i=1}^{n} \Phi(\|R_{\mu,a}(t_{i,m}) - R_{\mu,a}(t_{i-1,m})\|) \le v_{\Phi}(R_{\mu,a}; B).$$

Hence $v_{\Phi}(\mu; B) \le v_{\Phi}(R_{\mu,a}; B)$ for each $B \in \mathfrak{I}\llbracket a, b\rrbracket$ such that $B$ is not an atom of $\mu$ and the left endpoint of $B$ is not an atom of $\mu$ in $(a, b\rrbracket$ if $B$ is left-closed. For the converse inequality, let $B \in \mathfrak{I}\llbracket a, b\rrbracket$ and let $\kappa = \{t_i\}_{i=0}^{n}$ be a point partition of $B$. Then $R_{\mu,a}(t_i) - R_{\mu,a}(t_{i-1}) = \mu((t_{i-1}, t_i])$ for $i = 2, \ldots, n$, and

$$R_{\mu,a}(t_1) - R_{\mu,a}(t_0) = \begin{cases} \mu([t_0, t_1]) & \text{if } t_0 = a \in \llbracket a, b\rrbracket, \\ \mu((t_0, t_1]) & \text{if } t_0 > a. \end{cases}$$

Thus $s_{\Phi}(R_{\mu,a}; \kappa) \le v_{\Phi}(\mu; B)$ for any partition $\kappa$ of $B$, and so $v_{\Phi}(R_{\mu,a}; B) \le v_{\Phi}(\mu; B)$ for any $B \in \mathfrak{I}\llbracket a, b\rrbracket$, proving ($a$). The proof of ($b$) is symmetric, and is therefore omitted. The proof of the proposition is complete. ☐

## 3.3 Elementary Properties

In this section we collect several properties of a point function $f\colon J \to X$ and its $\Phi$-variation defined by (3.1), and of an interval function $\mu\colon \mathfrak{I}(J) \to X$ and its $\Phi$-variation defined by (3.35).

*Point functions*

It is easy to see that for any $a < c < b$,

$$v_\Phi(f; [\![a, c]) + v_\Phi(f; [c, b]\!]) \le v_\Phi(f; [\![a, b]\!]). \tag{3.51}$$

Thus the function $[\![a, b]\!] \ni t \mapsto v_\Phi(f; [\![a, t]) =: G_{f,\Phi}(t)$ on a nonempty interval $[\![a, b]\!]$ is nondecreasing, and bounded provided $f$ has bounded $\Phi$-variation on $[\![a, b]\!]$. One can use the function $G_{f,\Phi}$ to characterize boundedness of $\Phi$-variation of $f$ as follows.

Let $G$ be a nondecreasing function from a nonempty interval $J$ into $\mathbb{R}$, and let $\phi$ be a function from $[0, \infty)$ into itself. Then $\|\Delta f\| \le \phi(\Delta G)$ on $J$ will mean that for any $s, t \in J$, $\|f(s) - f(t)\| \le \phi(|G(s) - G(t)|)$. Here is a characterization of functions in $\mathcal{W}_\Phi$:

**Proposition 3.32.** *For any $\Phi \in \mathcal{V}$, with inverse $\phi = \Phi^{-1}$, and function $f$ from an interval $[\![a, b]\!]$ into a Banach space $X$, $f \in \mathcal{W}_\Phi([\![a, b]\!]; X)$ if and only if there is a nondecreasing, bounded function $G$ on $[\![a, b]\!]$ such that $\|\Delta f\| \le \phi(\Delta G)$ on $[\![a, b]\!]$, and then $v_\Phi(f) \le \operatorname{Osc}(G; [\![a, b]\!])$. Specifically, we can take $G = G_{f,\Phi}$.*

*Proof.* We can assume that $a < b$. Let $f \in \mathcal{W}_\Phi$. Then $H := G_{f,\Phi}$ is nondecreasing and bounded. It is easily seen that if $s, t \in [\![a, b]\!]$ and $s \le t$, then

$$H(s) + \Phi(\|f(t) - f(s)\|) \ \le \ H(t),$$

so $\Phi(\|f(t) - f(s)\|) \le H(t) - H(s)$ and $\|\Delta f\| \le \phi(\Delta H)$ on $[\![a, b]\!]$.

Conversely if $\|\Delta f\| \le \phi(\Delta G)$ on $[\![a, b]\!]$ for some nondecreasing, bounded $G$, then for any partition $\kappa = \{x_i\}_{i=0}^n$ of $[\![a, b]\!]$,

$$s_\Phi(f; \kappa) \ \le \ \sum_{i=1}^n G(x_i) - G(x_{i-1}) \ \le \ \operatorname{Osc}(G; [\![a, b]\!]) < \infty.$$

The proposition is proved. □

By Proposition 3.32, for $\Phi \in \mathcal{V}$ with inverse $\phi = \Phi^{-1}$ and $a \le b$, a function $f$ has bounded $\Phi$-variation on $[\![a, b]\!]$ if and only if there is a nondecreasing, bounded function $G$ on $[\![a, b]\!]$ such that $\|\Delta f\| \le \phi(\Delta G)$ on $[\![a, b]\!]$. Clearly $f$ can be discontinuous at each point of $[\![a, b]\!]$ where $G$ is discontinuous. But we show next that it cannot have discontinuities other than of the first kind, that is, $f$ must be regulated (as defined in Section 2.1).

**Proposition 3.33.** *For any $\Phi \in \mathcal{V}$, each function $f$ with finite $\Phi$-variation is regulated.*

*Proof.* Let $v_\Phi(f;J) < \infty$ for some $\Phi \in \mathcal{V}$ and a nondegenerate interval $J$. It will be shown that for each $t \in \bar{J}$ which is not the right endpoint of $J$, $f$ has a right limit at $t$. Suppose $\inf_{\delta>0} \operatorname{Osc}(f;(t,t+\delta)) =: \epsilon > 0$, where for an open interval $(u,v) \subset J$,

$$\operatorname{Osc}(f;(u,v)) = \sup\{\|f(t) - f(s)\|: u < s \le t < v\}.$$

Take $t < u_1 < v_1 \in J$ such that $\|f(u_1) - f(v_1)\| > \epsilon/2$. Given $t < u_n < v_n < u_{n-1} < \cdots < u_1 < v_1$, choose $t < u_{n+1} < v_{n+1} < u_n$ such that $\|f(u_{n+1}) - f(v_{n+1})\| > \epsilon/2$. This can be done for all $n = 1, 2, \ldots$ . Then $v_\Phi(f;J) \ge n\Phi(\epsilon/2)$ for all $n$, a contradiction. So $f$ has a right limit at $t$. Likewise $f$ has a left limit at each $s \in \bar{J}$ which is not the left endpoint for $J$, so $f$ is regulated. □

Next we construct a continuous function $g$ of bounded $\Phi$-variation such that $f = g \circ G_{f,\Phi}$. Let $f \in \mathcal{W}_\Phi((a,b])$ with $a < b$. If $f(a)$ is defined, then clearly also $f \in \mathcal{W}_\Phi([a,b])$. If $f(a)$ is not originally defined, it can be defined as a limit $f(a) := \lim_{x \downarrow a} f(x)$, and then $v_\Phi(f;[a,b]) = v_\Phi(f;(a,b])$. Symmetrically, we can treat $f(b)$ for $f \in \mathcal{W}_\Phi([a,b))$ with $a < b$. Recall that $\mathcal{VM}$ is the set of all $\Phi \in \mathcal{V}$ such that $\Phi^{-1}$ is a modulus of continuity.

**Proposition 3.34.** *Let $a < b$ and $f \in \mathcal{W}_\Phi([a,b];X)$ for some $\Phi \in \mathcal{VM}$. Then for each nondecreasing function $\chi\colon [a,b] \to [0,\infty)$ such that $\chi(a) = \lim_{x\downarrow a} \chi(x) = 0$ and*

$$G_{f,\Phi}(t) - G_{f,\Phi}(s) \le \chi(t) - \chi(s) \qquad \text{for } s,t \in [a,b],\ s \le t,$$

*there exists a continuous function $g\colon [0,\chi(b)] \to X$ such that $f = g \circ \chi$ on $[a,b]$ and*

$$\|g(v) - g(u)\| \le 5\Phi^{-1}(v-u) \qquad \text{for } 0 \le u \le v \le \chi(b).$$

*Proof.* The range of $\chi$ is of the form $[0,\chi(b)] \setminus \cup_j I_j$ where each $I_j$ is an interval of the form $[\chi(t-),\chi(t))$, $(\chi(t-),\chi(t))$, $(\chi(t),\chi(t+))$, or $(\chi(t),\chi(t+)]$. Define $g(u)$ for $u$ in the range of $\chi$, that is, for $u = \chi(t)$, $t \in [a,b]$, by $g(u) := f(t)$. Then letting $\phi := \Phi^{-1}$ and using the bounds (from Proposition 3.32) $\|\Delta f\| \le \phi(\Delta G_{f,\Phi}) \le \phi(\Delta\chi)$, we have

$$\|g(u) - g(v)\| \;\le\; \phi(|u-v|) \tag{3.52}$$

for all $u$, $v$ in the range of $\chi$. If $\chi$ does not have a point $\chi(t-)$ or $\chi(t+)$ in its range then we can extend $g$ to each such point by continuity using (3.52). Then $g$ is defined on $[0,\chi(b)]$ except on countably many disjoint open intervals $(u_j,v_j)$. Extend $g$ to such intervals by

$$g(\alpha u_j + (1-\alpha)v_j) \;:=\; \alpha g(u_j) + (1-\alpha) g(v_j), \qquad 0 \;\le\; \alpha \;\le\; 1.$$

Then for $0 \le \alpha \le \beta \le 1$, using (3.31), we have

$$\begin{aligned}&\big\|g(\alpha u_j+(1-\alpha)v_j)-g(\beta u_j+(1-\beta)v_j)\big\|\\&\quad=\big\|(\beta-\alpha)\big(g(u_j)-g(v_j)\big)\big\|=(\beta-\alpha)\big\|g(u_j)-g(v_j)\big\|\\&\quad\le(\beta-\alpha)\phi(v_j-u_j)\le 2\phi\big((\beta-\alpha)(v_j-u_j)\big)\\&\quad=2\phi\big(|\alpha u_j+(1-\alpha)v_j-\beta u_j-(1-\beta)v_j|\big).\end{aligned}$$

Suppose $u_i<w<v_i\le u_j<z<v_j$. Then

$$\begin{aligned}|g(z)-g(w)|&\le|g(z)-g(u_j)|+|g(u_j)-g(v_i)|+|g(v_i)-g(w)|\\&\le 2\phi(z-u_j)+\phi(u_j-v_i)+2\phi(v_i-w)\le 5\phi(z-w)\end{aligned}$$

since $\phi$ is nondecreasing. For other cases we get a similar bound with a factor less than 5. So, for the extended $g$, the proposition is proved. □

Reverse inequalities to (3.51) do not hold without additional assumptions on the functions $\Phi$ or $f$ or the point $c$. Indeed, let $f$ be a function on $[0,2]$ with values $f(t):=0$ for $t\in[0,1)$, $f(1):=1$, and $f(t):=2$ for $t\in(1,2]$. Then for $1<p<\infty$, $v_p(f;[0,1])=v_p(f;[1,2])=1$, while $v_p(f;[0,2])=2^p>2$. For arbitrary $f$ and $c$ we have the following inequalities:

**Proposition 3.35.** *For a function $f\in\mathcal{W}_\Phi([\![a,b]\!];X)$ with $\Phi\in\mathcal{V}$, and for $a<c<b$, the following hold:*

*(a) if $\Phi$ is convex then*

$$\|f\|_{[\![a,b]\!],(\Phi)}\le\|f\|_{[\![a,c],(\Phi)}+\|f\|_{[c,b]\!],(\Phi)};\tag{3.53}$$

*(b) if $\Phi$ is convex and satisfies the $\Delta_2$ condition* (3.3) *then*

$$v_\Phi(f;[\![a,b]\!])\le(D/2)\big\{v_\Phi(f;[\![a,c])+v_\Phi(f;[c,b]\!])\big\},\tag{3.54}$$

*where $D\ge 2$ is as in* (3.3).

*(c) Always*

$$v_\Phi(f;[\![a,b]\!])\le v_\Phi(f;[\![a,c])+v_\Phi(f;[c,b]\!])+w,$$

*where $w:=\sup\{w(u,v)\colon\ u\in[\![a,c],\ v\in[c,b]\!]\}$ and*

$$w(u,v):=\Phi(\|f(v)-f(u)\|)-\Phi(\|f(v)-f(c)\|)-\Phi(\|f(c)-f(u)\|).$$

*Proof.* For part $(a)$, replacing $f$ by $f-f(c)$ without changing any of the seminorms $\|f\|_{J,(\Phi)}$, we can assume $f(c)=0$. Then $\|f\mathbf{1}_{[\![a,c]}\|_{[\![a,b]\!],(\Phi)}=\|f\|_{[\![a,c],(\Phi)}$, $\|f\mathbf{1}_{[c,b]\!]}\|_{[\![a,b]\!],(\Phi)}=\|f\|_{[c,b]\!],(\Phi)}$, and writing $f=f\mathbf{1}_{[\![a,c]}+f\mathbf{1}_{[c,b]\!]}$, (3.53) follows from subadditivity of $\|\cdot\|_{[\![a,b]\!],(\Phi)}$, proved in Theorem 3.7$(c)$.

Next assume in addition that $\Phi$ satisfies the $\Delta_2$ condition. Then for $s<c\le t$, by convexity and (3.3), we have

$$\begin{aligned}\Phi(\|f(t)-f(s)\|)&\le(1/2)\big[\Phi(2\|f(t)-f(c)\|)+\Phi(2\|f(c)-f(s)\|)\big]\\&\le(D/2)\big[\Phi(\|f(t)-f(c)\|)+\Phi(\|f(c)-f(s)\|)\big].\end{aligned}\tag{3.55}$$

For a partition $\kappa = \{t_i\}_{i=0}^n$ of $[\![a,b]\!]$, we can assume that $t_0 \le c \le t_n$, and so there is an index $i_0$ such that $t_{i_0-1} < c \le t_{i_0}$. Then applying the preceding bound to the $i_0$th term in the sum $s_\Phi(f;\kappa)$ and noting that $D/2 \ge 1$ by convexity gives the inequality (3.54).

Given a partition $\kappa = \{t_i\}_{i=0}^n$ of $[\![a,b]\!]$ and the index $i_0$ as before, $\kappa_1 := \{t_0,\dots,t_{i_0-1},c\}$ and $\kappa_2 := \{c,t_{i_0},\dots,t_n\}$ (omitting $c$ from $\kappa_2$ if $c = t_{i_0}$) are partitions of $[\![a,c]$ and $[c,b]\!]$, respectively. Then $(c)$ follows from the relation

$$s_\Phi(f;\kappa) = s_\Phi(f;\kappa_1) + s_\Phi(f;\kappa_2) + w(t_{i_0-1},t_{i_0}).$$

The proof of the proposition is complete. □

The following shows that bounded $\Phi$-variation on adjoining subintervals implies it on the closure of their union, although not providing as useful a bound on the larger interval as we have from the preceding proposition.

**Proposition 3.36.** *Let $\Phi \in \mathcal{V}$, $J := [a,b]$, and $f : [a,b] \to X$. Suppose there is a point partition $\kappa = \{u_i\}_{i=0}^m$ of $J$ such that $v_\Phi(f;(u_{i-1},u_i)) < \infty$ for each $i = 1,\dots,m$. Then $v_\Phi(f;J) < +\infty$.*

*Proof.* It is easily seen that $f$ is bounded on $J$. Let $\tau = \{t_j\}_{j=0}^n$ be any point partition of $J$. In the $\Phi$-variation sum $s_\Phi(f;\tau)$ there are terms such that for some $i$, $u_{i-1} < t_{j-1} < t_j < u_i$, and at most $2m$ other terms. Thus we have

$$s_\Phi(f;\tau) \;\le\; 2m\Phi(2\|f\|_{\sup,J}) + \sum_{i=1}^m v_\Phi(f;(u_{i-1},u_i)) \;<\; +\infty,$$

and the conclusion follows. □

Suppose a convex $\Phi \in \mathcal{V}$ does not satisfy the $\Delta_2$ condition. For $n = 1,2,\dots$, take $u_n > 0$ such that $\Phi(2u_n) > n\Phi(u_n)$. Let $f_n(t) := u_n t$ for $0 \le t \le 2$. Then by convexity, $v_\Phi(f_n;[0,1]) = v_\Phi(f_n;[1,2]) = \Phi(u_n)$ and $v_\Phi(f_n;[0,2]) = \Phi(2u_n)$. Thus the $\Delta_2$ condition is necessary in order for (3.54) to hold for some constant $D < \infty$, even for all linear functions $f$.

In Proposition 3.35(b), in the special case $\Phi(u) = u^p$, $u \ge 0$, with $1 \le p < \infty$, we have $D = 2^p$ and $D/2 = 2^{p-1}$. The following gives an extension to more than two intervals provided the values of $f$ at the common endpoints of adjoining intervals are all equal. That assumption is vacuous if there are just two intervals.

**Proposition 3.37.** *Let $f \in \mathcal{W}_p([\![a,b]\!];X)$ with $1 \le p < \infty$, let $a = c_0 < c_1 < \dots < c_m = b$ with $m \ge 2$, and let $f(c_1) = \dots = f(c_{m-1})$. Then*

$$v_p(f;[\![a,b]\!]) \le 2^{p-1}\sum_{j=1}^m v_p(f;[c_{j-1},c_j]\cap[\![a,b]\!]). \tag{3.56}$$

*Proof.* Subtracting a constant which does not change $v_p(f)$ on any interval, we can assume that $f(c_j) = 0$ for $j = 1, \ldots, m-1$. Let $A := \{c_j\}_{j=1}^{m-1}$ and let $\kappa = \{x_i\}_{i=0}^n$ be a point partition of $[\![a, b]\!]$. If $(x_{i-1}, x_i) \cap A =: A_i \neq \emptyset$, let $u_i := \min A_i$ and $v_i := \max A_i$. Then each $[x_{i-1}, u_i]$ and $[v_i, x_i]$ is a subinterval of some $[c_{j-1}, c_j] \cap [\![a, b]\!]$, with $f(u_i) = f(v_i) = 0$, and so by convexity of the function $[0, \infty) \ni u \mapsto u^p$,

$$\|f(x_i) - f(x_{i-1})\|^p \le 2^{p-1}\{\|f(x_i) - f(v_i)\|^p + \|f(u_i) - f(x_{i-1})\|^p\}.$$

If $(x_{i-1}, x_i) \cap A = \emptyset$ then $[x_{i-1}, x_i] \subset [c_{j-1}, c_j] \cap [\![a, b]\!]$ for some $j$. Since $2^{p-1} \ge 1$, it then follows that for the $p$-variation sum $s_p$ as defined after (3.2),

$$s_p(f; \kappa) \le 2^{p-1} \sum_{j=1}^m v_p(f; [c_{j-1}, c_j] \cap [\![a, b]\!]).$$

The proof of (3.56) is complete. □

The following shows that boundedness of the $\Phi$-variation depends only on the behavior of $\Phi$ near zero.

**Proposition 3.38.** *Let $\Phi, \Psi \in \mathcal{V}$ and suppose that for some $c > 0$, $\Phi(u) \ge \Psi(u)$ for $0 \le u \le c$. For a nonempty interval $J$ and $f\colon J \to X$, if $v_\Phi(f; J) < \infty$ then $v_\Psi(f; J) < \infty$.*

*Proof.* We can assume that $J$ is a nondegenerate interval. Let $f\colon J \to X$ be such that $v_\Phi(f; J) < \infty$ and let $\kappa = \{t_i\}_{i=0}^n$ be a partition of $J$. Then

$$\mathrm{card}\{i\colon \|f(t_i) - f(t_{i-1})\| > c\} \le v_\Phi(f)/\Phi(c) =: K. \tag{3.57}$$

By Proposition 3.33, $f$ is regulated and so it is bounded by Corollary 2.2. Letting $N := \mathrm{Osc}(f; J)$ it follows that $s_\Psi(f; \kappa) \le K\Psi(N) + v_\Phi(f) < \infty$. Since $\kappa$ is an arbitrary partition, the proof is complete. □

Taking $\Psi(u) = cu$ for $0 \le u \le c$, for some $c > 0$, in the preceding proposition gives the following:

**Corollary 3.39.** *Let $\Phi \in \mathcal{V}$ and suppose that $\liminf_{u \downarrow 0} \Phi(u)/u > 0$. For a nonempty interval $J$ and $f\colon J \to X$, if $v_\Phi(f; J) < \infty$ then $v_1(f; J) < \infty$, that is, $f$ has bounded variation.*

For a regulated function $f$ on a nondegenerate interval $J$, given a constant $c > 0$, there is a Young interval partition $\{(z_{j-1}, z_j)\}_{j=1}^m$ of $J$ such that the oscillation $\mathrm{Osc}(f; (z_{j-1}, z_j)) \le c$ for each $1 \le j \le m$ (Theorem 2.1). For functions of bounded $\Phi$-variation one can strengthen this property, replacing the oscillation by the $\Phi$-variation. We recall that by definition (3.1), the $\Phi$-variation over an open interval $(u, v)$ is given by

$$v_\Phi(f;(u,v)) := \sup\Big\{\sum_{i=1}^{n}\Phi(\|f(t_i)-f(t_{i-1})\|)\colon\ u<t_0<\cdots<t_n<v\Big\}.$$

**Proposition 3.40.** *Let $J$ be a nondegenerate interval, let $f \in \mathcal{W}_\Phi(J;X)$ for some $\Phi \in \mathcal{V}$, and let $c > 0$. Then there exists a Young interval partition $\{(z_{j-1}, z_j)\}_{j=1}^m$ of $J$ such that $v_\Phi(f;(z_{j-1},z_j)) \le c$ for each $1 \le j \le m$, and $m \le 1 + v_\Phi(f;J)/c$.*

*Proof.* Let $J = [\![a,b]\!]$ with $a < b$ and let $\rho(t) := v_\Phi(f;[\![a,t]\!])$ for $t \in [\![a,b]\!]$. By (3.51), it follows that

$$\begin{aligned} v_\Phi(f;(u,v)) &= \lim_{[t,s]\uparrow(u,v)} v_\Phi(f;[t,s]) \\ &\le \lim_{s\uparrow v}\rho(s) - \lim_{t\downarrow u}\rho(t) = \rho(v-) - \rho(u+) \end{aligned} \tag{3.58}$$

for any $a \le u < v \le b$. Therefore the conclusion will follow from the next lemma. □

**Lemma 3.41.** *Let $\rho$ be a nondecreasing function on $(a,b)$ with $\rho(a+) \ge 0$, and let $c > 0$. Then there exists a point partition $\{z_j\}_{j=0}^m$ of $[a,b]$ such that $\rho(z_j-) - \rho(z_{j-1}+) \le c$ for each $1 \le j \le m$, and $m \le 1 + \rho(b-)/c$.*

*Proof.* The lemma holds with $m = 1$ if $c \ge \rho(b-)$ or $\rho(a+) = \rho(b-)$. Suppose $\max(\rho(a+), c) < \rho(b-)$. Let $M$ be the smallest integer such that $cM \ge \rho(b-)$ and let $z_0 := a$. For each $n = 1, \ldots, M$, let $z_1^n := \inf\{t \in (z_0, b)\colon \rho(t) \ge nc \wedge \rho(b-)\}$ and let $n_1 := \min\{n = 1, \ldots, M-1\colon z_1^n > z_0\}$, or $n_1 := M$ if there is no such $n$. If $n_1 = M$ let $z_1 := b$. Then $\rho(a+) \ge (M-1)c$ and $\rho(b-) - \rho(a+) \le Mc - (M-1)c = c$. Thus the lemma holds with $m = 1$ in this case. If $n_1 < M$ let $z_1 := z_1^{n_1}$. It then follows that $\rho(z_1-) \le n_1 c$ and $\rho(z_0+) \ge (n_1 - 1)c$.

Suppose $z_j < b$ and $n_j < M$ are chosen such that $\rho(z_j-) \le n_j c$ and $\rho(z_{j-1}+) \ge (n_j - 1)c$. For each integer $n$ with $n_j < n \le M$, let $z_{j+1}^n := \inf\{t \in (z_j, b)\colon \rho(t) \ge nc \wedge \rho(b-)\}$ and let $n_{j+1} := \min\{n\colon n_j < n < M,\ z_{j+1}^n > z_j\}$, or $n_{j+1} := M$ if there is no such $n$. If $n_{j+1} = M$ let $z_{j+1} := b$. Otherwise let $z_{j+1} := z_{j+1}^{n_{j+1}}$ and proceed further. The recursion stops after at most $M$ steps with a partition $\{z_j\}_{j=0}^m$ of $[a,b]$. Thus $m \le M \le 1 + \rho(b-)/c$ and $\rho(z_j-) - \rho(z_{j-1}+) \le c$ for each $j = 1, \ldots, m$, proving the lemma. □

The four relations in the next lemma, with the $\Phi$-variation replaced by the oscillation and $\Phi$ on the right being the identity function, hold for a regulated function. Boundedness of the $\Phi$-variation again leads to stronger properties.

**Proposition 3.42.** *Let $f \in \mathcal{W}_\Phi([\![a,b]\!];X)$ for some $\Phi \in \mathcal{V}$ and $a < b$. If $u \in [\![a,b)$ then*

$$\lim_{v\downarrow u} v_\Phi(f;(u,v]) = 0 \quad \textit{and} \quad \lim_{v\downarrow u} v_\Phi(f;[u,v]) = \Phi(\|(\Delta^+ f)(u)\|). \tag{3.59}$$

*If* $u \in (a,b]\!]$ *then*

$$\lim_{v\uparrow u} v_\Phi(f;[v,u)) = 0 \quad \textit{and} \quad \lim_{v\uparrow u} v_\Phi(f;[v,u]) = \Phi(\|(\Delta^- f)(u)\|). \tag{3.60}$$

*Proof.* Let $u \in [\![a,b)$. Suppose that $\lim_{v\downarrow u} v_\Phi(f;(u,v]) =: \delta > 0$. Then there is a $v > u$ such that $v_\Phi(f;(u,v]) < (3/2)\delta$. On the other hand, one can choose partitions $\kappa_1 = \{t_i\}_{i=0}^n$ of $(u,v]$ and $\kappa_2 = \{s_j\}_{j=0}^m$ of $(u,t_0)$ such that $s_\Phi(f;\kappa_1) \geq (3/4)\delta$ and $s_\Phi(f;\kappa_2) \geq (3/4)\delta$. Thus we have a partition $\lambda := \kappa_1 \cup \kappa_2$ of $(u,v]$ with $s_\Phi(f;\lambda) \geq (3/2)\delta$. This contradiction proves the first relation in (3.59). The first relation in (3.60) holds by a symmetric proof.

Also due to symmetry between the second relations in (3.59) and (3.60), we prove only the former. Let $u \in [\![a,b)$. Since $f$ is regulated by Proposition 3.33, the right side $A := \Phi(\|(\Delta^+ f)(u)\|)$ of (3.59) is defined and $A \leq v_\Phi(f;[u,v])$, which decreases to some finite limit as $v \downarrow u$. Suppose the limit is $A+\delta$ for some $\delta > 0$. Hence one can find $w \in (u,b]$ such that $v_\Phi(f;[u,w]) \leq A+(5/4)\delta$, and for each $v \in (u,w]$,

$$\big|\Phi(\|f(v)-f(u)\|) - A\big| < (1/8)\delta. \tag{3.61}$$

On the other hand, there is a partition $\kappa = \{t_i\}_{i=0}^n$ of $[u,w]$ such that $s_\Phi(f;\kappa) > A + (7/8)\delta$. Thus by (3.61) with $v = t_1$, we have

$$\sum_{i=2}^n \Phi\big(\|f(t_i)-f(t_{i-1})\|\big) = s_\Phi(f;\kappa) - \Phi\big(\|f(t_1)-f(u)\|\big) > (3/4)\delta. \tag{3.62}$$

Doing the same for $t_1$ in place of $w$ we get a partition $\{s_j\}_{j=1}^m$ of $(u,t_1]$ such that

$$\sum_{j=2}^m \Phi\big(\|f(s_j)-f(s_{j-1})\|\big) > (3/4)\delta. \tag{3.63}$$

Thus $\lambda := \{u, s_1, \dots, s_m = t_1, \dots, t_n\}$ is a partition of $[u,w]$ such that

$$\begin{aligned} &s_\Phi(f;\lambda)\\ &= \Phi\big(\|f(s_1)-f(u)\|\big) + \sum_{j=2}^m \Phi\big(\|f(s_j)-f(s_{j-1})\|\big) + \sum_{i=2}^n \Phi\big(\|f(t_i)-f(t_{i-1})\|\big)\\ &> A - (1/8)\delta + (3/4)\delta + (3/4)\delta > A + (5/4)\delta, \end{aligned}$$

where the first inequality follows by (3.61) with $v = s_1$, (3.63), and (3.62). This contradicts $s_\Phi(f;\lambda) \leq v_\Phi(f;[u,w]) \leq A + (5/4)\delta$, proving the second relation in (3.59). The proof of Proposition 3.42 is complete. □

**Corollary 3.43.** *For $f \in \mathcal{W}_p([a,b];X)$ with $1 \le p < \infty$ and $a < b$, the following four statements hold:*

(*a*) $\lim_{c\uparrow b} \|f\|_{[a,c],(p)} = \|f\|_{[a,b),(p)}$ *and* $\|f\|_{[a,b],(p)} \le \|f\|_{[a,b),(p)} + \|\Delta^- f(b)\|$;
(*b*) $\lim_{c\downarrow a} \|f\|_{[c,b],(p)} = \|f\|_{(a,b],(p)}$ *and* $\|f\|_{[a,b],(p)} \le \|\Delta^+ f(a)\| + \|f\|_{(a,b],(p)}$;
(*c*) $\lim_{u\downarrow a, v\uparrow b} \|f\|_{[u,v],(p)} = \|f\|_{(a,b),(p)}$.
(*d*) *If* $\Delta^+ f(a) = \Delta^- f(b) = 0$, *then* $\|f\|_{[a,b],(p)} = \|f\|_{(a,b),(p)}$.

*Proof.* We first prove (*c*). Any point partition of $(a,b)$ is a partition of some $[u,v]$ where $a < u < v < b$, and so $\|f\|_{[u,v],(p)} \uparrow \|f\|_{(a,b),(p)}$ as $u \downarrow a$ and $v \uparrow b$. Thus (*c*) holds. For (*a*), we have likewise $\|f\|_{[a,c],(p)} \uparrow \|f\|_{[a,b),(p)}$ as $c \uparrow b$. By Proposition 3.35(*a*), we have for any $c \in (a,b)$,

$$0 \le \|f\|_{[a,b],(p)} - \|f\|_{[a,c],(p)} \le \|f\|_{[c,b],(p)}.$$

Now letting $c \uparrow b$, the second part of (*a*) follows from the first part and the second relation in (3.60). Then (*b*) follows by symmetry. Part (*d*) follows from the proofs of (*a*) and (*b*), completing the proof. □

If $f$ has bounded $\Phi$-variation on $J = [\![a,b]\!]$ and $G_{f,\Phi}(t) = v_\Phi(f; [\![a,t]\!])$, $t \in J$, then by (3.51),

$$v_\Phi(f;[u,v]) \le G_{f,\Phi}(v) - G_{f,\Phi}(u) \qquad \text{for any } u,v \in J,\ u \le v. \tag{3.64}$$

Therefore if $f$ on $J$ is discontinuous at some point $u$, so that $\Delta^+ f(u) \ne 0$ for some $u \in [\![a,b)$ or $\Delta^- f(u) \ne 0$ for some $u \in (a,b]\!]$, then $G_{f,\Phi}$ is also discontinuous at $u$ by Proposition 3.42. The next proposition shows that the converse implication holds.

**Proposition 3.44.** *Let $f \in \mathcal{W}_\Phi([\![a,b]\!];X)$ for some $\Phi \in \mathcal{V}$ and $a < b$. Then for $t \in [\![a,b)$ and $s \in (a,b]\!]$, $\Delta^+ G_{f,\Phi}(t) = 0$ or $\Delta^- G_{f,\Phi}(s) = 0$ if and only if respectively $\Delta^+ f(t) = 0$ or $\Delta^- f(s) = 0$.*

*Proof.* We will prove only the right-continuity equivalence because a proof of the left-continuity equivalence is symmetric. If for some $t \in [\![a,b)$, $\Delta^+ G_{f,\Phi}(t) = 0$, then $\Delta^+ f(t) = 0$ by (3.64) and (3.59). Conversely, suppose that $\Delta^+ f(t) = 0$ for some $t \in [\![a,b)$. By Proposition 3.33, $f$ is regulated, and hence $c := \mathrm{Osc}\,(f;[\![a,b]\!]) < \infty$. Since $\Phi$ is uniformly continuous on $[0,c]$, given $\epsilon > 0$ there is a $\delta > 0$ such that $\big|\Phi(v) - \Phi(u)\big| < \epsilon$ for any $u,v \in [0,c]$ with $|u-v| < \delta$. Since $f$ is right-continuous at $t$, by (3.59), there is a $t_0 \in (t,b]\!]$ such that $\|f(r) - f(t)\| < \delta$ and $v_\Phi(f;[t,r]) < \epsilon$ for each $t < r \le t_0$. Also, we have

$$\Big|\|f(s_2) - f(s_1)\| - \|f(t) - f(s_1)\|\Big| \le \|f(s_2) - f(t)\| < \delta$$

for any $s_1 \in [\![a,t]$ and $s_2 \in (t,t_0]$. Therefore $w < \epsilon$, where $w$ is the supremum of

$$\Phi(\|f(s_2)-f(s_1)\|)-\Phi(\|f(t)-f(s_1)\|)-\Phi(\|f(s_2)-f(t)\|)$$

over $s_1 \in [\![a,t]$ and $s_2 \in [t,t_0]$. By inequality $(c)$ of Proposition 3.35, it then follows that $G_{f,\Phi}(r) - G_{f,\Phi}(t) \le v_\Phi(f;[t,r]) + w < 2\epsilon$ for any $r \in (t,t_0]$, proving the proposition. □

**Lemma 3.45.** $\|f\|_{(p)}$ *is a nonincreasing function of* $p > 0$. *Moreover, if* $0 < q < p < \infty$ *then*

$$\|f\|_{(p)} \le \min\Big\{\|f\|_{(q)}, \operatorname{Osc}(f)^{(p-q)/p}\|f\|_{(q)}^{q/p}\Big\}. \tag{3.65}$$

*Proof.* The first part follows from the inequality $(\sum_i a_i^p)^{1/p} \le (\sum_i a_i^q)^{1/q}$ valid for $0 < q < p < \infty$ and any $a_i \ge 0$, which clearly holds if all $a_i = 0$. Otherwise, letting $r := p/q$, $b_i := a_i^q$, and $\lambda_i := b_i/\sum_j b_j$, we get $\sum_i \lambda_i^r \le 1$, which yields the conclusion. To show the second part let $\kappa$ be a partition of an interval $J$ where $f$ is defined. Writing $p = q + (p-q)$, it follows that

$$s_p(f;\kappa) \le \operatorname{Osc}(f;J)^{p-q} v_q(f;J),$$

and so (3.65) holds. □

The class of functions considered in the next proposition will be used to establish several properties of $p$-variation.

**Proposition 3.46.** *Let* $1 < p < \infty$, *let* $M > 1$ *be an even integer such that* $M^{1-(1/p)} > 4$, *let* $\{h_k\}_{k\ge1}$ *be a sequence of real numbers such that*

$$\sum_{k=1}^{\infty} |h_k| M^{-k/p} < \infty, \tag{3.66}$$

*and let*

$$f(t) = \sum_{k=1}^{\infty} h_k M^{-k/p} \sin(2\pi M^k t), \quad 0 \le t \le 1. \tag{3.67}$$

*Then* $f \in \mathcal{W}_p[0,1]$ *if and only if* $\sup_{k\ge1} |h_k| < \infty$.

The proposition will be proved using the following lemma, which is also useful for other purposes. Recall that, as mentioned after the definition (1.18), for $0 < \alpha \le 1$ we use "$\alpha$-Hölder" and "$\alpha$-Hölder continuous" synonymously.

**Lemma 3.47.** *Let* $1 < p < \infty$, *let* $M > 1$ *be an even integer, let* $\{h_k\}_{k\ge1}$ *be a sequence of real numbers such that* (3.66) *holds, and let* $f$ *be the function defined by* (3.67). *If* $\sup_{k\ge1}|h_k| < \infty$ *then* $f$ *is a* $(1/p)$*-Hölder continuous function on* $[0,1]$. *Moreover, if* $M^{1-(1/p)} > 4$, *and if there is an integer* $n > 1$ *such that* $\max_{1\le k<n}|h_k| \le |h_n|$, *then* $s_p(f;\kappa_n) \ge (4/3^p)|h_n|^p$ *for the partition* $\kappa_n = \{i/(4M^n)\}_{i=0}^{4M^n}$ *of* $[0,1]$.

*Proof.* By (3.66), the series (3.67) converges uniformly for $t \in [0,1]$. To prove $(1/p)$-Hölder continuity of $f$ let $\|h\| := \sup_{k\geq 1} |h_k| < \infty$ and let $0 \leq s < t \leq 1$. If $t - s < 1/M$ let $k_0 = k_0(t-s) \geq 1$ be the maximal integer in the set $\{k\colon M^k \leq 1/(t-s)\}$ and let $k_0 := 1$ otherwise. Using $|\sin x - \sin y| \leq |x-y|$ and summing two geometric progressions, it follows that

$$|f(t) - f(s)| \leq 2\pi(t-s)\|h\| \sum_{k=1}^{k_0} M^{k(1-1/p)} + 2\|h\| \sum_{k>k_0} M^{-k/p} \leq C(t-s)^{1/p}$$

for a finite constant $C$ depending only on $p$, $M$, and $\|h\|$. Choosing $M$ large enough so that $M^{1/p} > 2$ and $M^{1-(1/p)} > 2$ we could take $C = 4(\pi+1)\|h\|$.

To prove the second part of the conclusion, let $M^{1-(1/p)} > 4$ and let $n > 1$ be an integer such that $\max_{1\leq k<n} |h_k| \leq |h_n|$. For $k = 1,2,\ldots$ and $0 \leq t \leq 1$, let $f_k(t) := h_k M^{-k/p} \sin(2\pi M^k t)$ and let $t_i := i/(4M^n)$ for $i = 0,\ldots,4M^n$. For $i = 1,\ldots,4M^n$, we have $|f_n(t_i) - f_n(t_{i-1})| = |h_n| M^{-n/p}$. For $k > n$, since $M$ is an even integer, we have $f_k(t_i) = 0$ for each $i$. For $1 \leq k < n$, since $|\sin x - \sin y| \leq |x-y|$, we have $|f_k(t_i) - f_k(t_{i-1})| \leq (\pi/2)|h_k| M^{(k-n)-k/p}$ for $i = 1,\ldots,4M^n$. Thus for each $i = 1,\ldots,4M^n$,

$$\begin{aligned} |f(t_i) - f(t_{i-1})| &\geq |f_n(t_i) - f_n(t_{i-1})| - \sum_{k=1}^{n-1} |f_k(t_i) - f_k(t_{i-1})| \\ &\geq \bigl[1 - (\pi/2)(M^{1-1/p} - 1)^{-1}\bigr] |h_n| M^{-n/p} > 3^{-1} |h_n| M^{-n/p} \end{aligned}$$

by the choice of $M$, and so the bound $s_p(f;\kappa_n) > (4/3^p)|h_n|^p$ holds, proving the lemma. □

**Proof of Proposition 3.46.** If $\sup_{k\geq 1} |h_k| < \infty$ then $f$ has bounded $p$-variation by the first part of the conclusion of Lemma 3.47. Conversely, suppose that $f \in \mathcal{W}_p[0,1]$ but $\sup_{k\geq 1} |h_k| = \infty$. Then for any finite constant $C > 0$ there exists an integer $n > 1$ such that $|h_n| > C$ and $\max_{1\leq k<n} |h_k| \leq |h_n|$. By the second part of the conclusion of Lemma 3.47, we have $v_p(f;[0,1]) > 4(C/3)^p$, a contradiction since $C$ can be arbitrarily large. Thus $\sup_{k\geq 1} |h_k| < \infty$, proving the proposition. □

Let $\mathcal{CW}_p[0,1]$ denote the set of all continuous real-valued functions on $[0,1]$ with bounded $p$-variation. A class of continuous functions of the form (3.67) can be used to show that $\mathcal{CW}_p[0,1]$ is not separable with respect to the $p$-variation norm.

**Proposition 3.48.** *For* $1 < p < \infty$, $(\mathcal{CW}_p[0,1], \|\cdot\|_{[p]})$ *is nonseparable.*

*Proof.* For each $h := \{h_k\}_{k\geq 1}$ with $h_k \in \{-1,1\}$, let $f_h = \sum_{k\geq 1} f_{h,k}$ be the function defined by (3.67) with $M = 2$. By the first part of the conclusion of

Lemma 3.47, $f_h \in \mathcal{CW}_p[0,1]$ for any such $h$. We will show that the $p$-variation seminorm between any two different such functions cannot be smaller than a positive constant. Let $h \neq h' = \{h'_k\}_{k\geq 1}$. Let $n$ be the least $k$ such that $h_k \neq h'_k$, and let $\kappa_n := \{t_i = i2^{-n-2}\}_{i=0}^{2^{n+2}}$. If $n > 1$ then $f_{h,k} \equiv f_{h',k}$ for each $k = 1, \ldots, n-1$. If $1 \leq n < k$ then $f_{h,k}(t_i) = f_{h',k}(t_i) = 0$ for each $i = 0, \ldots, 2^{n+2}$. Thus for each $i$, we have

$$\begin{aligned}|(f_h - f_{h'})(t_i) - (f_h - f_{h'})(t_{i-1})| \\ = |(f_{h,n} - f_{h',n})(t_i) - (f_{h,n} - f_{h',n})(t_{i-1})| = 2^{1-(n/p)},\end{aligned}$$

and so $s_p(f_h - f_{h'}; \kappa_n) = 2^{2+p}$, proving the proposition. □

It is easy to see that the Hölder space $\mathcal{H}_\alpha[0,1]$ with $0 < \alpha \leq 1$ is non-separable, as follows. Letting $f_{t,\alpha}(x) := (x-t)^\alpha$ if $0 \leq t \leq x \leq 1$ and $:= 0$ otherwise, for $s \neq t$, we have $\|f_{s,\alpha} - f_{t,\alpha}\|_{(\mathcal{H}_\alpha)} = 1$. The natural injection $\mathcal{H}_{1/p} \to \mathcal{CW}_p$, $1 \leq p < \infty$, is a bounded linear operator but not a homeomorphism onto its range. In fact, $t \mapsto f_{t,1/p}$ is continuous from $[0,1]$ into $\mathcal{CW}_p$ but not continuous into $\mathcal{H}_{1/p}$. So, these examples do not imply Proposition 3.48.

*Interval functions*

Analogously to (3.51), for an $X$-valued additive interval function $\mu$ on a nonempty interval $J$, and for any subintervals $A$, $B$ of $J$ such that $A \prec B$ and $A \cup B$ is an interval, we have

$$v_\Phi(\mu; A) + v_\Phi(\mu; B) \leq v_\Phi(\mu; A \cup B). \tag{3.68}$$

As in the case of point functions, reverse inequalities do not hold without additional assumptions on the functions $\Phi$ and $\mu$ or sets $A, B$. Indeed, let $\mu$ be the additive interval function on $[0,1]$ equal to 1 on each singleton $\{1/3\}$ and $\{2/3\}$, and 0 on any interval containing neither point. Then for $1 < p < \infty$, $v_p(\mu; [0,1/2]) = v_p(\mu; (1/2,1]) = 1$, while $v_p(\mu; [0,1]) = 2^p > 2$.

Iterating (3.68), we have: for an additive interval function $\mu$ on $J$ and for any subintervals $A, A_1, \ldots, A_n$ of $J$ such that $A_1 \prec \cdots \prec A_n$ and $A_1 \cup \cdots \cup A_n \subset A$,

$$\sum_{i=1}^{n} v_\Phi(\mu; A_i) \leq v_\Phi(\mu; A). \tag{3.69}$$

In general, $v_\Phi(\mu)$ and $v_p(\mu)$ are not additive interval functions. By contrast, it is easily seen that if $\mu$ is additive and upper continuous, the interval function $\sigma_p(\mu; \cdot)$ defined by (3.40) is additive. We also have additivity for $v_p^*$ defined by (3.41):

**Proposition 3.49.** *Let $\mu$ be an additive upper continuous interval function on a nonempty interval $J$ having values in $X$. Then for $0 < p < \infty$, $v_p^*(\mu; \cdot)$ is also an additive interval function on $J$.*

*Proof.* Let $\{A, B\}$ be an interval partition of $J$. Then

$$v_p^*(\mu; J) \le \inf_{\mathcal{B} \sqsupset \{A,B\}} \sup_{\mathcal{A} \sqsupset \mathcal{B}} s_p(\mu; \mathcal{A}) \le v_p^*(\mu; A) + v_p^*(\mu; B) =: C.$$

Suppose $v_p^*(\mu; J) < C$. Then there is $\mathcal{B} \in \mathrm{IP}\,(J)$ such that $\sup_{\mathcal{A} \sqsupset \mathcal{B}} s_p(\mu; \mathcal{A}) < C$. Letting $\mathcal{B} \sqcap A := \{D_j \cap A\}_{j=1}^k$ if $\mathcal{B} = \{D_j\}_{j=1}^k$ and similarly for $\mathcal{B} \sqcap B$, it follows that

$$\sup_{\mathcal{A} \sqsupset \mathcal{B}} s_p(\mu; \mathcal{A}) \ge \sup_{\mathcal{C} \sqsupset \mathcal{B} \sqcap A} s_p(\mu; \mathcal{C}) + \sup_{\mathcal{D} \sqsupset \mathcal{B} \sqcap B} s_p(\mu; \mathcal{D}) \ge C,$$

where $\mathcal{C} \in \mathrm{IP}\,(A)$ and $\mathcal{D} \in \mathrm{IP}\,(B)$, a contradiction, proving $v_p^*(\mu; J) = C$. □

**Proposition 3.50.** *Let $\mu \in \mathcal{AI}_\Phi(J; X)$ where $\Phi \in \mathcal{V}$ and $J$ is a nonempty interval. Then the interval function $A \mapsto v_\Phi(\mu; A)$, $A \in \mathfrak{I}(J)$, is upper continuous at $\emptyset$.*

*Proof.* The conclusion can be obtained from Proposition 3.42 since for half-open intervals $A$, $v_\Phi(\mu; A) = v_\Phi(h; A)$ for suitable $h \in \mathcal{W}_\Phi(J; X)$ by Proposition 3.31. Instead, we give a direct proof. Clearly $v_\Phi(\mu; B) \le v_\Phi(\mu; A)$ for $B \subset A$. Suppose that $\lim_{n\to\infty} v_\Phi(\mu; B_n) = \delta > 0$ for some $B_n \downarrow \emptyset$. There exist $\{u, v_n \colon n \ge 1\}$ in $\bar{J}$ such that for all large enough $n$, either $B_n = [\![v_n, u)$ with $v_n \uparrow u$ or $B_n = (u, v_n]\!]$ with $v_n \downarrow u$. By symmetry assume that the former holds. Then $\lim_{n\to\infty} \sup_{v_n \le t < u} \|\mu([\![t, u))\| = 0$. If not then there exists a sequence $\{t_k\}$ such that $t_k \uparrow u$ and $\mu([\![t_k, u)) \not\to 0$, contradicting upper continuity of $\mu$ and so proving the claim. Thus there is an integer $n_0$ such that

$$v_\Phi(\mu; B_{n_0}) < (5/4)\delta \qquad \text{and} \qquad \sup_{v_{n_0} \le t < u} \|\mu([\![t, u))\| < \Phi^{-1}(\delta/4). \tag{3.70}$$

On the other hand, there is an interval partition $\mathcal{A}' = \{A_i\}_{i=1}^m$ of $B_{n_0}$ such that $s_\Phi(\mu; \mathcal{A}') > (3/4)\delta$. By the second relation in (3.70), $\Phi(\|\mu(A_m)\|) < \delta/4$. Thus $\sum_{i=1}^{m-1} \Phi(\|\mu(A_i)\|) > \delta/2$. Since for large enough $n$, $v_\Phi(\mu; A_m) \ge v_\Phi(\mu; B_n) \ge \delta$, there is an interval partition $\mathcal{A}''$ of $A_m$ such that $s_\Phi(\mu; \mathcal{A}'') > (3/4)\delta$. Then $\mathcal{A} := \mathcal{A}'' \cup \{A_i\}_{i=1}^{m-1}$ is an interval partition of $B_{n_0}$ such that $s_\Phi(\mu; \mathcal{A}) > (5/4)\delta$, contradicting the first relation in (3.70) and proving the proposition. □

An interval function need not be upper continuous even if it is additive and has bounded $p$-variation for some $p < \infty$:

**Example 3.51.** On $[0, 2]$, let $\mu$ be the interval function such that $\mu(A) = 0$ for any interval $A$ included in $[0, 1]$ or in $[d, 2]$ for any $d > 1$, and $\mu([\![s, t]\!]) = r \neq 0$ for $s \le 1 < t \le 2$, specifically $\mu((1, c)) = r$ for any $1 < c \le 2$. Then $\mu$ is additive and of bounded $p$-variation for any $p$, but not upper continuous at $\emptyset$.

The following fact for interval functions corresponds to Proposition 3.40 for point functions.

**Proposition 3.52.** *For $\Phi \in \mathcal{V}$, let $\mu \in \mathcal{AI}_\Phi(J;X)$ for a nondegenerate interval $J$, and let $c > 0$. There exists a Young interval partition $\{(z_{j-1}, z_j)\}_{j=1}^m$ of $J$ such that $v_\Phi(\mu;(z_{j-1},z_j)) \le c$ for each $1 \le j \le m$, and $m \le 1 + v_\Phi(\mu;J)/c$.*

*Proof.* Let $J = [\![a,b]\!]$, and let $\rho(t) := v_\Phi(\mu;[\![a,t])$ for $t \in J$. By (3.68), $\rho$ is a nondecreasing function on $J$. If $J$ is left-open then let $\rho(a) := \rho(a+)$, which is 0 since $A \mapsto v_\Phi(\mu;A)$, $A \in \mathfrak{I}[\![a,b]\!]$, is upper continuous at $\emptyset$ by Proposition 3.50. If $J$ is right-open then let $\rho(b) := \rho(b-)$. Let $s,t \in [a,b]$, and let $s < s_k < t_k < t$, $k = 1,2,\ldots$, be such that as $k \to \infty$, $s_k \downarrow s$ and $t_k \uparrow t$. Then

$$v_\Phi(\mu;(s,t)) = \lim_{k\to\infty} v_\Phi(\mu;(s_k,t_k]). \tag{3.71}$$

Indeed, let $\mathcal{A} = \{A_i\}_{i=1}^n$ be an interval partition of $(s,t)$ with $A_1 = (s,u_1]\!]$ and $A_n = [\![u_n,t)$. Since $\mu$ is upper continuous, by Proposition 2.6($d$), we have

$$\lim_{k\to\infty} \Phi(\|\mu(s_k,u_1]\!]\|) = \Phi(\|\mu(A_1)\|) \quad \text{and} \quad \lim_{k\to\infty} \Phi(\|\mu[\![u_n,t_k]\|) = \Phi(\|\mu(A_n)\|).$$

Letting $\mathcal{A}_k := \{(s_k,u_1]\!], A_2, \ldots, A_{n-1}, [\![u_n,t_k]\}$, we then have $s_\Phi(\mu;\mathcal{A}_k) \to s_\Phi(\mu;\mathcal{A})$ as $k \to \infty$, and $s_\Phi(\mu;\mathcal{A}_k) \le v_\Phi(\mu;(s_k;t_k])$. Thus (3.71) holds with "$\le$" instead of "$=$". Since the reverse inequality is clear, (3.71) is proved. By (3.68) and (3.71), it then follows that

$$v_\Phi(\mu;(s,t)) \le \lim_{k\to\infty} \big[\rho(t_k) - \rho(s_k)\big] = \rho(t-) - \rho(s+).$$

Since $s,t \in [a,b]$ are arbitrary, the conclusion follows from Lemma 3.41. □

It will next be shown that the $\Phi$-variation of an additive and upper continuous interval function can be calculated using only Young interval partitions. Recall that the set of all Young interval partitions of a nonempty interval $J$ is denoted by YP$(J)$.

**Lemma 3.53.** *Let $\Phi \in \mathcal{V}$ and let $J$ be a nondegenerate interval. For $\mu \in \mathcal{AI}(J;X)$, $v_\Phi(\mu;J) = \sup\{s_\Phi(\mu;\mathcal{A})\colon \mathcal{A} \in \mathrm{YP}(J)\}$.*

*Proof.* It is enough to show that the $\Phi$-variation sum $s_\Phi(\mu;\mathcal{A})$ for any interval partition $\mathcal{A}$ of $J$ can be approximated arbitrary closely by $s_\Phi(\mu,\mathcal{A}_n)$ for Young interval partitions $\mathcal{A}_n$. For notational simplicity and by symmetry, this will be done for $J = (a,b)$ and $\mathcal{A} = \{(a,u),[u,b)\}$ with $u \in (a,b)$. Let $a < u_n \uparrow u$ be such that $\mu(\{u_n\}) = 0$, which is possible since by Proposition 2.6($c$) $\mu$ can be non-zero only for at most countably many singletons in $J$. Then letting $\mathcal{A}_n := \{(a,u_n),\{u_n\},(u_n,b)\}$, and using also statements ($a$) and ($d$) of Proposition 2.6, we get that $s_\Phi(\mu;\mathcal{A}_n) \to s_\Phi(\mu;\mathcal{A})$ as $n \to \infty$, proving the lemma. □

The proof of the following lemma is the same as the proof of Lemma 3.45 and is omitted. Recall also that the oscillation Osc for an interval function $\mu$ on an interval $J$, defined in (2.7), equals $\sup_{A\subset J}\|\mu(A)\|$.

**Lemma 3.54.** *For any additive interval function $\mu$, $\|\mu\|_{(p)}$ is a nonincreasing function of $p>0$. Moreover, if $0<q<p<\infty$ then*

$$\|\mu\|_{(p)} \le \min\Big\{\|\mu\|_{(q)}, \mathrm{Osc}(\mu)^{(p-q)/p}\|\mu\|_{(q)}^{q/p}\Big\}. \tag{3.72}$$

Recall that the class of all additive upper continuous interval functions from $J$ to $X$ having $p^*$-variation is denoted by $\mathcal{AI}_p^*(J;X)$ (Definition 3.28).

**Lemma 3.55.** *If $1\le q<p<\infty$ then $\mathcal{AI}_q(J;X)\subset\mathcal{AI}_p^*(J;X)$.*

*Proof.* We can assume that $J$ is a nondegenerate interval. Let $\mu\in\mathcal{AI}_q(J;X)$ and let $\epsilon>0$. By Corollary 2.12, there exists a Young interval partition $\{(z_{j-1},z_j)\}_{j=1}^m$ of $J$ such that $\mathrm{Osc}(\mu;(z_{j-1},z_j))<\epsilon$ for each $j=1,\dots,m$. Then by (3.72) and (3.69), we have

$$\begin{aligned}\sum_{j=1}^m v_p(\mu;(z_{j-1},z_j)) &\le \max_{1\le j\le m}\mathrm{Osc}(\mu;(z_{j-1},z_j))^{p-q}\sum_{j=1}^m v_q(\mu;(z_{j-1},z_j))\\ &< \epsilon^{p-q}v_q(\mu).\end{aligned}$$

Since $\epsilon>0$ is arbitrary, $\mu$ has $p^*$-variation by Lemma 3.29. □

The following is an adaptation of Proposition 3.37 to interval functions.

**Proposition 3.56.** *Let $1\le p<\infty$, let $\mu\in\mathcal{I}(J;X)$ be additive, and let $\{B_j\}_{j=1}^m$ be an interval partition of $J$ such that $\mu(B_j)=0$ for each $j=2,\dots,m-1$. Then*

$$v_p(\mu;J)\le 2^{p-1}\sum_{j=1}^m v_p(\mu;B_j). \tag{3.73}$$

*Proof.* We can assume that $J$ is a nondegenerate interval. Let $\mathcal{A}=\{A_i\}_{i=1}^n$ be an interval partition of $J$. For each $i$, if $A_i\not\subset B_j$ for all $j$, then letting $j_1:=\min\{j\colon A_i\cap B_j\ne\emptyset\}$ and $j_2:=\max\{j\colon A_i\cap B_j\ne\emptyset\}$, we have $j_1<j_2$. Since $\mu$ is additive, $\mu(B_j)=0$ for $j_1<j<j_2$, and the function $u^p$, $u\ge0$, is convex, it then follows that

$$\|\mu(A_i)\|^p\le 2^{p-1}\Big\{\|\mu(A_i\cap B_{j_1})\|^p+\|\mu(A_i\cap B_{j_2})\|^p\Big\}.$$

Otherwise $A_i\subset B_j$ for some $j$ and $A_i$ is in a partition $\{B_j\cap A_r\}_{r=s}^t$ of $B_j$, where $B_j\cap A_r\ne\emptyset$ if and only if $s\le r\le t$. Since $2^{p-1}\ge1$, it then follows that

$$s_p(\mu;\mathcal{A}) \le 2^{p-1}\sum_{j=1}^{m} v_p(\mu;B_j),$$

proving (3.73). □

*Local p-variation*

As before let $X$ be a Banach space and let $J$ be a nonempty interval. Also let $0<p<\infty$. For a function $f\colon J\to X$, its *local p-variation* is defined by

$$v_p^*(f) := v_p^*(f;J) := \inf_{\lambda\in \mathrm{PP}(J)} \sup_{\kappa\supset\lambda} s_p(f;\kappa).$$

If $f$ has bounded $p$-variation then it is regulated by Proposition 3.33. Thus by Corollary 2.2, $f$ has at most countably many jumps and we can define the sum

$$\begin{aligned}\sigma_p(f) := \sigma_p(f;J) &:= \sum_J \{\|\Delta^- f\|^p + \|\Delta^+ f\|^p\}\\ &:= \sum_{x\in(a,b]\!]} \|\Delta^- f(x)\|^p + \sum_{x\in[\![a,b)} \|\Delta^+ f(x)\|^p.\end{aligned}$$

Note that if at some $x$, $\Delta^- f(x)$ and $\Delta^+ f(x)$ are both non-zero, this differs from $\sigma_p$ for the corresponding interval function $\mu$ defined by (2.2) and (3.40). The following always holds:

$$\sigma_p(f) \le v_p^*(f) \le v_p(f). \tag{3.74}$$

Next is a definition of the class of functions whose local $p$-variation comes only from jumps.

**Definition 3.57.** Let $X$ be a Banach space, let $J$ be a nonempty interval and let $1<p<\infty$. We will say that a function $f\colon J\to X$ has *$p^*$-variation* on $J$ and write $f\in\mathcal{W}_p^*(J;X)$ if $\sigma_p(f)=v_p^*(f)<+\infty$.

We show first that the $p^*$-variation of a right-continuous function $f$ agrees with the $p^*$-variation of the corresponding interval function $\mu_f$ defined by (2.2). For $1\le p<\infty$ and $a<b$, let $\mathcal{DW}_p([\![a,b]\!];X)$ be the set of all $f\in\mathcal{W}_p([\![a,b]\!];X)$ such that $f$ is right-continuous on $(a,b)$ and $f(a)=0$ if $a\in[\![a,b]\!]$, or $f(a+)=0$ otherwise. Then $\mathcal{DW}_p([\![a,b]\!];X)$ is the intersection of $\mathcal{W}_p([\![a,b]\!];X)$ and $\mathcal{D}([\![a,b]\!];X)$ defined before Corollary 2.11. That Corollary gives a one-to-one correspondence $f\leftrightarrow\mu_f$ between $f\in\mathcal{D}([\![a,b]\!];X)$ and $\mu_f$ in the set $\mathcal{AI}([\![a,b]\!];X)$ of all $X$-valued additive and upper continuous interval functions on $[\![a,b]\!]$. We will show that this correspondence preserves the $p$-variation and the local $p$-variation between corresponding point functions and interval functions.

**Proposition 3.58.** *Let $X$ be a Banach space, $J = [\![a,b]\!]$ with $a < b$, and $1 \le p < \infty$. For $f \in \mathcal{DW}_p(J;X)$, we have $v_p(f) = v_p(\mu_f;J)$, $v_p^*(f) = v_p^*(\mu_f;J)$, and $\sigma_p(f) = \sigma_p(\mu_f;J)$.*

*Proof.* Let $f \in \mathcal{DW}_p(J;X)$. Since $\mu_f(\{x\}) = \Delta^- f(x)$ for each $x \in (a,b]\!]$, $\mu_f(\{a\}) = \Delta^+ f(a)$ if $a \in J$, and $\Delta^+ f(x) = 0$ for $x \in (a,b)$, we have $\sigma_p(f) = \sigma_p(\mu_f;J)$. To show that $v_p(f) = v_p(\mu_f;J)$, let $\lambda = \{t_i\}_{i=0}^n$ be a point partition of $J$. Let $\mathcal{B}_\lambda$ be an interval partition of $J$ consisting of the intervals $(t_{i-1},t_i]$ for $i = 2,\dots,n$ if $n \ge 2$; either $[a,t_1]$ if $a \in J$, or $(a,t_0]$ and $(t_0,t_1]$ if $a \notin J$; and $(t_n,b)$ if $b \notin J$. Since $\mu_f((t_{i-1},t_i]) = f(t_i) - f(t_{i-1})$ for each $i$, except if $a = t_0 \in J$ when $f(t_1) - f(t_0) = \mu_f([a,t_1])$, it follows that $s_p(f;\lambda) \le s_p(\mu_f;\mathcal{B}_\lambda)$, and so $v_p(f) \le v_p(\mu_f;J)$. The converse inequality holds since for any interval partition $\mathcal{A}$ of $J$, the $p$-variation sum $s_p(\mu_f;\mathcal{A})$ can be approximated arbitrarily closely by $s_p(f;\kappa)$ for some point partitions $\kappa$ of $J$.

To prove the equality of local $p$-variations, let $\lambda = \{t_i\}_{i=0}^n$ be a point partition of $J$, and let $\mathcal{B}_\lambda$ be the interval partition of $J$ defined above. If $J = [a,b]$ then $s_p(f;\lambda) = s_p(\mu_f;\mathcal{B}_\lambda)$. Otherwise, since $\mu_f((a,t_0]) = f(t_0) - f(a+)$ if $a \notin J$ and $\mu_f((t_n,b)) = f(b-) - f(t_n)$ if $b \notin J$, the $p$-variation sum $s_p(\mu_f;\mathcal{B}_\lambda)$ can be approximated arbitrarily closely by $s_p(f;\kappa)$ corresponding to point partitions $\kappa$ of $J$ which are refinements of $\lambda$. Therefore we have

$$s_p(f;\lambda) \le s_p(\mu_f;\mathcal{B}_\lambda) \le \sup_{\kappa \supset \lambda} s_p(f;\kappa).$$

For an interval partition $\mathcal{A}$ which is a refinement of $\mathcal{B}_\lambda$, the $p$-variation sum $s_p(\mu_f;\mathcal{A})$ can be approximated arbitrarily closely by $s_p(f;\kappa)$ corresponding to point partitions $\kappa$ of $J$ which are refinements of $\lambda$, and so

$$s_p(f;\lambda) \le \sup_{\mathcal{A} \sqsupset \mathcal{B}_\lambda} s_p(\mu_f;\mathcal{A}) \le \sup_{\kappa \supset \lambda} s_p(f;\kappa).$$

If $\kappa \supset \lambda$ then $\mathcal{B}_\kappa \sqsupset \mathcal{B}_\lambda$. Therefore for any point partition $\lambda$ of $J$,

$$\sup_{\kappa \supset \lambda} s_p(f;\kappa) = \sup_{\mathcal{A} \sqsupset \mathcal{B}_\lambda} s_p(\mu_f;\mathcal{A}) \ge v_p^*(\mu_f;J),$$

and so $v_p^*(f) \ge v_p^*(\mu_f;J)$.

To prove the converse inequality it is enough to show that for any interval partition $\mathcal{B}$ of $J = [\![a,b]\!]$,

$$\sup_{\mathcal{A} \sqsupset \mathcal{B}} s_p(\mu_f;\mathcal{A}) \ge v_p^*(f). \tag{3.75}$$

Taking a refinement of $\mathcal{B}$ if necessary, we can and do assume that $\mathcal{B} = \{(t_{j-1},t_j)\}_{j=1}^m$ is a Young interval partition of $J$. Let $\epsilon > 0$. If $J = [a,b]\!]$ let $s_j \in (t_{j-1},t_j)$, $j = 1,\dots,m$, be close enough to $t_j$ so that $S(J) := S_1 = S_1(\{u_j\}_{j=1}^m) < \epsilon$ for any $u_j \in [s_j,t_j)$, $j = 1,\dots,m$, where

$$S_1 := \sum_{j:\ t_j \in (a,b]\!]} \Big\{\|\Delta^- f(t_j)\|^p - \|f(t_j) - f(u_j)\|^p\Big\} + \sum_{j=1}^m \|f(t_j-) - f(u_j)\|^p.$$

If $J = (a, b]\!]$ let $s_j \in (t_{j-1}, t_j)$, $j = 1, \dots, m$, be close enough to $t_j$ and $s_0 \in (t_0, s_1)$ close enough to $t_0$ so that $S(J) := S_2 = S_2(\{u_j\}_{j=0}^m) < \epsilon$ for any $u_j \in [s_j, t_j)$, $j = 1, \dots, m$, and $u_0 \in (t_0, s_0]$, where $S_2 := S_1 + \|f(u_0) - f(a+)\|^p$. Then $\lambda(J) := \{t_0, t_j, s_j\}_{j=1}^m \cap J$ if $J = [a, b]\!]$, or $\lambda(J) := \{s_0, t_j, s_j\}_{j=1}^m \cap J$ if $J = (a, b]\!]$, is a point partition of $J$. Let $\zeta = \{x_i\}_{i=0}^n$ be a refinement of $\lambda(J)$ such that $s_p(f; \zeta) + \epsilon \geq \sup_{\kappa \supset \lambda} s_p(f; \kappa)$. Let $\mathcal{A}$ be the interval partition of $J$ consisting of the intervals $(x_{i-1}, x_i]$ such that $x_i \neq t_j$ for all $j$, $(x_{i-1}, x_i)$ such that $x_i = t_j$ for some $j$, $(a, x_0)$ if $t_0 = a \notin J$, $(x_n, b)$ if $t_m = b \notin J$, and of the singletons $\{t_j\} \cap J$ (if nonempty) for $j = 0, 1, \dots, m$. Then $\mathcal{A}$ is a refinement of $\mathcal{B}$. Letting for $j = 1, \dots, m-1$, $u_j := x_{i-1}$ if $x_i = t_j$, either $u_m := x_n$ if $b \notin J$ or $u_m := x_{n-1}$ if $b \in J$, and $u_0 := x_0$ if $a \notin J$, it follows that $\left|s_p(\mu_f; \mathcal{A}) - s_p(f; \zeta)\right| \leq S(J) < \epsilon$, and so

$$\sup_{\mathcal{A} \sqsupset \mathcal{B}} s_p(\mu_f; \mathcal{A}) \geq s_p(f; \zeta) - \epsilon \geq \sup_{\kappa \supset \lambda} s_p(f; \kappa) - 2\epsilon.$$

Since $\epsilon > 0$ is arbitrary, (3.75) holds, proving the proposition. □

The next fact follows from the preceding proposition and complements Corollary 2.11.

**Corollary 3.59.** *Let $X$ be a Banach space, $J = [\![a, b]\!]$ with $a < b$, and $1 < p < \infty$. Then $f \mapsto \mu_f$ gives a one-to-one linear mapping of $\mathcal{DW}_p(J; X)$ onto $\mathcal{AI}_p(J; X)$, whose inverse is $\mu \mapsto R_{\mu,a}$.*

*Proof.* The maps $f \mapsto \mu_f$ and $\mu \mapsto R_{\mu,a}$ are one-to-one, linear, and inverses of each other between $\mathcal{D}(J; X)$ and $\mathcal{AI}(J; X)$ by Corollary 2.11. For $f \in \mathcal{DW}_p(J; X)$ we have $\mu_f \in \mathcal{AI}_p(J; X)$ by Proposition 3.58. For $\mu \in \mathcal{AI}_p(J; X)$ take $h = R_{\mu,a} \in \mathcal{D}(J; X)$, so that $\mu_h = \mu$. As shown early in the proof of Proposition 3.58, $v_p(h) \leq v_p(\mu; J)$, so $h \in \mathcal{DW}_p(J; X)$, completing the proof. □

Now we continue characterizing functions having $p^*$-variation. The following proposition is analogous to Lemma 3.29 for interval functions. It also shows that $\mathcal{W}_p^*$ is the closure in $\mathcal{W}_p$ of the class of step functions, which clearly belong to $\mathcal{W}_p^*$.

**Proposition 3.60.** *Let $J$ be a nondegenerate interval, $f \in \mathcal{R}(J; X)$, and $1 < p < \infty$. Then statements (a), (b), and (c) are equivalent, where*

*(a)* *$f$ has $p^*$-variation on $J$;*
*(b)* *for every $\epsilon > 0$, there exists a Young interval partition $\{(z_{j-1}, z_j)\}_{j=1}^m$ of $J$ such that*

$$\sum_{j=1}^m v_p(f; (z_{j-1}, z_j)) < \epsilon; \tag{3.76}$$

(c) *f is a limit in* $(\mathcal{W}_p, \|\cdot\|_{[p]})$ *of step functions.*

*Proof.* It is easy to see that for any point partition $\lambda = \{x_i\}_{i=0}^n$ of the closure $\bar{J}$, $\sup\{s_p(f;\kappa)\colon\ \kappa \in \mathrm{PP}\,(J),\ \kappa \supset \lambda \cap J\}$ is equal to $\sum_{i=1}^n v_p(f;[x_{i-1},x_i]\cap J)$, and so

$$v_p^*(f) = \inf\Big\{\sum_{i=1}^n v_p(f;[x_{i-1},x_i]\cap J)\colon\ \{x_i\}_{i=0}^n \in \mathrm{PP}\,(\bar{J})\Big\}. \tag{3.77}$$

$(a) \Rightarrow (b)$. Let $f$ have $p^*$-variation on $J$. By (3.77), given $\epsilon > 0$ there is a partition $\{z_j\}_{j=0}^m$ of $\bar{J}$ such that

$$\sum_{j=1}^m v_p(f;[z_{j-1},z_j]\cap J) < v_p^*(f) + \epsilon/2$$

and

$$\sum_{j\colon\ z_{j-1},z_j\in J} \Big\{\|\Delta^+ f(z_{j-1})\|^p + \|\Delta^- f(z_j)\|^p\Big\} > \sigma_p(f) - \epsilon/2.$$

By (3.51), for any $\{u_{j-1},v_j\}_{j=1}^m \subset (a,b)$ with $[u_{j-1},v_j] \subset (z_{j-1},z_j)$ for each $j$, we have

$$v_p(f;[u_{j-1},v_j]) \le v_p(f;[z_{j-1},z_j]\cap J) - v_p(f;[z_{j-1},u_{j-1}]\cap J) - v_p(f;[v_j,z_j]\cap J)$$

for $j = 1,\dots,m$. By Proposition 3.42 and Corollary 3.43(*c*), for each such $j$, letting $u_{j-1} \downarrow z_{j-1}$ and $v_j \uparrow z_j$, it follows that

$$\begin{aligned}
&\sum_{j=1}^m v_p(f;(z_{j-1},z_j))\\
&\qquad \le \sum_{j=1}^m \Big\{v_p(f;[z_{j-1},z_j]\cap J) - \|\Delta^+ f(z_{j-1})\|^p - \|\Delta^- f(z_j)\|^p\Big\}\\
&\qquad < v_p^*(f) + \epsilon/2 - \sigma_p(f) + \epsilon/2 = \epsilon
\end{aligned}$$

since $f$ has $p^*$-variation.

$(b) \Rightarrow (c)$. Given $\delta > 0$, take a Young interval partition $\{(z_{j-1},z_j)\}_{j=1}^m$ of $J$ satisfying (3.76) for $\epsilon = \delta^p$. To define a step function, let $y_j := (z_{j-1}+z_j)/2$ if $2 \le j \le m-1$, or if $j = 1$ and $a > -\infty$, or if $j = m$ and $b < +\infty$. If $j = 1$ and $a = -\infty$, let $y_1 := z_1 - 1$. If $j = m$ and $b = +\infty$, let $y_m := z_{m-1} + 1$. Let $g_j := f(z_{j-1}+)1_{(z_{j-1},y_j]} + f(z_j-)1_{(y_j,z_j)}$ for $j = 1,\dots,m$. Then for each $j = 1,\dots,m$, letting $A_j := (z_{j-1},z_j)$,

$$v_p(f - g_j;A_j) \le 2^p v_p(f;A_j)$$

by Theorem 3.7(*c*), since $v_p(g_j;A_j) \le v_p(f;A_j)$. A step function now is defined by

$$g := g_\delta := \sum_{j:\ z_j \in J} f(z_j)1_{\{z_j\}} + \sum_{j=1}^m g_j.$$

Clearly $\|f-g\|_{\sup} < \delta$. At each $z_j \in J$, $f-g$ is continuous (one-sidedly if $j=0$ or $m$) and equals 0. Thus by Proposition 3.37 and (3.76) with $\epsilon = \delta^p$,

$$\begin{aligned} v_p(f-g;J) &\le 2^{p-1}\sum_{j=1}^m v_p(f-g;[z_{j-1},z_j]\cap J) \\ &= 2^{p-1}\sum_{j=1}^m v_p(f-g_j;(z_{j-1},z_j)) \le 2^{2p-1}\delta^p. \end{aligned}$$

Hence $\|f-g_\delta\|_{[p]} \le 5\delta$ and $g_\delta \to f$ in $\mathcal{W}_p$ as $\delta \downarrow 0$, proving $(c)$.

$(c) \Rightarrow (a)$. We have $f \in \mathcal{W}_p$. Suppose that $f \notin \mathcal{W}_p^*$. Then there is a constant $C>0$ such that $v_p^*(f) - \sigma_p(f) \ge C$. Take $0<\epsilon<C$. There is a step function $g$ with $\|f-g\|_{(p)} < \epsilon^{1/p}$. Take a Young interval partition $\{(z_{j-1},z_j)\}_{j=1}^m$ of $J$ such that $g$ is constant on each $(z_{j-1},z_j)$. Then by (3.51),

$$\sum_{j=1}^m v_p(f;(z_{j-1},z_j)) = \sum_{j=1}^m v_p(f-g;(z_{j-1},z_j)) \le v_p(f-g;J) < \epsilon$$

(by the way proving $(b)$). Also, we have

$$\sum_{j=1}^m \Big\{\|\Delta_J^+ f(z_{j-1})\|^p + v_p(f;(z_{j-1},z_j)) + \|\Delta_J^- f(z_j)\|^p\Big\} \ge v_p^*(f). \tag{3.78}$$

To prove this, let $t_{0j} := z_{j-1} < t_{1j} < t_{2j} < t_{3j} := z_j$ for each $j$. Then by (3.77),

$$v_p^*(f) \le \sum_{j=1}^m \sum_{i=1}^3 v_p(f;J\cap[t_{i-1,j},t_{ij}]).$$

Let $t_{1j} \downarrow t_{0j}$ and $t_{2j} \uparrow t_{3j}$ for each $j$. Then in the limit we get (3.78) by Proposition 3.42 and Corollary 3.43$(c)$, recalling that $\Delta_J^+ f(a) = 0$ for $J = (a,b]\!]$ and $\Delta_J^- f(b) = 0$ for $J = [\![a,b)$ as defined before (2.1). Thus, we get

$$\sum_{j=1}^m v_p(f;(z_{j-1},z_j)) \ge v_p^*(f) - \sigma_p(f) \ge C > \epsilon,$$

a contradiction, proving $(a)$. The proof of Proposition 3.60 is complete. □

Here is an analogue of Lemma 3.55 for point functions.

**Lemma 3.61.** *If $1 \le q < p < \infty$ then $\mathcal{W}_q(J;X) \subset \mathcal{W}_p^*(J;X)$.*

*Proof.* Let $f \in \mathcal{W}_q(J;X)$ and let $\epsilon > 0$. By Proposition 3.33, $f$ is regulated on $J$. Thus by Theorem 2.1(*b*), there exists a Young interval partition $\{(z_{j-1}, z_j)\}_{j=1}^m$ of $J$ such that $\mathrm{Osc}(f;(z_{j-1}, z_j)) < \epsilon$ for each $j = 1, \ldots, m$. Then by (3.65) and (3.51), we have

$$\sum_{j=1}^m v_p(f;(z_{j-1}, z_j)) \le \max_{1\le j\le m} \mathrm{Osc}(f;(z_{j-1}, z_j))^{p-q} \sum_{j=1}^m v_q(f;(z_{j-1}, z_j))$$
$$< \epsilon^{p-q} v_q(f).$$

Since $\epsilon > 0$ is arbitrary, $f$ has $p^*$-variation by Proposition 3.60. □

Closely related to the local $p$-variation is a related quantity defined as follows. Let $0 < p < \infty$ and $f\colon [a,b] \mapsto X$. For each $\epsilon > 0$, let $v_p(f;\epsilon) := \sup\{s_p(f;\kappa)\colon |\kappa| \le \epsilon\}$, where $|\kappa|$ is the mesh of $\kappa \in \mathrm{PP}\,[a,b]$, and $\bar{v}_p(f) := \bar{v}_p(f;[a,b]) := \lim_{\epsilon\downarrow 0} v_p(f;\epsilon)$. Since $|\lambda| \le |\kappa|$ for $\lambda \supset \kappa$, we have $\sup\{s_p(f;\lambda)\colon \lambda \supset \kappa\} \le v_p(f;\epsilon)$ for each $\kappa \in \mathrm{PP}\,[a,b]$ such that $|\kappa| \le \epsilon$. This implies the first inequality in

$$v_p^*(f) \le \bar{v}_p(f) \le v_p(f), \tag{3.79}$$

while the second one is clear. The strict inequality $v_p^*(f) < \bar{v}_p(f)$ may hold if $f$ has jumps on both sides of the same point and $p > 1$. For example, if $f(x) := 0$ for $x \in [0, 1/2)$, $f(1/2) := 1/2$, and $f(x) := 1$ for $x \in (1/2, 1]$, then $v_p^*(f) = 2^{1-p} < 1 = \bar{v}_p(f)$ for each $p > 1$. However, equality holds under the following conditions.

**Lemma 3.62.** *Let $f\colon [a,b] \to X$ and $0 < p < \infty$. If for each $x \in (a,b)$, either $\Delta^- f(x) = 0$ or $\Delta^+ f(x) = 0$, then $v_p^*(f) = \bar{v}_p(f)$.*

*Proof.* By (3.79), it is enough to prove that $v_p^*(f) \ge \bar{v}_p(f)$. Suppose not. Choose $C$ such that $v_p^*(f) < C < \bar{v}_p(f)$. Then there exist a partition $\kappa = \{x_i\}_{i=0}^n$ of $[a,b]$ and a sequence $\{\kappa_m\}_{m\ge 1}$ of partitions of $[a,b]$ such that $|\kappa_m| \downarrow 0$ and

$$\sup\{s_p(f;\lambda)\colon \lambda \supset \kappa\} < C < s_p(f;\kappa_m) \tag{3.80}$$

for $m = 1, 2, \ldots$. Since $f$ has only one-sided jumps, $f(y_i^m) \to f(x_i)$ for each $i \in \{1, \ldots, n-1\}$ and some $y_i^m \in \kappa_m$ chosen in such a way that no $y \in \kappa_m$ is between $y_i^m$ and $x_i$. For each large enough $m$, let $\kappa_m'$ be the same as $\kappa_m$ except that $y_i^m$ is replaced by $x_i$ for each $i = 1, \ldots, n-1$. Then $|s_p(f;\kappa_m) - s_p(f;\kappa_m')| \to 0$ as $m \to \infty$. Since $\kappa_m' \supset \kappa$, this contradicts (3.80), proving the lemma. □

The following shows that the first inequality in (3.74) can be strict.

**Proposition 3.63.** *For $1 < p < \infty$, there is a function of bounded $p$-variation, in fact a $(1/p)$-Hölder continuous function, not having $p^*$-variation.*

*Proof.* Let $M > 1$ be an even integer such that $M^{1-(1/p)} > 4$, and let $f$ be the function (3.67) with $h_k \equiv 1$. Then $f$ is $(1/p)$-Hölder by the first part of the conclusion of Lemma 3.47, and so $f$ is in $\mathcal{W}_p[0,1]$ with $\sigma_p(f) = 0$. By the second part of the same conclusion, for each integer $n > 1$, $s_p(f;\kappa_n) > 4/3^p$ with $\kappa_n = \{i/(4M^n)\}_{i=0}^{4M^n}$. By Lemma 3.62, since mesh $|\kappa_n| \to 0$ as $n \to \infty$, $v_p^*(f) = \bar{v}_p(f) \geq 4/3^p$, and so $f \notin \mathcal{W}_p^*[0,1]$, proving the proposition. □

It is easy to see that $(\mathcal{W}_p^*[0,1], \|\cdot\|_{[p]})$, $1 < p < \infty$, is nonseparable: for $t, s \in [0,1]$ and $t \neq s$, we have $\left\|1_{[t,1]} - 1_{[s,1]}\right\|_{[p]} \geq \left\|1_{[t,1]} - 1_{[s,1]}\right\|_{\sup} = 1$.

## *3.4 A Necessary and Sufficient Condition for Integral Duality

In this section, for a pair of convex functions $\Phi$ and $\Psi$, necessary and sufficient conditions are given such that the refinement Young–Stieltjes integral $(RYS) \int f \cdot dh$ exists whenever $f$ and $h$ have bounded $\Phi$- and $\Psi$-variations, respectively (Theorem 3.75 below). Use of the $(RYS)$ integral avoids the need for hypotheses of no common (one-sided) discontinuities, but the more familiar Riemann–Stieltjes and refinement Riemann–Stieltjes integrals exist under the same conditions on $\Phi$ and $\Psi$ if there are no such discontinuities. In other words, the full Stieltjes integral $(S) \int f \cdot dh$ exists, and Theorem 3.75 is actually formulated in terms of the $(S)$ integral.

The conditions on $\Phi$ and $\Psi$ may appear somewhat complex or unintuitive (cf. Definitions 3.67 and 3.70). Theorem 3.78 gives a simpler sufficient condition, due to Beurling, but this condition is not necessary (Example 3.80).

*Lacunary sequences*

The necessary and sufficient integrability conditions are formulated in terms of increasing sequences $\{n_k\} = \{n_1, n_2, \dots\}$ of positive integers which may be finite or infinite.

Let $\xi$ be a modulus of continuity. Then $\xi$ is a continuous, nondecreasing, and subadditive function on $[0,\infty)$ with $\xi(0) = 0$. By (3.32), the limit $0 < S := \lim_{u \downarrow 0} u^{-1}\xi(u)$ exists. It can be finite or infinite, e.g. for $\xi(u) = u^{1/p}$, $p = 1$ or 2 respectively. Recall that functions in $\mathcal{V}$ are strictly increasing and continuous from $[0,\infty)$ onto itself, and so they have inverses with the same properties. Recall also that $\mathcal{VM}$ is the class of all functions $\Xi \in \mathcal{V}$ such that the inverse $\Xi^{-1}$ is a modulus of continuity. If $\Xi \in \mathcal{VM}$ with inverse $\xi$ then

$$v = o(\xi(v)) \qquad \text{as } v \downarrow 0 \tag{3.81}$$

if and only if

$$\Xi(u) = o(u) \qquad \text{as } u \downarrow 0. \tag{3.82}$$

We begin with the following class of lacunary sequences.

**Definition 3.64.** Let $\xi$ be a modulus of continuity. A finite or infinite increasing sequence $\{n_k\}$ of positive integers is called *$\xi$-lacunary* if for some $C < \infty$, the inequalities

$$\sum_{k=1}^{m} n_k \xi(n_k^{-1}) \le C n_m \xi(n_m^{-1}) \quad \text{and} \quad \sum_{k=m}^{\infty} \xi(n_k^{-1}) \le C\xi(n_m^{-1}) \tag{3.83}$$

hold for each $m = 1, 2, \ldots$ such that $n_m$ is defined.

Every finite sequence $\{n_k\}$ is $\xi$-lacunary for any $\xi$.

**Proposition 3.65.** *Let $\xi$ be a modulus of continuity. There exists an infinite $\xi$-lacunary sequence $\{n_k\}$ if and only if $\xi$ satisfies* (3.81).

*Proof.* Suppose (3.81) holds. Let $n_1 := 1$. Given increasing positive integers $n_1 < \cdots < n_m$, let $\delta_m := 1/\sum_{k=1}^{m} n_k \xi(1/n_k)$. There is an $n_{m+1}$ large enough so that $1/n_{m+1} \le \delta_m \xi(1/n_{m+1})$ by (3.81), and $\xi(1/n_{m+1}) \le \xi(1/n_m)/2$ since $\xi(v) \downarrow 0$ as $v \downarrow 0$. It follows inductively that the inequalities (3.83) both hold with $C = 2$ in each.

Conversely, suppose (3.81) does not hold. Then the limit $S > 0$ is finite. Suppose a $\xi$-lacunary sequence $\{n_k\}$ exists. Then in the left inequality in (3.83), as $m \to \infty$, the left side is asymptotic to $mS$ and the right side to $CS$, a contradiction, completing the proof. □

A sequence $\{n_k\}$ is called *lacunary in the sense of Hadamard* or *Hadamard-lacunary* if for some $\theta > 1$,

$$n_{k+1}/n_k \ge \theta \qquad \text{for each } k \ge 1. \tag{3.84}$$

If for some $0 < \alpha < 1$, $\xi(u) = u^\alpha$, $u \ge 0$, then each Hadamard-lacunary sequence is also $\xi$-lacunary. To state a partial converse, the supremum of the numbers $\theta$ for which the condition (3.84) holds will be called the *degree of lacunarity* of the given sequence $\{n_k\}$.

**Lemma 3.66.** *For each $\mu > 1$ and modulus of continuity $\xi$, any infinite $\xi$-lacunary sequence can be decomposed into a finite number of Hadamard-lacunary sequences in each of which the degree of lacunarity is at least $\mu$.*

*Proof.* Let $\eta = \{n_k\}$ be an infinite $\xi$-lacunary sequence. It is enough to prove the lemma for some value of $\mu > 1$. Indeed if $\mu$ is arbitrary and (3.84) holds for some $\theta > 1$, then let $l$ be the least positive integer such that $\theta^l \ge \mu$. Divide the sequence $\{n_k\}$ into $l$ sequences $\eta^i = \{n_{i+kl}\}_{k=0}^{\infty}$, $i = 1, \ldots, l$. Then each member of $\eta$ belongs to one and only one of the sequences $\eta^i$, and for each $i = 1, \ldots, l$, $n_{i+(k+1)l}/n_{i+kl} \ge \theta^l \ge \mu$ for $k = 0, 1, \ldots$, proving the claim. Now consider the sequence of intervals $\Delta(m) := [2^m, 2^{m+1})$, $m = 0, 1, \ldots$. There exists a positive integer $l$ such that each $\Delta(m)$ contains at most $l$ members of the sequence $\eta$. Suppose not, that is, there exists a subsequence $m_i \uparrow \infty$ such that each $\Delta(m_i)$ contains $l_i$ members of $\eta$ and $l_i \uparrow \infty$. Then letting $M_i := \max\{k\colon\ n_k \in \Delta(m_i)\}$, we have

$$\begin{aligned}\sum_{k=1}^{M_i} n_k\xi(n_k^{-1}) \geq \sum_{n_k\in\Delta(m_i)} n_k\xi(n_k^{-1}) &\geq l_i \min_{n_k\in\Delta(m_i)} n_k\xi(M_i^{-1})\\ &\geq 2^{-1}l_iM_i\xi(M_i^{-1})\end{aligned}$$

for each $i$. This contradiction shows that $\eta$ can be decomposed into $2l$ Hadamard-lacunary sequences each of which has degree of lacunarity at least 2, proving the lemma. □

**Definition 3.67.** Let $\phi$ and $\psi$ be moduli of continuity. We say that an increasing infinite sequence $\eta = \{n_k\}$ of positive integers is a $W(\phi, \psi)$*-sequence* if

$$W(\phi, \psi; \eta) := \inf\big\{W(\phi, \psi; k)\colon\ k \geq 1\big\} > 0,$$

where for each $k \geq 1$,

$$W(\phi, \psi; k) := \max\Big\{\frac{\phi(n_{k+1}^{-1})}{\phi(n_k^{-1})}, \frac{\psi(n_{k+1}^{-1})}{\psi(n_k^{-1})}, \frac{n_k\phi(n_k^{-1})}{n_{k+1}\phi(n_{k+1}^{-1})}, \frac{n_k\psi(n_k^{-1})}{n_{k+1}\psi(n_{k+1}^{-1})}\Big\}.$$

If (3.81) fails for $\xi = \phi$ or for $\xi = \psi$ then $W(\phi, \psi; \eta) > 0$ for any sequence $\eta = \{n_k\}$ since in this case for some constants $c > 0$ and $C < \infty$, $c \leq u^{-1}\xi(u) \leq C$ for all $0 < u \leq n_1^{-1}$ by (3.32). Next we define a class of sequences with a slightly stronger property than that of $W(\phi, \psi)$-sequences.

**Definition 3.68.** Let $\phi$ and $\psi$ be moduli of continuity. For each $k \geq 1$, let

$$W_o(\phi, \psi; k) := \max\Big\{\frac{\phi(n_{k+1}^{-1})}{\phi(n_k^{-1})}, \frac{n_k\psi(n_k^{-1})}{n_{k+1}\psi(n_{k+1}^{-1})}\Big\} \quad \text{if } k \text{ is odd} \tag{3.85}$$

and

$$W_e(\phi, \psi; k) := \max\Big\{\frac{\psi(n_{k+1}^{-1})}{\psi(n_k^{-1})}, \frac{n_k\phi(n_k^{-1})}{n_{k+1}\phi(n_{k+1}^{-1})}\Big\} \quad \text{if } k \text{ is even.} \tag{3.86}$$

We say that an increasing infinite sequence $\eta = \{n_k\}$ of positive integers is a $W'(\phi, \psi)$*-sequence* if $n_1 = 1$ and for some finite constants $c_o$ and $c_e$,

$$W_o(\phi, \psi; \eta) := \inf\big\{W_o(\phi, \psi; k)\colon\ k \text{ is odd}\big\} = 1/c_o \tag{3.87}$$

and

$$W_e(\phi, \psi; \eta) := \inf\big\{W_e(\phi, \psi; k)\colon\ k \text{ is even}\big\} = 1/c_e. \tag{3.88}$$

It is clear that each infinite $W'(\phi, \psi)$-sequence $\eta$ also is a $W(\phi, \psi)$-sequence and we have the bound $W(\phi, \psi; \eta) \geq 1/\max\{c_o, c_e\}$.

**Lemma 3.69.** *Let $\phi$ and $\psi$ be moduli of continuity, and let $\eta = \{n_k\}$ be an infinite $W(\phi, \psi)$-sequence with $C := W(\phi, \psi; \eta)$. Then for each $k \geq 1$ and any $u \in [n_{k+1}^{-1}, n_k^{-1}]$,*

$$u^{-1}\phi(u)\psi(u) \leq 2C^{-1} \max\left\{n_k\phi(n_k^{-1})\psi(n_k^{-1}), n_{k+1}\phi(n_{k+1}^{-1})\psi(n_{k+1}^{-1})\right\}. \quad (3.89)$$

*In particular, if $\lim_{k\to\infty} n_k\phi(n_k^{-1})\psi(n_k^{-1}) = 0$ then $\phi(u)\psi(u) = o(u)$ as $u \downarrow 0$. Moreover, if in addition $\phi$ and $\psi$ are concave then the factor 2 in (3.89) is unnecessary.*

*Proof.* Recall that for $\xi = \phi$ or $\psi$, $\xi(u) \uparrow$ and by (3.31), $v^{-1}\xi(v) \leq 2u^{-1}\xi(u)$ for $0 < u < v$. Let $n_{k+1}^{-1} \leq u \leq n_k^{-1}$ for some $k \geq 1$. Then letting $h_j := n_j^{-1}$ for each $j$, we have $\xi(u) \leq \xi(h_k)$ and $u^{-1}\xi(u) \leq 2n_{k+1}\xi(h_{k+1})$. To bound $u^{-1}\phi(u)\psi(u)$, since $W(\phi, \psi; k) \geq C$ for each $k$, we have four cases:

$$\begin{aligned} u^{-1}(\phi\psi)(u) &\leq 2n_{k+1}\phi(h_k)\psi(h_{k+1}) \\ &\leq \begin{cases} 2C^{-1}n_{k+1}(\phi\psi)(h_{k+1}) & \text{if } \phi(h_{k+1}) \geq C\phi(h_k), \\ 2C^{-1}n_k(\phi\psi)(h_k) & \text{if } n_k\psi(h_k) \geq Cn_{k+1}\psi(h_{k+1}), \end{cases} \end{aligned}$$

and

$$\begin{aligned} u^{-1}(\phi\psi)(u) &\leq 2n_{k+1}\phi(h_{k+1})\psi(h_k) \\ &\leq \begin{cases} 2C^{-1}n_k(\phi\psi)(h_k) & \text{if } n_k\phi(h_k) \geq Cn_{k+1}\phi(h_{k+1}), \\ 2C^{-1}n_{k+1}(\phi\psi)(h_{k+1}) & \text{if } \psi(h_{k+1}) \geq C\psi(h_k). \end{cases} \end{aligned}$$

Therefore (3.89) holds, proving the first part of the lemma. The last part of the lemma follows by Proposition 3.14. □

Next we define special sequences depending on moduli of continuity $\phi$ and $\psi$. They will be shown to be $\phi$- and $\psi$-lacunary for suitable $\phi$ and $\psi$ in Proposition 3.71.

**Definition 3.70.** Let $\phi$ and $\psi$ be moduli of continuity. Define a sequence $\eta_0 = \eta_0(\phi, \psi) = \{n_k\}$, finite or infinite, recursively as follows. Let $n_1 := 4^s$ for some integer $s \geq 0$. If $n_l$ has been defined for $1 \leq l \leq k$, where $k$ is an odd integer, then let

$$n_{k+1} := \min\left\{4^r\colon\, r \in \mathbb{N},\, 4\phi(4^{-r}) \leq \phi(n_k^{-1}),\, 4^r\psi(4^{-r}) \geq 4n_k\psi(n_k^{-1})\right\} \quad (3.90)$$

provided the right side is defined. If $k$ is an even integer then let

$$n_{k+1} := \min\left\{4^r\colon\, r \in \mathbb{N},\, 4\psi(4^{-r}) \leq \psi(n_k^{-1}),\, 4^r\phi(4^{-r}) \geq 4n_k\phi(n_k^{-1})\right\} \quad (3.91)$$

provided the right side is defined. The recursive construction ends at a finite $n_k$ whenever $n_{k+1}$ is undefined. Then we call $\{n_k\}_{k\geq 1}$ a *D'yačkov sequence* for $\phi$, $\psi$. If $s = 0$, so $n_1 = 1$, we call it *the* D'yačkov sequence $\eta_0(\phi, \psi)$.

A D'yačkov sequence $\{n_k\}$ is easily seen to be infinite if (3.81) holds for $\xi = \phi$ and $\xi = \psi$. Conversely, suppose that a D'yačkov sequence $\{n_k\}$ is infinite. Then by (3.31) and (3.90), if $k \geq 1$ is odd then

$$n_{k+1}\psi(n_{k+1}^{-1}) \geq 2n_{k-1}\psi(n_{k-1}^{-1}) \geq \cdots \geq 2^{(k+1)/2}\psi(1),$$

and so (3.81) holds for $\xi = \psi$. If $k \geq 2$ is even then $n_{k+1}\phi(n_{k+1}^{-1}) \geq 2^{k/2}\phi(1)$ by (3.31) and (3.91), showing that (3.81) holds for $\xi = \phi$. Therefore (3.90) and (3.91) are defined for each $k \geq 1$ if and only if (3.81) holds for $\xi = \phi$ and $\xi = \psi$.

The following shows that $\eta_0(\phi, \psi)$ is a $\phi$- and $\psi$-lacunary sequence, and it is a $W(\phi, \psi)$-sequence with $W(\phi, \psi; \eta_0) \geq 1/16$.

**Proposition 3.71.** *Let $\{n_k\}$ be a D'yačkov sequence for some concave moduli of continuity $\phi$, $\psi$ such that (3.81) holds for $\xi = \phi$ and $\xi = \psi$. Then for $m = 1, 2, \ldots,$ and $\xi = \phi$ or $\psi$,*

$$\sum_{k=1}^{m} n_k\xi(n_k^{-1}) \leq \frac{8}{3}n_m\xi(n_m^{-1}) \quad \text{and} \quad \sum_{k=m}^{\infty} \xi(n_k^{-1}) \leq \frac{8}{3}\xi(n_m^{-1}). \tag{3.92}$$

*Also, the D'yačkov sequence $\eta_0 = \{n_k\}$ is a $W'(\phi, \psi)$-sequence with $c_o \leq 16$ and $c_e \leq 16$.*

*Proof.* By the remark following Definition 3.70, the sequence $\eta_0(\phi, \psi) = \{n_k\}$ is infinite. From the definition of $\{n_k\}$, for $k$ odd, we have

$$\phi(n_{k+1}^{-1}) \leq 4^{-1}\phi(n_k^{-1}) \quad \text{and} \quad n_k\psi(n_k^{-1}) \leq 4^{-1}n_{k+1}\psi(n_{k+1}^{-1}).$$

Then by monotonicity of $\phi(u)$ and $u^{-1}\psi(u)$ (Proposition 3.14), the inequalities

$$\phi(n_{k+2}^{-1}) \leq 4^{-1}\phi(n_k^{-1}) \quad \text{and} \quad n_k\psi(n_k^{-1}) \leq 4^{-1}n_{k+2}\psi(n_{k+2}^{-1})$$

hold for all $k = 1, 2, \ldots$. The same inequalities hold with $\phi$ and $\psi$ interchanged. Thus for $\xi = \phi$ or $\psi$,

$$\sum_{k \geq m} \xi(n_k^{-1}) \leq \left[\xi(n_m^{-1}) + \xi(n_{m+1}^{-1})\right] \sum_{j=0}^{\infty} 4^{-j} \leq \frac{8}{3}\xi(n_m^{-1}),$$

proving the second inequality in (3.92). For the first one, we have

$$\sum_{k \leq m} n_k\xi(n_k^{-1}) \leq \left[n_m\xi(n_m^{-1}) + n_{m-1}\xi(n_{m-1}^{-1})\right] \sum_{j=0}^{\infty} 4^{-j} \leq \frac{8}{3}n_m\xi(n_m),$$

proving (3.92).

To show that $\{n_k\}$ is a $W'(\phi, \psi)$-sequence, we have by definition of $\{n_k\}$, if $k$ is odd, that either $\phi(4n_{k+1}^{-1}) > 4^{-1}\phi(n_k^{-1})$, and then by concavity (Proposition 3.14)

$$\phi(n_{k+1}^{-1}) \geq 4^{-1}\phi(4n_{k+1}^{-1}) > 16^{-1}\phi(n_k^{-1}),$$

or $4^{-1}n_{k+1}\psi(4n_{k+1}^{-1}) < 4n_k\psi(n_k^{-1})$, in which case since $\psi$ is increasing,

$$n_{k+1}\psi(n_{k+1}^{-1}) < 16n_k\psi(n_k^{-1}).$$

For $k$ even, we have the same alternative with $\phi$ and $\psi$ interchanged. Thus Proposition 3.71 is proved. □

*An inequality for Riemann–Stieltjes sums*

An inequality to be given in Proposition 3.73 will be used to prove existence of the full Stieltjes integral (Theorem 3.75). The proof is based on constructing a special partition of a discrete triangle formed by indices: for an integer $n \geq 1$, let

$$D_n := \{(i,j) \in \mathbb{N} \times \mathbb{N}\colon\ 1 \leq i \leq n,\ 1 \leq j \leq i\}. \tag{3.93}$$

**Lemma 3.72.** *Let $F$ and $H$ be nondecreasing functions on an interval $[a,b]$ with $a < b$, such that $F(a) = H(a) = 0$ and $0 < \chi(b) \leq 1$ for $\chi = F$ or $H$. Let $\{t_i\}_{i=0}^n$ be a partition of $[a,b]$ such that* $\max\{F(t_i) - F(t_{i-1}), H(t_i) - H(t_{i-1})\} > 0$ *for $i = 1,\ldots,n$, and let $\{n_k\}_{k\geq 1}$ be an increasing sequence of positive integers with $n_1 = 1$. Let $k_0 := \max\{k \geq 1\colon \min\{F(b),H(b)\} \leq n_k^{-1}\}$. Then there exists a doubly indexed sequence $R = \{R_{k,m}\colon k \geq 0,\ m \geq 1\}$ with the following properties: $R$ contains exactly $n$ nonempty sets. The nonempty $R_{k,m}$ have $k \geq k_0$, are disjoint, and are of the form*

$$R_{k,m} = R(r_1,r_2,r_3) := \{(i,j) \in D_n\colon\ r_1 \leq j \leq r_2 \leq i \leq r_3\} \tag{3.94}$$

*for some integers $r_l = r_l(k,m)$, $l = 1,2,3$, with $1 \leq r_1 \leq r_2 \leq r_3 \leq n$. Also,*

*(a) $D_n = \cup_{k\geq k_0}\cup_{m\geq 1} R_{k,m}$;*
*(b) if $k$ is odd and $R_{k,m}$ is nonempty then*

$$H(t_{r_3}) - H(t_{r_2-1}) \geq n_{k+1}^{-1} \qquad \text{and} \qquad F(t_{r_2}) - F(t_{r_1-1}) \leq n_k^{-1}; \tag{3.95}$$

*(c) if $k$ is even and $R_{k,m}$ is nonempty then*

$$F(t_{r_2}) - F(t_{r_1-1}) \geq n_{k+1}^{-1} \qquad \text{and} \qquad H(t_{r_3}) - H(t_{r_2-1}) \leq n_k^{-1}.$$

*Proof.* A sequence $\{R_{k,m}\colon k \geq k_0,\ m \geq 1\}$ with the stated properties will be constructed recursively. When the recursive construction ends, any $R_{k,m}$ not yet defined are defined to be empty. Let $R_{k,m} := \emptyset$ for $0 \leq k \leq k_0 - 1$ and $m \geq 1$. By definition of $k_0$,

$$1/n_{k_0+1} < \min\{F(b),H(b)\} \leq 1/n_{k_0}. \tag{3.96}$$

The sets $R_{k,m}$ will be defined recursively in $k \geq k_0$ and for such $k$, recursively in $m \geq 1$ until the recursion ends for that $k$, then until the recursion in $k$ ends.

Suppose that for some $k \geq k_0$, $R_{l,m}$ has been defined for each $0 \leq l \leq k-1$, as is true for $k = k_0$ with each $R_{l,m} = \emptyset$. Let $W_{k-1} := \cup_{l=0}^{k-1} \cup_{m \geq 1} R_{l,m}$. If the set $\{(i,i)\colon 1 \leq i \leq n\} \setminus W_{k-1}$ is empty, the recursive construction ends. Assume that this set is not empty. For $\chi = F$ or $\chi = H$, let $\Delta_i \chi := \chi(t_i) - \chi(t_{i-1})$, $i = 1, \ldots, n$. Then define $R_{k,m}$, $m = 1, 2, \ldots$, depending on whether $k$ is odd or even as follows:

(1) Suppose that $k$ is odd. Given $\nu = 1, \ldots, n$ and $k$ let

$$u(\nu, k) \;:=\; \min\{r : \; \nu < r \leq n, \; (r, r) \in W_{k-1}\},$$

or $u(\nu, k) \;:=\; n+1$ if the given set is empty. Then let $\nu_{k,1} := n+1$, and for $m = 1, 2, \ldots$, recursively let

$$\nu_{k,m+1} := \max\Big\{\nu\colon \; 1 \leq \nu \leq \mu := \min\{u(\nu,k), \nu_{k,m}\} - 1, \; (\nu,\nu) \notin W_{k-1} \\ \text{and } n_{k+1}^{-1} \leq \sum_{j=\nu}^{\mu} \Delta_j H\Big\} \tag{3.97}$$

if the right side is defined, and let $\nu_{k,m+1} \;:=\; 0$ otherwise. For each $m = 1, 2, \ldots$, if $\nu_{k,m+1} \geq 1$ then let $R_{k,m}$ be the set of points $(i,j) \in D_n \setminus W_{k-1}$ such that

$$1 \leq j \leq \nu_{k,m+1} \leq i \leq \nu_{k,m} - 1. \tag{3.98}$$

If $\nu_{k,m+1} = 0$ then let $R_{k,m} := \emptyset$ and the recursion in $m$ for the given $k$ ends.

(2) Suppose that $k$ is even. Given $\nu = 1, \ldots, n$ and $k$ let

$$v(\nu, k) \;:=\; \max\{r : \; 1 \leq r < \nu, \; (r, r) \in W_{k-1}\},$$

or $v(\nu, k) \;:=\; 0$ if the given set is empty. Then let $\nu_{k,1} := 0$, and for $m = 1, 2, \ldots$, recursively let

$$\nu_{k,m+1} := \min\Big\{\nu\colon \; \mu := \max\{v(\nu,k), \nu_{k,m}\} + 1 \leq \nu \leq n, \; (\nu,\nu) \notin W_{k-1}, \\ \text{and } n_{k+1}^{-1} \leq \sum_{i=\mu}^{\nu} \Delta_i F\Big\} \tag{3.99}$$

if the right side is defined, and let $\nu_{k,m+1} := n+1$ otherwise. For each $m = 1, 2, \ldots$, if $\nu_{k,m+1} \leq n$ then let $R_{k,m}$ be the set of points $(i,j) \in D_n \setminus W_{k-1}$ such that

$$\nu_{k,m} + 1 \leq j \leq \nu_{k,m+1} \leq i \leq n. \tag{3.100}$$

If $\nu_{k,m+1} = n+1$ then let $R_{k,m} := \emptyset$ and the recursion in $m$ for the given $k$ ends.

If $k = k_0$ is odd, then for all $\nu$, $u(\nu,k) = n+1$, and $\nu_{k,1} = n+1$. So in the definition of $\nu_{k,m+1}$ for $m = 1$, the condition on $\nu$ holds at least for $\nu = 1$ by (3.96), and $R_{k,1} \neq \emptyset$. Symmetrically, $R_{k,1} \neq \emptyset$ for $k = k_0$ even.

For $k \geq k_0$, we claim and will prove by induction on $k$ three statements $(i)$, $(ii)$, and $(iii)$ as follows:

$(i)$ The set $D_n \setminus W_{k-1}$ is a union of sets $\{(i,j)\colon\ q_1 \le j \le i \le q_2\}$.

$(ii)$ The nonempty sets $R_{k,m}$ are rectangles of the form (3.94), with $r_2 = \nu_{k,m+1}$,

$$r_1 = r_1(k,m) = 1 + \max\left\{s < r_2 :\ (s,s) \in R'_{k,m-1} \cup W_{k-1}\right\}, \quad (3.101)$$

or $r_1 = 1$ if no such $s$ exists, where $R'_{k,w} = \emptyset$ for $k$ odd or $w = 0$, and otherwise equals $R_{k,w}$; and

$$r_3 = r_3(k,m) = \min\left\{s > r_2 :\ (s,s) \in R''_{k,m-1} \cup W_{k-1}\right\} - 1, \quad (3.102)$$

or $r_3 = n$ if no such $s$ exists, where $R''_{k,w} = \emptyset$ for $k$ even or $w = 0$, and otherwise equals $R_{k,w}$.

$(iii)$ For $r = 1,\ldots,n$ let $Q_r = \{(i,j) :\ 1 \le j \le r \le i \le n\}$. There is a set $S(k) \subset \{1,\ldots,n\}$ such that $W_k = \bigcup_{r\in S(k)} Q_r$.

To begin the induction, for $k = k_0$, statements $(i)$ and $(ii)$ both simplify because $W_{k_0-1} = \emptyset$. Thus $(i)$ holds for one set with $q_1 = 1$ and $q_2 = n$. For $(ii)$, $R_{k_0,m}$, if nonempty, as it is at least for $m = 1$, is either a set (3.98) for $k_0$ odd, or a set (3.100) for $k_0$ even. In either case it has the form of a discrete rectangle $R(r'_1, r'_2, r'_3)$ as in (3.94) for some integers $1 \le r'_1 \le r'_2 \le r'_3 \le n$ depending on $m$ with upper left corner $(r'_2, r'_2)$ and $r'_2 = \nu_{k,m+1} = r_2$ as stated in $(ii)$. These rectangles are disjoint for different $m$. To verify that $r'_1 = r_1$ as in (3.101) and $r'_3 = r_3$ as in (3.102), note that $(s,s) \in R_{k,m-1}$ if and only if $m \ge 2$ and $s = \nu_{k,m}$ and apply the definitions of $R'_{k,m}$, $R''_{k,m}$, and $R_{k,m}$ for $k$ even and odd. So $(ii)$ holds for $k = k_0$. Now $(iii)$ for $k = k_0$ follows easily, with $S(k_0) = \{\nu_{k_0,m}\}_{m\ge 2}$.

Next for the induction step, given $k > k_0$ and assuming $(i)$, $(ii)$, and $(iii)$ hold with $k-1$ in place of $k$, they will be proved for $k$. We first prove $(i)$ for $k$. Applying $(iii)$ for $k-1$, we have $S(k-1) = \{n_{k-1,t}\}_{t=1}^{N(k-1)}$ for some $N(k-1)$ and $n_{k-1,1} < n_{k-1,2} < \cdots < n_{k-1,N(k-1)}$. Let $n_{k-1,0} := 0$ and $n_{k-1,N(k-1)+1} := n+1$. If $S(k-1)$ is all of $\{1,\ldots,n\}$, the recursion has ended and there is no problem. Otherwise, there exists some $t = 1,\ldots,N(k-1)+1$ such that $n_{k-1,t} - n_{k-1,t-1} \ge 2$. It is then easily seen that $(i)$ holds for $k$ with, for each such $t$, $(q_1, q_2) = (n_{k-1,t-1} + 1, n_{k-1,t} - 1)$. Thus we have $(i)$ for $k$. Suppose $(i)$ holds for a given value of $k > k_0$ and $(iii)$ for $k-1$. We want to prove $(ii)$ for $k$. If $R_{k,1} = \emptyset$ then $(ii)$ holds vacuously, so suppose some $R_{k,m}$ are nonempty. Let $k$ be odd. If $R_{k,m}$ is nonempty then $r_2 = r_2(k,m) = \nu_{k,m+1} \ge 1$ and there is a unique $t \in \{1,\ldots,N(k-1)+1\}$ such that $n_{k-1,t-1} < r_2 < n_{k-1,t}$. In this case by the definition of $R_{k,m}$ we will have $r_1(k,m)$ as in (3.94) equal to $n_{k-1,t-1} + 1$, which confirms (3.101) since $R'_{k,m-1} = \emptyset$. Also, $r_3(k,1) = n_{k-1,t} - 1$, agreeing with (3.102). For $m \ge 2$ we will have $r_3(k,m) = \min(n_{k-1,t}, \nu_{k,m}) - 1$ which also agrees with (3.102). Thus $(ii)$ holds for $k$ odd. A symmetric proof applies for $k$ even, so the induction step for $(ii)$ is done.

Next, we want to complete the induction step by proving $(iii)$ for a $k > k_0$, assuming $(i)$ and $(ii)$ for $k$ and $(iii)$ for $k-1$. Specifically, $(iii)$ will be proved for $S(k) = S(k-1) \cup T(k)$ where $T(k) := \{r_2(k,m)\}_{m\geq 1}$. Let $k$ be odd. If $\nu_{k,2} = 0$, then $S(k) = S(k-1)$ and there is no problem. Let $S(k,0) := S(k-1)$ and recursively if $\nu_{k,m+1} \geq 1$ let $S(k,m) := S(k,m-1) \cup \{\nu_{k,m+1}\}$; otherwise the recursion ends with $S(k,m)$ defined for a largest $m = m(k)$. We will prove by induction on $M = 0, 1, \ldots, m(k)$ that

$$U_{k,M} := W_{k-1} \cup \bigcup_{1\leq m\leq M} R_{k,m} = V_{k,M} := \bigcup_{r\in S(k,M)} Q_r. \tag{3.103}$$

For $M = 0$ this follows from $(iii)$ for $k-1$ as assumed. For the induction step, let $r_2 := r_2(k,M) = \nu_{k,M+1}$. Clearly $R_{k,M} \subset Q_{r_2}$, so $U_{k,M} \subset V_{k,M}$. To prove the converse inclusion, let $(i,j) \in Q_{r_2}$, i.e. $1 \leq j \leq r_2 \leq i \leq n$. Recalling that $k$ is odd, if $j < r_1 = r_1(k,M)$ then $j \leq r_1 - 1 = n_{k-1,t-1} \leq i \leq n$ by (3.101). Thus $(i,j) \in W_{k-1} \subset U_{k,M}$ by $(iii)$ for $k-1$, so assume $j \geq r_1$. By (3.102), if $i > r_3 = r_3(k,M)$ there are two possibilities. If $r_3 = n_{k-1,t} - 1$ then $j \leq n_{k-1,t} \leq i \leq n$ so again $(i,j) \in U_{k,0} \subset U_{k,M}$. Or if $r_3 = \nu_{k,M} - 1$, which implies $M \geq 2$, and $i > r_3$, then $j \leq \nu_{k,M} \leq i \leq n$ so

$$(i,j) \in Q_{\nu_{k,M}} = Q_{r_2(k,M-1)} \subset U_{k,M-1}$$

by the induction hypothesis in $M$, and $U_{k,M-1} \subset U_{k,M}$. So we are left with $r_1 \leq j \leq r_2 \leq i \leq r_3$. Then by $(ii)$ for $k$, $(i,j) \in R_{k,M} \subset U_{k,M}$, finishing the inductive proof of (3.103) for $k$ odd. There is a symmetric proof for $k$ even. For $M = m(k)$ we then infer $(iii)$ for $k$, completing the inductive proof of $(i)$, $(ii)$, and $(iii)$ up through the largest $k$ for which $R_{k,1}$ is nonempty, as desired.

We claim that each $(i,i) \in D_n$ is in $\cup_{l\geq 0} \cup_{m\geq 1} R_{l,m}$. Suppose $(r,r) \in D_n \setminus W_{k-1}$ for some $k \geq k_0$ and either $\Delta_r F \geq n_{k+1}^{-1}$ with $k$ even (Case 1) or $\Delta_r H \geq n_{k+1}^{-1}$ with $k$ odd (Case 2). In Case 1, $\nu_{k,m}$ is strictly increasing in $m$ and $\nu_{k,m} < r \leq n$ for $m = 1$, so there is a largest such $m$, with $\nu_{k,m} < r \leq \nu_{k,m+1}$. In fact by definition of $\nu_{k,m+1}$ it equals $r$, so $(r,r) \in R_{k,m}$. There is a symmetric proof in Case 2. Therefore each pair $(i,i) \in D_n$ is in some $R_{l,m}$, because $n_k \to \infty$ and either $\Delta_i F > 0$ or $\Delta_i H > 0$, and the claim is proved. At each stage of the construction, if $(i_0,i_0) \in R_{k,m}$ for some $k$ and $m$, then each point of the discrete rectangle $\{(i,j)\colon 1 \leq j \leq i_0 \leq i \leq n\}$ is in $R_{k',m'}$ for some $k' \leq k$ and some $m'$, by $(iii)$ as proved above. Therefore each $(i,j) \in D_n$ is in some $R_{k,m}$, and $(a)$ holds. Since each nonempty $R_{k,m}$ contains exactly one point of the diagonal $\{(r,r)\colon r = 1,\ldots,n\}$, there are exactly $n$ nonempty sets $R_{k,m}$.

For $(b)$, let $k$ be odd, and let $R_{k,m_0}$ be a nonempty rectangle having the form (3.94). Then the first inequality in (3.95) holds by definition of $R_{k,m_0}$, because for $m = m_0$, $r_3(k,m)$, as given by (3.102), equals $\min(u(r_2,k),\nu_{k,m}) - 1$, which is the upper limit of the sum in (3.97), defining $\nu_{k,m+1} = r_2(k,m)$.

The second inequality in (3.95), that is,

$$F(t_{r_2}) - F(t_{r_1-1}) \leq n_k^{-1}, \tag{3.104}$$

will be proved next. Indeed, if $k = k_0$ then (3.104) holds because $F(b) \le n_{k_0}^{-1}$ by definition of $k_0$. Suppose that $k > k_0$, so that $k-1$ is even and $k-1 \ge k_0$. If all $R_{k-1,m}$ are empty then in (3.99) for $k-1$ in place of $k$ and $m = 1$, the given set of $\nu$ is empty. Consider $\nu = r_2 = r_2(k, m_0)$. Then $1 \le \nu \le n$ and $(\nu, \nu) \notin W_{k-1}$. So the given inequality must fail for this $\nu$, and the contrary inequality must hold. Since $\nu_{k-1,1} = 0$, it then follows that $F(t_{r_2}) - F(t_{v(r_2,k-1)}) < n_k^{-1}$. By (3.101) with $R'_{k,m-1} = \emptyset$, we have

$$r_1(k, m_0) - 1 = \max\{s < r_2\colon\ (s, s) \in W_{k-1}\} \ge v(r_2, k-1),$$

and so (3.104) holds if all $R_{k-1,m}$ are empty. Thus suppose that not all $R_{k-1,m}$ are empty, e.g. $R_{k-1,1}$ is not empty. Let $m$ be the largest integer such that $\nu_{k-1,m} < r_2$. Then $r_1(k, m_0) - 1 \ge \nu_{k-1,m} = \max\{v(\nu_{k-1,m}, k-1), \nu_{k-1,m}\}$. Thus (3.104) holds by (3.99) for $k-1$ in place of $k$ and $\nu = r_2$, since $r_2 < \nu_{k-1,m+1}$. So part (*b*) of the lemma is proved. The proof of part (*c*) is symmetric and we omit it. This completes the proof of the lemma. □

For moduli of continuity $\phi, \psi$ and for an increasing sequence $\eta = \{n_k\}$ of positive integers, let

$$\Lambda(\phi, \psi; \eta) := \phi(1)\psi(1) + \sum_{k=1}^{\infty} n_k \phi(n_k^{-1})\psi(n_k^{-1}) \le +\infty. \tag{3.105}$$

For $\Xi \in \mathcal{CV}$ and for a Banach-space-valued function $g$ on an interval $[a, b]$ with bounded $\Xi$-variation on $[a, b]$, let $N_\Xi(g; [a, b]) := g/\|g\|_{[a,b];(\Xi)}$ if $\|g\|_{[a,b];(\Xi)} \ne 0$ and let $N_\Xi(g; [a, b]) := 0$ otherwise.

**Proposition 3.73.** *Assume* (1.14) *and* $a < b$. *Let* $\Phi, \Psi \in \mathcal{CV}$ *be such that* (3.82) *holds for* $\Xi = \Phi$ *and* $\Xi = \Psi$, *let* $f \in \mathcal{W}_\Phi([a, b]; X)$ *with* $\check{f} := N_\Phi(f; [a, b])$, *and let* $h \in \mathcal{W}_\Psi([a, b]; Y)$ *with* $\check{h} := N_\Psi(h; [a, b])$. *Let* $\eta = \{n_k\}_{k\ge1}$ *be an infinite* $W'(\Phi^{-1}, \Psi^{-1})$*-sequence such that* (3.87) *and* (3.88) *hold. Then for any* $u, v$ *with* $a \le u < v \le b$, *and any partition* $\{t_i\}_{i=0}^n$ *of* $[u, v]$,

$$\Big\| \sum_{i=1}^{n} [f(t_i) - f(t_0)] \cdot [h(t_i) - h(t_{i-1})] \Big\| \tag{3.106}$$

$$\le \theta \Lambda(\Phi^{-1}, \Psi^{-1}; \eta) \|f\|_{[a,b],(\Phi)} \|h\|_{[a,b],(\Psi)} \{c_e v_\Phi(\check{f}; [u, v]) + c_o v_\Psi(\check{h}; [u, v])\},$$

*where* $\theta \le 1$. *Moreover, if* $\Lambda(\Phi^{-1}, \Psi^{-1}; \eta) < \infty$ *then for any* $\epsilon > 0$ *there is a* $\delta > 0$ *such that* $\theta < \epsilon$ *provided* $0 < \min\{v_\Phi(\check{f}; [u, v]), v_\Psi(\check{h}; [u, v])\} < \delta$.

*Proof.* It is enough to prove the proposition when $\Lambda(\Phi^{-1}, \Psi^{-1}; \eta) < \infty$ and $\|f\|_{(\Phi)} = \|h\|_{(\Psi)} = 1$, in which case $f = \check{f}$ and $h = \check{h}$, if we agree that (3.106) holds when it becomes $0 \le 0 \cdot (+\infty)$. For $t \in [u, v]$, let $F(t) := v_\Phi(f; [u, t])$ and $H(t) := v_\Psi(h; [u, t])$. Then $F(u) = H(u) = 0$. Since the left side of

(3.106) is zero if $f$ or $h$ is constant on $[u,v]$, we can assume that $F(v)>0$ and $H(v)>0$. By $(b)$ of Theorem 3.7, we have $F(v)\le v_\Phi(f;[a,b])\le 1$ and $H(v)\le v_\Psi(h;[a,b])\le 1$. For $\chi = f$, $h$, $F$, or $H$, let $\Delta_i\chi := \chi(t_i)-\chi(t_{i-1})$ for $i=1,\dots,n$. Suppose that for some $1\le i\le n$, $f$ and $h$ are constant on $[t_{i-1},t_i]$, so $\Delta_i f$, $\Delta_i h$, $v_\Phi(f;[t_{i-1},t_i])$, and $v_\Psi(h;[t_{i-1},t_i])$ are all zero. Then, we can write $f$ and $h$ as $f_1\circ G$ and $h_1\circ G$ respectively where $G$ is a nondecreasing continuous function from $[a,b]$ onto itself, strictly increasing except on $[t_{i-1},t_i]$ where it is constant, and replace $t_k$ by $u_j$ with $G(u_j)=t_k$ where $j=k$ for $k=0,\dots,i-1$ and $j=k-1$ for $k=i+1,\dots,n$, so that $j=0,1,\dots,n-1$. Both sides of (3.106) are the same for $f_1,h_1$, and $\{u_j\}_{j=0}^{n-1}$ as for $f,h$, and $\{t_k\}_{k=0}^n$. Iterating this step, either we arrive at a degenerate interval, in which case the proposition holds, or at a case where for each $1\le i\le n$, either $\Delta_iF>0$ or $\Delta_iH>0$. We have to bound the norm of

$$U:=\sum_{i=1}^n\big[f(t_i)-f(t_0)\big]\cdot\big[h(t_i)-h(t_{i-1})\big]=\sum_{(i,j)\in D_n}\Delta_jf\cdot\Delta_ih,$$

with $D_n$ defined by (3.93). To this aim we apply Lemma 3.72 to $F$, $H$, $\{t_i\}_{i=0}^n$ and $\eta=\{n_k\}$ to obtain an integer $k_0\ge 1$ and a sequence $\{R_{k,m}\colon k\ge k_0,\ m\ge 1\}$ of sets with the properties stated in Lemma 3.72. By Lemma 3.72$(a)$, each $(i,j)\in D_n$ is in some $R_{k,m}$, and so $U=\sum_{k=k_0}^\infty U_k$, where for $r_l=r_l(k,m)$ as in (3.94),

$$U_k:=\sum_{m=1}^\infty\sum_{(i,j)\in R_{k,m}}\Delta_jf\cdot\Delta_ih=\sum_{m=1}^\infty\big[f(t_{r_2})-f(t_{r_1-1})\big]\cdot\big[h(t_{r_3})-h(t_{r_2-1})\big].$$

For any $u\le s<t\le v$, by (3.51), we have $\Phi(\|f(t)-f(s)\|)\le F(t)-F(s)$, and the analogous inequalities hold for the functions $\Psi$, $h$, and $H$. Then for $k\ge k_0$ such that not all $R_{k,m}$ are empty, we have

$$\begin{aligned}\|U_k\|&\le\sum_{m=1}^\infty\big\|f(t_{r_2})-f(t_{r_1-1})\big\|\,\big\|h(t_{r_3})-h(t_{r_2-1})\big\|\\&\le\sum_{m=1}^\infty\phi\big(F(t_{r_2})-F(t_{r_1-1})\big)\,\psi\big(H(t_{r_3})-H(t_{r_2-1})\big),\end{aligned}$$

where $\phi$ and $\psi$ are the inverses $\Phi^{-1}$, $\Psi^{-1}$, respectively. To bound this series first suppose that $k$ is odd. Since $\Phi,\Psi\in\mathcal{CV}$, it follows that $\phi$, $\psi$ are concave moduli of continuity, and so $\phi(u)\downarrow$ and $u^{-1}\psi(u)\uparrow$ as $u\downarrow 0$ by Proposition 3.14. By Lemma 3.72$(b)$ and since $\eta$ is a $W'(\Phi^{-1},\Psi^{-1})$-sequence, it then follows that

$$\begin{aligned}\|U_k\|&\le\phi(n_k^{-1})n_{k+1}\psi(n_{k+1}^{-1})\sum_{m=1}^\infty\big[H(t_{r_3})-H(t_{r_2-1})\big]\\&\le c_o\big\{n_k\phi(n_k^{-1})\psi(n_k^{-1})+n_{k+1}\phi(n_{k+1}^{-1})\psi(n_{k+1}^{-1})\big\}H(v)\qquad\text{by (3.87)}\end{aligned}$$

since the intervals $[t_{r_2-1}, t_{r_3}]$ are nonoverlapping for fixed $k$. Second, the same inequality with $F(v)$ instead of $H(v)$ and $c_e$ instead of $c_o$ holds when $k$ is even by (3.88) and Lemma 3.72($c$). Thus

$$\|U\| \le \{c_o H(v) + c_e F(v)\} \sum_{k=k_0}^{\infty} n_k \phi(n_k^{-1}) \psi(n_k^{-1})$$

$$= \theta \Lambda(\phi, \psi; \eta) \{c_e v_\Phi(f; [u,v]) + c_o v_\Psi(h; [u,v])\},$$

where

$$\theta := \Lambda(\phi, \psi; \eta)^{-1} \sum_{k=k_0}^{\infty} n_k \phi(n_k^{-1}) \psi(n_k^{-1}).$$

By the definition of $k_0$ in Lemma 3.72, $n_{k_0+1}^{-1} < \min\{v_\Phi(f; [u,v]), v_\Psi(h; [u,v])\}$. Since $\Lambda(\phi, \psi; \eta) < \infty$, given $\epsilon > 0$ there exists $\delta > 0$ such that $\theta < \epsilon$ provided $0 < \min\{v_\Phi(f; [u,v]), v_\Psi(h; [u,v]) < \delta$, since $k_0 + 1$ and so $k_0$ goes to $+\infty$ as $\delta \to 0$. The proposition is now proved. □

*Full Stieltjes integrability*

Recall Definition 2.41 of the full Stieltjes integral. Now we are ready to prove necessary and sufficient conditions for full Stieltjes integrability for all functions $f, h$ with bounded $\Phi$- and $\Psi$-variation respectively when $\Phi$ and $\Psi$ are convex.

**Definition 3.74.** We say that a pair of functions $\Phi, \Psi \in \mathcal{V}$ form a *Stieltjes pair* if, assuming (1.14), the full Stieltjes integral $(S)\int_a^b f{\cdot}dh$ exists for any $f \in \mathcal{W}_\Phi([a,b]; X)$ and $h \in \mathcal{W}_\Psi([a,b]; Y)$.

Due to bilinearity of the full Stieltjes integral (Theorem 2.72), if $\Phi, \Psi \in \mathcal{V}$ form a Stieltjes pair, then the full Stieltjes integral $(S)\int_a^b f{\cdot}dh$ exists for all $f \in \widetilde{\mathcal{W}}_\Phi([a,b]; X)$ and $h \in \widetilde{\mathcal{W}}_\Psi([a,b]; Y)$.

When $a = b$, the integrals exist by definition and are 0, so in the definition we could assume equivalently that $a < b$.

It will be shown in Corollary 3.77 that $\Phi, \Psi$ are a Stieltjes pair if and only if $\Psi, \Phi$ are. For the $RS$ and $RRS$ integrals and continuous real-valued functions, for example, this would follow from integration by parts, Theorem 2.80. For $f$ and $h$ with common one-sided discontinuities, however, as for the $RYS$ integral in Theorem 2.81, integration by parts is more complicated.

For a modulus of continuity $\phi$, an interval $J$, and a Banach space $X$, let $\mathcal{H}_\phi^0(J; X)$ be the set of all functions $f\colon J \to X$ such that

$$\sup_{t, t+u \in J} \big\|f(t+u) - f(t)\big\| = o(\phi(u)) \qquad \text{as } u \downarrow 0,$$

and let $\mathcal{H}_\phi^0(J) := \mathcal{H}_\phi^0(J; \mathbb{R})$. If $J = [a,b]$ let $\mathcal{H}_\phi^0[a,b] := \mathcal{H}_\phi^0(J)$.

**Theorem 3.75.** *The following conditions about $\Phi$ and $\Psi$ in $\mathcal{CV}$ with inverses $\phi$ and $\psi$, respectively, are equivalent:*

*(a) $\Phi$ and $\Psi$ form a Stieltjes pair;*
*(b) the integral $(RS)\int_0^{2\pi} f\,\mathrm{d}h$ exists for any $f \in \mathcal{H}_\phi^0[0,2\pi]$ and $h \in \mathcal{H}_\psi^0[0,2\pi]$;*
*(c) $\Lambda(\phi,\psi;\eta) < \infty$ for any $\phi$- and $\psi$-lacunary sequence $\eta$;*
*(d) $\Lambda(\phi,\psi;\eta) < \infty$ for some $W(\phi,\psi)$-sequence $\eta$;*
*(e) $\Lambda(\phi,\psi;\eta_0) < \infty$ for the D'yačkov sequence $\eta_0(\phi,\psi)$.*

*Proof.* We begin by showing that all five statements are true if (3.81) fails for $\phi$ or $\psi$. Suppose that $v \neq o(\xi(v))$ as $v \downarrow 0$ for $\xi = \phi$ or $\psi$. Then first, by the remark following Definition 3.70, the D'yačkov sequence $\eta_0(\phi,\psi)$ is finite, and hence $(e)$ holds.

Second, if as $u \downarrow 0$, $u \neq o(\phi(u))$, then $\phi(u)/u$ remains bounded by the remarks following Definition 3.67, and so $u^{-1}\phi(u)\psi(u) \to 0$ as $u \downarrow 0$ since $\psi(u) \downarrow 0$. The same holds, interchanging $\phi$ and $\psi$, if $u \neq o(\psi(u))$. Therefore there exists a sequence $\eta$ such that $\Lambda(\phi,\psi;\eta) < \infty$. Also, again by the remarks following Definition 3.67, each $\{n_k\}$ is a $W(\phi,\psi)$-sequence, and hence $(d)$ holds.

Third, $(c)$ holds because every $\phi$- and $\psi$-lacunary sequence is finite by Proposition 3.65. Finally, if $f$ and $h$ have bounded $\Phi$- and $\Psi$-variations respectively, then one of them is of bounded variation by Proposition 3.39, the other is regulated by Proposition 3.33, and the full Stieltjes integral $(S)\int_a^b f\cdot\mathrm{d}h$ exists in this case by Theorems 2.20, 2.17, and 2.42. So $(a)$ and $(b)$ hold. Therefore for the rest of the proof we can assume that (3.81) holds for $\xi = \phi$ and for $\xi = \psi$.

$(b) \Rightarrow (c)$. Assuming that $(c)$ fails to hold, we construct functions $f \in \mathcal{H}_\phi^0[0,2\pi]$ and $h \in \mathcal{H}_\psi^0[0,2\pi]$ such that the integral $(RS)\int_0^{2\pi} f\,\mathrm{d}h$ does not exist. Suppose that $\Lambda(\phi,\psi;\eta) = \infty$ for some sequence $\eta = \{n_k\}$ which is $\phi$- and $\psi$-lacunary. By Lemma 3.66, one can decompose $\eta$ into finitely many sequences $\eta' = \{n_k'\}$ lacunary in the sense of Hadamard such that for each of them, $20n_k' \le n_{k+1}'$ for all $k \ge 1$, and for at least one of them, $\sum_{k\ge1} n_k'\phi(1/n_k')\psi(1/n_k') = \infty$. Fix $\{n_k'\}$ satisfying the latter condition. For each $k \ge 1$, let $\nu_k := \min\{4^m \colon m \in \mathbb{N},\ n_k' \le 4^m\}$. Then $4^{-1}\nu_k < n_k' \le \nu_k$. We note for later reference that

$$\nu_{k+1} \ge n_{k+1}' \ge 20n_k' \ge 5\nu_k. \tag{3.107}$$

Taking $\chi = \phi$ and $\psi$, since $\chi$ is concave, we have $\chi(1/\nu_k) \le \chi(1/n_k') \le 4\chi(1/\nu_k)$ for each $k \ge 1$.

It will be shown that $\{\nu_k\}$ is $\phi$- and $\psi$-lacunary. By assumption $\eta = \{n_k\}$ is $\phi$-lacunary, and so (3.83) holds for some $C = C(\eta)$. We have that $\eta' = \{n_k'\}$ is a subsequence of $\eta$. Thus for any $m \ge 1$ there is an $l \ge 1$ such that $n_l = n_m'$ and

$$\sum_{j=1}^m n_j'\phi(n_j'^{-1}) \le \sum_{k=1}^l n_k\phi(n_k^{-1}) \le Cn_l\phi(n_l^{-1}) = Cn_m'\phi(n_m'^{-1}).$$

Therefore the first inequality in (3.83) holds with $\eta$ replaced by $\eta'$. Similarly the second inequality in (3.83) holds with $\eta$ replaced by $\eta'$, and so $\eta'$ is $\phi$-lacunary. Next using the inequalities between $\nu_k$ and $n'_k$ stated before and after (3.107), we have

$$\sum_{k=1}^{m} \nu_k \phi(\nu_k^{-1}) \leq 4 \sum_{k=1}^{m} n'_k \phi(n_k'^{-1}) \leq 4Cn'_m \phi(n_m'^{-1}) \leq 4^2 C\nu_m \phi(\nu_m^{-1})$$

for each $m \geq 1$. Similarly it follows that $\sum_{k=m}^{\infty} \phi(\nu_k^{-1}) \leq 4C\phi(\nu_m^{-1})$ for each $m \geq 1$. Thus $\{\nu_k\}$ is $\phi$-lacunary. The same holds with $\phi$ replaced by $\psi$, so $\{\nu_k\}$ is indeed $\phi$- and $\psi$-lacunary.

We also have since $\chi(1/\nu_k) \geq \chi(1/n'_k)/4$ for $\chi = \phi$ or $\psi$ and $\nu_k \geq n'_k$ for each $k$ that

$$\sum_{k \geq 1} \nu_k \phi(1/\nu_k)\psi(1/\nu_k) = \infty.$$

Let $c_k \downarrow 0$ be a sequence of positive numbers such that

$$\sum_{k \geq 1} c_k^2 \nu_k \phi(1/\nu_k)\psi(1/\nu_k) = \infty.$$

Then define two real-valued functions $f$ and $h$ on $[0, 2\pi]$ by

$$f(t) := \sum_{k=1}^{\infty} c_k \phi(\nu_k^{-1}) \cos(\nu_k t) \quad \text{and} \quad h(t) := \sum_{k=1}^{\infty} c_k \psi(\nu_k^{-1}) \sin(\nu_k t)$$

for $0 \leq t \leq 2\pi$. Both series are uniformly convergent by the second inequality in (3.83) for $\xi = \phi$ and $\xi = \psi$. We show next that the functions defined by the two series are in $\mathcal{H}_\phi^0[0, 2\pi]$ and $\mathcal{H}_\psi^0[0, 2\pi]$, respectively. For any $u \in (0, 1/\nu_1]$, let $m$ be the integer such that $1/\nu_{m+1} < u \leq 1/\nu_m$. Then

$$|f(t+u) - f(t)| \leq u \sum_{k \leq m} c_k \nu_k \phi(\nu_k^{-1}) + 2 \sum_{k > m} c_k \phi(\nu_k^{-1}) =: T_1 + T_2 \quad (3.108)$$

for each $u \in (0, 1/\nu_1]$ and $t \in [0, 2\pi - u]$. Given $\epsilon > 0$ choose an integer $k_0 \geq 1$ such that $c_k \leq \epsilon$ for each $k \geq k_0$. Then choose an integer $k_1 \geq k_0$ such that $\nu_{k_0} \phi(\nu_{k_0}^{-1}) \leq \epsilon \nu_k \phi(\nu_k^{-1})$ for each $k \geq k_1$, which is possible due to (3.81) for $\xi = \phi$. Thus if $u \in (0, 1/\nu_{k_1+1}]$, then in (3.108), $m > k_1 \geq k_0$ and by the first inequality in (3.83) twice, once for $m$ there equal to $k_0$ here,

$$T_1 \leq u\big[Cc_1 \nu_{k_0} \phi(\nu_{k_0}^{-1}) + \epsilon C \nu_m \phi(\nu_m^{-1})\big] \leq \epsilon(c_1 + 1)C\phi(u)$$

because $\nu_m \phi(\nu_m^{-1}) \leq \phi(u)/u$ by Proposition 3.14. Using the second inequality in (3.83), it follows that $T_2 \leq 2\epsilon C\phi(u)$. Therefore $f \in \mathcal{H}_\phi^0[0, 2\pi]$. Similarly it follows that $h \in \mathcal{H}_\psi^0[0, 2\pi]$.

To show that the integral $(RS) \int_0^{2\pi} f \, dh$ does not exist, for each positive integer $n$, consider the tagged partition $\tau_n = (\{t_j\}_{j=0}^{4\nu_n}, \{t_j\}_{j=1}^{4\nu_n})$, where $t_j := \pi j/(2\nu_n)$, $j = 0, \ldots, 4\nu_n$. Applying the identities

$$\begin{aligned}\sin 2u - \sin 2v &= 2\sin(u-v)\cos(u+v), \\ 2\cos u \cos v &= \cos(u-v) + \cos(u+v),\end{aligned} \tag{3.109}$$

where the first follows from $\cos\theta = (\mathrm{e}^{\mathrm{i}\theta} + \mathrm{e}^{-\mathrm{i}\theta})/2$, $\sin\theta = (\mathrm{e}^{\mathrm{i}\theta} - \mathrm{e}^{-\mathrm{i}\theta})/(2\mathrm{i})$, we have

$$\begin{aligned}&S_{RS}(f, \mathrm{d}h; \tau_n) \\ &= \sum_{j=1}^{4\nu_n} \sum_{k,l=1}^{\infty} c_k c_l \phi(\nu_k^{-1}) \psi(\nu_l^{-1}) 2 \cos\frac{\pi\nu_k j}{2\nu_n} \sin\frac{\pi\nu_l}{4\nu_n} \cos\frac{\pi(2j-1)\nu_l}{4\nu_n} \\ &= \sum_{k,l=1}^{\infty} c_k c_l \phi(\nu_k^{-1}) \psi(\nu_l^{-1}) \sin\frac{\pi\nu_l}{4\nu_n} \\ &\quad \times \sum_{j=1}^{4\nu_n} \Big\{ \cos\frac{2\pi(\nu_k - \nu_l)j + \pi\nu_l}{4\nu_n} + \cos\frac{2\pi(\nu_k + \nu_l)j - \pi\nu_l}{4\nu_n} \Big\}.\end{aligned}$$

Since each $\nu_k$ is an integer power of 4 and $\nu_{k+1} \geq 5\nu_k$ for every $k \geq 1$ by (3.107), each term with indices $k$ or $l > n$ equals zero. Computing the inner sum by the formula, valid for $0 < t < 2\pi$ and any $M = 1, 2, \ldots$,

$$\sum_{j=1}^{M} \cos(jt + s) = \frac{\sin((M + \frac{1}{2})t + s) - \sin(\frac{1}{2}t + s)}{2\sin\frac{1}{2}t}, \tag{3.110}$$

which again follows from $\cos\theta = (\mathrm{e}^{\mathrm{i}\theta} + \mathrm{e}^{-\mathrm{i}\theta})/2$, $\sin\theta = (\mathrm{e}^{\mathrm{i}\theta} - \mathrm{e}^{-\mathrm{i}\theta})/(2\mathrm{i})$, and then some geometric series, noting that $Mt = 4\nu_n t$ is an integer multiple of $2\pi$, we see that all terms vanish except those with $\nu_k - \nu_l$ for $k = l$, where (3.110) does not apply. Then applying the inequality $(\sin u)/u \geq 2/\pi$ for $0 < u \leq \pi/2$, we get that

$$S_{RS}(f, \mathrm{d}h; \tau_n) = \sum_{k=1}^{n} c_k^2 \phi(\nu_k^{-1}) \psi(\nu_k^{-1}) 2\nu_n \sin\frac{\pi\nu_k}{2\nu_n} \geq 2 \sum_{k=1}^{n} c_k^2 \nu_k \phi(\nu_k^{-1}) \psi(\nu_k^{-1})$$

for each $n \geq 1$. Since the last sum increases without bound as $n \to \infty$, the Riemann–Stieltjes integral for the functions $f$ and $h$ does not exist, proving the implication $(b) \Rightarrow (c)$.

$(c) \Rightarrow (d)$ holds since a D'yačkov sequence is both $\phi$-, $\psi$-lacunary and a $W(\phi, \psi)$-sequence by Proposition 3.71.

$(d) \Rightarrow (e)$. Let $\eta = \{n_k\}$ be a $W(\phi, \psi)$-sequence such that $\Lambda(\phi, \psi; \eta) < \infty$ and let $C := W(\phi, \psi; \eta)$. We have to show that $\Lambda(\phi, \psi; \eta_0) < \infty$ for the D'yačkov sequence $\eta_0(\phi, \psi) = \{m_l\}_{l=1}^{\infty}$. To this aim we use Lemma 3.69 for the sequence $\eta$. Let $I(k) := \{l \in \mathbb{N} \colon n_k \leq m_l < n_{k+1}\}$ for each $k \geq 1$. By Definition 3.70, for $\xi = \phi$ or $\psi$, $4\xi(m_{l+2}^{-1}) \leq \xi(m_l^{-1})$ and $m_{l+2}\xi(m_{l+2}^{-1}) \geq 4m_l\xi(m_l^{-1})$ for each $l \geq 1$. If $M :=$card $I(k) \geq 3$, let $l_1 = l_1(k)$ and $l_2 = l_2(k)$ be respectively the smallest and the largest integer in $I(k)$. Then

$$\frac{\xi(n_{k+1}^{-1})}{\xi(n_k^{-1})} \leq \frac{\xi(m_{l_2}^{-1})}{\xi(m_{l_1}^{-1})} \leq \Big(\frac{1}{4}\Big)^{\lfloor (M-1)/2 \rfloor},$$

where $\lfloor x \rfloor$ is the largest integer $\leq x$, and

$$\frac{n_k \xi(n_k^{-1})}{n_{k+1}\xi(n_{k+1}^{-1})} \leq \frac{m_{l_1}\xi(m_{l_1}^{-1})}{m_{l_2}\xi(m_{l_2}^{-1})} \leq \Big(\frac{1}{4}\Big)^{\lfloor (M-1)/2 \rfloor}.$$

Therefore card $I(k) \leq 2+\log_2(C^{-1})$ for each $k \geq 1$. Since $\phi$ and $\psi$ are concave, this together with Lemma 3.69 gives the bound

$$\sum_{l \geq l_1(1)} m_l \phi(m_l^{-1})\psi(m_l^{-1}) \leq 2C^{-1}(2+\log_2(C^{-1}))\Lambda(\phi,\psi;\eta) < \infty,$$

proving $(e)$.

$(e) \Rightarrow (a)$. Assuming (1.14) and $a < b$, let $f \in \mathcal{W}_\Phi([a,b];X)$ and $h \in \mathcal{W}_\Psi([a,b];Y)$. Due to bilinearity of the full Stieltjes integral (Theorem 2.72) and since if $f$ or $h$ is constant the integral exists, we can and do assume that $\|f\|_{(\Phi)} = \|h\|_{(\Psi)} = 1$. Proposition 3.73 and the Cauchy test will be used to show the existence of the refinement Riemann–Stieltjes and refinement Young–Stieltjes integrals. Next we show that it is enough to consider Riemann–Stieltjes and Young–Stieltjes sums based on tagged partitions with suitable tags. Recall the definition (2.16) of Young–Stieltjes sums $S_{YS}$.

**Lemma 3.76.** *Let $\Phi, \Psi \in \mathcal{V}$ with inverses $\phi$, $\psi$, respectively, be such that*

$$\phi(u)\psi(u) = o(u) \qquad \text{as } u \downarrow 0. \tag{3.111}$$

*Let $f \in \mathcal{W}_\Phi([a,b];X)$ and $h \in \mathcal{W}_\Psi([a,b];Y)$. Given $\epsilon > 0$, there is a partition $\lambda$ of $[a,b]$ such that $(a)$, $(b)$, and $(c)$ hold, where:*

*$(a)$ For any refinement $\kappa$ of $\lambda$ and for tagged Young partitions $\tau' = (\kappa, \xi')$ and $\tau'' = (\kappa, \xi'')$,*

$$R(\kappa,\xi',\xi'') := \big\|S_{YS}(f,\mathrm{d}h;\tau') - S_{YS}(f,\mathrm{d}h;\tau'')\big\| < \epsilon.$$

*$(b)$ For any refinement $\kappa = \{t_j\}_{j=0}^m$ of $\lambda$,*

$$T(\kappa,\{\tau_j'\},\{\tau_j''\}) := \sum_{j=1}^m \big\|S_{RS}(f_j,\mathrm{d}h_j;\tau_j') - S_{RS}(f_j,\mathrm{d}h_j;\tau_j'')\big\| < \epsilon,$$

*where for $j = 1,\dots,m$, $\tau_j' = (\kappa_j,\xi_j')$, $\tau_j'' = (\kappa_j,\xi_j'')$ are tagged partitions of $[t_{j-1},t_j]$ and for $\chi = f$ or $h$, $\chi_j := \chi$ on $(t_{j-1},t_j)$, $\chi_j(t_{j-1}) := \chi(t_{j-1}+)$ and $\chi_j(t_j) := \chi(t_j-)$.*

*$(c)$ If $f$ and $h$ have no common one-sided discontinuities then for any refinement $\kappa$ of $\lambda$ and for tagged partitions $\tau' = (\kappa,\xi')$, $\tau'' = (\kappa,\xi'')$,*

$$Q(\kappa,\xi',\xi'') := \big\|S_{RS}(f,\mathrm{d}h;\tau') - S_{RS}(f,\mathrm{d}h;\tau'')\big\| < \epsilon. \tag{3.112}$$

*Proof.* For $u \in [a,b]$, let $F(u) := v_{\Phi}(f;[a,u])$ and $H(u) := v_{\Psi}(h;[a,u])$. By (3.51), for any $a \le u < s' \le s'' < v \le b$, we have $\Phi(\|f(s') - f(s'')\|) \le F(v-) - F(u+)$. Thus $\|f(s') - f(s'')\| \le \phi(F(v-) - F(u+))$. Similarly, we have $\|h(v-) - h(u+)\| \le \psi(H(v-) - H(u+))$. Let $V := F + H$ and let $\epsilon_0 > 0$. Since $V$ is nondecreasing and finite on $[a,b]$, by Theorem 2.1($b$), there exists a partition $\lambda = \{z_l\}_{l=0}^k$ of $[a,b]$ such that

$$V(z_l-) - V(z_{l-1}+) < \epsilon_0 \qquad \text{for each } l = 1, \dots, k. \tag{3.113}$$

For ($a$), let $\kappa = \{t_i\}_{i=0}^n$ be a refinement of $\lambda$. Let $\xi' = \{s_i'\}_{i=1}^n$ and $\xi'' = \{s_i''\}_{i=1}^n$ be sets of tags with $s_i', s_i'' \in (t_{i-1}, t_i)$ for $i = 1, \dots, n$. By (3.113), it follows that

$$\begin{aligned} R(\kappa, \xi', \xi'') &\le \sum_{i=1}^n \|f(s_i') - f(s_i'')\| \, \|h(t_i-) - h(t_{i-1}+)\| \\ &\le \sum_{i=1}^n \phi\big(V(t_i-) - V(t_{i-1}+)\big)\psi\big(V(t_i-) - V(t_{i-1}+)\big) \\ &\le V(b) \sup_{0<u\le\epsilon_0} u^{-1}\phi(u)\psi(u). \end{aligned}$$

Therefore statement ($a$) holds by assumption (3.111).

For ($b$), let $\kappa = \{t_j\}_{j=0}^m$ be a refinement of $\lambda$, and let $\kappa_j = \{t_{ji}\}_{i=0}^{n_j}$ be a partition of $[t_{j-1}, t_j]$ for $j = 1, \dots, m$. Let $\xi_j' = \{s_{ji}'\}_{i=1}^{n_j}$ and $\xi_j'' = \{s_{ji}''\}_{i=1}^{n_j}$ be two sets of tags for $\kappa_j$, and let $u_{ji} := \min\{V(t_{ji}), V(t_j-)\} - \max\{V(t_{j,i-1}), V(t_{j-1}+)\}$ for $i = 1, \dots, n_j$ and $j = 1, \dots, m$. Then each $u_{ji}$ is less than $\epsilon_0$ by (3.113) since $\kappa$ is a refinement of $\lambda$. As for $f$ and $h$, by (3.51), for $i = 1, \dots, n_j$, we have that each $\|f_j(s_{ji}') - f_j(s_{ji}'')\|$ is bounded by $\phi(u_{ji})$, and each $\|h_j(t_{ji}) - h_j(t_{j,i-1})\|$ is bounded by $\psi(u_{ji})$. Then we have

$$\begin{aligned} T(\kappa, \{\tau_j'\}, \{\tau_j''\}) &\le \sum_{j=1}^m \sum_{i=1}^{n_j} \|f_j(s_{ji}') - f_j(s_{ji}'')\| \, \|h_j(t_{ji}) - h_j(t_{j,i-1})\| \\ &\le \sum_{j=1}^m \sum_{i=1}^{n_j} \phi(u_{ji})\psi(u_{ji}) \le V(b) \sup_{0<u\le\epsilon_0} u^{-1}\phi(u)\psi(u). \end{aligned}$$

Now ($b$) also follows from the assumption (3.111).

To prove ($c$) suppose that $f$ and $h$ have no common one-sided discontinuities. Then by Proposition 3.44, one can choose $\mu = \{u_{l-1}, v_l \colon l = 1, \dots, k\} \subset (a,b)$ such that for each $l = 1, \dots, k$, we have $z_{l-1} < u_{l-1} < v_l < z_l$,

$$\min\big\{\phi\big(F(u_{l-1}) - F(z_{l-1})\big), \psi\big(H(u_{l-1}) - H(z_{l-1})\big)\big\} < \epsilon_0/k,$$

and

$$\min\{\phi(F(z_l)-F(v_l)),\psi(H(z_l)-H(v_l))\}<\epsilon_0/k.$$

Let $\lambda' := \lambda\cup\mu$. Then $(c)$ will be proved, and $(a)$ and $(b)$ still hold, for $\lambda'$ in place of $\lambda$. Let $\kappa=\{t_i\}_{i=0}^n$ be a refinement of $\lambda'$, and let $I$ be the set of indices $i\in\{1,\dots,n\}$ such that either $t_i\in\{z_1,\dots,z_k\}$ or $t_{i-1}\in\{z_0,\dots,z_{k-1}\}$. For $G=F$, $H$ or $V$, let $\Delta_iG:=G(t_i)-G(t_{i-1})$ for $i=1,\dots,n$. Let $\xi'=\{s_i'\}_{i=1}^n$ and $\xi''=\{s_i''\}_{i=1}^n$ be two sets of tags for $\kappa$. Then we have

$$\begin{aligned}
&Q(\kappa,\xi',\xi'')\\
&\quad\le\sum_{i=1}^n\|f(s_i')-f(s_i'')\|\,\|h(t_i)-h(t_{i-1})\|\le\sum_{i=1}^n\phi(\Delta_iF)\psi(\Delta_iH)\\
&\quad\le 2k(\epsilon_0/k)\max\{\phi(F(b)),\psi(H(b))\}+\sum_{i\notin I}\phi(\Delta_iV)\psi(\Delta_iV)\\
&\quad\le 2\epsilon_0\max\{\phi(F(b)),\psi(H(b))\}+V(b)\sup_{0<u\le\epsilon_0}u^{-1}\phi(u)\psi(u),
\end{aligned}\tag{3.114}$$

where the last inequality holds by (3.113). This together with assumption (3.111) proves statement $(c)$ of the lemma. □

Now continuing the proof of Theorem 3.75, by Proposition 3.71, the D'yačkov sequence $\eta_0(\phi,\psi)$ is a $W'(\phi,\psi)$-sequence with $c_o=c_e=16$ (cf. (3.87) and (3.88)). Since $\Lambda(\phi,\psi,\eta_0)<\infty$, by Lemma 3.69, it follows that (3.111) holds. Suppose that $f$ and $h$ have no common one-sided discontinuities. For a tagged partition $\tau$ of the form $(\{t_j\}_{j=0}^m,\{t_{j-1}\}_{j=1}^m)$, consider a tagged refinement of the form $\tau'=(\{s_i\}_{i=0}^n,\{s_i\}_{i=1}^n)$. Thus for each $j=0,\dots,m$, there is an $i(j)\in\{0,\dots,n\}$ such that $s_{i(j)}=t_j$. Then recalling that $\|f\|_{(\Phi)}=\|h\|_{(\Psi)}=1$ and using Proposition 3.73 with $[u,v]=[t_{j-1},t_j]$ for each $j$, we have for some $\theta_j\in(0,1]$,

$$\begin{aligned}
&\big\|S_{RS}(f,\mathrm{d}h;\tau')-S_{RS}(f,\mathrm{d}h;\tau)\big\|\\
&\quad\le\sum_{j=1}^m\Big\|\sum_{i=i(j-1)+1}^{i(j)}\big[f(s_i)-f(s_{i(j-1)})\big]\cdot\big[h(s_i)-h(s_{i-1})\big]\Big\|\\
&\quad\le 16\Lambda(\phi,\psi;\eta_0)\sum_{j=1}^m\theta_j\Big\{v_\Phi(f;[t_{j-1},t_j])+v_\Psi(h;[t_{j-1},t_j])\Big\}\\
&\quad\le 32\Lambda(\phi,\psi;\eta_0)\max_{1\le j\le m}\theta_j,
\end{aligned}\tag{3.115}$$

where the last inequality holds by iterating (3.51), and then (3.9). By the second part of Proposition 3.73, the right side of (3.115) can be made arbitrarily small provided given $\epsilon>0$ one can choose a partition $\{t_j\}_{j=0}^m$ of $[a,b]$ such that for each $j=1,\dots,m$,

$$\min\left\{v_\Phi(f;[t_{j-1},t_j]),v_\Psi(h;[t_{j-1},t_j])\right\}<\epsilon. \tag{3.116}$$

So, let $\epsilon>0$ be given. By Proposition 3.40, there exists a partition $\{z_l\}_{l=0}^k$ of $[a,b]$ such that for each $l=1,\dots,k$,

$$\max\left\{v_\Phi(f;(z_{l-1},z_l)),v_\Psi(h;(z_{l-1},z_l))\right\}<\epsilon. \tag{3.117}$$

Since $f$ and $h$ have no common one-sided discontinuities, by Proposition 3.42, there exists $\mu=\{u_{l-1},v_l\colon l=1,\dots,k\}\subset(a,b)$ such that for each $l=1,\dots,k$, $z_{l-1}<u_{l-1}<v_l<z_l$, we have

$$\min\left\{v_\Phi(f;[z_{l-1},u_{l-1}]),v_\Psi(h;[z_{l-1},u_{l-1}])\right\}<\epsilon,$$

and

$$\min\left\{v_\Phi(f;[v_l,z_l]),v_\Psi(h;[v_l,z_l])\right\}<\epsilon.$$

Therefore if $\{t_j\}_{j=0}^m$ is a refinement of $\lambda\cup\mu$, then (3.116) holds for each $j=1,\dots,m$. Again using (3.115) and Lemma 3.76($c$), we conclude that the Riemann–Stieltjes sums for $f$ and $h$ satisfy the Cauchy test under refinement, and hence $(RRS)\int_a^b f{\cdot}dh$ exists provided $f$ and $h$ have no common one-sided discontinuities.

Now let $f$ and $h$ be any functions in $\mathcal{W}_\Phi([a,b];X)$ and $\mathcal{W}_\Psi([a,b];Y)$, respectively, except that $\|f\|_{(\Phi)}=\|h\|_{(\Psi)}=1$ as before. Let $\tau=(\{t_j\}_{j=0}^m,\{s_j\}_{j=1}^m)$ be a Young tagged partition of $[a,b]$. For each $j=1,\dots,m$, let $\tau_j=(\{t_{ji}\}_{i=0}^{n_j},\{s_{ji}\}_{i=1}^{n_j})$ be a Young tagged partition of $[t_{j-1},t_j]$. Then $\cup_{j=1}^m\tau_j$ is a Young tagged refinement of $\tau$. Also for each $j=1,\dots,m$, let $f_j$ be a function defined on $[t_{j-1},t_j]$ so that it equals $f$ on $(t_{j-1},t_j)$, $f_j(t_{j-1}):=f(t_{j-1}+)$, and $f_j(t_j):=f(t_j-)$. Likewise for each $j=1,\dots,m$, define $h_j$ on $[t_{j-1},t_j]$. Then

$$S_{YS}(f,dh;\tau)-S_{YS}(f,dh;\cup_j\tau_j)=\sum_{j=1}^m\left[f(s_j)-f(t_{j-1}+)\right]{\cdot}[h(t_j-)-h(t_{j-1}+)]$$

$$+\sum_{j=1}^m\left\{f_j(t_{j-1}){\cdot}[h_j(t_j)-h_j(t_{j-1})]-S_{YS}(f_j,dh_j;\tau_j)\right\}=:R+T. \tag{3.118}$$

We will show that the norm of $R+T$ is small for suitable partitions $\tau$. At this stage no assumptions have been made on the partitions $\tau$ or $\tau_j$, $j=1,\dots,m$. A partition $\tau$ will be chosen near the end of the proof. The norm of $R$ will be made small using part ($a$) of Lemma 3.76. To bound the norm of $T$ will be less easy. To this aim Proposition 3.73 and part ($b$) of Lemma 3.76 will be used after decomposing each term of $T$ into three parts as follows. Let $j\in\{1,\dots,m\}$, and let $\kappa_j=\{u_{j,i-1},v_{ji}\}_{i=1}^{n_j}\subset[t_{j-1},t_j]$ be a partition of $[t_{j-1},t_j]$ such that

$$t_{j-1}=u_{j0}<s_{j1}<v_{j1}<t_{j1},\ \ t_{j,n_j-1}<u_{j,n_j-1}<s_{j,n_j}<v_{j,n_j}=t_j, \tag{3.119}$$

and

$$t_{j,i-1} < u_{j,i-1} < s_{ji} < v_{ji} < t_{ji} \quad \text{for} \quad i = 2, \dots, n_j - 1. \tag{3.120}$$

Let $\xi'_j = \{s_{ji}, t_{ji}\}_{i=1}^{n_j-1} \cup \{s_{j,n_j}\}$ and $\xi''_j = \{v_{ji}, u_{ji}\}_{i=1}^{n_j-1} \cup \{v_{j,n_j}\}$. Then $\tau'_j = (\kappa_j, \xi'_j)$ and $\tau''_j = (\kappa_j, \xi''_j)$ are tagged partitions of $[t_{j-1}, t_j]$, and

$$\begin{aligned} f_j(t_{j-1})\cdot[h_j(t_j) - h_j(t_{j-1})] &- S_{YS}(f_j, \mathrm{d}h_j; \tau_j) \\ = \big\{f_j(t_{j-1})\cdot[h_j(t_j) &- h_j(t_{j-1})] - S_{RS}(f_j, \mathrm{d}h_j; \tau''_j)\big\} \\ +\big\{S_{RS}(f_j, \mathrm{d}h_j; \tau''_j) &- S_{RS}(f_j, \mathrm{d}h_j; \tau'_j)\big\} \\ +\big\{S_{RS}(f_j, \mathrm{d}h_j; \tau'_j) &- S_{YS}(f_j, \mathrm{d}h_j; \tau_j)\big\} =: T_{j1} + T_{j2} + T_{j3}. \end{aligned}$$

For $T_{j3}$ suppose that in (3.119) and (3.120) $u_{ji} \downarrow t_{ji}$ and $v_{ji} \uparrow t_{ji}$ for $i = 1, \dots, n_j - 1$. Then since $\Delta^+ h_j(t_{j-1}) = h(t_{j-1}+) - h(t_{j-1}+) = 0$,

$$f_j(s_{j1})\cdot\big[h_j(v_{j1}) - h_j(t_{j-1})\big] \to f_j(s_{j1})\cdot\big[h_j(t_{j1}-) - h_j(t_{j-1}+)\big].$$

For $i = 2, \dots, n_j - 1$, we have

$$f_j(s_{ji})\cdot\big[h_j(v_{ji}) - h_j(u_{j,i-1})\big] \to f_j(s_{ji})\cdot\big[h_j(t_{ji}-) - h_j(t_{j,i-1}+)\big],$$

$$f_j(t_{ji})\cdot\big[h_j(u_{ji}) - h_j(v_{ji})\big] \to f_j(t_{ji})\cdot\big[h_j(t_{ji}+) - h_j(t_{ji}-)\big].$$

Since $\Delta^- h_j(t_j) = h(t_j-) - h(t_j-) = 0$,

$$f_j(s_{j,n_j})\cdot\big[h_j(t_j) - h_j(u_{j,n_j-1})\big] \to f_j(s_{j,n_j})\cdot\big[h_j(t_j-) - h_j(t_{j,n_j-1}+)\big].$$

Thus we have that each $T_{j3} \to 0$ as $u_{ji} \downarrow t_{ji}$ and $v_{ji} \uparrow t_{ji}$ for $i = 1, \dots, n_j - 1$. Now Proposition 3.73 with $\eta = \eta_0$ can be applied $m$ times, once to each sum $T_{j1}$ formed by the functions $f_j$ and $h_j$, with $[u, v] = [t_{j-1}, t_j]$. Since $v_\Phi(f_j; [t_{j-1}, t_j]) = v_\Phi(f; (t_{j-1}, t_j))$ and $v_\Psi(h_j; [t_{j-1}, t_j]) = v_\Psi(h; (t_{j-1}, t_j))$, we then have as in (3.115) for some $\theta_j \in (0, 1]$,

$$\begin{aligned} \sum_{j=1}^{m} \|T_{j1}\| &\le 16\Lambda(\phi, \psi; \eta_0) \sum_{j=1}^{m} \theta_j \Big\{v_\Phi(f; (t_{j-1}, t_j)) + v_\Psi(h; (t_{j-1}, t_j))\Big\} \\ &\le 32\Lambda(\phi, \psi; \eta_0) \max_{1\le j\le m} \theta_j. \end{aligned} \tag{3.121}$$

Now we are ready to specify partitions $\tau$ for which the norm of $R + T$ in (3.118) is small. Let $\epsilon > 0$. By the second part of Proposition 3.73, there exists $\delta > 0$ such that in (3.121), $\max_j \theta_j < \epsilon$ for any partition $\{t_j\}_{j=0}^m$ of $[a, b]$ such that

$$\max_{1\le j\le m} v_\Psi(h_j; [t_{j-1}, t_j]) = \max_{1\le j\le m} v_\Psi(h; (t_{j-1}, t_j)) < \delta. \tag{3.122}$$

By Proposition 3.40, there exists a partition $\lambda$ of $[a, b]$ such that (3.122) holds for any refinement $\{t_j\}_{j=0}^m$ of $\lambda$. We can and do assume that ($a$) and ($b$) of Lemma 3.76 hold for $\lambda$. Let $\tau = (\{t_j\}_{j=0}^m, \{s_j\}_{j=1}^m)$ be a Young tagged partition of $[a, b]$ which is a refinement of $\lambda$, and let $\tau_j = (\{t_{ji}\}_{i=0}^{n_j}, \{s_{ji}\}_{i=1}^{n_j})$ be a Young tagged partition of $[t_{j-1}, t_j]$ for $j = 1, \ldots, m$. For each $j = 1, \ldots, m$ and $i = 1, \ldots, n_j - 1$, letting $u_{ji} \uparrow t_{ji}$ and $v_{ji} \downarrow t_{ji}$, it follows that $\sum_{j=1}^m \|T_{j3}\| < \epsilon$ for some $\kappa_j = \{u_{j,i-1}, v_{ji}\}_{i=1}^{n_j}$ satisfying (3.119) and (3.120). Then $\|R\| < \epsilon$ for $R$ in (3.118) and $\sum_{j=1}^m \|T_{j2}\| < \epsilon$ by $(a)$ and $(b)$ of Lemma 3.76, respectively. Summing the bounds, we have

$$\begin{aligned}\big\|S_{YS}(f, \mathrm{d}h; \tau) - S_{YS}(f, \mathrm{d}h; \cup_j \tau_j)\big\| &\le \|R\| + \sum_{j=1}^m \{\|T_{j1}\| + \|T_{j2}\| + \|T_{j3}\|\} \\ &\le \epsilon + 32\epsilon \Lambda(\phi, \psi; \eta_0) + \epsilon + \epsilon.\end{aligned}$$

Since $\epsilon$ is arbitrary, $(RYS)\int_a^b f \cdot dh$ exists by the Cauchy test under refinement, proving the implication $(e) \Rightarrow (a)$. Since $\mathcal{H}_\phi^0[a,b] \subset \mathcal{W}_\Phi[a,b]$ (by Proposition 3.32 with $G$ the identity function), and likewise for $\psi$ and $\Psi$, clearly $(a)$ implies $(b)$, so the proof of Theorem 3.75 is complete. □

Notice that in Theorem 3.75, condition $(c)$ (or $(d)$) is symmetric in $\Phi$ and $\Psi$. Thus we have the following:

**Corollary 3.77.** *$\Phi$ and $\Psi$ in $\mathcal{CV}$ are a Stieltjes pair if and only if $\Psi$ and $\Phi$ are.*

## 3.5 Sufficient Conditions for Integrability

Let $\Phi, \Psi \in \mathcal{V}$ have inverses $\phi$, $\psi$ respectively which are moduli of continuity, so that $\Phi, \Psi \in \mathcal{VM}$. For $t > 0$, let

$$\begin{aligned}I_B(t) := I_B(\phi, \mathrm{d}\psi; t) &:= (RS)\int_{0+}^t \frac{\phi(u)}{u}\,\mathrm{d}\psi(u) \\ &:= \lim_{\epsilon \downarrow 0} (RS)\int_\epsilon^t \frac{\phi(u)}{u}\,\mathrm{d}\psi(u) \le +\infty.\end{aligned} \tag{3.123}$$

Note that the integral clearly exists for each $\epsilon \in (0, t)$ and increases as $\epsilon \downarrow 0$. Finiteness of $I_B(1)$ (or equivalently, of $I_B(t)$ for any $t > 0$) will be called *Beurling's condition*, since Beurling [14] made use of it. Since $\phi(u) \ge 0$ and $\psi$ is increasing, $I_B(t) = (LS)\int_0^t u^{-1}\phi(u)\,\mathrm{d}\psi(u)$ if either is finite. Specifically, the integrals from $\epsilon$ to $t$ are equal by Propositions 2.13, 2.18, 2.27 and Corollary 2.29, and as $\epsilon \downarrow 0$ we have monotone convergence for the $LS$ integral.

**Theorem 3.78.** *Let $\Phi, \Psi \in \mathcal{CV}$ with inverses $\phi$, $\psi$ respectively. If the Beurling condition $I_B(\phi, \mathrm{d}\psi; 1) < \infty$ holds then $\Phi$ and $\Psi$ form a Stieltjes pair, and $\phi(u)\psi(u) = o(u)$ as $u \downarrow 0$.*

*Proof.* To prove the first part of the conclusion we show that Theorem 3.75($d$) applies, that is, there is a $W(\phi, \psi)$-sequence $\eta$ such that $\Lambda(\phi, \psi; \eta) < \infty$. Recursively choose $n_1 := 1$, and for each $k \geq 1$ given $n_k$, let $n_{k+1}$ be the smallest integer $n$ such that $n \geq n_k$ and $\psi(1/n) \leq (1/2)\psi(1/n_k)$. Then $\eta := \{n_k\}$ is a $W(\phi, \psi)$-sequence (Definition 3.67) since for each $k \geq 1$, $n_{k+1} \leq 2(n_{k+1} - 1)$ and so by subadditivity of $\psi$,

$$\psi\Big(\frac{1}{n_{k+1}}\Big) \geq \psi\Big(\frac{1}{2(n_{k+1}-1)}\Big) \geq \frac{1}{2}\psi\Big(\frac{1}{n_{k+1}-1}\Big) > \frac{1}{4}\psi\Big(\frac{1}{n_k}\Big).$$

Using (3.31), for each $k \geq 1$, we have

$$\int_{n_{k+1}^{-1}}^{n_k^{-1}} \frac{\phi(u)}{u}\, \mathrm{d}\psi(u) \geq \frac{n_k}{2}\phi\Big(\frac{1}{n_k}\Big)\Big[\psi\Big(\frac{1}{n_k}\Big) - \psi\Big(\frac{1}{n_{k+1}}\Big)\Big] \geq \frac{n_k}{4}\phi\Big(\frac{1}{n_k}\Big)\psi\Big(\frac{1}{n_k}\Big).$$

Summing over $k$ it follows that $\Lambda(\phi, \psi; \eta) < \infty$, proving the first part of the conclusion. Moreover, $\lim_{k\to\infty} n_k\phi(n_k^{-1})\psi(n_k^{-1}) = 0$, and so the second part follows from Lemma 3.69, proving the theorem. □

Let $c, \beta, \gamma \in \mathbb{R}$ with $c > 0$. The function $g = g_{c,\beta,\gamma}$ from $(0, \mathrm{e}^{-\mathrm{e}}]$ into $\mathbb{R}$ defined by

$$g(x) = c\Big(\log\frac{1}{x}\Big)^{\beta}\Big(\log\log\frac{1}{x}\Big)^{\gamma} \tag{3.124}$$

will be called a *simple logarithmic function.* A finite linear combination of simple logarithmic functions with real coefficients will be called a *logarithmic function.* The following considerations hold for more general functions, but the given functions will suffice for present purposes.

**Lemma 3.79.** *Let $c, \alpha, \beta, \gamma \in \mathbb{R}$ with $c > 0$ and $\alpha \geq 0$. Let $g = g_{c,\beta,\gamma}$ be the simple logarithmic function (3.124) and let $f(x) := x^{\alpha}g(x)$ for $0 < x \leq \mathrm{e}^{-\mathrm{e}}$ and $f(0) := 0$. Then there exists a $\delta = \delta(\alpha, \beta, \gamma) > 0$ with $\delta \leq \mathrm{e}^{-\mathrm{e}}$ such that*

(*a*) *$f$ is convex on $[0, \delta]$ if either $\alpha > 1$, or $\alpha = 1$ and $\beta < 0$;*
(*b*) *$f$ is convex on $(0, \delta]$ (unbounded near 0) if $\alpha = 0$ and $\beta > 0$;*
(*c*) *$f$ is concave on $[0, \delta]$ if $0 < \alpha < 1$, or $\alpha = 0$ and $\beta < 0$, or $\alpha = 1$ and $\beta > 0$.*

*Moreover, if $\alpha > 1$, $\beta \leq 0$, and $\gamma \leq 0$, then $f$ is convex on the entire interval $[0, \mathrm{e}^{-\mathrm{e}}]$.*

*Proof.* Clearly $f$ is continuous on $(0, \mathrm{e}^{-\mathrm{e}}]$. It is continuous from the right at 0 unless $\alpha = 0$ and either $\beta > 0$ or $\beta = 0$ and $\gamma \geq 0$. There is a logarithmic function $g_1$ such that the derivative $g'(x) = x^{-1}g(x)g_1(x)$ for $0 < x \leq \mathrm{e}^{-\mathrm{e}}$ and $g_1(x) = o(1)$ as $x \downarrow 0$. It follows that if $\alpha > 0$ and $\alpha \neq 1$, then for all $0 < x < \mathrm{e}^{-\mathrm{e}}$,

$$f''(x) = x^{\alpha-2}g(x)\Big[\alpha(\alpha-1) + g_2(x)\Big],$$

where $g_2$ is a logarithmic function and $g_2(x) = o(1)$ as $x \downarrow 0$. If $\alpha > 1$, $\beta \leq 0$, and $\gamma \leq 0$, then $g_2$ is nonnegative on $(0, \mathrm{e}^{-\mathrm{e}}]$, and so the last conclusion follows.

To prove $(a)$, $(b)$, and $(c)$, if $\alpha > 0$ and $\alpha \neq 1$, then $f''(x) \sim \alpha(\alpha - 1)x^{\alpha-2}g(x)$ as $x \downarrow 0$, and the conclusions follow if $\alpha > 1$ or $0 < \alpha < 1$. If $\alpha = 0$, then $f \equiv g$ on $(0, \delta]$ and

$$g''(x) = \frac{c\beta}{x^2}\Big(\log\frac{1}{x}\Big)^{\beta-1}\Big(\log\log\frac{1}{x}\Big)^{\gamma}\big[1 + o(1)\big]$$

as $x \downarrow 0$. If $\beta < 0$, $f$ is continuous at 0. The corresponding conclusions follow. If $\alpha = 1$, then

$$f''(x) = 2g'(x) + xg''(x) = -\frac{c\beta}{x}\Big(\log\frac{1}{x}\Big)^{\beta-1}\Big(\log\log\frac{1}{x}\Big)^{\gamma}\big[1 + o(1)\big]$$

as $x \downarrow 0$. Conclusions $(a)$, $(b)$, and $(c)$ are now proved. □

A function $f$ as defined in the preceding lemma can be restricted to an interval $[0, \delta]$ and then extended to $[0, \infty)$ by letting

$$f(x) := \begin{cases} 0 & \text{if } x = 0, \\ x^{\alpha}g(x) & \text{if } 0 < x \leq \delta, \\ f(\delta) + (x - \delta)f'(\delta-) & \text{if } x > \delta. \end{cases} \tag{3.125}$$

Then $f$ is convex or concave on $[0, \infty)$ if it is on $[0, \delta]$, and likewise for $(0, \infty)$ and $(0, \delta]$ (if $\alpha = 0$ and $\beta > 0$).

The converse to Theorem 3.78 does not hold by the following.

**Example 3.80.** Let $\beta_0 > 0$ and $\beta_1 > 0$. Applying Lemma 3.79, take numbers $\delta(0, -\beta_0, 0)$ and $\delta(1, \beta_1, 0)$. Let $x_0 := \min(\mathrm{e}^{-\beta_1}, \delta(0, -\beta_0, 0), \delta(1, \beta_1, 0)) > 0$. For $0 < x \leq x_0$, let

$$\phi(x) := \Big(\log\frac{1}{x}\Big)^{-\beta_0} \quad \text{and} \quad \psi(x) := x\Big(\log\frac{1}{x}\Big)^{\beta_1}.$$

Notice that $\psi'(x) = |\log x|^{\beta_1} - \beta_1|\log x|^{\beta_1 - 1} \sim |\log x|^{\beta_1}$ as $x \downarrow 0$. Also, $\psi'(x) > 0$ for $0 < x < \mathrm{e}^{-\beta_1}$. Extending $\phi$ and $\psi$ to $[0, \infty)$ by (3.125), both functions are continuous, concave, and increasing. Thus Beurling's condition $I_B(\phi, \mathrm{d}\psi; 1) < \infty$ holds if and only if $\beta_0 > \beta_1 + 1$. The condition $\Lambda(\phi, \psi; \eta_0) < \infty$ for the D'yačkov sequence (Definition 3.70) $\eta_0 = \{n_k\}$ reduces to $\sum_{k=1}^{\infty}[\log n_k]^{\beta_1 - \beta_0} < \infty$. The D'yačkov sequence is lacunary in the sense of Hadamard with $\theta = 4$ as in (3.84), and so $n_k \geq 4^k$ for all $k \geq 1$. On the other hand, by (3.90), $\log n_{k+2} \geq 4^{1/\beta_0}\log n_k$ for all $k \geq 1$, so the series converges if and only if $\beta_0 > \beta_1$. Notice that the preceding condition holds if and only if $I_B(\psi, \mathrm{d}\phi; 1) < \infty$, that is, Beurling's condition holds with $\phi$ and $\psi$ interchanged.

Next, pairs $(\phi, \psi)$ such that Beurling's condition holds both for $(\phi, \psi)$ and for $(\psi, \phi)$ will be characterized. For functions $\phi$ and $\psi$ from $[0, \infty)$ into $[0, \infty)$, let

$$\Theta(\phi, \psi) := \sum_{k=1}^{\infty} \phi(k^{-1})\psi(k^{-1}) \le +\infty. \tag{3.126}$$

**Proposition 3.81.** *Let $\phi$ and $\psi$ be moduli of continuity.*

*(i) The following five conditions are equivalent:*
*(a) $\Theta(\phi, \psi) < \infty$;*
*(b) $\sum_{k=0}^{\infty} 2^k \phi(1/2^k)\psi(1/2^k) < \infty$;*
*(c) $\sum_{k=1}^{\infty} 2^k \phi(1/2^k)\psi(1/2^{k-1}) < \infty$;*
*(d) $\sum_{k=1}^{\infty} 2^k \phi(1/2^{k-1})\psi(1/2^k) < \infty$;*
*(e) $I_B(\phi, \mathrm{d}\psi; 1) < \infty$ and $I_B(\psi, \mathrm{d}\phi; 1) < \infty$.*

*(ii) If for some $\delta > 0$,*

$$\psi(1/2^k) \le (1-\delta)\psi(1/2^{k-1}) \tag{3.127}$$

*for $k = 1, 2, \ldots$, and $I_B(\phi, \mathrm{d}\psi; 1) < \infty$, then (a)–(e) hold.*

*(iii) There exist pairs $\phi$, $\psi$ for which $I_B(\psi, \mathrm{d}\phi; 1) < \infty$ but (a)–(e) do not hold, specifically $\phi$ and $\psi$ in Example* 3.80 *with $\beta_0 - 1 \le \beta_1 < \beta_0$.*

*Proof.* For $(i)$, clearly $(c)$ implies $(b)$ since $\psi$ is nondecreasing. The sum in $(b)$ equals

$$\sum_{j=1}^{\infty} 2^{j-1}\phi(1/2^{j-1})\psi(1/2^{j-1}),$$

whose convergence implies $(c)$ since $\phi$ is nondecreasing. So $(b)$ and $(c)$ are equivalent. Similarly, interchanging $\phi$ with $\psi$, it follows that $(b)$ and $(d)$ are equivalent.

For the sum in $(a)$ we have

$$\Theta(\phi, \psi) = \sum_{k=0}^{\infty} \sum_{n=2^k}^{2^{k+1}-1} \phi(1/n)\psi(1/n),$$

from which it follows easily that

$$\sum_{j=1}^{\infty} 2^{j-1}\phi(2^{-j})\psi(2^{-j}) = \sum_{k=0}^{\infty} 2^k \phi(2^{-k-1})\psi(2^{-k-1})$$
$$\le \Theta(\phi, \psi) \le \sum_{k=0}^{\infty} 2^k \phi(2^{-k})\psi(2^{-k}),$$

so $(a)$ is equivalent to $(b)$.

For $(b) \Rightarrow (e)$, by monotonicity of $\phi$, we have

$$I_B(\phi, \mathrm{d}\psi; 1) = \sum_{k=1}^{\infty} \int_{2^{-k}}^{2^{1-k}} \frac{\phi(u)}{u}\,\mathrm{d}\psi(u) \le 2\sum_{k=1}^{\infty} 2^{k-1}\phi(1/2^{k-1})\psi(1/2^{k-1}) < \infty.$$

Similarly, interchanging $\phi$ and $\psi$, it follows that $I_B(\psi, \mathrm{d}\phi; 1) < \infty$.

For $(e) \Rightarrow (b)$, let $0 < \epsilon < 1$. Applying (2.91) with $F(u) := u^{-1}$, and then applying (2.93) twice, first with $h \equiv 1$ (integration by parts) and second with $h(u) := u^{-1}$, we have

$$\begin{aligned}(RS)\int_{\epsilon}^{1} \frac{\phi(u)\psi(u)}{u^2}\,\mathrm{d}u &= -(RS)\int_{\epsilon}^{1} \phi(u)\psi(u)\,\mathrm{d}\Big(\frac{1}{u}\Big) \\ &= -\frac{\phi(u)\psi(u)}{u}\Big|_{\epsilon}^{1} + (RS)\int_{\epsilon}^{1} \frac{\phi(u)}{u}\,\mathrm{d}\psi(u) + (RS)\int_{\epsilon}^{1} \frac{\psi(u)}{u}\,\mathrm{d}\phi(u).\end{aligned} \tag{3.128}$$

By Theorem 3.78, we also have $(\phi\psi)(u) = o(u)$ as $u \downarrow 0$. Thus letting $\epsilon \downarrow 0$ on the right side of (3.128), we get that the limit of the left side exists and is finite. Also, with definitions as in (3.123), we have

$$\begin{aligned}&\phi(1)\psi(1) + (RS)\int_{0+}^{1} \frac{\phi(u)\psi(u)}{u^2}\,\mathrm{d}u \\ &\quad = (RS)\int_{0+}^{1} \frac{\phi(u)}{u}\,\mathrm{d}\psi(u) + (RS)\int_{0+}^{1} \frac{\psi(u)}{u}\,\mathrm{d}\phi(u).\end{aligned}$$

By monotonicity of $\phi$ and $\psi$, for each $k \ge 1$, we have

$$(RS)\int_{2^{-k}}^{2^{1-k}} \frac{\phi(u)\psi(u)}{u^2}\,du \ge 2^{k-2}\phi(1/2^k)\psi(1/2^k).$$

Summing over $k$, it follows that $(b)$ holds, and hence all five conditions are equivalent.

For $(ii)$, if (3.127) holds and $I_B(\phi, \mathrm{d}\psi; 1) < \infty$, then

$$\begin{aligned}&I_B(\phi, \mathrm{d}\psi; 1) \\ &\quad = \sum_{k=1}^{\infty} (RS)\int_{2^{-k}}^{2^{1-k}} \frac{1}{u}\phi(u)\,\mathrm{d}\psi(u) \ge \sum_{k=1}^{\infty} 2^{k-1}\phi(2^{-k})[\psi(2^{1-k}) - \psi(2^{-k})] \\ &\quad \ge \frac{\delta}{2}\sum_{k=1}^{\infty} 2^k\phi(2^{-k})\psi(2^{1-k}),\end{aligned}$$

implying $(c)$ and thus $(a)$–$(e)$.

For $(iii)$, taking $\phi$ and $\psi$ in Example 3.80, we have $I_B(\psi, \mathrm{d}\phi; 1) < \infty$ if and only if $\beta_1 < \beta_0$, while $I_B(\phi, \mathrm{d}\psi; 1) < \infty$ if and only if $\beta_1 < \beta_0 - 1$. So Proposition 3.81 is proved. □

Summing up: the equivalent conditions $(a)$–$(e)$ of Proposition 3.81$(i)$ clearly imply Beurling's condition $I_B(\phi, \mathrm{d}\psi; 1) < \infty$, which by Theorem 3.78 implies the equivalent conditions $(a)$ through $(e)$ of Theorem 3.75, but in neither case do the converse implications hold in general. Also, we have the following:

**Corollary 3.82.** *Assume* (1.14). *Let $\Phi, \Psi \in \mathcal{CV}$ with inverses $\phi$, $\psi$, respectively, and let $\Theta(\phi, \psi) < \infty$. The full Stieltjes integral $(S)\int_a^b f \cdot \mathrm{d}h$ exists if $f \in \mathcal{W}_\Phi([a,b]; X)$ and $h \in \mathcal{W}_\Psi([a,b]; Y)$.*

## 3.6 Love–Young Inequalities

The Love–Young inequality gives a bound for an integral in terms of a suitable $p$-variation norm of the integrand and $q$-variation seminorm of the integrator. Namely by Corollary 3.91 below in this section, if $f \in \mathcal{W}_p[a,b]$, $h \in \mathcal{W}_q[a,b]$ with $1 \le p < \infty$, $1 \le q < \infty$, and $p^{-1} + q^{-1} > 1$, then the full Stieltjes integral $(S)\int_a^b f\,\mathrm{d}h$ exists and we have the inequality

$$\left|(S)\int_a^b f\,\mathrm{d}h\right| \le K\|f\|_{[p]}\|h\|_{(q)}, \tag{3.129}$$

with the constant $K = \zeta(p^{-1} + q^{-1})$. Here $\zeta(r) := \sum_{n=1}^{\infty} n^{-r}$, $r > 1$, is the Riemann zeta function.

L. C. Young in 1936 [244] first published an inequality of the form (3.129), with the constant $1 + K$ in place of $K$. He acknowledged that E. R. Love had contributed to finding a proof, hence the name "Love–Young inequality." As Young pointed out, inequality (3.129) formally resembles the Hölder inequality (1.4). In contrast with the Hölder inequality, in the Love–Young inequality (3.129) it is not possible to take $p^{-1} + q^{-1} = 1$, as Proposition 3.104 will show. However, L. C. Young in 1938 [247] extended (3.129) to $f \in \mathcal{W}_\Phi[a,b]$ and $h \in \mathcal{W}_\Psi[a,b]$ with $\Phi, \Psi \in \mathcal{V}$ having inverses $\phi$, $\psi$, respectively, such that $\Theta(\phi, \psi)$, defined by (3.126), is finite, for a suitable constant $K$. Theorem 3.89 below gives such an inequality when $\Phi$ and $\Psi$ are convex.

*A Love–Young inequality for sums*

Recall that $\mathcal{CV}$ is the subclass of convex functions in $\mathcal{V}$. We start with a Love–Young inequality for Riemann–Stieltjes sums. Such sums are defined for functions over nondegenerate intervals $[a,b]$, and so we assume that $a < b$.

**Theorem 3.83.** *Assume* (1.14). *Let $\Phi, \Psi \in \mathcal{CV}$ with inverses $\phi$, $\psi$, respectively, and let $f\colon [c,d] \to X$ and $h\colon [a,b] \to Y$. Then for any partition $\{t_i\}_{i=0}^n$ of $[a,b]$, and for any $s_0 = c \le s_1 \le \cdots \le s_n = d$ and $\nu \in \{0, 1, \ldots, n\}$,*

$$\Big\| \sum_{i=1}^{n} f(s_i)\cdot\big[h(t_i) - h(t_{i-1})\big] - f(s_\nu)\cdot\big[h(b) - h(a)\big]\Big\| \le \sum_{k=1}^{\infty} \phi\Big(\frac{v_\Phi(f;[c,d])}{k}\Big)\psi\Big(\frac{v_\Psi(h;[a,b])}{k}\Big). \tag{3.130}$$

**Remark 3.84.** The theorem will be applied when $t_{i-1} \le s_i \le t_i$ for $i = 1, \dots, n$ and $a = c < d \le b$, so that we have Riemann–Stieltjes sums, and $v_\Phi(f;[c,d]) \le v_\Phi(f;[a,b])$.

*Proof.* We can assume that $c < d$. Let $\kappa := (\{s_i\}_{i=0}^n, \{t_i\}_{i=0}^n)$. Let $\Delta_i f := f(s_i) - f(s_{i-1})$ and $\Delta_i h := h(t_i) - h(t_{i-1})$ for $i = 1, \dots, n$. Also, let $S(\kappa) := \sum_{i=1}^n f(s_i)\cdot\Delta_i h$. Then we have

$$S(\kappa) = f(c)\cdot[h(b) - h(a)] + \sum_{1\le i\le j\le n} \Delta_i f\cdot\Delta_j h \tag{3.131}$$

because $\sum_{i=1}^j \Delta_i f = f(s_j) - f(c)$. On the other hand, since $f(s_j) = f(d) - \sum_{j<i\le n} \Delta_i f$, we also have

$$S(\kappa) = f(d)\cdot[h(b) - h(a)] - \sum_{1\le j<i\le n} \Delta_i f\cdot\Delta_j h. \tag{3.132}$$

Thus we have representations of the left side of (3.130) in the cases $\nu = 0$ or $n$. Now let $\nu \in \{1, \dots, n-1\}$. Then (3.131) for $(\{s_j\}_{j=\nu}^n, \{t_j\}_{j=\nu}^n)$ gives

$$\sum_{i=\nu+1}^{n} f(s_i)\cdot\Delta_i h = f(s_\nu)\cdot[h(b) - h(t_\nu)] + \sum_{\nu<i\le j\le n} \Delta_i f\cdot\Delta_j h. \tag{3.133}$$

Also, (3.132) for $(\{s_i\}_{i=0}^\nu, \{t_i\}_{i=0}^\nu)$ gives

$$\sum_{i=1}^{\nu} f(s_i)\cdot\Delta_i h = f(s_\nu)\cdot[h(t_\nu) - h(a)] - \sum_{1\le j<i\le \nu} \Delta_i f\cdot\Delta_j h. \tag{3.134}$$

Adding (3.133) and (3.134) gives the representation

$$S(\kappa) - f(s_\nu)\cdot[h(b) - h(a)] = -\sum_{1\le j<i\le\nu} \Delta_i f\cdot\Delta_j h + \sum_{\nu<i\le j\le n} \Delta_i f\cdot\Delta_j h \tag{3.135}$$

of the quantity we are aiming to bound. Therefore it is enough to bound the norm of the right side of (3.135). To this aim we replace $\Delta_i f$ by an element $x_i \in X$ and $\Delta_j h$ by an element $y_j \in Y$. Then, applying the following lemma, the proof of the Love–Young inequality for sums will be complete. □

Let $x = (x_1, \dots, x_n)$, $x_i \in X$, and $y = (y_1, \dots, y_n)$, $y_i \in Y$. For $\Phi, \Psi \in \mathcal{V}$, let

$$v_\Phi(x) := \max\Big\{\sum_{j=1}^{m} \Phi\Big(\Big\|\sum_{i=\theta(j-1)+1}^{\theta(j)} x_i\Big\|\Big) \colon 1 \le m \le n \qquad (3.136)$$
$$0 = \theta(0) < \theta(1) < \cdots < \theta(m) = n\Big\},$$

and let $v_\Psi(y)$ be defined similarly. Then the following holds:

**Lemma 3.85.** *As in Theorem 3.83, assume* (1.14), *and let $\Phi, \Psi \in \mathcal{CV}$ with inverses $\phi, \psi$, respectively. Then for any finite sequences $x = (x_1, \dots, x_n)$, $x_i \in X$, and $y = (y_1, \dots, y_n)$, $y_i \in Y$, and for any integer $\nu \in \{0, \dots, n\}$,*

$$\Big\| - \sum_{1 \le j < i \le \nu} x_i \cdot y_j + \sum_{\nu < i \le j \le n} x_i \cdot y_j \Big\| \le \sum_{k=1}^{\infty} \phi\Big(\frac{v_\Phi(x)}{k}\Big)\psi\Big(\frac{v_\Psi(y)}{k}\Big). \qquad (3.137)$$

*Proof.* Given two ordered $m$-tuples $\xi = \{\xi_1, \dots, \xi_m\}$ and $\eta = \{\eta_1, \dots, \eta_m\}$ of elements of $X$ and $Y$, respectively, and an integer $r \in \{0, \dots, m\}$, let

$$S_L(\xi, \eta; r) := \sum_{0 \le j < i \le r} \xi_i \cdot \eta_j = \sum_{1 \le j < i \le r} \xi_i \cdot \eta_j, \quad S_R(\xi, \eta; r, m) := \sum_{r < i \le j \le m} \xi_i \cdot \eta_j,$$

where $\eta_0 := 0$, $S_L(\xi, \eta; 0) := 0$ and $S_R(\xi, \eta; m, m) := 0$. Also let

$$U_m(\xi, \eta; r) := -S_L(\xi, \eta; r) + S_R(\xi, \eta; r, m).$$

We have to bound the norm of $U_n(x, y; \nu)$. To this aim we replace a pair $(x_l, x_{l+1})$ of elements in $x$ by a single element $x_l + x_{l+1}$, and a pair $(y_{l-1}, y_l)$ of elements in $y$ by a single element $y_{l-1} + y_l$, and show that for new sequences $x'$, $y'$ both of length $n-1$, and a new $\nu'$ equal to either $\nu$ or $\nu - 1$, the norm of $U_n(x, y; \nu) - U_{n-1}(x', y'; \nu')$ does not exceed the $(n-1)$st term of the right side of (3.137) if $\nu \in \{1, \dots, n\}$, or the $n$th term if $\nu = 0$. Repeating this reduction recursively we will get the desired bound.

To begin, for any $1 \le l \le n$, let $y_0 := x_{n+1} := 0$ and let

$$x'_j := x_j \quad \text{for} \quad 1 \le j \le l-1, \quad x'_l := x_l + x_{l+1}, \; x'_j := x_{j+1} \quad \text{for} \quad l < j < n,$$
$$y'_j := y_j \quad \text{for} \quad 0 \le j < l-1, \; y'_{l-1} := y_{l-1} + y_l, \; y'_j := y_{j+1} \quad \text{for} \quad l \le j < n$$

(where $x_{n+1} := 0$ is used when $l = n$). Then $x' := (x'_1, \dots, x'_{n-1})$ and $y' := (y'_1, \dots, y'_{n-1})$ are two sequences of elements of $X$ and $Y$, respectively, depending on $l$. If $1 \le l < \nu$ then $S_R(x', y'; \nu - 1, n - 1) = S_R(x, y; \nu, n)$ and

$$
\begin{aligned}
S_L(x', y'; \nu - 1) \\
&= \sum_{0<i\le\nu-1} x'_i\cdot(y'_0 + \cdots + y'_{i-1}) = \sum_{1<i<l} x_i\cdot(y_1 + \cdots + y_{i-1}) \\
&\quad + (x_l + x_{l+1})\cdot(y_1 + \cdots + y_l) + \sum_{l<i\le\nu-1} x_{i+1}\cdot(y_1 + \cdots + y_i) \\
&= x_l\cdot y_l + \sum_{1<i\le\nu} x_i\cdot(y_1 + \cdots + y_{i-1}) = x_l\cdot y_l + S_L(x, y; \nu).
\end{aligned}
$$

If $\nu < l \le n$ then $S_L(x', y'; \nu) = S_L(x, y; \nu)$ and

$$
\begin{aligned}
S_R(x', y'; \nu, n - 1) \\
&= \sum_{\nu<i\le n-1} x'_i\cdot(y'_i + \cdots + y'_{n-1}) = \sum_{\nu<i<l} x_i\cdot(y_i + \cdots + y_n) \\
&\quad + (x_l + x_{l+1})\cdot(y_{l+1} + \cdots + y_n) + \sum_{l<i<n} x_{i+1}\cdot(y_{i+1} + \cdots + y_n) \\
&= -x_l\cdot y_l + \sum_{\nu<i\le n} x_i\cdot(y_i + \cdots + y_n) = -x_l\cdot y_l + S_R(x, y; \nu, n).
\end{aligned}
$$

Let $\nu' := \nu'(\nu, l) := \nu - 1$ if $l < \nu$, and let $\nu' := \nu$ if $l > \nu$. Then

$$
U_n(x, y; \nu) - U_{n-1}(x', y'; \nu') = x_l\cdot y_l \tag{3.138}
$$

for $l \in \{1, \ldots, n\} \setminus \{\nu\}$ and $\nu \in \{0, \ldots, n\}$. We choose a particular value of $l$ using the following fact:

**Lemma 3.86.** *Let $\Phi, \Psi \in \mathcal{CV}$ with inverses $\phi, \psi$, respectively. Given two ordered $m$-tuples $u = \{u_1, \ldots, u_m\}$ and $v = \{v_1, \ldots, v_m\}$ of nonnegative real numbers, there is an index $r$, $1 \le r \le m$, such that*

$$
u_r\, v_r \le \phi\Big(\frac{1}{m}\sum_{j=1}^m \Phi(u_j)\Big)\psi\Big(\frac{1}{m}\sum_{j=1}^m \Psi(v_j)\Big). \tag{3.139}
$$

*Proof.* Take the least $r$ such that $u_r\, v_r = \min\{u_j\, v_j \colon j = 1, \ldots, m\}$. Using the inequality between geometric and arithmetic means, we have

$$
\begin{aligned}
u_r\, v_r \le \Big(\prod_{j=1}^m u_j\Big)^{1/m}\Big(\prod_{j=1}^m v_j\Big)^{1/m} &\le \phi\Big(\Phi\Big(\frac{1}{m}\sum_{j=1}^m u_j\Big)\Big)\psi\Big(\Psi\Big(\frac{1}{m}\sum_{j=1}^m v_j\Big)\Big) \\
&\le \phi\Big(\frac{1}{m}\sum_{j=1}^m \Phi(u_j)\Big)\psi\Big(\frac{1}{m}\sum_{j=1}^m \Psi(v_j)\Big)
\end{aligned}
$$

by convexity, proving the lemma. □

Now continuing the proof of Lemma 3.85, apply (3.139) to $u = \{\|x_l\|: l \in \{1,\dots,n\}\setminus\{\nu\}\}$ and $v = \{\|y_l\|: l \in \{1,\dots,n\}\setminus\{\nu\}\}$ with $m = n-1$ if $\nu \ge 1$ or to $u = \{\|x_l\|\}_{l=1}^n$ and $v = \{\|y_l\|\}_{l=1}^n$ with $m = n$ if $\nu = 0$. This gives an index $r \in \{1,\dots,n\}\setminus\{\nu\}$ such that $\|x_r \cdot y_r\| \le C_{n,\nu}(v_\Phi(x), v_\Psi(y))$, where

$$C_{m,\zeta}(s,t) := \begin{cases} \phi(s/m)\psi(t/m) & \text{if } \zeta = 0, \\ \phi(s/(m-1))\psi(t/(m-1)) & \text{if } \zeta \ge 1, \end{cases} \tag{3.140}$$

for any $s, t \ge 0$, $m \in \{3,\dots,n\}$, and $\zeta \in \{0,\dots,n\}$. For $l = r$, let $x^{(n-1)} := x'$, $y^{(n-1)} := y'$, and $\nu^{(n-1)} := \nu'$. Then by (3.138)

$$\|U_n(x, y; \nu)\| \le C_{n,\nu}(v_\Phi(x), v_\Psi(y)) + \|U_{n-1}(x^{(n-1)}, y^{(n-1)}; \nu^{(n-1)})\|.$$

Let $\nu^{(n)} := \nu$. Notice that $\nu^{(n-1)} = 1$ if $\nu = 1$ and $\nu^{(n-1)} = 0$ if $\nu = 0$. Applying the same argument to $x^{(n-1)}$, $y^{(n-1)}$, and $\nu^{(n-1)}$ instead of $x$, $y$, and $\nu$, one gets a similar inequality for the sum $U_{n-1}(x^{(n-1)}, y^{(n-1)}; \nu^{(n-1)})$ in terms of a sum $U_{n-2}(x^{(n-2)}, y^{(n-2)}; \nu^{(n-2)})$ of the same kind and a term $C_{n-1,\nu^{(n-1)}}(v_\Phi(x^{(n-1)}), v_\Psi(y^{(n-1)}))$. By the definitions, we have $v_\Phi(x^{(n-1)}) \le v_\Phi(x)$ and $v_\Psi(y^{(n-1)}) \le v_\Psi(y)$. Proceeding in this way we obtain sums $U_m(x^{(m)}, y^{(m)}; \nu^{(m)})$ for $m = n-1,\dots,2$ and the bound

$$\|U_n(x, y; \nu)\| \le \sum_{j=3}^{n} C_{j,\nu^{(j)}}(v_\Phi(x), v_\Psi(y)) + \|U_2(x^{(2)}, y^{(2)}; \nu^{(2)})\|. \tag{3.141}$$

To bound the last term consider two cases depending on whether $\nu = 0$ or $\nu \ge 1$. From the definition of $\nu'$, we have $\nu^{(j)} = 0$ for all $j$ if $\nu = 0$, and $\nu^{(j)} \ge 1$ for all $j \ge 2$ if $\nu \ge 1$. In the first case, $\nu^{(2)} = 0$. Then applying the preceding argument again we obtain for $l \in \{1, 2\}$,

$$U_1((x^{(2)})', (y^{(2)})'; (\nu^{(2)})') = S_R((x^{(2)})', (y^{(2)})'; 0, 1) = (x^{(2)})'_1 \cdot (y^{(2)})'_1$$

and by (3.138),

$$U_2(x^{(2)}, y^{(2)}; \nu^{(2)}) - U_1((x^{(2)})', (y^{(2)})'; (\nu^{(2)})') = x_l^{(2)} \cdot y_l^{(2)}.$$

Thus applying (3.139) for $l = r \in \{1, 2\}$, we get the bound

$$\|U_2(x^{(2)}, y^{(2)}; \nu^{(2)})\| \le \phi\Big(\frac{v_\Phi(x)}{2}\Big)\psi\Big(\frac{v_\Psi(y)}{2}\Big) + \phi(v_\Phi(x))\psi(v_\Psi(y)).$$

In the second case,

$$U_2(x^{(2)}, y^{(2)}; \nu^{(2)}) = \begin{cases} -S_L(x^{(2)}, y^{(2)}; 2) = -x_2^{(2)} \cdot y_1^{(2)} & \text{if } \nu^{(2)} = 2, \\ S_R(x^{(2)}, y^{(2)}; 1, 2) = x_2^{(2)} \cdot y_2^{(2)} & \text{if } \nu^{(2)} = 1. \end{cases}$$

Then the bound

$$\left\|U_2(x^{(2)}, y^{(2)}; \nu^{(2)})\right\| \le \phi\big(v_\Phi(x)\big)\psi\big(v_\Psi(y)\big)$$

holds. Inserting these bounds and (3.140) into (3.141) we obtain the desired inequality (3.137). □

So, the proof of Theorem 3.83 (a Love–Young inequality for sums) is completed. The following corollary of it will be used later on.

**Corollary 3.87.** *Assume* (1.14). *Let* $f \in \mathcal{W}_q([a,b];X)$ *and* $h \in \mathcal{W}_p([a,b];Y)$ *with* $1 \le p, q < \infty$, $p^{-1}+q^{-1} > 1$. *For any tagged partition* $\tau$ *of* $[a,b]$ *and any* $s \in [a,b]$, *we have*

$$\left\|S_{RS}(f, \mathrm{d}h; \tau) - f(s)\cdot[h(b) - h(a)]\right\| \le K_{p,q}\|f\|_{(q)}\|h\|_{(p)}, \tag{3.142}$$

*where* $K_{p,q} = 1 + \zeta(p^{-1}+q^{-1})$, *and* $K_{p,q} = \zeta(p^{-1}+q^{-1})$ *if in addition,* $s$ *is a tag of* $\tau$. *Moreover, we have*

$$\left\|S_{RS}(f, \mathrm{d}h; \tau)\right\| \le \zeta(p^{-1}+q^{-1})\|f\|_{[q]}\|h\|_{(p)}. \tag{3.143}$$

*Proof.* Since (3.143) is a simple consequence of (3.142) we prove only (3.142). We use Theorem 3.83 with $\Phi(u) \equiv u^q$, $\Psi(u) \equiv u^p$, $u \ge 0$, $1 \le p, q < \infty$, and $p^{-1}+q^{-1} > 1$. Let $\tau = (\{t_i\}_{i=0}^n, \{s_i\}_{i=1}^n)$ be a tagged partition of $[a,b]$, let $s \in [a,b]$, and let $s_0 := a$. Then by (3.130) with $a = c < d = b$ and $\nu = 0$, we have

$$\begin{aligned}
&\|S_{RS}(f, \mathrm{d}h; \tau) - f(s)\cdot[h(b) - h(a)]\| \\
&\qquad \le \|S_{RS}(f, \mathrm{d}h; \tau) - f(s_0)\cdot[h(b) - h(a)]\| + \|f(s_0) - f(s)\|\|h(b) - h(a)\| \\
&\qquad \le \zeta(p^{-1}+q^{-1})\|f\|_{(q)}\|h\|_{(p)} + \|f\|_{(q)}\|h\|_{(p)}
\end{aligned}$$

and the first conclusion holds. If $s \in \{s_i\}_{i=1}^n$ then (3.142) with $K_{p,q} = \zeta(p^{-1}+q^{-1})$ follows from (3.130) alone, proving the corollary. □

Next is a Love–Young inequality for Kolmogorov sums defined by (2.22) and approximating the Kolmogorov integral.

**Proposition 3.88.** *Assume* (1.14). *Let* $\Phi, \Psi \in \mathcal{CV}$ *with inverses* $\phi$, $\psi$, *respectively. For a nonempty interval* $J$, *let* $f \in \mathcal{W}_\Phi(J;X)$ *and* $\mu \in \mathcal{AI}_\Psi(J;Y)$. *Then for any tagged Young interval partition* $\mathcal{T}$ *of* $J$ *and any tag* $s \in J$ *of* $\mathcal{T}$,

$$\left\|S_{YS}(f, \mathrm{d}\mu; J, \mathcal{T}) - f(s)\cdot\mu(J)\right\| \le \sum_{k=1}^{\infty}\phi\Big(\frac{v_\Phi(f;J)}{k}\Big)\psi\Big(\frac{v_\Psi(\mu;J)}{k}\Big). \tag{3.144}$$

*Proof.* We can assume that $J$ is nondegenerate. First suppose that $J = [\![a,b]\!]$ for some $-\infty < a < b < \infty$. Let $\mathcal{T}$ be a tagged Young interval partition of $J$, let $s$ be a tag of $\mathcal{T}$, and let $h := R_{\mu,a}$ be the function defined by

(2.3) corresponding to the interval function $\mu$. Given $\epsilon > 0$, since $\mu$ is upper continuous, by Lemma 2.24, one can find an interval $[c, d] \subset J$ and a tagged partition $\tau$ of $[c, d]$ such that $s$ is a tag of $\tau$ and the norm of the difference

$$\big[S_{YS}(f, \mathrm{d}\mu; J, \mathcal{T}) - f(s)\cdot\mu(J)\big] - \big[S_{RS}(f, \mathrm{d}h; \tau) - f(s)\cdot[h(d) - h(c)]\big]$$

is less than $\epsilon$. By Theorem 3.83, it then follows that

$$\big\|S_{YS}(f, \mathrm{d}\mu; J, \mathcal{T}) - f(s)\cdot\mu(J)\big\| \leq \epsilon + \sum_{k=1}^{\infty} \phi\Big(\frac{v_{\Phi}(f; [c, d])}{k}\Big)\psi\Big(\frac{v_{\Psi}(h; [c, d])}{k}\Big).$$

By Proposition 3.31(a), $v_{\Psi}(h; [c, d]) \leq v_{\Psi}(\mu; [c, d]) \leq v_{\Psi}(\mu)$. Since $\epsilon > 0$ is arbitrary, the conclusion follows if $J$ is a finite interval.

Let $J = \mathbb{R}$, let $\mathcal{T} = (\{(t_{i-1}, t_i)\}_{i=1}^{n}, \{s_i\}_{i=1}^{n})$ be a tagged Young interval partition of $J$, let $s$ be a tag of $\mathcal{T}$, and let $\epsilon > 0$. If $\mathrm{Osc}(f) = 0$ there is no problem, so assume $\mathrm{Osc}(f) > 0$. Since $\mu$ is upper continuous, there are $u, v \in \mathbb{R}$ such that

$$\max\big\{\|\mu\|_{(-\infty,u],\sup}, \|\mu\|_{[v,\infty),\sup}\big\} \leq \epsilon/\mathrm{Osc}(f).$$

Let $a := s_1 - 1$ if $s_1 \leq u$ and $a := u$ otherwise. Also let $b := s_n + 1$ if $s_n \geq v$ and $b := v$ otherwise. Let $\mathcal{T}'$ be the tagged Young interval partition of $(a, b)$ obtained from $\mathcal{T}$ by replacing the interval $(-\infty, t_1)$ with $(a, t_1)$ and replacing the interval $(t_{n-1}, \infty)$ with $(t_{n-1}, b)$. Then $s \in (a, b)$, $s$ is a tag of $\mathcal{T}'$, and

$$\begin{aligned} &S_{YS}(f, \mathrm{d}\mu; J, \mathcal{T}) - f(s)\cdot\mu(J) \\ &\quad = S_{YS}(f, \mathrm{d}\mu; (a, b), \mathcal{T}') - f(s)\cdot\mu((a, b)) \\ &\qquad + [f(s_1) - f(s)]\cdot\mu((-\infty, a]) + [f(s_n) - f(s)]\cdot\mu([b, \infty)). \end{aligned}$$

The sum of the norms of the last two terms on the right is less than $2\epsilon$ by the choice of $u$ and $v$. Applying the bound obtained in the first part of the proof to the first difference on the right, the conclusion of the proposition follows again since $\epsilon > 0$ is arbitrary. The case when $J$ is a half-line can be treated similarly and therefore is omitted. The proof of the proposition is now complete. □

*Love–Young inequalities for the full Stieltjes integral*

Next is a Love–Young inequality, named for the special case (3.129), for the full Stieltjes integral defined by Definition 2.41. Recall the definition (3.126) of $\Theta(\phi, \psi)$.

**Theorem 3.89.** *Assume* (1.14). *Let $\Phi, \Psi \in \mathcal{CV}$ with inverses $\phi$, $\psi$, respectively, and let $f \in \mathcal{W}_{\Phi}([a, b]; X)$ and $h \in \mathcal{W}_{\Psi}([a, b]; Y)$. If $\Theta(\phi, \psi) < \infty$ then the full Stieltjes integral $(S)\int_a^b f\cdot\mathrm{d}h$ exists and the inequality*

$$\Big\|(S)\int_a^b [f-f(s)]\cdot dh\Big\| \le \sum_{k=1}^{\infty}\phi\Big(\frac{v_\Phi(f)}{k}\Big)\psi\Big(\frac{v_\Psi(h)}{k}\Big) \tag{3.145}$$

*holds for any* $s\in[a,b]$.

*Proof.* We can assume that $a<b$. The full Stieltjes integral $(S)\int_a^b f\cdot dh$ exists by Corollary 3.82. Let $s\in[a,b]$. If $(S)\int_a^b f\cdot dh = (RRS)\int_a^b f\cdot dh$ then (3.145) follows from (3.130) with $a=c<d\le b$, approximating the integral by Riemann–Stieltjes sums. Otherwise $(S)\int_a^b f\cdot dh = (RYS)\int_a^b f\cdot dh$ and (3.145) again follows from (3.130) because by Proposition 2.18, each Young–Stieltjes sum including a tag $s=s_\nu$ can be approximated arbitrarily closely by Riemann–Stieltjes sums also including a tag $s$. This completes the proof of the theorem. □

**Corollary 3.90.** *Assume* (1.14). *Let* $\Phi,\Psi\in\mathcal{CV}$, *and let* $f\in\widetilde{\mathcal{W}}_\Phi([a,b];X)$, $h\in\widetilde{\mathcal{W}}_\Psi([a,b];Y)$. *If* $\Theta(\Phi^{-1},\Psi^{-1})<\infty$ *then the full Stieltjes integral* $(S)\int_a^b f\cdot dh$ *exists and the inequality*

$$\Big\|(S)\int_a^b f\cdot dh\Big\| \le K\|f\|_{[\Phi]}\|h\|_{(\Psi)} \tag{3.146}$$

*holds with the constant* $K:=\max\{\Psi^{-1}(1),\Theta(\Phi^{-1},\Psi^{-1})\}$.

*Proof.* The existence of $(S)\int_a^b f\cdot dh$ follows from Theorem 3.89. It is enough to prove (3.146) for $a<b$. Due to bilinearity of the full Stieltjes integral (Theorem 2.72), we can assume $\|f\|_{(\Phi)}=\|h\|_{(\Psi)}=1$. Then by (3.9) we have $v_\Phi(f)\le 1$ and $v_\Psi(h)\le 1$. Thus $\|h(b)-h(a)\|\le\Psi^{-1}(1)$ by the second part of Theorem 3.7(*b*), and (3.146) follows from (3.145). □

The following is the special case of Theorem 3.89 and Corollary 3.90 when $\Psi(v)\equiv v^p$ and $\Phi(u)\equiv u^q$ with $p\ge 1$, $q\ge 1$, and $p^{-1}+q^{-1}>1$. Recall the Riemann zeta function $\zeta(r)=\sum_{n=1}^{\infty}n^{-r}$ for $r>1$.

**Corollary 3.91.** *Assume* (1.14). *Let* $f\in\mathcal{W}_q([a,b];X)$ *and* $h\in\mathcal{W}_p([a,b];Y)$ *for* $p\ge 1$, $q\ge 1$, *and* $p^{-1}+q^{-1}>1$. *Then the full Stieltjes integral* $(S)\int_a^b f\cdot dh$ *exists and we have the inequalities*

$$\Big\|(S)\int_a^b [f-f(s)]\cdot dh\Big\| \le K_{p,q}\|f\|_{(q)}\|h\|_{(p)} \tag{3.147}$$

*for any* $s\in[a,b]$, *where* $K_{p,q}:=\zeta(p^{-1}+q^{-1})$, *and*

$$\Big\|(S)\int_a^b f\cdot dh\Big\| \le K_{p,q}\|f\|_{[q]}\|h\|_{(p)}. \tag{3.148}$$

The next theorem extends Corollary 3.91 to the indefinite integral $I_S(f, \mathrm{d}h)$ defined by

$$I_S(f, \mathrm{d}h)(x) := (S)\int_a^x f{\cdot}\mathrm{d}h, \qquad x \in [a, b].$$

**Theorem 3.92.** *Assume* (1.14). *Let* $f \in \mathcal{W}_q([a, b]; X)$ *and* $h \in \mathcal{W}_p([a, b]; Y)$ *for* $p \geq 1$, $q \geq 1$, *and* $p^{-1} + q^{-1} > 1$. *Then the indefinite integral* $I_S(f, \mathrm{d}h)$ *exists, has bounded p-variation, and satisfies*

$$\big\|I_S(f, \mathrm{d}h)\big\|_{(p)} \leq K_{p,q}\|f\|_{[q]}\|h\|_{(p)},$$

*where* $K_{p,q} := \zeta(p^{-1} + q^{-1})$.

*Proof.* We can assume that $a < b$. The indefinite integral $I_S(f, \mathrm{d}h)$ exists by Corollary 3.91. Let $\kappa = \{t_i\}_{i=0}^n$ be a partition of $[a, b]$ and $K := K_{p,q}$. Then by additivity of the full Stieltjes integral (Theorem 2.73), by (3.148), and iterating (3.51), we have

$$\begin{aligned} s_p(I_S(f, \mathrm{d}h); \kappa) &= \sum_{i=1}^n \Big\|(S)\int_{t_{i-1}}^{t_i} f{\cdot}\mathrm{d}h\Big\|^p \\ &\leq K^p\|f\|_{[q]}^p \sum_{i=1}^n v_p(h; [t_{i-1}, t_i]) \leq K^p\|f\|_{[q]}^p\|h\|_{(p)}^p. \end{aligned}$$

Taking the supremum over $\kappa$ and $p$th roots, the proof of the theorem is complete. □

*Love–Young inequalities for the Kolmogorov integral*

Recall that the Kolmogorov integral was defined in Section 2.3, and that $\mathcal{AI}_\Phi(J; Y)$, $\Phi \in \mathcal{V}$, is the set of all additive upper continuous interval functions of bounded $\Phi$-variation on a nonempty interval $J$ having values in the Banach space $Y$ (Definition 3.16).

**Theorem 3.93.** *Assume* (1.14). *Let* $\Phi, \Psi \in \mathcal{CV}$ *with inverses* $\phi$, $\psi$, *respectively, be such that* $\Theta(\phi, \psi) < \infty$. *For a nonempty interval* $J$, *let* $f \in \mathcal{W}_\Phi(J; X)$ *and* $\mu \in \mathcal{AI}_\Psi(J; Y)$. *Then the Kolmogorov integral* $\oint_J f{\cdot}\mathrm{d}\mu$ *exists and the inequality*

$$\Big\|\oint_J [f - f(s)]{\cdot}\mathrm{d}\mu\Big\| \leq \sum_{k=1}^\infty \phi\Big(\frac{v_\Phi(f; J)}{k}\Big)\psi\Big(\frac{v_\Psi(\mu; J)}{k}\Big) \tag{3.149}$$

*holds for any* $s \in J$.

To prove the theorem we use the following:

**Lemma 3.94.** *Let $\Phi, \Psi \in \mathcal{CV}$ with inverses $\phi$, $\psi$, respectively, be such that $\Theta(\phi, \psi) < \infty$. For any $A, B \in (0, 1/2]$, we have*

$$\sum_{k=1}^{\infty} \phi\Big(\frac{A}{k}\Big)\psi\Big(\frac{B}{k}\Big) \le 4\Theta(\phi, \psi) \max\{A, B\}.$$

*Proof.* We have

$$\infty > \Theta(\phi, \psi) = \sum_{i=1}^{\infty} \sum_{k=2^{i-1}}^{2^i - 1} \phi\Big(\frac{1}{k}\Big)\psi\Big(\frac{1}{k}\Big) \ge \sum_{i=1}^{\infty} 2^{i-1} \phi(2^{-i})\psi(2^{-i}).$$

Take the integer $m \ge 1$ such that $2^{-m-1} < \max\{A, B\} \le 2^{-m}$. Then using the preceding lower bound for $\Theta(\phi, \psi)$, it follows that

$$\begin{aligned} \sum_{k=1}^{\infty} \phi\Big(\frac{A}{k}\Big)\psi\Big(\frac{B}{k}\Big) &= \sum_{i=0}^{\infty} \sum_{k=2^i}^{2^{i+1}-1} \phi\Big(\frac{A}{k}\Big)\psi\Big(\frac{B}{k}\Big) \\ &\le \frac{1}{2^m} \sum_{i=0}^{\infty} 2^{i+m} \phi\Big(\frac{1}{2^{i+m}}\Big)\psi\Big(\frac{1}{2^{i+m}}\Big) \le 4\Theta(\phi, \psi) \max\{A, B\}, \end{aligned}$$

proving the lemma. □

**Proof of Theorem 3.93.** We can assume that $J$ is nondegenerate. First suppose that $J = [\![a, b]\!]$ for some $-\infty < a < b < \infty$. Let $h := R_{\mu,a}$ be defined by (2.3), and let $\tilde{h}$, $\tilde{f}$ be defined by (2.23). Then $v_\Psi(\tilde{h}; [a, b]) = v_\Psi(h; J) \le v_\Psi(\mu; J) < \infty$ by Proposition 3.31(*a*). Also, we have $v_\Phi(\tilde{f}; [a, b]) \le 2\Phi(\|f\|_{\sup}) + v_\Phi(f; J) < \infty$. Then the integral $(RYS) \int_a^b \tilde{f} \cdot d\tilde{h}$ exists by Corollary 3.82, and so the Kolmogorov integral $\oint_J f \cdot d\mu$ exists by Corollary 2.26. Let $s \in J$. Then (3.149) follows by the Love–Young inequality (3.144) because one can approximate the Kolmogorov integral $\oint_J f \cdot d\mu$ arbitrarily closely by Kolmogorov sums $S_{YS}(f, d\mu; J, \mathcal{T})$ based on $\mathcal{T}$ having $s$ as a tag.

Now let $J = \mathbb{R}$ and $\epsilon \in (0, 2\Theta(\phi, \psi)]$. Using Propositions 3.42 and 3.50, choose $a, b \in \mathbb{R}$ such that $a < b$ and

$$4\Theta(\phi, \psi) \max\{v_\Phi(f; A), v_\Psi(\mu; A)\} \le \epsilon$$

holds when $A = (-\infty, a)$ or $A = (b, \infty)$. By Lemma 3.94, we then have for each such $A$,

$$\sum_{k=1}^{\infty} \phi\Big(\frac{v_\Phi(f; A)}{k}\Big)\psi\Big(\frac{v_\Psi(\mu; A)}{k}\Big) \le \epsilon.$$

By the first part of the proof, there is a tagged Young interval partition $(\mathcal{A}, \xi)$ of $[a, b]$ such that for any tagged refinement $\mathcal{B}$ of $\mathcal{A}$,

$$\|S_{YS}(f, \mathrm{d}\mu; [a,b], (\mathcal{A}, \xi)) - S_{YS}(f, \mathrm{d}\mu; [a,b], \mathcal{B})\| \le \epsilon.$$

Let $\mathcal{T}_0$ be a tagged Young interval partition $\mathcal{A}_0 := \{(-\infty, a), \mathcal{A}, (b, \infty)\}$ with tags $s \in (-\infty, a)$, $\xi$, and $t \in (b, \infty)$, and let $\mathcal{T}$ be a tagged Young interval partition of $J$ which is a refinement of $\mathcal{A}_0$. Let $\mathcal{T}_k$, $k = 1, 2, 3$, be the restriction of $\mathcal{T}$ to $(-\infty, a)$, $[a, b]$, and $(b, \infty)$, respectively. Then by the Love–Young inequality (3.144) for Kolmogorov sums over the intervals $(-\infty, a)$ and $(b, \infty)$, it follows that

$$\begin{aligned} &\|S_{YS}(f, \mathrm{d}\mu; J, \mathcal{T}) - S_{YS}(f, \mathrm{d}\mu; J, \mathcal{T}_0)\| \\ &\quad \le \|S_{YS}(f, \mathrm{d}\mu; (-\infty, a), \mathcal{T}_1) - f(s)\cdot\mu((-\infty, a))\| \\ &\qquad + \|S_{YS}(f, \mathrm{d}\mu; [a,b], \mathcal{T}_2) - S_{YS}(f, \mathrm{d}\mu; [a,b], (\mathcal{A}, \xi))\| \\ &\qquad + \|S_{YS}(f, \mathrm{d}\mu; (b, \infty), \mathcal{T}_3) - f(t)\cdot\mu((b, \infty))\| \le 3\epsilon. \end{aligned}$$

Thus the Kolmogorov integral $\fint_J f\cdot\mathrm{d}\mu$ exists by the Cauchy test. Now (3.149) for any $s \in J$ follows by the same Love–Young inequality (3.144) because one can approximate the Kolmogorov integral $\fint_J f\cdot\mathrm{d}\mu$ arbitrarily closely by Kolmogorov sums $S_{YS}(f, \mathrm{d}\mu; J, \mathcal{U})$ based on $\mathcal{U}$ having $s$ as a tag. The case when $J$ is a half-line can be treated similarly and therefore is omitted. The proof of Theorem 3.93 is complete. □

The following is the special case of Theorem 3.93 when $\Psi(v) \equiv v^p$ and $\Phi(u) \equiv u^q$ with $p \ge 1$, $q \ge 1$ and $p^{-1} + q^{-1} > 1$.

**Corollary 3.95.** *Assume* (1.14). *For $p \ge 1$ and $q \ge 1$ such that $p^{-1}+q^{-1} > 1$ and for a nonempty interval $J$, let $\mu \in \mathcal{AI}_p(J; Y)$ and $f \in \mathcal{W}_q(J; X)$. Then the Kolmogorov integral $\fint_J f\cdot\mathrm{d}\mu$ exists. Moreover, for any nonempty interval $A \subset J$ and any $s \in A$,*

$$\Big\|\fint_A [f - f(s)]\cdot\mathrm{d}\mu\Big\| \le K_{p,q}\|f\|_{A,(q)}\|\mu\|_{A,(p)}, \tag{3.150}$$

*where $K_{p,q} := \zeta(p^{-1} + q^{-1})$, and*

$$\Big\|\fint_A f\cdot\mathrm{d}\mu\Big\| \le K_{p,q}\|f\|_{A,[q]}\|\mu\|_{A,(p)}. \tag{3.151}$$

By Theorem 2.21, if the Kolmogorov integral $\fint_J f\cdot\mathrm{d}\mu$ exists on a nonempty interval $J$, then

$$\fint f\cdot\mathrm{d}\mu := \Big\{A \mapsto \fint_A f\cdot\mathrm{d}\mu\colon\ A \in \mathfrak{I}(J)\Big\} \tag{3.152}$$

is an additive interval function on $J$. Next we show that it is upper continuous and has bounded $p$-variation under suitable conditions. For $1 \le p < \infty$, let $\mathcal{Q}_p := [1, p/(p-1))$ if $p > 1$, and $\mathcal{Q}_1 := \{+\infty\}$. To include the case $q = \infty$, we write $\mathcal{W}_\infty(J; X)$ for the class of all regulated $X$-valued functions on $J$, and $\|\cdot\|_{A,[\infty]} := \|\cdot\|_{A,\sup}$ for nonempty $A \in \mathfrak{I}(J)$.

**Proposition 3.96.** *Assume* (1.14). *For a nonempty interval $J$, let $\mu \in \mathcal{AI}_p(J;Y)$ for $1 \le p < \infty$, and $f \in \mathcal{W}_q(J;X)$ for $q \in \mathcal{Q}_p$. Then $\fint f \cdot d\mu$ is an additive upper continuous interval function on $J$ of bounded p-variation, and for each nonempty $A \in \mathfrak{I}(J)$,*

$$\Big\| \fint f \cdot d\mu \Big\|_{A,(p)} \le K_{p,q} \|f\|_{A,[q]} \|\mu\|_{A,(p)}, \tag{3.153}$$

*where $K_{p,q} := \zeta(p^{-1} + q^{-1})$ if $p > 1$, and $K_{p,q} := 1$ if $p = 1$.*

*Proof.* We can assume that $A \subset J$ is a nondegenerate interval. The result holds clearly if $f \equiv 0$, so we can assume that $\|f\|_{\sup} > 0$. First let $p = 1$. If $J = [a, b]$ then the integral $\fint_J f \cdot d\mu$ is defined by Theorem 2.20 and Corollary 2.26. Let $J = (a, b)$ and $\epsilon > 0$. By Proposition 3.50, there are $c, d \in (a, b)$ such that $c < d$ and

$$v_1(\mu; (a, c]) + v_1(\mu; [d, b)) \le \epsilon / \|f\|_{\sup}.$$

Since the integral is defined over $[c, d]$ by the preceding references, there is an interval partition $\mathcal{A}$ of $(a, b)$ such that the norm of any difference between two Kolmogorov sums based on tagged refinements of $\mathcal{A}$ is less than $2\epsilon$. So the integral $\fint_J f \cdot d\mu$ is defined by the Cauchy test. The case when $J$ is half-open can be treated similarly and therefore is omitted. Thus the interval function (3.152) is defined on $J$ and is additive by the second part of Theorem 2.21. By Proposition 2.22 it is upper continuous. Applying the triangle inequality to Kolmogorov sums, we get

$$\Big\| \fint_A f \cdot d\mu \Big\| \le \|f\|_{A,\sup} \|\mu\|_{A,(1)}, \tag{3.154}$$

which also holds if $A$ is a singleton $\{x\} \subset J$. To prove (3.153), let $\mathcal{A} = \{A_i\}_{i=1}^n$ be a Young interval partition of $A$. Then by (3.154) and (3.69), we have

$$s_1(\fint f \cdot d\mu; \mathcal{A}) \le \|f\|_{A,\sup} \sum_{i=1}^n v_1(\mu; A_i) \le \|f\|_{A,\sup} \|\mu\|_{A,(1)}.$$

This together with Lemma 3.53 implies (3.153) with $p = 1$. If $p > 1$ the proof is the same except that now the integral $\fint_J f \cdot d\mu$ exists by the preceding corollary, and (3.151) is used instead of (3.154). The proof of the proposition is complete. □

### *Optimality of Love–Young inequalities*

Recall that $\mathcal{CV}$ is the class of all continuous increasing convex functions $\Phi\colon [0, \infty) \to [0, \infty)$ such that $\Phi(0) = 0$ and $\Phi(u) > 0$ for $u > 0$. In Section *3.4, necessary and sufficient conditions were given for $\Phi$ and $\Psi$ in $\mathcal{CV}$ to be a Stieltjes pair. The conditions were rather complex. One may ask whether in the special case $\Phi(u) \equiv u^p$, $u \ge 0$, for $1 < p < \infty$, simpler conditions could be found. If $\Psi(u) \equiv u^q$ for $u \geq 0$ and $1 < q < \infty$ also, then we know that $(\Phi, \Psi)$ form a Stieltjes pair if $p^{-1} + q^{-1} > 1$. They do not if $p^{-1} + q^{-1} = 1$. In fact, Proposition 3.104 will show that for any $p, q > 1$ with $p^{-1}+q^{-1} = 1$ there exist $f \in \mathcal{W}_p^*[0,1] \cap \mathcal{H}_{1/p}[0,1]$ and $g \in \mathcal{W}_q^*[0,1] \cap \mathcal{H}_{1/q}[0,1]$ such that $\int_0^1 f dg$ does not exist in any of the senses we have been considering. On the other hand, there are some necessity results giving $g \in \mathcal{W}_q$ under some conditions, e.g. Corollary 3.109. Thus one might ask whether there is some $\Psi \in \mathcal{CV}$ not of the form $u^q$, perhaps $u^q$ times some logarithmic factors, giving a Love–Young inequality as in (3.149) provided that $(RYS) \int f dh$ exists for all $f \in \mathcal{W}_p$. The next proposition shows that for some such $h$ there is no such $\Psi$.

Also, we write $\xi(u) \sim \eta(u)$ as $u \downarrow 0$ or $u \uparrow \infty$ if $\xi(u)/\eta(u) \to 1$ as $u \downarrow 0$ or $u \uparrow \infty$, respectively.

**Proposition 3.97.** *Let $1 < p, q < \infty$, $p^{-1} + q^{-1} = 1$. There is a right-continuous function $h \in \mathcal{W}_q^*[0,1]$ such that $h(0) = 0$ and $(RYS) \int_0^1 f\, dh$ exists for each $f \in \mathcal{W}_p[0,1]$, but there is no function $\Psi \in \mathcal{CV}$ such that*

$$h \in \mathcal{W}_\Psi[0,1] \quad \text{and} \quad \sum_{n=1}^{\infty} n^{-1/p} \Psi^{-1}(1/n) \; < \; \infty. \tag{3.155}$$

**Remark 3.98.** With $\Phi(u) \equiv u^p$ for $u \geq 0$, $\sum_{n=1}^{\infty} n^{-1/p}\Psi^{-1}(1/n) = \Theta(\Phi, \Psi)$. By Proposition 3.81 and Theorem 3.78, $\Theta(\Phi, \Psi) < \infty$ is a sufficient condition for $\Phi, \Psi$ to be a Stieltjes pair. The condition is not necessary by Example 3.80.

*Proof.* Let $\xi_q(u) := u^q \ln(1/u)^{3/2}$ for $0 < u \leq e^{-2}$ and $\xi_q(0) := 0$. Then one can check that $\xi_q$ has a derivative $\xi_q'$ on $(0, e^{-2})$ which extends to be continuous and strictly positive on $(0, e^{-2}]$. Define $\xi_q$ for $e^{-2} < u < +\infty$ so that it is linear there, and $C^1$ and strictly increasing on $(0, \infty)$, namely,

$$\xi_q(u) \; := \; \xi_q(e^{-2}) + \xi_q'(e^{-2}-)(u - e^{-2}), \qquad u > e^{-2}.$$

Let $\xi_q^{-1}$ be the inverse function of $\xi_q$. Then $\xi_q^{-1}(u) \sim (u/[\ln(1/u)]^{3/2})^{1/q}$ as $u \downarrow 0$. Let $h$ be the function on $[0,1]$ with value $\xi_q^{-1}(1/k)$ on the interval $[1/(2k+1), 1/(2k))$, $k = 1, 2, \ldots$, and 0 elsewhere. Then $\{\xi_q^{-1}(1/k) \colon k \geq 1\} \in \ell_q$ and $h \in \mathcal{W}_q^*[0,1]$ by Proposition 3.60(c) since $v_q(h; [0,\epsilon]) \leq 2\sum_{2k+1 \geq 1/\epsilon} 1/[k(\ln k)^{3/2}] \to 0$ as $\epsilon \downarrow 0$. Clearly, $h$ is right-continuous. Also, the following holds for the local $q$-variation:

$$v_q^*(h; [0,1]) \; = \; \sum_{(0,1)} |\Delta^- h|^q \; = \; \sum_{k=1}^{\infty} 2\big[\xi_q^{-1}(1/k)\big]^q \leq C \sum_{k=1}^{\infty} 1/\big[k(\ln k)^{3/2}\big] \; < \; \infty$$

for some $C < \infty$. To prove the existence of $(RYS) \int_a^b f\, dh$, let $f \in \mathcal{W}_p[0,1]$. For $m \geq 1$, let $\{x_i\}_{i=0}^n$ be a refinement of the partition $\{0, \{(2k+1)^{-1}, (2k)^{-1} \colon k = 1, \ldots, m\}, 1\}$ of $[0,1]$, and let $i(m) \in \{1, \ldots, n-2\}$ be such that $0 = x_0 < x_1 < \cdots < x_{i(m)} = (2m+1)^{-1}$. For a tagged Young partition $\tau = (\{x_i\}_{i=0}^n, \{y_i\}_{i=1}^n)$, a sum of some of the terms of the Young–Stieltjes sum $S_{YS}(f, \mathrm{d}h; \tau)$ defined by (2.16) is

$$U_{m,\tau} := \sum_{i=i(m)}^{n-1} \Big\{ [f\Delta^{\pm}h](x_i) + f(y_{i+1})[h(x_{i+1}-) - h(x_i+)] \Big\}.$$

For $i \geq i(m)$, $\Delta^{\pm}h(x_i) = \Delta^{-}h(x_i) = \xi_q^{-1}(1/k)$ if $x_i = 1/(2k+1)$, or $-\xi_q^{-1}(1/k)$ if $x_i = 1/(2k)$, both for $k = 1, \ldots, m$, and $\Delta^{\pm}h(x_i) = 0$ otherwise. We also have $h(x_{i+1}-) - h(x_i+) \equiv 0$ because $h$ is constant on the open interval $(x_i, x_{i+1})$, which is either included in some $[(2k+1)^{-1}, (2k)^{-1})$ or else $h \equiv 0$ on the interval. Thus

$$U_{m,\tau} = \sum_{k=1}^{m} \xi_q^{-1}(1/k)\big[f(1/(2k+1)) - f(1/(2k))\big].$$

Since $[f\Delta^{+}h](0) = [f\Delta^{-}h](1) = 0$, the sum of the remaining terms of $S_{YS}(f, \mathrm{d}h; \tau)$ is

$$V_{m,\tau} := \sum_{i=1}^{i(m)} f(y_i)\big[h(x_i-) - h(x_{i-1}+)\big] + \sum_{i=1}^{i(m)-1} f(x_i)\big[h(x_i+) - h(x_i-)\big].$$

We can assume that in $V_{m,\tau}$, for each $k > m$, $x_i \in ((2k+1)^{-1}, (2k)^{-1})$ for at most one value of $i$. Indeed, if $x_{i-1} \leq (2k+1)^{-1} < x_i < \cdots < x_{i+l} < (2k)^{-1} \leq x_{i+l+1}$ for some $l \geq 1$, then $\Delta^{\pm}h(x_j) = 0$ for $j = i, \ldots, i+l$ and $h(x_j-) - h(x_{j-1}+) = 0$ for $j = i+1, \ldots, i+l$. Thus we can replace the summands for $j = i+1, \ldots, i+l$ by $f(y_{i+l+1})[h(x_{i+l+1}) - h(x_i)]$, preserving the value of the sum $V_{m,\tau}$. Then we have, recalling that $h$ is right-continuous,

$$\begin{aligned} V_{m,\tau} &= \sum_{i=1}^{i(m)-1} h(x_i)\big[f(x_i) - f(y_{i+1})\big] \\ &\quad + \sum_{i=1}^{i(m)-1} h(x_i-)\big[f(y_i) - f(x_i)\big] + f(y_{i(m)})\big[h(x_{i(m)}-) - h(x_{i(m)-1})\big], \end{aligned} \tag{3.156}$$

and $V_{m,\tau} \to 0$ as $m \to \infty$ uniformly for all $x_i \leq (2m+1)^{-1}$. Indeed, the last single term tends to 0 as $m \to \infty$ because $f$ is bounded and $h(x) \to 0$ as $x \downarrow 0$. By definition of $h$, we have the bound

$$\max\Big\{ \sum_{i<i(m)} |h(x_i)|^q, \sum_{i<i(m)} |h(x_i-)|^q \Big\} \leq 2 \sum_{k>m} \big[\xi_q^{-1}(1/k)\big]^q,$$

and for $f$ we have

$$\max\Big\{ \sum_{i<i(m)} |f(x_i) - f(y_{i+1})|^p, \sum_{i<i(m)} |f(y_i) - f(x_i)|^p \Big\} \leq v_p(f; [0,1]) < \infty.$$

Thus one can apply Hölder's inequality (1.4) to show that the two sums in (3.156) tend to 0 as $m \to \infty$ uniformly for all $x_i \le (2m+1)^{-1}$. Therefore $f$ is $RYS$ integrable with respect to $h$ over $[0,1]$, and

$$(RYS)\int_0^1 f\,\mathrm{d}h \;=\; \sum_{k=1}^{\infty} \xi_q^{-1}(1/k)\big[f(1/(2k+1)) - f(1/(2k))\big],$$

where the sum converges absolutely.

Suppose that there is a $\Psi \in \mathcal{CV}$ such that (3.155) holds. Since $h \in \mathcal{W}_\Psi[0,1]$, we have

$$\begin{aligned}\infty \;&>\; \sum_{k=1}^{\infty} \Psi\circ\xi_q^{-1}(1/k) \\ &\ge\; \sum_{i=1}^{\infty} 2^{-i}\mathrm{card}\big\{k \ge 1\colon\ 2^{-i} \le \Psi\circ\xi_q^{-1}(1/k) < 2^{1-i}\big\}.\end{aligned} \tag{3.157}$$

For each $i \ge 0$, let $N_i := \mathrm{card}\{k \ge 1\colon \Psi\circ\xi_q^{-1}(1/k) \ge 2^{-i}\}$. Then for $i$ large enough so that $N_i \ge 1$, we have $1/(N_i+1) < \xi_q\circ\Psi^{-1}(2^{-i}) \le 1/N_i$. Thus letting $M_i := 1/\Psi^{-1}(2^{-i})$ for $i = 1, 2, \ldots$, it follows that $N_i$ is asymptotic to $1/\xi_q\circ\Psi^{-1}(2^{-i}) = M_i^q/\ln(M_i)^{3/2}$ as $i \uparrow \infty$. Moreover, by (3.157), we have

$$\infty \;>\; \sum_{i=1}^{\infty} 2^{-i}(N_i - N_{i-1}) \;\ge\; -N_0/2 + \sum_{i=1}^{\infty} 2^{-1-i}N_i.$$

Hence $\sum_{i=1}^{\infty} A_i < \infty$ with $A_i := M_i^q/[2^i \ln(M_i)^{3/2}]$. On the other hand, using convergence of the series in (3.155) and Proposition 3.81(i)(b), we get that $\sum_{i=1}^{\infty} B_i < \infty$ with $B_i := 2^{i/q}/M_i$. There is a finite $i_0$ such that $M_i \ge \mathrm{e}$ for $i \ge i_0$. Since $q > 1$, it follows that

$$\begin{aligned}M \;:=\; \sum_{i=i_0}^{\infty} \big[\ln(M_i)\big]^{-3/4} \;&\le\; \sum_{i=i_0}^{\infty} \big[\ln(M_i)\big]^{-3/[2(q+1)]} \\ &=\; \sum_{i=i_0}^{\infty} A_i^{1/(q+1)} B_i^{q/(q+1)} < \infty\end{aligned}$$

by Hölder's inequality (1.4). Hence $M_i \ge \exp\{((i-i_0)/M)^{4/3}\}$ for $i > i_0$. Thus

$$\limsup_{i\to\infty} 2^i \ln(M_i)^{3/2}/M_i^q \;\le\; \limsup_{i\to\infty} 2^i/M_i = 0.$$

So $A_i \to \infty$ as $i \to \infty$, and the series $\sum_i A_i$ diverges, a contradiction. □

**Note:** if right (or left) continuity is omitted from Proposition 3.97 then the examples of Proposition 3.108 would apply, making the statement less interesting.

Next is a formulation of Proposition 3.97 for interval function integrators $\mu$ rather than point functions $h$.

**Corollary 3.99.** *Let $1 < p,q < \infty$, $p^{-1} + q^{-1} = 1$. There is an interval function $\mu \in \mathcal{AI}_q^*[0,1]$ such that $⨍_{[0,1]} f\,\mathrm{d}\mu$ exists for each $f \in \mathcal{W}_p[0,1]$, but there is no function $\Psi \in \mathcal{CV}$ such that*

$$\mu \in \mathcal{AI}_\Psi[0,1] \quad \textit{and} \quad \sum_{n=1}^{\infty} n^{-1/p}\Psi^{-1}(1/n) \; < \; \infty.$$

*Proof.* Let $\mu$ be the interval function $\mu_h$ on $[0,1]$ defined by (2.2) with $h\colon [0,1] \to \mathbb{R}$ from the conclusion of Proposition 3.97. Then $\mu$ is non-zero, additive, and upper continuous by Theorem 2.8(*c*) with $h \in \mathcal{R}[0,1]$ and $h(0) = 0$. Moreover, $\mu \in \mathcal{AI}_q^*[0,1]$ by Lemma 3.29 and Propositions 3.60 and 3.30, using right-continuity of $h$. Thus by Corollary 2.26, $⨍_{[0,1]} f\,\mathrm{d}\mu = (RYS)\int_0^1 f\,\mathrm{d}h$ exists for each $f \in \mathcal{W}_p[0,1]$. By Proposition 3.30 once again, the $\Phi$-variation of $\mu$ is bounded or not depending on whether the $\Phi$-variation of $h$ is bounded or not. The conclusion then follows from Proposition 3.97. □

### *Substitution rules*

An equality between integrals $\int f\,\mathrm{d}(\int g\,\mathrm{d}\nu)$ and $\int fg\,\mathrm{d}\nu$, if it holds under some conditions, is often called a *substitution rule.* We start with one such rule for the refinement Riemann–Stieltjes integral.

**Proposition 3.100.** *Let $h \in \mathcal{W}_p[a,b]$ and $f,g \in \mathcal{W}_q[a,b]$ for some $p \geq 1$, $q \geq 1$ with $p^{-1} + q^{-1} > 1$. Suppose that the pairs $(h,g)$ and $(h,f)$ have no common one-sided discontinuities on $[a,b]$. Then $g$ and $fg$ are RRS integrable with respect to $h$, $f$ is RRS integrable with respect to the indefinite RRS integral $I_{RRS}(g,\mathrm{d}h)(y) := (RRS)\int_a^y g\,\mathrm{d}h$, $y \in [a,b]$, and*

$$(RRS)\int_a^b f\,dI_{RRS}(g,\mathrm{d}h) \;=\; (RRS)\int_a^b fg\,\mathrm{d}h\,. \tag{3.158}$$

*Proof.* We can assume that $a < b$. The *RRS* integrability of $g$ and $fg$ with respect to $h$ follows from Corollary 3.91 because $fg \in \mathcal{W}_q$ by Corollary 3.9. The indefinite integral $I(g,\mathrm{d}h) := I_{RRS}(g,\mathrm{d}h)$ is in $\mathcal{W}_p$ by Theorem 3.92. The functions $f$ and $I(g,\mathrm{d}h)$ have no common discontinuities on the same side at the same point by Theorem 2.75. Hence $f$ is *RRS* integrable with respect to $I(g,\mathrm{d}h)$ by Corollary 3.91. To prove (3.158) let $\tau = (\{x_i\}_{i=0}^n, \{y_i\}_{i=1}^n)$ be a tagged partition of $[a,b]$. Let $J_i := [x_{i-1},x_i]$ for $i = 1,\dots,n$. Since the *RRS* integral is additive for adjacent intervals by Theorem 2.73, we have

$$\begin{aligned}&\sum_{i=1}^n f(y_i)\big[I(g,\mathrm{d}h)(x_i) - I(g,\mathrm{d}h)(x_{i-1})\big] \\ &\quad = \sum_{i=1}^n f(y_i)g(y_i)\,\big[h(x_i) - h(x_{i-1})\big] + \sum_{i=1}^n f(y_i)(RRS)\int_{x_{i-1}}^{x_i} [g - g(y_i)]\,\mathrm{d}h.\end{aligned} \tag{3.159}$$

Let $p_1 > p$ and $q_1 > q$ be such that $p_1^{-1} + q_1^{-1} > 1$. By Lemma 3.45, $h \in \mathcal{W}_{p_1}[a,b]$ and $g \in \mathcal{W}_{q_1}[a,b]$. Then by the Love–Young inequality (3.147) it follows that for a finite constant $K = K_{p_1,q_1}$,

$$\begin{aligned}\sum_{i=1}^{n} |f(y_i)| \Big| (RRS) \int_{x_{i-1}}^{x_i} [g - g(y_i)] \, \mathrm{d}h \Big| &\le K \|f\|_{\sup} \sum_{i=1}^{n} \|h\|_{J_i,(p_1)} \|g\|_{J_i,(q_1)} \\ &\le \max_{1\le i\le n} \Big( (\mathrm{Osc}(h; J_i))^{(p_1-p)/p_1} \wedge (\mathrm{Osc}(g; J_i))^{(q_1-q)/q_1} \Big) \\ &\quad \times K \|f\|_{\sup} \Big( \mathrm{Osc}(h)^{(p_1-p)/p_1} \vee \mathrm{Osc}(g)^{(q_1-q)/q_1} \Big) \|h\|_{(p)}^{p/p_1} \|g\|_{(q)}^{q/q_1} . \end{aligned} \tag{3.160}$$

The last inequality follows by applying first (3.65), then writing a product $AB$ as $(A \wedge B)(A \vee B)$, using Hölder's inequality (1.4), and finally iterating (3.51). By Theorem 2.1(*b*), given any $\epsilon > 0$, there is a partition $\{u_j\}_{j=0}^m$ of $[a,b]$ such that $\mathrm{Osc}(g; (u_{j-1}, u_j)) < \epsilon$ and $\mathrm{Osc}(h; (u_{j-1}, u_j)) < \epsilon$ for each $j = 1, \ldots, m$. For $j = 1, \ldots, m$ take any $v_j \in (u_{j-1}, u_j)$. Let $n = 2m$ and $\{x_i\}_{i=0}^n = \{u_j\}_{j=0}^m \cup \{v_j\}_{j=1}^m$. Then for each $i = 1, \ldots, n$,

$$\min(\mathrm{Osc}(g; [x_{i-1}, x_i]), \mathrm{Osc}(h; [x_{i-1}, x_i])) < \epsilon,$$

since one endpoint of $[x_{i-1}, x_i]$ is in the interior of an interval $(u_{j-1}, u_j)$, and at the other, at least one of the two functions $g$ and $h$ is continuous from the side within $[x_{i-1}, x_i]$. It follows that one can make the right side of (3.160) arbitrarily small by choosing an appropriate $\tau$, and the same will hold for any refinement of $\tau$. Thus taking the limit on both sides of (3.159) under refinements of $\tau$, the relation (3.158) follows. This concludes the proof of Proposition 3.100. □

The next substitution rule, for the Riemann–Stieltjes integral, extends Proposition 2.86 (in which $p = 1$) to functions having bounded $p$-variation with $p > 1$. Recall that $\mathcal{Q}_p = [1, p/(p-1))$ if $p > 1$, $\mathcal{Q}_1 = \{+\infty\}$ and $\mathcal{W}_\infty(J; X) = \mathcal{R}(J; X)$, the class of all $X$-valued regulated functions defined on $J$.

**Proposition 3.101.** *For a Banach space $X$, $1 \le p < \infty$, and $q \in \mathcal{Q}_p$ let $h \in \mathcal{W}_p[a,b]$, $g \in \mathcal{W}_q([a,b]; X)$, and $f \in \mathcal{W}_q[a,b]$ be such that the two pairs $(h, g)$ and $(h, f)$ have no common discontinuities, and $gf\colon [a,b] \to X$ is the function defined by pointwise multiplication. Then the following Riemann–Stieltjes integrals (including those in integrands) are defined and*

$$(RS) \int_a^b \mathrm{d}I_{RS}(g, \mathrm{d}h) \cdot f = (RS) \int_a^b gf \cdot \mathrm{d}h = (RS) \int_a^b g \cdot \mathrm{d}I_{RS}(\mathrm{d}h, f), \tag{3.161}$$

*where $I_{RS}(g, \mathrm{d}h)(t) := (RS) \int_a^t g \cdot \mathrm{d}h \in X$ and $I_{RS}(\mathrm{d}h, f)(t) := (RS) \int_a^t f \, \mathrm{d}h$ for $t \in [a,b]$, and $\cdot$ denotes the natural bilinear mapping $X \times \mathbb{R} \to X$.*

*Proof.* We can assume that $a < b$ and $p > 1$. Since the pairs $(g, h)$ and $(f, h)$ have no common discontinuities, the indefinite integrals $I_{RS}(g, \mathrm{d}h)$ and $I_{RS}(\mathrm{d}h, f)$ exist and are in $\mathcal{W}_p([a, b]; X)$ and $\mathcal{W}_p[a, b]$, respectively, by Theorems 3.92 and 2.42. The discontinuities of $I_{RS}(g, \mathrm{d}h)$ and $I_{RS}(\mathrm{d}h, f)$ are subsets of those of $h$ by Theorem 2.75, and so the pairs $(I_{RS}(g, \mathrm{d}h), f)$ and $(g, I_{RS}(\mathrm{d}h, f))$ have no common discontinuities. Thus the leftmost and rightmost definite integrals in (3.161) exist by Corollary 3.91 and again by Theorem 2.42. Since $gf \in \mathcal{W}_q([a, b]; X)$ by Theorem 3.8 with $k = 2$ and the functions $gf$ and $h$ have no common discontinuities, the middle integral in (3.161) exists for the same reasons as the other two integrals. We will prove only the first equality in (3.161) since a proof of the second one is symmetric. Let $\tau = (\{t_i\}_{i=0}^n, \{s_i\}_{i=1}^n)$ be a tagged partition of $[a, b]$. Then

$$\big\|S_{RS}(\mathrm{d}I_{RS}(g, \mathrm{d}h), f; \tau) - S_{RS}(gf, \mathrm{d}h; \tau)\big\| \leq \|f\|_{\sup} R(\tau),$$

where

$$R(\tau) := \sum_{i=1}^{n} \Big\|(RS)\int_{t_{i-1}}^{t_i} g{\cdot}\mathrm{d}h - g(s_i){\cdot}\big[h(t_i) - h(t_{i-1})\big]\Big\|.$$

As in the proof of Proposition 3.100, by the Love–Young inequality (3.147) with $p_1 > p$ and $q_1 > q$ such that $p_1^{-1} + q_1^{-1} > 1$, and with the finite constant $K = K_{p_1,q_1}$, we have the bound

$$\begin{aligned}
R(\tau) &\leq K \sum_{i=1}^{n} \|h\|_{J_i,(p_1)} \|g\|_{J_i,(q_1)} \\
&\leq K \max_{1 \leq i \leq n} \Big( \big(\mathrm{Osc}(h; J_i)\big)^{(p_1-p)/p_1} \wedge \big(\mathrm{Osc}(g; J_i)\big)^{(q_1-q)/q_1} \Big) \\
&\quad \times \Big( \mathrm{Osc}(h; [a, b])^{(p_1-p)/p_1} \vee \mathrm{Osc}(g; [a, b])^{(q_1-q)/q_1} \Big) \|h\|_{(p)}^{p/p_1} \|g\|_{(q)}^{q/q_1},
\end{aligned}$$

where $J_i = [t_{i-1}, t_i]$ for $i = 1, \ldots, n$. By Theorem 2.1(*b*), given any $\epsilon > 0$, there is a partition $\{u_j\}_{j=0}^m$ of $[a, b]$ such that $\mathrm{Osc}(g; (u_{j-1}, u_j)) < \epsilon$ and $\mathrm{Osc}(h; (u_{j-1}, u_j)) < \epsilon$ for each $j = 1, \ldots, m$. Let $\delta$ be the minimum of $u_j - u_{j-1}$ for $j = 1, \ldots, m$ and let $\{t_i\}_{i=1}^n$ be a partition of $[a, b]$ with mesh less than $\delta$. Then for each $i = 1, \ldots, n$,

$$\min(\mathrm{Osc}(g; [t_{i-1}, t_i]), \mathrm{Osc}(h; [t_{i-1}, t_i])) < 2\epsilon,$$

since each $[t_{i-1}, t_i]$ contains at most one $u_j$ and either $g$ or $h$ is continuous at $u_j$. It follows that $R(\tau) \to 0$ as mesh $|\tau| \downarrow 0$. The proof of Theorem 3.101 is complete. □

A substitution rule for the Kolmogorov integral of functions with values in a Banach algebra will be given in Section 4.6 (Proposition 4.40).

## 3.7 Existence of Extended Riemann–Stieltjes Integrals

The Love–Young inequality, Corollary 3.91, showed existence of integrals $(RYS)\int_a^b f{\cdot}dg$ for $f \in \mathcal{W}_p([a,b];X)$ and $g \in \mathcal{W}_p([a,b];Y)$ if $p,q \geq 1$ and $p^{-1}+q^{-1} > 1$, and gave a bound (3.147) for such integrals. A main fact in this section will be Proposition 3.104, showing that such integrals need not exist if $p^{-1}+q^{-1}=1$, for any of the several extended Stieltjes integrals we have considered. In order to include the central Young integral among those which do not exist, we will next prove a proposition and corollary about this integral.

By Theorem 2.51($a$), for regulated functions, if their refinement Young–Stieltjes integral exists then so does the central Young integral, and the two are equal. Thus Theorem 3.89 gives sufficient conditions for existence of the central Young integral. By Proposition 2.52, there is a pair of functions $f,h$ for which $\int f\,dh$ exists in the central Young sense but not in the refinement Young–Stieltjes sense. The next fact gives a sufficient condition for the two integrals to agree.

For any interval $A \subset J$, Banach space $(X,\|\cdot\|)$, and $1 \leq q < \infty$, let

$$\ell^q(A,J;X) := \Big\{ f = \sum_i f_i 1_{\{\xi_i\}} \in c_0(A,J;X) \colon \sum_i \|f_i\|^q < \infty \Big\},$$

where $c_0(A,J;X)$ is defined in Definition 2.9. Also, recall that $\mathcal{W}_\infty([a,b];Y) := \mathcal{R}([a,b];Y)$.

**Proposition 3.102.** *Assume* (1.14) *and* $a<b$. *Let* $1 \leq p,q \leq \infty$ *with* $q^{-1}+p^{-1}=1$, *let* $h \in \mathcal{W}_p([a,b];Y)$, *and let* $f \in \mathcal{R}([a,b];X)$ *be such that* $\Delta^- f \in \ell^q((a,b),[a,b];X)$ *if* $q<\infty$. *Then* $(RYS)\int_a^b f{\cdot}dh$ *exists if and only if* $(CY)\int_a^b f{\cdot}dh$ *does, and the two integrals are equal.*

*Proof.* The proof is given only for the case $1<q<\infty$ because proofs for the cases $q=1$ and $q=\infty$ are similar. We can assume that $\|h\|_{(p)}>0$. By Theorem 2.51 it suffices to prove that the refinement Young–Stieltjes integral exists if the central Young integral does. Suppose that $(CY)\int_a^b f{\cdot}dh$ exists. Then by Proposition 2.48, $(RYS)\int_a^b f_-^{(a,b)}{\cdot}dh$ exists, where $f_-^{(a,b)}$ is defined by (2.46). Thus by Lemma 2.49, it is enough to prove that $(RYS)\int_a^b \Delta^-_{(a,b)} f{\cdot}dh$ exists, where $\Delta^-_{(a,b)} f = f - f_-^{(a,b)}$, using the definitions before (2.1) and in (2.46). By Hölder's inequality, the sum $\sum_{(a,b)} \Delta^- f{\cdot}\Delta^\pm h$ converges absolutely and so unconditionally in $Z$. Let $\epsilon>0$. Then there exists a finite set $\lambda_1 \subset (a,b)$ such that

$$\Big\| \sum_{x\in\mu} [\Delta^- f{\cdot}\Delta^\pm h](x) - \sum_{(a,b)} \Delta^- f{\cdot}\Delta^\pm h \Big\| < \epsilon$$

for any set $\mu \subset (a,b)$ including $\lambda_1$. Since $\Delta^- f \in \ell^q((a,b),[a,b];X)$ and $h \in \mathcal{W}_p([a,b];Y)$, there exists a finite set $\lambda_2 \subset (a,b)$ such that

$$\sum_{y\in\nu} \|\Delta^- f(y)\|^q < \Big(\epsilon/\|h\|_{(p)}\Big)^q$$

for any set $\nu \subset (a,b)$ disjoint from $\lambda_2$. Then $\lambda := \{a,b\}\cup\lambda_1\cup\lambda_2$ is a partition of $[a,b]$. Let $\tau = (\{t_i\}_{i=0}^n, \{s_i\}_{i=1}^n)$ be a tagged Young point partition of $[a,b]$, where $\{t_i\}_{i=0}^n$ is a refinement of $\lambda$. Recalling the definition (2.16) of Young–Stieltjes sum, by Hölder's inequality, we then have

$$\begin{aligned}
&\Big\|S_{YS}(\Delta^-_{(a,b)}f, h; \tau) - \sum_{(a,b)} \Delta^- f\cdot\Delta^{\pm} h\Big\| \\
&\quad\le \Big\|\sum_{i=1}^{n-1}[\Delta^- f\cdot\Delta^{\pm}h](t_i) - \sum_{(a,b)} \Delta^- f\cdot\Delta^{\pm} h\Big\| \\
&\quad\quad + \Big\|\sum_{i=1}^{n} \Delta^- f(s_i)\cdot[h(t_i-) - h(t_{i-1}+)]\Big\| \\
&\quad\le \epsilon + \Big(\sum_{i=1}^{n} \|\Delta^- f(s_i)\|^q\Big)^{1/q} \|h\|_{(p)} < 2\epsilon.
\end{aligned}$$

Therefore the integral $(RYS)\int_a^b \Delta^-_{(a,b)}f\cdot dh$ exists, and the proof is complete. □

The next fact follows from Proposition 3.102.

**Corollary 3.103.** *Let $1 \le p,q \le \infty$ with $p^{-1}+q^{-1}=1$, let $f \in \mathcal{W}_q([a,b];X)$, and let $h \in \mathcal{W}_p([a,b];Y)$. Then $(CY)\int_a^b f\cdot dh = (RYS)\int_a^b f\cdot dh$ if either side is defined.*

The following shows that integrals $\int_a^b f\cdot dg$ need not exist in any of several senses for $f \in \mathcal{W}_p$ and $g \in \mathcal{W}_q$ with $p^{-1}+q^{-1}=1$. Here it will be shown that $f$ can have further regularity properties, the Hölder property $f \in \mathcal{H}_{1/p}$ and $f \in \mathcal{W}_p^*$, and likewise for $q$, although of course $f \notin \mathcal{W}_r$ for any $r < p$. Here the two integrals of most interest may be the $RYS$ integral, since we have used it the most, and the Henstock–Kurzweil $HK$ integral, since it appears to be a very general integral. The Lebesgue–Stieltjes integral is very general as regards integrands $f$, but not integrators $g$, and so its nonexistence is not surprising.

**Proposition 3.104.** *If $p^{-1}+q^{-1}=1$ and $1 < p,q < \infty$, then there exist continuous functions $f \in \mathcal{W}_p^*[0,1]\cap\mathcal{H}_{1/p}[0,1]$ and $g \in \mathcal{W}_q^*[0,1]\cap\mathcal{H}_{1/q}[0,1]$ such that the integral $(A)\int_0^1 f\,dg$ does not exist for $A = RS, RRS, RYS, CY, HK$, or $LS$.*

*Proof.* For $j = 1,2,\dots$ and $t \in [0,1]$, let $\xi_j(t) := 2^{-j/p}\sin(2^{j+1}\pi t)$ and $\eta_j(t) := 2^{-j/q}(1-\cos(2^{j+1}\pi t))$. Then $\xi_j$ and $\eta_j$ are both periodic of period $1/2^j$. For $i = 1,2,\dots$, let $a_i := 1-2^{-i}$ and $J_i := [a_{i-1}, a_i]$, where

$a_0 := 0$. If $j \geq i$, then $\xi_j$ and $\eta_j$ are 0 at both endpoints of $J_i$, and $J_i$ includes $2^{j-i}$ complete periods of both functions. Also, if $j \geq i$ and $k \geq i$, then the Riemann–Stieltjes integral $(RS)\int_{a_{i-1}}^{a_i} \xi_j \, d\eta_k = 0$ for $j \neq k$ and

$$(RS)\int_{a_{i-1}}^{a_i} \xi_j d\eta_j = 2\pi \int_{a_{i-1}}^{a_i} \sin^2(2^{j+1}\pi t)\, dt = \pi/2^i.$$

Let $m_i := 2^i \geq i$ for all $i$, and on $J_i$, let $f_i := \sum_{j=m_i+1}^{m_{i+1}} \xi_j$ and $g_i := \sum_{j=m_i+1}^{m_{i+1}} \eta_j$. Then $(RS)\int_{a_{i-1}}^{a_i} f_i \, dg_i = \pi$ for all $i$. Let $f := f_i$ and $g := g_i$ on $J_i$ for each $i$ and $f(1) := g(1) := 0$. Then $f$ and $g$ are well defined and continuous at each $a_i$ for $i \geq 1$ since $\xi_j(a_i) = \eta_j(a_i) = 0$ for $j > \min(2^{i+1}, 2^i) = 2^i > i$. At $a_0 = 0$, $f = f_1$ and $g = g_1$ are right-continuous.

Let $\alpha := 1/p$. It will be shown that each $f_i$ is Hölder of order $\alpha$. For any $s, t \in [0,1]$ with $s \neq t$ we have $2^{-k-1} < |s-t| \leq 2^{-k}$ for some unique nonnegative integer $k$. Then using (3.109) and $|\sin x| \leq \min\{1, |x|\}$ for each real $x$,

$$\begin{aligned}|f_i(s) - f_i(t)| &\leq \sum_{j\geq 1} |\xi_j(s) - \xi_j(t)| \leq 2\pi|s-t|\sum_{j\leq k} 2^{(1-\alpha)j} + 2\sum_{j>k} 2^{-j\alpha} \\ &\leq \left[\frac{4\pi}{2^{1-\alpha}-1} + \frac{2}{1-2^{-\alpha}}\right]2^{-k\alpha} \leq C(\alpha)|s-t|^\alpha,\end{aligned}$$

where for $k = 0$ the sum over the empty set of indices is zero, and $C(\alpha) < +\infty$ depends only on $\alpha$. So $f_i$ is Hölder of order $\alpha = 1/p$. Since $f_i = 0$ at the endpoints of $J_i$ for each $i$, it follows that $f$ is also $\alpha$-Hölder, with $|f(s) - f(t)| \leq 2C(\alpha)(t-s)^\alpha$ for $0 \leq s \leq t < 1$, and since $f(1) = 0$, the same holds for $t = 1$. Likewise, $g$ is Hölder of order $1/q$. Thus by (1.7), $f \in \mathcal{W}_p[0,1]$ and $g \in \mathcal{W}_q[0,1]$. To show that $f \in \mathcal{W}_p^*[0,1]$, let $\epsilon > 0$. By Proposition 3.42, there is a $u \in (0,1)$ such that $v_p(f;[u,1]) < \epsilon/2$. Since $f$ is smooth on each interval $J_i$, and so is in $\mathcal{W}_1[0,u]$, $f \in \mathcal{W}_p^*[0,u]$ by Lemma 3.61. Thus by implication $(a) \Rightarrow (b)$ of Proposition 3.60, there is a partition $\{z_j\}_{j=0}^{m-1}$ of $[0,u]$ such that $\sum_{j=1}^{m-1} v_p(f;(z_{j-1}, z_j)) < \epsilon/2$. Letting $z_m := 1$, it then follows that $\sum_{j=1}^{m} v_p(f;(z_{j-1}, z_j)) < \epsilon$, and so $f \in \mathcal{W}_p^*[0,1]$ by the reverse implication $(b) \Rightarrow (a)$ of Proposition 3.60. Similarly it follows that $g \in \mathcal{W}_q^*[0,1]$.

We have that $(RS)\int_0^u f \, dg$ exists for each $u \in [0,1)$. Suppose that $(A)\int_0^1 f \, dg$ exists for $A = RS$, $RRS$, $RYS$, or $HK$. By Theorem 2.75 and property IV, defined by (2.72), we have

$$\lim_{u\uparrow 1}(A)\int_0^u f \, dg = (A)\int_0^1 f \, dg.$$

Also we have that $(A)\int_{a_{i-1}}^{a_i} f \, dg = \pi$ for each $i$ by construction if $A = RS$, or by Propositions 2.13 and 2.18 and Theorem 2.69 if $A = RRS$, $RYS$, or $HK$, respectively, a contradiction. Thus $(A)\int_0^1 f \, dg$ does not exist for $A = RS$, $RRS$, $RYS$, or $HK$. Also this integral does not exist for $A = CY$ by Corollary

3.103 and for $A = LS$ since neither $g$ nor $f$ is of bounded 1-variation (if either were, the $RS$ integral would exist by Theorems 2.17 and 2.42). The proof of Proposition 3.104 is complete. □

*The Henstock–Kurzweil integral*

We begin with necessary conditions for existence of $HK$ integrals.

**Proposition 3.105.** *Let* $1 < p, q < \infty$ *with* $p^{-1} + q^{-1} = 1$, *and let* $h$ *be a real-valued function on* $[a, b]$, $a < b$, *which is either right-continuous or left-continuous at each point. If* $(HK) \int_a^b f\,\mathrm{d}h$ *exists for each* $f \in \mathcal{W}_p^*[a, b]$, *then* $h \in \mathcal{W}_q[a, b]$.

To prove this fact we first prove two lemmas.

**Lemma 3.106.** *If* $h \notin \mathcal{W}_q[a, b]$ *with* $a < b$ *then:*

(*i*) *For at least one* $c \in [a, b]$, $v_q(h; J) = +\infty$ *for all intervals* $J$ *with* $c \in J \subset [a, b]$ *and* $J$ *relatively open in* $[a, b]$.
(*ii*) *If* $a < c < b$ *then either for all intervals* $J = [c, d]$, $c < d \leq b$, *or for all intervals* $J = [d, c]$, $a \leq d < c$, *we have* $v_q(h; J) = +\infty$ *for all such* $J$.

*Proof.* If (*i*) fails, then we have a covering of $[a, b]$ by finitely many relatively open intervals $J_i$ on each of which $v_q(h; J_i) < +\infty$, a contradiction. Statement (*ii*) then follows. □

**Lemma 3.107.** *If* $h \notin \mathcal{W}_q[a, b]$ *with* $a < b$ *then there is a* $c \in [a, b]$ *and a sequence* $\{u_j\}_{j \geq 1} \subset [a, b]$ *with* $u_j \uparrow c$ *or* $u_j \downarrow c$ *such that* $\sum_{j=1}^{\infty} |h(u_j) - h(u_{j-1})|^q = +\infty$.

*Proof.* Let $c$ be as in the preceding lemma, part (*i*). If $c \in [a, b)$ and $v_q(h; [c, d]) = +\infty$ for all $c < d \leq b$, then for a given such $d$ and some $u$, $c < u < d$, $v_q(h; [u, d]) > 1$ and $v_q(h; [c, u]) = +\infty$, so we can iterate. A similar iteration leads to the conclusion of the lemma if $c \in (a, b]$ and $v_q(h; [d, c]) = +\infty$ for all $a < d < c$. □

**Proof of Proposition 3.105.** Suppose $h \notin \mathcal{W}_q[a, b]$. Then by the preceding lemma and by symmetry, we can assume that $\sum_{j=1}^{\infty} |h(u_j) - h(u_{j-1})|^q = +\infty$ for some $u_j \downarrow c \in [a, b)$. Taking a subsequence, we can assume that $b_j := h(u_{j-1}) - h(u_j) \neq 0$ for all $j$. Replacing $u_j$ by $u_j + \delta_j$ for small enough $\delta_j > 0$ if $h$ is right-continuous at $u_j$, or replacing $u_j$ by $u_j - \delta_j$ for small enough $\delta_j > 0$ if $h$ is left-continuous at $u_j$, we can assume that $h$ is continuous at $u_j$ for all $j = 0, 1, \ldots$. There exists a sequence $\{a_j \colon j \geq 1\}$ in $\ell^p$ such that $a_j b_j > 0$ for all $j \geq 1$ and

$$\sum_{j=1}^{\infty} a_j b_j = +\infty. \tag{3.162}$$

To see this, since $b := \{b_j\}_{j\geq 1} \notin \ell^q$, it follows from well-known facts in functional analysis that there is some $\{a_j\}_{j\geq 1} \in \ell^p$ such that $\sum_{j=1}^{\infty} a_j b_j$ does not converge, e.g. [53, Theorems 6.4.1, 6.5.1]. The same will hold for some $a_j \neq 0$ for all $j$ and then for $a_j$ having the same sign as $b_j$, still with $\{a_j\}_{j\geq 1} \in \ell^p$. Let $f$ be a function on $[a,b]$ such that $f = a_j$ on $(u_j, u_{j-1})$ for each $j = 1, 2, \ldots$ and $f = 0$ elsewhere. Given $u \in (u_k, u_{k-1}]$ for some $k \geq 1$, let $f_u$ be the function on $[a,b]$ equal to $f$ on $[a,c] \cup (u,b]$ and zero on $(c,u]$. Then $f_u$ is a step function and $v_p(f - f_u) \leq 2\sum_{j\geq k} |a_j|^p \to 0$ as $u \downarrow c$, and so $f \in \mathcal{W}_p^*[a,b]$ by Proposition 3.60(c). For $c < u < b$ such that $c < u_{k+1} < u \leq u_k < \cdots < u_0 \leq b$,

$$(HK)\int_u^b f\,dh = a_{k+1}[h(u_k) - h(u)] + \sum_{j=1}^{k} a_j[h(u_{j-1}) - h(u_j)]. \tag{3.163}$$

Indeed, the integral on the left side exists by Proposition 2.56. By Lemma 2.57 applied recursively, given $\epsilon > 0$ there is a gauge function $\delta(\cdot)$ on $[u,b]$ such that for each $\delta$-fine tagged partition $\tau = (\{x_i\}_{i=0}^n, \{y_i\}_{i=1}^n)$ of $[u,b]$, the Riemann–Stieltjes sum $S_{RS}(f, dh; \tau)$ differs by at most $\epsilon$ from the integral on the left side of (3.163), and each $u_j$, $j = 0, \ldots, k$, is a tag $y_{i(j)}$ for some $i(j) \in \{1, \ldots, n\}$. Thus

$$S_{RS}(f, dh; \tau) = a_{k+1}[h(x_{i(k)-1}) - h(u)] + \sum_{j=1}^{k} a_j[h(x_{i(j-1)-1}) - h(x_{i(j)})].$$

Since $h$ is continuous at each $u_j$ and bounded by Proposition 2.60, the Riemann–Stieltjes sum can be made arbitrarily close to the sum on the right side of (3.163) by further refining $\delta$-fine tagged partitions. Likewise, $h(x_{i(k)-1})$ can be made to approach $h(u_k)$. Therefore (3.163) holds. By Proposition 2.58 and (3.162), in (3.163) letting $u \downarrow c$, it follows that $(HK)\int_c^b f\,dh = +\infty$. This contradiction proves the proposition. □

The following shows that the left- or right-continuity of the integrator in Proposition 3.105 cannot simply be omitted.

**Proposition 3.108.** *Let $h$ be a real-valued function on $[a,b]$ with $a < b$. If for some $u_k \downarrow c \in [a,b)$, $u_1 < b$, $h(u_k) \to 0$ (however slowly) and $h(x) = 0$ for all other $x \in [a,b]$, then $(RYS)\int_a^b f\,dh = (HK)\int_a^b f\,dh = 0$ for any $f\colon [a,b] \mapsto \mathbb{R}$.*

*Proof.* Given $\epsilon > 0$, define a gauge function $\delta(\cdot)$ on $[a,b]$ so that if $x \notin F := \{c\} \cup \{u_k\}_{k\geq 1}$ then $[x - \delta(x), x + \delta(x)]$ does not intersect $F$. Define each $\delta(u_k)$ so that $u_{k+1} < u_k - \delta(u_k)$ for all $k \geq 1$ and $u_k + \delta(u_k) < u_{k-1}$ for $k \geq 2$. Finally define $\delta(c)$ so that $|f(c)h(x)| < \epsilon$ for $c \leq x \leq c + \delta(c)$. Let $\tau(\delta) =$

$(\{x_i\}_{i=0}^n, \{y_i\}_{i=1}^n)$ be a $\delta$-fine tagged partition of $[a,b]$. By definition of $\delta(\cdot)$, if $x_i = y_i = u_k$ for some $k$, then $y_{i+1} = u_k$. Without changing the Riemann–Stieltjes sum $S_{RS}(f, \mathrm{d}h; \tau(\delta))$, we combine two adjoining intervals into one whenever they have a common tag. Then $h(x_i) - h(x_{i-1}) = 0$ except possibly if $y_i = c$ when $|f(c)[h(x_i) - h(x_{i-1})]| = |f(c)h(x_i)| < \epsilon$. Thus $|S_{RS}(f, \mathrm{d}h; \tau)| < \epsilon$ for any $\delta$-fine tagged partition $\tau$ of $[a,b]$ and the conclusion follows for the $HK$ integral. Since $h_-^{(a)} \equiv h_+^{(b)} \equiv 0$ on $[a,b]$ the Young–Stieltjes sums $S_{YS}(f, \mathrm{d}h; \tau)$ are all 0 by (2.16), and so the conclusion holds for the $RYS$ integrals. □

**Corollary 3.109.** *Let $p$, $q$, and $h$ be as in Proposition* 3.105. *If* $(RYS)\int_a^b f\,\mathrm{d}h$ *exists for each $f \in \mathcal{W}_p^*[a,b]$, then $h \in \mathcal{W}_q[a,b]$.*

*Proof.* In the proof of Proposition 3.105, the integral on the left side of (3.163) also exists as an $RYS$ integral with the same value because $h$ is a step function over each interval $[u,b]$ with $c < u < b$. By (3.162), $\lim_{u\downarrow c}(RYS)\int_u^b f\,\mathrm{d}h = +\infty$. On the other hand, by property IV (Theorem 2.75), the indefinite $RYS$ integral is a regulated function, and hence bounded. This contradiction proves that $h \in \mathcal{W}_q[a,b]$. □

## 3.8 Convolution and Related Integral Transforms

Suppose that the basic assumption (1.14) holds. For an interval $J = [\![a,b]\!]$ and $y \in \mathbb{R}$, let $J - y := [\![a-y, b-y]\!]$, with $-\infty - y := -\infty$ and $\infty - y := \infty$. Let $\mu$ and $\nu$ be upper continuous and additive interval functions on $\mathbb{R}$ with values in $X$ and $Y$, respectively. Letting $f_J(y) := \mu(J - y)$, $y \in \mathbb{R}$, define the interval function $\mu{*}\nu$ on $\mathbb{R}$ by

$$\mu{*}\nu(J) := \oint_{\mathbb{R}} \mu(J-y)\cdot\mathrm{d}\nu(y) \equiv \oint_{\mathbb{R}} f_J\cdot\mathrm{d}\nu$$

provided the Kolmogorov integral is defined for every interval $J \subset \mathbb{R}$. We will say that $\mu{*}\nu$ is the *convolution* of $\mu$ and $\nu$. The following relates the $p$-variation properties of the three interval functions $\mu$, $\nu$, and $\mu{*}\nu$.

**Theorem 3.110.** *Let $p, q \in [1,\infty)$ and $p^{-1} + q^{-1} > 1$. Assuming* (1.14), *let $\mu \in \mathcal{AI}_p(\mathbb{R}; X)$ and let $\nu \in \mathcal{AI}_q(\mathbb{R}; Y)$. For $r$ given by $1/r = 1/p + 1/q - 1$, we have $\mu{*}\nu \in \mathcal{AI}_r(\mathbb{R}; Z)$ and*

$$\big\|\mu{*}\nu\big\|_{(r)} \le C\|\mu\|_{(p)}\|\nu\|_{(q)},$$

*for some finite constant $C = C(p,q)$.*

Thus $(\mu,\nu) \mapsto \mu{*}\nu$ is a bounded bilinear operator from $\mathcal{AI}_p \times \mathcal{AI}_q$ into $\mathcal{AI}_r$ if $1 \le p, q, r < \infty$ and $1/p + 1/q - 1/r = 1$. The proof of the theorem is given below. It is a consequence of an analogous result for an integral transform of a more general kind, but for finite intervals, defined as follows. Let $G$ be a $Y$-valued regulated function defined on $[a,b]$, and let $H$ be an $X$-valued function on $[c,d]\times[a,b]$. Define an *integral transform* $F\equiv F(H,\mathrm{d}G)$ on $[c,d]$ with values in $Z$ by

$$F(x) := (RYS)\int_a^b H(x,y)\cdot\mathrm{d}G(y) \equiv (RYS)\int_a^b H(x,\cdot)\cdot\mathrm{d}G \qquad (3.164)$$

provided the refinement Young–Stieltjes integral exists for each $x\in[c,d]$.

An inequality bounding the $\gamma$-variation of the integral transform $F$ is proved in this section for a suitable $\gamma\in[1,\infty)$. Hypotheses on $H$ for this bound are expressed in the following terms. For $1\le\beta,p<\infty$, $a<b$, and $c<d$, let $\mathcal{W}_{\beta,p}(X)=\mathcal{W}_{\beta,p}([c,d]\times[a,b];X)$ be the class of functions $H\colon [c,d]\times[a,b]\to X$ such that $A_{\beta,\sup}(H)<\infty$ and $B_{\sup,p}(H)<\infty$, where

$$A_{\beta,\sup}(H) := A_{\beta,\sup}(H;[c,d],[a,b]) := \sup_{y\in[a,b]}\|H(\cdot,y)\|_{[c,d],(\beta)}, \qquad (3.165)$$

$$B_{\sup,p}(H) := B_{\sup,p}(H;[c,d],[a,b]) := \sup_{x\in[c,d]}\|H(x,\cdot)\|_{[a,b],(p)}. \qquad (3.166)$$

It is convenient first to bound the $\gamma$-variation of the integral transform $K := F(H-H(\cdot,a),\mathrm{d}G)$. Thus the function $K$ is defined by

$$K(x) = (RYS)\int_a^b \big[H(x,y)-H(x,a)\big]\cdot\mathrm{d}G(y), \qquad c\le x\le d,$$

provided the refinement Young–Stieltjes integral is defined for each $x$.

**Theorem 3.111.** *Let $p,q,\beta\in[1,\infty)$ and $p^{-1}+q^{-1}>1$. Assuming (1.14), $a<b$, and $c<d$, let $G\in\mathcal{W}_q([a,b];Y)$ and let $H\in\mathcal{W}_{\beta,p}([c,d]\times[a,b];X)$. Then the function $K$ is defined and for some finite constant $C=C(p,q,\beta)$, the bound*

$$\big\|K\big\|_{(\gamma)} \le CA_{\beta,\sup}(H)^{\beta/\gamma}B_{\sup,p}(H)^{1-(\beta/\gamma)}\big\|G\big\|_{(q)} \qquad (3.167)$$

*holds with $\gamma$ given by*

$$1/\gamma \ := \ (p/\beta)(1/p+1/q-1). \qquad (3.168)$$

Before starting to prove the theorem we give two corollaries. The first is a bound for the $\gamma$-variation seminorm of the function $F$ defined by (3.164) under the assumptions of Theorem 3.111. By the definition of $\gamma$ and since $q\ge1$, we have $\beta/\gamma=1+p(q^{-1}-1)\le1$, and so $\gamma\ge\beta$. By Lemma 3.45, we have

$$\|H(\cdot,a)\|_{(\gamma)} \le \|H(\cdot,a)\|_{(\beta)} \le A_{\beta,\sup}(H).$$

Since $F=K+H(\cdot,a)\cdot[G(b)-G(a)]$, and since $\|\cdot\|_{(\gamma)}$ is a seminorm, the following holds:

**Corollary 3.112.** *Under the assumptions of Theorem* 3.111 *and with the same constant* $C$, *we have the bound*

$$\begin{aligned}\|F\|_{(\gamma)} \leq & C A_{\beta,\sup}(H)^{\beta/\gamma} B_{\sup,p}(H)^{1-(\beta/\gamma)} \|G\|_{(q)} \\ & + A_{\beta,\sup}(H)\|G(b) - G(a)\|.\end{aligned} \tag{3.169}$$

A second corollary is a bound for the $r$-variation seminorm of a convolution $H{*}G$ defined on $\mathbb{R}$ by $(H{*}G)(x) = (RYS) \int_a^b H(x-y){\cdot}\mathrm{d}G(y)$ provided the refinement Young–Stieltjes integral exists for each $x \in \mathbb{R}$. Here we have in mind the case that $G$ is defined on $\mathbb{R}$, with $G(y) \equiv G(a)$ for $y \leq a$ and $G(y) \equiv G(b)$ for $y \geq b$. Then $\int_{-\infty}^{\infty} H(x-y){\cdot}\mathrm{d}G(y)$ is naturally defined as $(RYS) \int_a^b H(x-y){\cdot}\mathrm{d}G(y)$.

**Corollary 3.113.** *Let* $p, q \in [1, \infty)$ *and* $p^{-1} + q^{-1} > 1$. *Assuming* (1.14) *and* $a < b$, *let* $G \in \mathcal{W}_q([a,b];Y)$ *and let* $H \in \mathcal{W}_p(\mathbb{R};X)$. *For* $r$ *given by* $1/r = 1/p + 1/q - 1$, *we have* $H{*}G \in \mathcal{W}_r(\mathbb{R};Z)$ *and*

$$\|H{*}G\|_{\mathbb{R},(r)} \leq (1+C)\|H\|_{\mathbb{R},(p)}\|G\|_{[a,b],(q)}, \tag{3.170}$$

*where* $C = C(p,q,p)$ *is the constant in* (3.167).

*Proof.* Corollary 3.112 will be applied when $\beta = p$ and $\gamma = r$. For $-\infty < c < d < \infty$, let $\bar{H}(x,y) = H(x-y)$ for $x \in [c,d]$ and $y \in [a,b]$. Then $\bar{H} \in \mathcal{W}_{p,p}([c,d] \times [a,b];X)$ with $A_{p,\sup}(\bar{H}) \leq \|H\|_{\mathbb{R},(p)}$ and $B_{\sup,p}(\bar{H}) \leq \|H\|_{\mathbb{R},(p)}$. Given a point partition $\kappa = \{x_i\}_{i=0}^n$ of $\mathbb{R}$, by (3.169) applied to $F(\bar{H}, \mathrm{d}G)$ on $[c,d] = [x_0, x_n]$, we have

$$s_r(H{*}G;\kappa)^{1/r} \leq \|F\|_{[x_0,x_n],(r)} \leq (1+C)\|H\|_{\mathbb{R},(p)}\|G\|_{[a,b],(q)},$$

and the conclusion follows since $\kappa$ is an arbitrary partition of $\mathbb{R}$. □

*Convolution of interval functions*

Before proving Theorem 3.110 we give conditions under which the convolution of interval functions is well defined and upper continuous. For this we need a convergence theorem for Kolmogorov integrals.

**Lemma 3.114.** *Let* $1 \leq p, q < \infty$ *and* $p^{-1} + q^{-1} > 1$. *Assuming* (1.14), *for a nonempty interval* $J$, *let* $\mu \in \mathcal{AI}_q(J;Y)$ *and let* $\{f_k\}_{k\geq 1} \subset \mathcal{W}_p(J;X)$ *be such that* $\sup_k \|f_k\|_{(p)} < \infty$ *and* $f_k \to 0$ *as* $k \to \infty$ *on a dense set containing all atoms of* $\mu$. *Then* $⨍_J f_k{\cdot}\mathrm{d}\mu \to 0$ *as* $k \to \infty$.

*Proof.* We can assume that $J$ is nondegenerate. The Kolmogorov integral $⨍_J f_k{\cdot}\mathrm{d}\mu$ exists for each $k$ by Corollary 3.95. Let $\epsilon > 0$ and let $r > q$ be such that $p^{-1} + r^{-1} > 1$. By Lemma 3.55, $\mu$ has $r^*$-variation, and so by Lemma 3.29, there is a Young interval partition $\{(z_{j-1}, z_j)\}_{j=1}^m$ of $J$ such that

$$\sum_{j=1}^{m} v_r(\mu; (z_{j-1}, z_j)) < \epsilon. \tag{3.171}$$

For each $j = 1, \ldots, m$, let $t_j \in A_j := (z_{j-1}, z_j)$ be such that $f_k(t_j) \to 0$ as $k \to \infty$. For each $k \geq 1$, by additivity of the Kolmogorov integral (Theorem 2.21), we have

$$\fint_J f_k \cdot \mathrm{d}\mu = \sum_{z_j \in J} f_k(z_j) \cdot \mu(\{z_j\}) + \sum_{j=1}^{m} f_k(t_j) \cdot \mu(A_j) + \sum_{j=1}^{m} \fint_{A_j} [f_k - f_k(t_j)] \cdot \mathrm{d}\mu.$$

The first two sums on the right converge to zero as $k \to \infty$. For the last sum, by a Love–Young inequality (3.147) and applying Hölder's inequality (1.4), it follows that

$$\begin{aligned}
&\sum_{j=1}^{m} \Big\| \fint_{A_j} [f_k - f_k(t_j)] \cdot \mathrm{d}\mu \Big\| \\
&\quad \leq K_{p,r} \sum_{j=1}^{m} \|f_k\|_{A_j,(p)} \|\mu\|_{A_j,(r)} \\
&\quad \leq K_{p,r} \Big( \sum_{j=1}^{m} v_p(f_k; A_j) \Big)^{1/p} \Big( \sum_{j=1}^{m} v_r(\mu; A_j) \Big)^{1/r} \\
&\quad \leq \epsilon^{1/r} K_{p,r} \sup_k \|f_k\|_{(p)},
\end{aligned}$$

where the last inequality holds by (3.51) applied recursively, and (3.171). Since $\epsilon > 0$ is arbitrary, the conclusion follows. □

The following gives conditions for the existence of the convolution of interval functions.

**Proposition 3.115.** *Let $1 \leq p, q < \infty$ and $p^{-1} + q^{-1} > 1$. If $\mu \in \mathcal{AI}_p(\mathbb{R}; X)$ and $\nu \in \mathcal{AI}_q(\mathbb{R}; Y)$ then the convolution $\mu * \nu$ is a well-defined additive upper continuous interval function on $\mathbb{R}$ with values in $Z$.*

*Proof.* Let $f_x(y) := \mu((-\infty, x] - y)$ for $x, y \in \mathbb{R}$. For any partition $\kappa = \{t_i\}_{i=0}^n$ of $\mathbb{R}$, we have

$$s_p(f_x; \kappa) = \sum_{i=1}^{n} \big\| \mu((x - t_i, x - t_{i-1}]) \big\|^p \leq v_p(\mu; \mathbb{R}) < \infty.$$

Thus the bound $v_p(f_x; \mathbb{R}) \leq v_p(\mu; \mathbb{R})$ holds for any $x \in \mathbb{R}$. In particular, the integral $\fint_{\mathbb{R}} f_x \cdot \mathrm{d}\nu$ exists by Corollary 3.95. Similarly it follows that the function $y \mapsto \mu(J - y)$, $y \in \mathbb{R}$, has bounded $p$-variation uniformly in the interval

$J \subset \mathbb{R}$, and so the convolution $\mu * \nu$ is well defined. To show its additivity note that if $A, B \in \mathfrak{I}(\mathbb{R})$ are disjoint and $A \cup B \in \mathfrak{I}(\mathbb{R})$ then for each $y \in \mathbb{R}$, $A - y, B - y \in \mathfrak{I}(\mathbb{R})$ are disjoint and $A \cup B - y = (A - y) \cup (B - y) \in \mathfrak{I}(\mathbb{R})$. Then $\mu * \nu$ is additive since $\mu$ is additive and the Kolmogorov integral is linear with respect to the integrand by Theorem 2.72 (Property I) and Corollary 2.26. To show upper continuity of $\mu * \nu$, for $x \in \mathbb{R}$, let $g := R_{\mu * \nu, -\infty}$, that is,

$$g(x) = \mu * \nu((-\infty, x]) = \oint_{\mathbb{R}} \mu((-\infty, x - y]) \cdot d\nu(y)$$

for $x \in \mathbb{R}$. For any $x' < x''$, we have

$$g(x'') - g(x') = \oint_{\mathbb{R}} \mu((x', x''] - y) \cdot d\nu(y) \to 0$$

as $x', x'' \downarrow x$ or $x', x'' \uparrow x$ for any $x \in \mathbb{R}$ by Lemma 3.114 since $\mu$ is upper continuous and the set of all atoms of $\nu$ is at most countable by Proposition 2.6(*c*). Thus $g$ is a regulated function on $\mathbb{R}$ with values in $Z$. Moreover, for each $x \in \mathbb{R}$,

$$\oint_{\mathbb{R}} \mu((-\infty, x) - y) \cdot d\nu(y) - g(x-) = \lim_{z \uparrow x} \oint_{\mathbb{R}} \mu([z, x) - y) \cdot d\nu(y) = 0$$

by Lemma 3.114 again. Similarly $g(x+) = g(x)$ for each $x \in \mathbb{R}$. It then follows that the function $R_{\mu * \nu, -\infty} = g$ on $\mathbb{R}$ satisfies statement $(f)$ of Theorem 2.6, and so $\mu * \nu$ is upper continuous. The proof of the proposition is complete. □

Next, Theorem 3.110 will be proved assuming Theorem 3.111 and Corollary 3.112.

**Proof of Theorem 3.110.** The convolution $\mu * \nu$ is well defined by Proposition 3.115. To prove the stated bound let $\epsilon > 0$. By Proposition 3.115 again, the convolution interval function $\mu * \nu$ is upper continuous, and so there is $0 < M < \infty$ such that

$$\max \left\{ \|\mu * \nu\|_{(-\infty, -M], \sup}, \|\mu * \nu\|_{[M, \infty), \sup} \right\} < \epsilon.$$

Let $\mathcal{A} = \{A_i\}_{i=1}^n$ be an interval partition of $\mathbb{R}$. In taking the supremum of $s_r(\mu * \nu; \mathcal{A})$ over all such partitions we can assume $n \geq 4$ by upper continuity. Let $A_1 = (-\infty, u]\!]$ and $A_n = [\![v, \infty)$ for some $u < v$. Let $a := \min\{-M, u\} - 1$ and $b := \max\{M, v\} + 1$. Then

$$s_r(\mu * \nu; \mathcal{A}) \leq 2^r \epsilon^r + 2^{r-1} v_r(\mu * \nu; (a, b)), \tag{3.172}$$

as can be seen for $\rho := \mu * \nu$ by writing

$$\|\rho(A_1)\|^r = \|\rho((-\infty, a]) + \rho((a, u]\!])\|^r \leq 2^{r-1}(\epsilon^r + \|\rho((a, u]\!])\|^r)$$

and similarly for $A_n$, while $\cup_{j=2}^{n-1} A_j \subset [\![u, v]\!] \subset (a, b)$.

Let $\theta$ be an increasing homeomorphism from $(a,b)$ onto $\mathbb{R}$. For $x \in \mathbb{R}$, let $H(x,y) := \mu((-\infty,x] - \theta(y))$ if $y \in (a,b)$, $H(x,a) := H(x,a+) = \mu(\mathbb{R})$, and $H(x,b) := H(x,b-) = 0$. For any $y \in [a,b]$ and partition $\{x_i\}_{i=0}^n$ of $[a,b]$,

$$\sum_{i=1}^{n} \|H(x_i,y) - H(x_{i-1},y)\|^p = \sum_{i=1}^{n} \|\mu((x_{i-1},x_i] - \theta(y))\|^p \le v_p(\mu).$$

For each $x \in [a,b]$ and partition $\{y_j\}_{j=0}^m$ of $[a,b]$,

$$\sum_{j=1}^{m} \|H(x,y_j) - H(x,y_{j-1})\|^p = \sum_{j=1}^{m} \|\mu((x-\theta(y_j), x-\theta(y_{j-1})])\|^p \le v_p(\mu).$$

Thus $H \in \mathcal{W}_{p,p}([a,b]\times[a,b];X)$, $A_{p,\sup}(H) \le \|\mu\|_{(p)}$, and $B_{\sup,p}(H) \le \|\mu\|_{(p)}$. Let $\nu^\theta(J) := \nu(\theta(J))$ for intervals $J \subset (a,b)$ and let $\nu^\theta(\{a\}) := \nu^\theta(\{b\}) := 0$. Then $\|\nu^\theta\|_{[a,b],(q)} \le \|\nu\|_{(q)}$, and so $\nu^\theta \in \mathcal{AI}_q([a,b];Y)$. Also, let $G(y) := R_{\nu^\theta,a}(y) = \nu^\theta([a,y])$ for $y \in [a,b]$. Thus $\|G\|_{(q)} \le \|\nu\|_{(q)}$ by Proposition 3.31($a$). By a change of variables in the Kolmogorov integral (Proposition 2.30), we then have

$$\begin{aligned} R_{\mu*\nu,-\infty}(x) &= \oint_{(a,b)} \mu((-\infty,x] - \theta(y))\cdot \mathrm{d}\nu^\theta(y) \\ &= (RYS)\int_a^b H(x,y)\cdot \mathrm{d}G(y) =: F(x) \end{aligned}$$

for $x \in [a,b]$, where the next to last equality holds by Corollary 2.26 and since $\nu^\theta(\{a\}) = \nu^\theta(\{b\}) = 0$. This together with Proposition 3.31($a$) and Corollary 3.43($d$) give that $\|\mu*\nu\|_{(a,b),(r)} = \|R_{\mu*\nu,-\infty}\|_{(a,b),(r)} = \|F\|_{[a,b],(r)}$. The conclusion now follows from Corollary 3.112 with $\beta = p$, $\gamma = r$, and the bound (3.172), since the interval partition $\mathcal{A}$ of $\mathbb{R}$ and $\epsilon > 0$ are arbitrary. □

*Integral transforms*

To prepare for a proof of Theorem 3.111 we consider periodic functions. Assume that $H\colon \mathbb{R} \to X$ and $G\colon \mathbb{R} \to Y$ are periodic with period $\Lambda > 0$, and thus so is the function $z \mapsto H(z+u)\cdot G(z+v)\colon \mathbb{R} \to Z$, for any $u, v \in \mathbb{R}$. Let

$$\begin{aligned} I &:= \frac{1}{\Lambda}(RS)\int_0^\Lambda H\cdot \mathrm{d}G \quad \text{and} \\ F(z) &:= \frac{1}{\Lambda}(Bo)\int_{[0,\Lambda]} H(y-z)\cdot G(y)\,\mathrm{d}y \end{aligned} \tag{3.173}$$

for each real number $z$, provided the Riemann–Stieltjes integral and the Bochner integral with respect to Lebesgue measure (Definition 2.31) are defined.

**Lemma 3.116.** *Assuming* (1.14), *let* $H\colon \mathbb{R} \to X$ *and* $G\colon \mathbb{R} \to Y$ *be periodic with period* $\Lambda > 0$. *Let* $G$ *be continuous, let* $H$ *be bounded and* $\lambda$*-measurable for Lebesgue measure* $\lambda$, *and let the integral* $I$ *exist. Identifying* $L(\mathbb{R}, Z)$ *with* $Z$, $I$ *is the derivative of* $F$ *at* 0.

*Proof.* By the change of variables $y \mapsto y/\Lambda$, it is enough to prove the lemma for $\Lambda = 1$. It is enough to prove that

$$\lim_{t\downarrow 0} \big\|F(t) - F(0) - tI\big\|/t = 0 = \lim_{t\downarrow 0} \big\|F(0) - F(-t) - tI\big\|/t. \tag{3.174}$$

We prove only the first equality since the proof of the second one is similar. Denoting by $m := m(t) := \lfloor t^{-1} \rfloor$ the largest integer $\le t^{-1}$, for $y \in [0,1]$ and $t \in (0,1)$, let

$$U_t(y) := \sum_{i=1}^{m} H(y + (i-1)t)\cdot[G(y + it) - G(y + (i-1)t)]. \tag{3.175}$$

By periodicity, we have for each $t \in (0,1)$,

$$\begin{aligned} F(t) - F(0) &= (Bo)\int_{[0,1]} H(y)\cdot[G(y + t) - G(y)]\,\mathrm{d}y \\ &= (Bo)\int_{[0,1]} m^{-1} U_t(y)\,\mathrm{d}y. \end{aligned} \tag{3.176}$$

Next we compare $U_t(y)$ with a Riemann–Stieltjes sum for the integral $I$. Let $y \in [0,1]$, let $t \in (0,1)$, and let $k = k(y,t)$ be the minimal integer such that $y + kt > 1$. Then $k \in \{1, \ldots, m+1\}$. First suppose that $1 < k < m$. Define a tagged partition $\tau_k(y,t) = (\{t_j\}_{j=0}^{m}, \{s_j\}_{j=1}^{m})$ of $[0,1]$ by letting

$$s_j := \begin{cases} y + (j + k - 1)t - 1 & \text{for } j = 1, \ldots, m-k, \\ y + (j - m + k - 1)t & \text{for } j = m-k+1, \ldots, m, \end{cases}$$

$t_j := s_{j+1}$ for $j = 1, \ldots, m-1$, $t_0 := 0$, and $t_m := 1$. Then $t_{m-k} = y$, the mesh of $\tau_k(y,t)$ is $\le 2t$, and by periodicity, we have

$$U_t(y) = \sum_{j=1}^{m-k} H(y+(j+k-1)t-1)\cdot[G(y+(j+k)t-1) - G(y+(j+k-1)t-1)]$$

$$+ \sum_{j=m-k+1}^{m} H(y+(j-m+k-1)t)\cdot[G(y+(j-m+k)t) - G(y+(j-m+k-1)t)].$$

The $j$th term in the last display equals $H(s_j)\cdot[G(t_j) - G(t_{j-1})]$ except for $j = 1, m-k$, and $m$. Thus

$$\begin{aligned} U_t(y) = S_{RS}(H, \mathrm{d}G; \tau_k(y,t)) &+ H(y + kt)\cdot[G(0) - G(y + kt)] \\ &+ H(y + (m-1)t)\cdot[G(y + mt) - G(y)] \\ &+ H(y + (k-1)t)\cdot[G(y + kt) - G(1)] \end{aligned}$$

for any $y \in [0,1]$ and $t \in (0,1)$ such that $1 < k < m$. Let $\epsilon > 0$. There exists a $\delta > 0$ with $\delta < \epsilon/2$ such that for each $u \in [0,1]$, $\|G(u) - G(u+s)\| < \epsilon$ if $-\delta \le s \le \delta$ and $\|I - S_{RS}(H, \mathrm{d}G; \tau)\| < \epsilon$ if the mesh of a tagged partition $\tau$ is less than $2\delta$. Since the mesh of $\tau_k(y,t)$ is less than $2\delta$ when $0 < t < \delta$, it then follows that $\|U_t(y) - I\| < \epsilon + 3\epsilon\|H\|_{\sup}$ for any such $t$ and $y \in [0,1]$. Using relation (3.176), we then have

$$\begin{aligned} \frac{\|F(t) - F(0) - tI\|}{t} &\le (Bo)\int_{[0,1]} \frac{\|U_t(y) - I\|}{tm}\,dy + \Big\|I\Big(1 - \frac{1}{tm}\Big)\Big\| \\ &\le 2\epsilon[1 + 3\|H\|_{\sup}] + \epsilon\|I\| \end{aligned} \tag{3.177}$$

for any $0 < t \le \min\{\delta, 1/2\}$.

Now let $y \in [0,1]$ and $t \in (0,1)$ be such that $k = k(y,t) = 1$. By periodicity, we have

$$\begin{aligned} U_t(y) &= H(y) \cdot [G(y+t) - G(y)] \\ &\quad + \sum_{j=1}^{m-1} H(y + jt - 1)\cdot[G(y + (j+1)t - 1) - G(y + jt - 1)]. \end{aligned} \tag{3.178}$$

Let $t_{j-1} := s_j := y + jt - 1$ if $j = 2, \dots, m-1$, $s_1 := y + t - 1$, $s_m := t_{m-1} := y$, $t_0 := 0$ and $t_m := 1$. Then $\tau_1(y,t) := (\{t_j\}_{j=0}^m, \{s_j\}_{j=1}^m)$ is a tagged partition of $[0,1]$ with the mesh less than $2t$. Each term of (3.178) equals $H(s_j)\cdot[G(t_j) - G(t_{j-1})]$ except for the first term and the terms for $j = 1$ and $m - 1$. Thus

$$\begin{aligned} U_t(y) &= S_{RS}(H, \mathrm{d}G; \tau_1(y,t)) + H(y+t)\cdot[G(0) - G(y+t)] \\ &\quad + H(y)\cdot[G(y+t) - G(1)] + H(y + (m-1)t)\cdot[G(y + mt) - G(y)]. \end{aligned}$$

Lastly let $k = m$ or $k = m+1$. Let $s_j := y + (j-1)t$ if $j = 1, \dots, m$, $t_0 := 0$, $t_j := s_{j+1}$ if $j = 1, \dots, m-1$ and $t_m := 1$. Then $\tau_m(y,t) := (\{t_j\}_{j=0}^m, \{s_j\}_{j=1}^m)$ is a tagged partition of $[0,1]$. Since $y + (m-1)t \le 1$, in this case we have $y \le 2t$ and the mesh of $\tau_m(y,y)$ is less than $3t$. Each term of (3.175) equals $H(s_i)\cdot[G(t_i) - G(t_{i-1})]$ except for $i = 1$ or $m$. Thus

$$\begin{aligned} U_t(y) &= S_{RS}(H, \mathrm{d}G; \tau_m(y,t)) + H(y)\cdot[G(0) - G(y)] \\ &\quad + H(y + (m-1)t)\cdot[G(y + mt) - G(1)]. \end{aligned}$$

As in the case $1 < k < m$, given $\epsilon > 0$ there exists a $\delta > 0$ such that (3.177) holds for any $0 < t \le \min\{\delta, 1/2\}$. The first equality in (3.174) is proved, and so the lemma is proved. □

Assuming $H$, $G$, and $F$ to be as in (3.173), for each $k = 0, 1, 2, \dots$, let

$$F_k := \frac{2^k}{\Lambda}\big[F(2^{-k}\Lambda) - F(0)\big] = \frac{2^k}{\Lambda^2}(Bo)\int_{[0,\Lambda]} \big[H(y - 2^{-k}\Lambda) - H(y)\big]\cdot G(y)\,dy,$$

provided the Bochner integrals are defined. Notice that $F_0 \equiv 0$ by periodicity. The next lemma gives a relation between consecutive terms of the sequence $\{F_k\}_{k\geq 1}$.

**Lemma 3.117.** *For each $k \geq 1$, we have the equality*

$$F_k - F_{k-1} = \frac{2^{k-1}}{\Lambda^2}(Bo)\int_{[0,\Lambda]} \big[H(y-2^{-k}\Lambda) - H(y)\big]\cdot\big[G(y) - G(y+2^{-k}\Lambda)\big]\,dy.$$

*Proof.* Let $k \geq 1$. By periodicity, we have

$$\frac{\Lambda^2}{2^k}F_k = (Bo)\int_{[0,\Lambda]} \big[H(y) - H(y+2^{-k}\Lambda)\big]\cdot G(y+2^{-k}\Lambda)\,dy \tag{3.179}$$

and

$$\frac{\Lambda^2}{2^{k-1}}F_{k-1} = (Bo)\int_{[0,\Lambda]} \big[H(y-2^{-k}\Lambda) - H(y+2^{-k}\Lambda)\big]\cdot G(y+2^{-k}\Lambda)\,dy. \tag{3.180}$$

In the representation $F_k - F_{k-1} = F_k/2 + F_k/2 - F_{k-1}$, using (3.179) for the first $F_k/2$ and (3.180) for $F_{k-1}$, the desired equality follows, completing the proof. □

The next lemma will be used to bound each $\|F_k - F_{k-1}\|$.

**Lemma 3.118.** *Let $1 \leq p < \infty$, let $\Lambda > 0$, and let $H\colon \mathbb{R} \to X$ be $\lambda$-measurable for Lebesgue measure $\lambda$. If the $p$-variation of $H$ over any interval of length $\Lambda$ never exceeds $K$, then for each integer $n \geq 1$,*

$$\frac{1}{\Lambda}\int_{[0,\Lambda]} \|H(t+\Lambda/n) - H(t)\|^p\,dt \leq K/n. \tag{3.181}$$

*Proof.* Let $h(t) := H(\Lambda t)$ for each real number $t$. Then $h\colon \mathbb{R} \to X$ is a $\lambda$-measurable function with $p$-variation $v_p(h;[t,t+1]) = v_p(H;[\Lambda t, \Lambda t+\Lambda]) \leq K$ for each real number $t$. Thus using a change of variables in (3.181) we can suppose that $\Lambda = 1$. Let $n \geq 1$. Then by a change of variables again, we have

$$\begin{aligned}\int_{[0,1]} \|H(t+1/n) - H(t)\|^p\,dt &= \sum_{i=1}^n \int_{A_{in}} \|H(t+1/n) - H(t)\|^p\,dt\\ &= \int_{[0,1/n]} \sum_{i=1}^n \|H(t+i/n) - H(t+(i-1)/n)\|^p\,dt\\ &\leq \int_{[0,1/n]} v_p(H;[t,t+1])\,dt \leq K/n,\end{aligned}$$

where $A_{in} := [(i-1)/n, i/n]$ for $i = 1, \ldots, n$. The proof of Lemma 3.118 is complete. □

For $H\colon [c,d]\times[0,\Lambda]\mapsto X$ with $\Lambda>0$, and $G\colon \mathbb{R}\mapsto Y$ let

$$I(x) := I_\Lambda(x) := \frac{1}{\Lambda}(RS)\int_0^\Lambda H(x,y)\cdot\mathrm{d}G(y), \qquad x\in[c,d], \tag{3.182}$$

provided the Riemann–Stieltjes integral is defined for each $x$.

**Proposition 3.119.** *Let $p,q,\beta\in[1,\infty)$, $p^{-1}+q^{-1}>1$, $\alpha:=1/q$, $c<d$, and $\Lambda>0$. Let $H\in\mathcal{W}_{\beta,p}([c,d]\times[0,\Lambda];X)$ be such that $H(\cdot,0)\equiv H(\cdot,\Lambda)\equiv 0$, and let $G\colon\mathbb{R}\mapsto Y$ be periodic with period $\Lambda$ and be $\alpha$-Hölder continuous on $[0,\Lambda]$. Then the function $I$ is defined and there exists a finite constant $C=C(p,q,\beta)$ such that for $\gamma\geq 1$ given by (3.168), we have*

$$\begin{aligned}&\|I\|_{[c,d],(\gamma)} \\ &\quad\leq C A_{\beta,\sup}(H;[c,d],[0,\Lambda])^{\beta/\gamma} B_{\sup,p}(H;[c,d],[0,\Lambda])^{1-(\beta/\gamma)}\|G\|_{[0,\Lambda],(\mathcal{H}_\alpha)}.\end{aligned} \tag{3.183}$$

*Proof.* It is straightforward to check that since $\beta\geq 1$, we have $0<1/\gamma\leq 1$ in (3.168) and so $1\leq\gamma<\infty$.

Since $G$ is continuous and $p^{-1}+q^{-1}>1$, the function $I(x)$, $x\in[c,d]$, is well defined by Corollary 3.91 and Theorem 2.42. We extend $H(x,\cdot)$ from $[0,\Lambda]$ to $\mathbb{R}$ by periodicity. Let $y$ be a real number and let $m$ be the integer such that $y\leq m\Lambda<y+\Lambda$. Then by periodicity and since the $p$-variation seminorm is subadditive over intervals (Proposition 3.35($a$)), for each $x\in[c,d]$,

$$\begin{aligned}\|H(x,\cdot)\|_{[y,y+\Lambda],(p)} &\leq \|H(x,\cdot)\|_{[(m-1)\Lambda,m\Lambda],(p)} + \|H(x,\cdot)\|_{[m\Lambda,(m+1)\Lambda],(p)} \\ &= 2\|H(x,\cdot)\|_{[0,\Lambda],(p)} \leq 2B_{\sup,p}(H;[c,d],[0,\Lambda]).\end{aligned}$$

Likewise, for each $y$, $\|G\|_{[y,y+\Lambda],(\mathcal{H}_\alpha)}\leq 2\|G\|_{[0,\Lambda],(\mathcal{H}_\alpha)}$. If $\|G\|_{[0,\Lambda],(\mathcal{H}_\alpha)}=0$ then (3.183) holds. Moreover, (3.183) also holds if $B_{\sup,p}(H;[c,d],[0,\Lambda])=0$ because $G(\Lambda)=G(0)$. Since the inequality (3.183) is homogeneous, we can and do assume that

$$B_{\sup,p}(H;[c,d],[0,\Lambda])=1 \quad\text{and}\quad \|G\|_{[0,\Lambda],(\mathcal{H}_\alpha)}=1, \tag{3.184}$$

and so for each $y$,

$$B_{\sup,p}(H;[c,d],[y,y+\Lambda])\leq 2 \quad\text{and}\quad \|G\|_{[y,y+\Lambda],(\mathcal{H}_\alpha)}\leq 2. \tag{3.185}$$

In particular, $G$ and $H(x,\cdot)$, $x\in[c,d]$, are regulated, and so bounded measurable functions. For each $k=0,1,\ldots$ and each $x\in[c,d]$, let

$$\begin{aligned}F_k(x) &:= \frac{2^k}{\Lambda^2}(Bo)\int_{[0,\Lambda]}\big[H(x,y-2^{-k}\Lambda)-H(x,y)\big]\cdot G(y)\,\mathrm{d}y \\ &= \frac{2^k}{\Lambda^2}(Bo)\int_{[0,\Lambda]}H(x,y)\cdot\big[G(y+2^{-k}\Lambda)-G(y)\big]\,\mathrm{d}y.\end{aligned} \tag{3.186}$$

By Lemma 3.116, $I(x) = \lim_{k\to\infty} F_k(x)$ for each $x \in [c,d]$. By Lemma 3.117, (3.185), Hölder's inequality, and Lemma 3.118, respectively, for each $k \geq 1$ and $x \in [c,d]$, the following inequalities hold:

$$\begin{aligned}&\|F_k(x) - F_{k-1}(x)\| \\ &\quad \leq \frac{2^{k-1}}{\Lambda^2}\int_{[0,\Lambda]} \|H(x,y-2^{-k}\Lambda) - H(x,y)\|\,\|G(y) - G(y+2^{-k}\Lambda)\|\,dy \\ &\quad \leq \Lambda^{\alpha-1}2^{k(1-\alpha)}\Big(\frac{1}{\Lambda}\int_{[0,\Lambda]} \|H(x,y-2^{-k}\Lambda) - H(x,y)\|^p\,dy\Big)^{1/p} \\ &\quad \leq \Lambda^{\alpha-1}2^{1-(k\beta)/(p\gamma)},\end{aligned}$$

because $\alpha + 1/p - 1 = \beta/(p\gamma)$ by (3.168). Thus for each $k \geq 0$ and $x \in [c,d]$, we get the bound

$$\|I(x) - F_k(x)\| \leq \Lambda^{\alpha-1}\sum_{i=k+1}^{\infty} 2^{1-(i\beta)/(p\gamma)} = C_1 2^{-(k\beta)/(p\gamma)}, \tag{3.187}$$

where $C_1 = 2\Lambda^{\alpha-1}/(2^{\beta/(p\gamma)} - 1)$. Let $k \geq 1$ and $c \leq s < t \leq d$. By (3.186), (3.185) for $G$, and Hölder's inequality, we get

$$\begin{aligned}\|F_k(t) - F_k(s)\| &\leq \frac{2^k}{\Lambda^2}\int_{[0,\Lambda]} \|H(t,y) - H(s,y)\|\,\|G(y+2^{-k}\Lambda) - G(y)\|\,dy \\ &\leq \Lambda^{\alpha-1}2^{k(1-\alpha)+1}\Delta_\beta(s,t),\end{aligned} \tag{3.188}$$

where

$$\Delta := \Delta_\beta(s,t) := \Big(\frac{1}{\Lambda}\int_{[0,\Lambda]} \|H(t,y) - H(s,y)\|^\beta\,dy\Big)^{1/\beta}.$$

By considering three cases, we will show that for any $c \leq s < t \leq d$,

$$\|I(t) - I(s)\| \leq C_2\Delta^{\beta/\gamma}, \tag{3.189}$$

where $C_2 := 2\Lambda^{\alpha-1}[2^{1/p} + 2/(2^{\beta/(p\gamma)} - 1)]$. Assuming this is true, and applying it for a given partition $\kappa = \{t_i\}_{i=0}^n$ of $[c,d]$, it follows that

$$s_\gamma(I;\kappa)^{1/\gamma} \leq C_2\Big(\frac{1}{\Lambda}\int_{[0,\Lambda]} s_\beta(H(\cdot,y);\kappa)\,dy\Big)^{1/\gamma} \leq C_2 A_{\beta,\mathrm{sup}}(H;[c,d],[0,\Lambda])^{\beta/\gamma}.$$

Due to (3.184), the proof will be complete.

To prove (3.189), let $c \leq s < t \leq d$. First suppose that $1 < \Delta < \infty$. Then (3.189) follows from (3.187) with $k = 0$ because $2C_1 < C_2$. Second suppose that $0 < \Delta \leq 1$ and choose $k \geq 1$ so that $2^{-k/p} < \Delta \leq 2^{-(k-1)/p}$. By (3.187), it then follows that

$$\|I(x) - F_k(x)\| \leq C_1\Delta^{\beta/\gamma}$$

for all $x \in [c,d]$. Also by (3.188) and (3.168), we have

$$\|F_k(t) - F_k(s)\| \le \Lambda^{\alpha-1}2^{2-\alpha}\Delta^{\beta/\gamma}.$$

Then (3.189) is a consequence of the last two inequalities because $C_2 = 2C_1 + \Lambda^{\alpha-1}2^{1+(1/p)}$. Third suppose that $\Delta = 0$. Since $F_k \to I$ pointwise, and by (3.188),

$$\|I(t) - I(s)\| = \lim_{k\to\infty} \|F_k(t) - F_k(s)\| = 0.$$

Thus (3.189) holds in this case also, proving the proposition. □

Next we show that one can remove the periodicity condition on $G$ in the preceding proposition by increasing the constant $C$ in (3.183).

**Proposition 3.120.** *Under the hypotheses of Proposition* 3.119 *except for periodicity of* $G$, *the conclusion holds with a possibly larger constant* $C = C(p,q,\beta,\Lambda)$ *in* (3.183).

*Proof.* Define $\tilde{G}$ on $[0,\Lambda+1]$ by

$$\tilde{G}(y) := \begin{cases} G(y) & \text{if } y \in [0,\Lambda], \\ G(\Lambda) + [G(0) - G(\Lambda)](y-\Lambda) & \text{if } y \in (\Lambda,\Lambda+1]. \end{cases}$$

Then for $u \in [0,\Lambda]$ and $v \in [\Lambda,\Lambda+1]$, we have

$$\begin{aligned} \|\tilde{G}(v) - \tilde{G}(u)\| &\le \|\tilde{G}(v) - G(\Lambda)\| + \|G(\Lambda) - G(u)\| \\ &\le \|G(\Lambda) - G(0)\||v-\Lambda| + \|G\|_{(\mathcal{H}_\alpha)}|\Lambda - u|^\alpha \\ &\le \|G\|_{(\mathcal{H}_\alpha)}\Lambda^\alpha|v-\Lambda|^\alpha + \|G\|_{(\mathcal{H}_\alpha)}|\Lambda-u|^\alpha \\ &\le \|G\|_{(\mathcal{H}_\alpha)}|v-u|^\alpha(\Lambda^\alpha + 1). \end{aligned}$$

Thus $\|\tilde{G}\|_{(\mathcal{H}_\alpha)} \le (1+\Lambda^\alpha)\|G\|_{(\mathcal{H}_\alpha)}$. Also define $\tilde{H}$ on $[c,d]\times[0,\Lambda+1]$ by

$$\tilde{H}(x,y) := \begin{cases} H(x,y) & \text{if } (x,y) \in [c,d]\times[0,\Lambda], \\ 0 & \text{if } (x,y) \in [c,d]\times(\Lambda,\Lambda+1]. \end{cases}$$

Then $A_{\beta,\sup}(\tilde{H};[c,d],[0,\Lambda+1]) = A_{\beta,\sup}(H;[c,d],[0,\Lambda])$ and

$$B_{\sup,p}(\tilde{H};[c,d],[0,\Lambda+1]) = B_{\sup,p}(H;[c,d],[0,\Lambda])$$

since $H(x,\Lambda) = 0$ for all $x \in [c,d]$. For each $x \in [c,d]$, let

$$\tilde{I}(x) := \frac{1}{\Lambda+1}(RS)\int_0^{\Lambda+1} \tilde{H}(x,y)\cdot d\tilde{G}(y) = \frac{\Lambda}{\Lambda+1}I(x).$$

We can extend $\tilde{G}$ by periodicity outside the interval $[0,\Lambda+1]$. Since $\tilde{H}$ and $\tilde{G}$ satisfy all the conditions of Proposition 3.119 except for constant factors (e.g. $\Lambda+1$ in place of $\Lambda$), we can apply it to $\tilde{I}$. Then we get

$$\|I\|_{[c,d],(\gamma)} = \frac{\Lambda+1}{\Lambda}\|\tilde{I}\|_{[c,d],(\gamma)} \leq \frac{C(\Lambda+1)(\Lambda^{\alpha}+1)}{\Lambda} \times \tag{3.190}$$
$$\times A_{\beta,\sup}(H;[c,d],[0,\Lambda])^{\beta/\gamma} B_{\sup,p}(H;[c,d],[0,\Lambda])^{1-(\beta/\gamma)}\|G\|_{[0,\Lambda],(\mathcal{H}_\alpha)},$$

proving the proposition. □

Secondly we show that one can also remove the assumption $H(\cdot,\Lambda)\equiv 0$ in Proposition 3.119, again by increasing the constant $C$ in (3.183).

**Proposition 3.121.** *Under the hypotheses of Proposition* 3.119 *except for periodicity of $G$ and $H(\cdot,\Lambda)\equiv 0$, the conclusion holds with possibly a still larger constant $C$ in* (3.183).

*Proof.* Let $\tilde{H}$ be the same as $H$ except that $H(\cdot,\Lambda)$ is replaced by 0. The value of the integral $I$ does not change when $H$ is replaced by $\tilde{H}$ since $G$ is continuous. Since $B_{\sup,p}(\tilde{H}) \leq 2B_{\sup,p}(H)$ and $A_{\beta,\sup}(\tilde{H}) \leq A_{\beta,\sup}(H)$, (3.183) with a further factor $2^{1-(\beta/\gamma)}$ times $(\Lambda+1)(\Lambda^{\alpha}+1)/\Lambda$ follows from Proposition 3.120, specifically (3.190). □

Thirdly we show that for a continuous function $G$, in the preceding statement the $\alpha$-Hölder seminorm can be replaced by the $q$-variation seminorm.

**Proposition 3.122.** *Let $p,q,\beta \in [1,\infty)$, $p^{-1}+q^{-1}>1$, $c<d$, and $\Lambda>0$. Let $H \in \mathcal{W}_{\beta,p}([c,d]\times[0,\Lambda];X)$ be such that $H(\cdot,0)\equiv 0$, and let $G \in \mathcal{W}_q([0,\Lambda];Y)$ be continuous. Then the function $I$ in* (3.182) *is defined and there exists a finite constant $M = M(p,q,\beta,\Lambda)$ such that for $\gamma$ given by* (3.168), *we have*

$$\|I\|_{[c,d],(\gamma)} \tag{3.191}$$
$$\leq MA_{\beta,\sup}(H;[c,d],[0,\Lambda])^{\beta/\gamma} B_{\sup,p}(H;[c,d],[0,\Lambda])^{1-(\beta/\gamma)}\|G\|_{[0,\Lambda],(q)}.$$

*Proof.* Since $G$ is continuous and $p^{-1}+q^{-1}>1$, the function $I(x)$, $x\in[c,d]$, is well defined by Corollary 3.91 and Theorem 2.42. If $\|G\|_{[0,\Lambda],(q)}=0$ then (3.191) holds. Thus we can assume that $\|G\|_{[0,\Lambda],(q)} = (\Lambda/2)^{1/q}$. For each $y\in[0,\Lambda]$, let

$$\theta(y) := v_q(G;[0,y]) + y/2.$$

Also because $G$ is continuous, by Corollary 3.43, $\theta$ is a continuous increasing function on $[0,\Lambda]$, $\theta(0)=0$, and $\theta(\Lambda)=\Lambda$. Let $\phi$ be the inverse of $\theta$, and for each $y\in[0,\Lambda]$, let $\tilde{G}(y) := G(\phi(y))$ and $\tilde{H}(\cdot,y) := H(\cdot,\phi(y))$. For any $0\leq s<t\leq\Lambda$, we have

$$\begin{aligned}\|\tilde{G}(t)-\tilde{G}(s)\|^q &\leq v_q(G;[0,\phi(t)]) - v_q(G;[0,\phi(s)])\\ &= t-s-(1/2)[\phi(t)-\phi(s)] \leq t-s.\end{aligned}$$

Thus for $\alpha := 1/q$, $\|\tilde{G}\|_{[0,\Lambda],(\mathcal{H}_\alpha)} \le 1 = (2/\Lambda)^\alpha \|G\|_{[0,\Lambda],(q)}$. Also, we have $\tilde{H}(\cdot,0) \equiv H(\cdot,0) \equiv 0$, $A_{\beta,\sup}(\tilde{H}) \le A_{\beta,\sup}(H)$, $B_{\sup,p}(\tilde{H}) \le B_{\sup,p}(H)$, and for each $x \in [c,d]$,

$$I(x) = \frac{1}{\Lambda}(RS)\int_0^\Lambda \tilde{H}(x,y)\cdot d\tilde{G}(y)$$

by a change of variables property for the $RS$ integral (Proposition 2.77). Since all the conditions of Proposition 3.121 hold for $\tilde{G}$ and $\tilde{H}$, we get that (3.191) holds with $M = (2/\Lambda)^{1/q}C$ for the $C$ from Proposition 3.121. □

For a finite sequence $y = \{y_j\}_{j=1}^m$ in a normed space $Y$ and $1 \le q < \infty$, let $v_q(y) := v_\Phi(y;\{1,m\})$ as defined by (3.136) with $\Phi(u) = u^q$, $u \ge 0$.

**Lemma 3.123.** *For a Banach space $Y$, let $\{c_0,\dots,c_m\} \subset Y$, let $\lambda = \{z_j\}_{j=0}^m$ be a point partition of $[a,b]$ with $a < b$, and let $g\colon [a,b] \to Y$ be the function such that for $t \in [z_{j-1},z_j]$ and $j = 1,\dots,m$,*

$$g(t) = c_{j-1} + (c_j - c_{j-1})(t - z_{j-1})/(z_j - z_{j-1}).$$

*Then for any $q \ge 1$,*

$$v_q(g;[a,b]) \le 3^{q-1} v_q(\{c_j - c_{j-1}\}_{j=1}^m). \tag{3.192}$$

*For $Y = \mathbb{R}$ the factor $3^{q-1}$ in (3.192) is unnecessary.*

*Proof.* Let $q \ge 1$ and let $\kappa = \{t_i\}_{i=0}^n$ be a point partition of $[a,b]$. If $\kappa$ is a refinement of $\lambda$ then

$$\begin{aligned} s_q(g;\kappa) &= \sum_{j=1}^m \|c_j - c_{j-1}\|^q \sum_{t_i \in (z_{j-1},z_j]} \Big(\frac{t_i - t_{i-1}}{z_j - z_{j-1}}\Big)^q \\ &\le \sum_{j=1}^m \|c_j - c_{j-1}\|^q. \end{aligned} \tag{3.193}$$

Otherwise, let $A$, $B$, and $C$ be the sets of values of $i$ such that $(t_{i-1},t_i)$ contains no $z_j$, just one $z_j$, or more than one $z_j$, respectively. Let $\Delta_i := \|g(t_i) - g(t_{i-1})\|$ for each $i$. For $i \in B$ and $t_{i-1} < z_j < t_i$,

$$\Delta_i \le \|g(t_i) - g(z_j)\| + \|g(z_j) - g(t_{i-1})\|.$$

For $i \in C$, let $j_i$ and $J_i$ be the smallest and largest $j$ respectively such that $t_{i-1} < z_j < t_i$. Then

$$\Delta_i \le \|g(t_i) - g(z_{J_i})\| + \|g(z_{J_i}) - g(z_{j_i})\| + \|g(z_{j_i}) - g(t_{i-1})\|.$$

Jensen's inequality gives $(a+b+c)^q \le 3^{q-1}(a^q + b^q + c^q)$ for any $a,b,c \ge 0$. Thus $s_q(g;\kappa) \le 3^{q-1}S$, where $S$ is obtained from $\sum_{j=1}^m \|c_j - c_{j-1}\|^q$ by, for each $i \in C$ such that $J_i \geq j_i + 2$, deleting all the terms with $j = j_i + 1, \ldots, J_i$ and inserting the one term $\|c_{J_i} - c_{j_i}\|^q$. Thus $S \leq v_q(\{c_j - c_{j-1}\}_{j=1}^m)$ and the inequality (3.192) follows.

Now suppose that $Y = \mathbb{R}$. Again let $\kappa = \{t_i\}_{i=0}^n$ be a point partition of $[a, b]$. Since $u^q + v^q \leq (u + v)^q$ for any $u, v \geq 0$, we can assume that the consecutive increments $c_j - c_{j-1}$, $c_{j+1} - c_j$ of $g$ have opposite signs for $j = 1, \ldots, m-1$. Then if $g(t_{i-1}) < g(t_i) > g(t_{i+1})$, we can assume that $g$ has a local maximum at $t_i$, since otherwise $t_i$ could be replaced by such a point, increasing $s_q(g; \kappa)$. So $g(t_i) = c_j$ for some $j$. Likewise if $g(t_{i-1}) > g(t_i) < g(t_{i+1})$ then $g(t_i) = c_j$ for some $j$ since we can take a relative minimum of $g$. It then follows that $s_q(g; \kappa) = \sum_{r=1}^k |c_{j_r} - c_{j_{r-1}}|^q$ for some $0 = j_0 < j_1 < \cdots < j_k \leq m$, proving (3.192) with no factor $3^{q-1}$ in this case. The proof of the lemma is complete. $\square$

**Proof of Theorem 3.111.** Since $p^{-1} + q^{-1} > 1$, the function $K(x)$, $x \in [c, d]$, is well defined by Corollary 3.91. By the change of variables $y \mapsto (y-a)/(b-a)$, $y \in [a, b]$, it is enough to prove the theorem for $[a, b] = [0, 1]$ (Proposition 2.77). Since $A_{\beta,\sup}(H - H(\cdot, 0)) \leq 2A_{\beta,\sup}(H)$ and $B_{\sup,p}(H - H(\cdot, 0)) = B_{\sup,p}(H)$, we can and do assume that $H(\cdot, 0) \equiv 0$. To prove the theorem we will approximate each $K(x)$ by a similar integral with a continuous integrator and apply Proposition 3.122.

Let $\epsilon > 0$ and let $\kappa = \{x_j\}_{j=0}^m$ be a point partition of $[c, d]$. For each $j = 1, \ldots, m$, there exists a point partition $\lambda_j$ of $[0, 1]$ such that

$$\begin{aligned} &\Big\|(RYS)\int_0^1 \big[H(x_j, y) - H(x_{j-1}, y)\big]\cdot dG(y)\Big\|^\gamma \\ &\qquad \leq \big\|S_{YS}\big(H(x_j, \cdot) - H(x_{j-1}, \cdot), dG; \tau_j\big)\big\|^\gamma + \epsilon/m \end{aligned}$$

for each Young tagged point partition $\tau_j$ of $[0, 1]$ which is a refinement of $\lambda_j$, where $S_{YS}$ is the Young–Stieltjes sum defined by (2.16). Let $\{t_i\}_{i=0}^n$ be a simultaneous refinement of all $m$ partitions $\lambda_j$ and let $s_i := (t_{i-1} + t_i)/2$ for $i = 1, \ldots, n$. Then $\tau = (\{t_i\}_{i=0}^n, \{s_i\}_{i=1}^n)$ is a tagged Young point partition of $[0, 1]$ such that

$$s_\gamma(K; \kappa) \leq \epsilon + \sum_{j=1}^m \big\|K(x_j; \tau) - K(x_{j-1}; \tau)\big\|^\gamma, \tag{3.194}$$

where $K(x; \tau) := S_{YS}(H(x, \cdot), dG; \tau)$ for $x \in [c, d]$. Let $\mu := \{u_{i-1}, v_i\}_{i=1}^n$ be such that $t_{i-1} < u_{i-1} < s_i < v_i < t_i$ for $i = 1, \ldots, n$, $v_0 := 0$, and $u_n := 1$. Also, let $w_0 := 0 = t_0$, $w_{2i-1} := s_i$, and $w_{2i} := t_i$ for $i = 1, \ldots, n$. Then $\tau(\mu) := (\{v_i, u_i\}_{i=0}^n, \{w_i\}_{i=0}^{2n})$ is a tagged partition of $[0, 1]$. Define the function $g$ on $[0, 1]$ by

$$g(t) := G(u) + [G(v) - G(u)](t - u)/(v - u), \qquad t \in [u, v],$$

for each $[u,v]=[u_{i-1},v_i]$, $i=1,\dots,n$, and each $[u,v]:=[v_i,u_i]$, $i=0,\dots,n$. Also define the function $h$ on $[c,d]\times[0,1]$ by letting for each $x\in[c,d]$,

$$h(x,y):=\begin{cases} H(x,s_i) \text{ if } y\in[u_{i-1},v_i) \text{ for some } i=1,\dots,n, \\ H(x,t_i) \text{ if } y\in[v_i,u_i) \text{ for some } i=0,\dots,n,\end{cases}$$

and $h(x,1):=H(x,1)$. Notice that $h(\cdot,0)=H(\cdot,0)\equiv 0$. It is easy to verify that $A_{\beta,\sup}(h)\le A_{\beta,\sup}(H)$ and $B_{\sup,p}(h)\le B_{\sup,p}(H)$. Also for each $x\in[c,d]$, we have

$$S_{RS}(H(x,\cdot),\mathrm{d}G;\tau(\mu))=(RS)\int_0^1 h(x,y)\cdot\mathrm{d}g(y)=:I_{RS}(x;\tau(\mu)).$$

For each $x\in[c,d]$, letting $u_{i-1}\downarrow t_{i-1}$ and $v_i\uparrow t_i$ for each $i=1,\dots,n$, it follows that $I_{RS}(x;\tau(\mu))\to K(x;\tau)$. Thus one can choose $\tau(\mu)$ such that for the given $\kappa=\{x_j\}_{j=0}^m$,

$$\sum_{j=1}^m \|K(x_j;\tau)-K(x_{j-1};\tau)\|^\gamma\le\epsilon+\sum_{j=1}^m\|I_{RS}(x_j;\tau(\mu))-I_{RS}(x_{j-1};\tau(\mu))\|^\gamma.$$

By Lemma 3.123, $\|g\|_{(q)}\le 3^{1-(1/q)}\|G\|_{(q)}$. Applying (3.194) and Proposition 3.122 with $\Lambda=1$, $H=h$, and $G=g$ to $I_{RS}(\cdot;\tau(\mu))$, it follows that

$$\begin{aligned} s_\gamma(K;\kappa)^{1/\gamma}&\le 2\epsilon^{1/\gamma}+MA_{\beta,\sup}(h)^{\beta/\gamma}B_{\sup,p}(h)^{1-(\beta/\gamma)}\|g\|_{(q)}\\ &\le 2\epsilon^{1/\gamma}+3^{1-(1/q)}MA_{\beta,\sup}(H)^{\beta/\gamma}B_{\sup,p}(H)^{1-(\beta/\gamma)}\|G\|_{(q)}.\end{aligned}$$

Since $\epsilon>0$ and $\kappa$ are arbitrary, (3.167) holds. The proof of Theorem 3.111 is complete. □

## 3.9 Notes

*Notes on Section 3.1.* L. C. Young [246] suggested the notion of $\Phi$-variation and attributed the introduction of the power case $\Phi(u)\equiv u^p$ to Norbert Wiener. Wiener [241] actually defined $\bar{v}_p(f)$ as in (3.79). We do not know of a definition of $v_p$, as opposed to $\bar{v}_p$, before that of Marcinkiewicz [156].

Theorem 3.7 is due to Musielak and Orlicz [176, pp. 32–33]. It is analogous to corresponding results for the Luxemburg norm on Orlicz spaces.

The case of Corollary 3.9 for $p$-variation, when $\Phi(u)\equiv u^p$ for $u\ge 0$ for some $p$ with $p\ge 1$, was noted by G. Krabbe [124], [125].

The definition (3.17) of the complementary function and Proposition 3.11 are due to Birnbaum and Orlicz [18]. W. H. Young [253] discovered the inequality (3.21) and proved it under additional conditions on $P$ and $Q$.

Proposition 3.13 is a special case of Theorem 3 of Luxemburg [148, Section II.2], where a notion of complementary function allowing infinite values was used. The example following Proposition 3.12 and showing sharpness of the inequalities (3.24) was suggested by a similar example in [148, Section II.2, p. 49] showing sharpness of (3.27).

*Notes on Section 3.2.* The $\Phi$-variation of an additive set function was defined by I. Kluvánek [118] as follows. A set function is any function from a family $\mathcal{Q}$ of sets into a Banach space. It is called *additive* if whenever $A$ and $B$ are disjoint sets in $\mathcal{Q}$ such that $A \cup B \in \mathcal{Q}$, we have $\mu(A \cup B) = \mu(A) + \mu(B)$. A family $\mathcal{Q}$ of subsets of a set $\Omega$, containing the empty set, is called a *semiring* if for any $A$ and $B$ in $\mathcal{Q}$, we have $A \cap B \in \mathcal{Q}$ and $A \setminus B = \cup_{1 \le j \le n} C_j$ for some finite $n$ and disjoint $C_j \in \mathcal{Q}$. A family $\mathcal{P}$ of disjoint sets in $\mathcal{Q}$ is called a *partition* if for every $A \in \mathcal{Q}$, the subfamily $\{B \in \mathcal{P}\colon\ B \cap A \neq \emptyset\}$ of $\mathcal{P}$ is finite. The set of all partitions is denoted by $\Pi = \Pi(\mathcal{Q})$. Let $\Phi \in \mathcal{CV}$ be such that $\lim_{t \downarrow 0} \Phi(t)/t = 0$ and $\lim_{t \to \infty} \Phi(t)/t = \infty$. Let $(X, \|\cdot\|)$ be a Banach space, let $\mathcal{Q}$ be a semiring of subsets in a set $\Omega$, and let $\mu\colon\ \mathcal{Q} \to X$ be an additive set function. For $A \in \mathcal{Q}$ and a partition $\mathcal{P} \in \Pi$, let

$$s_\Phi(\mu, \mathcal{P}; A) := \sum_{B \in \mathcal{P}} \Phi\big(\|\mu(A \cap B)\|\big).$$

For a set of partitions $\Delta \subset \Pi$ and $A \in \mathcal{Q}$, let

$$v_\Phi(\mu, \Delta; A) := \sup\big\{s_\Phi(\mu, \mathcal{P}; A)\colon\ \mathcal{P} \in \Delta\big\}.$$

Then [118, p. 108] the $\Phi$-variation of a set function $\mu$ with respect to the family of partitions $\Delta$ is the set function $A \mapsto v_\Phi(\mu, \Delta; A)$, $A \in \mathcal{Q}$. The definition reduces to Definition 3.16 in case $\mathcal{Q} = \Delta$ is the semiring of all subintervals of a nonempty interval $J$ and $\mu$ is an additive interval function on $J$.

*Notes on Section 3.3.* The quantity $\bar{v}_p(f)$ related to the local $p$-variation was introduced by N. Wiener [241] and later used by many authors. The local $p$-variation $v_p^*(f)$ first appeared in Love and Young [147, p. 29].

*Notes on Section *3.4.* Theorem 3.75 extends Theorem 2 of D'yačkov [60], which proved equivalence of (*b*) through (*e*) and the part of (*a*) about Riemann–Stieltjes integrability. The main ingredient in the proof, an inequality for Riemann–Stieltjes sums (Proposition 3.73), is also due to D'yačkov [60].

Rubinšteĭn [197] introduced the class of $\xi$-lacunary sequences (as in Definition 3.64) in connection with his work on Fourier series of functions in Hölder classes.

*Notes on Section 3.6.* L. C. Young [244] proved (3.129) with a constant $K = 1 + \zeta(p^{-1} + q^{-1})$ when the full Stieltjes integral is replaced by the $Y_0$ integral defined in the notes to Section 2.5 (see Section 2.11). He also noted that the first, unpublished, proof of an inequality for finite sums, Theorem 3.83 for $\Phi(u) = u^p$, $\Psi(u) = u^q$, $u \ge 0$, for real-valued $f$ and $h$ on the same interval, was due to E. R. Love. Given such an inequality for finite sums, it seems that the main remaining difficulty in obtaining an inequality for integrals is to define an extension of the Riemann–Stieltjes integral to functions possibly having common (even one-sided) discontinuities, as Young did with the $Y_0$ integral.

In fact Love (unpublished) and L. C. Young [244] proved somewhat more, namely that a term $\|f\|_{(p)}\|h\|_{(q)}$ on the right side of an inequality such as (3.142) can be replaced by a supremum $\sup_\kappa (s_p(f;\kappa)^{1/p} s_q(h;\kappa)^{1/q})$, where $\kappa$ ranges over partitions of the given interval.

Love [144, p. 290] wrote in 1993: "L. C. Young (1936), partly in conjunction with the present writer, showed that Riemann–Stieltjes integration could reach out into areas inaccessible to Lebesgue–Stieltjes integration." There were previously known cases of such a phenomenon. For example, when $f$ is of bounded variation and $h$ is continuous but of unbounded variation, $(RS)\int_a^b f\,\mathrm{d}h$ exists and can be evaluated by integration by parts, whereas the Lebesgue–Stieltjes integral as it stands does not exist. But Love and Young found a much broader class of cases.

A year after L. C. Young's 1936 paper [244], a related joint paper by Love and Young appeared [146]. More recently, Love in 1993 [144] modified L. C. Young's proof to give the constant in (3.129), namely $K = \zeta(p^{-1} + q^{-1})$ as opposed to $1 + K$.

L. C. Young in 1938 [247, Theorem 5.1] proved that if $f \in \mathcal{W}_\Phi[a,b]$ and $h \in \mathcal{W}_\Psi[a,b]$ with $\Phi, \Psi \in \mathcal{V}$ having inverses $\phi$, $\psi$, respectively, such that $\Theta(\phi,\psi) < \infty$ defined by (3.126), then the integral $(RYS)\int_a^b f\,\mathrm{d}h$ exists and for any $s \in [a,b]$, the inequality

$$\Big|(RYS)\int_a^b [f - f(s)]\,\mathrm{d}h\Big| \le 20\sum_{k=1}^{\infty}\phi\Big(\frac{v_\Phi(f)}{k}\Big)\psi\Big(\frac{v_\Psi(h)}{k}\Big) \tag{3.195}$$

holds. Our Theorem 3.89 gives a conclusion of the form (3.195) under the additional hypothesis that $\Phi$ and $\Psi$ are convex but without the constant 20.

A function $\Phi \in \mathcal{V}$ is *log-convex* if

$$\Phi(u^\alpha v^\beta) \;\le\; \alpha\Phi(u) + \beta\Phi(v) \quad \text{for } u, v > 0 \text{ and } \alpha,\beta \ge 0,\ \alpha + \beta = 1.$$

Leśniewicz [138] By the inequality $u^\alpha v^\beta \le \alpha u + \beta v$ (Lemma 5.1.3 in [53]), if $\Phi \in \mathcal{V}$ is convex then it is log-convex. A function $\Phi \in \mathcal{V}$ is log-convex if and only if $\Phi(u) = \chi(\ln u)$ for $u > 0$, where $\chi$ is a convex function on $\mathbb{R}$. Related to Theorem 3.89 is Theorem 4.01 of Leśniewicz and Orlicz [139]: for log-convex functions $\Phi, \Psi \in \mathcal{V}$ with inverses $\phi, \psi$, and real-valued functions $f \in \mathcal{W}_\Phi[a,b]$, $h \in \mathcal{W}_\Psi[a,b]$, if their Riemann–Stieltjes integral is defined then

$$\Big|(RS)\int_a^b f\,\mathrm{d}h\Big| \le \phi(v_\Phi(f))\psi(v_\Psi(h)) + \sum_{k=1}^{\infty}\phi\Big(\frac{v_\Phi(f)}{k}\Big)\psi\Big(\frac{v_\Psi(h)}{k}\Big).$$

Lemma 3.86 can easily be proved for log-convex functions $\Phi, \Psi \in \mathcal{V}$. That lemma is the only part of the proof of Theorem 3.83 where convexity is used, so the theorem would follow for log-convex $\Phi$ and $\Psi$ in $\mathcal{V}$. Examples of log-convex functions in $\mathcal{V}$ that are not convex are given by $\Psi(u) = u^p$ for $0 < p < 1$. For such $\Psi$, however, a function in $\mathcal{W}_\Psi$ must have bounded variation (cf. Corollary 3.39), and so the existence of the full Stieltjes integral with suitable bounds holds whenever the other function is regulated (cf. Theorem 2.84).

*Notes on Section 3.7.* L. C. Young [244] gave the counterexamples in Proposition 3.104 (except that he did not consider the $HK$ integral). Leśniewicz and Orlicz [139] modified this example to show that the series condition (3.126) is best possible for convex functions $\Phi$ and $\Psi$ satisfying the $\Delta_2$ and $\nabla_2$ growth conditions: for some constants $c, d > 1$ and $u_0 > 0$, $\Phi(2u) \le c\Phi(u)$ and $2\Phi(u) \le d^{-1}\Phi(du)$ for $0 \le u \le u_0$, respectively, and the same for $\Psi$ instead of $\Phi$.

*Notes on Section 3.8.* L. C. Young [245, pp. 458–459] gave a proof of an earlier version of Theorem 3.111 under certain periodicity and continuity assumptions. His result when applied to the convolution $F{*}G$ yields (3.170) with $p^{-1} + q^{-1} - r^{-1} > 1$. The conclusion fails for $p^{-1} + q^{-1} - r^{-1} < 1$. Theorem 3.111 for real-valued and periodic functions is due to F. W. Gehring [74, Theorem 3.2].

Corollary 3.113 is a slight extension of a statement of Gehring [74, Theorem 4.1.4]. L. C. Young [248, Theorem 7.8] generalized Corollary 3.113 to functions with bounded $\Phi$-variation using a different extended Riemann–Stieltjes integral. There is a corresponding theorem in A. Zygmund [258, (1.26) on p. 39] for a convolution $(F{*}G)(x) := (2\pi)^{-1} \int_0^{2\pi} F(x-y)\,dG(y)$ defined in the Riemann–Stieltjes sense for each $x$ such that the functions $F(x-\cdot)$ and $G$ have no discontinuities in common, when $F(x+2\pi) - F(x)$ is constant for $x \in \mathbb{R}$, $F$ has bounded variation on $[0, 2\pi]$, and $G$ satisfies similar conditions.

More general results than Lemma 3.118 for real-valued functions are due to J. Marcinkiewicz [156, Theorem 4] and F. W. Gehring [74, Theorem 1.3.3].

# 4

# Banach Algebras

Throughout this chapter $\mathbb{K}$ denotes either the field $\mathbb{R}$ of real numbers or the field $\mathbb{C}$ of complex numbers.

**Definition 4.1.** A vector space $\mathbb{A}$ over $\mathbb{K}$ is called an *algebra over* $\mathbb{K}$ if it is equipped with a binary operation, referred to as multiplication and denoted by juxtaposition, from $\mathbb{A} \times \mathbb{A}$ to $\mathbb{A}$ such that for $x, y, z \in \mathbb{A}$ and $r \in \mathbb{K}$,

(*a*) $x(yz) = (xy)z$;
(*b*) $(x + y)z = xz + yz$, $x(y + z) = xy + xz$;
(*c*) $r(xy) = (rx)y = x(ry)$.

An algebra $\mathbb{A}$ is *commutative* if for all $x, y \in \mathbb{A}$,

(*d*) $xy = yx$.

An algebra $\mathbb{A}$ is *unital* if there exists an *identity* element $\mathbb{I} \in \mathbb{A}$ such that $\mathbb{I} \neq 0$ and for all $x \in \mathbb{A}$,

(*e*) $\mathbb{I}x = x\mathbb{I} = x$.

If $\mathbb{K} = \mathbb{R}$, $\mathbb{A}$ is called a *real algebra*, and if $\mathbb{K} = \mathbb{C}$, a *complex algebra*. A subset of an algebra $\mathbb{A}$ is a *subalgebra* of $\mathbb{A}$ if it is an algebra with respect to the operations of $\mathbb{A}$.

It is clear that if $\mathbb{A}$ is a unital algebra, then the identity element $\mathbb{I}$ is unique.

**Definition 4.2.** An element $x$ of a unital algebra $\mathbb{A}$ is said to be *invertible* with *inverse* $y$ if $yx = xy = \mathbb{I}$.

If $y$ and $z$ are two inverses of $x$, then $y = yxz = z$, so the inverse is unique. Let $x^{-1} := y$.

**Definition 4.3.** Let $S$ be a nonempty set and $\mathbb{A}$ a unital algebra over $\mathbb{K}$. The *pointwise operations* on $S$ are defined as follows. For $\mathbb{A}$-valued functions $f, g$ on $S$, a scalar $r \in \mathbb{K}$, and any $s \in S$, let

R.M. Dudley and R. Norvaiša, *Concrete Functional Calculus*, Springer Monographs in Mathematics, DOI 10.1007/978-1-4419-6950-7_4,

$$(rf)(s) = rf(s), \qquad (f+g)(s) = f(s) + g(s),$$
$$(fg)(s) = f(s)g(s), \qquad \mathbb{1}(s) := \mathbb{1}.$$

The set $\mathbb{A}^S$ of all $\mathbb{A}$-valued functions on $S$ with pointwise operations is an algebra with identity $\mathbb{1}(\cdot)$. An *algebra of* $\mathbb{A}$*-valued functions* on $S$ is a subalgebra of $\mathbb{A}^S$. An algebra of $\mathbb{K}$-valued functions will be called an *algebra of functions.*

For $S$ and $\mathbb{A}$ as in the preceding definition, let $\emptyset \neq E \subset S$ with $E \neq S$ and $\mathbb{V} := \{f \in \mathbb{A}^S \colon f = 0 \text{ on } S \setminus E\}$. Then $\mathbb{V}$ is a unital algebra of $\mathbb{A}$-valued functions on $S$ with identity $1_E \cdot \mathbb{1}$, but contains no constant functions (also if $\mathbb{A} = \mathbb{K}$). If $\mathbb{A}$ is commutative, then so is any algebra of $\mathbb{A}$-valued functions. Although $\mathbb{C}$ is an algebra over $\mathbb{R}$, for an algebra of $\mathbb{C}$-valued functions (which are not all $\mathbb{R}$-valued) we always take $\mathbb{K} = \mathbb{C}$.

## 4.1 Ideals and Normed Algebras

Let $\mathbb{A}$ be a commutative algebra over $\mathbb{K}$. Then an *ideal* in $\mathbb{A}$ is a subalgebra $I$ of $\mathbb{A}$ such that $xy \in I$ for all $x \in \mathbb{A}$ and $y \in I$. $I$ is a *proper* ideal if $I \neq \mathbb{A}$. A *maximal* ideal is a proper ideal which is not strictly included in any proper ideal.

**Theorem 4.4.** *Let* $\mathbb{A}$ *be a unital commutative algebra over* $\mathbb{K}$ *and* $I$ *a proper ideal in* $\mathbb{A}$*. Then* $I$ *is included in some maximal ideal.*

*Proof.* Proper ideals are partially ordered by inclusion. The identity $\mathbb{1}$ belongs to no proper ideal. The union of an inclusion-chain of proper ideals is an ideal and does not contain $\mathbb{1}$, so it is proper. The conclusion then follows from Zorn's lemma. □

A mapping $f$ from one algebra $\mathbb{A}$ over $\mathbb{K}$ to another $\mathbb{D}$ is called a *homomorphism* if $f(cx + y) = cf(x) + f(y)$ and $f(xy) = f(x)f(y)$ for all $x, y \in \mathbb{A}$ and $c \in \mathbb{K}$. For any commutative algebra $\mathbb{A}$ over $\mathbb{K}$ and ideal $I$ in $\mathbb{A}$, we can define cosets $x + I$ for $x \in \mathbb{A}$ by $x + I := \{x + y \colon y \in I\}$. Then $u + I = v + I$ if and only if $u - v \in I$. Let $\mathbb{A}/I$ be the set of all cosets $x + I$ for $x \in \mathbb{A}$. If we define algebra operations on $\mathbb{A}/I$ by

$$(x+I)+(u+I) := x+u+I, \qquad (x+I)\,(u+I) := xu+I, \qquad c(x+I) := cx+I,$$

for $x, u \in \mathbb{A}$ and $c \in \mathbb{K}$, then these operations are well defined because $I$ is an ideal. Thus $\mathbb{A}/I$ is an algebra over $\mathbb{K}$ and $x \mapsto x + I$ is a homomorphism from $\mathbb{A}$ onto $\mathbb{A}/I$.

**Theorem 4.5.** *Let* $\mathbb{A}$ *be a unital commutative algebra over* $\mathbb{K}$ *and* $L$ *a maximal ideal in* $\mathbb{A}$*. Then* $\mathbb{A}/L$ *is a unital commutative algebra in which all nonzero elements have inverses.*

*Proof.* Clearly $\mathbb{1} + L$ is the identity in $\mathbb{A}/L$. The 0 element is $L = 0 + L$. For any $x \in \mathbb{A}$, $x + L$ is a nonzero element in $\mathbb{A}/L$ if and only if $x \notin L$. Let $x \in \mathbb{A}$ and $x \notin L$. We need to show that $x + L$ has an inverse. Let $I := \{xy + u\colon y \in \mathbb{A},\ u \in L\}$. Then $I$ is an ideal strictly including $L$, so $I = \mathbb{A}$. Thus $xy + u = \mathbb{1}$ for some $y \in \mathbb{A}$ and $u \in L$, so $xy \in \mathbb{1} + L$ and $y + L$ is the desired inverse of $x + L$. □

**Definition 4.6.** A normed vector space $(\mathbb{B}, \|\cdot\|)$ over $\mathbb{K}$ will be called a *normed algebra* over $\mathbb{K}$ if $\mathbb{B}$ is an algebra over $\mathbb{K}$ and $(x, y) \mapsto xy$ is a bounded bilinear operator $\mathbb{B} \times \mathbb{B} \to \mathbb{B}$, so that for some $M < \infty$,

$$\|xy\| \le M\|x\|\,\|y\| \tag{4.1}$$

for all $x, y \in \mathbb{B}$.

If a normed algebra lacks an identity then one can be adjoined, as follows.

**Proposition 4.7.** *Any normed algebra $\mathbb{B}$ is included in a unital one.*

*Proof.* If $\mathbb{B}$ is not unital, let $\mathbb{U} := \mathbb{B} \times \mathbb{K}$ with multiplication $(x, z)(y, w) := (xy + zy + wx, zw)$ and norm $\|(x, z)\| := \|x\| + |z|$. Then $\mathbb{U}$ is a unital normed algebra with identity $\mathbb{1} = (0, 1)$. $\mathbb{B}$ is isometric as a normed algebra to $\{(x, 0)\colon x \in \mathbb{B}\} \subset \mathbb{U}$. □

As the following shows, it can be assumed that $M$ in (4.1) equals 1 and $\|\mathbb{1}\| = 1$.

**Theorem 4.8.** *Let $(\mathbb{B}, |||\cdot|||)$ be a unital normed algebra. Then there is a norm $\|\cdot\|$ on $\mathbb{B}$ equivalent to $|||\cdot|||$ such that $(\mathbb{B}, \|\cdot\|)$ is a normed algebra,*

$$\|xy\| \le \|x\|\,\|y\| \tag{4.2}$$

*for all $x, y \in \mathbb{B}$, and*

$$\|\mathbb{1}\| = 1. \tag{4.3}$$

*Proof.* Let (4.1) hold for $|||\cdot|||$ for some $M < \infty$. Define

$$\|x\| := \sup_{u \neq 0} |||xu|||/|||u|||.$$

It is easily seen that $\|\cdot\|$ is a norm and that (4.2) and (4.3) hold for $\|\cdot\|$. By (4.1) we have $\|x\| \le M|||x|||$ for all $x$. Conversely, $\|x\| \ge |||x|||/|||\mathbb{1}|||$, so $\|\cdot\|$ and $|||\cdot|||$ are equivalent. □

**Definition 4.9.** *A Banach algebra* is a normed algebra $(\mathbb{B}, \|\cdot\|)$, complete for $\|\cdot\|$, such that (4.2) holds and if $\mathbb{B}$ is unital, (4.3) also holds. If $\mathbb{B}$ is an algebra and $\|\cdot\|$ is a norm on it, then $\|\cdot\|$ is called a *Banach algebra norm* iff $(\mathbb{B}, \|\cdot\|)$ is a Banach algebra.

**Example 4.10.** Let $X$ be a Banach space over $\mathbb{K}$. The linear space $\mathbb{B}_X := L(X, X)$ of all bounded linear mappings from $X$ into itself with pointwise addition and scalar multiplication, and with the product defined by *composition* $(ST)x = S(Tx)$, $x \in X$, and the operator norm (1.17), is a unital Banach algebra over $\mathbb{K}$.

Let $S$ be a nonempty set and $\mathbb{A}$ a unital algebra. A subset $E$ of $\mathbb{A}^S$ *separates points* of $S$ if for each $s, t \in S$ with $s \neq t$, there exists $f \in E$ with $f(s) \neq f(t)$, and $E$ *strongly separates points of* $S$ if, further, for each $s \in S$, there exists $f \in E$ with $f(s) \neq 0$. Note that, if $E$ is an algebra of ($\mathbb{K}$-valued) functions on $S$ which strongly separates points of $S$ and if $s, t \in S$ with $s \neq t$, then there exists $f \in E$ with $f(s) = 0$ and $f(t) = 1$.

**Definition 4.11.** If $S$ is a nonempty set, $\mathbb{A}$ is a unital algebra, $\mathbb{B} \subset \mathbb{A}^S$ is an algebra of $\mathbb{A}$-valued functions on $S$, and $\|\cdot\|$ is a norm on $\mathbb{B}$, then $(\mathbb{B}, \|\cdot\|)$ will be called a *Banach algebra of* $\mathbb{A}$*-valued functions on* $S$ iff $\mathbb{B}$ strongly separates points of $S$ and $(\mathbb{B}, \|\cdot\|)$ is a Banach algebra. If $\mathbb{A} = \mathbb{K}$ then $(\mathbb{B}, \|\cdot\|)$ will be called a *Banach algebra of functions.*

**Remark 4.12.** Because of the strong separation, a Banach algebra of $\mathbb{A}$-valued functions $\mathbb{B}$ is unital iff it contains as identity $\mathbb{I}$ the constant function $\mathbb{I}(\cdot)$. Then, by Definition 4.9, the norm satisfies $\|\mathbb{I}(\cdot)\| = 1$. A Banach algebra of ($\mathbb{K}$-valued) functions is unital iff it contains all constant functions.

For example, Theorem 3.7 and Corollary 3.9 imply that the Banach space $\widetilde{\mathcal{W}}_\Phi(J)$ with the norm $\|\cdot\|_{[\Phi]}$ (Definition 3.5) of real-valued functions is a Banach algebra of functions. The Banach algebra property holds when the functions have values in any Banach algebra:

**Proposition 4.13.** *Let* $\Phi \in \mathcal{CV}$, *let* $J$ *be a nondegenerate interval, and let* $\mathbb{A}$ *be a real unital Banach algebra. Then* $(\widetilde{\mathcal{W}}_\Phi(J; \mathbb{A}), \|\cdot\|_{[\Phi]})$ *is a real unital Banach algebra of* $\mathbb{A}$*-valued functions.*

*Proof.* By Theorem 3.7, $\widetilde{\mathcal{W}}_\Phi(J; \mathbb{A})$ is a Banach space with the norm $\|\cdot\|_{[\Phi]}$. By (4.2), for $x, y \in \mathbb{A}$, $(x, y) \mapsto xy$ is a bilinear and 1-bounded mapping from $\mathbb{A} \times \mathbb{A}$ into $\mathbb{A}$. Thus by Theorem 3.8, the pointwise product $fg$ of two functions $f, g \in \widetilde{\mathcal{W}}_\Phi(J; \mathbb{A})$ also is in $\widetilde{\mathcal{W}}_\Phi(J; \mathbb{A})$ and $\|fg\|_{[\Phi]} \leq \|f\|_{[\Phi]}\|g\|_{[\Phi]}$. Since $\mathbb{A}$ is unital, the constant function $\mathbb{I}(\cdot)$ equal to $\mathbb{I}$ on $J$ is the identity for $\widetilde{\mathcal{W}}_\Phi(J; \mathbb{A})$ and $\|\mathbb{I}(\cdot)\|_{[\Phi]} = 1$ by (4.3) and since $\|\mathbb{I}(\cdot)\|_{(\Phi)} = 0$, proving the corollary. □

Another example of a real Banach algebra of functions is the Banach space of all bounded real-valued $\alpha$-Hölder functions with $0 < \alpha \le 1$ and pointwise operations on a nondegenerate interval. This will be a special case of the next theorem. Let $X$, $Y$ be normed spaces, let $U$ be a subset of $X$ with more than one element, and let $0 < \alpha \le 1$. Recall that $\mathcal{H}_\alpha(U;Y)$ is the class of all $\alpha$-Hölder functions $f\colon U \to Y$ with the seminorm $\|f\|_{(\mathcal{H}_\alpha)} = \|f\|_{U,(\mathcal{H}_\alpha)}$ defined by (1.18). $\mathcal{H}_{\alpha,\infty}(U;Y)$ denotes the class of all bounded functions $f$ in $\mathcal{H}_\alpha(U;Y)$ with the norm

$$\|f\|_{\mathcal{H}_\alpha} := \|f\|_{U,\mathcal{H}_\alpha} := \|f\|_{U,\sup} + \|f\|_{U,(\mathcal{H}_\alpha)}, \tag{4.4}$$

where $\|f\|_{\sup} = \|f\|_{U,\sup}$ is the sup norm. By the next theorem, $\mathcal{H}_{\alpha,\infty}(U;Y)$ with this norm also has the Banach algebra property if $Y$ does.

**Proposition 4.14.** *Let $0 < \alpha \le 1$, let $U$ be a subset of a normed space $X$, and let $\mathbb{A}$ be a real unital Banach algebra. Then $(\mathcal{H}_{\alpha,\infty}(U;\mathbb{A}), \|\cdot\|_{\mathcal{H}_\alpha})$ is a real unital Banach algebra of $\mathbb{A}$-valued functions.*

*Proof.* It is easy to check that $\mathcal{H}_{\alpha,\infty}(U;\mathbb{A})$ is a Banach space with the norm $\|\cdot\|_{\mathcal{H}_\alpha}$. Let $f, g \in \mathcal{H}_{\alpha,\infty}(U;\mathbb{A})$. For any $s, t \in U$ we have

$$\begin{aligned}\|(fg)(s) - (fg)(t)\| &\le \|f(s)\|\|g(s) - g(t)\| + \|g(t)\|\|f(s) - f(t)\| \\ &\le \|f\|_{\sup}\|g(s) - g(t)\| + \|g\|_{\sup}\|f(s) - f(t)\|.\end{aligned}$$

Then

$$\|fg\|_{(\mathcal{H}_\alpha)} \le \|f\|_{\sup}\|g\|_{(\mathcal{H}_\alpha)} + \|g\|_{\sup}\|f\|_{(\mathcal{H}_\alpha)}.$$

Since $\|fg\|_{\sup} \le \|f\|_{\sup}\|g\|_{\sup}$, we have that $\|fg\|_{\mathcal{H}_\alpha} \le \|f\|_{\mathcal{H}_\alpha}\|g\|_{\mathcal{H}_\alpha}$. Since $\mathbb{A}$ is unital, the constant function $\mathbb{1}(\cdot)$ equal to $\mathbb{1}$ on $U$ is the identity for $\mathcal{H}_\alpha(U;\mathbb{A})$ and $\|\mathbb{1}(\cdot)\|_{\mathcal{H}_\alpha} = 1$ by (4.3) and since $\|\mathbb{1}(\cdot)\|_{(\mathcal{H}_\alpha)} = 0$. □

## 4.2 The Spectral Radius

Let $\mathbb{B}$ be a Banach algebra over $\mathbb{K}$. For any $x \in \mathbb{B}$, the *spectral radius* is defined by

$$r(x) := \inf_{n \ge 1} \|x^n\|^{1/n}, \tag{4.5}$$

where $n$ ranges over positive integers, and $x^n := x_1 x_2 \cdots x_n$ with $x_j = x$ for $j = 1, \ldots, n$ is the $n$th power of $x$. Clearly for any $c \in \mathbb{K}$ and $x \in \mathbb{B}$,

$$r(cx) = |c| r(x). \tag{4.6}$$

In general, $r(\cdot)$ is not necessarily a seminorm, e.g. if $\mathbb{B}$ is the algebra of $2 \times 2$ matrices, $x = \left(\begin{smallmatrix} 0 & 1 \\ 0 & 0 \end{smallmatrix}\right)$, and $y = \left(\begin{smallmatrix} 0 & 0 \\ 1 & 0 \end{smallmatrix}\right)$, then $r(x) = r(y) = 0$ but $r(x + y) = 1$.

**Theorem 4.15.** *For any Banach algebra* $\mathbb{B}$ *and* $x \in \mathbb{B}$,

$$r(x) = \lim_{n\to\infty} \|x^n\|^{1/n}.$$

*Proof.* For each $n \geq 1$, let $a_n := \|x^n\|^{1/n}$. Then clearly $\liminf_{n\to\infty} a_n \geq r(x)$. Conversely, for any positive integers $n$ and $k$,

$$a_{nk} \leq \left(\|x^n\|^k\right)^{1/nk} = a_n,$$

and for $r = 0, 1, \ldots, n-1$,

$$a_{nk+r} \leq \left(a_n^{nk}\|x\|^r\right)^{1/(nk+r)}.$$

Letting $k \to +\infty$ while $n$ is fixed, we get $\limsup_{m\to\infty} a_m \leq a_n$. Taking the infimum in $n$ gives $\limsup_{m\to\infty} a_m \leq r(x)$, and the conclusion follows. □

**Theorem 4.16.** *In any unital Banach algebra* $\mathbb{B}$,

(*a*) *if* $x \in \mathbb{B}$ *and* $r(x) < 1$ *(and so if* $\|x\| < 1$*), then* $\mathbb{I} + x$ *has an inverse, given by*

$$(\mathbb{I} + x)^{-1} = \sum_{k=0}^{\infty} (-1)^k x^k, \tag{4.7}$$

*where* $x^0 := \mathbb{I}$ *and the series converges absolutely. In particular,* $\|y^{-1} - \mathbb{I}\| \leq \epsilon/(1-\epsilon)$ *whenever* $y \in \mathbb{B}$ *and* $\|y - \mathbb{I}\| \leq \epsilon < 1$.

(*b*) *the set* $U$ *of all invertible elements is open. In particular, if* $x \in U$ *and* $\|y\| < 1/\|x^{-1}\|$, *then* $x + y \in U$ *and*

$$(x + y)^{-1} = \sum_{k=0}^{\infty} (-1)^k (x^{-1}y)^k x^{-1}. \tag{4.8}$$

*Proof.* By the definition of spectral radius, $r(x) \leq \|x\|$ for all $x$, so suppose $r(x) < 1$. Then the series (4.7) is absolutely convergent in $\mathbb{B}$ by the root test. For any $n = 1, 2, \ldots$,

$$(\mathbb{I} + x)\Big(\sum_{k=0}^{n} (-1)^k x^k\Big) = \mathbb{I} - (-x)^{n+1} \to \mathbb{I}$$

as $n \to \infty$, and the two factors on the left commute, proving (*a*).

For (*b*), let $x \in U$ and let $y \in \mathbb{B}$ be such that $\|y\| < 1/\|x^{-1}\|$. By part (*a*), $\mathbb{I} + x^{-1}y$ is invertible. Then $z := x + y = x(\mathbb{I} + x^{-1}y)$ has an inverse because $x^{-1}$ and $(\mathbb{I} + x^{-1}y)^{-1}$ exist, and $z^{-1} = (\mathbb{I} + x^{-1}y)^{-1}x^{-1}$. Thus $U$ is open and (4.8) follows from (4.7). □

**Definition 4.17.** For any complex unital Banach algebra $\mathbb{B}$ and $x \in \mathbb{B}$, the *spectrum* $\sigma(x)$ of $x$ is the set of all $z \in \mathbb{C}$ such that $z\mathbb{1} - x$ is not invertible.

If $z \in \mathbb{C} \setminus \sigma(x)$ then $(z\mathbb{1} - x)^{-1}$ exists and is called the *resolvent* of $x$ at $z$. For any $z, w \in \mathbb{C} \setminus \sigma(x)$, the resolvent equation

$$(z\mathbb{1} - x)^{-1} - (w\mathbb{1} - x)^{-1} = -(z - w)(z\mathbb{1} - x)^{-1}(w\mathbb{1} - x)^{-1} \tag{4.9}$$

holds and is a consequence of

$$\begin{aligned}(z\mathbb{1} - x)^{-1} &= (z\mathbb{1} - x)^{-1}(w\mathbb{1} - x)(w\mathbb{1} - x)^{-1} \\ &= (z\mathbb{1} - x)^{-1}\big[(w - z)\mathbb{1} + (z\mathbb{1} - x)\big](w\mathbb{1} - x)^{-1} \\ &= (w - z)(z\mathbb{1} - x)^{-1}(w\mathbb{1} - x)^{-1} + (w\mathbb{1} - x)^{-1}.\end{aligned}$$

To state some properties of the spectrum, recall that holomorphic functions with values in a Banach space are defined in Definition 2.94.

**Theorem 4.18.** *For any complex unital Banach algebra* $\mathbb{B}$ *and* $x \in \mathbb{B}$,

(*a*) *the spectrum* $\sigma(x)$ *is a compact subset of* $\{z \in \mathbb{C}\colon |z| \le r(x)\}$;
(*b*) *the function* $z \mapsto (z\mathbb{1} - x)^{-1}$ *is holomorphic from* $\mathbb{C} \setminus \sigma(x)$ *into* $\mathbb{B}$;
(*c*) $\sigma(x)$ *is nonempty.*

*Proof.* The set of invertible elements is open by Theorem 4.16(*b*), and $z \mapsto z\mathbb{1} - x$ is continuous from $\mathbb{C}$ into $\mathbb{B}$, so $\sigma(x)$ is closed. If $|z| > r(x)$, then $r(x/z) < 1$ by (4.6), so $\mathbb{1} - x/z$ is invertible by Theorem 4.16(*a*), and

$$(z\mathbb{1} - x)^{-1} = (1/z)(\mathbb{1} - x/z)^{-1}. \tag{4.10}$$

So $\sigma(x) \subset \{z \in \mathbb{C}\colon |z| \le r(x)\}$ and (*a*) is proved.

For (*b*) let $w \in \mathbb{C} \setminus \sigma(x)$, $x_w := w\mathbb{1} - x$, and let $z \in \mathbb{C}$ be such that $|z - w| < 1/\|x_w^{-1}\|$. By Theorem 4.16(*b*), $z \in \mathbb{C} \setminus \sigma(x)$ and

$$(z\mathbb{1} - x)^{-1} = ((z - w)\mathbb{1} + x_w)^{-1} = \sum_{k=0}^{\infty}(z - w)^k(-1)^k x_w^{-k-1},$$

where $x^{-n} := (x^{-1})^n$, $n = 1, 2, \ldots$. Since $|z - w| < 1/\|x_w^{-1}\|$, the series converges absolutely in $\mathbb{B}$, so (*b*) is proved.

To prove (*c*), suppose $\sigma(x) = \emptyset$. So (for $z = 0$) $x^{-1}$ exists. By the Hahn–Banach theorem, let $L$ be a continuous linear functional on the Banach space $(\mathbb{B}, \|\cdot\|)$ with $L(x^{-1}) = 1$. For any $z \in \mathbb{C}$ let $g(z) := L((z\mathbb{1} - x)^{-1})$. By part (*b*), $g$ is an entire holomorphic function from $\mathbb{C}$ into $\mathbb{C}$. By (4.7) and (4.10), $g(z) \to 0$ as $|z| \to \infty$, so $g$ is bounded and by Liouville's theorem (see e.g. [42, Theorem 9.11.1]), $g$ is constant and so $g \equiv 0$, but $g(0) = 1$, a contradiction, proving (*c*). The proof of Theorem 4.18 is complete. □

**Theorem 4.19** (Mazur–Gelfand). *A unital Banach algebra $\mathbb{B}$ over $\mathbb{C}$ in which every non-zero element is invertible is equal to $\{z\mathbb{1}: z \in \mathbb{C}\}$, and so $\mathbb{B}$ is isomorphic to the complex numbers.*

*Proof.* Let $x \in \mathbb{B}$. If for each $z \in \mathbb{C}$, $z\mathbb{1} - x$ is nonzero and so invertible, $\sigma(x)$ is empty, contrary to Theorem 4.18($c$). The theorem is proved. □

**Theorem 4.20.** *For any unital complex Banach algebra $\mathbb{B}$ and $x \in \mathbb{B}$,*

$$r(x) = \sup\{|z|: z \in \sigma(x)\}.$$

*Proof.* Since $\sigma(x) \neq \emptyset$ by Theorem 4.18($c$), $s(x) := \sup\{|z|: z \in \sigma(x)\} \geq 0$, and $s(x) \leq r(x)$ by Theorem 4.18($a$). In proving $r(x) \leq s(x)$, we can assume $r(x) > 0$. For any $z \in \mathbb{C}$ with $|z| < 1/s(x)$, $\mathbb{1} - zx$ is invertible, since it clearly is if $z = 0$, whereas for $z \neq 0$, $z^{-1}\mathbb{1} - x$ is invertible because $|z^{-1}| > s(x)$. The function $f$ with $f(z) := (\mathbb{1} - zx)^{-1}$ is holomorphic from $\{z: |z| < 1/s(x)\}$ into $\mathbb{B}$ by Theorem 4.18($b$). Thus by Propositions 2.100 and 2.96, the Taylor series of $f$ around 0 exists and converges absolutely on the whole disk $|z| < 1/s(x)$ (the whole plane, if $s(x) = 0$). For $|z|$ small enough, say $< 1/r(x)$, by Theorem 4.16($a$), the Taylor series is $\sum_{k=0}^{\infty} z^k x^k$, so $h_k = x^k$ in (2.94). Thus for $0 < t < 1/s(x)$, $\limsup_{k\to\infty} t^k\|x^k\| = 0$, so $t \limsup_{k\to\infty} \|x^k\|^{1/k} = tr(x) \leq 1$. It follows that $r(x)/s(x) \leq 1$, so $r(x) \leq s(x)$, finishing the proof. □

**Theorem 4.21.** *In any commutative unital Banach algebra $\mathbb{A}$,*

($a$) *the closure $\bar{I}$ of any ideal $I$ is an ideal;*
($b$) *the closure of a proper ideal is a proper ideal;*
($c$) *any maximal ideal is closed.*

*Proof.* If $x_n, y_n \in I$, $x_n \to x \in \bar{I}$, $y_n \to y \in \bar{I}$, $c \in \mathbb{K}$, and $u \in \mathbb{A}$, then $cx_n \to cx$, $ux_n \to ux$, $x_n y_n \to xy$, and $x_n + y_n \to x + y$, so ($a$) holds.

A proper ideal contains no invertible elements. By Theorem 4.16($a$), elements in an open neighborhood of $\mathbb{1}$ are all invertible and so belong neither to $I$ nor to $\bar{I}$, so ($b$) holds.

If $M$ is a maximal ideal then $M \subset \bar{M}$, a proper ideal by ($b$), so $M = \bar{M}$ and ($c$) holds. □

## 4.3 Characters

**Definition 4.22.** Let $\mathbb{B}$ be a Banach algebra over $\mathbb{K} = \mathbb{C}$. Then a function $\phi$ from $\mathbb{B}$ into $\mathbb{C}$ is called a *character* if (*a*) $\phi$ is an algebra homomorphism, that is, for all $x, y \in \mathbb{B}$ and $c \in \mathbb{C}$,

$$\phi(cx + y) = c\phi(x) + \phi(y) \quad \text{and} \quad \phi(xy) = \phi(x)\phi(y);$$

(*b*) if $\mathbb{B}$ is unital, $\phi(\mathbb{1}) = 1$.

Let $\mathcal{M}(\mathbb{B})$ denote the set of all characters of $\mathbb{B}$.

**Remark 4.23.** In the definition of character, (*b*) is equivalent to the assumption that if $\mathbb{B}$ is unital, $\phi$ is not identically 0.

Recall that for any normed space $(X, \|\cdot\|)$ and any continuous linear functional $\phi$ from $X$ into $\mathbb{K}$, the dual norm is defined by $\|\phi\|' := \sup\{|\phi(x)|\colon x \in X, \|x\| = 1\}$.

**Theorem 4.24.** *Any character $\phi$ on a unital complex Banach algebra $\mathbb{B}$ is continuous and satisfies $\|\phi\|' = 1$. Also, $|\phi(x)| \leq r(x)$ for each $x$.*

*Proof.* Let $L := \phi^{-1}(0)$. Then for any $x \in \mathbb{B}$ having an inverse $x^{-1}$, $x \notin L$. Thus if $y \in \mathbb{B}$ and $r(y) < 1$, $\mathbb{1} + y \notin L$ by Theorem 4.16(*a*). In other words, $\phi(y) \neq -1$. Then $|\phi(y)| < 1$, since if $|\phi(y)| \geq 1$ then $\phi(cy) = -1$ for $c = -1/\phi(y)$ with $|c| \leq 1$, and so by (4.6), $r(cy) < 1$.

It follows that $|\phi(x)| \leq 1$ whenever $r(x) \leq 1$, and so whenever $\|x\| \leq 1$ by the definition (4.5) of $r(x)$. By homogeneity (4.6), $|\phi(x)| \leq r(x)$ for all $x$. □

Let $X$ be a vector space over $\mathbb{K}$ and $M$ a linear subspace. Then the quotient space $X/M$ is defined as the set of all cosets $x + M := \{x + u\colon u \in M\}$ for all $x \in X$, with $x + M = y + M$ if and only if $x - y \in M$. If $x, y \in X$ and $c \in \mathbb{K}$ let $(x + M) + (y + M) := x + y + M$ and $c(x + M) := cx + M$. It is easily checked that these cosets are well defined, and with these operations, $X/M$ is a vector space over $\mathbb{K}$.

If $(X, \|\cdot\|)$ is a normed linear space, let $\|x+M\| := \inf\{\|x+u\|\colon u \in M\}$. We then have the following.

**Proposition 4.25.** *Let $(X, \|\cdot\|)$ be a normed linear space and $M$ a closed linear subspace. Then*

(*a*) *$(X/M, \|\cdot\|)$ is a normed linear space;*
(*b*) *if $X$ is complete (a Banach space), so is $X/M$.*

*Proof.* For any $x, y \in X$ and $u, v \in M$ we have $\|x+y+u+v\| \leq \|x+u\|+\|y+v\|$, so $\|x + y + M\| \leq \|x + M\| + \|y + M\|$. For any $c \in \mathbb{K}$ with $c \neq 0$ we have $\|cx+cu\| = |c|\|x+u\|$ and $\|cx+v\| = |c|\|x+(v/c)\|$, so $\|c(x+M)\| = |c|\|x+M\|$. For $c = 0$, we have $\|0 + M\| = 0$. Thus $\|\cdot\|$ on $X/M$ is a seminorm.

If $x \notin M$ but $\|x + M\| = 0$, then $\|x + u_n\| \to 0$ for some $u_n \in M$, so $-u_n \to x$, a contradiction since $M$ is closed. Thus $\|\cdot\|$ on $X/M$ is a norm, proving (*a*).

Suppose $X$ is complete and let $\{x_n+M\}_{n\geq 1}$ be a Cauchy sequence in $X/M$. For each $k = 1, 2, \ldots$, there is an $n_k$ such that $\|(x_n + M) - (x_m + M)\| < 2^{-k}$ for $n \geq m \geq n_k$. We can assume $n_1 < n_2 < \cdots$. Recursively, a sequence $y_n \in x_n + M$ will be defined. Let $y_n := x_n$ for $n = 1, \ldots, n_1$. Given $y_n \in x_n + M$ for $n \leq n_k$ for some $k$, let $n_k < m \leq n_{k+1}$. Since $\|(x_m + M) - (y_{n_k} + M)\| < 2^{-k}$, take $y_m := x_m + u$ for some $u \in M$ such that $\|y_m - y_{n_k}\| < 2^{-k}$. This completes the recursive definition of $y_n$. For each $k$, we have $\|y_n - y_{n_k}\| < 2^{1-k}$ for all $n \geq n_k$. Thus $\{y_n\}_{n\geq 1}$ is a Cauchy sequence in $X$ and $y_n \to y$ for some $y \in X$. Then $x_n + M \to y + M$, proving ($b$) and completing the proof of Proposition 4.25. □

**Theorem 4.26.** *Let $\mathbb{B}$ be a commutative unital Banach algebra over $\mathbb{C}$ and $I$ a maximal ideal in $\mathbb{B}$. Then for each $x \in \mathbb{B}$, $x = z\mathbb{1} + u$ for some unique $z \in \mathbb{C}$ and $u \in I$, and $\mathbb{B}/I$ with the multiplication $(x + I)(y + I) := xy + I$ is a Banach algebra isometric to $\mathbb{C}$ via the correspondence $z\mathbb{1} + I \leftrightarrow z$.*

*Proof.* Recall that a maximal ideal is closed (Theorem 4.21($c$)) and the multiplication in $\mathbb{B}/I$ is well defined (before Theorem 4.5). $(\mathbb{B}/I, \|\cdot\|)$ is a Banach space by Proposition 4.25. For any $x, y \in \mathbb{B}$ and $u, v \in I$ we have $(x+u)(y+v) = xy+w$ with $w := xv+uy+uv \in I$, so $\|xy+I\| \leq \|x+I\|\|y+I\|$. $\mathbb{B}/I$ is unital, with identity $\mathbb{1} + I$. If $\|\mathbb{1} + u\| < 1$ then $-u$ is invertible by Theorem 4.16($b$), so $u \notin I$, recalling that a maximal ideal is proper. Thus $\|\mathbb{1} + I\| = 1$, and $\mathbb{B}/I$ is a Banach algebra.

If $x \notin I$, suppose $x + I$ is not invertible. Let $L := \{xy + u\colon\ y \in \mathbb{B}, u \in I\}$. Then $L$ is an ideal. If $xy + u = \mathbb{1}$ then $(x + I)(y + I) = \mathbb{1} + I$, a contradiction. So $L$ is a proper ideal, but it strictly includes $I$, again a contradiction. So every nonzero element of $\mathbb{B}/I$ is invertible. The conclusions of the theorem follow from Theorem 4.19. □

**Corollary 4.27.** *For any commutative unital Banach algebra $\mathbb{B}$ over $\mathbb{C}$, there is a 1-to-1 correspondence between maximal ideals $I$ in $\mathbb{B}$ and characters $\phi$ of $\mathbb{B}$, given by $I = \phi^{-1}(0)$, $\phi(z\mathbb{1} + u) = z$ for all $u \in I$ and $z \in \mathbb{C}$.*

**Definition 4.28.** Let $\mathbb{B}$ be a commutative unital Banach algebra $\mathbb{B}$ over $\mathbb{C}$. For $x \in \mathbb{B}$ and $\phi \in \mathcal{M}(\mathbb{B})$, let $\widehat{x}(\phi) := \phi(x)$. Then $\widehat{x}$ called the *Gelfand transform* of $x$.

On the set of characters $\mathcal{M}(\mathbb{B})$ of a Banach algebra $\mathbb{B}$, the *Gelfand topology* $\tau$ is the weak-star topology of pointwise convergence on $\mathbb{B}$, or equivalently, pointwise convergence on the closed unit ball $B_1 := \{x \in \mathbb{B}\colon\ \|x\| \leq 1\}$. Since a pointwise limit of characters is a character, and each character takes $B_1$ into $\{z \in \mathbb{C}\colon\ |z| \leq 1\}$ by Theorem 4.24, the Gelfand topology is compact by Tychonoff's theorem (e.g. [53, Theorem 2.2.8]). It is Hausdorff since distinct characters are unequal at some point.

If $\mathbb{B}$ is a commutative unital Banach algebra $\mathbb{B}$ over $\mathbb{C}$ then we always assume that $\mathcal{M}(\mathbb{B})$ is equipped with the Gelfand topology. Then $\widehat{x}$ is a continuous complex-valued function on $\mathcal{M}(\mathbb{B})$ for each $x \in \mathbb{B}$. Also, the mapping $x \mapsto \widehat{x}$ is an algebra homomorphism of $\mathbb{B}$ into $C(\mathcal{M}(\mathbb{B}); \mathbb{C})$, the Banach algebra of continuous complex-valued functions on $\mathcal{M}(\mathbb{B})$.

**Theorem 4.29.** *For any unital commutative complex Banach algebra $\mathbb{B}$ over $\mathbb{C}$ and $x \in \mathbb{B}$, the range* $\operatorname{ran}\widehat{x}$ *equals $\sigma(x)$.*

*Proof.* If $z \in \operatorname{ran}\widehat{x}$, suppose $z \notin \sigma(x)$. Then $z\mathbb{1} - x$ has an inverse $v$. Thus $(z - \widehat{x})\widehat{v} \equiv 1$, but at $\phi$ such that $\widehat{x}(\phi) = z$, this gives a contradiction, so $\operatorname{ran}\widehat{x} \subset \sigma(x)$.

Conversely let $z \in \sigma(x)$, so $z\mathbb{1} - x$ is not invertible. Then $L := \{(z\mathbb{1} - x)y \colon y \in \mathbb{B}\}$ is a proper ideal in $\mathbb{B}$. By Theorem 4.4, it is included in some maximal ideal $I$. By Corollary 4.27, let $\phi$ be a character with $I = \phi^{-1}(0)$. Then for $y = \mathbb{1}$ we get $\widehat{x}(\phi) = \phi(x) = z$, so $z \in \operatorname{ran}\widehat{x}$, completing the proof. □

**Corollary 4.30.** *For any commutative unital Banach algebra $\mathbb{B}$ over $\mathbb{C}$ and $x \in \mathbb{B}$, we have $r(x) = \|\widehat{x}\|_{\sup}$.*

*Proof.* This follows from Theorems 4.29 and 4.20. □

## 4.4 Holomorphic Functions of Banach Algebra Elements

Let $f$ be a complex valued holomorphic function on a disk $B(z_0, r) \subset \mathbb{C}$ for some $z_0 \in \mathbb{C}$ and $r > 0$ with a Taylor expansion around $z_0$

$$f(z) = \sum_{k=0}^{\infty} a_k (z - z_0)^k$$

for each $z \in B(z_0, r)$. Then $r \leq 1/\limsup_{k\to\infty} |a_k|^{1/k}$, and the series converges absolutely by Proposition 2.96. We have $a_k = f^{(k)}(z_0)/k!$ for each $k$ by Proposition 2.100. Moreover, for any unital complex Banach algebra $\mathbb{B}$, the series

$$f(x) := \sum_{k=0}^{\infty} a_k (x - z_0\mathbb{1})^k, \tag{4.11}$$

where $(x - z_0\mathbb{1})^0 := \mathbb{1}$, converges absolutely in $\mathbb{B}$ whenever the spectral radius $r(x - z_0\mathbb{1})$ is less than $r$. We will extend definition (4.11) using Cauchy formulas for curves in the complex plane.

Recall that contour integrals of suitable Banach-valued functions were defined in Section 2.10. For a curve $\zeta(\cdot) \colon [a, b] \to \mathbb{C}$ with $a < b$, the *reversed* curve will mean the curve $\eta(t) := \zeta(a + b - t)$ for $t \in [a, b]$. We have

$$\oint_{\eta(\cdot)} f(\eta)\,\mathrm{d}\eta = -\oint_{\zeta(\cdot)} f(\zeta)\,\mathrm{d}\zeta$$

by a change of variables for $RS$ integrals (Proposition 2.77 and Theorem 2.42). A *chain* will be a formal linear combination $C = \sum_{i=1}^{k} n_i\zeta_i$, where $\zeta_i(\cdot)$ are curves and $n_i \in \mathbb{Z}$. Let $-\zeta$ be the reversed curve of $\zeta$ for any $\zeta$, and let

$$\oint_C f(\zeta)\,\mathrm{d}\zeta := \sum_{i=1}^{k} n_i \oint_{\zeta_i(\cdot)} f(\zeta_i)\,\mathrm{d}\zeta_i,$$

where if $n_i < 0$ and $\eta_i(\cdot) = -\zeta_i(\cdot)$, then $n_i \oint_{\zeta_i(\cdot)} f(\zeta_i)\,\mathrm{d}\zeta_i = -n_i \oint_{\eta_i(\cdot)} f(\eta_i)\,\mathrm{d}\eta_i$. So $n_i\zeta_i = (-n_i)(-\zeta_i)$ and the notation is consistent. Also in Section 2.10, the winding number $w(\zeta(\cdot), z)$ of a closed curve $\zeta(\cdot)$ around a point $z$ not in the range of $\zeta(\cdot)$ is defined just after Definition 2.92. The winding number $w(C, z)$ of a chain $C = \sum_{i=1}^{k} n_i\zeta_i$, where $\zeta_i$ are piecewise $C^1$ curves, with respect to a point $z \notin \sum_{i=1}^{k} \operatorname{ran}(\zeta_i)$ is defined as $w(C,\zeta) := \sum_{i=1}^{k} n_i w(\zeta_i(\cdot), z)$. The definition is the extension of the conclusion of Proposition 2.93 to chains. The property "simply connected" is characterized in Theorem 2.105.

**Theorem 4.31.** *Let $\mathbb{B}$ be a unital Banach algebra over $\mathbb{C}$ and $x \in \mathbb{B}$. Suppose $U$ is a connected and simply connected open set in $\mathbb{C}$ and $K$ is a compact set such that $\sigma(x) \subset K \subset U$. Let $\zeta(\cdot)$ be a piecewise $C^1$ closed curve with range $\operatorname{ran}(\zeta) \subset U \setminus K$ such that $w(\zeta(\cdot), z) = 1$ for all $z \in K$. Let $f$ be a holomorphic function from $U$ into $\mathbb{C}$ and let*

$$f(x;\zeta(\cdot)) := \frac{1}{2\pi\mathrm{i}} \oint_{\zeta(\cdot)} f(\zeta)(\zeta\mathbb{I} - x)^{-1}\,\mathrm{d}\zeta. \tag{4.12}$$

*Then*

(*a*) *the integral $f(x;\zeta(\cdot))$ is well defined in $\mathbb{B}$, and does not depend on $\zeta(\cdot)$ with the given properties;*
(*b*) *if $U = B(z_0, r)$ for some $z_0 \in \mathbb{C}$ and $r > 0$ then the definitions* (4.11) *and* (4.12) *agree;*
(*c*) *if $\mathbb{B}$ is commutative, then for $F(x) := f(x;\zeta(\cdot))$, we have for the Gelfand transforms that $\widehat{F(x)} \equiv f \circ \widehat{x}$ on $\mathcal{M}(\mathbb{B})$.*

*Proof.* For ($a$), on $U \setminus \sigma(x)$, which includes $\operatorname{ran}(\zeta)$, both $f$ and, by Theorem 4.18($b$), $z \mapsto (z\mathbb{I} - x)^{-1}$, are holomorphic, thus continuous by Proposition 2.96. So the integrand in (4.12) is bounded and continuous on $\operatorname{ran}(\zeta)$ and the integral $f(x,\zeta(\cdot))$ is defined. Let $\zeta_1(\cdot)$ and $\zeta_2(\cdot)$ be two curves both satisfying the hypotheses of Theorem 4.31. Take any continuous linear form $L \in \mathbb{B}'$. Then $\zeta \mapsto L((\zeta\mathbb{I} - x)^{-1})$ is holomorphic from $\mathbb{C} \setminus \sigma(x)$ into $\mathbb{C}$. Consider the chain $C = \zeta_1 - \zeta_2$, and let $V := U \setminus K$. Then $V$ is no longer simply connected, but $\zeta_1$ and $\zeta_2$ both take values in $V$. We have $w(C, z) = 1 - 1 = 0$ for each $z \in K$ by hypothesis. For $z \notin U$ we have $w(\zeta_i(\cdot), z) = 0$ for $i = 1, 2$ by

Theorem 2.105. Thus $w(C, z) = 0$ for all $z \notin V$, and $C$ is homologous to 0 in $V$ as defined in Ahlfors [1, 2d ed., §4.4.4, Definition 2]. So we can apply the Cauchy integral theorem in the form given by Ahlfors [1, 2d ed., §4.4.4, Theorem 18] to $\zeta \mapsto f(\zeta)L((\zeta\mathbb{I} - x)^{-1})$, which is holomorphic on $V$. Thus by Proposition 2.78,

$$\begin{aligned} L(f(x;\zeta_1(\cdot))) &= \frac{1}{2\pi \mathrm{i}} \oint_{\zeta_1(\cdot)} f(\zeta)L((\zeta\mathbb{I} - x)^{-1})\,\mathrm{d}\zeta \\ &= \frac{1}{2\pi \mathrm{i}} \oint_{\zeta_2(\cdot)} f(\zeta)L((\zeta\mathbb{I} - x)^{-1})\,\mathrm{d}\zeta = L(f(x;\zeta_2(\cdot))). \end{aligned} \tag{4.13}$$

By the Hahn–Banach theorem it follows that $f(x;\zeta_1(\cdot)) = f(x;\zeta_2(\cdot))$, proving (*a*).

For (*b*), it is easily checked that for any $x \in \mathbb{B}$ and $z_0 \in \mathbb{C}$, $\sigma(z - z_0\mathbb{I}) = \sigma(x) - z_0 := \{z - z_0\colon\ z \in \sigma(x)\}$. Since $\sigma(x) \subset U = B(z_0, r)$, by Theorem 4.20 it then follows that $r(x - z_0\mathbb{I}) < r$. Let $r(x - z_0\mathbb{I}) < t < r$ and let $\zeta(\theta) = z_0 + t\mathrm{e}^{\mathrm{i}\theta}$, $0 \le \theta \le 2\pi$. For each $\zeta = \zeta(\theta) \in \operatorname{ran}(\zeta)$, by (4.6) and Theorem 4.16(*a*), $\mathbb{I} - (\zeta - z_0)^{-1}(x - z_0\mathbb{I})$ is invertible and

$$(\zeta\mathbb{I} - x)^{-1} = \frac{1}{\zeta - z_0}\left(\mathbb{I} - \frac{x - z_0\mathbb{I}}{\zeta - z_0}\right)^{-1} = \sum_{k=0}^{\infty} \frac{(x - z_0\mathbb{I})^k}{(\zeta - z_0)^{k+1}},$$

where the series converges absolutely in $\mathbb{B}$ and uniformly on $\operatorname{ran}(\zeta)$. Integrating term by term and using the Cauchy formulas for $f$ and its derivatives (Proposition 2.97 and Theorem 2.98), it follows that

$$f(x;\zeta(\cdot)) = \frac{1}{2\pi \mathrm{i}} \oint_{\zeta(\cdot)} f(\zeta) \sum_{k=0}^{\infty} \frac{(x - z_0\mathbb{I})^k}{(\zeta - z_0)^{k+1}}\,\mathrm{d}\zeta = \sum_{k=0}^{\infty} \frac{f^{(k)}(z_0)}{k!}(x - z_0\mathbb{I})^k,$$

which equals $f(x)$ as defined by (4.11).

For (*c*), let $\phi \in \mathcal{M}(\mathbb{B})$. For any $z \notin \sigma(x)$, we find that $\phi((z\mathbb{I} - x)^{-1}) = 1/(z - \phi(x))$. Since the range of $\zeta$ is included in $U \setminus \sigma(x)$, interchanging $\phi$ with $\oint_{\zeta(\cdot)}$ as in (4.13), we have

$$\widehat{F(x)}(\phi) = \phi(F(x)) = \frac{1}{2\pi \mathrm{i}} \oint_{\zeta(\cdot)} \frac{f(\zeta)\,\mathrm{d}\zeta}{\zeta - \phi(x)} = f(\phi(x)) = (f \circ \widehat{x})(\phi)$$

by the Cauchy integral formula (Theorem 2.107) since $\phi(x) \in \sigma(x)$ by Theorem 4.29, so $w(\zeta(\cdot), \phi(x)) = 1$ by assumption. Thus (*c*) follows and the theorem is proved. □

Curves satisfying the hypotheses of Theorem 4.31 always exist:

**Lemma 4.32.** *Let $U$ be a connected, simply connected, nonempty open set in $\mathbb{C}$, and let $K$ be a compact subset of $U$. Then there is a $C^\infty$ simple closed curve $\zeta(\theta)$, $0 \le \theta \le 2\pi$, with range* $\operatorname{ran}(\zeta) \subset U \setminus K$, *$\zeta'(\theta) \ne 0$ for all $\theta$, and winding number $w(\zeta(\cdot), z) = 1$ for all $z \in K$.*

*Proof.* If $U = \mathbb{C}$ then a circle with center at zero and sufficiently large radius will satisfy the conclusion. Therefore we can suppose that $U \neq \mathbb{C}$. By the Riemann mapping theorem (e.g. Greene and Krantz [86, p. 344] or Ahlfors [1, p. 222]), let $T$ be a one-to-one, holomorphic function from $U$ onto the unit disk $D := B(0,1)$. Then $T'(z) \neq 0$ for all $z \in U$ and the inverse function $T^{-1}$ is holomorphic from $D$ onto $U$ (e.g. Stein and Shakarchi [224, p. 206]). The image $T(K)$ is compact in $D$, and so $\rho := \sup\{|z|\colon z \in T(K)\} < 1$. Let $r \in (\rho, 1)$ and $\zeta(\theta) := T^{-1}(re^{i\theta})$, $0 \le \theta \le 2\pi$. Then $\zeta(\cdot)$ is a $C^\infty$ simple closed curve in $U$, where $\zeta'(\theta) \neq 0$ for all $\theta$ by the chain rule and since for each $z \in D$, $(T^{-1})'(z) = 1/T'(T^{-1}(z)) \neq 0$, with range $C := \operatorname{ran}(\zeta)$ and $\zeta'(0) = \zeta'(2\pi)$. We will show that $\zeta(\cdot)$ satisfies the conclusion of the lemma.

Let $D_1 := B(0,r)$, $D_2 := \{z \in \mathbb{C}\colon r < |z| < 1\}$ and $U_i := T^{-1}(D_i)$, $i = 1, 2$. Then $\mathbb{C}$ is the disjoint union $U_1 \cup C \cup U_2 \cup U^c$ of nonempty sets. The set $U_1$ is connected, open, and simply connected because $D_1$ has these properties (e.g. by Theorem 2.105($d$)). By the Jordan curve theorem, proved for piecewise $C^1$ simple closed curves in Stein and Shakarchi [224, Appendix B, Theorem 2.2], the complement of the range $C$ of $\zeta(\cdot)$ consists of two disjoint connected open sets $V_1$ and $V_2$, where $V_1$ is bounded and simply connected and $V_2$ is neither. We will show that $U_1 = V_1$.

We have $U^c \subset C^c = V_1 \cup V_2$. We claim that $V_1 \subset U$. Suppose $U^c \cap V_1 \neq \emptyset$ and choose $p \in U^c \cap V_1$. Let $F_p$ be the component of $p$ in $U^c$, given e.g. by Theorem 2.102. Then $F_p$ is closed in the relative topology of $U^c$, and so $F_p \subset V_1$ since $V_1$ and $V_2$ are open and disjoint. But then $F_p$ is bounded, contradicting the fact that $U$ is simply connected, e.g. Theorem 2.105($c$). Thus the claim $V_1 \subset U$ $(= U_1 \cup C \cup U_2)$ holds. Since $V_1 \cap C = \emptyset$, we have $V_1 \subset U_1 \cup U_2$. Since $V_1$ is connected and $U_i$, $i = 1, 2$, are disjoint and open, $V_1 \subset U_1$ or $V_1 \subset U_2$. Likewise $U_1 \subset V_1$ or $U_1 \subset V_2$. If $U_1 \subset V_2$ then $U_2 \subset V_1$ and $U_2 = V_1$ since $V_1 \subset U$. Now $U_1 \cup C = T^{-1}(\{z\colon |z| \le r\})$ is compact and $V_1 \cup C = V_2^c$ is closed and bounded, thus compact. It then follows that $U = (U_1 \cup C) \cup (V_1 \cup C)$ is compact, a contradiction. Thus $U_1 \subset V_1$, and so $V_1 \subset U_2$ is impossible, proving $V_1 \subset U_1$. Therefore we have $U_1 = V_1$.

Since $K \subset U_1$, by Theorem 2.2 in Appendix $B$ of Stein and Shakarchi [224, p. 351], $w(\zeta(\cdot), z) = k$ for all $z \in K$, where $k = \pm 1$. By reversing the direction of $\zeta(\cdot)$ if necessary, we can make $k = +1$, proving the lemma. □

**Note.** There is actually never a need to reverse the direction of $\zeta(\cdot)$ in the preceding proof, as follows. Let $T_{\mathbb{R}}$ be the transformation from an open set in $\mathbb{R}^2$ into $\mathbb{R}^2$ defined by $T$ from the Riemann mapping theorem and $z = x + iy \in U$, $(x, y) \in \mathbb{R}^2$. Then the Jacobian determinant of $T_{\mathbb{R}}$ at $(x, y)$ is easily shown to equal $|T'(z)|^2 > 0$. It follows that $T$ (or $T^{-1}$) preserves the winding number ($+1$ or $-1$) of a simple closed curve around points inside it, e.g. [178, Theorem VII.11.2, p. 198]. (The statement there, for transformations whose domain is the whole plane, extends to cover the present situation.)

By Theorem 4.31($a$) and Lemma 4.32, the following notation is well defined.

**Definition 4.33.** Let $\mathbb{B}$ be a unital Banach algebra over $\mathbb{C}$, let $x \in \mathbb{B}$, and let $U$ be a connected and simply connected open set in $\mathbb{C}$ such that $\sigma(x) \subset U$. For any holomorphic function $f\colon U \to \mathbb{C}$ define $f(x) := f(x;\zeta(\cdot)) \in \mathbb{B}$ by (4.12), where $\zeta(\cdot)$ is a piecewise $C^1$ simple closed curve with range $\operatorname{ran}(\zeta) \subset U \setminus \sigma(x)$ and winding number $w(\zeta(\cdot), z) = 1$ for all $z \in \sigma(x)$.

A mapping $(f, x) \mapsto f(x)$ for $f$ and $x$ satisfying the conditions of Definition 4.33 is sometimes called a *functional calculus*, e.g. Dales [36, p. 212].

First we apply this mapping to show that a composition $f \circ g$ is in a unital Banach algebra of functions $\mathbb{B}$ if $g \in \mathbb{B}$ and $f$ is a holomorphic function on a suitable open set including the spectrum of $g$.

**Proposition 4.34.** *Let $\mathbb{B}$ be a unital Banach algebra of $\mathbb{C}$-valued functions on a nonempty set $S$ and $g \in \mathbb{B}$. For a connected, simply connected open set $U \subset \mathbb{C}$ including the spectrum $\sigma(g)$, let $f$ be holomorphic on $U$. Then $f \circ g \in \mathbb{B}$ and $f \circ g = f(g)$ as defined in Definition* 4.33.

*Proof.* For each $s \in S$, the evaluation $e_s \in \mathcal{M}(\mathbb{B})$, and so $g(s) = \widehat{g}(e_s) \in \sigma(g) \subset U$ by Theorem 4.29. Then by Theorem 4.31($c$), for each $s \in S$, we have

$$f(g)(s) = \widehat{f(g)}(e_s) = f(\widehat{g}(e_s)) = (f \circ g)(s),$$

so $f(g) = f \circ g$, proving the proposition. □

Second we apply functional calculus to the exponential. Let $\mathbb{B}$ be a complex unital Banach algebra $\mathbb{B}$. Given $x \in \mathbb{B}$, let $U$ be a disk $B(0, R)$ in $\mathbb{C}$ such that $\sigma(x) \subset U$. The Taylor expansion around 0 of the exponential function $\exp(\cdot)\colon U \to \mathbb{C}$ gives the Taylor series $\sum_{k=0}^{\infty} z^k/k!$, which converges absolutely and uniformly. Let $\zeta(\theta) := re^{i\theta}$, $0 \le \theta \le 2\pi$, for some $r$ close enough to $R$. Then by Theorem 4.31($b$)

$$\exp(x) = \exp(x;\zeta(\cdot)) = \sum_{k=0}^{\infty} x^k/k!. \tag{4.14}$$

Here $\exp(x)$ is defined for any $x \in \mathbb{B}$ and the power series (4.14) converges absolutely and uniformly on any bounded set in $\mathbb{B}$.

**Lemma 4.35.** *If $\mathbb{B}$ is a unital Banach algebra, and $x$ and $y$ are elements of $\mathbb{B}$ which commute, then* $\exp(x+y) = \exp(x)\exp(y)$.

*Proof.* Let $x, y \in \mathbb{B}$ be such that $xy = yx$. For each positive integer $n$, let $C_n := \{0 \le i, j \le n\colon\ i + j \le n\}$. Using the binomial formula, we have

$$\sum_{k=0}^{n} \frac{(x+y)^k}{k!} = \sum_{k=0}^{n}\sum_{i=0}^{k} \frac{x^i y^{k-i}}{i!(k-i)!} = \sum_{i,j\in C_n} \frac{x^i y^j}{i!j!}$$

for each $n$. Therefore

$$\sum_{i=0}^{n} \frac{x^i}{i!} \sum_{j=0}^{n} \frac{y^j}{j!} - \sum_{k=0}^{n} \frac{(x+y)^k}{k!} = \sum_{i,j\in D_n} \frac{x^i y^j}{i!j!},$$

where $D_n := \{0 \le i,j \le n \colon i+j > n\}$. The norm of the right side is not greater than

$$\sum_{i,j\in D_n} \frac{\|x\|^i \|y\|^j}{i!j!} = \sum_{i=0}^{n} \frac{\|x\|^i}{i!} \sum_{j=0}^{n} \frac{\|y\|^j}{j!} - \sum_{k=0}^{n} \frac{(\|x\|+\|y\|)^k}{k!}$$
$$\to \mathrm{e}^{\|x\|}\mathrm{e}^{\|y\|} - \mathrm{e}^{\|x\|+\|y\|} = 0,$$

as $n \to \infty$, proving the conclusion. □

Next we will define the logarithm of a suitable Banach algebra element. To make the logarithm operation inverse to the exponential we will use the following result.

**Proposition 4.36.** *Let $\mathbb{B}$ be a unital Banach algebra over $\mathbb{C}$, let $x \in \mathbb{B}$, and let $U$ be a connected, simply connected, nonempty open set in $\mathbb{C}$ such that $\sigma(x) \subset U$. If $f \colon U \to \mathbb{C}$ and $g \colon \mathbb{C} \to \mathbb{C}$ are holomorphic functions on their domains then $(g \circ f)(x) = g(f(x))$.*

*Proof.* Since $\sigma(x)$ is compact by Theorem 4.18($a$), by Lemma 4.32 applied to $K = \sigma(x)$, there is a $C^\infty$ simple closed curve $\eta(\cdot)$ such that $\operatorname{ran}(\eta) \subset U \setminus \sigma(x)$ and $w(\eta(\cdot), z) = 1$ for all $z \in \sigma(x)$. By the Jordan curve theorem (e.g. Theorem 2.2 in Stein and Shakarchi [224, p. 351]), the complement of the range of $\eta(\cdot)$ consists of two disjoint connected open sets $V_1$ and $V_2$, where $V_1$ is bounded, simply connected and $w(\eta(\cdot), z) = 1$ if $z \in V_1$ and $= 0$ if $z \in V_2$. Thus $K := V_1 \cup \operatorname{ran}(\eta) = V_2^c$ is compact and $\sigma(x) \subset V_1 \subset K$. By Lemma 4.32 once again, there is a $C^\infty$ simple closed curve $\zeta(\cdot)$ such that $\operatorname{ran}(\zeta) \subset U \setminus K$ and $w(\zeta(\cdot), z) = 1$ for all $z \in \sigma(x)$. Let $h$ be another complex-valued holomorphic function on $U$. Then $f(x)$, $h(x)$ and $(fh)(x)$ in $\mathbb{B}$ are defined as in Definition 4.33. We claim that $(fh)(x) = f(x)h(x)$ in $\mathbb{B}$. To show this, by (4.12), we have

$$f(x)h(x) = \Big(\frac{1}{2\pi \mathrm{i}}\Big)^2 \oint_{\zeta(\cdot)} \oint_{\eta(\cdot)} f(\zeta)h(\eta)(\zeta\mathbb{1} - x)^{-1}(\eta\mathbb{1} - x)^{-1}\,\mathrm{d}\zeta\mathrm{d}\eta.$$

Using the resolvent equation (4.9), for each $\zeta, \eta \in \mathbb{C} \setminus \sigma(x)$, we have

$$(\zeta\mathbb{1} - x)^{-1}(\eta\mathbb{1} - x)^{-1} = (\eta - \zeta)^{-1}(\zeta\mathbb{1} - x)^{-1} + (\zeta - \eta)^{-1}(\eta\mathbb{1} - x)^{-1}.$$

By the Cauchy integral formula (Theorem 2.107), $\oint_{\eta(\cdot)} h(\eta)(\eta-\zeta)^{-1}\,\mathrm{d}\eta = 0$ for each $\zeta \in \operatorname{ran}(\zeta)$ since $\operatorname{ran}(\zeta) \subset V_2$, and so $w(\eta(\cdot),\zeta) = 0$ for each $\zeta \in \operatorname{ran}(\zeta)$. By the Cauchy integral formula once again, it then follows that

$$\begin{aligned} f(x)h(x) &= \Big(\frac{1}{2\pi\mathrm{i}}\Big)^2 \oint_{\eta(\cdot)} h(\eta)(\eta\mathbb{I}-x)^{-1}\Big[\oint_{\zeta(\cdot)} f(\zeta)(\zeta-\eta)^{-1}\,\mathrm{d}\zeta\Big]\,\mathrm{d}\eta \\ &= \frac{1}{2\pi\mathrm{i}} \oint_{\eta(\cdot)} h(\eta)f(\eta)(\eta\mathbb{I}-x)^{-1}\,\mathrm{d}\eta = (fh)(x), \end{aligned}$$

proving the claim. Using the claim and induction on $k$, it follows that for each positive integer $k$, $f^k(x) = f(x)^k$, and so

$$f(x)^k = \frac{1}{2\pi\mathrm{i}} \oint_{\zeta(\cdot)} f(\zeta)^k(\zeta\mathbb{I}-x)^{-1}\mathrm{d}\zeta.$$

Since $g$ is holomorphic on the whole plane, it has a Taylor expansion around 0, $g(z) = \sum_{k=0}^{\infty} a_k z^k$ for each $z \in \mathbb{C}$, which converges absolutely. Since the composition $g{\circ}f$ is holomorphic on $U$, applying (4.12) to $(g{\circ}f)(x)$, the Taylor expansion of $g$ at $z = f(\zeta)$, $\zeta \in \operatorname{ran}(\zeta)$, and then integrating term by term, we have

$$(g{\circ}f)(x) = \sum_{k=0}^{\infty} \frac{a_k}{2\pi\mathrm{i}} \oint_{\zeta(\cdot)} f(\zeta)^k(z\mathbb{I}-x)^{-1}\,\mathrm{d}\zeta = \sum_{k=0}^{\infty} a_k f(x)^k = g(f(x)),$$

where the last equality holds by Theorem 4.31($b$). The proof of the proposition is complete. □

Finally, we apply Definition 4.33 and Proposition 4.36 to define the logarithm. Let $g(w) = \mathrm{e}^w = \exp(w)$ from $\mathbb{C}$ into $\mathbb{C}$ and let $f(z) := \log z$ as a holomorphic function on $U := \{z \in \mathbb{C}\colon |z-1| < 1\}$ with $f(1) := \log 1 := 0$. Let $x \in \mathbb{B}$ satisfy $\|x - \mathbb{I}\| < 1$. To show that $\sigma(x) \subset U$ let $|z-1| \geq 1$. Then $z\mathbb{I} - x = (z-1)(\mathbb{I} + (z-1)^{-1}(\mathbb{I}-x))$ is invertible by Theorem 4.16($a$) since $\|(z-1)^{-1}(\mathbb{I}-x)\| < 1$, and so $z \notin \sigma(x)$, proving $\sigma(x) \subset U$. Thus we can define $f(x) = \log x$ and we have $\exp(\log(x)) = x$ for any such $x$ by Proposition 4.36. Moreover, by Theorem 4.31($b$), it follows that

$$\log x = \sum_{k=1}^{\infty} \frac{f^{(k)}(1)}{k!}(x-\mathbb{I})^k = -\sum_{k=1}^{\infty}(\mathbb{I}-x)^k/k, \tag{4.15}$$

and the series converges absolutely.

## 4.5 Complexification of Real Banach Algebras

We next see how a real Banach algebra can be embedded in a complex one.

**Definition 4.37.** Let $\mathbb{A}$ be an algebra over $\mathbb{R}$. Then the *complexification* $\mathbb{A}_{\mathbb{C}}$ of $\mathbb{A}$ is the set $\mathbb{A} \times \mathbb{A}$, with operations defined as follows: for $u, v, x, y \in \mathbb{A}$ and $a, b \in \mathbb{R}$, let

$$\begin{aligned}(u,v)+(x,y) &:= (u+x, v+y),\\ (u,v)\cdot(x,y) &:= (ux-vy, uy+vx),\\ (a+ib)(u,v) &:= (au-bv, av+bu).\end{aligned}$$

The following is straightforward to check:

**Proposition 4.38.** *For any algebra $\mathbb{A}$ over $\mathbb{R}$, the complexification $\mathbb{A}_{\mathbb{C}}$ as just defined is an algebra over $\mathbb{C}$. The map $u \mapsto (u,0)$ is 1-to-1 from $\mathbb{A}$ into $\mathbb{A}_{\mathbb{C}}$ and preserves real algebra operations. If $\mathbb{A}$ is unital with identity $\mathbb{1}_{\mathbb{A}}$, then so is $\mathbb{A}_{\mathbb{C}}$ with identity $\mathbb{1} := (\mathbb{1}_{\mathbb{A}}, 0)$.*

Via $x \leftrightarrow (x,0)$, we can view $\mathbb{A}$ as a real subalgebra of $\mathbb{A}_{\mathbb{C}}$, just as we view $\mathbb{R}$ as a subset of $\mathbb{C}$. Then we can write for each $(u,v) \in \mathbb{A}_{\mathbb{C}}$, $(u,v) = u+iv$, and each such $(u,v)$ can be written in multiple ways as a finite sum $\sum_{j=1}^n z_j x_j$ for $z_j \in \mathbb{C}$ and $x_j \in \mathbb{A}$.

**Theorem 4.39.** *Let $(\mathbb{A}, \|\cdot\|)$ be a real Banach algebra. For each $w$ in the complexification $\mathbb{A}_{\mathbb{C}}$, let*

$$|\!|\!|w|\!|\!| := \inf\Big\{\sum_{j=1}^n |z_j|\|x_j\| \colon w = \sum_{j=1}^n z_j x_j,\ z_j \in \mathbb{C},\ x_j \in \mathbb{A},\ n = 1, 2, \dots\Big\}.$$

*Then*

*(a) $|\!|\!|\cdot|\!|\!|$ is a Banach algebra norm on $\mathbb{A}_{\mathbb{C}}$;*
*(b) $\max\{\|u\|, \|v\|\} \le |\!|\!|(u,v)|\!|\!| \le \|u\| + \|v\|$ for any $(u,v) \in \mathbb{A}_{\mathbb{C}}$;*
*(c) $|\!|\!|(u,0)|\!|\!| = \|u\|$ for each $u \in \mathbb{A}$.*

*Proof.* If $\mathbb{A} = \{0\}$ there is no problem, so suppose $\mathbb{A} \neq \{0\}$. From the definition it is straightforward to check that $|\!|\!|\cdot|\!|\!|$ is a seminorm on the complex vector space $\mathbb{A}_{\mathbb{C}}$. Also, $|\!|\!|u+iv|\!|\!| \le \|u\| + \|v\|$, proving the right side of $(b)$.

If $|\!|\!|w|\!|\!| < 1$ for $w = (u,v) \in \mathbb{A}_{\mathbb{C}}$, then for some $n = 1, 2, \dots,$ $z_j \in \mathbb{C}$, and $x_j \in \mathbb{A}$ we have $w = \sum_{j=1}^n z_j x_j$ with $\sum_{j=1}^n |z_j|\|x_j\| < 1$. Let $z_j = a_j + ib_j$ for $a_j, b_j \in \mathbb{R}$. Then $u = \sum_{j=1}^n a_j x_j$ and $v = \sum_{j=1}^n b_j x_j$ with $\max\{\sum_{j=1}^n |a_j|\|x_j\|, \sum_{j=1}^n |b_j|\|x_j\|\} < 1$, so $\|u\| < 1$ and $\|v\| < 1$. It follows that for any $w = (u,v) \in \mathbb{A}_{\mathbb{C}}$, $\max\{\|u\|, \|v\|\} \le |\!|\!|w|\!|\!|$, proving the left side of $(b)$ and showing that $|\!|\!|\cdot|\!|\!|$ is a norm on $\mathbb{A}_{\mathbb{C}}$. Clearly, $(b)$ implies $(c)$. Also, if $\{w_n\}_{n=1}^\infty = \{(u_n, v_n)\}_{n=1}^\infty$ is a Cauchy sequence for $|\!|\!|\cdot|\!|\!|$, it is convergent by $(b)$, so $(\mathbb{A}_{\mathbb{C}}, |\!|\!|\cdot|\!|\!|)$ is a Banach space.

Let $w = \sum_{i=1}^m c_i x_i$ and $z = \sum_{j=1}^n d_j y_j$ for some $m$, $n$; $c_i, d_j \in \mathbb{C}$; and $x_i, y_j \in \mathbb{A}$. Then $wz = \sum_{i=1}^m \sum_{j=1}^n (c_i d_j)(x_i y_j)$ where

$$\sum_{i=1}^{m}\sum_{j=1}^{n}|c_jd_j|\|x_iy_j\| \le \Big(\sum_{i=1}^{m}|c_i|\|x_i\|\Big)\Big(\sum_{j=1}^{n}|d_j|\|y_j\|\Big).$$

Thus $\|\!|wz|\!\| \le \|\!|w|\!\|\,\|\!|z|\!\|$ and $\|\!|\cdot|\!\|$ is submultiplicative (4.2).

Suppose $\mathbb{A}_{\mathbb{C}}$ is unital and let $\mathbb{I} = (u, v) \neq (0, 0)$ be its identity. Then for any $x, y \in \mathbb{A}$, $xu - yv = ux - vy = x$. Letting $y = 0$ gives that $u$ is an identity $\mathbb{I}_{\mathbb{A}}$ of $\mathbb{A}$. Then $vy = yv = 0$ for all $y$, specifically $y = u$, gives $v = 0$. Thus $\mathbb{I} = (\mathbb{I}_{\mathbb{A}}, 0)$ and $\|\!|\mathbb{I}|\!\| = \|\mathbb{I}_{\mathbb{A}}\| = 1$, proving $(a)$, and so the theorem. $\square$

The complex Banach algebra $(\mathbb{A}_{\mathbb{C}}, \|\!|\cdot|\!\|)$ given by Theorem 4.39 will be called the *Banach algebra complexification* of $\mathbb{A}$.

Let $(\mathbb{B}, \|\cdot\|)$ be a Banach algebra of real-valued functions as defined in Definition 4.11. Then its Banach algebra complexification $(\mathbb{B}_{\mathbb{C}}, \|\!|\cdot|\!\|)$ can naturally be viewed as a Banach algebra of complex-valued functions, with $\mathbb{B}_{\mathbb{C}} = \{f + ig\colon\ f, g \in \mathbb{B}\}$.

## 4.6 A Substitution Rule for the Kolmogorov Integral

The Banach algebra property of $\mathcal{W}_p$ spaces of functions with values in a Banach algebra (Proposition 4.13) will be used to prove a substitution rule for the Kolmogorov integral, like those considered for Riemann–Stieltjes integrals at the end of Section 3.6. Recall that for $1 \le p < \infty$, $\mathcal{Q}_p := [1, p/(p-1))$ if $p > 1$, $\mathcal{Q}_1 := \{+\infty\}$, and $\mathcal{W}_\infty = \mathcal{R}$ is the class of regulated functions.

**Proposition 4.40.** *Assume that $\mathbb{B}$ is a Banach algebra over $\mathbb{K}$ and (1.14) holds with $Y = Z = \mathbb{B}$. Let $\alpha \in \mathcal{AI}_p([a,b];X)$ for $1 \le p < \infty$ and let $f, g \in \mathcal{W}_q([a,b];\mathbb{B})$ for $q \in \mathcal{Q}_p$. Then for $\mu(A) := \fint_A d\alpha{\cdot}g$, $A \in \mathfrak{I}[a,b]$, $\mu$ is an interval function in $\mathcal{AI}_p([a,b];\mathbb{B})$, and the two integrals $\fint_{[a,b]} d\mu\, f$, $\fint_{[a,b]} d\alpha{\cdot}(gf)$ exist and are equal.*

*Proof.* We can assume that $a < b$. By Proposition 3.96, we have that $\mu$ is defined and is in $\mathcal{AI}_p([a,b];\mathbb{B})$. By Proposition 4.13, $fg \in \mathcal{W}_q([a,b];\mathbb{B})$, and so the two integrals are defined. To prove that they are equal, let $\epsilon > 0$. Since $g$ is regulated, by Theorem 2.1, there exists a point partition $\{z_j\}_{j=0}^m$ of $[a,b]$ such that $\mathrm{Osc}(g;(z_{j-1},z_j)) < \epsilon$ for each $j = 1,\dots,m$. Let $\mathcal{T} = (\{(t_{i-1},t_i)\}_{i=1}^n, \{s_i\}_{i=1}^n)$ be a tagged Young interval partition of $[a,b]$ which is a tagged refinement of the Young interval partition $\lambda := \{(z_{j-1},z_j)\}_{j=1}^m$. Letting $A_i := (t_{i-1},t_i)$ for each $i$, we have

$$\begin{aligned}
&\big\|S_{YS}(d\mu, f;\mathcal{T}) - S_{YS}(d\alpha, gf;\mathcal{T})\big\|\\
&= \Big\|\sum_{i=1}^n \mu(A_i)f(s_i) - \alpha(A_i){\cdot}[g(s_i)f(s_i)]\Big\| \le \|f\|_{\sup}\sum_{i=1}^n \Big\|\fint_{A_i} d\alpha{\cdot}\big[g - g(s_i)\big]\Big\|.
\end{aligned}$$

If $p = 1$, then using (3.69), it follows that

$$\sum_{i=1}^{n} \Big\| \fint_{A_i} \mathrm{d}\alpha \cdot [g - g(s_i)] \Big\| \le \max_{1 \le i \le n} \mathrm{Osc}(g; A_i) \sum_{i=1}^{n} v_1(\alpha; A_i) < \epsilon v_1(\alpha).$$

If $p > 1$ then for $q' \in (q, p/(p-1))$, by the Love–Young inequality (3.150) and Lemma 3.45, we have

$$\sum_{i=1}^{n} \Big\| \fint_{A_i} \mathrm{d}\alpha \cdot [g - g(s_i)] \Big\| \le K_{p,q'} \sum_{i=1}^{n} \|g\|_{A_i,(q')} \|\alpha\|_{A_i,(p)}$$

$$\le \epsilon^{1-(q/q')} K_{p,q'} \sum_{i=1}^{n} v_q(g; A_i)^{1/q'} v_p(\alpha; A_i)^{1/p} \le \epsilon^{1-(q/q')} K_{p,q'} v_q(g)^{1/q'} \|\alpha\|_{(p)},$$

where the last inequality holds by Hölder's inequality (1.4) and by (3.69). Since $\mathcal{T}$ is an arbitrary tagged Young interval partition of $[a, b]$ which is a refinement of $\lambda$, and since $\epsilon > 0$ is arbitrary, we have that $\fint_{[a,b]} \mathrm{d}\mu\, f = \fint_{[a,b]} \mathrm{d}\alpha \cdot (gf)$, proving the proposition. □

## 4.7 Notes

*Notes on Section 4.1.* There are numerous books giving expositions of the basic theory of Banach algebras. For example, the fact that a unit can be adjoined if needed (Proposition 4.7) is given in Larsen [133, Theorem 1.1.5]. Often the multiplication in a Banach algebra is taken to satisfy $\|xy\| \le \|x\|\|y\|$ for all $x, y$ by definition, e.g. [216, p. 209]. Theorem 4.8 showed that this inequality can be obtained by renorming. The theorem holds also when $(x, y) \mapsto xy$ need only be continuous in $x$ for each $y$ and in $y$ for each $x$, as I. M. Gelfand proved; see e.g. Theorem 1.3.1 of Larsen [133].

*Notes on Section 4.2.* Belfi and Doran [13] write: "commutative Banach algebras enjoy a remarkably complete and beautiful structure theory due, in large part, to the efforts of a single man, I. M. Gelfand." A basic paper was Gelfand [76]. An earlier related announcement was Gelfand [75]. A book on the theory of commutative Banach algebras is Gelfand, Raikov, and Shilov [77]. According to Belfi and Doran [13], Theorem 4.19 was announced by Mazur [162], with a first published proof by Gelfand [76].

*Notes on Section 4.3.* Proposition 4.25 is a classical fact in functional analysis. It was stated e.g. by Dunford and Schwartz [57, pp. 88–89]. They say that it was first discovered and published by Banach in the first, 1931, edition in Polish [7] of his classic book *Théorie des opérations linéaires* [8] published in French in 1932. Meanwhile, Hausdorff [89] in 1932 independently also discovered the quotient space by a closed linear subspace.

In Banach [8] we have not found quotient spaces very explicitly. In Chapitre I, §2, Banach considers left and right cosets of subsets (not necessarily subgroups) of a not necessarily abelian group with a topology. In Chapitre IV, §3, Lemme, he considers an element $y_0$ in a normed linear space $E$ at distance $d$ from a linear subspace $G$ [or its closure $\bar{G}$]. [Here $d = d(y_0) =: d(y_0 + G)$ defines the quotient seminorm on $E/G$, but we did not find this formulation in [8].]

The other facts in Section 4.3 were found by Gelfand ([76], [77]).

*Notes on Section 4.4.* Theorem 4.31 is also due to Gelfand ([76] and [77, I.§6, Theorem 1])

*Notes on Section 4.5.* Theorem 4.39 is essentially Proposition 3 in §13 of Bonsall and Duncan [22]. The norm $|||\cdot|||$ is the same as in Theorem 1.3.5(*ii*) in Ruston [200], where it is shown to be the largest among all admissible norms, that is, those satisfying Theorem 4.39(*b*). Ruston [200, p. 221] and Kirwan [117] give references to earlier results on complexification of real normed spaces. On the norm $|||\cdot|||$ as a tensor product norm see also Section 5.4.

*Notes on Section 4.6.* Proposition 4.40 had many predecessors with interval functions $\alpha$, $\mu$ replaced by point functions (e.g. Propositions 2.86, 3.100, and 3.101). In its present form, the substitution rule is stated here for the first time as far as we know.

# 5

# Derivatives and Analyticity in Normed Spaces

All of the vector spaces that we consider in this chapter will be defined over a field $\mathbb{K}$ which will be either the field $\mathbb{R}$ of real numbers or the field $\mathbb{C}$ of complex numbers.

Let $X$ and $Y$ be normed spaces with the norm on each denoted by $\|\cdot\|$, and let $U$ be an open subset of $X$. Let $L(X,Y)$ denote the space of all bounded linear operators from $X$ into $Y$. A function (nonlinear operator) $F\colon U \to Y$ is called (Fréchet) *differentiable* or has a (Fréchet) *derivative at* $u \in U$ if there exists an $L_u \in L(X,Y)$ such that for each $x \in X$ with $u + x \in U$,

$$(\Delta F)(x) := F(u+x) - F(u) = L_u x + \mathrm{Rem}_F(u,x), \tag{5.1}$$

where $\|\mathrm{Rem}_F(u,x)\| = o(\|x\|)$ as $\|x\| \to 0$. It follows that $\|(\Delta F)(x)\| = O(\|x\|)$ as $\|x\| \to 0$, and so $F$ is continuous at $u$. The linear operator $L_u$ is easily seen to be unique, is called the *derivative* of $F$ *at* $u$, and is denoted by $DF(u)$. If the function $F$ is differentiable at each $u \in U$ then we say that $F$ is differentiable on $U$, and we call the map $D^1F := DF\colon U \to L(X,Y)$ the *first derivative of* $F$. The difference $\mathrm{Rem}_F(u,x) = (\Delta F)(x) - L_u x$ is called the *remainder* in the differentiation of $F$.

We will recall some basic facts about differentiation. The first one concerns differentiability of composition mappings.

**Theorem 5.1** (Chain rule). *Let $X, Y, Z$ be three normed spaces, $U$ an open neighborhood of $u \in X$, $F$ a mapping of $U$ into $Y$, $v = F(u)$, $V$ an open neighborhood of $v$ in $Y$, and $G$ a mapping of $V$ into $Z$. If $F$ is differentiable at $u$ and $G$ is differentiable at $v$, then the composition mapping $G{\circ}F$ defined by $(G{\circ}F)(u) := G(F(u))$ is differentiable at $u$ and*

$$D(G{\circ}F)(u) = DG(v){\circ}DF(u).$$

*Proof.* For each $x \in X$ with $u + x \in U$, using the notation (5.1), we have

$$R_{G\circ F}(u,x) := (G{\circ}F)(u+x) - (G{\circ}F)(u) - (DG(v){\circ}DF(u))x$$

R.M. Dudley and R. Norvaiša, *Concrete Functional Calculus*, Springer Monographs in Mathematics, DOI 10.1007/978-1-4419-6950-7_5,

$$= [G(v + (\Delta F)(x)) - G(v) - DG(v)(\Delta F)(x)] + DG(v)\mathrm{Rem}_F(u, x).$$

Since $G$ is differentiable at $v = F(u)$ and $F$ is continuous at $u$, we have

$$\|G(v + (\Delta F)(x)) - G(v) - DG(v)(\Delta F)(x)\| = o(\|(\Delta F)(x)\|)$$

as $\|x\| \to 0$. Since $F$ is differentiable at $u$, we have $\|(\Delta F)(x)\| = O(\|x\|)$ and $\|\mathrm{Rem}_F(u, x)\| = o(\|x\|)$ as $\|x\| \to 0$. Thus $\|R_{G\circ F}(u, x)\| = o(\|x\|)$ as $\|x\| \to 0$, proving the theorem. □

**Remark 5.2.** A bounded linear operator $A$ from $Y$ to $Z$, or $B$ from $X$ to $Y$, is clearly Fréchet differentiable everywhere and equal to its derivative. Thus in Theorem 5.1 we have $D(A\circ F)(u) = (A\circ DF)(u)$, and $D(G\circ B)(u) = (DG\circ B)(u)$ if $B(u) \in V$.

The second fact is a variant of the mean value theorem, which gives a bound for the remainder in a differentiation. For $u, u+x \in X$, the line segment joining $u$ and $u + x$ is the set $[u, u + x] := \{u + tx\colon\ 0 \le t \le 1\}$.

**Theorem 5.3** (Mean value theorem). *Let $X, Y$ be two normed spaces over $\mathbb{K}$, $U$ an open subset of $X$, and $F$ a differentiable mapping of $U$ into $Y$. If the segment $[u, u + x]$ is included in $U$, then*

$$\big\|F(u + x) - F(u)\big\| \le \|x\| \sup_{v\in[u,u+x]} \big\|DF(v)\big\|, \tag{5.2}$$

$$\big\|F(u + x) - F(u) - DF(u)x\big\| \le \|x\| \sup_{v\in[u,u+x]} \big\|DF(v) - DF(u)\big\|. \tag{5.3}$$

*Proof.* To prove (5.2), let $\phi(t) := F(u + tx)$ for each $t \in \mathbb{R}$. By the chain rule, the function $\phi$ is differentiable from an open neighborhood $V$ of $[0, 1]$ into $Y$ and $D\phi(t) = DF(u + tx)x$ for each $t \in [0, 1]$. By the Hahn–Banach theorem (Corollary 6.1.5 in [53]), there is a continuous functional $f \in Y'$ (linear over $\mathbb{K}$) such that $\|f\|' = 1$ and

$$|f(\phi(1) - \phi(0))| = \|\phi(1) - \phi(0)\| = \|F(u + x) - F(u)\|.$$

The derivative of $f$ is the constant mapping $Df(y) = f$ for each $y \in Y$. Then by the chain rule again, the composition $f\circ\phi$ is differentiable on $V \subset \mathbb{R}$ into $\mathbb{K}$ and $D(f\circ\phi)(t) = (f\circ DF(u + tx))x$ for each $t \in [0, 1]$. Thus by the ordinary mean value theorem there is a $t \in [0, 1]$ such that $f(\phi(1)) - f(\phi(0)) = D(f\circ\phi)(t)$, and so

$$|f(\phi(1)) - f(\phi(0))| \le \sup_{t\in[0,1]} |(f\circ DF(u + tx))x| \le \|x\| \sup_{v\in[u,u+x]} \|DF(v)\|,$$

proving (5.2).

The proof of (5.3) is the same except that now we choose a continuous linear functional $f \in Y'$ with unit norm so that

$$|f(\phi(1)-\phi(0)-D\phi(0))| = \|\phi(1)-\phi(0)-D\phi(0)\| = \|F(u+x)-F(u)-DF(u)x\|.$$

Then again by the ordinary mean value theorem there is a $t \in [0,1]$ such that

$$\begin{aligned}\big|f(\phi(1)) - f(\phi(0)) - f(D\phi(0))\big| &= \big|f\circ DF(u+tx)x - f\circ DF(u)x\big| \\ &\le \|x\| \sup_{v\in[u,u+x]} \big\|DF(v) - DF(u)\big\|,\end{aligned}$$

proving the theorem. □

## 5.1 Polynomials and Power Series

Finite and infinite Taylor series can be defined for suitable functions from one normed space to another. To treat such series we will first need some multilinear algebra.

A function $A$ from a Cartesian product $V_1 \times \cdots \times V_k$ of $k$ vector spaces into a vector space $V_0$ is called *$k$-linear* if for each $j = 1, \dots, k$, $A(v_1, \dots, v_k)$ is linear in $v_j$ for any fixed values of $v_i$, $i \neq j$. If $V_j = (V_j, \|\cdot\|_j)$ are normed spaces for $j = 0, 1, \dots, k$, $k \ge 1$, then a $k$-linear map $A\colon V_1 \times \cdots \times V_k \to V_0$ is called *bounded* if

$$\|A\| := \sup\big\{\|A(v_1,\dots,v_k)\|_0\colon \|v_j\|_j \le 1,\ j = 1,\dots,k\big\} < \infty. \tag{5.4}$$

This agrees with the usual definition of operator norm (1.17) if $k = 1$.

Let $\mathcal{M}_k(V_1, \dots, V_k; V_0)$ be the set of all bounded $k$-linear maps from $V_1 \times \cdots \times V_k$ into $V_0$ with the norm defined by (5.4). It is easily seen that $\mathcal{M}_k(V_1, \dots, V_k; V_0)$ with this norm is a Banach space provided $(V_0, \|\cdot\|_0)$ is.

**Theorem 5.4.** *Let $(V_j, \|\cdot\|_j)$, $j = 0, 1, \dots, k$, be normed spaces over $\mathbb{K}$ and let $L$ be a $k$-linear map from $V_1 \times \cdots \times V_k$ into $V_0$. Then the following are equivalent:*

*(a) $L$ is jointly continuous everywhere;*
*(b) $L$ is jointly continuous at $(0, 0, \dots, 0)$;*
*(c) $L$ is bounded.*

*Proof.* $(a) \Rightarrow (b)$ is clear. Assume $(b)$. Then for some $\delta > 0$, $\|x_j\|_j \le \delta$ for $j = 1, \dots, k$ implies $\|L(x_1, \dots, x_k)\|_0 \le 1$. It follows by $k$-linearity that for any $y_j \in V_j$, $j = 1, \dots, k$, $\|L(y_1, \dots, y_k)\|_0 \le \delta^{-k}\|y_1\| \cdots \|y_k\|$, so $\|L\| \le \delta^{-k}$ and $(c)$ follows.

Assume $(c)$ with $\|L\| = M < \infty$. To prove $(a)$, take any $x_j \in V_j$ and $y_j \in V_j$ for $j = 1, \dots, k$. Then

$$\begin{aligned}L(y_1,\dots,y_k) - L(x_1,\dots,x_k) &= L(y_1 - x_1, y_2, \dots, y_k) \\ +L(x_1, y_2 - x_2, y_3, \dots, y_k) &+ \cdots + L(x_1, \dots, x_{k-1}, y_k - x_k),\end{aligned}$$

and so

$$\|L(y_1,\ldots,y_k) - L(x_1,\ldots,x_k)\|_0 \le M \sum_{j=1}^{k} \|x_j - y_j\|_j \prod_{i\neq j} \max\{\|x_i\|_i, \|y_i\|_i\}.$$

Thus for fixed $x_1,\ldots,x_k$, as $y_j \to x_j$ for each $j$, the right side tends to zero, and so $L(y_1,\ldots,y_k) \to L(x_1,\ldots,x_k)$, proving $(a)$ and the theorem. □

Let $A$ be a $k$-linear map from $V_1 \times \cdots \times V_k$ into $V_0$, and let $1 \le m < k$. For any $v_j \in V_j$, $j = 1,\ldots,m$, a mapping $A(v_1,\ldots,v_m)$ from $V_{m+1} \times \cdots \times V_k$ into $V_0$ is defined by

$$A(v_1,\ldots,v_m)(v_{m+1},\ldots,v_k) := A(v_1,\ldots,v_m,v_{m+1},\ldots,v_k). \tag{5.5}$$

**Proposition 5.5.** *Let $1 \le m < k$, let $V_j = (V_j, \|\cdot\|_j)$, $j = 0,1,\ldots,k$, be normed spaces, and let $\mathcal{M}_{k,(m)} := \mathcal{M}_m(V_1,\ldots,V_m;\mathcal{M}_{k-m}(V_{m+1},\ldots,V_k;V_0))$. For $v_j \in V_j$, $j = 1,\ldots,m$, and $A \in \mathcal{M}_k := \mathcal{M}_k(V_1,\ldots,V_k;V_0)$, define $L_{m,k}(A)(v_1,\ldots,v_m) \in \mathcal{M}_{k-m}(V_{m+1},\ldots,V_k;V_0)$ by*

$$L_{m,k}(A)(v_1,\ldots,v_m) := A(v_1,\ldots,v_m).$$

*Then $L_{m,k}$ is a linear isometry from $\mathcal{M}_k$ onto $\mathcal{M}_{k,(m)}$.*

*Proof.* For any $A \in \mathcal{M}_k$ and $v_j \in V_j$, $j = 1,\ldots,m$, $L_{m,k}(A)(v_1,\ldots,v_m)$ is clearly a $(k-m)$-linear map of $V_{m+1} \times \cdots \times V_k$ into $V_0$ with

$$\big\|L_{m,k}(A)(v_1,\ldots,v_m)\big\| \le \|A\| \prod_{j=1}^{m} \|v_j\|_j.$$

Thus $L_{m,k}(A) \in \mathcal{M}_{k,(m)}$ and $\|L_{m,k}(A)\| \le \|A\|$. The map $L_{m,k}$ is clearly linear.

Conversely, let $B \in \mathcal{M}_{k,(m)}$ and for any $v_j \in V_j$, $j = 1,\ldots,k$, let $A(B)(v_1,\ldots,v_k) := B(v_1,\ldots,v_m)(v_{m+1},\ldots,v_k)$. Then from the definitions, clearly $A(B)$ is $k$-linear from $V_1 \times \cdots \times V_k$ into $V_0$ with $\|A(B)\| \le \|B\| < \infty$, so $A(B) \in \mathcal{M}_k$. Evidently $L_{m,k}(A(B)) = B$, so $L_{m,k}$ and $B \mapsto A(B)$ are inverses of one another and are linear isometries, completing the proof. □

Let $L$ be a $k$-linear function from a Cartesian product $V_1 \times \cdots \times V_k$ of $k$ vector spaces into a vector space $V_0$. If $V_1 = \cdots = V_k = X$, then $L$ is defined on the $k$-fold product set $X^k := V_1 \times \cdots \times V_k$ and is called *symmetric* if $L(x_{\pi(1)},\ldots,x_{\pi(k)}) = L(x_1,\ldots,x_k)$ for any permutation $\pi$ of $\{1,\ldots,k\}$ and any $x_1,\ldots,x_k \in X$. For normed spaces $X$, $Y$ and an integer $k \ge 1$, let $L(^kX,Y) := \mathcal{M}_k(V_1,\ldots,V_k;Y)$, where $V_1 = V_2 = \cdots = V_k = X$. Thus $L(^1X,Y) = L(X,Y)$, with the usual operator norm (1.17). Let $L_s(^kX,Y)$

denote the vector subspace of symmetric elements of $L(^kX,Y)$. For $k=0$ let $L_s(^0X,Y):=Y$. Any $L\in L(^kX,Y)$ has a *symmetrization* given by $L^{(s)}=L$ for $k=0$ or 1 and for $k\geq 2$,

$$L^{(s)}(x_1,\dots,x_k):=\frac{1}{k!}\sum_{\pi}L(x_{\pi(1)},\dots,x_{\pi(k)}),$$

where the sum is over all permutations $\pi$ of $\{1,2,\dots,k\}$. Clearly $L^{(s)}$ is symmetric and if $L$ is symmetric, then $L^{(s)}=L$. For any $L\in L(^kX,Y)$,

$$\left\|L^{(s)}\right\|\leq\|L\|. \tag{5.6}$$

For $L\in L(^kX,Y)$, let

$$Lx_1^{\otimes k_1}\otimes\cdots\otimes x_m^{\otimes k_m}:=L(x_1,\dots,x_1,x_2,\dots,x_2,x_3,\dots,x_{m-1},x_m,\dots,x_m)$$

where $x_j$ appears $k_j$ times on the right for $j=1,\dots,m$ and $k_j$ are nonnegative integers for $j=1,\dots,m$ such that $k_1+\cdots+k_m=k$ and $m=1,2,\dots$. For each symmetric $k$-linear mapping $L$, $k\geq 2$, and any integer $m\geq 1$, we have the following multinomial formula:

$$L(x_1+\cdots+x_m)^{\otimes k}=\sum\frac{k!}{k_1!\cdots k_m!}Lx_1^{\otimes k_1}\otimes\cdots\otimes x_m^{\otimes k_m}, \tag{5.7}$$

where $\sum$ denotes summation over all $m$-tuples $(k_1,\dots,k_m)$ of nonnegative integers satisfying $k_1+\cdots+k_m=k$. As a special case, when $m=2$, we have the binomial formula:

$$L(x+y)^{\otimes k}=Lx^{\otimes k}+\sum_{l=1}^{k-1}\binom{k}{l}Lx^{\otimes k-l}\otimes y^{\otimes l}+Ly^{\otimes k}. \tag{5.8}$$

Let $X$ and $Y$ be vector spaces. For $k=1,2,\dots$, a mapping $P\colon X\to Y$ is called a *$k$-homogeneous polynomial* if there exists a $k$-linear mapping $L$ from $X^k$ into $Y$ such that $Px=L(x,\dots,x)$ for all $x\in X$. Then let $P=\tilde{L}$. If $M=L^{(s)}$ then clearly $\tilde{M}=\tilde{L}$. Thus, in the definition of $k$-homogeneous polynomial, we can equivalently take $L$ to be symmetric. To show that the map $L\mapsto\tilde{L}$ is one-to-one for $L$ symmetric, and find its inverse, one can use the following polarization identity:

**Theorem 5.6** (Polarization). *Let $L\colon X^k\to Y$ be a symmetric $k$-linear map. Then*

$$L(x_1,\dots,x_k)=\frac{1}{k!2^k}\sum{}'e_1\cdots e_kL(e_1x_1+\cdots+e_kx_k)^{\otimes k}, \tag{5.9}$$

*where $\sum'$ denotes summation over all $e_i\in\{-1,1\}$, $i=1,\dots,k$. Thus, if $\tilde{L}\equiv\tilde{M}$ for two symmetric $k$-linear maps $X^k\to Y$ then $L\equiv M$.*

*Proof.* By the multinomial formula (5.7), we have

$$L(e_1x_1 + \cdots + e_kx_k)^{\otimes k} = \sum{}'' \frac{k!}{n_1!\cdots n_k!} L(e_1x_1)^{\otimes n_1} \otimes \cdots \otimes (e_kx_k)^{\otimes n_k}$$
$$= \sum{}'' \frac{k!}{n_1!\cdots n_k!} e_1^{n_1}\cdots e_k^{n_k} Lx_1^{\otimes n_1} \otimes \cdots \otimes x_k^{\otimes n_k},$$

where $\sum''$ denotes summation over all $k$-tuples $(n_1, \dots, n_k)$ of nonnegative integers satisfying $n_1 + \cdots + n_k = k$. Denoting the right side of (5.9) by $A$, we then have

$$A = \frac{1}{2^k}\sum{}'' \frac{1}{n_1!\cdots n_k!} a(n_1, \dots, n_k) Lx_1^{\otimes n_1} \otimes \cdots \otimes x_k^{\otimes n_k},$$

where $a(n_1, \dots, n_k) = \sum' e_1^{n_1+1} \cdots e_k^{n_k+1}$. If some $n_i = 0$ then $a(n_1, \dots, n_k) = 0$. Thus $a(n_1, \dots, n_k) \neq 0$ if and only if $n_1 = \cdots = n_k = 1$. Since $a(1, \dots, 1) = 2^k$, $A = L(x_1, \dots, x_k)$. The last conclusion follows easily, proving the theorem. □

By the preceding polarization theorem, a symmetric $k$-linear map $L$ from $X^k$ into $Y$ is uniquely defined by its values on the diagonal of $X^k$. The one-to-one relation between $L$ and the $k$-homogeneous polynomial $P = \tilde{L}$ will be assumed and used frequently in what follows.

A function $P$ from $X$ into $Y$ will be called a *polynomial* iff for some finite $k$, $P = \sum_{j=0}^{k} P_j$ where for $j = 1, \dots, k$, $P_j$ is a $j$-homogeneous polynomial from $X$ into $Y$ and $P_0(x) \equiv y_0$ for some $y_0 \in Y$.

If $X$ and $Y$ are normed spaces and $L$ is a continuous $k$-linear map from $X$ into $Y$, then clearly the $k$-homogeneous polynomial $\tilde{L}$ is continuous. Conversely, if $\tilde{L}$ is continuous, then by the polarization theorem 5.6, $L$ is continuous, and so bounded by Theorem 5.4.

Let $P^k(X, Y)$ be the vector space of all continuous $k$-homogeneous polynomials from $X$ to $Y$. A norm on $P^k(X, Y)$ is given by

$$\|P\| := \sup\{\|Px\| \colon \|x\| \leq 1\} = \inf\{C \geq 0 \colon \|Px\| \leq C\|x\|^k \text{ for all } x \in X\}.$$

It is clear that if $L \in L_s({}^kX, Y)$, then $\tilde{L} \in P^k(X, Y)$ and $\|\tilde{L}\| \leq \|L\|$. A converse bound follows from the next theorem:

**Theorem 5.7.** *The mapping $L_s({}^kX, Y) \ni L \mapsto \tilde{L} \in P^k(X, Y)$ is a vector space isomorphism and a homeomorphism of the first onto the second space. Moreover,*

$$\|\tilde{L}\| \leq \|L\| \leq \frac{k^k}{k!}\|\tilde{L}\|. \tag{5.10}$$

*Proof.* It is enough to prove the second inequality in (5.10) when $k > 1$. By Theorem 5.6, we have

$$\|L(x_1,\dots,x_k)\| \le \frac{1}{k!2^k}\sum{}' \|\tilde{L}\|\big(\|e_1x_1\|+\cdots+\|e_kx_k\|\big)^k,$$

where $\sum'$ denotes summation over all $e_i \in \{-1,1\}$, $i=1,\dots,k$, and has $2^k$ terms. Thus if $\|x_i\|\le 1$ for $i=1,\dots,k$, we obtain the second inequality in (5.10), completing the proof. □

The constant $k^k/k!$ in (5.10) is best possible. Let $X := L^1$ be the Banach space of Lebesgue integrable functions over $[0,1]$, let $Y := \mathbb{R}$, and let $L \in L_s(^kX,Y)$ be defined by

$$L(x_1,\dots,x_k) := \sum_{\pi}\int_{[0,1/k]} x_{\pi(1)}(t)\,\mathrm{d}t\cdots\int_{[(k-1)/k,1]} x_{\pi(k)}(t)\,\mathrm{d}t,$$

where the sum is over all permutations of $\pi$ of $\{1,\dots,k\}$. Kopeć and Musielak [123] proved for this case that $\|L\| = (k^k/k!)\|\tilde{L}\|$. It is well known that the mapping $L_s(^kX,Y)\ni L\mapsto \tilde{L}\in P^k(X,Y)$ is an isometry if $X=H$ is a real Hilbert space and $Y$ is a Banach space, as shown in Corollary 5.23($e$) below.

**Definition 5.8.** Let $X$ and $Y$ be Banach spaces, and let $P^0(X,Y)$ be the set of all constant functions from $X$ into $Y$. A *power series* from $X$ to $Y$ around $u\in X$ is a series of the form $\sum_{k\ge 0}P_u^k(x-u)$, where $P_u^k\in P^k(X,Y)$ for each $k=0,1,\dots$. The largest $r\in[0,\infty]$ such that the series converges in $Y$ for all $x\in X$ with $\|x-u\|<r$ is called the *radius of convergence* of the power series. Let $U$ be an open subset of $X$. A mapping $F\colon U\to Y$ has a *power series expansion* around a point $u\in U$ if there exists a power series $\sum_{k\ge 0}P_u^k(x-u)$ from $X$ to $Y$ around $u$ which converges to $F(x)$ for $x$ in a ball $B(u,r)\subset U$ for some $r>0$.

If $P$ is any polynomial from $X$ into $Y$ and $u$ is a fixed element of $X$, then the function $x\mapsto P(x-u)$ is also a polynomial, by Theorem 5.6 (polarization) and the binomial formula (5.8).

Power series with radius of convergence equal to 0 will not be of interest. If two power series $\sum_{k\ge 0}P_u^k(x-u)$ and $\sum_{k\ge 0}Q_u^k(x-u)$ converge absolutely and have the same sum on a neighborhood of $u$, then $P_u^k=Q_u^k$ for $k=0,1,\dots$. This is a consequence of the following theorem.

**Theorem 5.9.** *If the power series $\sum_{k\ge 0}P_u^k(x-u)$ from $X$ to $Y$ around $u\in X$ converges absolutely and its sum is equal to zero for $x$ in some neighborhood of $u$, then $P_u^k\equiv 0$ for $k=0,1,\dots$.*

*Proof.* Taking $x=u$ we get $P_u^0\equiv 0$. By the assumption, there is an $r>0$ such that the power series $\sum_{k\ge 1}P_u^k(x-u)$ from $X$ to $Y$ around $u$ converges absolutely to zero for all $x\in X$ such that $\|x-u\|\le r$. Let $y\in X$ be arbitrary and non-zero. For any $t\in\mathbb{K}$ such that $|t|\le r/\|y\|$, letting $x:=ty+u$, we have $\|x-u\|=|t|\,\|y\|\le r$, and so

$$0 = \sum_{k \geq 1} P_u^k(x-u) = \sum_{k \geq 1} P_u^k(y) t^k.$$

By Lemma 2.95 applied to $h_k := P_u^k(y)$ and $\delta := r/\|y\|$, it follows that $P_u^k(y) = 0$ for $k = 1, 2, \ldots$. Since $y \in X$ is arbitrary, the proof of the theorem is complete. □

For a given power series $\sum_{k \geq 0} P_u^k(x-u)$ from $X$ to $Y$ around $u \in X$, a number $\rho = \rho_u := \rho_{u,\mathrm{U}}$, $0 \leq \rho \leq +\infty$, is said to be the *radius of uniform convergence* of the series if $\rho$ is the supremum of all $r$, $0 \leq r < +\infty$, such that the power series converges uniformly on the closed ball $\bar{B}(u,r)$ with center at $u$ and radius $r > 0$. Here in the expanded notation $\rho_{u,\mathrm{U}}$, $u$ is the point around which the expansion is taken, and "U" indicates "uniform."

**Theorem 5.10.** *For a normed space $X$ and a Banach space $Y$, the radius of uniform convergence $\rho_u$ of the power series $\sum_{k \geq 0} P_u^k(x-u)$ around $u$ from $X$ to $Y$ is given by the Cauchy–Hadamard formula:*

$$\frac{1}{\rho_u} = \limsup_{k \to \infty} \|P_u^k\|^{1/k}. \tag{5.11}$$

*Moreover, if $\rho_u > 0$ then for each $0 < r < \rho_u$, $\sum_{k \geq 0} \|P_u^k\| r^k < \infty$.*

*Proof.* Let $\alpha := \limsup_{k \to \infty} \|P_u^k\|^{1/k}$. First suppose that $\alpha = +\infty$. Let $r > 0$. Since $\alpha > 1/r$, there is a sequence $k(j) \to \infty$ as $j \to \infty$ such that for some $y_j \in X$ with $\|y_j\| \leq 1$, $\|P_u^{k(j)}(y_j)\| > r^{-k(j)}$ for all $j$. For each $j \geq 1$, letting $x_j := u + r y_j$, it follows that $\|P_u^{k(j)}(x_j - u)\| > 1$ and $x_j \in \bar{B}(u,r)$. Since $r > 0$ is arbitrary, $\rho_u = 0$, that is, (5.11) holds in the first case.

Second suppose that $0 < \alpha < +\infty$. Let $0 < r < 1/\alpha$. Then

$$\limsup_{k \to \infty} (\|P_u^k\| r^k)^{1/k} = \alpha r < 1.$$

By the root test on series of nonnegative numbers, it follows that the series $\sum_{k \geq 1} \|P_u^k\| r^k$ converges. Therefore the series $\sum_{k \geq 0} P_u^k(x-u)$ converges uniformly for $x \in \bar{B}(u,r)$, and so $1/\alpha \leq \rho_u$. If $r > 1/\alpha$ then one can show as in the first case that the power series does not converge uniformly on $\bar{B}(u,r)$, proving (5.11) in the second case.

Finally suppose that $\alpha = 0$. Let $0 < r < \infty$ and let $\epsilon := 1/(2r)$. In this case there is an integer $k_0$ such that $\|P_u^k\| \leq \epsilon^k$ for each $k \geq k_0$. If $x \in \bar{B}(u,r)$ then $\|P_u^k(x-u)\| \leq (r\epsilon)^k = 2^{-k}$ for all $k \geq k_0$. Thus the power series converges uniformly on $\bar{B}(u,r)$. Since $r$ is arbitrary, $\rho_u = +\infty$, completing the proof of (5.11) in all cases. □

**Corollary 5.11.** *Let $X$ be a normed space and let $Y$ be a Banach space. For the power series $\sum_{k\geq 0} P_u^k(x-u)$ from $X$ to $Y$ around $u \in X$, the following are equivalent:*

(*a*) $\sum_{k\geq 0} P_u^k(x-u)$ *converges uniformly on some closed ball $\bar{B}(u,r)$, $r>0$;*
(*b*) $\limsup_{k\to\infty} \|P_u^k\|^{1/k} < \infty$;
(*c*) *there exist $C>0$ and $c>0$ such that $\|P_u^k\| \leq Cc^k$ for $k=0,1,\ldots$.*

*Proof.* By Theorem 5.10, (*a*) implies (*b*) because $0 < r \leq \rho_u$. Since clearly (*c*) follows from (*b*), it is enough to show that (*c*) implies (*a*). To this aim let $r>0$ be such that $rc<1$. Then for $x \in \bar{B}(u,r)$,

$$\sum_{k\geq 0} \|P_u^k(u-x)\| \leq C\sum_{k\geq 0}(rc)^k < \infty,$$

and so (*a*) holds, proving the corollary. □

For a power series of complex numbers its radii of absolute convergence and uniform convergence are the same. This is no longer true for power series in an infinite-dimensional space, as the following example shows. Recall that $\ell^p$ spaces were defined under the heading "Classes of measurable functions" in Section 1.4.

**Example 5.12.** Let $X = \ell^2$ and $Y = \mathbb{K} = \mathbb{R}$. For $x = \{x_j\}_{j\geq 1} \in \ell^2$, let $P_0^k(x) := (x_k)^k$ for $k = 1,2,\ldots$. Then for each $k \geq 1$, $P_0^k \in P^k(\ell^2,\mathbb{R})$ and $\|P_0^k\| = 1$, and so $\rho_0 = 1$ although the power series $F(x) := \sum_{k\geq 1} P_0^k(x)$ converges absolutely for all $x \in \ell^2$, so that its radius of convergence is $+\infty$.

Let $\mathbb{B}$ be a Banach algebra and let $h \in \mathbb{B}$. For each $k = 1,2,\ldots$ define the $k$-linear map $M^k := M^k[h]$ from $\mathbb{B}\times\cdots\times\mathbb{B}$ ($k$ times) into $\mathbb{B}$ by

$$M^k[h](x_1,\ldots,x_k) := hx_1x_2\cdots x_k \tag{5.12}$$

for $x_1,\ldots,x_k \in \mathbb{B}$. Let $S^k = S^k[h]$ be the symmetrization of $M^k[h]$, and let $P^k = P^k[h]$ be the $k$-homogeneous polynomial defined by $P^k := (S^k)^\frown$. Then $P^k x = hx^k$ for $x \in \mathbb{B}$.

**Lemma 5.13.** *Let $\mathbb{B}$ be a unital Banach algebra and $h \in \mathbb{B}$. For each $k = 1,2,\ldots$, we have $\|P^k[h]\| = \|S^k[h]\| = \|M^k[h]\| = \|h\|$.*

**Remark 5.14.** In this case $P^k = (S^k)^\frown$ but in (5.10), $k^k/k!$ can be replaced by 1.

*Proof.* Clearly, since $P^k = (S^k)^\frown$, by the left inequality in (5.10) and by (5.6), $\|P^k\| \leq \|S^k\| \leq \|M^k\| \leq \|h\|$. Conversely, taking $x_1 = \cdots = x_k = x = \mathbb{1}$ we get $\|P^k[h]\| = \|S^k[h]\| = \|M^k[h]\| = \|h\|$. □

Let $\mathbb{B}$ be a Banach algebra. Then a $\mathbb{B}$-*power series* will mean a series

$$\sum_{k=0}^{\infty} h_k x^k \tag{5.13}$$

for $x \in \mathbb{B}$ and $h_k \in \mathbb{B}$ for all $k$, where if $\mathbb{B}$ is unital, $x^0 := \mathbb{1}$, or if it is not, the $k = 0$ term is omitted ($h_0 = 0$). It is a power series from $\mathbb{B}$ into $\mathbb{B}$ of the form $\sum_{k\geq 0} P^k[h_k]x^k$ with $P^k[h_k] \in P^k(\mathbb{B}, \mathbb{B})$ as defined before Lemma 5.13.

**Remark 5.15.** The notion of $\mathbb{B}$-power series will usually be applied when the coefficients $h_k$ in (5.14) all belong to the *center* $\{h \in \mathbb{B} : hx = xh$ for all $x \in \mathbb{B}\}$, for example, if $\mathbb{B}$ is commutative or if all $h_k$ are constant multiples of the identity $\mathbb{1}$.

**Theorem 5.16.** *For any Banach algebra* $\mathbb{B}$, *sequence* $\{h_k\}_{k\geq 0} \subset \mathbb{B}$, *and* $x \in \mathbb{B}$, *if the spectral radius* $r(x)$ *satisfies*

$$r(x) < \rho := 1/\limsup_{k\to\infty} \|h_k\|^{1/k},$$

*then the* $\mathbb{B}$-*power series* (5.13) *converges absolutely.*

*Proof.* Let $r(x) < \sigma < \tau < \rho$. Then by Theorem 4.15, for $k$ large enough, $\|x^k\|^{1/k} < \sigma$, while $\|h_k\|^{1/k} < 1/\tau$. Thus $\|h_k x^k\|^{1/k} < \sigma/\tau < 1$ for $k$ large enough, and the theorem follows by the root test on series of nonnegative numbers. □

It will follow from Theorem 6.17 and Corollary 6.23 that if $1 \leq p < \infty$ and $J$ is a nonempty interval, then for any $f$ in the Banach algebra $\mathcal{W}_p(J)$, the spectral radius $r(f)$ equals $\|f\|_{\sup}$, and the same holds for Banach algebras of Hölder functions.

As stated before Theorem 2.102, a topological space $X$ is said to be *connected* if the only subsets of $X$ which are both open and closed are the empty set $\emptyset$ and the set $X$ itself. A subset $U$ of a topological space $X$ is *connected* if it is connected with its relative topology.

**Definition 5.17.** Let $\mathbb{B}$ be a Banach algebra and $U$ a nonempty connected open subset of $\mathbb{B}$. A function $F$ from $U$ into $\mathbb{B}$ will be said to have a $\mathbb{B}$-*Taylor expansion* around $u \in U$ if there are an $r > 0$ and a sequence $\{h_k\} \subset \mathbb{B}$ such that $F$ is given by a $\mathbb{B}$-power series, called the $\mathbb{B}$-*Taylor series* of $F$ around $u$,

$$F(x) = \sum_{k=0}^{\infty} h_k (x-u)^k, \tag{5.14}$$

for each $x \in U$ with $\|x - u\| < r$. A function $F$ from $U$ into $\mathbb{B}$ will be called $\mathbb{B}$-*analytic* on $U$ if $F$ has a $\mathbb{B}$-Taylor expansion around each point of $U$.

If $F$ is $\mathbb{B}$-analytic on $U$ then its $\mathbb{B}$-Taylor series around each point of $U$ is unique by Theorem 5.9, since convergence of a $\mathbb{B}$-power series on a set $B(0, r)$ yields its absolute convergence there due to the next theorem.

**Theorem 5.18.** *If $\mathbb{B}$ is a unital Banach algebra and $\{h_k\}_{k=0}^{\infty} \subset \mathbb{B}$, then the following are equivalent:*

*(a) for some $r \in (0, +\infty]$, the $\mathbb{B}$-power series $\sum_{k=0}^{\infty} h_k x^k$ converges in $\mathbb{B}$ for all $x$ with $\|x\| < r$;*
*(b) $M := \limsup_{k\to\infty} \|h_k\|^{1/k} < \infty$.*

*If (a) and (b) hold then the largest $r$ for which (a) holds is $\rho := 1/M$, and for any $0 < s < \rho$, the series in (a) converges absolutely and uniformly for $\|x\| \leq s$.*

**Remark 5.19.** The Hilbert space $\ell^2$ in Example 5.12 is a Banach algebra under pointwise operations of sequences but does not have an identity. In it, for $h_k = e_k$ with $(e_k)_j = \delta_{jk} = 1_{j=k}$ for $k \geq 1$, $h_0 = 0$, we have $r = +\infty$ in $(a)$ but $M = 1$ in $(b)$ so $\rho = 1$.

*Proof.* Recall that the norm on a Banach algebra $\mathbb{B}$ with identity $\mathbb{I}$ can be and has been chosen so that $\|\mathbb{I}\| = 1$ (see Theorem 4.8 and Definition 4.9).

$(a) \Rightarrow (b)$: Let $x = t\mathbb{I}$ for any $t$ with $0 < t < r$. Then the series $\sum_{k=0}^{\infty} t^k h_k$ converges in $\mathbb{B}$. Thus $\sup_k t^k \|h_k\| < \infty$, and so taking $k$th roots implies $M < \infty$.

$(b) \Rightarrow (a)$: For each $k$, let $P^k\colon x \mapsto h_k x^k$. By Lemma 5.13, $\|P^k\| = \|h_k\|$. Thus $(b)$ implies $(a)$ with $r = \rho$ by Theorem 5.10, and the absolute and uniform convergence for $\|x\| \leq s$ follows for $s < \rho$.

Let $t > \rho$ and $x = t\mathbb{I}$, so $\|x\| = t$. Then $\|h_k x^k\| = t^k \|h_k\|$. Let $\rho < s < t$. Then $\|h_k\|^{1/k} > 1/s$ for infinitely many $k$, so $t^k\|h_k\| > t^k/s^k$, which is unbounded. Thus $\sum_{k=0}^{\infty} h_k x^k$ does not converge and $\rho$ is the largest $r$ for which $(a)$ holds, proving the theorem. □

**Example 5.20.** Let $\mathbb{B}$ be a Banach algebra containing an element $y \neq 0$ which is *nilpotent*, meaning that $y^n = 0$ for some $n \geq 2$. Specifically, let $M(2, \mathbb{R})$ be the algebra of $2 \times 2$ real matrices with usual matrix multiplication and $y = \left(\begin{smallmatrix} 0 & 1 \\ 0 & 0 \end{smallmatrix}\right)$. Then the power series $\sum_{k=0}^{\infty} h_k y^k$ converges for arbitrary $h_k \in \mathbb{B}$, but the spectral radius $r(y)$ is 0. Since $y$ can be multiplied by any scalar in $\mathbb{K}$, its norm can be taken arbitrarily large, e.g., larger than the radius $\rho_0$ of uniform convergence of the power series $\sum_{k=0}^{\infty} h_k x^k$. Thus, unlike the case of analytic functions of one real or complex variable, divergence of a power series $\sum a_k x^k$ for some $x = x_0$ does not imply its divergence for all $x$ with $\|x\| > \|x_0\|$.

Next we show that if a power series around $u \in X$ has a positive radius of uniform convergence then its sum has a locally uniformly convergent power series expansion around each point $v$ in some neighborhood of $u$. To define the power series around $v$ the following is used. Let $L$ be a symmetric $m$-linear mapping and let $k$ be an integer such that $0 < k < m$. For $(x_1, \ldots, x_{m-k}) \in X^{m-k}$, the mapping $L(x_1, \ldots, x_{m-k})$ defined by (5.5) is a symmetric $k$-linear mapping from $X^k$ to $Y$. Thus the mapping

$$X \ni x \mapsto L(x_1, \ldots, x_{m-k})x^{\otimes k} \in Y$$

is a $k$-homogeneous polynomial. If $x_1 = \cdots = x_{m-k} = x$ for some $x$, then we write $Lx^{\otimes m-k} := L(x_1, \ldots, x_{m-k}) \in L_s({}^kX, Y)$.

**Theorem 5.21.** *For a normed space $X$, a Banach space $Y$, and $u \in X$, let $\sum_{m\geq 0} P_u^m(x-u)$ be a power series around $u$ from $X$ to $Y$ whose radius of uniform convergence is $\rho_u > 0$, and let*

$$F(x) := \sum_{m\geq 0} P_u^m(x-u) \tag{5.15}$$

*for each $x \in B(u, \rho_u)$. For each $m \geq 1$, let $L_u^m \in L_s({}^mX, Y)$ be such that $P_u^m = (L_u^m)\tilde{}$, and let*

$$\overline{\rho} := \overline{\rho}_{\{L_u^m\}_{m\geq 1}} := 1/\limsup_{m\to\infty} \|L_u^m\|^{1/m}. \tag{5.16}$$

*Then $\overline{\rho} \geq \mathrm{e}^{-1}\rho_u$, and for each $v \in B(u, \overline{\rho})$, (a), (b), and (c) hold, where*

*(a) for each $k \geq 1$,*

$$P_v^k := P_u^k + \sum_{m>k} \binom{m}{k} L_u^m(v-u)^{\otimes m-k} \in P^k(X, Y); \tag{5.17}$$

*(b) the power series $\sum_{k\geq 0} P_v^k(x-v)$ with $P_v^0 \equiv F(v)$ has radius of uniform convergence $\rho_v \geq \overline{\rho} - \|v-u\|$, and*

$$\sum_{k=1}^{\infty} \|P_v^k(x-v)\| \leq \sum_{m=1}^{\infty} \|L_u^m\| r^m < \infty$$

*for all $x$ such that $\|x-v\| \leq r - \|v-u\|$ with $r < \overline{\rho}$;*

*(c) for each $x \in X$ such that $\|x-v\| < \overline{\rho} - \|v-u\|$,*

$$F(x) = \sum_{k\geq 0} P_v^k(x-v). \tag{5.18}$$

**Remark 5.22.** Timothy Nguyen has proved that the bound $\overline{\rho} \geq \mathrm{e}^{-1}\rho_u$ can be improved to $\overline{\rho} \geq \mathrm{e}^{-1/2}\rho_u$: see the Notes. Although the bound in Theorem 5.7 is sharp as shown in references given after it, the sharpness pertains to $m$ distinct elements of $X$ for each $m$, where in the present case one only needs to consider two elements of $X$ at a time, such as $u$ and $v$ in (5.17). The equality $\overline{\rho} = \rho_u$ holds for some infinite-dimensional $X$ and $Y$; see Corollary 5.23. Indeed we do not know at this writing any example where it fails. If $\rho_u = +\infty$, the factor $\mathrm{e}^{-1}$ or $\mathrm{e}^{-1/2}$ makes no difference.

Cena [31] calls the quantity $\overline{\rho}$ the "radius of restricted convergence" of a power series. We do not know whether it may have other names.

*Proof.* By Stirling's formula, we have $\lim_{m\to\infty} m/(m!)^{1/m} = \mathrm{e}$. Thus using (5.10) and (5.11) it follows that

$$\frac{1}{\overline{\rho}} = \limsup_{m\to\infty} \|L_u^m\|^{1/m} \le \limsup_{m\to\infty} \Big(\frac{m^m}{m!}\|P_u^m\|\Big)^{1/m} = \frac{\mathrm{e}}{\rho_u},$$

and so $\overline{\rho} \ge \mathrm{e}^{-1}\rho_u$.

Let $v \in B(u, \overline{\rho})$ and let $r$ be such that $\overline{\rho} > r > \|v-u\|$. By the root test on series of nonnegative numbers, it follows that $\sum_{m\ge 1}\|L_u^m\|r^m < \infty$. Thus for any $x \in X$ such that $\|x-v\| \le r - \|v-u\|$, as in part (*b*) or, for a suitable $r$, in part (*c*) of the statement, we have

$$\begin{aligned} &\sum_{m=1}^{\infty}\sum_{k=1}^{m}\binom{m}{k}\|L_u^m\|\|v-u\|^{m-k}\|x-v\|^k \\ &\qquad \le \sum_{m=1}^{\infty}\|L_u^m\|(\|v-u\|+r-\|v-u\|)^m < \infty. \end{aligned} \tag{5.19}$$

Interchanging the order of summation on the left side, it follows that

$$\sum_{k=1}^{\infty}\Bigg(\sum_{m=k}^{\infty}\binom{m}{k}\|L_u^m\|\|v-u\|^{m-k}\Bigg)\|x-v\|^k < \infty. \tag{5.20}$$

Therefore for each $k \ge 1$, taking $x \ne v$, $\sum_{m=k}^{\infty}\binom{m}{k}L_u^m(v-u)^{\otimes m-k}$ converges in $L_s({}^kX, Y)$ to some $Q_v^k$. Thus the series defining $P_v^k$ in (5.17) converges absolutely and $P_v^k w = Q_v^k w^{\otimes k}$ for all $w \in X$ and $k \ge 1$, so (*a*) is proved.

It follows that

$$\sum_{k=1}^{\infty}\|P_v^k(x-v)\| \le \sum_{m=1}^{\infty}\|L_u^m\|r^m < \infty$$

for any $x$ such that $\|x-v\| \le r - \|v-u\|$ with $r < \overline{\rho}$. The power series $\sum_{k\ge 0}P_v^k(x-v)$ converges uniformly for $\|x-v\| \le r - \|v-u\|$. Since $r$ can be taken arbitrarily close to $\overline{\rho}$, the radius of uniform convergence is $\rho_v \ge \overline{\rho} - \|v-u\|$, proving (*b*).

For (*c*), let $x \in X$ be such that $\|x-v\| < \overline{\rho} - \|v-u\|$. By the binomial formula (5.8), for each integer $m \ge 1$, we have

$$\begin{aligned} L_u^m(x-u)^{\otimes m} &= L_u^m(v-u+x-v)^{\otimes m} \\ &= L_u^m(v-u)^{\otimes m} + \sum_{k=1}^{m-1}\binom{m}{k}L_u^m(v-u)^{\otimes m-k}\otimes(x-v)^{\otimes k} + L_u^m(x-v)^{\otimes m}, \end{aligned}$$

where the sum over the empty set of indices is zero. Since $x \in B(u, \overline{\rho}) \subset B(u, \rho_u)$, $F(x)$ is defined and equals

$$F(x) = F(u) + \sum_{m=1}^{\infty} \Big\{ P_u^m(v-u) + \sum_{k=1}^{m-1} \binom{m}{k} L_u^m (v-u)^{\otimes m-k} \otimes (x-v)^{\otimes k} + P_u^m(x-v) \Big\}.$$

The series, as a double series in $m$ and $k$, converges absolutely by (5.19). Thus the order of summation can be reversed. Also, (5.15) with $x = v$ shows that the sum of terms with $k = 0$ is $F(v)$. So $F(x)$ equals

$$F(v) + \sum_{k=1}^{\infty} \Big\{ P_u^k(x-v) + \Big( \sum_{m=k+1}^{\infty} \binom{m}{k} L_u^m (v-u)^{\otimes m-k} \Big) (x-v)^{\otimes k} \Big\}.$$

This and (5.17) imply (5.18) with $P_v^0 \equiv F(v)$, proving $(c)$ and the theorem. □

**Corollary 5.23.** *Under the hypotheses of Theorem* 5.21, *statements* $(a)$, $(b)$, *and* $(c)$ *hold for each* $v \in B(u, \overline{\rho})$ *with* $\overline{\rho} = \rho_u$ *in the following cases*:

$(d)$ *If* $X = Y = \mathbb{B}$ *is a unital Banach algebra and* (5.15) *is a* $\mathbb{B}$*-power series.*
$(e)$ *If* $X$ *is a Hilbert space over* $\mathbb{K}$ *and* $Y$ *is a Banach space over* $\mathbb{K}$.

*Proof.* For $(d)$, the power series (5.15) is given by a $\mathbb{B}$-power series (5.13) for some sequence $\{h_m\}_{m \geq 0}$. Using notation as defined before Lemma 5.13, we have for each $m \geq 1$, $L_u^m = M^m[h_m]$ and $P_u^m = P^m[h_m]$. Thus for the $\overline{\rho}$ and $\rho_u$ defined by (5.16) and (5.11), respectively, we have $\overline{\rho} = \rho_u$ by Lemma 5.13.

For $(e)$, it is enough to prove that $\|L\| = \|\tilde{L}\|$ for any mapping $L \in L_s({}^kX, Y)$ and $k = 2, 3, \ldots$. First suppose that $X$ is finite-dimensional. To begin, let $k = 2$. By compactness take $x, u \in X$ with $\|x\| = \|u\| = 1$ and $\|L\| = \|L(x, u)\|$. Since $L(x, u) = \frac{1}{4}[\tilde{L}(x+u) - \tilde{L}(x-u)]$, we have

$$\begin{aligned} \|L\| &= \|L(x,u)\| \leq \frac{1}{4}[\|\tilde{L}(x+u)\| + \|\tilde{L}(x-u)\|] \leq \frac{1}{4}\|\tilde{L}\|[\|x+u\|^2 + \|x-u\|^2] \\ &\leq \frac{1}{4}\|L\|[\|x+u\|^2 + \|x-u\|^2] = \frac{1}{2}\|L\|(\|x\|^2 + \|u\|^2) = \|L\| \end{aligned}$$

using the parallelogram law in Hilbert space, and all inequalities must be equalities, in particular $\|\tilde{L}\| = \|L\|$ as desired for $k = 2$. Also, in the three expressions with coefficient $\frac{1}{4}$, since the inequalities hold between the first terms of each, and the second terms, the inequalities for the first terms and the second terms must each be equalities, and hence

$$\|\tilde{L}(x \pm u)\| = \|L\|\|x \pm u\|^2. \tag{5.21}$$

For $k \geq 3$, let $S := \{x \in X : \|x\| \leq 1\}$. By compactness, take $y_1, \ldots, y_k \in S$ such that $\|L\| = \|L(y_1, \ldots, y_k)\|$. For some $a \in S$, none of the inner products $\langle a, y_j \rangle$ is 0 for $j = 1, \ldots, k$ since for each $j$, $\{a \in S : \langle a, y_j \rangle = 0\}$ is nowhere dense in $S$. Replacing $y_j$ by $-y_j$ as needed, we can assume that that $\langle a, y_j \rangle > 0$ for each $j$. For some $\epsilon > 0$, the compact set

$$A := \{x = (x_1, \ldots, x_k) \in S^k : \langle a, x_j \rangle \geq \epsilon, \ j = 1, \ldots, k, \ \text{ and } \\ \|L(x_1, \ldots, x_k)\| = \|L\|\}$$

is nonempty. Clearly, for each $x \in A$, $\|x_j\| = 1$ for all $j = 1, \ldots, k$. At some $(u_1, \ldots, u_k) \in A$, $\sum_{j=1}^k \langle a, x_j \rangle$ attains its maximum on $A$. It will be enough to show that $u_p = \pm u_q$ for all $p, q = 1, \ldots, k$. Suppose that $u_p \pm u_q \neq 0$ for some $p, q$. Define $v_j := u_j$ for $j \neq p, q$ and $v_j := (u_p + u_q)/\|u_p + u_q\|$ for $j = p, q$. We have $\|L(v_1, \ldots, v_k)\| = \|L\|$ by (5.21) for the 2-linear map $(x_p, x_q) \mapsto L(x)$ where $x_j = u_j$ for $j \neq p, q$ are fixed. Then $\sum_{j=1}^k \langle a, v_j \rangle > \sum_{j=1}^k \langle a, u_j \rangle$ because $\|u_p + u_q\| < 2$ (using the Hilbert space assumption). This contradicts the choice of $u_j$ and finishes the proof when $X$ is finite-dimensional.

Now let $X$ be infinite-dimensional. We want to show that for any $a_1, \ldots, a_k$ in $S$, $\|L(a_1, \ldots, a_k)\| \leq \|\tilde{L}\|$. Restricting to the finite-dimensional space spanned by $a_1, \ldots, a_k$, this reduces to the previous case. The corollary is proved. □

We finish this section with a special case of Theorem 3.8 applied to $L(^kX, Y)$-valued functions. It will be used in Section 6.6 in relation to higher order derivatives of a Nemytskii operator acting between $\mathcal{W}_p$ spaces.

**Corollary 5.24.** *Let $k = 1, 2, \ldots$, let $X$, $Y$ be Banach spaces, let $J$ be a nondegenerate interval and let $\Phi \in \mathcal{CV}$. For $F \in \widetilde{\mathcal{W}}_\Phi(J; L(^kX, Y))$ and $g_1, \ldots, g_k \in \widetilde{\mathcal{W}}_\Phi(J; X)$, let $h(s) := F(s)(g_1(s), \ldots, g_k(s))$, $s \in J$. Then $h \in \widetilde{\mathcal{W}}_\Phi(J; Y)$ and*

$$\|h\|_{[\Phi]} \leq \|F\|_{[\Phi]} \|g_1\|_{[\Phi]} \cdots \|g_k\|_{[\Phi]}.$$

*Proof.* For $A \in L(^kX, Y)$ and $x_1, \ldots, x_k \in X$, let $L(A, x_1, \ldots, x_k) := A(x_1, \ldots, x_k)$. Then $L$ is a $(k+1)$-linear and 1-bounded mapping from $L(^kX, Y) \times X^k$ into $Y$ as defined before Theorem 3.8. The conclusion then follows from Theorem 3.8 applied to $L$. □

## 5.2 Higher Order Derivatives and Taylor Series

Throughout this section let $X$ and $Y$ be normed spaces and let $U$ be a nonempty open subset of $X$. Let $L^1(X, Y) := L(X, Y)$, the normed space of bounded linear operators from $X$ into $Y$, with the usual operator norm (1.17). Recursively, for $k = 1, 2, \ldots$, given a normed space $L^k(X,Y)$, let $L^{k+1}(X,Y) := L(X, L^k(X,Y))$ with the operator norm.

For $k \geq 2$, we say that $F\colon U \to Y$ has a $k$th (Fréchet) *derivative* or is (Fréchet) *differentiable of order* $k$ *at* $u$ if $F$ has a $(k-1)$st derivative $D^{k-1}F(x)$ at each point $x$ of some neighborhood of $u$, and the mapping $D^{k-1}F$ with values in $L^{k-1}(X,Y)$ is differentiable at $u$. Then $D^kF(u)$, the $k$th derivative of $F$ at $u$, is defined as the derivative of $D^{k-1}F$ at $u$. Thus $D^kF(u)$ is in $L^k(X,Y)$. If $F$ is differentiable of order $k$ at $u$ for each $u \in U$ then we say that $F$ is *differentiable of order* $k$ *on* $U$.

The space $L^k(X,Y)$, which occurs naturally in the definition of $D^kF(u)$, can be identified conveniently with the space $L(^kX,Y)$ of $k$-linear mappings defined previously. Specifically, let $\Phi_{(k)}\colon L^k(X,Y) \to L(^kX,Y)$ be the natural isomorphism defined by

$$\begin{aligned}\Phi_{(k)}(A)(x_1,\ldots,x_k) &:= A(x_1)(x_2)\cdots(x_k)\\ &:= \big[\cdots\big[A(x_1)\big](x_2)\cdots\big](x_k)\end{aligned} \tag{5.22}$$

for each $A \in L^k(X,Y)$. It is easily seen by induction on $k$ that $\Phi_{(k)}$ is an isometry for the norm on $L(^kX,Y)$ defined by (5.4) and the operator norm on $L^k(X,Y)$. For $k \geq 2$, let $F$ be differentiable of order $k$ at $u \in U$. Then the *$k$th differential at* $u$ is defined by $\mathrm{d}^kF(u) := \Phi_{(k)}(D^kF(u))$. If $F$ is differentiable of order $k$ on $U$ then the $k$th differential $\mathrm{d}^kF$ is a mapping from $U$ into $L(^kX,Y)$, that is, for $u \in U$, $\mathrm{d}^kF(u)$ is a $k$-linear mapping such that

$$\mathrm{d}^kF(u)(x_1,\ldots,x_k) = D^kF(u)(x_1)\cdots(x_k) \tag{5.23}$$

for any $x_1,\ldots,x_k \in X$. For $k = 1$ and $u \in U$, we let $\mathrm{d}F(u) := \mathrm{d}^1F(u) := DF(u)$ whenever $F$ is differentiable at $u$.

The following gives a condition for existence of the $k$th derivative in terms of differentials.

**Proposition 5.25.** *Let $k \geq 2$, let $F\colon U \to Y$ be differentiable of order $k-1$ on $U$, let $L \in L(^kX,Y)$, and let $u \in U$. Then the following two statements are equivalent:*

*(a) $F$ is differentiable of order $k$ at $u$ and $\mathrm{d}^kF(u) = L$;*
*(b) the function $x \mapsto f(x) := \mathrm{d}^{k-1}F(x)$, $x \in U$, is differentiable at $u$ and $Df(u)(x) = L(x,\ldots)$ for each $x \in X$.*

*Moreover, whenever (a) or (b) holds, we have $\|Dd^{k-1}F(u)\| = \|d^kF(u)\|$.*

*Proof.* (a) $\Rightarrow$ (b). We have $f(x) = (\Phi_{(k-1)} \circ D^{k-1}F)(x)$, where $\Phi_{(k-1)}$ is continuous and linear, and so its derivative $D\Phi_{(k-1)}$ equals $\Phi_{(k-1)}$. Thus by the chain rule in the form of Theorem 5.1, $f$ is differentiable at $u$ and for each $x \in X$, we have

$$Df(u)(x) = \Phi_{(k-1)}\Big(D^k F(u)(x)\Big) = D^k F(u)(x)(\cdot)\cdots(\cdot) = L(x,\cdots),$$

where the last equality holds by (5.23) and since $\mathrm{d}^k F(u) = L$, proving $(b)$.

$(b) \Rightarrow (a)$. Let $\Phi^{-1}_{(k-1)}$ be the inverse of $\Phi_{(k-1)}$, and so $D^{k-1}F(x) = \left(\Phi^{-1}_{(k-1)} \circ \mathrm{d}^{k-1}F\right)(x)$ for $x \in U$. Again since $\Phi^{-1}_{(k-1)}$ is continuous and linear, by the chain rule (Theorem 5.1), $D^k F(u)$ is defined and for each $x \in X$, we have

$$D^k F(u)(x) = \Phi^{-1}_{(k-1)}\Big(Df(u)(x)\Big) = \Phi^{-1}_{(k-1)}(L(x,\ldots)). \tag{5.24}$$

Thus for each $x \in X$, we have

$$\mathrm{d}^k F(u)(x,\ldots) = \Phi_{(k)}\Big(D^k F(u)\Big)(x,\ldots) = \Phi_{(k-1)}\Big(D^k F(u)(x)\Big) = L(x,\ldots),$$

proving $(a)$.

Finally, let $(b)$ hold. Then using Proposition 5.5 with $m = 1$, $V_1 = \cdots = V_k = X$, and $V_0 = Y$, we have

$$\|Df(u)\|_{L(X,L({}^{k-1}X,Y))} = \sup\big\{\|L(x,\ldots)\|_{L({}^{k-1}X,Y)} \colon \|x\| \le 1\big\} = \|L\|_{L({}^kX,Y)}.$$

The proof of the proposition is complete. □

**Corollary 5.26.** *Let $k \ge 2$, let $F\colon U \to Y$ be differentiable of order $k-1$ on $U$ and for some $x_2,\ldots,x_k \in X$, let $g(x) := \mathrm{d}^{k-1}F(x)(x_2,\ldots,x_k)$, $x \in U$. For $u \in U$, if $F$ is differentiable of order $k$ at $u$ then $g\colon U \to Y$ is differentiable at $u$ with the derivative*

$$\mathrm{d}g(u)(x) = \mathrm{d}^k F(u)(x, x_2, \ldots, x_k), \qquad x \in X.$$

*Proof.* Let $f\colon U \to L({}^{k-1}X, Y)$ be the function defined in Proposition 5.25$(b)$. Let $T\colon L({}^{k-1}X,Y) \to Y$ be the evaluation operator defined by $T(A) := A(x_2,\ldots,x_k)$ for $A \in L({}^{k-1}X,Y)$. Then $T$ is a bounded linear operator and $g = T \circ f$. Thus using the chain rule (Remark 5.2) and $(a) \Rightarrow (b)$ of Proposition 5.25, it follows that $g$ is differentiable at $u$ and for each $x \in X$, we have

$$(Dg)(u)(x) = T(Df(u)(x)) = \mathrm{d}^k F(u)(x, x_2, \ldots, x_k),$$

proving the corollary. □

Further, we will prove that the $k$th differential, whenever it exists, is a symmetric $k$-linear mapping, that is, $\mathrm{d}^k F(u) \in L_s({}^kX, Y)$. The following is known as the Schwarz theorem.

**Theorem 5.27.** *For each $k \ge 2$, if $F\colon U \to Y$ is differentiable of order $k$ at $u \in U$ then the $k$-linear mapping $\mathrm{d}^k F$ is symmetric.*

*Proof.* The proof is by induction. First let $k = 2$. Since $F$ is twice differentiable at $u$, $F$ is differentiable on an open ball $B(u,r) \subset U$ for some $r > 0$. For $x_1, x_2 \in X$ such that $\|x_1\| < r/2$ and $\|x_2\| < r/2$, let

$$\Delta(x_1,x_2) := F(u+x_1+x_2) - F(u+x_1) - F(u+x_2) + F(u).$$

Next we show that the difference between $\mathrm{d}^2F(u)(x_1,x_2)$ and $\Delta(x_1,x_2)$ is $o((\|x_1\|+\|x_2\|)^2)$ as $\|x_1\|+\|x_2\| \to 0$. For $x \in X$ such that $\|x-u\| < r/2$, let $G_{x_1}(x) := F(x+x_1) - F(x)$. Then $G_{x_1}$ is a differentiable mapping of $B(u,r/2)$ into $Y$. By (5.3) in the mean value theorem, we have for any $C \in L(X,Y)$,

$$\begin{aligned}\big\|G_{x_1}(u+x_2) &- G_{x_1}(u) - DG_{x_1}(u)x_2\big\| \\ &\le \|x_2\| \sup_{0\le t\le 1} \big\|DG_{x_1}(u+tx_2) - DG_{x_1}(u)\big\| \\ &\le \|x_2\| \sup_{0\le t\le 1} \big\|(DG_{x_1}(u+tx_2) - C) - (DG_{x_1}(u) - C)\big\| \\ &\le 2\|x_2\| \sup_{0\le t\le 1} \big\|DG_{x_1}(u+tx_2) - C\big\|.\end{aligned}$$

Take $C = D^2F(u)(x_1)$. Recall that $\mathrm{d}^2F(u)(x_1,x_2) = D^2F(u)(x_1)(x_2) = C(x_2)$ by (5.23) with $k = 2$. Thus

$$\begin{aligned}\big\|\Delta(x_1,x_2) &- \mathrm{d}^2F(u)(x_1,x_2)\big\| \\ &\le \big\|G_{x_1}(u+x_2) - G_{x_1}(u) - DG_{x_1}(u)x_2\big\| \\ &\quad + \big\|\big[DG_{x_1}(u) - D^2F(u)(x_1)\big](x_2)\big\| \\ &\le 3\|x_2\| \sup_{0\le t\le 1} \big\|DG_{x_1}(u+tx_2) - D^2F(u)(x_1)\big\|.\end{aligned} \tag{5.25}$$

For each $t \in [0,1]$, since $D^2F(u)(\cdot)\colon X \to L(X,Y)$ is linear, we have

$$\begin{aligned}DG_{x_1}&(u+tx_2) - D^2F(u)(x_1) \\ &= \big[DF(u+x_1+tx_2) - DF(u) - D^2F(u)(x_1+tx_2)\big] \\ &\quad - \big[DF(u+tx_2) - DF(u) - D^2F(u)(tx_2)\big].\end{aligned} \tag{5.26}$$

Let $\epsilon > 0$. Since $DF$ is differentiable at $u$ there is a $\delta < r/4$ such that

$$\big\|DF(u+x) - DF(u) - D^2F(u)x\| < \epsilon\|x\|$$

for each $x \in X$ with $\|x\| < \delta$. Thus by (5.25) and (5.26), we get the bound

$$\big\|\Delta(x_1,x_2) - \mathrm{d}^2F(u)(x_1,x_2)\big\| < 3\epsilon\|x_2\|\big(\|x_1\| + 2\|x_2\|\big)$$

whenever $\|x_1\| + \|x_2\| < \delta$. Since $\Delta(x_1,x_2) = \Delta(x_2,x_1)$, interchanging $x_1$ and $x_2$ in the preceding bound, it follows that

$$\big\|\Delta(x_1,x_2) - \mathrm{d}^2F(u)(x_2,x_1)\big\| < 3\epsilon\|x_1\|\big(\|x_2\| + 2\|x_1\|\big)$$

whenever $\|x_1\| + \|x_2\| < \delta$. The preceding two bounds yield the inequality

$$\big\|\mathrm{d}^2F(u)(x_1, x_2) - \mathrm{d}^2F(u)(x_2, x_1)\big\| < 6\epsilon\big(\|x_1\| + \|x_2\|\big)^2 \tag{5.27}$$

for each $x_1, x_2 \in X$ such that $\|x_1\| + \|x_2\| < \delta$. Since $\mathrm{d}^2F(u)$ is bilinear, inequality (5.27) is homogeneous of order 2, and so it holds for arbitrary $x_1, x_2 \in X$. Thus letting $\epsilon \downarrow 0$, $\mathrm{d}^2F(u)(x_1, x_2) = \mathrm{d}^2F(x_2, x_1)$ for any $x_1, x_2 \in X$, proving the conclusion for $k = 2$.

Now suppose that $k > 2$ and $\mathrm{d}^{k-1}F(v)$ is symmetric for all $v \in B(u, r)$ for some $r > 0$. Thus for $x_2, \ldots, x_k \in X$, we have for any permutation $\pi$ of $\{2, \ldots, k\}$,

$$\mathrm{d}^{k-1}F(v)(x_{\pi(2)}, \ldots, x_{\pi(k)}) = \mathrm{d}^{k-1}F(v)(x_2, \ldots, x_k).$$

Then by Corollary 5.26, we have

$$\mathrm{d}^kF(u)(x_1, x_{\pi(2)}, \ldots, x_{\pi(k)}) = \mathrm{d}^kF(u)(x_1, x_2, \ldots, x_k) \tag{5.28}$$

for any $x_1, x_2, \ldots, x_k \in X$ and any permutation $\pi$ of $\{2, \ldots, k\}$. Let $G(x) := \mathrm{d}^{k-2}F(x)(x_3, \ldots, x_k)$ for $x \in U$ and $x_3, \ldots, x_k \in X$. The first derivative $DG\colon U \to L(X, Y)$ exists by Corollary 5.26 and is given by

$$(DG)(x)(w) = d^{k-1}F(x)(w, x_3, \ldots, x_k)$$

for any $x \in U$ and $w \in X$. We claim that the second derivative $D^2G(u)$ exists and satisfies, for any $x_1, x_2 \in X$,

$$D^2G(u)(x_1)(x_2) = \mathrm{d}^2G(u)(x_1, x_2) = \mathrm{d}^kF(u)(x_1, x_2, x_3, \ldots, x_k).$$

Indeed, since $F$ is differentiable of order $k$ at $u$, letting $M := \prod_{j=3}^k \|x_j\|$,

$$\begin{aligned}
&\|DG(x) - DG(u) - d^kF(u)(x - u, \cdot, x_3, \ldots, x_k)\|_{L(X,Y)} \\
&\quad \le M\|d^{k-1}F(x) - d^{k-1}F(u) - d^kF(u)(x - u, \ldots)\|_{L(^{k-1}X,Y)} = o(\|x - u\|)
\end{aligned}$$

as $x \to u$ by $(a) \Rightarrow (b)$ of Proposition 5.25, proving the claim.

By the first part of the proof, we have

$$\begin{aligned}
\mathrm{d}^kF(u)(x_1, x_2, x_3, \ldots, x_k) &= \mathrm{d}^2G(u)(x_1, x_2) = \mathrm{d}^2G(u)(x_2, x_1) \\
&= \mathrm{d}^kF(u)(x_2, x_1, x_3, \ldots, x_k).
\end{aligned} \tag{5.29}$$

Now the symmetry of $\mathrm{d}^kF(u)$ follows from (5.28) and (5.29). The proof of the theorem is complete. □

A mapping $F$ is called a $C^k$ *function on* $U \subset X$ if the derivatives of $F$ through order $k$ all exist on $U$ and are continuous. A mapping $F$ is called a $C^\infty$ *function on* $U \subset X$ if it is a $C^k$ function for each $k$.

**Theorem 5.28.** *For a normed space $X$, a Banach space $Y$, and $u \in X$, let $\sum_{m\geq 0} P_u^m(x-u)$ be a power series around $u$ from $X$ to $Y$ whose radius of uniform convergence is $\rho_u > 0$, and let $F(x)$ be its sum for $x \in B(u,\rho_u)$. For each $m \geq 1$, let $L_u^m \in L_s({}^m X, Y)$ be such that $P_u^m = (L_u^m)\check{}$, and let $\overline{\rho}$ be defined by (5.16). Then $F$ is a $C^\infty$ function on $B(u,\overline{\rho})$, and for each $v \in B(u,\overline{\rho})$ and $n \geq 1$,*

$$\mathrm{d}^n F(v) = n! L_u^n + \sum_{m \geq n+1} m(m-1)\cdots(m-n+1) L_u^m (v-u)^{\otimes m-n}. \quad (5.30)$$

*In particular, $L_u^n = d^n F(u)/n!$.*

*Proof.* The proof is by induction. Let $v \in B(u,\overline{\rho})$. By Theorem 5.21, for each $x \in B(v, \overline{\rho} - \|v-u\|)$, we have $F(x) = F(v) + \sum_{k\geq 1} P_v^k(x-v)$, where each $P_v^k$ is given by (5.17) and the power series $\sum_{k\geq 0} P_v^k(x-v)$ has radius of uniform convergence $\rho_v \geq \overline{\rho} - \|v-u\|$. If $0 < r < \overline{\rho} - \|v-u\|$ then by Theorem 5.10, the series $\sum_{k\geq 1} \|P_v^k\| r^k$ converges. Letting $C_1 := \sup\{\|P_v^k\| r^k \colon k \geq 2\} < \infty$, if $\|x-v\| < r$, we have

$$\begin{aligned}\big\|F(x) - F(v) - P_v^1(x-v)\big\| &\leq \sum_{k\geq 2} \|P_v^k\|\,\|x-v\|^k \\ &\leq \frac{C_1\|x-v\|^2}{r(r-\|x-v\|)} = o(\|x-v\|)\end{aligned}$$

as $\|x-v\| \to 0$. Thus $F$ is differentiable at $v$ and its derivative $D^1F(v)$ is given by (5.17) with $k=1$. Since $v \in B(u,\overline{\rho})$ is arbitrary, $F$ is differentiable on $B(u,\overline{\rho})$ and (5.30) holds with $n=1$.

Suppose that for some $n \geq 1$, $F$ has an $n$th differential on $B(u,\overline{\rho})$ and (5.30) holds. Recalling the definition (5.5), for each integer $m \geq 1$ and $(x_1,\dots,x_m) \in X^m$, let

$$\begin{aligned}&K_u^{m,n}(x_1,\dots,x_m) \\ &\quad := (m+n)(m+n-1)\cdots(m+1) L_u^{m+n}(x_1,\dots,x_m) \in L({}^n X, Y).\end{aligned} \quad (5.31)$$

Let $K_u^m := K_u^{m,n}$ since $n$ is fixed for the present. It is clear that each $K_u^m$ is a bounded symmetric $m$-linear mapping from $X^m$ to $L({}^n X, Y)$. Let $\tilde{K}_u^m := (K_u^m)\check{}$ be the $m$-homogeneous polynomial corresponding to $K_u^m$. By the Cauchy–Hadamard formula (Theorem 5.10), we have

$$\limsup_{m\to\infty} \big\|\tilde{K}_u^m\big\|^{1/m} \leq \limsup_{m\to\infty} \big\{(m+n)\cdots(m+1)\|P_u^{m+n}\|\big\}^{1/m} = \frac{1}{\rho_u}.$$

Again by Theorem 5.10, it follows that the power series $\sum_{m\geq 1} \tilde{K}_u^m(x-u)$ around $u$ from $X$ to $L({}^n X, Y)$ has radius of uniform convergence $\geq \rho_u$. By (5.30) with $x=u$, $\mathrm{d}^n F(u) = n! L_u^n$, and so by a change of index of summation,

$$\begin{aligned} \mathrm{d}^n F(x) - \mathrm{d}^n F(u) &= \sum_{m>n} m(m-1)\cdots(m-n+1) L_u^m (x-u)^{\otimes m-n} \\ &= \sum_{m\geq 1} \tilde{K}_u^m (x-u). \end{aligned}$$

Let $v \in B(u, \overline{\rho})$. By Theorem 5.21 with $\mathrm{d}^n F$ in place of $F$ and $L({}^n X, Y)$ in place of $Y$, for each $x \in B(v, \overline{\rho} - \|v-u\|)$, we have $\mathrm{d}^n F(x) = \sum_{k\geq 0} Q_v^k (x-v)$, where $Q_v^0 \equiv \mathrm{d}^n F(v)$,

$$Q_v^k = \tilde{K}_u^k + \sum_{m>k} \binom{m}{k} K_u^m (v-u)^{\otimes m-k} \in P^k(X, L({}^n X, Y))$$

for each $k \geq 1$, and the power series $\sum_{k\geq 0} Q_v^k (x-v)$ has radius of uniform convergence $\rho_v \geq \overline{\rho} - \|v-u\|$. In particular, for the term with $k=1$ and for each $x \in X$, we have by (5.31)

$$Q_v^1(x) = \tilde{K}_u^1(x) + \sum_{m>1} m K_u^m (v-u)^{\otimes m-1}(x) = L_v^{n+1}(x, \cdots) \in L({}^n X, Y),$$

where $L_v^{n+1} \in L({}^{n+1} X, Y)$ is defined by

$$\begin{aligned} &(n+1)! L_u^{n+1} + \sum_{m\geq 2} m(m+1)\cdots(m+n) L_u^{m+n} (v-u)^{\otimes m-1} \\ &\qquad = (n+1)! L_u^{n+1} + \sum_{k\geq n+2} k(k-1)\cdots(k-n) L_u^k (v-u)^{\otimes k-n-1}. \end{aligned}$$

As in the case $n=1$, for $0 < r < \overline{\rho} - \|v-u\|$, letting $C_n := \sup\{\|Q_v^k\| r^k \colon k \geq 2\} < \infty$, if $\|x-v\| < r$, it follows that

$$\left\|\mathrm{d}^n F(x) - \mathrm{d}^n F(v) - Q_v^1(x-v)\right\| \leq \sum_{k\geq 2} \|Q_v^k\| \, \|x-v\|^k \leq \frac{C_n \|x-v\|^2}{r(r - \|x-v\|)},$$

and the right side is $o(\|x-v\|)$ as $\|x-v\| \to 0$. Thus by Proposition 5.25, $F$ is differentiable of order $n+1$ at $v$ and $\mathrm{d}^{n+1} F(v) = L_v^{n+1}$. Since $v \in B(u, \overline{\rho})$ is arbitrary, $\mathrm{d}^n F$ is differentiable on $B(u, \overline{\rho})$ and (5.30) holds with $n$ replaced by $n+1$. Since differentiable functions are continuous, it then follows that each $\mathrm{d}^n F$ is continuous. By induction, $F$ is a $C^\infty$ function on $B(u, \overline{\rho})$. The proof of Theorem 5.28 is complete. □

Let $X$ and $Y$ be Banach spaces and $U$ a connected open subset of $X$. A mapping $F \colon U \to Y$ *has a Taylor expansion around a point* $u \in U$ if there exists a power series $\sum_{k\geq 0} P_u^k (x-u)$ around $u$ from $X$ to $Y$ which converges to $F$ uniformly on a ball $B(u, r) \subset U$ for some $r > 0$. By Theorem 5.10, the power series converges absolutely on a neighborhood of $u$, and so it is unique by Theorem 5.9. Moreover, by Theorem 5.28, for $x \in B(u, \overline{\rho})$, where $\overline{\rho} \geq \mathrm{e}^{-1}r$,

$$F(x) = F(u) + \sum_{k\geq 1} \frac{1}{k!} \mathrm{d}^k F(u)(x-u)^{\otimes k}. \tag{5.32}$$

The right side will be called the *Taylor series* of $F$ around $u$. We have $P_u^k = (\mathrm{d}^k F(u))\tilde{\ }/k!$, and so the Taylor series has a form as in the one-dimensional case. A mapping $F\colon U \to Y$ is *analytic* on $U$ if $F$ has a Taylor expansion around each point of $U$.

A sum of a power series around $u$ with $\rho_u > 0$ is analytic on $B(u, \overline{\rho})$ for some $\overline{\rho} > 0$, as follows from Theorem 5.21:

**Corollary 5.29.** *For a normed space $X$, a Banach space $Y$, and $u \in X$, let $\sum_{m\geq 0} P_u^m(x-u)$ be a power series around $u$ from $X$ to $Y$ whose radius of uniform convergence is $\rho_u > 0$, and let $F(x)$ be its sum for $x \in B(u, \rho_u)$. Then $F$ is analytic on $B(u, \overline{\rho})$, where $\overline{\rho} \geq \mathrm{e}^{-1}\rho_u$.*

The following compares analyticity with $\mathbb{B}$-analyticity as in Definition 5.17.

**Theorem 5.30.** *Let $U$ be a nonempty connected open subset of a unital Banach algebra $\mathbb{B}$. Then*

(*a*) *any $\mathbb{B}$-analytic function $F$ from $U$ into $\mathbb{B}$ is analytic from $U$ into $\mathbb{B}$, and for each $u \in U$, its $\mathbb{B}$-Taylor series* (5.14) *is its Taylor series around $u$;*
(*b*) *if* (5.14) *holds whenever $\|x-u\| < r$, then for the radius $\rho_u$ of uniform convergence, $\rho_u \geq r$, and $F$ has a $\mathbb{B}$-Taylor series around $x$ with $\rho_x \geq \rho_u - \|x-u\|$.*

*Proof.* For each $k$, $X \ni x \mapsto h_k x^k = P^k[h_k](x)$ is a $k$-homogeneous polynomial as defined before Lemma 5.13, and so (5.14) is a power series around $u$ from $\mathbb{B}$ to $\mathbb{B}$. By Theorem 5.18, (5.14) converges uniformly on some neighborhood of $u$, and so (*a*) holds.

For (*b*), by Lemma 5.13 we have $\|P^k[h_k]\| = \|h_k\|$. Thus $\rho_u \geq r$ by Theorems 5.18 and 5.10, proving the first part of (*b*). The second part follows from Theorem 5.21(*b*) and Corollary 5.23(*d*), proving the theorem. □

### *Analytic continuation*

Here we prove that an analytic function from an interval in $\mathbb{R}$ into a Banach space can be extended to a holomorphic function on an open subset of the complex plane. We begin with several lemmas.

**Lemma 5.31.** *Let $Y$ be a Banach space and let $\sum_{k\geq 0} c_k t^k$ be a power series from $\mathbb{R}$ to $Y$ around 0, identifying $P^k(\mathbb{R}, Y)$ with $Y$ for $k = 0, 1, \ldots$, where for $c \in Y$, $P \in P^k(\mathbb{R}, Y)$ is defined by $P(t) = t^k c$. Suppose that the power series converges to $g(t)$ on a ball $B(0, r)$ with radius $r > 0$. If there is a $c_k \neq 0$ then there is a $0 < q < r$ such that $g(t) \neq 0$ for $0 < |t| < q$.*

*Proof.* Let $k$ be the minimal integer such that $c_k \neq 0$. Then $g(t) = t^k(c_k + \cdots + c_{k+n}t^n + \cdots)$ and the series $\sum_{n\geq 0} c_{k+n}t^n$ converges, say to $h(t)$, for $t \in B(0,r)$. The sum $h$ is continuous by the proof of Proposition 2.96 with $\mathbb{C}$ replaced by $\mathbb{R}$. Since $h(0) = c_k \neq 0$, there is a $0 < q < r$ such that $h(t) \neq 0$ for $t \in B(0,q)$, proving the lemma. □

**Lemma 5.32.** *Let $X$, $Y$ be Banach spaces, and let $f$ be an analytic function from a convex open set $U \subset X$ into $Y$. If $f$ is equal to 0 on a nonempty open subset of $U$ then $f \equiv 0$ on $U$.*

*Proof.* Let $V$ be an open subset of $U$ such that $f = 0$ on $V$, let $v \in V$, and let $u$ be any point in $U$. Since the segment joining $v$ and $u$ is included in $U$, the function $g(t) := f(v + t(u - v))$, $t \in [0,1]$, is defined and analytic on an open interval including $[0,1]$. Let $D$ be the set of all $t \in [0,1]$ such that $g(s) = 0$ for $0 \leq s \leq t$, and let $d := \sup\{t \in [0,1] \colon t \in D\}$. By assumption, $d > 0$. Suppose $d < 1$. Since $g$ has a Taylor expansion around $d$, there is a Taylor series $\sum_{k\geq 0} c_k(t-d)^k$ which converges to $g(t)$ for $t \in B(d,\epsilon)$ and some $\epsilon > 0$. If all $c_k = 0$ then $g(t) = 0$ for $t < d + \epsilon$, a contradiction. If there is a $c_k \neq 0$ then $g(t) \neq 0$ for some $0 < t < d$ by Lemma 5.31, again a contradiction. Thus $d = 1$, and so $f(x) = 0$, proving the lemma. □

**Theorem 5.33.** *Let $X, Y$ be Banach spaces, let $U$ be a connected open subset of $X$, and let $f$ be an analytic function from $U$ into $Y$. If for some $u \in U$, $f$ is zero on a neighborhood of $u$, then $f \equiv 0$ on $U$.*

*Proof.* Let $A$ be the interior of the set of all $x \in U$ such that $f(x) = 0$. Then $A$ is open and nonempty by assumption. We prove that $A$ is also closed in $U$, and so equal to $U$ since $U$ is connected. Let $u \in U$ be a limit point of $A$. Since $f$ has a Taylor expansion around $u$, there is an open ball $B(u,r) \subset U$ with center $u$ and radius $r > 0$, so that $f$ is analytic on $B(u,r)$. The intersection $B(u,r) \cap A$ includes another open ball $V$ such that $f(x) = 0$ for $x \in V$. By Lemma 5.32, $f(x) = 0$ for $x \in B(u,r)$, and so $u \in A$, proving the theorem. □

**Theorem 5.34.** *Let $X$, $Y$ be complex Banach spaces, and let $R$ be a real vector subspace of $X$ such that the complex subspace $R_c := R + \mathrm{i}R$ is dense in $X$. If $f$ and $g$ are analytic on a connected open set $U$ in $X$, and $f = g$ on some nonempty set $V \subset U \cap R$, relatively open in $R$, then $f = g$ on $U$.*

For the proof we need the following lemma:

**Lemma 5.35.** *Let $R$ be as in the preceding theorem, and let for some $k \geq 0$, $P \in P^k(X,Y)$ be such that $P$ vanishes on $R$. Then $P \equiv 0$.*

*Proof.* We can assume that $k \geq 2$. Let $L \in L_s({}^k X, Y)$ be such that $P = \tilde{L}$, that is, $Px = Lx^{\otimes k}$ for all $x \in X$. By the binomial formula (5.8), we have

$$P(x + \mathrm{i}y) = L(x + \mathrm{i}y)^{\otimes k} = \sum_{l=0}^{k} \mathrm{i}^l \binom{k}{l} Lx^{\otimes k-l} \otimes y^{\otimes l}$$

for $x, y \in X$. Since $P = 0$ on $R$, by the polarization Theorem 5.6, $L = 0$ on $R^k$, and so $P(x + \mathrm{i}y) = 0$ for all $x, y \in R$. Thus $P = 0$ on the dense set $R_c$ in $X$. Since $P$ is continuous, $P \equiv 0$, proving the lemma. □

**Proof of Theorem 5.34.** It is enough to prove the theorem with $g \equiv 0$ on $U$. Let $u \in V$, and let $\sum_{k \geq 0} P_u^k$ be a power series representation of $f$ on an open ball $B(u, \rho) \subset X$ for some $\rho > 0$. By Theorem 5.10, the series converges absolutely on $B(u, \rho)$. Since $f = 0$ on $V \subset R$, by Theorem 5.9, $P_u^k(x) = 0$ for all $x \in R$ and $k = 0, 1, \ldots$. By Lemma 5.35, $P_u^k(x) = 0$ for all $x \in R_c = R + \mathrm{i}R$ and $k = 0, 1, \ldots$, and so $f$ is zero in a neighborhood of $u$ in $X$. This proves that $f \equiv 0$ on $U$ by Theorem 5.33. The proof of Theorem 5.34 is complete. □

Now we are ready to prove that an analytic function on an open interval extends to a holomorphic function on an open subset of $\mathbb{C}$.

**Theorem 5.36.** *Let $Y$ be a complex Banach space, let $U$ be a nonempty open interval in $\mathbb{R}$, and let $f$ be an analytic function from $U$ into $Y$. Then there are an open connected set $V \subset \mathbb{C}$ such that $V \cap \mathbb{R} = U$ and a holomorphic mapping $g$ from $V$ into $Y$ such that $g = f$ on $V$.*

*Proof.* For each $k = 1, 2, \ldots$, we identify $P^k(\mathbb{R}, Y)$ with $Y$ as in Lemma 5.31. Let $u \in U$. There is a sequence $\{c_k\}_{k \geq 0} \subset Y$ and $r > 0$ such that $f(x) = \sum_{k=0}^{\infty} c_k (x - u)^k$ and (by Theorem 5.10, or directly) the series converges absolutely and uniformly for $x \in \mathbb{R}$ and $|x - u| < r$. Let $A_u := \{z \in \mathbb{C} \colon\ |z - u| < r\}$ and $Q_u := A_u \cap \mathbb{R}$. Then the power series $\sum_{k=0}^{\infty} c_k (z - u)^k$ converges absolutely and uniformly on $A_u$ by Theorem 5.10. Let $g_u$ be its sum on $A_u$. If $u, v \in U$ are such that $A_u \cap A_v \neq \emptyset$ then also $Q_u \cap Q_v = (A_u \cap A_v) \cap \mathbb{R} \neq \emptyset$ and we have $g_u = g_v = f$ on $Q_u \cap Q_v$. Since $A_u \cap A_v$ is connected, by Theorem 5.34, $g_u = g_v$ on $A_u \cap A_v$. Take $V := \cup_{u \in U} A_u$ and define $g$ to be equal to $g_u$ on each $A_u$. Clearly $V$ is connected. Since $g$ has a Taylor expansion around each point of $V$ by Corollary 5.23($e$), $g$ is analytic on $V$ and agrees with $f$ on $U$. The proof is complete. □

### *Uniformly entire mappings*

A function of one complex variable is called entire if it is analytic (holomorphic) on the complex plane. It then has a Taylor series around any point, converging on the whole plane, uniformly on bounded sets. This is no longer true for power series on an infinite-dimensional space as the following shows.

**Example 5.37.** As in Example 5.12, let $F\colon \ell^2 \to \mathbb{R}$ be defined by $F(x) := \sum_{k\geq 1}(x_k)^k$ for each $x = \{x_j\}_{j\geq 1} \in \ell^2$. For any $u = \{u_j\}_{j\geq 1} \in \ell^2$, by the binomial formula, we have

$$F(x) = F(u+x-u) = \sum_{k\geq 1}(u_k + x_k - u_k)^k = \sum_{k=1}^{\infty}\sum_{j=0}^{k}\binom{k}{j}u_k^{k-j}(x_k-u_k)^j.$$

For any fixed $u$, the double series converges absolutely and uniformly for $\|x-u\| \leq \rho$ provided $\sum_{k\geq 1}(|u_k|+\rho)^k < \infty$. The latter holds if and only if $\rho < 1$. Then the series can be rearranged so that $F(x) = F(u) + \sum_{j\geq 1} P_u^j(x-u)$, where for each $j \geq 1$,

$$P_u^j(x-u) := \sum_{k=j}^{\infty}\binom{k}{j}u_k^{k-j}(x_k-u_k)^j.$$

This gives the Taylor series of $F$ around $u$ with radius of uniform convergence $\rho_u \geq 1$. For $l = 1, 2, \ldots$, letting $x_k^l := u_k + 1$ if $k = l$ and $x_k^l := u_k$ if $k \neq l$, we have that $x^l = \{x_k^l\}_{k\geq 1} \in \ell^2$ with $\|x^l - u\| = 1$ and $\sum_{j\geq l} P_u^j(x^l - u) = P_u^l(x^l-u) = 1$. Thus the power series $\sum_{j\geq 1} P_u^j(x-u)$ does not converge uniformly on the closed ball $\bar{B}(u,1)$, and so $\rho_u = 1$ for any $u \in \ell^2$.

A mapping $F\colon X \to Y$ will be called *uniformly entire* on $X$ if $F$ is analytic on $X$ with infinite radius of uniform convergence of its Taylor series around any point of $X$. It is equivalent, by Theorem 5.21, that its Taylor series around 0 has infinite radius of uniform convergence.

*Composition of analytic functions*

We will prove that a composition of analytic functions between Banach spaces is analytic. Let $X$ and $Y$ be Banach spaces and for $k = 0, 1, \ldots$ let $P^k \in P^k(X,Y)$ be a continuous $k$-homogeneous polynomial. Let $\mathcal{P} := \{P^k\}_{k=0}^{\infty}$ and $\rho_{\mathcal{P}} := 1/\limsup_{k\to\infty}\|P^k\|^{1/k}$, which by (5.11) is the radius of uniform convergence of the power series $\sum_{k=0}^{\infty} P^k(x)$ around 0, or of $\sum_{k=0}^{\infty} P^k(x-u)$ around $u$ for any $u \in X$. Also let $L^k \in L_s(^kX, Y)$ be a continuous symmetric $k$-linear map for each $k = 1, 2, \ldots$, let $L^0$ be any element of $Y$, and let $\mathcal{L} := \{L_k\}_{k=0}^{\infty}$. Let $\overline{\rho}_{\mathcal{L}} := 1/\limsup_{k\to\infty}\|L^k\|^{1/k}$, as in (5.16). If $P^k = (L^k)^{\widehat{}}$ for each $k \geq 1$ and $P^0 \equiv L^0 \in Y$ then a power series around any $u \in X$ can be written in two equivalent ways, $\sum_{k\geq 0} P^k(x-u) \equiv \sum_{k\geq 0} L^k(x-u)^k$, where $L^0(x-u)^0 := L^0 \in Y$. Its radius of uniform convergence is $\rho_{\mathcal{P}}$. We have $\|P^k\| \leq \|L^k\|$ for all $k$, so $\overline{\rho}_{\mathcal{L}} \leq \rho_{\mathcal{P}}$. Conversely it was shown in Theorem 5.21 that $\overline{\rho}_{\mathcal{L}} \geq \rho_{\mathcal{P}}/\mathrm{e}$.

Here is a fact about composition of analytic functions.

**Theorem 5.38.** *Let $X$, $Y$, and $Z$ be three Banach spaces. Let $\xi \in X$ with $\xi \in U$, a connected open set, and let $F$ be an analytic function from $U$ into a connected open set $V \subset Y$. Let $F$ have the power series around $\xi$*

$$F(x) = \sum_{k=0}^{\infty} P^k(x - \xi)$$

*where $P^k \in P^k(X,Y)$ and the series has radius of uniform convergence $\rho_{\mathcal{P}} > 0$. Let $\eta := F(\xi) \in Y$. Let $G$ be an analytic function from $V$ into $Z$, with power series around $\eta$ given by*

$$G(y) = \sum_{j=0}^{\infty} B^j(y - \eta)^j$$

*where $B^j \in L_s({}^jY, Z)$ for each $j$, and let this series also have radius of uniform convergence $\rho_{\mathcal{Q}} > 0$ where $\mathcal{Q} = \{(B^j)^{\frown}\}_{j \geq 0}$. Then for some $r$ with $0 < r < \rho_{\mathcal{P}}$, there exists an $M$ with $0 < M < \overline{\rho}_{\mathcal{B}}$ for $\mathcal{B} = \{B^j\}_{j \geq 0}$ such that*

$$\sum_{k=0}^{\infty} \|P^k\| r^k \leq M. \tag{5.33}$$

*For any such $r$ and $M$, $F$ restricted to a ball $B(\xi, r)$ takes values in another ball $B(\eta, M)$, the composition $G \circ F$ from $B(\xi, r)$ into $Z$ is analytic, and its power series around $\xi$ has radius of uniform convergence at least $r$.*

*If $F$ and $G$ are uniformly entire from $X$ into $Y$ and $Y$ into $Z$ respectively, then $G \circ F$ is uniformly entire from $X$ into $Z$.*

*Proof.* Let $\phi(u) := F(\xi + u) - F(\xi)$ for all $u \in X$ such that $\xi + u \in U$. Then $\phi$ is defined and analytic on a neighborhood of 0 and has the power series $\phi(u) = \sum_{k=1}^{\infty} P^k(u)$ around 0. Let $\psi(v) := G(\eta + v) - G(\eta)$ for all $v \in Y$ such that $\eta + v \in V$. Then $\psi$ is defined and analytic on a neighborhood of 0 in $Y$ and has the power series $\psi(v) = \sum_{j=1}^{\infty} B^j v^j$ around 0. The hypotheses hold for $\phi, \psi, 0, 0$ in place of $F, G, \xi, \eta$ respectively, with the same $P^k$ for $k \geq 1$, the same $B^j$ for $j \geq 1$, and so the same $\rho_{\mathcal{P}}$ and $\overline{\rho}_{\mathcal{B}}$. Suppose the conclusion holds also. We would then have a power series $(\psi \circ \phi)(u) = \sum_{l=1}^{\infty} \pi^l(u)$ for some $l$-homogeneous polynomials $\pi^l$ from $X$ into $Z$, where the series has convergence properties as stated. Substituting, we would then get

$$G(F(x)) \;=\; G(F(\xi)) + \sum_{l=1}^{\infty} \pi^l(x - \xi)$$

for $\|x - \xi\| < r$, with the same polynomials $\pi^l$ for $l \geq 1$. Thus we can assume that $\xi = 0$ in $X$ and $\eta = 0$ in $Y$.

Next, here is a lemma for power series in real variables.

**Lemma 5.39.** *Let $r > 0$ and $a_l \geq 0$ be real numbers and set $f(u) = \sum_{l=1}^{\infty} a_l u^l$. Assume that the series converges for all real $u$ with $|u| \leq r$. Then for any $j = 2, 3, \ldots$, $f(u)^j = \sum_{l=1}^{\infty} c_{j,l} u^l$ where all $c_{j,l} \geq 0$ and the series also converges for $|u| \leq r$. We have a finite sum*

$$c_{j,l} = \sum \left\{ \prod_{m=1}^{j} a_{l_m} : l_m \geq 1, \sum_{m=1}^{j} l_m = l \right\}. \tag{5.34}$$

The lemma is straightforward to prove (the sum has at most $l^j$ terms). We return to the proof of the theorem.

Since $0 < r < \rho_{\mathcal{P}}$, the series $f(r) := \sum_{k=1}^{\infty} \|P^k\| r^k$ converges. Recall that $\overline{\rho}_{\mathcal{B}} \geq \rho_{\mathcal{Q}}/\mathrm{e} > 0$. We have $f(r) \downarrow 0$ as $r \downarrow 0$. Thus for any $M$ with $0 < M < \overline{\rho}_{\mathcal{B}}$, there exists $r > 0$ for which $f(r) \leq M$. Fix such an $M$ and $r$.

We have for $\|x\| \leq r$ that $F(x) = \sum_{k=1}^{\infty} P^k(x)$ is uniformly and absolutely convergent, with the norms of terms dominated termwise by the terms of the series of nonnegative numbers $\sum_{k=1}^{\infty} \|P^k\| r^k \leq M$.

The series $\sum_{j=1}^{\infty} B^j y^j$ is absolutely convergent uniformly for $y$ with $\|y\| \leq M$, in particular for $y = F(x)$ with $\|x\| \leq r$. Moreover, in each term $B^j F(x)^j$, if we substitute in $F(x) = \sum_k P^k(x)$, we will get, due to the boundedness of $B^j$, a convergent infinite series of terms of the form $B^j(P^{i_1}(x), \ldots, P^{i_j}(x))$, such that the sum of the norms of all these terms is bounded above by $\|B^j\| M^j$. Further, when we sum over $j$, we again get a convergent sum of norms because $M < \overline{\rho}_{\mathcal{B}}$. Thus we get by dominated convergence a series converging absolutely to $G(F(x))$ in $Z$. In the whole sum, for any given finite degree $l$ in $x$, there are only finitely many terms of that degree, since such terms occur only for $j \leq l$ and $k \leq l$. Specifically, for each $l \geq 1$ let

$$\pi^l(x) := \sum_{j=1}^{l} \sum \left\{ B^j(P^{i_1}(x), \ldots, P^{i_j}(x)) : i_m \geq 1 \text{ for all } m, \sum_{m=1}^{j} i_m = l \right\}.$$

This is a finite sum (although with many terms for large $l$), and so a polynomial. For each $j$, its terms correspond one-to-one to terms in (5.34). It is the sum of all terms homogeneous of degree $l$ in the series for $G(F(x))$. Each term in the sum has norm at most $\|B^j\| \prod_{m=1}^{j} \|P^{i_m}\| r^l$. Now we apply Lemma 5.39 to the function $f(u) = \sum_{l=1}^{\infty} \|P^l\| u^l$, so that $f(r) \leq M$ and for each $j = 1, 2, \ldots$, $f(r)^j \leq M^j$. Then in the lemma,

$$c_{j,l} = \sum \left\{ \prod_{m=1}^{j} \|P^{i_m}\| : i_m \geq 1 \text{ for all } m, \sum_{m=1}^{j} i_m = l \right\}.$$

This implies the bound $\|\pi^l\| \leq \sum_{j=1}^{l} \|B^j\| c_{j,l}$. We want to show that $\sum_{l=1}^{\infty} \|\pi^l\| r^l < \infty$. For each fixed $j$, we have by Lemma 5.39 that $\sum_{l=1}^{\infty} c_{j,l} r^l = f(r)^j \leq M^j$. Then, $\sum_j \|B^j\| M^j < \infty$ because $0 < M < \overline{\rho}_{\mathcal{B}}$. So indeed, $\sum_{l=1}^{\infty} \|\pi^l\| r^l < \infty$. Thus the series $G(F(x)) = \sum_{l=1}^{\infty} \pi^l(x)$ is absolutely and uniformly convergent for $\|x\| \le r$, with values in $Z$. The conclusions follow. □

## 5.3 Taylor's Formulas

Here we prove a Taylor's formula with integral remainder (Theorem 5.42) for a smooth function on an open subset $U$ of a Banach space $X$ with values in another Banach space $Y$. Then a criterion for analyticity, Theorem 5.43, will follow. In this section for functions $f\colon [0,1] \to Y$ and $\theta\colon [0,1] \to \mathbb{R}$, the integral

$$\int_0^1 f(t)\,\mathrm{d}\theta(t) := (RS)\int_0^1 f\cdot\mathrm{d}\theta$$

is defined as an element of $Y$ using the natural bilinear mapping $B\colon Y \times \mathbb{R} \to Y$, provided the Riemann–Stieltjes integral is defined.

**Lemma 5.40.** *Let $X$, $Y$ be two Banach spaces, $U$ an open subset of $X$, and $F$ a continuously differentiable mapping of $U$ into $Y$. If the segment joining $u$ and $u + x$ is included in $U$, then*

$$F(u+x) - F(u) = \int_0^1 DF(u+tx)x\,\mathrm{d}t.$$

*Proof.* Let $\phi(t) := u + tx \in U$ for all $t$ in an open interval $V \subset \mathbb{R}$ including $[0,1]$, and let $G$ be the composition $F \circ \phi$, defined from $V$ into $Y$. By the chain rule for differentiation of a composition (Theorem 5.1), $DG(t) = DF(u + tx) \circ D\phi(t)$ for each $t \in V$, and so $G$ is a continuously differentiable function on $V$. By (5.3) in the mean value theorem, for $0 \le s < t \le 1$,

$$\big\|G(t) - G(s) - DG(s)(t-s)\big\| \le (t-s) \sup_{s \le v \le t} \big\|DG(s) - DG(v)\big\|.$$

Since $DG$ is continuous on $[0,1]$, given $\epsilon > 0$ there is a $\delta > 0$ such that $\mathrm{Osc}(DG; [s,t]) < \epsilon$ whenever $t - s < \delta$. Therefore using telescoping sums, it follows that

$$\|G(1) - G(0) - S_{RS}(DG, \mathrm{d}\theta; \tau)\| < 2\epsilon$$

for each tagged partition $\tau$ of $[0,1]$ with mesh $|\tau| < \delta$, where $\theta(t) \equiv t$. Thus

$$F(u+x) - F(u) = G(1) - G(0) = \int_0^1 DG(t)\,\mathrm{d}t = \int_0^1 DF(u+tx)x\,\mathrm{d}t,$$

as desired. □

First, we prove the Taylor formula in the case when $X = \mathbb{R}$ and $U$ is an open interval. Let $F$ be a function from $U$ to $Y$, differentiable of order $n$ at $u \in U$. Since $X = \mathbb{R}$, for each $k = 1, \ldots, n$, we identify $L_s({}^k\mathbb{R}, Y)$ with $Y$ via $y \leftrightarrow L^k[y]$ defined by $L^k[y](t_1, \ldots, t_k) := t_1 \cdots t_k y$. So there exists $h_k(u) \in Y$ such that $\mathrm{d}^k F(u) = L^k[h_k(u)]$. In this case, we let

$$\mathrm{d}^k F(u) := \mathrm{d}^k F(u)(1, \ldots, 1) = h_k(u). \tag{5.35}$$

The *Taylor polynomial* $T_n(\cdot)$ of $F$ around $u$ of order $n$ is defined by

$$T_n(x) = F(u) + \sum_{k=1}^{n} \frac{1}{k!} \mathrm{d}^k F(u) x^k$$

for each $x \in \mathbb{R}$. In this case Taylor's formula with an integral remainder is given by the following theorem.

**Theorem 5.41.** *Let $U$ be an open interval in $\mathbb{R}$, let $Y$ be a Banach space, and for some integer $n \geq 1$, let $F$ be a $C^n$ function on $U$ with values in $Y$. Then for any pair of numbers $u$ and $u + x$ in $U$, we have*

$$F(u + x) - T_n(x) = \int_0^1 \big[\mathrm{d}^n F(u + tx) - \mathrm{d}^n F(u)\big] x^n \, \mathrm{d}\theta_n(t), \tag{5.36}$$

*where $\theta_n(t) := -(1 - t)^n / n!$ for $t \in [0, 1]$.*

*Proof.* The Riemann–Stieltjes integral in (5.36) exists since the integrand is continuous and $\theta_n$ is smooth. The proof of the equality in (5.36) will be by finite induction in $k = 1, \ldots, n$. Suppose that $k = 1$. By Lemma 5.40, it follows that

$$\begin{aligned} F(u + x) - T_1(x) &= F(u + x) - F(u) - \mathrm{d}^1 F(u) x \\ &= \int_0^1 \big[\mathrm{d}^1 F(u + tx) - \mathrm{d}^1 F(u)\big] x \, \mathrm{d}t. \end{aligned}$$

Thus, (5.36) holds with $n = 1$. Suppose that (5.36) holds for some integer $1 \leq k < n$ instead of $n$. We then have (recalling Theorem 2.80 on integration by parts)

$$
\begin{aligned}
F(u+x) &- T_k(x) \\
&= \int_0^1 \big[\mathrm{d}^k F(u+tx) - \mathrm{d}^k F(u)\big] x^k \,\mathrm{d}\theta_k(t) \\
&= \int_0^1 \int_0^t (D(\mathrm{d}^k F))(u+sx)\,\mathrm{d}s\,\mathrm{d}\theta_k(t) x^{k+1} \quad \text{by Lemma 5.40} \\
&= \theta_k \int_0^{(\cdot)} (D(\mathrm{d}^k F))(u+sx)\,\mathrm{d}s\Big|_0^1 \qquad \text{integrating by parts} \\
&- \int_0^1 \theta_k(t)\,\mathrm{d}\Big(\int_0^t (D(\mathrm{d}^k F))(u+sx)\,\mathrm{d}s\Big)\, x^{k+1} \quad \text{e.g. (2.77)} \\
&= - \int_0^1 \theta_k(t) (D(\mathrm{d}^k F))(u+tx)\,\mathrm{d}t\, x^{k+1} \qquad \text{by Proposition 2.86} \\
&= \int_0^1 (\mathrm{d}^{k+1} F)(u+tx) x^{k+1}\,\mathrm{d}\theta_{k+1}(t) \qquad \text{by Proposition 5.25}
\end{aligned}
$$

since $\theta_{k+1}(t) = -\int_0^t \theta_k(s)\,\mathrm{d}s$ for $t \in [0,1]$. It then follows that

$$
\begin{aligned}
F(u+x) - T_{k+1}(x) &= F(u+x) - T_k(x) - \int_0^1 \mathrm{d}^{k+1} F(u) x^{k+1}\,\mathrm{d}\theta_{k+1}(t) \\
&= \int_0^1 \big[\mathrm{d}^{k+1} F(u+tx) - \mathrm{d}^{k+1} F(u)\big] x^{k+1}\,\mathrm{d}\theta_{k+1}(t).
\end{aligned}
$$

Thus (5.36) holds with $k+1$ instead of $n$ whenever it holds with $k$ instead of $n$. The proof is complete by induction. □

Let $X, Y$ be two Banach spaces, $U$ an open subset of $X$, and for a positive integer $n$, $G\colon U \to Y$ a function differentiable of order $n$ at $u \in U$. Let $T_n(\cdot)$ be the *Taylor polynomial* of $G$ around $u$ of order $n$, that is, for each $x \in X$,

$$
T_n(x) = G(u) + \sum_{k=1}^n \frac{1}{k!} \mathrm{d}^k G(u) x^{\otimes k}.
$$

The following extends the Taylor formula with an integral remainder of the preceding theorem to a Banach space $X$.

**Theorem 5.42.** *Let $X$, $Y$ be two Banach spaces, let $U$ be an open subset of $X$, and for some integer $n \geq 1$, let $G$ be a $C^n$ function on $U$ with values in $Y$. If the segment joining $u$ and $u+x$ is included in $U$, then*

$$
G(u+x) - T_n(x) = \int_0^1 \big[\mathrm{d}^n G(u+tx) - \mathrm{d}^n G(u)\big] x^{\otimes n}\,\mathrm{d}\theta_n(t), \tag{5.37}
$$

*where $\theta_n(t) := -(1-t)^n/n!$ for $t \in [0,1]$.*

*Proof.* The Riemann–Stieltjes integral in (5.37) exists since the integrand is continuous and $\theta_n$ is smooth. Let $\phi_x(t) := u + tx \in U$ for all $t$ in an open interval $V$ including $[0, 1]$. To prove (5.37) we apply (5.36) to $F(t) := G(u + tx) = G \circ \phi_x(t)$, with $t \in V$. By the chain rule (Theorem 5.1), using finite induction in $k = 1, \ldots, n$ and (5.35), it follows that $F$ is a $C^n$ function on $V$ and

$$\mathrm{d}^k F(t) = D^k F(t)(1) \cdots (1) = D^k G(u + tx)(x) \cdots (x) = \mathrm{d}^k G(u + tx) x^{\otimes k}$$

for $k = 1, \ldots, n$. Thus by (5.36), it follows that

$$\begin{aligned} G(u + x) = F(1) &= F(0) + \sum_{k=1}^{n} \frac{\mathrm{d}^k F(0)}{k!} + \int_0^1 \big[\mathrm{d}^n F(t) - \mathrm{d}^n F(0)\big]\, \mathrm{d}\theta_n(t) \\ &= T_n(x) + \int_0^1 \big[\mathrm{d}^n G(u + tx) - \mathrm{d}^n G(u)\big] x^{\otimes n}\, \mathrm{d}\theta_n(t), \end{aligned}$$

proving (5.37). The proof of Theorem 5.42 is complete. □

Next, necessary and sufficient conditions will be established for a function to have a Taylor expansion around a point. A condition related to condition (*b*) below was used in Corollary 5.11 to characterize uniform convergence of a power series.

**Theorem 5.43.** *Let $X$, $Y$ be two Banach spaces and let $U$ be a connected open subset of $X$. A function $F \colon U \to Y$ has a Taylor expansion around $u \in U$ if and only if there exist constants $c \geq 0$ and $r > 0$ such that*

(*a*) *$F$ is a $C^\infty$ function on the open ball $B(u, r)$;*
(*b*) *$\|\mathrm{d}^n F(x)\| \leq c^n n!$ for $x \in B(u, r)$ and $n = 1, 2, \ldots$.*

*Then the Taylor series converges uniformly and absolutely on $B(u, \rho)$ whenever $0 < \rho < 1/c$ and $\rho \leq r$.*

*Proof.* First suppose that $F$ has a Taylor expansion around $u \in U$. Therefore there is a power series around $u$ from $X$ to $Y$ whose radius of uniform convergence is $\rho_u > 0$ and $F(x)$ is its sum for $x \in B(u, \rho_u)$. Let $\overline{\rho}$ be defined by (5.16), and let $r := \overline{\rho}/2$. Then $r \geq \rho_u/(2\mathrm{e}) > 0$ by Theorem 5.21 and (*a*) follows by Theorem 5.28. By statements (*b*) and (*c*) of Theorem 5.21, for each $v \in B(u, r)$, there is a power series $\sum_{n \geq 0} P_v^n(x - v)$ converging to $F(x)$ for each $x \in X$ such that $\|x - v\| < \overline{\rho} - \|v - u\|$, and having radius of uniform convergence $\rho_v \geq \overline{\rho} - r = r$. By Theorem 5.28 applied to the power series $\sum_{n \geq 0} P_v^n(x - v)$ for $v = u$, $(\mathrm{d}^n F(v))\check{} = n! P_v^n$ on $X$ for each $n \geq 1$. Therefore using (5.10), (5.11), and Stirling's formula, for each such $v$,

$$\limsup_{n \to \infty} \Big(\frac{\|\mathrm{d}^n F(v)\|}{n!}\Big)^{1/n} \leq \limsup_{n \to \infty} \Big(\frac{n^n \|P_v^n\|}{n!}\Big)^{1/n} = \frac{\mathrm{e}}{\rho_v} \leq \frac{2\mathrm{e}}{\overline{\rho}} < \infty.$$

Thus there is a finite constant $c$ such that $\|\mathrm{d}^n F(v)\|/n!)^{1/n} \le c$ for all $n \ge 1$ and $v \in B(u,r)$, and so $(b)$ follows.

To prove the converse implication let $(a)$ and $(b)$ hold for some constants $c \ge 0$ and $r > 0$. We can assume that $c > 0$ since otherwise $F$ is a constant function. By $(a)$ and Theorem 5.42, for each integer $n \ge 1$, Taylor's formula with an integral remainder (5.37) holds with $U = B(u,r)$. Let $n \ge 1$ and $\|x\| < r$. By Proposition 2.13, using the bound (2.84) for the $RS$ integral, since $v_1(\theta_n) = 1/n!$, we have

$$\big\|F(u+x) - T_n(x)\big\| \le \frac{6\|x\|^n}{n!} \sup_{0\le t\le 1} \big\|\mathrm{d}^n F(u+tx)\big\| \le 6(rc)^n$$

for each $n \ge 1$. For any $\rho$ such that $\rho c < 1$ and $\rho \le r$, it follows that the power series $F(u) + \sum_{n\ge 1}(\mathrm{d}^n F(u)/n!)x^{\otimes n}$ converges to $F(u+x)$ absolutely and uniformly on $B(u,\rho)$, proving the "if" part and the last conclusion, and hence proving the theorem. □

Let $X$, $Y$ be two Banach spaces, $U$ an open subset of $X$, and $G\colon U \to Y$ a function differentiable of order $n$ at $u \in U$. For $x \in X$ such that $u + x \in U$, the *remainder in Taylor's expansion* of $G$ around $u$ of order $n$ is defined by

$$\mathrm{Rem}_G^n(u,x) := G(u+x) - T_n(x) = G(u+x) - G(u) - \sum_{k=1}^{n} \frac{\mathrm{d}^k G(u) x^{\otimes k}}{k!}.$$

If $G$ is a $C^n$ function on $U$, then the Taylor formula with an integral remainder holds by Theorem 5.42, which implies that $\|\mathrm{Rem}_G^n(u,x)\| = o(\|x\|^n)$ as $\|x\| \to 0$. The next theorem gives the same order of smallness but requires only differentiability of order $n$ at $u$.

**Theorem 5.44.** *Let $X$, $Y$ be two Banach spaces, let $u$ be an element of an open subset $U$ of $X$, and let $G\colon U \to Y$ be a function differentiable of order $n$ at $u$. Then $\|\mathrm{Rem}_G^n(u,x)\| = o(\|x\|^n)$ as $\|x\| \to 0$.*

*Proof.* If $n = 1$, then the conclusion holds by the definition of Fréchet differentiability of $G$ at $u$. Suppose that $n > 1$ and that the statement holds when $n$ is replaced by some $m$, $1 \le m < n$. It will be proved to hold for $m+1$. Let $r > 0$ be such that $B(u,r) \subset U$ and $G$ is differentiable of order $m$ on $B(u,r)$. For each $x \in B(0,r)$, let $R(x) := \mathrm{Rem}_G^{m+1}(u,x) \in Y$. It will be shown that $R$ is differentiable on $B(0,r)$ with derivative $DR(x) = L(x) \in L(X,Y)$ for $x \in B(0,r)$, where

$$L(x)(\cdot) := DG(u+x)(\cdot) - DG(u)(\cdot) - \sum_{k=2}^{m+1} \frac{1}{(k-1)!} \mathrm{d}^k G(u) x^{\otimes k-1} \otimes (\cdot).$$

Let $x \in B(0,r) \setminus \{0\}$. For each $y \in B(0,r)$ such that $x + y \in B(0,r)$, we have

$$
\begin{aligned}
&\|R(x+y) - R(x) - L(x)(y)\| \\
&\quad \le \|G(u+x+y) - G(u+x) - DG(u+x)(y)\| \\
&\qquad + \sum_{k=2}^{m+1} \frac{1}{k!} \big\| \mathrm{d}^k G(u)\{(x+y)^{\otimes k} - x^{\otimes k} - kx^{\otimes k-1} \otimes y\} \big\|.
\end{aligned}
$$

The sum on the right side is of order $O(\|y\|^2) = o(\|y\|)$ as $\|y\| \to 0$ by the binomial formula (5.8), and the first term on the right side is of order $o(\|y\|)$ due to differentiability of $G$ at $u+x$. Thus $R$ is differentiable on $B(0,r)$ with the derivative $DR = L$.

Up through the next display, all derivatives $D, \dots, D^k$ and differentials $d, \dots, d^{k-1}$ are with respect to the variable $u$. Derivatives $D$ can be interchanged with evaluations at points $x$ or $x, x_1, x_2, \dots, x_{k-1}$ by the chain rule, Remark 5.2, since the evaluations are bounded linear operators. We claim that $d^k G(u)(x, \dots) = d^{k-1}(DG(u)(x))$ holds in $L({}^{k-1}X, Y)$ for each $x \in X$ and $k = 2, \dots, m+1$. Indeed for any $x_1, \dots, x_k$ in $X$, using the definitions of $d^k G(u)$ in (5.23), $D^k G(u)$, and (5.22), we have

$$
\begin{aligned}
&d^k G(u)(x_1, \dots, x_k) \\
&\quad = D^k G(u)(x_1)(x_2) \cdots (x_k) = D[\cdots D[DG(u)(x_1)](x_2) \cdots](x_k) \\
&\quad = D^{k-1}[DG(u)(x_1)](x_2) \cdots (x_k) = d^{k-1}[DG(u)(x_1)](x_2, \dots, x_k),
\end{aligned}
$$

proving the claim. Therefore we have $L(x) = \mathrm{Rem}^m_{DG}(u, x)$. Since $DG$ is differentiable of order $m$ at $u$, $\|DR(x)\| = o(\|x\|^m)$ as $\|x\| \to 0$ by the induction assumption. By (5.2) in the mean value theorem, for $x \in B(0,r) \setminus \{0\}$, we have

$$
\begin{aligned}
\|\mathrm{Rem}^{m+1}_G(u, x)\| &= \|R(x)\| = \|R(x) - R(0)\| \\
&\le \|x\| \sup_{t \in [0,1]} \|DR(tx)\| = o(\|x\|^{m+1})
\end{aligned}
$$

as $\|x\| \to 0$, proving the conclusion with $m$ replaced by $m+1$. The theorem is proved by induction. □

## 5.4 Tensor Products of Banach Spaces

Given Banach spaces $X$ and $Y$ over $\mathbb{K}$, we have the Banach space $\mathcal{M}(X,Y) := \mathcal{M}_2(X, Y; \mathbb{K})$ of bounded bilinear maps from $X \times Y$ into $\mathbb{K}$, defined just before Theorem 5.4 with the norm (5.4). For each $x \in X$, $y \in Y$, and $L \in \mathcal{M}(X,Y)$, let $(x \otimes y)(L) := L(x,y)$. Then clearly $x \otimes y$ is in the dual Banach space $\mathcal{M}(X,Y)'$ with the dual norm $\|x \otimes y\|' \le \|x\|\|y\|$. Define the *tensor product* $X \otimes Y$ as the linear subspace of $\mathcal{M}(X,Y)'$ spanned by all such $x \otimes y$, i.e. as the set of all finite sums $\sum_{i=1}^n x_i \otimes y_i$, $x_i \in X$, $y_i \in Y$, $n = 1, 2, \dots$.

Beside the most obvious norm from the present approach, namely the dual norm $\|\cdot\|'$ in $\mathcal{M}(X,Y)'$, other norms can be defined on $X \otimes Y$. One is what is called the projective tensor product norm: for each $v \in X \otimes Y$, let

$$\|v\|_\pi := \inf\Big\{\sum_{i=1}^n \|x_i\|\|y_i\|\colon v = \sum_{i=1}^n x_i \otimes y_i,\ x_i \in X,\ y_i \in Y,\ n = 1, 2, \dots\Big\}. \tag{5.38}$$

It is easily checked that $\|v\|' \le \|v\|_\pi$ for all $v \in X \otimes Y$ and that $\|\cdot\|_\pi$ is a seminorm, thus a norm since $\|\cdot\|'$ is on $\mathcal{M}(X,Y)'$.

## 5.5 Notes

M. Fréchet defined Fréchet differentiability in 1925 ([68] and [69]) and proved the chain rule for it. A. E. Taylor [230], in a survey of Fréchet's work in analysis, discussed historical precursors of Fréchet's work.

The chain rule and the mean value theorem (5.2) are Theorems 8.2.1 and 8.5.4 in Dieudonné [42], respectively.

*Notes on Section 5.1.* The polarization identity of Theorem 5.6 is Theorem 4.6 in Chae [32]. Bounds for norms of $k$-homogeneous polynomials as in Theorem 5.7 are given in Theorem 4.13 in [32]. Uniqueness of power series as in Theorem 5.9 is given in Theorem 11.10 in [32]. The Cauchy–Hadamard formula of Theorem 5.10 is Theorem 11.5 in [32].

A. Mazur and W. Orlicz in 1935 [163] defined polynomials of degree $k$ in real normed spaces. A historical discussion of different concepts of polynomials in real and complex normed spaces is given in Taylor [229].

An early proof of Corollary 5.23($e$) was by Banach in 1938 [9]. Several other authors either discovered the fact independently, perhaps even earlier, or gave other proofs. The proof given is from Bochnak and Siciak [20]. They attribute it to S. Łojasiewicz.

*Notes on Section 5.2.* The higher order derivatives of mappings between normed spaces are elements of $L(X, L(X,Y)), L(X, L(X, L(X,Y))), \dots$. Usually (e.g. [42], [32]) these derivatives are identified with multilinear mappings using the natural isomorphism between spaces $L^k(X,Y)$ and $L({}^kX, Y)$ defined by (5.22). Alternatively, the $k$th differential can be defined directly by induction as follows. Let $X, Y$ be Banach spaces, $U$ an open subset of $X$, and $F\colon U \to Y$. Let $d^1F := DF$ where it is defined, as usual. Suppose that for some $k \ge 1$, the $k$th differential of $F$ is defined in a neighborhood of $x \in U$ as a $k$-linear mapping $\mathrm{d}^kF(y) \in L({}^kX, Y)$. Then the $(k+1)$st differential of $F$ at $x$ is defined if there exists a $(k+1)$-linear bounded mapping $L$ from $X \times \cdots \times X$ ($k+1$ times) into $Y$ such that

$$\begin{aligned}\mathrm{d}^kF(x + x_{k+1})(x_1, \dots, x_k) &- \mathrm{d}^kF(x)(x_1, \dots, x_k)\\ &= L(x_1, \dots, x_{k+1}) + R(x; x_1, \dots, x_{k+1})\end{aligned}$$

for all $x_1, \ldots, x_k \in X$, $x_{k+1}$ in a neighborhood of $x$ and

$$\sup\left\{\|R(x, x_1, \ldots, x_{k+1})\|: \|x_1\| = \cdots = \|x_k\| = 1\right\} = o(\|x_{k+1}\|)$$

as $x_{k+1} \to 0$. Then the $(k+1)$st differential of $F$ at $x$ is defined to be $\mathrm{d}^{k+1}F(x) := L \in L(^{(k+1)}X, Y)$. In Zeidler [257, Proposition 4.19] it is proved that the definition of the $k$th differential $\mathrm{d}^k F(x)$ just given is equivalent to the one given by (5.23).

The Schwarz theorem on symmetry of differentials given by Theorem 5.27 is Theorem 7.9 in [32]. Differentiability of power series given by Theorem 5.28 is Theorem 11.12 in [32]. The principle of analytic continuation as in Theorem 5.36 is Theorem 9.4.5 in [42].

A. Alexiewicz and W. Orlicz [2] gave a definition of analytic function from an open set $U$ in a real Banach space $X$ into another Banach space $Y$ essentially as above in and around (5.32). Taylor [229, §4] discussed earlier notions of analyticity in real and complex Banach spaces. More on higher-order derivatives in normed spaces can be found in the books Chae [32] and Dieudonné [42].

*Notes on Section 5.3.* The Taylor formula with an integral remainder in Theorem 5.42 is Theorem 8.14.3 in [42]. The form of Taylor's theorem as in Theorem 5.44 is sometimes given W. H. Young's name. His expansion theorem in section 13 of [252] has this form when $X = Y = \mathbb{R}$. For the case of Banach spaces see Theorem 5.6.3 in [30].

Many authors write or assume that the composition $G \circ F$ of analytic functions between (connected) open sets in Banach spaces is analytic provided that the range of $F$ is included in the domain of $G$. Lemma 5.39 for power series from $\mathbb{R}$ into $\mathbb{R}$ with nonnegative coefficients is surely well known and very likely has been known for a century or more.

In the literature, often the Banach space $X$ is a Hilbert space, in which case the radius $\rho_u$ of uniform convergence equals $\overline{\rho}$ defined by (5.16) as shown in Corollary 5.23($e$). Following Cena [31], we call $\overline{\rho}$ the radius of restricted convergence. Also often, $\mathbb{K} = \mathbb{C}$. But for general real Banach spaces, we do not know whether Theorems 5.21 or 5.38 could have been formulated correctly before the possibility that $\overline{\rho} \neq \rho_u$ was first noticed, apparently by L. Nachbin in the late 1960s according to Chae [32, p. 44]. T. Nguyen [179] proves that in Theorem 5.21, the factor $\mathrm{e}^{-1}$ can be improved to $\mathrm{e}^{-1/2}$.

Cena [31, Definition 12] notes, without details, that a power series around $u$ represents an analytic function in the ball $B(u, \overline{\rho})$, as in our Theorem 5.21. Cena [31, Theorem 14] states a theorem on composition of analytic functions between real Banach spaces, which is as close as we could find to our Theorem 5.38 in the literature. We did not do a full search. Cena does not give a proof, but the references and discussion suggest an approach via a fact like our lemma 5.39.

*Notes on Section 5.4.* Schatten [205] in 1950 gave apparently the first systematic book-length exposition on tensor products of normed spaces. He cites earlier papers by J. von Neumann and himself separately and jointly. The projective tensor product norm (5.38) is defined in Schatten [205, pp. 36–37].

The norm $\|\cdot\|$ defined in Theorem 4.39 on the complexification of a real Banach algebra is the special case of the projective tensor product norm $\|\cdot\|_\pi$ defined by (5.38) where $X = \mathbb{C}$, but considered as a real Banach space, i.e. as the 2-dimensional real Hilbert space $\ell_2^2$, and $Y$ is the real Banach algebra $\mathbb{A}$.

# 6

# Nemytskii Operators on Some Function Spaces

## 6.1 Overview

For a nonempty set $S$, Banach spaces $X$ and $Y$ over a field $\mathbb{K} = \mathbb{R}$ or $\mathbb{C}$, and a nonempty open set $U \subset X$, let $\mathbb{G}$, $\mathbb{F}$, and $\mathbb{H}$ be vector spaces of functions acting from $S$ into $X$, from $U$ into $Y$, and from $S$ into $Y$, respectively. Usually, but not always, $\mathbb{F}$, $\mathbb{G}$, and $\mathbb{H}$ will be normed spaces. Given a function $(u, s) \mapsto \psi(u, s)$ from $U \times S$ into $Y$, and a function $g\colon S \to U$, the *Nemytskii operator* $N_\psi$ is defined by

$$N_\psi(g)(s) := (N_\psi g)(s) := \psi(g(s), s), \qquad s \in S. \tag{6.1}$$

Other authors call such an operator a superposition operator, e.g. Appell and Zabrejko [3]. We use the term Nemytskii operator (as many others have) partly to distinguish it from the two-function composition operator $(F, G) \mapsto F \circ G$ to be treated in Chapter 8. Recall also that for a linear operator $A$ we write $Ax := A(x)$. We will often apply this rule also when $A$ is a Nemytskii operator.

**Definition 6.1.** For a set $V \subset \mathbb{G}$ such that each $g \in V$ has values in $U$, we say that $N_\psi$ *acts* from $V$ into $\mathbb{H}$ if $N_\psi g \in \mathbb{H}$ for each $g \in V$.

Some of the questions we consider in this chapter and the next are under what conditions a Nemytskii operator acts from $V$ into $\mathbb{H}$, is continuous, Fréchet differentiable, or analytic.

A Nemytskii operator $N_\psi$ will be called *autonomous* if $\psi(u, s) \equiv F(u)$ for a function $F$ from $U$ into $Y$. Then $N_F g := N_\psi g \equiv F \circ g$. An autonomous Nemytskii operator is nonlinear unless $F$ is linear. Fréchet differentiability of the operator $V \ni g \mapsto F \circ g \in \mathbb{H}$ at $g = G$ is equivalent to that of $g \mapsto F \circ (G + g)$ at $g = 0$ in $\mathbb{G}$. This is equivalent to Fréchet differentiability of the Nemytskii operator $N_\phi$ at $g = 0$ for $\phi(u, s) := F(G(s) + u)$, $(u, s) \in U \times S$.

In the present chapter, $S$ will be any nonempty set or an interval. It will be a measure space in Chapter 7. Often $U = X = Y = \mathbb{K} = \mathbb{R}$ or $\mathbb{C}$. $\mathbb{G}$ will be

R.M. Dudley and R. Norvaiša, *Concrete Functional Calculus*, Springer Monographs in Mathematics, DOI 10.1007/978-1-4419-6950-7_6,

a space of bounded functions with sup norm in Section 6.2 and a $\mathcal{W}_p$ space in Section 6.4. For $\psi(\cdot,\cdot)$ also depending on $s$, Nemytskii operators $N_\psi$ will be considered on a Banach algebra of real-valued bounded functions in Section 6.3 and on $\mathcal{W}_p$ spaces in Section 6.5. This chapter is concerned with spaces of functions as opposed to equivalence classes for a measure. Chapter 7 will treat Nemytskii operators from spaces $L^s$ into $L^p$ on a measure space.

## 6.2 Remainders in Differentiability

In this section we give a simple bound for remainders in differentiability of an autonomous Nemytskii operator $N_F$ acting between spaces of bounded functions, provided the derivative of $F$ is a Hölder function. Hölder classes were already introduced in Section 1.4. Here is some more related notation.

Let $X, Y$ be Banach spaces over a field $\mathbb{K}$, let $B$ be a subset of $X$ with more than one element, and let $0 < \alpha \le 1$. Recall that $\mathcal{H}_\alpha(B;Y)$ is the space of all $\alpha$-Hölder functions $F\colon B \to Y$ with the seminorm $\|F\|_{(\mathcal{H}_\alpha)} = \|F\|_{B,(\mathcal{H}_\alpha)}$ defined by (1.18). As in (4.4), $\mathcal{H}_{\alpha,\infty}(B;Y)$ denotes the space of all bounded functions $F$ in $\mathcal{H}_\alpha(B;Y)$ with the norm

$$\|F\|_{\mathcal{H}_\alpha} := \|F\|_{B,\mathcal{H}_\alpha} := \|F\|_{\sup} + \|F\|_{(\mathcal{H}_\alpha)}, \tag{6.2}$$

where $\|F\|_{\sup} = \|F\|_{B,\sup} = \sup\{\|F(x)\|\colon x \in B\}$ is the supremum norm. Since completeness is straightforward to check, we have the following:

**Proposition 6.2.** *For any Banach spaces $X$ and $Y$, set $B \subset X$ containing more than one point, and $0 < \alpha \le 1$, $(\mathcal{H}_{\alpha,\infty}(B;Y), \|\cdot\|_{\mathcal{H}_\alpha})$ is a Banach space.*

To define a local $\alpha$-Hölder property let $U$ be a nonempty open subset of $X$ and let $\delta > 0$. Let $U_\delta$ be the set of all $x \in U$ such that $\|x\| \le 1/\delta$ and $y \in U$ whenever $\|x - y\| < \delta$. Then $U_\delta$ is closed and bounded (possibly empty), and $U_\delta \uparrow U$ as $\delta \downarrow 0$. Note that $U_\delta = \{x\colon \|x\| \le 1/\delta\}$ if $U = X$.

The following is straightforward to prove:

**Proposition 6.3.** *Let $U$ be an open subset of a Banach space and let $K$ be compact with $K \subset U$. Then for some $k = 1, 2, \ldots$, $K \subset B_k := U_{1/k}$.*

**Definition 6.4** (of $\mathcal{H}_\alpha^{\mathrm{loc}}$). Let $0 < \alpha \le 1$ and let $X$, $Y$ be Banach spaces. For a nonempty open set $U \subset X$, let $U_\delta$, $\delta > 0$, be as just defined. For $m = 1, 2, \ldots$, let $B_m := B_m(U) := U_{1/m}$. We say that a function $F\colon U \to Y$ is *$\alpha$-Hölder locally on $U$* if it is $\alpha$-Hölder on $B_m$ for all sufficiently large $m$. The class of all such functions is denoted by $\mathcal{H}_\alpha^{\mathrm{loc}}(U;Y)$. We write $F \in \mathcal{H}_\alpha^{\mathrm{loc}}$ if $U = X = Y = \mathbb{K}$.

Note that if $U = X$, $F$ is $\alpha$-Hölder locally on $X$ if and only if it is $\alpha$-Hölder on each bounded set in $X$.

**Definition 6.5** (of $\mathcal{H}_{1+\alpha}$ and $\mathcal{H}_{1+\alpha}^{\mathrm{loc}}$). Let $0 < \alpha \le 1$, let $X$, $Y$ be Banach spaces, and let $U$ be an open subset of $X$. For a function $F\colon U \to Y$, we write $F \in \mathcal{H}_{1+\alpha}(U;Y)$ if $F$ is Fréchet differentiable on $U$ and its derivative $DF$ is in $\mathcal{H}_\alpha(U;L(X,Y))$, that is, (1.18) holds with $F$ and $Y$ replaced by $DF$ and $L(X,Y)$, respectively. Also, we write $F \in \mathcal{H}_{1+\alpha}^{\mathrm{loc}}(U;Y)$ if $F$ is Fréchet differentiable on $U$ and $DF \in \mathcal{H}_\alpha^{\mathrm{loc}}(U;L(X,Y))$. If $U = X = Y = \mathbb{K}$, we write $\mathcal{H}_{1+\alpha} = \mathcal{H}_{1+\alpha}(\mathbb{K};\mathbb{K})$ and $\mathcal{H}_{1+\alpha}^{\mathrm{loc}} = \mathcal{H}_{1+\alpha}^{\mathrm{loc}}(\mathbb{K};\mathbb{K})$.

Note that if $F\colon U \to Y$ is in $\mathcal{H}_\alpha^{\mathrm{loc}}(U;Y)$ or in $\mathcal{H}_{1+\alpha}^{\mathrm{loc}}(U;Y)$ then $F$ is bounded on each nonempty subset $B_m(U) \subset U$.

Recall that as defined at the beginning of Section 6.1 we have a nonempty set $S$, Banach spaces $X$ and $Y$, and a nonempty open set $U \subset X$. Also, $\mathbb{G}$, $\mathbb{F}$, and $\mathbb{H}$ are vector spaces, which we now assume are normed spaces, of functions acting from $S$ into $X$, from $U$ into $Y$, and from $S$ into $Y$, respectively. For $F \in \mathbb{F}$ and $G \in \mathbb{G}$, if the autonomous Nemytskii operator $N_F$ is Fréchet differentiable at $G$ from $\mathbb{G}$ into $\mathbb{H}$, the remainder in the differentiability is defined by (5.1), that is,

$$\mathrm{Rem}_{N_F}(G,g) = N_F(G+g) - N_F(G) - (DN_F(G))g. \tag{6.3}$$

Thus $\mathrm{Rem}_{N_F}(G,g) \in \mathbb{H}$ for each $g \in \mathbb{G}$, and $\|\mathrm{Rem}_{N_F}(G,g)\| = o(\|g\|)$ as $\|g\| \to 0$. We will see that for $X = Y = \mathbb{K}$ the derivative $DN_F(G)$, if it exists, has the following form.

**Definition 6.6.** For an integer $k \ge 1$ and for vector spaces $\mathbb{G}_1, \dots, \mathbb{G}_k, \mathbb{H}$ of $\mathbb{K}$-valued functions on a set $S$, let $T$ be an operator from the Cartesian product $\mathbb{G}_1 \times \cdots \times \mathbb{G}_k$ into $\mathbb{H}$, and let $h$ be a $\mathbb{K}$-valued function on $S$. We say that $T$ is the *$k$-linear multiplication operator $M^k[h]$ induced by $h$* if

$$M^k[h](g_1,\dots,g_k) := hg_1g_2\cdots g_k = T(g_1,\dots,g_k) \in \mathbb{H}$$

for all $g_j \in \mathbb{G}_j$, $j = 1,\dots,k$. If $k = 1$, the linear operator $M^1[h]$ will be written as $M[h]$.

If $\mathbb{G}_1 = \cdots = \mathbb{G}_k = \mathbb{H} = \mathbb{B}$, a Banach algebra of functions, the $k$-linear multiplication operator $M^k[h]$ is defined and bounded for any $h \in \mathbb{B}$, as in any Banach algebra by (5.12).

The following rather easy differentiability fact will give bounds on sup norms of remainders for autonomous Nemytskii operators on open subsets of $\ell^\infty$ spaces. Recall that for a Banach space $X$, $\ell^\infty(S;X)$ is the space of all bounded $X$-valued functions on the set $S$ with the supremum norm. We will consider functions $G$ whose ranges are "well inside" an open set $U$ in the sense that $\mathrm{ran}\,(G) \subset B_m$ for some $m$.

**Proposition 6.7.** *Let $S$ be a nonempty set, let $X$, $Y$ be Banach spaces, let $U$ be an open subset of $X$, and let $F \in \mathcal{H}_{1+\alpha}^{\mathrm{loc}}(U;Y)$ for some $0 < \alpha \le 1$. Let $G \in \ell^\infty(S;X)$ be such that $ran(G) \subset B_m$ for some $m$. Then there is* *a neighborhood of zero $V$ in $\ell^\infty(S;X)$ such that the autonomous Nemytskii operator $N_F$ acts from $G+V$ into $\ell^\infty(S;Y)$, is Fréchet differentiable at $G$ from $\ell^\infty(S;X)$ into $\ell^\infty(S;Y)$, with derivative $g \mapsto (s \mapsto DF(G(s))g(s))$, and the remainder bound*

$$\|\mathrm{Rem}_{N_F}(G,g)\|_{\sup} \leq C\|g\|_{\sup}^{1+\alpha} \tag{6.4}$$

*holds for all $g \in V$, where $C := \|DF\|_{B_{2m},(\mathcal{H}_\alpha)}$. If in addition $X = Y = \mathbb{K}$ then $DN_F(G)$ is the linear multiplication operator $M[F'{\circ}G]$ from $\ell^\infty(S)$ into itself.*

*Proof.* Let $V$ be the set of all $g \in \ell^\infty(S;X)$ such that $\|g\|_{\sup} < 1/(2m)$. It is easy to see that $\mathrm{ran}\,(G+rg) \subset B_{2m} \subset U$ for each $g \in V$ and $r \in [0,1]$. Since $F$ is bounded on $B_{2m}$, $\|N_F(G+g)\|_{\sup} \leq \|F\|_{B_{2m},\sup} < \infty$ for each $g \in V$, and so $N_F$ acts from $G+V$ into $\ell^\infty(S;Y)$.

To show Fréchet differentiability of $N_F$ at $G$, for each $g \in \ell^\infty(S;X)$, let $Hg$ be the function $s \mapsto DF(G(s))g(s)$, $s \in S$. Then for each $s \in S$, $\|(Hg)(s)\| \leq \|DF\|_{B_m,\sup}\|g\|_{\sup}$, and so $H$ is a bounded linear operator from $\ell^\infty(S;X)$ into $\ell^\infty(S;Y)$. Since $F$ is continuously differentiable on $U$, by Lemma 5.40 we have for some constant $C < \infty$, any $s \in S$, and $g \in V$,

$$\begin{aligned}&\big\|\{N_F(G+g) - N_F(G) - Hg\}(s)\big\| \\ &\quad = \Big\|\int_0^1 \big[DF(G(s)+rg(s)) - DF(G(s))\big]g(s)\,\mathrm{d}r\Big\| \leq C\|g\|_{\sup}^{1+\alpha}\end{aligned}$$

since $DF \in \mathcal{H}_\alpha(B_{2m};L(X,Y))$. Thus $N_F$ is Fréchet differentiable at $G$ with the derivative $DN_F(G) = H$ and (6.4) holds.

If in addition $X = Y = \mathbb{K}$ then for each $s \in S$, $F'(G(s)) = DF(G(s))1$, and so the value $DF(G(s))g(s)$ is the product $F'(G(s))$ times $g(s)$. Since $\|F'{\circ}G\|_{\sup} \leq \|DF\|_{B_m,\sup} < \infty$, $DN_F(G) = H$ is the linear multiplication operator $M[F'{\circ}G]$, proving the proposition. □

## 6.3 Higher Order Differentiability and Analyticity

In this section Nemytskii operators are treated on Banach spaces of real-valued functions, then on Banach algebras of bounded $\mathbb{K}$-valued functions. We begin with necessary conditions for higher order Fréchet differentiability and analyticity of Nemytskii operators. The conditions for a Nemytskii operator $N_\psi$ are expressed in terms of higher order derivatives (more precisely, differentials as defined by (5.23)) of the function $\psi$, defined as follows. Recall the Banach space $L(^nX,Y)$ of bounded $n$-linear functions from $X^n$ into $Y$ defined in Section 5.1.

For functions $\psi$ used in defining Nemytskii operators $N_\psi$ we will be considering the $n$th Fréchet differential $\psi_u^{(n)}$ with respect to $u$ in $\psi(u,s)$, with $s$

held fixed. For $n = 1$ and $X = \mathbb{R}$, we could write in partial derivative notation $\psi_u^{(1)}(u, s) \equiv \partial\psi(u, s)/\partial u$. Notions of partial derivative can also be defined for Fréchet derivatives. We will not consider $\partial\psi(u, s)/\partial s$, however.

**Definition 6.8.** Let $S$ be a nonempty set, let $X$, $Y$ be Banach spaces over $\mathbb{K}$, let $U$ be a nonempty open subset of $X$, and let $\psi\colon U \times S \to Y$. Let $\psi_u^{(0)} \equiv \psi$ and $L(^0X, Y) \equiv Y$. For $n = 1, 2 \ldots$, $v \in U$ and $s \in S$, we will say that $\psi$ is (Fréchet) $u$-differentiable of order $n$ at $(v, s)$ if the function $u \mapsto f(u) := \psi(u, s)$, $u \in U$, is (Fréchet) differentiable of order $n$ at $v$, and then we let $\psi_u^{(n)}(v, s) := d^n f(v) \in L(^nX, Y)$ be the $n$th differential at $(v, s)$. Thus, $\psi$ is *Fréchet u-differentiable* at $(v, s)$ if it is Fréchet $u$-differentiable of order 1 there. Also, for $\psi$ to be Fréchet $u$-differentiable of order $n$ on a subset $W \subset U \times S$ will mean that for each $(v, s) \in W$, $(u, s) \in W$ for $u$ in a neighborhood of $v$ and $\psi$ is Fréchet $u$-differentiable of order $n$ at $(v, s)$. Let $g$ be a function on $S$ with range in $U$. Then we will say $\psi$ is Fréchet $u$-differentiable of order $n$ *on the graph of* $g$ or, briefly, *at* $g$ if $\psi$ is Fréchet $u$-differentiable of order $n$ at $(g(s), s)$ for all $s \in S$.

For $X = Y = \mathbb{K}$ we identify $\psi_u^{(n)}(v, s)$ with its value at $1^{\otimes n}$ and say that $\psi$ is $u$-differentiable of order $n$ at $(v, s)$. Therefore $\psi$ is $u$-differentiable of order $n + 1$ at $(v, s)$ if $\psi_u^{(n)}(u, s)$ is defined for $u$ in a neighborhood of $v$ and the partial derivative $\psi_u^{(n+1)}(v, s) := \partial\psi_u^{(n)}(u, s)/\partial u\big|_{u=v}$ exists.

Next we show that Fréchet differentiability of order $n$ of the Nemytskii operator $N_\psi$ at a function $G$ implies Fréchet $u$-differentiability of order $n$ of $\psi$ at $G$, and their differentials are closely related.

**Proposition 6.9.** *Let $S$ be a nonempty set, let $X$, $Y$ be Banach spaces, and let $\mathbb{G}$ and $\mathbb{H}$ be Banach spaces of $X$-valued and $Y$-valued functions on $S$, respectively, such that $\mathbb{G}$ contains the constant functions $x1(\cdot)$, $x \in X$, and $x \mapsto x1(\cdot)$ is a bounded operator from $X$ into $\mathbb{G}$. Assume that for each $s \in S$, the evaluation $h \mapsto h(s)$ (which is clearly linear) is a bounded operator from $\mathbb{H}$ into $Y$. Let $U$ be a nonempty open subset of $X$, and let $\psi\colon U \times S \to Y$. Let $G \in \mathbb{G}$ be such that there is a neighborhood $V$ of $G$ in $\mathbb{G}$ with* $\operatorname{ran}(g) \subset U$ *for each $g \in V$. For $k \geq 1$, if the Nemytskii operator $N_\psi$ acts from $V$ into $\mathbb{H}$ and is Fréchet differentiable of order $k$ at $G$, then $\psi$ is Fréchet $u$-differentiable of order $k$ at $G$ and the $k$th differential $\mathrm{d}^k N_\psi(G) \in L(^k\mathbb{G}, \mathbb{H})$ is the $k$-linear operator*

$$\mathrm{d}^k N_\psi(G)(g_1, \ldots, g_k)(s) = \psi_u^{(k)}(G(s), s)(g_1(s), \ldots, g_k(s)) \tag{6.5}$$

*for each $s \in S$ and any $g_1, \ldots, g_k \in \mathbb{G}$. If in addition $X = Y = \mathbb{K}$ then $\psi$ is $u$-differentiable of order $k$ at $G$ and the $k$th differential $\mathrm{d}^k N_\psi(G) \in L(^k\mathbb{G}, \mathbb{H})$ is the $k$-linear multiplication operator $M^k[N_\phi(G)]$ for $\phi = \psi_u^{(k)}$, that is,*

$$\mathrm{d}^k N_\psi(G)(g_1, \ldots, g_k)(s) = \psi_u^{(k)}(G(s), s) g_1(s) \cdots g_k(s) \tag{6.6}$$

*for each $s \in S$ and any $g_1, \ldots, g_k \in \mathbb{G}$.*

*Proof.* Let $k = 1$. For any $x \in X$, we have $x1(\cdot) \in \mathbb{G}$ and $\|x1(\cdot)\|_{\mathbb{G}} \to 0$ as $\|x\| \to 0$. Thus if $\|x\|$ is small enough then $G + x1(\cdot) \in V$ and the remainder in the differentiation of $N_\psi$ has values

$$\mathrm{Rem}_{N_\psi}(G, x1(\cdot))(s) = \psi(G(s) + x, s) - \psi(G(s), s) - (DN_\psi(G)x1(\cdot))(s)$$

for each $s \in S$. Since $x \mapsto x1(\cdot)$ is a bounded operator from $X$ into $\mathbb{G}$ and $\|x1(\cdot)\|_{\mathbb{G}} > 0$ if $x \neq 0$, we have $\|\mathrm{Rem}_{N_\psi}(G, x1(\cdot))\|_{\mathbb{H}} = o(\|x\|)$ as $\|x\| \to 0$. Since the evaluation operators are also bounded, we then have for each $s \in S$,

$$\|\psi(G(s) + x, s) - \psi(G(s), s) - (DN_\psi(G)x1(\cdot))(s)\| = o(\|x\|)$$

as $\|x\| \to 0$. Thus for each $s \in S$, $\psi$ is Fréchet $u$-differentiable at $(G(s), s)$ and $\psi_u^{(1)}(G(s), s) \in L(X, Y)$ is the composition $x \mapsto x1(\cdot) \mapsto DN_\psi(G)x1(\cdot) \mapsto (DN_\psi(G)x1(\cdot))(s) \in Y$, $x \in X$. If $X = Y = \mathbb{K}$, then for each $s \in S$, $\psi$ is $u$-differentiable at $(G(s), s)$ and $\psi_u^{(1)}(G(s), s) = (DN_\psi(G)1(\cdot))(s) \in \mathbb{K}$. To show (6.5) and (6.6) for $k = 1$, let $g \in \mathbb{G}$ and $s \in S$. For each $t \in \mathbb{K}$ such that $G(s) + tg(s) \in U$, let $\phi(t) := \psi(G(s) + tg(s), s)$. Then $\phi$ is defined in a neighborhood of 0 and as $t \to 0$,

$$\|\phi(t) - \phi(0) - (DN_\psi(G)tg)(s)\| = \|\mathrm{Rem}_{N_\psi}(G, tg)(s)\| = o(\|tg\|_{\mathbb{G}}) = o(|t|).$$

On the other hand, $\phi$ is the composition of the functions $t \mapsto G(s) + tg(s)$ and $u \mapsto \psi(u, s)$. Thus by the chain rule (Theorem 5.1), $\phi'(0) = \psi_u^{(1)}(G(s), s)g(s)$ and

$$\|\phi(t) - \phi(0) - t\psi_u^{(1)}(G(s), s)g(s)\| = o(|t|)$$

as $|t| \to 0$. Due to uniqueness of the derivative $\phi'(0)$, we have $(DN_\psi(G)g)(s) = \psi_u^{(1)}(G(s), s)g(s)$ for each $g \in \mathbb{G}$ and $s \in S$, proving (6.5) and (6.6) for $k = 1$.

We will show by induction on $k$ that the conclusions hold for all functions $\Gamma$, in place of $G$, satisfying the hypotheses on $G$. Suppose that the conclusion holds for $k$ with $k \geq 1$ and $N_\psi$ is Fréchet differentiable of order $k+1$ at $G$. By definition of the $(k+1)$st differential, $\mathrm{d}^k N_\psi(\Gamma)$ is defined for all $\Gamma$ in some neighborhood of $G$. By the induction assumption, $\psi_u^{(k)}(\Gamma(s), s)$ is defined for all such $\Gamma$ and all $s \in S$. Since $\|(G + x1(\cdot)) - G\|_{\mathbb{G}} = \|x1(\cdot)\|_{\mathbb{G}} \to 0$ as $x \to 0$ in $X$, it then follows that for each $s \in S$, $f(u) := \psi_u^{(k)}(u, s) \in L(^k X, Y)$ is defined for all $u$ in a neighborhood $W$ of $G(s)$. For any $s \in S$, any $x \in X$ such that $G(s) + x \in W$, and any $x_1, \ldots, x_k$ in the closed unit ball $\bar{B}(0, 1) \subset X$, by (6.5) and Proposition 5.25($b$) applied to $F = N_\psi$, we have

$$\begin{aligned}
&\big\|\big[f(G(s) + x) - f(G(s))\big](x_1, \ldots, x_k) - D\mathrm{d}^k N_\psi(G)(x1)(x_1 1, \ldots, x_k 1)(s)\big\| \\
&\quad = \big\|\mathrm{Rem}_{\mathrm{d}^k N_\psi}(G, x1)(x_1 1, \ldots, x_k 1)(s)\big\| \\
&\quad \leq \sup\Big\{\mathrm{Rem}_{d^k N_\psi}(G, x1)(g_1, \ldots, g_k)(s) \colon \|g_1\| \leq \|1\|, \ldots, \|g_k\| \leq \|1\|\Big\} \\
&\quad = o(\|x1\|) = o(\|x\|) \qquad \text{as } \|x\| \to 0,
\end{aligned}$$

where $1 \equiv 1(\cdot)$. For $x_1, \ldots, x_{k+1} \in X$ and $s \in S$, let

$$\psi_u^{(k+1)}(G(s), s)(x_1, \ldots, x_{k+1}) := d^{k+1} N_\psi(G)(x_1 1(\cdot), \ldots, x_{k+1} 1(\cdot))(s),$$

and so $\psi_u^{(k+1)}(G(s), s) \in L({}^{k+1}X, Y)$. By Proposition 5.25(*b*) once again, $Dd^k N_\psi(G)g = d^{k+1} N_\psi(G)(g, \ldots)$ for all $g \in \mathbb{G}$, and so $Df(G(s))x = \psi_u^{(k+1)}(G(s), s)(x, \ldots)$ for all $x \in X$ and $s \in S$. By Proposition 5.25(*a*) now applied to $F = \psi(\cdot, s)$, it then follows that $\psi$ is Fréchet $u$-differentiable of order $k+1$ at $G$ and (6.5) holds with $k+1$ instead of $k$, setting $x_j := g_j(s)$, $j = 1, \ldots, k+1$. In the case $X = Y = \mathbb{K}$, for each $s \in S$, it follows that $\psi$ is $u$-differentiable of order $k+1$ at $G$ with $\psi_u^{(k+1)}(G(s), s) = d^{k+1} N_\psi(G)(1(\cdot), \ldots, 1(\cdot))(s)$. Thus the conclusion of the proposition holds for any $k \geq 1$ by induction. □

The following shows that a power series expansion of a Nemytskii operator $N_\psi$ around $G$ (as defined in Definition 5.8) has a simple form and implies a Taylor expansion of $\psi(u, s)$ around $u = G(s)$ for each $s$ (as defined before (5.32)). As before, we write $x^0 := 1(\cdot)$ if $x \in \mathbb{K}^S$ or $x^0 := \mathbb{I}$ if $x$ is in a unital Banach algebra.

**Theorem 6.10.** *Let $\mathbb{G}$ and $\mathbb{H}$ be Banach spaces of functions on a set $S$ such that $\mathbb{G}$ contains the constant functions on $S$. Let $\psi\colon \mathbb{K} \times S \to \mathbb{K}$, and let $V$ be an open subset of $\mathbb{G}$. Suppose that for some $G \in V$, the Nemytskii operator $N_\psi$ acts from $V$ into $\mathbb{H}$ and has a power series expansion around $G$ with radius of (not necessarily uniform) convergence $r > 0$. Then there is a sequence $\{h_k\}_{k\geq 0} \subset \mathbb{H}$ such that for each $g \in \mathbb{G}$ with $\|g\|_{\mathbb{G}} < r$, the series $\sum_{k\geq 0} h_k g^k$ converges in $\mathbb{G}$, and for all $s \in S$,*

$$N_\psi(G+g)(s) = N_\psi(G)(s) + \sum_{k=1}^{\infty} h_k(s) g(s)^k. \tag{6.7}$$

*We have*

$$\limsup_{k\to\infty} \|h_k\|_{\mathbb{H}}^{1/k} < \infty. \tag{6.8}$$

*Moreover, $\psi(u, s)$ has a Taylor expansion in $u$ around $u = G(s)$ for each $s \in S$ and $h_k(s) = \psi_u^{(k)}(G(s), s)/k!$ for all $k \geq 1$ and all $s \in S$.*

*Proof.* We can assume that $B(G, r) \subset V$. By assumption, for each $k \geq 1$, there is a $k$-homogeneous polynomial $P_G^k$ from $\mathbb{G}$ into $\mathbb{H}$ such that

$$N_\psi(G+g) = N_\psi(G) + \sum_{k=1}^{\infty} P_G^k(g)$$

and the series converges in $\mathbb{H}$ for each $g \in B(0, r)$. Let $h_0 := N_\psi(G)$. Recalling that the constant function $1 = 1(\cdot)$ is in $\mathbb{G}$, let $h_k := P_G^k(1) \in \mathbb{H}$ for each $k \geq 1$.

For any $g \in \mathbb{G}$ and $s \in S$ such that $c := g(s) \neq 0$, there is a $\delta_0 > 0$ such that $\|\delta g\|_{\mathbb{G}} < r$ and $\|\delta c1\|_{\mathbb{G}} < r$ for each $0 \leq \delta \leq \delta_0$. Since each $P_G^k$ is a $k$-homogeneous polynomial, we have

$$\begin{aligned} N_\psi G(s) &+ \sum_{k=1}^{\infty} P_G^k(g)(s)\delta^k \\ &= N_\psi(G+\delta g)(s) = \psi(G(s)+\delta c, s) = N_\psi(G+\delta c1)(s) \\ &= h_0(s) + \sum_{k=1}^{\infty} h_k(s)(\delta c)^k = \sum_{k=0}^{\infty} h_k(s)\delta^k g(s)^k. \end{aligned}$$

All these power series converge in $\mathbb{K}$ absolutely, and due to the uniqueness of such series (Lemma 2.95), $P_G^k(g)(s) = h_k(s)g(s)^k$ for all $g \in \mathbb{G}$ and all $s \in S$, also when $g(s) = 0$, proving the first conclusion. To prove (6.8), for some $u \neq 0$ in $\mathbb{K}$, taking $g$ as the constant function $u$, the series $\sum_{k=1}^{\infty} u^k h_k$ converges in $\mathbb{H}$. Then $\sup_k |u|^k \|h_k\| < \infty$. Taking $k$th roots, (6.8) follows. To prove the last conclusion, for each $s \in S$ and some $\delta > 0$, if $|v| < \delta$ then

$$\psi(G(s)+v, s) = N_\psi(G+v1)(s) = \sum_{k=0}^{\infty} h_k(s)v^k,$$

and so the function $u \mapsto \psi(u,s)$ has a Taylor expansion around $u = G(s)$ for each $s \in S$. A (scalar-valued) power series is an analytic (holomorphic) function inside its circle of convergence, and so $h_k(s) = \psi_u^{(k)}(G(s),s)/k!$ for all $k \geq 1$ and all $s \in S$ by Proposition 2.100, proving the theorem. □

The power series expansion (6.7) has the form of (5.14), which gives the following:

**Corollary 6.11.** *If in addition to the hypotheses of Theorem* 6.10, $V = \mathbb{G} = \mathbb{H} = \mathbb{B}$, *a unital Banach algebra of functions, then* (6.7) *is the* $\mathbb{B}$*-Taylor expansion of* $N_\psi$ *around* $G$ *and* $N_\psi$ *is* $\mathbb{B}$*-analytic on* $\{h \in \mathbb{B}\colon \|h - G\| < r\}$.

*Proof.* Under the present hypotheses, the power series (6.7) satisfies Definition 5.17 of a $\mathbb{B}$-Taylor expansion. Thus $N_\psi$ is $\mathbb{B}$-analytic on $\{h \in \mathbb{B}\colon \|h-G\| < r\}$ by Theorem 5.30($b$). □

Using Theorem 6.10 once again we show next that analyticity of a Nemytskii operator $N_\psi$ implies analyticity of the function $u \mapsto \psi(u,\cdot)$.

**Corollary 6.12.** *Let* $\mathbb{G}$ *and* $\mathbb{H}$ *be Banach spaces of functions on a set* $S$ *such that* $\mathbb{G}$ *contains the constant functions on* $S$ *and for some* $K < \infty$, $\|\cdot\|_{\sup} \leq K\|\cdot\|_{\mathbb{G}}$ *on* $\mathbb{G}$. *Let* $\psi\colon \mathbb{K}\times S \to \mathbb{K}$, *and let* $V := \{g \in \mathbb{G}\colon \|g\|_{\sup} < M\}$ *for some* $M > 0$. *If the Nemytskii operator* $N_\psi$ *acts from* $V$ *into* $\mathbb{H}$ *and is* *analytic on $V$, then the function $u \mapsto \psi(u,\cdot) \in \mathbb{H}$ is analytic on the open ball $B(0,M) \subset \mathbb{K}$. Moreover, for each $u \in B(0,M)$, there is an $r > 0$ such that*

$$\psi(u+x,\cdot) = \sum_{k=0}^{\infty} \psi_u^{(k)}(u,\cdot)\frac{x^k}{k!}$$

*and the series converges in $\mathbb{H}$ for all $x \in B(0,r)$.*

*Proof.* The set $V$ is open in $\mathbb{G}$ since the sup norm is continuous on $\mathbb{G}$. Since $V$ contains the constant functions $u1(\cdot)$ with $|u| < M$ and $N_\psi$ maps $V$ into $\mathbb{H}$, $\psi(u,\cdot) \in \mathbb{H}$ for each $u \in B(0,M)$. The desired analyticity follows from Theorem 6.10 applied also to each $G = u1(\cdot)$. The second part of the conclusion follows from (6.7) applied to $g = x1(\cdot)$. □

For the next result recall the definitions of spectral radius $r(\cdot)$ given by (4.5) and (Banach) algebra of functions, Definitions 4.3 and 4.11.

**Theorem 6.13.** *Let $(\mathbb{B}, \|\cdot\|)$ be a unital Banach algebra of $\mathbb{K}$-valued functions on a set $S$. For each $s \in S$ and $f \in \mathbb{B}$, let $e_s(f) := f(s)$. Then, with $\mathbb{K} = \mathbb{C}$ except that (b) also holds for $\mathbb{K} = \mathbb{R}$,*

*(a) $e_s$ is a character on $\mathbb{B}$ for each $s \in S$;*
*(b) for all $f \in \mathbb{B}$, $\|f\|_{\sup} \le r(f) \le \|f\|$;*
*(c) $\|f\|_{\sup} = r(f)$ if the closure of $\operatorname{ran}(f)$ equals $\sigma(f)$;*
*(d) if $S$ is a topological space and the functions in $\mathbb{B}$ are continuous, then $s \mapsto e_s$ is continuous from $S$ into $\mathcal{M}(\mathbb{B})$ with Gelfand topology.*

*Proof.* By Remark 4.12, the identity in $\mathbb{B}$ is the constant function $1(\cdot)$. Then $(a)$ follows directly. By Theorem 4.24 and (4.5), for each $f \in \mathbb{B}$ and $s \in S$,

$$|f(s)| = |e_s(f)| \le r(f) \le \|f\|,$$

so $(b)$ follows for $\mathbb{K} = \mathbb{C}$.

If $\mathbb{K} = \mathbb{R}$, let $(\mathbb{B}_{\mathbb{C}}, \|\!|\cdot|\!\|)$ be the Banach algebra complexification of $\mathbb{B}$ given by Theorem 4.39, which is also unital. We have the identification $f \leftrightarrow (f,0)$ of $\mathbb{B}$ with a real subalgebra of $\mathbb{B}_{\mathbb{C}}$, where $\|f\| = \|\!|(f,0)|\!\|$ by Theorem 4.39$(c)$. For all $n = 1, 2, \ldots$, we have $\|f^n\| = \|\!|(f^n,0)|\!\| = \|\!|(f,0)^n|\!\|$. Thus $r(f) = r((f,0))$ and by the case $\mathbb{K} = \mathbb{C}$,

$$\|f\|_{\sup} = \|(f,0)\|_{\sup} \le r((f,0)) \le \|\!|(f,0)|\!\| = \|f\|,$$

proving $(b)$ for $\mathbb{K} = \mathbb{C}$.

For $(c)$ suppose $\overline{\operatorname{ran}(f)} = \sigma(f)$. Note that $\sup\{|z|\colon z \in \overline{\operatorname{ran}(f)}\} = \|f\|_{\sup}$, and so $\|f\|_{\sup} = r(f)$ holds by Theorem 4.20. For $(d)$, if $s(i) \to s$ in $S$, then $f(s(i)) \to f(s)$ for all $f \in \mathbb{B}$, so $e_{s(i)} \to e_s$ as stated. □

It follows that if $\|f_n - f\| \to 0$ in a unital Banach algebra $\mathbb{B}$ of functions, then $f_n \to f$ not only pointwise, as assumed in Proposition 6.9, but uniformly.

**Remark 6.14.** Any Banach space $(X, \|\cdot\|)$ becomes a Banach algebra with the trivial multiplication defined by $xy = 0$ for all $x, y \in X$. Of course such an algebra is not unital. It has $r(x) = 0$ for each $x \in X$. $X$ may be a space of real- or complex-valued functions which need not be bounded. Thus in Theorem 6.13($b$), the hypothesis that $\mathbb{B}$ is unital cannot simply be removed.

**Proposition 6.15.** *Let $\mathbb{B}$ be a unital Banach algebra of functions on a set $S$. For a function $h\colon S \to \mathbb{K}$ and $k = 1, 2, \ldots$, $(g_1, \ldots, g_k) \mapsto hg_1 \cdots g_k$ defines a bounded $k$-linear operator from $\mathbb{B} \times \cdots \times \mathbb{B}$ into $\mathbb{B}$ if and only if $h \in \mathbb{B}$.*

*Proof.* "If" is obvious, and "only if" follows from setting $g_1 = g_2 = \cdots = g_k = \mathbb{1} = 1(\cdot)$. ☐

If the operator defined in the preceding proposition is bounded then it is the operator $M^k[h]$ defined by (5.12).

**Definition 6.16.** If $\mathbb{B}$ is a unital Banach algebra of functions on a set $K$, where $K$ is a compact Hausdorff space and $\mathbb{B} \subset C(K; \mathbb{C})$, then $\mathbb{B}$ will be called a *Banach function algebra* on $K$. A Banach function algebra $\mathbb{B}$ on $K$ is called *natural* if the evaluation map $s \mapsto e_s$ takes $K$ onto $\mathcal{M}(\mathbb{B})$.

In the preceding definition, recall that by definition of Banach algebras of functions (Definition 4.11), $\mathbb{B}$ strongly separates points of $K$. Thus the evaluation map $s \mapsto e_s$ is 1-to-1 as well as onto. Also the map $s \mapsto e_s$ is continuous since $\mathbb{B} \subset C(K; \mathbb{C})$. Thus it is a homeomorphism (e.g. [53, Theorem 2.2.11]).

Recall that any unital Banach algebra of functions on $S$ contains the constant functions and its identity element $\mathbb{1} = 1(\cdot)$ (Remark 4.12). Before giving examples of natural Banach algebras of functions (Corollary 6.23 below) we prove next some important properties of such algebras.

**Theorem 6.17.** ($a$) *If $\mathbb{B}$ is any Banach algebra of $\mathbb{C}$-valued functions on a set $S$ and $f \in \mathbb{B}$, then* $\mathrm{ran}\,(f) \subset \mathrm{ran}\,(\widehat{f})$.

($b$) *For any natural Banach function algebra $\mathbb{B}$ on $K$, a compact Hausdorff space, and any $f \in \mathbb{B}$,* $\mathrm{ran}\,(f) = \mathrm{ran}\,(\widehat{f}) = \sigma(f)$. *Also* $r(f) = \|f\|_{K,\mathrm{sup}}$.

*Proof.* Part ($a$) is clear since for each $s \in S$, the evaluation $e_s$ is a character of $\mathbb{B}$. The first statement in ($b$) then follows from the definitions and Theorem 4.29. For the second one, by Corollary 4.30, $r(f) = \|\widehat{f}\|_{\mathrm{sup}} = \|f\|_{K,\mathrm{sup}}$, proving the theorem. ☐

Recall Definition 4.28 of the Gelfand transform. A commutative unital complex Banach algebra $(\mathbb{A}, \|\cdot\|)$ is called *semisimple* if the Gelfand transform $x \mapsto \widehat{x}$ is 1-to-1 from $\mathbb{A}$ into $C(\mathcal{M}(\mathbb{A}), \mathbb{C})$. Then $x \mapsto \widehat{x}$ is an algebra isomorphism, in other words a 1-to-1 function preserving addition and multiplication.

A commutative unital complex Banach algebra need not be semisimple, as the following shows:

**Example 6.18.** Consider the set $\mathbb{A}$ of $2 \times 2$ matrices of the form

$$T = \begin{pmatrix} a & b \\ 0 & a \end{pmatrix}$$

for any $a, b \in \mathbb{C}$, with usual addition and multiplication of matrices. Then $\mathbb{A}$ is a unital commutative algebra over $\mathbb{C}$. It becomes a Banach algebra with the usual matrix (operator) norm $\|T\| := \sup\{\|Tx\|\colon x \in \mathbb{C}^2, \|x\| \le 1\}$. Clearly, $\mathbb{A}$ has just one proper, hence maximal, ideal, composed of the matrices $T$ with $a = 0$. Thus by Corollary 4.27, there is only one character $\phi$, with $\phi(T) = a$ for all such $T$, $a$. Clearly $\mathbb{A}$ is not semisimple.

**Proposition 6.19.** *Let $(\mathbb{A}, \|\cdot\|_{\mathbb{A}})$ be a commutative unital complex Banach algebra. Then the range $\widehat{\mathbb{A}}$ of the Gelfand transform with the norm $\|\widehat{x}\|_{\widehat{\mathbb{A}}} := \|x\|_{\mathbb{A}}$ is a Banach function algebra on $\mathcal{M}(\mathbb{A})$. If $\mathbb{A}$ is semisimple, then $\widehat{\mathbb{A}}$ is natural.*

*Proof.* The set $\mathcal{M}(\mathbb{A})$ of all characters on $\mathbb{B}$ with Gelfand topology is a compact Hausdorff space as was stated after Definition 4.28. The set $\widehat{\mathbb{A}} = \{\widehat{f}\colon f \in \mathbb{A}\}$ always separates points of $\mathcal{M}(\mathbb{A})$ since for any $\phi \neq \psi$ in $\mathcal{M}(\mathbb{A})$, $\phi(f) \neq \psi(f)$ for some $f \in \mathbb{A}$, so $\widehat{f}(\phi) \neq \widehat{f}(\psi)$. Also, $\mathbb{A}$ strongly separates points of $\mathcal{M}(\mathbb{A})$ since for each $\phi \in \mathcal{M}(\mathbb{A})$, $\widehat{\mathbb{I}}(\phi) = \phi(\mathbb{I}) = 1 \neq 0$. Since each $\widehat{f}$ is a continuous complex-valued function on $\mathcal{M}(\mathbb{A})$, $\widehat{\mathbb{A}}$ is a Banach function algebra on $\mathcal{M}(\mathbb{A})$ with the norm $\|\cdot\|_{\widehat{\mathbb{A}}}$ and identity $\widehat{\mathbb{I}} = 1(\cdot)$.

For each $\phi \in \mathcal{M}(\widehat{\mathbb{A}})$ and $f \in \mathbb{A}$, let $\widetilde{\phi}(f) := \phi(\widehat{f})$. If $\mathbb{A}$ is semisimple, then $\widetilde{\phi} \in \mathcal{M}(\mathbb{A})$ since the Gelfand transform is an algebra homomorphism and a 1-to-1 mapping onto its range, and so $\widehat{\mathbb{A}}$ is natural, proving the proposition. $\square$

Next the class $\mathcal{W}_p(J; X)$ of all $X$-valued functions of bounded $p$-variation on an interval $J$ will be imbedded into the set $C(K; X)$ of continuous functions on a compact Hausdorff space $K$. Partitions and the class $\mathcal{W}_p$ for any linearly ordered set were defined in Section 1.4. For an interval $J = [a, b]$ with $-\infty < a < b < \infty$, let

$$J^{\pm} := \{a, a+, b-, b\} \cup \bigcup_{a<x<b} \{x-, x, x+\}. \tag{6.9}$$

On $J^{\pm}$ we put the linear ordering:

$$a < a+ < x- < x < x+ < y- < y < y+ < b- < b$$

whenever $a < x < y < b$. $J^{\pm}$ has an *interval topology* with a subbase given by all sets $\{\xi\colon \xi < \eta\}$ and $\{\xi\colon \xi > \eta\}$ for all $\eta \in J^{\pm}$. Then $J^{\pm}$ with the interval topology is a compact, non-metrizable, Hausdorff space. Each regulated function $f$ on $J$, and thus each function $f \in \mathcal{W}_p(J)$, has a unique extension to a function $\tilde{f}$ on $J^{\pm}$ which is continuous on $J^{\pm}$ for the interval topology. Let $I\colon \mathcal{R}(J) \to C(J^{\pm})$ be the mapping defined by $I(f) := \tilde{f}$.

**Proposition 6.20.** *Let $X$ be a Banach space over $\mathbb{K}$ and let $J$ be a nondegenerate closed interval.*

(*a*) *For each $f \in \mathcal{R}(J;X)$, the closure $\overline{\operatorname{ran}(f)} = \operatorname{ran}\tilde{f}$ and is compact.*
(*b*) *The mapping $I$ is an isometry between $\mathcal{R}(J;X)$ and $C(J^{\pm};X)$, both with supremum norm.*
(*c*) *for $1 \le p < \infty$, the mapping $I$ when restricted to $\mathcal{W}_p(J;X)$ is an isometry between $\mathcal{W}_p(J;X)$ and $\mathcal{W}_p(J^{\pm};X)$.*
(*d*) *The spectra $\sigma(f)$ and $\sigma(\tilde{f})$ are equal for each $f \in \mathcal{R}(J;\mathbb{C})$ and for each $f \in \mathcal{W}_p(J;\mathbb{C})$ with $1 \le p < \infty$.*

*Proof.* In (*a*), clearly $\overline{\operatorname{ran}(f)} = \operatorname{ran}\tilde{f}$, and so $\|f\|_{J,\sup} = \|\tilde{f}\|_{J^{\pm},\sup}$ for any regulated function $f\colon J \to X$. Since $I$ takes $\mathcal{R}(J;X)$ onto $C(J^{\pm};X)$, (*b*) holds. Since $J^{\pm}$ is compact, so is $\operatorname{ran}\tilde{f} = \overline{\operatorname{ran}(f)}$ and (*a*) is proved.

For (*c*), for any such $f$, we have $v_p(f;J) \le v_p(\tilde{f};J^{\pm})$. For the converse inequality, let $\kappa = \{x_i\}_{i=0}^n$ be a partition of $J^{\pm}$. If for some $x \in J$ and $i$, $x_i = x-$ or $x_i = x+$, $\tilde{f}(x_i)$ can be approximated arbitrarily closely by $f(y_i)$ with $y_i \in J$. Taking $y_i := x_i$ for all other $i$, one can make a partition $\lambda = \{y_i\}_{i=0}^n$ of $J$ so that $s_p(f;\lambda)$ is arbitrarily close to $s_p(\tilde{f};\kappa)$, proving the converse inequality, and so $v_p(f;J) = v_p(\tilde{f};J^{\pm})$. Thus when restricted to $\mathcal{W}_p(J;X)$, $I$ is an isometry from $\mathcal{W}_p(J;X)$ onto $\mathcal{W}_p(J^{\pm};X)$, proving (*c*).

For (*d*), let $f \in \mathcal{W}_p(J;\mathbb{C})$. If $z \in \mathbb{C} \setminus \sigma(f)$ then let $g := (z1(\cdot) - f)^{-1} \in \mathcal{W}_p(J;\mathbb{C})$ by Definition 4.17 of $\sigma(\cdot)$. Clearly $\tilde{g} = (z1(\cdot) - \tilde{f})^{-1} \in \mathcal{W}_p(J^{\pm};\mathbb{C})$, and so $z \in \mathbb{C} \setminus \sigma(\tilde{f})$. Conversely, if $z \in \mathbb{C} \setminus \sigma(\tilde{f})$ then the restriction of $(z1(\cdot) - \tilde{f})^{-1} \in \mathcal{W}_p(J^{\pm};\mathbb{C})$ to $J$ is the inverse of $z1(\cdot) - f$ in $\mathcal{W}_p(J;\mathbb{C})$, and so $z \in \mathbb{C} \setminus \sigma(f)$, proving (*d*) and the proposition. □

For a complex-valued function $f = g + ih$ where $g$ and $h$ are real-valued, the *complex conjugate* is $\bar{f} = g - ih$. A set (vector space, algebra, etc.) $\mathcal{F}$ of complex-valued functions is called *self-adjoint* if $\bar{f} \in \mathcal{F}$ whenever $f \in \mathcal{F}$. Here are characterizations of natural algebras.

**Theorem 6.21.** *Let $\mathbb{B}$ be a Banach function algebra on $K$. Then*

(*a*) *$\mathbb{B}$ is natural if and only if for any $n = 1, 2, \ldots,$ and $f_1, \ldots, f_n \in \mathbb{B}$ such that $\bigcap_{j=1}^n f_j^{-1}(0) = \emptyset$, there exist $g_1, \ldots, g_n \in \mathbb{B}$ with $\sum_{j=1}^n g_j f_j = 1(\cdot)$;*

(*b*) *if $\mathbb{B}$ is self-adjoint, then it is natural if and only if for every $f \in \mathbb{B}$ with $f^{-1}(0) = \emptyset$, $1/f \in \mathbb{B}$.*

*Proof.* For (*a*), to prove "only if," let $\mathbb{B}$ be natural, $f_j \in \mathbb{B}$, and $\bigcap_{j=1}^n f_j^{-1}(0) = \emptyset$. Let $L := \{\sum_{j=1}^n g_j f_j : g_j \in \mathbb{B}\}$. Then $L$ is an ideal. If $L = \mathbb{B}$ then $1(\cdot) \in L$ as desired. Otherwise, $L$ is a proper ideal and is included in some maximal ideal $M$ by Theorem 4.4. By Corollary 4.27, there is a character $\phi$ of $\mathbb{B}$ with $M = \phi^{-1}(0)$. Since $\mathbb{B}$ is natural, there is some $s \in K$ with $\phi(f) = f(s)$ for all $f \in M$. Thus $s \in \bigcap_{j=1}^n f_j^{-1}(0)$, a contradiction.

To prove "if" in (*a*), suppose $\phi \in \mathcal{M}(\mathbb{B})$ is a character which is not of the form $e_s(f) = f(s)$ for any $s \in K$. Thus by Corollary 4.27 the maximal ideal $M = \phi^{-1}(0)$ is different from $M_s := \{f \in \mathbb{B} : f(s) = 0\}$ for each $s \in K$. So $M \not\subset M_s$ for each $s$ since $M$ is maximal. For each $s \in K$, take $h_s \in \mathbb{B}$ such that $\phi(h_s) = 0$ and $h_s(s) \neq 0$. The open sets $h_s^{-1}(\mathbb{C} \setminus \{0\})$ cover $K$, so there is a finite subcover. Thus there exist finitely many $f_i = h_{s_i}$, $i = 1, \ldots, n$, such that $\bigcap_{i=1}^n f_i^{-1}(0) = \emptyset$ and $\phi(f_i) = 0$ for each $i$. Take $g_1, \ldots, g_n \in \mathbb{B}$ with $\sum_{i=1}^n g_i f_i = 1(\cdot)$. Since $\phi$ is a character, we get $\phi(1(\cdot)) = 0$, a contradiction, so (*a*) is proved.

For (*b*), to prove "if," let $f_1, \ldots, f_n \in \mathbb{B}$ with $\bigcap_{j=1}^n f_j^{-1}(0) = \emptyset$. Then $f := \sum_{j=1}^n |f_j|^2 \in \mathbb{B}$ since $\mathbb{B}$ is self-adjoint. Thus $f^{-1}(0) = \emptyset$ so $1/f \in \mathbb{B}$. Let $g_j := \bar{f}_j/f \in \mathbb{B}$ for each $j$. Then $\sum_{j=1}^n g_j f_j = 1(\cdot)$, so by part (*a*), $\mathbb{B}$ is natural.

Conversely, if $\mathbb{B}$ is natural and $f^{-1}(0) = \emptyset$, let $n = 1$ and $f_1 = f$. Then (*a*) gives $g_1 = 1/f \in \mathbb{B}$ (self-adjointness is not needed in this direction), finishing the proof of Theorem 6.21. $\square$

Recall the class $\mathcal{H}_\alpha(B; Y)$ of all $\alpha$-Hölder functions from $B$ to $Y$ defined in Section 1.4. The following is easy to prove:

**Lemma 6.22.** *Let $X, Y$ be normed spaces over $\mathbb{K}$, let $B$ be a nonempty subset of $X$ and let $F \in \mathcal{H}_\beta(B; Y)$ for some $0 < \beta \le 1$. Let $f$ be a function on a nonempty interval $J$ with range in $B$. Then*

(*a*) *if $0 < p < \infty$ and $f \in \mathcal{W}_p(J; X)$, then $F \circ f \in \mathcal{W}_{p/\beta}(J; Y)$ and*

$$\|F \circ f\|_{(p/\beta)} \le \|F\|_{B,(\mathcal{H}_\beta)} \|f\|_{(p)}^\beta;$$

(*b*) *if $0 < \alpha \le 1$ and $f \in \mathcal{H}_\alpha(J; X)$, then $F \circ f \in \mathcal{H}_{\alpha\beta}(J; Y)$ and*

$$\|F \circ f\|_{(\mathcal{H}_{\alpha\beta})} \le \|F\|_{B,(\mathcal{H}_\beta)} \|f\|_{(\mathcal{H}_\alpha)}^\beta.$$

*Proof.* We can assume that $J$ is nondegenerate. First let $0 < p < \infty$ and $f \in \mathcal{W}_p(J; X)$. For any $s, t \in J$ we have

$$\begin{aligned} \|F(f(t)) - F(f(s))\|^{p/\beta} &\le \left[\|F\|_{B,(\mathcal{H}_\beta)} \|f(t) - f(s)\|^\beta\right]^{p/\beta} \\ &= \|F\|_{B,(\mathcal{H}_\beta)}^{p/\beta} \|f(t) - f(s)\|^p. \end{aligned} \tag{6.10}$$

Summing over any point partition of the interval $J$, then taking the supremum over partitions gives $v_{p/\beta}(F\circ f) \le \|F\|_{B,(\mathcal{H}_\beta)}^{p/\beta} v_p(f)$, proving $(a)$.

Now let $0 < \alpha \le 1$ and $f \in \mathcal{H}_\alpha(J;X)$. Then again for any $s,t \in J$, we have

$$\begin{aligned}\|F(f(t)) - F(f(s))\| &\le \|F\|_{B,(\mathcal{H}_\beta)}\|f(t) - f(s)\|^\beta \\ &\le \|F\|_{B,(\mathcal{H}_\beta)}\|f\|_{(\mathcal{H}_\alpha)}^\beta |t-s|^{\alpha\beta},\end{aligned}$$

proving $(b)$. The lemma is proved. □

**Corollary 6.23.** *Let $J := [a,b]$ with $-\infty < a < b < +\infty$. In each of the following cases, $\mathbb{A}$ is a natural Banach function algebra:*

*(a) $\mathbb{A} = C(K;\mathbb{C})$ with the norm $\|\cdot\|_{\sup}$, where $K$ is a compact Hausdorff space;*
*(b) for $0 < \alpha \le 1$, the Hölder space $\mathbb{A} = \mathcal{H}_{\alpha,\infty}(J;\mathbb{C})$ with the norm $\|\cdot\|_{\mathcal{H}_\alpha}$;*
*(c) the space $\mathbb{A} = C^n(J)$ of $C^n$ $\mathbb{C}$-valued functions $f$ on $J$, namely, $C^n$ functions on $(a,b)$ such that for $j = 0,1,\ldots,n$, with $f^{(0)} := f$, $f^{(j)}$ extends to a continuous function on $J$, with the norm*

$$\|f\|_{C^n} := \sum_{k=0}^n \frac{\|f^{(k)}\|_{\sup}}{k!};$$

*(d) for $1 \le p < \infty$, $\mathbb{A} = \mathcal{W}_p(J^\pm;\mathbb{C})$ with the norm $\|\cdot\|_{[p]}$.*

*Proof.* Each $\mathbb{A}$ given is a unital, self-adjoint algebra of continuous functions on a compact Hausdorff space. The given norms are Banach algebra norms, clearly for $(a)$, for $(b)$ by Proposition 4.14, and for $(d)$ by Proposition 4.13. For $(c)$, $\|\cdot\|_{C^n}$ is a Banach algebra norm because

$$\begin{aligned}\|fg\|_{C^n} &= \sum_{k=0}^n \frac{\|(fg)^{(k)}\|_{\sup}}{k!} \le \sum_{k=0}^n \frac{1}{k!}\sum_{j=0}^k \binom{k}{j}\|f^{(j)}\|_{\sup}\|g^{(k-j)}\|_{\sup} \\ &\le \Big[\sum_{k=0}^n \frac{\|f^{(k)}\|_{\sup}}{k!}\Big]\Big[\sum_{i=0}^n \frac{\|g^{(i)}\|_{\sup}}{i!}\Big] = \|f\|_{C^n}\|g\|_{C^n}.\end{aligned}$$

Thus each $\mathbb{A}$ with its given norm is a Banach function algebra. To show that each $\mathbb{A}$ is natural, by Theorem 6.21$(b)$ it suffices to check that $1/f \in \mathbb{A}$ for each $f \in \mathbb{A}$ with $f^{-1}(0) = \emptyset$. For $(a)$ this is immediate. The function $z \mapsto 1/z$ is uniformly Lipschitz on the complement of any neighborhood of 0 in $\mathbb{C}$. This implies $(b)$ and $(d)$ by Lemma 6.22 with $\beta = 1$. Also, $z \mapsto 1/z$ is $C^\infty$ with each derivative uniformly bounded on the complement of each neighborhood of 0. For derivatives through the $n$th, this implies $(c)$, completing the proof. □

**Remark 6.24.** The Banach algebras of functions $\mathcal{R}(J;\mathbb{C})$ of regulated functions with supremum norm, or $\mathcal{W}_p(J;\mathbb{C})$ with its usual norm, do not consist of continuous functions on $J$. So they are not Banach function algebras, hence not natural. But they are isometric as Banach algebras to the natural algebras $C(J^{\pm};\mathbb{C})$ and $\mathcal{W}_p(J^{\pm};\mathbb{C})$, respectively. In other words, the maximal ideal spaces of $\mathcal{R}(J;\mathbb{C})$ and of $\mathcal{W}_p(J;\mathbb{C})$ are both given by $J^{\pm}$.

**Example 6.25.** Here is an example of a non-natural Banach function algebra. Let $D := \{z \in \mathbb{C}\colon |z| \le 1\}$. Let $\mathcal{A}(D)$ be the Banach algebra of continuous $\mathbb{C}$-valued functions on $D$ which are holomorphic on its interior, with norm $\|\cdot\|_{\sup}$ and pointwise operations. Then $\mathcal{A}(D)$ is a Banach function algebra but is not self-adjoint. For each $f \in \mathcal{A}(D)$ let $T(f)$ be its restriction to $[-1,1]$. Let $\mathbb{A} := \{T(f)\colon f \in \mathcal{A}(D)\}$. Now $f \mapsto T(f)$ is 1-to-1, since $T(f)$ determines all the derivatives $f^{(n)}(0)$ and thus the Taylor series of $f$ around 0, which converges to it for $|z| < 1$ and thus determines $f$ uniquely by continuity. So $\|T(f)\| := \|f\|_{\sup}$ is a well-defined norm on $\mathbb{A}$. $T$ is an algebra isomorphism and an isometry of $\mathcal{A}(D)$ onto $\mathbb{A}$. Clearly $\mathbb{A}$ contains the constants and strongly separates points of $[-1,1]$. For each $z = x + iy \in \mathbb{C}$ with $x, y \in \mathbb{R}$, let $\bar{z} := x - iy$. To see that $\mathbb{A}$ is self-adjoint, note that for each $f \in \mathcal{A}(D)$ we have $f^* \in \mathcal{A}(D)$ where $f^*(z) := \bar{f}(\bar{z})$ for all $z \in D$. (If $f(z) = \sum_{n=0}^{\infty} a_n z^n$ for $|z| < 1$, then $f^*(z) = \sum_{n=0}^{\infty} \bar{a}_n z^n$.) On $[-1,1]$, $f^* = \bar{f}$, so $\mathbb{A}$ is self-adjoint. Evidently $\mathbb{A}$ is not natural, since evaluations at points of $D$ not in $[-1,1]$ give characters of $\mathcal{A}(D)$ and thus of $\mathbb{A}$ not given by points of $[-1,1]$.

For a vector space of ($\mathbb{K}$-valued) functions $\mathbb{B}$ and a nonempty open set $U \subset \mathbb{K}$, let $\mathbb{B}^{[U]}$ be the set of all $f \in \mathbb{B}$ such that the closure of the range $\operatorname{ran}(f)$ is included in $U$. Note that if $\mathbb{B}$ is a Banach function algebra then each such $\operatorname{ran}(f)$ is already compact and closed. But for a general Banach algebra of functions such as $\mathcal{W}_p(J)$, $\operatorname{ran}(f)$ need not be closed.

**Proposition 6.26.** *Let $\mathbb{B}$ be a unital Banach algebra of $\mathbb{C}$-valued functions on a set $S$ such that for each $f \in \mathbb{B}$, the closure of* $\operatorname{ran}(f)$ *equals $\sigma(f)$. If $U$ is a connected open set in $\mathbb{C}$, then $\mathbb{B}^{[U]}$ is a connected open set in $\mathbb{B}$.*

**Remark 6.27.** If $\mathbb{B}$ is a natural Banach function algebra then the hypothesis holds by Theorem 6.17($b$). If $\mathbb{B} = \mathcal{W}_p(J;\mathbb{C})$ with $1 \le p < \infty$ and $J = [a,b]$, then $\overline{\operatorname{ran}(f)} = \operatorname{ran}(I(f))$ and $\sigma(f) = \sigma(I(f))$ for each $f \in \mathcal{W}_p(J;\mathbb{C})$, where $I$ is the isometry between $\mathcal{W}_p(J;\mathbb{C})$ and $\mathcal{W}_p(J^{\pm};\mathbb{C})$ of Proposition 6.20($c$). Therefore the hypothesis holds again because $\mathcal{W}_p(J^{\pm};\mathbb{C})$ is a natural Banach function algebra by Corollary 6.23($d$).

Let $(\mathbb{A}, \tau)$ be a topological vector space of real-valued functions, i.e. addition and multiplication by scalars are both jointly continuous. Let $U$ be a connected open set in $\mathbb{R}$, i.e. an open interval. Then $\mathbb{A}^{[U]}$ is evidently convex and thus connected for $\tau$. If $\mathbb{A}$ is a unital Banach algebra of $\mathbb{R}$-valued functions then $\mathbb{A}^{[U]}$ is open by Theorem 6.13($b$) and the following proof.

*Proof.* For any set $A \subset \mathbb{C}$ and $z \in \mathbb{C}$, let $d(z, A) := \inf\{|z - w| : w \in A\}$. For $\delta > 0$, let $A^\delta := \{z : d(z, A) < \delta\}$. Let $f \in \mathbb{B}^{[U]}$ and let $A := \overline{\mathrm{ran}\,(f)}$. By Theorem 6.13($b$), $f$ is bounded, and so $A$ is a compact set in $\mathbb{C}$. Since $U$ is open, there is an $\epsilon > 0$ such that $A^\epsilon \subset U$. Let $g \in B(f, \epsilon/3) \subset \mathbb{B}$. By Theorem 6.13($b$) again, $\|f - g\|_{\sup} \leq \|f - g\| < \epsilon/3$ and so $\overline{\mathrm{ran}\,(g)} \subset A^\epsilon \subset U$, proving that $\mathbb{B}^{[U]}$ is an open set in $\mathbb{B}$.

Connectedness of $\mathbb{B}^{[U]}$ is clear if $U = \mathbb{C}$. Let $U = D = B(0, 1) \subset \mathbb{C}$. Then $\mathbb{B}^{[D]}$ is connected because it is convex and the segment joining two functions in $\mathbb{B}^{[D]}$ is a continuous mapping from $[0, 1]$ into $\mathbb{B}^{[D]}$. Let $U$ be a connected open set in $\mathbb{C}$ and let $T$ be a Riemann mapping from $U$ onto $D$, namely a 1-to-1 holomorphic function from $U$ onto $D$, where the inverse mapping $T^{-1}$ is also holomorphic from $D$ onto $U$ (e.g. Ahlfors [1, p. 222]). For each $f \in \mathbb{B}^{[D]}$, the spectrum $\sigma(f) = \overline{\mathrm{ran}\,(f)}$ is included in $D$, and so $N_{T^{-1}}f = T^{-1} \circ f \in \mathbb{B}$ by Proposition 4.34. Also, $\overline{\mathrm{ran}\,(T^{-1} \circ f)} \subset U$ since $T^{-1}(\overline{\mathrm{ran}\,(f)})$ is a compact subset of $U$. To show continuity of the autonomous Nemytskii operator $N_{T^{-1}}$ from $\mathbb{B}^{[D]}$ into $\mathbb{B}^{[U]}$ we need the following:

**Lemma 6.28.** *If $f_n$, $n \geq 0$, are functions from a set $S$ into a metric space $(X, d)$, $f_n \to f_0$ uniformly on $S$ as $n \to \infty$, and each $A_n := \overline{\mathrm{ran}\,(f_n)}$ is a compact set in $X$, then $A := \cup_{n \geq 0} A_n$ is compact in $X$.*

*Proof.* It suffices to show that any sequence $\{x_j\}_{j=1}^\infty \subset A$ has a convergent subsequence. If infinitely many $x_j$ belong to one $A_n$ this is clear. Thus taking a subsequence suppose that for some $n_j \to \infty$ and $s_j \in S$, $d(x_j, f_{n_j}(s_j)) < 1/j$, where $\sup_{s \in S} d(f_n(s), f_0(s)) < 1/j$ for each $n \geq n_j$. Let $y_j := f_0(s_j) \in A_0$. Taking a further subsequence, we can assume that $y_j \to y$ for some $y \in A_0$. Then $x_j \to y$ also, proving the lemma. □

To show continuity of $N_{T^{-1}}$ let $f_n$, $n \geq 0$, be functions in $\mathbb{B}^{[D]}$ such that $f_n \to f_0$ in $\mathbb{B}$ as $n \to \infty$. By Theorem 6.13($b$) again, $f_n \to f_0$ uniformly on $S$, and so the union $A$ of all $\overline{\mathrm{ran}\,(f_n)}$, $n \geq 0$, is a compact set in $D$ by the preceding lemma. By Lemma 4.32 there is a $C^\infty$ simple closed curve $\zeta(\theta)$, $0 \leq \theta \leq 2\pi$, with range $\mathrm{ran}\,(\zeta) \subset D \setminus A$ and winding number $w(\zeta(\cdot), z) = 1$ for all $z \in A$. Thus by a classical Cauchy integral formula, e.g. Theorem 2.107 for $X = \mathbb{C}$, for each $n \geq 0$ and $s \in S$, we have

$$T^{-1}(f_n(s)) = \frac{1}{2\pi \mathrm{i}} \oint_{\zeta(\cdot)} \frac{T^{-1}(\zeta)\, \mathrm{d}\zeta}{\zeta - f_n(s)}.$$

Since $\sigma(f_n) = \overline{\mathrm{ran}\,(f_n)} \subset A$, by Theorem 4.18($b$), the function $z \mapsto (z1(\cdot) - f_n)^{-1}$ is holomorphic from $\mathbb{C} \setminus A$ into $\mathbb{B}$. Thus it is continuous by Proposition 2.96, and so bounded on $\mathrm{ran}\,(\zeta)$. Therefore there is a finite number $M$ such that $\|(z1(\cdot) - f_n)^{-1}\| \leq M$ for each $z \in \mathrm{ran}\,(\zeta)$ and $n \geq 0$. It then follows that there is a finite constant $C$ such that

$$\|T^{-1}\circ f_n - T^{-1}\circ f_0\| = \frac{1}{2\pi}\Big\| \oint_{\zeta(\cdot)} \frac{T^{-1}(\zeta)(f_n - f_0)\,\mathrm{d}\zeta}{(\zeta 1(\cdot) - f_n)(\zeta 1(\cdot) - f_0)} \Big\| \le C\|f_n - f_0\|$$

for each $n \ge 1$. Thus $N_{T^{-1}} f_n \to N_{T^{-1}} f_0$ in $\mathbb{B}$ as $n \to \infty$, and so $N_{T^{-1}}$ is continuous from $\mathbb{B}^{[D]}$ onto $\mathbb{B}^{[U]}$, proving connectedness of $\mathbb{B}^{[U]}$ and the proposition. □

**Theorem 6.29.** *Let $\mathbb{B}$ be a unital Banach algebra of $\mathbb{C}$-valued functions on a set $S$ such that for each $f \in \mathbb{B}$, we have $\overline{\mathrm{ran}\,(f)} = \sigma(f)$, let $U$ be a connected and simply connected open set in $\mathbb{C}$, and let $F$ be a holomorphic function from $U$ into $\mathbb{C}$. Then the autonomous Nemytskii operator $N_F$ acts from $\mathbb{B}^{[U]}$ into $\mathbb{B}$ and is $\mathbb{B}$-analytic on $\mathbb{B}^{[U]}$.*

*Proof.* The set $\mathbb{B}^{[U]}$ is open and connected by Proposition 6.26. If $f \in \mathbb{B}^{[U]}$ then $\sigma(f) = \overline{\mathrm{ran}\,(f)}$ is a compact subset of $U$. Thus $N_F f = F\circ f \in \mathbb{B}$ for each $f \in \mathbb{B}^{[U]}$ by Proposition 4.34.

To prove $\mathbb{B}$-analyticity of $N_F$ on $\mathbb{B}^{[U]}$ let $f \in \mathbb{B}^{[U]}$. Since $K := \overline{\mathrm{ran}\,(f)}$ is a compact subset of $U$, by Lemma 4.32 there is a $C^\infty$ simple closed curve $\zeta(\theta)$, $0 \le \theta \le 2\pi$, with range $\mathrm{ran}\,(\zeta) \subset U \setminus K$ and winding number $w(\zeta(\cdot), z) = 1$ for all $z \in K$. Thus by a classical Cauchy integral formula, e.g. Theorem 2.107 for $X = \mathbb{C}$, for each $k = 1, 2, \ldots$ and $x \in S$, we have

$$F^{(k)}(f(x)) = \frac{k!}{2\pi \mathrm{i}} \oint_{\zeta(\cdot)} \frac{F(\zeta)\,\mathrm{d}\zeta}{(\zeta - f(x))^{k+1}}.$$

Since $\sigma(f) = K$, by Theorem 4.18(*b*), the function $z \mapsto (z1(\cdot) - f)^{-1}$ is holomorphic from $\mathbb{C} \setminus K$ into $\mathbb{B}$. Thus it is continuous by Proposition 2.96, and so bounded on $\mathrm{ran}\,(\zeta)$. Therefore there is a finite number $M$ such that $\|(z1(\cdot) - f)^{-1}\| \le M$ for each $z \in \mathrm{ran}\,(\zeta)$, and so $\|(z1(\cdot) - f)^{-k-1}\| \le M^{k+1}$ for each $z \in \mathrm{ran}\,(\zeta)$ and $k = 1, 2, \ldots$. It follows that there is a finite constant $C$ such that for each $k = 1, 2, \ldots$, $\|h_k\| \le C\|F\|_{\mathrm{ran}\,(\zeta),\sup} M^{k+1}$, where

$$h_k := (F^{(k)}\circ f)/k! = \frac{1}{2\pi \mathrm{i}} \oint_{\zeta(\cdot)} \frac{F(\zeta)\,\mathrm{d}\zeta}{(\zeta 1(\cdot) - f)^{k+1}} \in \mathbb{B}.$$

Then $\limsup_{k\to\infty} \|h_k\|^{1/k} \le M$, and by the root test, the $\mathbb{B}$-power series $\sum_{k\ge 1} h_k g^k$ converges in $\mathbb{B}$ for each $g \in \mathbb{B}$ with $\|g\| < 1/M$. Let $h_0 := N_F f$. Since $F$ has an ordinary Taylor expansion around $f(x)$, for each $x \in S$,

$$F(f(x) + z) = \sum_{k\ge 0} h_k(x) z^k$$

for each $z$ with $|z| < r(x)$ and some $r(x) > 0$. Since $\|h_k\|_{S,\sup} \le \|h_k\|$ for each $k$, there is an $r > 0$ such that $r(x) \ge 1/M$ for all $x$. Thus the $\mathbb{B}$-power series $\sum_{k\ge 0} h_k g^k$ converges in $\mathbb{B}$ to $N_F(f + g)$ for each $g \in \mathbb{B}$ such that $f + g \in \mathbb{B}^{[U]}$

and $\|g\|_{S,\sup} \le \|g\| < \min\{r, 1/M\}$, and so it is the $\mathbb{B}$-Taylor series of $N_F$ around $f$, proving the theorem. □

Let $\mathbb{B}$ be a self-adjoint unital Banach algebra of $\mathbb{C}$-valued functions and let $\mathbb{B}_{\mathbb{R}}$ be the set of all $\mathbb{R}$-valued functions in $\mathbb{B}$. Then $\mathbb{B}_{\mathbb{R}}$ is a real Banach algebra of $\mathbb{R}$-valued functions with the norm $\|\cdot\|_{\mathbb{B}}$. For example, if $\mathbb{B} = \mathcal{W}_p(J;\mathbb{C})$ with $1 \le p < \infty$ then $\mathbb{B}_{\mathbb{R}} = \mathcal{W}_p(J;\mathbb{R}) \equiv \mathcal{W}_p(J)$.

If $W = (-M, M)$ for some $0 < M \le \infty$, then we have $\mathbb{B}_{\mathbb{R}}^{[W]} = \{f \in \mathbb{B}_{\mathbb{R}} \colon \|f\|_{\sup} < M\}$ since $\overline{\operatorname{ran}(f)} \subset W$ if and only if $\|f\|_{\sup} < M$. By Theorem 6.13($b$) and Remark 6.27, the set $\mathbb{B}_{\mathbb{R}}^{[W]}$ is connected and open.

**Corollary 6.30.** *Let $\mathbb{B}$ be a self-adjoint unital Banach algebra of $\mathbb{C}$-valued functions on a set $S$ such that $\operatorname{ran}(f) = \sigma(f)$ for each $f \in \mathbb{B}$, and let $W := (-M, M)$ for some $0 < M \le \infty$. For $F\colon \mathbb{R} \to \mathbb{R}$, the following statements are equivalent:*

($a$) *The autonomous Nemytskii operator $N_F\colon g \mapsto F \circ g$ acts from $\mathbb{B}_{\mathbb{R}}^{[W]}$ into $\mathbb{B}$, thus into $\mathbb{B}_{\mathbb{R}}$, and is analytic on $\mathbb{B}_{\mathbb{R}}^{[W]}$.*

($b$) *$F$ is analytic on $W$.*

*Proof.* Let $F\colon \mathbb{R} \to \mathbb{R}$. Suppose ($a$) holds. By Theorem 6.13($b$), $\|\cdot\|_{\sup} \le \|\cdot\|_{\mathbb{B}}$. Since $\mathbb{B}$ contains the constant functions, by Corollary 6.12 with $\mathbb{G} = \mathbb{H} = \mathbb{B}_{\mathbb{R}}$, $V = \mathbb{B}_{\mathbb{R}}^{[W]}$ and $\psi \equiv F$, ($b$) holds.

Conversely, suppose that ($b$) holds. By analytic continuation in the form of Theorem 5.36, there is a connected open set $U \subset \mathbb{C}$ such that $U \cap \mathbb{R} = W$ and a holomorphic extension of $F$ from $W$ into $U$, which will be denoted also by $F$. Thus $N_F$ acts from $\mathbb{B}_{\mathbb{R}}^{[W]}$ into $\mathbb{B}$ and is analytic on $\mathbb{B}_{\mathbb{R}}^{[W]}$ by Theorem 6.29, proving the corollary. □

Now we return to non-autonomous Nemytskii operators $N_\psi$. Under some conditions, we will establish their analyticity on open sets $\mathbb{B}_{\mathbb{R}}^{[W]}$. We begin by giving conditions for a Taylor expansion of $N_\psi$ around a fixed function in $\mathbb{B}_{\mathbb{R}}$. For the last conclusion, recall that the $n$-linear multiplication operator $M^n[h]$ was defined in Definition 6.6.

**Proposition 6.31.** *Let $\mathbb{B}$ be a self-adjoint unital Banach algebra of $\mathbb{C}$-valued functions on a set $S$ such that $\operatorname{ran}(f) = \sigma(f)$ for each $f \in \mathbb{B}$. Let $\psi\colon \mathbb{R} \times S \to \mathbb{R}$ be $u$-differentiable of order $n$ everywhere on $\mathbb{R} \times S$ for each $n \ge 1$. Let $V$ be a connected open set in $\mathbb{B}_{\mathbb{R}}$ and let $f \in V$. If the Nemytskii operator $N_\psi$ acts from $V$ into $\mathbb{B}_{\mathbb{R}}$ then the following statements are equivalent:*

($a$) *there are $c \ge 0$ and $r > 0$ such that for each $n = 1, 2, \ldots$, $\sup\{|\psi_u^{(n)}(f(s)+t, s)|\colon s \in S\} \le c^n n!$ for $t \in \mathbb{R}$ with $|t| \le r$, $N_{\psi_u^{(n)}}(f) \in \mathbb{B}_{\mathbb{R}}$ and*

$$\limsup_{k\to\infty} \left\| N_{\psi_u^{(k)}}(f)/k! \right\|_{\mathbb{B}}^{1/k} < \infty;$$

(*b*) *there is an* $r > 0$ *such that* $N_\psi(f+g)$ *is given by the* $\mathbb{B}_{\mathbb{R}}$*-Taylor series* $\sum_{k=0}^{\infty} N_{\psi_u^{(k)}}(f)g^k/k!$, *which converges absolutely for each* $g \in \mathbb{B}_{\mathbb{R}}$ *such that* $\|g\|_{\sup} < r$;

(*c*) $N_\psi$ *on* $V$ *has a Taylor expansion around* $f$.

*If* (*c*) *holds then there is an* $r > 0$ *such that for* $n = 1, 2, \ldots$ *and* $g \in B(f,r) \cap \mathbb{B}_{\mathbb{R}}$, *the nth differential* $\mathrm{d}^n N_\psi(g) \in L({}^n\mathbb{B}_{\mathbb{R}}, \mathbb{B}_{\mathbb{R}})$ *is the n-linear multiplication operator* $M^n[h]$ *induced by the function* $h := N_{\psi_u^{(n)}}(g) \in \mathbb{B}_{\mathbb{R}}$.

*Proof.* (*a*) $\Rightarrow$ (*b*). Let $h_0 := N_\psi f$ and for $k = 1, 2, \ldots$, let $h_k := N_{\psi_u^{(k)}}(f)/k!$. Then each $h_k$ is in $\mathbb{B}_{\mathbb{R}}$, and $T := \limsup_{k\to\infty} \|h_k\|_{\mathbb{B}}^{1/k} < \infty$. For each $g \in \mathbb{B}_{\mathbb{R}}$ such that $\|g\|_{\sup} < 1/T$, the $\mathbb{B}_{\mathbb{R}}$-power series $\sum_{k=0}^{\infty} h_k g^k$ converges in $\mathbb{B}_{\mathbb{R}}$ absolutely by Theorem 5.16 since the spectral radius $r(g)$ equals $\|g\|_{\sup}$ by Theorem 6.13(*c*). Let $\rho \in \mathbb{R}$ be such that $0 < \rho < 1/c$ and $\rho \le r$. By Theorem 5.43 for $X = Y = \mathbb{R}$, $F(u) := \psi(u,s)$, and $U := U_\rho := \{f(s) + t\colon t \in \mathbb{R}, |t| < \rho\} \subset \mathbb{R}$, for each fixed $s \in S$ and $t \in \mathbb{R}$ with $|t| < \rho$, we have that the Taylor series

$$\psi(f(s)+t, s) = F(f(s)+t) = \sum_{k=0}^{\infty} F^{(k)}(f(s))\frac{t^k}{k!} = \sum_{k=0}^{\infty} h_k(s)t^k$$

converges absolutely on $B(f(s),\rho) = U$. Thus $N_\psi(f+g)(s) = (\sum_{k=0}^{\infty} h_k g^k)(s)$ for each $s \in S$ and each $g \in \mathbb{B}_{\mathbb{R}}$ with $\|g\|_{\sup} < \min\{1/c, r\}$, proving (*b*).

Statement (*c*) follows from (*b*) by Theorem 5.30(*a*). Suppose that (*c*) holds. Then by Theorem 5.28, there exists a constant $\rho > 0$ such that $N_\psi$ is a $C^\infty$ function on $B(f,\rho)$. By Theorem 6.13(*b*), $\|\cdot\|_{\sup} \le \|\cdot\|_{\mathbb{B}}$. Since $\mathbb{B}$ contains the constant functions, by Proposition 6.9, for each $n = 1, 2, \ldots$ and each $g \in B(f,\rho) \subset \mathbb{B}_{\mathbb{R}}$, the $n$th differential $\mathrm{d}^n N_\psi(g)$ is the multiplication operator $M^n[h]$ from $\mathbb{B}_{\mathbb{R}}^n$ into $\mathbb{B}_{\mathbb{R}}$ induced by the function $h = N_{\psi_u^{(n)}}(g) \in \mathbb{B}_{\mathbb{R}}$, proving the last conclusion of the theorem. Since $\|\cdot\|_{\sup} \le \|\cdot\|_{\mathbb{B}}$ and $\|N_{\psi_u^{(n)}}(g)\|_{\mathbb{B}} \le \|\mathrm{d}^n N_\psi(g)\|$ for $g \in B(f,r) \cap \mathbb{B}_{\mathbb{R}}$ and $n = 1, 2, \ldots$, (*a*) follows from (*c*) by Theorem 5.43. The proof of Proposition 6.31 is complete. $\square$

**Theorem 6.32.** *Let* $\mathbb{B}$ *be a self-adjoint unital Banach algebra of* $\mathbb{C}$*-valued functions on a set* $S$ *such that* $\operatorname{ran}(f) = \sigma(f)$ *for each* $f \in \mathbb{B}$. *Let* $W := (-M, M)$ *for some* $0 < M \le \infty$ *and let* $\psi\colon W \times S \to \mathbb{R}$ *be u-differentiable of order n everywhere on* $W \times S$ *for each* $n \ge 1$. *If the Nemytskii operator* $N_\psi$ *acts from* $\mathbb{B}_{\mathbb{R}}^{[W]}$ *into* $\mathbb{B}_{\mathbb{R}}$ *then for each* $u \in W$, $\psi(u,\cdot) \in \mathbb{B}_{\mathbb{R}}$ *and the following statements are equivalent*:

(*a*) *for each closed interval* $B \subset W$, *there is a* $c \ge 0$ *such that for* $k = 1, 2, \ldots$,

$$\left\|\psi_u^{(k)}\right\|_{B\times S,\sup} \le c^k k!, \tag{6.11}$$

*and for each* $f \in \mathbb{B}_{\mathbb{R}}^{[W]}$, $N_{\psi_u^{(n)}}(f) \in \mathbb{B}_{\mathbb{R}}$ *for* $n = 1, 2, \ldots$, *and*

$$\limsup_{k\to\infty} \left\|N_{\psi_u^{(k)}}(f)/k!\right\|_{\mathbb{B}}^{1/k} < \infty;$$

(*b*) $N_\psi$ *is* $\mathbb{B}_{\mathbb{R}}$*-analytic on* $\mathbb{B}_{\mathbb{R}}^{[W]}$ *and for each* $f \in \mathbb{B}_{\mathbb{R}}^{[W]}$, $\sum_{k\geq 0} N_{\psi_u^{(k)}}(f)g^k/k!$ *is the* $\mathbb{B}_{\mathbb{R}}$*-Taylor expansion around* $f$;

(*c*) $N_\psi$ *is analytic on* $\mathbb{B}_{\mathbb{R}}^{[W]}$.

*Proof.* First, $\mathbb{B}$ is unital, so $\psi(u,\cdot) = N_\psi(u1_S(\cdot)) \in \mathbb{B}_{\mathbb{R}}$ for $u \in W$. Next to show that ($a$) implies ($b$), let $f \in \mathbb{B}_{\mathbb{R}}^{[W]}$, let $M_0$ be such that $\|f\|_{\sup} < M_0 < M$, let $B := [-M_0, M_0]$, and let $r := M_0 - \|f\|_{\sup}$. Thus for each $t \in \mathbb{R}$ such that $|t| \leq r$ and for $n = 1, 2, \ldots$, we have

$$\sup_{s\in K} \left|\psi_u^{(n)}(f(s)+t, s)\right| \leq \left\|\psi_u^{(n)}\right\|_{B\times S,\sup} \leq c^n n!.$$

By the implication ($a$) $\Rightarrow$ ($b$) of Proposition 6.31, there is a $\rho > 0$ such that $\sum_{k\geq 0} N_{\psi_u^{(k)}}(f)g^k/k!$ is a $\mathbb{B}_{\mathbb{R}}$-power series which converges (absolutely) to $N_\psi(f+g)$ for each $g \in \mathbb{B}_{\mathbb{R}}$ such that $\|g\|_{\sup} < \rho$, and so for each $g \in B(0,\rho)\cap\mathbb{B}_{\mathbb{R}}$ by Theorem 6.13($b$). Thus it is the $\mathbb{B}_{\mathbb{R}}$-Taylor expansion of $N_\psi$ around $f$. Since $f \in \mathbb{B}_{\mathbb{R}}^{[W]}$ is arbitrary, ($b$) is proved.

Statement ($c$) follows from ($b$) by Theorem 5.30($a$). To prove that ($c$) implies ($b$), let ($c$) hold and let $B$ be a closed finite interval included in $W$. By Corollary 6.12 for $\mathbb{G} = \mathbb{H} = \mathbb{B}_{\mathbb{R}}$ and so $V = \mathbb{B}_{\mathbb{R}}^{[W]}$, the function $u \mapsto \psi(u,\cdot) \in \mathbb{B}_{\mathbb{R}}$, $u \in W$, is analytic on $W$. By analytic continuation as in Theorem 5.36, there are a connected open set $U \subset \mathbb{C}$ such that $U\cap\mathbb{R} = W$ and a holomorphic extension of the function from $W$ to $U$, which will be denoted by $z \mapsto \psi(z,\cdot) \in \mathbb{B}$. Let $\epsilon > 0$ be such that $D := \{z = u + w \in \mathbb{C}\colon |w| \leq \epsilon, u \in B\} \subset U$. By Proposition 2.96, $z \mapsto \psi(z,\cdot) \in \mathbb{B}$ is continuous on $U$, and so $\|\psi(z,\cdot)\|_{\mathbb{B}}$ is bounded for $z \in D$. Let $u \in B$ and let $\zeta_u(t) := u + (\epsilon/2)e^{2\pi it}$ for $t \in [0,1]$. By Cauchy integral formulas for derivatives as in Theorem 2.98 for $f(z) = \psi(z,\cdot)$, $z \in B(u,\epsilon)$, and for each $n = 1, 2, \ldots$, we have

$$\psi_u^{(n)}(u,\cdot) = \frac{n!}{2\pi i}\oint_{\zeta_u(\cdot)} \frac{\psi(\zeta,\cdot)\,d\zeta}{(\zeta-u)^{n+1}} = \frac{n!}{(\epsilon/2)^n}\int_0^1 \frac{\psi(\zeta_u(t),\cdot)\,dt}{e^{2\pi int}}.$$

Thus for each $u \in B$ and each $n = 1, 2, \ldots$,

$$\left\|\psi_u^{(n)}(u,\cdot)\right\|_{S,\sup} \leq \left\|\psi_u^{(n)}(u,\cdot)\right\|_{\mathbb{B}} \leq \frac{n!}{(\epsilon/2)^n}\sup_{z\in D}\left\|\psi(z,\cdot)\right\|_{\mathbb{B}} \leq c^n n!,$$

where $c := 2C/\epsilon$ and $C := \max\{1, \sup_{z\in D}\|\psi(z,\cdot)\|_{\mathbb{B}}\}$, proving (6.11). The rest of ($a$) follows by the implication ($c$) $\Rightarrow$ ($a$) of Proposition 6.31 applied to each $f \in \mathbb{B}_{\mathbb{R}}^{[W]}$. The proof of Theorem 6.32 is complete. □

## 6.4 Autonomous Nemytskii Operators on $\mathcal{W}_p$ Spaces

In this section, the autonomous Nemytskii operator $N_F$ is considered between the spaces $\mathbb{G} = W_p(J;X)$ and $\mathbb{H} = W_q(J;Y)$ with $p \le q$. In particular, we give conditions for $N_F$ to act and be Fréchet differentiable.

*Necessary and sufficient acting conditions from $\mathcal{W}_p$ into $\mathcal{W}_q$*

We will prove that for $X = Y = \mathbb{R}$ and for a closed interval $J$, the autonomous Nemytskii operator $N_F$ acts from $\mathcal{W}_p(J)$ into $\mathcal{W}_q(J)$ with $0 < p \le q < \infty$ if and only if $F$ is locally $\alpha$-Hölder with $\alpha = p/q$ (Corollary 6.36). But first here is an example which shows that one cannot take $F \in \mathcal{W}_1$ and get $N_F f \in \mathcal{W}_p$ even for $f$ very smooth.

**Example 6.33.** Let $f(x) := \mathrm{e}^{-1/x}\sin(x)$ for $x > 0$ and $f(x) := 0$ for $x \le 0$. The function $f$ is $C^\infty$, with all derivatives 0 at 0. For $F := 1_{[0,1]}$, the composition $F \circ f$ has no right limit at 0, and so is not regulated. Thus it is not of bounded $p$-variation for any $p < \infty$ by Proposition 3.33.

Now we show that a local $\alpha$-Hölder property (Definition 6.4) is a sufficient condition for an autonomous Nemytskii operator to act from a suitable subset of $\mathcal{W}_p$ into $\mathcal{W}_q$ with $0 < p \le q < \infty$ and $\alpha = p/q$, as well as to be bounded from $\mathcal{W}_p$ into $\mathcal{W}_q$. A nonlinear operator from a subset $E$ of one normed space into another will be called *bounded* if it takes bounded subsets of $E$ into bounded sets. Recall that for a nonempty open set $U \subset X$, $\mathcal{W}_p^{[U]}(J;X)$ is the set of all $f \in \mathcal{W}_p(J;X)$ such that the closure of the range $\operatorname{ran}(f)$ is included in $U$. Also recall Definition 6.4 of the class $\mathcal{H}_\alpha^{\mathrm{loc}}(U;Y)$ and of the sets $B_m(U) = U_{1/m}$ with $m = 1, 2, \ldots$.

**Proposition 6.34.** *Let $0 < \alpha \le 1$, $0 < p < \infty$ and let $J := [a,b]$ with $a < b$. Let $X$, $Y$ be Banach spaces, let $U$ be a nonempty open subset of $X$, and let $F \in \mathcal{H}_\alpha^{\mathrm{loc}}(U;Y)$. Then $\mathcal{W}_p^{[U]}(J;X)$ is open and for each $f \in \mathcal{W}_p^{[U]}(J;X)$, $N_F f \in \mathcal{W}_{p/\alpha}(J;Y)$. Also, if $U = X$ then the autonomous Nemytskii operator $N_F$ is bounded from $\mathcal{W}_p(J;X)$ into $\mathcal{W}_{p/\alpha}(J;Y)$.*

*Proof.* Let $f \in \mathcal{W}_p^{[U]}(J;X)$. By Proposition 6.20(*a*), the closure of $\operatorname{ran}(f)$ is a compact subset of $U$. Since $\|\cdot\|_{J,\sup} \le \|\cdot\|_{J,[p]}$, it follows that a ball $B(f,\epsilon)$ is included in $\mathcal{W}_p^{[U]}(J;X)$ for some $\epsilon > 0$ just as at the beginning of the proof of Proposition 6.26, and so $\mathcal{W}_p^{[U]}(J;X)$ is open. By Proposition 6.3, $\operatorname{ran}(f) \subset B_m(U)$ for some $m \ge 1$. Then $N_F f \in \mathcal{W}_{p/\alpha}(J;Y)$ by the first part of Lemma 6.22(*a*) with $B = B_m(U)$ and $\beta = \alpha$. If $U = X$ and $E$ is a bounded subset of $\mathcal{W}_p(J;X)$ then $E \subset B_m(X)$ for some $m$ and $\{N_F f\colon f \in E\}$ is a bounded set in $\mathcal{W}_{p/\alpha}(J;Y)$ by the second part of Lemma 6.22(*a*) with $B = B_m(X)$ and $\beta = \alpha$, proving the proposition. □

In spite of the simplicity of the proofs of Lemma 6.22($a$) and Proposition 6.34, in order that $F \circ f \in \mathcal{W}_{p/\alpha}(J;\mathbb{R})$ for all $f \in \mathcal{W}_p(J;\mathbb{R})$, it is necessary that $F \in \mathcal{H}_\alpha^{\mathrm{loc}}$, as will follow from the next theorem, taking $\Psi(y) = y^{p/\alpha}$ and $\Phi(x) = x^p$. Recall that a function $\Phi\colon [0,\infty) \to [0,\infty)$ belongs to the class $\mathcal{V}$ if it is strictly increasing, continuous, unbounded, and 0 at 0.

**Theorem 6.35** (Ciemnoczołowski and Orlicz). *Let $J := [a,b]$ with $a < b$, let $\Phi, \Psi \in \mathcal{V}$, and let $\Psi$ satisfy the $\Delta_2$ condition* (3.3). *Suppose that a function $F\colon \mathbb{R} \to \mathbb{R}$ is such that $v_\Psi(F \circ f; J) < \infty$ for each $f \in \mathcal{W}_\Phi(J)$. Then for every $M > 0$ there is a finite constant $K$ such that for any $u, v \in [-M, M]$,*

$$\Psi(|F(u) - F(v)|) \le K\Phi(|u - v|). \tag{6.12}$$

*Proof.* First we prove that $F$ is bounded over each interval $[-M, M]$. Suppose not. Then for some $M > 0$, there exists a sequence $\{u_i\}_{i\ge1} \subset [-M, M]$ such that $|F(u_i)| \to +\infty$ as $i \to \infty$. Due to compactness of $[-M, M]$, by passing to a subsequence if necessary, we can and do assume that for some $u_0 \in [-M, M]$ we have $|u_i - u_0| \downarrow 0$ and $\Phi(2|u_i - u_0|) \le 2^{-i}$ for each $i = 1, 2, \ldots$. For a sequence $\{t_i\}_{i\ge1}$ of distinct points of $J$, let $f$ be a function on $J$ such that $f(t_i) = u_i$ for $i = 1, 2, \ldots$, and $f(t) = u_0$ for $t \in J \setminus \{t_i\}_{i\ge1}$. Then for any point partition $\kappa$ of $J$, we have

$$s_\Phi(f;\kappa) \le 2\sum_{i\ge1} \Phi(|u_i - u_0|) + 2\sum_{i\ge1} \sup_{j>i} \Phi(|u_i - u_j|) \le 4\sum_{i=1}^{\infty} \Phi(2|u_i - u_0|) \le 4,$$

and so $f \in \mathcal{W}_\Phi(J)$. On the other hand, we have

$$v_\Psi(F \circ f; J) \ge \Psi(|F(f(t_i)) - F(f(t_1))|) = \Psi(|F(u_i) - F(u_1)|) \to +\infty$$

as $i \to \infty$, a contradiction, proving that $F$ is bounded on every bounded interval.

To prove (6.12) suppose that it does not hold for some $M > 0$. For each $u, v \in [-M, M]$ with $v > u$, let $k(u, v)$ be the ratio of $\Psi(|F(v) - F(u)|)$ to $\Phi(v - u)$. Therefore there exists a sequence $\{[u_i, v_i]\}_{i\ge1}$ of subintervals of $J$ such that

$$k(u_i, v_i) \to +\infty \tag{6.13}$$

as $i \to \infty$. Since $F$ is bounded on $[-M, M]$, we have that $v_i - u_i \to 0$ as $i \to \infty$. For each $i \ge 1$, let $w_i := (u_i + v_i)/2$. Due to compactness of $[-M, M]$, taking a subsequence if necessary, we can and do assume that there is a $w_0 \in [-M, M]$ such that $w_i \to w_0$ as $i \to \infty$. Let $I := \{i \ge 1 : w_0 \in [u_i, v_i]\}$ and let $N$ be the cardinality of $I$. First suppose that $N < \infty$. By symmetry we can and do assume that there are infinitely many intervals $[u_i, v_i]$ to the right of $w_0$. Recursively select from this set of intervals a subsequence $\{[u_i', v_i']\}_{i\ge1}$ such that for $i = 1, 2, \ldots$:

($a$) $w_0 < \cdots < u_{i+1}' < v_{i+1}' < u_i' < v_i' < \cdots$ ;

(*b*) $\Phi(v_i' - w_0) \le 2^{-i-1}$;
(*c*) $k(u_i', v_i') \ge 2^{2i+1}$.

By (*b*), we have that $2^i\Phi(v_i' - u_i') < 1/2$ for each $i \ge 1$. Thus for each $i \ge 1$, letting $m_i$ be the largest integer $\le (2^i\Phi(v_i' - u_i'))^{-1}$, we have that

$$1/2^{i+1} < m_i\Phi(v_i' - u_i') \le 1/2^i. \tag{6.14}$$

Let $\{[s_i, t_i)\}_{i\ge1}$ be a sequence of nonempty disjoint subintervals of $J$, and for each $i \ge 1$, let $\kappa_i = \{s_{ij}, t_{ij}\}_{j=1}^{m_i}$ be a point partition of $[s_i, t_i)$, so that $s_i = s_{i1} < t_{i1} < \cdots < s_{im_i} < t_{im_i} < t_i$. Let $f$ be a function on $J$ such that

$$f(t) = \begin{cases} v_i' & \text{if } t = t_{ij} \text{ for some } i, j, \\ u_i' & \text{if } t \in [s_i, t_i) \setminus \{t_{ij}\}_{j=1}^{m_i} \text{ for some } i \ge 1, \\ w_0 & \text{if } t \in J \setminus \cup_i [s_i, t_i). \end{cases} \tag{6.15}$$

Then by (*b*) and the right side of (6.14), for any point partition $\kappa$ of $J$, we have

$$s_\Phi(f;\kappa) \le 2\sum_{i\ge1} m_i\Phi(v_i' - u_i') + 2\sum_{i\ge1} \Phi(v_i' - w_0) \le 3, \tag{6.16}$$

and so $f \in \mathcal{W}_\Phi(J)$. On the other hand by (*c*) and by the left side of (6.14), we have for each $i$

$$\begin{aligned} v_\Psi(F\circ f; J) &\ge \sum_{j=1}^{m_i} \Psi(|F(f(t_{ij})) - F(f(s_{ij}))|) = m_i\Psi(|F(v_i') - F(u_i')|) \\ &= k(u_i', v_i')m_i\Phi(v_i' - u_i') \ge 2^{2i+1}/2^{i+1} = 2^i \to \infty \end{aligned} \tag{6.17}$$

as $i \to \infty$, a contradiction.

Now suppose that $N = +\infty$. Then for infinitely many $i \ge 1$, we have either $w_0 \in (u_i, v_i)$, or $w_0 = u_i$, or $w_0 = v_i$. Since the proof in each case is similar, we assume that $u_i < w_0 < v_i$ for each $i \ge 1$. For each $i \ge 1$, let $S_i := \max\{k(u_i, w_0), k(w_0, v_i)\}$. Then by the $\Delta_2$ condition (3.3) for $\Psi$, we have

$$\begin{aligned} \Psi(|F(v_i) - F(u_i)|) &\le \Psi(2|F(v_i) - F(w_0)|) + \Psi(2|F(w_0) - F(u_i)|) \\ &\le DS_i\big[\Phi(v_i - w_0) + \Phi(w_0 - u_i)\big] \le 2DS_i\Phi(v_i - u_i). \end{aligned}$$

Thus for each $i \ge 1$, it follows that $S_i \ge k(u_i, v_i)/(2D)$. By (6.13), we can recursively select a sequence $\{[u_i', v_i']\}_{i\ge1}$ such that for each $i \ge 1$, either $k(w_0, v_i') \ge 2^{2i+1}$ or $k(u_i', w_0) \ge 2^{2i+1}$. By symmetry, we assume that the first case holds. Taking a subsequence if necessary, we assume that $\Phi(v_i' - w_0) \le 1/2^{i+1}$ for each $i \ge 1$. Thus for each $i \ge 1$, letting $m_i$ be the largest integer $\le (2^i\Phi(v_i' - w_0))^{-1}$ we have that (6.14) holds with $w_0$ instead of $u_i'$. Let $f$ be the function on $J$ defined by (6.15) with $w_0$ instead of $u_i'$ for each $i \ge 1$. Then for any point partition $\kappa$ of $J$, (6.16) holds with all $u_i'$ replaced by $w_0$, and so

$f \in \mathcal{W}_\Phi(J)$. On the other hand, (6.17) holds with all $u_i'$ replaced by $w_0$, again a contradiction, proving (6.12). □

Proposition 6.34 and Theorem 6.35 imply the following:

**Corollary 6.36.** *Let $0 < p \le q < \infty$, $\alpha := p/q$, and $J := [a,b]$ with $a < b$. For a function $F\colon \mathbb{R} \to \mathbb{R}$, the autonomous Nemytskii operator $N_F$ acts from $\mathcal{W}_p(J)$ into $\mathcal{W}_q(J)$ if and only if $F \in \mathcal{H}_\alpha^{\mathrm{loc}}$.*

*Fréchet differentiability and remainder bounds*

We will give a bound for the remainder $\mathrm{Rem}_{N_F}$, as defined in (6.3), in the Fréchet differentiability of the autonomous Nemytskii operator $N_F$ acting from $\mathcal{W}_p(J;X)$ into $\mathcal{W}_r(J;Y)$ with $p < r$. Recall Definition 6.5 of the class $\mathcal{H}_{1+\alpha}^{\mathrm{loc}}(U;Y)$.

**Theorem 6.37.** *Let $0 < \alpha \le 1$, $1 \le p < \alpha r < \infty$ and let $J := [a,b]$ with $a < b$. For Banach spaces $X$, $Y$, let $U$ be a nonempty open subset of $X$, let $F \in \mathcal{H}_{1+\alpha}^{\mathrm{loc}}(U;Y)$, and let $G \in \mathcal{W}_p^{[U]}(J;X)$. Then there is a neighborhood $V$ of zero in $\mathcal{W}_p(J;X)$ such that the autonomous Nemytskii operator $N_F$ acts from $G+V$ into $\mathcal{W}_r(J;Y)$, and is Fréchet differentiable at $G$ with derivative $g \mapsto (s \mapsto (DF)(G(s))g(s))$, and there is a finite constant $C := C(F,G,V,\alpha,p,r)$ such that the remainder bound*

$$\big\|\mathrm{Rem}_{N_F}(G,g)\big\|_{[r]} \le C\|g\|_{\sup}^{\alpha-p/r}\|g\|_{[p]} \tag{6.18}$$

*holds for each $g \in V$. We can take $C := \|DF\|_{B_{2m},\mathcal{H}_\alpha}[E\|G\|_{(p)}^{p/r} + 2E + 1]$ where for some $m \ge 1$, $G$ takes values in $B_m := B_m(U)$, $V := \{g \in \mathcal{W}_p(J;X)\colon \|g\|_{[p]} < 1/(2m)\}$, $E := C(\alpha^{-1}, (1-\alpha+p/r)^{-1}, p\alpha^{-1})$, and $C(\cdot,\cdot,\cdot)$ is the constant in (3.167). If in addition $X = Y = \mathbb{K}$ then the derivative $DN_F(G)$ is a linear multiplication operator $M[F'\circ G]$.*

The proof is based on bounds, obtained in Theorem 3.111, of the $\gamma$-variation seminorm of the integral transform $K_H := F(H - H(\cdot,0), \mathrm{d}G)$ on $[a,b]$ with $G(t) \equiv t$ on $[0,1]$, defined by (3.164), and so

$$K_H(s) = \int_0^1 H(s,t)\,\mathrm{d}t - H(s,0), \qquad a \le s \le b. \tag{6.19}$$

We will take a function $H$ and a number $\gamma = r \in [1,\infty)$ as in the next lemma.

**Lemma 6.38.** *Let $0 < \alpha \le 1$, $1 \le p \le \alpha r < \infty$, $J := [a,b]$ with $a < b$. For Banach spaces $X$, $Y$, let $U$ be a nonempty open subset of $X$ and let $F \in \mathcal{H}_{1+\alpha}^{\mathrm{loc}}(U;Y)$. Let $G, g \in \mathcal{W}_p(J;X)$ and $B := B_m(U)$ for some $m \ge 1$ be such that $G(s) + tg(s) \in B$ for all $s \in J$ and $t \in [0,1]$. Then for the integral transform $K := K_H$ with*

$$H(s,t) := DF(G(s)+tg(s))g(s), \qquad (s,t) \in J \times [0,1], \tag{6.20}$$

*we have the bound*

$$\|K\|_{[r]} \le \|DF\|_{B,\mathcal{H}_\alpha}\Big\{E\|G\|_{(p)}^{p/r}+(1+E)\|g\|_{[p]}^{p/r}+E\Big\}\|g\|_{[p]}\|g\|_{\sup}^{\alpha-p/r}, \tag{6.21}$$

*where* $E := C(\alpha^{-1},(1-\alpha+p/r)^{-1},p\alpha^{-1})$ *and* $C(\cdot,\cdot,\cdot)$ *is the constant in* (3.167).

*Proof.* Let $\widetilde{p}$ and $\widetilde{q}$ respectively be the parameters called $p$ and $q$ in Theorem 3.111, let the function called $G$ there be here $\tilde{G}(t) \equiv t$, $0 \le t \le 1$, and apply the theorem to $\beta := p/\alpha$, $\widetilde{p} := 1/\alpha$, and $\widetilde{q} := 1/(1-\alpha+p/r) \ge 1$. Then $\widetilde{p}^{-1}+\widetilde{q}^{-1} = 1+p/r > 1$, $\gamma = r$, and $\|\tilde{G}\|_{(\widetilde{q})} \le \|\tilde{G}\|_{(1)} = 1$. To bound $A_{\beta,\sup}(H;J\times[0,1])$ defined by (3.165), let $\kappa = \{s_i\}_{i=0}^n$ be a point partition of $J$ and $t \in [0,1]$. Then using the Minkowski inequality (1.5), the fact that $DF \in \mathcal{H}_\alpha(B;L(X,Y))$, and letting $f := G+tg$, we have the bound

$$\begin{aligned}
s_\beta(H(\cdot,t);\kappa)^{1/\beta} &\le \Big(\sum_{i=1}^n \big\|[DF(f(s_i))-DF(f(s_{i-1}))]g(s_i)\big\|^\beta\Big)^{1/\beta}\\
&\quad+\Big(\sum_{i=1}^n \big\|DF(f(s_{i-1}))[g(s_i)-g(s_{i-1})]\big\|^\beta\Big)^{1/\beta}\\
&\le \|DF\|_{B,(\mathcal{H}_\alpha)}\{\|G\|_{(p)}^\alpha+\|g\|_{(p)}^\alpha\}\|g\|_{\sup}+\|DF\|_{B,\sup}\|g\|_{(p)}\\
&\le \|DF\|_{B,\mathcal{H}_\alpha}\{\|G\|_{(p)}^\alpha+\|g\|_{(p)}^\alpha+1\}\|g\|_{[p]}.
\end{aligned}$$

Since $\kappa \in \mathrm{PP}\,(J)$ and $t \in [0,1]$ are arbitrary, the right side gives a bound for $A_{\beta,\sup}(H;J\times[0,1])$. Now for each point partition $\lambda = \{t_j\}_{j=0}^m$ of $[0,1]$ and $s \in J$, we have the bound

$$\begin{aligned}
s_{\widetilde{p}}(H(s,\cdot);\lambda)^{1/\widetilde{p}} &\le \|DF\|_{B,(\mathcal{H}_\alpha)}\|g\|_{\sup}^{1+\alpha}\Big(\sum_{j=1}^m t_j-t_{j-1}\Big)^\alpha\\
&\le \|DF\|_{B,(\mathcal{H}_\alpha)}\|g\|_{[p]}\|g\|_{\sup}^\alpha.
\end{aligned}$$

Again since $\lambda \in \mathrm{PP}\,[0,1]$ and $s \in J$ are arbitrary, the right side gives a bound for $B_{\sup,\widetilde{p}}(H;J\times[0,1])$ defined by (3.166). Therefore by (3.167), since $\beta/\gamma = p/(\alpha r) \le 1$, it follows that

$$\|K\|_{(r)} \le C\|DF\|_{B,\mathcal{H}_\alpha}\Big\{\|G\|_{(p)}^{p/r}+\|g\|_{(p)}^{p/r}+1\Big\}\|g\|_{[p]}\|g\|_{\sup}^{\alpha-p/r}.$$

Also, for each $s \in J$, we have $\|K(s)\| \le \|DF\|_{B,(\mathcal{H}_\alpha)}\|g\|_{\sup}^{1+\alpha}$. Therefore (6.21) holds. □

**Proof of Theorem 6.37.** By Propositions 6.20($a$) and 6.3, there is a positive integer $m$ such that $\mathrm{ran}\,(G) \subset B_m = B_m(U)$. Let $V$ be the set of all $g \in$

$\mathcal{W}_p(J;X)$ such that $\|g\|_{[p]} < 1/(2m)$. It is easy to see that $\operatorname{ran}(G+tg) \subset B_{2m} = B_{2m}(U) \subset U$ for each $g \in V$ and $t \in [0,1]$. Since $F$ is Lipschitz on $B_{2m}$ and $p < r$, $N_F$ acts from $G+V$ into $\mathcal{W}_r(J;Y)$ by Proposition 6.34.

To show Fréchet differentiability of $N_F$, for each $g \in \mathcal{W}_p(J;X)$ let $Lg$ be the function $s \mapsto (DF(G(s)))g(s)$, $s \in J$. To show that $Lg$ has bounded $r$-variation let $\kappa = \{t_i\}_{i=0}^n$ be a point partition of $J$. Then using the Minkowski inequality (1.5), we have

$$\begin{aligned} s_r(Lg;\kappa)^{1/r} &\le \Big(\sum_{i=1}^n \big\|[DF(G(t_i)) - DF(G(t_{i-1}))]g(t_i)\big\|^r\Big)^{1/r} \\ &\quad + \Big(\sum_{i=1}^n \big\|DF(G(t_{i-1}))[g(t_i) - g(t_{i-1})]\big\|^r\Big)^{1/r} \\ &\le \|DF\|_{B_m,(\mathcal{H}_\alpha)}\|G\|_{(\alpha r)}^{\alpha}\|g\|_{\sup} + \|DF\|_{B_m,\sup}\|g\|_{(r)}. \end{aligned}$$

Since $\kappa$ is arbitrary and $p < \alpha r \le r$, it follows that $Lg \in \mathcal{W}_r(J;Y)$, and so $L$ is a bounded linear operator from $\mathcal{W}_p(J;X)$ into $\mathcal{W}_r(J;Y)$. Since $F$ is continuously differentiable on $U$ and since the segment joining $G(s)$ and $(G+g)(s)$ is included in $U$ for each $s \in J$, by Lemma 5.40, we have

$$\begin{aligned} &\big[N_F(G+g) - N_F(G) - Lg\big](s) \\ &\qquad = \int_0^1 \big[DF(G(s)+tg(s)) - DF(G(s))\big]g(s)\,dt = K_H(s), \end{aligned}$$

where $K_H$ is defined by (6.19) and (6.20). By Lemma 6.38 with $B = B_{2m}$, since now $p/r < \alpha$, $\|K_H\|_{[r]} = o(\|g\|_{[p]})$ as $\|g\|_{[p]} \to 0$, and so $N_F$ is Fréchet differentiable at $G$. Since $\|g\|_{[p]} < 1/2 < 1$ for $g \in V$, the remainder bound (6.18) follows from (6.21) with $C := \|DF\|_{B_{2m},\mathcal{H}_\alpha}[E\|G\|_{(p)}^{p/r} + 2E + 1]$, proving Theorem 6.37. □

On Fréchet differentiability of autonomous Nemytskii operators we also have the following, recalling Definition 6.5 with here $\mathbb{K} = \mathbb{R}$:

**Theorem 6.39.** *Let $0 < \alpha \le 1$, $1 \le p < \alpha r < \infty$ and $J := [a,b]$, where $-\infty < a < b < \infty$. Then for any $F \in \mathcal{H}_{1+\alpha}$, $N_F$ is Fréchet differentiable from $\mathcal{W}_p(J)$ into $\mathcal{W}_r(J)$, with derivative $DN_F(G)(g) = F'(G)g$ for any $G, g \in \mathcal{W}_p(J)$. There is a constant $M := M(\alpha,p,r) < \infty$ such that for each $G \in \mathcal{W}_p(G)$, $F \in \mathcal{H}_{1+\alpha}$, and $g \in \mathcal{W}_p(J)$,*

$$\begin{aligned} &\big\|\operatorname{Rem}_{N_F}(G,g)\big\|_{[r]} \\ &\qquad \le \|F'\|_{(\mathcal{H}_\alpha)}\Big\{M\|G\|_{(p)}^{p/r} + (1+M)\|g\|_{[p]}^{p/r}\Big\}\|g\|_{\sup}^{\alpha-p/r}\|g\|_{[p]}. \end{aligned}$$

**Sketch of proof.** The differentiability, with the given derivative, follows from the remainder bound as in the proof of Proposition 6.7. The statement is clearly similar to that of Theorem 6.37. There we assume $F \in \mathcal{H}_{1+\alpha}^{\rm loc}(U;Y)$, here $F \in \mathcal{H}_{1+\alpha}(\mathbb{R};\mathbb{R})$. There we had $DF(G(s)+tg(s))]g(s)$, which in the present scalar case becomes the product of the functions $F'(G(s)+tg(s))$ and $g(s)$. The $r$-variation norm of the product can be bounded using the Banach algebra property of $\mathcal{W}_r(J)$. So we need only consider $F'(G(s)+tg(s))$. The proof then is similar to but easier than those of Lemma 6.38 and Theorem 6.37 and is omitted. □

An operator $N$ acting from a normed space $(W_1, \|\cdot\|_1)$ into a normed space $(W_2, \|\cdot\|_2)$ is *globally Lipschitz* if there is a finite constant $L$ such that for each $x, y \in W_1$,

$$\big\|N(y) - N(x)\big\|_2 \le L\big\|y - x\big\|_1, \tag{6.22}$$

or *locally Lipschitz* if for each $0 < M < \infty$ there is a finite constant $L = L(M)$ such that (6.22) holds for all $x, y \in W_1$ having norms bounded by $M$. The next theorem gives a local Lipschitz property of the autonomous Nemytskii operator $N_F$ for functions $F$ in the class $\mathcal{H}_{1+\alpha}^{\rm loc}(X;Y)$ (Definition 6.5). Theorem 6.68 extends the local Lipschitz property to non-autonomous Nemytskii operators. The global Lipschitz property for Nemytskii operators is treated in Theorem 6.70.

**Theorem 6.40.** *Let $1 \le p \le q < \infty$, $\alpha := p/q$, and $J := [a,b]$ with $a < b$. For Banach spaces $X$ and $Y$, let $F \in \mathcal{H}_{1+\alpha}^{\rm loc}(X;Y)$. Then the autonomous Nemytskii operator $N_F$ acts from $\mathcal{W}_p(J;X)$ into $\mathcal{W}_q(J;Y)$. Also, for any $0 < R < \infty$ there are finite constants $C$ and $D$ such that for any $f, g \in \mathcal{W}_p(J;X)$ with sup norms bounded by $R$ and for any nondegenerate interval $A \subset J$,*

$$\|N_F f - N_F g\|_{A,\sup} \le C\|f-g\|_{A,\sup}, \tag{6.23}$$

$$\|N_F f - N_F g\|_{A,(q)} \le C\|f-g\|_{A,(q)} + D\|f-g\|_{A,\sup}\|g\|_{A,(p)}^{\alpha}. \tag{6.24}$$

*Moreover, if $F \in \mathcal{H}_{1+\alpha}(X;Y)$ and $DF$ is bounded then* (6.23) *and* (6.24) *hold with $C = \|DF\|_{\sup}$ and $D = \|DF\|_{(\mathcal{H}_\alpha)}$ for any $f, g \in \mathcal{W}_p(J;X)$.*

*Proof.* Since $F \in \mathcal{H}_{1+\alpha}^{\rm loc}(X;Y)$, its derivative $DF$ is in $\mathcal{H}_{\alpha}^{\rm loc}(X;L(Y;X))$ and so $DF$ is bounded on bounded sets. Thus by (5.2) in the mean value theorem, $F$ is locally Lipschitz, and so $F \in \mathcal{H}_1^{\rm loc}(X;Y) \subset \mathcal{H}_\alpha^{\rm loc}(X;Y)$. Hence the autonomous Nemytskii operator $N_F$ acts from $\mathcal{W}_p(J;X)$ into $\mathcal{W}_p(J;Y) \subset \mathcal{W}_q(J;Y)$ by Proposition 6.34 with $U = X$.

Letting $B := \{x\colon \|x\| \le 3R\}$, we will show that (6.23) and (6.24) hold with $C := \|DF\|_{B,\sup}$ and $D = \|DF\|_{B,(\mathcal{H}_\alpha)}$. By (5.2) in the mean value theorem, for each $s \in A$ we have

$$\big\|F(f(s)) - F(g(s))\big\| \le \|DF\|_{B,\sup}\|f-g\|_{A,\sup},$$

and so (6.23) holds. To prove (6.24) let $a \le s < t \le b$. By (5.2) in the mean value theorem again, we have

$$S := \big\|F(f(t)) - F(g(t) + (f-g)(s))\big\| \le \|DF\|_{B,\sup}\|(f-g)(t) - (f-g)(s)\|,$$

since the segment joining $f(t)$ and $g(t) + (f-g)(s)$ is in $B$. Since $DF \in \mathcal{H}_\alpha(B; L(X,Y))$, and so $F$ is continuously Fréchet differentiable, by two applications of Lemma 5.40, it follows that

$$\begin{aligned} T &:= \big\|\big[F(g(t) + (f-g)(s)) - F(g(t))\big] - \big[F(f(s)) - F(g(s))\big]\big\| \\ &= \Big\|\int_0^1 \Big[DF(g(t)+r(f-g)(s)) - DF(g(s)+r(f-g)(s))\Big]\Big((f-g)(s)\Big)\,dr\Big\| \\ &\le \|DF\|_{B,(\mathcal{H}_\alpha)}\|g(t)-g(s)\|^\alpha\|f(s)-g(s)\|. \end{aligned}$$

Then we have

$$\begin{aligned} R(s,t) := \big\|(F\circ f - F\circ g)(t) - (F\circ f - F\circ g)(s)\big\| &\le S + T \\ &\le C\|(f-g)(t) - (f-g)(s)\| + D\|g(t)-g(s)\|^\alpha\|f(s)-g(s)\|. \end{aligned}$$

Let $\kappa = \{t_i\}_{i=0}^n$ be a point partition of $A$. Applying the Minkowski inequality (1.5), it follows that

$$\begin{aligned} s_q(F\circ f - F\circ g;\kappa)^{1/q} &= \Big(\sum_{i=1}^n R(t_{i-1},t_i)^q\Big)^{1/q} \\ &\le Cs_q(f-g;\kappa)^{1/q} + D\|f-g\|_{A,\sup}s_p(g;\kappa)^{1/q} \\ &\le C\|f-g\|_{A,(q)} + D\|f-g\|_{A,\sup}\|g\|_{A,(p)}^\alpha. \end{aligned}$$

Since $\kappa$ is an arbitrary partition of $A$, (6.24) holds, completing the proof of the theorem. □

## 6.5 Nemytskii Operators on $\mathcal{W}_p$ Spaces

In this section, the Nemytskii operator $N_\psi$, where $\psi = \psi(u,s)$ can depend on $s$, is considered between the spaces $\mathbb{G} = \mathcal{W}_p(J;X)$ and $\mathbb{H} = \mathcal{W}_q(J;Y)$.

*Acting and boundedness conditions from $\mathcal{W}_p$ into $\mathcal{W}_q$*

For $X = Y = \mathbb{R}$, when does the Nemytskii operator $N_\psi$ act from $\mathcal{W}_p(J)$ into $\mathcal{W}_q(J)$ with $0 < p \le q < \infty$ and $\alpha := p/q$? If $\psi(u,s) \equiv F(u)$, then a necessary and sufficient condition is that $F$ is $\alpha$-Hölder on bounded sets, by Corollary 6.36. For an arbitrary $\psi$, an $\alpha$-Hölder condition in $u$ for each $s$ is not necessary, as the following example shows:

**Example 6.41.** For a linearly ordered set $S$, let $T$ be a finite subset of $S$, and let $\psi = \psi(u,s)$ be such that $\psi(u,s) \equiv 0$ if $s \in S \setminus T$ and $\psi(u,s)$ is arbitrary if $s \in T$. Then $N_\psi(g) \in \mathcal{W}_q(S)$ for each $q > 0$ and any function $g\colon S \to \mathbb{R}$.

For an arbitrary $\psi$, we use an $\alpha$-Hölder condition uniformly in $s$ as follows:

**Definition 6.42** (of $\mathcal{UH}_\alpha$ and $H_\alpha$). Let $0 < \alpha \leq 1$ and let $S$ be a nonempty set. For Banach spaces $X$ and $Y$, let $B$ be a subset of $X$ with more than one element and let $\psi\colon B \times S \to Y$. We say that $\psi$ is *$s$-uniformly $\alpha$-Hölder on $B$*, or on $B \times S$, or $\psi \in \mathcal{UH}_\alpha(B \times S; Y)$, if there exists a finite constant $H$ such that

$$\|\psi(u,s) - \psi(v,s)\| \leq H\|u - v\|^\alpha \tag{6.25}$$

for all $u, v \in B$ and all $s \in S$. Let $H_\alpha(\psi) := H_\alpha(\psi; B \times S, Y)$ be the minimal $H \geq 0$ such that (6.25) holds. We write $\psi \in \mathcal{UH}_\alpha(B \times S)$ if $X = Y = \mathbb{K}$.

If $\psi\colon X \times S \to Y$ is such that $B \ni u \mapsto \psi(u, \cdot) =: f(u) \in \ell^\infty(S;Y)$, then the preceding condition $\psi \in \mathcal{UH}_\alpha(B \times J; Y)$ holds if and only if $f\colon B \to \ell^\infty(S;Y)$ has the $\alpha$-Hölder property. Clearly, if $\psi(u,s) \equiv F(u)$ then $\psi \in \mathcal{UH}_\alpha(B \times S; Y)$ if and only if $F \in \mathcal{H}_\alpha(B;Y)$ and $H_\alpha(\psi) = \|F\|_{(\mathcal{H}_\alpha)}$. To define a local version of a uniform $\alpha$-Hölder condition recall Definition 6.4 and the sequence of sets $B_m(U)$, $m = 1, 2, \ldots$, defined there given an open set $U$.

**Definition 6.43** (of $\mathcal{UH}_\alpha^{\rm loc}$). Let $\alpha$, $S$, $X$, and $Y$ be as in Definition 6.42, let $U$ be a nonempty open subset of $X$, and let $\psi\colon U \times S \to Y$. We say that $\psi$ is *$s$-uniformly $\alpha$-Hölder locally on $U$*, or $\psi \in \mathcal{UH}_\alpha^{\rm loc}(U \times S; Y)$, if it is $s$-uniformly $\alpha$-Hölder on $B_m(U)$ for all sufficiently large $m$. We write $\psi \in \mathcal{UH}_\alpha^{\rm loc}(U \times S)$ if $X = Y = \mathbb{K}$.

We show next that if the Nemytskii operator $N_\psi$ is bounded from $\mathcal{W}_p$ into $\mathcal{W}_q$, then for $\psi \equiv \psi(u,s)$, the $(p/q)$-Hölder condition in $u$ is necessary for all but countably many $s$. Recall that a nonlinear operator from one normed space $E$ into another is called bounded if it takes bounded subsets of $E$ into bounded sets.

**Theorem 6.44.** *Let $1 \leq p \leq q < \infty$, $\alpha := p/q$, $J := [a,b]$ with $a < b$, and $\psi\colon \mathbb{R} \times J \to \mathbb{R}$. Suppose the Nemytskii operator $N_\psi$ is bounded from $\mathcal{W}_p(J)$ into $\mathcal{W}_q(J)$. Then $\psi \equiv \eta + \xi$, where*

(*a*) $\eta \in \mathcal{UH}_\alpha^{\rm loc}(\mathbb{R} \times J)$.
(*b*) *for some countable set $E \subset J$, $\xi(u,s) \equiv 0$ for $s \notin E$.*
(*c*) *we have $(u,s) \mapsto \psi(u, s+) \in \mathcal{UH}_\alpha^{\rm loc}(\mathbb{R} \times [a,b))$ and $(u,s) \mapsto \psi(u,s-) \in \mathcal{UH}_\alpha^{\rm loc}(\mathbb{R} \times (a,b])$.*
(*d*) *for each $u$ and for $f = \psi$, $\eta$, or $\xi$, $f(u,\cdot) \in \mathcal{W}_q(J)$. For $u$ bounded, $f(u,\cdot)$ are bounded in $\mathcal{W}_q(J)$.*
(*e*) *if for a dense set of values of $u$, $\psi(u,\cdot)$ is right-continuous on $[a,b)$ and left-continuous at $b$, then we can take $\xi \equiv 0$.*
(*f*) *$N_\eta$ and $N_\xi$ are also bounded operators from $\mathcal{W}_p(J)$ into $\mathcal{W}_q(J)$.*

*Proof.* On $\mathcal{W}_r(J)$ we have the equivalent norm $|\!|\!| f |\!|\!|_{[r]} := \max\{\|f\|_{(r)}, \|f\|_{\sup}\}$, with $|\!|\!| f |\!|\!|_{[r]} \leq \|f\|_{[r]} \leq 2|\!|\!| f |\!|\!|_{[r]}$ for all $f$. Then for any $0 < C < \infty$, $N_\psi$ acts from $\{h\colon \|h\|_{[p]} \le C\}$ into $\{f\colon \|f\|_{(q)} \le M\}$ for some $M := M(C) < \infty$. For any $C > 0$ let

$$H := H(C) := 2^{2\alpha+1+(1/q)}M. \tag{6.26}$$

We will first show that for some countable set $T \subset J$, and $\psi_T = \psi$ restricted to $\mathbb{R} \times (J \setminus T)$,

$$\psi_T \in \mathcal{UH}_\alpha^{\mathrm{loc}}(\mathbb{R} \times (J \setminus T)). \tag{6.27}$$

Then (6.27) will also hold for any countable $E \supset T$ in place of $T$. Suppose (6.27) fails for all countable $T \subset J$. Then for some positive integer $k$, there is no countable set $T$ such that $\psi_T \in \mathcal{UH}_\alpha([-k,k] \times (J \setminus T))$. Indeed, otherwise for each $k = 1,2,\ldots,$ there is a countable $T(k) \subset J$ such that $\psi_{T(k)} \in \mathcal{UH}_\alpha([-k,k] \times (J \setminus T(k)))$. Then letting $T := \cup_{k\ge1} T(k)$, (6.27) holds with $T$ countable, a contradiction. Therefore for each $m = 1,2,\ldots,$ the set $\mathcal{E}_m := \mathcal{E}_{m,k}$ of all $s \in J$ such that for some $u,v \in [-k,k]$, $\big|\psi(u,s) - \psi(v,s)\big| > m|u-v|^\alpha$ is uncountable. For any $u,v \in [-k,k]$ there are points $u_l, v_l \in [-k,k]$, $1 \le l \le 10k$, such that $\max_l |u_l - v_l| \le 1/5$ and

$$|\psi(u,s) - \psi(v,s)| \le 10k \max_l |\psi(u_l,s) - \psi(v_l,s)|$$

for any $s \in S$. Therefore for each $m = 1,2,\ldots$ and for all $s \in \mathcal{E}_{m,k}$ there are $u,v \in [-k,k]$ such that $|u-v| \le 1/5$ and $\big|\psi(u,s) - \psi(v,s)\big| > (m/(10k))|u-v|^\alpha$. Take $C = k$, fix $m \ge 10kH$ with $H$ defined by (6.26), and let $\mathcal{F} := \mathcal{E}_{m,k}$. For $r = 5,6,\ldots,$ let $\mathcal{F}_r$ be the set of all $s \in \mathcal{F}$ such that $u$ and $v$ can be chosen with $1/(2r) < |u-v| \le 1/r$. Since $\mathcal{F} = \cup_{r=5}^\infty \mathcal{F}_r$, some $\mathcal{F}_r$ is uncountable, and we can assume $\mathcal{F} = \mathcal{F}_r$. Let $\delta := 1/(2r) < 1/9$. We can assume $u = y_s$ and $v = z_s$ chosen so that $\delta < z_s - y_s \le 2\delta$ for all $s$. There is a finite set $F \subset [-C,C]$ such that each point of $[-C,C]$ is within $\delta/3$ of some point in $F$. Thus for some $u \in F$, there is an uncountable set $A_u \subset J$ with $y_s \le u \le z_s$ for all $s \in A_u$. For an $n > (2M/H)^q/\delta^p$, take $a < s_1 < \cdots < s_n < b$ with $s_j \in A_u$ for $j = 1,\ldots,n$. Let $s_0 := a$ and $g(s_0) := u$. Recursively, suppose we have defined $g(s_0),\ldots,g(s_{j-1})$ for $j \le n$, where $g(s_i) = z_{s_i}$ or $y_{s_i}$ for each $i$ with $1 \le i < j$, so that $|g(s_i) - u| \le 2\delta$. Since $|\psi(z_{s_j},s_j) - \psi(y_{s_j},s_j)| > H\delta^\alpha$, we can choose $g(s_j) = y_{s_j}$ or $z_{s_j}$ such that $\Delta_j := |\psi(g(s_j),s_j) - \psi(g(s_{j-1}),s_{j-1})| > H\delta^\alpha/2$. In either case, $|g(s_j) - g(s_{j-1})| \le 4\delta$. Thus $g(s_j)$ are defined for $j = 0,1,\ldots,n$. For each $r = 1,\ldots,n$, we have $\sum_{j=1}^r \Delta_j^q \ge r(H/2)^q\delta^p$. There is a least $r \le n$ such that $r(H/2)^q\delta^p > M^q$, and so $2r(4\delta)^p > 1$ by (6.26). Then $r \ge 2$ and

$$\sum_{j=1}^r \big|N_\psi(g)(s_j) - N_\psi(g)(s_{j-1})\big|^q = \sum_{j=1}^r \Delta_j^q > M^q. \tag{6.28}$$

Also, we have $(r/2)(H/2)^q\delta^p \le M^q$, and so using (6.26) and $C = k \ge 1$,

$$\sum_{j=1}^r |g(s_j) - g(s_{j-1})|^p \le r(4\delta)^p \le 1 \le C^p. \tag{6.29}$$

Let $h(s_j) := g(s_j)$ for $j = 0, \ldots, r$. Let $h$ be linear on each interval $[s_{j-1}, s_j]$, $j = 1, \ldots, r$, and $h(s) := h(s_r) = g(s_r)$ for $s \in [s_r, b]$. Any $p$-variation sum $s_p(h; \kappa)$ for $h$ with a point partition $\kappa$ of $[a, b]$ is dominated by one in which adjoining increments of the same sign are combined, so we can assume that adjoining increments are of opposite signs. Then we can assume that the points of $\kappa$ are local maxima or minima of $h$ and that $\kappa \subset \{s_j\}_{j=0}^r$. Each $g(s_j)$ is within $2\delta$ of $u$, so $v_p(h) \le r(4\delta)^p \le 1$ by (6.29), and $\|h\|_{(p)} \le 1 \le C$. Since $\|h\|_{\sup} \le C$ we have $\|h\|_{[p]} \le C$, but $\|N_\psi(h)\|_{(q)} > M$ by (6.28), a contradiction. It follows that (6.27) holds for a countable set $T \subset J$.

By taking any constant function $u1(\cdot)$ in $\mathcal{W}_p(J)$ we get $\psi(u, \cdot) \in \mathcal{W}_q(J)$, which is $(d)$ for $f = \psi$, and so $\psi(u, \cdot)$ is regulated by Proposition 3.33. Therefore for $u \in \mathbb{Q}$ (rational), $\psi(u, \cdot)$ is continuous except at most for $s$ in a countable set $E_u \subset J$. Let $E := T \cup \bigcup_{u \in \mathbb{Q}} E_u \cup \{b\}$. Then $E$ is countable.

For every $u \in \mathbb{R}$ and $a \le s < b$, the limit $\psi(u, s+) = \lim_{t \downarrow s} \psi(u, t)$ exists. The limit can be taken through $t \notin E$ since $E$ is countable. For $t \notin E$, the functions $\psi(\cdot, t)$ are uniformly $\alpha$-Hölder, thus uniformly equicontinuous, on every bounded interval $[-k, k]$, $k = 1, 2, \ldots$. So as $t \downarrow s$ through $t \notin E$, $\psi(\cdot, t)$ converges uniformly on $[-k, k]$ to $\psi(\cdot, s+)$. Thus $(u, s) \mapsto \psi(u, s+) \in \mathcal{UH}_\alpha([-k, k] \times [a, b))$. Moreover, if $s \notin E$ then $\psi(u, s+) = \psi(u, s)$ for all $u \in \mathbb{Q}$ and so for all $u$ since both functions are $\alpha$-Hölder.

For all $u \in \mathbb{R}$, let $\eta(u, s) := \psi(u, s+)$ for $a \le s < b$, and let $\eta(u, b) := \lim_{t \uparrow b} \psi(u, t)$. Then $(a)$ holds for $\eta$. Let $\xi(u, s) := \psi(u, s) - \eta(u, s)$ for all $u \in \mathbb{R}$ and $s \in J$. Then $(b)$ holds for $\xi$. The first conclusion of $(c)$ is already proved. The proof of the second one is symmetric. Next, $(d)$ holds for $f = \eta$ since taking right or left limits cannot increase the $q$-variation. Since $\mathcal{W}_q$ is a vector space, $(d)$ also holds for $f = \xi$. Boundedness of $f(u, \cdot)$ for $u$ bounded holds by the hypotheses, so $(d)$ is proved. For $(e)$, replacing $\mathbb{Q}$ by the given dense set, the same proof gives $\eta \equiv \psi$ so $\xi \equiv 0$.

To prove $(f)$, for $\|h\|_{\sup} \le \|h\|_{[p]} \le S < \infty$, we have

$$\|N_\eta(h)\|_{\sup} \le \sup\{|\eta(u, s)|\colon\ s \in J,\ |u| \le S\} < \infty$$

by $(d)$. Any $q$-variation sum

$$\sum_{i=1}^{n} \left|\eta(h(s_i), s_i) - \eta(h(s_{i-1}), s_{i-1})\right|^q$$

can be approximated arbitrarily closely by a sum

$$\sum_{i=1}^{n} \left|\psi(h(s_i), t_i) - \psi(h(s_{i-1}), t_{i-1})\right|^q$$

for $t_i \notin E$, $t_i$ close enough to $s_i$ and $t_i > s_i$ except if $s_n = b$, when $t_n < s_n$, and $t_1 < t_2 < \cdots < t_n$. For any such $\{t_i\}_{i=1}^n$, there is a $g \in \mathcal{W}_p(J)$ with $g(t_i) = h(s_i)$ for $i = 1, \ldots, n$ and $\|g\|_{[p]} \le \|h\|_{[p]}$. It follows that for any $S < \infty$,

$$\sup\left\{\|N_\eta(h)\|_{(q)}\colon \|h\|_{[p]} \le S\right\} \le \sup\left\{\|N_\psi(g)\|_{(q)}\colon \|g\|_{[p]} \le S\right\} < \infty.$$

So $(f)$ holds for $N_\eta$, and since $N_\xi \equiv N_\psi - N_\eta$, also for $N_\xi$. The theorem is proved. □

**Remark 6.45.** For $\psi(u,s) \equiv F(u)$ not depending on $s$ the preceding theorem gives that $F$ is $\alpha$-Hölder on bounded intervals. This also follows from the case $\Psi(y) = y^q$ and $\Phi(y) = y^p$ of Theorem 6.35, which assumes that $N_F$ acts (without assuming it is bounded) from $\mathcal{W}_\Phi$ to $\mathcal{W}_\Psi$. In this case the acting condition implies boundedness by Proposition 6.34.

Example 6.41 shows that if the set $E$ in $(b)$ is finite, $\psi(\cdot,s)$ for $s \in E$ can be arbitrary bounded functions. Specifically, let $E$ have cardinality $n < \infty$, let $|\psi(u,s)| \le M < \infty$ for any $s \in E$ and $u \in \mathbb{R}$, and let $\psi(u,s) = 0$ whenever $s \in J \setminus E$. Then for an arbitrary function $g\colon J \to \mathbb{R}$ we have $N_\psi g \in \mathcal{W}_q(J)$ with $\|N_\psi g\|_{\sup} \le M$ and $\|N_\psi g\|_{(q)} \le 2n^{1/q}M$. So $N_\psi$ is bounded in the unusually strong sense that it is bounded uniformly on any space of real-valued functions, not only on bounded sets for some norm. In this case $\eta \equiv 0$ and $\xi \equiv \psi$.

The following shows that $\psi(u,\cdot)$ is regulated uniformly for $u$ in bounded intervals whenever the Nemytskii operator $N_\psi$ acts from $\mathcal{W}_1(J)$ into $\mathcal{R}(J)$. Recall that $\mathcal{W}_\infty(J)$ is the space $\mathcal{R}(J)$ of real-valued regulated functions on $J$ and $\ell^\infty(J)$ is the space of real-valued bounded functions on $J$, both with supremum norm.

**Proposition 6.46.** *Let $1 \le p \le \infty$, $1 \le q \le \infty$, $J := [a,b]$ with $a < b$, and $\psi\colon \mathbb{R} \times J \to \mathbb{R}$. Suppose the Nemytskii operator $N_\psi$ acts from $\mathcal{W}_p(J)$ into $\mathcal{W}_q(J)$. Then*

*(a) $\psi(u,\cdot) \in \mathcal{W}_q(J)$ for each $u \in \mathbb{R}$;*
*(b) for each $k = 1,2,\ldots$, $\psi$ is bounded on $[-k,k] \times J$;*
*(c) for each $C > 0$, letting $\psi_C$ be $\psi$ restricted to $[-C,C] \times J$, the function $s \mapsto \psi_C(\cdot,s)$ is in $\mathcal{R}(J;\ell^\infty([-C,C]))$.*

*Proof.* Since $\mathcal{W}_1(J) \subset \mathcal{W}_p(J)$ for $p \ge 1$ by Lemma 3.45, we can assume $p = 1$. For each $u \in \mathbb{R}$, $\psi(u,\cdot) = N_\psi(u1(\cdot)) \in \mathcal{W}_q(J)$ since $u1(\cdot) \in \mathcal{W}_1(J)$, proving $(a)$. To prove $(b)$, suppose it fails for some $k$. For each $j = 1,2,\ldots$, choose $u_j \in [-k,k]$ and $t_j \in J$ with $|\psi(u_j,t_j)| > j$. Taking subsequences and by symmetry, we may and do assume that $u_j \downarrow u_0 \in [-k,k]$ and $t_j \uparrow s \in J$. Let $h(a) := u_1$, $h(t_j) := u_j$ for $j \ge 1$, let $h$ be linear on each interval $[t_{j-1},t_j]$, where $t_0 := a$, and let $h(t) := u_0$ for $t \in [s,b]$. Then $h$ is nonincreasing and bounded, so $h \in \mathcal{W}_1(J)$, but $(N_\psi h)(t_j)$ are unbounded, a contradiction. Thus $(b)$ holds.

To prove $(c)$, each $\psi_C(\cdot,s)$ is bounded on $[-C,C]$ by $(b)$. By $(a)$ and Proposition 3.33, $\psi(u,\cdot)$ is regulated for each $u \in \mathbb{R}$. By symmetry it is enough to prove that for each $s \in (a,b]$ and $0 < C < \infty$, $\psi(u,t)$ converges to $\psi(u,s-)$

as $t \uparrow s$ uniformly for $|u| \leq C$. Suppose not. Then for some $0 < C < \infty$ and $s \in (a, b]$ there are $\epsilon > 0$, $a < t_j \uparrow s$, and $|u_j| \leq C$ such that $|\psi(u_j, t_j) - \psi(u_j, s-)| \geq 2\epsilon$. Taking a subsequence, we can assume that $u_j \to u_0$ for some $u_0 \in [-C, C]$, and then by symmetry that $u_j \downarrow u_0$. For each $j$, take $s_j \in (t_j, s)$ such that $|\psi(u_j, s_j) - \psi(u_j, s-)| < \epsilon$ and so $|\psi(u_j, t_j) - \psi(u_j, s_j)| > \epsilon$. Taking a subsequence of $\{t_j\}$, we can assume that $t_j < s_j < t_{j+1}$ for all $j$. Let $g(t_j) := g(s_j) := u_j$ for all $j \geq 1$ and $g(a) := u_1$. Let $g$ be linear on each interval between adjacent values where it is defined and $g \equiv u_0$ on $[s, b]$. Then $g$ is nonincreasing, so it is in $\mathcal{W}_1(J)$. Also, for each $j = 1, 2, \ldots$ we have that $|(N_\psi g)(t_j) - (N_\psi g)(s_j)| > \epsilon$. So $N_\psi g(s-)$ does not exist and $N_\psi g$ is not regulated, contradicting the acting hypothesis. □

We will next define some conditions related to boundedness of the Nemytskii operator $N_\psi$ from $\mathcal{W}_p$ into $\mathcal{W}_q$ with $p \leq q$. For $0 < p < \infty$ and $x = (x_1, \ldots, x_n)$, $x_i \in X$, let $w_p(x) := v_p(\{x_i - x_{i-1}\}_{i=2}^n)^{1/p} + \max_i \|x_i\|$, where

$$v_p(\{x_i - x_{i-1}\}_{i=2}^n) := \max\Big\{\sum_{j=1}^m \big\|x_{\theta(j)} - x_{\theta(j-1)}\big\|^p \colon 1 \leq m < n, \quad 1 = \theta(0) < \theta(1) < \cdots < \theta(m) = n\Big\}. \tag{6.30}$$

Clearly, $v_p(\{x_i - x_{i-1}\}_{i=2}^n) = v_\Phi(\{x_i - x_{i-1}\}_{i=2}^n)$ with $\Phi(u) \equiv u^p$, $u \geq 0$, defined by (3.136). For $0 < \alpha \leq 1$, let $\psi \in \mathcal{UH}_\alpha(\mathbb{R} \times J)$, $q := p/\alpha$, and $f \in \mathcal{W}_p(J)$. Using the Minkowski inequality (1.5), for any point partition $\kappa = \{s_i\}_{i=0}^n$ of $J$, we have

$$s_q(N_\psi f; \kappa)^{1/q} \leq H_\alpha(\psi)\|f\|_{(p)}^\alpha + \Big(\sum_{i=1}^n \big|\psi(f(s_i), s_i) - \psi(f(s_i), s_{i-1})\big|^q\Big)^{1/q}$$

and $w_p(\{f(s_i)\}_{i=1}^n) \leq \|f\|_{[p]}$. Therefore $N_\psi f \in \mathcal{W}_q(J)$ if $\psi$ also satisfies the condition formulated next.

**Definition 6.47** (of $\mathcal{W}_{\alpha,q}$, $W_{\alpha,q}$, and $\mathcal{HW}_{\alpha,q}$). Let $0 < \alpha \leq 1$, let $0 < q < \infty$ and let $J$ be a nondegenerate interval. For Banach spaces $X$, $Y$, let $B$ be a nonempty subset of $X$ and let $\psi \equiv \psi(u, s) \colon B \times J \to Y$. We say that $\psi$ is in the class $\mathcal{W}_{\alpha,q}(B \times J; Y)$ if for each $0 \leq K < \infty$, there exists a finite constant $W = W(K)$ such that

$$s_q(\psi; \mu, \kappa) := \sum_{i=1}^n \|\psi(u_i, s_i) - \psi(u_i, s_{i-1})\|^q \leq W^q \tag{6.31}$$

for all partitions $\kappa = \{s_i\}_{i=0}^n$ of $J$ and for all finite sets $\mu = \{u_i\}_{i=1}^n \subset B$ such that $w_{\alpha q}(\mu) \leq K$. Let $W_{\alpha,q}(\psi, K) = W_{\alpha,q}(\psi, K; B \times J, Y)$ be the minimal $W \geq 0$ such that (6.31) holds for a given $K$.

Suppose in addition that $B$ has more than one element. We say that $\psi$ is in the class $\mathcal{HW}_{\alpha,q}(B \times J; Y)$ if

$(a)$ $\psi$ is $s$-uniformly $\alpha$-Hölder on $B$, or $\psi \in \mathcal{UH}_\alpha(B \times J; Y)$ (Definition 6.42);
$(b)$ $\psi \in \mathcal{W}_{\alpha,q}(B \times J; Y)$.

In other words, $\mathcal{HW}_{\alpha,q}(B \times J; Y) = \mathcal{UH}_\alpha(B \times J; Y) \cap \mathcal{W}_{\alpha,q}(B \times J; Y)$. We write $\psi \in \mathcal{HW}_{\alpha,q}(B \times J)$ if $X = Y = \mathbb{K}$.

We will next compare the conditions $\psi \in \mathcal{UH}_\alpha(B \times J; Y)$, $\psi \in \mathcal{W}_{\alpha,q}(B \times J; Y)$ and $\psi \in \mathcal{HW}_{\alpha,q}(B \times J; Y)$ with the following two properties of $\psi$:

$$u \mapsto \psi(u, \cdot) \in \mathcal{H}_\alpha(B; \mathcal{W}_q(J; Y)), \tag{6.32}$$

$$s \mapsto \psi(\cdot, s) \in \mathcal{W}_q(J; \mathcal{H}_{\alpha,\infty}(B; Y)). \tag{6.33}$$

In (6.33), if $B$ is bounded, $\mathcal{H}_{\alpha,\infty}(B; Y) = \mathcal{H}_\alpha(B; Y)$ as a set, but $\mathcal{H}_{\alpha,\infty}(B; Y)$ has the norm $\|\cdot\|_{\mathcal{H}_\alpha}$ and $\mathcal{H}_\alpha(B; Y)$ only a seminorm.

**Proposition 6.48.** *Let $0 < \alpha \le 1$, $1 \le q < \infty$, $J := [a, b]$ with $a < b$, let $X, Y$ be Banach spaces, and let $B$ be a bounded set in $X$ with more than one element. For $\psi\colon B \times J \to Y$ the following hold:*

*$(a)$ if (6.32) holds then $\psi \in \mathcal{UH}_\alpha(B \times J; Y)$;*
*$(b)$ if (6.33) holds then $\psi \in \mathcal{HW}_{\alpha,q}(B \times J; Y)$.*

*Proof.* For $(a)$, assuming that (6.32) holds, it follows directly that $\psi \in \mathcal{UH}_\alpha(B \times J; Y)$. For $(b)$ let (6.33) hold. The closure of the range of $s \mapsto \psi(\cdot, s)$ is compact in $\mathcal{H}_{\alpha,\infty}(B; Y)$ by Propositions 6.2 and 6.20$(c)$, and so it is bounded, showing that $\psi \in \mathcal{UH}_\alpha(B \times J; Y)$. Clearly (6.33) implies $\psi \in \mathcal{W}_{\alpha,q}(B \times J; Y)$. Thus (6.33) implies $\psi \in \mathcal{HW}_{\alpha,q}(B \times J; Y)$, proving the proposition. □

**Remark 6.49.** In the preceding proposition, clearly $(a)$ also holds if in (6.32) $\mathcal{W}_q(J; Y)$ is replaced by $\ell^\infty(J; Y)$, and $(b)$ also holds if in (6.33) $\mathcal{H}_{\alpha,\infty}(B; Y)$ is replaced by $\ell^\infty(B; Y)$.

The following example shows that $\psi \in \mathcal{HW}_{\alpha,q}(B \times J; Y)$ does not imply (6.33), that is, the converse to $(b)$ in the preceding proposition does not hold.

**Example 6.50.** Let $Y := \mathbb{R}$ and $X = \mathbb{H}$, a separable, infinite-dimensional Hilbert space with orthonormal basis $\{e_j\}_{j=1}^\infty$. For $j = 1, 2, \ldots$ and $u \in B := \{x\colon \|x\| \le 1\} \subset \mathbb{H}$, let $F_j(u) := \max\{0, 1 - 2\|e_j - u\|\}$. Let $J := [0, 1]$ and for $s \in J$, let $h_j(s) := \max\{0, 1 - 2^{j+1}|s - (1/2^j)|\}$, $j = 1, 2, \ldots$. Then for $(u, s) \in B \times J$, let $\psi(u, s) := \sum_{j=1}^\infty F_j(u) h_j(s) \in \mathbb{R}$. The sum converges since for any $u$, $F_j(u) \ne 0$ for at most one value of $j$, and so $\|\psi\|_{\sup} = 1$. If for some $u, v \in B$, $F_j(u) \ne 0$, $F_i(v) \ne 0$, and $i \ne j$, then $\|u - v\| \ge \sqrt{2} - 1$. Thus if $\|u - v\| < \sqrt{2} - 1$ and $s \in J$ then either $\psi(u, s) - \psi(v, s) = 0$, or for some $j$,

$$|\psi(u, s) - \psi(v, s)| = |F_j(u) - F_j(v)||h_j(s)| \le 2\|u - v\|,$$

and so $\|\psi(\cdot, s)\|_{\mathcal{H}_1} \le 1 + 2/(\sqrt{2} - 1)$ for all $s$ and $\psi \in \mathcal{UH}_1(B \times J; \mathbb{R}) \subset \mathcal{UH}_\alpha(B \times J; \mathbb{R})$ for any $\alpha \in (0, 1]$. Let $\alpha \in (0, 1]$, $1 \le q < \infty$, and $p := \alpha q$.

To show that $\psi \in \mathcal{W}_{\alpha,q}(B \times J;\mathbb{R})$, let $\kappa = \{s_i\}_{i=0}^n$ be a partition of $J$, let $0 < K < \infty$, and let $\mu = \{u_i\}_{i=1}^n \subset B$ be such that $w_p(\mu) \le K$. Then we have at most finitely many values of $j$, say $N$, such that $F_j(u_i) \ne 0$ for some $i$. Let $I$ be the set of such $j$. Since $w_p(\mu) \le K$, $K^p \ge (N-1)(\sqrt{2}-1)^p$, and so $N \le 1 + (K/(\sqrt{2}-1))^p$. Using the Minkowski inequality (1.5), it then follows that

$$s_q(\psi;\mu,\kappa)^{1/q} \le \sum_{j\in I}\Big(\sum_{i:\ \|u_i-e_j\|\le 1/2} |F_j(u_i)|^q |h_j(s_i)-h_j(s_{i-1})|^q\Big)^{1/q}$$
$$\le 2(1+(K/(\sqrt{2}-1))^p),$$

since $v_1(h_j) \le 2$ for each $j$, proving $\psi \in \mathcal{W}_{\alpha,q}(B \times J;\mathbb{R})$. But letting $e_0 := 0$, $s_i := 1/2^{n-i}$, $i = 0,1,\dots,n$, for any $n = 1,2,\dots$ and $0 < q < \infty$ we have

$$n - 1 = \sum_{i=1}^n |\psi(e_{n-i},s_i) - \psi(e_{n-i},s_{i-1})|^q \le \sum_{i=1}^n \|\psi(\cdot,s_i) - \psi(\cdot,s_{i-1})\|_{\sup}^q.$$

Thus the mapping $s \mapsto \psi(\cdot,s)$ is not in $\mathcal{W}_q(J;\ell^\infty(B;\mathbb{R}))$, and so $s \mapsto \psi(\cdot,s) \notin \mathcal{W}_q(J;\mathcal{H}_{\alpha,\infty}(B;\mathbb{R}))$ for each $\alpha \in (0,1]$, although it takes values in $\mathcal{H}_\alpha(B;\mathbb{R}) = \mathcal{H}_{\alpha,\infty}(B;\mathbb{R})$.

Next, local versions of the conditions in Definition 6.47 will be defined on $U \times J$ for an open set $U \subset X$, using the sequence of closed, bounded sets $B_m := U_{1/m} \uparrow U$ as $m \to \infty$ defined before Proposition 6.3. The $\mathcal{W}_{\alpha,q}^{\mathrm{loc}}$ condition on $\psi \in \mathcal{UH}_\alpha^{\mathrm{loc}}$ will be shown to be sufficient (Proposition 6.54(*b*)) for the Nemytskii operator $N_\psi$ to be bounded from $\mathcal{W}_p$ into $\mathcal{W}_q$ with $q = \alpha p$. Proposition 6.57 will show that a weaker condition is not sufficient.

**Definition 6.51** (of $\mathcal{W}_{\alpha,q}^{\mathrm{loc}}$ and $\mathcal{HW}_{\alpha,q}^{\mathrm{loc}}$). Let $\alpha$, $q$, $J$, $X$, and $Y$ be as in Definition 6.47. Let $U$ be a nonempty open subset of $X$ and let $\psi\colon U \times S \to Y$. We say that $\psi$ is in the class $\mathcal{W}_{\alpha,q}^{\mathrm{loc}}(U \times J;Y)$ or in the class $\mathcal{HW}_{\alpha,q}^{\mathrm{loc}}(U \times J;Y)$ if for each sufficiently large $m$, $\psi$ is, respectively, in the class $\mathcal{W}_{\alpha,q}(B_m \times J;Y)$ or in the class $\mathcal{HW}_{\alpha,q}(B_m \times J;Y)$. We write $\psi \in \mathcal{W}_{\alpha,q}^{\mathrm{loc}}(U \times J)$ and $\psi \in \mathcal{HW}_{\alpha,q}^{\mathrm{loc}}(U \times J)$ if $X = Y = \mathbb{K}$.

Recalling the definitions of $H_\alpha(\cdot;B \times J,Y)$ and $W_{\alpha,q}(\cdot,K;B \times J,Y)$ in Definitions 6.42 and 6.47, respectively, the following is easy to check:

**Proposition 6.52.** *Let $1 \le q < \infty$, $\alpha \in (0,1]$, let $J$ be a nondegenerate interval, and for Banach spaces $X,Y$, let $B$ be a subset of $X$ with more than one element. Then $H_\alpha(\cdot;B \times J,Y)$, and $W_{\alpha,q}(\cdot,K;B \times J,Y)$ for each $K \in [0,\infty)$, are seminorms.*

The two seminorms for $\psi$ just introduced will be used to give sufficient conditions for acting and boundedness of the Nemytskii operator $N_\psi$, in parts (*a*) and (*b*) of the next fact. Conversely, part (*c*) shows that finiteness of $W_{\alpha,q}$ is necessary given that of $H_\alpha$. See further Theorem 6.58.

**Lemma 6.53.** *Let $1 \le p \le q < \infty$, $\alpha := p/q$, $J := [a,b]$ with $a < b$ and let $A$ be a nondegenerate subinterval of $J$. For Banach spaces $X$ and $Y$, let $B$ be a subset of $X$ with more than one element and let $\psi\colon B \times J \to Y$. Then the following hold:*

*(a) for $g \in \mathcal{W}_p(J;X)$, if the range of $g$ is included in $B$, then*

$$\big\|N_\psi g\big\|_{A,(q)} \le H_\alpha(\psi; B\times A, Y)\|g\|_{A,(p)}^\alpha + W_{\alpha,q}(\psi, \|g\|_{A,[p]}; B\times A, Y);$$

*(b) for $g \in \mathcal{W}_p(J;X)$, if $\operatorname{ran}(g) \subset B$, $v \in B$ and $t \in A$, then*

$$\begin{aligned}\big\|N_\psi g\big\|_{A,\sup} &\\ &\le \|\psi\|_{B\times A,\sup}\\ &\le H_\alpha(\psi; B\times A, Y)\sup_{u\in B}\|u-v\|^\alpha + W_{\alpha,q}(\psi, \|v\|; \{v\}\times A, Y) + \|\psi(v,t)\|;\end{aligned}$$

*(c) for $0 \le K < \infty$,*

$$\begin{aligned}&W_{\alpha,q}(\psi, K; B\times A, Y)\\ &\quad\le H_\alpha(\psi; B\times A, Y)K^\alpha + \sup\{\|N_\psi g\|_{A,(q)}\colon \|g\|_{A,[p]} \le K, \operatorname{ran}(g) \subset B\}.\end{aligned}$$

*Proof.* In proving each inequality we can assume that its right side is finite since otherwise there is nothing to prove. If $g$ is constant, the proof will show that the inequality in $(a)$ holds with $\infty \cdot 0$ replaced by $0$ if $H_\alpha(\psi; B\times A, Y) = +\infty$. To prove $(a)$, let $\kappa = \{s_i\}_{i=0}^n$ be a partition of $A$. Then $w_p(\{g(s_{i-1})\}_{i=1}^n) \le \|g\|_{A,[p]}$. Thus by the Minkowski inequality (1.5), (6.25) and (6.31), we have

$$\begin{aligned}s_q(N_\psi g;\kappa)^{1/q} &\le \Big(\sum_{i=1}^n \Big[\|\psi(g(s_i),s_i) - \psi(g(s_{i-1}),s_i)\|\\ &\qquad\qquad + \|\psi(g(s_{i-1}),s_i) - \psi(g(s_{i-1}),s_{i-1})\|\Big]^q\Big)^{1/q}\\ &\le H_\alpha(\psi; B\times A, Y)\Big(\sum_{i=1}^n \|g(s_i) - g(s_{i-1})\|^p\Big)^{1/q}\\ &\quad + \Big(\sum_{i=1}^n \|\psi(g(s_{i-1}),s_i) - \psi(g(s_{i-1}),s_{i-1})\|^q\Big)^{1/q}\\ &\le H_\alpha(\psi; B\times A, Y)\|g\|_{A,(p)}^\alpha + W_{\alpha,q}(\psi, \|g\|_{A,[p]}; B\times A, Y).\end{aligned}$$

Since $\kappa$ is an arbitrary partition of $A$, $(a)$ follows.

To prove $(b)$, for $g \in \mathcal{W}_p(J;X)$ with $\operatorname{ran}(g) \subset B$, for any $u, v \in B$ and $s, t \in J$, we have

$$\begin{aligned}\big\|\psi(u,s)\big\| &\le \|\psi(u,s) - \psi(v,s)\| + \|\psi(v,s) - \psi(v,t)\| + \|\psi(v,t)\|\\ &\le H_\alpha(\psi; B\times A, Y)\sup_{u\in B}\|u-v\|^\alpha + W_{\alpha,q}(\psi, \|v\|; \{v\}\times A, Y) + \|\psi(v,t)\|.\end{aligned}$$

Since the first inequality in $(b)$ is immediate, $(b)$ holds.

For $(c)$, let $\kappa = \{s_i\}_{i=0}^n$ be a partition of $A$, and let $\mu = \{u_i\}_{i=1}^n \subset B$ be such that $w_p(\mu) \le K$. For such $\{u_i\}$ there is a $g \in \mathcal{W}_p(J;X)$ with $g(s_i) = u_i$ for all $i$, $\operatorname{ran}(g) \subset \mu \subset B$, and $\|g\|_{A,[p]} \le K$. Then applying the Minkowski inequality (1.5), (6.25), and (6.31), where $s_q(\psi;\mu,\kappa)$ is defined, we have

$$\begin{aligned}
&s_q(\psi;\mu,\kappa)^{1/q}\\
&\le \Big(\sum_{i=1}^n \Big[\|(N_\psi g)(s_i)-(N_\psi g)(s_{i-1})\|+\|\psi(g(s_{i-1}),s_{i-1})-\psi(g(s_i),s_{i-1})\|\Big]^q\Big)^{\frac{1}{q}}\\
&\le \big\|N_\psi g\big\|_{A,(q)} + H_\alpha(\psi;B\times A,Y)\|g\|_{A,(p)}^\alpha\\
&\le \sup\big\{\|N_\psi g\|_{A,(q)}\colon\ \|g\|_{A,[p]} \le K,\ \operatorname{ran}(g)\subset B\big\} + H_\alpha(\psi;B\times A,Y)K^\alpha.
\end{aligned}$$

Since $\kappa$ and $\mu$ are arbitrary, $(c)$ holds, completing the proof of the lemma. □

Recall that for a nonempty open set $U \subset X$, the sequence $B_m(U) = U_{1/m} \uparrow U$ as $m \to \infty$ is defined before Proposition 6.3 and $\mathcal{W}_p^{[U]}(J;X)$ is the set of all $g \in \mathcal{W}_p(J;X)$ such that the closure of the range $\operatorname{ran}(g)$ is included in $U$. Also recall that a nonlinear operator from a subset $E$ of one normed space into another is called bounded if it takes bounded subsets of $E$ into bounded sets.

**Proposition 6.54.** *Let $1 \le p \le q < \infty$, $\alpha := p/q$, and $J := [a,b]$ with $a < b$. For Banach spaces $X$ and $Y$, let $\psi\colon X \times J \to Y$. Then the following hold:*

*(a) if $\psi \in \mathcal{UH}_\alpha^{\rm loc}(U\times J;Y)$ for a nonempty open subset $U$ of $X$, then (i) for each $0 \le K < \infty$ and each $m = 1,2,\dots,$*

$$\sup\big\{\|N_\psi g\|_{[q]}\colon\ \|g\|_{[p]} \le K,\ \operatorname{ran}(g)\subset B_m(U)\big\} < \infty \tag{6.34}$$

*if and only if (ii) $\psi \in \mathcal{W}_{\alpha,q}^{\rm loc}(U\times J;Y)$. In particular, $N_\psi$ acts from $\mathcal{W}_p^{[U]}(J;X)$ into $\mathcal{W}_q(J;Y)$ provided $\psi \in \mathcal{HW}_{\alpha,q}^{\rm loc}(U\times J;Y)$;*

*(b) if $\psi \in \mathcal{UH}_\alpha^{\rm loc}(X\times J;Y)$ then $N_\psi$ is a bounded operator from $\mathcal{W}_p(J;X)$ into $\mathcal{W}_q(J;Y)$ if and only if $\psi \in \mathcal{W}_{\alpha,q}^{\rm loc}(X\times J;Y)$;*

*(c) if $B$ is a bounded subset of $X$ having more than one element and if $\psi \in \mathcal{UH}_\alpha(B\times J;Y)$, then $N_\psi$ is a bounded operator from the set $\{g \in \mathcal{W}_p(J;X)\colon \operatorname{ran}(g)\subset B\}$ into $\mathcal{W}_q(J)$ if and only if $\psi \in \mathcal{W}_{\alpha,q}(B\times J;Y)$.*

*Proof.* For $(a)$, to show $(ii)$ implies $(i)$, (6.34) follows from $(a)$ and $(b)$ of Lemma 6.53, provided $\psi \in \mathcal{W}_{\alpha,q}^{\rm loc}(U\times J;Y)$. Conversely, suppose that $(i)$ holds. Then by Lemma 6.53$(c)$, for any $K \ge 0$ and $m = 1,2,\dots$, $W_{\alpha,q}(\psi,K;B_m(U)\times J,Y) < \infty$, and so $\psi \in \mathcal{W}_{\alpha,q}^{\rm loc}(U\times J;Y)$, proving the first part of $(a)$. To prove the second part let $g \in \mathcal{W}_p^{[U]}(J;X)$. By Propositions 6.20$(a)$ and 6.3, ran$(g) \subset B_m(U)$ for some $m$. Then $N_\psi g \in \mathcal{W}_q(J;Y)$ by the first part of $(a)$. The proof of $(a)$ is complete.

To prove $(b)$, for $U = X$ and each $m$, $B_m = B_m(X) = \{x\colon \|x\| \le m\}$. Given a bounded set $D$ in $\mathcal{W}_p(J;X)$, there is an $m$ such that ran$(g) \subset B_m$ for each $g \in D$. Applying $(a)$ and $(b)$ of Lemma 6.53 to $B = B_m$ and $v = 0 \in B_m$, we have that there is a finite constant $C = C(\psi, D, m)$ such that $\|N_\psi g\|_{[q]} \le C$ for each $g \in D$ provided $\psi \in \mathcal{W}_{\alpha,q}^{\mathrm{loc}}(X \times J;Y)$. The converse implication holds by Lemma 6.53$(c)$ applied to each $B = B_m$, proving $(b)$.

The proof of $(c)$ is the same as the proof of $(b)$, replacing $B_m$ by $B$ and 0 by any fixed point $v$ of $B$. The proof of Proposition 6.54 is complete. □

**Remark 6.55.** In relation to Proposition 6.54$(b)$, recall that by Definition 6.47, for a subset $B$ of $X$ with more than one element, $\mathcal{HW}_{\alpha,q}(B \times J;Y) = \mathcal{UH}_\alpha(B \times J;Y) \cap \mathcal{W}_{\alpha,q}(B \times J;Y)$ and so $\mathcal{HW}_{\alpha,q}^{\mathrm{loc}}(B \times J;Y) = \mathcal{UH}_\alpha^{\mathrm{loc}}(B \times J;Y) \cap \mathcal{W}_{\alpha,q}^{\mathrm{loc}}(B \times J;Y)$.

We next consider a condition weaker than $\psi \in \mathcal{W}_{\alpha,q}^{\mathrm{loc}}(X \times J;Y)$. Namely, condition (6.31) restricted to singletons $\mu$ means that $\psi(u,\cdot)$ must have bounded $q$-variation uniformly for $u$ in bounded sets. Proposition 6.57 will show that in the case $p = q = \alpha = 1$, boundedness of the 1-variation of $\psi(u,\cdot)$ uniformly in $u$ and an $s$-uniform Lipschitz condition on $\psi(\cdot, s)$ are not sufficient for $N_\psi$ to act from $\mathcal{W}_1$ into $\mathcal{W}_1$. In that direction we first prove the following.

**Lemma 6.56.** *For any $n = 1, 2, \ldots$, there exists a real-valued function $\psi = \psi_n$ on $[0,1] \times [0,1]$ such that*

*(a) for any $s, u, v \in [0,1]$, $|\psi(u,s) - \psi(v,s)| \le |u - v|$;*
*(b) for any $u \in [0,1]$, $\|\psi(u,\cdot)\|_{(1)} \le 2$;*
*(c) for $g(s) \equiv s$, $\|N_\psi g\|_{(1)} \ge n$;*
*(d) $\psi(u,s) = 0$ if $s$ or $u = 0$ or $1$, and $|\psi(u,s)| \le 1/(4n+2)$ for all $u, s \in [0,1]$.*

*Proof.* Let $n$ be a positive integer, and let $M := M_n := (2n+1)^2$. For $j = 0, 1, \ldots, M$, let $s_j := j/M$. Define $\psi$ on $[0,1] \times [0,1]$, for $u \in [0,1]$, letting $\psi(u, s_j) := (-1)^j \max\{0, (4n+2)^{-1} - |u - s_j|\}$ if $j = n+2, \ldots, M-n-2$, $\psi(u, s_j) := 0$ if $j = 0, 1, \ldots, n+1$ or $j = M-n-1, \ldots, M$, and letting $\psi(u,\cdot)$ be linear on each interval $[s_{j-1}, s_j]$, $j = 1, \ldots, M$. To check $(a)$ notice first that it holds when $s = s_j$ for each $j$, and so it holds for all $0 \le s \le 1$.

To check $(b)$ let $u \in [0,1]$. By the piecewise linearity of $\psi(u,\cdot)$, we have

$$\|\psi(u,\cdot)\|_{(1)} \le \sum_{j=1}^{M} |\psi(u, s_j) - \psi(u, s_{j-1})|.$$

For each $j$, we have $\psi(u, s_j) = 0$ if $|u - s_j| \ge 1/(4n+2)$, and so $\psi(u, s_j) \ne 0$ for at most $2n+1$ values of $j$. Thus in the sum there are at most $2n+3$ non-zero terms, each no larger than $(2n+1)^{-1}$, giving a sum $\le 2$, proving $(b)$.

For $(c)$ where $g$ is the identity function, we have

$$\|N_\psi g\|_{(1)} \geq \sum_{j=n+2}^{M-n-1} \big|\psi(s_j, s_j) - \psi(s_{j-1}, s_{j-1})\big| \geq 2n - 1 \geq n,$$

proving $(c)$.

We have $(d)$ for $s = 0$ or $1$ by the definition of $\psi$, and

$$\psi(0, s_{n+2}) = \psi(1, s_{M-n-2}) = 0.$$

It follows that $\psi(0, s) = \psi(1, s) = 0$ for $s = s_j$ for all $j$, and then all $s$, so $(d)$ and the lemma are proved. □

Now we can prove that boundedness of the 1-variation of $\psi(u, \cdot)$ uniformly in $u$ and an $s$-uniform Lipschitz condition on $\psi(\cdot, s)$ are not acting conditions from $\mathcal{W}_1$ into $\mathcal{W}_1$ for the Nemytskii operator $N_\psi$.

**Proposition 6.57.** *There exists a real-valued function $\psi$ on $[0, 1] \times [0, 1]$ such that $(a)$ and $(b)$ of Lemma 6.56 hold and $\|N_\psi g\|_{(1)} = +\infty$ where $g(s) \equiv s$, $0 \leq s \leq 1$.*

*Proof.* For $n = 1, 2, \ldots$, let $\psi_n$ be functions on $[0, 1] \times [0, 1]$ satisfying all four properties of Lemma 6.56. For $j = 1, 2, \ldots$, let $p_j := 1/[j(1 + j)]$. Let $t_0 := 0$ and for $n \geq 1$, let $t_n := \sum_{j=1}^n p_j \uparrow 1$ as $n \to \infty$. Let $C_n := [t_{n-1}, t_n] \times [t_{n-1}, t_n]$. Define a function $\psi$ on $[0, 1] \times [0, 1]$ to be 0 outside $\cup_{n=1}^\infty C_n$, and on $C_n$ let $\psi(t_{n-1} + p_n u, t_{n-1} + p_n s) := p_n \psi_n(u, s)$ for $0 \leq u \leq 1$ and $0 \leq s \leq 1$.

To prove $(a)$ of Lemma 6.56 for $\psi$, let $s = t_{n-1} + p_n s_n$ for some $s_n$, $0 \leq s_n \leq 1$. For $t_{n-1} \leq u = t_{n-1} + p_n u_n \leq v = t_{n-1} + p_n v_n \leq t_n$,

$$\begin{aligned}\big|\psi(u, s) - \psi(v, s)\big| &= p_n \big|\psi_n(u_n, s_n) - \psi_n(v_n, s_n)\big| \\ &\leq p_n |u_n - v_n| = p_n |u - v|/p_n = |u - v|.\end{aligned}$$

So $(a)$ holds on $C_n$ for each $n$. Since $\psi$ is 0 on the boundary of each $C_n$ and outside their union, $(a)$ holds on $[0, 1] \times [0, 1]$.

For $(b)$ of Lemma 6.56, notice that $\psi(u, 1-) = 0$ for each $u \in [0, 1]$. Using additivity over adjoining intervals of the 1-variation, we then have

$$\|\psi(u, \cdot)\|_{(1)} = \sum_{n \geq 1} p_n \|\psi_n(u, \cdot)\|_{(1)} \leq 2$$

for each $u \in [0, 1]$, proving $(b)$. Actually $\|\psi_n(u, \cdot)\|_{(1)} > 1$ implies $s_{n-1} < u < s_n$, which can hold for only one value of $n$.

For $g(s) \equiv s$, using $(c)$ of Lemma 6.56, we have

$$\big\|N_\psi g\big\|_{(1)} \geq \sum_{n \geq 1} p_n \|N_{\psi_n} g\|_{(1)} \geq \sum_{n \geq 1} \frac{n}{n(n+1)} = +\infty.$$

The proof of Proposition 6.57 is complete. □

The main result of this section thus far is the following sufficient condition for acting and boundedness of a Nemytskii operator $N_\psi$ on Banach-valued functions. Part $(b)$ gives necessity for real-valued functions under a hypothesis on $\psi$.

**Theorem 6.58.** *Let $1 \le p \le q < \infty$, $\alpha := p/q$, and $J := [a,b]$ with $a < b$. For Banach spaces $X$ and $Y$, let $\psi\colon X \times J \to Y$.*

*$(a)$ If $\psi \in \mathcal{HW}^{\mathrm{loc}}_{\alpha,q}(X \times J; Y)$ then the Nemytskii operator $N_\psi$ acts from $\mathcal{W}_p(J;X)$ into $\mathcal{W}_q(J;Y)$ and is bounded.*

*$(b)$ Suppose that $X = Y = \mathbb{R}$ and for a dense set of values of $u$, $\psi(u,\cdot)$ is right-continuous on $[a,b)$ and left-continuous at $b$. If the Nemytskii operator $N_\psi$ acts and is bounded from $\mathcal{W}_p(J)$ into $\mathcal{W}_q(J)$ then $\psi \in \mathcal{HW}^{\mathrm{loc}}_{\alpha,q}(\mathbb{R} \times J)$.*

*Proof.* For $(a)$, $N_\psi$ acts from $\mathcal{W}_p(J;X)$ into $\mathcal{W}_q(J;Y)$ by the second part of Proposition 6.54$(a)$ with $U = X$. By Proposition 6.54$(b)$, $N_\psi$ is bounded.

For $(b)$, by Theorem 6.44$(e)$, $\psi \in \mathcal{UH}^{\mathrm{loc}}_\alpha(\mathbb{R} \times J)$. Thus $\psi \in \mathcal{HW}^{\mathrm{loc}}_{\alpha,q}(\mathbb{R} \times J)$ by the implication $(i) \Rightarrow (ii)$ of Proposition 6.54$(a)$ with $U = X = Y = \mathbb{R}$. □

*Continuity and Lipschitz properties from $\mathcal{W}_p$ into $\mathcal{W}_q$*

The continuity of a Nemytskii operator $N_\psi$ will be proved using the following conditions on $\psi$, which in turn use Definition 6.47 of $\mathcal{W}_{\alpha,q}(B\times J;Y)$, Definition 6.8 of $u$-differentiability of $\psi$, and the definition of $w_p(x)$ for $x = (x_1,\dots,x_n)$ before (6.30).

**Definition 6.59** (of $\mathcal{CW}_{0,q}$ and $\mathcal{HW}_{0,\alpha,q}$). Let $0 < \alpha \le 1$, let $0 < q < \infty$, and let $J$ be a nondegenerate interval. For Banach spaces $X$, $Y$ and for a nonempty open set $U \subset X$, let $\psi \equiv \psi(u,s)\colon U \times J \to Y$ and let $B$ be a nonempty subset of $U$. We say that $\psi$ is in the class $\mathcal{CW}_{0,q}(B \times J;Y)$ if

$(a)$ $\|\psi\|_{B\times J,\sup} < \infty$ and $\psi$ is $s$-uniformly continuous on $B$, that is, for each $\epsilon > 0$ there is a $\delta > 0$ such that $\|\psi(u,s) - \psi(v,s)\| < \epsilon$ whenever $u,v \in B$, $\|u - v\| < \delta$ and $s \in J$;

$(b)$ $\psi \in \mathcal{W}_{1,q}(B \times J;Y)$.

We say that $\psi$ is in the class $\mathcal{HW}_{0,\alpha,q}(B\times J;Y)$ for $\alpha = 1$ if $0 \in B$, $\psi$ is Fréchet $u$-differentiable on $U \times J$ with derivative $\psi^{(1)}_u \in \mathcal{CW}_{0,q}(B \times J; L(X,Y))$, and $\psi(0,\cdot) \in \mathcal{W}_q(J;Y)$, or for $\alpha \in (0,1)$, if the following hold:

$(c)$ $\sup\{\|\psi(u,s) - \psi(v,s)\|\colon\ s \in J,\ u,v \in B,\ \|u - v\| < \delta\} = o(\delta^\alpha)$ as $\delta \downarrow 0$;

$(d)$ $\psi \in \mathcal{W}_{\alpha,q}(B \times J;Y)$, and for each $0 \le K < \infty$ and $\epsilon > 0$ there exists a $\delta > 0$ such that

$$\Delta_q(\psi;\mu_1,\mu_2,\kappa) := \sum_{i=1}^{n} \big\|\psi(u_i,s_i)-\psi(v_i,s_i)-\psi(u_i,s_{i-1})+\psi(v_i,s_{i-1})\big\|^q < \epsilon^q$$

for all partitions $\kappa = \{s_i\}_{i=0}^n$ of $J$ and for all sets $\mu_1 = \{u_i\}_{i=1}^n \subset B$, $\mu_2 = \{v_i\}_{i=1}^n \subset B$ such that $w_{\alpha q}(\mu_1) \le K$, $w_{\alpha q}(\mu_2) \le K$, and $\max_i \|u_i - v_i\| < \delta$.

We write $\psi \in \mathcal{CW}_{0,q}(B\times J)$ and $\psi \in \mathcal{HW}_{0,\alpha,q}(B\times J)$ if $X = Y = \mathbb{K}$.

Next a local version of these conditions will be defined on $U \times J$ for an open set $U \subset X$. Recall that for a nonempty open set $U \subset X$, the sequence of closed, bounded sets $B_m := B_m(U) := U_{1/m} \uparrow U$ as $m \to \infty$ is defined before Proposition 6.3.

**Definition 6.60** (of $\mathcal{HW}_{0,\alpha,q}^{\mathrm{loc}}$). Let $\alpha$, $q$, $J$, $X$, $Y$, $U$, and $\psi$ be as in Definition 6.59 and let $0 \in U$ if $\alpha = 1$. We say that $\psi$ is in the class $\mathcal{HW}_{0,\alpha,q}^{\mathrm{loc}}(U \times J;Y)$ if for each sufficiently large $m$, $\psi$ is in the class $\mathcal{HW}_{0,\alpha,q}(B_m \times J;Y)$. We write $\psi \in \mathcal{HW}_{0,\alpha,q}^{\mathrm{loc}}(U\times J)$ if $X = Y = \mathbb{K}$.

The following conditions will be shown to imply a Lipschitz property for a Nemytskii operator, and will be used in Section 6.6 concerning higher order differentiability on $\mathcal{W}_p$ spaces.

**Definition 6.61** (of $\mathcal{HW}_{n+\alpha,q}$ and $\mathcal{HW}_{n+\alpha,q}^{\mathrm{loc}}$). Let $0<\alpha\le 1$ and $0<q<\infty$. Let $J$ be a nondegenerate interval. For Banach spaces $X$, $Y$, let $U$ be an open subset of $X$, let $B$ be a subset of $U$ containing 0, and let $n$ be a positive integer. We say that $\psi\colon U\times J\to Y$ is in the class $\mathcal{HW}_{n+\alpha,q}(B\times J;Y)$ if $\psi$ is Fréchet $u$-differentiable of order $n$ on $U \times J$ with derivatives such that for $k = 0,\dots,n-1$,

$$\psi_u^{(k)}(0,\cdot) \in \mathcal{W}_q(J;L(^kX,Y)) \quad \text{and} \quad \psi_u^{(n)} \in \mathcal{HW}_{\alpha,q}(B\times J;L(^nX,Y)), \tag{6.35}$$

where $L(^0X,Y) \equiv Y$. Let $0 \in U$, let $B_m := U_{1/m}$, $m = 1,2,\dots$, be the sequence defined before Proposition 6.3, and let $m_0 \ge 1$ be the least integer such that $0 \in B_{m_0}$. We say that $\psi$ is in the class $\mathcal{HW}_{n+\alpha,q}^{\mathrm{loc}}(U\times J;Y)$ if for each $m > m_0$, $\psi \in \mathcal{HW}_{n+\alpha,q}(B_m\times J;Y)$. Let $\mathcal{HW}_{n+\alpha,q}(B\times J) := \mathcal{HW}_{n+\alpha,q}(B\times J;\mathbb{R})$ and $\mathcal{HW}_{n+\alpha,q}^{\mathrm{loc}}(U\times J) := \mathcal{HW}_{n+\alpha,q}^{\mathrm{loc}}(U\times J;\mathbb{R})$.

**Lemma 6.62.** *Let $0<\alpha\le 1$, $1\le q<\infty$, and $J := [a,b]$ with $a<b$. For Banach spaces $X$ and $Y$, let $B$ be a nonempty bounded and convex subset of $X$ such that $0 \in B$, let $U$ be an open subset of $X$ such that $B \subset U$, and let $\psi\colon U\times J\to Y$ be in the class $\mathcal{HW}_{n+\alpha,q}(B\times J;Y)$ for some positive integer $n$. Then for $k=0,\dots,n-1$,*

$$\psi_u^{(k)} \in \mathcal{HW}_{1,q}(B\times J;L(^kX,Y)). \tag{6.36}$$

*Moreover,* (6.36) *holds with $k=0$ if $\psi \in \mathcal{HW}_{0,1,q}(B\times J;Y)$.*

*Proof.* To prove (6.36) for $k = n-1$, let $u, v \in B$. Since $B$ is convex the segment $[u, v]$ joining $u$ and $v$ is included in $B$. Then by (5.2) in the mean value theorem, we have for each $s \in J$,

$$\|\psi_u^{(n-1)}(v,s) - \psi_u^{(n-1)}(u,s)\| \le \|\psi_u^{(n)}\|_{B\times J,\sup}\|u-v\|, \tag{6.37}$$

since $\|D\psi_u^{(n-1)}(\cdot,s)\| = \|\psi_u^{(n)}(\cdot,s)\|$ for each $s$ by Proposition 5.25. Moreover, since $B$ is bounded, $\|\psi_u^{(n)}\|_{B\times J,\sup} < \infty$ by the second inequality in Lemma 6.53($b$) applied to $\psi = \psi_u^{(n)}$. Thus $\psi_u^{(n-1)}$ is $s$-uniformly Lipschitz on $B$, that is, condition ($a$) in Definition 6.47 holds for $\psi = \psi_u^{(n-1)}$ with $\alpha = 1$.

To prove $\psi_u^{(n-1)} \in \mathcal{W}_{1,q}(B \times J; L(^{n-1}X, Y))$, let $0 \le K < \infty$, let $\mu = \{u_i\}_{i=1}^n \subset B$ be such that $w_q(\mu) \le K$, and let $\kappa = \{s_i\}_{i=0}^n$ be a partition of $J$. Recall Definition 6.47, including the definition (6.31) of $s_q(\eta; \mu, \kappa)$. Since for each $s \in J$, $\psi_u^{(n-1)}(\cdot, s)$ is a continuously Fréchet differentiable function on $U$, by Proposition 5.25 and using the fact that $Lu := L(u)(\cdots) = L(u,\cdots) \in L(^{n-1}X, Y)$ for $L \in L(^nX, Y)$ (see (5.5) and Proposition 5.5 with $k = n$ and $m = 1$), we have $D\psi_u^{(n-1)}(rv,s)(v) = \psi_u^{(n)}(rv,s)v \in L(^{n-1}X,Y)$ for each $r \in [0,1]$ and $v \in B$. By Lemma 5.40, by the Minkowski inequality (1.5) and by Jensen's inequality applied to integrals, we have

$$\begin{aligned} s_q(\psi_u^{(n-1)}; \mu, \kappa)^{1/q} &= \Big(\sum_{i=1}^n \Big\| \int_0^1 \Big[\psi_u^{(n)}(ru_i, s_i) - \psi_u^{(n)}(ru_i, s_{i-1})\Big] u_i \, dr \\ &\qquad + \psi_u^{(n-1)}(0, s_i) - \psi_u^{(n-1)}(0, s_{i-1})\Big\|^q\Big)^{1/q} \\ &\le K\Big(\int_0^1 \sum_{i=1}^n \|\psi_u^{(n)}(ru_i, s_i) - \psi_u^{(n)}(ru_i, s_{i-1})\|^q \, dr\Big)^{1/q} \\ &\quad + s_q(\psi_u^{(n-1)}(0,\cdot); \kappa)^{1/q} \\ &\le KW_{1,q}(\psi_u^{(n)}, K) + \|\psi_u^{(n-1)}(0,\cdot)\|_{(q)} \end{aligned}$$

since $\max_i \|u_i\| \le K$ and $w_q(\{ru_i\}_{i=1}^n) \le K$ for each $r \in [0,1]$. Thus $W_{1,q}(\psi_u^{(n-1)}, K; B \times J, L(^{n-1}X, Y)) < \infty$ for any $0 \le K < \infty$, and so (6.36) holds for $k = n-1$. Proceeding in the same way for the lower derivatives we conclude that (6.36) holds for $k = 0, \ldots, n-1$.

Finally let $\psi \in \mathcal{HW}_{0,1,q}(B \times J; Y)$. Then $\|\psi_u^{(1)}\|_{B\times J,\sup} < \infty$ by Definition 6.59($a$), and so $\psi$ is $s$-uniformly Lipschitz on $B$ by (6.37) with $n = 1$. Since for each $s \in J$, $\psi(\cdot, s)$ is a continuously Fréchet differentiable function on $U$, and since $\psi_u^{(1)} \in \mathcal{W}_{1,q}(B \times J; L(X, Y))$ by Definition 6.59($b$), the preceding argument for $k = 0$ yields (6.36) with $k = 0$, proving the lemma. □

Now we are ready to prove continuity properties of the Nemytskii operator for $\psi$ in one of the classes in Definition 6.60.

**Theorem 6.63.** *Let $1 \le p \le q < \infty$, $\alpha := p/q$, and $J := [a, b]$ with $a < b$. For Banach spaces $X$ and $Y$, let $\psi \in \mathcal{HW}^{\rm loc}_{0,\alpha,q}(X \times J; Y)$. Then the Nemytskii operator $N_\psi$ is uniformly continuous on bounded sets from $\mathcal{W}_p(J; X)$ into $\mathcal{W}_q(J; Y)$.*

*Proof.* By Definition 6.60 with $U = X$, the class $\mathcal{HW}^{\rm loc}_{0,\alpha,q}(X \times J; Y)$ is defined with respect to the sequence of balls $B_m := B_m(X) = \{x\colon \|x\| \le m\}$ for $m = 1, 2, \ldots$, and so $\psi \in \mathcal{HW}_{0,\alpha,q}(B_m \times J; Y)$ for each $m$. If $\alpha \in (0, 1)$, by Definition 6.59 with $B = B_m$, $m = 1, 2, \ldots$, its condition $(c)$ implies that $\psi$ is $s$-uniformly $\alpha$-Hölder on $B_m$, or $\psi \in \mathcal{UH}_\alpha(B_m \times J; Y)$ according to Definition 6.42. Thus $\psi \in \mathcal{UH}^{\rm loc}_\alpha(X \times J; Y)$ by Definition 6.43. Also, condition $(d)$ of Definition 6.59 implies that $\psi \in \mathcal{W}_{\alpha,q}(B_m \times J; Y)$ for each $m$ as defined in Definition 6.47. Thus $\psi \in \mathcal{W}^{\rm loc}_{\alpha,q}(X \times J; Y)$ by Definition 6.51. Therefore $\psi \in \mathcal{W}^{\rm loc}_{\alpha,q} \cap \mathcal{UH}^{\rm loc}_\alpha$, and so $N_\psi$ acts from $\mathcal{W}_p(J; X)$ into $\mathcal{W}_q(J; Y)$ by Proposition 6.54$(b)$ if $\alpha \in (0, 1)$. For $\alpha = 1$, so $p = q$, by Definition 6.59, $\psi$ is Fréchet $u$-differentiable everywhere on $X \times J$ with derivative $\psi^{(1)}_u \in \mathcal{CW}_{0,p}(B_m \times J; L(X, Y))$ for each $m$, and $\psi(0, \cdot) \in \mathcal{W}_p(J; Y)$. Thus by the second part of Lemma 6.62, it follows that $\psi \in \mathcal{HW}_{1,p}(B_m \times J; Y)$ for each $m$, and so $\psi \in \mathcal{HW}^{\rm loc}_{1,p}(X \times J; Y)$ according to Definition 6.51. Thus the Nemytskii operator $N_\psi$ acts from $\mathcal{W}_p(J; X)$ into $\mathcal{W}_p(J; Y)$ for $\alpha = 1$, again by Proposition 6.54$(b)$.

To prove the uniform continuity of $N_\psi$ on bounded sets let $1 \le M < \infty$ and $\epsilon \in (0, 1]$. Let $K := M + 1$. We will show that there are finite constants $C = C(\psi, M)$ and $\delta > 0$ such that if $G \in \mathcal{W}_p(J; X)$ with $\|G\|_{[p]} \le M$, then

$$\|N_\psi(G + g) - N_\psi(G)\|_{[q]} \le \epsilon C \tag{6.38}$$

for each $g \in \mathcal{W}_p(J; X)$ with $\|g\|_{[p]} \le \delta$. Let $B := \{x \in X\colon \|x\| \le K\}$. First suppose that $\alpha = 1$. Since $\psi \in \mathcal{HW}^{\rm loc}_{0,1,q}(X \times J; Y)$, $\psi$ is Fréchet $u$-differentiable everywhere on $X \times J$ with $\psi^{(1)}_u \in \mathcal{CW}_{0,p}(B \times J; Y)$, and so by Definition 6.59, there is a $\delta \in (0, \epsilon)$ such that $\|\psi^{(1)}_u(u, y) - \psi^{(1)}_u(v, y)\| < \epsilon$ whenever $u, v \in B$, $\|u - v\| \le \delta$, and $y \in J$. Let $g \in \mathcal{W}_p(J; X)$ be such that $\|g\|_{[p]} \le \delta$. For each $s \in J$, since the line segment joining $G(s)$ and $(G + g)(s)$ is included in $B$, by (5.2) in the mean value theorem, we have

$$\big\|N_\psi(G+g)(s) - N_\psi(G)(s)\big\| \le \|\psi^{(1)}_u\|_{B\times J,\sup}\|g\|_{\sup} < \epsilon\|\psi^{(1)}_u\|_{B\times J,\sup}. \tag{6.39}$$

To bound the $p$-variation seminorm of $N_\psi(G + g) - N_\psi(G)$, for any $a \le s < t \le b$, let

$$\begin{aligned} T(s,t) &:= \big[N_\psi(G + g) - N_\psi(G)\big](t) - \big[N_\psi(G + g) - N_\psi(G)\big](s) \qquad (6.40)\\ &= [\psi((G + g)(t), t) - \psi((G + g)(s), t)] - [\psi(G(t), t) - \psi(G(s), t)]\\ &\quad +[\psi((G + g)(s), t) - \psi(G(s), t)] - [\psi((G + g)(s), s) - \psi(G(s), s)]. \end{aligned}$$

Since $\psi^{(1)}_u(\cdot, s)$ and $\psi^{(1)}_u(\cdot, t)$ are continuous on $B$, by Lemma 5.40, we have

$$
\begin{aligned}
T(s,t) = \int_0^1 \Big[&\psi_u^{(1)}((G+g)(s) + r[(G+g)(t) - (G+g)(s)], t) \\
&-\psi_u^{(1)}(G(s) + r[G(t) - G(s)], t)\Big][G(t) - G(s)]\,\mathrm{d}r \\
&+ \int_0^1 \psi_u^{(1)}((G+g)(s)+r[(G+g)(t)-(G+g)(s)], t)[g(t)-g(s)]\,\mathrm{d}r \\
&+ \int_0^1 \Big[\psi_u^{(1)}((G+rg)(s), t) - \psi_u^{(1)}((G+rg)(s), s)\Big] g(s)\,\mathrm{d}r.
\end{aligned}
$$

For each $r \in [0,1]$ we have $u := (G+g)(s) + r[(G+g)(t) - (G+g)(s)] \in B$, $v := G(s) + r[G(t) - G(s)] \in B$, and $\|u - v\| \le \|g\|_{[p]} \le \delta$. Therefore

$$
\begin{aligned}
\|T(s,t)\| \le \epsilon\|G(t) - G(s)\| &+ \|\psi_u^{(1)}\|_{B\times J,\sup}\|g(t) - g(s)\| + \\
&+ \|g\|_{\sup} \int_0^1 \Big\|\psi_u^{(1)}((G+rg)(s), t) - \psi_u^{(1)}((G+rg)(s), s)\Big\|\,\mathrm{d}r.
\end{aligned}
$$

Let $\kappa = \{s_i\}_{i=0}^n$ be a partition of $J$. For any $r \in [0,1]$, letting $x_i := (G + rg)(s_{i-1})$ for $i = 1, \ldots, n$, we have $w_p(\{x_i\}_{i=1}^n) \le \|G\|_{[p]} + \|g\|_{[p]} \le K$. Since $\|g\|_{[p]} \le \epsilon$ and $\|G\|_{[p]} \le M$, using the Minkowski inequality (1.5), Jensen's inequality for integrals, and Definition 6.47 for $W_{1.p}$, it then follows that

$$
\begin{aligned}
\Big(\sum_{i=1}^n \|T(s_{i-1}, s_i)\|^p\Big)^{1/p} &\le \epsilon\|G\|_{(p)} + \|\psi_u^{(1)}\|_{B\times J,\sup}\|g\|_{(p)} + \|g\|_{\sup}\times \\
\times\Big(\int_0^1 \sum_{i=1}^n &\Big\|\psi_u^{(1)}((G+rg)(s_{i-1}), s_i) - \psi_u^{(1)}((G+rg)(s_{i-1}), s_{i-1})\Big\|^p\,\mathrm{d}r\Big)^{1/p} \\
\le \epsilon\big\{M + \|\psi_u^{(1)}\|_{B\times J,\sup} &+ W_{1,p}(\psi_u^{(1)}, K; B\times J, L(X,Y))\big\},
\end{aligned}
$$

where $W_{1,p}(\psi_u^{(1)}, K; \ldots)$ is finite because $\psi_u^{(1)} \in \mathcal{CW}_{0,p}(B \times J; L(X,Y)) \subset \mathcal{W}_{1,p}(B\times J; L(X,Y))$ by Definition 6.59($b$). Letting $C := M + 2\|\psi_u^{(1)}\|_{B\times J,\sup} + W_{1,p}(\psi_u^{(1)}, K; B \times J, L(X,Y))$, and recalling that $K = M + 1$ and $B = \{x\colon \|x\| \le K\}$, so that $C$ depends only on $\psi$ and $M$, the preceding bound and (6.39) yield (6.38) in the case $\alpha = 1$.

Now suppose that $\alpha \in (0,1)$. By Definition 6.59($c$), there is a $\delta_1 > 0$ such that $\|\psi(u,s) - \psi(v,s)\| < \epsilon\|u - v\|^\alpha$ whenever $u, v \in B$, $\|u - v\| < \delta_1$, and $s \in J$. By Definition 6.59($d$), there is a $\delta_2 > 0$ such that $\Delta_q(\psi; \mu_1, \mu_2, \kappa) < \epsilon^q$ whenever $\kappa$ is a partition of $J$, $\mu_1 = \{u_i\}_{i=1}^n \subset B$, and $\mu_2 = \{v_i\}_{i=1}^n \subset B$ are such that $w_p(\mu_1) \le K$, $w_p(\mu_2) \le K$, and $\max_i \|u_i - v_i\| < \delta_2$. Let $\delta := \min\{\delta_1, \delta_2, 1\}$. By Proposition 3.40 with $\Phi(u) = u^p$ and $c = (\delta/2)^p$, there exists a partition $\{z_j\}_{j=0}^m$ of $J$ such that $m \le 1 + 2^p v_p(G)/\delta^p \le 1 + (2M/\delta)^p$ and $\mathrm{Osc}\,(G, (z_{j-1}, z_j)) \le \delta/2$ for each $j = 1, \ldots, m$. Let $g \in \mathcal{W}_p(J)$ be such that $\|g\|_{[p]}^\alpha \le \delta^p/(2(\delta^p + (2M)^p)) < \delta/2 < 1$. If $z_{j-1} < s < t < z_j$ for some $j \in \{1, \ldots, m\}$ then $\|(G+g)(t) - (G+g)(s)\| < \delta_1$ and $\|G(t) - G(s)\| < \delta_1$, and so by (6.40), we have

$$\|T(s,t)\| \le 2\epsilon\|G(t)-G(s)\|^\alpha + \epsilon\|g(t)-g(s)\|^\alpha \\ +\big\|\psi((G+g)(s),t)-\psi(G(s),t)-\psi((G+g)(s),s)+\psi(G(s),s)\big\|.$$

Let $\kappa = \{s_i\}_{i=0}^n$ be a partition of $J$ and for each $j = 1,\dots,m$, let either $I_j := \{i\colon z_{j-1} < s_{i-1} < s_i < z_j\}$ or $I_j := \emptyset$ if there is no such $i$. Let $\mu_1 = \{(G+g)(s_{i-1})\}_{i=1}^n$ and $\mu_2 = \{G(s_{i-1})\}_{i=1}^n$. Then recalling that $K = M+1$, we have $w_p(\mu_1) \le K$, $w_p(\mu_2) \le K$, and $\max_i \|(G+g)(s_{i-1}) - G(s_{i-1})\| \le \|g\|_{\sup} < \delta_2$. Using the Minkowski inequality (1.5), and again recalling Definition 6.59($d$) for $\Delta_q$, we have

$$\Big(\sum_{j=1}^m \sum_{i\in I_j} \|T(s_{i-1},s_i)\|^q\Big)^{1/q} \le 2\epsilon\|G\|_{(p)}^\alpha + \epsilon\|g\|_{(p)}^\alpha + \Delta_q(\psi;\mu_1,\mu_2,\kappa)^{1/q} \\ \le 2\epsilon(M+1),$$

where the sum over the empty set is defined as 0. Since for each $s \in J$, $\|(G+g)(s) - G(s)\| \le \|g\|_{\sup} < \delta_1$, if $i \in \{1,\dots,n\}\setminus \cup_{j=1}^m I_j$ then we have by (6.40) $\|T(s_{i-1},s_i)\| < 2\epsilon\|g\|_{\sup}^\alpha$. Since there are no more than $m$ such indices $i$ and $\|g\|_{\sup}^\alpha \le \delta^p/(2(\delta^p + (2M)^p)) \le 1/(2m)$, we get the bound

$$s_q(N_\psi(G+g) - N_\psi(G);\kappa)^{1/q} \le 2\epsilon m\|g\|_{\sup}^\alpha + 2\epsilon(M+1) \le \epsilon(2M+3).$$

Again since $\|g\|_{\sup} < \delta_1$, we have

$$\big\|N_\psi(G+g)(s) - N_\psi(G)(s)\big\| < \epsilon\|g\|_{\sup}^\alpha \le \epsilon$$

for each $s \in J$. Therefore (6.38) holds with $C = 2M+4$ when $\alpha \in (0,1)$, proving the theorem. □

In Definitions 6.59 and 6.60 for $\alpha = 1$ and $U = X = Y = \mathbb{R}$, if $\psi(u,s) \equiv F(u)$ not depending on $s$, then Definition 6.59($b$) holds with $W_{1,q}(\psi,K;B\times J,\mathbb{R}) \equiv 0$ for each $B = B_m(\mathbb{R}) = [-m,m]$, $m = 1,2,\dots$, and $0 \le K < \infty$ (Definition 6.47), and so $\psi \in \mathcal{HW}_{0,1,q}^{\mathrm{loc}}(\mathbb{R}\times J;\mathbb{R})$ for some, or equivalently all, $q$ with $1 \le q < \infty$ if and only if $F$ is a $C^1$ function on $\mathbb{R}$. Thus we can apply the preceding theorem to obtain the following:

**Corollary 6.64.** *Let $1 \le p < \infty$ and $J := [a,b]$ with $a < b$. If $F$ is a real-valued $C^1$ function on $\mathbb{R}$ then the autonomous Nemytskii operator $N_F$ is uniformly continuous on bounded sets from $\mathcal{W}_p(J)$ into $\mathcal{W}_p(J)$.*

In Definitions 6.59 and 6.60 for $\alpha \in (0,1)$ and $U = X = Y = \mathbb{R}$, if $\psi(u,s) \equiv F(u)$ not depending on $s$, then Definition 6.59($d$) holds with $W_{1,q}(\psi,K;B\times J,\mathbb{R}) \equiv 0$ for each $B = B_m(\mathbb{R}) = [-m,m]$, $m = 1,2,\dots$, and $0 \le K < \infty$ (Definition 6.47), and $\Delta_q \equiv 0$. Therefore such $\psi \in \mathcal{HW}_{0,\alpha,q}^{\mathrm{loc}}(\mathbb{R}\times J;\mathbb{R})$ for some, or equivalently all, $q$ with $1 \le q < \infty$ if and only if for each $m = 1,2,\dots,$

$$\sup\{|F(u)-F(v)|:\ u,v\in[-m,m],\ |u-v|<\delta\}=o(\delta^\alpha) \tag{6.41}$$

as $\delta\downarrow 0$. Now we can apply Theorem 6.63 to a function $\psi(u,s)\equiv F(u)$ in the class $\mathcal{HW}^{\rm loc}_{0,\alpha,q}(\mathbb{R}\times J;\mathbb{R})$ with $\alpha\in(0,1)$ to obtain local uniform continuity of the autonomous Nemytskii operator $N_F$:

**Corollary 6.65.** *Let $1\le p<q<\infty$, $\alpha:=p/q$, and $J:=[a,b]$ with $a<b$. If $F$ is a real-valued function on $\mathbb{R}$ such that (6.41) holds for each $m=1,2,\dots$, then the autonomous Nemytskii operator $N_F$ is uniformly continuous on bounded sets from $\mathcal{W}_p(J)$ into $\mathcal{W}_q(J)$.*

In Corollary 6.64, the operator $N_F$ is not necessarily uniformly continuous on bounded sets if $F$ is only a Lipschitz function:

**Proposition 6.66.** *For $F(u):=|u|$, $J:=[0,2\pi]$, and $1\le p<\infty$, the autonomous Nemytskii operator $N_F$ is not uniformly continuous on bounded sets from $\mathcal{W}_p(J)$ into itself; in fact, there is a bounded sequence $\{G_n\}\subset\mathcal{W}_p(J)$ such that for the sequence of constant functions $g_n\equiv n^{-1/p}$, $F\circ(G_n+g_n)-F\circ G_n$ does not converge to 0 in $\mathcal{W}_p(J)$.*

*Proof.* Let $G_n(s):=n^{-1/p}\cos(ns)$ for $s\in J$. Consider the partition $\kappa_n$ of $J$ consisting of the points $s_j:=j\pi/(2n)$ for $j=0,1,\dots,4n$. Then along $\kappa_n$, $n^{1/p}G_n$ runs through the sequence of values $1,0,-1,0,1,0,-1,\dots,-1,0,1$. Considering only the even values of $j$ we get all the relative maxima and minima of $G_n$ and find that $v_p(G_n)=2^{1+p}$ for each $n$, so $\|G_n\|_{[p]}$ are bounded. If $j=4i-1$ for $i=1,\dots,n$, we have $G_n(s_j)=0$, $G_n(s_j)+g_n=n^{-1/p}$, and $G_n(s_{j-1})=-n^{-1/p}$, so

$$|F(G_n(s_j))+g_n)-F(G_n(s_j))-F(G_n(s_{j-1})+g_n)+F(G_n(s_{j-1}))| \;=\; 2n^{-1/p}$$

for each such $j$. Thus $v_p(F\circ(G_n+g_n)-F\circ G_n)\ge 2^p$, which does not converge to 0 as $n\to\infty$. □

**Remark 6.67.** In fact, in the above proof $\|G_n\|_{\mathcal{H}_{1/p}}$ are bounded and $\|g_n\|\to 0$ in any normed space of real functions on $[0,2\pi]$ containing the constants.

Under some conditions on $\psi$ we will prove a local Lipschitz property of a Nemytskii operator $N_\psi$, which in turn will be used to prove existence and uniqueness of solutions of nonlinear integral equations in Chapter 10. Recall Definition 6.61 of the class $\mathcal{HW}^{\rm loc}_{n+\alpha,q}$. Here we will take $n=1$. For a function $\psi(u,s)\equiv F(u)$, the condition $\psi\in\mathcal{HW}^{\rm loc}_{1+\alpha,q}(X\times J;Y)$ is equivalent to $F\in\mathcal{H}^{\rm loc}_{1+\alpha}(X;Y)$ (Definition 6.5), and so the next result extends Theorem 6.40.

**Theorem 6.68.** *Let $1\le p\le q<\infty$, $\alpha:=p/q$, and $J:=[a,b]$ with $a<b$. For Banach spaces $X$ and $Y$, let $\psi\in\mathcal{HW}^{\rm loc}_{1+\alpha,q}(X\times J;Y)$. Then the Nemytskii*

*operator $N_\psi$ acts from $\mathcal{W}_p(J;X)$ into $\mathcal{W}_q(J;Y)$. Also, for each $0 < R < \infty$, if $f, g \in \mathcal{W}_p(J;X)$ and $3\max\{\|f\|_{[p]}, \|g\|_{[p]}\} \le R$ then for any nondegenerate subinterval $A \subset J$,*

$$\big\|N_\psi f - N_\psi g\big\|_{A,\sup} \le C\|f-g\|_{A,\sup} \tag{6.42}$$

*and*

$$\big\|N_\psi f - N_\psi g\big\|_{A,(q)} \le C\|f-g\|_{A,(q)} + D\|f-g\|_{A,\sup}\big[\|g\|_{A,(p)}^{\alpha} + 1\big], \tag{6.43}$$

*where $C := H_1(\psi; B_R \times J, Y)$ with $B_R := \{x \in X\colon \|x\| \le R\}$ and $D := H_\alpha(\psi_u^{(1)}; B_R \times J, Y) + W_{\alpha,q}(\psi_u^{(1)}, R; B_R \times J, Y)$.*

*Proof.* Each $B_m = B_m(X)$ in the sequence $\{B_m\}_{m=1}^\infty$ in the definition of $\mathcal{HW}_{1+\alpha,q}^{\mathrm{loc}}(X \times J, Y)$ is a ball with center zero and radius $m$, and so it is bounded and convex with $0 \in B_m$. Thus by Lemma 6.62 with $n = 1$, we have $\psi \in \mathcal{HW}_{1,q}(B_m \times J; Y)$ for each $m$, and so $N_\psi$ acts from $\mathcal{W}_p(J;X)$ into $\mathcal{W}_q(J;Y)$ by Proposition 6.54(*b*) and Remark 6.55. Let $0 < R < \infty$, let $f, g \in \mathcal{W}_p(J;X)$ be such that $3\max\{\|f\|_{[p]}, \|g\|_{[p]}\} \le R$, and let $A$ be a nondegenerate subinterval of $J$. Let $\lceil R\rceil$ be the minimal integer $m$ such that $m \ge R$. Then $C \le H_1(\psi; B_{\lceil R\rceil} \times J, Y) < \infty$, $\mathrm{ran}\,(f) \cup \mathrm{ran}\,(g) \subset B_R$, and for each $s \in A$,

$$\big\|(N_\psi f - N_\psi g)(s)\big\| = \big\|\psi(f(s), s) - \psi(g(s), s)\big\| \le C\|f-g\|_{A,\sup},$$

and so (6.42) holds. To prove (6.43) let $s, t \in A$ and let

$$T_1(s,t) := \big\|\psi(f(t), t) - \psi(g(t) + (f-g)(s), t)\big\|.$$

Clearly, $\psi(\cdot, s)$ and $\psi(\cdot, t)$ are continuously Fréchet differentiable functions on $X$. Then by Lemma 5.40, we have

$$\begin{aligned}
&\Big|\big\|(N_\psi f - N_\psi g)(t) - (N_\psi f - N_\psi g)(s)\big\| - T_1(s,t)\Big| \\
&\quad \le \Big\|\big[\psi(g(t) + (f-g)(s), t) - \psi(g(t), t)\big] - \big[\psi(f(s), s) - \psi(g(s), s)\big]\Big\| \\
&\quad \le \|(f-g)(s)\|\Big\|\int_0^1 \Big[\psi_u^{(1)}(g(t) - g(s) + \xi(r,s), t) - \psi_u^{(1)}(\xi(r,s), s)\Big]\,\mathrm{d}r\Big\| \\
&\quad \le \|f-g\|_{A,\sup}\big[T_2(s,t) + T_3(s,t)\big],
\end{aligned}$$

where $\xi(r,s) := g(s) + r(f-g)(s)$ for each $r \in [0,1]$,

$$T_2(s,t) := \int_0^1 \Big\|\psi_u^{(1)}(g(t) - g(s) + \xi(r,s), t) - \psi_u^{(1)}(\xi(r,s), t)\Big\|\,\mathrm{d}r,$$

and

$$T_3(s,t) := \int_0^1 \Big\|\psi_u^{(1)}(\xi(r,s), t) - \psi_u^{(1)}(\xi(r,s), s)\Big\|\,\mathrm{d}r.$$

Let $\kappa = \{s_i\}_{i=0}^n$ be a partition of $A$. Then by the Minkowski inequality (1.5), we have

$$\begin{aligned} s_q\big(N_\psi f - N_\psi g;\kappa\big)^{1/q} &\le \Big(\sum_{i=1}^n T_1(s_{i-1},s_i)^q\Big)^{1/q} \\ &\quad + \|f-g\|_{A,\sup}\sum_{k=2}^3\Big(\sum_{i=1}^n T_k(s_{i-1},s_i)^q\Big)^{1/q}. \end{aligned} \tag{6.44}$$

Since ran $(f)$ and ran $(g)+$ran $(f-g)$ are subsets of $B_R$ and $\psi \in \mathcal{HW}_{1,q}(B_R \times J;Y) \subset \mathcal{UH}_1(B_R \times J;Y)$ (Definition 6.42 and Remark 6.55 for $\alpha = 1$ and $B = B_R$), we have

$$\sum_{i=1}^n T_1(s_{i-1},s_i)^q \le C^q\|f-g\|_{A,(q)}^q.$$

Let $D_1 := H_\alpha(\psi_u^{(1)}; B_R \times J, Y)$. By Jensen's inequality and the $s$-uniform $\alpha$-Hölder property (6.25) for $\psi = \psi_u^{(1)}$ and $B = B_R$, we have

$$\begin{aligned} &\sum_{i=1}^n T_2(s_{i-1},s_i)^q \\ &\quad\le \sum_{i=1}^n \int_0^1 \Big\|\psi_u^{(1)}(g(s_i)-g(s_{i-1})+\xi(r,s_{i-1}),s_i) - \psi_u^{(1)}(\xi(r,s_{i-1}),s_i)\Big\|^q\,dr \\ &\quad\le \sum_{i=1}^n D_1^q\big\|g(s_i)-g(s_{i-1})\big\|^p \le D_1^q\|g\|_{A,(p)}^{q\alpha}, \end{aligned}$$

since ran $(g)+r\cdot$ran $(f-g) \subset B_R$ for each $r \in [0,1]$. Also, by the definition of $W_{\alpha,q}(\psi_u^{(1)}, R)$ following (6.31) with $\psi_u^{(1)}$ in place of $\psi$ and $B = B_R$, we have

$$\begin{aligned} &\sum_{i=1}^n T_3(s_{i-1},s_i)^q \\ &\quad\le \sum_{i=1}^n \int_0^1 \Big\|\psi_u^{(1)}(\xi(r,s_{i-1}),s_i) - \psi_u^{(1)}(\xi(r,s_{i-1}),s_{i-1})\Big\|^q\,dr \\ &\quad\le W_{\alpha,q}(\psi_u^{(1)}, R; B_R \times J, Y)^q, \end{aligned}$$

since $w_p(\{\xi(r,s_{i-1})\}_{i=1}^n) \le \|g+r(f-g)\|_{A,[p]} \le R$ for each $r \in [0,1]$. Letting $D_2 := W_{\alpha,q}(\psi_u^{(1)}, R; B_R \times J, Y)$, we have $D = D_1 + D_2$. The inequality (6.43) now follows by (6.44), proving the theorem. □

A local Lipschitz property of a Nemytskii operator now easily follows:

**Corollary 6.69.** *Under the hypotheses of Theorem 6.68, for each $0 < M < \infty$ there is a finite constant $L = L(M)$ such that the bound*

$$\left\|N_\psi f - N_\psi g\right\|_{[q]} \le L\|f - g\|_{[p]}$$

*holds for each $f$ and $g$ in $\mathcal{W}_p(J;X)$ having norms bounded by $M$.*

An entire holomorphic function $f$ from $\mathbb{C}$ into $\mathbb{C}$ is always locally Lipschitz. But it is globally Lipschitz if and only if it is linear, $f(z) \equiv a+bz$, by Liouville's theorem for $f'$. Analogously the following fact, due to Matkowski and Miś [161] for $q = p = 1$, shows that for a Nemytskii operator $N_\psi$ to be globally Lipschitz from $\mathcal{W}_p$ into $\mathcal{W}_q$ requires $\psi = \psi(u,s)$ to be of a very special linear form in $u$, except perhaps for $s$ in a countable set.

**Theorem 6.70.** *Let $1 \le p \le q < \infty$, $J := [a,b]$ with $a < b$, and $\psi\colon \mathbb{R}\times J \to \mathbb{R}$. Suppose that the Nemytskii operator $N_\psi$ acts from $\mathcal{W}_p(J)$ into $\mathcal{W}_q(J)$ and there is a constant $L < \infty$ such that for each $f, g \in \mathcal{W}_p(J)$,*

$$\left\|N_\psi(f) - N_\psi(g)\right\|_{[q]} \le L\|f - g\|_{[p]}. \tag{6.45}$$

*Then there are functions $\eta, \xi\colon \mathbb{R}\times J \to \mathbb{R}$ such that $\psi \equiv \eta+\xi$ and a countable set $T \subset J$ such that:*

*(a) $\xi(u,s) = 0$ for all $s \in J\setminus T$ and $u \in \mathbb{R}$;*
*(b) there are functions $\alpha, \beta \in \mathcal{W}_q(J)$ such that $\eta(u,s) = \alpha(s) + \beta(s)u$ for all $u \in \mathbb{R}$ and $s \in J$;*
*(c) for $f = \psi, \eta$ or $\xi$, $f$ is $s$-uniformly Lipschitz on $\mathbb{R}$ and for each $u \in \mathbb{R}$, $f(u,\cdot) \in \mathcal{W}_q(J)$ and so is regulated;*
*(d) for each $u \in \mathbb{R}$, $\xi(u,s-) = 0$ for $a < s \le b$ and $\xi(u,s+) = 0$ for $a \le s < b$;*
*(e) the Nemytskii operators $N_\eta$, $N_\xi$ are globally Lipschitz.*

*Proof.* Since $\mathcal{W}_p(J)$ contains the constant functions, the function $\psi(u,\cdot)$ is in $\mathcal{W}_q(J)$ for each $u \in \mathbb{R}$, and so it is regulated by Proposition 3.33. Using (6.45) for two constant functions $u1(\cdot)$ and $v1(\cdot)$, we get that $|\psi(u,s) - \psi(v,s)| \le L|u - v|$ for any $u, v \in \mathbb{R}$ and $s \in J$. Let $\eta(u,s) := \psi(u,s-)$ if $s \in (a,b]$ and $\eta(u,a) := \psi(0,a)$ for each $u \in \mathbb{R}$, and let $\xi := \psi - \eta$. Clearly $\eta$ is $s$-uniformly Lipschitz on $\mathbb{R}$ and for each $u \in \mathbb{R}$, $\eta(u,\cdot) \in \mathcal{W}_q(J)$. The same is true for $\xi$ by linearity, and so $(c)$ holds.

To prove $(b)$, define two functions $f_k$, $k = 1,2$, as follows. Let $\{c_k, d_k\}$, $k = 1,2$, be real numbers, $t \in (a,b]$, and let $\{t_i\}_{i=0}^{2n+1}$ be a partition of $[a,t]$. For each $s \in [a,b]$ and $k = 1,2$, let

$$f_k(s) := \begin{cases} d_k \text{ if } s \notin \{t_{2i}\}_{i=1}^n, \\ c_k \text{ if } s \in \{t_{2i}\}_{i=1}^n. \end{cases}$$

Clearly, $f_1, f_2 \in \mathcal{W}_p(J)$. Also, we have

$$\begin{aligned}\|f_1 - f_2\|_{[p]} &\le \big|(f_1 - f_2)(a)\big| + 2\|f_1 - f_2\|_{(p)}\\ &= |d_1 - d_2| + 2(2n)^{1/p}|c_1 - c_2 - d_1 + d_2|.\end{aligned}$$

Thus by (6.45), it follows that

$$\begin{aligned}&\Big(\sum_{i=1}^{n} |\psi(c_1, t_{2i}) - \psi(c_2, t_{2i}) - \psi(d_1, t_{2i-1}) + \psi(d_2, t_{2i-1})|^q\Big)^{1/q}\\ &\qquad \le \big\|N_\psi(f_1) - N_\psi(f_2)\big\|_{(q)} \le L\Big\{|d_1 - d_2| + 2(2n)^{1/p}|c_1 - c_2 - d_1 + d_2|\Big\}.\end{aligned}$$

In the left side of the preceding inequalities, letting $t_1 \to t_{2n+1} = t$, we get the bound

$$\begin{aligned}n^{1/q}\Big|\psi(c_1, t-) - \psi(c_2, t-) - \psi(d_1, t-) + \psi(d_2, t-)\Big|\\ \le L\Big\{|d_1 - d_2| + 2(2n)^{1/p}\big|c_1 - c_2 - d_1 + d_2\big|\Big\}.\end{aligned}$$

For any real numbers $v, w$, taking $c_1 := v + w$, $c_2 := v$, $d_1 := w$, and $d_2 := 0$, we then have

$$\Big|\psi(v + w, t-) - \psi(v, t-) - \psi(w, t-) + \psi(0, t-)\Big| \le Ln^{-1/q}|w|.$$

Letting $n \to \infty$ on the right side, it follows that the left side is zero. This shows that the function $\ell_t(u) := \psi(u, t-) - \psi(0, t-)$, $u \in \mathbb{R}$, is additive. By $(c)$, $\ell_t$ is continuous, hence linear. Since $\psi(1, \cdot) - \psi(0, \cdot) \in \mathcal{W}_q(J)$ there is a function $\beta \in \mathcal{W}_q(J)$ such that $\beta(a) = 0$ and for $t \in (a, b]$, $\ell_t(u) = \beta(t)u$, $u \in \mathbb{R}$. Letting $\alpha(t) := \psi(0, t-)$ for $t \in (a, b]$ and $\alpha(a) := \psi(0, a)$, we have the representation $\eta(u, s) = \alpha(s) + \beta(s)u$ for all $u \in \mathbb{R}$ and $s \in J$, proving $(b)$.

To prove $(a)$ and $(d)$, for each $u \in \mathbb{Q}$ (rational), $\psi(u, \cdot)$ is continuous except at most for $s$ in a countable set $T_u \subset J$. Let $T := \bigcup_{u \in \mathbb{Q}} T_u$. Then $T$ is a countable subset of $J$. For each $u \in \mathbb{Q}$ and $s \notin T$, $\xi(u, s) = 0$. By $(c)$, $\xi$ is $s$-uniformly Lipschitz on $\mathbb{R}$, so $\xi(\cdot, s) \equiv 0$ for each $s \in J \setminus T$, proving $(a)$. Also by $(c)$, $\xi(u, \cdot)$ is regulated for each $u \in \mathbb{R}$, proving $(d)$.

For $(e)$, since $\mathcal{W}_q(J)$ is a Banach algebra by Corollary 3.9 and $\|\cdot\|_{[q]} \le \|\cdot\|_{[p]}$ by Lemma 3.45, for each $f, g \in \mathcal{W}_p(J)$, we have

$$\big\|N_\eta f - N_\eta g\big\|_{[q]} \le \|\beta(f - g)\|_{[q]} \le \|\beta\|_{[q]}\|f - g\|_{[p]}.$$

Thus the Nemytskii operator $N_\eta$ is globally Lipschitz, and so is the Nemytskii operator $N_\xi$ by linearity. The proof of the theorem is complete. □

The following example shows that $\psi(\cdot, s)$ can in fact be non-affine for $s$ in a countable set.

**Example 6.71.** Let $1 \le p < \infty$, let $Q := \mathbb{Q} \cap [0, 1] = \{r_0, r_1, \dots\}$ be the set of rational numbers in $[0, 1]$ with $r_0 = 0$, and for each $u \in \mathbb{R}$, let

$$\xi(u,s) := \begin{cases} 2^{-k}\sin u \text{ if } s = r_k \text{ for some } k \geq 0, \\ 0 \qquad\qquad \text{if } s \in [0,1]\setminus Q. \end{cases} \tag{6.46}$$

Then for any partition $\kappa = \{t_i\}_{i=0}^n$ of $[0,1]$ and for any function $f \in \mathcal{W}_p[0,1]$, we have

$$s_p(N_\xi f;\kappa) \leq 2^p \sum_{i=0}^{n} |\xi(f(t_i),t_i)|^p \leq 2^p \sum_{k=0}^{\infty} 2^{-kp}|\sin f(r_k)| \leq 4^p/(2^p-1).$$

Thus the Nemytskii operator $N_\xi$ acts from $\mathcal{W}_p[0,1]$ into itself. To check the global Lipschitz condition for $N_\xi$, let $f,g \in \mathcal{W}_p[0,1]$ and again let $\kappa$ be a partition of $[0,1]$. Then we have

$$s_p(N_\xi f - N_\xi g;\kappa) \leq 2^p \sum_{r_k\in\kappa} 2^{-kp}\big|\sin f(r_k) - \sin g(r_k)\big|^p \leq 4^p\|f-g\|_{\sup}^p,$$

and so $\|N_\xi f - N_\xi g\|_{(p)} \leq 4\|f-g\|_{\sup}$. Since for any $s \in [0,1]$,

$$\big|(N_\xi f - N_\xi g)(s)\big| \leq \big|\sin f(s) - \sin g(s)\big| \leq \|f-g\|_{\sup},$$

it then follows that $\|N_\xi f - N_\xi g\|_{[p]} \leq 5\|f-g\|_{[p]}$, that is, a global Lipschitz condition holds for the Nemytskii operator $N_\xi$ generated by the function (6.46).

If a function $\psi = \psi(u,s)$ is left- or right-continuous in $s$ then global Lipschitz conditions from $\mathcal{W}_p$ into $\mathcal{W}_q$ for the Nemytskii operator $N_\psi$ can be characterized as follows:

**Corollary 6.72.** *Let $1 \leq p \leq q < \infty$, let $J := [a,b]$ with $a < b$, and let $\psi = \psi(u,s)\colon \mathbb{R}\times J \to \mathbb{R}$ be either left-continuous in $s \in (a,b]$ or right-continuous in $s \in [a,b)$. Assume that the Nemytskii operator $N_\psi$ acts from $\mathcal{W}_p(J)$ into $\mathcal{W}_q(J)$. There exists a constant $L < \infty$ such that* (6.45) *holds for each $f,g \in \mathcal{W}_p(J)$ if and only if there are functions $\alpha,\beta \in \mathcal{W}_q(J)$ such that $\psi(u,s) = \alpha(s) + \beta(s)u$ for all $s \in (a,b]$ or $s \in [a,b)$ respectively and $u \in \mathbb{R}$.*

*Proof.* The "if" part holds because $\mathcal{W}_q(J)$ is a Banach algebra by Corollary 3.9 and $\|\cdot\|_{[q]} \leq \|\cdot\|_{[p]}$ by Lemma 3.45. The "only if" part holds in case each $\psi(u,\cdot)$ is left-continuous on $(a,b]$ since then in the proof of Theorem 6.70, $\psi(u,s) = \eta(u,s)$ for $a < s \leq b$ and we apply part (*b*) of the theorem. The other case follows by symmetry. □

## 6.6 Higher Order Differentiability and Analyticity on $\mathcal{W}_p$ Spaces

Let $M > 0$, $W := (-M,M) \subset \mathbb{R}$, and let $\psi = \psi(u,s)$ be a real-valued function defined for $(u,s) \in W\times[a,b]$. Assuming $\psi = \psi(u,s)$ analytic with respect to $u$ on $W$ for each $s$, and under further conditions, we will prove that the Nemytskii operator $N_\psi$ is analytic as a mapping acting from $\mathcal{W}_p^{[W]}$ into $\mathcal{W}_p$ with $p \geq 1$ (see Corollary 6.78 below for $M = +\infty$, or Corollary 6.79 for general $M > 0$). We notice that analytic Nemytskii operators over $\mathcal{W}_p$ do not reduce to polynomials as do analytic Nemytskii operators over $L_p$-spaces (see Proposition 7.59 below). The present situation is more similar to the case when $N_\psi$ acts on a Banach space $C$ of bounded continuous $\mathbb{R}$-valued functions. By Theorem 6.8 of Appell and Zabrejko [3], for $\psi(u,s) \equiv F(u)$, the Nemytskii operator $N_F$ is analytic on the space $C$ if and only if $F$ is analytic from $\mathbb{R}$ into $\mathbb{R}$.

We will give a characterization of the $C^k$ property for the Nemytskii operator from $\mathcal{W}_p$ to $\mathcal{W}_q$. Recall that for Banach spaces $X$, $Y$ and an open set $U \subset X$, a mapping $F$ from $U$ into $Y$ is called $C^k$ on $U$ if the Fréchet derivatives of $F$ through order $k$ all exist on $U$ and are continuous. Before we state the characterization, we will find the norm of a $k$-linear multiplication operator (Definition 6.6).

**Lemma 6.73.** *Let $1 \leq p \leq q < \infty$, let $J$ be a nondegenerate interval, and let $k$ be a positive integer. Then $h \in \mathcal{W}_q(J)$ if and only if the $k$-linear multiplication operator $M^k[h]$ acts and is bounded from the Cartesian product $\mathcal{W}_p(J) \times \cdots \times \mathcal{W}_p(J)$ ($k$ times) into $\mathcal{W}_q(J)$. Moreover, in either case, $\|M^k[h]\| = \|h\|_{[q]}$.*

*Proof.* Let $h \in \mathcal{W}_q(J)$. By the Banach algebra property of $\mathcal{W}_q(J)$ (Corollary 3.9) and since $\|\cdot\|_{[q]} \leq \|\cdot\|_{[p]}$ (Lemma 3.45), for the norm on $L({}^k\mathcal{W}_p, \mathcal{W}_q)$ defined by (5.4), we have

$$\|M^k[h]\| = \sup\Big\{\big\|hg_1\cdots g_k\big\|_{[q]}\colon \|g_i\|_{[p]} \leq 1,\ i = 1,\dots,k\Big\} \leq \|h\|_{[q]},$$

and so $M^k[h]$ is a bounded operator as stated. The converse implication and the equality of the norms follow by taking $M^k[h](g_1,\dots,g_k)$ with $g_1 = \cdots = g_k = 1(\cdot) \in \mathcal{W}_p(J)$. □

The special case $p = q$ of the preceding lemma is also a special case of Lemma 5.13 on Banach algebras.

**Theorem 6.74.** *Let $1 \leq p \leq q < \infty$, $J := [a,b]$ with $a < b$ and $\psi\colon \mathbb{R} \times J \to \mathbb{R}$. Then for $k = 1, 2, \dots,$ the following two statements are equivalent:*

*(a) the Nemytskii operator $N_\psi$ is a $C^k$ mapping from $\mathcal{W}_p(J)$ into $\mathcal{W}_q(J)$;*
*(b) the function $\psi$ is $u$-differentiable of order $k$ on $\mathbb{R} \times J$ and each Nemytskii operator $N_{\psi_u^{(j)}}$ for $j = 0, 1, \dots, k$ is a continuous mapping from $\mathcal{W}_p(J)$ into $\mathcal{W}_q(J)$.*

*When (a) or (b) holds, for each $s \in J$, $u \mapsto \psi(u,s)$ is a $C^k$ function on $\mathbb{R}$.*

**Remark 6.75.** Let $\psi(u,s) \equiv \eta(u,s) + \xi(s)$. Then for any $k \geq 1$, $\psi_u^{(k)}(u,s) = \eta_u^{(k)}(u,s)$ if either side exists. But it can happen that the Nemytskii operator $N_\eta$ is a $C^k$ mapping from $\mathcal{W}_p(J)$ into $\mathcal{W}_q(J)$ while the Nemytskii operator $N_\psi$ does not act from $\mathcal{W}_p(J)$ into $\mathcal{W}_q(J)$, if $\xi(\cdot) \notin \mathcal{W}_q(J)$. Thus, in $(b)$ of Theorem 6.74, it is not sufficient to take $j = k$ (e.g. for $k = 1$).

*Proof.* Let $k$ be a positive integer and suppose that $(a)$ holds. Since $N_\psi$ is Fréchet differentiable of order $k$ at each constant function, by Proposition 6.9 with $X = Y = \mathbb{R}$, $S = J$, $\mathbb{G} = \mathcal{W}_p(J)$, and $\mathbb{H} = \mathcal{W}_q(J)$, $\psi$ is $u$-differentiable of order $k$ everywhere on $\mathbb{R} \times J$, and by (6.6), for each $G \in \mathcal{W}_p(J)$, the $k$th differential $\mathrm{d}^k N_\psi(G)$ is the $k$-linear multiplication operator $M^k[N_{\psi_u^{(k)}}(G)]$ in $L({}^k\mathcal{W}_p, \mathcal{W}_q)$. Thus by Lemma 6.73, the Nemytskii operator $N_{\psi_u^{(k)}}$ acts from $\mathcal{W}_p(J)$ into $\mathcal{W}_q(J)$. Also for any $f_1, f_2 \in \mathcal{W}_p(J)$,

$$\mathrm{d}^k N_\psi f_1 - \mathrm{d}^k N_\psi f_2 = M^k\big[N_{\psi_u^{(k)}} f_1 - N_{\psi_u^{(k)}} f_2\big].$$

Thus by the second part of Lemma 6.73, we have

$$\big\|\mathrm{d}^k N_\psi f_1 - \mathrm{d}^k N_\psi f_2\big\| = \big\|N_{\psi_u^{(k)}} f_1 - N_{\psi_u^{(k)}} f_2\big\|_{[q]}, \tag{6.47}$$

proving continuity of $N_{\psi_u^{(k)}}$, and so $(b)$ holds for $j = k$. Also $(b)$ must hold with $k$ replaced by $j$ for $j = 0, \ldots, k$, since by the definition of the $C^k$ property, $N_\psi$ is a $C^j$ mapping from $\mathcal{W}_p(J)$ into $\mathcal{W}_q(J)$.

The converse implication will be proved by induction. Suppose that $(b)$ holds with $k = 1$. For any $u, v \in \mathbb{R}$ and $s \in J$, we have

$$\big|\psi_u^{(1)}(u,s) - \psi_u^{(1)}(v,s)\big| \leq \big\|N_{\psi_u^{(1)}}(u1(\cdot)) - N_{\psi_u^{(1)}}(v1(\cdot))\big\|_{\sup}. \tag{6.48}$$

Since $N_{\psi_u^{(1)}}$ is a continuous mapping and $\|u1(\cdot) - v1(\cdot)\|_{[p]} = |u - v|$, the function $\psi_u^{(1)}(\cdot, s)$ is continuous for each $s \in J$. Let $G \in \mathcal{W}_p(J)$. For $\phi\colon \mathbb{R} \times J \to \mathbb{R}$, $f\colon J \to \mathbb{R}$, and $s \in J$, let

$$\Delta\phi(f;s) := \phi((G+f)(s), s) - \phi(G(s), s) = [N_\phi(G+f) - N_\phi(G)](s). \tag{6.49}$$

Then for $a \leq s < t \leq b$ and $g \in \mathcal{W}_p(J)$, by Lemma 5.40, we have

$$\begin{aligned}
&T(g;s,t)\\
&:= \Big[\Delta\psi(g;t) - \psi_u^{(1)}(G(t),t)g(t)\Big] - \Big[\Delta\psi(g;s) - \psi_u^{(1)}(G(s),s)g(s)\Big]\\
&= g(t)\int_0^1 \Delta\psi_u^{(1)}(rg;t)\,\mathrm{d}r - g(s)\int_0^1 \Delta\psi_u^{(1)}(rg;s)\,\mathrm{d}r \qquad (6.50)\\
&= [g(t)-g(s)]\int_0^1 \Delta\psi_u^{(1)}(rg;t)\,\mathrm{d}r + g(s)\int_0^1 \Big[\Delta\psi_u^{(1)}(rg;t) - \Delta\psi_u^{(1)}(rg;s)\Big]\,\mathrm{d}r\\
&=: T_1(g;s,t) + T_2(g;s,t).
\end{aligned}$$

Let $\kappa = \{x_i\}_{i=0}^n$ be a partition of $J$. Since $p \le q$, then we have

$$\Big(\sum_{i=1}^n \big|T_1(g;x_{i-1},x_i)\big|^q\Big)^{1/q} \le \Big(\sum_{i=1}^n \big|T_1(g;x_{i-1},x_i)\big|^p\Big)^{1/p}$$
$$\le \|g\|_{(p)} \sup_{0\le r\le 1} \big\|\Delta\psi_u^{(1)}(rg;\cdot)\big\|_{\sup}.$$

Also by Jensen's inequality for integrals, we have

$$\Big(\sum_{i=1}^n \big|T_2(g;x_{i-1},x_i)\big|^q\Big)^{1/q}$$
$$\le \|g\|_{\sup}\Big(\int_0^1 \sum_{i=1}^n \Big|\Delta\psi_u^{(1)}(rg;x_i) - \Delta\psi_u^{(1)}(rg;x_{i-1})\Big|^q\,\mathrm{d}r\Big)^{1/q}$$
$$\le \|g\|_{\sup} \sup_{0\le r\le 1} \big\|\Delta\psi_u^{(1)}(rg,\cdot)\big\|_{(q)}.$$

Thus by Minkowski's inequality (1.5), it follows that

$$\Big(\sum_{i=1}^n \big|T(g;x_{i-1},x_i)\big|^q\Big)^{1/q}$$
$$\le \|g\|_{(p)} \sup_{0\le r\le 1} \big\|\Delta\psi_u^{(1)}(rg;\cdot)\big\|_{\sup} + \|g\|_{\sup} \sup_{0\le r\le 1} \big\|\Delta\psi_u^{(1)}(rg;\cdot)\big\|_{(q)}.$$

$N_\psi$ acts from $\mathcal{W}_p(J)$ into $\mathcal{W}_q(J)$ by $(b)$ for $j=0$. Recalling the definition of $T$, since $\kappa$ is an arbitrary partition, the right side of the preceding inequality gives a bound for $\|N_\psi(G+g) - N_\psi(G) - N_{\psi_u^{(1)}}(G)g\|_{(q)}$. To bound the sup norm, again using Lemma 5.40, for each $s \in J$, we get as in (6.50)

$$\Big|N_\psi(G+g)(s) - N_\psi(G)(s) - N_{\psi_u^{(1)}}(G)(s)g(s)\Big|$$
$$= |g(s)|\Big|\int_0^1 \Delta\psi_u^{(1)}(rg;s)\,\mathrm{d}r\Big| \le \|g\|_{\sup} \sup_{0\le r\le 1} \big\|\Delta\psi_u^{(1)}(rg;\cdot)\big\|_{\sup}.$$

It then follows that

$$\big\|N_\psi(G+g) - N_\psi(G) - N_{\psi_u^{(1)}}(G)g\big\|_{[q]}$$
$$\le \|g\|_{(p)} \sup_{0\le r\le 1} \big\|\Delta\psi_u^{(1)}(rg;\cdot)\big\|_{\sup}$$
$$+\|g\|_{\sup}\Big[\sup_{0\le r\le 1} \big\|\Delta\psi_u^{(1)}(rg;\cdot)\big\|_{(q)} + \sup_{0\le r\le 1} \big\|\Delta\psi_u^{(1)}(rg;\cdot)\big\|_{\sup}\Big]$$
$$\le 2\|g\|_{[p]} \sup_{0\le r\le 1} \big\|N_{\psi_u^{(1)}}(G+rg) - N_{\psi_u^{(1)}}(G)\big\|_{[q]}.$$

The right side is $o(\|g\|_{[p]})$ as $\|g\|_{[p]} \to 0$ since $N_{\psi_u^{(1)}}$ is continuous at $G$ by assumption. Also by assumption, $N_{\psi_u^{(1)}}(G) \in \mathcal{W}_q(J)$, so the linear multiplication operator $M_{1,\psi,G} := M[N_{\psi_u^{(1)}}(G)]$ is bounded from $\mathcal{W}_p(J)$ into $\mathcal{W}_q(J)$

by Lemma 6.73 with $k = 1$. Thus for any $G \in \mathcal{W}_p(J)$, $N_\psi$ is differentiable at $G$ from $\mathcal{W}_p(J)$ to $\mathcal{W}_q(J)$ with derivative equal to $M_{1,\psi,G}$. Also since the Nemytskii operator $N_{\psi_u^{(1)}}$ is continuous, the Nemytskii operator $N_\psi$ is a $C^1$ mapping by (6.47) with $k = 1$, proving $(a)$ with $k = 1$.

Suppose that $(b)$ implies $(a)$ for $k = 1, \dots, n$ for some $n \geq 1$ and $(b)$ holds for $k = n + 1$. Then $N_\psi$ is a $C^n$ mapping from $\mathcal{W}_p(J)$ into $\mathcal{W}_q(J)$ since $(b)$ implies $(a)$ with $k = n$. Also for each $G \in \mathcal{W}_p(J)$, the $n$th differential $\mathrm{d}^n N_\psi(G) \in L(^n\mathcal{W}_p, \mathcal{W}_q)$ is the $n$-linear multiplication operator $M^n[N_{\psi_u^{(n)}}(G)]$ with $N_{\psi_u^{(n)}}(G) \in \mathcal{W}_q(J)$ by Proposition 6.9. By the implication $(b) \Rightarrow (a)$ with $k = 1$ already proved and by Proposition 6.9 with $k = 1$, both applied to $\psi_u^{(n)}$, for each $G \in \mathcal{W}_p(J)$, $N_{\psi_u^{(n)}} \colon \mathcal{W}_p(J) \to \mathcal{W}_q(J)$ is differentiable at $G$ and its derivative is $M[N_{\psi_u^{(n+1)}}(G)]$, with $N_{\psi_u^{(n+1)}}(G) \in \mathcal{W}_q(J)$. By Lemma 6.73 with $k = n$, we then have

$$\begin{aligned} &\big\|\mathrm{d}^n N_\psi(G+g) - \mathrm{d}^n N_\psi(G) - M^n[N_{\psi_u^{(n+1)}}(G)g]\big\|_{L(^n\mathcal{W}_p,\mathcal{W}_q)} \\ &\quad = \big\|N_{\psi_u^{(n)}}(G+g) - N_{\psi_u^{(n)}}(G) - N_{\psi_u^{(n+1)}}(G)g\big\|_{[q]}. \end{aligned}$$

These norms are $o(\|g\|_{[p]})$ as $\|g\|_{[p]} \to 0$ by the facts for $k = 1$ just mentioned, and so $\mathrm{d}^n N_\psi \colon \mathcal{W}_p \to L(^n\mathcal{W}_p, \mathcal{W}_q)$ is differentiable at $G \in \mathcal{W}_p(J)$ with the derivative $D\mathrm{d}^n N_\psi(G)g = M^n[N_{\psi_u^{(n+1)}}(G)g]$, $g \in \mathcal{W}_p(J)$. For each $G \in \mathcal{W}_p(J)$, we have by the implication $(b) \Rightarrow (a)$ in Proposition 5.25 applied to $L = M^{n+1}[N_{\psi_u^{(n+1)}}(G)]$ that $N_\psi$ is differentiable of order $n+1$ at $G$ and $\mathrm{d}^{n+1}N_\psi(G) = M^{n+1}[N_{\psi_u^{(n+1)}}(G)]$. Since $N_{\psi_u^{(n+1)}}$ is continuous, it follows that $N_\psi$ is a $C^{n+1}$ mapping from $\mathcal{W}_p(J)$ into $\mathcal{W}_q(J)$, proving $(a)$ with $k = n+1$. By induction, $(b) \Rightarrow (a)$ holds for each $k = 1, 2, \dots$. By (6.48) with $\psi_u^{(k)}$ in place of $\psi_u^{(1)}$ and since $N_{\psi_u^{(k)}}$ is continuous by $(b)$, it follows that $u \mapsto \psi_u^{(k)}(u, s)$ is continuous for each $s$, completing the proof of the theorem. □

The next statement is in preparation for giving sufficient conditions on $\psi$ for higher order differentiability of the Nemytskii operator $N_\psi$. Recall Definition 6.61 of the class $\mathcal{HW}_{n+\alpha,q}$.

**Lemma 6.76.** *Let $1 \leq p \leq q \leq r < \infty$, $\alpha := p/q$, $J := [a, b]$ with $a < b$, and let $n$ be a positive integer. For Banach spaces $X$, $Y$, let $B$ be a convex bounded set in $X$ containing $0$, let $V$ be an open set in $X$ including $B$, and let $\phi \colon V \times J \to Y$ be in the class $\mathcal{HW}_{n+\alpha,q}(B \times J; Y)$. Then for each $k \in \{1, \dots, n\}$, there exists a finite constant $C_k = C_k(\phi, B, p, r, \alpha)$ such that the bound*

$$\big\|\phi_u^{(k-1)}(g(\cdot), \cdot) - \phi_u^{(k-1)}(0, \cdot) - \phi_u^{(k)}(0, \cdot)(g(\cdot), \dots)\big\|_{[r]} \leq \|g\|_{[p]}\epsilon_k \tag{6.51}$$

*with*

$$\epsilon_k = \begin{cases} C_k \|g\|_{\sup}^{1-(q/r)}, & \text{if } k \in \{1,\dots,n-1\} \text{ and } n > 1, \\ C_n \|g\|_{\sup}^{\alpha-(p/r)}, & \text{if } k = n, \end{cases}$$

*holds for each* $g \in \mathcal{W}_p(J;X)$ *such that* $\|g\|_{[p]} \leq 1$ *and* $\operatorname{ran}(g) \subset B$.

*Proof.* Let $g \in \mathcal{W}_p(J;X)$ with $\operatorname{ran}(g) \subset B$. For $k = 1,\dots,n$, let $R_k(g) = R_k(g;\cdot)$ be the function on $J$ with values in $L({}^{k-1}X,Y)$ for each $s \in J$ defined by

$$R_k(g;s) := \phi_u^{(k-1)}(g(s),s) - \phi_u^{(k-1)}(0,s) - \phi_u^{(k)}(0,s)(g(s),\dots). \tag{6.52}$$

Let $h_k(s,t) := \phi_u^{(k)}(tg(s),s)(g(s),\dots) \in L({}^{k-1}X,Y)$ for $t \in [0,1]$ and $s \in J$. By Lemma 5.40, it follows that

$$R_k(g;s) = \int_0^1 \Big[h_k(s,t) - h_k(s,0)\Big]\, dt =: I_k(g)(s) \tag{6.53}$$

for $s \in J$, since $D\phi_u^{(k-1)}(0,s)(g(s)) = \phi_u^{(k)}(0,s)(g(s),\dots)$ by Proposition 5.25. To bound the $r$-variation seminorm of $I_k(g) = I_k(g)(\cdot)$, we will use the bounds for the $\gamma$-variation seminorm given by Theorem 3.111 of the integral transform $K(s) = F(h_k(s,\cdot) - h_k(s,0), d\bar{G}) = I_k(g)(s)$, $s \in J$, where $\bar{G}(t) \equiv t$, $t \in [0,1]$. To this aim we will bound $A_{\beta,\sup}(h_k) = A_{\beta,\sup}(h_k; J \times [0,1])$ and $B_{\sup,\bar{p}}(h_k) = B_{\sup,\bar{p}}(h_k; J \times [0,1])$ defined by (3.165) and (3.166), respectively, for $\beta = q$ and for suitable values of $\bar{p} \geq 1$.

If $n > 1$, then by Lemma 6.62, for each $k = 1,\dots,n-1$, $\phi_u^{(k)} \in \mathcal{HW}_{1,q}(B \times J; L({}^kX,Y))$, and so

$$H_1(\phi_u^{(k)}) < \infty \quad \text{and} \quad W_{1,q}(\phi_u^{(k)}, K) < \infty \tag{6.54}$$

for each $0 \leq K < \infty$, according to Definitions 6.42 and 6.47. Since $B$ is convex and contains 0, $tg(s) \in B$ for each $t \in [0,1]$ and $s \in J$. Then for any partition $\lambda = \{t_j\}_{j=0}^m$ of $[0,1]$ and $s \in J$, we have

$$\begin{aligned} s_1(h_k(s,\cdot);\lambda) &\leq \|g\|_{\sup} \sum_{j=1}^m \Big\|\phi_u^{(k)}(t_j g(s),s) - \phi_u^{(k)}(t_{j-1}g(s),s)\Big\| \\ &\leq H_1(\phi_u^{(k)})\|g\|_{\sup}^2, \end{aligned}$$

which gives a bound for $B_{\sup,1}(h_k)$. By Lemma 3.45, for each $s \in J$, $\|h_k(s,\cdot)\|_{(\bar{p})} \leq \|h_k(s,\cdot)\|_{(1)} \leq B_{\sup,1}(h_k)$, and so we have

$$B_{\sup,\bar{p}}(h_k) \leq B_{\sup,1}(h_k) \leq H_1(\phi_u^{(k)})\|g\|_{\sup}^2 \tag{6.55}$$

for each $k = 1,\dots,n-1$ and $\bar{p} \geq 1$. For $n \geq 1$ and $k = n$, since $\phi_u^{(n)} \in \mathcal{HW}_{\alpha,q}(B \times J; L({}^nX;Y))$ by definition (6.35), we have

$$H_\alpha(\phi_u^{(n)}) < \infty \quad \text{and} \quad W_{\alpha,q}(\phi_u^{(n)}, K) < \infty \tag{6.56}$$

for each $0 \le K < \infty$, according to Definition 6.47 again. Similarly, for any partition $\lambda$ of $[0,1]$ and $s \in J$, we get the bound

$$s_{q/p}(h_n(s,\cdot);\lambda) \le (H_\alpha(\phi_u^{(n)}))^{q/p}\|g\|_{\sup}^{(q/p)+1}$$

since $p = \alpha q$, and so the bound

$$B_{\sup,q/p}(h_n) \le H_\alpha(\phi_u^{(n)})\|g\|_{\sup}^{1+\alpha} \tag{6.57}$$

holds. Next we will apply the second property in (6.54) with $K = \|g\|_{[q]}$. For any partition $\kappa = \{s_i\}_{i=0}^n$ of $J$ and $t \in [0,1]$, using the Minkowski inequality (1.5) and (6.31), it follows that for $n > 1$ and $k = 1, \dots, n-1$,

$$\begin{aligned}
&s_q(h_k(\cdot,t);\kappa)^{1/q} \\
&\quad \le \Big(\sum_{i=1}^n \Big\|[\phi_u^{(k)}(tg(s_i),s_i) - \phi_u^{(k)}(tg(s_{i-1}),s_i)](g(s_i),\dots)\Big\|^q\Big)^{1/q} \\
&\qquad + \Big(\sum_{i=1}^n \Big\|[\phi_u^{(k)}(tg(s_{i-1}),s_i) - \phi_u^{(k)}(tg(s_{i-1}),s_{i-1})](g(s_i),\dots)\Big\|^q\Big)^{1/q} \\
&\qquad + \Big(\sum_{i=1}^n \Big\|\phi_u^{(k)}(tg(s_{i-1}),s_{i-1})(g(s_i) - g(s_{i-1}),\dots)\Big\|^q\Big)^{1/q} \\
&\quad \le \big[H_1(\phi_u^{(k)})\|g\|_{(q)} + W_{1,q}(\phi_u^{(k)},\|g\|_{[q]})\big]\|g\|_{\sup} + \|\phi_u^{(k)}\|_{B\times J,\sup}\|g\|_{(q)},
\end{aligned}$$

since $w_q(\{tg(s_{i-1})\}_{i=1}^n) \le \|g\|_{[q]}$ for each $t \in [0,1]$. For each $k = 1, \dots, n-1$, letting $D_k := H_1(\phi_u^{(k)}) + W_{1,q}(\phi_u^{(k)},1) + \|\phi_u^{(k)}\|_{B\times J,\sup}$, the bound

$$A_{q,\sup}(h_k) \le D_k\|g\|_{[q]} \tag{6.58}$$

holds if $\|g\|_{[q]} \le 1$. If $n \ge 1$ and $k = n$ then similarly to the preceding step, except that now (6.56) with $K = \|g\|_{[p]}$ is used, we have

$$\begin{aligned}
&s_q(h_n(\cdot,t);\kappa)^{1/q} \\
&\quad \le \big[H_\alpha(\phi_u^{(n)})\|g\|_{(p)}^\alpha + W_{\alpha,q}(\phi_u^{(n)},\|g\|_{[p]})\big]\|g\|_{\sup} + \|\phi_u^{(n)}\|_{B\times J,\sup}\|g\|_{(q)}.
\end{aligned}$$

Thus letting $D_n := H_\alpha(\phi_u^{(n)}) + W_{\alpha,q}(\phi_u^{(n)},1) + \|\phi_u^{(n)}\|_{B\times J,\sup}$, the bound

$$A_{q,\sup}(h_n) \le D_n\|g\|_{[q]} \tag{6.59}$$

holds if $\|g\|_{[p]} \le 1$. Denote by $\bar{p}$ and $\bar{q}$ respectively the parameters called $p$ and $q$ in Theorem 3.111. If $n > 1$ and $k = 1, \dots, n-1$, then apply the theorem with $\bar{p} = \bar{q} := 2 - q/r \in [1,2)$ and $\beta := q$. Thus $1/\gamma = (\bar{p}/\beta)(1/\bar{p} + 1/\bar{q} - 1) = 1/r$ and $\|\bar{G}\|_{[0,1],(\bar{q})} \le \|\bar{G}\|_{[0,1],(1)} = 1$ by Lemma 3.45. Since $\bar{p} \ge 1$ and $\beta/\gamma = q/r$, by (3.167), (6.55), and (6.58), we have

$$\begin{aligned}
\big\|I_k(g)\big\|_{(r)} &\le CA_{q,\sup}(h_k)^{q/r}B_{\sup,1}(h_k)^{1-(q/r)} \\
&\le E_k(\phi,B,p,q,r)\|g\|_{\sup}^{1-(q/r)}\|g\|_{[q]}
\end{aligned} \tag{6.60}$$

if $\|g\|_{[q]} \le 1$, where $E_k(\phi, B, p, q, r) := CD_k^{q/r} H_1(\phi_u^{(k)})^{1-(q/r)}$. If $k = n$ then apply Theorem 3.111 with $\bar{q} := 1/(1 + p(1/r - 1/q))$, $\bar{p} := q/p$, and $\beta := q$. Thus $1/\bar{p} + 1/\bar{q} = 1 + p/r > 1$ and $1/\gamma = (\bar{p}/\beta)(1/\bar{p} + 1/\bar{q} - 1) = 1/r$. Since $\bar{p} \ge 1$, $\bar{q} \ge 1$, and $\beta/\gamma = q/r$, by (3.167), (6.57), and (6.59), we have

$$\begin{aligned} \big\|I_n(g)\big\|_{(r)} &\le CA_{q,\sup}(h_n)^{q/r} B_{\sup,q/p}(h_n)^{1-(q/r)} \\ &\le E_n(\phi, B, p, q, r)\|g\|_{\sup}^{(p/q)-(p/r)}\|g\|_{[q]} \end{aligned} \tag{6.61}$$

if $\|g\|_{[p]} \le 1$, where $E_n(\phi, B, p, q, r) := CD_n^{q/r} H_\alpha(\phi_u^{(n)})^{1-(q/r)}$. To bound the supremum norm of $I_k(g)$ for $k \in \{1, \ldots, n-1\}$ with $n > 1$, since $\phi_u^{(k)}$ is $s$-uniformly Lipschitz (the first property in (6.54)), it follows that

$$\|I_k(g)(s)\| \le \|g\|_{\sup} \int_0^1 \Big\|\phi_u^{(k)}(tg(s), s) - \phi_u^{(k)}(0, s)\Big\|\, dt \le H_1(\phi_u^{(k)})\|g\|_{\sup}^2 \tag{6.62}$$

for each $s \in J$. For $k = n$, since $\phi_u^{(n)}$ is $s$-uniformly $\alpha$-Hölder (the first property in (6.56)) with $\alpha = p/q$, we have

$$\|I_n(g)(s)\| \le \|g\|_{\sup} \int_0^1 \Big\|\phi_u^{(n)}(tg(s), s) - \phi_u^{(n)}(0, s)\Big\|\, dt \le H_\alpha(\phi_u^{(n)})\|g\|_{\sup}^{1+\alpha} \tag{6.63}$$

for each $s \in J$. Let $M := \sup\{\|x\|\colon x \in B\} < \infty$. Then for $k \in \{1, \ldots, n-1\}$ with $n > 1$, using (6.60) and (6.62), it follows that

$$\begin{aligned} \|I_k(g)\|_{[r]} &= \|I_k(g)\|_{\sup} + \|I_k(g)\|_{(r)} \\ &\le \big[H_1(\phi_u^{(k)})M^{q/r} + E_k(\phi, B, p, q, r)\big]\,\|g\|_{\sup}^{1-(q/r)}\|g\|_{[q]} \end{aligned}$$

since $\|g\|_{\sup} \le M$. Similarly, using (6.61) and (6.63), it follows that

$$\|I_n(g)\|_{[r]} \le \big[H_{p/q}(\phi_u^{(n)})M^{p/r} + E_n(\phi, B, p, q, r)\big]\,\|g\|_{\sup}^{(p/q)-(p/r)}\|g\|_{[q]}$$

since $\|g\|_{\sup} \le M$. This together with (6.52) and (6.53) implies (6.51) since $\|g\|_{[q]} \le \|g\|_{[p]}$ by Lemma 3.45, completing the proof of Lemma 6.76. □

Recall that for a nonempty open set $U \subset X$, $\mathcal{W}_p^{[U]}(J; X)$ is the set of all $g \in \mathcal{W}_p(J; X)$ such that the closure of the range $\operatorname{ran}(g)$ is included in $U$.

**Theorem 6.77.** *Let $0 < \alpha \le 1 \le p < \alpha r < \infty$ and $J := [a, b]$ with $a < b$. For Banach spaces $X$, $Y$, let $U$ be an open subset of $X$ with $0 \in U$ and let $n$ be a positive integer. Let $\psi \in \mathcal{HW}^{\rm loc}_{n+\alpha,p/\alpha}(U \times J; Y)$ and $G \in \mathcal{W}_p^{[U]}(J; X)$. Then the Nemytskii operator $N_\psi$ acts from $\mathcal{W}_p^{[U]}(J; X)$ into $\mathcal{W}_p(J; Y)$ and is Fréchet differentiable of order $n$ at $G$ as a mapping with values in $\mathcal{W}_r(J; Y)$. Moreover, there exist constants $C = C(\psi, G, U, p, r, \alpha) < +\infty$ and $\delta = \delta(G, U) > 0$ such that for the remainder in the differentiation of $d^{n-1}N_\psi$,*

$$\big\|\mathrm{Rem}_{d^{n-1}N_\psi}(G,g)\big\| \le C\|g\|_{[p]}^{1+\alpha-(p/r)} \tag{6.64}$$

*for each* $g \in \mathcal{W}_p(J;X)$ *with* $\|g\|_{[p]} \le \delta$.

*Proof.* Let $q := p/\alpha < r$. By Definition 6.61, $\psi \in \mathcal{HW}_{n+\alpha,q}(B_m \times J;Y)$ for each $m > m_0$. Thus $\psi_u^{(n)} \in \mathcal{HW}_{\alpha,q}^{\mathrm{loc}}(U \times J; L(^nX,Y))$ and by Lemma 6.62, $\psi_u^{(k)} \in \mathcal{HW}_{1,q}^{\mathrm{loc}}(U \times J; L(^kX,Y))$ for $k = 0,\dots,n-1$, as defined in Definition 6.51. By Proposition 6.54($a$), for each $h \in \mathcal{W}_p^{[U]}(J;X)$,

$$N_{\psi_u^{(n)}}(h) \in \mathcal{W}_q(J;L(^nX,Y)) \quad \text{and} \quad N_{\psi_u^{(k)}}(h) \in \mathcal{W}_p(J;L(^kX,Y)) \tag{6.65}$$

for $k = 0,1,\dots,n-1$. In particular, it then follows that $N_\psi$ acts from $\mathcal{W}_p^{[U]}(J;X)$ into $\mathcal{W}_p(J;Y)$. To prove higher order differentiability of $N_\psi$ we will use Proposition 5.25 with $k$-linear mappings $L_{k,\psi}$ defined as follows. For a nonempty open set $V \subset X$, $\phi \in \mathcal{HW}_{n+\alpha,q}^{\mathrm{loc}}(V \times J;Y)$, $k \in \{1,\dots,n\}$, $h \in \mathcal{W}_p^{[V]}(J;X)$, and $g_1,\dots,g_k \in \mathcal{W}_p(J;X)$, let

$$\Big(L_{k,\phi}(h)(g_1,\dots,g_k)\Big)(s) := \phi_u^{(k)}(h(s),s)(g_1(s),\dots,g_k(s)) \in Y \tag{6.66}$$

for $s \in J$. Then $L_{k,\psi}(h)$ is a $k$-linear mapping from the $k$-fold product $\mathcal{W}_p(J;X) \times \cdots \times \mathcal{W}_p(J;X)$ into $Y$-valued functions on $J$. (The $\mathcal{W}_p$ property has not yet been used.) For each $k \in \{1,\dots,n\}$, using Corollary 5.24 with $F = N_{\psi_u^{(k)}}(h)$ and $\Phi(u) \equiv u^r$, $u \ge 0$, (6.65), and since $\|\cdot\|_{[r]} \le \|\cdot\|_{[q]} \le \|\cdot\|_{[p]}$ by Lemma 3.45, it follows that

$$\|L_{k,\psi}(h)(g_1,\dots,g_k)\|_{[r]} \le \big\|N_{\psi_u^{(k)}}(h)\big\|_{[r]}\|g_1\|_{[p]}\cdots\|g_k\|_{[p]} < \infty.$$

Thus for each $k \in \{1,\dots,n\}$ and $h \in \mathcal{W}_p^{[U]}(J;X)$, $L_{k,\psi}(h)$ is in $L(^k\mathcal{W}_p,\mathcal{W}_r)$.

Now the differentiability of order $n$ will be proved by induction. By Propositions 6.20($a$) and 6.3, $\mathrm{ran}\,(G) \subset B_m(U)$ for some $m \ge 1$. Let $\delta := \delta(G,U) := 1/(4m)$ and let $W := \{g \in \mathcal{W}_p(J;X)\colon \|g\|_{[p]} < 2\delta\}$. Then $G + W \subset \mathcal{W}_p^{[U]}(J;X)$. Let $\mathrm{d}^0N_\psi := N_\psi$, $L_{0,\psi}(h) := N_\psi(h)$, and $L(^0\mathcal{W}_p,\mathcal{W}_r) := \mathcal{W}_r$. Suppose that for some $k \in \{1,\dots,n\}$, the $(k-1)$st differential $\mathrm{d}^{k-1}N_\psi$ is defined on $G+W$. Then $\mathrm{d}^{k-1}N_\psi(h) = L_{k-1,\psi}(h)(\cdot,\cdots,\cdot)$ for all $h \in G+W$ by definition if $k=1$ and by Proposition 6.9 if $k > 1$. Let $V := B(0,2\delta) \subset X$ and $\phi(u,s) := \psi(G(s)+u,s)$ for $(u,s) \in V \times J$. By the chain rule (Theorem 5.1) we have $\phi_u^{(k)}(u,s) = \psi_u^{(k)}(G(s)+u,s)$ for $k = 1,\dots,n$, $s \in J$ and $u \in V$. For $g \in \mathcal{W}_p(J;X)$, let

$$D_k(g) := \begin{cases} L_{1,\psi}(G)(g) & \text{if } k=1,\\ L_{k,\psi}(G)(g,\cdot,\dots,\cdot) & \text{if } k>1,\end{cases} = \begin{cases} L_{1,\phi}(0)(g) & \text{if } k=1,\\ L_{k,\phi}(0)(g,\cdot,\dots,\cdot) & \text{if } k>1.\end{cases}$$

We apply Lemma 6.76 to $\phi\colon V \times J \to Y$ with $B := \{x\colon \|x\| \le \delta\}$. Let $g \in \mathcal{W}_p(J;X)$ be such that $\|g\|_{[p]} \le \delta$. If $k=1$ then by (6.51) we have

$$\big\|\mathrm{d}^{k-1}N_\psi(G+g)-\mathrm{d}^{k-1}N_\psi(G)-D_k(g)\big\|\leq\|g\|_{[p]}\epsilon_1,$$

and the right side is of order $o(\|g\|_{[p]})$ as $\|g\|_{[p]}\to 0$ since $r>q$. If $k\geq 2$ then using Corollary 5.24 with $F=R_k(g)$ defined by (6.52), and $\Phi(u)\equiv u^r$, $u\geq 0$, we have

$$\begin{aligned}
&\big\|\mathrm{d}^{k-1}N_\psi(G+g)-\mathrm{d}^{k-1}N_\psi(G)-D_k(g)\big\|_{L({}^{k-1}\mathcal{W}_p,\mathcal{W}_r)}\\
&\quad=\big\|L_{k-1,\phi}(g)-L_{k-1,\phi}(0)-D_k(g)\big\|_{L({}^{k-1}\mathcal{W}_p,\mathcal{W}_r)}\\
&\quad\leq\sup\big\{\|R_k(g)\|_{[r]}\|g_1\|_{[r]}\cdots\|g_{k-1}\|_{[r]}\colon\ \|g_1\|_{[p]}\leq 1,\dots,\|g_{k-1}\|_{[p]}\leq 1\big\}\\
&\quad\leq\|R_k(g)\|_{[r]},
\end{aligned}$$

since $\|\cdot\|_{(r)}\leq\|\cdot\|_{(p)}$ by Lemma 3.45. By (6.51), the right side is of order $o(\|g\|_{[p]})$ as $\|g\|_{[p]}\to 0$ since $r>q$. Therefore $\mathrm{d}^{k-1}N_\psi$ is differentiable at $G$ with derivative $D(\mathrm{d}^{k-1}N_\psi(h))\big|_{h=G}g=D_k(g)$, and so $N_\psi$ is Fréchet differentiable of order $k$ at $G$ by definition if $k=1$ and by Proposition 5.25 $(b)\Rightarrow(a)$ with $L=L_{k,\psi}(G)$ if $k>1$. By induction, $N_\psi$ is differentiable of order $n$ at $G$ as a mapping with values in $\mathcal{W}_r(J;Y)$ and (6.64) holds with $C(\psi,G,U,p,r,\alpha):=C_n(\phi,B,p,r,\alpha)$ by (6.51). The proof of Theorem 6.77 is complete. □

Next we apply results from the end of Section 6.3, concerning analyticity of Nemytskii operators acting from a real Banach algebra $\mathbb{B}_\mathbb{R}$ to itself, where $\mathbb{B}$ is a self-adjoint complex Banach algebra of functions and $\mathbb{B}_\mathbb{R}$ is the set of all real-valued elements of $\mathbb{B}$. Here we consider the case $\mathbb{B}=\mathcal{W}_p(J;\mathbb{C})$ with $1\leq p<\infty$ and $J=[a,b]$, and so $\mathbb{B}_\mathbb{R}=\mathcal{W}_p(J;\mathbb{R})\equiv\mathcal{W}_p(J)$. By Remark 6.27, $\mathbb{B}$ is a unital Banach algebra of $\mathbb{C}$-valued functions on $J$ such that $\overline{\operatorname{ran}(f)}=\sigma(f)$ for each $f\in\mathbb{B}$. Then the following statement is a special case of Proposition 6.31.

**Corollary 6.78.** *Let $1\leq p<\infty$, $J:=[a,b]$ with $a<b$, and let $V$ be a connected open set in $\mathcal{W}_p(J)$. Let $\psi\colon\mathbb{R}\times J\to\mathbb{R}$ be u-differentiable of order $n$ everywhere on $\mathbb{R}\times J$ for each $n\geq 1$. Let the Nemytskii operator $N_\psi$ act from $V$ into $\mathcal{W}_p(J)$, and let $G\in V$. Then $N_\psi$ on $V$ has a Taylor expansion around $G$ if and only if*

*(a) there are $c\geq 0$ and $r>0$ such that for each $k=1,2,\dots,$*

$$\sup\big\{|\psi_u^{(k)}(G(x)+t,x)|\colon\ x\in J,\ t\in\mathbb{R},\ |t|\leq r\big\}\leq c^k k!;$$

*(b) $N_{\psi_u^{(k)}}(G)\in\mathcal{W}_p(J)$ for all $k\geq 1$, and $\sup_{k\geq 1}\big\|N_{\psi_u^{(k)}}(G)/k!\big\|_{[p]}^{1/k}<\infty$.*

The next statement is a special case of Theorem 6.32.

**Corollary 6.79.** *Let $1\leq p<\infty$, $J:=[a,b]$ with $a<b$, and $W:=(-M,M)$ for some $0<M\leq\infty$. Let $\psi\colon W\times J\to\mathbb{R}$ be u-differentiable of order $n$*

*everywhere on* $W \times J$ *for each* $n \geq 1$*, and suppose that the Nemytskii operator* $N_\psi$ *acts from* $\mathcal{W}_p^{[W]}(J)$ *into* $\mathcal{W}_p(J)$*. Then* $N_\psi$ *is analytic on* $\mathcal{W}_p^{[W]}(J)$ *if and only if*

(*a*) *for each closed interval* $B \subset W$*, there is a* $c \geq 0$ *such that for* $k = 1, 2, \ldots,$

$$\|\psi_u^{(k)}\|_{B \times J, \sup} \leq c^k k!;$$

(*b*) *statement* (*b*) *of Corollary* 6.78 *holds for each* $G \in \mathcal{W}_p^{[W]}(J)$.

The last statement in this section is a special case of Corollary 6.30.

**Corollary 6.80.** *Let* $F\colon \mathbb{R} \to \mathbb{R}$ *and* $1 \leq p < \infty$*. Then the autonomous Nemytskii operator* $N_F$*, as a mapping from* $\mathcal{W}_p$ *into* $\mathcal{W}_p$*, is analytic on* $\mathcal{W}_p$ *if and only if* $F$ *is analytic on* $\mathbb{R}$.

## 6.7 Notes

We are aware of little previous literature on the specific topics of this chapter, e.g. higher order differentiability of Nemytskii operators, and such operators on $\mathcal{W}_p$ spaces, except for the few references mentioned in the following notes. There is much more literature about Nemytskii operators on other domain spaces, e.g. $L^p$ spaces; see Chapter 7. The notes to Chapter 7 mention several other classes of domain spaces.

*Notes on Section 6.3.* Dales [36, §4.1] treats natural Banach algebras. Theorem 6.21 is Proposition 4.1.5 of Dales [36]; his Notes do not say who first proved it.

*Notes on Section 6.5.* Theorem 6.35 essentially is the main result of Ciemnoczołowski and Orlicz [33]. Proposition 6.57 seems to contradict Appell and Zabrejko [3, p. 174, Theorem 6.12]. Theorem 6.70 with $q = p = 1$ is due to Matkowski and Miś [161].

# 7

# Nemytskii Operators on $L^p$ Spaces

In this chapter, the Nemytskii operator $N_\psi$ is considered between the spaces $\mathbb{G} = L^s$ and $\mathbb{H} = L^p$ with $s, p \geq 1$. Thus, it will be considered as an operator between spaces of equivalence classes of measurable functions on a measure space.

Throughout this chapter $(\Omega, \mathcal{S}, \mu)$ is a measure space with $\mu(\Omega) > 0$. As before, "$\mu$-almost everywhere" is abbreviated "a.e. $(\mu)$." For $1 \leq p \leq \infty$, $\mathcal{L}^p := \mathcal{L}^p(\Omega, \mathcal{S}, \mu) := \mathcal{L}^p(\Omega, \mathcal{S}, \mu; \mathbb{R})$ denotes the set of all real-valued functions $f$ on $\Omega$ such that $f$ is $\mu$-measurable, that is, measurable for the completion of $\mu$, and $\|f\|_p < \infty$ where for $1 \leq p < \infty$,

$$\|f\|_p := \Big( \int |f(\omega)|^p \, d\mu(\omega) \Big)^{1/p} \qquad \text{and} \qquad \|f\|_\infty := \text{ess.sup}\,|f|$$

(cf. (1.22) and (1.23) with $\|f\|_{\mathbb{R}} := |f|$). $L^p := L^p(\Omega, \mathcal{S}, \mu) := L^p(\Omega, \mathcal{S}, \mu; \mathbb{R})$ with norm $\|\cdot\|_p$ is the Banach space of all equivalence classes of functions in $\mathcal{L}^p(\Omega, \mathcal{S}, \mu; \mathbb{R})$ for equality a.e. $(\mu)$.

In this chapter, the set $S$ in the definition (6.1) of the Nemytskii operator $N_\psi$ will be $\Omega$. Also in this chapter, $U = X = Y = \mathbb{R}$, and so $\psi$ is a function $(u, \omega) \mapsto \psi(u, \omega)$ from $\mathbb{R} \times \Omega$ into $\mathbb{R}$.

## 7.1 Acting, Boundedness, and Continuity Conditions

Given a function $\psi$ from $\mathbb{R} \times \Omega$ into $\mathbb{R}$, for the Nemytskii operator $N_\psi$ to act from $L^s$ into $L^p$, it first of all must preserve measurability of functions.

*Preserving measurability*

A function $\psi\colon \mathbb{R} \times \Omega \to \mathbb{R}$ will be called *bimeasurable* or *$\mathcal{S}$-bimeasurable* if for each measurable function $f\colon \Omega \to \mathbb{R}$, $N_\psi(f)$ is measurable. Next some properties of $\psi$ are stated and shown to imply its bimeasurability.

R.M. Dudley and R. Norvaiša, *Concrete Functional Calculus*, Springer Monographs in Mathematics, DOI 10.1007/978-1-4419-6950-7_7, 

**Definition 7.1.** Let $A \in \mathcal{S}$. Then $A$ will be called *hereditarily $\mathcal{S}$-measurable*, or $A \in \mathcal{S}_h$, if for every $C \subset A$, $C \in \mathcal{S}$.

**Remark 7.2.** Any subset of a set in $\mathcal{S}_h$ is also in $\mathcal{S}_h$. So is any countable union of sets in $\mathcal{S}_h$. If $\omega \in \Omega$ and the singleton $\{\omega\}$ is in $\mathcal{S}$ then $\{\omega\} \in \mathcal{S}_h$. If $(\Omega, \mathcal{S}, \mu)$ is complete and $\mu(A) = 0$ then $A \in \mathcal{S}_h$. Thus if both $(\Omega, \mathcal{S}, \mu)$ is complete and all singletons are in $\mathcal{S}$, then $A \in \mathcal{S}_h$ if $A = B \cup C$ where $\mu(B) = 0$ and $C$ is countable. If $\mu$ is a finite or $\sigma$-finite measure on $\Omega = [0, 1]$ then under some set-theoretic assumptions there are no sets in $\mathcal{S}_h$ other than such $B \cup C$ [53, Appendix C].

The following condition on a function $\psi$ will be shown to be sufficient for its bimeasurability. It says that for some $N \in \mathcal{S}_h$, $\psi$ restricted to $(u, \omega)$ such that $\omega \notin N$ is jointly measurable.

**Definition 7.3.** Let $\mathcal{B}$ be the Borel $\sigma$-algebra on $\mathbb{R}$. A function $\psi = \psi(u, \omega)$, $(u, \omega) \in \mathbb{R} \times \Omega$, is called a *Shragin function* or is said to satisfy *Shragin's condition* if there is a set $N \in \mathcal{S}_h$ such that for any $A \in \mathcal{B}$,

$$\psi^{-1}(A) \setminus (\mathbb{R} \times N) = \{(u, \omega) \in \mathbb{R} \times N^c\colon\ \psi(u, \omega) \in A\} \in \mathcal{B} \otimes \mathcal{S}, \tag{7.1}$$

where $\mathcal{B} \otimes \mathcal{S}$ is the product $\sigma$-algebra on $\mathbb{R} \times \Omega$. If in the Shragin condition, in addition, $\mu(N) = 0$, then the *strong* or *$\mu$-strong Shragin condition* will be said to hold and a $\psi$ will be called a *strong* or *$\mu$-strong Shragin function.*

If $(\Omega, \mathcal{S}, \mu)$ is complete, then a function $\psi\colon\ \mathbb{R} \times \Omega \to \mathbb{R}$ is $\mu$-strong Shragin if and only if (7.1) holds for some $\mu$-null set $N \subset \Omega$.

**Proposition 7.4.** *A Shragin function on $\mathbb{R} \times \Omega$ is bimeasurable.*

*Proof.* Let $\psi$ be a Shragin function on $\mathbb{R} \times \Omega$. Let $f\colon\ \Omega \to \mathbb{R}$ be measurable, $A \in \mathcal{B}$, and let $N$ be a hereditarily $\mathcal{S}$-measurable set for which Shragin's condition (7.1) holds. The function $\hat{f}\colon\ \Omega \to \mathbb{R} \times \Omega$ defined by $\hat{f}(\omega) := (f(\omega), \omega)$ is measurable for the $\sigma$-algebra $\mathcal{B} \otimes \mathcal{S}$ on $\mathbb{R} \times \Omega$. Since $\omega \notin N$ if and only if $\hat{f}(\omega) \notin \mathbb{R} \times N$, we have that

$$\{\omega \in \Omega \setminus N\colon\ \psi(f(\omega), \omega) \in A\} = \{\omega \in \Omega\colon\ \hat{f}(\omega) \in \psi^{-1}(A) \setminus (\mathbb{R} \times N)\} \in \mathcal{S}$$

by Shragin's condition. Also, $\{\omega \in N\colon\ \psi(f(\omega), \omega) \in A\} \in \mathcal{S}$ since $N \in \mathcal{S}_h$. The proof of the proposition is complete. □

**Definition 7.5.** For a measurable space $(X, \mathcal{S})$, a set $A \subset X$ is called *universally measurable* (*u.m.*) iff it is measurable for the completion of every finite measure $\mu$ on $(X, \mathcal{S})$. If $(Y, \mathcal{B})$ is another measurable space, then a function $f$ from $X$ into $Y$ is called *universally measurable* (*u.m.*) iff $f^{-1}(B)$ is u.m. for all $B \in \mathcal{B}$. If $X$ and $Y$ are separable metric spaces, then unless otherwise specified, $\mathcal{S}$ and $\mathcal{B}$ will be taken as the Borel $\sigma$-algebras in the respective spaces.

The collection $\mathcal{U}_{\mathcal{S}}$ of all u.m. sets for $\mathcal{S}$ is easily seen to form a $\sigma$-algebra. Note that if $(\Omega, \mathcal{S})$ is a measurable space and $(\Omega, \mathcal{S}, \mu)$ is complete for some finite measure $\mu$, then $\mathcal{U}_{\mathcal{S}} = \mathcal{S}$. Also, if $(\Omega, \mathcal{S}, \mu)$ is a complete, $\sigma$-finite measure space, then there is a finite measure $\nu$ on $\mathcal{S}$ equivalent to $\mu$, so $\mathcal{U}_{\mathcal{S}} = \mathcal{S}$ in this case also. If $X$ is any uncountable complete separable metric space, e.g. $X = \mathbb{R}$, and $\mathcal{S}$ is the Borel $\sigma$-algebra, then there exist u.m. sets which are not Borel measurable, e.g. [53, §13.2].

The space of all Shragin functions on $\mathbb{R} \times \Omega$ is linear, as is the space of all strong Shragin functions.

**Proposition 7.6.** *If $(\Omega, \mathcal{S}, \mu)$ is complete and $\sigma$-finite and if $\psi$ is a $\mu$-strong Shragin function on $\mathbb{R} \times \Omega$, then the following are equivalent:*

*(a) $N_\psi f = 0$ a.e. $(\mu)$ for all $\mu$-measurable functions $f\colon \Omega \to \mathbb{R}$;*
*(b) for some $p$, $1 \le p \le \infty$, and some $r > 0$, $N_\psi f = 0$ a.e. $(\mu)$ for all $f \in \mathcal{L}^p(\Omega, \mathcal{S}, \mu)$ such that $\|f\|_p < r$;*
*(c) for $\mu$-almost all $\omega \in \Omega$, $\psi(u, \omega) = 0$ for all $u \in \mathbb{R}$.*

*Proof.* Implications $(c) \Rightarrow (a) \Rightarrow (b)$ are immediate. Suppose that $(b)$ holds for some $p \in [1, \infty]$ and let $D := \{(u, \omega)\colon \psi(u, \omega) \neq 0\}$. Since $\psi$ is a strong Shragin function there is a $\mu$-null set $N \subset \Omega$ such that $D \setminus (\mathbb{R} \times N) \in \mathcal{B} \otimes \mathcal{S}$. According to Sainte-Beuve's selection theorem (e.g. [52, Theorem 5.3.2] for $\mathcal{F} := Y := \mathbb{R}$ with Borel $\sigma$-algebra and $T$ the identity $\mathbb{R} \to \mathbb{R}$, $X = \psi$ restricted to $\mathbb{R} \times N^c$), the projection $E := \{\omega\colon (u, \omega) \in D \setminus (\mathbb{R} \times N)$ for some $u \in \mathbb{R}\}$ is (universally) measurable in $\Omega$ and there is a (universally) measurable function $\eta\colon E \to \mathbb{R}$ such that $(\eta(\omega), \omega) \in D \setminus (\mathbb{R} \times N)$ for all $\omega \in E$. If $\mu(E) > 0$ then by $\sigma$-finiteness there is a $C \in \mathcal{S}$ with $C \subset E$, $0 < \mu(C) < \infty$, $|\eta| \le M$ on $C$ for some $M < \infty$, and $\|\eta 1_C\|_p < r$. Let $\eta_C := \eta 1_C$. Thus $\eta_C \in \mathcal{L}^p(\Omega, \mathcal{S}, \mu)$ with norm less than $r$, and so $N_\psi \eta_C = 0$ a.e. $(\mu)$. It then follows that $C$ is a $\mu$-null set, a contradiction, so $\mu(E) = 0$. For $\omega \in \Omega \setminus (N \cup E)$, $\psi(u, \omega) = 0$ for all $u \in \mathbb{R}$, proving $(c)$. The proof of the proposition is complete. □

**Definition 7.7.** For two bimeasurable functions $\psi$ and $\phi$ on $\mathbb{R} \times \Omega$, we write $\psi \simeq \phi$ or equivalently $\psi \simeq_\mu \phi$ if $N_\psi f = N_\phi f$ a.e. $(\mu)$ for each measurable function $f$. For $\psi$ again bimeasurable and for a Borel measurable function $h\colon \Omega \to \mathbb{R}$, we write $\psi \simeq h$ if $\psi \simeq \hat{h}$ for $\hat{h}(u, \omega) := h(\omega)$, $(u, \omega) \in \mathbb{R} \times \Omega$.

**Remark 7.8.** If $\psi$ and $\phi$ are strong Shragin functions, then so is $\psi - \phi$, and by Proposition 7.6, $\psi \simeq \phi$ if and only if for $\mu$-almost all $\omega \in \Omega$, $\psi(u, \omega) = \phi(u, \omega)$ for all $u \in \mathbb{R}$.

*Acting and boundedness in the autonomous case*

Let $F$ be a function from $\mathbb{R}$ into $\mathbb{R}$ and let $N_F$ be the autonomous Nemytskii operator defined by $N_F f = F \circ f$. Given a complete and finite measure space $(\Omega, \mathcal{S}, \mu)$, $1 \le s \le \infty$, and $1 \le p < \infty$, Theorem 7.13$(a)$ will show that $N_F$ acts from $L^s = L^s(\Omega, \mathcal{S}, \mu)$ into $L^p = L^p(\Omega, \mathcal{S}, \mu)$ if $F$ is universally measurable and a condition on the growth of $F$ holds, defined as follows (where $\beta$ will be $s/p$). For $0 < \beta < \infty$, let $\mathcal{G}_\beta$ be the class of all functions $F$ from $\mathbb{R}$ into itself satisfying the $\beta$ *growth condition*:

$$\|F\|_{\mathcal{G}_\beta} := \sup_u |F(u)|/(1 + |u|^\beta) < \infty. \tag{7.2}$$

Let $\mathcal{G}_\infty$ be the class of all functions from $\mathbb{R}$ into itself which are bounded on bounded intervals, i.e. $F \in \mathcal{G}_\infty$ if and only if $\sup_{|x| \le M} |F(x)| < \infty$ for each $M$ with $0 < M < \infty$. A norm $\|\cdot\|_{\mathcal{G}_\infty}$ is not defined.

For $0 < \beta < \infty$, let $\Gamma_\beta$ be the class of functions $F$ defined Lebesgue $\lambda$-almost everywhere from $\mathbb{R}$ into $\mathbb{R}$ such that

$$\|F\|_{\Gamma_\beta} := \inf\{K\colon\ |F(u)| \le K(1 + |u|^\beta) \text{ a.e. } (\lambda)\} < \infty. \tag{7.3}$$

For $\beta = +\infty$, $F \in \Gamma_\infty$ will mean that for each $M$ with $0 < M < \infty$, $F$ restricted to $[-M, M]$ is in $\mathcal{L}^\infty([-M, M], \lambda)$. For the next fact recall the class $\mathcal{H}_1^{\mathrm{loc}}$ of functions that are 1-Hölder locally on $\mathbb{R}$, as defined in Definition 6.4.

**Lemma 7.9.** *Let $F \in \mathcal{H}_1^{\mathrm{loc}}$ be such that for some $0 < \beta < \infty$, $F' \in \Gamma_\beta$, with $K := \|F'\|_{\Gamma_\beta}$. Then $|F'(u)| \le K(1 + |u|^\beta)$ for all $u$ such that $F'(u)$ is defined.*

*Proof.* Let $t, u \in \mathbb{R}$ and $\epsilon > 0$. Suppose $t \ne u$ and $|t| \le |u| + \epsilon$. Then $F(t) - F(u) = \int_u^t F'(u)\,du$, where the integral is well defined although $F'(u)$ is defined only a.e. $(\lambda)$ (e.g. Theorem 8.18 and Example 8.20 in [198]), so

$$|F(t) - F(u)| \le |t - u| \cdot K(1 + |u + \epsilon|^\beta).$$

If $F'(u)$ is defined, then letting $t \to u$ and $\epsilon \downarrow 0$, by definition of derivative the conclusion follows. □

Recall that an *atom* is a set $A \in \mathcal{S}$ (not necessarily a singleton) with $\mu(A) > 0$ such that for all $C \in \mathcal{S}$, $\mu(A \cap C) = 0$ or $\mu(A)$. If there are no atoms, then $(\Omega, \mathcal{S}, \mu)$ is called *nonatomic*. It is known that if $(\Omega, \mathcal{S}, \mu)$ is a nonatomic finite measure space then for any $A \in \mathcal{S}$ and $0 < c < \mu(A)$ there is a $C \in \mathcal{S}$ with $C \subset A$ and $\mu(C) = c$ (Appendix, Proposition A.1).

We will also need $N_F$ to preserve measurability of functions. Evidently, if $F$ is Borel measurable and $u(\cdot)$ is measurable from $\Omega$ into $\mathbb{R}$, then so is $F \circ u$. For $\psi(u, \omega) \equiv F(u)$, the function $\psi(\cdot, \cdot)$ is a Shragin function if and only if $F$ is Borel measurable. We next define a regularity condition for a measure space $(X, \mathcal{S}, \mu)$. For any function $T$ from $X$ to another set $Y$, let $\mathcal{S}_T := \{A \subset Y\colon\ T^{-1}(A) \in \mathcal{S}\}$. Then $\mathcal{S}_T$ is a $\sigma$-algebra of subsets of $Y$ and $\mu \circ T^{-1}$ is a measure defined on each $A \in \mathcal{S}_T$ by $(\mu \circ T^{-1})(A) := \mu(T^{-1}(A))$. If $(Y, \mathcal{B})$ is a measurable space and $T$ is measurable for $\mathcal{S}$ and $\mathcal{B}$, then $\mathcal{B} \subset \mathcal{S}_T$, where the inclusion may be strict.

**Definition 7.10.** A measure space $(X, \mathcal{S}, \mu)$ is called *perfect* iff for every measurable function $T$ from $X$ into $\mathbb{R}$ and every set $A \subset \mathbb{R}$ such that $T^{-1}(A)$ is measurable for the completion of $\mu$, $A$ is measurable for the completion of the restriction of the measure $\mu \circ T^{-1}$ to the Borel sets in $\mathbb{R}$.

The following shows that some commonly occurring measure spaces are perfect.

**Proposition 7.11.** *Let $(X, d)$ be a complete separable metric space, let $\mathcal{S}$ be its Borel $\sigma$-algebra, and let $\mu$ be a finite measure on $\mathcal{S}$. Then $(X, \mathcal{S}, \mu)$ is perfect.*

*Proof.* Let $T$ be Borel measurable from $X$ into $\mathbb{R}$ and let $A \subset \mathbb{R}$ be such that $T^{-1}(A)$ is measurable for the completion of $\mu$, i.e. there are Borel sets $B_i \subset X$, $i = 1, 2$, with $B_1 \subset T^{-1}(A) \subset B_2$ and $\mu(B_2 \setminus B_1) = 0$. Each $T[B_i] = \{T(x) \colon x \in B_i\}$ is u.m. in $\mathbb{R}$ [53, §13.2]. Thus for some Borel sets $C_i \subset \mathbb{R}$, $C_1 \subset T[B_1] \subset A \subset T[B_2] \subset C_2$ with $(\mu \circ T^{-1})(C_2 \setminus C_1) = 0$, so $A$ is measurable for the completion of $\mu \circ T^{-1}$, which finishes the proof. □

**Example 7.12.** For $X = [0, 1]$, $\mathcal{S}$ the Borel $\sigma$-algebra, and $\lambda$ = Lebesgue measure on $(X, \mathcal{S})$, let $E$ be a non-measurable set and extend $\lambda$ to a measure $\bar{\lambda}$ on the smallest $\sigma$-algebra $\mathcal{S}_E$ including $\mathcal{S}$ and containing $E$ [53, §3.4, Problems 3-4]. Considering the identity $T$ from $X$ into $\mathbb{R}$ and $A = E$, we see that $(X, \mathcal{S}_E, \bar{\lambda})$ is not perfect.

Here is a fact on acting and boundedness conditions for $N_F$ from $L^s$ to $L^p$ for $1 \le p < \infty$ and $1 \le s \le \infty$. (Recall that an operator between Banach spaces is called bounded if it takes bounded sets into bounded sets.) If $g$, $h$, $F \circ g$, and $F \circ h$ are all measurable, then $g = h$ a.e. $(\mu)$ implies that $F \circ g = F \circ h$ a.e. $(\mu)$. Thus if $N_F$ acts from $\mathcal{L}^s$ into $\mathcal{L}^p$, it will naturally define an operator from $L^s$ into $L^p$.

**Theorem 7.13.** *Let $(\Omega, \mathcal{S}, \mu)$ be complete and finite, let $F$ be a function from $\mathbb{R}$ into $\mathbb{R}$, let $1 \le p < \infty$ and $1 \le s \le \infty$. Then (a) always holds, (b) holds when $(\Omega, \mathcal{S}, \mu)$ is nonatomic, and (c) holds when $(\Omega, \mathcal{S}, \mu)$ is nonatomic and perfect, where*

(a) *If $F \in \mathcal{G}_{s/p}$ and $F$ is u.m., then $N_F$ acts from $L^s = L^s(\Omega, \mathcal{S}, \mu)$ into $L^p$ and is a bounded operator.*
(b) *If $N_F$ acts from $V$ into $L^p$ then $F \in \mathcal{G}_{s/p}$, where $V$ is a nonempty open subset of $L^s$ for $s < \infty$ or $V = L^\infty$ for $s = +\infty$.*
(c) *$N_F$ acts from $L^s$ into $L^p$ if and only if both $F \in \mathcal{G}_{s/p}$ and $F$ is u.m.*

**Remark 7.14.** By Proposition 7.11, the characterization in part (c) applies when $\Omega = [0, 1]$, $\mu$ is Lebesgue measure, and $\mathcal{S}$ is the $\sigma$-algebra of Lebesgue measurable sets.

*Proof.* (*a*): Let $F \in \mathcal{G}_{s/p}$ be u.m. and $g \in \mathcal{L}^s$. Then for any Borel set $B \subset \mathbb{R}$, $(F\circ g)^{-1}(B) = g^{-1}(F^{-1}(B)) \in \mathcal{S}$ because $F^{-1}(B)$, being u.m. in $\mathbb{R}$, must be measurable for the completion of $\mu \circ g^{-1}$ on the Borel sets of $\mathbb{R}$, and $\mathcal{S}$ is complete for $\mu$ by assumption. So $F\circ g$ is a measurable function. If $s < \infty$, then we have

$$\left(\int |F(g(\omega))|^p \mathrm{d}\mu(\omega)\right)^{1/p} \le \|F\|_{\mathcal{G}_{s/p}} \left[\int (1 + |g(\omega)|^{s/p})^p \mathrm{d}\mu(\omega)\right]^{1/p}$$
$$\le \|F\|_{\mathcal{G}_{s/p}} \left[\|1(\cdot)\|_p + \|g\|_s^{s/p}\right] < \infty.$$

If $g \in \mathcal{L}^\infty$ and $\|g\|_\infty \le K$ for some $K < \infty$, then

$$\|F\circ g\|_p \le \sup_{|x|\le K} |F(x)|\mu(\Omega)^{1/p} < \infty.$$

Thus $N_F$ acts from $\mathcal{L}^s$ into $\mathcal{L}^p$ and defines a bounded operator from $L^s$ into $L^p$, proving (*a*).

(*b*): Let $s < \infty$ and let $N_F$ act from a neighborhood $V$ of some $G \in \mathcal{L}^s$ into $\mathcal{L}^p$. Suppose that $F \notin \mathcal{G}_{s/p}$. Then for some $x_n \in \mathbb{R}$ and all $n = 1, 2, \ldots$,

$$|F(x_n)| > n(1 + |x_n|^{s/p}). \tag{7.4}$$

We will find some numbers $\mu_n > 0$ and disjoint sets $A_n \in \mathcal{S}$ with $\mu(A_n) = \mu_n$ such that

$$\sum_n \mu_n |x_n|^s < +\infty, \tag{7.5}$$

$$\sum_n \mu_n |F(x_n)|^p = +\infty. \tag{7.6}$$

If $\sup_n |x_n| < \infty$, let $\mu_n := C/n^2$ for a constant $C > 0$. Or if the $x_n$ are unbounded, then taking a subsequence, we can assume that $|x_n| \ge 1$ for all $n$, and let $\mu_n := C/(|x_n|^s n^2)$ for a constant $C > 0$. Then in either case (7.5) holds and since $p \ge 1$, by (7.4), also (7.6) holds. Taking $C$ small enough, we have $\sum_n \mu_n < \mu(\Omega)$ and since $\mu$ is nonatomic, disjoint $A_n \in \mathcal{S}$ with $\mu(A_n) = \mu_n$ exist by Proposition A.1 and induction. Then $h := \sum_n x_n 1_{A_n} \in \mathcal{L}^s$ by (7.5). Let $C_k := \bigcup_{n\ge k} A_n$ for $k = 1, 2, \ldots$. Let $G_k := h$ on $C_k$ and $G_k := G$ elsewhere. Then $G_k \in V$ for $k$ large enough, but

$$\int_\Omega |F\circ G_k|^p \,\mathrm{d}\mu \ge \int_{C_k} |F\circ h|^p \,\mathrm{d}\mu = +\infty$$

by (7.6), a contradiction, proving that $F \in \mathcal{G}_{s/p}$ if $s < \infty$. Now let $s = +\infty$ and $N_F$ act from $L^\infty$ into $L^p$. Suppose that $F \notin \mathcal{G}_\infty$. Then there is a sequence $\{x_n\}$ of real numbers such that $\sup_n |x_n| < \infty$ and $|F(x_n)| > n$ for all $n = 1, 2, \ldots$. Let $\mu_n := C/n^2$ for all $n = 1, 2, \ldots$, where the constant $C > 0$ is such that

$\sum_n \mu_n < \mu(\Omega)$, and let $A_n \in \mathcal{S}$ be disjoint sets with $\mu(A_n) = \mu_n$. Then (7.6) holds and $h := \sum_n x_n 1_{A_n} \in \mathcal{L}^\infty$. It follows that $\int |F \circ h|^p \, d\mu = +\infty$, a contradiction, proving $F \in \mathcal{G}_\infty$, and hence (*b*).

(*c*): By (*a*) and (*b*), it is enough to prove that $F$ is u.m. provided $N_F$ acts from $\mathcal{L}^s$ into $\mathcal{L}^p$. Taking a constant times $\mu$, we can assume that $\mu$ is a probability measure. Let $0 < M < \infty$. Each measurable function $g$ from $\Omega$ into $[-M, M]$ is in $\mathcal{L}^s$. Since $\mu$ is nonatomic, for any probability measure $Q$ on the Borel sets of $[-M, M]$, there is a measurable $g$ with $\mu \circ g^{-1} = Q$ on the Borel sets by Proposition A.7. Since $F \circ g \in \mathcal{L}^p$, $F \circ g$ must be measurable. Thus for each Borel set $B \subset \mathbb{R}$, $g^{-1}(F^{-1}(B))$ is measurable. So in $\mathbb{R}$, $(\mu \circ g^{-1})(F^{-1}(B))$ is defined via the completion of the restriction $Q$ of $\mu \circ g^{-1}$ to the Borel sets, since $(\Omega, \mathcal{S}, \mu)$ is perfect. Thus $F^{-1}(B)$ is $Q$-measurable. Since $Q$ is an an arbitrary probability measure on the Borel sets of $[-M, M]$, $F^{-1}(B) \cap [-M, M]$ is u.m. for all $M$. Letting $M \to +\infty$ we get that $F^{-1}(B)$ is u.m., so $F$ is u.m., proving (*c*) and the theorem. □

Next, here is a criterion for $N_F$ to act from $L^s$ into $L^\infty$.

**Proposition 7.15.** *Let $(\Omega, \mathcal{S}, \mu)$ be a complete, $\sigma$-finite, nonatomic measure space, let $F: \mathbb{R} \to \mathbb{R}$ be universally measurable, and let $1 \leq s < \infty$. Then $N_F$ acts from $L^s$ into $L^\infty$ if and only if $\|F\|_{\sup} < +\infty$.*

*Proof.* As in the proof of Theorem 7.13(*a*), for each $g \in \mathcal{L}^s$, $F \circ g$ is a measurable function. Thus if $\|F\|_{\sup} < +\infty$ then $N_F g \in \mathcal{L}^\infty$, proving the "if" part. To prove "only if," suppose $\|F\|_{\sup} = +\infty$. Then there is a sequence $\{u_k\}_{k \geq 1} \subset \mathbb{R}$ such that $|F(u_k)| \to +\infty$. Take $B \in \mathcal{S}$ with $0 < \mu(B) < +\infty$. Since $\mu$ is nonatomic, by Proposition A.1 for $\mu$ restricted to $B$ or subsets of it and induction on $k$, for each $k$, take disjoint $A(k) \in \mathcal{S}$ such that $A(k) \subset B$ and $0 < \mu(A(k)) \leq \mu(B)/[2^k(1 + |u_k|^s)]$. Let $g := \sum_{k=1}^\infty u_k 1_{A(k)}$. Then clearly $g \in L^s$ but $N_F g \notin L^\infty$, a contradiction, proving the proposition. □

### *Continuity*

Necessary and sufficient conditions for the Nemytskii operator $N_\psi$ to act from $L^s$ into $L^p$ continuously will be given in Theorem 7.19 if $\psi$ is a strong Shragin function. It is not hard to show (cf. the proof of Proposition 7.18 below) that a Nemytskii operator $N_\psi$ is continuous on $L^0$ assuming the following condition on $\psi$. A function $\psi: \mathbb{R} \times \Omega \to \mathbb{R}$ is called a *Carathéodory function* (for $\mu$) or is said to satisfy the *Carathéodory condition* (for $\mu$) if $\psi(\cdot, \omega)$ is continuous on $\mathbb{R}$ for almost all $\omega \in \Omega$, and $\psi(u, \cdot)$ is $\mu$-measurable on $\Omega$ for all $u \in \mathbb{R}$. On complete measure spaces $(\Omega, \mathcal{S}, \mu)$, the Carathéodory condition implies the strong Shragin condition, as the following shows:

**Proposition 7.16.** *If $(\Omega, \mathcal{S}, \mu)$ is complete, then any Carathéodory function $\psi$ on $\mathbb{R} \times \Omega$ is a strong Shragin function.*

*Proof.* Let $\psi$ be a Carathéodory function on $\mathbb{R} \times \Omega$ and let $N \subset \Omega$ be a $\mu$-null set such that for each $\omega \in N^c$, $\psi(\cdot, \omega)$ is continuous on $\mathbb{R}$. Since $(\Omega, \mathcal{S}, \mu)$ is complete, $N \in \mathcal{S}_h$. For $\delta > 0$ and $(u, \omega) \in \mathbb{R} \times N^c$, let

$$\psi_\delta(u, \omega) := \sup\{\psi(v, \omega)\colon\ v \in \mathbb{Q},\ |u - v| < \delta\}.$$

For $\delta > 0$ and any $c \in \mathbb{R}$, we have

$$\{(u, \omega) \in \mathbb{R} \times N^c\colon\ \psi_\delta(u, \omega) > c\} = \bigcup_{v \in \mathbb{Q}} \Big[(v - \delta, v + \delta) \times \{\omega \in N^c\colon\ \psi(v, \omega) > c\}\Big].$$

Since $\psi$ is a Carathéodory function, it follows that $\psi_\delta^{-1}((c, \infty)) \setminus (\mathbb{R} \times N) \in \mathcal{B} \otimes \mathcal{S}$ for each $c \in \mathbb{R}$ and $\delta > 0$, and so $\psi_\delta$ is a strong Shragin function for each $\delta > 0$. Also, letting $\psi_n := \psi_{\delta_n}$ for some $\delta_n \downarrow 0$, it follows that for each $(u, \omega) \in \mathbb{R} \times N^c$, we have $\psi(u, \omega) = \lim_{n \to \infty} \psi_n(u, \omega)$. Thus $\psi(\cdot, \cdot)$ is jointly measurable on $\mathbb{R} \times N^c$ (e.g. [53, Theorem 4.2.2]), so $\psi$ is a strong Shragin function as stated. □

The converse to the preceding proposition will be shown to hold if in addition the Nemytskii operator $N_\psi$ is continuous from $L^\infty$ into $L^0$. Recall that for a finite measure space $(\Omega, \mathcal{S}, \mu)$, $L^0 = L^0(\Omega, \mathcal{S}, \mu)$ is the space of all $\mu$-equivalence classes of real-valued $\mu$-measurable functions on $\Omega$, with the topology of convergence in $\mu$-measure, metrized by a metric $d_0$ (Section 1.4).

**Proposition 7.17.** *Let $(\Omega, \mathcal{S}, \mu)$ be finite and complete, and let $\psi$ be a strong Shragin function on $\mathbb{R} \times \Omega$ such that the Nemytskii operator $N_\psi$ is continuous from $L^\infty(\Omega, \mathcal{S}, \mu)$ into $L^0(\Omega, \mathcal{S}, \mu)$. Then $\psi$ is a Carathéodory function.*

*Proof.* By Proposition 7.4, $\psi(u, \cdot) = N_\psi(u1(\cdot))$ is measurable for all $u \in \mathbb{R}$. Thus it is sufficient to prove that for $\mu$-almost all $\omega \in \Omega$, the function $\psi(\cdot, \omega)$ is uniformly continuous on any interval $[-c, c]$ with $0 < c < \infty$. For $C := [-c, c]$, let $A = A(C)$ be the set of all $\omega \in \Omega$ such that $\psi(\cdot, \omega)$ is not uniformly continuous on $C$. For $k, n = 1, 2, \ldots$, let $C_n := \{(u, v) \in C \times C\colon\ |u - v| \le 1/n\}$,

$$A_{k,n} := \{\omega \in \Omega\colon\ |\psi(u, \omega) - \psi(v, \omega)| > 1/k \text{ for some } (u, v) \in C_n\},$$

and $A_k := \cap_{n \ge 1} A_{k,n}$. Then $A_k \uparrow A$ as $k \to \infty$. To show measurability of $A_{k,n}$ fix any $k$ and $n$. Let

$$B_{k,n} := \{(\omega, u, v) \in \Omega \times C_n\colon\ |\psi(u, \omega) - \psi(v, \omega)| > 1/k\}.$$

Since $\psi$ is a strong Shragin function, there is a $\mu$-null set $N$ such that $B_{k,n} \setminus (N \times \mathbb{R}^2)$ belongs to the $\sigma$-algebra $\mathcal{S} \otimes \mathcal{B}(\mathbb{R}^2)$, where $\mathcal{B}(\mathbb{R}^2)$ denotes the $\sigma$-algebra of Borel subsets of $\mathbb{R}^2$. According to Sainte-Beuve's selection theorem (e.g. [52, Theorem 5.3.2] for $\mathcal{F} := Y := C_n \subset \mathbb{R}^2$ with Borel $\sigma$-algebra and $T$ the identity $\mathbb{R}^2 \to \mathbb{R}^2$, $X(\omega, u, v) = \psi(u, \omega) - \psi(v, \omega)$ restricted to $N^c \times \mathbb{R}^2$), the projection

$$D_{k,n} := \{\omega \in \Omega\colon\ (\omega, u, v) \in B_{k,n} \setminus (N \times \mathbb{R}^2) \text{ for some } (u,v) \in C_n\}$$

is (universally) measurable in $\Omega$ and there are (universally) measurable functions $u_{k,n}(\cdot)$ and $v_{k,n}(\cdot)$ from $D_{k,n}$ into $C$ such that $(\omega, u_{k,n}(\omega), v_{k,n}(\omega)) \in B_{k,n}$ for all $\omega \in D_{k,n}$. Since $N$ is a $\mu$-null set and $A_{k,n} = (A_{k,n} \cap N) \cup D_{k,n}$, each $A_{k,n}$ is measurable, and so are the sets $A_k$ and $A$.

Suppose that for some interval $C = [-c, c]$ with $0 < c < \infty$, the set $A$ has positive $\mu$-measure. Then there is an integer $k \geq 1$ such that $\mu(A_k) > 0$. Fix this $k$. For each $\omega \in D := \bigcap_{n \geq 1} D_{k,n} \supset A_k \setminus N$, by compactness, there is a $u \in C$ such that for every neighborhood $W$ of $u$, $u_{k,n}(\omega) \in W$ for infinitely many values of $n$. Let $A_*$ be the set of all such pairs $(\omega, u) \in D \times C$. Then

$$A_* = \bigcap_{m \geq 1} \bigcup_{n \geq m} \bigcup_{r \in \mathbb{Q}} \{(\omega, u)\colon\ |u_{k,n}(\omega) - r| < 1/m,\ |r - u| < 1/m\}, \qquad (7.7)$$

and so $A_* \in \mathcal{S} \otimes \mathcal{B}$. The projection of $A_*$ into $\Omega$ equals $D$. By Sainte-Beuve's selection theorem again there is a measurable function $\eta\colon D \to C$ such that $(\omega, \eta(\omega)) \in A_*$ for all $\omega \in D$. For $m = 1, 2, \ldots$, let

$$\begin{aligned} E_{k,m} := \{(\omega, U, V) \in D \times C \times C\colon\ &|U - \eta(\omega)| < 2/m,\ |V - \eta(\omega)| < 3/m, \\ &\text{and } |\psi(U, \omega) - \psi(V, \omega)| > 1/k\}. \end{aligned}$$

Then $E_{k,m} \in \mathcal{S} \otimes \mathcal{B}(\mathbb{R}^2)$. For each $\omega \in D$ and $m$, there are $U$ and $V$ in $C = [-c, c]$ with $(\omega, U, V) \in E_{k,m}$, by (7.7) with $u = \eta(\omega)$, $U = u_{k,n}(\omega)$ for some $n \geq m$, and $V = v_{k,n}(\omega)$. Thus by Sainte-Beuve's selection theorem again, there are measurable functions $U_m$ and $V_m$ on $D$ such that $(\omega, U_m(\omega), V_m(\omega)) \in E_{k,m}$ for all $\omega \in D$. Let $g_m := 1_D U_m(\cdot)$, $h_m := 1_D V_m(\cdot)$ and $\eta_D := 1_D \eta$. Then $\eta_D \in \mathcal{L}^\infty$, $g_m \to \eta_D$, and $h_m \to \eta_D$ in $L^\infty$ as $m \to \infty$, and for each $m$, $|N_\psi g_m - N_\psi h_m| > 1/k$ on the set $D$. But since $N_\psi$ is continuous on $L^\infty$, $N_\psi g_m - N_\psi h_m$ converges to zero in $\mu$-measure as $m \to \infty$, a contradiction, proving the proposition. □

**Proposition 7.18.** *Let $(\Omega, \mathcal{S}, \mu)$ be finite and complete, and let $\psi$ be a strong Shragin function on $\mathbb{R} \times \Omega$. The Nemytskii operator $N_\psi$ is continuous from $L^0(\Omega, \mathcal{S}, \mu)$ into itself if and only if $\psi$ is a Carathéodory function.*

*Proof.* The "only if" part of the conclusion follows from Proposition 7.17. To prove the "if" part, let $\psi$ be a Carathéodory function. Suppose that $N_\psi$ is not continuous on $L^0$. Then there is a sequence $\{f_n\} \subset \mathcal{L}^0$ converging in $\mu$-measure to an $f \in \mathcal{L}^0$ and such that for some $\epsilon > 0$, $d_0(N_\psi f_n, N_\psi f) \geq \epsilon$ for each $n \geq 1$. By a theorem of F. Riesz (e.g. [53, Theorem 9.2.1]) one can extract a subsequence $\{f_{n'}\}$ which converges to $f$ a.e. $(\mu)$. Thus the subsequence $N_\psi f_{n'}$ converges to $N_\psi f$ a.e. $(\mu)$ as $n \to \infty$, and also in $\mu$-measure, a contradiction, proving continuity of $N_\psi$ on $L^0$. The proof of the proposition is complete. □

**Theorem 7.19.** *Let $(\Omega, \mathcal{S}, \mu)$ be finite and complete, let $\psi$ be a strong Shragin function on $\mathbb{R} \times \Omega$, and let the Nemytskii operator $N_\psi$ act from $L^s = L^s(\Omega, \mathcal{S}, \mu)$ into $L^p$ for some $1 \le s, p < \infty$. Then $N_\psi$ is continuous if and only if $\psi$ is a Carathéodory function.*

*Proof.* Since convergence in $L^p$ implies convergence in $\mu$-measure, the "only if" part of the conclusion follows by Proposition 7.17. To prove the "if" part, let $\psi$ be a Carathéodory function. First suppose that $N_\psi 0 = 0$ and that $N_\psi$ is not continuous at zero. Then there is a sequence $\{f_n\} \subset \mathcal{L}^0$ such that

$$\sum_{n=1}^{\infty} \|f_n\|_s^s < \infty \tag{7.8}$$

and for some finite constant $C > 0$,

$$\int_\Omega |N_\psi f_n|^p \, \mathrm{d}\mu \ge C \qquad \forall n \ge 1. \tag{7.9}$$

Further, by induction, one can construct a sequence $\{\epsilon_k\}$ of positive numbers, a sequence $\{A_k\}$ of measurable subsets of $\Omega$, and a subsequence $\{f_{n_k}\}$ such that for each $k \ge 1$, (1) $0 < \epsilon_{k+1} < \epsilon_k/2$, (2) $\mu(A_k) \le \epsilon_k$, (3)

$$\int_{A_k} |N_\psi f_{n_k}|^p \, \mathrm{d}\mu \ge (2/3)C, \tag{7.10}$$

and (4) for each $B \in \mathcal{S}$ such that $\mu(B) \le 2\epsilon_{k+1}$,

$$\int_B |N_\psi f_{n_k}|^p \, \mathrm{d}\mu \le (1/3)C. \tag{7.11}$$

Indeed, let $\epsilon_1 := \mu(\Omega)$, $A_1 := \Omega$, and $f_{n_1} := f_1$. Suppose that $\epsilon_k$, $A_k$ and $f_{n_k}$ are chosen for some $k \ge 1$. Since $\|1_B N_\psi f_{n_k}\|_p \to 0$ as $\mu(B) \to 0$, there is an $\epsilon_{k+1}$ satisfying (4) and (1). Since convergence in $L^s$ implies convergence in $\mu$-measure, by Proposition 7.18, there is an integer $n_{k+1} > n_k$ such that $\mu(A_{k+1}) < \epsilon_{k+1}$, where

$$A_{k+1} := \{\omega \in \Omega \colon \big|(N_\psi f_{n_{k+1}})(\omega)\big| > (C/3\mu(\Omega))^{1/p}\}.$$

By (7.9), it then follows that

$$\int_{A_{k+1}} \big|N_\psi f_{n_{k+1}}\big|^p \, \mathrm{d}\mu = \int_\Omega \big|N_\psi f_{n_{k+1}}\big|^p \, \mathrm{d}\mu - \int_{\Omega \setminus A_{k+1}} \big|N_\psi f_{n_{k+1}}\big|^p \, \mathrm{d}\mu \ge (2/3)C,$$

and so (2) and (3) hold with $k$ replaced by $k+1$, completing the inductive construction of $\{\epsilon_k\}$, $\{A_k\}$, and $\{f_{n_k}\}$. For $k = 1, 2, \ldots$, let $B_k := A_k \setminus \cup_{i>k} A_i$. Then for each $k \ge 1$,

$$\mu(A_k \setminus B_k) \le \mu\Big(\bigcup_{i>k} A_i\Big) \le \sum_{i>k} \epsilon_i < 2\epsilon_{k+1}. \tag{7.12}$$

Define a function $g\colon \Omega \to \mathbb{R}$ by $g(\omega) := f_{n_k}(\omega)$ if $\omega \in B_k$ for some $k \ge 1$, and $g(\omega) := 0$ otherwise. Then $g$ is measurable and by (7.8),

$$\int_\Omega |g|^s \, d\mu = \sum_{k\ge1} \int_{B_k} |f_{n_k}|^s \, d\mu \le \sum_{k\ge1} \|f_{n_k}\|_s^s < \infty,$$

and so $g \in L^s$. However, for each $k \ge 1$, by (7.10), (7.11), and (7.12), we have

$$\begin{aligned}\int_{B_k} |N_\psi g|^p \, d\mu &= \int_{B_k} |N_\psi f_{n_k}|^p \, d\mu \\ &= \int_{A_k} |N_\psi f_{n_k}|^p \, d\mu - \int_{A_k \setminus B_k} |N_\psi f_{n_k}|^p \, d\mu \ge C/3.\end{aligned}$$

Since the sets $B_k$ are disjoint, it follows that $\int_\Omega |N_\psi g|^p \, d\mu = +\infty$, contradicting the acting condition for $N_\psi$. Thus $N_\psi$ must be continuous at zero provided $N_\psi 0 = 0$.

Now let $\psi$ be any Carathéodory function and let $G \in L^s$. For $(u, \omega) \in \mathbb{R} \times \Omega$, let

$$\phi(u, \omega) := \psi(G(\omega) + u, \omega) - \psi(G(\omega), \omega).$$

Then clearly $N_\phi$ acts from $L^s$ into $L^p$ and $N_\phi 0 = 0$. For any $u \in \mathbb{R}$ and $\omega \in \Omega$ let $f_u(\omega) := u1(\omega) \equiv u$. Then $f_u \in L^s$, so $N_\phi(f_u) \in L^p$ and $\phi(u, \cdot) = N_\phi(f_u)(\cdot)$ is measurable. It is easily seen that $\phi$ is continuous in $u$ for each $\omega$ such that $\psi(\cdot, \omega)$ is, so $\phi$ is a Carathéodory function. By Proposition 7.16, $\phi$ is a strong Shragin function. By the preceding part of the proof, $N_\phi$ is continuous at zero. Thus $N_\psi$ is continuous at $G$. The proof of the theorem is complete. □

**Corollary 7.20.** *Let $(\Omega, \mathcal{S}, \mu)$ be finite, complete, and nonatomic, let $\psi$ be a Carathéodory function, and let $1 \le p, s < \infty$. If the Nemytskii operator $N_\psi$ acts from $L^s$ into $L^p$, then it is bounded, and there is a constant $C = C_\psi > 0$ such that $\|N_\psi f\|_p \le \|N_\psi 0\|_p + C\|f\|_s^{s/p}$ for each $f \in L^s$.*

*Proof.* Let $\phi(u, \omega) := \psi(u, \omega) - \psi(0, \omega)$, $(u, \omega) \in \mathbb{R} \times \Omega$. Then for any $f \in L^s$, $N_\psi f = N_\phi f + \psi(0, \cdot)$, and $\psi(0, \cdot) \in \mathcal{L}^p$, so it will suffice to prove the result for $\phi$ in place of $\psi$, or in other words to assume that $\psi(0, \cdot) \equiv 0$. By Theorem 7.19, $N_\psi$ is continuous. Therefore there is an $r > 0$ such that $\|N_\psi g\|_p \le 1$ for each $g \in L^s$ with $\|g\|_s \le r$. By Proposition A.9, for $f \in L^s$ with $\|f\|_s > r$, there exists a partition $\{A_i\}_{i=0}^n$ of $\Omega$ into measurable sets such that $\int_{A_i} |f|^s \, d\mu = \|f\|_s^s/(n+1) \in (r^s/2, r^s]$ for $i = 0, \dots, n$. Then

$$\|N_\psi f\|_p^p = \sum_{i=0}^n \|N_\psi(1_{A_i} f)\|_p^p \le n + 1 \le (2/r^s)\|f\|_s^s.$$

Thus the conclusion holds with $C = (2/r^s)^{1/p}$. □

A Nemytskii operator acting from an open subset of $L^s$ into $L^p$ extends to all of $L^s$ by the next proposition.

**Proposition 7.21.** *Let $(\Omega, \mathcal{S}, \mu)$ be nonatomic, let $1 \le p, s < \infty$, and let $V$ be a nonempty open set in $L^s(\Omega, \mathcal{S}, \mu)$. For the Nemytskii operator $N_\psi$ the following hold:*

*(a) if $N_\psi$ acts from $V$ into $L^p$, then it acts from $L^s$ into $L^p$;*
*(b) if in addition $N_\psi$ is continuous from $V$ into $L^p$, then it is continuous from $L^s$ into $L^p$.*

*Proof.* For $(a)$, let $G \in V$ and let $r > 0$ be such that $G + B_r \subset V$, where $B_r = \{f \in L^s\colon\ \|f\|_s < r\}$. For $(u, \omega) \in \mathbb{R} \times \Omega$, let

$$\phi(u, \omega) := \psi(G(\omega) + u, \omega) - \psi(G(\omega), \omega). \tag{7.13}$$

Then $N_\phi$ acts from $B_r$ into $L^p$. Let $f \in L^s$ be such that $f \notin G + B_r$, and so $\|f - G\|_s \ge r$. By Proposition A.9 with $h = f - G$ and $c = r/2$, there is a partition $\{A_i\}_{i=0}^k$ of $\Omega$ into measurable sets such that $\|(f-G)1_{A_i}\|_s \le r/2$ for $i = 0, \dots, k$. Since $N_\phi(f - G) = \sum_{i=0}^k N_\phi((f - G)1_{A_i})$, we have $N_\phi(f - G) \in L^p$, and so $N_\psi f = N_\phi(f - G) + N_\psi G \in L_p$, proving $(a)$.

For $(b)$, as for $(a)$, let $G \in V$, let $r > 0$ be such that $G + B_r \subset V$, and let $\phi$ be defined by (7.13). Then $N_\phi$ is continuous from $B_r$ into $L^p$. Let $f \in L^s$ be such that $f \notin G+B_r$ and for a sequence $\{f_n\} \subset L^s$, $f_n \to f$ in $L^s$. Again, let a partition $\{A_i\}_{i=0}^k$ of $\Omega$ into measurable sets be such that $\|(f-G)1_{A_i}\|_s \le r/2$ for $i = 0, \dots, k$ (Proposition A.9). Thus $\|(f_n - G)1_{A_i}\|_s < r$ for each $i$ and all sufficiently large $n$. Since

$$\|N_\phi(f_n - G) - N_\phi(f - G)\|_p^p = \sum_{i=0}^k \|N_\phi((f_n - G)1_{A_i}) - N_\phi((f - G)1_{A_i})\|_p^p \to 0$$

as $n \to \infty$, it follows that $N_\psi f_n - N_\psi f = N_\phi(f_n - G) - N_\phi(f - G) \to 0$ in $L^p$ as $n \to \infty$. The proof of the proposition is complete. □

Let $F$ be a Borel measurable function from $\mathbb{R}$ into $\mathbb{R}$ such that the autonomous Nemytskii operator $N_F$ acts from $L^s$ into $L^p$. By Theorem 7.19, $N_F$ is (globally) continuous if and only if $F$ is continuous. The following gives a local continuity condition.

**Proposition 7.22.** *Let $(\Omega, \mathcal{S}, \mu)$ be complete and finite, and let $1 \le s, p < \infty$. Let $G \in \mathcal{L}^s$ and let $F$ be universally measurable from $\mathbb{R}$ into $\mathbb{R}$, continuous a.e. for $\mu{\circ}G^{-1}$, and in $\mathcal{G}_{s/p}$. Then $N_F$ is continuous at $G$ from $L^s$ into $L^p$.*

*Proof.* By Theorem 7.13$(a)$, $N_F$ acts from $L^s$ into $L^p$. To show its continuity at $G$, let $g_n \to 0$ in $L^s$. Taking a subsequence if necessary, we can assume that $g_n \to 0$ a.e. $(\mu)$. Then $f_n := N_F(G + g_n) - N_F(G) \to 0$ as $n \to \infty$ a.e. $(\mu)$

since $F$ is continuous at $G(\omega)$ for $\mu$-almost all $\omega \in \Omega$. By e.g. Theorem III.3.6 in [57], it is enough to show that $\{|f_n|^p\}_{n\geq 1}$ are uniformly integrable, that is, for each $\epsilon > 0$ there is a $\delta > 0$ such that if $\mu(A) < \delta$ then $\int_A |f_n|^p \, d\mu < \epsilon$ for each $n \geq 1$. Using the $\beta$ growth condition (7.2) with $\beta = s/p$, it follows that

$$\int_A |f_n|^p \, d\mu = \|[N_F(G+g_n) - N_F(G)]1_A\|_p^p$$

$$\leq 4^{p-1}\|F\|_{\mathcal{G}_{s/p}}^p \Big\{\mu(A) + \int_A |G+g_n|^s \, d\mu\Big\} + 2^{p-1}\int_A |N_F(G)|^p \, d\mu$$

for any $A \in \mathcal{S}$. Since $G + g_n \to G$ in $L^s$ as $n \to \infty$, and so $\{|G+g_n|^s\}_{n\geq 1}$ are uniformly integrable, the right side tends to zero as $\mu(A) \to 0$ uniformly in $n$, proving the proposition. □

Recall that Proposition 7.15 gave a criterion for $N_F$ to act from $L^s$ into $L^\infty$ of a suitable measure space, namely that $F$ be bounded. The following shows that for $N_F$ not only to act but to be continuous into $L^\infty$ is much more restrictive:

**Proposition 7.23.** *Let $(\Omega, \mathcal{S}, \mu)$ be a $\sigma$-finite, nonatomic measure space, let $F$ be a function from $\mathbb{R}$ into $\mathbb{R}$, and let $1 \leq s < \infty$. Then $N_F$ both acts and is continuous from $L^s$ into $L^\infty$ if and only if $F$ is a constant.*

*Proof.* "If" is clear. To prove "only if," suppose $F$ is not a constant. Then $F(a) \neq F(b)$ for some $a, b$. Take $B$ with $0 < \mu(B) < +\infty$ and let $G := a1_B \in L^s$. Take $A(k) \subset B$, $A(k) \in \mathcal{S}$, with $0 < \mu(A(k)) \to 0$ as $k \to +\infty$, as is possible by Proposition A.1. Let $g_k := (b-a)1_{A(k)}$. Then as $k \to \infty$, $g_k \to 0$ in $L^s$ but $F \circ (G + g_k)$ does not converge to $F \circ G$ in $L^\infty$. So, either $N_F$ does not act from $L^s$ into $L^\infty$, or if it acts, it is not continuous, proving the proposition. □

## 7.2 Hölder Properties

For $0 < \alpha \leq 1$, recall Definition 6.42 of a uniformly $\alpha$-Hölder function $\psi(u, s)$ specialized to the case when $S$ is a singleton, $B = L_s$, and $Y = L_p$, where $1 \leq s, p \leq \infty$. For a given $G \in L_s$, we say that a Nemytskii operator $N_\psi$ acting from $L_s$ into $L_p$ is *$\alpha$-Hölder at $G$* from $L_s$ into $L_p$ if there are a neighborhood $V$ of 0 in $L_s$ and a finite constant $L$ such that $\|N_\psi(G+g) - N_\psi G\|_p \leq L\|g\|_s^\alpha$ for each $g \in V$. In other words, $N_\psi$ is $\alpha$-Hölder at $G$ if $\|N_\psi(G+g) - N_\psi G\|_p = O(\|g\|_s^\alpha)$ as $\|g\|_s \to 0$.

It will be shown that the autonomous Nemytskii operator $N_F$ has a Hölder property from spaces $L^s$ to $L^p$ for $F \in \mathcal{W}_p(\mathbb{R})$, but only at suitable functions $G$ which will be characterized. For $1 \leq p < \infty$, let $\gamma(p, s) := 1 + s/[p(1+s)]$ for $1 \leq s < \infty$ and $\gamma(p, \infty) := 1 + 1/p$.

**Theorem 7.24.** *Let $(\Omega, \mathcal{S}, \mu)$ be a finite measure space, let $J$ be a nondegenerate interval (bounded or unbounded), and let $G$ be a measurable function on $\Omega$ with values in $J$. Then the following are equivalent:*

$(a)$ *there is a $C < \infty$ such that for some $p \in [1, \infty)$ and $s \in [1, \infty]$,*

$$\|N_F(G+g) - N_F(G)\|_p \leq C\|F\|_{(p)}\|g\|_s^{\gamma(p,s)-1} \tag{7.14}$$

*for all $F \in \mathcal{W}_p(J; \mathbb{R})$ and $g \in L^s(\Omega, \mathcal{S}, \mu)$ such that $G + g$ has values in $J$;*

$(a')$ *$(a)$ holds for all $p \in [1, \infty)$ and $s \in [1, \infty]$;*

$(b)$ *$G \in \mathcal{D}_\lambda$, that is, there is an $M < \infty$ such that $\mu \circ G^{-1}$ has a density with respect to Lebesgue measure $\lambda$ on $\mathbb{R}$ bounded by $M$, so that $(\mu \circ G^{-1})((c, d]) \leq M(d - c)$ for $-\infty < c < d < \infty$.*

*Moreover, the constants $C$ in $(a)$ or $(a')$ can be chosen to depend only on $\mu(\Omega)$ and $M$ in $(b)$, or also on $p$; we can take $M \geq 1$ and then $C = C_1$ or $C_2$ where $C_1 := 2(\max\{\mu(\Omega), 6M + 1\})^{1/p} \leq C_2 := 2\max\{\mu(\Omega), 6M + 1\}$.*

**Remark.** Proposition 7.28 will show that in (7.14) the power $\gamma(p, s) - 1$ on $\|g\|_s$, the power 1 on $\|F\|_{(p)}$, and the seminorm $\|\cdot\|_{(p)}$ itself are all optimal. $(b)$ implies that $\mu$ is nonatomic by Proposition A.4; thus, so must $(a)$. In fact, if $\mu \circ G^{-1}$ has an atom at a point where $F$ has a jump, then $g \mapsto N_F(G + g)$ is not continuous at $g = 0$ (Remark 7.26 below).

*Proof.* $(b)$ implies $(a')$: fix $p \in [1, \infty)$ and $s \in [1, \infty]$. Let $g \in L^s(\Omega, \mathcal{S}, \mu)$ be such that $G + g$ has values in $J$, let $\delta := \|g\|_s$, $\beta := s/(1 + s)$, or $\beta := 1$ if $s = +\infty$, $K := \mu(\Omega)$, and $F \in \mathcal{W}_p(J; \mathbb{R})$. Subtracting a constant from $F$, we can assume that $\sup F = -\inf F$ and so that $\|F\|_{\sup} \leq \|F\|_{(p)}/2$. We can assume that $\delta \leq 1$, since otherwise (7.14) holds using $C \geq K^{1/p}$. Let $A := \{\omega \in \Omega : |g(\omega)| \leq \delta^\beta\}$. Then $\mu(\Omega \setminus A) \leq \delta^{s(1-\beta)} = \delta^\beta$ if $s < \infty$, or $\mu(\Omega \setminus A) = 0$ if $s = +\infty$. We can assume that $M \geq 1$ in $(b)$. Let $\Delta := \Delta(F, G, g) := N_F(G + g) - N_F(G)$. It follows that

$$\int_{\Omega \setminus A} |\Delta|^p \mathrm{d}\mu \leq \|F\|_{(p)}^p \delta^\beta. \tag{7.15}$$

Let $k$ be a positive integer such that $1/(2k) \leq \delta^\beta \leq 1/k$. For fixed $k$ we can decompose $\mathbb{R}$ into disjoint intervals $I_{k,j} := [(j-1)/k, j/k)$, $j \in \mathbb{Z}$. Let $\mathbb{Z}_0$ be the set of all $j \in \mathbb{Z}$ such that $A(k, j) := A \cap G^{-1}(I_{k,j})$ is nonempty. Then

$$\int_A |\Delta|^p \mathrm{d}\mu \leq \sum_{j \in \mathbb{Z}_0} (\sup_{A(k,j)} |\Delta|)^p \mu(A(k, j)).$$

Choose $s_{k,j} \in A(k, j)$ for each $j \in \mathbb{Z}_0$ such that $|\Delta(s_{k,j})| \geq \sup_{A(k,j)} |\Delta|/2$. Then $|g(s_{k,j})| \leq \delta^\beta \leq 1/k$. For $i = 0, 1, 2$ let $\mathbb{Z}_{0,i}$ be the set of all integers $j \in \mathbb{Z}_0$ such that $j \equiv i \bmod 3$. For each $i$, the intervals with endpoints $(G + g)(s_{k,j})$ and $G(s_{k,j})$ for all $j \in \mathbb{Z}_{0,i}$ are disjoint. Thus

$$\int_A |\Delta|^p \mathrm{d}\mu \le 2^p \frac{3M}{k} \|F\|_{(p)}^p \le 6M\delta^\beta \|2F\|_{(p)}^p.$$

Combining this with (7.15) and taking $p$th roots we get $\|\Delta\|_p \le C\|F\|_{(p)}\|g\|_s^{\beta/p}$ for $C$ as given at the end of the theorem. So $(a')$ is proved.

Clearly $(a')$ implies $(a)$. Lastly, to show that $(a)$ implies $(b)$, assume $(a)$. If $(b)$ fails, then for $n = 1, 2, \ldots,$ there exist $c_n < d_n$ in $\mathbb{R}$ with $q_n := (\mu\circ G^{-1})((c_n, d_n]) > n(d_n - c_n)$. Let $g_n := 2(c_n - d_n)$ on $G^{-1}((c_n, d_n])$ and $g_n := 0$ elsewhere, and let $F_n := 1_{[c_n,\infty)}$. Then for all $n$, $\|F_n\|_{(p)} = 1$ and $\|F_n\circ(G + g_n) - F_n\circ G\|_p = q_n^{1/p}$. For $s < \infty$, $\|g_n\|_s = 2(d_n - c_n)q_n^{1/s}$. Letting $\delta_n := d_n - c_n > 0$ gives $q_n > n\delta_n$, while by $(a)$, $q_n^{1/p} \le C(2\delta_n)^{s/(p(1+s))}q_n^{1/(p(1+s))}$, giving $q_n^{s/(1+s)} \le C^p(2\delta_n)^{s/(1+s)}$ and so $q_n \le 2C^{p(s+1)/s}\delta_n$, a contradiction for $n$ large. For $s = \infty$, $q_n^{1/p} \le C \cdot 1 \cdot (2\delta_n)^{1/p}$ so $q_n \le 2C^p\delta_n$ while $q_n > n\delta_n$, again a contradiction for $n$ large. So $(b)$ holds and the theorem is proved. □

The next fact will be used in Theorem 8.9 below to prove differentiability of a two-function composition operator $(F, g) \mapsto F\circ(G+g)$, where both $\|F\|_{(p)} \to 0$ and $\|g\|_s \to 0$.

**Corollary 7.25.** *If $(a)$ in Theorem 7.24 holds then also for any $a > 1$, $b > 1$ with $a^{-1} + b^{-1} = 1$ and $\gamma := \gamma(p, s)$,*

$$\|N_F(G + g) - N_F(G)\|_p \le C\Big(a^{-1}\|F\|_{(p)}^a + b^{-1}\|g\|_s^{(\gamma-1)b}\Big),$$

$$\|N_F(G + g) - N_F(G)\|_p \le C\left(\frac{1}{\gamma}\|F\|_{(p)}^\gamma + \left(1 - \frac{1}{\gamma}\right)\|g\|_s^\gamma\right)$$

*for each $F \in \mathcal{W}_p(J; \mathbb{R})$ and $g \in L^s(\Omega, \mathcal{S}, \mu)$ such that $G$ and $G+g$ have values in $J$.*

*Proof.* For any $u, v \ge 0$ we have $uv \le a^{-1}u^a + b^{-1}v^b$ by (3.20), and the first conclusion follows from (7.14). Then $a := \gamma$, $b = \gamma/(\gamma - 1)$ gives the second. □

The following example shows that the nonatomicity of $\mu\circ G^{-1}$ in Theorem 7.24$(b)$ is essential, not only for the given Hölder property, but for continuity of $N_F$.

**Remark 7.26.** Let $G \in \mathcal{L}^\infty(\Omega, \mathcal{S}, \mu)$ and suppose $\mu\circ G^{-1}$ has an atom at $y$, meaning that $\mu(G^{-1}(\{y\})) > 0$. Then if $F(y+)$ exists but $F(y+) \ne F(y)$, the autonomous Nemytskii operator $N_F$ is not continuous at $G$ on $L^\infty$; to see this let $g_n(\omega) \equiv 1/n$, so $\|g_n\|_\infty \to 0$, and on $G^{-1}(\{y\})$, $N_F(G) \equiv F(y)$, while $N_F(G + g_n) \to F(y+)$. Similarly if $F(y-) \ne F(y)$, let $g_n(\omega) \equiv -1/n$.

**Remark 7.27.** Let $G(x) = x$, $0 \le x \le 1$, and $F_1(x) := 1_{\{x \le 1/2\}}$. Let $J(n)$ be the interval $[1/2, 1/2 + 1/n)$, and $g_n := -n^{-1} 1_{J(n)}$, $n = 2, 3, \ldots$. Then for $1 \le p < \infty$ and $\mu$ equal to Lebesgue measure on $\Omega = [0,1]$, $\|N_{F_1}(G + g_n) - N_{F_1}(G)\|_p = n^{-1/p}$ and $\|F_1\|_{(p)} = 1$. Also $\|g_n\|_s = n^{-(1+s)/s}$ for $1 \le s < \infty$, $\|g_n\|_{\sup} = 1/n$, $\|g_n\|_{(r)} = 2^{1/r}/n$, and so $\|g_n\|_{[r]} = (1 + 2^{1/r})/n$, $1 \le r < \infty$. We can also take multiples $F = tF_1$ with $t \to 0$.

The last remark gives the following:

**Proposition 7.28.** *If $(\mathbb{F}, \|\cdot\|)$ is a normed space of real-valued functions on $[0,1]$ containing $F_1$ of Remark 7.27, and for some finite constants $C > 0$, $\lambda > 0$, and $\zeta > 0$,*

$$\|N_F(G + g_n) - N_F(G)\|_p \le C\|F\|^{\lambda}\|g_n\|_s^{\zeta}$$

*for $F$, $G$, and $g_n$ as in Remark 7.27, then $\lambda \le 1$ and $\zeta \le s/(p(1+s))$ for $1 \le s < \infty$, $\zeta \le 1/p$ for $s = \infty$. Thus the powers $\lambda = 1$ in Theorem 7.24 (a) and $\zeta = s/(p(1+s))$ for $s < \infty$, $\zeta = 1/p$ for $s = \infty$, are separately optimal, and no norm on $F$ allows a better exponent than $\|\cdot\|_{(p)}$ does.*

On the other hand, the exponents for $s = \infty$ are not improved if we replace $\|\cdot\|_\infty$ by a stronger $r$-variation norm $\|g\|_{[r]}$, $1 \le r < \infty$.

By Theorem 7.24, for $F \in \mathcal{W}_p(\mathbb{R})$ and $\alpha = s/[p(1+s)]$, the autonomous Nemytskii operator $N_F$ is $\alpha$-Hölder from $L^s$ into $L^p$ at suitable elements of $L^s$. We will show in Proposition 7.30 that $N_F$ is $\alpha$-Hölder from $L^s$ into $L^p$ with $\alpha = s/p \le 1$ provided $F \in \mathcal{H}_\alpha(\mathbb{R}; \mathbb{R})$. Although for a bounded nondegenerate interval $J$, $\mathcal{H}_\alpha(J; \mathbb{R}) \subset \mathcal{W}_{1/\alpha}(J; \mathbb{R})$ for $0 < \alpha \le 1$, $\mathcal{W}_p(\mathbb{R})$ does not include any space $\mathcal{H}_\alpha(\mathbb{R}; \mathbb{R})$.

Let $M_k(f_1, \ldots, f_k) := f_1 \cdots f_k$ be the function defined by the pointwise multiplication of real-valued functions $f_1, \ldots, f_k$ on $\Omega$. The following bound for the $L_p$ norm of $M_k(f_1, \ldots, f_k)$ extends Hölder's inequality and will be used to prove $\alpha$-Hölder properties of Nemytskii operators and several times later for other purposes.

**Lemma 7.29.** *Let $(\Omega, \mathcal{S}, \mu)$ be $\sigma$-finite and $L^r := L^r(\Omega, \mathcal{S}, \mu)$. For any $k = 2, 3, \ldots$, if $p_j \in [1, \infty)$ for $j = 1, \ldots, k$ and $s := \sum_{j=1}^k 1/p_j \le 1$, then for $p := 1/s$,*

$$\|M_k(f_1, \ldots, f_k)\|_p \le \|f_1\|_{p_1} \cdots \|f_k\|_{p_k} \tag{7.16}$$

*and $(f_1, \ldots, f_k) \mapsto M_k(f_1, \ldots, f_k)$ is a bounded $k$-linear map from $L^{p_1} \times \cdots \times L^{p_k}$ into $L^p$ with norm $\|M_k\| = 1$.*

*Proof.* We have $p < p_j$ for all $j$. For $k = 2$, by Hölder's inequality (e.g. [53, Theorem 5.1.2] with $p = sp_1$ and $q = sp_2$), if $f \in \mathcal{L}^{p_1}$ and $g \in \mathcal{L}^{p_2}$, then

$$\int_\Omega |fg|^p \, d\mu \le \Big(\int_\Omega |f|^{p_1} \, d\mu\Big)^{p/p_1} \Big(\int_\Omega |g|^{p_2} \, d\mu\Big)^{p/p_2},$$

and so (7.16) holds. For the norm defined by (5.4), we then have

$$\|M_2\| = \sup\{\|fg\|_p\colon \ \|f\|_{p_1} \leq 1,\ \|g\|_{p_2} \leq 1\} \leq 1.$$

By induction on $k$ it follows that (7.16) and $\|M_k\| \leq 1$ hold for all $k$.

Conversely, take any $h \in \mathcal{L}^1$ with $\|h\|_1 = 1$ and let $f_j := |h|^{1/p_j}$ for $j = 1, \ldots, k$. Then $\|f_j\|_{p_j} = 1$ for all $j$ and $\|M_k(f_1, \ldots, f_k)\|_p = 1$, so $\|M_k\| = 1$, proving the lemma. □

The following statement gives sufficient conditions for an autonomous Nemytskii operator to be $\alpha$-Hölder from $L^s$ into $L^p$ with $1 \leq s \leq p < \infty$.

**Proposition 7.30.** *Let $(\Omega, \mathcal{S}, \mu)$ be complete and finite and let $1 \leq s \leq p < \infty$. If $F \in \mathcal{H}_\alpha(\mathbb{R};\mathbb{R})$ with $0 < \alpha \leq s/p$, then the autonomous Nemytskii operator $N_F$ is $\alpha$-Hölder from $L^s(\Omega, \mathcal{S}, \mu)$ into $L^p(\Omega, \mathcal{S}, \mu)$.*

*Proof.* Clearly $F \in \mathcal{G}_\alpha$ and $F$ is universally measurable. Thus $N_F$ acts from $L^s$ into $L^p$ by Theorem 7.13(*a*) since $\alpha \leq s/p$. For some $K < \infty$, we have $|F(x) - F(y)| \leq K|x-y|^\alpha$ for all $x, y \in \mathbb{R}$. Then for any $G$ and $g$ in $L^s(\Omega, \mathcal{S}, \mu)$, by Lemma 7.29 with $k = 2$ if $\alpha < s/p$, or directly if $\alpha = s/p$,

$$\int_\Omega \left|F{\circ}(G+g) - F{\circ}G\right|^p d\mu \leq K^p \int_\Omega |g|^{p\alpha}\, d\mu \leq M\|g\|_s^{p\alpha},$$

where $M := K^p \mu(\Omega)^{1-(p\alpha/s)}$, which gives the conclusion. □

The upper bound $s/p$ for $\alpha$ in the last proposition cannot be made larger, as Corollary 7.34 will show. But first we show that for a specific $F \in \mathcal{H}_\alpha$, the autonomous Nemytskii operator $N_F$ is $\beta$-Hölder at suitable elements of $L^s$ for a $\beta = \beta(\alpha, s, p)$ with $\alpha < \beta < s/p \leq 1$. As defined in the Appendix, $\mathcal{D}_\lambda = \mathcal{D}_\lambda(\Omega, \mathcal{S}, \mu)$ is the set of all functions $G \in \mathcal{L}^0(\Omega, \mathcal{S}, \mu)$ such that $\mu{\circ}G^{-1}$ has a bounded density $\xi_G$ with respect to Lebesgue measure $\lambda$ on $\mathbb{R}$.

**Proposition 7.31.** *Let $(\Omega, \mathcal{S}, \mu)$ be complete and finite, $1 \leq s \leq p < \infty$, and let $F_\alpha(u) \equiv |u|^\alpha$ with $0 < \alpha \leq s/p$. If $G \in \mathcal{D}_\lambda \cap \mathcal{L}^s$ then the autonomous Nemytskii operator $N_{F_\alpha}$ acts from $L^s$ into $L^p$ and there is a constant $K < \infty$ such that for each $g \in L^s$ with $\|g\|_s \leq 1$,*

$$\|N_{F_\alpha}(G+g) - N_{F_\alpha}(G)\|_p \leq K\|g\|_s^\beta, \tag{7.17}$$

*where $\beta := [\alpha + (1/p)]s/(1+s) \in [\alpha, s/p]$.*

*Proof.* If $\alpha = s/p$ then $\beta = \alpha$ and the conclusion follows from Proposition 7.30. Suppose that $\alpha < s/p$, and so $\alpha < \beta < s/p \leq 1$. Since $F_\alpha \in \mathcal{G}_\alpha \subset \mathcal{G}_{s/p}$, $N_{F_\alpha}$ acts from $L^s$ into $L^p$ by Theorem 7.13(*a*). For each $g \in L^s$ with $\|g\|_s \leq 1$, let $\Delta(g) := F_\alpha{\circ}(G+g) - F_\alpha{\circ}G$. Let $\rho := s/(1+s)$ and $\epsilon := \epsilon(g) := \|g\|_s^\rho$. Thus $\beta = \rho[\alpha + (1/p)]$. Also, let $A_1 := \{\omega\colon\ |G(\omega)| \leq \epsilon\}$, $A_2 := A_1^c \cap \{\omega\colon\ |g(\omega)| >$

$\epsilon/2\}$, and $A_3 := A_1^c \cap \{\omega\colon |g(\omega)| \le \epsilon/2\}$. Thus the left side of (7.17) does not exceed the sum of $T_i := \|\Delta(g)1_{A_i}\|_p$, $i = 1, 2, 3$.

Let $M := \|\xi_G\|_\infty < \infty$. Then $\mu(A_1) = \int_{\{|x|\le\epsilon\}} \xi_G(x)\,\mathrm{d}x \le 2M\epsilon = 2M\|g\|_s^\rho$. Since $|F_\alpha(u) - F_\alpha(v)| \le |u - v|^\alpha$ for all $u, v \in \mathbb{R}$, it then follows by Lemma 7.29 with $k = 2$ if $\alpha < s/p$, or directly if $\alpha = s/p$, that

$$T_1 \le \Big(\int_\Omega |g|^{\alpha p} 1_{A_1}\,\mathrm{d}\mu\Big)^{1/p} \le \|g\|_s^\alpha \mu(A_1)^{(1/p)-(\alpha/s)} \le K_1\|g\|_s^\beta, \tag{7.18}$$

where $K_1 := (2M)^{(1/p)-(\alpha/s)}$. For $T_2$, again, since $\epsilon = \|g\|_s^\rho$, we have

$$\mu(A_2) \le \mu(\{\omega\colon |g(\omega)| > \epsilon/2\}) \le (2/\epsilon)^s \|g\|_s^s = 2^s\|g\|_s^\rho.$$

Thus as in (7.18), we get the bound

$$T_2 \le \Big(\int_\Omega |g|^{\alpha p} 1_{A_2}\,\mathrm{d}\mu\Big)^{1/p} \le \|g\|_s^\alpha \mu(A_2)^{(1/p)-(\alpha/s)} \le K_2\|g\|_s^\beta,$$

where $K_2 := 2^{(s/p)-\alpha}$. To bound $T_3$, by (5.2) in the mean value theorem, for each $x, y \in \mathbb{R}$ such that $|x| > \epsilon$ and $|y| \le \epsilon/2$, we have

$$\big|F_\alpha(x+y) - F_\alpha(x)\big| \le |y| \sup_{0\le t\le 1} |F_\alpha'(x+ty)|$$

where $|x + ty| > \epsilon/2$, so the sup is finite. It then follows that for each $\omega \in A_3$,

$$\begin{aligned}\big|\Delta(g)(\omega)\big| &\le \alpha|g(\omega)| \sup_{0\le t\le 1} |G(\omega) + tg(\omega)|^{\alpha-1} \\ &\le (\epsilon/2)^{1-(s/p)}|g(\omega)|^{s/p}\big[|G(\omega)| - \epsilon/2\big]^{\alpha-1}\end{aligned}$$

since $s \le p$. Since $\epsilon = \|g\|_s^\rho \le 1$, we have

$$\begin{aligned}T_3 &\le \Big(\frac{\epsilon}{2}\Big)^{1-(s/p)}\Big[\Big(\int_{A_3\cap\{|G|>1\}} |g|^s 2^{p(1-\alpha)}\mathrm{d}\mu\Big)^{1/p} \\ &\qquad + \Big(\int_{A_3\cap\{|G|\le 1\}} |g|^s (\epsilon/2)^{p(\alpha-1)}\,\mathrm{d}\mu\Big)^{1/p}\Big] \\ &\le (\epsilon/2)^{1-(s/p)}\|g\|_s^{s/p}\Big[2^{1-\alpha} + (\epsilon/2)^{\alpha-1}\Big] \\ &= 2^{(s/p)-\alpha}\Big[\|g\|_s^{\rho[1+(1/p)]} + \|g\|_s^\beta\Big] \le K_3\|g\|_s^\beta,\end{aligned}$$

where $K_3 := 2K_2$. Summing the bounds of $T_i$, $i = 1, 2, 3$, and letting $K := K_1 + K_2 + K_3$, it follows that (7.17) holds whenever $\|g\|_s \le 1$. The proof of the proposition is complete. □

For the Nemytskii operator $N_{F_\alpha}$, the order of the remainder bound (7.17) cannot be improved in general, as the following shows.

**Proposition 7.32.** *Let* $(\Omega, \mathcal{S}, \mu) = ([0,1], \mathcal{B}, \lambda)$, $1 \le s, p < \infty$, $F_\alpha(u) \equiv |u|^\alpha$, $u \in \mathbb{R}$, *with* $0 < \alpha < \infty$, $G(x) \equiv x$, $x \in [0,1]$, *and* $\beta > 0$. *If the Nemytskii operator* $N_{F_\alpha}$ *acts from* $L^s$ *into* $L^p$ *and* $\|N_{F_\alpha}(G+g) - N_{F_\alpha}(G)\|_p = O(\|g\|_s^\beta)$ *as* $\|g\|_s \to 0$, *then* $\alpha \le s/p$ *and* $\beta \le [\alpha + (1/p)]s/(1+s) \in [\alpha, s/p]$.

*Proof.* By Theorem 7.13($b$), it follows that $\alpha \le s/p$, and so as in the last proof $[\alpha + (1/p)]s/(1+s) \in [\alpha, s/p]$. For $g \in L^s$, let $\Delta(g) := F_\alpha \circ (G+g) - F_\alpha \circ G$. For $\delta > 0$, let $g_\delta(x) := -x 1_{[0,\delta]}(x)$, $x \in [0,1]$. Then $\|g_\delta\|_s = \delta^{1+(1/s)}/(1+s)^{1/s}$ and $|\Delta(g_\delta)| = F_\alpha 1_{[0,\delta]}$. Thus there is a constant $C = C(\alpha, s, p) < \infty$ such that for each $\delta > 0$,

$$\|\Delta(g_\delta)\|_p = \delta^{\alpha+(1/p)}/(1+\alpha p)^{1/p} = C\|g_\delta\|_s^{[\alpha+(1/p)]s/(1+s)}.$$

Since $\|g_\delta\|_s \to 0$ as $\delta \downarrow 0$, the stated conclusion follows. □

For $(u, x) \in \mathbb{R} \times [0,1]$, letting $\phi_\alpha(u,x) := |x+u|^\alpha - |x|^\alpha$, $\phi_\alpha$ is a Shragin function on $\mathbb{R} \times [0,1]$. By Theorem 7.13($a$), if $\alpha \le s/p$ then $N_{F_\alpha}$ and so $N_{\phi_\alpha}$ acts from $L^s([0,1], \mathcal{B}, \lambda)$ into $L^p([0,1], \mathcal{B}, \lambda)$. Thus by the preceding proposition, if $\alpha \le s/p$ and $\|N_{\phi_\alpha} g\|_p = O(\|g\|_s^\beta)$ as $\|g\|_s \to 0$, then $\beta \le s/p$. This fact holds for any non-degenerate Nemytskii operator acting from $L^s$ into $L^p$, as the following shows.

**Proposition 7.33.** *Let* $(\Omega, \mathcal{S}, \mu)$ *be nonatomic and finite,* $1 \le s, p < \infty$, *and* $0 < \alpha < \infty$. *Let a Shragin function* $\phi$ *be such that* $\phi \not\equiv 0$, *and let the Nemytskii operator* $N_\phi$ *act from* $L^s$ *into* $L^p$. *Then the following hold:*

($a$) *if* $\|N_\phi g\|_p = O(\|g\|_s^\alpha)$ *as* $\|g\|_s \to 0$, *then* $\alpha \le s/p$;
($b$) *if* $\|N_\phi g\|_p = o(\|g\|_s^\alpha)$ *as* $\|g\|_s \to 0$, *then* $\alpha < s/p$.

*Proof.* Let $f \in \mathcal{L}^0$ be such that $N_\phi f \ne 0$ on a set of positive $\mu$-measure. Then for some constant $C > 0$, the set $A := \{\omega \colon |\phi(f(\omega), \omega)| \ge C\}$ has positive $\mu$-measure. We can and do assume that for some $M < \infty$, $|f| \le M$ on $A$. Since $\mu$ is nonatomic, by Proposition A.1, for each $n = 1, 2, \ldots$, there are $A_n \in \mathcal{S}$ with $A_n \subset A$ and $0 < \mu_n := \mu(A_n) \to 0$ as $n \to \infty$. For each $n = 1, 2, \ldots$, let $g_n := f 1_{A_n}$. Then $\|g_n\|_s \le M\mu_n^{1/s}$ and $\|N_\phi(g_n)\|_p \ge C\mu_n^{1/p}$. Thus for ($a$) we have $\mu_n^{1/p} = O(\mu_n^{\alpha/s})$ as $n \to \infty$, and so $\alpha \le s/p$. For ($b$) we have $\mu_n^{1/p} = o(\mu_n^{\alpha/s})$ as $n \to \infty$, and so $\alpha < s/p$. The proof of the proposition is complete. □

**Corollary 7.34.** *Let* $(\Omega, \mathcal{S}, \mu)$ *be nonatomic, finite, and complete,* $1 \le s \le p < \infty$, *and* $0 < \alpha \le 1$. *Let* $F$ *be a Borel measurable non-constant function. If the autonomous Nemytskii operator* $N_F$ *is* $\alpha$*-Hölder at some* $G \in L^s = L^s(\Omega, \mathcal{S}, \mu)$ *from* $L^s$ *into* $L^p$, *then* $\alpha \le s/p$.

*Proof.* Let $G \in L^s$ be such that $N_F$ is $\alpha$-Hölder at $G$. For $(u,\omega) \in \mathbb{R} \times \Omega$, let $\phi(u,\omega) := F(G(\omega)+u) - F(G(\omega))$. Then $\phi$ is jointly measurable on $\mathbb{R} \times \Omega$, and so it is a $\mu$-strong Shragin function on $\mathbb{R} \times \Omega$. Since $F$ is not constant, for each $\omega \in \Omega$, $\phi(u,\omega) \neq 0$ for some $u \in \mathbb{R}$, and so $\phi \not\simeq 0$ by Proposition 7.6. Since $\|N_\phi g\|_p = O(\|g\|_s^\alpha)$ as $\|g\|_s \to 0$, the conclusion follows from Proposition 7.33($a$). □

**Proposition 7.35.** *Let $(\Omega, \mathcal{S}, \mu)$ be nonatomic, finite, and complete, $1 \le s < p < \infty$, $s/p < \alpha \le 1$, and let $\psi$ be a Shragin function on $\mathbb{R}\times\Omega$. The Nemytskii operator $N_\psi$ is $\alpha$-Hölder at some $G \in \mathcal{L}^s = \mathcal{L}^s(\Omega, \mathcal{S}, \mu)$ from $\mathcal{L}^s$ into $\mathcal{L}^p$ if and only if there is a function $h \in \mathcal{L}^p$ such that $\psi \simeq h$.*

*Proof.* The "if" part is obvious. To prove the "only if" part suppose that $N_\psi$ is $\alpha$-Hölder at $G \in \mathcal{L}^s$ from $\mathcal{L}^s$ into $\mathcal{L}^p$. For $(u,\omega) \in \mathbb{R} \times \Omega$, let $\phi(u,\omega) := \psi(G(\omega)+u,\omega) - \psi(G(\omega),\omega)$. Then $\phi$ is a Shragin function on $\mathbb{R} \times \Omega$ and $\|N_\phi g\|_p = O(\|g\|_s^\alpha)$ as $\|g\|_s \to 0$. Since $\alpha > s/p$, by Proposition 7.33($a$), $N_\phi f = 0$ a.e. $(\mu)$ for each $f \in \mathcal{L}^0$. Let $h := N_\psi G \in \mathcal{L}^p$ and $f \in \mathcal{L}^0$. Then $N_\psi f \equiv N_\phi(f-G) + h = h$ a.e. $(\mu)$, proving the proposition. □

## 7.3 Differentiability

Proposition 7.35 will be used to show that only degenerate Nemytskii operators acting from $L^s$ into $L^p$ with $1 \le s < p < \infty$ are Fréchet differentiable. Recall the remainder in the differentiability as defined in (5.1) for a Nemytskii operator $N_\psi$:

$$\mathrm{Rem}_{N_\psi}(G,g) = N_\psi(G+g) - N_\psi(G) - (DN_\psi(G))g. \tag{7.19}$$

**Corollary 7.36.** *Let $(\Omega, \mathcal{S}, \mu)$ be nonatomic, finite, and complete. For $1 \le s < p < \infty$, the following statements hold:*

($a$) *if $\psi$ is a Shragin function on $\mathbb{R} \times \Omega$ and the Nemytskii operator $N_\psi$ is Fréchet differentiable at some $G \in L^s$ from $L^s$ into $L^p$ then there is a function $h \in L^p$ such that $\psi \simeq h$.*

($b$) *if $F$ is a Borel function on $\mathbb{R}$ and the autonomous Nemytskii operator $N_F$ is Fréchet differentiable at some $G \in L^s$ from $L^s$ into $L^p$ then $F$ must be a constant function.*

*Proof.* Let $N$ be either $N_\psi$ or $N_F$. Since $N$ is Fréchet differentiable at some $G$ from $L^s$ into $L^p$ and using (7.19), we have

$$\big\|N(G+g) - N(G)\big\|_p \le \big\|\mathrm{Rem}_N(G,g)\big\|_p + \|(DN(G))g\|_p = O(\|g\|_s)$$

as $\|g\|_s \to 0$, and so $N$ is 1-Hölder at $G$. Thus ($a$) follows by Proposition 7.35.

For $(b)$ let $\psi(u,\omega) \equiv F(u)$. Then $\psi$ is a $\mu$-strong Shragin function on $\mathbb{R}\times\Omega$ (with $N = \emptyset$). By part $(a)$, $N_\psi f = h$ for each $f \in L^0$. By Proposition 7.6 applied to $\phi(u,\omega) := \psi(u,\omega) - h(\omega)$, $(u,\omega) \in \mathbb{R}\times\Omega$, we have for $\mu$-almost all $\omega$, $\psi(u,\omega) = h(\omega)$ for each $u \in \mathbb{R}$. Since $\psi(u,\omega) \equiv F(u)$, $h$ must equal a constant $c$ a.e. $(\mu)$ and $F \equiv c$, proving $(b)$. □

In the case $1 \le s = p < \infty$, $\psi$ must be affine in $u$ to give a Fréchet differentiable operator $N_\psi$ from $L^s$ to $L^p$, as will be shown in Proposition 7.39. But first the derivative linear operator in case of differentiability of a Nemytskii operator $N_\psi$ will be shown to be a multiplication operator $M[h]$ induced by some function $h \in \mathcal{L}^r$, where for $1 \le p \le s \le \infty$ with $p < \infty$,

$$r := r(s,p) := \begin{cases} +\infty & \text{if } s = p, \\ (sp)/(s-p) & \text{if } p < s < \infty, \\ p & \text{if } s = +\infty. \end{cases} \tag{7.20}$$

Recall that for a function $h\colon \Omega \to \mathbb{R}$, the multiplication operator $M[h]$ from $L^s$ into $L^p$ induced by $h$ takes each $g \in L^s$ to the pointwise product $M[h](g) = hg \in L^p$.

**Lemma 7.37.** *Let $(\Omega,\mathcal{S},\mu)$ be $\sigma$-finite, let $1 \le p \le s \le \infty$ with $p < \infty$, and let $h\colon \Omega \to \mathbb{R}$ be $\mu$-measurable. Then the following three statements are equivalent:*

*(a) $h \in \mathcal{L}^{r(s,p)}$;*
*(b) $h$ induces a bounded linear multiplication operator $M[h]$ from $L^s$ into $L^p$;*
*(c) for any $f \in \mathcal{L}^s$, $hf \in \mathcal{L}^p$.*

*If these conditions hold, then the operator norm of $M[h]$ in (b) is $\|h\|_r$.*

*If $1 \le s < p < \infty$ and $(\Omega,\mathcal{S},\mu)$ is nonatomic, then (b) and (c) are equivalent to $h = 0$ in $L^0$.*

*Proof.* To show that $(a)$ implies $(b)$, first let $s < \infty$. If $p < s$ then for any $f \in \mathcal{L}^s$, $fh$ is $\mu$-measurable and we have

$$\int |hf|^p d\mu = \int |f|^p |h|^p d\mu \le \Big(\int |f|^s d\mu\Big)^{p/s} \Big(\int |h|^r d\mu\Big)^{p/r} < \infty \tag{7.21}$$

by Lemma 7.29 with $k = 2$. Taking $p$th roots, we see that $M[h]$ is a bounded linear operator from $\mathcal{L}^s$ into $\mathcal{L}^p$. If $p = s$ and $h \in \mathcal{L}^\infty$, then for any $f \in \mathcal{L}^p$, we have $(\int |hf|^p\, d\mu)^{1/p} \le \|h\|_\infty \|f\|_p$. Similarly if $s = +\infty$ and so $r = p$, then for any $f \in \mathcal{L}^\infty$, we have $(\int |hf|^p\, d\mu)^{1/p} \le \|h\|_p \|f\|_\infty$. Thus $(b)$ follows from $(a)$.

Clearly $(b)$ implies $(c)$. Suppose that $(c)$ holds. To prove $(a)$, first let $s < +\infty$. Let $f_n \to f$ in $\mathcal{L}^s$ and $f_n h \to g$ in $\mathcal{L}^p$. Taking a subsequence, we can assume that $f_n \to f$ $\mu$-almost everywhere and thus that $f_n h \to g$ a.e. $(\mu)$, so $fh = g$ a.e. $(\mu)$. Thus the linear multiplication operator $M[h]$ from $\mathcal{L}^s$ into $\mathcal{L}^p$

is closed, and by the closed graph theorem it is bounded, that is, its operator norm

$$M := \sup\{\|fh\|_p \colon \|f\|_s \le 1\} \tag{7.22}$$

is finite. For any $g \in \mathcal{L}^{s/p}$, let $f := |g|^{1/p}$. Then $f \in \mathcal{L}^s$ and

$$\begin{aligned}\Big|\int g|h|^p d\mu\Big| \;&\le\; \int |g||h|^p d\mu \;=\; \int |fh|^p d\mu \\ &\le\; M^p\Big(\int |f|^s d\mu\Big)^{p/s} \;=\; M^p\Big(\int |g|^{s/p} d\mu\Big)^{p/s}.\end{aligned}$$

Thus $g \mapsto \int g|h|^p d\mu$ is a bounded linear functional on $\mathcal{L}^{s/p}$. By the Riesz representation theorem, it is of the form $g \mapsto \int g\psi\, d\mu$ for some function $\psi \in \mathcal{L}^{r/p}$, or $\psi \in \mathcal{L}^\infty$ for $s = p$, unique up to equality a.e. $(\mu)$. If $\mu(\{\omega\colon \psi(\omega) \ne |h(\omega)|^p\}) > 0$ let $A \in \mathcal{S}$ with $0 < \mu(A) < +\infty$ and $\psi \ne |h|^p$ on $A$. By symmetry we can assume $\psi > |h|^p$ on $A$. We have $1_A \in \mathcal{L}^{s/p}$ and $\int \psi 1_A\, d\mu > \int |h|^p 1_A\, d\mu$, a contradiction. Thus $\psi = |h|^p$ a.e. $(\mu)$ and so $h \in \mathcal{L}^r$, proving $(a)$ when $s < +\infty$. If $s = +\infty$ then $(a)$ also holds since $f := 1(\cdot) \in \mathcal{L}^\infty$, $r = p$, and $h = hf \in \mathcal{L}^r$.

To prove that $M = \|h\|_r$ holds suppose that the three conditions hold. If $p < s < +\infty$, by (7.21) and (7.22), we have $M \le \|h\|_r$. Conversely, given $h \in \mathcal{L}^r$, let $f := |h|^{r/s}$. Then $f \in \mathcal{L}^s$, $\|f\|_s = \|h\|_r^{r/s}$, and $\|fh\|_p = \|h\|_r^{r/p}$, so $\|fh\|_p/\|f\|_s = \|h\|_r$ and $M = \|h\|_r$. If $s = p$, then $M \le \|h\|_\infty$ and we get $M = \|h\|_\infty$ by considering functions $f_n := 1_{\{|h| > \|h\|_\infty - 1/n\}}$, $n \to \infty$. If $s = +\infty$, then $p = r$ and $M = \|h\|_r$ by (7.22), since $\|1(\cdot)\|_\infty = 1$. So the equality of norms is proved.

Clearly, if $h = 0$ a.e. $(\mu)$ then $(b)$ and $(c)$ hold. Conversely, suppose $1 \le s < p < \infty$, $(\Omega, \mathcal{S}, \mu)$ is nonatomic, and $(c)$ holds for a given $h \in \mathcal{L}^0(\Omega, \mathcal{S}, \mu)$, not equal to 0 a.e. $(\mu)$. Then for some $\delta > 0$ and set $A \in \mathcal{S}$ with $0 < \mu(A) < \infty$, $|h| \ge \delta$ on $A$ and so $(c)$ holds with $h$ replaced by $\delta 1_A$ and thus by $1_A$. Let $\mu_A(E) := \mu(E \cap A)$ for all $E \in \mathcal{S}$. Then clearly $\mu_A$ is a nonatomic measure. By Proposition A.5 in the Appendix, there is a measurable function $G$ from $\Omega$ into $[0,1]$ such that $\mu_A \circ G^{-1}$ is $\mu(A)$ times Lebesgue measure on $[0,1]$. We can set $G(\omega) = 1$ for all $\omega$ in the set $\Omega \setminus A$ of $\mu_A$-measure 0. Let $f(\omega) := G(\omega)^{-1/p} 1_A(\omega)$. Then $f \in L^s$ but $f \notin L^p$, a contradiction. The lemma is proved. □

**Proposition 7.38.** *Let $(\Omega, \mathcal{S}, \mu)$ be $\sigma$-finite and $1 \le p \le s \le \infty$ with $p < \infty$. Let $G$ be a $\mu$-measurable real-valued function on $\Omega$ and let $\psi$ be a Shragin function on $\mathbb{R} \times \Omega$. If the Nemytskii operator $N_\psi$ is Fréchet differentiable at $G$ from $L^s(\Omega, \mathcal{S}, \mu)$ into $L^p(\Omega, \mathcal{S}, \mu)$ then the derivative operator $D_{\psi,G} := (DN_\psi)(G)(\cdot)$ is the multiplication operator $M[h]$ induced by some function $h \in \mathcal{L}^r(\Omega, \mathcal{S}, \mu)$, where $r$ is defined by (7.20).*

*Proof.* $D_{\psi,G}$ is a bounded linear operator from $L^s$ into $L^p$. Let $A \in \mathcal{S}$ with $\mu(A) < \infty$. Let $h_A := D_{\psi,G}(1_A) \in \mathcal{L}^p$. Then as $t \to 0$,

$$\|(N_\psi(G+t1_A)-N_\psi G)/t-h_A\|_p \;\to\; 0. \tag{7.23}$$

It follows that $h_A(\omega)=0$ for $\mu$-almost all $\omega\notin A$. If $B\subset A$, we have $h_B(\omega)=h_A(\omega)$ for $\mu$-almost all $\omega\in B$. Since $\mu$ is $\sigma$-finite, it follows that there is a $\mu$-measurable function $h\colon\,\Omega\to\mathbb{R}$ such that we can take $h_A=h1_A$ for all $A\in\mathcal{S}$ with $\mu(A)<\infty$. Taking $t$ along the sequence $1/n$ in (7.23), and by $\sigma$-finiteness again, there is a sequence $n_k\uparrow+\infty$ such that

$$n_k[\psi(G(\omega)+{n_k}^{-1},\omega)-\psi(G(\omega),\omega)] \;\to\; h(\omega) \tag{7.24}$$

for $\mu$-almost all $\omega$ as $k\to\infty$. We can redefine $h(\omega)$ as the limit of the left side if it exists and 0 otherwise.

Let $g$ be any $\mu$-simple function on $\Omega$, that is, $g=\sum_{i=1}^n c_i1_{A_i}$ where $c_i\in\mathbb{R}$, $A_i\in\mathcal{S}$, and $\mu(A_i)<\infty$. Then since $D_{\psi,G}$ is linear, we have $D_{\psi,G}(g)=hg$.

Now take any $g\in\mathcal{L}^s(\Omega,\mathcal{S},\mu)$ and take $\mu$-simple $g_n\to g$ in $\mathcal{L}^s$. We have $D_{\psi,G}(g_n)=hg_n$ for all $n$ and $D_{\psi,G}(g_n)\to D_{\psi,G}(g)$ in $\mathcal{L}^p$. Taking a subsequence, we can assume that $g_n\to g$ and $D_{\psi,G}(g_n)=hg_n\to D_{\psi,G}(g)$ hold $\mu$-almost everywhere. Thus $D_{\psi,G}(g)=hg$ a.e. $(\mu)$. It follows that $D_{\psi,G}(g)=M[h](g)\in L^p$ for each $g\in L^s$, and so $h\in\mathcal{L}^r$ by $(b)\Rightarrow(a)$ of Lemma 7.37. □

Now we can show that for $N_\psi$ to be Fréchet differentiable from an $L^p$ space to itself, $\psi$ must have an affine form in $u$, up to equivalence in the sense of Definition 7.7.

**Proposition 7.39.** *Let $(\Omega,\mathcal{S},\mu)$ be nonatomic, finite, and complete, let $1\le p<\infty$, and let $\psi$ be a Shragin function. If the Nemytskii operator $N_\psi$ is Fréchet differentiable at some $G\in\mathcal{L}^p(\Omega,\mathcal{S},\mu)$ from $L^p$ into $L^p$, then for some $\xi\in\mathcal{L}^p$ and $h\in\mathcal{L}^\infty$, $\psi\backsimeq\eta$ where $\eta(u,\omega):=\xi(\omega)+h(\omega)u$, $(u,\omega)\in\mathbb{R}\times\Omega$.*

*Proof.* Suppose that $N_\psi$ is Fréchet differentiable at $G\in\mathcal{L}^p$ from $L^p$ into $L^p$. By Proposition 7.38 with $s=p$, there is an $h\in\mathcal{L}^\infty$ such that $(DN_\psi G)(f)=hf$ a.e. $(\mu)$ for each $f\in\mathcal{L}^p$. For $(u,\omega)\in\mathbb{R}\times\Omega$, let $\phi(u,\omega):=\psi(G(\omega)+u,\omega)-\psi(G(\omega),\omega)-h(\omega)u$. Then $\phi$ is a Shragin function on $\mathbb{R}\times\Omega$ and $\|N_\phi g\|_p=\|\mathrm{Rem}_{N_\psi}(G,g)\|_p=o(\|g\|_p)$ as $\|g\|_p\to 0$. By Proposition 7.33($b$) with $\alpha=1$ and $s=p$, we must have $\phi\backsimeq 0$, and so $N_\phi f=0$ a.e. $(\mu)$ for each $f\in\mathcal{L}^0$. Let $\xi:=N_\psi G-hG\in\mathcal{L}^p$ and $f\in\mathcal{L}^0$. Then $N_\psi f\equiv N_\phi(f-G)+(N_\psi G-hG)+hf=\xi+hf=N_\eta f$ a.e. $(\mu)$, proving the proposition. □

Proposition 7.41 will give sufficient conditions for Fréchet differentiability of an autonomous Nemytskii $N_F$ operator from $L^s$ into $L^p$ with $p<s$. The following gives necessary conditions. Let $F$ be a u.m. function from $\mathbb{R}$ into $\mathbb{R}$ and let $\nu$ be a finite Borel measure on $\mathbb{R}$. Then $F$ will be said to be *differentiable in $\nu$-measure* if there exists a measurable function $\eta$ from $\mathbb{R}$ into $\mathbb{R}$

such that $(F(y+t)-F(y))/t \to \eta(y)$ in $\nu$-measure as $t \to 0$, and then we will write $D_{(\nu)}F = \eta$, defined up to equality a.e. $(\nu)$.

**Theorem 7.40.** *Let $(\Omega, \mathcal{S}, \mu)$ be $\sigma$-finite and complete. Let $G\colon \Omega \to \mathbb{R}$ be $\mu$-measurable and let $F\colon \mathbb{R} \to \mathbb{R}$ be Borel measurable. Let $1 \le p < s \le \infty$. If the autonomous Nemytskii operator $N_F$ is Fréchet differentiable at $G$ from $L^s(\Omega, \mathcal{S}, \mu)$ into $L^p(\Omega, \mathcal{S}, \mu)$, then:*

*(a) the derivative operator $D_{F,G} := (DN_F)(G)(\cdot)$ must be the multiplication operator $M[h]$ induced by some function $h \in \mathcal{L}^r(\Omega, \mathcal{S}, \mu)$, where $h$ can be taken as $\eta \circ G$ for some Borel measurable function $\eta$, and $r$ is defined by (7.20);*
*(b) if $\mu$ is finite and $\nu$ is the image measure $\mu \circ G^{-1}$ on $\mathbb{R}$, then $F$ must be differentiable in $\nu$-measure with $\eta = D_{(\nu)}F$.*

*Proof.* Since $F$ is Borel, $\psi(u, \omega) \equiv F(u)$ is a Shragin function, and so by Proposition 7.38, the derivative operator $D_{F,G}$ is multiplication by some function $h \in \mathcal{L}^r$. As in (7.24), there is a sequence $n_k \uparrow +\infty$ such that

$$h(\omega) = \lim_{k\to\infty} n_k \big[F(G(\omega) + n_k^{-1}) - F(G(\omega))\big]$$

for $\mu$-almost all $\omega$. We can take $h(\omega)$ to be the limit if it exists and 0 otherwise. Then $h$ is $G^{-1}(\mathcal{B}) := \{G^{-1}(B)\colon B \in \mathcal{B}\}$ measurable. It then follows that $h = \eta \circ G$ for some Borel measurable function $\eta$, e.g. by [53, Theorem 4.2.8] with $X = \Omega$, $Y = \mathbb{R}$, $T = G$, and $f = h$. So $(a)$ is proved.

For $(b)$, if $\mu$ is finite and $X$ is the identity function on $\mathbb{R}$, then it follows easily that $\gamma \mapsto F\circ(X+\gamma)$ is Fréchet differentiable at $\gamma = 0$ from $\mathcal{L}^s(\mathbb{R}, \mathcal{B}, \nu)$ to $\mathcal{L}^p(\mathbb{R}, \mathcal{B}, \nu)$, setting $g := \gamma \circ G$ and applying the image measure theorem, e.g. [53, Theorem 4.1.11], twice. Then, to see that $F$ is differentiable in $\nu$-measure with $D_{(\nu)}F = \eta$, let $\gamma \to 0$ through constants. Theorem 7.40 is proved. □

A natural question is whether, under the conditions of the previous theorem, the ordinary derivative $F'$ exists $\nu$-almost everywhere. Examples will be given in Proposition 7.63 of Section 7.5 showing that it may not.

Recall the classes $\mathcal{G}_\beta$ and $\Gamma_\beta$ of functions satisfying growth conditions, defined in (7.2) and in (7.3), respectively, and the classes $\mathcal{H}_1^{\mathrm{loc}}$ of functions that are 1-Hölder locally on $\mathbb{R}$, as defined in Definition 6.4. We will say that a function $f$ defined almost everywhere for $\mu$ is in $\mathcal{L}^r(\Omega, \mathcal{S}, \mu)$ if $f = g$ a.e. $(\mu)$ for some $g \in \mathcal{L}^r(\Omega, \mathcal{S}, \mu)$.

**Proposition 7.41.** *Let $(\Omega, \mathcal{S}, \mu)$ be a finite complete measure space, and let $1 \le p < s < \infty$. Let $F \in \mathcal{H}_1^{\mathrm{loc}}$, $F' \in \Gamma_{(s/p)-1}$ and let $G \in \mathcal{L}^s(\Omega, \mathcal{S}, \mu)$. If $F'{\circ}G$ exists a.e. $(\mu)$, then $F'{\circ}G \in \mathcal{L}^r(\Omega, \mathcal{S}, \mu)$ with $r$ defined by (7.20), the autonomous Nemytskii operator $N_F$ is Fréchet differentiable at $G$ from $L^s$ into $L^p$, and $DN_F(G) = M[F'{\circ}G]$.*

The proof of this proposition will use the following measurability fact.

**Lemma 7.42.** *Let $F$ be a real-valued Borel function on $\mathbb{R}$, and let $D$ be the set of all $x \in \mathbb{R}$ such that $F$ is differentiable at $x$. Then $D$ is u.m. (universally measurable) and for each $m = 1, 2, \ldots$, there is a u.m. function $D \ni x \mapsto \delta(x, m) > 0$ such that for each $t \in \mathbb{R}$ with $|t| < \delta(x, m)$,*

$$\big|F(x+t) - F(x) - F'(x)t\big| < |t|/m.$$

*Proof.* For $x, s, t \in \mathbb{R}$ such that $s \neq 0$ and $t \neq 0$, let

$$Q(F; x, s, t) := \Big|\frac{F(x+t) - F(x)}{t} - \frac{F(x+s) - F(x)}{s}\Big|.$$

For $k = 1, 2, \ldots$, and $m = 1, 2, \ldots$, let

$$\begin{aligned} B_{k,m} &:= B_{k,m}(F) \\ &:= \big\{(x, s, t) \in \mathbb{R}^3 \colon 0 < |s| < 1/k,\ 0 < |t| < 1/k,\ Q(F; x, s, t) > 1/m\big\}. \end{aligned}$$

Since $F$ is Borel measurable, each $B_{k,m}$ is a Borel set in $\mathbb{R}^3$. For $y := (x, s, t) \in \mathbb{R}^3$, let $\pi_1(y) := x$ and let $A_{k,m} := \{\pi_1(y) \colon y \in B_{k,m}\}$. Then each $A_{k,m}$ is analytic and so a u.m. set in $\mathbb{R}$ (e.g. [53, Theorem 13.2.6]), and $A_{k,m}$ decreases as $k$ increases. The complement $A_{k,m}^c$, that is the set of all $x \in \mathbb{R}$ such that $0 < |s| < 1/k$ and $0 < |t| < 1/k$ implies $Q(F; x, s, t) \leq 1/m$, also is a u.m. set. Now $F$ is differentiable at $x$ if and only if $x \in D := \bigcap_{m=1}^{\infty} \bigcup_{k=1}^{\infty} A_{k,m}^c$. Thus $D$ is a u.m. set. For any $x \in D$ and each $m = 1, 2, \ldots$, there is a least $k = k(x, m)$ such that $x \in A_{k,m}^c$. The function $D \ni x \mapsto k(x, m)$ is u.m. for each $m$ since $k(x, m) = k$ if and only if $x \in A_{k,m}^c \cap A_{k-1,m}$, where $A_{0,m} := \mathbb{R}$, for $k = 1, 2, \ldots$. Letting $\delta(x, m) := 1/k(x, m)$ for $x \in D$ and $m = 1, 2, \ldots$ completes the proof of the lemma. □

**Proof of Proposition 7.41.** Since $F \in \mathcal{H}_1^{\text{loc}}$, it is locally absolutely continuous and is an indefinite integral of $F'$ (e.g. Theorem 8.18 and Example 8.20 in [198]). Thus $F \in \mathcal{G}_{s/p}$, and so $N_F$ acts from $L^s$ into $L^p$ by Theorem 7.13(*a*). Notice that $r = (sp)/(s-p) > p \geq 1$ and $s/r = (s/p) - 1$. Since there is a $K < \infty$ such that $|F' \circ G| \leq K(1(\cdot) + |G|^{s/r})$ whenever $F' \circ G$ is defined, by Lemma 7.9, and so a.e. $(\mu)$, by the Minkowski inequality for integrals (e.g. [53, Theorem 5.1.5]), we have

$$\big\|F' \circ G\big\|_r \leq K \|1(\cdot) + |G|^{s/r}\|_r \leq K\big(\mu(\Omega)^{1/r} + \|G\|_s^{s/r}\big) < \infty.$$

Thus $M[F' \circ G]$ is a bounded linear operator from $L^s$ into $L^p$ by Lemma 7.37. Let $N$ be a $\mu$-null set such that $F'(G(\omega))$ exists for each $\omega \in \Omega \setminus N$. For $u \in \mathbb{R}$, let

$$\psi(u, \omega) := F(G(\omega) + u) - F(G(\omega)) - F'(G(\omega))u$$

if $\omega \in \Omega \setminus N$ and $\psi(u,\omega) := 0$ if $\omega \in N$. Then the Nemytskii operator $N_\psi$ acts from $L^s$ into $L^p$. To prove that $N_F$ is Fréchet differentiable at $G$ with derivative $M[F'{\circ}G]$ we need to show that

$$\|N_\psi g\|_p = o(\|g\|_s) \quad \text{as} \quad \|g\|_s \to 0. \tag{7.25}$$

Let $\epsilon = 1/m$ for some $m = 1, 2, \ldots$, and let $\delta(\cdot, m)$ be the function from Lemma 7.42. If $\omega \in \Omega \setminus N$, let $\delta_G(\omega) := \delta(G(\omega), m)$, and let $\delta_G(\omega) := 1$ if $\omega \in N$. Then for each $\omega \in \Omega$, $|\psi(u,\omega)| \le \epsilon|u|$ whenever $|u| < \delta_G(\omega)$. For each $g \in L^s$, let $A := A(g) := \{\omega\colon |g(\omega)| < \delta_G(\omega)\}$ and write $N_\psi(g) = R_1(g) + R_2(g) + R_3(g)$, where $R_1(g) := N_\psi(g)1_A$, $R_2(g) := (F'{\circ}G)g1_{A^c}$ and $R_3(g) := [F{\circ}(G+g) - F{\circ}G]1_{A^c}$. For $R_1$, we have

$$\|R_1(g)\|_p \le \epsilon\|g\|_p \le \epsilon\|g\|_s\mu(\Omega)^{1/r} \tag{7.26}$$

by Lemma 7.37. For $R_2$, using the same lemma we have

$$\|R_2(g)\|_p \le \|g\|_s\|(F'{\circ}G)1_{A^c}\|_r = o(\|g\|_s) \tag{7.27}$$

as $\|g\|_s \to 0$, because in this case $\mu(A^c) \to 0$, and so $\|(F'{\circ}G)1_{A^c}\|_r \to 0$ by dominated convergence. Turning to $R_3$, for each $\omega \in \Omega$, since $t \mapsto F(G(\omega) + tg(\omega))$ is Lipschitz and so absolutely continuous on $0 \le t \le 1$, we have

$$F(G(\omega) + g(\omega)) - F(G(\omega)) = g(\omega)\int_0^1 F'((G+tg)(\omega))\,dt.$$

Thus by Lemma 7.37, we have

$$\big\|R_3(g)\big\|_p \le \|g\|_s\Big(\int_{A^c}\Big|\int_0^1 F'{\circ}(G+tg)\,dt\Big|^r d\mu\Big)^{1/r}. \tag{7.28}$$

Applying Hölder's inequality (e.g. Lemma 7.29 with $k = 2$), the $\beta$ growth condition with $\beta = s/r = (s/p) - 1$ and the Minkowski inequality for integrals twice, we get

$$\begin{aligned}
\Big(\int_{A^c}&\Big|\int_0^1 F'{\circ}(G+tg)\,dt\Big|^r d\mu\Big)^{1/r}\\
&\le \Big(\int_0^1\int_{A^c}\big|F'{\circ}(G+tg)\big|^r d\mu\,dt\Big)^{1/r}\\
&\le K\Big(\int_{A^c}\Big|1(\cdot) + (|G| + |g|)^{s/r}\Big|^r d\mu\Big)^{1/r}\\
&\le K\Big\{\mu(A^c)^{1/r} + \big(\|G1_{A^c}\|_s + \|g\|_s\big)^{s/r}\Big\} = o(1)
\end{aligned}$$

as $\|g\|_s \to 0$, where $K := \|F'\|_{\mathcal{G}_{s/r}}$. This together with (7.28) yields that $\|R_3(g)\|_p = o(\|g\|_s)$ as $\|g\|_s \to 0$. Letting $\epsilon \downarrow 0$, by (7.26) and (7.27), we see that (7.25) holds, finishing the proof. □

**Proposition 7.43.** *Let $F$ be a u.m. function from $\mathbb{R}$ into $\mathbb{R}$ and $\nu$ a finite Borel measure on $\mathbb{R}$.*

*(a) If the derivative $F'(x)$ exists for $\nu$-almost all $x$, then a derivative in measure $D_{(\nu)}F$ exists and equals $F'(x)$ for $\nu$-almost all $x$.*
*(b) If $\nu$ has an atom at a point $x$, so that $\nu(\{x\}) > 0$, and $D_{(\nu)}F$ exists, then $F'(x)$ exists and equals $D_{(\nu)}F(x)$.*

*Proof.* For $(a)$, the set $\mathcal{D}_F := \{x\colon\ F'(x) \text{ exists}\}$ is u.m. by Lemma 7.42 and $\nu(\mathbb{R} \setminus \mathcal{D}_F) = 0$. Let $\eta(x) := F'(x)$ for $x \in \mathcal{D}_F$ and $\eta(x) := 0$ otherwise. Then since pointwise convergence a.e. $(\nu)$ implies convergence in $\nu$-measure, $\eta$ is a derivative of $F$ in $\nu$-measure, and any other function $\rho$ satisfying the definition of $D_{(\nu)}F$ must equal $\eta$ and therefore $F'(x)$ for $\nu$-almost all $x$.

For $(b)$, let $\nu(\{x\}) > 0$ and $\eta = D_{(\nu)}F$. Convergence in measure implies pointwise convergence at an atom, so $F'(x)$ must exist and equal $D_{(\nu)}F(x)$. □

**Corollary 7.44.** *If $F\colon \mathbb{R} \to \mathbb{R}$ is everywhere differentiable, then for every finite Borel measure $\nu$ on $\mathbb{R}$, $D_{(\nu)}F$ exists and can be taken equal to $F'$.*

*Proof.* Since $F'(x)$ exists for all $x$, $F$ is continuous and so Borel measurable and u.m. The rest follows from Proposition 7.43$(a)$. □

**Proposition 7.45.** *Let $(\Omega, \mathcal{S}, \mu)$ be finite and complete, let $1 \le p < s < \infty$, and let $F\colon \mathbb{R} \to \mathbb{R}$ be differentiable everywhere. The following statements are equivalent:*

*(a) $F' \in \mathcal{G}_{(s/p)-1}$;*
*(b) the autonomous Nemytskii operator $N_F$ is everywhere differentiable from $L^s$ into $L^p$ and $DN_F(G) = M[F'{\circ}G]$ for each $G \in L^s$.*

*Proof.* Assuming $(a)$, since $F'$ is locally bounded, $F \in \mathcal{H}_1^{\text{loc}}$ by the mean value theorem of calculus. Since $\mathcal{G}_\beta \subset \varGamma_\beta$ and $F'{\circ}G$ is defined a.e. $(\mu)$ for each $G \in L^s$, $(b)$ follows from Proposition 7.41.

Suppose that $(b)$ holds. Then for $r = sp/(s-p)$, $F'{\circ}G \in L^r$ for each $G \in L^s$ by Lemma 7.37, and so $F' \in \mathcal{G}_{(s/p)-1}$ by Theorem 7.13$(b)$ with $r$ instead of $p$. Thus $(a)$ holds, proving the proposition. □

**Proposition 7.46.** *Let $(\Omega, \mathcal{S}, \mu)$ be finite, complete, and nonatomic, and $1 \le p < s < \infty$. Let a Borel measurable $F\colon \mathbb{R} \to \mathbb{R}$ be such that the autonomous Nemytskii operator $N_F$ acts from $L^s$ into $L^p$. If $N_F$ is Fréchet differentiable on a nonempty open set $V$, then $F$ is everywhere differentiable on $\mathbb{R}$, it is locally Lipschitz with $F' \in \mathcal{G}_{(s/p)-1}$, $N_F$ is Fréchet differentiable on all of $L^s$, and the derivative operator $DN_F(G)(\cdot)$ at any $G \in L^s$ is the multiplication operator $M[F'{\circ}G](\cdot)$ induced by the function $F'{\circ}G$, which is in $L^r$ with $r$ defined by (7.20).*

*Proof.* To show that $F$ is differentiable, let $u \in \mathbb{R}$ and let $G \in V$. For $A \in \mathcal{S}$ with $\mu(A) > 0$, let $f : \Omega \to \mathbb{R}$ be such that $f = u$ on $A$ and $f = G$ elsewhere. Since $\mu$ is nonatomic, by Proposition A.1, one can choose $A$ such that $f \in V$, and so $N_F$ is Fréchet differentiable at $f$. Since $\mu \circ f^{-1}$ has an atom at $u$, $F'(u)$ exists by Theorem 7.40 and Proposition 7.43($b$). Since $u$ is arbitrary, $F$ is differentiable everywhere. By Theorem 7.40 and Corollary 7.44, for each $f \in V$, the derivative operator $DN_F(f)$ equals $M[F' \circ f]$ and $F' \circ f \in L^r$ with $r = (sp)/(s-p)$ as in (7.20). Thus by Theorem 7.13($b$) with $r$ instead of $p$, $F' \in \mathcal{G}_{(s/p)-1}$.

By Proposition 7.21($a$), $N_F$ acts from $L^s$ into $L^p$. To show that $N_F$ is Fréchet differentiable on all of $L^s$, let $G \in V$ and let $\rho > 0$ be such that $G + B_\rho \subset V$, where $B_\rho = \{f \in L^s : \|f\|_s < \rho\}$. For $(u, \omega) \in \mathbb{R} \times \Omega$, let

$$\phi(u, \omega) := F(G(\omega) + u) - F(G(\omega)).$$

Then $\phi$ is a Shragin function on $\mathbb{R} \times \Omega$ and the Nemytskii operator $N_\phi$ acts from $L^s$ into $L^p$. Recalling Definition 6.8 of $\phi_u^{(1)}$, we have

$$\begin{aligned} &N_\phi(g_0 + g) - N_\phi(g_0) - N_{\phi_u^{(1)}}(g_0)g \\ &\quad = N_F(G + g_0 + g) - N_F(G + g_0) - F' \circ (G + g_0)g \end{aligned} \tag{7.29}$$

for any $g_0, g \in L^s$. Let $g_0 \in B_\rho$ and let $g \in L^s$ be such that $\|g\| < \rho - \|g_0\|$, and so $g_0 + g \in B_\rho$. Then $G + g_0 \in V$, $G + g_0 + g \in V$, and so by (7.29)

$$\begin{aligned} &\|N_\phi(g_0 + g) - N_\phi(g_0) - N_{\phi_u^{(1)}}(g_0)g\|_p \\ &\quad = \|\mathrm{Rem}_{N_F}(G + g_0, g)\|_p = o(\|g\|_s) \end{aligned} \tag{7.30}$$

as $\|g\|_s \to 0$, where the remainder $\mathrm{Rem}_{N_F}$ is defined as in (7.19). Thus $N_\phi$ is Fréchet differentiable at $g_0 \in B_\rho$ and the derivative operator $DN_\phi(g_0)(\cdot)$ is the multiplication operator from $L^s$ into $L^p$ induced by $N_{\phi_u^{(1)}}(g_0) = F' \circ (G + g_0)$. To show that the derivative operator $DN_F$ extends from $V$ to all of $L^s$, let $f \in L^s$ be such that $f \notin G + B_\rho$, and so $\|f - G\|_s \geq \rho$. By Proposition A.9 with $h = f - G$ and $c = \rho/2$, there exists a partition $\{A_i\}_{i=0}^m$ of $\Omega$ into measurable sets such that $\|(f - G)1_{A_i}\|_s \leq \rho/2$ for $i = 0, \ldots, m$. For $g \in L^s$ such that $\|g\|_s < \rho/2$, by (7.29) with $g_0 = f - G$, we have

$$\begin{aligned} &\|N_F(f + g) - N_F(f) - (F' \circ f)g\|_p^p \\ &\quad = \|N_\phi((f - G) + g) - N_\phi(f - G) - N_{\phi_u^{(1)}}(f - G)g\|_p^p \\ &\quad = \sum_{i=0}^m \|\mathrm{Rem}_{N_\phi}((f - G)1_{A_i}, g1_{A_i})\|_p^p = o(\|g\|_s^p) \end{aligned}$$

as $\|g\|_s \to 0$ by (7.30) with $g_0 = (f - G)1_{A_i}$ and with $g1_{A_i}$ instead of $g$. Thus $N_F$ is Fréchet differentiable on all of $L^s$ and the derivative operator $DN_F(f)(\cdot)$ at $f \in L^s$ is the multiplication operator $M[F' \circ f](\cdot)$ induced by the function $F'\circ f \in L^r$ with $r$ defined by (7.20). The proof of the proposition is complete. □

**Corollary 7.47.** *Let $(\Omega, \mathcal{S}, \mu)$ be finite, complete, and nonatomic, and $1 \le p < s < \infty$. Let a Borel measurable $F\colon \mathbb{R} \to \mathbb{R}$ be such that the autonomous Nemytskii operator $N_F$ acts from $L^s$ into $L^p$. The following statements are equivalent:*

*(a) $N_F$ is Fréchet differentiable on a nonempty open set $V$ in $L^s$;*
*(b) $N_F$ is Fréchet differentiable everywhere on $L^s$, $F$ is differentiable everywhere on $\mathbb{R}$, and for each $G \in L^s$, $DN_F(G) = M[F'\circ G]$;*
*(c) $F$ is differentiable everywhere on $\mathbb{R}$ and $F' \in \mathcal{G}_{(s/p)-1}$.*

*Proof.* Statement $(b)$ and thus $(a)$ follow from $(c)$ by the implication $(a) \Rightarrow (b)$ of Proposition 7.45. By Proposition 7.46, $(a)$ implies $(b)$. Then $(b)$ implies $(c)$ since $(b)$ implies $(a)$ in Proposition 7.45, proving the corollary. □

By Theorem 7.24 and Proposition 7.28, the autonomous Nemytskii operator $N_F$ has a Hölder property at $G$ of optimal order $s/[p(1+s)]$ from $L^s$ into $L^p$, uniformly for $\|F\|_{(p)} \le 1$, if and only if $G \in \mathcal{D}_\lambda$, that is, $\mu\circ G^{-1}$ has a bounded density with respect to Lebesgue measure $\lambda$. Such $G$ are dense in $L^s(\Omega, \mathcal{S}, \mu)$ for any nonatomic measure space $(\Omega, \mathcal{S}, \mu)$ by Proposition A.6. One might ask whether Proposition 7.46 extends to $F$ such that $N_F$ is Fréchet differentiable at all such $G$. The answer is negative: $F$ need not be locally Lipschitz, as the following proposition shows for $\alpha < 1$.

**Proposition 7.48.** *Let $(\Omega, \mathcal{S}, \mu)$ be finite, complete, and nonatomic, let $1 \le p < s < \infty$, let $F_\alpha(u) \equiv |u|^\alpha$ with $0 < \alpha \le 1$, and let $G \in \mathcal{D}_\lambda \cap \mathcal{L}^s$. If $\alpha > 1 + (1/s) - (1/p)$, then the autonomous Nemytskii operator $N_{F_\alpha}$ is Fréchet differentiable at $G$ from $L^s$ into $L^p$, and there is a constant $K < \infty$ such that for each $g \in L^s$ with $\|g\|_s \le 1$,*

$$\|\mathrm{Rem}_N(G, g)\|_p \le K\|g\|_s^\beta, \tag{7.31}$$

*where $N = N_{F_\alpha}$ and $\beta := [\alpha + (1/p)]s/(1+s) > 1$.*

*Proof.* Since $F_\alpha \in \mathcal{G}_\alpha \subset \mathcal{G}_{s/p}$, $N_{F_\alpha}$ acts from $L^s$ into $L^p$ by Theorem 7.13$(a)$. For $u \ne 0$, $|F'_\alpha(u)| = \alpha|u|^{\alpha-1}$, and $F'_\alpha(u)$ is undefined for $u = 0$. Let $\xi_G$ be a density of $\mu\circ G^{-1}$ and let $M := \|\xi_G\|_\infty < \infty$. If $\alpha < 1$, by the image measure theorem, we have for any $r > 0$,

$$\begin{aligned}\int_\Omega |F'_\alpha\circ G|^r \,d\mu &= \alpha^r \int_{\mathbb{R}} |x|^{(\alpha-1)r}\xi_G(x)\,dx \\ &\le \alpha^r\mu(\{\omega\colon |G(\omega)| \ge 1\}) + \alpha^r M\int_{\{|x|\le 1\}} |x|^{(\alpha-1)r}\,dx.\end{aligned}$$

For $r := (sp)/(s-p)$, $(\alpha-1)r > -1$, and the right side is finite since $\mu$ is finite. If $\alpha = 1$, then the left side is finite since $\mu$ is finite and $F_1' = \pm 1(\cdot)$. Thus $F_\alpha' \circ G \in \mathcal{L}^r$, and so $\mathcal{D}(g) := (F_\alpha' \circ G)g \in L^p$ for each $g \in L^s$ by Lemma 7.37. Also, for each $g \in L^s$, let $\Delta(g) := F_\alpha \circ (G+g) - F_\alpha \circ g$ and $R(g) := \Delta(g) - \mathcal{D}(g)$. We will show that for a finite constant $K < \infty$, $\|R(g)\|_p \le K\|g\|_s^\beta$ for each $g \in L^s$ such that $\|g\|_s \le 1$. From this bound it will follow that $N_{F_\alpha}$ is Fréchet differentiable at $G$ from $L^s$ into $L^p$, the derivative operator $DN_{F_\alpha}(G)(g)$ equals $\mathcal{D}(g)$, and the stated bound (7.31) holds.

We have $1/p = (1/r) + (1/s)$ and $r \ge p \ge 1$. Let $\rho := s/(1+s)$ and $\epsilon := \epsilon(g) := \|g\|_s^\rho \le 1$. Thus $\beta = \rho[\alpha + (1/p)]$. Also, let

$$\begin{cases} A := \{\omega\colon\ |G(\omega)| \le \epsilon\}, \\ B := \{\omega\colon\ |G(\omega)| > \epsilon,\ |g(\omega)| > \epsilon/2\}, \\ C := \{\omega\colon\ |G(\omega)| > \epsilon,\ |g(\omega)| \le \epsilon/2\}, \end{cases} \quad \text{and} \quad \begin{cases} T_1 := \|\Delta(g)1_A\|_p, \\ T_2 := \|\mathcal{D}(g)1_A\|_p, \\ T_3 := \|\Delta(g)1_B\|_p, \\ T_4 := \|\mathcal{D}(g)1_B\|_p, \\ T_5 := \|R(g)1_C\|_p. \end{cases}$$

Thus $\|R(g)\|_p \le \sum_{i=1}^5 T_i$. To bound $T_1$, we have $\mu(A) = \int_{\{|x|\le\epsilon\}} \xi_G(x)\,dx \le 2M\epsilon = 2M\|g\|_s^\rho$, where $M = \|\xi_G\|_\infty$. Since $|F_\alpha(u) - F_\alpha(v)| \le |u-v|^\alpha$ for all $u, v \in \mathbb{R}$, by Lemma 7.29 with $k = 2$ it then follows that

$$T_1 \le \Big(\int_\Omega |g|^{\alpha p} 1_A\,d\mu\Big)^{1/p} \le \|g\|_s^\alpha \mu(A)^{(1/p)-(\alpha/s)} \le K_1\|g\|_s^\beta, \tag{7.32}$$

where $K_1 := (2M)^{(1/p)-(\alpha/s)}$. For $T_2$ and $\alpha < 1$, since $\|\xi_G\|_\infty = M$, we have

$$\int_A |F_\alpha' \circ G|^r\,d\mu = \alpha^r \int_{\{|x|\le\epsilon\}} |x|^{r(\alpha-1)} \xi_G(x)\,dx \le \frac{2\alpha^r M}{r(\alpha-1)+1} \|g\|_s^{[r(\alpha-1)+1]\rho}.$$

Thus by Lemma 7.29 with $k = 2$, we have

$$T_2 \le \|g\|_s \big\|(F_\alpha' \circ G)1_A\big\|_r \le K_2\|g\|_s^\beta$$

for a constant $K_2 = K_2(\alpha, r, M)$. The same holds when $\alpha = 1$. For $T_3$, since $\epsilon = \|g\|_s^\rho$, we have

$$\mu(B) \le \mu(\{\omega\colon\ |g(\omega)| > \epsilon/2\}) \le (2/\epsilon)^s \|g\|_s^s = 2^s\|g\|_s^\rho. \tag{7.33}$$

Thus as in (7.32), we get the bound

$$T_3 \le \Big(\int_\Omega |g|^{\alpha p} 1_B\,d\mu\Big)^{1/p} \le \|g\|_s^\alpha \mu(B)^{(1/p)-(\alpha/s)} \le K_3\|g\|_s^\beta,$$

where $K_3 := 2^{(s/p)-\alpha}$. For $T_4$ and $\alpha < 1$, we have

$$\int_B |F_\alpha' \circ G|^r d\mu = \alpha^r \int_{B\cap\{|G|<1\}} |G|^{r(\alpha-1)}\,d\mu + \alpha^r \int_{B\cap\{|G|\ge1\}} |G|^{r(\alpha-1)}\,d\mu$$

$$\leq \alpha^r \Big[ \epsilon^{r(\alpha-1)} + 1 \Big] \mu(\{\omega\colon\, |g(\omega)| > \epsilon/2\}) \leq \alpha^r 2^s \Big[ \|g\|_s^{[r(\alpha-1)+1]\rho} + \|g\|_s^{\rho} \Big]$$

by the second inequality and the equality in (7.33). Thus by Lemma 7.29 with $k = 2$, it follows that

$$T_4 \leq \|g\|_s \big\| (F'_\alpha \circ G) 1_B \big\|_r \leq \alpha 2^{s/r} \Big[ \|g\|_s^{1+[\alpha-1+(1/r)]\rho} + \|g\|_s^{1+\rho/r} \Big] \leq K_4 \|g\|_s^{\beta}$$

since $\|g\|_s \leq 1$ and $1 + \rho/r > \beta$, where $K_4 := \alpha 2^{1+s/r}$. For $\alpha = 1$, since $|F'_1| = 1(\cdot)$, by Lemma 7.29 with $k = 2$ and (7.33), we have

$$T_4 \leq \|g\|_s \mu(B)^{1/r} \leq 2^{s/r} \|g\|_s^{1+\rho/r} \leq K_4 \|g\|_s^{\beta},$$

where $K_4 := 2^{s/r}$. Finally, for $T_5$, by Taylor's theorem with remainder, for each $x, y \in \mathbb{R}$ such that $|x| > \epsilon$ and $|y| \leq \epsilon/2$, we have

$$\big| F_\alpha(x+y) - F_\alpha(x) - F'_\alpha(x) y \big| \leq \frac{y^2}{2} \sup_{0 \leq t \leq 1} \big| F''_\alpha(x + ty) \big|.$$

For $\alpha = 1$, the left side is zero, and so we can assume that $\alpha < 1$. It then follows that for each $\omega \in C$,

$$\begin{aligned} |R(g)(\omega)| &\leq \frac{\alpha(1-\alpha)}{2} \big[ g(\omega) \big]^2 \sup_{0 \leq t \leq 1} \big| G(\omega) + t g(\omega) \big|^{\alpha-2} \\ &\leq \frac{\alpha(1-\alpha)\epsilon^2}{8} \big[ |G(\omega)| - \epsilon/2 \big]^{\alpha-2}. \end{aligned}$$

Since $\epsilon = \|g\|_s^{\rho} \leq 1$ and $\|\xi_G\|_\infty = M$, we have

$$\begin{aligned} &\Big( \int_{\{\epsilon \leq |G| < 1\}} \big[ |G| - \epsilon/2 \big]^{p(\alpha-2)} \, d\mu \Big)^{1/p} \\ &\quad = \Big( \int_{\{\epsilon \leq |x| \leq 1\}} \big[ |x| - \epsilon/2 \big]^{p(\alpha-2)} \xi_G(x) \, dx \Big)^{1/p} \\ &\quad \leq \Big( \frac{2^{(2-\alpha)p} M}{p(2-\alpha) - 1} \Big)^{1/p} \|g\|_s^{\rho[\alpha-2+(1/p)]}. \end{aligned}$$

Thus, splitting the integration in $T_5$ over $C \cap \{|G| \geq 1\}$ and $C \cap \{|G| < 1\}$, it follows that

$$T_5 \leq \frac{\alpha(1-\alpha)}{8} \Big[ 2^{\alpha-2} \mu(\Omega)^{1/p} \|g\|_s^{2\rho} + \Big( \frac{2^{(2-\alpha)p} M}{p(2-\alpha) - 1} \Big)^{1/p} \|g\|_s^{\beta} \Big] \leq K_5 \|g\|_s^{\beta},$$

since $\|g\|_s \leq 1$ and $2\rho > \beta$, for a suitable constant $K_5$. Summing the bounds of $T_i$, $i = 1, \ldots, 5$, it follows that for $K := \sum_{i=1}^5 K_i$, $\|R(g)\|_p \leq K \|g\|_s^{\beta}$ whenever $\|g\|_s \leq 1$. The proof of the proposition is complete. □

The following shows that the order of the remainder bound (7.31) cannot be improved in general.

**Proposition 7.49.** *Let $(\Omega, \mathcal{S}, \mu) = ([0,1], \mathcal{B}, \lambda)$, $1 \le p < s < \infty$, $F_\alpha(u) \equiv |u|^\alpha$, $u \in \mathbb{R}$, with $0 < \alpha < \infty$ and $G(x) \equiv x$, $x \in [0,1]$. If the autonomous Nemytskii operator $N_{F_\alpha}$ is Fréchet differentiable at $G$ from $L^s$ into $L^p$, and there are constants $K < \infty$, $0 < \rho \le 1$ such that (7.31) holds for all $g \in L^s$ with $\|g\|_s \le \rho$, then $\beta \le [\alpha + (1/p)]s/(1+s)$.*

*Proof.* By Theorem 7.40, we have that the derivative operator $(DN_{F_\alpha})(G) = M[(D_{(\lambda)}F_\alpha)\circ G]$, where we take $D_{(\lambda)}F_\alpha(u) = \alpha|u|^\alpha/u$ for $u \neq 0$ by Proposition 7.43($a$). Thus for each $g \in L^s$, we have

$$R(g) := \mathrm{Rem}_{N_{F_\alpha}}(G,g) = F_\alpha\circ(G+g) - F_\alpha\circ G - [(D_{(\lambda)}F_\alpha)\circ G]g.$$

For $\delta > 0$, let $g_\delta(x) := -2x1_{[0,\delta]}(x)$, $x \in [0,1]$. Then $\|g_\delta\|_s = 2\delta^{1+(1/s)}/(1+s)^{1/s}$ and $|R(g_\delta)| = 2\alpha F_\alpha 1_{[0,\delta]}$. Thus there is a constant $C = C(\alpha, s, p) < \infty$ such that for each $\delta > 0$,

$$\|R(g_\delta)\|_p = 2\alpha\delta^{\alpha+(1/p)}/(1+\alpha p)^{1/p} = C\|g_\delta\|_s^{[\alpha+(1/p)]s/(1+s)}.$$

Since $\|g_\delta\|_s \to 0$ as $\delta \downarrow 0$, the stated conclusion follows. □

The next fact shows that the sufficient lower bound for $\alpha$ in Proposition 7.48 is necessary:

**Proposition 7.50.** *Let $(\Omega, \mathcal{S}, \mu)$ be nonatomic, finite, and complete, let $1 \le p < s < \infty$, and let $F_\alpha(u) \equiv |u|^\alpha$, $u \in \mathbb{R}$, for some $\alpha > 0$. Let $G \in \mathcal{D}_\lambda \cap \mathcal{L}^s$ be such that for some $C > 0$ and $\delta > 0$, $\xi_G(x) \ge C$ for all $x \in [-\delta, \delta]$. Such $G$ exist. If the autonomous Nemytskii operator $N_{F_\alpha}$ is Fréchet differentiable at $G$ from $L^s$ into $L^p$, then $1 + (1/s) - (1/p) < \alpha \le s/p$.*

*Proof.* Functions $G$ satisfying the hypotheses exist by Proposition A.5. The interval $(1 + (1/s) - (1/p), s/p]$ is nonempty if and only if $s > p$. Since $N_{F_\alpha}$ acts from $L^s$ into $L^p$ by Proposition 7.31 and $F_\alpha \notin \mathcal{G}_\beta$ for $\beta < \alpha$, we have $\alpha \le s/p$ by Theorem 7.13($c$). If $\alpha \ge 1$, then the proof is complete. Suppose that $\alpha < 1$. By Theorem 7.40, the derivative operator $(DN_{F_\alpha})(G)$ equals $M[(D_{(\lambda)}F_\alpha)\circ G]$, where $D_{(\lambda)}F_\alpha(u) = \alpha|u|^\alpha/u$ if $u \neq 0$ by Proposition 7.43($a$). Letting $r := (sp)/(s-p)$, by Lemma 7.37, $(D_{(\lambda)}F_\alpha)\circ G \in L^r$. Thus by the image measure theorem, we have

$$\infty > \int_\Omega |D_{(\lambda)}F_\alpha\circ G|^r \, d\mu \ge \alpha^r \int_{|x|\le\delta} |x|^{(\alpha-1)r}\xi_G(x)\, dx \ge 2C\alpha^r \int_0^\delta x^{(\alpha-1)r}\, dx,$$

and the conclusion follows. □

The following is a consequence of Propositions 7.48 and 7.50:

**Corollary 7.51.** *Let* $(\Omega, \mathcal{S}, \mu)$ *be nonatomic, finite, and complete, let* $1 \leq p < s < \infty$, *and let* $F_\alpha(u) \equiv |u|^\alpha$, $u \in \mathbb{R}$, *for some* $0 < \alpha < 1$. *The following statements are equivalent:*

(*a*) *for each* $G \in \mathcal{D}_\lambda \cap \mathcal{L}^s$, *the autonomous Nemytskii operator* $N_{F_\alpha}$ *is Fréchet differentiable at* $G$ *from* $L^s$ *into* $L^p$;
(*b*) $[\alpha + (1/p)]s/(1+s) > 1$.

The following applies if $F$ is the distribution function $P((-\infty, x])$ of a probability measure $P$ with support in $[a, b]$ and density in $\mathcal{H}_\beta([a, b]; \mathbb{R})$. Recall Definition 6.5 of the class $\mathcal{H}_{1+\beta}$.

**Proposition 7.52.** *Let* $(\Omega, \mathcal{S}, \mu)$ *be a finite measure space,* $-\infty < a < b < \infty$, $1 \leq p < s < \infty$, *and*

$$\beta := \frac{s(1+p)}{p(1+s)} - 1 = \frac{s-p}{p(1+s)}.$$

*Let* $F\colon \mathbb{R} \to \mathbb{R}$ *be a function whose restriction to* $[a, b]$ *is in* $\mathcal{H}_{1+\beta}([a, b]; \mathbb{R})$, $F(x) = F(a)$ *for all* $x \leq a$, *and* $F(x) = F(b)$ *for all* $x \geq b$. *Let* $G$ *be a measurable function from* $\Omega$ *into* $[a, b]$ *such that for some* $K < \infty$,

$$(\mu\circ G^{-1})\big([a, a+x] \cup [b-x, b]\big) \leq Kx \tag{7.34}$$

*for all* $x \geq 0$. *Then the autonomous Nemytskii operator* $N_F$ *acts from* $L^s$ *into* $L^p$, *it is Fréchet differentiable at* $G$, *and the remainder in the differentiation has the bound*

$$\|\mathrm{Rem}_{N_F}(G, g)\|_p = O(\|g\|_s^{1+\beta}) \tag{7.35}$$

*as* $\|g\|_s \to 0$.

*Proof.* Clearly $0 < \beta < 1$. Since $F$ is bounded and $\mu$ is finite, $N_F$ acts from $L^s$ into $L^p$. To prove its Fréchet differentiability at $G$ we can assume $a = 0$ and $b = 1$, possibly changing $K$ in (7.34). If $h$ is such that $G + h$ has values in $[0, 1]$ then by the mean value theorem of calculus

$$|\mathrm{Rem}_{N_F}(G, h)| = |F'\circ(G + th) - F'\circ G|\cdot|h| \leq \|F'\|_{(\mathcal{H}_\alpha)}|h|^{1+\beta}$$

for a variable $t$ with $0 \leq t \leq 1$. Then by Hölder's inequality (e.g. Lemma 7.29 with $k = 2$),

$$\int_\Omega |\mathrm{Rem}_{N_F}(G, h)|^p \, d\mu \leq \|F'\|_{(\mathcal{H}_\alpha)}^p \mu(\Omega)^{1-p(1+\beta)/s} \|h\|_s^{(1+\beta)p}. \tag{7.36}$$

Now let $g$ be any function in $L^s(\Omega, \mathcal{S}, \mu)$. Let $A := \{G + g \leq 0\}$, $B := \{0 < G + g < 1\}$, and $C := \{G + g \geq 1\}$. Let $g_1 := 1_A\cdot(G + g)$, $g_3 := 1_C\cdot(G + g - 1)$, and $g_2 := g - g_1 - g_3$. Then we have $G + g_2 = \max\{0, \min\{G + g, 1\}\}$, $\|g_1\|_s \leq \|g\|_s$, and $\|g_3\|_s \leq \|g\|_s$. Thus by (7.36) for $h = g_2$,

$$\|\mathrm{Rem}_{N_F}(G,g_2)\|_p \le C_1\|g_2\|_s^{1+\beta} \le 3^{1+\beta}C_1\|g\|_s^{1+\beta} \tag{7.37}$$

for some constant $C_1$. We have $G+g_2 = G+g-g_1 \equiv 0$ on $A$ and $G+g_2 = G+g-g_3 \equiv 1$ on $C$. Thus $F\circ(G+g) \equiv F\circ(G+g_2)$, on $A$ since $F(y)=F(0)$ for $y \le 0$, on $C$ since $F(y)=F(1)$ for $y \ge 1$, and on $B$ since $g_2 = g$ there. It follows that

$$\mathrm{Rem}_{N_F}(G,g) = \mathrm{Rem}_{N_F}(G,g_2) - (F'\circ G)\cdot(g_1+g_3). \tag{7.38}$$

Again by Hölder's inequality,

$$\int_\Omega |g_1|^p\,\mathrm{d}\mu = \int_A (-G-g)^p\,\mathrm{d}\mu \le \|g\|_s^p\mu(A)^{1-p/s}. \tag{7.39}$$

For any $\delta > 0$, let $m := \sup\{\mu(A)\colon\ \|g\|_s \le \delta\}$. On $A$ we have $0 \ge -G \ge g$. If we replace $g$ by $-G$ on $A$, then $\|g\|_s$ is not increased and $A$ is unchanged. Thus in finding $m$ we can assume $g=-G$ on $A$. Then, $m$ is attained by $\mu(A)$ under the constraint $\int_A G^s\,\mathrm{d}\mu \le \delta^s$ when $A$ is a set such that for some $t \ge 0$, $\{G<t\} \subset A \subset \{G \le t\}$ by the Neyman–Pearson lemma (e.g. Lehmann [137, Theorem 3.1, p. 74]), applied to $\mu$ and $G^s\mu$ normalized to be probability measures. Let $\mu_A(E) := \mu(A\cap E)$ for all $E \in \mathcal{S}$, $\nu := \mu_A\circ G^{-1}$, and $L(x) := \nu([0,x])$. Then $\int_A G^s\,\mathrm{d}\mu = \int_0^t x^s\,\mathrm{d}L(x)$ and $L(t)=\mu(A)$. We can assume that $t$ is the (unique) smallest number for which the latter holds. For all $x \ge 0$, we have $L(x) \le Kx$ by (7.34) and so $L(x) \le \min\{Kx,\mu(A)\}$. Integrating by parts, and then writing $\int_0^t = \int_0^{\mu(A)/K} + \int_{\mu(A)/K}^t$, it follows that

$$\|g\|_s^s \ge \int_0^t x^s\,\mathrm{d}L(x) = t^s\mu(A) - \int_0^t L(x)sx^{s-1}\,\mathrm{d}x \ge \mu(A)^{1+s}/K^s(1+s).$$

Thus $\mu(A) \le C_2\|g\|_s^{s/(1+s)}$ for some constant $C_2$. By (7.39) we then have the bound

$$\|g_1\|_p \le C_2^{(s-p)/ps}\|g\|_s^{1+\beta}.$$

The function $g_3$ and the set $C$ can be treated symmetrically, interchanging $G$ with $1-G$ and $g$ with $-g$, giving the same bound. Thus

$$\|(F'\circ G)(g_1+g_3)\|_p \le 2\|F'\|_{\sup}C_2^{(s-p)/sp}\|g\|_s^{1+\beta}.$$

This together with (7.37) and (7.38) yields Fréchet differentiability with the remainder bound (7.35), proving the proposition. □

## 7.4 Higher Order Differentiability and Finite Taylor Series

The $k$th order differential of an autonomous Nemytskii operator from $L^s$ to $L^p$ will be shown under some conditions to be a $k$-linear multiplication operator $M^k[h]$ induced by a function $h \in \mathcal{L}^{r_k}$, where for $1 \le p \le kp \le s < \infty$,

$$r_k := r_k(s,p) := (sp)/(s-kp), \tag{7.40}$$

defined as $+\infty$ if $kp = s$. (This was already done when $k = 1$ in Definition 7.20 of $r(s,p) \equiv r_1(s,p)$.)

Recall that for a function $h\colon \Omega \to \mathbb{R}$ and a positive integer $k$, if for any $g_1,\ldots,g_k \in L^s$, the pointwise product $M^k[h](g_1,\ldots,g_k) = hg_1\cdots g_k$ is in $L^p$, then $M^k[h]$ is the $k$-linear multiplication operator from $(L^s)^k := L^s \times \cdots \times L^s$ ($k$ times) into $L^p$ induced by $h$. The following extends Lemma 7.37 to $k$-linear multiplication operators.

**Lemma 7.53.** *Let $(\Omega,\mathcal{S},\mu)$ be $\sigma$-finite, let $h\colon \Omega \to \mathbb{R}$ be $\mu$-measurable, let $k$ be a positive integer, and let $1 \le p \le kp \le s < \infty$. Then the following three statements are equivalent:*

(*a*) *for any $g_1,\ldots,g_k \in L^s$, $hg_1\cdots g_k \in L^p$;*
(*b*) *$h$ induces a bounded $k$-linear multiplication operator $M^k[h]$ from $(L^s)^k$ into $L^p$;*
(*c*) *$h \in L^{r_k}$ with $r_k$ defined by* (7.40).

*If these conditions hold, then the operator norm of $M^k[h]$ in* (*b*) *is $\|h\|_{r_k}$.*

*If $1 \le s < kp$ and $(\Omega,\mathcal{S},\mu)$ is nonatomic, then* (*a*) *and* (*b*) *are equivalent to $h = 0$ in $L^0$.*

*Proof.* Clearly ($b$) implies ($a$). To show that ($a$) implies ($c$), take any $f \in L^{s/k}$. Let $g_1 = g_2 = \cdots = g_k = |f|^{1/k} \in L^s$. By ($a$), $h|f| \in L^p$, and so $hf \in L^p$. It follows by Lemma 7.37 with $s/k$ in place of $s$ that $h \in L^{r_k}$, i.e. ($c$) holds.

To show that ($c$) implies ($b$), we apply Lemma 7.29 with $k+1$ instead of $k$, $p_j = s$ for $j = 1,\ldots,k$, and $p_{k+1} = r_k$. By (7.16), it follows that

$$\|M^k[h]\| = \sup\bigl\{\|M^k[h](g_1,\ldots,g_k)\|_p\colon\ \|g_j\|_s \le 1,\ j = 1,\ldots,k\bigr\} \le \|h\|_{r_k},$$

completing the equivalence proof. To prove the reverse inequality let $h \in L^{r_k}$ and $g_j := |h|^{r_k/s}/\|h\|_{r_k}^{r_k/s}$ for $j = 1,\ldots,k$. Then $g_j \in L^s$, $\|g_j\|_s = 1$, and $\|M^k[h](g_1,\ldots,g_k)\|_p = \|h\|_{r_k}$, completing the proof for $kp \le s$.

If $kp > s$ and $(\Omega,\mathcal{S},\mu)$ is nonatomic, then $h = 0$ in $L^0$ clearly implies ($a$) and ($b$). Conversely, suppose ($a$) holds. Then as in the case $kp \le s$, we have $fh \in L^p$ for any $f \in L^{s/k}$. Applying the last statement of Lemma 7.37 with $s/k < p$ in place of $s$, we get $h = 0$ in $L^0$, finishing the proof of the lemma. $\square$

The next fact extends Proposition 7.46 to $k$th differentials.

**Proposition 7.54.** *Let $(\Omega,\mathcal{S},\mu)$ be finite, complete, and nonatomic, $k$ a positive integer and $1 \le p \le kp < s < \infty$. Let a Borel measurable function $F\colon \mathbb{R} \to \mathbb{R}$ be such that the autonomous Nemytskii operator $N_F$ acts from $L^s$ into $L^p$. If $N_F$ is $k$ times Fréchet differentiable on a nonempty open set $V$, then $F$ is $k$ times differentiable on $\mathbb{R}$, $N_F$ is $k$ times Fréchet differentiable on* *all of $L^s$, and for each $f \in L^s$, the kth differential $\mathrm{d}^k N_F(f)$ is the bounded $k$-linear multiplication operator from $(L^s)^k$ into $L^p$ induced by the function $F^{(k)} \circ f$, which is in $L^{r_k}$ with $r_k$ defined by (7.40).*

The proof of this proposition is by induction on $m$, $1 \le m \le k$. First, we prove a lemma giving a part of the induction step, which also extends the hypothesis to allow $f$ to be fixed and for that $f$, $(k-1)p < s \le kp$, but shows that if $s < kp$ then $d^k N_F(f) = 0$. The lemma will also be used later in Proposition 7.59, which will show that if $N_F$ is $k$ times differentiable with $k \ge sp$ at some $G \in L^s$ then $F$ is a polynomial of degree $\le k$ (or $k-1$, if $s/p$ is not an integer).

**Lemma 7.55.** *Let $(\Omega, \mathcal{S}, \mu)$ be finite, complete, and nonatomic, and let $1 \le p \le mp < s < \infty$ for some positive integer $m$. Let a Borel measurable function $F\colon \mathbb{R} \to \mathbb{R}$ be such that the autonomous Nemytskii operator $N_F$ acts from $L^s$ into $L^p$. If $N_F$ is $m+1$ times Fréchet differentiable at $f \in L^s$ and there is a Borel measurable function $G$ such that $\mathrm{d}^m N_F(g) = M^m[G \circ g]$ for each $g \in V$, a neighborhood of $f$, then there is a function $h_f \in \mathcal{L}^0(\Omega, \mathcal{S}, \mu)$ such that $\mathrm{d}^{m+1} N_F(f) = M^{m+1}[h_f]$. If $s < (m+1)p$ then $h_f = 0$ in $L^0$, so $d^{m+1} N_F(f) = 0$.*

*Proof.* By assumption, the $(m+1)$st differential $L := \mathrm{d}^{m+1} N_F(f)$ exists and is in $L({}^{m+1}L^s, L^p)$. By Proposition 5.25, the function $g \mapsto \mathrm{d}^m N_F(g) = M^m[G \circ g]$, $g \in V$, is differentiable at $f$ and $D(M^m[G \circ f])(g_0) = L(g_0, \dots)$ for each $g_0 \in L^s$. Thus for each $g_0, g_1, \dots, g_m \in L^s$ and $t \in \mathbb{R}$ such that $f + tg_0 \in V$, we have

$$\big\| t^{-1}\big[N_G(f + tg_0) - N_G(f)\big] g_1 \cdots g_m - L(g_0, g_1, \dots, g_m) \big\|_p \to 0 \qquad (7.41)$$

as $t \to 0$. It will be shown by induction on $i = 1, 2, \dots, m+1$ that for any fixed $g_0, g_1, \dots, g_{m-i} \in L^s$ there is a $\mu$-measurable function $h = h(g_0, g_1, \dots, g_{m-i})(\cdot)$ such that for the $i$-linear multiplication operator $M^i[h]$ and any $g_{m-i+1}, \dots, g_m$ in $L^s$ we have

$$L(g_0, g_1, \dots, g_m) = M^i[h](g_{m-i+1}, \dots, g_m). \qquad (7.42)$$

For $i = m+1$ none of $g_0, \dots$ will be fixed and $h$ will be just one function, giving the conclusion of the lemma.

For $i = 1$, let $g_0, g_1, \dots, g_{m-1} \in L^s$ be fixed and for each $g_m \in L^s$, let $T(g_m) := L(g_0, g_1, \dots, g_m)$. Let $A \in \mathcal{S}$ with $\mu(A) > 0$ and $g_m = 1_A$. By (7.41), it follows that $T(1_A)(\omega) = 0$ for $\mu$-almost all $\omega \notin A$. If $B \in \mathcal{S}$ and $B \subset A$, then $T(1_B)(\omega) = T(1_A)(\omega)$ for $\mu$-almost all $\omega \in B$. Since $\mu$ is finite, there is a $\mu$-measurable function $h = h(g_0, g_1, \dots, g_{m-1})(\cdot)\colon \Omega \to \mathbb{R}$, namely $h = T(1(\cdot))$, such that $T(1_A) = 1_A h$ for all $A \in \mathcal{S}$ with $\mu(A) > 0$. In particular for $A = \Omega$, taking $t$ along the sequence $1/n$ in (7.41) and $g_m = 1(\cdot)$, there is a sequence $n_j \uparrow +\infty$ such that

$$n_j\big[G(f(\omega)+n_j^{-1}g_0(\omega))-G(f(\omega))\big]g_1\cdots g_{m-1}\to h(\omega)$$

for $\mu$-almost all $\omega$ as $j\to\infty$. We can redefine $h(\omega)$ as the limit of the left side if it exists and 0 otherwise.

If $g_m$ is a $\mu$-simple function, since $T(\cdot)$ is linear, we have $T(g_m)=hg_m$. For any $g_m\in L^s$, there exist $\mu$-simple $g_{m,j}\to g_m$ in $L^s$ as $j\to\infty$. We have $T(g_{m,j})=hg_{m,j}$ and $T(g_{m,j})\to T(g_m)$ in $L^p$ as $j\to\infty$. Taking a subsequence, we can assume that $g_{m,j}\to g_m$ and $T(g_{m,j})=hg_{m,j}\to T(g_m)$ $\mu$-almost everywhere as $j\to\infty$. Thus $L(g_0,g_1,\dots,g_m)=T(g_m)=h(g_0,g_1,\dots,g_{m-1})g_m$ a.e. $(\mu)$ for any $g_m\in L^s$, and so $T(\cdot)$ is the multiplication operator from $L^s$ into $L^p$ induced by the function $h$. By Proposition 5.5, the operator norm $\|T(\cdot)\|$ is bounded by $\|L\|\|g_0\|_s\|g_1\|_s\cdots\|g_{m-1}\|_s$. Since $p<s$, by Lemma 7.53 with $k=1$, $h\in L^{r_1}$ with $r_1=sp/(s-p)$, and $\|T(\cdot)\|=\|h\|_{r_1}$. Thus (7.42) is proved for $i=1$.

For the induction step, given (7.42) for some $i$ with $1\le i\le m$, we want to prove it for $i+1$ in place of $i$. We can write the right side of (7.42) as $h(g_0,\dots,g_{m-i})g_{m-i+1}\cdots g_m\in L^p$. Since $i\le m$ and $mp<s$ we have by Lemma 7.53 that $h(g_0,\dots,g_{m-i})\in L^{r_i(s,p)}$. We need to show that for any fixed $g_0,\dots,g_{m-i-1}\in L^s$ (or with none of $g_0,\dots$ fixed if $i=m$), we can write $h(g_0,\dots,g_{m-i})=h(g_0,\dots,g_{m-i-1})g_{m-i}$ for some measurable function $h(g_0,\dots,g_{m-i-1})(\cdot)$, or simply a function $h$ if $i=m$. As in the $i=1$ case, for any $g_{m-i}\in L^s$, let $T(g_{m-i})(\cdot):=h(g_0,\dots,g_{m-i})\in L^{r_i(s,p)}$, and let $g_{m-i}=1_A$ for any $A\in\mathcal{S}$ with $\mu(A)>0$. By (7.41) and (7.42) we have $T(1_A)(\omega)=0$ for $\mu$-almost all $\omega\notin A$. Also, $(m+1)$-linearity of $L$ and (7.42) imply that $T$ is linear from $L^s$ into $L^{r_i(s,p)}$. Thus, the proof as for $i=1$ gives that $h(g_0,\dots,g_{m-i-1})$ exists and (7.42) holds for $i+1$ in place of $i$. For $i=m+1$ we get that there is a $\mu$-measurable function $h_f:=h\colon\Omega\to\mathbb{R}$ such that $L(g_0,g_1,\dots,g_m)=h_fg_0g_1\cdots g_m$ a.e. $(\mu)$ for any $g_0,g_1,\dots,g_m\in L^s$. If $s<(m+1)p$ then $h_f=0$ in $L^0$ by the last statement in Lemma 7.53 for $k=m+1$. Since $f\in V$ is arbitrary, the proof of the lemma is complete. □

**Proof of Proposition 7.54.** If $k=1$, then the conclusion follows from Proposition 7.46. Suppose that $k>1$ and the conclusion holds for some $m$ in place of $k$ with $1\le m<k$. It will be proved to hold for $m+1$. By the induction assumption, (1) $F^{(m)}$ exists everywhere on $\mathbb{R}$, and (2) for each $f\in L^s$, the $m$th differential $\mathrm{d}^mN_F(f)$ is the bounded $m$-linear multiplication operator from $(L^s)^m$ into $L^p$ induced by the function $F^{(m)}\circ f\in\mathcal{L}^{r_m}$. Thus for $f\in L^s$, $\mathrm{d}^mN_F(f)(g_1,\dots,g_m)=N_{F^{(m)}}(f)g_1\cdots g_m$ for each $g_1,\dots,g_m\in L^s$.

To prove (1) and (2) with $m+1$ in place of $m$, let $f\in V$. Since $N_F$ is $m+1$ times differentiable on $V$, by Lemma 7.55 with $G=F^{(m)}$, there is an $h_f\in\mathcal{L}^0(\Omega,\mathcal{S},\mu)$ such that $\mathrm{d}^{m+1}N_F(f)=M^{m+1}[h_f]$. Since $(m+1)p\le kp<s$, the function $h_{m+1}(f):=h_f\in\mathcal{L}^{r_{m+1}}$ by the implication $(b)\Rightarrow(c)$ of Lemma 7.53 with $k$ replaced by $m+1$. By Proposition 5.25 with $k=m+1$, the function $\tilde f\mapsto\mathrm{d}^mN_F(\tilde f)$, $\tilde f\in V$, is differentiable at $f$ and $D\mathrm{d}^mN_F(f)g=M^{m+1}[h_{m+1}(f)](g,\dots)$ for each $g\in L^s$. Since

$$r_{m+1} = sr_m/(s - r_m) \tag{7.43}$$

the linear multiplication operator $M[h_{m+1}(f)]$ is bounded from $L^s$ into $L^{r_m}$, and for each $g \in L^s$, the $m$-linear multiplication operator $M^m[h_{m+1}(f)g]$ is bounded from $(L^s)^m$ into $L^p$. Therefore by Lemma 7.53 with $k = m$, for each $g \in L^s$ such that $f + g \in V$, we have

$$\begin{aligned} &\|N_{F^{(m)}}(f+g) - N_{F^{(m)}}(f) - h_{m+1}(f)g\|_{r_m} \\ &\quad = \left\|\mathrm{d}^m N_F(f+g) - \mathrm{d}^m N_F(f) - D\mathrm{d}^m N_F(f)g\right\|_{L({}^m L^s, L^p)} = o(\|g\|_s) \end{aligned}$$

as $\|g\|_s \to 0$. Thus the autonomous Nemytskii operator $N_{F^{(m)}}$ is differentiable from $V$ into $L^{r_m}$ and its derivative at $f \in V$ is the bounded multiplication operator induced by the function $h_{m+1}(f)$. Next we will apply Proposition 7.46 with $r_m$ and $F^{(m)}$ instead of $p$ and $F$, respectively, as we may, recalling that $(m+1)p \le kp < s$ and so $r_m < s$. It follows that $F^{(m+1)}$ exists everywhere on $\mathbb{R}$ and for each $f \in L^s$, $h_{m+1}(f) = F^{(m+1)} \circ f$, proving (1) and (2) with $m+1$ in place of $m$.Therefore the conclusion of the proposition holds for $m+1$ instead of $k$. The induction argument is then complete, and the proposition is proved. □

In particular, the preceding proposition concludes that $F^{(k)} \circ f$ is in $L^{r_k}$ for each $f \in L^s$. Using Theorem 7.13($b$) with $r_k = sp/(s - kp)$ instead of $p$, and recalling the growth-condition classes $\mathcal{G}_\beta$ defined in (7.2), we get the following:

**Corollary 7.56.** *Under the hypotheses of Proposition* 7.54, $F^{(k)} \in \mathcal{G}_{(s/p)-k}$.

The necessary condition just stated is also sufficient:

**Proposition 7.57.** *Let $(\Omega, \mathcal{S}, \mu)$ be finite and complete, $k$ a positive integer, and $1 \le p \le kp < s < \infty$. If $F\colon \mathbb{R} \to \mathbb{R}$ is $k$ times differentiable everywhere on $\mathbb{R}$ and its $k$th derivative $F^{(k)}$ is in $\mathcal{G}_{(s/p)-k}$, then the autonomous Nemytskii operator $N_F$ is $k$ times Fréchet differentiable from $L^s$ into $L^p$ and $\mathrm{d}^k N_F(f) = M^k[F^{(k)} \circ f]$ for each $f \in L^s$.*

*Proof.* If $k = 1$, then the conclusion holds by Proposition 7.45. Suppose that $k > 1$. Since $F^{(k)}$ is locally bounded, $F^{(k-1)} \in \mathcal{H}_1^{\mathrm{loc}}$ by the mean value theorem of calculus, so it is locally absolutely continuous and is an indefinite integral of $F^{(k)}$ (e.g. Theorem 8.18 and Example 8.20 in [198]). Thus $F^{(k-1)} \in \mathcal{G}_{(s/p)-(k-1)}$. Iterating, we find that

$$F^{(j)} \in \mathcal{G}_{(s/p)-j} \qquad \text{for } j = 0, 1, \ldots, k, \tag{7.44}$$

where $F^{(0)} \equiv F$.

Now suppose that the conclusion holds with $k$ replaced by some $m$, $1 \le m < k$. It will be proved to hold for $m + 1$. By the induction assumption, $N_F$

is $m$ times Fréchet differentiable from $L^s$ into $L^p$, and the $m$th differential $\mathrm{d}^m N_F(f)$ equals $M^m[F^{(m)}\circ f] \in L(^m L^s, L^p)$ for each $f \in L^s$. Thus $F^{(m)}\circ f \in L^{r_m}$ by Lemma 7.53. Let $G \in L^s$. Recall that $r_j = (sp)/(s - jp)$ for $jp < s$. Since $(m+1)p \le kp < s$, we have $(s/p) - (m+1) = s/r_{m+1}$ and (7.43). By (7.44) with $j = m+1$ and Theorem 7.13(*a*) with $r_{m+1}$ instead of $p$, it follows that $F^{(m+1)}\circ G \in L^{r_{m+1}}$, and so $M^{m+1}[F^{(m+1)}\circ G]$ is a bounded $(m+1)$-linear multiplication operator from $(L^s)^{m+1}$ into $L^p$ by Lemma 7.53. Also for each $g \in L^s$, $(F^{(m+1)}\circ G)g \in L^{r_m}$ and $M^m[(F^{(m+1)}\circ G)g]$ is a bounded $m$-linear multiplication operator from $(L^s)^m$ into $L^p$ by Lemma 7.37 with $r_m$ instead of $p$ and by Lemma 7.53 with $k = m$, respectively. By Proposition 5.25, it is enough to prove that the function $f \mapsto \mathrm{d}^m N_F(f)$, $f \in L^s$, is differentiable at $G$ and $D\mathrm{d}^m N_F(G)(g) = M^{m+1}[F^{m+1}\circ G](g, \dots)$ for each $g \in L^s$. For $g \in L^s$, let

$$\begin{aligned} R(g) &:= \mathrm{d}^m N_F(G+g) - \mathrm{d}^m N_F(G) - M^{m+1}[F^{(m+1)}\circ G](g, \dots) \\ &= M^m[F^{(m)}\circ(G+g) - F^{(m)}\circ G - (F^{(m+1)}\circ G)g] \in L(^m L^s, L^p). \end{aligned}$$

By the norm statement in Lemma 7.53 with $k = m$, the operator norm of $R(g)$ is

$$\|R(g)\| = \|F^{(m)}\circ(G+g) - F^{(m)}\circ G - (F^{(m+1)}\circ G)g\|_{r_m}.$$

The right side is $o(\|g\|_s)$ as $\|g\|_s \to 0$ by Proposition 7.45 applied to $F = F^{(m)}$ and $p = r_m$, using (7.44) with $j = m+1$ and the identity $(s/p) - (m+1) = (s/r_m) - 1$. The proof of the proposition is complete by induction. □

**Proposition 7.58.** *Let $(\Omega, \mathcal{S}, \mu)$ be finite, complete, and nonatomic, $k$ a positive integer, and $1 \le p \le kp < s < \infty$. Let a Borel measurable $F\colon \mathbb{R} \to \mathbb{R}$ be such that the autonomous Nemytskii operator $N_F$ acts from $L^s$ into $L^p$, and let $V$ be a nonempty open subset of $L^s(\Omega, \mathcal{S}, \mu)$. Then the following statements are equivalent:*

(*a*) *$N_F$ is $C^k$ on $V$;*
(*b*) *$N_F$ is $C^k$ on $L^s$;*
(*c*) *$F$ is $C^k$ on $\mathbb{R}$ and $N_{F^{(k)}}$ acts and is continuous from $L^s$ into $L^{r_k}$ with $r_k$ defined by* (7.40).

*Proof.* First suppose that (*a*) holds. By Proposition 7.54, the $k$th derivative $F^{(k)}$ exists on $\mathbb{R}$, and for each $f \in L^s$, the $k$th differential $\mathrm{d}^k N_F(f)$ exists and is the bounded $k$-linear multiplication operator from $(L^s)^k$ into $L^p$ induced by the function $F^{(k)}\circ f \in L^{r_k}$. Thus the Nemytskii operator $N_{F^{(k)}}$ acts from $L^s$ into $L^{r_k}$. By Lemma 7.53, the operator norm of the $k$-linear multiplication operator from $(L^s)^k$ into $L^p$ induced by a function $h \in L^{r_k}$ is equal to $\|h\|_{r_k}$. Thus for any $f_1, f_2 \in L^s$, we have

$$\begin{aligned}\|N_{F^{(k)}}(f_1) - N_{F^{(k)}}(f_2)\|_{r_k} \\ &= \sup\left\{\left\|\left[F^{(k)}\circ f_1 - F^{(k)}\circ f_2\right]g_1\cdots g_k\right\|_p:\ \|g_1\|_s \le 1,\ldots,\|g_k\|_s \le 1\right\} \\ &= \|\mathrm{d}^k N_F(f_1) - \mathrm{d}^k N_F(f_2)\|.\end{aligned}$$

Since $\mathrm{d}^k N_F$ is continuous on $V$, it then follows that $N_{F^{(k)}}$ is continuous on $V$. By Proposition 7.21($b$), $N_{F^{(k)}}$ is continuous from $L^s$ into $L^{r_k}$, proving the second part of ($c$). The first part follows from Theorem 7.19, since $F^{(k)}$ is Borel measurable and $\psi(u,\omega) \equiv F^{(k)}(u)$ is a Carathéodory function if and only if $F^{(k)}$ is continuous.

Suppose that ($c$) holds. By Theorem 7.13 with $r_k$ in place of $p$ and $F^{(k)}$ in place of $F$, since $N_{F^{(k)}}$ acts from $L^s$ to $L^{r_k}$, we have $F^{(k)} \in \mathcal{G}_{s/r_k} = \mathcal{G}_{(s/p)-k}$. Thus by Proposition 7.57, $N_F$ is $k$ times Fréchet differentiable from $L^s$ into $L^p$, with $d^k N_F(G) = M^k[F^{(k)}\circ G]$ for each $G \in L^s$. By ($c$), $N_{F^{(k)}}$ is continuous from $L^s$ into $L^{r_k}$. By the norm equality in Lemma 7.53, it follows that $d^k N_F(\cdot)$ is continuous on $L^s$, so that ($b$) holds. Since clearly ($b$) implies ($a$), the proof of the proposition is complete. □

In the preceding proposition it was assumed that $kp < s$. If $kp > s$ then $r_k$ in part ($c$) is not defined. If $kp = s$, then $r_k = +\infty$, and the condition in part ($c$), if it holds, implies by Proposition 7.23 that $F^{(k)}$ is a constant, in other words, that $F$ is a polynomial of degree at most $k$. The following proposition confirms this and proves that $F$ is a polynomial also if $N_F$ is $k$ times differentiable, even at one point, for a $k \ge s/p$. Thus such differentiability is very restrictive.

**Proposition 7.59.** *Let $(\Omega, \mathcal{S}, \mu)$ be complete, nonatomic, and finite, let $1 \le p \le s < \infty$, and let $j$ and $k$ be respectively the largest and the smallest integers such that $j \le s/p \le k$. Let $F$ be a Borel function on $\mathbb{R}$ such that the autonomous Nemytskii operator $N_F$ acts from $L^s$ into $L^p$. If $N_F$ is $k$ times Fréchet differentiable at a point $G$ then $F$ is a polynomial of degree at most $j$.*

*Proof.* If $k = 1$ then $p = s$ and $j = 1$, and the conclusion follows from Proposition 7.39. Thus we can assume that $k > 1$. Let $r > 0$ be such that $N_F$ is $k-1$ times differentiable on $G + B_r$, where $B_r = \{g \in L^s\colon \|g\|_s < r\}$. For $g \in B_r$, the remainder in Taylor's expansion of $N_F$ around $G$ of order $k$ is defined by

$$R_k(G,g) := \mathrm{Rem}^k_{N_F}(G,g) = N_F(G+g) - N_F(G) - \sum_{i=1}^{k} \frac{1}{i!}\mathrm{d}^i N_F(G) g^{\otimes i}.$$

Thus $R_k(G,g) \in L^p$ for each $g \in B_r$. For $i = 1,\ldots,k-1$, since $ip < s$, by Proposition 7.54, $F^{(i)}$ exists and $\mathrm{d}^i N_F(G) = M^i[F^{(i)}\circ G]$. By Lemma 7.55 with $m = k-1$, there exists $h_G \in \mathcal{L}^0(\Omega,\mathcal{S},\mu)$ such that $\mathrm{d}^k N_F(G) =$

$M^k[h_G]$. Let $a_i(\omega) := (i!)^{-1}F^{(i)}\circ G(\omega)$ for $i = 1, \ldots, k-1$, and let $a_k(\omega) := (k!)^{-1}h_G(\omega)$. For $(u, \omega) \in \mathbb{R} \times \Omega$, let

$$\psi(u,\omega) := F(G(\omega)+u) - F(G(\omega)) - \sum_{i=1}^{k} a_i(\omega)u^i. \tag{7.45}$$

Then $\psi$ is a strong Shragin function on $\mathbb{R} \times \Omega$ and the Nemytskii operator $N_\psi$ acts from $B_r$ into $L^p$ since $N_\psi(g) = R_k(G, g)$ for each $g \in B_r$. By Theorem 5.44, $\|N_\psi(g)\|_p = o(\|g\|_s^k)$ as $\|g\|_s \to 0$. If for some $g \in B_r$, $N_\psi(g) \neq 0$ on a set of positive $\mu$-measure, then by Proposition 7.33($b$), we must have $kp < s$. This contradiction yields that $N_\psi(g) = 0$ for each $g \in B_r$. Therefore by Proposition 7.6, there is a $\mu$-null set $N$ such that for each $\omega \in N^c$, $\psi(u, \omega) = 0$ for all $u \in \mathbb{R}$. Thus taking an $\omega_0 \in N^c$, it follows that

$$F(u) = F(G(\omega_0)) + \sum_{i=1}^{k} a_i(\omega_0)(u - G(\omega_0))^i$$

for each $u \in \mathbb{R}$. The proof is complete if $j = k$, in other words if $s/p$ is an integer.

Let $j < k$. Then $kp > s$ and $j = k - 1$. In this case, for $g \in B_r$, consider the remainder $R_j(G, g) := \mathrm{Rem}^j_{N_F}(G, g)$ in Taylor's expansion of $N_F$ around $G$ of order $j$. Also, for $(u, \omega) \in \mathbb{R} \times \Omega$, let $\psi(u, \omega)$ be defined by (7.45) without the last term. Then again, $\psi$ is a strong Shragin function on $\mathbb{R} \times \Omega$ and the Nemytskii operator $N_\psi$ acts from $B_r$ into $L^p$ since $N_\psi(g) = R_j(G, g)$ for each $g \in B_r$. By Theorem 5.44, $\|N_\psi(g)\|_p = O(\|g\|_s^k)$ as $\|g\|_s \to 0$. If for some $g \in B_r$, $N_\psi(g) \neq 0$ on a set of positive $\mu$-measure, then by Proposition 7.33($a$), we must have $k \leq s/p$. This contradiction yields that $N_\psi(g) = 0$ for each $g \in B_r$. Again by Proposition 7.6, it follows that $F$ is a polynomial of degree at most $j$. The proposition is now proved. $\square$

We will give bounds for the remainders in Taylor expansions of the autonomous Nemytskii operator $g \mapsto N_F(G + g)$. For an open set $U \subset \mathbb{R}$, $n = 1, 2, \ldots$, and $0 < \alpha \leq 1$, we write $F \in \mathcal{H}_{n+\alpha}(U; \mathbb{R})$ if a function $F\colon U \to \mathbb{R}$ is $C^n$ on $U$ and its $n$th derivative $F^{(n)}$ is in $\mathcal{H}_\alpha(U; \mathbb{R})$. If $n = 1$ this agrees with Definition 6.5 for $X = Y = \mathbb{R}$.

**Proposition 7.60.** *Let $(\Omega, \mathcal{S}, \mu)$ be a complete, finite measure space. Let $1 \leq p < s < \infty$, $0 < \alpha \leq 1$, and a positive integer $n$ be such that $(n+\alpha)p \leq s$. Let $F \in \mathcal{H}_{n+\alpha}(\mathbb{R}; \mathbb{R})$ and let $G \in L^s(\Omega, \mathcal{S}, \mu)$. Then the autonomous Nemytskii operator $N_F$ acts from $L^s(\Omega, \mathcal{S}, \mu)$ into $L^p(\Omega, \mathcal{S}, \mu)$, it is $n$ times Fréchet differentiable at $G$, and the remainder in the Taylor series around $G$ has the bound*

$$\Big\| N_F(G+g) - N_F(G) - \sum_{k=1}^{n} \frac{F^{(k)}\circ G}{k!} g^k \Big\|_p \leq K \|g\|_s^{n+\alpha}, \tag{7.46}$$

*where* $K = \mu(\Omega)^{(1/p)-(n+\alpha)/s}\|F^{(n)}\|_{(\mathcal{H}_\alpha)}/n!$, *for all* $g \in L^s$.

*Proof.* Since for each $k = 1, \ldots, n$, $F^{(k-1)}$ is an indefinite integral of $F^{(k)}$, $F^{(n)} \in \mathcal{H}_\alpha$, and $n+\alpha \le s/p$, it follows that $F$ satisfies the $\beta$ growth condition (7.2) with $\beta = s/p$. By Theorem 7.13, it then follows that $N_F$ acts from $L^s$ into $L^p$, proving the first part of the conclusion. By Taylor's formula with an integral remainder (e.g. Theorem 5.41), for each $(u, \omega) \in \mathbb{R} \times \Omega$, we have

$$\begin{aligned} \phi(u,\omega) &:= F(G(\omega)+u) - F(G(\omega)) - \sum_{k=1}^{n} \frac{F^{(k)}(G(\omega))}{k!} u^k \\ &= u^n \int_0^1 \Big[F^{(n)}(G(\omega)+tu) - F^{(n)}(G(\omega))\Big]\, d\theta_n(t), \end{aligned} \tag{7.47}$$

where $\theta_n(t) = -(1-t)^n/n!$ for $0 \le t \le 1$. Let $1 \le k \le n$. Since $(n+\alpha)p \le s$, $F^{(k)}$ satisfies the $\beta$ growth condition with $\beta = (s-kp)/p = s/r_k$, and so by Theorem 7.13 the Nemytskii operator $N_{F^{(k)}}$ acts from $L^s$ into $L^{r_k}$. By Lemma 7.29 with $k+1$ in place of $k$, $p_1 = \cdots = p_k = s$ and $p_{k+1} = r_k$, for $g \in L^s$ and each $1 \le k \le n$, we then have

$$\int_\Omega |(F^{(k)} \circ G) g^k|^p \, d\mu \le \|N_{F^{(k)}} G\|_{r_k}^p \|g\|_s^{kp} < \infty.$$

Thus $N_\phi$ acts from $L^s$ into $L^p$. Since $F^{(n)} \in \mathcal{H}_\alpha$, for each $(u,\omega) \in \mathbb{R} \times \Omega$,

$$|\phi(u,\omega)| \le |u|^{n+\alpha} \|F^{(n)}\|_{(\mathcal{H}_\alpha)}/n!.$$

Therefore by Lemma 7.29 with $k = 2$, we have

$$\int_\Omega |N_\phi g|^p \, d\mu \le \big[\|F^{(n)}\|_{(\mathcal{H}_\alpha)}/n!\big]^p \mu(\Omega)^{1-p(n+\alpha)/s} \|g\|_s^{p(n+\alpha)},$$

and the conclusion follows. The proof of the proposition is complete. □

**Remark 7.61.** The assumption $(n+\alpha)p \le s$ in the preceding proposition cannot be improved in general. Indeed, under the hypotheses of Proposition 7.60 suppose in addition that $(\Omega, \mathcal{S}, \mu)$ is nonatomic and $F(u) \ne F(0) + \sum_{k=1}^n (k!)^{-1} F^{(k)}(0) u^k$ for some $u \in \mathbb{R}$. Thus $\phi \not\equiv 0$, where $\phi$ is the Shragin function defined by (7.47) with $G \equiv 0$. If (7.46) holds for $G \equiv 0$, then by Proposition 7.33($a$) we must have $n + \alpha \le s/p$.

The remainder bound $O(\|g\|_s^{n+\alpha})$ in (7.46) is of best possible order under the conditions of Proposition 7.60, as the following shows (for $n = 1$).

**Proposition 7.62.** *Let* $0 < \alpha < 1$ *and* $1 \le p < p(1+\alpha) \le s < \infty$. *Then there are a function* $F \in \mathcal{H}_{1+\alpha}(\mathbb{R};\mathbb{R})$, *a finite measure space* $(\Omega, \mathcal{S}, \mu)$, *and a* $G \in L^s(\Omega, \mathcal{S}, \mu)$ *such that if for some* $\beta > 1$,

$$\|\mathrm{Rem}_{N_F}(G, g)\|_p = O(\|g\|_s^\beta)$$

*as* $\|g\|_s \to 0$, *then* $\beta \le 1 + \alpha$.

*Proof.* Let $(\Omega, \mathcal{S}, \mu)$ be $[0,1]$ with Lebesgue measure $\lambda$ and $G(x) := x$, $0 \le x \le 1$. Let $M > 1$ be a large enough integer so that

$$\max\{1/(M^\alpha - 1), \pi/(M^{1-\alpha} - 1)\} < 1/150, \tag{7.48}$$

and let

$$F(u) := (2\pi)^{-1} \sum_{n=1}^{\infty} M^{-n(1+\alpha)} \sin(2\pi M^n u), \qquad u \in \mathbb{R}.$$

Then

$$F'(u) = \sum_{n=1}^{\infty} M^{-n\alpha} \cos(2\pi M^n u), \qquad u \in \mathbb{R}.$$

We have $F' \in \mathcal{H}_\alpha$ by the proof of Proposition 3.104. Thus $F \in \mathcal{H}_{1+\alpha}$ and so the autonomous Nemytskii operator $N_F$ is Fréchet differentiable at $G$ from $L^s$ to $L^p$ by Proposition 7.41. By the fundamental theorem of calculus for $g \in \mathcal{L}^s([0,1], \lambda)$ and $x \in [0,1]$, we have

$$\mathrm{Rem}_{N_F}(G, g)(x) = g(x) \int_0^1 \big[F'(x + tg(x)) - F'(x)\big]\, dt.$$

For $k = 1, 2, \ldots$, and $x \in [0,1]$, let $g_k(x) := rM^{-k} - x$, where $r$ is the nearest integer to $xM^k$, or the smaller of two equally close integers. The values of $|g_k|$ are uniformly distributed over $[0, M^{-k}/2]$, so

$$\|g_k\|_s = \Big(2M^k \int_0^{M^{-k}/2} y^s\, dy\Big)^{1/s} = C(s) M^{-k}$$

for a constant $C(s)$ not depending on $M$ or $k$. For $k, n = 1, 2, \ldots$, and $x, t \in [0,1]$, let

$$\Delta(k, n, x, t) := \cos(2\pi M^n(x + tg_k(x))) - \cos(2\pi M^n x).$$

Then for $k = 1, 2, \ldots$ and $x \in [0,1]$, we have

$$\mathrm{Rem}_{N_F}(G, g_k)(x) = g_k(x) \sum_{n=1}^{\infty} M^{-n\alpha} \int_0^1 \Delta(k, n, x, t)\, dt. \tag{7.49}$$

Using the definition of $g_k$ and the periodicity of the cosine, it follows that

$$\Delta(k, k, x, t) = \cos(2\pi M^k (1 - t) g_k(x)) - \cos(2\pi M^k g_k(x)) \ge 0,$$

where the last inequality holds since $t \in [0,1]$ and $2\pi M^k g_k(x) \in [-\pi, \pi]$ for $x \in [0,1]$. If $|g_k(x)| \ge M^{-k}/4$, which occurs on a set $A \subset [0,1]$ with $\lambda(A) = 1/2$, and if $t \ge 3/4$, then $\Delta(k, k, x, t) \ge 2^{-1/2}$. Thus for the $p$th norm of the $k$th term of the sum on the right side of (7.49), we have the bound

$$\left\| g_k M^{-k\alpha} \int_0^1 \Delta(k,k,x,t)\,\mathrm{d}t \right\|_p \geq M^{-k(1+\alpha)} 2^{-(9/2)-(1/p)} \geq M^{-k(1+\alpha)}/50.$$

For the sum of the terms with $n > k$, we have since $\|g_k\|_{\sup} = M^{-k}/2$,

$$\begin{aligned}\left\| g_k \sum_{n=k+1}^{\infty} M^{-n\alpha} \int_0^1 \Delta(k,n,x,t)\,\mathrm{d}t \right\|_p &\leq M^{-k-(k+1)\alpha}(1-M^{-\alpha})^{-1} \\ &\leq M^{-k(1+\alpha)}/(M^{\alpha}-1).\end{aligned}$$

For the sum of the terms with $n < k$, we have

$$\begin{aligned}\left\| g_k \sum_{n=1}^{k-1} M^{-n\alpha} \int_0^1 \Delta(k,n,x,t)\,\mathrm{d}t \right\|_p &\leq \left\| g_k \sum_{n=1}^{k-1} M^{-n\alpha}(2\pi M^n)\cdot M^{-k} \right\|_p \\ &\leq \pi M^{-2k} \sum_{n=1}^{k-1} M^{n(1-\alpha)} \leq \pi M^{-k(1+\alpha)}/(M^{1-\alpha}-1).\end{aligned}$$

By choice of $M$ (7.48), we have

$$\|\mathrm{Rem}_{N_F}(G, g_k)\|_p \geq M^{-k(1+\alpha)}/150 \geq C\|g_k\|_s^{1+\alpha}$$

for a constant $C > 0$. Since $\|g_k\|_s \to 0$ as $k \to \infty$, the conclusion follows. □

## 7.5 Examples where $N_F$ Is Differentiable and $F$ Is Not

Here are examples showing that when an autonomous Nemytskii operator $N_F$ is Fréchet differentiable at one $G$, the derivative of $F$ in measure as in Theorem 7.40($b$) may not be an ordinary derivative. Let $\mathcal{B}$ be the Borel $\sigma$-algebra in $\mathbb{R}$. For a measurable function $h \geq 0$ and a measure $\mu$, we define a measure $h\mu$ by $(h\mu)(A) := \int_A h\,\mathrm{d}\mu$, $A \in \mathcal{B}$.

**Proposition 7.63.** *There exist a probability measure $\nu$ on $\mathcal{B}$, a compact set $K \subset I := [0,1]$ with $\nu(K) = 1$, and a continuous function $F$ from $\mathbb{R}$ into $\mathbb{R}$ such that for the identity function $X(x) \equiv x$ from $\mathbb{R}$ into $\mathbb{R}$, and any $p, s$ with $1 \leq p < s < \infty$, the autonomous Nemytskii operator $N_F$ is Fréchet differentiable at $X$, in other words, $g \mapsto F \circ (X + g)$ is Fréchet differentiable at $g = 0$, from $\mathcal{L}^s(I, \mathcal{B}, \nu)$ into $\mathcal{L}^p(I, \mathcal{B}, \nu)$ with derivative 0, but $F'$ exists nowhere on $K$. Moreover, $F$, $K$, and $\nu$ can be chosen so that for Lebesgue measure $\lambda$, either (a) or (b) holds, where*

(*a*) *$\lambda(K) = 1/2$, $F$ is $\alpha$-Hölder for each $\alpha \in (0,1)$, and*

$$\nu = 2 \cdot 1_K \lambda; \tag{7.50}$$

*(b)* $\lambda(K) = 0$ *and* $F$ *is Lipschitz, with* $|F(x)-F(y)| \le 2|x-y|$ *for all* $x, y \in \mathbb{R}$.

**Remark**. The $F$'s constructed in the proof will be differentiable everywhere outside of $K$. In part $(b)$, $F$, being Lipschitz, is necessarily differentiable almost everywhere for $\lambda$.

*Proof.* Numbers $b_i \equiv 1 - a_i \in (0, 1/2]$ will be defined for $i = 0, 1, \ldots$ as follows. For part $(b)$ we just take $a_i \equiv b_i \equiv 1/2$, $i = 0, 1, \ldots$. For part $(a)$ and $i = 0, 1, \ldots$, let

$$a_i := \frac{(i+1)/(i+2)}{(i+2)/(i+3)} = \frac{(i+1)(i+3)}{(i+2)^2}, \qquad b_i = 1 - a_i = 1/(i+2)^2. \quad (7.51)$$

Then for any $k = 0, 1, \ldots$ and $n > k$,

$$a_k a_{k+1} \cdots a_{n-1} = \frac{(k+1)/(k+2)}{(n+1)/(n+2)} \to \frac{k+1}{k+2} \quad (7.52)$$

as $n \to \infty$, so

$$\text{in part } (a), \quad \prod_{i=0}^{\infty} a_i = \frac{1}{2}. \quad (7.53)$$

We will define the set $K$ as the intersection of a decreasing sequence of sets $K_i$, $i = 0, 1, \ldots$, each of which is a finite union of nondegenerate closed intervals $K_{ij}$, $j = 1, \ldots, M(i)$, for some $M(i)$. The lexicographic ordering of ordered pairs of positive integers is defined by $(i, j) \prec (i', j')$ if and only if $i < i'$ or $i = i'$ and $j < j'$.

First, let $K_{01} := [0, 1]$ and $M(0) := 1$. Closed intervals $K_{ij}$ will be defined recursively in $i$ and $j$, $i = 0, 1, \ldots$, $j = 1, \ldots, M(i)$. Suppose that $K_{ij}$, $j = 1, \ldots, M(i)$, are defined for some $i \ge 0$. For a positive integer $m(i, j)$ defined below, each $K_{ij}$ will be a union of $2^{m(i,j)}$ nonoverlapping intervals $I_{ijr} = [a_{ijr}, b_{ijr}]$, $r = 1, \ldots, 2^{m(i,j)}$, of equal length

$$b_{ijr} - a_{ijr} = h_{ij} := \lambda(K_{ij})/2^{m(i,j)}. \quad (7.54)$$

Let $c_{ijr} := a_{ijr} + b_i h_{ij}/2$ and $d_{ijr} := b_{ijr} - b_i h_{ij}/2$. Then

$$K_{ijr} := [c_{ijr}, d_{ijr}] \quad (7.55)$$

is a subinterval of $I_{ijr}$ with

$$\lambda(K_{ijr}) = a_i \lambda(I_{ijr}) = a_i h_{ij} \ge h_{ij}/2 \quad (7.56)$$

since $b_i \le 1/2$ for all $i$. The intervals $K_{i+1,s}$ are defined as the intervals $K_{ijr}$ for all $j = 1, \ldots, M(i)$ and $r = 1, \ldots, 2^{m(i,j)}$ where $s < s'$ if and only if $(j, r) \prec (j', r')$. For $i = 0, 1, \ldots$ let

$$M(i+1) := \sum_{j=1}^{M(i)} 2^{m(i,j)}. \quad (7.57)$$

Thus $K_{i+1,s}$ for $s = 1, \ldots, M(i+1)$ will be defined once $m(i,j)$ are defined.
Now to define $m(i,j)$ and so $h_{ij}$ recursively, we begin with

$$m(0,1) := 1 \quad \text{and} \qquad h_{01} = \lambda(K_{01})/2^{m(0,1)} = 1/2. \tag{7.58}$$

If $K_{ij}$, $m(i,j)$, and so $h_{ij}$ are defined for some $i \geq 0$ and for all $j = 1, \ldots, M(i)$, as they are for $i = 0$, then all the sets $K_{i+1,j}$ are defined as in (7.55) and just after (7.56). Let $m(i+1,1)$ be the smallest positive integer such that

$$h_{i+1,1} = \lambda(K_{i+1,1})/2^{m(i+1,1)} < (h_{i,M(i)}/(i+2))^{i+2} < h_{i,M(i)}/2. \tag{7.59}$$

For each $j = 1, \ldots, M(i) - 1$, choose $m(i+1, j+1)$ as the smallest positive integer such that

$$h_{i+1,j+1} = \lambda(K_{i+1,j+1})/2^{m(i+1,j+1)} < (h_{i+1,j}/(i+1))^{i+1}. \tag{7.60}$$

Then, iterating (7.60) and (7.59), the recursive definition of all $m(i,j)$, $M(i)$ by (7.57), $K_{ij}$, and $h_{ij}$ is complete. It follows from (7.58), (7.59), and (7.60) that for all $i = 0, 1, \ldots$,

$$h_{i+1,1} \leq h_{i,1}/2 \leq 1/2^{i+2}, \quad \text{so} \quad \sum_{n=0}^{\infty} h_{n1} \leq 1. \tag{7.61}$$

It follows that $h_{ij} > h_{i'j'}$ whenever both are defined and $(i,j) \prec (i',j')$, i.e. $h_{0,1} > h_{1,1}$ and for all $i = 1, 2, \ldots$,

$$h_{i1} > h_{i2} > \cdots > h_{i,M(i)} > h_{i+1,1} > 0. \tag{7.62}$$

For $i = 0, 1, \ldots$, let $K_i := \bigcup_{j=1}^{M(i)} K_{ij}$, a finite union of closed intervals. Then $\lambda(K_0) = 1$ and $\lambda(K_{i+1}) = a_i \lambda(K_i)$ for all $i \geq 0$, with $K_0 \supset K_1 \supset K_2 \supset \cdots$. Let

$$K := \bigcap_{i=0}^{\infty} K_i. \tag{7.63}$$

Then $K$ is a compact, perfect set with $\lambda(K) = \Pi_{i=0}^{\infty} a_i = 0$ in part $(b)$, while in part $(a)$, $\lambda(K) = 1/2$ by (7.53). For part $(a)$, $\nu$ is defined by (7.50). For part $(b)$, we define $\nu(K_{01}) := 1$ and recursively let $\nu(K_{ijr}) := \nu(K_{ij})/2^{m(i,j)}$ for all $j = 1, \ldots, M(i)$ and $r = 1, \ldots, 2^{m(i,j)}$ (these equations are also true, by the way, in part $(a)$). Then in part $(b)$, $\nu$ extends to a unique Borel probability measure on $\mathbb{R}$ with $\nu(K) = 1$. We have

$$\begin{gathered} \text{in part } (a), \text{ for each } n \geq 1 \text{ and } 1 \leq j \leq M(n), \\ \nu(K_{nj}) < 2\lambda(K_{nj}) < 2h_{n-1,1} \leq 1/2^{n-1} \end{gathered} \tag{7.64}$$

by (7.50), (7.56), and (7.61).

We will use the following fact, which holds for the probability measure $\nu$ in each of parts $(a)$ and $(b)$.

**Lemma 7.64.** *For any $i = 2, 3, 4, \ldots$ and $r = 1, \ldots, M(i)$, let the interval $K_{ir}$ be $[a, b]$ with $a < b$ and let $J$ be a subinterval of $K_{ir}$ with a common endpoint, $J = [a, x]$ or $[x, b]$ for some $x \in K_{ir}$. Then*

$$\nu(J)/\nu(K_{ir}) \leq 2\lambda(J)/\lambda(K_{ir}). \tag{7.65}$$

*Proof.* For part $(a)$, by (7.55), (7.56) iterated, and (7.52), we have for each $i \geq 1$,

$$\lambda(K_{ir} \cap K) = \lambda(K_{ir})i/(i+1) \geq \lambda(K_{ir})/2,$$

while by (7.50), $\nu(K_{ir}) = 2\lambda(K_{ir} \cap K)$ and $\nu(J) \leq 2\lambda(J)$, so the conclusion follows. The rest of the proof will be for part $(b)$. For any Borel set $B$ and $i = 0, 1, \ldots$, let

$$\nu_i(B) := \sum_{j=1}^{M(i)} \lambda(B \cap K_{ij}) / \sum_{j=1}^{M(i)} \lambda(K_{ij}). \tag{7.66}$$

Then each $\nu_i$ is a Borel probability measure on $I$. We claim and will show by induction that:

$$\text{in case } (b), \quad \nu(K_{ij}) = 2^i \lambda(K_{ij}) \tag{7.67}$$

for all $i = 0, 1, \ldots$ and $j = 1, \ldots, M(i)$. For $i = 0$ the claim holds. Let $i \geq 1$ and suppose the claim holds for $i-1$ in place of $i$. For each $j = 1, \ldots, M(i)$ we have $K_{ij} = K_{i-1,u,v}$ for some $u = 1, \ldots, M(i-1)$ and $v = 1, \ldots, m(i-1, u)$. Then

$$\begin{aligned}
\nu(K_{ij}) &= \nu(K_{i-1,u})/2^{m(i-1,u)} && \text{by definition of } \nu \\
&= 2^{i-1}\lambda(K_{i-1,u})/2^{m(i-1,u)} && \text{by induction hypothesis} \\
&= 2^{i-1}\lambda(I_{i-1,u,v}) && \text{by (7.54) with } i-1 \text{ in place of } i \\
&= 2^i \lambda(K_{ij}) && \text{by (7.56) and since } a_{i-1} = 1/2,
\end{aligned}$$

proving (7.67) by induction. Iterating the first equation in (7.56), it follows that $\sum_{j=1}^{M(i)} \lambda(K_{ij}) = a_{i-1} \cdots a_0 = 2^{-i}$, and so

$$\nu_i(K_{ij}) = \nu(K_{ij}) \text{ for each } j. \tag{7.68}$$

The sequence $\{\nu_i\}_{i \geq 0}$ converges weakly to $\nu$, which has a continuous distribution function. Thus it suffices to prove (7.65) for each $\nu_k$, $k \geq i$, in place of $\nu$. Let $k = i + q$ for $q = 0, 1, 2, \ldots$. On $K_{ir}$, $\nu_i = 2^i\lambda$ by (7.66) and (7.67), and $\nu_{i+q}(K_{ir}) = \nu_i(K_{ir})$. Suppose $J = [a, x]$ for $x \in [a, b] = K_{ir}$. We can assume $x > a$ and want to prove that for $q = 0, 1, 2, \ldots$,

$$G_q(x) := \nu_{i+q}([a, x])/(2^i(x - a)) \leq 2. \tag{7.69}$$

For $q = 0$, $G_q(x) \equiv 1$. For $q = 1$, we will show that $G_1(x) \leq 4/3$ for all $x$. $K_{ir}$ is decomposed into $2^{m(i,r)}$ intervals $I_{irp}$ of equal length $h_{ir}$. For each $p$, $K_{irp}$

is the middle half of $I_{irp}$. The Radon–Nikodym derivative (density) $d\nu_{i+1}/d\nu_i$ equals 2 on each $K_{irp}$ and is 0 on the complementary intervals $I_{irp}\setminus K_{irp}$. It is easily seen that $G_1 = 1$ at the endpoints and midpoints of the intervals $I_{irp}$, except that for $p = 1$, $G_1 = 0$ on an interval $(a, a+\delta)$ for some $\delta > 0$. The midpoints of $I_{irp}$ and $K_{irp}$ are the same. Let $u$ be the midpoint of $K_{irp}$. Then on $K_{irp}$ we have $G_1(x) = [u-a+2(x-u)]/(x-a)$ since $u-a+(x-u) \equiv x-a$. Since $u > a$ it follows that $G_1$ is increasing on each $K_{irp}$ and attains a relative maximum at its right-hand endpoint $d_{irp}$. Then

$$G_1(d_{irp}) = \frac{ph_{ir}}{\left(p-\frac{1}{4}\right)h_{ir}} = p/\left(p-\frac{1}{4}\right),$$

which is maximized for $p = 1, \ldots, 2^{m(i,r)}$ when $p = 1$, where it equals $4/3$, so $G_1(x) \le 4/3$ for $a < x \le b$. Applying the same proof with $i+1$, $G_2/G_1$, and $2(x-u)$ in place of $i$, $G_1$, and $x-u$ respectively gives that $G_2(x) \le (4/3)G_1(x)$ for $a < x \le b$, and hence

$$G_2(x) \le (4/3)^2, \quad a < x \le b. \tag{7.70}$$

For each $q = 0, 1, \ldots,$ $\nu_{i+q}$ is concentrated on intervals $K_{i+q,\xi}$, on each of which it has density $2^q$ with respect to $\nu_i$, or $2^{i+q}$ with respect to $\lambda$. Let $h_q(x) := (x-a)G_q(x)$ for each $x$. Then for each $q \ge 1$, $h_q'(x) = 0$ or $2^q$ wherever the derivative is defined (as it is except on a finite set). We have $\mathrm{d}G_q(x)/\mathrm{d}x = ((x-a)h_q'(x) - h_q(x))/(x-a)^2 > 0$ if and only if $x$ is in the interior of $K_{i+q,\xi}$ for some $\xi$ and $G_q(x) = h_q(x)/(x-a) < h_q'(x)$, which equals $2^q$ on such an interior.

Now, the bound

$$G_q(x) = h_q(x)/(x-a) < 2^{q-1} \tag{7.71}$$

for each $q \ge 2$ will be proved by induction on $q$. We have $d\nu_{q+1}/d\nu_q \le 2$ everywhere, and so $h_{q+1}(x) \le 2h_q(x)$ for all $x \in [a, b]$. For $q = 2$, $h_2(x) < 2(x-a)$ by (7.70). So (7.71) holds for all $q = 2, 3, \ldots$ and $a < x \le b$ as claimed, finishing the induction.

It follows that $G_q$ is increasing on each $K_{i+q,\xi}$ and has a relative maximum at its right-hand endpoint. Suppose that $q \ge 2$ and the absolute maximum of $G_q$ is at a point $B$, which must be the right endpoint of some $K_{i+q,\xi}$. Then $K_{i+q,\xi} \subset K_{i+q-1,u}$ for some $u$. Let $h := h_{i+q-1,u} = 2\lambda(K_{i+q,\xi})$.

It will be shown that $\xi$ is the largest $\eta$ such that $K_{i+q,\xi} \subset K_{i+q-1,u}$. Suppose not. The right endpoint of $K_{i+q,\xi+1}$ is $B+h$. Let $\Delta := h_q(B+h) - h_q(B) = \nu_{i+q}(K_{i+q,\xi+1})/2^i = 2^q\lambda(K_{i+q,\xi+1})$ [by (7.67) and (7.68)] $= 2^{q-1}h$. To show that $G_q(B+h) > G_q(B)$ is equivalent to showing that $(B-a)\Delta > hh_q(B)$, or $h_q(B) < (B-a)\cdot 2^{q-1}$, or $G_q(B) < 2^{q-1}$, which is true by (7.71). So $G_q(B+h) > G_q(B)$, which gives a contradiction, and $\xi$ is the maximal $\eta$ such that $K_{i+q,\eta} \subset K_{i+q-1,u}$. Let $B_q := B$ and let $C_{q-1}$ be the right endpoint of $K_{i+q-1,u}$. Then $h_{q-1}(C_{q-1}) = h_q(C_{q-1}) = h_q(B_q)$ and $C_{q-1} - B_q = h/4$.

Let $C := C_{q-1} - a$ and $D := B_q - a$. Then $D/C = 1 - (h/(4C))$. We have $K_{i+q-1,u} \subset K_{i+q-2,s}$ for some $s$ and by (7.59) since $i \geq 2$ and $q \geq 2$, so $i + q - 2 \geq 2$ and

$$C \geq \lambda(K_{i+q-1,u}) = h_{i+q-2,s}/2 \geq 2^{2(i+q)-1} h_{i+q-1,1}.$$

Since $h = h_{i+q-1,u} \leq h_{i+q-1,1}$, it follows that $h/(4C) \leq 1/2^{2(i+q)+1}$.

Now $1/(1-z) \leq 1 + 2z$ for $0 \leq z \leq 1/2$, so

$$\sup_x G_q(x) = G_q(B_q) = \frac{h_q(B_q)}{B_q - a} = \frac{h_{q-1}(C_{q-1})}{D} = \frac{C}{D} G_{q-1}(C_{q-1})$$

$$\leq \frac{C}{D} \sup_x G_{q-1}(x) \leq (1 + 4^{-q}) \sup_x G_{q-1}(x).$$

Using this for $q \geq 3$, the bound (7.70), and since $1 + t \leq e^t$ for $t \geq 0$, we have

$$\sup_{q \geq 1} \sup_x G_q(x) \leq (4/3)^2 \exp(1/(64(1 - 4^{-1}))) < 2,$$

proving (7.69) and Lemma 7.64 for intervals $J = [a, x]$. A proof for intervals $J = [x, b]$ is similar. □

Next the function $F$ will be defined. For any $b$ with $0 < b < 1$ let $G(\cdot; b)$ be the trapezoidal function on $\mathbb{R}$ defined by

$$G(x; b) := \begin{cases} 0 & \text{if } x < 0, \\ x/b & \text{if } 0 \leq x \leq b/2, \\ 1/2 & \text{if } b/2 < x \leq 1 - b/2, \\ (1-x)/b & \text{if } 1 - b/2 < x \leq 1, \\ 0 & \text{if } x > 1. \end{cases}$$

For each $i = 0, 1, \ldots$, $j = 1, \ldots, M(i)$, and $r = 1, \ldots, 2^{m(i,j)}$, let

$$H_{ijr}(x) := h_{ij} G((x - a_{ijr})/h_{ij}; b_i), \quad H_{i+1} := \sum_{k=1}^{M(i)} \sum_{s=1}^{2^{m(i,k)}} H_{iks}. \tag{7.72}$$

Each $H_{ijr}$ is a continuous, piecewise linear function with support $I_{ijr}$. We have $H'_{ijr} = 0$ or $\pm 1/b_i$ except at the four points where right and left derivatives have different values. For $k = 1, 2, \ldots$, let $F_k := \sum_{i=1}^{k} H_i$. Then each $H_i$ and $F_k$ is piecewise linear. All functions $H_{ijr}$, $H_i$, and $F_k$ are nonnegative and are 0 outside $[0, 1]$. For each $i$ and each $x \in [0, 1]$, there is at most one ordered pair $(j, r)$ such that $H_{ijr}(x) > 0$. Also, for the finite union $U_i$ of open intervals where $H_{i+1}$ has a non-zero derivative, $U_i \subset K_i \setminus K_{i+1}$, so the $U_i$ are disjoint,

$$U_i \cap U_k = \emptyset, \quad i \neq k. \tag{7.73}$$

For any set $S$ and bounded real-valued function $f$ on $S$, recall that the *oscillation* of $f$ on $S$ is defined by $\mathrm{Osc}(f;S) = (\sup - \inf)(\{f(x) : x \in S\})$ and that $\|f\|_{\sup} := \sup\{|f(x)| : x \in S\}$. If $f \geq 0$ and $f(x) = 0$ for some $x \in S$ then clearly $\mathrm{Osc}(f;S) = \|f\|_{\sup}$. For any $b \in (0,1)$, $\|G(\cdot;b)\|_{\sup} = 1/2$, from the definition of $G$. It follows from the definitions of $H_{ijr}$ (7.72) and $K_{ijr}$ (7.55) that for each $i = 0,1,\ldots$, $j = 1,\ldots,M(i)$, and $r = 1,2,\ldots,2^{m(i,j)}$,

$$\|H_{ijr}\|_{\sup} = h_{ij}/2, \quad \mathrm{Osc}(H_{ijr}, K_{ijr}) = 0. \tag{7.74}$$

We have for each $i = 0,1,\ldots$, by (7.61), and (7.62) for $i+1$ in place of $i$,

$$\|H_{i+1}\|_{\sup} \;=\; \|H_{i11}\|_{\sup} \;=\; h_{i1}/2 \;\leq\; 1/2^{i+2}. \tag{7.75}$$

Also $\mathrm{Osc}(H_{i+1}; K_{i+1,s}) = 0$ for all $i \geq 0$ and all $s$, or in other words $\mathrm{Osc}(H_i; K_{i,u}) = 0$ for all $i \geq 1$ and all $u$. Since each $K_{i+1,s} = K_{ijr} \subset K_{ij}$ for some $j$ and $r$, it follows that for $1 \leq u \leq i$ and all $s = 1,\ldots,M(i)$,

$$\mathrm{Osc}(H_u; K_{i,s}) \;=\; 0. \tag{7.76}$$

From (7.75) it follows that the series

$$F(x) \;:=\; \sum_{i\geq 1} H_i(x), \tag{7.77}$$

or equivalently the sequence $\{F_k\}_{k\geq 1}$, converges absolutely and uniformly on $\mathbb{R}$ to a function $F$, which is 0 outside $[0,1]$.

For each $i = 1,2,\ldots$ and $j = 1,\ldots,M(i)$, by (7.76) and (7.74), we have

$$\mathrm{Osc}(F;K_{ij}) \leq \sum_{k>i} \mathrm{Osc}(H_k;K_{ij}) \leq h_{ij}/2 + \sum_{k>i+1} \|H_k\|_{\sup}.$$

For each $k > i+1$, by (7.75), (7.59), and (7.62), it follows that $\|H_k\|_{\sup} = h_{k-1,1}/2$ and $\|H_{k+1}\|_{\sup} < \|H_k\|_{\sup}/2$. Also, $h_{i+1,1} < h_{ij}/2$, and so

$$\sum_{k>i+1} \|H_k\|_{\sup} \leq 2^{-2} h_{ij} \sum_{k\geq 0} 2^{-k} = h_{ij}/2.$$

Therefore for each $i = 1,2,\ldots$ and $j = 1,\ldots,M(i)$, we have the bound

$$\mathrm{Osc}(F;K_{ij}) \leq h_{ij} \leq h_{i1}. \tag{7.78}$$

It will be shown that $F'$ exists nowhere on $K$. Let $t \in K$. Then for each $i = 0,1,\ldots$, there are some unique $j$, $r$, and $s$ with $t \in K_{ijr} = K_{i+1,s}$ and so $u_i := a_{ijr} < c_{ijr} \leq t \leq d_{ijr} < v_i := b_{ijr}$. We have $\mathrm{Osc}(H_k; I_{ijr}) = 0$ for all $k = 1,\ldots,i$ by (7.76) since $[u_i,v_i] = I_{ijr} \subset K_{ij}$. Also, $H_{i+1}(t) = H_{ijr}(t) = h_{ij}/2$ by (7.72), while $H_{i+1}(u_i) = H_{i+1}(v_i) = 0$. For $k > i+1$ we have $H_k(u_i) = H_k(v_i) = 0$ since $u_i, v_i \notin K_{i+1}$. Thus $F(u_i) = F(v_i)$ and $F(t) - F(u_i) = F(t) - F(v_i) \geq h_{ij}/2$, while $\max(t-u_i, v_i - t) < h_{ij}$, so

$$(F(t) - F(u_i))/(t - u_i) \geq 1/2, \quad (F(v_i) - F(t))/(v_i - t) \leq -1/2.$$

As $i \to \infty$, $u_i = a_{ijr} \uparrow t$ and $v_i = b_{ijr} \downarrow t$, so $F'(t)$ does not exist.

In part (*b*), since $b_i \equiv 1/2$, we have by (7.72) that $H_i'(x) = 0$ or $\pm 2$ everywhere except at finitely many points where $H_i$ has left and right derivatives from among the same three values. Recall that the sets where $H_i' \neq 0$ are disjoint (7.73). It follows from (7.77) that $|F(x) - F(y)| \leq 2(y - x)$ for $0 \leq x \leq y \leq 1$. Since $F(x) \equiv 0$ for all $x \notin [0,1]$, we have $|F(x) - F(y)| \leq 2|x - y|$ for all real $x, y$.

In both parts (*a*) and (*b*) we have Hölder conditions as follows.

**Lemma 7.65.** *For each $\alpha \in (0,1)$ $F$ is $\alpha$-Hölder, i.e. there is a $C_\alpha < +\infty$ such that for all real $x$ and $y$, $|F(x) - F(y)| \leq C_\alpha |x - y|^\alpha$.*

*Proof.* In part (*b*), we have for $0 \leq x \leq y \leq 1$ that $|F(x) - F(y)| \leq 2|x - y| \leq 2|x - y|^\alpha$. Since $F \equiv 0$ on $(-\infty, 0]$ and on $[1, \infty)$, the same Hölder condition holds for all $x, y \in \mathbb{R}$. In part (*a*), by (7.72) and the definition of $G$, each $H_{ijr}$ is piecewise linear, with slope $\pm 1/b_i = \pm(i+2)^2$ by (7.51) on two intervals each of length $b_i h_{ij}/2$, and otherwise constant. If $|x - y| \leq b_i h_{ij}/2$ then

$$|H_{ijr}(x) - H_{ijr}(y)| \ \leq \ |x - y|/b_i \ = \ |x-y|^\alpha |x-y|^{1-\alpha}/b_i \ \leq \ |x - y|^\alpha h_{ij}^{1-\alpha}/b_i^\alpha.$$

By (7.62) this is $\leq |x - y|^\alpha h_{i1}^{1-\alpha}/b_i^\alpha \leq \gamma_i |x - y|^\alpha$ where by (7.61) and (7.51) we can take $\gamma_i := (i+2)^{2\alpha}/2^{i(1-\alpha)}$.

Or if $|x - y| > b_i h_{ij}/2$ then by (7.62), (7.72), and (7.74),

$$|H_{ijr}(x) - H_{ijr}(y)| \leq h_{ij}/2 \ = \ h_{ij}^{1-\alpha} h_{ij}^\alpha/2 \ \leq \ h_{i1}^{1-\alpha}|x - y|^\alpha/b_i^\alpha \ \leq \ \gamma_i |x - y|^\alpha$$

again and so in both cases. Then by (7.72), $|H_{i+1}(x) - H_{i+1}(y)| \leq \gamma_i |x - y|^\alpha$ for $0 \leq x \leq y \leq 1$. It follows that $|F(x) - F(y)| \leq C_\alpha |x - y|^\alpha$ for $C_\alpha := \sum_i \gamma_i = \sum_i (i+2)^{2\alpha}/2^{i(1-\alpha)} < \infty$. □

Let $F$, $\nu$, $K_{ij}$, $h_{ij}$, and $M(i)$ be as defined in the proof of Proposition 7.63 so far. We still need to prove that $N_F$ is Fréchet differentiable, with derivative 0, at the identity function. This will be done by the following:

**Lemma 7.66.** *Let $1 \leq p < s < \infty$. Let $g_n \to 0$ in $\mathcal{L}^s(\mathbb{R}, \nu)$. For any function $f : \mathbb{R} \to \mathbb{R}$ and $x \in \mathbb{R}$, let $\Delta_n f(x) := f(x + g_n(x)) - f(x)$. Then $\|\Delta_n F\|_p = o(\|g_n\|_s)$ as $n \to \infty$.*

*Proof.* It will be shown that it suffices to consider sequences $g_n$ such that for all $n$,

$$h_{n+1,1} \ < \ \|g_n\|_s \ \leq \ h_{n,1}. \tag{7.79}$$

To prove this, note that if the conclusion fails, there are an $\epsilon > 0$ and a subsequence $g_{n(k)}$ such that $\|\Delta_{n(k)} F\|_p \geq \epsilon \|g_{n(k)}\|_s$ for all $k$. Taking a further subsequence, we can assume that for each $n$, there is at most one $k$ such that $h_{n+1,1} < \|g_{n(k)}\|_s \le h_{n,1}$. This subsequence can be filled out by way of constants so that there is exactly one $k$ for each $n$, and (7.79) holds.

We claim that for a sequence $B_n$ of Borel subsets of $[0,1]$,

$$\int_{B_n} |\Delta_n F|^p \, d\nu = o(\|g_n\|_s^p) \qquad \text{as } n \to \infty, \tag{7.80}$$

provided that for some $\delta > 0$,

$$\nu(B_n) \; = \; o(\|g_n\|_s^\delta) \qquad \text{as } n \to \infty. \tag{7.81}$$

To prove the claim, let $B_n$ be such a sequence. By Lemma 7.65, $F$ is $\alpha$-Hölder for any $0 < \alpha < 1$. Thus by Lemma 7.29 with $k = 2$,

$$\int_{B_n} |\Delta_n F|^p d\nu \; \le \; C_\alpha^p \int_{B_n} |g_n|^{p\alpha} d\nu \; \le \; C_\alpha^p \|g_n\|_s^{p\alpha} \nu(B_n)^{(s-p\alpha)/s}. \tag{7.82}$$

To show that the latter product is $o(\|g_n\|_s^p)$ for some $\alpha \in (0,1)$, it will suffice if

$$\nu(B_n) \; = \; o(\|g_n\|_s^{p(1-\alpha)s/(s-p\alpha)}).$$

For this it will suffice in turn if (7.81) holds for some $\delta > 0$ since if so, we can take $\alpha \in (0,1)$ close enough to 1 so that $p(1-\alpha)s/(s-p\alpha) < \delta$. So the claim is proved. Moreover, for part $(b)$, (7.80) holds provided

$$\nu(B_n) \to 0 \qquad \text{as } n \to \infty. \tag{7.83}$$

This follows from (7.82) with $\alpha = 1$ since $F$ is a Lipschitz function in this case.

For a sequence of numbers $\gamma_n \to 0$ let $A_n \; := \; \{x \in [0,1] : \; |g_n(x)| > \gamma_n\}$. Then $\nu(A_n) \le \|g_n\|_s^s/\gamma_n^s$, which will be $o(\|g_n\|_s^\delta)$ for a given $\delta > 0$ if $\|g_n\|_s = o(\gamma_n^{s/(s-\delta)})$. This holds for $\gamma_n := (n\|g_n\|_s)^{(s-\delta)/s}$, which goes to 0 as $n \to \infty$ by (7.79) and (7.61). Taking $\delta := s/5$ and $B_n = A_n$, by (7.81), it follows that (7.80) holds.

Let for each $n = 1, 2, \ldots$, $C_n \; := \; [0,1] \setminus A_n$ and $g_n^{(1)} \; := \; g_n 1_{C_n}$. Then

$$\|g_n^{(1)}\|_{\sup} \; \le \; (n\|g_n\|_s)^{4/5}. \tag{7.84}$$

Now it will suffice to prove (7.80) with $C_n$ in place of $B_n$, namely

$$\int_{C_n} |\Delta_n F|^p \, d\nu = o(\|g_n\|_s^p) \qquad \text{as } n \to \infty. \tag{7.85}$$

To continue the proof of Lemma 7.66, here is a technical lemma.

**Lemma 7.67.** *For $n \ge 3$, let $i = i(n) = n$ or $n+1$, and let $1 \le v = v_n \le q = q_n \le M(i)$ be integers. Suppose that*

$$h_{i,v} \; = \; o(\|g_n\|_s) \;\; \textit{as} \;\; n \to \infty, \tag{7.86}$$

*and* $\{h_n\}_{n\geq 3}$ *is a sequence such that* $0 < h_n \leq h_{n-1,1}$,

$$\lambda(K_{iq}) \geq h_n/2, \tag{7.87}$$

*and*

$$\|g_n^{(1)}\|_{\sup} \leq h_n^2. \tag{7.88}$$

*Then* (7.80) *holds for the sequence* $B_n = C_n \cap \bigcup_{j=v}^q K_{ij}$.

*Proof.* Let $n \geq 3$. For $v \leq j \leq q$, let $I_{ij}$ be a subinterval of $K_{ij}$ with endpoints distant by $\|g_n^{(1)}\|_{\sup}$ from those of $K_{ij}$. The assumption $h_n \leq h_{n-1,1}$ implies by (7.61) that $h_n < 1/4$. By (7.87), (7.62), and (7.88), $I_{ij}$ exists and is nondegenerate. Then $J_{ij} := K_{ij} \setminus I_{ij}$ is a union of two intervals, with $\lambda(J_{ij}) = 2\|g_n^{(1)}\|_{\sup}$.

If $v \leq j \leq q$ and $x \in C_n \cap I_{ij}$ then $x + g_n(x) = x + g_n^{(1)}(x) \in K_{ij}$, and so

$$|\Delta_n F(x)| = |F(x + g_n^{(1)}(x)) - F(x)| \leq \mathrm{Osc}(F; K_{ij}) \leq h_{ij}$$

by (7.78). Thus by (7.62), $|\Delta_n F(x)| \leq h_{i,v}$ for each $x \in U_n := C_n \cap \bigcup_{v\leq j\leq q} I_{ij}$, and hence (7.80) holds for the sequence $B_n = U_n$ by (7.86).

So, let $F_n := \bigcup_{v\leq j\leq q} J_{ij}$. Then in part $(a)$,

$$\nu(F_n) \leq 4\sum_{j=v}^{q} \|g_n^{(1)}\|_{\sup} = 4\|g_n^{(1)}\|_{\sup} \sum_{j=v}^{q} \lambda(K_{ij})/\lambda(K_{ij}) \leq 8\|g_n^{(1)}\|_{\sup}/h_n$$

by (7.87). In part $(b)$, since each of the two intervals forming $J_{ij}$ has a common endpoint with $K_{ij}$, by Lemma 7.64 and (7.87) we have

$$\nu(F_n) = \sum_{j=v}^{q} \nu(J_{ij}) \leq 8\sum_{j=v}^{q} \|g_n^{(1)}\|_{\sup} h_n^{-1} \nu(K_{ij}) \leq 8\|g_n^{(1)}\|_{\sup}/h_n.$$

Thus in both parts $(a)$ and $(b)$, by (7.84) and (7.88), we have

$$\nu(F_n) \leq 8h_n^{-1} \min\{h_n^2, (n\|g_n\|_s)^{4/5}\} \leq \epsilon_n \|g_n\|_s^{1/5},$$

where $\epsilon_n := 8n^{2/5}\|g_n\|_s^{1/5}$ if $h_n^2 \leq (n\|g_n\|_s)^{4/5}$ and $\epsilon_n := 8n^{1/5}h_n^{1/2}$ if $(n\|g_n\|_s)^{4/5} < h_n^2$. By (7.79) and (7.61), we have $\|g_n\|_s \leq 2^{-n-1}$. By the assumption $h_n \leq h_{n-1,1}$ and (7.61), $h_n \leq 2^{-n}$. Thus $\nu(F_n) = o(\|g_n\|_s^{1/5})$ as $n \to \infty$, and so (7.80) holds for $B_n = F_n$ by (7.81) with $\delta = 1/5$. Thus Lemma 7.67 is proved. □

We define $K_{n,M(n)+1} := K_{n+1,1}$ and $h_{n,M(n)+1} := h_{n+1,1}$. The latter is consistent with (7.60) for $j = M(n)$ by (7.59).

**Lemma 7.68.** *For $n = 2, 3, \ldots$ and $j = j_n = 1, \ldots, M(n)$, suppose that $h_{n,j+1} \le \|g_n\|_s \le h_{n,j}$ and $U_n = K_{n,j}$ or $K_{n,j+1}$. Then (7.80) holds for the sequence $B_n = U_n \cap C_n$.*

*Proof.* For part $(b)$, the conclusion holds by (7.83) since $\nu(U_n) \to 0$ as $n \to \infty$.

For part $(a)$, if $\lambda(U_n) > 2\|g_n^{(1)}\|_{\sup}$ let the interval $W_n \subset U_n$ have endpoints distant by $\|g_n^{(1)}\|_{\sup}$ from those of $U_n$ and $V_n := U_n \setminus W_n$. Otherwise let $W_n = \emptyset$ and $V_n = U_n$. In either case, to show that (7.80) holds for the sequence $B_n = V_n$, it suffices to verify (7.81) for $\delta = 1/5$ and $B_n = V_n$. We have

$$\nu(V_n) \le 2\lambda(V_n) \le 4\|g_n^{(1)}\|_{\sup} \le 4(n\|g_n\|_s)^{4/5}$$

by (7.84). Now $(n\|g_n\|_s)^{4/5} = o(\|g_n\|_s^{1/5})$, or equivalently $n^4\|g_n\|_s^3 \to 0$ as $n \to \infty$, by (7.79) and (7.61), proving (7.81) in this case.

If $x \in W_n \cap C_n$ then $x + g_n(x) = x + g_n^{(1)}(x) \in U_n$. Thus by (7.76), $\Delta_n H_k(x) = 0$ for each $k \le n$, and so

$$|\Delta_n F(x)| = |F(x + g_n^{(1)}(x)) - F(x)| \le |\Delta_n H_{n+1}(x)| + \sum_{k>n+1} \|H_k\|_{\sup} \tag{7.89}$$

for each $x \in W_n \cap C_n$. By (7.75), (7.61), and (7.79),

$$S_n := \sum_{r=n+2}^{\infty} \|H_r\|_{\sup} \le h_{n+1,1} \le \|g_n\|_s. \tag{7.90}$$

Thus since $\nu(U_n) \to 0$ as $n \to \infty$ by (7.64),

$$\int_{W_n} S_n^p d\nu \le \nu(U_n) h_{n+1,1}^p = o(\|g_n\|_s^p). \tag{7.91}$$

It remains to bound $\Delta_n H_{n+1}$ on $W_n \cap C_n$. First let $U_n = K_{n,j+1}$. If $x \in W_n \cap C_n$ then again $x + g_n(x) = x + g_n^{(1)}(x) \in U_n$ and

$$|\Delta_n H_{n+1}(x)| = |H_{n+1}(x + g_n^{(1)}(x)) - H_{n+1}(x)| \le h_{n,j+1}/2$$

if $j < M(n)$ by (7.74), or if $j = M(n)$, then $\Delta_n H_{n+1}(x) = 0$. In either case, $|\Delta_n F(x)| \le h_{n,j+1}$ by (7.89), (7.90), and (7.59). Since $h_{n,j+1} \le \|g_n\|_s$, it then follows by (7.64) that

$$\int_{W_n \cap C_n} |\Delta_n F|^p d\nu \le h_{n,j+1}^p \nu(U_n) \le \|g_n\|_s^p / 2^{n-1} = o(\|g_n\|_s^p).$$

Now let $U_n = K_{n,j}$. Then note that by (7.72), $|H_{n+1}(x) - H_{n+1}(y)| \le |x-y|/b_n$ for all $x, y$. It follows by Hölder's inequality (Lemma 7.29 with $k = 2$) that

$$\int_{W_n} |\Delta_n H_{n+1}|^p d\nu \le \int_{W_n} |g_n|^p d\nu / b_n^p \le \|g_n\|_s^p \nu(W_n)^{(s-p)/s} / b_n^p.$$

This is $o(\|g_n\|_s^p)$ because $\nu(W_n)^{(s-p)/s} = o((n+2)^{-3p})$ and $(n+2)^{-3p} = o(b_n^p)$ by (7.64) and (7.51), with bounds not depending on $j$. By (7.89), we have on $W_n \cap C_n$ that $|\Delta_n F|^p \le 2^p[|\Delta_n H_{n+1}|^p + S_n^p]$. Lemma 7.68 follows by (7.91). □

Next, to prove (7.85) and so Lemma 7.66, for each $n$ we have three possible cases.

$$\text{In Case 1, for some } \rho = 1, \dots, M(n) - 2, \quad h_{n,\rho+2} < \|g_n\|_s \le h_{n,\rho+1}. \tag{7.92}$$

$$\text{In Case 2,} \quad h_{n+1,1} < \|g_n\|_s \le h_{n,M(n)}. \tag{7.93}$$

$$\text{In Case 3,} \quad h_{n,2} < \|g_n\|_s \le h_{n,1}. \tag{7.94}$$

The three cases include all possibilities for each $n$ by (7.79) and (7.62). It suffices to consider $n \ge 3$. The sequence $\{g_n\}_{n\ge 3}$ can be decomposed into three subsequences $\{g_{n_{ij}}\}_{j\ge 1}$, $i = 1, 2, 3$, such that Case $i$ holds for $n = n_{ij}$ for all $j$. It suffices to prove Lemma 7.66 for each subsequence. Filling out the subsequences with constants, it suffices to prove (7.85) assuming that (7.79) and one of the three Cases holds for all $n$.

For the proof in Case 1, first apply Lemma 7.67 with $i = n+1$, $v = 1$, $q = q(n, \rho)$, the largest $k$ such that $K_{n+1,k} \subset K_{n,\rho}$, and $h_n := h_{n,\rho} \le h_{n,1} \le h_{n-1,M(n-1)}$ by (7.62). Then (7.86) holds by (7.92), (7.59), and (7.62); (7.87) holds since by (7.56), $\lambda(K_{n+1,q}) \ge h_{n,\rho}/2 = h_n/2$; and (7.88) holds by (7.84), (7.92), and (7.60). Thus Lemma 7.67 applies and gives relation (7.80) for the sequence $B_n = C_n \cap K \cap \bigcup_{t=1}^{\rho} K_{n,t} \subset C_n \cap \bigcup_{j=1}^{q(n,\rho)} K_{n+1,j}$.

Relation (7.80) for the sequences $B_n = C_n \cap K_{n,\rho+1}$ and $B_n = C_n \cap K_{n,\rho+2}$ holds by Lemma 7.68 for $j = \rho + 1$.

In Case 1, it remains to treat the set $F_{n,\rho} := \bigcup_{j=\rho+3}^{M(n)} K_{n,j}$. For this set apply Lemma 7.67 with $i = n$, $v = \rho+3$, $q = M(n)$, and $h_n := h_{n-1,M(n-1)} \le h_{n-1,1}$ by (7.62). Note that (7.86) holds by (7.92) and (7.60); (7.87) holds since $\lambda(K_{n,M(n)}) \ge h_{n-1,M(n-1)}/2$ by (7.56); and (7.88) holds by (7.84), (7.92), (7.62), and (7.59). Thus relation (7.80) holds for $B_n = C_n \cap F_{n,\rho}$. Since $\nu(K) = 1$ and $C_n \cap K \subset C_n \cap [(K \cap \bigcup_{t=1}^{\rho} K_{n,t}) \cup K_{n,\rho+1} \cup K_{n,\rho+2} \cup F_{n,\rho}]$, (7.85) holds in Case 1.

In Case 2 (7.93), first Lemma 7.67 will be applied with $i = n+1$, $v = 2$, $q$ the largest $k$ such that $K_{n+1,k} \subset K_{n,M(n)-1}$, and $h_n := h_{n,M(n)-1} < h_{n-1,1}$. Then (7.86) holds by (7.93) and (7.60); (7.87) holds since $\lambda(K_{n+1,q}) \ge h_{n,M(n)-1}/2$ by (7.56); and (7.88) holds by (7.84), (7.93), and (7.60). Thus relation (7.80) holds for $B_n = C_n \cap \bigcup_{j=2}^{q} K_{n+1,j}$.

The sets $K_{n,M(n)}$ and $K_{n+1,1}$ are treated by Lemma 7.68 with $j = M(n)$. Each $x \in K$ is either in some $K_{n+1,j} \subset K_{n,r}$ for $j \ge 2$ and $r \le M(n) - 1$, or in $K_{n,M(n)}$ or in $K_{n+1,1}$. So (7.85) holds in Case 2.

For Case 3 (7.94), let $L_n := \bigcup_{p\geq 3} K_{n,p}$. To apply Lemma 7.67, let $i = n$, $v = 3$, $q = M(n)$, and $h_n := h_{n-1,M(n-1)} < h_{n-1,1}$ by (7.62). Then (7.86) holds by (7.94) and (7.60); (7.87) holds, i.e. $\lambda(K_{n,M(n)}) \geq h_{n-1,M(n-1)}/2$, by (7.56); and (7.88) holds by (7.84), (7.94), and (7.59). Thus relation (7.80) follows for $B_n = C_n \cap L_n$.

The sets $K_{n,1}$ and $K_{n,2}$ are treated by Lemma 7.68 with $j = 1$. Then (7.85) holds in Case 3, and so the proof of Lemma 7.66 is complete. □

Thus the proof of Proposition 7.63 is also done. □

## 7.6 Notes

*Notes on Section 7.1.* The Shragin measurability condition was given by Shragin [218]. In [220] Shragin also used a different version of this condition, which we call the strong Shragin condition. Proposition 7.4 is a special case of Theorem 2 of Shragin [219]. Proposition 7.6 is close to Lemma 1.5 in [3]. Proposition 7.17 is a special case of Theorem 2 of Shragin [220, Theorem 2]. A related result when a function $\psi$ is just bimeasurable follows from Theorem 2 of Ponosov [188]. Namely for such a $\psi$, the Nemytskii operator $N_\psi$ is continuous on $L^0$ if and only if $\psi \simeq \phi$ for some Carathéodory function $\phi$.

Appell and Zabrejko [3, Chapter 3] treat Nemytskii operators between $L^p$ spaces, which they call Lebesgue spaces. Most of Theorem 7.13 is a corollary of their Theorem 3.1, which treats operators $N_\psi$ for functions $\psi$ of two variables and allows atoms. For both reasons the statement becomes more complicated. If $\mu$ is nonatomic the growth condition on $\psi$ is that

$$|\psi(u,x)| \leq a(x) + b|u|^{s/p} \tag{7.95}$$

for some constant $b$ and function $a(\cdot) \in L^p$. Krasnosel'skiĭ [127] proved that (7.95) is necessary for $N_\psi$ to act on $L^s$ into $L^p$ for $1 \leq p < s < \infty$, if $\psi(\cdot,\cdot)$ is a Carathéodory function. As Krasnosel'skii points out, (7.95) had been given earlier, by Vaĭnberg [232], who showed that for a Carathéodory function, it is sufficient for $N_\psi$ not only to act but to be continuous from $L^s$ into $L^p$. In the autonomous case $\psi(u,x) \equiv F(u)$ the Carathéodory condition reduces to continuity of $F$, which is not at all necessary for $N_F$ to act from $L^s$ into $L^p$, although it is necessary for continuity by Theorem 7.19.

We do not know previous references for the (easy) extension to the case that $N_F$ is defined only on a nonempty open subset of $L^s$ as in Theorem 7.13(*b*) and the extension to universal measurability in (*a*) and part (*c*) of Theorem 7.13, other than remarks in [54, Part III] about measurability.

*Notes on Section 7.2.* Theorem 7.24 is an extension of [55, Theorem 3.3], where $G$ was unnecessarily assumed to be bounded. The present proof is shorter and is self-contained. A predecessor of this fact was [50, Theorem 2.2], which also included Proposition 7.28.

*Notes on Section 7.3.* The fact that a Carathéodory function $\psi$ must be affine whenever the Nemytskii operator $N_\psi$ taking $L^p$ into $L^p$, $1 \le p < \infty$, is Fréchet differentiable at some function (Proposition 7.39) was proved by Vaĭnberg [233, pp. 91–92] in the case $p = 2$. The idea of his proof was similar to Lemma 7.33(*b*) as shown more explicitly when $p = 2$ by Durdil [58, Theorem 11]. Apparently Corollary 7.36 was known to Appell and Zabrejko [4, Theorem 3.12]. Theorem 7.40 is due to Wang Sheng-Wang [237].

*Notes on Section 7.4.* Proposition 7.59 shows that if the autonomous Nemytskii operator $N_F$ acts from $L^s$ into $L^p$ and is $k$ times Fréchet differentiable at a point $G$, with $k \ge s/p$, then $F$ is a polynomial of degree $\le s/p$. This may be considered as a refinement of the known fact that an analytic Nemytskii operator acting from $L^s$ into $L^p$ must be a polynomial (e.g. Theorem 2.20 and Theorem 3.16 in [3]). The authors of [3, p. 116] say that "[t]his fact is rather disappointing, in view of the usefulness of Lebesgue spaces in applications, and may be the reason for the fact that analyticity properties of superposition operators in other than Lebesgue spaces have not been studied ...". In contrast, Nemytskii operators $N_\psi$ or $N_F$ acting on certain Banach algebras of functions (Theorems 6.29, and 6.32), specifically acting on $\mathcal{W}_p$ spaces (Corollaries 6.78, 6.79, and 6.80), are analytic for $F$ analytic, or for $(u, x) \mapsto \psi(u, x)$, analytic in $u$ with suitable bounds with respect to $x$.

*Notes on Section 7.5.* The counterexample, Proposition 7.63, and its proof are given here for the first time. It shows that Theorem 2.1 of [54, Part I] is not correct. The proof of the latter relied on the sentence after (2.56) in [3], which apparently is also incorrect.

Appell and Zabrejko [3] consider Nemytskii operators also on Orlicz spaces, symmetric spaces ($\mu$-ideal spaces whose norm may be defined by means of nondecreasing rearrangement of measurable functions), spaces of continuous functions, functions of bounded variation, Hölder spaces, spaces of smooth functions, and Sobolev spaces.

Runst and Sickel [199, Chapter 5] consider Nemytskii operators on spaces of "Besov–Triebel–Lizorkin type." Such spaces of generalized functions (tempered distributions) on $\mathbb{R}^k$ are defined by way of Fourier transforms, partitions of unity, $L^p$ properties with respect to Lebesgue measure, and $\ell^q$ properties of sequences. They include several known classes of spaces such as Sobolev spaces but not $\mathcal{W}_p$ spaces. Runst and Sickel consider mainly autonomous operators $f \mapsto G \circ f$ or $(f_1, \ldots, f_n) \mapsto (x \mapsto G(f_1(x), \ldots, f_n(x)))$ and to a lesser extent the case where $G$ can depend separately on $x$. In a long section [199, §5.3] $G$ is assumed to be $C^\infty$; in [199, §5.4], $G(y) = |y|^\mu$, which is nonsmooth at 0 for $\mu$ not an even integer.

# 8

# Two-Function Composition

## 8.1 Overview; General Remarks

Recall that composition of two functions $F$ and $G$ is defined by

$$(F\circ G)(s) := F(G(s)), \tag{8.1}$$

where $s \in S$, $G\colon S \to U$, and $F\colon U \to Y$. In this chapter $Y$ will be a Banach space and $U$ will be an open subset of a Banach space $X$. The set $S$ will be an interval in Section 8.5. Sometimes, as in Section 8.3, a measure $\mu$ is given on a $\sigma$-algebra of subsets of $S$. Often $X = Y = \mathbb{K} = \mathbb{R}$ or $\mathbb{C}$.

By *two-function composition operator* we mean the operator

$$TC:\ (F,G) \mapsto TC(F,G)\ :=\ F\circ G,$$

where $F\circ G$ is defined by (8.1). In Chapters 6 and 7, the composition $F\circ G$ is treated for fixed $F$ and is called in this case the autonomous Nemytskii operator $N_F$, and so $TC(F,G) = N_F G$.

For $S$, $U$, $X$, and $Y$ as before, let $\mathbb{G}$, $\mathbb{F}$, and $\mathbb{H}$ be Banach spaces of functions, or for $\mathbb{G}$ and $\mathbb{H}$ of $\mu$-equivalence classes, acting from $S$ into $X$, from $U$ into $Y$, and from $S$ into $Y$, respectively. For a set $W \subset \mathbb{F}\times\mathbb{G}$ such that $g$ has values in $U$ for each $(f,g) \in W$, we say that $TC$ *acts* from $W$ into $\mathbb{H}$ if $TC(f,g) \in \mathbb{H}$ for each $(f,g) \in W$. Here $TC(f,g)$ is defined when $f$ is an actual function, not an equivalence class for a fixed $\sigma$-finite measure $\nu$ on $U$. If $g = g_1$ a.e. $(\mu)$ ($\mu$-almost everywhere) for two $\mu$-measurable functions $g, g_1\colon S \to U$ and $f$ is Borel measurable from $U$ into $Y$, then clearly also $f\circ g = f\circ g_1$ a.e. $(\mu)$. But for two functions $f$ and $f_1$, we will have $f\circ g = f_1\circ g$ a.e. $(\mu)$ if and only if $f = f_1$ a.e. $(\mu\circ g^{-1})$, a measure on $U$ which depends on $g$. Suppose that $\mathbb{G}$ contains all constant functions $g(s) \equiv x$, $s \in S$, for each $x \in U$, or the equivalence classes of such functions. If $f$ and $f_1$ are two Borel measurable functions defined on $U$ and $f\circ g = f_1\circ g$ (a.e. $(\mu)$ if $\mathbb{G}$ is a space of $\mu$-equivalence classes) for all $g \in \mathbb{G}$, then $f(x) = f_1(x)$ for all $x \in U$, i.e. $f = f_1$. So in

R.M. Dudley and R. Norvaiša, *Concrete Functional Calculus*, Springer Monographs in Mathematics, DOI 10.1007/978-1-4419-6950-7_8,

$$\mathbb{F} \times \mathbb{G} \ni (f, g) \mapsto TC(f, g) = f \circ g \in \mathbb{H} \tag{8.2}$$

we will take $f$ in a space $\mathbb{F}$ of functions, not of equivalence classes.

**Remark 8.1.** We may assume that $TC$ acts from $\mathbb{F} \times V$ into $\mathbb{H}$ for a nonempty open set $V \subset \mathbb{G}$. Indeed, let $W$ be a nonempty open set in $\mathbb{F} \times \mathbb{G}$ such that $TC$ acts from $W$ into $\mathbb{H}$. Take any nonempty open sets $O \subset \mathbb{F}$ and $V \subset \mathbb{G}$ such that $O \times V \subset W$. Fix $F \in O$. Then $(F + f) \circ G \in \mathbb{H}$ for all $f$ in a neighborhood of 0 in $\mathbb{F}$ and all $G \in V$. Taking $f = 0$ and subtracting, we get that $f \circ G \in \mathbb{H}$. But then we can multiply $f$ by an arbitrary constant, so that in fact $TC$ acts from $\mathbb{F} \times V$ into $\mathbb{H}$. Let $V_W$ be the union of all open $V \subset \mathbb{G}$ such that $O \times V \subset W$ for some nonempty open $O \subset \mathbb{F}$. Then $TC$ acts from $\mathbb{F} \times V_W$ into $\mathbb{H}$ and $\mathbb{F} \times V_W \supset W$. So if $TC$ acts on an open set $W \subset \mathbb{F} \times \mathbb{G}$, as will be needed to define continuity and differentiability at a point $(F, G)$, we may as well assume $W = \mathbb{F} \times V$ for an open $V \subset \mathbb{G}$. On the other hand, restricting $g$ to a neighborhood of 0 can actually make a difference in whether $f \circ (G + g)$ is defined at all, or is in $\mathbb{H}$, for example if $\mathbb{G}$ is a space of functions with supremum norm. However, if $\mathbb{G}$ is an $L^p$ space and $F$ is such that $F \circ G \in \mathbb{H}$ for all $G$ in a nonempty open set $V \subset \mathbb{G}$, then $F \circ G \in \mathbb{H}$ for all $G \in \mathbb{G}$ under some conditions (cf. Proposition 7.21($a$)).

We will consider some cases in which $TC$ acts between some function spaces and is continuous or differentiable at certain $(F, G)$. Continuity or differentiability of $TC$ at $(F, G)$ is clearly equivalent to that of

$$(f, g) \mapsto (F + f) \circ (G + g) \tag{8.3}$$

at $f = g = 0$. $TC$ will be considered when $G$ and $G + g$ take values in $U$.

For fixed $G$ ($g \equiv 0$), the operator $C_{\circ G}\colon F \mapsto F \circ G$ is linear. So if it is a bounded linear operator from a Banach space $\mathbb{F}$ of functions on $U$ to a Banach space $\mathbb{H}$ of functions on $S$, then it is everywhere Fréchet differentiable with derivative $DC_{\circ G} \equiv C_{\circ G}$ and all higher order derivatives identically 0. Thus the operators $C_{\circ G}$ are treated straightforwardly with regard to the kinds of differentiability considered in this book. Such operators have their own interest as linear operators. Singh and Manhas [221] consider operators $C_{\circ G}$ followed by multiplication by a suitable function, especially when the domain of functions $F$ is an $L^p$ space, and list over 400 references. Shapiro [217] considers the case where $S$ is an open set in the complex plane $\mathbb{C}$ and both $F$ and $G$ are holomorphic, giving nearly 150 references. Rosenthal [194] reviewed both books.

## 8.2 Differentiability of Two-Function Composition in General

Let $\mathbb{F}$, $\mathbb{G}$, and $\mathbb{H}$ be Banach spaces and let $V \subset \mathbb{G}$ be an open set such that the two-function composition operator $TC$ acts from $\mathbb{F} \times V$ into $\mathbb{H}$. Fréchet differentiability of $TC$ at a point can be viewed as joint Fréchet differentiability as follows. There is a natural 1-to-1 correspondence between $A \in L(\mathbb{F} \times \mathbb{G}; \mathbb{H})$ and $(B, C) \in L(\mathbb{F}; \mathbb{H}) \times L(\mathbb{G}; \mathbb{H})$, given by $A(f, g) = B(f) + C(g)$ for all $f \in \mathbb{F}$ and $g \in \mathbb{G}$. Thus the operator $TC$ is Fréchet differentiable at $(F, G) \in \mathbb{F} \times V$ from $\mathbb{F} \times V$ into $\mathbb{H}$ if and only if there exist bounded linear partial derivative operators $D_1TC(F, G)(\cdot) \in L(\mathbb{F}; \mathbb{H})$ and $D_2TC(F, G)(\cdot) \in L(\mathbb{G}; \mathbb{H})$ such that

$$\|\mathrm{Rem}_{TC}(F, G)(f, g)\|_{\mathbb{H}} = o(\|f\|_{\mathbb{F}} + \|g\|_{\mathbb{G}})$$

as $f \to 0$ in $\mathbb{F}$ and $g \to 0$ in $\mathbb{G}$, where for each $(f, g) \in \mathbb{F} \times \mathbb{G}$, the remainder in the differentiation defined by (5.1) is

$$\begin{aligned}\mathrm{Rem}_{TC}(F, G)(f, g) := &\ \mathrm{Rem}_{TC}((F, G), (f, g)) \\ = &\ TC(F + f, G + g) - TC(F, G) - D_1TC(F, G)(f) \\ &-D_2TC(F, G)(g).\end{aligned} \tag{8.4}$$

If so, then $D_1TC(F, G)(\cdot)$ must be $C_{\circ G} : f \mapsto f \circ G$, and $D_2TC(F, G)(\cdot)$ must be the Fréchet derivative $DN_F(G)$ of the autonomous Nemytskii operator $N_F$, as will be seen in the next theorem. Recall that if $N_F$ is differentiable at $G$, the remainder in the differentiation as defined by (6.3) is $\mathrm{Rem}_{N_F}(G, g) = F \circ (G + g) - F \circ G - ((DN_F)(G))(g)$. The theorem also gives a representation of $\mathrm{Rem}_{TC}(F, G)$ in which one term is $\mathrm{Rem}_{N_F}(G, g)$.

For a nonempty open set $V \subset \mathbb{G}$, a family of autonomous Nemytskii operators $\{N_f \colon f \in \mathcal{F}\}$ acting from $V$ into $\mathbb{H}$ is equicontinuous at $G \in V$ if for each $\epsilon > 0$ there is a $\delta > 0$ such that $\|N_f(G + g) - N_f(G)\|_{\mathbb{H}} < \epsilon$ for all $g \in \mathbb{G}$ such that $G + g \in V$ and $\|g\|_{\mathbb{G}} < \delta$, and all $f \in \mathcal{F}$. Here is a characterization of differentiability of the two-function composition operator $TC$ at a point.

**Theorem 8.2.** *Let $X$ and $Y$ be Banach spaces. Let $\mathbb{G} = (\mathbb{G}, \|\cdot\|_{\mathbb{G}})$ and $\mathbb{H} = (\mathbb{H}, \|\cdot\|_{\mathbb{H}})$ be non-zero Banach spaces of $X$- and $Y$-valued functions respectively or $\mu$-equivalence classes of them on a nonempty set $S$. Let $U$ be a nonempty open subset of $X$. Let $\mathbb{F} = (\mathbb{F}, \|\cdot\|_{\mathbb{F}})$ be a Banach space of functions from $U$ into $Y$. Let $F \in \mathbb{F}$ and $G \in \mathbb{G}$. Let $V$ be a neighborhood of $G$ in $\mathbb{G}$ such that all functions in $V$ take values in $U$. Suppose $TC$ acts from $\mathbb{F} \times V$ into $\mathbb{H}$. Then $TC$ is Fréchet differentiable at $(F, G)$ from $\mathbb{F} \times V$ into $\mathbb{H}$ if and only if the following three conditions all hold:*

*(a) the autonomous Nemytskii operator $N_F$ acts from $V$ into $\mathbb{H}$ and is Fréchet differentiable at $G$;*
*(b) for each $f \in \mathbb{F}$, the autonomous Nemytskii operator $N_f$ is continuous at $G$ from $V$ into $\mathbb{H}$, and the set $\{N_f \colon f \in \mathbb{F},\ \|f\|_{\mathbb{F}} \le 1\}$ is equicontinuous at $G$;*
*(c) $C_{\circ G} = TC(\cdot, G)$ is a bounded linear operator from $\mathbb{F}$ into $\mathbb{H}$.*

*If so, then the partial derivatives are $D_1TC(F, G) = C_{\circ G}$ and $D_2TC(F, G) = DN_F(G)$. For $(f, g) \in \mathbb{F} \times \mathbb{G}$,*

$$\mathrm{Rem}_{TC}(F,G)(f,g) = \mathrm{Rem}_{N_F}(G,g) + f\circ(G+g) - f\circ G. \tag{8.5}$$

*If, in addition, for some* $\alpha > 0$,

$$\|\mathrm{Rem}_{N_F}(G,g)\|_{\mathbb{H}} = O(\|g\|_{\mathbb{G}}^{1+\alpha}) \tag{8.6}$$

*as* $g \to 0$ *in* $\mathbb{G}$ *and for some* $\beta > 1$ *and* $\gamma > 1$,

$$\|f\circ(G+g) - f\circ G\|_{\mathbb{H}} = O(\|f\|_{\mathbb{F}}^{\beta} + \|g\|_{\mathbb{G}}^{\gamma}) \tag{8.7}$$

*as* $f \to 0$ *in* $\mathbb{F}$ *and* $g \to 0$ *in* $\mathbb{G}$, *then*

$$\|\mathrm{Rem}_{TC}(F,G)(f,g)\|_{\mathbb{H}} = O\left(\|f\|_{\mathbb{F}}^{\beta} + \|g\|_{\mathbb{G}}^{\min\{\gamma,1+\alpha\}}\right), \tag{8.8}$$

*as* $f \to 0$ *in* $\mathbb{F}$ *and* $g \to 0$ *in* $\mathbb{G}$.

*Proof.* To prove "only if," assume that $TC$ is Fréchet differentiable at $(F, G)$. Then there exist $B \in L(\mathbb{F}, \mathbb{H})$ and $C \in L(\mathbb{G}, \mathbb{H})$ such that as $f \to 0$ in $\mathbb{F}$ and $g \to 0$ in $\mathbb{G}$,

$$\|(F+f)\circ(G+g) - F\circ G - B(f) - C(g)\|_{\mathbb{H}} = o(\|f\|_{\mathbb{F}} + \|g\|_{\mathbb{G}}). \tag{8.9}$$

Specializing to $f = 0$, we get that

$$\|F\circ(G+g) - F\circ G - C(g)\|_{\mathbb{H}} = o(\|g\|_{\mathbb{G}}) \tag{8.10}$$

as $\|g\|_{\mathbb{G}} \to 0$. Thus $N_F$ is Fréchet differentiable at $G$ with derivative $C$, proving $(a)$. Next, letting $g = 0$ in (8.9), we get that $B = C_{\circ G}$, so $(c)$ must hold. Combining (8.9) and (8.10) gives as $\|f\|_{\mathbb{F}} + \|g\|_{\mathbb{G}} \to 0$,

$$\|f\circ(G+g) - f\circ G\|_{\mathbb{H}} = o(\|f\|_{\mathbb{F}} + \|g\|_{\mathbb{G}}). \tag{8.11}$$

Given any $\epsilon > 0$, take $\delta > 0$ such that if $\|f\|_{\mathbb{F}} + \|g\|_{\mathbb{G}} < \delta$ and $G + g \in V$ then

$$\|f\circ(G+g) - f\circ G\|_{\mathbb{H}} < \epsilon(\|f\|_{\mathbb{F}} + \|g\|_{\mathbb{G}}).$$

Take any $f \in \mathbb{F}$. If $0 < \|g\|_{\mathbb{G}} < \delta/(1 + \|f\|_{\mathbb{F}})$ and $G + g \in V$ (such $g$ exist), let $\tilde{f} := \|g\|_{\mathbb{G}} f$. Then

$$\|\tilde{f}\circ(G+g) - \tilde{f}\circ G\|_{\mathbb{H}} < \epsilon(1 + \|f\|_{\mathbb{F}})\|g\|_{\mathbb{G}}.$$

Thus $\|f\circ(G+g) - f\circ G\|_{\mathbb{H}} < \epsilon(1+\|f\|_{\mathbb{F}})$. If $\|g\|_{\mathbb{G}} = 0$ then $\|f\circ(G+g) - f\circ G\|_{\mathbb{H}} = 0$ by the acting condition. Thus $N_f$ is indeed continuous at $G$ and the set of all $N_f$ with $\|f\|_{\mathbb{F}} \le 1$ is equicontinuous at $G$, since for all such $f$, $\|g\|_{\mathbb{G}} < \delta/2$ and $G + g \in V$ imply $\|f\circ(G+g) - f\circ G\|_{\mathbb{H}} < 2\epsilon$, proving $(b)$ and thus "only if."

To prove "if," suppose $(a)$, $(b)$, and $(c)$ hold. By $(a)$, a $C \in L(\mathbb{G}, \mathbb{H})$ satisfying (8.10) exists. By $(c)$, $B := C_{\circ G}$ is a bounded linear operator. To prove (8.9) for these $B$ and $C$ we now just need to prove (8.11). If $\|f\|_{\mathbb{F}} = 0$

then $\|f\circ(G+g) - f\circ G\|_{\mathbb{H}} = 0$ by the acting condition. If $\|f\|_{\mathbb{F}} > 0$ let $\tilde{f} := f/\|f\|_{\mathbb{F}}$. Then (*b*) implies that for every $\epsilon > 0$ there is a $\delta > 0$, not depending on $f$ by equicontinuity, such that if $\|g\|_{\mathbb{G}} < \delta$ and $G + g \in V$ then for any $t \geq 0$, $\|t\tilde{f}\circ(G+g) - t\tilde{f}\circ G\|_{\mathbb{H}} \leq \epsilon t$. For $t = \|f\|_{\mathbb{F}}$ this gives

$$\|f\circ(G+g) - f\circ G\|_{\mathbb{H}} \leq \epsilon\|f\|_{\mathbb{F}} \leq \epsilon(\|f\|_{\mathbb{F}} + \|g\|_{\mathbb{G}}),$$

which implies (8.11). So $TC$ is Fréchet differentiable at $(F, G)$, proving "if." The partial derivatives $D_1$ and $D_2$ are as stated. Then (8.5) follows directly from definition (8.4), and (8.8) follows from (8.5), (8.6), and (8.7). The proof of the theorem is complete. □

**Corollary 8.3.** *In Theorem* 8.2, (*b*) *can be replaced by*

(*b'*) $\|f\circ(G+g) - f\circ G\|_{\mathbb{H}} = o(\|f\|_{\mathbb{F}} + \|g\|_{\mathbb{G}})$ *as* $\|f\|_{\mathbb{F}} + \|g\|_{\mathbb{G}} \to 0$.

*Proof.* To see that (*b*) implies (*b'*), the statement holds when $f = 0$. For $f \neq 0$ let $\tilde{f} := f/\|f\|_{\mathbb{F}}$. Then

$$\|f\circ(G+g) - f\circ G\|_{\mathbb{H}} = \|f\|_{\mathbb{F}}\|\tilde{f}\circ(G+g) - \tilde{f}\circ G\|_{\mathbb{H}} = o(\|f\|_{\mathbb{F}} + \|g\|_{\mathbb{G}})$$

as $\|f\|_{\mathbb{F}} + \|g\|_{\mathbb{G}} \to 0$, since $\|\tilde{f}\circ(G+g) - \tilde{f}\circ G\|_{\mathbb{H}} \to 0$ as $\|g\|_{\mathbb{G}} \to 0$ uniformly in $f$ by (*b*), and noting that $G + g \in V$ for $\|g\|_{\mathbb{G}}$ small enough, proving (*b'*).

Conversely, suppose that (*a*), (*b'*), and (*c*) all hold. Then we will show that $TC$ is Fréchet differentiable at $(F, G)$ from $\mathbb{F} \times V$ into $\mathbb{H}$ with the stated partial derivatives $D_1TC$ and $D_2TC$. The given $D_1TC$ and $D_2TC$ are bounded linear operators between the appropriate Banach spaces by (*c*) and (*a*) respectively. Thus it remains to show that the expression on the right in (8.5) is $o(\|f\|_{\mathbb{F}} + \|g\|_{\mathbb{G}})$ as $\|f\|_{\mathbb{F}} + \|g\|_{\mathbb{G}} \to 0$, which is true by (*a*) and (*b'*). The corollary is proved. □

## 8.3 Measure Space Domains

In this section we will consider the two-function composition operator (8.2) where $\mathbb{G}$ and $\mathbb{H}$ are spaces of $\mu$-equivalence classes of functions, that is, spaces included in $L^0(\mu;\mathbb{K})$ such as $L^p(\mu;\mathbb{K})$, where $\mu$ is a finite measure on a set $\Omega = S$ and $\mathbb{K} = \mathbb{C}$ or $\mathbb{R}$.

Let $(\Omega, \mathcal{S}, \mu)$ be a finite measure space. As defined in Section 1.4, $L^0(\mu;\mathbb{K}) = L^0(\Omega, \mathcal{S}, \mu;\mathbb{K})$ is the space of all $\mu$-equivalence classes of $\mathbb{K}$-valued $\mu$-measurable functions on $\Omega$, with the topology of convergence in $\mu$-measure, metrized by a metric $d_0$. As usual, equivalence classes will be represented by individual functions in them, and equations or inequalities will be understood to hold $\mu$-almost everywhere.

**Definition 8.4.** Let $\mathbb{G} = (\mathbb{G}, \|\cdot\|)$ be a normed space and $\mathbb{G} \subset L^0(\mu;\mathbb{K})$. We will say that $\mathbb{G}$ is a *$\mu$-pre-ideal space* if for all $g \in \mathbb{G}$ and $f \in L^0(\mu;\mathbb{K})$ with $|f| \leq |g|$ we have $f \in \mathbb{G}$ and $\|f\| \leq \|g\|$. A $\mu$-pre-ideal space $\mathbb{G}$ which is complete (a Banach space) will be called a *$\mu$-ideal space.* A $\mu$-(pre)-ideal space $\mathbb{G}$ will be called *full* if $L^\infty(\mu;\mathbb{K}) \subset \mathbb{G}$ and whenever for a sequence $\{g_n\}_{n\geq 1} \subset L^\infty$ with $\sup_n \|g_n\|_\infty < \infty$ and $d_0(g_n, 0) \to 0$ we have $\|g_n\| \to 0$.

For example, the spaces $L^p(\mu;\mathbb{K})$ are $\mu$-ideal spaces for $1 \leq p \leq \infty$ and full for $1 \leq p < \infty$.

**Proposition 8.5.** *If $(\Omega, \mathcal{S}, \mu)$ is a finite measure space and $\mathbb{G} = (\mathbb{G}, \|\cdot\|)$ is a $\mu$-pre-ideal space, then whenever $g_n, g \in \mathbb{G}$ and $\|g_n - g\| \to 0$ we have $d_0(g_n, g) \to 0$.*

*Proof.* We can assume $g = 0$. If $d_0(g_n, 0) \not\to 0$ then taking a subsequence, we can assume that for some $\epsilon > 0$, $\mu(A(n)) > \epsilon$ for all $n$, where $A(n) := \{\omega : |g_n(\omega)| > \epsilon\}$. Let $h_n := \epsilon 1_{A(n)}$. Then $|h_n| \leq |g_n|$ and $h_n$ is $\mu$-measurable, so $h_n \in \mathbb{G}$ and $1_{A_n} \in \mathbb{G}$ with $\|h_n\| \leq \|g_n\| \to 0$. Thus $\|1_{A(n)}\| \to 0$. Taking a further subsequence, we can assume that $\|1_{A(n)}\| \leq 1/2^n$ for all $n$. For any $k \leq n$ let $B(k,n) := \bigcup_{j=k}^n A(j)$. Then $1_{B(k,n)} \leq \sum_{j=k}^n 1_{A(j)}$, so $1_{B(k,n)} \in \mathbb{G}$ and $\|1_{B(k,n)}\| \leq 2^{1-k}$. Let $B(k) := \bigcup_{j=k}^\infty A(j)$. Then $\mu(B(k)) \geq \mu(A(k)) > \epsilon$ for all $k$. We have $B(k) \downarrow B$ as $k \to \infty$ where $\mu(B) \geq \epsilon$. For each $k = 1, 2, \ldots$, there is an $n(k)$ large enough so that $\mu(B(k) \setminus C(k)) < \epsilon/3^k$ for $C(k) := B(k, n(k))$. Let $A := \bigcap_{k=1}^\infty C(k)$. Then $A = B \setminus \bigcup_k (B(k) \setminus C(k))$, so $\mu(A) > \epsilon - \epsilon/2 = \epsilon/2$. We have $1_A \in \mathbb{G}$ since $1_A \leq 1_{C(1)} = 1_{B(1,n(1))} \in \mathbb{G}$ and $1_A \in L^0$, and for each $k$, $\|1_A\| \leq \|1_{C(k)}\| \leq 2^{1-k}$. Thus $1_A = 0$ in $\mathbb{G}$ but not in $L^0(\mu;\mathbb{K})$, a contradiction since $\mathbb{G} \subset L^0$. This proves the proposition. □

The next fact follows directly from the definition of "full":

**Proposition 8.6.** *If $(\Omega, \mathcal{S}, \mu)$ is a finite measure space and $\mathbb{H} = (\mathbb{H}, \|\cdot\|)$ is a full $\mu$-ideal space, then the injection $I : L^\infty(\mu) \to \mathbb{H}$ is bounded, with norm $\|I\| := \|I\|_{L^\infty,\mathbb{H}} < \infty$.*

For example, if $\mathbb{H} = L^p(\Omega, \mathcal{S}, \mu)$, with $1 \leq p < \infty$, then $\mathbb{H}$ is a full $\mu$-ideal space with $\|I\| = \mu(\Omega)^{1/p}$.

As one possible class of Banach spaces to serve as $\mathbb{F}$ in (8.2) we consider the Banach spaces of bounded functions defined as follows. Let $\ell^\infty_{\mathcal{U}}(\mathbb{K})$ be the space of all uniformly bounded, universally measurable functions $f$ from $\mathbb{K}$ into itself, with the supremum norm $\|\cdot\|_{\sup}$. (Since $\mathbb{K}$ is a separable metric space, recall that universal measurability means that for every Borel set $B \subset \mathbb{K}$, $f^{-1}(B)$ is measurable for the completion of every finite measure $\mu$ on the Borel sets of $\mathbb{K}$.) Then $(\ell^\infty_{\mathcal{U}}(\mathbb{K}), \|\cdot\|_{\sup})$ is a Banach space over $\mathbb{K}$. Recall that $C_b(\mathbb{K}) := C_b(\mathbb{K}, \mathbb{K})$ denotes the space of all bounded continuous functions from $\mathbb{K}$ into itself. Then $C_b(\mathbb{K})$ is a Banach subspace of $\ell^\infty_{\mathcal{U}}(\mathbb{K})$.

Here are some sufficient conditions for acting, boundedness, and continuity of the two-function composition operator $TC$.

**Theorem 8.7.** *Let $(\Omega, \mathcal{S}, \mu)$ be a finite measure space, $\mathbb{G}$ a Banach subspace of $L^0(\mu)$, and $\mathbb{H}$ a full $\mu$-ideal space. Then*

*(a) $TC$ acts from $\ell^\infty_{\mathcal{U}}(\mathbb{K}) \times \mathbb{G}$ into $L^\infty(\mu)$.*
*(b) If $B \subset \ell^\infty_{\mathcal{U}}(\mathbb{K})$ is bounded, then $TC$ acts from $B \times \mathbb{G}$ into a bounded subset of $L^\infty(\mu)$.*
*(c) If $F \in \ell^\infty_{\mathcal{U}}(\mathbb{K})$ and $G \in \mathbb{G}$ are such that the autonomous Nemytskii operator $N_F$ is continuous at $G$ from $\mathbb{G}$ into $\mathbb{H}$, then $TC$ is jointly continuous at $(F, G)$ from $\ell^\infty_{\mathcal{U}}(\mathbb{K}) \times \mathbb{G}$ into $\mathbb{H}$.*
*(d) If $F \in C_b(\mathbb{K})$ and $\mathbb{G}$ is a $\mu$-ideal space, then for any $G \in \mathbb{G}$, $TC$ is jointly continuous at $(F, G)$ from $\ell^\infty_{\mathcal{U}}(\mathbb{K}) \times \mathbb{G}$ into $\mathbb{H}$.*

*Proof.* The proof of $(a)$ is as in that of Theorem 7.13$(a)$. Then that of $(b)$ is immediate. For $(c)$, $TC$ acts from $\ell^\infty_{\mathcal{U}}(\mathbb{K}) \times \mathbb{G}$ into $\mathbb{H} = (\mathbb{H}, \|\cdot\|)$ since $L^\infty(\mu) \subset \mathbb{H}$. Let $f_n \to 0$ in $\ell^\infty_{\mathcal{U}}(\mathbb{K})$ and $G_n \to G$ in $\mathbb{G}$. Then

$$\|(F + f_n) \circ G_n - F \circ G\| \le \|(F + f_n) \circ G_n - F \circ G_n\| + \|F \circ G_n - F \circ G\|.$$

The first term equals $\|f_n \circ G_n\|$, which is at most $\|I\| \|f_n\|_{\sup} \to 0$ by Proposition 8.6 and since $\|f_n \circ G_n\|_\infty \le \|f_n\|_{\sup}$. The second term goes to 0 by continuity of $N_F$ at $G$. So $(c)$ is proved.

For $(d)$, if $F \in C_b(\mathbb{K})$ and $G_n \to G$ in $\mathbb{G}$, then $d_0(G_n, G) \to 0$ by Proposition 8.5. Thus $d_0(F \circ G_n, F \circ G) \to 0$. The functions $F \circ G_n \in \mathbb{H}$ are uniformly bounded. Thus by the definition of full $\mu$-ideal space applied to $F \circ G_n - F \circ G$, $\|F \circ G_n - F \circ G\| \to 0$. So $N_F$ is continuous at (the arbitrary) $G \in \mathbb{G}$ from $\mathbb{G}$ into $\mathbb{H}$, proving $(d)$ and the theorem. □

**Remark 8.8.** Suppose $\mathbb{G}$ contains some $g$ with $\mu(g^{-1}(\{t\})) > 0$ for some $t \ne 0$, as is true whenever $\mathbb{G}$ is a non-zero $\mu$-ideal space. Then by considering scalar multiples of $g$, we see that continuity of $F$ in part $(d)$ of the last theorem is necessary for that of $N_F$.

Let $(\Omega, \mathcal{S}, \mu)$ be a finite measure space and $1 \le p < \infty$. We will investigate conditions on Banach spaces $\mathbb{F}$, $\mathbb{G}$ such that $TC$ acts from $\mathbb{F} \times \mathbb{G}$ into $\mathbb{H} := L^p(\mu) = L^p(\Omega, \mathcal{S}, \mu)$, and further conditions under which $TC$ is bounded, continuous at a point, and differentiable at a point. Here $\mathbb{G}$ will be a space of real-valued $\mathcal{S}$-measurable functions on $\Omega$ or of $\mu$-equivalence classes of such functions, and $\mathbb{F}$ will be a space of functions from $\mathbb{R}$ into $\mathbb{R}$, as noted early in Section 8.1.

Suppose that $TC$ acts from $\mathbb{F} \times \mathbb{G}$ into $L^p(\mu)$. If $F \in \mathbb{F}$, then the Nemytskii operator $N_F$ must act from $\mathbb{G}$ into $L^p(\mu)$. If $F \in \mathbb{F}$ and $g, h \in \mathbb{G}$ with $g = h$ a.e. $(\mu)$ then $F \circ g = F \circ h$ a.e. $(\mu)$. So we may as well take $\mathbb{G}$ as a space of $\mu$-equivalence classes of measurable functions.

To apply Theorem 7.13, we will take $\mathbb{G} = L^s(\Omega, \mathcal{S}, \mu)$, where $1 \le s \le \infty$ and $(\Omega, \mathcal{S}, \mu)$ is complete, nonatomic, and perfect. Then by Theorem 7.13($c$), $N_F$ acts from $L^s(\mu)$ into $L^p(\mu)$, $1 \le p < \infty$, if and only if $F$ satisfies the growth condition $F \in \mathcal{G}_{s/p}$ and is universally measurable. Under those conditions, $N_F$ is a bounded nonlinear operator.

For Fréchet differentiability of $TC$ at $(F, G) \in \mathbb{F} \times L^s(\mu)$ we need to check the conditions of Theorem 8.2 for $X = Y = U = \mathbb{R}$. Only condition ($b$) involves both $f$ and $g$. To deal with it we can apply Theorem 7.24, if we take $\mathbb{F} = \mathcal{W}_p(\mathbb{R})$. By Proposition 7.28, no other choice of $\mathbb{F}$ containing a particular function with a simple jump can give a smaller bound for the order of the norm of the remainder term $\|f\circ(G+g) - f\circ G\|_p$.

Recall that $\mathcal{D}_\lambda = \mathcal{D}_\lambda(\Omega, \mathcal{S}, \mu)$ is the set of all $G \in \mathcal{L}^0(\Omega, \mathcal{S}, \mu)$ for which $\mu\circ G^{-1}$ has a bounded density with respect to Lebesgue measure $\lambda$ on $\mathbb{R}$; see also Appendix A. Theorem 7.24 showed that the condition $G \in \mathcal{D}_\lambda$ is both necessary and sufficient for a sharp Hölder condition on an operator $N_F$, in the present case on $N_{F'}$ for a remainder bound. Also recall that $\mathcal{H}_{1+\alpha} = \mathcal{H}_{1+\alpha}(\mathbb{R}; \mathbb{R})$ is defined in Definition 6.5.

**Theorem 8.9.** *Let $(\Omega, \mathcal{S}, \mu)$ be a complete and finite measure space, let $1 \le p < s < \infty$ and $0 < \alpha < 1$ be such that $(1+\alpha)p \le s$. Let $G \in L^s(\Omega, \mathcal{S}, \mu) \cap \mathcal{D}_\lambda$ and $F \in \mathcal{W}_p(\mathbb{R}) \cap \mathcal{H}_{1+\alpha}$. Then the two-function composition operator $TC$ is Fréchet differentiable at $(F, G)$ from $\mathcal{W}_p(\mathbb{R}) \times L^s(\Omega, \mathcal{S}, \mu)$ into $L^p(\Omega, \mathcal{S}, \mu)$. Also, for any $b$ with $p(1+s)/s < b \le (1+\alpha)p(1+s)/s$ and $a := b/(b-1)$, the remainder in the differentiation has the bound* (8.8) *with $\beta = a$ and $\gamma = sb/[p(1+s)]$.*

*Proof.* We apply Theorem 8.2 for $X = Y = U = \mathbb{R}$, $\mathbb{F} = \mathcal{W}_p(\mathbb{R})$, $V = \mathbb{G} = L^s(\Omega, \mathcal{S}, \mu)$, and $\mathbb{H} = L^p(\Omega, \mathcal{S}, \mu)$. For each $f \in \mathcal{W}_p(\mathbb{R})$, $F + f \in \mathcal{W}_p(\mathbb{R})$ is bounded and so satisfies the growth condition $\mathcal{G}_{s/p}$. Also, $F + f$ is Borel by Corollary 2.2, hence universally measurable. Thus by Theorem 7.13($a$), $TC$ acts from $\mathcal{W}_p(\mathbb{R}) \times L^s(\Omega, \mathcal{S}, \mu)$ into $L^p(\Omega, \mathcal{S}, \mu)$. Condition ($a$) and (8.6) hold by Proposition 7.60 with $n = 1$. Condition ($b$) holds by Theorem 7.24. Condition ($c$) holds since $C_{\circ G}$ is bounded and linear $\mathcal{W}_p \to L^\infty \to L^p$, and (8.7) holds by Corollary 7.25. □

**Remark 8.10.** In Theorem 8.9, if $b > (1+\alpha)p(1+s)/s$, then $\min\{1+\alpha, \gamma\} = 1+\alpha$ in (8.8), while $\beta$ would become smaller. Thus there is no advantage in choosing such $b$.

**Remark 8.11.** Let $(\mathbb{F}, \|\cdot\|)$ be a Banach space of bounded Borel measurable functions from $\mathbb{R}$ into $\mathbb{R}$ with $\|\cdot\| \ge \|\cdot\|_{\sup}$, such as $\mathbb{F} = \mathcal{W}_p(\mathbb{R})$. Then for any function $G\colon \Omega \to \mathbb{R}$, measurable for the completion of $\mu$, the map $f \mapsto f\circ G$ is a bounded linear operator from $\mathbb{F}$ into $L^\infty(\mu)$ and so into $L^p(\mu)$. Thus condition ($c$) of Theorem 8.2 will hold.

The following result complements Theorem 8.9 in case the functions $F$ and $G$ satisfy the hypotheses of Proposition 7.52 and

$$\alpha = \frac{s(1+p)}{p(1+s)} - 1 = \frac{s-p}{p(1+s)}. \tag{8.12}$$

**Theorem 8.12.** *Let $(\Omega, \mathcal{S}, \mu)$ be a complete, nonatomic, finite measure space, $-\infty < a < b < \infty$, $1 \le p < s < \infty$, and let $\alpha$ be defined by (8.12). Let $F$ be a function from $\mathbb{R}$ into $\mathbb{R}$ whose restriction to $[a, b]$ is in $\mathcal{H}_{1+\alpha}([a, b]; \mathbb{R})$, $F(x) = F(a)$ for all $x \le a$, and $F(x) = F(b)$ for all $x \ge b$, and let $G$ be a measurable function in $\mathcal{D}_\lambda(\Omega, \mathcal{S}, \mu)$ with values in $[a, b]$. Then the two-function composition operator $TC$ acts from $\mathcal{W}_p \times L^s(\Omega, \mathcal{S}, \mu)$ into $L^p(\Omega, \mathcal{S}, \mu)$ and is Fréchet differentiable at $(F, G)$. The remainder in the differentiation has the bound (8.8) with $\gamma = sb/[p(1+s)]$ and $\beta = b/(b-1)$ for any $b$ with $p(1+s)/s < b \le 1+p$.*

*Proof.* The proof is the same as for Theorem 8.9 except that now in Theorem 8.2, condition $(a)$ and (8.6) hold by Proposition 7.52. □

## 8.4 Spaces with Norms Stronger than Supremum Norms

In this section we will consider the two-function composition operator (8.2) where $\mathbb{G}$ and $\mathbb{H}$ are normed spaces with norms stronger than supremum norms.

Let $\mathbb{B}_1 = (\mathbb{B}_1, \|\cdot\|_1)$ and $\mathbb{B}_2 = (\mathbb{B}_2, \|\cdot\|_2)$ be two Banach spaces over a field $\mathbb{K} = \mathbb{R}$ or $\mathbb{C}$ of $\mathbb{K}$-valued functions defined on a nonempty set $S$. Let $\mathbb{B}_2/\mathbb{B}_1$ be the set of all functions $h\colon S \to \mathbb{K}$ such that $hg \in \mathbb{B}_2$ for each $g \in \mathbb{B}_1$ and $g \mapsto hg$ is a bounded linear operator from $\mathbb{B}_1$ into $\mathbb{B}_2$. For each $h \in \mathbb{B}_2/\mathbb{B}_1$, let

$$\|h\|_{2/1} := \sup\bigl\{\|hg\|_2\colon\ g \in \mathbb{B}_1,\ \|g\|_1 \le 1\bigr\}.$$

Then $\mathbb{B}_2/\mathbb{B}_1$ equipped with the norm $\|\cdot\|_{2/1}$ is a normed vector space over $\mathbb{K}$. If $\mathbb{B}_1$ contains the constant functions $x1(\cdot)$, $x \in \mathbb{K}$, then $\mathbb{B}_2/\mathbb{B}_1 \subset \mathbb{B}_2$ and the embedding of $\mathbb{B}_2/\mathbb{B}_1$ into $\mathbb{B}_2$ is a bounded operator.

**Remark 8.13.** The boundedness of $g \mapsto hg$ in the definition of the set $\mathbb{B}_2/\mathbb{B}_1$ is not automatic as the following shows. Let $\mathbb{B}_1 = \mathbb{B}_2$ be an infinite-dimensional real Banach space with a Hamel basis $\{e_i\}_{i\in I}$. Thus $\mathbb{B}_1$ is the set of all sums $\sum_{i\in I} x_i e_i$ such that $x_i \in \mathbb{R}$ and $x_i = 0$ for all but finitely many $i \in I$. Let $S := I$. Then $\mathbb{B}_1$ is represented as a Banach space of $\mathbb{R}$-valued functions $x\colon i \mapsto x_i$ on $S$ (namely, the set of all functions from $I$ to $\mathbb{R}$, zero except at most on finite sets). For any $h\colon I \to \mathbb{R}$, $hx := \sum_{i\in I} h(i)x_i \in \mathbb{B}_2 = \mathbb{B}_1$, but $x \mapsto hx$ is not a bounded operator if $h$ is unbounded or since, in general, $x \mapsto x_i$ is not a bounded linear functional.

Let $U$ be a nonempty open set in $\mathbb{K}$. Recall that for a vector space of $\mathbb{K}$-valued functions $\mathbb{B}$, $\mathbb{B}^{[U]}$ is the set of all $f \in \mathbb{B}$ such that the closure of the range $\operatorname{ran}(f)$ is included in $U$. Let $(Y(U), \|\cdot\|)$ be a normed space of $\mathbb{K}$-valued functions defined on $U$. Later, we will take $Y(U)$ to be the space $\mathcal{H}_{1+p/q}(-M, M)$ with $0 < M < \infty$. The following is a characterization of differentiability of the two-function composition operator (8.2) at a point $(F, G)$, where $F \in \mathbb{F} = Y(U)$, $G \in \mathbb{B}_1^{[U]} \subset \mathbb{B}_1 = \mathbb{G}$, and $\mathbb{H} = \mathbb{B}_2$.

**Theorem 8.14.** *Let $\mathbb{B}_1 = (\mathbb{B}_1, \|\cdot\|_1)$ and $\mathbb{B}_2 = (\mathbb{B}_2, \|\cdot\|_2)$ be Banach spaces over $\mathbb{K}$ of $\mathbb{K}$-valued functions on a nonempty set $S$ with pointwise operations such that $\mathbb{B}_1 \subset \mathbb{B}_2$, $\|\cdot\|_2 \leq \|\cdot\|_1$ on $\mathbb{B}_1$, $\|\cdot\|_{\sup} \leq \|\cdot\|_2$ on $\mathbb{B}_2$, and $\mathbb{B}_1$ contains the constant functions $x1(\cdot)$, $x \in \mathbb{K}$. For a nonempty open set $U \subset \mathbb{K}$, let $G \in \mathbb{B}_1^{[U]}$ and $F \in Y(U)$. Suppose that $TC$ acts from $Y(U) \times \mathbb{B}_1^{[U]}$ into $\mathbb{B}_2$. Then $TC$ is Fréchet differentiable at $(F, G)$ if and only if the following three conditions hold:*

(*a*) *$F$ is differentiable at $G(x)$ for each $x \in S$, $F'{\circ}G \in \mathbb{B}_2/\mathbb{B}_1$, and as $g \to 0$ in $\mathbb{B}_1$,*

$$\|F{\circ}(G+g) - F{\circ}G - (F'{\circ}G)g\|_2 = o(\|g\|_1); \tag{8.13}$$

(*b*) *for each $f \in Y(U)$, the autonomous Nemytskii operator $N_f$ is continuous at $G$ from $\mathbb{B}_1^{[U]}$ into $\mathbb{B}_2$, and the set $\{N_f \colon f \in Y(U), \|f\| \leq 1\}$ is equicontinuous at $G$;*

(*c*) *$C_{\circ G} = TC(\cdot, G)$ is a bounded linear operator from $Y(U)$ into $\mathbb{B}_2$.*

*Proof.* Since $\|\cdot\|_{\sup} \leq \|\cdot\|_1$ on $\mathbb{B}_1$, $B_1^{[U]}$ is an open set in $\mathbb{B}_1$. We apply Theorem 8.2 for $X = Y = \mathbb{K}$, $\mathbb{F} = Y(U)$, $\mathbb{G} = \mathbb{B}_1$, $V = \mathbb{B}_1^{[U]}$, and $\mathbb{H} = \mathbb{B}_2$. Since these choices in (*b*) and (*c*) of Theorem 8.2 yield (*b*) and (*c*) of Theorem 8.14, one needs only to check the equivalence of the conditions (*a*).

Suppose that (*a*) of Theorem 8.2 holds in this case. To show that the current (*a*) holds, we use Proposition 6.9 with $X = Y = \mathbb{K}$, $\mathbb{G} = \mathbb{B}_1$, $V = \mathbb{B}_1^{[U]}$, $\mathbb{H} = \mathbb{B}_2$, $\psi(u, x) \equiv F(u)$, and $k = 1$. Clearly, $x \mapsto x1(\cdot)$ is a bounded operator from $\mathbb{K}$ into $\mathbb{B}_1$. Since $\|\cdot\|_{\sup} \leq \|\cdot\|_2$ on $\mathbb{B}_2$, for each $s \in S$, the evaluation $h \mapsto h(s)$ is a bounded operator from $\mathbb{B}_2$ into $\mathbb{K}$. Since the autonomous Nemytskii operator $N_F$ acts from $V = \mathbb{B}_1^{[U]}$ into $\mathbb{B}_2$ and is Fréchet differentiable at $G$, by the second part of Proposition 6.9, $F$ is differentiable at $G(x)$ for each $x \in S$ and the derivative operator $DN_F(G)$ is the linear multiplication operator $M[F'{\circ}G]$, and so $F'{\circ}G \in \mathbb{B}_2/\mathbb{B}_1$. Thus (8.13) holds since the left side is the norm of the remainder $\mathrm{Rem}_{N_F}(G, g)$. Since the current condition (*a*) clearly implies condition (*a*) of Theorem 8.2, the proof is complete. □

**Corollary 8.15.** *In Theorem 8.14, (b) can be replaced by*

(*b'*) $\|f \circ (G+g) - f \circ G\|_2 = o(\|f\| + \|g\|_1)$ *as* $\|f\| + \|g\|_1 \to 0$.

*Proof.* In Theorem 8.14, $(a)$ is equivalent to condition $(a)$ of Theorem 8.2, while $(b)$ and $(c)$ are the same as conditions $(b)$ and $(c)$ of Theorem 8.2. Thus the conclusion holds by Corollary 8.3. □

Next consider two-function composition $TC\colon \mathbb{F}\times\mathbb{G}\to\mathbb{H}$ with $\mathbb{G}=\mathbb{H}=(\ell^\infty(S),\|\cdot\|_{\sup})$ and $\mathbb{F}=(\mathcal{H}_{\alpha,\infty},\|\cdot\|_{\mathcal{H}_\alpha})$ for some $\alpha\in(0,1]$, as defined before (6.2). Under some conditions we will show its Fréchet differentiability at $(F,G)$, where $G\colon S\to[c,d]\subset(a,b)$ and $F\colon [a,b]\to\mathbb{R}$.

**Proposition 8.16.** *Let $S$ be a nonempty set, let $J$ be a nonempty open interval (bounded or unbounded), and let $0<\alpha,\beta\le 1$. Let $F\in\mathcal{H}_{1+\beta}(J;\mathbb{R})$, let $G\in\ell^\infty(S)$, and let $V$ be a neighborhood of $G$ in $\ell^\infty(S)$ such that all functions in $V$ take values in $J$. Then $TC$ acts from $\mathcal{H}_{\alpha,\infty}(J;\mathbb{R})\times V$ into $\ell^\infty(S)$, it is Fréchet differentiable at $(F,G)$ with the remainders*

$$\big\|\mathrm{Rem}_{N_F}(G,g)\big\|_{\sup}=O(\|g\|_{\sup}^{1+\beta}) \tag{8.14}$$

*as $\|g\|_{\sup}\to 0$,*

$$\big\|f\circ(G+g)-f\circ G\big\|_{\sup}\le\|f\|_{\mathcal{H}_\alpha}\|g\|_{\sup}^{\alpha}, \tag{8.15}$$

*and*

$$\big\|\mathrm{Rem}_{TC}(F,G)(f,g)\big\|_{\sup}=O\left(\|f\|_{\mathcal{H}_\alpha}^{1+\alpha}+\|g\|_{\sup}^{1+\min\{\beta,\alpha\}}\right) \tag{8.16}$$

*as $\|f\|_{\mathcal{H}_\alpha}+\|g\|_{\sup}\to 0$.*

*Proof.* Clearly $TC$ acts from $\mathcal{H}_{\alpha,\infty}(J;\mathbb{R})\times V$ into $\ell^\infty(S)$. We will apply Theorem 8.2 for $X=Y=\mathbb{R}$, $U=J$, $\mathbb{F}=\mathcal{H}_{\alpha,\infty}(J;\mathbb{R})$, and $\mathbb{G}=\mathbb{H}=\ell^\infty(S)$. By hypothesis, there is a $\delta>0$ such that $\|G\|_{\sup}\le 1/\delta$ and $G(s)+u\in J$ for each $s\in S$ and $u\in\mathbb{R}$ with $|u|\le\delta$. Recalling the definition of $U_\delta$ before Proposition 6.3, the range $\mathrm{ran}\,(G)$ is included in $U_{1/m}=: B_m$ for some integer $m$ with $1/m\le\delta$. Thus Fréchet differentiability of $N_F$ at $G$ and (8.14) follow from Proposition 6.7 with $\alpha$ there replaced by $\beta$. Thus $(a)$ of Theorem 8.2 holds, and $(c)$ clearly holds. Relation (8.15) is immediate from the definition of $\|\cdot\|_{(\mathcal{H}_\alpha)}\le\|\cdot\|_{\mathcal{H}_\alpha}$. Since $(b)$ of Theorem 8.2 follows from (8.15), $TC$ is Fréchet differentiable at $(F,G)$ by the first part of Theorem 8.2. By inequality (3.20) of W. H. Young applied to the right side of (8.15), (8.7) holds for $\beta=\gamma=1+\alpha$. Thus (8.16) follows from (8.8), proving the proposition. □

**Remark 8.17.** Suppose we are given $F\in\mathcal{H}_{1+\beta}([c,d];\mathbb{R})$ with $c<d$ and want to apply Proposition 8.16 only when $G+g$ also takes values in $[c,d]$. This can occur for example when $G$ and $G+g$ are probability distribution functions, with $c=0,d=1$. Then, we can set $a:=c-1$, $b:=d+1$, extend $F'$ to be constant on $[a,c]$ and on $[d,b]$, and extend each $f\in\mathcal{H}_\alpha([c,d];\mathbb{R})$ to an element of $\mathcal{H}_\alpha([a,b];\mathbb{R})$ without increasing $\|f\|_{(\mathcal{H}_\alpha)}$ or $\|f\|_{\sup}$, e.g. [53, Prop. 11.2.3] for the metric $d(x,y):=|x-y|^\alpha$, $x,y\in\mathbb{R}$.

## 8.5 Two-Function Composition on $\mathcal{W}_p$ Spaces

In this section we consider the two-function composition operator (8.2), where $\mathbb{F}$ is a normed space of functions, $\mathbb{G} = \mathcal{W}_p$ and $\mathbb{H} = \mathcal{W}_q$ with $1 \le p \le q < \infty$.

By Corollary 6.36, $F{\circ}G \in \mathcal{W}_q$ for all $G \in \mathcal{W}_p$ if and only if $F$ satisfies a Hölder condition of order $p/q$ locally. This gives the following acting conditions for the $TC$ operator.

**Proposition 8.18.** *Let $1 \le p \le q < \infty$, $\alpha := p/q$, and $J := [a, b]$ with $a < b$. Then the two-function composition operator $TC$ acts from $\mathcal{H}_\alpha^{\rm loc} \times \mathcal{W}_p(J)$ into $\mathcal{W}_q(J)$.*

Generally we consider operators acting on Banach spaces or at least normed spaces, so we should note that in the preceding proposition, $\mathcal{H}_\alpha^{\rm loc}$ is not a normed space.

We will show that the joint Fréchet differentiability of the two-function composition operator with $\mathcal{W}_r$, $r > q$, instead of $\mathcal{W}_q$, holds whenever $F$ and $f$ have derivatives in the Hölder space $\mathcal{H}_\alpha$ with $\alpha = p/q$. Recall that classes of such functions, denoted by $\mathcal{H}_{1+\alpha} \subset \mathcal{H}_{1+\alpha}^{\rm loc}$, are defined in Definition 6.5.

Theorem 8.14 characterizes Fréchet differentiability of the two-function composition operator at a point. The following corollary will be applied to show that condition $(a)$ of Theorem 8.14 holds. It follows from Theorem 6.77 taking $n = 1$, $X = Y = \mathbb{R}$, $U = (-M, M)$ with $0 < M < \infty$, and $\psi(u, s) \equiv F(u)$ not depending on $s$.

**Corollary 8.19.** *Let $1 \le p \le q < r < \infty$, $\alpha := p/q$, $J := [a, b]$ with $a < b$, and $W := (-M, M)$ with $0 < M < \infty$. Let $G \in \mathcal{W}_p^{[W]}(J)$ and $F \in \mathcal{H}_{1+\alpha}^{\rm loc}(W)$. Then the autonomous Nemytskii operator $N_F$ acts from $\mathcal{W}_p^{[W]}(J)$ into $\mathcal{W}_p(J)$ and is Fréchet differentiable at $G$ from $\mathcal{W}_p^{[W]}(J)$ into $\mathcal{W}_r(J)$. Moreover, there exist constants $C = C(F, G, M, p, q, r) < +\infty$ and $\delta = \delta(G, M) > 0$ such that for the remainder in the differentiation of $N_F$,*

$$\begin{aligned} \|{\rm Rem}_{N_F}(G, g)\|_{[r]} &= \|F{\circ}(G + g) - F{\circ}G - (F'{\circ}G)g\|_{[r]} \\ &\le C\|g\|_{[p]}^{1+\alpha-(p/r)} \end{aligned} \tag{8.17}$$

*for each $g \in \mathcal{W}_p(J)$ such that $\|g\|_{[p]} \le \delta$.*

Now we are ready to state bounds for the remainder ${\rm Rem}_{TC}$ in the joint differentiability of the two-function composition operator, given by (8.4) and (8.5). Recalling the definitions (6.2), for a nonempty open set $U \subset \mathbb{R}$ and $\alpha \in (0, 1]$, let $\mathcal{H}_{1+\alpha,\infty}(U)$ be the set of all bounded $C^1$ functions $f$ such that $f' \in \mathcal{H}_{\alpha,\infty}(U)$. With the norm $\|f\|_{1+\alpha,\infty} := \|f\|_{U,\mathcal{H}_1} + \|f'\|_{U,\mathcal{H}_\alpha}$, $\mathcal{H}_{1+\alpha,\infty}(U)$ is a Banach space.

**Theorem 8.20.** *Let $1 \le p \le q < r < \infty$, $\alpha := p/q$, $J := [a, b]$ with $a < b$, and $W := (-M, M)$ with $0 < M < \infty$. Let $G \in \mathcal{W}_p^{[W]}(J)$ and $F \in \mathcal{H}_{1+\alpha,\infty}(W)$.*

*Then the two-function composition operator $TC$ is Fréchet differentiable at $(F, G)$ from $\mathcal{H}_{1+\alpha,\infty}(W) \times \mathcal{W}_p^{[W]}(J)$ into $\mathcal{W}_r(J)$. Moreover, there exist constants $C = C(F, G, M, p, q, r) < +\infty$ and $\delta = \delta(G, M) > 0$ such that*

$$\begin{aligned}\|\mathrm{Rem}_{TC}(F,G)(f,g)\|_{[r]} &= \big\|(F+f)\circ(G+g) - F\circ G - f\circ G - (F'\circ G)g\big\|_{[r]} \\ &\leq C\|g\|_{[p]}\big(\|f\|_{1+\alpha,\infty} + \|g\|_{[p]}^{\alpha-(p/r)}\big)\end{aligned} \tag{8.18}$$

*for each $f \in \mathcal{H}_{1+\alpha,\infty}(W)$ and for each $g \in \mathcal{W}_p(J)$ such that $\|g\|_{[p]} \leq \delta$.*

**Remark 8.21.** The bound (8.18), as well as the bound (8.17), holds when $q = r$. In this case, however, the stated Fréchet differentiability may not hold.

**Proof of Theorem 8.20.** By Corollary 8.19, $TC$ acts from $\mathcal{H}_{1+\alpha,\infty}(W) \times \mathcal{W}_p^{[W]}(J)$ into $\mathcal{W}_p(J)$. We apply Theorem 8.14 with $S = J$, $\mathbb{B}_1 = \mathcal{W}_p(J)$, $\mathbb{B}_2 = \mathcal{W}_r(J)$, $U = W$, and $Y(U) = \mathcal{H}_{1+\alpha,\infty}(W)$. Since $p \leq q < r$, we have $\mathcal{W}_p(J) \subset \mathcal{W}_q(J) \subset \mathcal{W}_r(J)$ and $\|\cdot\|_{[r]} \leq \|\cdot\|_{[q]} \leq \|\cdot\|_{[p]}$. By Proposition 6.34 with $\alpha = 1$, $U = W$, and $Y = X = \mathbb{R}$, for $f \in \mathcal{H}_{1,\infty}(W)$ we have

$$\|f\circ G\|_{(r)} \leq \|f\circ G\|_{(q)} \leq \|f\|_{W,(\mathcal{H}_1)}\|G\|_{(q)} \leq \|f\|_{W,(\mathcal{H}_1)}\|G\|_{(p)}.$$

Also, since $\|f\circ G\|_{\sup} \leq \|f\|_{W,\sup}$, we have

$$\|TC(f,G)\|_{[r]} \leq \|f\|_{W,\mathcal{H}_1} \max\{1, \|G\|_{(p)}\}.$$

Thus $(c)$ of Theorem 8.14 holds. We have $\|G\|_{\sup} < M$. So, for $g \in \mathcal{W}_p(J)$ such that $\|g\|_{[p]} < \delta_1(G, M) := M - \|G\|_{\sup}$, $G + g \in \mathcal{W}_p^{[W]}(J)$. Then for $f \in \mathcal{H}_{1+\alpha,\infty}(W)$, by Theorem 6.40 with $X = Y = \mathbb{R}$, it follows that

$$\begin{aligned}\big\|f\circ(G+g) - f\circ G\big\|_{[q]} &\leq \|f'\|_{W,\mathcal{H}_\alpha}\|g\|_{[q]} \max\{1, \|G\|_{(p)}^\alpha\} \\ &\leq \|f\|_{1+\alpha,\infty}\|g\|_{[p]} \max\{1, \|G\|_{(p)}^\alpha\}.\end{aligned} \tag{8.19}$$

Thus the autonomous Nemytskii operator $N_f$ is continuous at $G$ from $\mathcal{W}_p^{[W]}(J)$ into $\mathcal{W}_r(J)$, and the set of all functions $N_f$ with $f \in \mathcal{H}_{1+\alpha,\infty}(W)$ and $\|f\|_{1+\alpha,\infty} \leq 1$ is equicontinuous at $G$, so $(b)$ of Theorem 8.14 holds. Since $(a)$ of Theorem 8.14 follows from Corollary 8.19, the desired Fréchet differentiability holds. Taking a minimum if necessary, we can assume that $\delta = \delta(G, M)$ in the second part of Corollary 8.19 is less than or equal to $\delta_1(G, M)$. By (8.5), the bound (8.18) for the remainder follows from (8.19) and (8.17). The proof of Theorem 8.20 is complete. □

## 8.6 Notes

*Notes on Section 8.2.* In the literature outside of statistics, where we found that the two-function composition operator $(f, g) \mapsto (F + f)\circ(G + g)$ has been considered, $f$ has been assumed differentiable at least once: Brokate and Colonius [26], Gray [85], Hartung and Turi [88]; regarding $C^k$ spaces for $k \geq 1$ and Sobolev spaces see Ebin and Marsden [61, p. 108], who give earlier references, and on $C^k$ spaces, also Garay [73], whom we thank for pointing out his paper and some other references. On the $C^\infty$ case, which apparently has been much studied in connection with infinite-dimensional Lie groups, cf. Milnor [171]. On the composition operator for holomorphic functions, cf. Stevenson [225], [226]; or for linear operators, Dieudonné [42, (8.3.1) p. 148]. Thus it was a striking innovation by Reeds [192] to take $f$ non-differentiable and indeed discontinuous, in the space $D[0,1]$ of right-continuous functions with left limits on $[0,1]$. Reeds and, following his lead, Fernholz [65] proved (compact) differentiability of the two-function composition operator.

*Notes on Section 8.3.* Appell and Zabrejko treat Nemytskii operators on ideal Banach spaces, which we call $\mu$-ideal, in [3, Chapter 2]. What we call $\mu$-pre-ideal spaces and $\mu$-ideal spaces have been called respectively *Köthe function spaces* and *Banach function spaces* by many authors, e.g. in the books [215, p. 127] and [170, p. 115]. The notion of Banach function space apparently first appeared in the 1955 thesis of W. A. J. Luxemburg [148] and was developed in a series of 19 notes, the first 13 with A. C. Zaanen, beginning with [149]. See also [256, p. 252]. The names "pre-ideal" and "ideal" spaces were apparently coined by Zabrejko. We have adopted the terms (but added "$\mu$-") because there are many Banach spaces of functions such as spaces of continuous, Hölder, or differentiable functions or $\mathcal{W}_p$ spaces which are not $\mu$-ideal spaces; on the other hand, $\mu$-ideal spaces are spaces of equivalence classes of functions depending on $\mu$. Proposition 8.5 is well known and appears e.g. as [215, Prop. 13.2] and [170, Prop. 2.6.3].

# 9

# Product Integration

As we saw in Chapter 1, the matrix-valued product integral first arose as a way of representing the solutions of initial value problems for linear ordinary differential equations

$$df(t)/dt = C(t) \cdot f(t)$$

and for linear integral equations

$$f(t) = f(a) + \int_a^t C(s)f(s)\, ds,$$

where $-\infty < a \leq t < \infty$, $f$ has values in $\mathbb{R}^k$, and $C$ is a $k \times k$ matrix-valued function. This representation has a long history, going back to the 1880s. It originated in work of Volterra [235]. He defined the product integral with respect to $C$ over an interval $[a, t]$ with $a < t$ to be the limit

$$f(t) := \lim_{\tau} \big[I + C(s_n)(t_n - t_{n-1})\big] \cdots \big[I + C(s_1)(t_1 - t_0)\big] f(a), \tag{9.1}$$

as the mesh $|\tau|$ of tagged partitions $\tau = (\{t_i\}_{i=0}^n, \{s_i\}_{i=1}^n)$ of $[a, t]$ tends to zero, where $I$ is the $k \times k$ identity matrix. Almost at the same time, Peano [183] suggested a representation of the solution as a series of iterated integrals of the form

$$f(t) := f(a) + \sum_{k=1}^{\infty} \int_{a<s_1<s_2<\cdots<s_k<t} C(s_k) \cdots C(s_2)C(s_1)f(a) ds_k \cdots ds_2 ds_1$$

for each $t > a$, which is often considered as a series representation of the product integral (9.1). On the history of the product integral generally, see also the notes at the end of this chapter.

In this book, we consider the product integral with respect to interval functions rather than point functions. In Theorem 9.51 we will give a Taylor series expansion for the nonlinear operator induced by the product integral

R.M. Dudley and R. Norvaiša, *Concrete Functional Calculus*, Springer Monographs in Mathematics, DOI 10.1007/978-1-4419-6950-7_9,

acting between suitable Banach spaces of interval functions of bounded $p$-variation for $p < 2$. Then, Corollary 9.52 will show that this Taylor expansion has a form analogous to Peano's series.

In this chapter, we consider functions with values in a Banach algebra $\mathbb{B}$ with norm $\|\cdot\|$, and assume that $\mathbb{B}$ is unital, that is, it contains an identity $\mathbb{I}$. The multiplication from $\mathbb{B} \times \mathbb{B}$ into $\mathbb{B}$ will be written $(x, y) \mapsto xy$ rather than $x{\cdot}y$. Likewise, in integrals the dots between integrands and integrators will be omitted. In Section 9.12 we consider a Banach space $Y$ and the Banach algebra $\mathbb{B} = \mathbb{B}_Y$ of bounded linear operators from $Y$ into itself (which is unital). There we treat some $Y$-valued functions, and in some integrals, dots appear.

## 9.1 Multiplicative Interval Functions and $\Phi$-Variation

Let $J$ be a nonempty interval in $\mathbb{R}$, let $Y$ be a Banach space, and let $\mathcal{I}(J; Y)$ be the class of all interval functions defined on the set $\mathfrak{I}(J)$ of subintervals of $J$ with values in $Y$ as defined in Section 2.1. There, we considered the properties of additivity and upper continuity, and proved several facts. The $\Phi$-variation of such functions was introduced and examined in Sections 3.2 and 3.3. If the Banach space $Y$ is a Banach algebra $\mathbb{B}$, one can define what will be called a multiplicative interval function. The product integral defined next will be the main example of such a function. Recall the definitions in Section 1.4. Also recall that a nondegenerate interval is one containing more than one point.

For an interval function $\mu$ on $J$ with values in a Banach algebra $\mathbb{B}$, for a nonempty interval $A \subset J$, and for an interval partition $\mathcal{A} = \{A_i\}_{i=1}^n$ of $A$, let

$$P(\mu; A, \mathcal{A}) := \prod_{i=1}^{n} \big(\mathbb{I} + \mu(A_i)\big) = \big(\mathbb{I} + \mu(A_n)\big) \cdots \big(\mathbb{I} + \mu(A_1)\big). \tag{9.2}$$

Since $\mathbb{B}$ may not be commutative, the product sign is used below with the prescribed order. In (1.12) we saw this order in solving linear ordinary differential equations. Some authors use the reverse order.

**Definition 9.1.** Let $\mu \in \mathcal{I}([a, b]; \mathbb{B})$ with $-\infty < a \le b < +\infty$. We say that the *product integral* $\prod(\mathbb{I} + d\mu)$ with respect to $\mu$ is defined on $[a, b]$ if for each nonempty $A \in \mathfrak{I}[a, b]$ the limit

$$\prod_A(\mathbb{I} + d\mu) := \lim_{\mathcal{A}} P(\mu; A, \mathcal{A}) \tag{9.3}$$

exists under refinements of partitions $\mathcal{A}$ of $A$. We define $\prod_{\emptyset}(\mathbb{I} + d\mu) := \mathbb{I}$.

Thus the product integral $\prod(\mathbb{I} + d\mu)\colon\ A \mapsto \prod_A(\mathbb{I} + d\mu)$, $A \in \mathfrak{I}[a, b]$, if it exists, is a $\mathbb{B}$-valued interval function on $[a, b]$.

**Remark 9.2.** The product integral is of most interest when $\mu$ is additive. But additivity is by no means necessary for existence of the product integral, as can be seen as follows. Let $a < b$ and let $\mu$ be an interval function on $[a, b]$ having a product integral. For example, let $\mu$ be additive, specifically, $\mu \equiv 0$. Let $0 < c < b - a$. Let $\nu$ be an interval function on $[a, b]$ equal to $\mu$ on all intervals of length less than $c$ and defined arbitrarily on other subintervals of $[a, b]$. Then $\nu$ has a product integral equal to that of $\mu$, since in taking refinements of partitions we will eventually have only intervals of length less than $c$. Clearly, $\nu$ will not be additive nor bounded in general.

**Definition 9.3.** An interval function $\rho$ on $[a, b]$ with values in a Banach algebra $\mathbb{B}$ will be called *multiplicative* if $\rho(\emptyset) = \mathbb{1}$ and for any $A, B \in \mathfrak{I}[a, b]$ such that $A \cup B \in \mathfrak{I}[a, b]$ and $A \prec B$ if both intervals are nonempty, we have

$$\rho(A \cup B) = \rho(B)\rho(A). \tag{9.4}$$

If instead of (9.4), we have $\rho(A \cup B) = \rho(A)\rho(B)$ then $\rho$ will be called $*$-multiplicative.

**Remark 9.4.** The order of multiplication in (9.4) is as in the solution of (forward) initial value problems; see (1.12).

The product integral (9.3), if it exists, is a multiplicative interval function, as we will show in Proposition 9.27. Simple examples of multiplicative interval functions can be given using invertible point functions as follows. Recall that $[a, b]^{\pm} = \{a, a+, b-, b\} \cup \bigcup_{a<t<b}\{t-, t, t+\}$ as defined in (6.9). Let $h$ be a regulated function on $[a, b]$ with values in $\mathbb{B}$. Then $h$ extends naturally to a function $\tilde{h}$ on $[a, b]^{\pm}$ (see Proposition 6.20). If each $x$ in $\{\tilde{h}(t)\colon\ t \in [a, b]^{\pm}\} \subset \mathbb{B}$ has an inverse $x^{-1}$, then we say that $h$ is an *invertible function*, and define the *reciprocal function* $h^{\mathrm{inv}}$ on $[a, b]^{\pm}$ by $h^{\mathrm{inv}}(u) := (h(u))^{-1}$ for each $u \in [a, b]^{\pm}$. By Theorem 4.16($b$), the reciprocal function $h^{\mathrm{inv}}$ is regulated whenever defined. That is, for $a \le s < t \le b$, we have

$$\lim_{u \downarrow s} h^{\mathrm{inv}}(u) = h^{\mathrm{inv}}(s+) \quad \text{and} \quad \lim_{u \uparrow t} h^{\mathrm{inv}}(u) = h^{\mathrm{inv}}(t-). \tag{9.5}$$

Moreover, by Theorem 4.16($a$), if $\|x - \mathbb{1}\| < 1$ then $x$ has an inverse $x^{-1}$. Thus if $\|h(t) - \mathbb{1}\| < 1$ for each $t \in [a, b]^{\pm}$ then $h$ is an invertible function.

If $h$ is an invertible function on $[a, b]$, let $\alpha_h := \alpha_{h,[a,b]}$ be the interval function on $[a, b]$ defined on open intervals or singletons $\subset [a, b]$ by

$$\begin{aligned}
\alpha_{h,[a,b]}((s,t)) &:= h(t-)h(s+)^{\mathrm{inv}} \quad \text{for } a \le s < t \le b,\\
\alpha_{h,[a,b]}(\{t\}) &:= h(t+)h(t-)^{\mathrm{inv}} \quad \text{for } a < t < b,\\
\alpha_{h,[a,b]}(\{a\}) &:= h(a+)h(a)^{\mathrm{inv}} \quad \text{and} \quad \alpha_{h,[a,b]}(\{b\}) := h(b)h(b-)^{\mathrm{inv}}
\end{aligned}$$

if $a < b$, and $\alpha_{h,[a,a]}(\{a\}) := \alpha_{h,[a,a]}(\emptyset) := \mathbb{1}$ if $a = b$. Then $\alpha_h$ extends uniquely to a multiplicative interval function on $[a, b]$.

Here we consider upper continuity and $\Phi$-variation of an arbitrary multiplicative interval function. Recall Definition 2.3 of upper continuous interval functions and Definition 3.16 of the $p$-variation of an interval function.

Next is an example of a multiplicative interval function which is neither upper continuous at $\emptyset$ nor has bounded $p$-variation for any $0 < p < \infty$.

**Example 9.5.** As in Example 2.4, let $(\mathbb{B}, \|\cdot\|)$ be either the Banach space $\ell^\infty[0,1]$ with the supremum norm or the Banach space $L^\infty([0,1],\lambda)$ with the essential supremum norm, where $\lambda$ is Lebesgue measure. In both cases $\mathbb{B}$ is a Banach algebra under pointwise multiplication. Also, $\mathbb{B}$ is commutative and unital with the identity $\mathbb{I}$ being the function equal to 1 everywhere on $[0,1]$. For an interval $A \subset [0,1]$, let $\rho(A) := \mathbb{I} + 1_A$. Thus $\rho$ is an interval function from $\mathfrak{I}[0,1]$ into $\mathbb{B}$. If intervals $A, B \subset [0,1]$, $A \prec B$, and $A \cup B$ is again an interval, then

$$\rho(B)\rho(A) = (\mathbb{I} + 1_B)(\mathbb{I} + 1_A) = \mathbb{I} + 1_B + 1_A = \rho(A \cup B).$$

Since also $\rho(\emptyset) = \mathbb{I}$, the interval function $\rho$ is multiplicative. Let $A_k \downarrow \emptyset$ and $A_k \neq \emptyset$ if $\mathbb{B} = \ell^\infty[0,1]$, or $\lambda(A_k) > 0$ if $\mathbb{B} = L^\infty([0,1],\lambda)$. Then $\|\rho(A_k)\| = 2 \not\to 1 = \|\mu(\emptyset)\|$, so $\rho$ is not upper continuous at $\emptyset$. Let $\mathcal{A} = \{A_i\}_{i=1}^n$ be an interval partition of $[0,1]$, with $\lambda(A_i) > 0$ in the case $\mathbb{B} = L^\infty([0,1],\lambda)$, and $0 < p < \infty$. Then

$$s_p(\rho; \mathcal{A}) = \sum_{i=1}^{n} \|1_{A_i}\|^p = n.$$

Since $n$ is arbitrary, $v_p(\rho; [0,1]) = +\infty$.

As for additive interval functions, the following three propositions characterize multiplicative upper continuous interval functions. Given an interval function $\mu$ on $[a,b]$ define the functions $R_{\mu,a}$ and $L_{\mu,b}$ on $[a,b]$ respectively by

$$R_{\mu,a}(t) := \begin{cases} \mu(\emptyset) & \text{if } t = a, \\ \mu([a,t]) & \text{if } t \in (a,b], \end{cases} \quad \text{and} \quad L_{\mu,b}(t) := \begin{cases} \mu([t,b]) & \text{if } t \in [a,b), \\ \mu(\emptyset) & \text{if } t = b. \end{cases}$$

The function $R_{\mu,a}$ on $[a,b]$ agrees with the function $R_{\mu,a}$ on $[\![a,b]\!]$ defined by (2.3). We say that an interval function $\mu$ on $J$ is *bounded* if $\sup\{\|\mu(A)\| \colon A \in \mathfrak{I}(J)\} < \infty$. The following statement is similar to parts of Proposition 2.6.

**Proposition 9.6.** *Let $\rho$ be a $\mathbb{B}$-valued bounded and multiplicative interval function on $J = [a,b]$. Then the following three statements are equivalent:*

*(a) $\rho$ is upper continuous;*
*(b) $\rho$ is upper continuous at $\emptyset$;*
*(c) $\rho(A_n) \to \mathbb{I}$ whenever open intervals $A_n \downarrow \emptyset$ and*

$$\operatorname{card}\{t \in J \colon \|\rho(\{t\}) - \mathbb{I}\| > \epsilon\} < \infty \quad \textit{for all } \epsilon > 0. \tag{9.6}$$

*Any of (a), (b), or (c) implies the following two statements:*

(d) $R_{\rho,a}$ *is regulated,* $R_{\rho,a}(t-)=\rho([a,t))$ *for* $t\in(a,b]$*, and* $R_{\rho,a}(t+)=\rho([a,t])$ *for* $t\in[a,b)$*;*
(e) $L_{\rho,b}$ *is regulated,* $L_{\rho,b}(t-)=\rho([t,b])$ *for* $t\in(a,b]$*, and* $L_{\rho,b}(t+)=\rho((t,b])$ *for* $t\in[a,b)$*.*

*Moreover, statement (b) holds if either of the following two statements holds:*

(d′) *statement (d) holds and* $R_{\rho,a}$ *is an invertible function;*
(e′) *statement (e) holds and* $L_{\rho,b}$ *is an invertible function.*

*Proof.* (a) ⇔ (b): clearly, (a) implies (b). For (b) ⇒ (a), let intervals $A_n\downarrow A$. Then $A_n=B_n\cup A\cup C_n$ for intervals $B_n\prec A\prec C_n$, with $B_n\downarrow\emptyset$ and $C_n\downarrow\emptyset$. So $\rho(A_n)=\rho(C_n)\rho(A)\rho(B_n)\to\rho(A)$ by continuity of the product in $\mathbb{B}$.

(b) ⇒ (c). The first part of statement (c) is clear. For the second part, suppose that there exist an $\epsilon>0$ and an infinite sequence $\{t_j\colon j\geq 1\}$ of distinct points of $J$ such that

$$\|\rho(\{t_j\})-\mathbb{I}\|>\epsilon\qquad\text{for all } j\geq 1. \tag{9.7}$$

Then there exist $t\in J$ and a subsequence $\{t_j'\colon j\geq 1\}$ such that either $t_j'\downarrow t$ or $t_j'\uparrow t$ as $j\to\infty$. In the first case, by multiplicativity of $\rho$, we have

$$\rho((t,t_j'])-\mathbb{I}=\rho(\{t_j'\})\,[\rho((t,t_j'))-\mathbb{I}]+\rho(\{t_j'\})-\mathbb{I}.$$

Due to the boundedness assumption, $\sup_j\|\rho(\{t_j\})\|<\infty$. Thus the left side and the first term on the right side tend to zero because $(t,t_j']\downarrow\emptyset$ and $(t,t_j')\downarrow\emptyset$. Hence $\rho(\{t_j'\})\to\mathbb{I}$ as $j\to\infty$. This contradicts (9.7). Therefore (9.6) holds in the first case. A proof in the second case is symmetric.

For (c) ⇒ (b), let intervals $A_n\downarrow\emptyset$. Then for some $t$ and $n_0$, either for all $n\geq n_0$, $A_n=(t,s_n]\!]$ for some $s_n\in(t,b)$, or for all $n\geq n_0$, $A_n=[\![s_n,t)$ for some $s_n\in(a,t)$. Using continuity of the multiplication in $\mathbb{B}$, in each of the two cases, $\rho(A_n)\to\mathbb{I}$ follows by (c).

For (b) ⇒ (d), since $\rho$ is bounded and multiplicative, we have for $a<t\leq b$,

$$R_{\rho,a}(s)-\rho([a,t))=\big[\mathbb{I}-\rho((s,t))\big]\rho([a,s])\to 0$$

as $s\uparrow t$, and for $a\leq t<b$,

$$R_{\rho,a}(s)-\rho([a,t])=\big[\rho((t,s])-\mathbb{I}\big]\rho([a,t])\to 0$$

as $s\downarrow t$. Thus $R_{\rho,a}$ is regulated and (d) holds. The implication (b) ⇒ (e) follows by a symmetric argument.

For (d′) ⇒ (b), let intervals $A_n\downarrow\emptyset$. Then for some $t$ and $n_0$, either for all $n\geq n_0$ $A_n$ is left-open at $t\in[a,b)$, or for all $n\geq n_0$ $A_n$ is right-open at $t\in(a,b]$. If $A_n=(t,s_n]\!]$ for $a\leq t<s_n<b$, then by multiplicativity of $\rho$, we have

$$\rho(A_n)=\rho([a,s_n]\!])(R_{\rho,a})^{\rm inv}(t+)=\begin{Bmatrix}R_{\rho,a}(s_n+)(R_{\rho,a})^{\rm inv}(t+)\text{ if }A_n=(t,s_n],\\ R_{\rho,a}(s_n-)(R_{\rho,a})^{\rm inv}(t+)\text{ if }A_n=(t,s_n),\end{Bmatrix}$$

$$\to R_{\rho,a}(t+)(R_{\rho,a})^{\rm inv}(t+)=\mathbb{I}\qquad\text{as } n\to\infty.$$

If $A_n = [\![s_n, t)$ for $a < s_n < t \le b$, then again by multiplicativity of $\rho$ and by (9.5) with $h = R_{\rho,a}$, we have

$$\rho(A_n)$$
$$= \left\{ \begin{array}{l} \rho([a,t))(R_{\rho,a})^{\mathrm{inv}}(s_n-) \text{ if } A_n = [s_n,t), \\ \rho([a,t))(R_{\rho,a})^{\mathrm{inv}}(s_n+) \text{ if } A_n = (s_n,t), \end{array} \right\} \to R_{\rho,a}(t-)(R_{\rho,a})^{\mathrm{inv}}(t-) = \mathbb{1}$$

as $n \to \infty$, proving $(b)$. The implication $(e') \Rightarrow (b)$ follows by a symmetric argument. The proof of Proposition 9.6 is now complete. □

We will say that an interval function $\rho$ on $J$ is *nondegenerate* if for any $A_1, A_2, \ldots \in \mathfrak{I}(J)$ such that $A_n \downarrow \emptyset$, $\rho(A_n)$ is invertible for all sufficiently large $n$. (This is unrelated to the notion of nondegenerate interval.) Every element $x \in \mathbb{B}$ such that $\|x - \mathbb{1}\| < 1$ is invertible (Theorem 4.16$(a)$). Thus a multiplicative interval function upper continuous at $\emptyset$ is nondegenerate. The following together with Proposition 9.6 completes the analogy to Proposition 2.6.

**Proposition 9.7.** *Let $\rho$ be a multiplicative interval function on $[a,b]$. Then $(a) \Rightarrow (c) \Rightarrow (b)$, and $(b) \Rightarrow (a)$ if, in addition, $\rho$ is nondegenerate, where*

*$(a)$ $\rho$ is upper continuous at $\emptyset$;*
*$(b)$ $\rho(A_n) \to \rho(A)$ whenever intervals $A_n \uparrow A$;*
*$(c)$ $\rho(A_n) \to \rho(A)$ whenever intervals $A_n \to A \neq \emptyset$.*

*Proof.* For $(a) \Rightarrow (c)$, let intervals $A_n \to A \neq \emptyset$. If $A = (u,v)$ then $\{u\} \prec A_n \prec \{v\}$ for $n$ large enough. For such $n$, there are intervals $C_n$ and $D_n$ with $\{u\} \prec C_n \prec A_n \prec D_n \prec \{v\}$ and $C_n \cup A_n \cup D_n = A$. Also for such $n$, we have either $C_n = \emptyset$ or $C_n = (u, \cdot]\!]$, and either $D_n = \emptyset$ or $D_n = [\![\cdot, v)$. Clearly $C_n \to \emptyset$. If $N := \{n\colon\ C_n \neq \emptyset\}$ is infinite, there is a function $j \mapsto n(j)$ onto $N$ such that $C_{n(j)} \downarrow \emptyset$. Thus $\rho(C_n) \to \mathbb{1}$. Similarly $\rho(D_n) \to \mathbb{1}$. Each $x_n \in \mathbb{B}$ such that $\|x_n - \mathbb{1}\| < 1$ has an inverse $x_n^{-1}$ and $x_n^{-1} \to \mathbb{1}$ as $x_n \to \mathbb{1}$ (Theorem 4.16). Therefore for all large enough $n$, $\rho(A_n) = \rho(D_n)^{-1}\rho(A)\,\rho(C_n)^{-1} \to \rho(A)$ as $n \to \infty$. Similar arguments work for the other cases $A = (u,v]$, $[u,v)$ and $[u,v]$.

Clearly, $(c)$ implies $(b)$. For $(b) \Rightarrow (a)$, let intervals $A_n \downarrow \emptyset$. Then for some $t$ and $n_0$, either for all $n \ge n_0$, $A_n = (t, s_n]\!]$ for some $s_n \in (t,b)$, or for all $n \ge n_0$, $A_n = [\![s_n, t)$ for some $s_n \in (a,t)$. Suppose that $A_n = (t, s_n]\!]$ for all $n \ge n_0 \ge 1$. Since $\rho$ is nondegenerate we can assume that $\rho(A_{n_0})$ is invertible. Then $A_{n_0} = A_{n_0+k} \cup B_k$ for some intervals $B_k = [\![s_{n_0+k}, s_{n_0}]\!]$, $k \ge 1$, with $B_k \uparrow A_{n_0}$. Since the set $U$ of all invertible elements $x$ in $\mathbb{B}$ is open and $x \mapsto x^{-1}$ is a continuous function on $U$, $\rho(B_k)^{-1} \to \rho(A_{n_0})^{-1}$ as $k \to \infty$ (Theorem 4.16$(b)$). Since $\rho$ is multiplicative, $\rho(A_{n_0+k}) = \rho(B_k)^{-1}\rho(A_{n_0}) \to \mathbb{1}$ as $k \to \infty$. A similar argument works in the case $A_n = [\![\cdot, t)$ for some $t \in (a,b]$. The proof of Proposition 9.7 is complete. □

Let $\rho$ be a $\mathbb{B}$-valued interval function on $J$. If for each $A \in \mathfrak{I}(J)$, $\rho(A)$ has an inverse $(\rho(A))^{-1}$, then we say that $\rho$ is an *invertible interval function*, and define the *reciprocal interval function* $\rho^{\mathrm{inv}}$ on $J$ by $\rho^{\mathrm{inv}}(A) := (\rho(A))^{-1}$ for each $A \in \mathfrak{I}(J)$.

**Proposition 9.8.** *Let $\rho$ be a bounded multiplicative interval function on $[a,b]$. Then the following two statements are equivalent:*

*(a) $\rho$ is upper continuous at $\emptyset$ and invertible;*
*(b) statements (d) and (e) of Proposition 9.6 hold, and the functions $R_{\rho,a}$ and $L_{\rho,b}$ are both invertible.*

*Moreover, if (a) or (b) holds then the reciprocal interval function $\rho^{\mathrm{inv}}$ of $\rho$ is $*$-multiplicative,*

$$\rho^{\mathrm{inv}}([s,t]) \tag{9.8}$$
$$= \begin{cases} R_{\rho,a}(s-)(R_{\rho,a})^{\mathrm{inv}}(t+) = (L_{\rho,b})^{\mathrm{inv}}(s-)L_{\rho,b}(t+) & \text{if } a < s \le t < b, \\ (R_{\rho,a})^{\mathrm{inv}}(t+) = (L_{\rho,b})^{\mathrm{inv}}(a)L_{\rho,b}(t+) & \text{if } a = s \le t < b, \\ R_{\rho,a}(s-)(R_{\rho,a})^{\mathrm{inv}}(b) = (L_{\rho,b})^{\mathrm{inv}}(s-) & \text{if } a < s \le t = b; \end{cases}$$

$$\rho^{\mathrm{inv}}((s,t]) \tag{9.9}$$
$$= \begin{cases} R_{\rho,a}(s+)(R_{\rho,a})^{\mathrm{inv}}(t+) = (L_{\rho,b})^{\mathrm{inv}}(s+)L_{\rho,b}(t+) & \text{if } a \le s < t < b, \\ R_{\rho,a}(s+)(R_{\rho,a})^{\mathrm{inv}}(b) = (L_{\rho,b})^{\mathrm{inv}}(s+) & \text{if } a \le s < t = b; \end{cases}$$

$$\rho^{\mathrm{inv}}([s,t)) \tag{9.10}$$
$$= \begin{cases} R_{\rho,a}(s-)(R_{\rho,a})^{\mathrm{inv}}(t-) = (L_{\rho,b})^{\mathrm{inv}}(s-)L_{\rho,b}(t-) & \text{if } a < s < t \le b, \\ (R_{\rho,a})^{\mathrm{inv}}(t-) = (L_{\rho,b})^{\mathrm{inv}}(a)L_{\rho,b}(t-) & \text{if } a = s < t \le b; \end{cases}$$

*and*

$$\rho^{\mathrm{inv}}((s,t)) = R_{\rho,a}(s+)(R_{\rho,a})^{\mathrm{inv}}(t-) = (L_{\rho,b})^{\mathrm{inv}}(s+)L_{\rho,b}(t-) \tag{9.11}$$

*if $a \le s < t \le b$.*

*Proof.* $(a) \Rightarrow (b)$: since $\rho$ is upper continuous at $\emptyset$, statements $(d)$ and $(e)$ of Proposition 9.6 hold. To show that $R_{\rho,a}$ is an invertible function, let $x$ be in the closure of the set $\{R_{\rho,a}(t)\colon\ a \le t \le b\}$. Due to compactness of $[a,b]$ and by statement $(d)$ of Proposition 9.6, there exists $t \in [a,b]$ such that for some sequence $\{t_k\}_{k\ge1} \subset [a,b]$, $t = \lim_k t_k$ and either $x = R_{\rho,a}(t) = \rho([a,t])$, or $x = R_{\rho,a}(t+) = \rho([a,t])$ with $a \le t < b$, or $x = R_{\rho,a}(t-) = \rho([a,t))$ with $a \le t < b$. Thus $x$ has an inverse $x^{-1} = \rho^{\mathrm{inv}}([a,t]\!])$, and since $x$ is arbitrary, $R_{\rho,a}$ is an invertible function. A symmetric argument yields that $L_{\rho,b}$ is also an invertible function, completing the proof of $(b)$.

$(b) \Rightarrow (a)$: the interval function $\rho$ is upper continuous at $\emptyset$ by the last part of Proposition 9.6. Let $R^{\mathrm{inv}}$ be the reciprocal function of $R_{\rho,a}$, and let $L^{\mathrm{inv}}$

be the reciprocal function of $L_{\rho,b}$. To show that $\rho((s,t])$ has an inverse for $a \le s < t \le b$, let

$$x_r := R_{\rho,a}(s+)(R^{\mathrm{inv}})^{(b)}_{+}(t) \quad \text{and} \quad x_l := L^{\mathrm{inv}}(s+)\rho((t,b]),$$

where $f^{(b)}_{+} = f(t+)$ if $t \in [a,b)$ or $= f(b)$ if $t = b$. Due to multiplicativity of $\rho$ and by statements $(d)$ and $(e)$ of Proposition 9.6, we have

$$\begin{aligned} x_l &= L^{\mathrm{inv}}(s+)\rho((t,b])\rho([a,t])(R^{\mathrm{inv}})^{(b)}_{+}(t) \\ &= L^{\mathrm{inv}}(s+)\rho((s,b])\rho([a,s])(R^{\mathrm{inv}})^{(b)}_{+}(t) = x_r, \\ \rho((s,t])x_r &= \rho((s,t])\rho([a,s])(R^{\mathrm{inv}})^{(b)}_{+}(t) = (R_{\rho,a})^{(b)}_{+}(t)(R^{\mathrm{inv}})^{(b)}_{+}(t) = \mathbb{I}, \end{aligned}$$

and

$$x_l\rho((s,t]) = L^{\mathrm{inv}}(s+)\rho((t,b])\rho((s,t]) = L^{\mathrm{inv}}(s+)L_{\rho,b}(s+) = \mathbb{I}.$$

Thus the inverse $(\rho((s,t]))^{-1}$ exists and equals $x_r = x_l$. Similarly it follows that inverses $(\rho(A))^{-1}$ exist for $A = [s,t)$, $A = (s,t)$, and $A = [s,t]$, and so the reciprocal interval function $\rho^{\mathrm{inv}}$ is defined on $[a,b]$ with values given by (9.8), (9.9), (9.10), and (9.11). Thus $(a)$ holds. Moreover, if intervals $A, B$ satisfy $A \prec B$ and $A \cup B$ is an interval then $\rho^{\mathrm{inv}}(A \cup B) = \rho^{\mathrm{inv}}(A)\,\rho^{\mathrm{inv}}(B)$, and so $\rho^{\mathrm{inv}}$ is $*$-multiplicative. □

**Proposition 9.9.** *If a $\mathbb{B}$-valued interval function $\rho$ is multiplicative and upper continuous on $[a,b]$ then $\rho$ is bounded.*

*Proof.* Suppose not. Recall that the oscillation Osc for interval functions, defined in (2.7), equals the supremum norm. By compactness of $[a,b]$ and multiplicativity of $\rho$, there are $t \in [a,b]$ and a sequence of closed intervals $\{A_k\}_{k\ge1}$ such that $A_k \downarrow \{t\}$ and $\mathrm{Osc}(\rho; A_k) = +\infty$ for each $k \ge 1$. Let $\{B_k\}_{k\ge1}$ be a sequence of intervals such that $\|\rho(B_k)\| \to +\infty$ and $B_k \subset A_k$ for each $k \ge 1$. If infinitely many $B_k$ contain $t$ or have $t$ as an endpoint, then taking a subsequence, we can assume that $B_k \downarrow \{t\}$ or $B_k \downarrow \emptyset$, contradicting upper continuity. So we can assume that $t < c_k \le d_k$ and $B_k = [\![c_k, d_k]\!] \ne \emptyset$ for each $k \ge 1$. Again by upper continuity, we have $\rho((t,c_k]\!]) \to \mathbb{I}$ and $\rho((t,d_k]\!]) \to \mathbb{I}$ as $k \to \infty$. By Theorem 4.16, $\rho((t,c_k]\!])$ are invertible for all sufficiently large $k$ and $\rho((t,c_k]\!])^{-1} \to \mathbb{I}$ as $k \to \infty$, and so by multiplicativity, $\rho(B_k) = \rho((t,d_k]\!])\rho((t,c_k]\!])^{-1} \to \mathbb{I}$ as $k \to \infty$, a contradiction, proving the proposition. □

Multiplicative interval functions do not form a linear space: if $\rho$ is multiplicative then for a constant $c \ne 1$, $c\rho$ is not since $c\rho(\emptyset) = c\mathbb{I} \ne \mathbb{I}$. This is one reason why the nonlinear operator induced by the product integral and defined on additive interval functions is considered in most of the rest of this chapter as acting from a subspace of the Banach space of interval functions of bounded $p$-variation into itself, for $p < 2$.

Recall (Section 3.1) that $\mathcal{V}$ is the class of all functions $\Phi\colon [0,\infty) \to [0,\infty)$ which are strictly increasing, continuous, unbounded, and 0 at 0, and $\mathcal{CV}$ is the class of those $\Phi \in \mathcal{V}$ which are convex. Let $\Phi \in \mathcal{CV}$ and let $J$ be a nonempty interval. The $\Phi$-variation for interval functions is defined in Section 3.2 by (3.35), and the set of all interval functions on $J$ with values in a Banach algebra $\mathbb{B}$ and having bounded $\Phi$-variation is denoted by $\mathcal{I}_{\Phi}(J;\mathbb{B})$. Then $\widetilde{\mathcal{I}}_{\Phi}(J;\mathbb{B})$ is the set of all interval functions $\mu \in \mathcal{I}(J;\mathbb{B})$ such that $c\mu \in \mathcal{I}_{\Phi}(J;\mathbb{B})$ for some $c > 0$. By Proposition 3.22, $\mu \mapsto \|\mu\|_{(\Phi)} = \sup\{\inf\{c > 0\colon v_{\Phi}(\mu/c; A) \le 1\}\colon A \in \mathfrak{I}(J)\}$ for $\mu \in \widetilde{\mathcal{I}}_{\Phi}(J;\mathbb{B})$ is a seminorm, which we call the $\Phi$-variation seminorm. The next fact follows from Proposition 3.20 since $\mu(\emptyset) = \mathbb{I}$ for $\mu$ multiplicative.

**Corollary 9.10.** *For $\Phi \in \mathcal{CV}$ and $\mu \in \widetilde{\mathcal{I}}_{\Phi}([a,b];\mathbb{B})$ multiplicative, $\|\mu\|_{(\Phi)} = 0$ if and only if $\mu \equiv \mathbb{I}$.*

## 9.2 Product Integrals for Real-Valued Interval Functions

Let $\mu$ be an additive real-valued interval function on $[a,b]$. For a nonempty interval $A \subset [a,b]$, and for an interval partition $\mathcal{A} = \{A_i\}_{i=1}^{n}$ of $A$, recall the product $P(\mu; A, \mathcal{A})$ defined by (9.2). Also recall that the product integral $\prod_A(1+\mathrm{d}\mu)$ with respect to $\mu$ over a nonempty interval $A$ is defined and equals $C \in \mathbb{R}$ if for any $\epsilon > 0$, there exists an interval partition $\mathcal{B}$ of $A$ such that $|P(\mu; A, \mathcal{A}) - C| < \epsilon$ for any interval partition $\mathcal{A}$ of $A$ which is a refinement of $\mathcal{B}$. If $A$ is the empty set the product integral is defined by $\prod_A(1+\mathrm{d}\mu) = 1$. We say that the product integral $\prod(1+\mathrm{d}\mu)$ exists on $[a,b]$ if it is defined over each interval $A \subset [a,b]$. We further say that the product integral $\prod(1+\mathrm{d}\mu)$ exists and is non-zero on $[a,b]$ if for each interval $A \in \mathfrak{I}[a,b]$, $\prod_A(1+\mathrm{d}\mu)$ is defined and non-zero.

The main result of this section is the next theorem, which gives necessary and sufficient conditions for the product integral with respect to an additive and upper continuous interval function to exist and be non-zero. Recall (Definition 3.28) that an additive and upper continuous interval function $\mu$ has $2^*$-variation on $[a,b]$, or $\mu \in \mathcal{AI}_2^*[a,b] \equiv \mathcal{AI}_2^*([a,b];\mathbb{R})$, if $\sigma_2(\mu;[a,b]) = v_2^*(\mu;[a,b]) < +\infty$ (as defined in (3.40) and (3.41)).

**Theorem 9.11.** *Let $\mu$ be a real-valued additive and upper continuous interval function on $[a,b]$. The product integral $\prod(1+\mathrm{d}\mu)$ exists and is non-zero on $[a,b]$ if and only if $\mu \in \mathcal{AI}_2^*[a,b]$ and $\mu(\{t\}) \neq -1$ for all $t \in [a,b]$.*

Before proving the theorem we note that the characterization of additive $\mu$ such that $\prod(1+\mathrm{d}\mu)$ exists and is non-zero may fail if $\mu$ is not upper continuous at $\emptyset$, as the following shows.

**Example 9.12.** Let $\mu$ be the additive interval function on $[0,2]$ defined in Example 3.51, which has bounded $p$-variation for all $p \in (0,\infty)$ but is not upper continuous at $\emptyset$. If $A$ is an interval included in $[0,1]$ or in $[d,2]$ for any $d>1$, then $\prod_A(1+\mathrm{d}\mu)=1$. If $A$ is not such an interval then $A$ includes $(1,c)$ for some $1<c\le 2$, and so $\prod_A(1+\mathrm{d}\mu)=1+r$. The product integral $\prod(1+\mathrm{d}\mu)$ is a multiplicative interval function on $[0,2]$ which has bounded $p$-variation for any $p$, but is not upper continuous at $\emptyset$. Also, $\mu(\{t\})=0\neq -1$ for each $t\in[0,2]$, but $\prod_{[0,2]}(1+\mathrm{d}\mu)=0$ for $r=-1$.

The proof of Theorem 9.11 starts with a series of auxiliary statements which together will prove the necessity part. Then we will prove the sufficiency part of the theorem.

By Taylor's theorem with remainder, for $u\in(-1,1)$,

$$\log(1+u)=u-\frac{\theta(u)}{2}u^2, \tag{9.12}$$

where

$$\theta(u)=-\frac{\mathrm{d}^2}{\mathrm{d}v^2}\log(1+v)=\frac{1}{(1+v)^2}$$

for some $v$ between 0 and $u$. Thus for $|u|\le\epsilon<1$, (9.12) holds with

$$\frac{1}{(1+\epsilon)^2}\le\theta(u)\le\frac{1}{(1-\epsilon)^2}. \tag{9.13}$$

Let $\mu$ be an additive and upper continuous interval function on $[a,b]$, and let $A\subset[a,b]$ be a nondegenerate interval, in other words, we recall, $A$ contains more than one point. By Corollary 2.12, there exists a Young interval partition $\mathcal{B}=\{(z_{j-1},z_j)\}_{j=1}^m$ of $A$ such that $\mathrm{Osc}(\mu;(z_{j-1},z_j))\le 1/2$ for all $j=1,\dots,m$. Let $B:=\{z_j\}_{j=0}^m\cap A$ and let $\mathcal{A}=\{A_i\}_{i=1}^n$ be an interval partition of $A$ which is a refinement of the Young interval partition $\mathcal{B}$, and let $I=I(\mathcal{A},\mathcal{B})$ be the set of all indices $i\in\{1,\dots,n\}$ such that $B\cap A_i=\emptyset$. Then using the notation (9.2), we have

$$\begin{aligned}P(\mu;A,\mathcal{A})&=\exp\Big\{\sum_{i\in I}\mu(A_i)-\frac{1}{2}\sum_{i\in I}\theta_i\mu(A_i)^2\Big\}\prod_{t\in B}(1+\mu(\{t\}))\\&=\exp\Big\{\mu(A)-\frac{1}{2}\sum_{i\in I(\mathcal{A},\mathcal{B})}\theta_i\mu(A_i)^2\Big\}\prod_{t\in B}(1+\mu(\{t\}))\exp\{-\mu(\{t\})\},\end{aligned} \tag{9.14}$$

where by (9.13), $4/9\le\theta_i:=\theta(\mu(A_i))\le 4$ for $i\in I(\mathcal{A},\mathcal{B})$.

**Lemma 9.13.** *Let $\mu$ be an additive and upper continuous interval function on $[a,b]$. If $\prod_{[a,b]}(1+\mathrm{d}\mu)$ exists and is non-zero then $v_2(\mu)<\infty$.*

*Proof.* We can assume that $a<b$. There exist $\epsilon>0$ and a Young interval partition $\mathcal{B}=\{(z_{j-1},z_j)\}_{j=1}^m$ of $[a,b]$ such that $\mathrm{Osc}(\mu;(z_{j-1},z_j))\le 1/2$ for

$j = 1, \ldots, m$, and $|P(\mu; [a,b], \mathcal{A})| \geq \epsilon$ for each interval partition $\mathcal{A}$ of $[a,b]$ which is a refinement of $\mathcal{B}$. Then (9.14) holds with $A = [a,b]$. Thus we have

$$\begin{aligned}\epsilon &\leq \big|P(\mu; [a,b], \mathcal{A})\big| \\ &\leq \exp\Big\{\mu([a,b]) - \frac{2}{9} \sum_{i \in I(\mathcal{A},\mathcal{B})} \mu(A_i)^2\Big\} \prod_{t \in B} \big|1 + \mu(\{t\})\big| \exp\big\{ - \mu(\{t\})\big\}\end{aligned}$$

for any refinement $\mathcal{A}$ of $\mathcal{B}$. If $v_2(\mu) = +\infty$, then by (3.39), for some $j$, $v_2(\mu; (z_{j-1}, z_j)) = +\infty$. Then by choosing $\mathcal{A}$, $\sum_{i \in I(\mathcal{A},\mathcal{B})} \mu(A_i)^2$ can be made arbitrarily large, and so $|P(\mu; [a,b], \mathcal{A})|$ is arbitrarily small. This contradiction proves $v_2(\mu) < \infty$. □

The following example shows that the assumption in Lemma 9.13 that the product integral is non-zero is essential.

**Example 9.14.** Let $\mu$ be the interval function on $[0,1]$ defined as follows: for $k = 2, 3, \ldots$, $\mu(\{1/(2k-1)\}) := 1/\sqrt{k}$, $\mu(\{1/(2k)\}) := -1/\sqrt{k}$, and for any interval $A \subset [0,1]$,

$$\mu(A) := \sum_{k=3}^{\infty} \mu(\{1/k\}) 1_A(1/k).$$

For an interval $A \subset [0,1]$, if $0 \notin \bar{A}$ then $\mu(A)$ is a finite sum. Also, if $(0,\epsilon) \subset A$ for some $0 < \epsilon < 1$, then $\mu(A)$ is a conditionally convergent sum, and for each $n \geq 2$, $\mu((0, 1/(2n-1)]) = 0$ and $\mu((0, 1/(2n)]) = -1/\sqrt{n}$. Clearly, $\mu$ is additive and upper continuous. It has bounded $p$-variation on $[0,1]$ if and only if $p > 2$. The product integral with respect to $\mu$ exists, and $\prod_A (1 + d\mu) = 0$ whenever $(0,\epsilon) \subset A$ for some $0 < \epsilon < 1$ because $\prod_{j \geq k} (1 - j^{-1}) = 0$ for all $k$.

In this example $\prod(1 + d\mu)$ is degenerate (not nondegenerate, in the sense defined before Proposition 9.7), and it is not upper continuous although $\mu$ is.

Let $\xi$ be a real-valued function defined on a nonempty set $T$. The product $\prod_T \xi$ will be said to converge *unconditionally* to a number $C$ if $C \neq 0$ and for each $\epsilon > 0$ there is a finite set $B \subset T$ such that

$$\Big|\prod_{t \in D} \xi(t) - C\Big| < \epsilon \tag{9.15}$$

for each finite subset $D$ of $T$ including $B$. We have the following:

**Lemma 9.15.** *Let $\xi$ be a function from a nonempty set $T$ into $\mathbb{R}$. Then the following two statements are equivalent:*

(*a*) *For every $\delta > 0$ there is a finite set $B \subset T$ such that for every finite $E \subset T \setminus B$, $|1 - \prod_{t \in E} \xi(t)| < \delta$.*

(*b*) *The sum $\sum_T (1 - \xi)$ converges absolutely, that is, $\xi(t) = 1$ except for at most countably many $t \in T$ and $\sum_{t \in T} |\xi(t) - 1| < \infty$.*

*Moreover, the product* $\prod_T \xi$ *converges unconditionally if and only if both* $(a)$ *holds and* $\xi(s) \neq 0$ *for all* $s \in T$.

*Proof.* If $|\xi(t) - 1| > 1/2$ for infinitely many $t \in T$ then both $(a)$ and $(b)$ fail, so we can assume that $|\xi(t) - 1| \leq 1/2$ except for $t$ in a finite set $F$. Let $x_i = \xi(t_i) - 1$ for some $t_i \in T \setminus F$. If $|x| \leq 1/2$ then $\mathrm{e}^{-2|x|} \leq 1 + x \leq \mathrm{e}^{|x|}$. Thus if $0 < \delta \leq 1/2$ and $\sum_i |x_i| < \delta$ then

$$1 - 2\delta \leq \mathrm{e}^{-2\delta} \leq \prod_i (1 + x_i) \leq \mathrm{e}^{\delta} \leq 1 + 2\delta.$$

Letting $\delta \downarrow 0$, it follows that $(b)$ implies $(a)$. Conversely, if $\sum_i |x_i| = +\infty$ then, taking a subsequence, we have $|x_i| \leq 1/2$ for all $i$ and either:

(*i*) $x_i > 0$ for all $i$ and $\sum_i x_i = +\infty$, in which case $\infty = \exp\{\sum_{i \geq j} x_i/2\} = \prod_{i \geq j}(1 + x_i)$ for any $j$,

or

(*ii*) $x_i < 0$ for all $i$ and $\sum_i x_i = -\infty$, in which case $0 = \exp\{\sum_{i \geq j} x_i\} = \prod_{i \geq j}(1 + x_i)$ for any $j$.

In either case, $(a)$ fails, so $(a)$ implies $(b)$ and the two are equivalent.

Now suppose $\prod_T \xi$ converges unconditionally to some $C \neq 0$. Then $\xi(s) \neq 0$ for all $s \in T$, since otherwise (9.15) is violated for $\epsilon = |C|/2$ and $D = B \cup \{s\}$. It suffices to prove $(a)$ for $0 < \delta < 1$. Let $\epsilon := |C|\delta/5$. Take a finite set $B \subset T$ such that (9.15) holds for the given $\epsilon$. To show that $(a)$ holds for the same $B$ and $\delta$ as defined, suppose it fails for some finite $E \subset T$. Let $\prod_A := \prod_{t \in A} \xi(t)$ for any finite $A \subset T$. Apply (9.15) to $D = B$ and to $D = B \cup E$. We get $|\prod_B - C| < \epsilon$ and $|\prod_{B \cup E} - C| < \epsilon$. Thus

$$2\epsilon > \Big|\prod_{B \cup E} - \prod_B\Big| = \Big|\prod_B\Big| \cdot \Big|\prod_E - 1\Big| \geq \frac{|C|\delta}{2} = \frac{5\epsilon}{2},$$

a contradiction. So $(a)$ holds.

Conversely, suppose $(a)$ holds and $\xi(t) \neq 0$ for all $t \in T$. From the proof of equivalence of $(a)$ and $(b)$ it follows that $\sum_{t \in T \setminus F} \log \xi(t)$ converges absolutely to a limit $L$. It then follows that $\prod_{t \in T} \xi(t)$ converges unconditionally to $C = \mathrm{e}^L \prod_{t \in F} \xi(t) \neq 0$. □

For any real-valued interval function $\mu$ on $[a, b]$ and nonempty subset $T \subset [a, b]$, let

$$\gamma(\mu; T) := \prod_T (1 + \mu)\mathrm{e}^{-\mu} := \prod_{t \in T} (1 + \mu(\{t\}))\mathrm{e}^{-\mu(\{t\})} \tag{9.16}$$

provided the product converges unconditionally, and $\gamma(\mu; \emptyset) := 1$.

**Lemma 9.16.** *Let $\mu$ be an upper continuous, additive interval function on $[a,b]$ such that $\sigma_2(\mu;[a,b]) < \infty$ and $\mu(\{t\}) \neq -1$ for all $t \in [a,b]$. Then $\gamma(\mu;T)$ as defined by (9.16) exists for all sets $T \subset [a,b]$ and is bounded, with $\inf\{|\gamma(\mu;T)|\colon T \subset [a,b]\} > 0$. For any disjoint sets $D, E \subset [a,b]$,*

$$\gamma(\mu;D)\gamma(\mu;E) = \gamma(\mu;D \cup E). \tag{9.17}$$

*Thus, restricted to $\mathfrak{I}[a,b]$, $\gamma(\mu;\cdot)$ is a multiplicative and bounded interval function.*

*Proof.* By Taylor's theorem with remainder, for $|u| \le 1/2$,

$$\xi(u) := (1+u)\mathrm{e}^{-u} = 1 - \frac{\theta(u)}{2}u^2, \quad \text{where} \quad \frac{1}{2\sqrt{\mathrm{e}}} \le \theta(u) \le \frac{3\sqrt{\mathrm{e}}}{2}. \tag{9.18}$$

Since $\mu$ is upper continuous, by Proposition 2.6(*c*), there is an at most countable set $C \subset [a,b]$ such that $\mu(\{t\}) \neq 0$ just for $t \in C$ and a finite set $F \subset C$ such that $|\mu(\{t\})| \le 1/2$ for all $t \in [a,b] \setminus F$. Thus $\xi(\mu(\{t\})) = 1$ for all $t \in [a,b] \setminus C$ and

$$\sum_{t\in[a,b]\setminus F} \bigl|1 - \xi(\mu(\{t\}))\bigr| \le (3\sqrt{\mathrm{e}}/4)\sigma_2(\mu) < \infty. \tag{9.19}$$

Therefore the product (9.16) converges unconditionally for any nonempty set $T \subset [a,b]$ by Lemma 9.15, and (9.17) holds by the definition of product. By (9.18), (9.19), and Lemma 9.15, for any set $D \subset [a,b] \setminus F$, $0 < \gamma(\mu;D) \le 1$. Thus by (9.17), for each set $T \subset [a,b]$, we have $|\gamma(\mu;T)| \le \max_{E\subset F} |\gamma(\mu;E)| < \infty$ and

$$\bigl|\gamma(\mu;T)\bigr| \ge \gamma(\mu;[a,b]\setminus F) \min_{E\subset F} \bigl|\gamma(\mu;E)\bigr| > 0.$$

The results for intervals follow, proving the lemma. □

For the next statement recall that $s_2(\mu;\mathcal{A}) = \sum_{i=1}^n \mu(A_i)^2$ for any interval partition $\mathcal{A} = \{A_i\}_{i=1}^n$ of $[a,b]$.

**Lemma 9.17.** *Let $\mu$ be an additive and upper continuous interval function on $[a,b]$. If $\prod_{[a,b]}(1+\mathrm{d}\mu)$ exists and is non-zero then the limit*

$$\lim_{\mathcal{A}} s_2(\mu;\mathcal{A}) \tag{9.20}$$

*exists under refinements of partitions $\mathcal{A}$ of $[a,b]$.*

*Proof.* We can assume that $a < b$. By Corollary 2.12, it follows that given $\epsilon \in (0,1/2]$, there exists a Young interval partition $\mathcal{B} = \{(z_{j-1},z_j)\}_{j=1}^m$ of $[a,b]$ such that $\mathrm{Osc}(\mu;(z_{j-1},z_j)) < \epsilon$ for $j = 1,\ldots,m$. Since $\prod_{[a,b]}(1+\mathrm{d}\mu) \neq 0$, $\mu(\{t\}) \neq -1$ for $a \le t \le b$. By Lemma 9.13, $v_2(\mu) < \infty$, and so by Lemma 9.16, the interval function $\gamma(\mu)$ given by (9.16) is well defined. Let $B := \{z_j\}_{j=0}^m$, and taking a refinement of $\mathcal{B}$ and thus of $B$, we can assume that the set $B$ is large enough so that, by the method of proof of Lemma 9.15,

$$G := \big|\log|\gamma(\mu;[a,b])| - \log|\gamma(\mu;B)|\big| < \epsilon.$$

Since $\prod_{[a,b]}(1+\mathrm{d}\mu)$ exists and is non-zero, we can also assume that

$$H := \Big|\log\Big|\prod_{[a,b]}(1+\mathrm{d}\mu)\Big| - \log\big|P(\mu;[a,b],\mathcal{A})\big|\Big| < \epsilon$$

for any interval partition $\mathcal{A}$ of $[a,b]$ which is a refinement of $\mathcal{B}$. Letting

$$C := \mu([a,b]) + \log|\gamma(\mu;[a,b])| - \log\Big|\prod_{[a,b]}(1+\mathrm{d}\mu)\Big|,$$

which does not depend on $\mathcal{B}$, $\mathcal{A}$, or $\epsilon$, it then follows by (9.14) with $A=[a,b]$ that

$$\Big|\frac{1}{2}\sum_{i\in I(\mathcal{A},\mathcal{B})}\theta_i\mu(A_i)^2 - C\Big| \le G+H < 2\epsilon$$

for any interval partition $\mathcal{A}=\{A_i\}_{i=1}^n$ of $[a,b]$ which is a refinement of $\mathcal{B}$. Since $|\mu(A_i)|\le\epsilon$ for each $i\in I(\mathcal{A},\mathcal{B})$, we have by (9.13),

$$2(1-\epsilon)^2(C-2\epsilon) < \sum_{i\in I(\mathcal{A},\mathcal{B})}\mu(A_i)^2 < 2(1+\epsilon)^2(C+2\epsilon).$$

Therefore for any two interval partitions $\mathcal{A}'=\{A_i'\}$ and $\mathcal{A}''=\{A_i''\}$ of $[a,b]$ which are refinements of $\mathcal{B}$, we have

$$\big|s_2(\mu;\mathcal{A}')-s_2(\mu;\mathcal{A}'')\big| = \Big|\sum_{i\in I(\mathcal{A}',\mathcal{B})}\mu(A_i')^2 - \sum_{i\in I(\mathcal{A}'',\mathcal{B})}\mu(A_i'')^2\Big| < 8\epsilon(1+\epsilon^2+C).$$

Since $\epsilon>0$ is arbitrary, the proof is complete. □

By Lemma 3.27 with $p=2$, the limit (9.20) exists if and only if $\sigma_2^*(\mu;[a,b]) = v_2^*(\mu;[a,b]) < +\infty$. Since $\sigma_2^*(\mu;[a,b]) = \sigma_2(\mu;[a,b])$ by Proposition 3.26, it follows then that $\mu\in\mathcal{AI}_2^*[a,b]$. Now we are ready to finish the proof of Theorem 9.11.

**Proof of Theorem 9.11.** We can assume that $a<b$. Since $\mu$ is upper continuous and additive, it is bounded by Corollary 2.12. First suppose that $\prod(1+\mathrm{d}\mu)$ exists and is non-zero on $[a,b]$, in particular on each singleton $\{t\}$, $t\in[a,b]$. Thus $\mu(\{t\})\ne -1$ for each $t\in[a,b]$ by the definition of product integral. The interval function $\mu$ has $2^*$-variation, by Lemmas 3.27 and 9.17 and Proposition 3.26. Thus the "only if" part holds.

Now suppose that $\mu \in \mathcal{AI}_2^*[a,b]$ is such that $\mu(\{t\}) \neq -1$ for all $t \in [a,b]$. Let $A \subset [a,b]$ be a nondegenerate interval and let $\epsilon \in (0,1/4]$. By Lemma 3.25, $\sigma_2(\mu;A) < \infty$, and so by Lemma 9.16, $\gamma(\mu;T)$ given by (9.16) is well defined for any subset $T \subset A$, (9.17) holds for any disjoint subsets $D, E$ of $T$, and

$$0 < C_1 := \inf\{|\gamma(\mu;B)|\colon B \subset A\} \le C_2 := \sup\{|\gamma(\mu;B)|\colon B \subset A\} < \infty.$$

Therefore by definition of unconditionally convergent product, there exists a finite set $B_0 \subset A$ such that

$$\bigl|1 - \gamma(\mu; A \setminus B)\bigr| \le |\gamma(\mu;A) - \gamma(\mu;B)|/C_1 < \epsilon \tag{9.21}$$

for any finite $B$ such that $B_0 \subset B \subset A$. Since $\mu \in \mathcal{AI}_2^*[a,b]$, by Lemma 3.29 with $p = 2$, there exists a Young interval partition $\mathcal{B} = \{(z_{j-1}, z_j)\}_{j=1}^m$ of $A$ such that $B := A \cap \{z_j\}_{j=0}^m \supset B_0$ and

$$\sum_{j=1}^m v_2(\mu; (z_{j-1}, z_j)) < \epsilon. \tag{9.22}$$

Let $\mathcal{A} = \{A_i\}_{i=1}^n$ be an interval partition of $A$ which is a refinement of $\mathcal{B}$, and let $I = I(\mathcal{A},\mathcal{B})$ be the set of all indices $i \in \{1,\dots,n\}$ such that $A_i \cap B = \emptyset$. By (9.22), for each $i \in I$, $|\mu(A_i)| \le \sqrt{\epsilon} \le 1/2$. Thus by (9.14), we have

$$\begin{aligned} &P(\mu; A, \mathcal{A}) - \mathrm{e}^{\mu(A)}\gamma(\mu;A) \\ &\quad = \mathrm{e}^{\mu(A)}\gamma(\mu;B)\Bigl[\exp\Bigl\{-\frac{1}{2}\sum_{i\in I(\mathcal{A},\mathcal{B})} \theta_i \mu(A_i)^2\Bigr\} - \gamma(\mu; A\setminus B)\Bigr], \end{aligned}$$

where $\theta_i = \theta(\mu(A_i))$ for $i \in I(\mathcal{A},\mathcal{B})$. Since $|1 - \mathrm{e}^{-x}| \le |x|\mathrm{e}^{|x|}$ for any $x \in \mathbb{R}$, by (9.13) and (9.22), it follows that

$$\begin{aligned} &\Bigl|1 - \exp\Bigl\{-\frac{1}{2}\sum_{i\in I(\mathcal{A},\mathcal{B})} \theta_i \mu(A_i)^2\Bigr\}\Bigr| \\ &\qquad \le C \sum_{i\in I(\mathcal{A},\mathcal{B})} \mu(A_i)^2 \le C\sum_{j=1}^m v_2(\mu;(z_{j-1},z_j)) < C\epsilon \end{aligned}$$

where $C := 2\exp\{2v_2(\mu;A)\}$. This together with (9.21) gives the bound

$$\bigl|P(\mu;A,\mathcal{A}) - \mathrm{e}^{\mu(A)}\gamma(\mu;A)\bigr| \le \mathrm{e}^{\|\mu\|_{\sup}} C_2(C+1)\epsilon \tag{9.23}$$

for any refinement $\mathcal{A}$ of the Young interval partition $\mathcal{B}$ of $A$. Since $\epsilon > 0$ is arbitrary, the product integral $\prod_A(1+\mathrm{d}\mu)$ is defined and equal to $\mathrm{e}^{\mu(A)}\gamma(\mu;A)$ if $A \subset [a,b]$ is a nondegenerate interval. The same conclusion clearly holds if $A$ is a singleton. So for each $A \in \mathfrak{I}[a,b]$, $\prod_A(1+\mathrm{d}\mu) \neq 0$ if and only if

$\gamma(\mu; A) \neq 0$, which is true whenever the product is defined, proving the "if" part. The proof of Theorem 9.11 is complete. □

The preceding proof of Theorem 9.11 also gives the value of the product integral over an interval. In particular, recalling the notation (9.2) and Definition 9.1, the bound (9.23) implies the following:

**Corollary 9.18.** *Let $\mu$ be as in Theorem 9.11. If the product integral with respect to $\mu$ exists and is non-zero on $[a, b]$, then for any interval $A \subset [a, b]$,*

$$\prod_A (1 + \mathrm{d}\mu) = \mathrm{e}^{\mu(A)} \prod_A (1 + \mu)\mathrm{e}^{-\mu},$$

*where the product converges unconditionally for nonempty $A$, and equals 1 for $A = \emptyset$.*

Let $\mu$ be as in Theorem 9.11 and let $\mu(\{t\}) \neq -1$ for all $t \in [a, b]$. By that theorem, for $\mu \notin \mathcal{AI}_2^*[a, b]$, if the product integral with respect to $\mu$ exists it must be zero. Example 9.14 gives a $\mu \in \mathcal{AI}_p[a, b] := \mathcal{AI}_p([a, b]; \mathbb{R})$ for all $p > 2$ such that $\prod_{[a,b]}(1 + \mathrm{d}\mu) = 0$. The following shows that for $p = 2$ this cannot happen. Propositions 3.63 and 3.58 and Corollary 3.59 give examples of interval functions satisfying the hypothesis.

**Proposition 9.19.** *Let $\mu$ be in $\mathcal{AI}_2(J)$ but not in $\mathcal{AI}_2^*(J)$ for $J = [a, b]$ with $a < b$, and assume that $\mu(\{t\}) \neq -1$ for all $t \in J$. Then $\prod_J(1 + \mathrm{d}\mu)$ does not exist.*

*Proof.* By the assumptions, Proposition 3.26, and Definition 3.28, $\sigma_2(\mu; J) = \sigma_2^*(\mu; J) < v_2^*(\mu; J)$. Let $c := (v_2^* - \sigma_2)(\mu; J)$ and let $\epsilon > 0$. By Corollary 2.12, there exists a Young interval partition $\{(z_{j-1}, z_j)\}_{j=1}^m$ of $J$ such that $\mathrm{Osc}(\mu; (z_{j-1}, z_j)) < 1/2$ for $j = 1, \ldots, m$. By Lemma 9.16, the product $\gamma(\mu; J)$ is not 0. Moreover, letting $B := \{z_j\}_{j=0}^m$, we can assume that

$$(1 - \epsilon)\big|\gamma(\mu; J)\big| \leq \big|\gamma(\mu; B)\big| \leq (1 + \epsilon)\big|\gamma(\mu; J)\big|$$

and $0 \leq \sigma_2(\mu; J) - \sum_{t \in B} \mu(\{t\})^2 < \epsilon$. Let $\mathcal{B}$ be a refinement of $\{(z_{j-1}, z_j)\}_{j=1}^m$. By (3.42), $\mathcal{B}$ has a refinement $\mathcal{A}_1$ such that $s_2(\mu; \mathcal{A}_1) < \sigma_2(\mu; J) + \epsilon$. By (9.14), since $\theta_i \leq 4$ for $i \in I(\mathcal{A}_1, \mathcal{B})$ and $\sum_{i \in I(\mathcal{A}_1, \mathcal{B})} \mu(A_i)^2 = s_2(\mu; \mathcal{A}_1) - \sum_{t \in B} \mu(\{t\})^2$, it follows that

$$\big|P(\mu; J, \mathcal{A}_1)\big| > \mathrm{e}^{\mu(J)}\big|\gamma(\mu; J)\big|(1 - \epsilon)\exp\{-4\epsilon\}.$$

On the other hand, by (3.41), there is another refinement $\mathcal{A}_2$ of $\mathcal{B}$ such that $s_2(\mu; \mathcal{A}_2) > v_2^*(\mu; J) - \epsilon = c + \sigma_2(\mu; J) - \epsilon$. Again by (9.14), since $\theta_i \geq 4/9$ for $i \in I(\mathcal{A}_2, \mathcal{B})$, it follows that

$$\big|P(\mu; J, \mathcal{A}_2)\big| < \mathrm{e}^{\mu(J)}\big|\gamma(\mu; J)\big|(1 + \epsilon)\exp\{-(2/9)(c - \epsilon)\}.$$

Thus, letting $\epsilon \downarrow 0$, $\prod_J(1 + \mathrm{d}\mu)$ does not exist. □

But the results of this section for real-valued interval functions do not extend even to 3-dimensional Banach algebras, as the following proposition and example will show.

**Proposition 9.20.** *Let $\mathbb{B}$ be a Banach algebra, let $J$ be a nondegenerate interval, and let $\mu \in \mathcal{I}(J; \mathbb{B})$ be additive and such that*

$$\mu(B)\mu(A) = 0 \qquad \text{if } A, B \in \mathfrak{I}(J) \text{ and } A \cap B = \emptyset. \tag{9.24}$$

*Then for any interval $C \subset J$, the product integral $\prod_C(\mathbb{I}+\mathrm{d}\mu)$ exists and equals $\mathbb{I} + \mu(C)$.*

*Proof.* Since $\mu$ is additive, it is easily seen that for any partition $\mathcal{A}$ of $C$, the product $P(\mu; C, \mathcal{A})$ equals $\mathbb{I} + \mu(C)$, and the conclusion follows. □

If $\mu$ is an additive interval function satisfying the conditions of Proposition 9.20, then $\mu$ has certain properties if and only if $\prod(\mathbb{I} + \mathrm{d}\mu)$ does, for example, bounded $p$-variation for any $p \in (0, \infty)$.

**Example 9.21.** Let $\mathbb{B}$ be the Banach algebra of $2 \times 2$ upper triangular real matrices with usual addition and multiplication, and the matrix (operator) norm $\|T\| := \sup\{\|Tx\|: x \in \mathbb{R}^2, \|x\| \leq 1\}$. Let $J$ be a nondegenerate interval, and let $\nu$ be an additive real-valued interval function on $J$. For $A \in \mathfrak{I}(J)$, let

$$\mu(A) := \begin{pmatrix} 0 & \nu(A) \\ 0 & 0 \end{pmatrix}.$$

Then $\mu$ is an additive interval function on $J$ with values in $\mathbb{B}$ and $\mu(A)\mu(B) = 0$ for any $A$ and $B$, so (9.24) holds. Thus the product integral $\prod(\mathbb{I}+\mathrm{d}\mu)$ exists and equals $\mathbb{I} + \mu \neq 0$.

The last example shows that Theorem 9.11 for the one-dimensional Banach algebra $\mathbb{R}$ does not extend to a 3-dimensional Banach algebra $\mathbb{B}$. Let $h(x) := (1/\log(2/x))\cos(\pi/x)$ for $x \in (0, 1]$ and $h(0) := 0$. Then $\nu = \mu_h$ is upper continuous by Theorem 2.7 and does not have bounded $p$-variation for any $p \in (0, +\infty)$ since $\sum_j |\nu((1/(j+1), 1/j])|^p = +\infty$.

In other examples, in contrast to Theorem 9.11, $\mu$ is not upper continuous. In Example 2.5, $\mu$ is a real-valued additive interval function for which the functions $L_{\mu,a}$ and $R_{\mu,a}$ are regulated but $\mu$ is not upper continuous. Here is an easily described example where $\mathbb{B}$ is infinite-dimensional. The condition (9.24) is satisfied also by the additive interval function $\mu(A) = 1_A$, $A \in \mathfrak{I}[0, 1]$, which is neither upper continuous nor has bounded $p$-variation for any $0 < p < \infty$ when considered with values either in $\ell^\infty[0, 1]$ or in $L^\infty([0, 1], \lambda)$ (as in Example 2.4). By Proposition 9.20 the multiplicative interval function $A \mapsto \mathbb{I} + 1_A$, $A \in \mathfrak{I}[0, 1]$, of Example 9.5 is the product integral with respect to $\mu$.

## 9.3 Nonexistence for $p > 2$

Recall that $\mathcal{AI}_p([a,b];\mathbb{B})$ is the set of all additive, upper continuous interval functions on $[a,b]$ with values in the Banach algebra $\mathbb{B}$ having bounded $p$-variation (Definition 3.16). Most of this chapter, beginning with Section 9.4, will give positive results about product integrals for interval functions in $\mathcal{AI}_p$ with $p < 2$. The next fact, and Theorem 9.11, give some reasons why $p < 2$ is needed. It will be shown that for $p > 2$ the product integral may not exist. Let $M_2(\mathbb{R})$ be the Banach algebra of $2 \times 2$ real matrices with the matrix (operator) norm.

**Proposition 9.22.** *There exists an interval function $\mu \in \mathcal{AI}_p([0,1];\mathbb{B})$ for all $p \in (2,\infty)$ such that the product integral $\prod(\mathbb{1} + \mathrm{d}\mu)$ does not exist if*

*(a)* $\mathbb{B} = M_2(\mathbb{R})$;
*(b)* $\mathbb{B}$ *is the* 1*-dimensional complex Banach algebra* $\mathbb{C}$*, the field of complex numbers*;
*(c)* $\mathbb{B} = \mathbb{R}$.

*Proof.* For $(a)$ let $c_k > 0$ be any sequence of real numbers with $\sum_k c_k^2 = +\infty$ but $\sum_k c_k^p < \infty$ for all $p > 2$, such as $c_k = 1/\sqrt{k}$. Define a $2 \times 2$ matrix-valued interval function $\mu$ on $[0,1]$ as follows. For a nonempty interval $J \subset [0,1]$, let $\mu(J) := \sum_{k=1}^{\infty} \mu_k(J)$ where $\mu_k(J) = c_k T[1_J(1/(2k-1)) - 1_J(1/(2k))]$ and $T = \left(\begin{smallmatrix} 0 & 1 \\ -1 & 0 \end{smallmatrix}\right)$, and $\mu(\emptyset) := 0$. For a given interval $J$, clearly $\mu_k(J) \neq 0$ for at most two values of $k$, so the sum defining $\mu$ converges. Here $\mu$ is an additive, upper continuous interval function. For $k = 1, 2, \ldots$, let $B_{2k-1} := [1/(2k), 1/(2k-1)]$ and $B_{2k} := (1/(2k+1), 1/(2k))$. Then $v_p(\mu; B_{2k-1}) = 2c_k^p$ and $v_p(\mu; B_{2k}) = 0$ for each $k$. By Proposition 3.56 for $\{B_j\}_{j=1}^{2n-1}$, $v_p(\mu; [1/(2n), 1]) \le 2^p \sum_{k=1}^{\infty} c_k^p$ for each integer $n \ge 1$. Since $\mu$ is upper continuous it follows that $v_p(\mu; [0,1]) \le \sup_n v_p(\mu; [1/(2n), 1]) < \infty$. Thus $\mu \in \mathcal{AI}_p([0,1]; M_2(\mathbb{R}))$ for all $p > 2$. Any interval partition of $[0,1]$, for each sufficiently large $k \ge 1$, has a refinement $\mathcal{A}$ which contains intervals $A_i$, $i = 1, \ldots, 2k+1$, such that $1/j \in A_{2k+2-j}$ for $j = 1, \ldots, 2k$ and $1/j \in A_1$ for all $j > 2k$. Then

$$P(\mu; [0,1], \mathcal{A}) = \prod_{r=1}^{2k+1} (\mathbb{1} + \mu(A_r)) = \prod_{i=1}^{k} (\mathbb{1} - c_i^2 T^2) = \prod_{i=1}^{k} (1 + c_i^2)\mathbb{1}$$

because on all other intervals in $\mathcal{A}$, $\mu$ is 0. These products evidently diverge as $k \to \infty$.

For the complex case $(b)$ we can just replace the matrix $T$ by the number $\mathrm{i} = \sqrt{-1}$.

In case $(c)$ let $f(x) := \sqrt{x}\cos(\pi/x)$ for $0 < x \le 1$ and $f(0) := 0$. Then $f$ is continuous. Let $\mu([\![a,b]\!]) := f(b) - f(a)$ for $0 \le a \le b \le 1$. Then $\mu \in \mathcal{AI}_p[0,1]$ for any $p > 2$ by (3.7) with $\Phi(u) := u^p$ and $g(t) := \sqrt{t}$ and by Proposition 3.30. But $\mu \notin \mathcal{AI}_2[0,1]$, since $\sum_j \mu((a_{j+1}, a_j])^2 = +\infty$ for

$a_j := 1/j$, $j = 1, 2, \ldots$. Given any interval partition $\mathcal{B} := \{B_i\}_{i=1}^k$ of $[0,1]$, if $B_1 = \{0\}$ then renumber the intervals so that $B_0 := \{0\}$ and in any case $B_1 = [\![0, b]\!]$ for some $b > 0$.

However the partition $B_2, \ldots, B_k$ of $[\![b, 1]$ is refined, the product over the resulting intervals will approach a bounded non-zero limit, namely $e^{f(1)-f(b)}$, by Corollary 9.18.

Consider two kinds of partitions of $B_1$, each giving a kind of refinement of $\mathcal{B}$. In one kind, subdivide $B_1$ into intervals with endpoints 0, $b$, and $1/r$ for $r = 2m-1, 2m, \ldots, 2n+1$ for some positive integers $m < n$ with $1/(2m-1) < b$. The product of $1+\mu$ over the resulting partition of $B_1$ will have a bounded factor $1 + f(1/(2n+1))$ and a factor

$$\prod_{j=m}^{n}\left(1+\frac{1}{\sqrt{2j}}+\frac{1}{\sqrt{2j+1}}\right)\left(1-\frac{1}{\sqrt{2j}}-\frac{1}{\sqrt{2j-1}}\right) \leq \prod_{j=m}^{n}\left(1-\frac{1}{2j}\right) \to 0$$

as $n \to \infty$. Thus the infimum of the given products is 0.

In another kind of partition, subdivide $B_1$ by first choosing a smallest endpoint $c > 0$ and then dividing $[c, b]\!]$ arbitrarily finely. Then by Corollary 9.18 again the products over partitions of $[c, b]\!]$ approach $e^{f(b)-f(c)}$, and $1+f(c)$ remains bounded.

Thus the products over the resulting subintervals of $B_1$ will approach a bounded non-zero limit. Since the limits for the two kinds of refinements are different, the product integral does not exist. □

## 9.4 Inequalities for Finite Products

As the examples from the end of Section 9.2 show, product integrals in a general Banach algebra require a correspondingly general treatment, which we begin to prepare now.

Let $x_1, \ldots, x_n$ be arbitrary elements in $\mathbb{B}$, and let

$$\prod_{j=1}^{n}(\mathbb{1}+x_j) := (\mathbb{1}+x_n)\cdots(\mathbb{1}+x_1) = \mathbb{1} + \sum_{k=1}^{n}\sum_{1\leq j_1<\cdots<j_k\leq n} x_{j_k}\cdots x_{j_1}.$$

For $k = 1, \ldots, n$, letting

$$H_k(m) := \sum_{j=k}^{m} x_j\, H_{k-1}(j-1), \qquad m = k, \ldots, n, \tag{9.25}$$

where $H_0(m) := \mathbb{1}$ for $m = 0, 1, \ldots, n$, it then follows that

$$\prod_{j=1}^{n}(\mathbb{1}+x_j) = \mathbb{1} + \sum_{k=1}^{n} H_k(n). \tag{9.26}$$

For $0 < p < \infty$ and $x = (x_1, \ldots, x_n)$, $x_i \in \mathbb{B}$, recall the notation

$$v_p(x) = \max\Big\{\sum_{j=1}^{m}\Big\|\sum_{i=\theta(j-1)+1}^{\theta(j)} x_i\Big\|^p : 1 \le m \le n, \tag{9.27}$$
$$0 = \theta(0) < \theta(1) < \cdots < \theta(m) = n\Big\}.$$

Clearly, $v_p(x) = v_\Phi(x)$ with $\Phi(r) \equiv r^p$, $r \ge 0$, defined by (3.136) and (6.30). Next is the main result of this section:

**Theorem 9.23.** *Let $1 \le p < 2$ and let $x = (x_1, \ldots, x_n)$. For each $k = 1, \ldots, n$,*

$$\big\|H_k(n)\big\| \le C_p^k \Big(\frac{v_p(x)^k}{k!}\Big)^{1/p}, \tag{9.28}$$

*where $C_p = 1 + 4^{1/p}\zeta(2/p)$ and $\zeta(r) = \sum_{k=1}^{\infty} k^{-r}$ for $r > 1$.*

Before proving the theorem we give an application of these bounds. Recall (1.20) that for $1 \le p < \infty$, $\|\mu\|_{J,(p)} = v_p(\mu; J)^{1/p}$. For products $P(\mu; J, \mathcal{A})$ as defined by (9.2) we have the following inequalities.

**Corollary 9.24.** *Let $1 \le p < 2$, let $a < b$, and let $\mu \in \mathcal{I}_p([a,b]; \mathbb{B})$ be additive. Let $C_p$ be the constant from (9.28) and let*

$$K := K(\mu, p) := 1 + \sum_{k=1}^{\infty} C_p^k \Big(\frac{v_p(\mu; [a,b])^k}{k!}\Big)^{1/p}. \tag{9.29}$$

*Then for each nonempty interval $J \subset [a,b]$ and for any interval partition $\mathcal{A}$ of $J$, we have*

$$\big\|P(\mu; J, \mathcal{A})\big\| \le K, \tag{9.30}$$

$$\big\|P(\mu; J, \mathcal{A}) - \mathbb{I}\big\| \le C_p K \|\mu\|_{J,(p)}, \tag{9.31}$$

$$\big\|P(\mu; J, \mathcal{A}) - \mathbb{I} - \mu(J)\big\| \le C_p^2 K \|\mu\|_{J,(p)}^2. \tag{9.32}$$

*Proof.* Let $\mathcal{A} = \{A_i\}_{i=1}^n$ be an interval partition of $J$. Then letting $x := (\mu(A_1), \ldots, \mu(A_n))$, we have $v_p(x) \le v_p(\mu; J) \le v_p(\mu; [a,b])$, and the corollary follows from Theorem 9.23 and (9.26), where for (9.32), $H_1(n) = \mu(J)$. □

Recall that for a regulated function $f$ on $[a,b]$ we defined the function $f_-^{(a)}$ on $[a,b]$ with values $f_-^{(a)}(t) = f(t-)$ if $t \in (a,b]$ or $= f(a)$ if $t = a$. To prove Theorem 9.23 we will use step functions defined recursively by

$$f_k(t) := (RRS)\int_0^t \mathrm{d}h\, (f_{k-1})_-^{(0)}, \qquad t \in [0,1], \tag{9.33}$$

where $h$ is a right-continuous step function depending on the $n$-tuple $x$, $f_0 \equiv \mathbb{I}$, and $k = 1, \ldots, n$. Then the proof of Theorem 9.23 will be based on the inequalities in the next two lemmas.

**Lemma 9.25.** *Let $\rho$ be a nondecreasing function on $[a, b]$ with $a < b$, let $\{t_i\}_{i=0}^n$ be a point partition of $[a, b]$ with $n \geq 2$, and let $k$ be a positive integer. Then there exists an index $i \in \{1, \ldots, n-1\}$ such that*

$$\left[\rho^k(t_i) - \rho^k(t_{i-1})\right]\left[\rho(t_{i+1}) - \rho(t_i)\right] \leq \frac{4}{(n-1)^2}\Big(\frac{\rho^{k+1}(b) - \rho^{k+1}(a)}{k+1}\Big).$$

*Proof.* Letting $\Delta_i \chi := \chi(t_i) - \chi(t_{i-1})$ for a function $\chi$, for each $i = 1, \ldots, n-1$, we have

$$\begin{aligned} \Delta_i \rho^k \Delta_{i+1}\rho &= k \int_{\rho(t_{i-1})}^{\rho(t_i)} u^{k-1}\, du\, \Delta_{i+1}\rho \\ &\leq k \int_{\rho(t_{i-1})}^{\rho(t_i)} u^{(k-1)/2}\, du\, \rho(t_i)^{(k-1)/2}\left[\rho(t_{i+1}) - \rho(t_i)\right] \qquad (9.34) \\ &\leq k \int_{\rho(t_{i-1})}^{\rho(t_i)} u^{(k-1)/2}\, du \int_{\rho(t_i)}^{\rho(t_{i+1})} u^{(k-1)/2}\, du. \end{aligned}$$

Then we get

$$\begin{aligned} \min_{1\leq i\leq n-1}\{\Delta_i\rho^k\Delta_{i+1}\rho\} &= \Big(\min_{1\leq i\leq n-1}\sqrt{\Delta_i\rho^k\Delta_{i+1}\rho}\Big)^2 \\ &\leq \Big\{\frac{1}{n-1}\sum_{i=1}^{n-1}\sqrt{\Delta_i\rho^k\Delta_{i+1}\rho}\Big\}^2 . \end{aligned}$$

By (9.34) and using the inequality $2\sqrt{uv} \leq u + v$ for $u, v \geq 0$, we can continue the above inequality with

$$\begin{aligned} &\leq \Big\{\frac{1}{n-1}\frac{\sqrt{k}}{2}\Big(\int_{\rho(a)}^{\rho(t_{n-1})} u^{(k-1)/2}\, du + \int_{\rho(t_1)}^{\rho(b)} u^{(k-1)/2}\, du\Big)\Big\}^2 \\ &\leq \Big\{\frac{\sqrt{k}}{n-1}\int_{\rho(a)}^{\rho(b)} u^{(k-1)/2}\, du\Big\}^2 = \Big\{\frac{2\sqrt{k}}{n-1}\Big(\frac{\rho^{(k+1)/2}(b) - \rho^{(k+1)/2}(a)}{k+1}\Big)\Big\}^2 \\ &\leq \frac{4}{(n-1)^2}\Big(\frac{\rho^{k+1}(b) - \rho^{k+1}(a)}{k+1}\Big), \end{aligned}$$

where in the last step, the inequality $(d-c)^2 < d^2 - c^2$ for $0 < c < d$ was used. The proof is complete. □

The next inequality will be used to bound the $p$-variation of the functions $f_k$ defined by (9.33).

**Lemma 9.26.** *Let $1 \le p < 2$, let $a < b$, and let $h \in \mathcal{W}_p([a,b];\mathbb{B})$ be right-continuous on $[a,b)$. Also, let $f \in \mathcal{W}_p([a,b];\mathbb{B})$ be right-continuous on $(a,b)$ and $f(a) = 0$. Assume that for a nondecreasing function $\rho$ on $[a,b]$ with $\rho(a) \ge 0$ and for a positive integer $k$,*

$$\|f(t) - f(s)\| \le C_p^k \Big(\frac{\rho^k(t) - \rho^k(s)}{k!}\Big)^{1/p} \tag{9.35}$$

*and*

$$\|h(t) - h(s)\| \le \big(\rho(t) - \rho(s)\big)^{1/p} \tag{9.36}$$

*for each $a \le s < t \le b$, where $C_p$ is the constant from (9.28). Then the indefinite integral $I(f,\mathrm{d}h)(t) := (RRS)\int_a^t \mathrm{d}h\, f_-^{(a)}$, $t \in [a,b]$, exists and for any $a \le c < d \le b$,*

$$\big\|I(f,\mathrm{d}h)\big\|_{[c,d],(p)} \le C_p^{k+1}\Big(\frac{\rho^{k+1}(d) - \rho^{k+1}(c)}{(k+1)!}\Big)^{1/p}.$$

*Proof.* Since $h$ and $f_-^{(a)}$ have no common one-sided discontinuities, the indefinite integral $I(f,\mathrm{d}h)$ exists by Definition 2.41 of the full Stieltjes integral and by Theorem 3.92. Let $\widetilde{h} := [(k!)^{1/p}/C_p^k]\, h$. By the additivity property of $I(f,\mathrm{d}\widetilde{h})$ (Corollary 2.91), we have for $a \le u < v \le b$,

$$I(f,\mathrm{d}\widetilde{h})(v) - I(f,\mathrm{d}\widetilde{h})(u) = (RRS)\int_u^v \mathrm{d}\widetilde{h}\, f_-^{(u)}.$$

Thus it is enough to prove that for any $a \le u < v \le b$,

$$\Big\|(RRS)\int_u^v \mathrm{d}\widetilde{h}\, f_-^{(u)}\Big\| \le [1 + 4^{1/p}\zeta(2/p)]\Big(\frac{\rho^{k+1}(v) - \rho^{k+1}(u)}{k+1}\Big)^{1/p}. \tag{9.37}$$

Let $a \le u < v \le b$ and let $\epsilon > 0$. There exists a point partition $\lambda = \{t_i\}_{i=0}^n$ of $[u,v]$ with $n \ge 2$ such that for each $s_i \in (t_{i-1}, t_i]$, $i = 1,\dots,n$, if $u > a$, or for each $s_i \in (t_{i-1}, t_i]$, $i = 2,\dots,n$, and $s_1 = a$ if $u = a$,

$$\Big\|(RRS)\int_u^v \mathrm{d}\widetilde{h}\, f_-^{(u)} - \sum_{i=1}^n \big[\widetilde{h}(t_i) - \widetilde{h}(t_{i-1})\big] f_-^{(u)}(s_i)\Big\| < \epsilon.$$

Let $S(\lambda) := \sum_{i=1}^n [\widetilde{h}(t_i) - \widetilde{h}(t_{i-1})] f(t_{i-1})$. Since $f$ is right-continuous on $(a,b)$, letting $s_i \downarrow t_{i-1}$ for each $i$ such that $s_i > t_{i-1}$, it follows that

$$\Big\|(RRS)\int_u^v \mathrm{d}\widetilde{h}\, f_-^{(u)} - S(\lambda)\Big\| \le \epsilon.$$

Let $\lambda^i := \lambda \setminus \{t_i\}$ for each $i \in \{1,\dots,n-1\}$. Then by (9.36) and (9.35), we have

$$\|S(\lambda)-S(\lambda^i)\| \le \|\widetilde{h}(t_{i+1})-\widetilde{h}(t_i)\|\,\|f(t_i)-f(t_{i-1})\|$$
$$\le \big\{[\rho(t_{i+1})-\rho(t_i)]\,[\rho^k(t_i)-\rho^k(t_{i-1})]\big\}^{1/p}.$$

By Lemma 9.25, there exists an index $i \in \{1,\dots,n-1\}$ such that

$$\|S(\lambda)-S(\lambda^i)\| \le \Big\{\frac{4}{(n-1)^2}\Big(\frac{\rho^{k+1}(v)-\rho^{k+1}(u)}{k+1}\Big)\Big\}^{1/p}.$$

Deleting further points according to Lemma 9.25, we obtain the bound

$$\Big\|(RRS)\int_u^v \mathrm{d}\widetilde{h}\, f_-^{(u)}\Big\| \le 4^{1/p}\Big(\sum_{m=2}^{\infty}\frac{1}{(m-1)^{2/p}}\Big)\Big(\frac{\rho^{k+1}(v)-\rho^{k+1}(u)}{k+1}\Big)^{1/p}$$
$$+\,\|[\widetilde{h}(v)-\widetilde{h}(u)]f(u)\|+\epsilon. \tag{9.38}$$

Since $f(a)=0$, by (9.35), $\|f(u)\|^p \le [C_p^{kp}/k!]\rho^k(u)$. Then by (9.36), we have

$$\|[\widetilde{h}(v)-\widetilde{h}(u)]f(u)\|^p \le [\rho(v)-\rho(u)]\rho^k(u) \le \int_{\rho(u)}^{\rho(v)} t^k\,\mathrm{d}t = \frac{\rho^{k+1}(v)-\rho^{k+1}(u)}{k+1}.$$

Letting $\epsilon \downarrow 0$ in (9.38), it then follows that (9.37) holds, proving the lemma. □

Now we are ready to prove the inequalities (9.28).

**Proof of Theorem 9.23.** Let $t_j := j/n$, $j=0,1,\dots,n$, and let $h$ be the right-continuous $\mathbb{B}$-valued step function on $[0,1]$ defined by

$$h(t) := \begin{cases} 0 & \text{if } t\in[0,t_1),\\ \sum_{i=1}^{j} x_i & \text{if } t\in[t_j,t_{j+1}),\ j=1,\dots,n-1,\\ \sum_{i=1}^{n} x_i & \text{if } t=t_n. \end{cases}$$

For this $h$, and $f_0 := \mathbb{I}$ on $[0,1]$, define $\mathbb{B}$-valued functions $f_k$, $k=1,\dots,n$, on $[0,1]$ by (9.33). Since $h$ and $(f_{k-1})_-^{(0)}$ have no common one-sided discontinuities and $h$ is a step function, the integral (9.33) exists by Theorem 2.17. Each $f_k$, $k=1,\dots,n$, is a right-continuous step function equal to 0 on $[0,t_k)$, constant on $[t_j,t_{j+1})$, $j=k,\dots,n-1$, and having jumps at $t_j$, $j=k,\dots,n$, given by

$$\Delta^- f_k(t_j)=\Delta^- h(t_j)f_{k-1}(t_j-)=x_j f_{k-1}(t_j-).$$

Thus for $m=k,\dots,n$,

$$f_k(t_m)=\sum_{j=k}^{m}\Delta^- f_k(t_j)=\sum_{j=k}^{m} x_j f_{k-1}(t_j-).$$

Since $f_1\equiv h$, it then follows from (9.25) that $f_k(t_m)=H_k(m)$ for $k=1,\dots,n$ and $m=k,\dots,n$. Also since $x_i=h(t_i)-h(t_{i-1})$ for $i=1,\dots,n$, we have

$v_p(x) = v_p(h; [0,1])$. Since $\|H_k(n)\| = \|f_k(1)\| \leq \|f_k\|_{[0,1],(p)}$, it is enough to prove that for $k = 1, \ldots, n$,

$$\|f_k\|_{[s,t],(p)} \leq C_p^k \Big( \frac{v_p(h; [0,t])^k - v_p(h; [0,s])^k}{k!} \Big)^{1/p} \tag{9.39}$$

for each $0 \leq s < t \leq 1$. Since $f_1 \equiv h$ and $C_p \geq 1$, (9.39) holds for $k = 1$ by (3.51). Suppose (9.39) holds for some $1 \leq k < n$. Applying Lemma 9.26 to $f = f_k$ and $\rho := v_p(h; [0, \cdot])$, we conclude that (9.39) holds with $k$ replaced by $k+1$. Thus by induction, (9.39) holds for $k = 1, \ldots, n$, proving Theorem 9.23. □

## 9.5 The Product Integral

Here we will prove that the product integral with respect to an additive interval function $\mu$ with values in a Banach algebra exists provided $\mu$ is upper continuous and has bounded $p$-variation for some $1 \leq p < 2$. First we show that the product integral, if it exists, is a multiplicative interval function. For this, recalling Remark 9.2, $\mu$ need not be additive.

**Proposition 9.27.** *Let $\mu$ be an interval function on $[a,b]$ with values in a Banach algebra $\mathbb{B}$ such that $\mu(\emptyset) = 0$. The product integral $\prod(\mathbb{I} + \mathrm{d}\mu)$, if it exists as an interval function on $[a,b]$, is multiplicative.*

*Proof.* Let $A, B \in \mathfrak{I}[a,b]$ be such that $A \cup B \in \mathfrak{I}[a,b]$. To prove multiplicativity we can assume that $a < b$, and that the intervals $A, B$ are nonempty and $A \prec B$. Since $\prod(\mathbb{I} + \mathrm{d}\mu)$ exists over each of the intervals $A$, $B$ and $A \cup B$, there exist nested sequences $\mathcal{A}_A^k$, $\mathcal{A}_B^k$, and $\mathcal{A}_{A\cup B}^k$ of interval partitions of $A$, $B$, and $A \cup B$, respectively, such that for $C = A$, $C = B$, and $C = A \cup B$,

$$\prod_C (\mathbb{I} + \mathrm{d}\mu) = \lim_{k\to\infty} P(\mu; C, \mathcal{A}_C^k).$$

Moreover, taking refinements of partitions if necessary, we can assume that for each $k \geq 1$, $\mathcal{A}_{A\cup B}^k = \mathcal{A}_A^k \cup \mathcal{A}_B^k$. Therefore for each $k \geq 1$, we have

$$P(\mu; A \cup B, \mathcal{A}_{A\cup B}^k) = P(\mu; B, \mathcal{A}_B^k) P(\mu; A, \mathcal{A}_A^k).$$

Taking the limit on both sides as $k \to \infty$, it follows that

$$\prod_{A\cup B} (\mathbb{I} + \mathrm{d}\mu) = \prod_B (\mathbb{I} + \mathrm{d}\mu) \prod_A (\mathbb{I} + \mathrm{d}\mu).$$

Since $A$ and $B$ with the stated properties are arbitrary, the product integral $\prod(\mathbb{I} + \mathrm{d}\mu)$ is a multiplicative interval function on $[a,b]$, as claimed. □

Now we are ready to prove the main result of this section. Recall that $\mathcal{AI}_p([a,b];\mathbb{B})$ is the class of all $\mathbb{B}$-valued additive and upper continuous interval functions on $[a,b]$ with bounded $p$-variation.

**Theorem 9.28.** *Let $\mu \in \mathcal{AI}_p([a,b];\mathbb{B})$ with $1 \le p < 2$. The product integral $A \mapsto \prod_A(\mathbb{I}+\mathrm{d}\mu)$ is an upper continuous, bounded, and multiplicative interval function on $[a,b]$, and for each $A \in \mathfrak{I}[a,b]$,*

$$\Big\|\prod_A(\mathbb{I}+\mathrm{d}\mu)\Big\| \le K, \tag{9.40}$$

$$\Big\|\prod_A(\mathbb{I}+\mathrm{d}\mu)-\mathbb{I}\Big\| \le C_pK\|\mu\|_{A,(p)}, \tag{9.41}$$

$$\Big\|\prod_A(\mathbb{I}+\mathrm{d}\mu)-\mathbb{I}-\mu(A)\Big\| \le C_p^2K\|\mu\|_{A,(p)}^2, \tag{9.42}$$

*where $K=K(\mu,p)$ and $C_p$ are as defined in* (9.29) *and* (9.28), *respectively.*

*Proof.* To prove the convergence of some products we will use the algebraic identity

$$\prod_{j=1}^m x_j-\prod_{j=1}^m y_j=\sum_{i=1}^m\Big(\prod_{j=i+1}^{m+1}x_j\Big)(x_i-y_i)\Big(\prod_{j=0}^{i-1}y_j\Big), \tag{9.43}$$

valid by a telescoping sum for any elements $x_1,\dots,x_m$ and $y_1,\dots,y_m$ of a Banach algebra, where we set $x_{m+1}:=y_0:=\mathbb{I}$. We can assume that $a<b$. Let $A=(s,t)\subset[a,b]$ be open, and let $\epsilon>0$. By Proposition 3.52, there exists a Young interval partition $\{(z_{j-1},z_j)\}_{j=1}^m$ of $A$ such that

$$\max_{1\le j\le m} v_p(\mu;(z_{j-1},z_j))<\epsilon. \tag{9.44}$$

For $j=1,\dots,m$, let $B_{2j-1}:=(z_{j-1},z_j)$, and for $j=1,\dots,m-1$, let $B_{2j}:=\{z_j\}$. Let $\mathcal{A}=\{A_i\}_{i=1}^n$ be a refinement of $\mathcal{B}=\{B_k\}_{k=1}^{2m-1}$. For $j=1,\dots,m$, let $I(j)$ be the set of indices $i$ such that $A_i\subset(z_{j-1},z_j)$, and for $j=1,\dots,m-1$, let the index $i(j)$ be such that $A_{i(j)}=B_{2j}$. Also for $j=1,\dots,m$, let

$$U_j:=\begin{cases}\prod_{i=i(j)}^n\big[\mathbb{I}+\mu(A_i)\big] & \text{if } j=1,\dots,m-1,\\ \mathbb{I} & \text{if } j=m,\end{cases}$$

and

$$V_j:=\begin{cases}\prod_{k=1}^{2j-2}\big[\mathbb{I}+\mu(B_k)\big] & \text{if } j=2,\dots,m,\\ \mathbb{I} & \text{if } j=1.\end{cases}$$

To apply the algebraic identity (9.43), with $2m-1$ in place of $m$, let $x_{2r-1}=\prod_{i\in I(r)}(\mathbb{I}+\mu(A_i))$, $r=1,\dots,m$, $x_{2r}=\mathbb{I}+\mu(A_{i(r)})$, $r=1,\dots,m-1$, and $y_j=\mathbb{I}+\mu(B_j)$, $j=1,\dots,2m-1$. Then $x_{2r}=y_{2r}$, $r=1,\dots,m-1$, and we get

$$\begin{aligned}
&\big\|P(\mu;A,\mathcal{A})-P(\mu;A,\mathcal{B})\big\| \\
&\quad=\Big\|\sum_{j=1}^{m}U_j\Big(\prod_{i\in I(j)}[\mathbb{I}+\mu(A_i)]-\mathbb{I}-\mu(B_{2j-1})\Big)V_j\Big\| \\
&\quad\le\sum_{j=1}^{m}\|U_j\|\Big\|\prod_{i\in I(j)}[\mathbb{I}+\mu(A_i)]-\mathbb{I}-\mu(B_{2j-1})\Big\|\|V_j\| \\
&\quad\le K^3C_p^2\sum_{j=1}^{m}\|\mu\|_{(z_{j-1},z_j),(p)}^2\le\epsilon^{(2/p)-1}K^3C_p^2v_p(\mu;A)=:M(\epsilon),
\end{aligned}$$

where the next to last inequality holds by (9.30) and (9.32) and the last by (9.44) and (3.69), writing the exponent $2=(2-p)+p$. Since $\epsilon>0$ is arbitrary and $\mathcal{A}$ is any refinement of $\mathcal{B}\in \mathrm{IP}\,(A)$, $\prod_A(\mathbb{I}+\mathrm{d}\mu)$ exists provided $A$ is open. If $A=[s,t]\subset[a,b]$ is closed then take the refinement $\mathcal{A}'=\{s\}\cup\mathcal{A}\cup\{t\}$ of the partition $\mathcal{B}'=\{s\}\cup\mathcal{B}\cup\{t\}$ with $\mathcal{A}$ and $\mathcal{B}$ as for the open interval $(s,t)$. Then we have

$$\begin{aligned}
&P(\mu;[s,t],\mathcal{A}')-P(\mu;[s,t],\mathcal{B}') \\
&\quad=\big[\mathbb{I}+\mu(\{t\})\big]\big[P(\mu;(s,t),\mathcal{A})-P(\mu;(s,t),\mathcal{B})\big]\big[\mathbb{I}+\mu(\{s\})\big],
\end{aligned}$$

and the norm of the left side is less than or equal to $K^2M(\epsilon)$ by (9.30). Since the same bound holds if $A$ is a half-open interval, the product integral $\prod(\mathbb{I}+\mathrm{d}\mu)$ is defined on $\mathfrak{I}[a,b]$.

Approximating the product integral over an interval $A$ by finite products $P(\mu;A,\mathcal{A})$, the inequalities (9.40), (9.41), and (9.42) follow respectively from the inequalities (9.30), (9.31), and (9.32). Then the product integral interval function is bounded by (9.40), and it is upper continuous by (9.41) and by Propositions 3.50 and 9.6. The proof of Theorem 9.28 is complete. □

## 9.6 The Strict Product Integral

Given an interval function $\mu$, a condition on a multiplicative interval function $\alpha$ in relation to $\mu$ will be defined, which if it holds for a bounded $\mu$, implies that $\alpha$ is the product integral with respect to $\mu$, as will be shown in Theorem 9.34.

**Definition 9.29.** Let $\mu$ be a $\mathbb{B}$-valued interval function on $[a,b]$. A $\mathbb{B}$-valued multiplicative interval function $\alpha$ on $[a,b]$ will be called a *multiplicative transform of* $\mu$ on $[a,b]$ if for any $\epsilon>0$ there is a Young interval partition $\mathcal{B}$ of $[a,b]$ such that for each refinement $\{A_i\}_{i=1}^n$ of $\mathcal{B}$,

$$\sum_{i=1}^{n}\|\alpha(A_i)-\mathbb{I}-\mu(A_i)\|<\epsilon. \tag{9.45}$$

If the product integral $\beta(A) := \prod_A(\mathbb{I} + \mathrm{d}\mu)$ exists for each interval $A \subset [a,b]$ and (9.45) holds for $\alpha = \beta$, then the product integral will be called *strict* and will be denoted by $\widehat{\mu} := \beta = \prod(\mathbb{I} + \mathrm{d}\mu)$.

The product integral $\prod(\mathbb{I} + \mathrm{d}\mu)$, if it exists, is unique and a multiplicative interval function by Proposition 9.27, and so the strict product integral $\widehat{\mu}$ is well defined if it exists.

Let $\mu$ be an additive interval function on $[a,b]$ satisfying the conditions of Proposition 9.20, that is, $\mu(A)\mu(B) = 0$ for disjoint intervals $A, B \subset [a,b]$. It is straightforward that $\alpha := \mathbb{I} + \mu$ is a multiplicative interval function and (9.45) holds, and so Example 9.21 gives examples of strict product integrals.

It is easy to see that if $\alpha$ is a multiplicative transform of $\mu$ then for a singleton $\{t\}$, $t \in [a,b]$, $\alpha(\{t\}) = \mathbb{I} + \mu(\{t\})$. Thus if $\mu$ has no atoms in $[a,b]$ then $\alpha(\{t\}) = \mathbb{I}$ for each $t \in [a,b]$. Also, letting $\delta_t$, $t \in [a,b]$, be the interval function on $[a,b]$ such that $\delta_t(A) = \mathbb{I}$ if $t \in A$ and $= 0$ otherwise, $\delta_t$ has the strict product integral $\mathbb{I} + \delta_t$. The following seems to be obvious, but needs a little proof.

**Lemma 9.30.** *If a $\mathbb{B}$-valued interval function $\mu$ on $[a,b]$ has a multiplicative transform $\alpha$ and $A \subset [a,b]$ is a nonempty interval such that $\mu(B) = 0$ for every interval $B \subset A$, then $\alpha(A) = \mathbb{I}$. In particular, if $\mu = \sum_{i=1}^n x_i\delta_{t_i}$ on $[a,b]$ for some $\{x_i\}_{i=1}^n \subset \mathbb{B}$ and $\{t_i\}_{i=1}^n \subset [a,b]$ with $a \le t_1 < t_2 < \cdots < t_n \le b$, then*

$$\alpha = \prod_{i=1}^{n}(\mathbb{I} + x_i\delta_{t_i}) = \prod(\mathbb{I} + \mathrm{d}\mu).$$

*Proof.* Given $\epsilon > 0$, there is an interval partition $\{A_i\}_{i=1}^n$ of $A$ such that $\sum_{i=1}^n \epsilon_i < \epsilon$ where $\epsilon_i := \|\alpha(A_i) - \mathbb{I}\|$ for each $i$. Then $\prod_{i=1}^n(1+\epsilon_i) \le \exp\{\sum_{i=1}^n \epsilon_i\} < \mathrm{e}^{\epsilon}$. By multiplicativity and (9.43),

$$\|\alpha(A) - \mathbb{I}\| = \Big\|\Big(\prod_{i=1}^{n}\alpha(A_i)\Big) - \mathbb{I}\Big\| \le \epsilon\mathrm{e}^{\epsilon}.$$

Letting $\epsilon \downarrow 0$, the first conclusion follows. Applying it to each interval $(t_{i-1}, t_i)$, $i = 2, \ldots, n$, the interval $(a, t_1)$ if $a < t_1$, $(t_n, b)$ if $t_n < b$, and any subintervals of these intervals, gives the second conclusion. □

**Proposition 9.31.** *Let $\mu$ be an interval function with $\mu(\emptyset) = 0$. Any multiplicative transform $\alpha$ of $\mu$, if it exists, is upper continuous at $\emptyset$ if and only if $\mu$ is upper continuous at $\emptyset$.*

*Proof.* Let $\mu$ be an interval function on $[a,b]$, and let $\alpha$ be a multiplicative transform of $\mu$. Given $\epsilon > 0$ there is an interval partition $\{B_j\}_{j=1}^m$ of $[a,b]$ such that for any interval $A \subset [a,b]$,

$$\max_{1\le j\le m} \|\alpha(A\cap B_j) - \mathbb{I} - \mu(A\cap B_j)\| < \epsilon.$$

This holds because the sets $A\cap B_j$ and $B_j \setminus A$ that are nonempty give a refinement of $\{B_j\}_{j=1}^m$, where the sets are either empty, nonempty intervals, or for one value of $j$, $B_j \setminus A$ may be a union of two nonempty intervals. Let intervals $A_k \downarrow \emptyset$. Then there is $j \in \{1,\dots,m\}$ such that for all sufficiently large $k$, $A_k \subset B_j$, and so

$$\|\alpha(A_k) - \mathbb{I}\| < \epsilon + \|\mu(A_k)\| \qquad \text{and} \qquad \|\mu(A_k)\| < \epsilon + \|\alpha(A_k) - \mathbb{I}\|.$$

Letting $k \to \infty$, the conclusion follows. □

If a multiplicative transform $\alpha$ exists for a bounded interval function $\mu$, then $\alpha$ is the product integral $\prod(\mathbb{I} + d\mu)$, as will be shown in Theorem 9.34. But a multiplicative transform may not exist while the product integral does, and so a product integral need not be strict, as follows.

**Example 9.32.** Let $\mu$ be the additive, upper continuous interval function on $[0,1]$ of Example 9.14. Then $\mu$ does not have a multiplicative transform, as follows. Suppose $\alpha$ is a multiplicative transform of $\mu$. Then for any $0 < s < 1$, by Lemma 9.30, $\alpha$ restricted to $[s,1]$ is equal to the product integral with respect to the interval function $\sum_{3\le k\le 1/s} \mu(\{1/k\})\delta_{\{1/k\}}$. Hence for any $t > 0$, $\alpha([s,t)) \to 0$ as $s \downarrow 0$. By Proposition 9.31, $\alpha$ is upper continuous at $\emptyset$, and so by the implication $(a) \Rightarrow (b)$ of Proposition 9.7, $\alpha((0,t)) = 0$ for each $t > 0$. On the other hand, by upper continuity at $\emptyset$ again, $\alpha((0,t)) \to 1$ as $t \downarrow 0$, a contradiction.

In the next fact, the converse part may not hold if $\mu$ is not additive, by Remark 9.2.

**Proposition 9.33.** *Let $\mu$ be an interval function having a multiplicative transform $\alpha$. If $\mu$ is bounded, then so is $\alpha$. Conversely, if $\mu$ is additive, then if $\alpha$ is bounded, so is $\mu$.*

*Proof.* Let $\mu$ and $\alpha$ be defined on $[a,b]$. By definition of multiplicative transform, there is an interval partition $\mathcal{B} = \{B_j\}_{j=1}^m$ of $[a,b]$ such that for any interval partition $\{A_i\}_{i=1}^n$ of $[a,b]$ which is a refinement of $\mathcal{B}$,

$$\max_{1\le i\le n} \|\alpha(A_i) - \mathbb{I} - \mu(A_i)\| \le \sum_{i=1}^n \|\alpha(A_i) - \mathbb{I} - \mu(A_i)\| \le 1.$$

Let $\|\mu\|_{\sup} < \infty$, and let $A \in \mathfrak{I}[a,b]$. By multiplicativity of $\alpha$, we then have

$$\|\alpha(A)\| \le \prod_{j=1}^m \|\alpha(A\cap B_j)\| \le \prod_{j=1}^m \left(2 + \|\mu(A\cap B_j)\|\right) \le \left(2 + \|\mu\|_{\sup}\right)^m.$$

Thus $\|\alpha\|_{\sup} < \infty$. Conversely, let $\mu$ be additive, let $\|\alpha\|_{\sup} < \infty$, and let $A \in \mathfrak{I}[a,b]$. We have similarly

$$\|\mu(A)\| \leq \sum_{j=1}^{m} \|\mu(A \cap B_j)\| \leq \sum_{j=1}^{m} (1 + \|\alpha(A \cap B_j) - \mathbb{I}\|) \leq m(\|\alpha\|_{\sup} + 2).$$

Thus $\|\mu\|_{\sup} < \infty$, completing the proof. □

Next we show that for a bounded interval function $\mu$, if a multiplicative transform of $\mu$ exists then it is a strict product integral of $\mu$.

**Theorem 9.34.** *If $\mu$ is a bounded interval function on $[a,b]$ having a multiplicative transform $\alpha$, then the product integral $\prod_A(\mathbb{I} + \mathrm{d}\mu)$ exists for any $A \in \mathfrak{I}[a,b]$ and equals $\alpha(A)$. Thus, the product integral is strict and is the unique multiplicative transform of $\mu$.*

*Proof.* By Proposition 9.33, $\alpha$ is bounded. Let $M := \|\alpha\|_{\sup} < \infty$. Then $M \geq 1$ since $\alpha(\emptyset) = \mathbb{I}$. For any $0 < \delta < 1$, let $\epsilon = \delta/(7M^2)$. For any $A \in \mathfrak{I}[a,b]$, there is an interval partition $\mathcal{B}$ of $A$ such that for any refinement $\{A_i\}_{i=1}^n$ of $\mathcal{B}$, (9.45) holds. Let $\mathcal{A} = \{A_i\}_{i=1}^n$ be such a refinement. Let $x_i := \alpha(A_i)$ and $y_i := \mathbb{I} + \mu(A_i)$ for $i = 1, \ldots, n$. Let $z_i := y_i - x_i$ and $\epsilon_i := \|y_i - x_i\|$, so that $\sum_{i=1}^n \epsilon_i < \epsilon$. Let

$$S := \prod_{i=1}^{n}(x_i + z_i) - \prod_{i=1}^{n} x_i = y_n \cdots y_2 y_1 - x_n \cdots x_2 x_1 = P(\mu; A, \mathcal{A}) - \alpha(A).$$

Then $S = \sum_T w_T$, where $T$ runs over the $2^n - 1$ nonempty subsets of $\{1, \ldots, n\}$ and $w_T = u_n \cdots u_2 u_1$ where $u_i = z_i$ for $i \in T$ and $u_i = x_i$ for $i \notin T$. For each $T$, $\|w_T\|$ will be bounded by a product of terms, including $\epsilon_i$ for each $i \in T$. Let a *gap* in $T$ be a nonempty set $G \subset \{1, \ldots, n\} \setminus T$ of consecutive integers, $G = \{i, i+1, \ldots, j\}$, maximal in the sense that $i = 1$ or $i - 1 \in T$ and $j = n$ or $j + 1 \in T$. For each such gap $G$ we use the bound $\|x_j \cdots x_{i+1} x_i\| = \|\alpha(A_i \cup \cdots \cup A_j)\| \leq M$. Thus $\|w_T\| \leq M^{\gamma(T)} \prod_{i \in T} \epsilon_i$, where $\gamma(T)$ is the number of different gaps for $T$. We have $\gamma(T) \leq \lfloor (n+1)/2 \rfloor$, where $\lfloor x \rfloor$ is the largest integer $\leq x$. Thus

$$\|S\| \leq \sum_{k=0}^{\lfloor (n+1)/2 \rfloor} M^k U_k, \qquad \text{where} \qquad U_k := \sum_{\{T:\ \gamma(T)=k\}} \prod_{i \in T} \epsilon_i.$$

Recall that each $T$ is nonempty. Then $U_0 = \epsilon_1 \epsilon_2 \cdots \epsilon_n < \epsilon^n$, $U_1 < \epsilon_1 + \epsilon_n + \epsilon^2 + \epsilon^3 + \cdots < \epsilon/(1-\epsilon) < 2\epsilon$, and for $k \geq 2$,

$$U_k \leq \sum_{r=k-1}^{\lfloor (n+1)/2 \rfloor} \sum_{1 \leq i_1 < i_2 < \cdots < i_r \leq n} \epsilon_{i_1} \epsilon_{i_2} \cdots \epsilon_{i_r} \leq \sum_{r=k-1}^{\infty} \epsilon^r = \epsilon^{k-1}/(1-\epsilon) < 2\epsilon^{k-1}.$$

Thus

$$\|S\| \le \epsilon^n + 2M\epsilon + 2M\sum_{k=1}^{\infty}(M\epsilon)^k \le 3M\epsilon + 2M^2\epsilon/(1-M\epsilon) \le 7M^2\epsilon = \delta.$$

Letting $\delta \downarrow 0$, it follows that $\prod_A(\mathbb{I} + \mathrm{d}\mu)$ exists and equals $\alpha(A)$. □

The following example shows that there is a bounded additive interval function $\mu$ such that $\mu$ has a strict product integral, but $\mu$ neither is upper continuous nor has bounded $p$-variation for any $0 < p < \infty$.

**Example 9.35.** Let $H = L^2([0,1],\lambda)$ be the Hilbert space of functions which are square integrable with respect to Lebesgue measure $\lambda$ on $[0,1]$, and let $\mathbb{B}_H := L(H,H)$ be the Banach algebra of bounded linear operators from $H$ into itself (see Example 4.10). For each interval $J \subset [0,1]$, let

$$\mu(J)\colon\ f \mapsto 1_J f =: \mu(J)(f), \qquad f \in H.$$

Then $\mu$ is an additive interval function on $[0,1]$ with values in $\mathbb{B}_H$. It is easy to see that $\mu(A) = 0$ if $A$ is a singleton or empty and $\|\mu(A)\| = 1$ otherwise, if $A$ is a nondegenerate interval. Thus $\mu$ is bounded, not upper continuous (in operator norm), and has bounded $p$-variation for no $0 < p < \infty$. Moreover, $\mu(A)\mu(B) = 0$ if $A, B \in \mathfrak{I}[0,1]$ and $A \cap B = \emptyset$. So by Proposition 9.20 the product integral with respect to $\mu$ exists, and for any interval $A \subset [0,1]$,

$$\prod_A(\mathbb{I} + \mathrm{d}\mu) = \mathbb{I} + \mu(A) = \mathbb{I} + m_{1_A},$$

where $m_g$ is multiplication by $g$. By Theorem 9.34, it follows that the product integral is strict.

It will be convenient to consider a reversed form of $p$-variation of interval functions as follows. For any $1 \le p < \infty$, nondegenerate interval $A$, interval function $\alpha$ on $A$ with values in $\mathbb{B}$, and interval partition $\mathcal{A} = \{A_i\}_{i=1}^n$ of $A$, let

$$s_p^{\leftarrow}(\alpha;\mathcal{A}) := \sum_{j=1}^{n}\left\|\alpha\left(\bigcup_{i=j}^{n} A_i\right) - \alpha\left(\bigcup_{i=j+1}^{n} A_i\right)\right\|^p \tag{9.46}$$

where the union over the empty set of indices is empty. Let $v_p^{\leftarrow}(\alpha;A) := \sup_{\mathcal{A}} s_p^{\leftarrow}(\alpha;\mathcal{A})$ where the supremum is over all interval partitions of $A$. If $\alpha$ is additive, then clearly $v_p^{\leftarrow}(\alpha;A) = v_p(\alpha;A)$. Let $\mathcal{I}_p^{\leftarrow}(J;\mathbb{B})$ be the set of all interval functions $\alpha$ on $J$ with values in $\mathbb{B}$ such that $\sup_{A\in\mathfrak{I}(J)} v_p^{\leftarrow}(\alpha;A) < \infty$. Then on $\mathcal{I}_p^{\leftarrow}(J;\mathbb{B})$, it is easily seen that

$$\|\alpha\|_{(p)}^{\leftarrow} := \|\alpha\|_{J,(p)}^{\leftarrow} := \sup_{A\in\mathfrak{I}(J)} v_p^{\leftarrow}(\alpha;A)^{1/p} \tag{9.47}$$

is a seminorm and $\|\alpha\|_{[p]}^{\leftarrow} := \|\alpha\|_{J,[p]}^{\leftarrow} := \|\alpha\|_{(p)}^{\leftarrow} + \|\alpha\|_{\sup}$ is a norm. Further, for each $p$ let $\overline{\mathcal{I}}_p(J;\mathbb{B}) := \mathcal{I}_p(J;\mathbb{B}) \cap \mathcal{I}_p^{\leftarrow}(J;\mathbb{B})$ with the seminorm

$$\|\alpha\|_{\overline{(p)}} := \|\alpha\|_{A,\overline{(p)}} := \max(\|\alpha\|_{(p)}, \|\alpha\|_{(p)}^{\leftarrow}) \tag{9.48}$$

and the norm $\|\alpha\|_{\overline{[p]}} := \|\alpha\|_{A,\overline{[p]}} := \|\alpha\|_{\overline{(p)}} + \|\alpha\|_{\sup}$. It is easily seen that $(\mathcal{I}_p^{\leftarrow}(J;\mathbb{B}), \|\cdot\|_{[p]}^{\leftarrow})$ and $(\overline{\mathcal{I}}_p(J;\mathbb{B}), \|\cdot\|_{\overline{[p]}})$ are Banach spaces, the first by symmetry with $\mathcal{I}_p$ and the second by intersecting two Banach spaces of (interval) functions with the maximum of their norms.

**Theorem 9.36.** *For $\mu \in \mathcal{AI}_p([a,b];\mathbb{B})$, $1 \le p < 2$, the strict product integral $\widehat{\mu}$ with respect to $\mu$ exists,*

$$\|\widehat{\mu}\|_{\sup} \le K, \quad \text{and} \quad \|\widehat{\mu}\|_{\overline{(p)}} \le C_p K \|\widehat{\mu}\|_{\sup} \|\mu\|_{(p)}, \tag{9.49}$$

*where $K := K(\mu,p)$ is defined by* (9.29) *and $C_p$ just after* (9.28).

*Proof.* Let $\mu \in \mathcal{AI}_p([a,b];\mathbb{B})$ for some $1 \le p < 2$. Then the interval function $\mathfrak{I}[a,b] \ni A \mapsto \prod_A(\mathbb{1} + d\mu)$ is defined and multiplicative by Theorem 9.28. To check the condition of Definition 9.29, we can assume that $a < b$. Let $\epsilon > 0$. By Proposition 3.52, there exists a Young partition $\mathcal{B} = \{(z_{j-1}, z_j)\}_{j=1}^m$ of $[a,b]$ such that $v_p(\mu;(z_{j-1},z_j)) < \epsilon$ for each $j = 1,\dots,m$. Let $\{A_i\}_{i=1}^n$ be an interval partition of $[a,b]$ which is a refinement of $\mathcal{B}$, and let $I$ be the set of all indices $i \in \{1,\dots,n\}$ such that $A_i \cap \{z_j\}_{j=0}^m = \emptyset$. Then $\prod_{A_i}(\mathbb{1}+d\mu) = \mathbb{1}+\mu(A_i)$ for each $i \in \{1,\dots,n\} \setminus I$, and so by the inequality (9.42) of Theorem 9.28,

$$\begin{aligned}
\sum_{i=1}^n \Big\| \prod_{A_i}(\mathbb{1} + d\mu) - \mathbb{1} - \mu(A_i)\Big\| &\le K C_p^2 \sum_{i\in I} \|\mu\|_{A_i,(p)}^2 \\
&\le K C_p^2 \max_j \|\mu\|_{(z_{j-1},z_j),(p)}^{2-p} \sum_{i\in I} v_p(\mu;A_i) \\
&\le K C_p^2 \epsilon^{(2/p)-1} v_p(\mu;[a,b]).
\end{aligned}$$

In the last inequality, (3.69) was used. Thus $\widehat{\mu} = \prod(\mathbb{1}+d\mu)$, and the first bound in (9.49) follows from (9.40). To prove the second bound, let $A \in \mathfrak{I}[a,b]$ be nonempty and let $\mathcal{A} = \{A_i\}_{i=1}^n$ be an interval partition of $A$. To bound $\|\widehat{\mu}\|_{(p)}$, let $B_i = \cup_{j=1}^i A_j$ for $i = 1,\dots,n$, and let $B_0 := \emptyset$. By multiplicativity of $\widehat{\mu}$, (9.41) and (3.69) again, we have

$$\begin{aligned}
s_p(\widehat{\mu};\mathcal{A}) = \sum_{i=1}^n \big\|\widehat{\mu}(B_i) - \widehat{\mu}(B_{i-1})\big\|^p &\le \|\widehat{\mu}\|_{\sup}^p \sum_{i=1}^n \big\|\widehat{\mu}(A_i) - \mathbb{1}\big\|^p \\
&\le \|\widehat{\mu}\|_{\sup}^p C_p^p K^p v_p(\mu;A).
\end{aligned}$$

For $\|\widehat{\mu}\|_{(p)}^{\leftarrow}$, define instead $B_i = \bigcup_{j=i}^n A_i$ and in place of $B_{i-1}$ put $B_{i+1}$ with $B_{n+1} := \emptyset$. Then a symmetric argument gives the same bound for $s_p^{\leftarrow}$ as for $s_p$.

Since $\mathcal{A}$ is an arbitrary partition of $A$ and $\mu$ is additive, $v_p(\mu; A) \leq v_p(\mu; [a, b])$, so the second bound in (9.49) follows, completing the proof. □

Next we define a transform which for suitable interval functions is an inverse of the product integral operator. Let $\nu$ be an interval function on $[a, b]$ with values in a Banach algebra $\mathbb{B}$. For a nonempty interval $A \subset [a, b]$ and for an interval partition $\mathcal{A} = \{A_i\}_{i=1}^n$ of $A$, let

$$S(\nu; A, \mathcal{A}) := \sum_{i=1}^{n} \big[\nu(A_i) - \mathbb{I}\big].$$

**Definition 9.37.** Let $\nu$ be a $\mathbb{B}$-valued interval function on $[a, b]$. We say that an interval function $\check{\nu}$ (pronounced "nu check") is the *additive transform* of $\nu$ if $\check{\nu}(\emptyset) = 0$ and for each nonempty $A \in \mathfrak{I}[a, b]$,

$$\check{\nu}(A) = \lim_{\mathcal{A}} S(\nu; A, \mathcal{A}),$$

provided the limit exists under refinements of partitions $\mathcal{A}$ of $A$.

Clearly $\check{\nu}$, if it exists, is unique and an additive interval function. Let $\mu$ be an additive interval function on $[a, b]$ which has a strict product integral $\widehat{\mu}$. Then it is easy to see that the additive transform of $\widehat{\mu}$ exists and $(\widehat{\mu})\check{} = \mu$.

In an alternative notation, let $\mathcal{L}(\nu) := \check{\nu}$. If $\mu$ is an additive interval function on $[a, b]$ such that the product integral with respect to $\mu$ exists, let $\mathcal{P}(\mu) := \prod(\mathbb{I} + \mathrm{d}\mu)$.

**Corollary 9.38.** *Let $\mu \in \mathcal{AI}_p([a, b]; \mathbb{B})$, $1 \leq p < 2$. Then $\mathcal{L} \circ \mathcal{P}(\mu) = \mu$ and $\mathcal{P} \circ \mathcal{L}(\widehat{\mu}) = \widehat{\mu}$.*

## 9.7 Commutative Banach Algebras

Throughout this section, let $\mathbb{A}$ be a commutative Banach algebra. We will extend the sufficiency part of Theorem 9.11 to $\mathbb{A}$-valued interval functions. Specifically we show that if a suitable additive upper continuous interval function $\mu$ has values in $\mathbb{A}$ and has no atoms, then the product integral with respect to $\mu$ is strict and equals the exponential of $\mu$. If $\mu$ has atoms then we will see that the product integral of $\mu$ is strict and is the exponential of $\mu$ multiplied by a product of $\mathbb{A}$-valued functions.

Let $\xi$ be an $\mathbb{A}$-valued function defined on a nonempty set $T$. The product $\prod_T \xi$ will be said to converge *unconditionally* to an element $C \in \mathbb{A}$ if $C \neq 0$ and for each $\epsilon > 0$ there is a finite set $B \subset T$ such that $\| \prod_{t \in D} \xi(t) - C\| < \epsilon$ for each finite subset $D$ of $T$ including $B$.

Recall the exponential function $\exp(x)$, $x \in \mathbb{B}$, defined by (4.14) when $\mathbb{B}$ is a complex Banach algebra. If $\mathbb{A}$ is a real Banach algebra the exponential function is defined by the series representation on the right side of (4.14) using the complexification of $\mathbb{A}$ given by Definition 4.37 and Proposition 4.38. For an $\mathbb{A}$-valued interval function $\mu$ on $[a,b]$ let $\xi(t) := (\mathbb{I}+\mu(\{t\}))\exp\{-\mu(\{t\})\}$, $t \in [a,b]$, for a nonempty subset $T \subset [a,b]$ let

$$\gamma(\mu;T) := \prod_{T}(\mathbb{I}+\mu)\mathrm{e}^{-\mu} := \prod_{t\in T}\xi(t) \tag{9.50}$$

provided the product converges unconditionally, and let $\gamma(\mu;\emptyset) := \mathbb{I}$. Recall the definition (3.40) of $\sigma_2(\mu;T)$ for any set $T \subset [a,b]$.

**Lemma 9.39.** *Let $\mu$ be an upper continuous and additive $\mathbb{A}$-valued interval function on $[a,b]$ such that $\sigma_2(\mu;[a,b]) < \infty$ and $\mu(\{t\}) \neq -\mathbb{I}$ for each $t \in [a,b]$. Then $\gamma(\mu) = \{\gamma(\mu;A)\colon A \in \mathfrak{I}[a,b]\}$ is a well-defined $\mathbb{A}$-valued multiplicative interval function such that*

$$\|\gamma(\mu)\|_{\sup} \leq \exp\{4\sigma_2(\mu;[a,b])\}. \tag{9.51}$$

*Moreover, for each $A \in \mathfrak{I}[a,b]$,*

$$\|\mathbb{I} - \gamma(\mu;A)\| \leq 2^{-1}\sigma_2(\mu;A)\|\gamma(\mu)\|_{A,\sup}\mathrm{e}^{\|\mu\|_{A,\sup}}. \tag{9.52}$$

*Proof.* First we show that the product (9.50) is well defined for a set $T \subset [a,b]$ such that $\|\mu(\{t\})\| \leq 1/2$ for each $t \in T$. Let $B$ be a finite subset of $T$. Since $\exp\big(\log(u)\big) = u$ for any $u \in \mathbb{A}$ such that $\|\mathbb{I} - u\| < 1$, by Lemma 4.35 and since any finite product converges unconditionally we have

$$\xi(t) = \exp\{\log(\mathbb{I}+\mu(\{t\}))\}\exp\{-\mu(\{t\})\} = \exp\{\eta(t)\}$$

for each $t \in T$, where $\eta(t) := \log(\mathbb{I}+\mu(\{t\})) - \mu(\{t\})$. Using again Lemma 4.35, it follows that $\gamma(\mu;B) = \exp\{\sum_B \eta\}$ for any finite subset $B$ of $T$. By the series representation of the logarithm function, $\|\log(\mathbb{I}+u) - u\| \leq \|u\|^2$ if $\|u\| \leq 1/2$. Thus $\sum_B \|\eta\| \leq \sigma_2(\mu;B)$ for any subset $B \subset T$, and so the sum $\sum_T \eta$ converges absolutely in $\mathbb{A}$. Since the exponential is continuous, it then follows that the product (9.50) is well defined and the bound

$$\|\gamma(\mu;T)\| \leq \exp\big\{\sigma_2(\mu;T)\big\} \tag{9.53}$$

holds for any set $T \subset [a,b]$ such that $\|\mu(\{t\})\| \leq 1/2$ for each $t \in T$.

Now let $A = (s,t)$ be an open subinterval of $[a,b]$. By Proposition 2.6(*c*) since $\mu$ is upper continuous and additive, the set $\{z \in A\colon \|\mu(\{z\})\| > 1/2\}$ is finite, and so we can assume that it is a set $\{z_j\}_{j=1}^{m-1}$ for some $m \geq 2$. Let $z_0 := s$, $z_m := t$, $A_j := (z_{j-1},z_j)$ for $j = 1,\ldots,m$, $V_1 := \mathbb{I}$, and $V_k := \prod_{j=1}^{k-1}\xi(z_j)\gamma(\mu;A_j)$ for $k = 2,\ldots,m$. Using the series representation of the exponential function, for each $j = 1,\ldots,m-1$, we have

$$\|\xi(z_j)\| \leq \exp\{2\|\mu(\{z_j\})\|\} \leq \exp\{4\|\mu(\{z_j\})\|^2\} \tag{9.54}$$

since $2\|\mu(\{z_j\})\| > 1$. Since $\mu(\{t\}) \neq -\mathbb{I}$ for $t \in [a,b]$, $V := V_m\gamma(\mu; A_m) \neq 0$. To show that $\gamma(\mu; A) = V$, let $\epsilon > 0$. By the first part of the proof, for each $j = 1, \ldots, m$, there is a finite set $B_j \subset A_j$ such that

$$\|\gamma(\mu; D) - \gamma(\mu; A_j)\| < \epsilon \tag{9.55}$$

for each finite set $D$ such that $B_j \subset D \subset A_j$. Let $T$ be a finite set such that $T_0 := \{z_j\}_{j=1}^{m-1} \cup (\cup_{j=1}^m B_j) \subset T \subset A$. Then letting $T_j := T \cap A_j$ for $j = 1, \ldots, m$, and applying the algebraic identity (9.43) with $2m - 1$ in place of $m$, we have

$$\begin{aligned}\|\gamma(\mu; T) - V\| &= \Big\| \prod_{j=1}^{m-1} \xi(z_j) \prod_{j=1}^{m} \gamma(\mu; T_j) - \prod_{j=1}^{m-1} \xi(z_j) \prod_{j=1}^{m} \gamma(\mu; A_j) \Big\| \\ &\leq \sum_{j=1}^{m} \|\gamma(\mu; T \cap [z_j, t))\| \big\|\gamma(\mu; T_j) - \gamma(\mu; A_j)\big\| \|V_j\| \\ &< \epsilon m \exp\{8\sigma_2(\mu; A)\}.\end{aligned}$$

The last bound holds by (9.53), (9.54), and (9.55) since $B_j \subset T_j \subset A_j$ for each $j$. Since $\epsilon > 0$ is arbitrary and $T$ is any finite set including $T_0$, $\gamma(\mu; A)$ is well defined provided $A$ is an open subinterval of $[a, b]$.

If $A = [s,t] \subset [a,b]$ with $\|\mu(\{s\})\| > 1/2$ and $\|\mu(\{t\})\| > 1/2$, then instead of $T$ and $V$ as above take $T' := \{s\} \cup T \cup \{t\}$ and $V' := \xi(t)V\xi(s)$. Then the difference $\gamma(\mu; T') - V'$ equals $\xi(t)\big[\gamma(\mu; T) - V\big]\xi(s)$ and its norm is bounded as in the last display (with now $A$ closed rather than open). Since the same bound holds in all other cases, the interval function $\gamma(\mu)$ is defined on $[a,b]$. Clearly $\gamma(\mu)$ is multiplicative and (9.51) holds due to (9.53) and (9.54).

Finally to prove (9.52), let $A \in \mathfrak{I}[a,b]$ and let $T = \{t_1, \ldots, t_m\} \subset A$. Applying the algebraic identity (9.43) with $x_j := \mathbb{I}$ and $y_j := \xi(t_j)$, we have

$$\mathbb{I} - \gamma(\mu; T) = \sum_{j=1}^{m} \big[\mathbb{I} - \xi(t_j)\big]\gamma(\mu; T_j),$$

where $T_j := \{t_1, \ldots, t_{j-1}\}$ if $j = 2, \ldots, m$ and $T_1 := \emptyset$. For any $t \in [a,b]$, using the series representation of the exponential function, it follows that

$$\|\mathbb{I} - \xi(t)\| \leq 2^{-1}\|\mu(\{t\})\|^2 \mathrm{e}^{\|\mu(\{t\})\|}.$$

Therefore we have the bound

$$\|\mathbb{I} - \gamma(\mu; T)\| \leq 2^{-1}\sigma_2(\mu; A)\|\gamma(\mu)\|_{A,\sup} \mathrm{e}^{\|\mu\|_{A,\sup}}.$$

Since $T$ is any finite subset of $A$, (9.52) holds, proving the lemma. □

The following extends the sufficiency part of Theorem 9.11 and Corollary 9.18. Recall that $\mathcal{AI}_p^*([a,b];\mathbb{A})$, $0 < p < \infty$, is the class of all $\mathbb{A}$-valued additive upper continuous interval functions on $[a,b]$ having $p^*$-variation as defined in Definition 3.28.

**Theorem 9.40.** *Let $\mu \in \mathcal{AI}_2^*([a,b];\mathbb{A})$ be such that $\mu(\{t\}) \neq -\mathbb{I}$ for each $t \in [a,b]$. Then the strict product integral $\widehat{\mu}$ exists and for any interval $A \subset [a,b]$,*

$$\widehat{\mu}(A) = \exp\{\mu(A)\}\gamma(\mu; A). \tag{9.56}$$

*Proof.* We can assume that $a < b$. By Definition 3.28, $\sigma_2(\mu;[a,b]) < \infty$, and so the right side of (9.56) is defined for each $A \in \mathfrak{I}[a,b]$ by Lemma 9.39. Let $\alpha(A)$ be the right side of (9.56). By Lemmas 4.35 and 9.39, $\alpha = \{\alpha(A)\colon A \in \mathfrak{I}[a,b]\}$ is an $\mathbb{A}$-valued multiplicative interval function on $[a,b]$. Since $\mu$ is bounded, by Theorem 9.34, it is enough to prove that $\alpha$ is a multiplicative transform of $\mu$. To check whether the condition of Definition 9.29 holds, let $\epsilon \in (0,1/2]$. Since $\mu \in \mathcal{AI}_2^*$, by Lemma 3.29, there exists a Young interval partition $\mathcal{B} := \{(z_{j-1},z_j)\}_{j=1}^m$ of $[a,b]$ such that $\sum_{j=1}^m v_2(\mu;(z_{j-1},z_j)) < \epsilon$. Let $\mathcal{A} = \{A_i\}_{i=1}^n$ be an interval partition of $[a,b]$ which is a refinement of $\mathcal{B}$, and let $I$ be the set of all indices $i \in \{1,\dots,n\}$ such that $A_i \cap \{z_j\}_{j=0}^m = \emptyset$. For each $i \in \{1,\dots,n\} \setminus I$, $A_i = \{z_j\}$ for some $j \in \{0,\dots,m\}$, and so $\alpha(A_i) = \mathbb{I} + \mu(A_i)$, again using the multiplicative property of the exponential function (Lemma 4.35). Using the series representation of the exponential function, we have

$$\|\mathbb{I} + u - \exp(u)\| \le 2^{-1}\|u\|^2 e^{\|u\|} \quad \text{and} \quad \|\exp(u)\| \le e^{\|u\|}$$

for each $u \in \mathbb{A}$. This and the bound (9.52) yield

$$\begin{aligned}
\sum_{i=1}^n \|\mathbb{I} + \mu(A_i) - \alpha(A_i)\| &= \sum_{i\in I} \|\mathbb{I} + \mu(A_i) - \alpha(A_i)\| \\
&\le \sum_{i\in I} \|\mathbb{I} + \mu(A_i) - \exp\{\mu(A_i)\}\| + \sum_{i\in I} \|\exp\{\mu(A_i)\}\|\|\mathbb{I} - \gamma(\mu;A_i)\| \\
&\le 2^{-1}e^{\|\mu\|_{\sup}} \sum_{i\in I} \|\mu(A_i)\|^2 + 2^{-1}\|\gamma(\mu)\|_{\sup} e^{2\|\mu\|_{\sup}} \sum_{i\in I} \sigma_2(\mu;A_i) \\
&\le C\sum_{j=1}^m v_2(\mu;(z_{j-1},z_j)) < \epsilon C,
\end{aligned}$$

where $C := 2^{-1}\exp\{2\|\mu\|_{\sup}\}(1 + \|\gamma(\mu)\|_{\sup})$ and (3.69) was used in the last line. Thus $\alpha$ is a multiplicative transform of $\mu$, proving the theorem. □

## 9.8 Integrals with Two Integrands

In the next section we will extend the algebraic identity (9.43) to a class of multiplicative transforms. To prepare for that, in this section we define and examine an integral with two integrands, as in $\int f\, \mathrm{d}\mu\, g$. Such an integral is needed and may differ from $\int f g\, \mathrm{d}\mu$ when $\mathbb{B}$ is noncommutative. Two-integrand integrals are used to express the Fréchet derivative and the Taylor series representation of the product integral operator in Proposition 9.63 and Theorem 9.51 below, respectively (the proposition is proved before the proof of the theorem is completed).

*Definitions of integrals*

Let $f, g$ be $\mathbb{B}$-valued functions on $[a,b]$ and let $\mu$ be an additive interval function on $[a,b]$. Recall from Section 1.4 the notion of a tagged Young interval partition. For a tagged Young interval partition $\mathcal{T} = (\{(t_{i-1}, t_i)\}_{i=1}^n, \{s_i\}_{i=1}^n)$ of a nonempty interval $A \subset [a,b]$, let

$$S_{YS}(f, \mathrm{d}\mu, g; A, \mathcal{T}) := \sum_{i=1}^n f(s_i)\mu((t_{i-1}, t_i))g(s_i) + \sum_{i=0}^n f(t_i)\mu(\{t_i\} \cap A)g(t_i).$$

For an interval $A \subset [a,b]$, the Kolmogorov integral with two integrands $⨍_A f\, \mathrm{d}\mu\, g$ is defined as 0 if $A = \emptyset$ or if $A$ is nonempty, as the limit

$$⨍_A f\, \mathrm{d}\mu\, g := \lim_{\mathcal{T}} S_{YS}(f, \mathrm{d}\mu, g; A, \mathcal{T})$$

if it exists in the refinement sense. It is not at all surprising, and holds by Proposition 2.25, that for an upper continuous additive interval function $\mu$ on $[a,b]$ and an interval $A \in \mathfrak{I}[a,b]$, the Kolmogorov integral with two integrands $⨍_A f\, \mathrm{d}\mu\, \mathbb{I}$ exists if and only if the Kolmogorov integral $⨍_A f\, \mathrm{d}\mu$ also does, and then the two are equal since $S_{YS}(f, \mathrm{d}\mu, \mathbb{I}; A, \mathcal{T})$ equals the sum $S_{YS}(f, \mathrm{d}\mu; A, \mathcal{T})$ defined by (2.22) for each tagged Young interval partition $\mathcal{T}$ of $A$. Moreover, if the Banach algebra $\mathbb{B}$ is commutative, then $S_{YS}(f, \mathrm{d}\mu, g; A, \mathcal{T}) = S_{YS}(fg, \mathrm{d}\mu; A, \mathcal{T})$ for each tagged Young interval partition $\mathcal{T}$ of $A$, and so $⨍_A f\, \mathrm{d}\mu\, g = ⨍_A fg\, \mathrm{d}\mu$ if either integral exists.

Similarly we will define the *RYS* integral with two integrands. Let $f, g, h$ be $\mathbb{B}$-valued functions on $[a,b]$. For a tagged Young point partition $\tau = (\{t_i\}_{i=0}^n, \{s_i\}_{i=1}^n)$ of $[a,b]$ with $a < b$ the Young–Stieltjes sum $S_{YS}(f, \mathrm{d}h, g; \tau)$ based on $\tau$ is defined by

$$S_{YS}(f, \mathrm{d}h, g; \tau) := \sum_{i=1}^n f(s_i)[h(t_i-) - h(t_{i-1}+)]g(s_i) + \sum_{i=0}^n f(t_i)\Delta^{\pm}_{[a,b]}h(t_i)g(t_i),$$

where $\Delta^{\pm}_{[a,b]}h$ is defined by (2.1). The *RYS* integral with two integrands $(RYS)\int_a^b f\, \mathrm{d}h\, g$ is defined as 0 if $a = b$, or as

$$(RYS)\int_a^b f\,\mathrm{d}h\,g := \lim_{\tau} S_{YS}(f, \mathrm{d}h, g; \tau)$$

if $a < b$, provided the limit exists in the refinement sense. The $RYS$ integral with two integrands extends the $RYS$ integral (with one integrand), in the sense that if either of the integrals $(RYS)\int_a^b f\,\mathrm{d}h$ and $(RYS)\int_a^b f\,\mathrm{d}h\,\mathbb{I}$ exists then clearly so does the other, with the same value.

Given a regulated function $h$ on $[a, b]$, there is an additive interval function $\mu_h := \mu_{h,[a,b]}$ corresponding to $h$ defined by (2.2). Let

$$\underset{a}{\overset{b}{⨍}} f\,\mathrm{d}h\,g := \underset{[a,b]}{⨍} f\,\mathrm{d}\mu_{h,[a,b]}\,g$$

provided the Kolmogorov integral with two integrands is defined.

**Proposition 9.41.** *Let $f, h, g$ be $\mathbb{B}$-valued regulated functions on $[a, b]$. Then $⨍_a^b f\,\mathrm{d}h\,g = (RYS)\int_a^b f\,\mathrm{d}h\,g$ if either side is defined.*

*Proof.* We can assume that $a < b$. Since there is a one-to-one correspondence between tagged Young partitions $\tau$ and tagged Young interval partitions $\mathcal{T}$ with $S_{YS}(f, \mathrm{d}h, g; \tau) = S_{YS}(f, \mathrm{d}\mu_h, g; [a, b], \mathcal{T})$, the conclusion follows from the definitions. □

Integrals with two integrands have properties extending those of integrals with one integrand, often straightforwardly. For one such property, the proof of the following theorem is similar to the proof of Theorem 2.21 and is omitted.

**Theorem 9.42.** *Let $f, g\colon [a, b] \to \mathbb{B}$ and let $\mu\colon \mathfrak{I}[a, b] \to \mathbb{B}$ be additive. For $A, A_1, A_2 \in \mathfrak{I}(J)$ such that $A = A_1 \cup A_2$ and $A_1 \cap A_2 = \emptyset$, $⨍_A f\,\mathrm{d}\mu\,g$ exists if and only if both $⨍_{A_1} f\,\mathrm{d}\mu\,g$ and $⨍_{A_2} f\,\mathrm{d}\mu\,g$ exist, and then*

$$\underset{A}{⨍} f\,\mathrm{d}\mu\,g = \underset{A_1}{⨍} f\,\mathrm{d}\mu\,g + \underset{A_2}{⨍} f\,\mathrm{d}\mu\,g.$$

*In particular, if the integral $⨍_{[a,b]} f\,\mathrm{d}\mu\,g$ is defined, then $\mathfrak{I}[a, b] \ni A \mapsto ⨍_A f\,\mathrm{d}\mu\,g$ is a $\mathbb{B}$-valued additive interval function on $[a, b]$.*

The next theorem implies that the operators

$$U \times I \times W \ni (f, \mu, g) \mapsto \underset{[a,b]}{⨍} f\,\mathrm{d}\mu\,g \in \mathbb{B}$$

and

$$U \times V \times W \ni (f, h, g) \mapsto \underset{a}{\overset{b}{⨍}} f\,\mathrm{d}h\,g \in \mathbb{B}$$

are trilinear for any point function spaces $U, V, W$ and interval function space $I$ on which they are defined.

**Theorem 9.43.** *Let $f, g\colon [a,b] \to \mathbb{B}$, let $\mu\colon \mathfrak{I}[a,b] \to \mathbb{B}$ be additive, and let $u_1, u_2 \in \mathbb{K}$. The following three statements mean that the left side exists provided the right side does, and then the equality holds:*

(*a*) *for $f_1, f_2\colon [a,b] \to \mathbb{B}$,*

$$\oint_{[a,b]} (u_1 f_1 + u_2 f_2)\,\mathrm{d}\mu\, g = u_1 \oint_{[a,b]} f_1\,\mathrm{d}\mu\, g + u_2 \oint_{[a,b]} f_2\,\mathrm{d}\mu\, g;$$

(*b*) *for $\mu_1, \mu_2\colon \mathfrak{I}[a,b] \to \mathbb{B}$,*

$$\oint_{[a,b]} f\,\mathrm{d}(u_1\mu_1 + u_2\mu_2)\, g = u_1 \oint_{[a,b]} f\,\mathrm{d}\mu_1\, g + u_2 \oint_{[a,b]} f\,\mathrm{d}\mu_2\, g;$$

(*c*) *for $g_1, g_2\colon [a,b] \to \mathbb{B}$,*

$$\oint_{[a,b]} f\,\mathrm{d}\mu\,(u_1 g_1 + u_2 g_2) = u_1 \oint_{[a,b]} f\,\mathrm{d}\mu\, g_1 + u_2 \oint_{[a,b]} f\,\mathrm{d}\mu\, g_2.$$

*Moreover, the preceding three statements hold when $\mu$ is replaced by $h\colon [a,b] \to \mathbb{B}$ and $\oint_{[a,b]}$ is replaced by $\oint_a^b$.*

*Proof.* Statement (*a*) follows from the equality

$$S_{YS}(u_1 f_1 + u_2 f_2, \mathrm{d}\mu, g; [a,b], \mathcal{T}) \\ = u_1 S_{YS}(f_1, \mathrm{d}\mu, g; [a,b], \mathcal{T}) + u_2 S_{YS}(f_2, \mathrm{d}\mu, g; [a,b], \mathcal{T}),$$

valid for any tagged Young interval partition $\mathcal{T}$ of $[a,b]$. The proof of the other statements is symmetric and is omitted. □

The same argument as in the preceding proof gives the following statement.

**Theorem 9.44.** *Let $f\colon [a,b] \to \mathbb{B}$, let $\mu\colon \mathfrak{I}[a,b] \to \mathbb{B}$ be additive, and let $u \in \mathbb{B}$. If $\oint_{[a,b]} f\,\mathrm{d}\mu$ exists then so does $\oint_{[a,b]} uf\,\mathrm{d}\mu$ and*

$$\oint_{[a,b]} uf\,\mathrm{d}\mu = u \oint_{[a,b]} f\,\mathrm{d}\mu. \tag{9.57}$$

*Love–Young inequalities*

Here we extend the Love–Young inequalities of Section 3.6 to Riemann–Stieltjes and Young–Stieltjes sums and integrals with two integrands. For a partition $\kappa = \{t_j\}_{j=0}^n$ of $[a,b]$, let

$$S(f, \mathrm{d}h, g; \kappa) := \sum_{j=1}^n f(t_j)\big[h(t_j) - h(t_{j-1})\big]g(t_j).$$

**Theorem 9.45.** *Let $\Phi, \Psi \in \mathcal{CV}$ with inverses $\phi$ and $\psi$, respectively, and let $f, g, h\colon [a,b] \to \mathbb{B}$ with $a < b$. Then for any partition $\kappa$ of $[a,b]$,*

$$\begin{aligned} &\big\|S(f, \mathrm{d}h, g; \kappa) - f(a)[h(b) - h(a)]g(a)\big\| \\ &\qquad \le \big\{\|f\|_{\sup} + \|g\|_{\sup}\big\} \sum_{k=1}^{\infty} \phi\Big(\frac{v_\Phi(f) + v_\Phi(g)}{k}\Big)\psi\Big(\frac{v_\Psi(h)}{k}\Big). \end{aligned} \tag{9.58}$$

*Moreover, if for $u \ge 0$, $\Phi(u) \equiv u^q$ and $\Psi(u) \equiv u^p$ with $1 \le p, q < \infty$ and $p^{-1} + q^{-1} > 1$, then instead of (9.58),*

$$\begin{aligned} &\big\|S(f, \mathrm{d}h, g; \kappa) - f(a)[h(b) - h(a)]g(a)\big\| \\ &\qquad \le \zeta(p^{-1} + q^{-1})\big\{\|f\|_{(q)}\|g\|_{\sup} + \|g\|_{(q)}\|f\|_{\sup}\big\}\|h\|_{(p)}, \end{aligned} \tag{9.59}$$

*where $\zeta(r) := \sum_{k \ge 1} k^{-r}$ for $r > 1$.*

*Proof.* Let $\kappa = \{t_j\}_{j=0}^n$ be a partition of $[a,b]$. Define a sequence $\{\Box_{ij}\}_{i,j=1}^n$ of (trilinear) operators $\Box_{ij}\colon (f, h, g) \to \mathbb{B}$ by

$$(\Box_{ij})(f, h, g) := f(t_i)(\Delta_j h)g(t_i) - f(t_{i-1})(\Delta_j h)g(t_{i-1}), \tag{9.60}$$

where $\Delta_j F := F(t_j) - F(t_{j-1})$ for $j = 1, \ldots, n$ and $F = f, g, h$. Then we have

$$S(f, \mathrm{d}h, g; \kappa) - f(a)\big[h(b) - h(a)\big]g(a) = \Big(\sum_{1 \le i \le j \le n} \Box_{ij}\Big)(f, h, g).$$

In the proof of (9.58), for $r = 1, \ldots, n$ and $x, y, w \ge 0$, let

$$C_r(x, y, w) := \big\{\|f\|_{\sup} + \|g\|_{\sup}\big\}\phi\Big(\frac{x + y}{r}\Big)\psi\Big(\frac{w}{r}\Big). \tag{9.61}$$

For the proof of (9.59), instead of (9.61), let

$$C_r(x, y, w) := \Big[\Big(\frac{x}{r}\Big)^{1/q}\|g\|_{\sup} + \Big(\frac{y}{r}\Big)^{1/q}\|f\|_{\sup}\Big]\Big(\frac{w}{r}\Big)^{1/p}. \tag{9.62}$$

Thus (9.58) and (9.59) will be proved once we show that

$$\Big\|\Big(\sum_{1 \le i \le j \le n} \Box_{ij}\Big)(f, h, g)\Big\| \le \sum_{r=1}^{n} C_r\big(v_\Phi(f), v_\Phi(g), v_\Psi(h)\big). \tag{9.63}$$

The proof is similar to the proof of Lemma 3.85. We proceed recursively through $k = n, n-1, \ldots, 2$, replacing the sequence $\{\Box_{ij}\}_{i,j=1}^n$ of operators by new sequences and choosing one among them to give the desired bound. For discrete intervals of indices $A, B \subset \{1, \ldots, k\}$ such as $\{l, l+1, \ldots, m\}$, suppose that trilinear operators $\xi_{uv}^{(k)}$, $u \in A$, $v \in B$, with values in $\mathbb{B}$, are defined on ordered triples $(f, h, g)$ of functions from $J$ into $\mathbb{B}$.

For $k = 2, \ldots, n$, we define $I_k := \{1, \ldots, k\} \times \{1, \ldots, k\}$ and

$$S_{A,B}^{(k)} := \sum_{u \in A, v \in B} \xi_{uv}^{(k)}, \tag{9.64}$$

where $\xi_{uv}^{(n)} := \square_{uv}$ as defined by (9.60) for $(u,v) \in I_n$, and $\{\xi_{uv}^{(k)}\}_{(u,v)\in I_k}$ will be defined recursively for other $k$ in the course of the proof. In each recursion step we will have operators $S_{C,D}^{(k)} = S_{A,B}^{(n)}$, recalling that $\xi_{ij}^{(n)} = \square_{ij}$, where $C$, $D$, $A$, and $B$ are discrete intervals $\{l, l+1, \ldots, m\}$. In that case, for $A = \{l, \ldots, m\} \subset \{1, \ldots, n\}$ and any $B$,

$$\begin{aligned} &S_{A,B}^{(n)}(f,h,g) \\ &\quad = f(t_m)\Big(\sum_{j\in B} \Delta_j h\Big) g(t_m) - f(t_{l-1})\Big(\sum_{j\in B} \Delta_j h\Big) g(t_{l-1}) \\ &\quad = \Big(\sum_{i\in A} \Delta_i f\Big)\Big(\sum_{j\in B} \Delta_j h\Big) g(t_m) + f(t_{l-1})\Big(\sum_{j\in B} \Delta_j h\Big)\Big(\sum_{i\in A} \Delta_i g\Big). \end{aligned}$$

Thus

$$\begin{aligned} &\Big\| S_{A,B}^{(n)}(f,h,g) \Big\| \\ &\quad \le \Big\{ \Big\| \sum_{i\in A} \Delta_i f \Big\| \|g\|_{\sup} + \Big\| \sum_{i\in A} \Delta_i g \Big\| \|f\|_{\sup} \Big\} \Big\| \sum_{j\in B} \Delta_j h \Big\|. \end{aligned} \tag{9.65}$$

Given a set $\zeta = \{\xi_{uv}\}_{(u,v)\in I_k}$, where $1 \le k \le n$, let

$$U_k(\zeta) := \sum_{1\le u\le v\le k} \xi_{uv}.$$

Let $z^{(n)} := z = \{\square_{ij}\}_{(i,j)\in I_n}$. We will choose recursively for each $k = n-1, \ldots, 2$ a sequence $z^{(k)} = \{\xi_{uv}^{(k)}\}_{(u,v)\in I_k}$ of operators

$$\xi_{uv}^{(k)} = S_{A(u,k),B(v,k)}^{(n)} \tag{9.66}$$

for some discrete intervals $A(u,k) \subset \{1, \ldots, n\}$, $B(v,k) \subset \{1, \ldots, n\}$. Then the following inequalities will be proved:

$$\begin{aligned} &\Big\| \Big( \sum_{1\le i\le j\le n} \square_{ij} \Big)(f,h,g) \Big\| = \big\| U_n(z)(f,h,g) \big\| \\ &\quad \le \sum_{r=k+1}^{n} C_r\big(v_\Phi(f), v_\Phi(g), v_\Psi(h)\big) + \big\| U_k(z^{(k)})(f,h,g) \big\| \end{aligned} \tag{9.67}$$

and

$$\big\| U_2(z^{(2)})(f,h,g) \big\| \le \sum_{r=1}^{2} C_r(v_\Phi(f), v_\Phi(g), v_\Psi(h)). \tag{9.68}$$

Sets $A(u,k)$ and $B(v,k)$ included in $\{1,2,\ldots,n\}$ for $(u,v) \in I_k$ and sets $A_{i,l,k}$ and $B_{j,l,k}$ included in $\{1,2,\ldots,n\}$ for $(i,j) \in I_{k-1}$ and $l \in \{1,\ldots,k\}$ will be defined first for $k=n$ and then recursively for $k=n-1,\ldots,2$. Let $A(u,n) := \{u\}$ and $B(v,n) := \{v\}$ for $u,v=1,\ldots,n$. To define $A_{i,l,k}$ and $B_{j,l,k}$ from $A(u,k)$ and $B(v,k)$ let for $(i,j) \in I_{k-1}$,

$$A_{i,l,k} := \begin{cases} A(i,k) & \text{if } 1 \le i \le l-1, \\ A(l,k) \cup A(l+1,k) & \text{if } i = l, \\ A(i+1,k) & \text{if } l < i < k, \end{cases} \tag{9.69}$$

and

$$B_{j,l,k} := \begin{cases} B(j,k) & \text{if } 1 \le j < l-1, \\ B(l-1,k) \cup B(l,k) & \text{if } j = l-1, \\ B(j+1,k) & \text{if } l \le j < k. \end{cases} \tag{9.70}$$

Define $\xi_{ij}^{(k,l)} := S^{(n)}_{A_{i,l,k},B_{j,l,k}}$ for $(i,j) \in I_{k-1}$. At this stage $A(i,k)$ and $B(i,k)$ are defined just for $k=n$. Then $z^{(k,l)} := \{\xi_{ij}^{(k,l)}\}_{(i,j)\in I_{k-1}}$ is a sequence of operators depending on $k$ and $l$. Suppose we have defined $z^{(k)}$ for a given $k=n,n-1,\ldots,3$ and thus $z^{(k,l)}$. We then have

$$\begin{aligned} U_{k-1}(z^{(k,l)}) &= \sum_{1\le i\le k-1}\sum_{j=i}^{k-1} \xi_{ij}^{(k,l)} = \sum_{1\le i<l} S^{(k)}_{\{i\},\{i,\ldots,k\}} \\ &\quad + S^{(k)}_{\{l,l+1\},\{l+1,\ldots,k\}} + \sum_{l<i<k} S^{(k)}_{\{i+1\},\{i+1,\ldots,k\}} \\ &= -S^{(k)}_{\{l\},\{l\}} + U_k(z^{(k)}) = -\xi_{ll}^{(k)} + U_k(z^{(k)}). \end{aligned}$$

Thus for $l \in \{1,\ldots,k\}$,

$$U_k(z^{(k)}) - U_{k-1}(z^{(k,l)}) = \xi_{ll}^{(k)}. \tag{9.71}$$

By (9.66), we have $\xi_{ll}^{(k)} = S^{(n)}_{A(l,k),B(l,k)}$. Thus by (9.65),

$$\begin{aligned} &\left\|\xi_{ll}^{(k)}(f,h,g)\right\| \qquad\qquad (9.72)\\ &\le \Big\{\Big\|\sum_{i\in A(l,k)} \Delta_i f\Big\|\|g\|_{\sup} + \Big\|\sum_{i\in A(l,k)} \Delta_i g\Big\|\|f\|_{\sup}\Big\}\Big\|\sum_{j\in B(l,k)} \Delta_j h\Big\|. \end{aligned}$$

We will choose a particular value of $l$ using Lemma 3.86. Let $\Delta'_j F := \sum_{i\in A(j,k)} \Delta_i F$ for $F=f$ or $g$ and $\Delta''_j h := \sum_{i\in B(j,k)} \Delta_i h$. First let $u_j := \max\{\|\Delta'_j f\|, \|\Delta'_j g\|\}$ and $v_j := \|\Delta''_j h\|$. Then (3.139) with $m=k$ gives us an index $l' \in \{1,\ldots,k\}$ such that

$$u_{l'}v_{l'} \le \phi\Big(\frac{1}{k}\sum_{j=1}^k \Phi(u_j)\Big)\psi\Big(\frac{1}{k}\sum_{j=1}^k \Psi(v_j)\Big). \tag{9.73}$$

The inequality $\Phi(u_j) \le \Phi(\|\Delta'_j f\|) + \Phi(\|\Delta'_j g\|)$ and (9.72) give the bound

$$\big\|(\xi^{(k)}_{l'l'})(f,h,g)\big\| \le C_k(v_\Phi(f), v_\Phi(g), v_\Psi(h)), \tag{9.74}$$

where $C_k$ is defined by (9.61). Or in the case when $\Phi(u) \equiv u^q$ and $\Psi(u) \equiv u^p$ for some $1 \le p, q < \infty$ with $p^{-1} + q^{-1} > 1$, to bound (9.72) we keep $v_l$ unchanged but let $u_l := \|\Delta'_l f\|\,\|g\|_{\sup} + \|\Delta'_l g\|\,\|f\|_{\sup}$. An application of Lemma 3.86 again gives us an index $l' \in \{1,\dots,k\}$ such that (9.73) holds with the new values of $\{u_j\}$. That is, in this case we have the bound

$$\begin{aligned}&\big\|(\xi^{(k)}_{l'l'})(f,h,g)\big\| \\ &\quad\le \Big(\frac{1}{k}\sum_{j=1}^k \big(\|\Delta'_j f\|\,\|g\|_{\sup} + \|\Delta'_j g\|\,\|f\|_{\sup}\big)^q\Big)^{1/q}\Big(\frac{1}{k}\sum_{j=1}^k \|\Delta''_j h\|^p\Big)^{1/p}.\end{aligned}$$

By the Minkowski inequality (1.5) with $q$ instead of $p$, and for each $j$, taking $a_j = \|\Delta'_j f\|\,\|g\|_{\sup}$ and $b_j = \|\Delta'_j g\|\,\|f\|_{\sup}$, (9.74) holds in this case with $C_k$ defined by (9.62). In either case set $A(u,k-1) := A_{u,l',k}$ and $B(v,k-1) := B_{v,l',k}$. With these definitions, (9.69) and (9.70), the recursive definition of these sets is completed. Thus $z^{(k)} = \{\xi^{(k)}_{uv}\}_{(u,v)\in I_k}$ are also defined recursively by (9.66) for $k = n, n-1, \dots, 2$. By (9.71) with $l = l'$ and (9.74), we have the bound

$$\big\|U_k(z)(f,h,g)\big\| \le C_k(v_\Phi(f), v_\Phi(g), v_\Psi(h)) + \big\|U_{k-1}(z^{(k-1)})(f,h,g)\big\|. \tag{9.75}$$

Putting $k = n$ in (9.75) gives (9.67) with $k$ there set equal to $n-1$. Applying (9.75) inductively, it follows that (9.67) holds with $k = 2$. To prove (9.68) the recursive construction starting with $z^{(2)}$ gives $z^{(1)} = \{\xi^{(1)}_{11}\} := z^{(2,l')}$ such that

$$U_1(z^{(1)}) = \xi^{(1)}_{11} \quad\text{and}\quad U_2(z^{(2)}) - U_1(z^{(1)}) = \xi^{(2)}_{ll}$$

for some $l = l' \in \{1,2\}$ as in (9.71). Thus applying (3.139) as in (9.73) together with (9.65) and the inequality $\|\Delta\chi\| \le \Xi^{-1}(v_\Xi(\chi))$, we get the bound as in (9.74),

$$\big\|U_2(z^{(2)})(f,h,g)\big\| \le C_2(v_\Phi(f), v_\Phi(g), v_\Psi(h)) + C_1(v_\Phi(f), v_\Phi(g), v_\Psi(h)).$$

Thus (9.68) holds, which together with (9.67) for $k = 2$ yields (9.63). The proof of Theorem 9.45 is complete. □

The preceding Love–Young inequalities were proved for Riemann–Stieltjes sums based on tagged partitions of a particular form. To bound Young–Stieltjes sums we need to extend these inequalities to Riemann–Stieltjes sums based on arbitrary tagged partitions.

**Corollary 9.46.** *Let $h \in \mathcal{W}_p([a,b];\mathbb{B})$ and $f,g \in \mathcal{W}_q([a,b];\mathbb{B})$ with $a<b$, $1 \le p,q < \infty$, $p^{-1}+q^{-1} > 1$. For any tagged partition $\tau$ of $[a,b]$, we have*

$$\big\|S_{RS}(f,\mathrm{d}h,g;\tau)\big\| \le K_{p,q}\|f\|_{[q]}\|g\|_{[q]}\|h\|_{(p)}, \tag{9.76}$$

*where $K_{p,q} := 1+\zeta(p^{-1}+q^{-1})$. Also, if $\mu \in \mathcal{AI}_p([a,b];\mathbb{B})$ then for any tagged Young interval partition $\zeta$ of $[a,b]$,*

$$\big\|S_{YS}(f,\mathrm{d}\mu,g;[a,b],\zeta)\big\| \le K_{p,q}\|f\|_{[q]}\|g\|_{[q]}\|\mu\|_{(p)}. \tag{9.77}$$

*Proof.* Let $\tau = (\{t_i\}_{i=0}^n, \{s_i\}_{i=1}^n)$ be a tagged partition of $[a,b]$. Without changing the value of the sum $S_{RS}(f,\mathrm{d}h,g;\tau)$ one can assume that $s_i \in \{t_{i-1},t_i\}$ for $i=1,\dots,n$, since for each $i$ such that $t_{i-1}<s_i<t_i$ one can refine the partition $\{t_i\}_{i=0}^n$ to contain $s_i$ and let $s_i$ be the tag for both $[t_{i-1},s_i]$ and $[s_i,t_i]$. Letting $\Delta_i\chi := \chi(t_i)-\chi(t_{i-1})$ for $\chi = f$, $g$, or $h$, we have

$$\begin{aligned}
&S_{RS}(f,\mathrm{d}h,g;\tau) - f(a)[h(b)-h(a)]g(a)\\
&\quad = \sum_{i=1}^n f(t_i)(\Delta_i h)g(t_i) - f(a)[h(b)-h(a)]g(a)\\
&\quad\quad + \sum_{i=1}^n \Big\{f(s_i)(\Delta_i h)g(s_i) - f(t_i)(\Delta_i h)g(t_i)\Big\} =: T_1+T_2.
\end{aligned}$$

Let $J_i := [t_{i-1},t_i]$ for $i=1,\dots,n$. By Hölder's inequality (1.4) and (3.51), we get

$$\begin{aligned}
\|T_2\| &\le \sum_{i=1}^n \Big\{\|\Delta_i f\|\|\Delta_i h\|\|g(t_i)\| + \|f(t_{i-1})\|\|\Delta_i h\|\|\Delta_i g\|\Big\}\\
&\le \|g\|_{\sup}\sum_{i=1}^n \|f\|_{J_i,(q)}\|h\|_{J_i,(p)} + \|f\|_{\sup}\sum_{i=1}^n \|g\|_{J_i,(q)}\|h\|_{J_i,(p)}\\
&\le \|g\|_{\sup}\Big(\sum_{i=1}^n v_q(f;J_i)\Big)^{1/q}\Big(\sum_{i=1}^n v_p(h;J_i)\Big)^{1/p}\\
&\quad + \|f\|_{\sup}\Big(\sum_{i=1}^n v_q(g;J_i)\Big)^{1/q}\Big(\sum_{i=1}^n v_p(h;J_i)\Big)^{1/p}\\
&\le \big\{\|f\|_{(q)}\|g\|_{\sup} + \|g\|_{(q)}\|f\|_{\sup}\big\}\|h\|_{(p)}.
\end{aligned}$$

To bound $\|T_1\|$ we use Theorem 9.45 with $\Phi(u) \equiv u^q$, $\Psi(u) \equiv u^p$, $u \ge 0$, $1 \le p,q < \infty$ and $p^{-1}+q^{-1} > 1$. Thus we have

$$\begin{aligned}
\big\|S_{RS}(f,\mathrm{d}h,g;\tau)\big\| &\le \|T_1\| + \|T_2\| + \|f(a)[h(b)-h(a)]g(a)\|\\
&\le K_{p,q}\big\{\|f\|_{(q)}\|g\|_{\sup} + \|g\|_{(q)}\|f\|_{\sup}\big\}\|h\|_{(p)}\\
&\quad + \|f\|_{\sup}\|h\|_{(p)}\|g\|_{\sup}\\
&\le K_{p,q}\|f\|_{[q]}\|g\|_{[q]}\|h\|_{(p)},
\end{aligned}$$

proving (9.76).

To prove (9.77), let $h = R_{\mu,a}$ as defined by (2.3). By Corollary 2.11, $\mu = \mu_h$ defined by (2.2). Let $\tau = (\{t_i\}_{i=0}^n, \{u_i\}_{i=1}^n)$ be a tagged Young point partition of $[a, b]$ and let $\zeta = ((t_{i-1}, t_i)_{i=1}^n, \{u_i\}_{i=1}^n)$ be the corresponding tagged Young interval partition. It follows that $S_{YS}(f, \mathrm{d}h, g; \tau) \equiv S_{YS}(f, \mathrm{d}\mu, g; [a, b], \zeta)$. Now Young–Stieltjes sums $S_{YS}(f, \mathrm{d}h, g; \tau)$ can be approximated arbitrarily closely by Riemann–Stieltjes sums $S_{RS}$, just as in Proposition 2.18. So (9.77) follows, proving the corollary. □

Next we get existence of some integrals with two integrands and bounds for them.

**Proposition 9.47.** *For $1 \le p, q < \infty$ such that $p^{-1} + q^{-1} > 1$, let $\mu \in \mathcal{AI}_p([a, b]; \mathbb{B})$ and let $f, g \in \mathcal{W}_q([a, b]; \mathbb{B})$. Then the integral $⨎_{[a,b]} f \,\mathrm{d}\mu\, g$ exists, and for each $B \in \mathfrak{I}[a, b]$,*

$$\Big\| ⨎_B f \,\mathrm{d}\mu\, g \Big\| \le K_{p,q} \|f\|_{B,[q]} \|g\|_{B,[q]} \|\mu\|_{B,(p)}, \tag{9.78}$$

*where $K_{p,q} := 1 + \zeta(p^{-1} + q^{-1})$.*

*Proof.* By Theorem 9.42, the integral in (9.78) exists provided it does when $B = [a, b]$. To prove this take $q_1 > q$ such that $1/p + 1/q_1 > 1$. Then $f, g \in \mathcal{W}_{q_1}^*$ by Lemma 3.61, so in proving existence of the integral we can assume $f, g \in \mathcal{W}_q^*$. Given $\epsilon > 0$, let $\delta = (\epsilon/5)^q$. By Proposition 3.60, take a Young interval partition $\kappa = \{(z_{j-1}, z_j)\}_{j=1}^n$ of $[a, b]$ such that $\sum_{j=1}^n v_q(f; (z_{j-1}, z_j)) < \delta$. We have another such partition for $g$, and we can take a common refinement of both for which the same bounds will hold by superadditivity of $v_q$ as an interval function, (3.51). We can assume that $\kappa$ is this common refinement. Let $y_j := (z_{j-1} + z_j)/2$ for each $j$. Let $f_1$ be a step function, as given by Proposition 3.60(c) and its proof, such that $f_1(z_j) = f(z_j)$ for each $j$ and $f_1$ has values $f(z_{j-1}+)$ on $(z_{j-1}, y_j]$ and $f(z_j-)$ on $(y_j, z_j)$, with $\|f - f_1\|_{[q]} < \epsilon$. Let $g_1$ be a step function defined from $g$ as $f_1$ is from $f$. Let $\alpha$ be the Young interval partition $\{(z_{j-1}, y_j), (y_j, z_j)\}_{j=1}^n$. Then $f_1$ and $g_1$ are constant on each interval in $\alpha$. Let $\zeta$ be any tagged refinement of $\alpha$. It follows that $S_{YS}(f_1, \mathrm{d}\mu, g_1; [a, b], \zeta) =: S(f_1, \mathrm{d}\mu, g_1; \zeta) = S(f_1, \mathrm{d}\mu, g_1; \alpha)$ for each $\zeta$. Write

$$\begin{aligned}
&\|S(f, \mathrm{d}\mu, g; \zeta) - S(f_1, \mathrm{d}\mu, g_1; \alpha)\| \\
&\quad = \|S(f, \mathrm{d}\mu, g; \zeta) - S(f_1, \mathrm{d}\mu, g_1; \zeta)\| \\
&\quad \le \|S(f, \mathrm{d}\mu, g; \zeta) - S(f_1, \mathrm{d}\mu, g; \zeta)\| + \|S(f_1, \mathrm{d}\mu, g; \zeta) - S(f_1, \mathrm{d}\mu, g_1; \zeta)\| \\
&\quad = T_1 + T_2,
\end{aligned}$$

say. Then by (9.77),

$$T_1 \le K_{p,q} \|f - f_1\|_{[q]} \|g\|_{[q]} \|\mu\|_{(p)} \le \epsilon K_{p,q} \|g\|_{[q]} \|\mu\|_{(p)}$$

and $T_2 \le \epsilon K_{p,q}\|f\|_{[q]}\|\mu\|_{(p)}$, where $\|f_1\|_{[q]} \le \|f\|_{[q]}$ since clearly $\|f_1\|_{\sup} \le \|f\|_{\sup}$ and $q$-variation sums for $f_1$ can be approximated as well as desired by such sums for $f$. Letting $\epsilon \downarrow 0$, it follows by the Cauchy test that $⨏_{[a,b]} f d\mu g$ exists.

Now we return to the given $q$. We omit the proof of inequality (9.78) because it is the same as the proof of the Love–Young inequality in Theorem 3.93 except that now we use (9.77) if $B$ is closed, and otherwise we first approximate the Young–Stieltjes sum over $B$ by Young–Stieltjes sums over closed subintervals of $B$ and use the resulting bound. □

Finally we bound the $p$-variation of an interval function defined by an integral:

$$⨏ f\, d\mu\, g := \Big\{ B \mapsto ⨏_B f\, d\mu\, g \colon\ B \in \mathfrak{I}[a,b] \Big\}.$$

**Corollary 9.48.** *For $1 \le p, q < \infty$ such that $p^{-1} + q^{-1} > 1$, let $\mu \in \mathcal{AI}_p([a,b];\mathbb{B})$ and let $f, g \in \mathcal{W}_q([a,b];\mathbb{B})$. Then $⨏ f\, d\mu\, g$ is in $\mathcal{AI}_p([a,b];\mathbb{B})$ and for each $J \in \mathfrak{I}[a,b]$,*

$$\big\| ⨏ f\, d\mu\, g \big\|_{J,(p)} \le K_{p,q}\|f\|_{J,[q]}\|g\|_{J,[q]}\|\mu\|_{J,(p)}, \tag{9.79}$$

*where $K_{p,q} := 1 + \zeta(p^{-1} + q^{-1})$.*

*Proof.* The interval function $⨏ f\, d\mu\, g$ is defined and additive due to Proposition 9.47 and Theorem 9.42. Also, it is upper continuous by (9.78) and Proposition 2.6 since $\mathfrak{I}[a,b] \ni A \mapsto v_p(\mu; A)$ is upper continuous at $\emptyset$ by Proposition 3.50. To prove (9.79), let $J \in \mathfrak{I}[a,b]$ and let $\mathcal{A} = \{A_i\}_{i=1}^n$ be an interval partition of $J$. By (9.78) and (3.69) with $\Phi(u) \equiv u^p$, it follows that

$$s_p(⨏ f\, d\mu\, g; \mathcal{A}) \le K_{p,q}^p \|f\|_{J,[q]}^p \|g\|_{J,[q]}^p v_p(\mu; J).$$

Since $\mathcal{A}$ is arbitrary, (9.79) holds, and so $⨏ f\, d\mu\, g$ is in $\mathcal{AI}_p([a,b];\mathbb{B})$. □

## 9.9 Duhamel's Formula

In this section the algebraic identity (9.43) is extended to a suitable class of multiplicative interval functions which includes the product integrals with respect to an additive upper continuous interval function of bounded $p$-variation for $1 \le p < 2$.

In the next theorem and hereafter, $⨏_A f(t)\, \gamma(dt)\, g(t) \equiv ⨏_A f\, d\gamma\, g$ for any point functions $f, g$ on $[a,b]$ and interval function $\gamma$ on $[a,b]$.

**Theorem 9.49.** *Let $\mu, \nu \in \mathcal{AI}_2^*([a,b];\mathbb{B})$, and suppose that $\mu$ and $\nu$ have strict product integrals $\widehat{\mu}$, $\widehat{\nu}$ respectively. Then for each $A \in \mathfrak{I}[a,b]$, the following two integrals exist, and*

$$\widehat{\mu}(A) - \widehat{\nu}(A) = \oint_A \widehat{\nu}((t,b] \cap A)\,(\mu - \nu)(\mathrm{d}t)\,\widehat{\mu}([a,t) \cap A) \tag{9.80}$$

$$= \oint_A \widehat{\mu}((t,b] \cap A)\,(\mu - \nu)(\mathrm{d}t)\,\widehat{\nu}([a,t) \cap A). \tag{9.81}$$

Either of the two relations (9.80) and (9.81) will be called Duhamel's formula. Before proving Theorem 9.49 we establish a relation between the $p$-variation of an additive interval function $\mu$ and a strict product integral $\widehat{\mu}$. Let $\Phi \in \mathcal{CV}$ and let $\pi$ be a $\mathbb{B}$-valued interval function on $J = [a,b]$. For any $c > 0$, let

$$w_\Phi(\pi, c) := w_\Phi(\pi, c; J) := \sup\Big\{ \sum_{i=1}^n \Phi\Big(\frac{\|\pi(A_i) - \pi(\emptyset)\|}{c}\Big) \colon \{A_i\}_{i=1}^n \in \mathrm{IP}\,(J)\Big\},$$

and let

$$V_\Phi(\pi) := V_\Phi(\pi; J) := \inf\{c > 0 \colon w_\Phi(\pi, c; J) \le 1\}, \tag{9.82}$$

or $+\infty$ if no such finite $c$ exists. For $\Phi(u) \equiv u^p$, $u \ge 0$, $1 \le p < \infty$, we write $V_p := V_\Phi$. Then

$$V_p(\pi) \;=\; \Big(\sup\Big\{ \sum_{i=1}^n \|\pi(A_i) - \pi(\emptyset)\|^p \colon \{A_i\}_{i=1}^n \in \mathrm{IP}\,(J)\Big\}\Big)^{1/p}. \tag{9.83}$$

It is easy to check that on interval functions $\pi$ for which it is finite, $V_p$ is a seminorm, with $V_p(\pi) = 0$ if and only if $\pi \equiv \pi(\emptyset)$, in other words, $\pi$ is constant.

Let $1 \le p < 2$ and let $\widehat{\mu}$ be the strict product integral of $\mu \in \mathcal{AI}_p(J;\mathbb{B})$, which exists and is in $\overline{\mathcal{I}}_p(J;\mathbb{B})$ by Theorems 9.36 and 9.28. For an interval partition $\{A_i\}_{i=1}^n$ of $J$, by (9.41) and (3.69), we have

$$\sum_{i=1}^n \big\|\widehat{\mu}(A_i) - \mathbb{1}\big\|^p \le C_p^p K^p v_p(\mu; J),$$

where $K = K(\mu, p)$ and $C_p$ are as defined in (9.29) and (9.28), respectively, and so

$$V_p(\widehat{\mu}) \le C_p K \|\mu\|_{(p)} < \infty. \tag{9.84}$$

Thus, in this section and the next, various bounds in terms of $V_p(\widehat{\mu})$ will yield bounds in terms of $K(\mu,p)$ (which is bounded for $\|\mu\|_{[p]}$ bounded) and $\|\mu\|_{(p)}$. To give a more precise characterization of $V_\Phi(\widehat{\mu})$, recall $\widetilde{\mathcal{I}}_\Phi$ and $\|\nu\|_{(\Phi)}$ for not necessarily additive interval functions $\nu$ as defined in Definition 3.18.

From the following lemma, we will actually use only the bound, for $1 \le p < 2$, $\|\widehat{\mu}\|_{\overline{(p)}} \le V_p(\widehat{\mu})\|\widehat{\mu}\|_{\sup}$.

**Lemma 9.50.** *Let $\mu$ be a $\mathbb{B}$-valued additive interval function on $[a,b]$, having a strict product integral $\widehat{\mu}$. For $\Phi \in \mathcal{CV}$, $\mu \in \widetilde{\mathcal{I}}_{\Phi}([a,b];\mathbb{B})$ if and only if $\max\{\|\widehat{\mu}\|_{\sup}, V_{\Phi}(\widehat{\mu})\} < \infty$, and if either of the two statements holds then $\|\widehat{\mu}\|_{(\Phi)} \le V_{\Phi}(\widehat{\mu})\|\widehat{\mu}\|_{\sup} < \infty$. If $1 \le p < \infty$ and $\Phi(x) = x^p$ for $0 \le x < \infty$ then we also have $\|\widehat{\mu}\|_{\overline{(p)}} \le V_p(\widehat{\mu})\|\widehat{\mu}\|_{\sup}$ and*

$$\|\widehat{\mu}\|_{\overline{[p]}} \le C_p(\widehat{\mu}) := (1 + V_p(\widehat{\mu}))\|\widehat{\mu}\|_{\sup}. \tag{9.85}$$

*Proof.* We can assume that $a < b$. There exists a Young interval partition $\mathcal{B} = \{(z_{j-1}, z_j)\}_{j=1}^{m}$ of $[a,b]$ such that if $\{B_l\}_{l=1}^{k}$ is a refinement of $\mathcal{B}$ then

$$\max_{1\le l\le k} \|\widehat{\mu}(B_l) - \mathbb{1} - \mu(B_l)\| \le \sum_{l=1}^{k} \|\widehat{\mu}(B_l) - \mathbb{1} - \mu(B_l)\| \le 1. \tag{9.86}$$

Let $\|\mu\|_{[\Phi]} < \infty$. Then $\|\widehat{\mu}\|_{\sup} < \infty$ by Proposition 9.33. To show that $V_{\Phi}(\widehat{\mu}) < \infty$ let $\{A_i\}_{i=1}^{n}$ be an interval partition of $[a,b]$. There are at most $m+1$ values of $i$ such that $A_i$ contains some $z_j$. For all other values of $i$, $A_i$ is included in some $(z_{j-1}, z_j)$. For $j = 1, \dots, m$, let $I(j) := \{i = 1, \dots, n\colon A_i \subset (z_{j-1}, z_j)\}$, and let $c > 2\max\{\|\mu\|_{(\Phi)}, \|\widehat{\mu}\|_{\sup} + 1 + \|\mu\|_{\sup}\}$. Then the intervals $\{A_i\colon i \in I(j),\, j = 1, \dots, m\}$ are members of a refinement of $\mathcal{B}$. Therefore using convexity of $\Phi$ twice and $\Phi(0) = 0$, we get

$$\begin{aligned}
&\sum_{j=1}^{m} \sum_{i\in I(j)} \Phi\Big(\frac{\|\widehat{\mu}(A_i) - \mathbb{1}\|}{c}\Big) \\
&\qquad \le \sum_{j=1}^{m} \sum_{i\in I(j)} \Big\{\frac{1}{2}\Phi\Big(\frac{2\|\widehat{\mu}(A_i) - \mathbb{1} - \mu(A_i)\|}{c}\Big) + \frac{1}{2}\Phi\Big(\frac{2\|\mu(A_i)\|}{c}\Big)\Big\} \qquad (9.87)\\
&\qquad \le \frac{\Phi(1)}{c} \sum_{j=1}^{m} \sum_{i\in I(j)} \|\widehat{\mu}(A_i) - \mathbb{1} - \mu(A_i)\| + \frac{1}{2} v_{\Phi}\Big(\frac{2\mu}{c}\Big).
\end{aligned}$$

Thus by (9.86), we have the bound

$$\sum_{i=1}^{n} \Phi\Big(\frac{\|\widehat{\mu}(A_i) - \mathbb{1}\|}{c}\Big) \le (m+1)\Phi\Big(\frac{\|\widehat{\mu}\|_{\sup} + 1}{c}\Big) + \frac{\Phi(1)}{c} + \frac{1}{2} v_{\Phi}\Big(\frac{2\mu}{c}\Big). \tag{9.88}$$

Since the interval partition $\{A_i\}_{i=1}^{n}$ is arbitrary, $w_{\Phi}(\widehat{\mu}, c) < \infty$ and hence, letting $c \to +\infty$ for $m$ fixed, $V_{\Phi}(\widehat{\mu}) < \infty$, proving the "only if" part. Conversely, let $\max\{\|\widehat{\mu}\|_{\sup}, V_{\Phi}(\widehat{\mu})\} < \infty$. Then $\|\mu\|_{\sup} < \infty$ by Proposition 9.33. To prove $\|\mu\|_{(\Phi)} < \infty$, let $\{A_i\}_{i=1}^{n}$ be an interval partition of $A \in \mathfrak{I}[a,b]$. As for (9.88), with $\widehat{\mu}(A_i) - \mathbb{1}$ and $\mu(A_i)$ in (9.87) interchanged, we get that

$$\sum_{i=1}^{n} \Phi\Big(\frac{\|\mu(A_i)\|}{c}\Big) \le (m+1)\Phi\Big(\frac{\|\mu\|_{\sup}}{c}\Big) + \frac{\Phi(1)}{c} + \frac{w_{\Phi}(\widehat{\mu}, c/2)}{2}$$

holds for any $c > 2\max\{V_\Phi(\widehat{\mu}), \|\widehat{\mu}\|_{\sup} + 1 + \|\mu\|_{\sup}\}$. For $c$ large enough the right side of the last display is less than 1 by continuity of $\Phi$ at 0 and the definition of $V_\Phi$. Since $\{A_i\}_{i=1}^n$ is arbitrary, $\|\mu\|_{(\Phi)} < \infty$, proving the first part of the conclusion.

To prove the second part let again $\max\{\|\widehat{\mu}\|_{\sup}, V_\Phi(\widehat{\mu})\} < \infty$, let $c > V_\Phi(\widehat{\mu})\|\widehat{\mu}\|_{\sup}$, and let $\{A_i\}_{i=1}^n$ be an interval partition of $A$. Then by multiplicativity of $\widehat{\mu}$, we have

$$\sum_{i=1}^n \Phi\Big(\Big\|\widehat{\mu}\Big(\bigcup_{j=1}^i A_j\Big) - \widehat{\mu}\Big(\bigcup_{j=1}^{i-1} A_j\Big)\Big\|/c\Big) \le \sum_{i=1}^n \Phi\Big(\frac{\|\widehat{\mu}(A_i) - \mathbb{1}\|\|\widehat{\mu}\|_{\sup}}{c}\Big) \le 1.$$

Since $c > V_\Phi(\widehat{\mu})\|\widehat{\mu}\|_{\sup}$ is arbitrary, $\|\widehat{\mu}\|_{(\Phi)} \le V_\Phi(\widehat{\mu})\|\widehat{\mu}\|_{\sup}$. If $\Phi(x) \equiv x^p$ then we can replace $\cup_{j=1}^i A_j$ by $\cup_{j=i}^n A_j$ and $\cup_{j=1}^{i-1} A_i$ by $\cup_{j=i+1}^n A_i$ ($= \emptyset$ for $i = n$) for each $i = 1, \dots, n$ in the last display, giving the same bound for $v_p^{\leftarrow}$ as for $v_p$. The last two statements follow, completing the proof of the lemma. □

Now we are ready to prove Duhamel's formula.

**Proof of Theorem 9.49.** Taking the negative of both sides of either of the two equalities (9.80) or (9.81), the other follows. Thus it is enough to prove (9.81). By Definition 9.29 of multiplicative transform, for any $x \in [a, b]$,

$$\widehat{\mu}(\{x\}) - \widehat{\nu}(\{x\}) = \mu(\{x\}) - \nu(\{x\}), \tag{9.89}$$

and so (9.81) holds with $A = \{x\}$. Thus we can assume that $a < b$ and $A \in \mathfrak{I}[a, b]$ is nondegenerate. First suppose that $A = [u, v] \subset [a, b]$. Let $\mathcal{T} = (\{(x_{i-1}, x_i)\}_{i=1}^n, \{y_i\}_{i=1}^n)$ be a tagged Young interval partition of $[u, v]$, and let $A_{2i} := \{x_i\}$ for $i = 0, 1, \dots, n$, $A_{2i-1} := (x_{i-1}, x_i)$ for $i = 1, \dots, n$. Then by multiplicativity, we have

$$\widehat{\mu}([u, v]) - \widehat{\nu}([u, v]) = \prod_{k=0}^{2n} \widehat{\mu}(A_k) - \prod_{k=0}^{2n} \widehat{\nu}(A_k) =: M.$$

Applying the algebraic identity (9.43) to the right side, it follows that

$$M = \sum_{i=1}^n \widehat{\mu}([x_i, v])\{\widehat{\mu}((x_{i-1}, x_i)) - \widehat{\nu}((x_{i-1}, x_i))\}\widehat{\nu}([u, x_{i-1}])$$

$$+ \sum_{i=0}^n \widehat{\mu}((x_i, v])\{\widehat{\mu}(\{x_i\}) - \widehat{\nu}(\{x_i\})\}\widehat{\nu}([u, x_i)).$$

Then by (9.89),

$$M = S_{YS}(\widehat{\mu}((\cdot, v]), \mathrm{d}(\mu - \nu), \widehat{\nu}([u, \cdot)); [u, v], \mathcal{T}) + R(\mathcal{T}),$$

where $R(\mathcal{T}) := \sum_{i=1}^{n} D_i$ and for $i = 1, \dots, n$,

$$D_i := \widehat{\mu}([x_i, v])\{\widehat{\mu}((x_{i-1}, x_i)) - \widehat{\nu}((x_{i-1}, x_i))\}\widehat{\nu}([u, x_{i-1}])$$
$$- \widehat{\mu}((y_i, v])\{\mu((x_{i-1}, x_i)) - \nu((x_{i-1}, x_i))\}\widehat{\nu}([u, y_i)).$$

To prove (9.81) for $A = [u, v]$ it is enough to show that

$$\lim_{\mathcal{T}} R(\mathcal{T}) = 0. \tag{9.90}$$

For any $i \in \{1, \dots, n\}$ we have

$$\begin{aligned} D_i &= \widehat{\mu}([x_i, v])\{\widehat{\mu}((x_{i-1}, x_i)) - \mathbb{I} - \mu((x_{i-1}, x_i))\}\widehat{\nu}([u, x_{i-1}]) \\ &\quad - \widehat{\mu}([x_i, v])\{\widehat{\nu}((x_{i-1}, x_i)) - \mathbb{I} - \nu((x_{i-1}, x_i))\}\widehat{\nu}([u, x_{i-1}]) \\ &\quad + \widehat{\mu}([x_i, v])\{\mu((x_{i-1}, x_i)) - \nu((x_{i-1}, x_i))\}\widehat{\nu}([u, x_{i-1}]) \\ &\quad - \widehat{\mu}((y_i, v])\{\mu((x_{i-1}, x_i)) - \nu((x_{i-1}, x_i))\}\widehat{\nu}([u, y_i)) \\ &=: U_1(i) - U_2(i) + U_3(i) - U_4(i). \end{aligned}$$

Let $\epsilon > 0$. By Lemma 3.29 with $p = 2$ and Definition 9.29, there exists a Young interval partition $\lambda = \{(z_{j-1}, z_j)\}_{j=1}^{m}$ of $[a, b]$ such that

$$S_1 := \sum_{j=1}^{m} v_2(\mu; (z_{j-1}, z_j)) < \epsilon, \qquad S_2 := \sum_{j=1}^{m} v_2(\nu; (z_{j-1}, z_j)) < \epsilon,$$

and for any Young interval partition $\mathcal{A} = \{A_i\}_{i=0}^{N}$ that is a refinement of $\lambda$,

$$S_3(\mathcal{A}) := \sum_{i=0}^{N} \|\widehat{\mu}(A_i) - \mathbb{I} - \mu(A_i)\| < \epsilon, \quad S_4(\mathcal{A}) := \sum_{i=0}^{N} \|\widehat{\nu}(A_i) - \mathbb{I} - \nu(A_i)\| < \epsilon.$$

(The sums can be restricted to the open intervals in $\mathcal{A}$ since the terms corresponding to singletons are 0.) Let $\mathcal{T} = (\{(x_{i-1}, x_i)\}_{i=1}^{n}, \{y_i\}_{i=1}^{n})$ be a tagged refinement of $\{(z_{j-1}, z_j)\}_{j=1}^{m}$, and $\mathcal{A} := \{(x_{i-1}, x_i)\}_{i=1}^{n}$. Then we have

$$\sum_{i=1}^{n} \|U_1(i)\| \le \|\widehat{\mu}\|_{\sup}\|\widehat{\nu}\|_{\sup} S_3(\mathcal{A}) < \epsilon\|\widehat{\mu}\|_{\sup}\|\widehat{\nu}\|_{\sup}.$$

Likewise, $\sum_{i=1}^{n} \|U_2(i)\| < \epsilon\|\widehat{\mu}\|_{\sup}\|\widehat{\nu}\|_{\sup}$. Let $\gamma := \mu - \nu$. For any $i \in \{1, \dots, n\}$, since $x_{i-1} < y_i < x_i$, by straightforward algebra we have

$$\begin{aligned} &U_4(i) - U_3(i) \\ &= \{\widehat{\mu}((y_i, v]) - \widehat{\mu}([x_i, v]) - \widehat{\mu}([x_i, v])\mu((y_i, x_i))\}\gamma((x_{i-1}, x_i))\widehat{\nu}([u, y_i)) \\ &\quad + \widehat{\mu}([x_i, v])\gamma((x_{i-1}, x_i))\{\widehat{\nu}([u, y_i)) - \widehat{\nu}([u, x_{i-1}]) - \nu((x_{i-1}, y_i))\widehat{\nu}([u, x_{i-1}])\} \\ &\quad + \widehat{\mu}([x_i, v])\mu((y_i, x_i))\gamma((x_{i-1}, x_i))\widehat{\nu}([u, y_i)) \\ &\quad + \widehat{\mu}([x_i, v])\gamma((x_{i-1}, x_i))\nu((x_{i-1}, y_i))\widehat{\nu}([u, x_{i-1}]). \end{aligned}$$

Using multiplicativity of $\widehat{\mu}$ and $\widehat{\nu}$, we continue the equality

$$\begin{aligned} U_4(i) - U_3(i) \\ &= \widehat{\mu}([x_i, v])\big\{\widehat{\mu}((y_i, x_i)) - \mathbb{I} - \mu((y_i, x_i))\big\}\gamma((x_{i-1}, x_i))\widehat{\nu}([u, y_i]) \qquad (9.91) \\ &\quad + \widehat{\mu}([x_i, v])\gamma((x_{i-1}, x_i))\big\{\widehat{\nu}((x_{i-1}, y_i)) - \mathbb{I} - \nu((x_{i-1}, y_i))\big\}\widehat{\nu}([u, x_{i-1}]) \\ &\quad + \widehat{\mu}([x_i, v])\mu((y_i, x_i))\big\{\mu((x_{i-1}, x_i)) - \nu((x_{i-1}, x_i))\big\}\widehat{\nu}([u, y_i]) \\ &\quad + \widehat{\mu}([x_i, v])\big\{\mu((x_{i-1}, x_i)) - \nu((x_{i-1}, x_i))\big\}\nu((x_{i-1}, y_i))\widehat{\nu}([u, x_{i-1}]). \end{aligned}$$

It is clear that

$$T_1 := \sum_{i=1}^{n} \big\|\widehat{\mu}((y_i, x_i)) - \mathbb{I} - \mu((y_i, x_i))\big\| < \epsilon,$$

and likewise

$$T_2 := \sum_{i=1}^{n} \big\|\widehat{\nu}((x_{i-1}, y_i)) - \mathbb{I} - \nu((x_{i-1}, y_i))\big\| < \epsilon.$$

Using Hölder's inequality (1.4) with $p = q = 2$, and then Minkowski's inequality (1.5) with $r = 2$, it follows that

$$\begin{aligned} T_3 &:= \sum_{i=1}^{n} \big\|\mu((y_i, x_i))\big\|\Big\{\big\|\mu((x_{i-1}, x_i))\big\| + \big\|\nu((x_{i-1}, x_i))\big\|\Big\} \\ &< \sqrt{\epsilon}\big\{\|\mu\|_{(2)} + \|\nu\|_{(2)}\big\} \end{aligned}$$

and

$$\begin{aligned} T_4 &:= \sum_{i=1}^{n} \big\|\nu((x_{i-1}, y_i))\big\|\Big\{\big\|\mu((x_{i-1}, x_i))\big\| + \big\|\nu((x_{i-1}, x_i))\big\|\Big\} \\ &< \sqrt{\epsilon}\big\{\|\mu\|_{(2)} + \|\nu\|_{(2)}\big\}. \end{aligned}$$

Now summing the norms of the right side of equality (9.91), we have

$$\begin{aligned} \sum_{i=1}^{n} \big\|U_4(i) - U_3(i)\big\| \\ &\le \|\widehat{\mu}\|_{\sup}\|\gamma\|_{\sup}\|\widehat{\nu}\|_{\sup}\big[T_1 + T_2\big] + \|\widehat{\mu}\|_{\sup}\|\widehat{\nu}\|_{\sup}\big[T_3 + T_4\big] \\ &< 2\epsilon\|\widehat{\mu}\|_{\sup}\|\gamma\|_{\sup}\|\widehat{\nu}\|_{\sup} + 2\sqrt{\epsilon}\|\widehat{\mu}\|_{\sup}\|\|\widehat{\nu}\|_{\sup}\big\{\|\mu\|_{(2)} + \|\nu\|_{(2)}\big\}. \end{aligned}$$

It follows from these bounds that (9.90) holds.

Now if $A = (u, v]$ then all calculations are the same except that $\widehat{\mu}(\{u\}) = \widehat{\nu}(\{u\}) = \mathbb{I}$ and $\mu(\{u\}) = \nu(\{u\}) = 0$. The proof is also similar to the above for the other two cases $A = [u, v)$ and $A = (u, v)$. The proof of Theorem 9.49 is complete. □

## 9.10 Smoothness of the Product Integral Operator

For $1 \le p < \infty$, recall that $\mathcal{I}_p(J;\mathbb{B})$ is the set of all interval functions on an interval $J$ with values in $\mathbb{B}$ having bounded $p$-variation, and $\mathcal{AI}_p(J;\mathbb{B})$ is the set of all additive and upper continuous interval functions in $\mathcal{I}_p(J;\mathbb{B})$. Integrals $⨍ \, \mathrm{d}\nu \, f$ may be written $⨍ \, \nu(\mathrm{d}s)f(s)$.

Let $1 \le p < 2$. If $\mu \in \mathcal{AI}_p([a,b];\mathbb{B})$ then the strict product integral $\widehat{\mu}$ of $\mu$ exists and is in $\overline{\mathcal{I}}_p([a,b];\mathbb{B})$ by Theorem 9.36. Thus we have defined the nonlinear operator $\mathcal{P}$:

$$\mathcal{AI}_p([a,b];\mathbb{B}) \ni \mu \mapsto \mathcal{P}(\mu) = \widehat{\mu} = \prod(\mathbb{I} + \mathrm{d}\mu) \in \overline{\mathcal{I}}_p([a,b];\mathbb{B}), \tag{9.92}$$

acting between the two Banach spaces. We call $\mathcal{P}$ the *product integral operator*. Investigation of smoothness of this operator in the sense of differential calculus on normed spaces (see Chapter 5) is the subject of the present section. We will prove that $\mathcal{P}$ has a Taylor expansion around each $\mu \in \mathcal{AI}_p([a,b];\mathbb{B})$, and that the Taylor series has infinite radius of uniform convergence. We will also show that the $k$th term of the Taylor series is defined by the interval function $Q^k_\mu(\nu) = \{Q^k_\mu(\nu)(A)\colon\ A \in \mathfrak{I}[a,b]\}$, where

$$Q^k_\mu(\nu)(A) \tag{9.93}$$
$$= \begin{cases} ⨍_A \widehat{\mu}((s,b]\cap A)\,\nu(\mathrm{d}s)\,\widehat{\mu}([a,s)\cap A) & \text{if } k=1,\\ ⨍_A \widehat{\mu}((s_1,b]\cap A)\,\nu(\mathrm{d}s_1)\,⨍_{[a,s_1)\cap A} \widehat{\mu}((s_2,s_1))\,\nu(\mathrm{d}s_2)\cdots & \\ \qquad \cdots ⨍_{[a,s_{k-1})\cap A} \widehat{\mu}((s_k,s_{k-1}))\,\nu(\mathrm{d}s_k)\,\widehat{\mu}([a,s_k)\cap A) & \text{if } k \ge 2. \end{cases}$$

In the latter iterated integral, we have $s_j \in A$ for $j = 1,\dots,k$, and so each interval $(s_j, s_{j-1}) \subset A$ and $(s_j,s_{j-1}) \cap A = (s_j,s_{j-1})$ for $j = 2,\dots,k$. Here the iterated integral is first done with respect to $s_k$ for each fixed $s_{k-1}$, then with respect to $s_{k-1}$ for each $s_{k-2}$, and so on. The main result to be proved in this section is the following theorem:

**Theorem 9.51.** *Let $\mathbb{B}$ be a Banach algebra, $a < b$, and $1 \le p < 2$. The product integral operator* (9.92) *is a uniformly entire mapping from $\mathcal{AI}_p([a,b];\mathbb{B})$ into $\overline{\mathcal{I}}_p([a,b];\mathbb{B})$. More specifically, for $\mu \in \mathcal{AI}_p([a,b];\mathbb{B})$, the following hold:*

(*a*) *for each integer $k \ge 1$ and $\nu \in \mathcal{AI}_p([a,b];\mathbb{B})$, relation* (9.93) *defines an interval function $Q^k_\mu(\nu) = \{Q^k_\mu(\nu)(A)\colon\ A \in \mathfrak{I}[a,b]\}$ in $\overline{\mathcal{I}}_p([a,b];\mathbb{B})$;*

(*b*) *for each integer $k \ge 1$, the mapping*

$$\mathcal{AI}_p := \mathcal{AI}_p([a,b];\mathbb{B}) \ni \nu \mapsto Q^k_\mu(\nu) \in \overline{\mathcal{I}}_p([a,b];\mathbb{B}) =: \overline{\mathcal{I}}_p$$

*is a $k$-homogeneous polynomial;*

(*c*) *the power series $\sum_{k\ge1} Q^k_\mu(\nu-\mu)$ from $\mathcal{AI}_p$ to $\overline{\mathcal{I}}_p$ around $\mu$ has infinite radius of uniform convergence and its sum is equal to $\mathcal{P}(\nu) - \mathcal{P}(\mu)$.*

The following is a special case of the preceding theorem: a Taylor expansion of the product integral operator around 0.

**Corollary 9.52.** *Let $\nu \in \mathcal{AI}_p([a,b];\mathbb{B})$ with $1 \le p < 2$ and $a < b$. Then for any interval $A \subset [a,b]$,*

$$\prod_A(\mathbb{I} + d\nu) = \mathbb{I} + \nu(A) + \sum_{k\ge 2} \fint_A \nu(ds_1) \fint_{[a,s_1)\cap A} \nu(ds_2) \cdots \fint_{[a,s_{k-1})\cap A} \nu(ds_k). \tag{9.94}$$

Before proving Theorem 9.51 we establish several auxiliary results. Let $\mu, \nu \in \mathcal{AI}_p([a,b];\mathbb{B})$ for some $1 \le p < 2$. By the Duhamel formula (9.80), for $A \in \mathfrak{I}[a,b]$, we have

$$\widehat{\mu+\nu}(A) - \widehat{\mu}(A) = \fint_A \widehat{\mu}\big((s,b]\cap A\big)\, \nu(ds)\, \widehat{\mu+\nu}\big([a,s)\cap A\big). \tag{9.95}$$

The Love–Young inequality (9.79), if applicable to the integral on the right side, would imply the Lipschitzian property of the nonlinear product integral operator $\mathcal{P}: \ \mu \mapsto \widehat{\mu}$. To justify its applicability the next two lemmas give bounds for the $p$-variation of the two integrands in (9.95).

**Lemma 9.53.** *For $1 \le p < 2$ and $a < b$, let $\alpha \in \mathcal{I}_p^{\leftarrow}([a,b];\mathbb{B})$. For nonempty $A \in \mathfrak{I}[a,b]$ and $t \in A$, let $f_A(t) := \alpha((t,b]\cap A)$. Then for any nonempty $A \in \mathfrak{I}[a,b]$,*

$$\|f_A\|_{A,[p]} \le \|\alpha\|_{A,[p]}^{\leftarrow}. \tag{9.96}$$

*In particular if $\alpha = \widehat{\mu}$ for $\mu \in \mathcal{AI}_p([a,b];\mathbb{B})$, then*

$$\|f_A\|_{A,[p]} \le \|\widehat{\mu}\|_{A,[p]}^{\leftarrow} \le \|\widehat{\mu}\|_{A,\overline{[p]}} \le C_p(\widehat{\mu}) \tag{9.97}$$

*where the right side is defined by (9.85).*

*Proof.* Statement (9.96) follows by the definition (9.47) of the reversed form of the $p$-variation. Then (9.97) follows from Lemma 9.50. □

**Lemma 9.54.** *Let $a < b$, let $\eta \in \mathcal{I}_q([a,b];\mathbb{B})$ with $1 \le q < \infty$, let $A \in \mathfrak{I}[a,b]$ be nonempty, and let $g_A(t) := \eta([a,t)\cap A)$, $t \in A$. Then $g_A \in \mathcal{W}_q(A;\mathbb{B})$,*

$$\|g_A\|_{A,(q)} \le \|\eta\|_{A,(q)} \qquad \text{and} \qquad \|g_A\|_{A,\sup} \le \|\eta\|_{A,\sup}. \tag{9.98}$$

*Proof.* The second inequality in (9.98) holds because $[a,t)\cap A$ is a subinterval of $A$. To prove the first one we can assume that $A$ is nondegenerate. Let $\kappa = \{t_i\}_{i=0}^n$ be a point partition of $A$ and let $A_i := [t_{i-1},t_i)\cap A$, $i = 1,\dots,n$. Then $\{A_i\}_{i=1}^n$ is an interval partition of $[t_0,t_n)\cap A \subset A$. So we have

$$s_q(g_A;\kappa) = \sum_{i=1}^n \left\| \eta\left(\bigcup_{j=1}^i A_j\right) - \eta\left(\bigcup_{j=1}^{i-1} A_j\right)\right\|^q \le \|\eta\|_{A,(q)}^q,$$

where a union over the empty set of indices is defined as the empty set. Thus the first inequality in (9.98) holds, proving the lemma. □

Next is a technical device to be used for bounding the remainder in differentiating the product integral operator. Its proof will use additivity of the interval function $A \mapsto \oint_A f(s)\nu(\mathrm{d}s)g(s)$ for fixed $f, g \in \mathcal{W}_p([a,b];\mathbb{B})$, which holds by Theorem 9.42 and Proposition 9.47. But in (9.99), the integrands depend on $A$, and $\beta$ itself is in general not additive (nor multiplicative, since $\beta(\emptyset) = 0 \neq \mathbb{I}$).

**Proposition 9.55.** *For $1 \le p < 2$ and $a < b$, let $\mu, \nu \in \mathcal{AI}_p([a,b];\mathbb{B})$ and let $\alpha \in \mathcal{I}_p([a,b];\mathbb{B})$. Then the interval function $\beta = \{\beta(A)\colon A \in \mathfrak{I}[a,b]\}$ defined by*

$$\beta(A) := \oint_A \widehat{\mu}\big((s,b]\cap A\big)\,\nu(\mathrm{d}s)\,\alpha\big([a,s)\cap A\big), \qquad A \in \mathfrak{I}[a,b], \tag{9.99}$$

*exists and we have the bound*

$$\|\beta\|_{[p]} \le K_p \|\widehat{\mu}\|_{\overleftarrow{[p]}}\big[2 + V_p(\widehat{\mu})\big]\,\|\nu\|_{(p)}\|\alpha\|_{[p]}, \tag{9.100}$$

*where $V_p$ is defined by (9.83) and $K_p := K_{p,p} = 1 + \zeta(2/p)$.*

*Or if instead $\alpha \in \mathcal{I}_p^{\leftarrow}([a,b];\mathbb{B})$, and we define*

$$\widetilde{\beta}(A) := \oint_A \alpha\big((s,b]\cap A\big)\,\nu(\mathrm{d}s)\,\widehat{\mu}\big([a,s)\cap A\big), \qquad A \in \mathfrak{I}[a,b], \tag{9.101}$$

*then $\widetilde{\beta}$ exists and*

$$\|\widetilde{\beta}\|_{\overleftarrow{[p]}} \le K_p \|\widehat{\mu}\|_{\overline{[p]}}\big[2 + V_p(\widehat{\mu})\big]\,\|\nu\|_{(p)}\|\alpha\|_{\overleftarrow{[p]}}. \tag{9.102}$$

*Proof.* The strict product integral $\widehat{\mu}$ is defined by Theorem 9.36. By Lemmas 9.50, 9.53, and 9.54, the point functions $\widehat{\mu}((t,b]\cap A)$, $t \in A$, and $\alpha([a,t)\cap A)$, $t \in A$, have bounded $p$-variation for each nonempty $A \in \mathfrak{I}[a,b]$. Therefore by Proposition 9.47 with $q = p$, the integral (9.99) is defined, and for each nonempty $A \in \mathfrak{I}[a,b]$,

$$\begin{aligned} &\Big\|\oint_A \widehat{\mu}((s,b]\cap A)\,\nu(\mathrm{d}s)\,\alpha([a,s)\cap A)\Big\| \\ &\qquad \le K_p \|\widehat{\mu}((\cdot,b]\cap A)\|_{A,[p]}\|\alpha([a,\cdot)\cap A)\|_{A,[p]}\|\nu\|_{A,(p)} \\ &\qquad \le K_p \|\widehat{\mu}\|_{\overleftarrow{[p]}}\|\alpha\|_{[p]}\|\nu\|_{A,(p)}, \end{aligned} \tag{9.103}$$

where the second inequality holds by (9.97) and (9.98). To bound the $p$-variation of $\beta$, let $D \in \mathfrak{I}[a,b]$ be nonempty, let $\mathcal{A} = \{A_i\}_{i=1}^n$ be an interval partition of $D$, and for each $i = 1, \ldots, n$, let $B_i := \cup_{j\le i} A_j$. Since $\widehat{\mu}$ is multiplicative and $[a,s)\cap B_i = [a,s)\cap D$ for each $s \in B_i$, $i = 1, \ldots, n$, we have for $i = 2, \ldots, n$,

$$\begin{aligned}\beta(B_i)-\beta(B_{i-1}) &= \fint_{A_i} \widehat{\mu}((s,b]\cap B_i)\,\nu(\mathrm{d}s)\,\alpha([a,s)\cap D)\\ &\quad+\fint_{B_{i-1}} \widehat{\mu}((s,b]\cap A_i)\widehat{\mu}((s,b]\cap B_{i-1})\,\nu(\mathrm{d}s)\,\alpha([a,s)\cap D)\\ &\quad-\fint_{B_{i-1}} \widehat{\mu}((s,b]\cap B_{i-1})\,\nu(\mathrm{d}s)\,\alpha([a,s)\cap D)\\ &= \fint_{A_i} \widehat{\mu}((s,b]\cap A_i)\,\nu(\mathrm{d}s)\,\alpha([a,s)\cap D)\\ &\quad+\Big[\widehat{\mu}(A_i)-\mathbb{I}\Big]\fint_{B_{i-1}} \widehat{\mu}((s,b]\cap B_{i-1})\,\nu(\mathrm{d}s)\,\alpha([a,s)\cap D).\end{aligned}$$

Since $\beta(\emptyset)=0$, by the Minkowski inequality (1.5), we have

$$\begin{aligned}&s_p(\beta;\mathcal{A})^{1/p}\\ &\quad=\Big(\|\beta(A_1)\|^p+\sum_{i=2}^{n}\|\beta(B_i)-\beta(B_{i-1})\|^p\Big)^{1/p}\\ &\quad\le\Big(\sum_{i=1}^{n}\Big\|\fint_{A_i}\widehat{\mu}((s,b]\cap A_i)\,\nu(\mathrm{d}s)\,\alpha([a,s)\cap D)\Big\|^p\Big)^{1/p}\\ &\qquad+\Big(\sum_{i=2}^{n}\|\widehat{\mu}(A_i)-\mathbb{I}\|^p\Big\|\fint_{B_{i-1}}\widehat{\mu}((s,b]\cap B_{i-1})\,\nu(\mathrm{d}s)\,\alpha([a,s)\cap D)\Big\|^p\Big)^{1/p}.\end{aligned}\tag{9.104}$$

As in the bound (9.103) with the point function $\alpha([a,t)\cap A)$, $t\in A$, replaced by the function $\alpha([a,t)\cap D)$, $t\in A\in\mathfrak{I}(D)$, we have

$$\Big\|\fint_{A}\widehat{\mu}((s,b]\cap A)\,\nu(\mathrm{d}s)\alpha([a,s)\cap D)\Big\|\le K_p\|\widehat{\mu}\|_{\overleftarrow{[p]}}\|\alpha\|_{[p]}\|\nu\|_{A,(p)}$$

for each $A\in\mathfrak{I}(D)$. Applying this bound to (9.104) yields

$$\begin{aligned}s_p(\beta;\mathcal{A})^{1/p}&\le K_p\|\widehat{\mu}\|_{\overleftarrow{[p]}}\|\alpha\|_{[p]}\Big(\sum_{i=1}^{n}v_p(\nu;A_i)\Big)^{1/p}\\ &\quad+K_p\|\widehat{\mu}\|_{\overleftarrow{[p]}}\|\alpha\|_{[p]}\|\nu\|_{(p)}\Big(\sum_{i=1}^{n}\|\widehat{\mu}(A_i)-\mathbb{I}\|^p\Big)^{1/p}\\ &\le K_p\|\widehat{\mu}\|_{\overleftarrow{[p]}}\|\alpha\|_{[p]}\big[1+V_p(\widehat{\mu})\big]\|\nu\|_{(p)}\quad\text{by (3.69) and (9.83).}\end{aligned}$$

Since $\mathcal{A}$ is an arbitrary interval partition of $D$, we get the bound

$$v_p(\beta;D)^{1/p}\le K_p\|\widehat{\mu}\|_{\overleftarrow{[p]}}\|\alpha\|_{[p]}\big[1+V_p(\widehat{\mu})\big]\|\nu\|_{(p)}$$

for each nonempty $D\in\mathfrak{I}[a,b]$. This and (9.103) yield (9.100).

A proof for (9.101) is symmetric, proving the proposition. □

We get as a byproduct the following general fact, in which the product integral does not appear.

**Theorem 9.56.** *For $1 \le p < 2$ and $a < b$, let $\nu \in \mathcal{AI}_p([a,b];\mathbb{B})$ and let $\alpha \in \mathcal{I}_p([a,b];\mathbb{B})$. Then the interval function $\beta = \{\beta(A)\colon A \in \mathfrak{I}[a,b]\}$ defined by*

$$\beta(A) := \fint_A \nu(\mathrm{d}s)\alpha\big(A \cap [a,s)\big) \tag{9.105}$$

*exists and satisfies*

$$\|\beta\|_{[p]} \le 2K_p \|\nu\|_{(p)} \|\alpha\|_{[p]}, \tag{9.106}$$

*where $K_p := K_{p,p} = 1 + \zeta(2/p)$. Thus $(\nu,\alpha) \mapsto \beta$ is a bounded bilinear operator from $\mathcal{AI}_p([a,b];\mathbb{B}) \times \mathcal{I}_p([a,b];\mathbb{B})$ into $\mathcal{I}_p([a,b];\mathbb{B})$.*

*Or if $\alpha \in \mathcal{I}_p^{\leftarrow}([a,b];\mathbb{B})$, then the interval function $\widetilde{\beta} = \{\widetilde{\beta}(A)\colon A \in \mathfrak{I}[a,b]\}$ defined by*

$$\widetilde{\beta}(A) := \fint_A \alpha((s,b] \cap A)\nu(\mathrm{d}s), \qquad A \in \mathfrak{I}[a,b], \tag{9.107}$$

*exists and satisfies*

$$\|\widetilde{\beta}\|_{[p]}^{\leftarrow} \le 2K_p \|\nu\|_{(p)} \|\alpha\|_{[p]}^{\leftarrow}. \tag{9.108}$$

*Thus $(\nu,\alpha) \mapsto \widetilde{\beta}$ takes $\mathcal{AI}_p([a,b];\mathbb{B}) \times \mathcal{I}_p^{\leftarrow}([a,b];\mathbb{B})$ into $\mathcal{I}_p^{\leftarrow}([a,b];\mathbb{B})$ and is a bounded bilinear operator.*

*Proof.* In either case we apply Proposition 9.55 to $\mu \equiv 0$, so that $\widehat{\mu} \equiv \mathbb{1}$. For (9.106) we apply (9.100) and for (9.108) we use (9.101). The conclusions follow. □

The following gives a bound for $V_p(\beta)$ when $\beta$ is an interval function defined by a suitable integral.

**Proposition 9.57.** *Let $a < b$, $1 \le p < 2$, $\alpha \in \mathcal{I}_p^{\leftarrow}([a,b];\mathbb{B})$, $\nu \in \mathcal{AI}_p([a,b];\mathbb{B})$ and $\eta \in \mathcal{I}_p([a,b];\mathbb{B})$. Then $\beta$ defined by*

$$\beta(A) := \fint_A \alpha((s,b] \cap A)\nu(\mathrm{d}s)\eta([a,s) \cap A)$$

*is an interval function with $\beta(\emptyset) = 0$ and*

$$\|\beta\|_{\sup} \le V_p(\beta) \le K_p \|\alpha\|_{[p]}^{\leftarrow} \|\eta\|_{[p]} \|\nu\|_{(p)},$$

*where $K_p := K_{p,p} := 1 + \zeta(2/p)$.*

*Proof.* From the definition it is clear that $\beta(\emptyset) = 0$. For any such interval function $\beta$, clearly $\|\beta\|_{\sup} \le V_p(\beta)$. We apply the Love–Young inequality (9.78) with $q = p$ for $f_A := \alpha((\cdot,b] \cap A)$, $g_A := \eta([a,\cdot) \cap A)$, and $\nu$ in place of $\mu$ there, and then apply Lemmas 9.53 and 9.54 for $f_A$ and $g_A$, respectively, giving

$$\|\beta(A)\| \le K_p \|f_A\|_{A,[p]} \|g_A\|_{A,[p]} \|\nu\|_{A,(p)} \le K_p \|\alpha\|_{\overleftarrow{[p]}} \|\eta\|_{[p]} \|\nu\|_{A,(p)}.$$

For any interval partition $\{A_i\}_{i=1}^n$ of $[a,b]$, using (3.69), we have

$$\sum_{i=1}^n \|\beta(A_i)\|^p \le K_p^p (\|\alpha\|_{\overleftarrow{[p]}})^p \|\eta\|_{[p]}^p \|\nu\|_{(p)}^p,$$

which also gives the same bound for $V_p(\beta)^p$, completing the proof. □

Let $1 \le p < 2$ and $\mu \in \mathcal{AI}_p([a,b];\mathbb{B})$. Then the strict product integral $\widehat{\mu}$ of $\mu$ exists and is in $\overline{\mathcal{I}}_p([a,b];\mathbb{B}) \subset \mathcal{I}_p([a,b];\mathbb{B})$ by Theorem 9.36. Let $L_\mu^0 := \widehat{\mu}$. For $k \ge 1$, $\nu_1, \dots, \nu_k \in \mathcal{AI}_p([a,b];\mathbb{B})$ and $A \in \mathfrak{I}[a,b]$, define recursively

$$\begin{aligned} &L_\mu^k(\nu_1,\dots,\nu_k)(A) \\ &\quad := \oint_A \widehat{\mu}((s,b]\cap A)\,\nu_k(ds)\,L_\mu^{k-1}(\nu_1,\dots,\nu_{k-1})([a,s)\cap A) \\ &\quad = \oint_A \widehat{\mu}((s_1,b]\cap A)\,\nu_k(ds_1) \oint_{[a,s_1)\cap A} \widehat{\mu}((s_2,s_1))\,\nu_{k-1}(ds_2)\cdots \\ &\qquad \cdots \oint_{[a,s_{k-1})\cap A} \widehat{\mu}((s_k,s_{k-1}))\,\nu_1(ds_k)\,\widehat{\mu}([a,s_k)\cap A) \end{aligned} \tag{9.109}$$

provided the Kolmogorov integrals exist. In the iterated integral we integrate first with respect to $s_k$, and then $s_{k-1},\dots,s_1$. If $\nu_1,\dots,\nu_k = \nu$ then $L_\mu^k(\nu,\dots,\nu) = Q_\mu^k(\nu)$ defined by (9.93). As in (9.93), for all $j = 1,\dots,k$ we have $s_j \in A$, thus for $j = 2,\dots,k$, $(s_j,s_{j-1}) \subset A$ and $(s_j,s_{j-1}) \cap A = (s_j,s_{j-1})$.

**Lemma 9.58.** *Let $\mu \in \mathcal{AI}_p([a,b];\mathbb{B})$, $1 \le p < 2$. Then for each $k \ge 1$ and any $\nu_1,\dots,\nu_k \in \mathcal{AI}_p([a,b];\mathbb{B})$, the interval function defined by $L_k := L_\mu^k(\nu_1,\dots,\nu_k) = \{L_\mu^k(\nu_1,\dots,\nu_k)(A)\colon A \in \mathfrak{I}[a,b]\}$ exists,*

$$\|L_k\|_{[p]} \le \Big\{K_p \|\widehat{\mu}\|_{\overleftarrow{[p]}} (2 + V_p(\widehat{\mu}))\Big\}^k \|\widehat{\mu}\|_{[p]} \|\nu_1\|_{(p)} \cdots \|\nu_k\|_{(p)}, \tag{9.110}$$

*and*

$$V_p(L_k) \le \Big\{K_p \|\widehat{\mu}\|_{\overleftarrow{[p]}}\Big\}^k (2 + V_p(\widehat{\mu}))^{k-1} \|\widehat{\mu}\|_{[p]} \|\nu_1\|_{(p)} \cdots \|\nu_k\|_{(p)}. \tag{9.111}$$

*Proof.* By the first part of Proposition 9.55 with $\alpha = L_\mu^0 = \widehat{\mu}$, the interval function $L_\mu^1(\nu_1) = \{L_\mu^1(\nu_1)(A)\colon A \in \mathfrak{I}[a,b]\}$ is defined and (9.110) holds with $k=1$. Assuming it holds for some $k \ge 1$ and applying (9.100) with $\alpha = L_k = L_\mu^k(\nu_1,\dots,\nu_k)$ it follows that (9.110) holds with $k+1$ in place of $k$, and so it holds for each $k \ge 1$ by induction. Also, for each $k \ge 1$ the interval function $L_\mu^k(\nu_1,\dots,\nu_k)$ is defined and is in $\mathcal{I}_p([a,b];\mathbb{B})$. For (9.111), by Proposition 9.57 and (9.110), for each $k \ge 1$ we have

$$V_p(L_k) \le K_p \|\widehat{\mu}\|_{\overleftarrow{[p]}} \|L_{k-1}\|_{[p]} \|\nu_k\|_{(p)} \tag{9.112}$$
$$\le \Big\{K_p \|\widehat{\mu}\|_{\overleftarrow{[p]}}\Big\}^k (2+V_p(\widehat{\mu}))^{k-1} \|\widehat{\mu}\|_{[p]} \|\nu_1\|_{(p)} \cdots \|\nu_k\|_{(p)},$$

proving the lemma. □

For $\mu$ as before in $\mathcal{AI}_p([a,b];\mathbb{B})$, let $Q^0_\mu(\nu) := \widehat{\mu}$ for any $\nu \in \mathcal{AI}_p$, and let $Q^k_\mu(\nu)$ be defined by (9.93) for each $k \ge 1$. For $\nu$ also in $\mathcal{AI}_p([a,b];\mathbb{B})$ let $\gamma_0(\mu,\nu) := \widehat{\mu+\nu} - \widehat{\mu}$. For each integer $n \ge 1$, let

$$\gamma_n(\mu,\nu) := \widehat{\mu+\nu} - \sum_{k=0}^{n} Q^k_\mu(\nu) = \gamma_0(\mu,\nu) - \sum_{k=1}^{n} Q^k_\mu(\nu). \tag{9.113}$$

By (9.95), we have an integral representation of $\gamma_0(\mu,\nu) = \{\gamma_0(\mu,\nu)(A)\colon\ A \in \mathfrak{I}[a,b]\}$. The following extends this fact to $\gamma_n(\mu,\nu)$.

**Lemma 9.59.** *Let $\mu,\nu \in \mathcal{AI}_p([a,b];\mathbb{B})$. For each $n \ge 1$ and any $A \in \mathfrak{I}[a,b]$,*

$$\gamma_n(A) = \fint_A \widehat{\mu}\big((s,b]\cap A\big)\,\nu(\mathrm{d}s)\,\gamma_{n-1}\big([a,s)\cap A\big). \tag{9.114}$$

*Proof.* Recall that $Q^n_\mu(\nu) = L^n_\mu(\nu,\dots,\nu)$ for each $n \ge 1$. Let $\gamma_n := \gamma_n(\mu,\nu)$ for each $n \ge 0$, and let $A \in \mathfrak{I}[a,b]$. By Theorem 9.43(*c*), (9.95), and (9.109) with $k=1$, we have

$$\gamma_1(A) = \gamma_0(A) - L^1_\mu(\nu)(A) = \fint_A \widehat{\mu}\big((s,b]\cap A\big)\,\nu(\mathrm{d}s)\,\gamma_0\big([a,s)\cap A\big).$$

Using Theorem 9.43(*c*) again, $\gamma_n = \gamma_{n-1} - L^n_\mu(\nu,\dots,\nu)$ for $n \ge 1$, and (9.109) for the induction step, it follows that (9.114) holds for each $n \ge 1$ by induction. □

The next lemma gives a technical device to bound $\|\rho\|_{\overline{[p]}}$ in terms of $V_p(\rho)$ for suitable non-additive interval functions $\rho$.

**Lemma 9.60.** *Let $\mu \in \mathcal{AI}_p([a,b];\mathbb{B})$, $B_1, B_2 \in \mathfrak{I}[a,b]$, $B_1 \cup B_2 \in \mathfrak{I}[a,b]$, and $B_1 \prec B_2$. For for each $n \ge 0$ and for any $\nu_1,\dots,\nu_n \in \mathcal{AI}_p([a,b];\mathbb{B})$,*

$$L^n_\mu(\nu_1,\dots,\nu_n)(B_1\cup B_2) = \sum_{k=0}^{n} L^k_\mu(\nu_{n-k+1},\dots,\nu_n)(B_2) L^{n-k}_\mu(\nu_1,\dots,\nu_{n-k})(B_1). \tag{9.115}$$

*Also, for each $n \ge 0$ and any $\nu \in \mathcal{AI}_p$, writing $\gamma_n := \gamma_n(\mu,\nu)$, we have*

$$\gamma_n(B_1\cup B_2) = \gamma_n(B_2)\widehat{\mu+\nu}(B_1) + \sum_{k=0}^{n} Q^k_\mu(\nu)(B_2)\gamma_{n-k}(B_1). \tag{9.116}$$

*Proof.* The proof is by induction. To prove (9.115) note that it holds when $n = 0$ by multiplicativity of $\widehat{\mu}$. Assume that (9.115) holds for some $n \geq 0$. Let $L_0 := L_\mu^0$ and $L_n := L_\mu^n(\nu_1, \ldots, \nu_n)$ for each $n \geq 1$. By additivity and trilinearity of $\displaystyle\ne\!\!\!\!\int$ (Theorems 9.42 and 9.43(*c*)), we then have $L_{n+1}(B_1 \cup B_2) = I_1 + I_2$ where by (9.109),

$$I_1 := \widehat{\mu}(B_2) \underset{B_1}{\ne\!\!\!\!\int} \widehat{\mu}((s,b] \cap B_1)\, \nu_{n+1}(ds)\, L_n([a,s) \cap B_1) = \widehat{\mu}(B_2) L_{n+1}(B_1)$$

and

$$I_2 := \underset{B_2}{\ne\!\!\!\!\int} \widehat{\mu}((s,b] \cap B_2)\, \nu_{n+1}(ds)\, L_n(B_1 \cup ([a,s) \cap B_2)).$$

The latter integrand in $I_2$, by induction hypothesis, equals

$$\sum_{k=0}^{n} L_k(\nu_{n-k+1}, \ldots, \nu_n)([a,s) \cap B_2) L_{n-k}(\nu_1, \ldots, \nu_{n-k})(B_1)$$

for $s \in B_2$, and since the $L_{n-k}$ factor does not depend on $s$ we get by (9.109),

$$I_2 = \sum_{k=0}^{n} L_{k+1}(\nu_{n-k+1}, \ldots, \nu_{n+1})(B_2) L_{n-k}(\nu_1, \ldots, \nu_{n-k})(B_1).$$

Thus (9.115) holds for $n+1$ in place of $n$ and for each $n \geq 0$ by induction.

To prove (9.116) note that it holds when $n = 0$ by multiplicativity of $\widehat{\mu}$ and $\widehat{\mu+\nu}$. Assume that (9.116) holds for some $n \geq 0$. Then using (9.115) with $n+1$ in place of $n$ and with $\nu_1 = \cdots = \nu_{n+1} = \nu$, we have

$$\begin{aligned}
&\gamma_{n+1}(B_1 \cup B_2) \\
&\quad = \gamma_n(B_1 \cup B_2) - Q_{n+1}(B_1 \cup B_2) \\
&\quad = \big[\gamma_n(B_2) - Q_{n+1}(B_2)\big]\widehat{\mu+\nu}(B_1) + Q_{n+1}(B_2)\big[\widehat{\mu+\nu}(B_1) - \widehat{\mu}(B_1)\big] \\
&\qquad + \sum_{k=0}^{n} Q_k(B_2)\big[\gamma_{n-k}(B_1) - Q_{n+1-k}(B_1)\big] \\
&\quad = \gamma_{n+1}(B_2)\widehat{\mu+\nu}(B_1) + \sum_{k=0}^{n+1} Q_k(B_2)\gamma_{n+1-k}(B_1),
\end{aligned}$$

and so (9.116) holds for each $n \geq 0$, proving the lemma. □

The following shows that the $n$-linear mapping $L_\mu^n$ defined by (9.109) acts from the $n$-fold product $(\mathcal{AI}_p)^n$ into $\overline{\mathcal{I}}_p$ and gives bounds for norms.

**Lemma 9.61.** *Let $\mu \in \mathcal{AI}_p([a,b];\mathbb{B})$, $1 \leq p < 2$. For each $k \geq 1$ and $\nu_1, \ldots, \nu_k \in \mathcal{AI}_p([a,b];\mathbb{B})$, let $L_k(\nu_1, \ldots, \nu_k) := L_\mu^k(\nu_1, \ldots, \nu_k)$ be the interval function defined recursively by* (9.109). *Then for each integer $n \geq 1$ and any $\nu_1, \ldots, \nu_n \in \mathcal{AI}_p$,*

$$
\begin{aligned}
&\|L_n(\nu_1,\ldots,\nu_n)\|_{\overleftarrow{[p]}} \\
&\quad\le D(\widehat{\mu})V_p(L_n(\nu_1,\ldots,\nu_n)) \qquad (9.117)\\
&\qquad+\sum_{k=1}^{n-1} V_p(L_k(\nu_{n-k+1},\ldots,\nu_n))V_p(L_{n-k}(\nu_1,\ldots,\nu_{n-k}))\\
&\quad\le \Big(K_p\|\widehat{\mu}\|_{\overleftarrow{[p]}}\Big)^n D(\widehat{\mu})^{n-2}\|\widehat{\mu}\|_{[p]}\Big(D(\widehat{\mu})^2+(n-1)\|\widehat{\mu}\|_{[p]}\Big)\prod_{i=1}^n\|\nu_i\|_{(p)} \qquad (9.118)
\end{aligned}
$$

*where* $K_p = 1+\zeta(2/p)$ *and* $D(\widehat{\mu}) := 1+V_p(\widehat{\mu})+\|\widehat{\mu}\|_{\sup}$.

*Proof.* Let $L_0 := L^0_\mu := \widehat{\mu}$ and $L_k := L_k(\nu_1,\ldots,\nu_k)$ for each $k\ge 1$. Let $n\ge 1$, let $D\in\mathfrak{I}[a,b]$ be nonempty, and let $\mathcal{A}=\{A_i\}_{i=1}^m\in \mathrm{IP}\,(D)$. To bound the $p$-variation sum $s_p(L_n;\mathcal{A})$, by (9.115), we have

$$
\begin{aligned}
L_n(B_1\cup B_2)-L_n(B_1) &= \big[\widehat{\mu}(B_2)-\mathbb{I}\big]L_n(B_1)\\
&\quad+\sum_{k=1}^{n} L_k(\nu_{n-k+1},\ldots,\nu_n)(B_2)L_{n-k}(\nu_1,\ldots,\nu_{n-k})(B_1)
\end{aligned}
$$

for $B_1 := \bigcup_{j=1}^{i-1}A_j$ and $B_2 := A_i$ for $i=1,\ldots,m$. As usual, the union over the empty set of indices is defined as empty. Using the Minkowski inequality (1.5), and the definition of $V_p(\cdot)$ for $\widehat{\mu}$ with $\widehat{\mu}(\emptyset)=\mathbb{I}$ and for $L_k$ with $L_k(\emptyset)=0$, it then follows that

$$
\begin{aligned}
s_p(L_n;\mathcal{A})^{1/p} &\le V_p(\widehat{\mu})\|L_n\|_{\sup}\\
&\quad+\sum_{k=1}^{n} V_p(L_k(\nu_{n-k+1},\ldots,\nu_n))\|L_{n-k}(\nu_1,\ldots,\nu_{n-k})\|_{\sup},
\end{aligned}
$$

and so

$$
\begin{aligned}
\|L_n\|_{[p]} &\le D(\widehat{\mu})V_p(L_n)\\
&\quad+\sum_{k=1}^{n-1} V_p(L_k(\nu_{n-k+1},\ldots,\nu_n))\|L_{n-k}(\nu_1,\ldots,\nu_{n-k})\|_{\sup}.
\end{aligned} \qquad (9.119)
$$

To bound the reversed $p$-variation we use the equality

$$
\begin{aligned}
L_n(B_1\cup B_2)-L_n(B_2) &= L_n(B_2)\big[\widehat{\mu}(B_1)-\mathbb{I}\big]\\
&\quad+\sum_{k=0}^{n-1} L_k(\nu_{n-k+1},\ldots,\nu_n)(B_2)L_{n-k}(\nu_1,\ldots,\nu_{n-k})(B_1)
\end{aligned}
$$

with $B_1 := A_i$ and $B_2 := \bigcup_{j>i}A_j$ for $i - 1,\ldots,m$. Similarly using the Minkowski inequality, it follows that

$$
\begin{aligned}
\|L_n\|_{\overleftarrow{[p]}} &\le D(\widehat{\mu})V_p(L_n)\\
&\quad+\sum_{k=1}^{n-1} \|L_k(\nu_{n-k+1},\ldots,\nu_n)\|_{\sup}V_p(L_{n-k}(\nu_1,\ldots,\nu_{n-k})).
\end{aligned}
$$

This together with (9.119) yields (9.117). Inserting (9.110) and (9.111) into the right side of (9.117) and using $2 + V_p(\widehat{\mu}) \le D(\widehat{\mu})$ gives (9.118), proving the lemma. □

Next we give bounds for the remainder $\gamma_n(\mu,\nu)$ defined by (9.113). In its proof we will use the fact that $\|\alpha\|_{\sup} \le \|\alpha(\emptyset)\| + V_p(\alpha)$ for every interval function $\alpha$.

**Lemma 9.62.** *Let $\mu,\nu \in \mathcal{AI}_p([a,b];\mathbb{B})$, $1 \le p < 2$. Then*

$$\|\gamma_0(\mu,\nu)\|_{\overline{[p]}} \le C(\mu,\nu)K_p\|\widehat{\mu}\|_{\overleftarrow{[p]}}\|\widehat{\mu+\nu}\|_{[p]}\|\nu\|_{(p)} \tag{9.120}$$

*and for each $n \ge 1$,*

$$\|\gamma_n(\mu,\nu)\|_{\overline{[p]}} \le C(\mu,\nu)V_p(\gamma_n) + \sum_{k=1}^{n} V_p(Q_\mu^k(\nu))V_p(\gamma_{n-k}) \tag{9.121}$$

$$\le \|\widehat{\mu+\nu}\|_{[p]}\Big(C(\mu,\nu)^2 + \|\widehat{\mu}\|_{[p]} + (n-1)C(\mu,\nu)\|\widehat{\mu}\|_{[p]}\Big) \times \times D(\widehat{\mu})^{n-1}\Big\{K_p\|\widehat{\mu}\|_{\overleftarrow{[p]}}\|\nu\|_{(p)}\Big\}^{n+1} \tag{9.122}$$

*where $K_p = 1 + \zeta(2/p)$, $C(\mu,\nu) = 2 + V_p(\widehat{\mu}) + V_p(\widehat{\mu+\nu})$, and $D(\widehat{\mu}) := 1 + V_p(\widehat{\mu}) + \|\widehat{\mu}\|_{\sup}$.*

*Proof.* For $k \ge 0$, let $Q_k := Q_\mu^k(\nu)$ and $\gamma_k := \gamma_k(\mu,\nu)$. Let $n \ge 0$, let $D \in \mathfrak{I}[a,b]$ be nonempty, and let $\mathcal{A} = \{A_i\}_{i=1}^m \in \mathrm{IP}\,(D)$. To bound the $p$-variation sum $s_p(\gamma_n;\mathcal{A})$ we use

$$\begin{aligned}\gamma_n(B_1 \cup B_2) &- \gamma_n(B_1)\\ &= \big[\widehat{\mu}(B_2) - \mathbb{I}\big]\gamma_n(B_1) + \gamma_n(B_2)\widehat{\mu+\nu}(B_1) + \sum_{k=1}^{n} Q_k(B_2)\gamma_{n-k}(B_1)\end{aligned}$$

with $B_1 := \bigcup_{j=1}^{i-1} A_j$ and $B_2 := A_i$ for $i = 1,\dots,m$, which holds by (9.116). The sum on the right side is zero when $n = 0$. As always, the union over the empty set of indices is defined as empty. Using the Minkowski inequality (1.5), and the definition of $V_p(\cdot)$ for $\widehat{\mu}$ with $\widehat{\mu}(\emptyset) = \mathbb{I}$ and for $\gamma_n$ with $\gamma_n(\emptyset) = 0$, it then follows that

$$s_p(\gamma_n;\mathcal{A})^{1/p} \le V_p(\widehat{\mu})\|\gamma_n\|_{\sup} + V_p(\gamma_n)\|\widehat{\mu+\nu}\|_{\sup} + \sum_{k=1}^{n} V_p(Q_k)\|\gamma_{n-k}\|_{\sup}.$$

This gives the bound

$$\|\gamma_n\|_{[p]} \le C(\mu,\nu)V_p(\gamma_n) + \sum_{k=1}^{n} V_p(Q_k)\|\gamma_{n-k}\|_{\sup}.$$

Next using (9.116) with $B_1 := A_i$ and $B_2 := \bigcup_{j>i} A_j$ for $i = 1, \ldots, m$, it follows that

$$\|\gamma_n\|_{\overleftarrow{[p]}} \le C(\mu,\nu)V_p(\gamma_n) + \sum_{k=1}^{n} \|Q_k\|_{\sup} V_p(\gamma_{n-k}),$$

and so if $n \ge 1$, (9.121) follows. Also if $n = 0$ we have from the last two displays

$$\|\gamma_0\|_{\overleftarrow{[p]}} \le C(\mu,\nu)V_p(\gamma_0). \tag{9.123}$$

To prove (9.120) we use the integral representation (9.95) for $\gamma_0$. By Proposition 9.57 with $\alpha = \widehat{\mu}$ and $\eta = \widehat{\mu+\nu}$, it follows that

$$V_p(\gamma_0) \le K_p\|\widehat{\mu}\|_{\overleftarrow{[p]}}\|\widehat{\mu+\nu}\|_{[p]}\|\nu\|_{(p)}. \tag{9.124}$$

Inserting this bound into the right side of (9.123) gives (9.120).

To prove (9.122) we use the integral representation (9.114) of $\gamma_k$. Also using (9.100) recursively and (9.120), it follows that for each $k \ge 1$,

$$\begin{aligned}
\|\gamma_k\|_{[p]} &\le K_p\|\widehat{\mu}\|_{\overleftarrow{[p]}}(2 + V_p(\widehat{\mu}))\|\nu\|_{(p)}\|\gamma_{k-1}\|_{[p]} \\
&\le \cdots \le \Big\{K_p\|\widehat{\mu}\|_{\overleftarrow{[p]}}(2 + V_p(\widehat{\mu}))\|\nu\|_{(p)}\Big\}^k \|\gamma_0\|_{[p]} \\
&\le C(\mu,\nu)\|\widehat{\mu+\nu}\|_{[p]}(2 + V_p(\widehat{\mu}))^k\Big\{K_p\|\widehat{\mu}\|_{\overleftarrow{[p]}}\|\nu\|_{(p)}\Big\}^{k+1}.
\end{aligned}$$

Again using (9.114) and Proposition 9.57 with $\alpha = \widehat{\mu}$ and $\eta = \gamma_{k-1}$, together with the preceding bound, it follows that the bound

$$\begin{aligned}
V_p(\gamma_k) &\le K_p\|\widehat{\mu}\|_{\overleftarrow{[p]}}\|\gamma_{k-1}\|_{[p]}\|\nu\|_{(p)} \\
&\le C(\mu,\nu)\|\widehat{\mu+\nu}\|_{[p]}(2 + V_p(\widehat{\mu}))^{k-1}\Big\{K_p\|\widehat{\mu}\|_{\overleftarrow{[p]}}\|\nu\|_{(p)}\Big\}^{k+1}
\end{aligned} \tag{9.125}$$

holds for each $k \ge 1$. Finally, by (9.111), for each $k \ge 1$ we have

$$V_p(Q_k) \le \Big\{K_p\|\widehat{\mu}\|_{\overleftarrow{[p]}}\|\nu\|_{(p)}\Big\}^k (2 + V_p(\widehat{\mu}))^{k-1}\|\widehat{\mu}\|_{[p]}. \tag{9.126}$$

Now we will prove (9.122) as follows. First we use (9.125) and (9.126) with $k = n$ together with (9.124) to get that

$$\begin{aligned}
&C(\mu,\nu)V_p(\gamma_n) + V_p(Q_n)V_p(\gamma_0) \\
&\quad\le \|\widehat{\mu+\nu}\|_{[p]}\Big(C(\mu,\nu)^2 + \|\widehat{\mu}\|_{[p]}\Big)D(\widehat{\mu})^{n-1}\Big\{K_p\|\widehat{\mu}\|_{\overleftarrow{[p]}}\|\nu\|_{(p)}\Big\}^{n+1},
\end{aligned}$$

since $2 + V_p(\widehat{\mu}) \le D(\widehat{\mu})$. Second if $n > 1$ we use (9.125) and (9.126) with $k \in \{1, \ldots, n-1\}$ to get the bound

$$\sum_{k=1}^{n-1} V_p(Q_\mu^k(\nu))V_p(\gamma_{n-k})$$
$$\leq (n-1)C(\mu,\nu)\|\widehat{\mu+\nu}\|_{[p]}\|\widehat{\mu}\|_{[p]}D(\widehat{\mu})^{n-2}\Big\{K_p\|\widehat{\mu}\|_{\overleftarrow{[p]}}\|\nu\|_{(p)}\Big\}^{n+1}.$$

Adding the preceding two bounds gives (9.122), completing the proof of the lemma. □

Next we show that $L_\mu^1$ defined by (9.109) with $k=1$ is the derivative of the product integral operator.

**Proposition 9.63.** *Under the conditions of Theorem* 9.51, *the product integral operator $\mathcal{P}$ is everywhere differentiable from $\mathcal{AI}_p$ into $\overline{\mathcal{I}}_p$, its derivative at $\mu \in \mathcal{AI}_p$ is given by $L_\mu^1(\nu) = Q_\mu^1(\nu)$, $\nu \in \mathcal{AI}_p$, and the remainder $\gamma_1 := \gamma_1(\mu,\nu)$ defined by* (9.113) *has the bound*

$$\begin{aligned}\|\gamma_1\|_{\overleftarrow{[p]}} &\leq C(\mu,\nu)V_p(\gamma_1)+V_p(L_\mu^1(\nu))V_p(\gamma_0)\\ &\leq \Big\{K_p\|\widehat{\mu}\|_{\overleftarrow{[p]}}\Big\}^2\|\widehat{\mu+\nu}\|_{[p]}\Big(C(\mu,\nu)^2+\|\widehat{\mu}\|_{[p]}\Big)\|\nu\|_{(p)}^2,\end{aligned} \tag{9.127}$$

*where $K_p = 1+\zeta(2/p)$ and $C(\mu,\nu) := 2+V_p(\widehat{\mu})+V_p(\widehat{\mu+\nu})$.*

*Proof.* Let $\mu \in \mathcal{AI}_p([a,b];\mathbb{B})$. By Lemma 9.58 with $k=1$ the interval function $L_\mu^1(\nu) = \{L_\mu^1(\nu)(A)\colon A \in \mathfrak{I}[a,b]\}$ is defined for each $\nu \in \mathcal{AI}_p([a,b];\mathbb{B})$. By Lemma 9.61 with $n=1$, $L_\mu^1(\cdot)$ maps $\mathcal{AI}_p$ into $\overline{\mathcal{I}}_p$ and is linear by Theorem 9.43($b$), which also holds for $[a,b]$ replaced by any interval $A$. By Lemma 9.62 with $n=1$, (9.127) holds. By (9.49) and (9.84), respectively,

$$\|\widehat{\mu+\nu}\|_{\overleftarrow{[p]}} \leq K+C_pK^2\|\mu+\nu\|_{(p)} \quad\text{and}\quad V_p(\widehat{\mu+\nu}) \leq C_pK\|\mu+\nu\|_{(p)} \tag{9.128}$$

where $K = K(\mu+\nu,p)$ and $C_p$ are defined by (9.29) and just after (9.28), respectively. By (9.29), $K(\mu+\nu,p)$ is bounded uniformly for $\nu$ in any bounded set in $\mathcal{AI}_p([a,b];\mathbb{B})$ and $\mu$ fixed. Thus so are $\|\widehat{\mu+\nu}\|_{\overleftarrow{[p]}}$ and $V_p(\widehat{\mu+\nu})$. It follows that as $\|\nu\|_{(p)} \to 0$, $\|\gamma_1(\mu,\nu)\|_{\overleftarrow{[p]}} = O(\|\nu\|_{(p)}^2) = o(\|\nu\|_{(p)})$. Thus $L_\mu^1$ is the derivative of the operator $\nu \mapsto \widehat{\mu+\nu} = \mathcal{P}(\mu+\nu)$ at $\nu=0$, proving the proposition. □

The conclusion of the next proposition implies that the product integral operator $\mathcal{P}$ defined by (9.92) is analytic.

**Proposition 9.64.** *Let $\mathbb{B}$ be a Banach algebra, $a<b$, $1 \leq p < 2$, and $\mu \in \mathcal{AI}_p([a,b];\mathbb{B})$. Then the statements,* ($a$), ($b$) *of Theorem* 9.51 *and the following statement* ($c'$) *hold:*

($c'$) *there exists* $T > 0$ *such that the power series* $\sum_{k\geq 1} Q_\mu^k(\nu - \mu)$ *converges to* $\mathcal{P}(\nu) - \mathcal{P}(\mu)$ *absolutely and uniformly for* $\|\nu - \mu\|_{[p]} \leq T$.

*Proof.* Let $\mu \in \mathcal{AI}_p([a,b];\mathbb{B})$. By Lemmas 9.58 and 9.61, for each $k \geq 1$ and $\nu_1, \ldots, \nu_k$, the interval function $L_\mu^k(\nu_1, \ldots, \nu_k) = \{L_\mu^k(\nu_1, \ldots, \nu_k)(A) \colon A \in \mathfrak{I}[a,b]\}$ is defined by (9.109) and is in $\overline{\mathcal{I}}_p$. Since $Q_\mu^k(\nu) = L_\mu^k(\nu, \ldots, \nu)$ for $\nu \in \mathcal{AI}_p$, ($a$) of Theorem 9.51 holds.

By Theorem 9.43($b$), extended to not necessarily closed intervals $A$, for each $k \geq 1$, $L_\mu^k$ is a $k$-linear mapping from the $k$-fold product $(\mathcal{AI}_p)^k$ into $\overline{\mathcal{I}}_p$. For $\nu_1, \ldots, \nu_k \in \mathcal{AI}_p$, let

$$L_\mu^{k,\mathrm{sym}}(\nu_1, \ldots, \nu_k) := (k!)^{-1} \sum_\pi L_\mu^k(\nu_{\pi(1)}, \ldots, \nu_{\pi(k)}), \tag{9.129}$$

where the sum is over all permutations $\pi$ of $\{1, \ldots, k\}$. Then $L_\mu^{k,\mathrm{sym}}$ is a $k$-linear symmetric mapping from the $k$-fold product $(\mathcal{AI}_p)^k$ into $\overline{\mathcal{I}}_p$. By (9.118) $L_\mu^{k,\mathrm{sym}}$ is a bounded $k$-linear symmetric mapping from $(\mathcal{AI}_p([a,b];\mathbb{B}))^k$ into $\overline{\mathcal{I}}_p([a,b];\mathbb{B})$ with norm

$$\|L_\mu^{k,\mathrm{sym}}\| \leq K_p^k \|\widehat{\mu}\|_{\overline{[p]}}^{k+1} D(\widehat{\mu})^{k-2} \Big\{ D(\widehat{\mu})^2 + (k-1)\|\widehat{\mu}\|_{[p]} \Big\},$$

where $K_p = 1 + \zeta(2/p)$ and $D(\widehat{\mu}) := 1 + V_p(\widehat{\mu}) + \|\widehat{\mu}\|_{\sup}$. Since for $\nu \in \mathcal{AI}_p$, $Q_\mu^1(\nu) = L_\mu^1(\nu)$ and $Q_\mu^k(\nu) = L_\mu^{k,\mathrm{sym}}(\nu, \ldots, \nu) = L_\mu^k(\nu, \ldots, \nu)$ for $k \geq 2$, each $Q_\mu^k$ is a $k$-homogeneous polynomial, proving ($b$) of Theorem 9.51.

For ($c'$) recall the definition (9.113) of the remainder $\gamma_n(\mu, \nu)$. By (9.122) there is a finite constant $C = C(p, \widehat{\mu}) := D(\widehat{\mu}) K_p \|\widehat{\mu}\|_{\overline{[p]}} \geq 1$ such that for each $n \geq 1$ and any $\nu \in \mathcal{AI}_p$,

$$\|\gamma_n(\mu, \nu)\|_{\overline{[p]}} \leq (n+1) C^{n+2} \|\widehat{\mu + \nu}\|_{[p]} (2 + V_p(\widehat{\mu}) + V_p(\widehat{\mu + \nu}))^2 \|\nu\|_{(p)}^{n+1}.$$

By (9.128) and the boundedness of $K(\mu + \nu, p)$ for $\nu$ bounded, as in the previous proof, we obtain that for $0 < r < 1/C$, $\|\gamma_n(\mu, \nu)\|_{[p]} \to 0$ as $n \to \infty$ uniformly for $\|\nu\|_{(p)} \leq r$, proving ($c'$). The proof of Proposition 9.64 is complete. □

Now we are ready to prove the main result of this section.

**Proof of Theorem 9.51.** Statements ($a$) and ($b$) hold by Proposition 9.64. It is enough to prove ($c$) for $\mu \equiv 0$. Indeed, assuming ($c$) to hold for the 0 interval function, let $\mu \in \mathcal{AI}_p([a,b];\mathbb{B})$ be arbitrary. By Theorem 5.21 with $X = \mathcal{AI}_p$, $Y = \overline{\mathcal{I}}_p$, $u = 0$, and $v = \mu$, for each integer $k \geq 1$ there exists a $k$-homogeneous polynomial $P_\mu^k$ such that the power series $\sum_{k\geq 1} P_\mu^k(\nu - \mu)$ has infinite radius of uniform convergence, and its sum is $\mathcal{P}(\nu) - \mathcal{P}(\mu)$. On the other hand, by Proposition 9.64, the power series $\sum_{k\geq 1} Q_\mu^k(\nu - \mu)$ converges uniformly and absolutely in a neighborhood of $\mu$, and its sum is also $\mathcal{P}(\nu) - \mathcal{P}(\mu)$. Due to uniqueness of power series (Theorem 5.9), $Q_\mu^k = P_\mu^k$ for each $k \geq 1$. Thus (c) holds provided it holds for $\mu \equiv 0$.

To prove (c) for $\mu \equiv 0$, we will first prove that for each $\nu \in \mathcal{AI}_p$, the bound

$$\left\|Q_0^k(\nu)\right\|_{[p]} \leq 2C_p^k \|\nu\|_{(p)}^k/(k!)^{1/p} \tag{9.130}$$

holds for each $k \geq 2$, where $C_p := 1 + 4^{1/p}\zeta(2/p)$. To prove (9.130) we use Lemma 9.26 inductively to bound the norms of the increments of the functions $f_k := R_{Q_k,a}$, $k \geq 2$, defined by (2.3), where $Q_k := Q_0^k(\nu)$. By Proposition 3.30,

$$v_p(Q_k; [a,b]) = v_p(f_k; [a,b]) \tag{9.131}$$

provided

$$\lim_{r \uparrow t} Q_k([a,r]) = Q_k([a,t)) \qquad \text{for } a < t \leq b \tag{9.132}$$

and

$$\lim_{r \downarrow t} Q_k([a,r]) = Q_k([a,t]) \qquad \text{for } a \leq t < b. \tag{9.133}$$

Let $Q_1 := \nu$. Then for each $k \geq 2$ and $A \in \mathfrak{I}[a,b]$, by (9.109) with $\widehat{\mu} \equiv \mathbb{I}$,

$$Q_k(A) = \displaystyle{\neq}\!\!\!\!\!\int_A \nu(\mathrm{d}s) Q_{k-1}([a,s) \cap A). \tag{9.134}$$

For any $a \leq r < t \leq b$, we have that

$$Q_k([a,t)) - Q_k([a,r]) = \displaystyle{\neq}\!\!\!\!\!\int_{(r,t)} \nu(\mathrm{d}s) Q_{k-1}([a,s)).$$

This Kolmogorov integral over an interval $B$ in place of $(r,t)$ defines an additive and upper continuous interval function by Corollary 9.48. Letting $r \uparrow t$, we obtain (9.132). Similarly it follows that (9.133) holds, and so (9.131) does. So we will next bound the $p$-variation of the point function $f_k$ for each $k \geq 2$.

Let $h$ be the $\mathbb{B}$-valued point function on $[a,b]$ defined by $h(t) := \nu([a,t])$ for $a \leq t \leq b$. Since $\nu$ is upper continuous, $h$ is right-continuous on $[a,b)$. Let $\rho$ be the nondecreasing function on $[a,b]$ defined by $\rho(t) := v_p(\nu; [a,t])$ if $a < t \leq b$, and $\rho(a) := 0$. By (3.68), it follows that for each $a \leq s < t \leq b$,

$$\|h(t) - h(s)\| \leq v_p(\nu; (s,t])^{1/p} \leq \big(\rho(t) - \rho(s)\big)^{1/p}.$$

Letting $f_1 := R_{Q_1,a} = R_{\nu,a}$, it follows similarly that for each $a \leq s < t \leq b$,

$$\|f_1(t) - f_1(s)\| \leq \big(\rho(t) - \rho(s)\big)^{1/p}.$$

For $k \geq 2$, since $Q_k(\{a\}) = 0 = Q_k(\emptyset)$, we have for each $a \leq t \leq b$,

$$f_k(t) = R_{Q_k,a}(t) = \displaystyle{\neq}\!\!\!\!\!\int_{[a,t]} \nu(\mathrm{d}s)\, Q_{k-1}([a,s)) = (RRS)\int_a^t \mathrm{d}h\,\big(f_{k-1}\big)_-^{(a)},$$

where the last equality holds by Proposition 2.90 due to (9.132) if $k \geq 3$ and due to Proposition 2.6($f$) if $k = 2$. Also for each $k \geq 1$, $f_k$ is right-continuous on $(a, b)$ due to (9.133) if $k \geq 2$ and due to Proposition 2.6($f$) if $k = 1$. By Lemma 9.26 with $k = 1$, we then have that for each $a \leq s < t \leq b$,

$$\|f_2(t) - f_2(s)\| \leq \|f_2\|_{[s,t],(p)} \leq C_p^2 \Big(\frac{\rho^2(t) - \rho^2(s)}{2}\Big)^{1/p}.$$

Further applying the same lemma inductively, for each $a \leq s < t \leq b$,

$$\|f_k(t) - f_k(s)\| \leq C_p^k \Big(\frac{\rho^k(t) - \rho^k(s)}{k!}\Big)^{1/p}$$

for $k = 2, 3, \ldots$. From this and (9.131) it follows that for each $k \geq 2$ and each $\nu \in \mathcal{AI}_p([a,b]; \mathbb{B})$, we have

$$v_p(Q_0^k(\nu); [a,b])^{1/p} \leq C_p \|\nu\|_{(p)}^k / (k!)^{1/p}. \tag{9.135}$$

Let $a \leq c < d \leq b$, let $\nu \in \mathcal{AI}_p([a,b]; \mathbb{B})$, and let $\widetilde{\nu}(A) := \nu(A \cap [\![c,d]\!])$ for each $A \in \mathfrak{I}[a,b]$. Clearly $\widetilde{\nu} \in \mathcal{AI}_p([a,b]; \mathbb{B})$ and $\|\widetilde{\nu}\|_{(p)} \leq \|\nu\|_{(p)}$. We claim that for each $k \geq 2$ and each $t \in [\![c,d]\!]$,

$$Q_0^k(\nu)([\![c,t]\!]) = Q_0^k(\widetilde{\nu})([a,t]\!]). \tag{9.136}$$

Indeed for any $t \in [\![c,d]\!]$, by (9.134) we have

$$\begin{aligned} Q_0^2(\nu)([\![c,t]\!]) &= \fint_{[\![c,t]\!]} \nu(\mathrm{d}s)\, \widetilde{\nu}([a,s) \cap [a,t]\!]) \\ &= \fint_{[a,t]\!]} \widetilde{\nu}(\mathrm{d}s)\, \widetilde{\nu}([a,s) \cap [a,t]\!]) = Q_0^2(\widetilde{\nu})([a,t]\!]). \end{aligned}$$

Thus (9.136) holds for $k = 2$. Assuming it holds for some $k \geq 2$, one can show similarly that it holds with $k$ replaced by $k+1$, and so (9.136) holds for each $k \geq 2$ by induction. Then by (9.135), for each $k \geq 2$,

$$v_p(Q_0^k(\nu); [\![c,d]\!])^{1/p} = v_p(Q_0^k(\widetilde{\nu}); [a,b])^{1/p} \leq C_p \frac{\|\widetilde{\nu}\|_{(p)}^k}{(k!)^{1/p}} \leq C_p \frac{\|\nu\|_{(p)}^k}{(k!)^{1/p}}.$$

Since $a \leq c < d \leq b$ are arbitrary, (9.130) follows.

Next we will show that for each $k = 0, 1, \ldots,$

$$\|Q_0^k\| = \sup\{\|Q_k\|_{\overline{[p]}} \colon \|\nu\|_{[p]} \leq 1\} \;\leq\; (k^3+1) K_p (2C_p)^k / k!^{1/p}. \tag{9.137}$$

For each $k \geq 4$, by (9.117) with $\nu_1 = \cdots = \nu_k = \nu$ and $\mu \equiv 0$, we have

$$\|Q_k\|_{\overline{[p]}} \leq 2V_p(Q_k) + 2\|\nu\|_{(p)} V_p(Q_{k-1}) + \sum_{i=2}^{k-2} V_p(Q_i) V_p(Q_{k-i}), \tag{9.138}$$

since $D(\widehat{\mu}) = 2$ in this case, and $V_p(Q_1) = \|\nu\|_{(p)}$. Next using (9.134) and Proposition 9.57 to bound the right side, it follows that

$$\|Q_k\|_{\overline{[p]}} \le 2K_p\|\nu\|_{(p)}\|Q_{k-1}\|_{[p]} + 2K_p\|\nu\|_{(p)}^2\|Q_{k-2}\|_{[p]} \\ + K_p^2\|\nu\|_{(p)}^2 \sum_{i=2}^{k-2} \|Q_{i-1}\|_{[p]}\|Q_{k-i-1}\|_{[p]}.$$

This together with (9.130) applied to each $\|Q_j\|_{[p]}$ gives the bound

$$\|Q_k\|_{\overline{[p]}} \le 4K_p^2 C_p^{k-1}\|\nu\|_{(p)}^k \Big\{\frac{2}{((k-2)!)^{1/p}} + \sum_{i=2}^{k-2} \frac{1}{((i-1)!(k-i-1)!)^{1/p}}\Big\}.$$

For any integers $r, s \ge 0$ we have $1/(r!s!) \le 2^{r+s}/(r+s)!$. It follows that

$$\|Q_k\|_{\overline{[p]}} \le 4K_p^2 C_p^{k-1}\|\nu\|_{(p)}^k \left\{\frac{2+(k-3)2^{k-2}}{(k-2)!^{1/p}}\right\} \le \frac{kK_p(2C_p)^k\|\nu\|_{(p)}^k}{(k-2)!^{1/p}},$$

which implies (9.137) for $k \ge 4$. For $k = 1, 2, 3$, (9.138) still holds where the sum on the right drops out; for $k = 2$, the factor of 2 in the middle term on the right is unnecessary; and for $k = 1$, the middle term also disappears, and we can put $(k-1)!$ in place of $(k-2)!$. For $k = 0$, $Q_0 \equiv \mathbb{I}$, so $\|Q_0\|_{\overline{[p]}} = 1$ and the given bound holds, so (9.137) is proved for all $k = 0, 1, \ldots$. Therefore

$$\limsup_{k\to\infty} \|Q_0^k\|^{1/k} = 0, \tag{9.139}$$

and the radius of uniform convergence is infinite by the Cauchy–Hadamard formula (Theorem 5.10).

The next step will be to show that for each $\nu \in \mathcal{AI}_p$,

$$\mathcal{P}(\nu) = \mathbb{I} + \sum_{k\ge1} Q_0^k(\nu). \tag{9.140}$$

To prove this we will apply a lemma, to be used in full generality in the next section. Recall that $\mathcal{Q}_p = [1, p/(p-1))$ if $p > 1$, $\mathcal{Q}_1 = \{+\infty\}$, and $\mathcal{W}_\infty([a,b];\mathbb{B}) = \mathcal{R}([a,b];\mathbb{B})$, the class of all regulated functions as defined before Proposition 3.96. Let $\mathcal{U}_p := \{q \in \mathcal{Q}_p : q \ge p\}$. Then $\mathcal{U}_p$ is nonempty for each $p$ with $1 \le p < 2$.

**Lemma 9.65.** *Let $\nu \in \mathcal{AI}_p([a,b];\mathbb{B})$ for $1 \le p < 2$, and let $q \in \mathcal{U}_p$. If $g \in \mathcal{W}_q([a,b];\mathbb{B})$ and either $g(t) = \oint_{(a,t)} \mathrm{d}\nu\, g$ for $a \le t \le b$, or $g(t) = \oint_{(a,t]} \mathrm{d}\nu\, g_-$ for $a \le t \le b$, then $g \equiv 0$ on $[a,b]$.*

*Proof.* We can assume that $a < b$. In either case, $g(a) = g(a+) = 0$ since $\oint \mathrm{d}\nu\, g$ and $\oint \mathrm{d}\nu\, g_-^{(a)}$ are additive upper continuous interval functions on $[a,b]$ by Proposition 3.96 and (1.16). First suppose that $g(t) = \oint_{(a,t)} \mathrm{d}\nu\, g$ for $a \le$

$t \le b$. For each $a \le t \le b$, since $\|g\|_{[a,t],[q]}$ is a nonincreasing function of $q$ (Lemma 3.45), then using Corollary 3.43($d$) and the Love–Young inequalities (3.151) and (3.153), we have

$$\|g\|_{[a,t],[q]} \le \|g\|_{[a,t],[p]} = \|\fint \mathrm{d}\nu\, g\|_{(a,t],[p]} \le 2K_{p,q}\|\nu\|_{(a,t],(p)}\|g\|_{(a,t],[q]},$$

where $K_{p,q} = \zeta(p^{-1}+q^{-1})$ if $p > 1$ and $K_{p,q} = 1$ if $p = 1$. If $2K_{p,q}\|\nu\|_{(a,t],(p)} < 1$, as is true for some $t > a$ by Proposition 3.50, then $\|g\|_{[a,t],[q]} = 0$, so $g \equiv 0$ on $[a,t]$. Let $u := \sup\{t \in [a,b]\colon\ g(s) = 0 \text{ for } a \le s \le t\}$. Then $u > a$. Since $g(t) - g(s) = \fint_{[s,t)} \mathrm{d}\nu\, g$ for $a \le s < t \le b$ and $\fint \mathrm{d}\nu\, g$ is an upper continuous interval function, $g$ is left-continuous, and so $g \equiv 0$ on $[a,u]$. If $u = b$ then we are done. If $u < b$, then for $u \le t \le b$, $g(t) = \fint_{(u,t)} \mathrm{d}\nu\, g$ by additivity of the integral (Theorem 9.42). Then by the same argument with $u$ in place of $a$, $g \equiv 0$ on some interval $[u,v]$ for $v > u$, contradicting the definition of $u$ and proving the lemma in the first case.

Now suppose that $g(t) = \fint_{(a,t]} \mathrm{d}\nu\, g_-$ for $a \le t \le b$. For $a \le t \le b$, since $\|g_-\|_{(a,t],[q]} \le \|g\|_{(a,t],[q]}$, as in the first case, we have

$$\|g\|_{[a,t],[q]} \le 2K_{p,q}\|\nu\|_{(a,t],(p)}\|g\|_{(a,t],[q]}.$$

Again, we conclude that $u := \sup\{t \in [a,b]\colon\ g(s) = 0 \text{ for } a \le s \le t\} > a$. As before, by additivity and upper continuity of the Kolmogorov integral $\fint \mathrm{d}\nu\, g_-^{(a)}$, we have

$$\lim_{s \uparrow u} \fint_{(s,u]} \mathrm{d}\nu\, g_- = \nu(\{u\})g(u-) = 0.$$

Thus $g(u) = g(u-) = 0$, and so $g \equiv 0$ on $[a,u]$. By the same argument as in the first case, $g \equiv 0$ on $[a,b]$, completing the proof of the lemma. □

Continuing the proof of Theorem 9.51, for each $n \ge 1$, let

$$\gamma_n := \gamma_n(0,\nu) = \mathcal{P}(\nu) - \mathbb{I} - \sum_{k=1}^{n} Q_0^k(\nu)$$

(see (9.113)). By (9.139), the series

$$\gamma_\infty := \mathcal{P}(\nu) - \mathbb{I} - \sum_{k=1}^{\infty} Q_0^k(\nu) = \lim_{n\to\infty} \gamma_n$$

converges in $\overline{\mathcal{I}}_p([a,b];\mathbb{B})$, absolutely and uniformly for $\nu$ in any bounded subset in $\mathcal{AI}_p([a,b];\mathbb{B})$. By Lemma 9.59 with $\mu \equiv 0$, we have

$$\gamma_n(A) = \fint_A \nu(\mathrm{d}s)\, \gamma_{n-1}([a,s) \cap A) \tag{9.141}$$

for any $n \geq 2$ and interval $A \subset [a, b]$. Letting $n \to \infty$, by the Love–Young inequality (9.78) and by Lemma 9.54 applied to $\eta = \gamma_{n-1} - \gamma_\infty$, we have

$$\gamma_\infty(A) = \fint_A \nu(\mathrm{d}s)\, \gamma_\infty([a, s) \cap A) \tag{9.142}$$

for any interval $A \subset [a, b]$. For $A = [\![r, t) \subset [a, b]$, letting $g(t) := \gamma_\infty([\![r, t))$, we have $g(t) = \fint_{[\![r,t)} \mathrm{d}\nu\, g$. Since $\gamma_\infty(\emptyset) = 0$ (by the definition of the Kolmogorov integral), it follows that $g(r) = \gamma_\infty(\emptyset) = 0$. Thus by additivity of the integral (Theorem 9.42), $g(t) = \fint_{(r,t)} \mathrm{d}\nu\, g$ for any $a \leq r \leq t \leq b$. By Lemma 9.65, it then follows that $\gamma_\infty([\![r, t)) = 0$ for any $a \leq r \leq t \leq b$. Then by additivity again,

$$\gamma_\infty([\![r, t]\!]) = \fint_{[\![r,t]\!]} \nu(\mathrm{d}s)\, \gamma_\infty([a, s) \cap [\![r, t)) = \gamma_\infty([\![r, t)) + \nu(\{t\})\gamma_\infty([\![r, t)) = 0,$$

giving $\gamma_\infty([\![r, t]\!]) = 0$ for any $a \leq r \leq t \leq b$. Thus $\gamma_\infty \equiv 0$ on $[a, b]$, and hence (9.140) holds. The proof of Theorem 9.51 is complete. □

## 9.11 Linear Integral Equations

In the integral equations in this section we either integrate over intervals open at their variable endpoint $t$, or take an integrand $f(t-)$ or $g(t+)$. To take the integral over $[a, t]$ in Theorem 9.66, for example, risks contradictions as in (1.8). In this section we give solutions, unique in suitable $\mathcal{W}_q$ spaces, of four types of linear integral equations with respect to an additive upper continuous interval function: ($a$) forward homogeneous, ($b$) backward homogeneous, ($c$) forward inhomogeneous and ($e$) backward inhomogeneous. The solutions are expressed in terms of product integrals.

The next theorem gives solutions of some forward homogeneous linear integral equations. Recall the notation $\mathcal{U}_p$, $p \geq 1$, given before Lemma 9.65.

**Theorem 9.66.** *Let $\mu \in \mathcal{AI}_p([a, b]; \mathbb{B})$ for $1 \leq p < 2$ and let $y \in \mathbb{B}$. Then the following two statements hold:*

($a$) *The integral equation*

$$f(t) = y + \fint_{[a,t)} \mathrm{d}\mu\, f, \qquad a \leq t \leq b, \tag{9.143}$$

*has the solution $f = f_1$ where $f_1(t) := \widehat{\mu}([a, t))y$ for $a \leq t \leq b$, $f_1 \in \mathcal{W}_p([a, b]; \mathbb{B})$, and $f_1$ is the unique solution in $\mathcal{W}_q([a, b]; \mathbb{B})$ for each $q \in \mathcal{U}_p$.*

($b$) *The integral equation*

$$f(t) = y + \fint_{(a,t]} \mathrm{d}\mu\, f_-, \qquad a \leq t \leq b, \tag{9.144}$$

*has the solution $f = f_2$ where $f_2(t) := \widehat{\mu}((a, t])y$ for $a \leq t \leq b$, $f_2 \in \mathcal{W}_p([a, b]; \mathbb{B})$, and $f_2$ is the unique solution in $\mathcal{W}_q([a, b]; \mathbb{B})$ for each $q \in \mathcal{U}_p$.*

*Proof.* By Theorems 9.28 and 9.36, the strict product integral $\widehat{\mu}$ exists and is an upper continuous, bounded, and multiplicative interval function on $[a,b]$ having bounded $p$-variation. By the Duhamel formula (9.80) with $\nu \equiv 0$ multiplied from the right by $y \in \mathbb{B}$, for each $A \in \mathfrak{I}[a,b]$,

$$\widehat{\mu}(A)y = y + \oint_A d\mu\, \widehat{\mu}([a,\cdot) \cap A)y. \tag{9.145}$$

We have $f_1 \in \mathcal{W}_p([a,b];\mathbb{B})$ by Lemma 9.54 with $A = [a,b]$. In (9.145), letting $A = [a,t)$, it follows that $f_1$ is a solution of (9.143). To prove the stated uniqueness, let $q \in \mathcal{U}_p$ and let $h \in \mathcal{W}_q([a,b];\mathbb{B})$ be such that (9.143) holds for $h$ in place of $f$. Then let $g := h - f_1$ on $[a,b]$. Since $f_1 \in \mathcal{W}_q([a,b];\mathbb{B})$ by Lemma 3.45, $g \in \mathcal{W}_q([a,b];\mathbb{B})$ also. By (9.143) with $h$ in place of $f$, and by (9.145) with $A = [a,t)$,

$$g(t) = \oint_{[a,t)} d\mu\, g, \qquad a \le t \le b. \tag{9.146}$$

Since $g(a) = h(a) - y = 0$, by additivity of the integral (Theorem 2.21), for each $t \in (a,b]$,

$$g(t) = \mu(\{a\})g(a) + \oint_{(a,t)} d\mu\, g = \oint_{(a,t)} d\mu\, g. \tag{9.147}$$

Hence $g \equiv 0$ on $[a,b]$ by Lemma 9.65, proving the uniqueness part of statement $(a)$.

To prove statement $(b)$, first we show that $f_2$ has bounded $p$-variation on $[a,b]$. By multiplicativity, we have for any $a \le s < t \le b$,

$$\|f_2(s) - \widehat{\mu}((a,t))y\| \le \|\widehat{\mu}\|_{\sup}\|y\|\|\mathbb{I} - \widehat{\mu}((s,t))\|, \tag{9.148}$$

and for any $a \le t < r \le b$,

$$\|f_2(r) - \widehat{\mu}((a,t])y\| \le \|\widehat{\mu}\|_{\sup}\|y\|\|\widehat{\mu}((t,r]) - \mathbb{I}\|. \tag{9.149}$$

Due to boundedness and upper continuity of $\widehat{\mu}$ the right sides tend to 0 as $s \uparrow t$ or $r \downarrow t$, respectively. Thus $(f_2)_-(t) = \widehat{\mu}((a,t))y$ for each $t \in (a,b]$ and $f_2(t+) = f_2(t)$ for each $t \in [a,b)$. Applying Proposition 3.30 with $J = (a,b]$ to $f_2$ restricted to $(a,b]$ and the interval function $\{\widehat{\mu}(A)y\colon\ A \in \mathfrak{I}(a,b]\}$, and then using Corollary 3.43$(b)$, we have that $f_2$ has bounded $p$-variation on $[a,b]$. Thus by (9.145) with $A = (a,t]$, it follows that $f_2$ is a solution of (9.144). To prove the uniqueness statement for it, let $q \in \mathcal{U}_p$ and let $h \in \mathcal{W}_q([a,b];\mathbb{B})$ be such that (9.144) holds for $h$ in place of $f$. Let $g := h - f_2$. Then $g \in \mathcal{W}_q([a,b];\mathbb{B})$ and instead of (9.146) and (9.147), now we have

$$g(t) = \oint_{(a,t]} d\mu\, g_-$$

for each $t \in [a,b]$. Again applying Lemma 9.65, it follows that $g \equiv 0$ on $[a,b]$, proving the uniqueness of the solution for statement $(b)$. The proof of Theorem 9.66 is complete. □

Let $\mathcal{MI}_q([a,b];\mathbb{B})$, $0<q<\infty$, be the set of all multiplicative, bounded, and upper continuous interval functions in $\mathcal{I}_q([a,b];\mathbb{B})$. The following will be used to solve inhomogeneous linear integral equations. Recall again the point function $R_{\nu,a}$ defined for an interval function $\nu$ by Definition 2.3.

**Corollary 9.67.** *Let $a<b$, let $\mu\in\mathcal{AI}_p([a,b];\mathbb{B})$ for $1\le p<2$, let $q\ge p$ be such that $p^{-1}+q^{-1}>1$, and let $\mathcal{D}_a:=\{[a,t),(a,t]\colon t\in[a,b]\}$. For $\alpha\in\mathcal{MI}_q([a,b];\mathbb{B})$, the following two statements are equivalent:*

*(a) $\alpha(A)=\widehat{\mu}(A)$ for each $A\in\mathcal{D}_a$;*
*(b) for each $A\in\mathcal{D}_a$,*

$$\alpha(A)=\mathbb{1}+\fint_A \mathrm{d}\mu\,\alpha([a,\cdot)\cap A). \tag{9.150}$$

*If in addition the function $R_{\widehat{\mu},a}$ on $[a,b]$ is invertible, then either of the above statements is equivalent to the following statement: (c) $\alpha=\widehat{\mu}$ on $[a,b]$.*

**Remark 9.68.** The following example shows that in general, statement $(c)$ of the preceding corollary does not follow from $(a)$ or $(b)$. Let $\mathbb{B}=\mathbb{R}$. Recall that $\delta_t$, $t\in[a,b]$, is the interval function on $[a,b]$ such that $\delta_t(A)=1$ if $t\in A$ and $=0$ otherwise. Let $a<u<v<b$, $\mu:=-\delta_u$, and $\nu:=-\delta_u-\delta_v$. Since $\mu,\nu\in\mathcal{AI}_1([a,b];\mathbb{R})$, by Theorem 9.36, the strict product integrals $\widehat{\mu}$ and $\widehat{\nu}$ are in $\mathcal{MI}_1([a,b];\mathbb{R})$. Also, by Lemma 9.30, $\widehat{\mu}=1-\delta_u$ and $\widehat{\nu}=(1-\delta_u)(1-\delta_v)$. Letting $\alpha:=\widehat{\nu}$, we have that $(a)$ and $(b)$ hold but not $(c)$.

*Proof.* As in the previous proof, $\widehat{\mu}$ exists and is in $\mathcal{MI}_p([a,b];\mathbb{B})$. Let $y=\mathbb{1}$ in Theorem 9.66, and let $\alpha\in\mathcal{MI}_q([a,b];\mathbb{B})$.

$(a)\Rightarrow(b)$. Relation (9.150) for $A=[a,t)$, $t\in[a,b]$, holds by Theorem 9.66$(a)$, and for $A=(a,t]$, $t\in[a,b]$, (9.150) holds by Theorem 9.66$(b)$ since

$$\alpha((a,t))=\lim_{s\uparrow t}\alpha((a,s])=\lim_{s\uparrow t}\widehat{\mu}((a,s])=\widehat{\mu}((a,t))$$

for each $t\in(a,b]$ by upper continuity of $\widehat{\mu}$ and $\alpha$ as in (9.148).

$(b)\Rightarrow(a)$. Relation (9.150) for $A=[a,t)$, $t\in[a,b]$, gives that the function $f:=\alpha([a,\cdot))$ is a solution of the integral equation (9.143). By Lemma 9.54 with $A=[a,b]$, $f\in\mathcal{W}_q([a,b];\mathbb{B})$. Thus $\alpha([a,t))=\widehat{\mu}([a,t))$ for all $t\in[a,b]$ by Theorem 9.66$(a)$. Let $g:=\alpha((a,\cdot])$. By upper continuity, as in (9.148), $g_-(t)=\alpha((a,t))$ for each $t\in(a,b]$. Thus (9.150) for $A=(a,t]$, $t\in[a,b]$, gives that $g$ is a solution of the integral equation (9.144). Once again as in (9.148) and (9.149), it follows that the hypotheses of Proposition 3.30 with $J=(a,b]$ hold for $g$ and $\alpha$ restricted to $(a,b]$. This together with Corollary 3.43$(b)$ yields $g\in\mathcal{W}_q([a,b];\mathbb{B})$, and so $\alpha((a,t])=\widehat{\mu}((a,t])$ for all $t\in[a,b]$ by Theorem 9.66$(b)$, proving $(a)$.

$(a)\Leftrightarrow(c)$. Clearly, $(c)$ implies $(a)$. For the converse implication, since $a<b$ and $[a,t)\downarrow\{a\}$ as $t\downarrow a$, $\alpha(\{a\})=\widehat{\mu}(\{a\})$ by upper continuity of $\alpha$ and $\widehat{\mu}$. Thus by multiplicativity, $R_{\alpha,a}(t)=\alpha([a,t])=\widehat{\mu}([a,t])=R_{\widehat{\mu},a}(t)$ for each $t\in(a,b]$, and $R_{\alpha,a}(a)=\alpha(\emptyset)=\mathbb{1}=\widehat{\mu}(\emptyset)=R_{\widehat{\mu},a}(a)$. By $(a)\Rightarrow(d)$ of

Proposition 9.6, $R_{\widehat{\mu},a} \equiv R_{\alpha,a}$ is a regulated function on $[a,b]$, and $\alpha([a,t)) = R_{\alpha,a}(t-) = R_{\widehat{\mu},a}(t-) = \widehat{\mu}([a,t))$ holds for each $t \in (a,b]$. Let $(R_{\widehat{\mu},a})^{\text{inv}}$ be the reciprocal function of $R_{\widehat{\mu},a}$, and thus of $R_{\alpha,a}$. So by multiplicativity, we have

$$\begin{aligned}\alpha([s,t]) &= \alpha([s,t])\alpha([a,s))(R_{\alpha,a})^{\text{inv}}(s-) = R_{\alpha,a}(t)(R_{\alpha,a})^{\text{inv}}(s-)\\ &= R_{\widehat{\mu},a}(t)(R_{\widehat{\mu},a})^{\text{inv}}(s-) = \widehat{\mu}([s,t])\widehat{\mu}([a,s))(R_{\widehat{\mu},a})^{\text{inv}}(s-) = \widehat{\mu}([s,t])\end{aligned}$$

for each $a < s \le t \le b$. Similarly it follows that $\alpha([\![s,t]\!]) = \widehat{\mu}([\![s,t]\!])$ for each $a \le s < t \le b$, proving $(c)$. The proof of Corollary 9.67 is complete. □

The next theorem gives a solution of a backward homogeneous linear integral equation. Its proof is omitted since it is similar to the proof of Theorem 9.66 except that here we use Duhamel's formula (9.81) instead of (9.80).

**Theorem 9.69.** *Let $\mu \in \mathcal{AI}_p([a,b];\mathbb{B})$ for $1 \le p < 2$ and let $y \in \mathbb{B}$. Then the following two statements hold:*

*(a) The integral equation*

$$g(t) = y + \oint_{(t,b]} g\,\mathrm{d}\mu, \qquad a \le t \le b, \tag{9.151}$$

*has the solution $g = g_1$ where $g_1(t) = y\widehat{\mu}((t,b])$ for $a \le t \le b$, $g_1 \in \mathcal{W}_p([a,b];\mathbb{B})$, and $g_1$ is the unique solution in $\mathcal{W}_q([a,b];\mathbb{B})$ for each $q \in \mathcal{U}_p$.*

*(b) The integral equation*

$$g(t) = y + \oint_{[t,b)} g_+\,\mathrm{d}\mu, \qquad a \le t \le b, \tag{9.152}$$

*has the solution $g = g_2$ where $g_2(t) = y\widehat{\mu}([t,b))$ for $a \le t \le b$, $g_2 \in \mathcal{W}_p([a,b];\mathbb{B})$, and $g_2$ is the unique solution in $\mathcal{W}_q([a,b];\mathbb{B})$ for each $q \in \mathcal{U}_p$.*

The next statement will also be used to solve inhomogeneous linear integral equations. Recall the point function $L_{\nu,b}$ defined for an interval function $\nu$ by Definition 2.3.

**Corollary 9.70.** *Let $a < b$, let $\mu \in \mathcal{AI}_p([a,b];\mathbb{B})$ for $1 \le p < 2$, let $q \ge p$ be such that $p^{-1} + q^{-1} > 1$, and let $\mathcal{D}_b := \{(t,b],[t,b)\colon t \in [a,b]\}$. For $\alpha \in \mathcal{MI}_q([a,b];\mathbb{B})$, the following two statements are equivalent:*

*(a) $\alpha(B) = \widehat{\mu}(B)$ for each $B \in \mathcal{D}_b$;*

*(b) for each $B \in \mathcal{D}_b$,*

$$\alpha(B) = \mathbb{1} + \oint_B \alpha((\cdot,b] \cap B)\,\mathrm{d}\mu. \tag{9.153}$$

*If in addition the function $L_{\widehat{\mu},b}$ is invertible then either of the above statements is equivalent to the following statement: (c) $\alpha = \widehat{\mu}$ on $[a,b]$.*

A proof of the corollary is based on arguments symmetric to those used to prove Corollary 9.67 and therefore is omitted.

Now we turn to inhomogeneous linear integral equations.

**Theorem 9.71.** *Let* $\mu, \nu \in \mathcal{AI}_p([a,b];\mathbb{B})$ *for* $1 \le p < 2$, *and let* $y \in \mathbb{B}$. *Then the following two statements hold:*

(*a*) *The integral equation*

$$f(t) = y + \oint_{[a,t)} \mathrm{d}\mu\, f + \nu([a,t)), \qquad a \le t \le b, \tag{9.154}$$

*has the solution* $f = f_3$, *in* $\mathcal{W}_p([a,b];\mathbb{B})$ *and unique in* $\mathcal{W}_q([a,b];\mathbb{B})$ *for each* $q \in \mathcal{U}_p$, *where*

$$f_3(t) := \widehat{\mu}([a,t))y + \oint_{[a,t)} \widehat{\mu}((\cdot,t))\,\mathrm{d}\nu, \qquad a \le t \le b. \tag{9.155}$$

(*b*) *The integral equation*

$$f(t) = y + \oint_{(a,t]} \mathrm{d}\mu\, f_- + \nu((a,t]), \qquad a \le t \le b, \tag{9.156}$$

*has the solution* $f = f_4$, *in* $\mathcal{W}_p([a,b];\mathbb{B})$ *and unique in* $\mathcal{W}_q([a,b];\mathbb{B})$ *for each* $q \in \mathcal{U}_p$, *where*

$$f_4(t) := \widehat{\mu}((a,t])y + \oint_{(a,t]} \widehat{\mu}((\cdot,t])\,\mathrm{d}\nu, \qquad a \le t \le b. \tag{9.157}$$

For the proof we will use an integral relation to be proved next by solving a homogeneous linear integral equation with respect to a suitable $2\times 2$ matrix-valued interval function.

**Lemma 9.72.** *Let* $a < b$, *let* $\mu, \nu \in \mathcal{AI}_p([a,b];\mathbb{B})$ *for* $1 \le p < 2$, *and let* $\gamma$ *be the interval function on* $[a,b]$ *defined by*

$$\gamma(A) := \oint_A \widehat{\mu}((\cdot,b] \cap A)\,\mathrm{d}\nu, \qquad A \in \mathfrak{I}[a,b]. \tag{9.158}$$

*Then for each* $A \in \mathcal{D}_a = \{(a,t], [a,t)\colon\ t \in [a,b]\}$, *we have*

$$\gamma(A) = \oint_A \mathrm{d}\mu\, \gamma([a,\cdot) \cap A) + \nu(A). \tag{9.159}$$

*Proof.* The interval function $\gamma$ is well defined by Proposition 9.55 with $\alpha \equiv \mathbb{I}$, which also gives the bound

$$\|\gamma\|_{[p]} \le K_p C_p(\widehat{\mu})\big[2 + V_p(\widehat{\mu})\big]\,\|\nu\|_{(p)}, \tag{9.160}$$

where $K_p := K_{p,p} = 1+\zeta(2/p)$, $C_p(\widehat{\mu}) := [1+V_p(\widehat{\mu})]\|\widehat{\mu}\|_{\sup}$, and $W_p$ is defined by (9.83). Then, by Proposition 9.55 with $\alpha = \gamma$, $\mu$ and $\nu$ interchanged, and then $\nu \equiv 0$, so that $\widehat{\nu} \equiv \mathbb{I}$, it follows that the Kolmogorov integral in (9.159) exists for each interval $A$. Let $\mathbb{B}^{\urcorner}$ be the set of all $2 \times 2$ matrices $\left(\begin{smallmatrix} x & y \\ 0 & z \end{smallmatrix}\right)$ with $x, y, z \in \mathbb{B}$. As for matrices with scalar elements, define addition of matrices elementwise, and let

$$\begin{pmatrix} x & y \\ 0 & z \end{pmatrix}\begin{pmatrix} u & v \\ 0 & w \end{pmatrix} := \begin{pmatrix} xu & xv + yw \\ 0 & zw \end{pmatrix} \quad \text{and} \quad \begin{pmatrix} x & y \\ 0 & z \end{pmatrix}\begin{pmatrix} u \\ v \end{pmatrix} := \begin{pmatrix} xu + yv \\ zv \end{pmatrix}.$$

Then $\mathbb{B}^{\urcorner}$ is a vector space. For $u, v \in \mathbb{B}$, let $\|\left(\begin{smallmatrix} u \\ v \end{smallmatrix}\right)\| := \max\{\|u\|, \|v\|\}$. Define a norm on $\mathbb{B}^{\urcorner}$ by

$$\left\|\begin{pmatrix} x & y \\ 0 & z \end{pmatrix}\right\| := \sup\left\{\left\|\begin{pmatrix} x & y \\ 0 & z \end{pmatrix}\begin{pmatrix} u \\ v \end{pmatrix}\right\| : u, v \in \mathbb{B}, \ \|u\| \le 1, \ \|v\| \le 1\right\}.$$

Then

$$\max\{\|x\|, \|y\|, \|z\|\} \le \left\|\begin{pmatrix} x & y \\ 0 & z \end{pmatrix}\right\| \le 2\max\{\|x\|, \|y\|, \|z\|\}.$$

Equipped with the identity $\mathbb{I}^{\urcorner} := \left(\begin{smallmatrix} \mathbb{I} & 0 \\ 0 & \mathbb{I} \end{smallmatrix}\right)$, $\mathbb{B}^{\urcorner}$ is a Banach algebra. Let $\xi := \left(\begin{smallmatrix} \mu & \nu \\ 0 & 0 \end{smallmatrix}\right)$. It is easy to verify that $\xi$ is an additive upper continuous interval function on $[a, b]$ with bounded $p$-variation, that is, $\xi \in \mathcal{AI}_p([a, b]; \mathbb{B}^{\urcorner})$. Also, let $\tilde{\xi} := \left(\begin{smallmatrix} \widehat{\mu} & \gamma \\ 0 & \mathbb{I} \end{smallmatrix}\right)$, another $\mathbb{B}^{\urcorner}$-valued interval function on $[a, b]$. The $p$-variation of $\tilde{\xi}$ is bounded because the $p$-variation of $\gamma$ is bounded by (9.160), and the $p$-variation of $\widehat{\mu}$ is bounded by Theorem 9.36. We next show that $\tilde{\xi}$ is multiplicative and $\tilde{\xi}(A) = \widehat{\xi}(A)$ for each $A \in \mathcal{D}_a$. To this aim, let nonempty $A, B \in \mathfrak{I}[a, b]$ be such that $A \prec B$ and $C := A \cup B$ is an interval. For $t \in A$, $(t, b] \cap C = ((t, b] \cap A) \cup B$, and so by multiplicativity, $\widehat{\mu}((t, b] \cap C) = \widehat{\mu}(B)\widehat{\mu}((t, b] \cap A)$. For $t \in B$, $(t, b] \cap C = (t, b] \cap B$. Thus by definition (9.158) of $\gamma$, by additivity of the Kolmogorov integral (Theorem 2.21), and by (9.57) with $u = \widehat{\mu}(B)$, where $\fint_A \widehat{\mu}((\cdot, b] \cap A) \mathrm{d}\nu = \gamma(A)$ exists, we have

$$\begin{aligned} \gamma(C) &= \fint_A \widehat{\mu}((\cdot, b] \cap C) \,\mathrm{d}\nu + \fint_B \widehat{\mu}((\cdot, b] \cap C) \,\mathrm{d}\nu \\ &= \widehat{\mu}(B) \fint_A \widehat{\mu}((\cdot, b] \cap A) \,\mathrm{d}\nu + \fint_B \widehat{\mu}((\cdot, b] \cap B) \,\mathrm{d}\nu \\ &= \widehat{\mu}(B)\gamma(A) + \gamma(B). \end{aligned}$$

Multiplying two matrices, it then follows that

$$\tilde{\xi}(B)\tilde{\xi}(A) = \begin{pmatrix} \widehat{\mu}(B)\widehat{\mu}(A) & \widehat{\mu}(B)\gamma(A) + \gamma(B) \\ 0 & \mathbb{I} \end{pmatrix} = \begin{pmatrix} \widehat{\mu}(C) & \gamma(C) \\ 0 & \mathbb{I} \end{pmatrix} = \tilde{\xi}(C).$$

Since $\gamma(\emptyset) = 0$, $\tilde{\xi}(\emptyset)$ is the identity in $\mathbb{B}^{\urcorner}$, and so $\tilde{\xi}$ is a multiplicative interval function on $[a, b]$. For each $a \le t < c \le b$, we have, letting $y := \mathbb{I}$ in Theorem 9.69, with $c$ in place of $b$ there,

$$\mathbb{I} + \fint_{A_{t,c}} \widehat{\mu}((\cdot, c] \cap A_{t,c})\,\mathrm{d}\mu = \widehat{\mu}(A_{t,c})$$

for $A_{t,c} = (t, c]$ by Theorem 9.69($a$), and for $A_{t,c} = [t, c)$ by Theorem 9.69($b$) since $\widehat{\mu}((s, c)) = \lim_{r\downarrow s} \widehat{\mu}([r, c))$ by multiplicativity and upper continuity. By definition of the Kolmogorov integral, we then get

$$\begin{aligned} &\mathbb{I}^{\urcorner} + \fint_{A_{t,c}} \tilde{\xi}((\cdot, c] \cap A_{t,c})\,\mathrm{d}\xi \\ &= \begin{pmatrix} \mathbb{I} + \fint_{A_{t,c}} \widehat{\mu}((\cdot, c] \cap A_{t,c})\,\mathrm{d}\mu & \fint_{A_{t,c}} \widehat{\mu}((\cdot, c] \cap A_{t,c})\,\mathrm{d}\nu \\ 0 & \mathbb{I} \end{pmatrix} \\ &= \begin{pmatrix} \widehat{\mu}(A_{t,c}) & \fint_{A_{t,c}} \widehat{\mu}((\cdot, c] \cap A_{t,c})\,\mathrm{d}\nu \\ 0 & \mathbb{I} \end{pmatrix} = \tilde{\xi}(A_{t,c}) \end{aligned} \tag{9.161}$$

for each $a \le t < c \le b$. By implication ($b$) $\Rightarrow$ ($a$) of Corollary 9.70 applied to the $\mathbb{B}^{\urcorner}$-valued interval functions $\xi$ and $\alpha = \tilde{\xi}$ restricted to $[a, c]$, we have $\tilde{\xi}(A_{t,c}) = \widehat{\xi}(A_{t,c})$ for each $a \le t < c \le b$. In particular, $\tilde{\xi}(A) = \widehat{\xi}(A)$ for each $A \in \mathcal{D}_a$. Now by implication ($a$) $\Rightarrow$ ($b$) of Corollary 9.67 applied to the $\mathbb{B}^{\urcorner}$-valued interval functions $\xi$ and $\alpha = \tilde{\xi}$, it follows that for each $A \in \mathcal{D}_a$,

$$\begin{aligned} \begin{pmatrix} \widehat{\mu}(A) & \gamma(A) \\ 0 & \mathbb{I} \end{pmatrix} = \tilde{\xi}(A) &= \mathbb{I}^{\urcorner} + \fint_A d\xi\, \tilde{\xi}([a, \cdot) \cap A) \\ &= \begin{pmatrix} \mathbb{I} + \fint_A \mathrm{d}\mu\, \widehat{\mu}([a, \cdot) \cap A) & \fint_A \mathrm{d}\mu\, \gamma([a, \cdot) \cap A) + \nu(A) \\ 0 & \mathbb{I} \end{pmatrix}. \end{aligned}$$

Thus (9.159) must hold for each $A \in \mathcal{D}_a$, completing the proof of the lemma. □

**Proof of Theorem 9.71.** If $a = b$ it is straightforward that all conclusions hold, with $f_3(a) = f_4(a) = y$. So we can assume that $a < b$. Let $\gamma$ be the interval function on $[a, b]$ defined by (9.158). Then for $f_3$ defined by (9.155), $f_3(t) = \widehat{\mu}([a, t))y + \gamma([a, t))$ for $t \in [a, b]$. We have $f_3 \in \mathcal{W}_p([a, b]; \mathbb{B})$ by Lemma 9.54 with $q = p$ and $A = [a, b]$ and since $\widehat{\mu}, \gamma \in \mathcal{I}_p([a, b]; \mathbb{B})$ by Theorem 9.36 and (9.160). Therefore by Theorem 9.66($a$) and by (9.159) with $A = [a, t)$, we have

$$\begin{aligned} \fint_{[a,t)} \mathrm{d}\mu\, f_3 &= \fint_{[a,t)} \mathrm{d}\mu\, \widehat{\mu}([a, \cdot))y + \fint_{[a,t)} \mathrm{d}\mu\, \gamma([a, \cdot)) \\ &= \widehat{\mu}([a, t))y - y + \gamma([a, t)) - \nu([a, t)) = f_3(t) - y - \nu([a, t)) \end{aligned}$$

for each $t \in [a, b]$, that is, $f_3$ indeed solves the integral equation (9.154). To show uniqueness, let $h \in \mathcal{W}_q([a, b]; \mathbb{B})$ be such that (9.154) holds for $h$ in place of $f$. Then letting $g := h - f_3$, we have $g(a) = h(a) - y = 0$ and

$$g(t) = \oint_{[a,t)} \mathrm{d}\mu\, g = \oint_{(a,t)} \mathrm{d}\mu\, g$$

for each $t \in [a,b]$ as in (9.147). By Lemma 9.65, it then follows that $g \equiv 0$ on $[a,b]$. Therefore (9.155) gives the unique solution of (9.154) in $\mathcal{W}_q$, proving $(a)$.

To prove $(b)$, for $f_4$ defined by (9.157), using (9.158) with $A = (a,t]$, we have $f_4(t) = \widehat{\mu}((a,t])y + \gamma((a,t])$ for $a \le t \le b$. First we show that

$$(f_4)_-(t) = \widehat{\mu}((a,t))y + \gamma((a,t)) \qquad \text{for } t \in (a,b]. \tag{9.162}$$

Since the strict product integral $\widehat{\mu}$ is an upper continuous, bounded, and multiplicative interval function, it follows by (9.148) with $y = \mathbb{I}$ that $\widehat{\mu}((a,s]) \to \widehat{\mu}((a,t))$ as $s \uparrow t$ for $t \in (a,b]$. For $\gamma$, by additivity of the Kolmogorov integral (Theorem 2.21), and by bringing a fixed element $u = \widehat{\mu}((s,t))$ of $\mathbb{B}$ outside an integral via (9.57), as we can since the integral defining $\gamma((a,s])$ exists,

$$\begin{aligned}\gamma((a,t)) &= \oint_{(a,s]} \widehat{\mu}((\cdot,s] \cup (s,t))\,\mathrm{d}\nu + \oint_{(s,t)} \widehat{\mu}((\cdot,t))\,\mathrm{d}\nu \\ &= \widehat{\mu}((s,t))\gamma((a,s]) + \gamma((s,t))\end{aligned} \tag{9.163}$$

for $a < s < t \le b$. Since $\widehat{\mu}$ is upper continuous at $\emptyset$, $\widehat{\mu}((s,t)) \to \mathbb{I}$ as $s \uparrow t$. By Lemma 9.53 with $f_{[a,u)} = \widehat{\mu}((\cdot,u))$, $\|\widehat{\mu}((\cdot,u))\|_{(s,u),[p]} \le C_p(\widehat{\mu})$ for each $a \le s < u \le b$, where the finite constant $C_p(\widehat{\mu})$ is defined just after (9.160). Therefore by the Love–Young inequality (9.78), for each $a \le s < t \le b$,

$$\|\gamma((s,t))\| = \Big\| \oint_{(s,t)} \widehat{\mu}((\cdot,t))\,\mathrm{d}\nu \Big\| \le K_p C_p(\widehat{\mu}) \|\nu\|_{(s,t),(p)},$$

where $K_p := K_{p,p} = 1 + \zeta(2/p)$. By Proposition 3.50, $\|\nu\|_{(s,t),(p)} \to 0$ as $s \uparrow t$, and so $\|\gamma((s,t))\| \to 0$ as $s \uparrow t$. Hence since $\|\gamma\|_{\sup} < \infty$ by (9.160), we get by (9.163) that

$$\gamma((a,t)) - \gamma((a,s]) = \big[\widehat{\mu}((s,t)) - \mathbb{I}\big]\gamma((a,s]) + \gamma((s,t)) \to 0 \tag{9.164}$$

as $s \uparrow t$ for $a < t \le b$, and so (9.162) holds. Now as for $(a)$, by Theorem 9.66$(b)$, and by (9.159) with $A = (a,t]$, we have

$$\begin{aligned}\oint_{(a,t]} \mathrm{d}\mu\,(f_4)_- &= \oint_{(a,t]} \mathrm{d}\mu\,\widehat{\mu}((a,\cdot))y + \oint_{(a,t]} \mathrm{d}\mu\,\gamma((a,\cdot)) \\ &= \widehat{\mu}((a,t])y - y + \gamma((a,t]) - \nu((a,t]) \;=\; f_4(t) - y - \nu((a,t])\end{aligned}$$

for each $t \in (a,b]$, that is, $f_4$ solves the integral equation (9.156). To show uniqueness, let $h \in \mathcal{W}_q([a,b];\mathbb{B})$ be such that (9.156) holds with $h$ in place of $f$. For $a \le t \le b$, let $g(t) := h(t) - \widehat{\mu}((a,t])y - \gamma((a,t])$. By (9.156) with $h$ in place of $f$, Theorem 9.66$(b)$ and (9.159) with $A = (a,t]$, we then have

$$g(t) = \fint_{(a,t]} \mathrm{d}\mu\big[h_- - \widehat{\mu}((a,\cdot))y - \gamma([a,\cdot))\big] = \fint_{(a,t]} \mathrm{d}\mu\, g_-$$

for each $a \le t \le b$. By Lemma 9.65, it follows that $g \equiv 0$ on $[a,b]$. Therefore (9.157) is the unique solution of (9.156) in $\mathcal{W}_q$. The proof of Theorem 9.71 is complete. □

We next have an analogous result for a backward inhomogeneous integral equation.

**Theorem 9.73.** *Let $\mu, \nu \in \mathcal{AI}_p([a,b];\mathbb{B})$ for $1 \le p < 2$, and let $y \in \mathbb{B}$. Then the following two statements hold:*

(*a*) *The integral equation*

$$g(t) = y + \fint_{(t,b]} g\,\mathrm{d}\mu + \nu((t,b]), \qquad a \le t \le b, \tag{9.165}$$

*has the solution $g = g_3$, in $\mathcal{W}_p([a,b];\mathbb{B})$ and unique in $\mathcal{W}_q([a,b];\mathbb{B})$ for each $q \in \mathcal{U}_p$, where*

$$g_3(t) := y\widehat{\mu}((t,b]) + \fint_{(t,b]} \mathrm{d}\nu\,\widehat{\mu}((t,\cdot)), \qquad a \le t \le b. \tag{9.166}$$

(*b*) *The integral equation*

$$g(t) = y + \fint_{[t,b)} g_+\,\mathrm{d}\mu + \nu([t,b)), \qquad a \le t \le b, \tag{9.167}$$

*has the solution $g = g_4$, in $\mathcal{W}_p([a,b];\mathbb{B})$ and unique in $\mathcal{W}_q([a,b];\mathbb{B})$ for each $q \in \mathcal{U}_p$, where*

$$g_4(t) := y\widehat{\mu}([t,b)) + \fint_{[t,b)} \mathrm{d}\nu\,\widehat{\mu}([t,\cdot)), \qquad a \le t \le b. \tag{9.168}$$

To prove this theorem we use the following lemma, analogous to Lemma 9.72. The rest of the proof is similar to the proof of Theorem 9.71 and so is omitted.

**Lemma 9.74.** *Let $a < b$, let $\mu, \nu \in \mathcal{AI}_p([a,b];\mathbb{B})$ for $1 \le p < 2$, and let $\gamma$ be the interval function on $[a,b]$ defined by*

$$\gamma(A) := \fint_A \mathrm{d}\nu\,\widehat{\mu}([a,\cdot) \cap A), \qquad A \in \mathfrak{I}[a,b]. \tag{9.169}$$

*Then for each $A \in \mathcal{D}_b = \{(t,b], [t,b) \colon t \in [a,b]\}$,*

$$\gamma(A) = \fint_A \gamma((\cdot,b] \cap A)\mathrm{d}\mu + \nu(A). \tag{9.170}$$

*Proof.* The interval function $\gamma$ is well defined by Proposition 9.55, with $(\mu, \nu, \alpha)$ there equal to $(0, \nu, \widehat{\mu})$ here, so that $\widehat{\mu} \equiv 1$ there. The rest of the proof of the lemma is essentially symmetric to that of Lemma 9.72 and only a few differences will be mentioned. Let $\mathbb{B}^{\llcorner}$ be the set of all $2 \times 2$ matrices $\left(\begin{smallmatrix} x & 0 \\ y & z \end{smallmatrix}\right)$ with $x, y, z \in \mathbb{B}$. Define algebra operations in $\mathbb{B}^{\llcorner}$ as for ordinary matrices, analogously to those in $\mathbb{B}^{\urcorner}$. Defining a norm as before, $\mathbb{B}^{\llcorner}$ is a unital Banach algebra. Let $\xi := \left(\begin{smallmatrix} \mu & 0 \\ \nu & 0 \end{smallmatrix}\right)$. Then $\xi \in \mathcal{AI}_p([a,b]; \mathbb{B}^{\llcorner})$. Also, for $\gamma$ now defined by (9.169), let $\tilde{\xi} := \left(\begin{smallmatrix} \widehat{\mu} & 0 \\ \gamma & \mathbb{I} \end{smallmatrix}\right)$ another $\mathbb{B}^{\llcorner}$-valued interval function on $[a, b]$. Although one cannot simply take transposes of matrices (see Remark 9.75), one can show in analogy to the proof of (9.161) that

$$\mathbb{I}^{\llcorner} + \oint_{A_{c,t}} d\xi \, \tilde{\xi}([c, \cdot) \cap A_{c,t}) = \tilde{\xi}(A_{c,t})$$

for each $A_{c,t} = (c, t]$ with $a \le c < t \le b$. By the implication $(b) \Rightarrow (a)$ of Corollary 9.67 applied to the $\mathbb{B}^{\llcorner}$-valued interval functions $\xi$ and $\alpha = \tilde{\xi}$ restricted to $[c, a]$, we have $\tilde{\xi}(A_{c,t}) = \widehat{\xi}(A_{c,t})$ for each $a \le c < t \le b$. In particular, $\tilde{\xi}(A) = \widehat{\xi}(A)$ for each $A \in \mathcal{D}_b$. Now using implication $(a) \Rightarrow (b)$ of Corollary 9.70, it follows that (9.170) holds for each $A \in \mathcal{D}_b$, completing the proof of the lemma. □

**Remark 9.75.** The reader might have noted that as a set of matrices, $\mathbb{B}^{\llcorner}$ as in the last proof is the set of all transposes $C^t$ of matrices $C \in \mathbb{B}^{\urcorner}$ in the proof of Lemma 9.72, where $C_{ij}^t := C_{ji}$ for all $i, j$. Unfortunately, however, the relation $(CD)^t = D^t C^t$, familiar for matrices with scalar elements and valid if $\mathbb{B}$ is commutative, reverses the direction of multiplication and is not even valid if $\mathbb{B}$ is not commutative unless one uses the reversed multiplication ($*$-multiplication) on one side. If $\mathbb{B}$ is commutative, even if $\mathbb{B} = \mathbb{K}$, two matrices in $\mathbb{B}^{\llcorner}$ (or two in $\mathbb{B}^{\urcorner}$) still do not commute in general.

The interval function $\xi^{\llcorner} := \xi$ as defined in the last proof is exactly the transpose of $\xi^{\urcorner} = \xi$ as in the proof of Lemma 9.72. But $\gamma^{\llcorner} := \gamma$ defined by (9.169) is not the same as $\gamma^{\urcorner} := \gamma$ defined by (9.158). So, $\tilde{\xi}$ defined in the proofs of Lemmas 9.72 and 9.74 are not actually transposes of each other.

Let $1 \le p < 2$, let $a < b$, and let $\mathbb{B}$ be a Banach algebra. By Theorem 9.71$(a)$, for $y \in \mathbb{B}$ and $\mu, \nu \in \mathcal{AI}_p([a,b]; \mathbb{B}) \equiv \mathcal{AI}_p$, the forward inhomogeneous linear integral equation (9.154) has a unique solution $f_3 \in \mathcal{W}_p([a,b]; \mathbb{B}) \equiv \mathcal{W}_p$. Define the *solution mapping* $\mathcal{S}^{\succ}_{[a,\cdot)}$ corresponding to the integral equation (9.154) by

$$\mathbb{B} \times \mathcal{AI}_p \times \mathcal{AI}_p \ni (y, \mu, \nu) \mapsto \mathcal{S}^{\succ}_{[a,\cdot)}(y, \mu, \nu) := f_3 \in \mathcal{W}_p. \tag{9.171}$$

Likewise by Theorems 9.71$(b)$ and 9.73, there exist *solution mappings* $\mathcal{S}^{\succ}_{(a,\cdot]}$, $\mathcal{S}^{\prec}_{(\cdot,b]}$, and $\mathcal{S}^{\prec}_{[\cdot,b)}$ corresponding to the integral equations (9.156), (9.165), and

(9.167), respectively. We show next that the four solution mappings are analytic.

**Theorem 9.76.** *Let $\mathbb{B}$ be a Banach algebra, $a < b$, and $1 \le p < 2$. Then the four solution mappings $\mathcal{S}^{\succ}_{[a,\cdot)}$, $\mathcal{S}^{\succ}_{(a,\cdot]}$, $\mathcal{S}^{\prec}_{(\cdot,b]}$ and $\mathcal{S}^{\prec}_{[\cdot,b)}$ corresponding to the integral equations* (9.154), (9.156), (9.165) *and* (9.167), *respectively, are uniformly entire.*

*Proof.* There are symmetric or similar proofs for each of the four mappings, so a detailed proof will be given only for the mapping (9.171). At the end of the proof we will indicate what changes are needed for the other three mappings.

By Theorem 5.38 it is enough to show that $\mathcal{S}^{\succ}_{[a,\cdot)}$ is a composition of uniformly entire mappings. The product integral operator $\mathcal{P}$ defined by (9.92) is a uniformly entire mapping from $\mathcal{AI}_p([a,b];\mathbb{B})$ into $\overline{\mathcal{I}}_p([a,b];\mathbb{B})$ by Theorem 9.51.

The map $\beta \mapsto \beta([a,\cdot))$ is a bounded linear operator with norm 1 from $\mathcal{I}_p([a,b];\mathbb{B})$ into $\mathcal{W}_p([a,b];\mathbb{B})$. For $y \in \mathbb{B}$ and $\mu \in \mathcal{AI}_p([a,b];\mathbb{B})$ let

$$F_1(y,\mu) \ := \ \widehat{\mu}([a,\cdot))y. \tag{9.172}$$

Then $F_1$ is uniformly entire from $\mathbb{B} \times \mathcal{AI}_p([a,b];\mathbb{B})$ into $\mathcal{W}_p([a,b];\mathbb{B})$ by composition of uniformly entire functions, Theorem 5.38. $F_1$ gives the first term on the right in (9.155). Most of the proof, to be done, will deal with the second term.

Let $\nu \in \mathcal{AI}_p([a,b];\mathbb{B})$. For a $\mathbb{B}$-valued interval function $\alpha$ on $[a,b]$, let

$$\chi(\alpha,\nu)(A) := \fint_A \alpha((\cdot,b]\cap A)\,\mathrm{d}\nu \tag{9.173}$$

provided the Kolmogorov integral is defined for each interval $A \subset [a,b]$. Thus if defined, $\chi := \chi(\alpha,\nu) := \{\chi(\alpha,\nu)(A)\colon\ A \in \mathfrak{I}[a,b]\}$ is a $\mathbb{B}$-valued interval function on $[a,b]$. The function $t \mapsto \chi(\widehat{\mu},\nu)([a,t))$, $t \in [a,b]$, gives the second term on the right in (9.155), and is the composition of the interval function $\chi(\widehat{\mu},\nu)$ with the map $t \mapsto [a,t)$ from $[a,b]$ into $\mathfrak{I}[a,b]$.

**Remark 9.77.** By the second part of Theorem 9.56, if $\alpha \in \mathcal{I}_p^{\leftarrow}$ then $\chi(\alpha,\nu) \in \mathcal{I}_p^{\leftarrow}$ and the map $(\alpha,\nu) \mapsto \chi(\alpha,\nu)$ is a bounded bilinear operator from $\mathcal{I}_p^{\leftarrow} \times \mathcal{AI}_p$ into $\mathcal{I}_p^{\leftarrow}$. By Proposition 9.55 with $\alpha \equiv 1\!\!1$ there, we have that $\chi(\widehat{\mu},\nu) \in \mathcal{I}_p$. But whether the map $\chi$ is defined on $\mathcal{I}_p \times \mathcal{AI}_p$ and if so, takes values in $\mathcal{I}_p$ and is a bounded bilinear operator, is not obvious (to us) in general. We will show (see (9.176) below) that there is a subspace $M_p \subset \overline{\mathcal{I}}_p$ such that $\chi$ takes $M_p \times \mathcal{AI}_p$ into $\mathcal{I}_p$. It turns out that $\widehat{\mu}$ and its Taylor polynomials and remainders, to be used in the proof, all are in $M_p$. On $M_p$ we will have a norm $\|\cdot\|_{\overline{[p]},W} \ge \|\cdot\|_{\overline{[p]}}$.

For an interval function $\alpha$ on $[a,b]$ and a nondegenerate interval $J \subset [a,b]$, let $W_p(\alpha) \equiv W_p(\alpha;J)$, where

$$W_p(\alpha) := \sup\Big(\sum_{i=2}^{n}\|\alpha((\cdot,b]\cap\bigcup_{j\le i}A_j) - \alpha((\cdot,b]\cap\bigcup_{j\le i-1}A_j)\|^p_{\bigcup_{j\le i-1}A_j,[p]}\Big)^{1/p},$$

the norm is for point functions restricted to $\bigcup_{j\le i-1}A_j$ in the $i$th term, and the supremum is taken over $\mathcal{A} = \{A_i\}_{i=1}^n \in \mathrm{IP}\,(\overline{J})$. Then $W_p(\cdot)$ is a seminorm and $\|\cdot\|^{\leftarrow}_{\overline{J},[p]} + W_p(\cdot)$ is a norm. If $\alpha = \widehat{\mu}$, using multiplicativity we have

$$\begin{aligned}
W_p(\widehat{\mu}) &= \sup_{\mathcal{A}}\Big(\sum_{i=2}^{n}\|[\widehat{\mu}(A_i) - \mathbb{I}]\widehat{\mu}((\cdot,b]\cap\bigcup_{j\le i-1}A_j)\|^p_{\bigcup_{j\le i-1}A_j,[p]}\Big)^{1/p}\\
&\le \sup_{\mathcal{A}}\Big(\sum_{i=2}^{n}\|\widehat{\mu}(A_i) - \mathbb{I}\|^p\|\widehat{\mu}((\cdot,b]\cap\bigcup_{j\le i-1}A_j)\|^p_{\bigcup_{j\le i-1}A_j,[p]}\Big)^{1/p}\\
&\le \sup_{\mathcal{A}}\Big(\sum_{i=2}^{n}\|\widehat{\mu}(A_i) - \mathbb{I}\|^p\Big)^{1/p}\sup_{A\in\mathfrak{I}(J)}\|\widehat{\mu}((\cdot,b]\cap A)\|_{A,[p]}\\
&\le V_p(\widehat{\mu})\sup_{A\in\mathfrak{I}(J)}\|\widehat{\mu}\|^{\leftarrow}_{\overline{A},[p]}
\end{aligned}$$

by definition of $V_p(\cdot)$, and by the first inequality in (9.97). Therefore for each nondegenerate $J \subset [a,b]$,

$$W_p(\widehat{\mu};J) \le V_p(\widehat{\mu};J)\|\widehat{\mu}\|^{\leftarrow}_{\overline{J},[p]} \le V_p(\widehat{\mu};[a,b])\|\widehat{\mu}\|^{\leftarrow}_{[p]} < \infty \tag{9.174}$$

by (9.84).

For boundedness of $\chi$ into $\mathcal{I}_p$ we will show that for any $\alpha \in \overline{\mathcal{I}}_p$,

$$v_p(\chi;J)^{1/p} \le K_p\|\alpha\|^{\leftarrow}_{\overline{J},[p]}\|\nu\|_{(p)} + K_pW_p(\alpha;J)\|\nu\|_{(p)}. \tag{9.175}$$

Indeed, let $B_1, B_2 \in \mathfrak{I}(J)$ be such that $B_1\cup B_2 \in \mathfrak{I}(J)$ and $B_1 \prec B_2$. Then

$$\chi(B_1\cup B_2) - \chi(B_1) = \chi(B_2) + \fint_{B_1}\big[\alpha((t,b]\cap(B_1\cup B_2)) - \alpha((t,b]\cap B_1)\big]\,\nu(\mathrm{d}t).$$

This also holds when $B_1 = \emptyset$. Taking $B_1 := \bigcup_{j\le i-1}A_j$, $B_2 := A_i$, and using the Minkowski inequality, we have since $\chi(\emptyset) = 0$,

$$s_p(\chi;\mathcal{A})^{1/p} \le V_p(\chi) + \Big(\sum_{i=2}^{n}\Big\|\fint_{B_1}\big[\alpha((\cdot,b]\cap(B_1\cup B_2)) - \alpha((\cdot,b]\cap B_1)\big]\,\mathrm{d}\nu\Big\|^p\Big)^{1/p}.$$

Then we apply Proposition 9.57 to the first term on the right side. To bound the second term, for $i = 2,\dots,n$, by the Love–Young inequality (9.78) with $q = p$, $f(t) := \alpha((t,b]\cap(B_1\cup B_2)) - \alpha((t,b]\cap B_1)$, $g \equiv \mathbb{I}$, $B := B_1 = \bigcup_{j\le i-1}A_j \in \mathfrak{I}[a,b]$, and $\nu$ in place of $\mu$, we have

$$\begin{aligned}
&\Big\|\fint_{B_1}\big[\alpha((\cdot,b]\cap(B_1\cup B_2)) - \alpha((\cdot,b]\cap B_1)\big]\,\mathrm{d}\nu\Big\|\\
&\qquad \le K_p\|\alpha((\cdot,b]\cap(B_1\cup B_2)) - \alpha((\cdot,b]\cap B_1)\|_{B_1,[p]}\|\nu\|_{(p)}.
\end{aligned}$$

Then taking $p$th powers and summing over $i = 2, \ldots, n$, we have

$$\Big(\sum_{i=2}^{n} \Big\| \fint_{B_1} [\alpha((\cdot, b] \cap (B_1 \cup B_2)) - \alpha((\cdot, b] \cap B_1)] \, d\nu \Big\|^p \Big)^{1/p}$$
$$\leq K_p \Big( \sum_{i=2}^{n} \|\alpha((\cdot, b] \cap (B_1 \cup B_2)) - \alpha((\cdot, b] \cap B_1)\|^p_{B_1,[p]} \Big)^{1/p} \|\nu\|_{(p)}$$
$$\leq K_p W_p(\alpha; J) \|\nu\|_{(p)},$$

and (9.175) follows.

Let $W_p(\alpha; \emptyset) := 0$. Since $\|\chi\|_{\sup} \leq V_p(\chi)$, by (9.175) we have

$$\|\chi\|_{[p]} \leq K_p \big[ 2\|\alpha\|_{\overline{[p]}} + \sup_{A \in \mathfrak{I}[a,b]} W_p(\alpha; A) \big] \|\nu\|_{(p)}. \tag{9.176}$$

For an interval function $\alpha$ let

$$\|\alpha\|_{\overline{[p]},W} := \|\alpha\|_{\overline{[p]}} + \sup_{A \in \mathfrak{I}[a,b]} W_p(\alpha; A) \ \leq \ +\infty$$

and let $M_p := \{\alpha \in \overline{\mathcal{I}}_p([a, b]; \mathbb{B}) \colon \|\alpha\|_{\overline{[p]},W} < \infty\}$. Then $\|\cdot\|_{\overline{[p]},W}$ is a norm on $M_p$ and $(\alpha, \nu) \mapsto \chi(\alpha, \nu)$ is bounded and bilinear from $M_p \times \mathcal{AI}_p$ into $\mathcal{I}_p$. Also, by (9.174) we have $\widehat{\mu} \in M_p$.

Let $\mathcal{AI}_p := \mathcal{AI}_p([a, b]; \mathbb{B})$ and likewise for $\mathcal{I}_p$ and $\overline{\mathcal{I}}_p$. We will show that $(\mu, \nu) \mapsto \chi(\widehat{\mu}, \nu)$ is uniformly entire from $\mathcal{AI}_p \times \mathcal{AI}_p$ into $\mathcal{I}_p$. We will find a Taylor expansion for $(\mu, \nu) \mapsto \chi(\widehat{\mu}, \nu)$ around $(0, 0)$. The following lemma will provide individual terms in the expansion. Recall the notation $Q_0^k$ defined by (9.93) with $Q_0^0 := \mathbb{1}$.

**Lemma 9.78.** *Let $1 \leq p < 2$. For each integer $k \geq 1$,*

*(a) the mapping $\chi_{(k)}$ defined by $\chi_{(k)}(\mu, \nu) := \chi(Q_0^k(\mu), \nu)$ is a $(k + 1)$-homogeneous polynomial from $\mathcal{AI}_p \times \mathcal{AI}_p$ into $\mathcal{I}_p$;*

*(b) for each $\nu \in \mathcal{AI}_p$, $\mu \mapsto \chi_{(k)}(\mu, \nu)$ is a $k$-homogeneous polynomial from $\mathcal{AI}_p$ into $\mathcal{I}_p$;*

*(c) for all $\mu$ and $\nu$ in $\mathcal{AI}_p$, and all $k \geq 1$,*

$$\|\chi_{(k)}(\mu, \nu)\|_{[p]} \leq 3K_p^4(k^7 + 3)(4eC_p)^k k!^{-1/p} \|\mu\|^k_{[p]} \|\nu\|_{(p)},$$

*where $K_p = 1 + \zeta(2/p)$ and $C_p = 1 + 4^{1/p}\zeta(2/p)$;*

*(d) for each $\mu \in \mathcal{AI}_p$, $\nu \mapsto \chi_{(k)}(\mu, \nu)$ is a bounded linear operator (homogeneous polynomial of degree 1) from $\mathcal{AI}_p$ into $\mathcal{I}_p$.*

*Proof.* First, (*b*) and (*c*) will be proved. For $k \geq 1$ and $\nu_1, \ldots, \nu_k \in \mathcal{AI}_p$, let $L_\mu^{k,\mathrm{sym}}(\nu_1, \ldots, \nu_k)$ be defined by (9.109) and (9.129) and let

$$M_\nu^k(\nu_1, \ldots, \nu_k) := \chi(L_0^{k,\mathrm{sym}}(\nu_1, \ldots, \nu_k), \nu) = \frac{1}{k!} \sum_\pi \chi(L_0^k(\nu_{\pi(1)}, \ldots, \nu_{\pi(k)}), \nu).$$

The mapping $M_\nu^k$ is $k$-linear and symmetric. It satisfies $\chi(Q_0^k(\mu),\nu) = M_\nu^k(\mu,\ldots,\mu)$. We will prove that $M_\nu^k$ is a mapping from the $k$-fold product $(\mathcal{AI}_p)^k$ into $\mathcal{I}_p$.

Let $k \geq 1$, $\nu_1,\ldots,\nu_k \in \mathcal{AI}_p$, $L_k := L_0^k(\nu_1,\ldots,\nu_k)$, $B_1, B_2 \in \mathfrak{I}[a,b]$, $B_1 \cup B_2 \in \mathfrak{I}[a,b]$, and $B_1 \prec B_2$. By (9.115) with $n = k$, and since $L_0^0 \equiv \mathbb{I}$, we have

$$L_k(B_1 \cup B_2) - L_k(B_1) = \sum_{l=1}^{k} L_0^l(\overline{\nu}_2)(B_2)L_0^{k-l}(\overline{\nu}_1)(B_1), \tag{9.177}$$

where $\overline{\nu}_2 := (\nu_{k-l+1},\ldots,\nu_k)$ and $\overline{\nu}_1 := (\nu_1,\ldots,\nu_{k-l})$. Let $A \in \mathfrak{I}[a,b]$. Let $\mathcal{A} = \{A_i\}_{i=1}^n \in \mathrm{IP}(A)$, $B_1 = \cup_{j<i}A_j$, and $B_2 = A_i$. Also, for each $x \in B_1$ let $B_{r,x} := B_r \cap (x,b]$ for $r = 1,2$ and note that (9.177) holds for $B_{r,x}$ in place of $B_r$, since its hypotheses do. By the Minkowski inequality (1.5), it follows that

$$\begin{aligned}
W_p(L_k;A) &\leq \sum_{l=1}^{k} \sup_{\mathcal{A}} \Big( \sum_{i=2}^{n} \|L_0^l(\overline{\nu}_2)(A_i)L_0^{k-l}(\overline{\nu}_1)((\cdot,b]\cap B_1)\|_{B_1,[p]}^p \Big)^{1/p} \\
&\leq \sum_{l=1}^{k} V_p(L_0^l(\overline{\nu}_2)) \sup_{B\in\mathfrak{I}(A)} \|L_0^{k-l}(\overline{\nu}_1)((\cdot,b]\cap B)\|_{B,[p]} \\
&\leq \sum_{l=1}^{k} V_p(L_0^l(\nu_{k-l+1},\ldots,\nu_k))\|L_0^{k-l}(\nu_1,\ldots,\nu_{k-l})\|_{\overleftarrow{[p]}}
\end{aligned}$$

by (9.96). Therefore we have

$$\sup_{A\in\mathfrak{I}[a,b]} W_p(L_k;A) \leq \sum_{l=1}^{k} V_p(L_0^l(\nu_{k-l+1},\ldots,\nu_k))\|L_0^{k-l}(\nu_1,\ldots,\nu_{k-l})\|_{\overleftarrow{[p]}}. \tag{9.178}$$

The bound (9.137) for $\|Q_0^k\|$ gives via Theorem 5.7 for the norm of a $k$-linear operator, into $\overline{\mathcal{I}}_p([a,b];\mathbb{B})$ in this case, and then Stirling's formula that for all $k = 0,1,\ldots,$

$$\|L_0^k\| \leq \frac{(k^3+1)K_p(2kC_p)^k}{k!^{(p+1)/p}} \leq \frac{K_p(k^3+1)(2\mathrm{e}C_p)^k}{k!^{1/p}}. \tag{9.179}$$

To bound $V_p(L_k)$ for $k \geq 1$, (9.112) gives in this case

$$V_p(L_k) \leq K_p\|L_{k-1}\|_{[p]}\|\nu_k\|_{(p)} \leq \frac{K_p^2(k^3+1)(2\mathrm{e}C_p)^k}{k!^{1/p}} \prod_{i=1}^{k} \|\nu_i\|_{(p)}. \tag{9.180}$$

Using (9.176) and (9.178), then (9.179) and (9.180), it follows that

$$\begin{aligned}
\|\chi(L_k,\nu)\|_{[p]} &\leq K_p \Big[ 2\|L_k\|_{\overline{[p]}} + \sup_{A\in\mathfrak{I}[a,b]} W_p(L_k;A) \Big] \|\nu\|_{(p)} \\
&\leq K_p\Big[2\|L_k\|_{\overline{[p]}} + V_p(L_k) + \sum_{l=1}^{k-1} V_p(L_0^l(\overline{\nu}_2))\|L_0^{k-l}(\overline{\nu}_1)\|_{\overleftarrow{[p]}}\Big]\|\nu\|_{(p)}.
\end{aligned}$$

Now by (9.180) applied to $l$ in place of $k$ and to $\overline{\nu}_2$, we have

$$V_p(L_0^l(\overline{\nu}_2)) \le K_p^2(l^3+1)(2\mathrm{e}C_p)^l\|\nu_{k-l+1}\|_{[p]}\cdots\|\nu_k\|_{[p]}/l!^{1/p},$$

and from (9.179) applied to $k-l$ in place of $k$ and to $\overline{\nu}_1$, we get

$$\|L_0^{k-l}(\overline{\nu}_1)\|_{\overline{[p]}} \le K_p((k-l)^3+1)(2\mathrm{e}C_p)^{k-l}\|\nu_1\|_{[p]}\cdots\|\nu_{k-l}\|_{[p]}/(k-l)!^{1/p}.$$

For each $l$ we have $1/(l!(k-l)!)^{1/p} \le 2^{k/p}/k!^{1/p}$. It can be verified that for $k>0$, $f(x) := (x^3+1)((k-x)^3+1)$ has $f(0)=f(k)=k^3+1$ and $f'(x)=3(k-2x)[x^2(k-x)^2-k]$, so $f'(0)=-3k^2=-f'(k)$. For $k\ge 3$, $f'$ has three simple zeros in $(0,k)$ of which the outer two are relative minima and $x=k/2$ is a relative maximum. It is the absolute maximum on $[0,k]$ for $k\ge 4$, and then

$$\sum_{l=1}^{k-1}(l^3+1)((k-l)^3+1) \le k\left(\left(\frac{k}{2}\right)^3+1\right)^2. \tag{9.181}$$

One can verify directly that the same bound holds for $k=1$ and 2. For $k=3$ it holds with a further factor of 3 on the right. From this and last four displays it follows that for $k\ge 1$

$$\begin{aligned}\|\chi(L_k,\nu)\|_{[p]} &\le K_p^4(2\mathrm{e}C_p)^k k!^{-1/p}\|\nu\|_{(p)}\|\nu_1\|_{[p]}\cdots\|\nu_k\|_{[p]}\times\\ &\quad\times\left[2(k^3+1)+(k^3+1)+3k2^{k/p}\left\{\left(\frac{k}{2}\right)^3+1\right\}^2\right].\end{aligned}$$

Routine calculations show that the quantity in square brackets is bounded above by $3k^72^k$ for $k\ge 4$ (it is quite easy to see that it is no larger in order of magnitude) and by $3(k^7+3)2^k$ for all $k\ge 1$. Thus for all $k\ge 1$,

$$\begin{aligned}\|M_\nu^k(\nu_1,\dots,\nu_k)\|_{[p]} &\le \frac{1}{k!}\sum_\pi\|\chi(L_0^k(\nu_{\pi(1)},\dots,\nu_{\pi(k)}),\nu)\|_{[p]}\\ &\le 3K_p^4(k^7+3)(4\mathrm{e}C_p)^k k!^{-1/p}\|\nu\|_{(p)}\|\nu_1\|_{[p]}\cdots\|\nu_k\|_{[p]}.\end{aligned}$$

Therefore the mapping $M_\nu^k$ takes the $k$-fold product $(\mathcal{AI}_p)^k$ into $\mathcal{I}_p$. Thus $(b)$ and $(c)$ follow, taking $\nu_1=\cdots=\nu_k=\mu$.

Clearly $\nu\mapsto M_\nu^k(\nu_1,\dots,\nu_k)$ is linear for fixed $\nu_1,\dots,\nu_k$. For $\mu_j$ and $\nu_j\in\mathcal{AI}_p$, $j=0,1,\dots,k$, let $\nu:=\mu_0$ and

$$Y^{k+1}((\mu_0,\nu_0),(\mu_1,\nu_1),\dots,(\mu_k,\nu_k)) := M_\nu^k(\nu_1,\dots,\nu_k).$$

Inequality $(c)$ implies that $Y^{k+1}$ is a bounded $(k+1)$-linear map from $(\mathcal{AI}_p\times\mathcal{AI}_p)^{k+1}$ into $\mathcal{I}_p$. It follows that

$$(\mu,\nu)\mapsto\chi(Q_0^k(\mu),\nu) = Y^{k+1}(\nu,\mu)^{\otimes(k+1)}$$

is a $(k+1)$-homogeneous polynomial, so $(a)$ holds. It follows that $(d)$ also holds, so the lemma is proved. □

Let $1 \le p < 2$ and $\mu, \nu \in \mathcal{AI}_p([a,b];\mathbb{B})$. Recall the notation $\gamma_n$ defined by (9.113). We have for each $n = 0, 1, \dots$ by Theorem 9.51,

$$\widehat{\mu} = \gamma_n(0,\mu) + \sum_{k=0}^{n} Q_0^k(\mu),$$

where each term is in $\overline{\mathcal{I}}_p$. For fixed $\mu$, the mapping $\nu \mapsto \chi(\widehat{\mu},\nu)$ is linear since the Kolmogorov integral is linear with respect to an integrator, and bounded from $\mathcal{AI}_p$ into $\mathcal{I}_p$ by (9.176) with $\alpha = \widehat{\mu}$ and (9.174).

By Lemma 9.78, $\chi(Q_0^k(\mu),\nu)$ is defined for each $k$. Then by bilinearity of the Kolmogorov integral and induction, for each integer $n \ge 1$ we have

$$\chi(\widehat{\mu},\nu) = \sum_{k=0}^{n} \chi(Q_0^k(\mu),\nu) + R_n(\mu,\nu), \tag{9.182}$$

where

$$R_n(\mu,\nu) = \chi(\gamma_n(0,\mu),\nu). \tag{9.183}$$

We need to find a suitable bound for $\|R_n(\mu,\nu)\|_{[p]}$.

For $\mu \in \mathcal{AI}_p$, recall $K(\mu,p)$ defined by (9.29), which is easily seen to be bounded if $\|\mu\|_{[p]}$ is bounded (although it grows rapidly as $\|\mu\|_{[p]}$ becomes large). We have $K(\mu,p) \ge 1$ for all $\mu$. Also recall $C_p$ as defined just after (9.28). By Theorem 9.36 we have $\|\widehat{\mu}\|_{\sup} \le K(\mu,p)$ and $\|\widehat{\mu}\|_{\overline{(p)}} \le C_p K(\mu,p)\|\widehat{\mu}\|_{\sup}\|\mu\|_{(p)}$. It follows that

$$\|\widehat{\mu}\|_{\overline{[p]}} \;\le\; N(\mu,p) \;:=\; K(\mu,p)[1 + C_p K(\mu,p)\|\mu\|_{(p)}], \tag{9.184}$$

which clearly also remains bounded for $\|\mu\|_{[p]}$ bounded. Also, $N(\mu,p) \ge 1$ for $1 \le p < 2$ and any $\mu$.

**Lemma 9.79.** *Let $\mu,\nu \in \mathcal{AI}_p([a,b];\mathbb{B})$, $1 \le p < 2$. There exists $n_0 = n_0(\mu,p)$ depending only on $\mu$ and $p$ such that for each integer $n \ge 1$ and $\gamma_n := \gamma_n(0,\mu)$,*

$$\begin{aligned}\|\chi(\gamma_n,\nu)\|_{[p]} \le B(n,\mu,\nu,p) &:= 4n_0 K_p^4 \|\nu\|_{(p)}[(n+n_0)^3+1]^2 \times \\ &\times \{2\mathrm{e}C_p(1+|\mu\|_{[p]})\}^{n+n_0}\frac{1}{n!^{1/p}}\left[1 + N(\mu,p) + n2^{(n+1)/p}\right].\end{aligned} \tag{9.185}$$

*Proof.* Let $n \ge 1$. By (9.176) with $\alpha = \gamma_n$, we have

$$\|\chi(\gamma_n,\nu)\|_{[p]} \le K_p\left[2\|\gamma_n\|_{\overline{[p]}} + \sup_{A \in \mathfrak{I}[a,b]} W_p(\gamma_n;A)\right]\|\nu\|_{(p)}. \tag{9.186}$$

By Theorem 9.51 we have

$$\gamma_n = \gamma_n(0,\mu) \equiv \sum_{k=n+1}^{\infty} Q_0^k(\mu), \tag{9.187}$$

where the sum converges in $\overline{\mathcal{I}}_p$. From (9.137) we get for all $k \geq 0$,

$$\|Q_0^k(\mu)\|_{\overline{[p]}} \leq T_k := T_k(\mu, p) := (k^3+1)K_p(2C_p\|\mu\|_{[p]})^k / k!^{1/p}. \tag{9.188}$$

It is easy to check that $(k+1)^3 + 1 \leq 2k^3$ for all $k \geq 4$. Recall that $\lfloor x \rfloor$ is the largest integer $\leq x$ and that $N(\mu, p) \geq 1$. For each $k \geq 4$ we have

$$T_{k+1}/T_k \leq 4C_p\|\mu\|_{[p]}/(k+1)^{1/p},$$

which is $\leq 1/2$ if $k \geq n_0$ defined by

$$n_0 := n_0(\mu, p) := \lfloor \max\{3 + N(\mu, p), (8C_p\|\mu\|_{[p]})^p\} \rfloor \geq 4, \tag{9.189}$$

where the $N(\mu, p)$ is included for later convenience. If $n \geq n_0(\mu, p)$ it then follows via domination by a geometric series with ratio $1/2$ and (9.188) that for $n \geq n_0(\mu, p)$,

$$\|\gamma_n(0,\mu)\|_{\overline{[p]}} \ \leq \ 2T_{n+1} \ = \ \frac{2K_p((n+1)^3+1)(2C_p\|\mu\|_{[p]})^{n+1}}{(n+1)!^{1/p}}. \tag{9.190}$$

Suppose on the other hand that $1 \leq n < n_0 = n_0(\mu, p)$. Let

$$S_n := S_n(\mu, p) := \sum_{k=n+1}^{n_0} T_k.$$

In this sum the number of terms is at most $n_0$ and each term, bounding each factor separately, satisfies

$$T_k \leq K_p(n_0^3 + 1)\left[2C_p(1 + \|\mu\|_{[p]})\right]^{n_0} / (n+1)!^{1/p}, \qquad n < k \leq n_0.$$

Thus $S_n$ is bounded above by $n_0$ times the latter bound. Further, applying (9.190) with $n$ there replaced by $n_0$, for which it holds, we get that if $0 \leq n < n_0$ then

$$\begin{aligned}
&\|\gamma_n\|_{\overline{[p]}} \leq S_n + 2T_{n_0+1} \\
&\leq \frac{n_0 K_p(n_0^3+1)\left[2C_p(1+\|\mu\|_{[p]})\right]^{n_0}}{(n+1)!^{1/p}} + \frac{2K_p((n_0+1)^3+1)\left[2C_p\|\mu\|_{[p]}\right]^{n_0+1}}{(n_0+1)!^{1/p}} \\
&\leq (n_0+1)K_p((n_0+1)^3+1)\left[2C_p(1+\|\mu\|_{[p]})\right]^{n_0+1} / (n+1)!^{1/p},
\end{aligned}$$

since $n_0 > 1$. Combining with (9.190) gives for all $n \geq 1$,

$$\|\gamma_n(0,\mu)\|_{\overline{[p]}} \ \leq \ 2n_0 K_p((n+n_0)^3+1)[2C_p(1+\|\mu\|_{[p]})]^{n+n_0}/(n+1)!^{1/p}. \tag{9.191}$$

By Lemma 9.59 with $\mu = 0$ there, and $\nu$ there equal to $\mu$ here, we have for all $A \in \mathfrak{I}[a,b]$ and $n \geq 1$,

$$\gamma_n(A) = \oint_A \mu(\mathrm{d}s)\gamma_{n-1}([a,s)\cap A).$$

It follows from Proposition 9.57 that for all $n \geq 2$,

$$V_p(\gamma_n) \leq K_p\|\mu\|_{(p)}\|\gamma_{n-1}(0,\mu)\|_{[p]}$$

and thus that

$$V_p(\gamma_n) \;\leq\; 2n_0K_p^2((n+n_0)^3+1)[2C_p(1+\|\mu\|_{[p]})]^{n+n_0}/n!^{1/p}. \tag{9.192}$$

To bound $W_p(\gamma_n;A)$, let $A \in \mathfrak{I}[a,b]$ and $\mathcal{A} = \{A_i\}_{i=1}^m \in \mathrm{IP}\,(A)$. For $B_1 := \bigcup_{j<i} A_j$ and $B_2 := A_i$, by (9.116), we have

$$\gamma_n(B_1\cup B_2) - \gamma_n(B_1) = \gamma_n(B_2)\widehat{\mu}(B_1) + \sum_{k=1}^n Q_k(B_2)\gamma_{n-k}(B_1), \tag{9.193}$$

where $Q_k := Q_0^k(\mu)$. As in the last proof, for each $x \in B_1$ let $B_{r,x} := B_r \cap (x,b]$ for $r = 1,2$ and note that (9.193) holds for $B_{r,x}$ in place of $B_r$, since its hypotheses do, with $B_{2,x} \equiv B_2 = A_i$. Thus by the Minkowski inequality (1.5) and then by (9.96), it follows that

$$\begin{aligned} W_p(\gamma_n;A) &\leq \sup_{\mathcal{A}}\Big(\sum_{i=2}^m \|\gamma_n(A_i)\|^p\Big)^{1/p} \sup_{B\in\mathfrak{I}(A)} \|\widehat{\mu}((\cdot,b]\cap B)\|_{B,[p]} \\ &\quad + \sum_{k=1}^n \sup_{\mathcal{A}}\Big(\sum_{i=2}^m \|Q_k(A_i)\|^p\Big)^{1/p} \sup_{B\in\mathfrak{I}(A)} \|\gamma_{n-k}((\cdot,b]\cap B)\|_{B,[p]} \\ &\leq V_p(\gamma_n)\|\widehat{\mu}\|_{\overleftarrow{[p]}} + \sum_{k=1}^n V_p(Q_k)\|\gamma_{n-k}\|_{\overleftarrow{[p]}}. \end{aligned}$$

Inserting this bound into (9.186), we have

$$\|\chi(\gamma_n,\nu)\|_{[p]} \leq K_p\Big[2\|\gamma_n\|_{\overline{[p]}} + V_p(\gamma_n)\|\widehat{\mu}\|_{\overleftarrow{[p]}} + \sum_{k=1}^n V_p(Q_k)\|\gamma_{n-k}\|_{\overleftarrow{[p]}}\Big]\|\nu\|_{(p)}. \tag{9.194}$$

We bound $V_p(Q_k)$ by (9.180), which gives

$$V_p(Q_k) \;\leq\; K_p^2(k^3+1)(2\mathrm{e}C_p)^k\|\mu\|_{[p]}^k/k!^{1/p}. \tag{9.195}$$

We can bound $\|\gamma_{n-k}\|_{\overleftarrow{[p]}}$ by (9.191) with $n-k$ in place of $n$ for $k<n$, giving

$$\|\gamma_{n-k}\|_{\overline{[p]}} \;\leq\; \frac{2n_0K_p((n-k+n_0)^3+1)[2C_p(1+\|\mu\|_{[p]})]^{n-k+n_0}}{(n-k+1)!^{1/p}}. \tag{9.196}$$

For $k = n$, we get

$$\|\gamma_0\|_{\overline{[p]}} = \|\widehat{\mu} - \mathbb{1}\|_{\overline{[p]}} \le \|\widehat{\mu}\|_{\overline{[p]}} + 1 \le N(\mu, p) + 1 \le n_0(\mu, p)$$

by (9.184) and the definition (9.189) of $n_0$. Thus (9.196) holds for $k = n$ also and so for all $k = 1, \ldots, n$.

As in previous proofs we have $1/(k!(n-k+1)!)^{1/p} \le 2^{(n+1)/p}/(n+1)!^{1/p}$. Thus the sum in (9.194) is bounded above by

$$2n_0 K_p^3 (2\mathrm{e}C_p\{1 + \|\mu\|_{[p]}\})^{n+n_0} 2^{(n+1)/p} n!^{-1/p} \sum_{k=1}^{n} (k^3 + 1)((n + n_0 - k)^3 + 1). \tag{9.197}$$

As in (9.181) and since $n_0 \ge 4$, the sum in (9.197) is bounded above by

$$n[\{(n + n_0)/2\}^3 + 1]^2 \tag{9.198}$$

for all $n \ge 1$. Inserting (9.198) in place of the sum in (9.197) and then (9.184), (9.197), and (9.192) into (9.194) gives for $n \ge 1$, $\|\chi(\gamma_n, \nu)\|_{[p]} \le B(n, \mu, \nu, p)$ as defined in (9.185). Thus (9.185) is proved. The proof of the lemma is complete. □

It follows from Lemma 9.79 that for each $\mu, \nu \in \mathcal{AI}_p$ and $B(n, \mu, \nu, p)$ as defined in (9.185), as $n \to \infty$,

$$\frac{B(n+1, \mu, \nu, p)}{B(n, \mu, \nu, p)} \sim \frac{2^{(p+1)/p}\mathrm{e}C_p\{1 + \|\mu\|_{[p]}\}}{(n+1)^{1/p}} \to 0.$$

Thus for any $\mu, \nu \in \mathcal{AI}_p$, $\|\chi(\gamma_n, \nu)\|_{[p]} \to 0$ as $n \to \infty$. From this, (9.182), and (9.183) it follows that

$$\chi(\widehat{\mu}, \nu) = \nu + \sum_{k=1}^{\infty} \chi(Q_0^k(\mu), \nu), \tag{9.199}$$

where the series converges absolutely in $\mathcal{I}_p$ for all $\mu$ and $\nu$ in $\mathcal{AI}_p$ and uniformly for $\|\mu\|_{[p]}$ and $\|\nu\|_{[p]}$ bounded. By Lemma 9.78, it is the Taylor expansion around $(0, 0)$ of the mapping $(\mu, \nu) \mapsto \chi(\widehat{\mu}, \nu)$ from $\mathcal{AI}_p \times \mathcal{AI}_p$ into $\mathcal{I}_p$. By the Cauchy–Hadamard formula (Theorem 5.10), it has infinite radius of uniform convergence, and so by Theorem 5.21 and in light of Lemma 9.78 the function $(\mu, \nu) \mapsto \chi(\widehat{\mu}, \nu)$ is uniformly entire from $\mathcal{AI}_p \times \mathcal{AI}_p$ into $\mathcal{I}_p$.

From this and composition of uniformly entire functions (Theorem 5.38), taking $\chi(\widehat{\mu}, \nu)$, then the bounded linear operator $\chi \mapsto \chi([a, t))$ from $\mathcal{I}_p$ into $\mathcal{W}_p$, we get that the second term in (9.155) is a uniformly entire function of $(\mu, \nu)$. The first term, in other words $F_1(y, \mu)$ as defined in (9.172), was seen there to be a uniformly entire function of $(y, \mu)$. Since the sum of two uniformly entire functions of $(y, \mu, \nu)$ is uniformly entire, the conclusion of the theorem for the solution (9.155) of the forward equation (9.154) follows.

The other forward equation (9.156) has a solution given by (9.157). To adapt the proof to that case we need to replace $[a, \cdot)$ by $(a, \cdot]$ in (9.172) and $[a, t)$ by $(a, t]$ in the preceding paragraph.

To prove that the solution mapping

$$\mathbb{B} \times \mathcal{AI}_p \times \mathcal{AI}_p \ni (y, \mu, \nu) \mapsto \mathcal{S}^{\prec}_{(\cdot, b]}(y, \mu, \nu) := g_3 \in \mathcal{W}_p$$

corresponding to the backward integral equation (9.165), and given by (9.166), is uniformly entire we make the following changes. Let $\nu \in \mathcal{AI}_p([a, b]; \mathbb{B})$. For an interval function $\alpha$ on $[a, b]$, let $\chi^{\leftarrow}(\nu, \alpha)(A) := \oint_A d\nu\, \alpha([a, \cdot) \cap A)$ provided the Kolmogorov integral is defined for each subinterval $A \subset [a, b]$. Then if defined, $\chi^{\leftarrow} \equiv \chi^{\leftarrow}(\nu, \alpha) = \{\chi^{\leftarrow}(\nu, \alpha)(A) \colon A \in \mathfrak{I}[a, b]\}$ is an interval function on $[a, b]$. The function $t \mapsto \chi^{\leftarrow}(\nu, \widehat{\mu})((t, b])$, $t \in [a, b]$, gives the second term on the right in (9.166), and is the composition of the interval function $\chi^{\leftarrow}(\nu, \widehat{\mu})$ with the map $t \mapsto (t, b]$ from $[a, b]$ into $\mathfrak{I}[a, b]$. Here the map $\beta \mapsto \beta((\cdot, b])$ is a bounded linear operator with norm 1 from $\mathcal{I}_p^{\leftarrow}([a, b]; \mathbb{B})$ into $\mathcal{W}_p([a, b]; \mathbb{B})$. Since the product integral operator is uniformly entire from $\mathcal{AI}_p$ into $\mathcal{I}_p^{\leftarrow}$, the function $(y, \mu) \mapsto y\widehat{\mu}((\cdot, b])$ is uniformly entire from $\mathbb{B} \times \mathcal{AI}_p$ into $\mathcal{W}_p$, and this gives the first term on the right in (9.166). To prove that the second term on the right in (9.166) is uniformly entire it is enough to show that the function $(\mu, \nu) \mapsto \chi^{\leftarrow}(\nu, \widehat{\mu})$ is uniformly entire from $\mathcal{AI}_p \times \mathcal{AI}_p$ into $\mathcal{I}_p^{\leftarrow}$.

For an interval function $\alpha$ on $[a, b]$ and a nondegenerate interval $J \subset [a, b]$, let $W_p^{\leftarrow}(\alpha) \equiv W_p^{\leftarrow}(\alpha; J)$, where

$$W_p^{\leftarrow}(\alpha) := \sup_{\mathcal{A}} \Big( \sum_{i=1}^{n-1} \|\alpha([a, \cdot) \cap \bigcup_{j \geq i} A_j) - \alpha([a, \cdot) \cap \bigcup_{j \geq i+1} A_j)\|^p_{\bigcup_{j \geq i+1} A_j, [p]} \Big)^{1/p},$$

the norm is for point functions restricted to $\bigcup_{j \geq i+1} A_j$ in the $i$th term, and the supremum is taken over $\mathcal{A} = \{A_i\}_{i=1}^n \in \mathrm{IP}(J)$. Let $W_p^{\leftarrow}(\nu, \alpha)(\emptyset) := 0$. Then as for (9.176), we obtain the bound

$$\|\chi^{\leftarrow}(\nu, \alpha)\|_{[p]}^{\leftarrow} \leq K_p \Big[ 2\|\alpha\|_{\overline{[p]}} + \sup_{J \in \mathfrak{I}[a, b]} W_p^{\leftarrow}(\alpha; J) \Big] \|\nu\|_{(p)}. \tag{9.200}$$

For an interval function $\alpha$ let

$$\|\alpha\|_{[p], W}^{\leftarrow} := \|\alpha\|_{\overline{[p]}} + \sup_{J \in \mathfrak{I}[a, b]} W_p^{\leftarrow}(\alpha; J) \ \leq \ +\infty$$

and let $M_p^{\leftarrow} := \{\alpha \in \overline{\mathcal{I}}_p([a, b]; \mathbb{B}) \colon \|\alpha\|_{[p], W}^{\leftarrow} < \infty\}$. Then $\|\cdot\|_{[p], W}^{\leftarrow}$ is a norm on $M_p^{\leftarrow}$ and $(\nu, \alpha) \mapsto \chi^{\leftarrow}(\nu, \alpha)$ is bounded and bilinear from $M_p^{\leftarrow} \times \mathcal{AI}_p$ into $\mathcal{I}_p^{\leftarrow}$. As in Lemma 9.78 we obtain that the mapping $\chi_{(k)}^{\leftarrow}$ defined by $\chi_{(k)}^{\leftarrow}(\mu, \nu) := \chi^{\leftarrow}(\nu, Q_0^k(\mu))$ is a $(k + 1)$-homogeneous polynomial from $\mathcal{AI}_p \times \mathcal{AI}_p$ into $\mathcal{I}_p^{\leftarrow}$

and the bound in Lemma 9.78($c$) holds when $\chi_{(k)}$ is replaced by $\chi^{\leftarrow}_{(k)}$. As for (9.182), we have

$$\chi^{\leftarrow}(\nu, \widehat{\mu}) = \sum_{k=0}^{n} \chi^{\leftarrow}(\nu, Q_0^k(\mu)) + \chi^{\leftarrow}(\nu, \gamma_n(0, \mu))$$

for each $n \geq 1$. We show that $\|\chi^{\leftarrow}(\nu, \gamma_n(0, \mu))\|^{\leftarrow}_{[p]}$ tends to zero as $n \to \infty$ rapidly enough as in Lemma 9.79 except that now we use the bound (9.200) instead of (9.176). Then it follows that the function $(\mu, \nu) \mapsto \chi^{\leftarrow}(\nu, \widehat{\mu})$ is uniformly entire from $\mathcal{AI}_p \times \mathcal{AI}_p$ into $\mathcal{I}_p^{\leftarrow}$ as desired.

For the other backward equation (9.167) with solution (9.168) we replace $(t, b]$ by $[t, b)$ and $(\cdot, b]$ by $[\cdot, b)$ in the above proof. The proof of the theorem is complete. □

## 9.12 Integral Equations for Banach-Space-Valued Functions

If $\mathbb{B}_2$ is a Banach algebra and a subset $\mathbb{B}_1 \subset \mathbb{B}_2$ is also a Banach algebra, with the same algebra operations and norm, then $\mathbb{B}_1$ will be called a Banach subalgebra of $\mathbb{B}_2$. Let $Y$ be a Banach space and let $\mathbb{B}_Y := L(Y, Y)$ be the Banach algebra of all bounded linear operators from $Y$ into itself, as in Example 4.10. Let $\mathbb{B}$ be a Banach subalgebra of $\mathbb{B}_Y$. Let $\mu$ be a $\mathbb{B}$-valued additive interval function on $[a, b]$ and let $y \in Y$. We will show that the linear integral equation for an unknown function $\phi\colon [a, b] \to Y$,

$$\phi(t) = y + \oint_{[a,t)} \mathrm{d}\mu\, \phi, \qquad a \leq t \leq b, \tag{9.201}$$

can be solved for $\phi$ given $y$ and $\mu \in \mathcal{AI}_p([a, b]; \mathbb{B})$, $1 \leq p < 2$, using the results of the preceding section.

By Theorem 9.66 with $y = \mathbb{I}$ there, the linear integral equation (9.143) has the solution $f_1(t) = \widehat{\mu}([a, t))$, $t \in [a, b]$, unique in $\mathcal{W}_q([a, b]; \mathbb{B}_Y)$ for $q \in \mathcal{U}_p$. It follows that $\phi(t) = f_1(t)y$ gives a solution of (9.201).

**Corollary 9.80.** *Let $Y$ be a Banach space, let $\mathbb{B}$ be a Banach subalgebra of $\mathbb{B}_Y$, let $\mu \in \mathcal{AI}_p([a, b]; \mathbb{B})$ for $1 \leq p < 2$, and let $y \in Y$. Then the following two statements hold:*

(*a*) *The integral equation*

$$\phi(t) = y + \oint_{[a,t)} \mathrm{d}\mu\, \phi, \qquad a \leq t \leq b, \tag{9.202}$$

*has the solution $\phi_{(1)}(t) = \widehat{\mu}([a, t))y$ for $a \leq t \leq b$, $\phi_{(1)} \in \mathcal{W}_p([a, b]; \mathbb{B})$, and $\phi_{(1)}$ is the unique solution in $\mathcal{W}_q([a, b]; \mathbb{B})$ for $q \in \mathcal{U}_p$.*

(*b*) *The integral equation*

$$\phi(t) = y + \fint_{(a,t]} \mathrm{d}\mu\,\phi_-, \qquad a \le t \le b, \tag{9.203}$$

*has the solution* $\phi_{(2)}(t) = \widehat{\mu}((a,t])y$ *for* $a \le t \le b$, $\phi_{(2)} \in \mathcal{W}_p([a,b];\mathbb{B})$, *and* $\phi_{(2)}$ *is the unique solution in* $\mathcal{W}_q([a,b];\mathbb{B})$ *for* $q \in \mathcal{U}_p$.

We already have shown the existence of a solution of the integral equation (9.202). Likewise we get a solution of (9.203). The uniqueness statements can be proved as in the proof of Theorem 9.66(*a*) except that here instead of Lemma 9.65 we use the following analogue under the present assumptions.

**Lemma 9.81.** *Let* $Y$ *be a Banach space and let* $\mathbb{B}$ *be a Banach subalgebra of* $\mathbb{B}_Y$. *Let* $\mu \in \mathcal{AI}_p([a,b];\mathbb{B})$ *for* $1 \le p < 2$, *and let* $q \in \mathcal{U}_p$. *If* $g \in \mathcal{W}_q([a,b];Y)$ *and either* $g(t) = \fint_{(a,t)} \mathrm{d}\mu\, g$ *for* $a \le t \le b$, *or* $g(t) = \fint_{(a,t]} \mathrm{d}\mu\, g_-$ *for* $a \le t \le b$, *then* $g \equiv 0$ *on* $[a,b]$.

We omit the proof of this lemma because it is the same as the proof of Lemma 9.65.

The next statement gives solutions of some forward inhomogeneous linear integral equations. The given functions $\phi_3$ and $\phi_4$ are solutions by way of Theorem 9.71, and the uniqueness statements follow from Lemma 9.81.

**Corollary 9.82.** *Let* $Y$ *be a Banach space, let* $\mathbb{B}$ *be a Banach subalgebra of* $\mathbb{B}_Y$, *let* $\nu \in \mathcal{AI}_p([a,b];Y)$, *let* $\mu \in \mathcal{AI}_p([a,b];\mathbb{B})$ *for* $1 \le p < 2$, *and let* $y \in Y$. *Then the following two statements hold:*

(*a*) *The integral equation*

$$\phi(t) = y + \fint_{[a,t)} \mathrm{d}\mu\,\phi + \nu([a,t)), \qquad a \le t \le b, \tag{9.204}$$

*has the solution* $\phi_3$, *in* $\mathcal{W}_p([a,b];Y)$ *and unique in* $\mathcal{W}_q([a,b];Y)$ *for* $q \in \mathcal{U}_p$, *where*

$$\phi_3(t) := \widehat{\mu}([a,t))y + \fint_{[a,t)} \widehat{\mu}((\cdot,t))\,\mathrm{d}\nu, \qquad a \le t \le b. \tag{9.205}$$

(*b*) *The integral equation*

$$\phi(t) = y + \fint_{(a,t]} \mathrm{d}\mu\,\phi_- + \nu((a,t]), \qquad a \le t \le b,$$

*has the solution* $\phi_4$, *in* $\mathcal{W}_p([a,b];Y)$ *and unique in* $\mathcal{W}_q([a,b];Y)$ *for* $q \in \mathcal{U}_p$, *where*

$$\phi_4(t) := \widehat{\mu}((a,t])y + \fint_{(a,t]} \widehat{\mu}((\cdot,t])\,\mathrm{d}\nu, \qquad a \le t \le b.$$

**Example 9.83.** Suppose the hypotheses of the preceding corollary hold with $\mathbb{B} = \mathbb{B}_Y$ and that for any two intervals $A, B \subset [a, b]$, $\mu(A)\mu(B) = \mu(B)\mu(A)$. Let $S$ be the range of $\mu$ in $\mathbb{B}_Y$. There is a smallest algebra $\mathbb{A}$ including $S$, namely the set of all finite linear combinations of finite products of elements of $S$, and $\mathbb{A}$ is commutative. The closure $\mathbb{B}$ of $\mathbb{A}$ in $\mathbb{B}_Y$ is a commutative Banach subalgebra of $\mathbb{B}_Y$, and we can apply the Corollary to $\mathbb{B}$. If $\mu(\{t\}) \neq -\mathbb{I}$ for all $t \in [a, b]$ then $\widehat{\mu}$ has an explicit form given by Theorem 9.40.

*Linear differential equations.*

Let $Y$ be a Banach space, and again let $\mathbb{B}_Y := L(Y, Y)$ be the Banach algebra of all bounded linear operators from $Y$ into itself. By analogy with stochastic differential equations (e.g. [108]), if $\mu \in \mathcal{AI}_p([a, b]; \mathbb{B}_Y)$ and $\nu \in \mathcal{AI}_p([a, b]; Y)$ with $1 \le p < 2$, we will say that the *formal linear differential equation*

$$\mathrm{d}\phi = \mathrm{d}\mu\, \phi + \mathrm{d}\nu \tag{9.206}$$

holds on $[a, b]$, where $\phi\colon\ [a, b] \to Y$, if and only if the integral equation (9.204) holds for $a \le t \le b$ with $y := \phi(a)$. For any $y \in Y$, such an equation has a solution $\phi = \phi_3$ with $\phi(a) = y$, unique in $\mathcal{W}_q$ for $q \in \mathcal{U}_p$, given by (9.205), where $\widehat{\mu}\colon\ \mathfrak{I}[a, b] \to \mathbb{B}_Y$.

Let $\lambda$ be Lebesgue measure, $g(\cdot) \in \mathcal{L}^1([a, b], \lambda; \mathbb{B}_Y)$, and $h \in \mathcal{L}^1([a, b], \lambda; Y)$. We will say that the linear differential equation

$$\frac{\mathrm{d}\phi}{\mathrm{d}t} = g\phi + h \tag{9.207}$$

holds on $[a, b]$ if $\phi$ is continuous from $[a, b]$ into $Y$, $g\phi + h \in \mathcal{L}^1([a, b], \lambda; Y)$, $\mathrm{d}\phi(t)/\mathrm{d}t$ exists in $Y$ for $\lambda$-almost all $t \in (a, b)$, and we have the Bochner integrals

$$\phi(t) - \phi(a) = (Bo)\int_{[a,t]} (g\phi + h)\, \mathrm{d}\lambda, \qquad \text{for } a \le t \le b. \tag{9.208}$$

The (a) continuity of $\phi$ and (b) existence of $\mathrm{d}\phi(t)/\mathrm{d}t$ equal to $g\phi + h$ for $\lambda$-almost all $t$ actually follow from the other assumptions, (a) from dominated convergence for Bochner integrals, and (b) by an extension of a theorem of Lebesgue to Banach-valued functions, given as Theorem 2.35.

**Theorem 9.84.** *Under the hypotheses of (9.207) and (9.208), let*

$$\mu(A) := I(g)(A) := (Bo) \int_A g(s)\, \mathrm{d}\lambda(s), \quad \nu(A) := I(h)(A) := (Bo) \int_A h(s)\, \mathrm{d}\lambda(s)$$

*for each $A \in \mathfrak{I}[a, b]$. Then $\mu \in \mathcal{AI}_1([a, b]; \mathbb{B}_Y)$ and $\nu \in \mathcal{AI}_1([a, b]; Y)$. For any $y \in Y$, the solution $\phi_3$ of (9.204) given by Corollary 9.82 is also a solution of (9.207), in other words of (9.208), with $\phi_3(a) = y$, and is the unique continuous solution $\phi$ with $\phi(a) = y$.*

*The mapping $\Phi_3$ defined by $\Phi_3(y, g, h) := \phi_3(\cdot)$ is a uniformly entire function from the Banach space $Y \times \mathcal{L}^1([a, b], \lambda; \mathbb{B}_Y) \times \mathcal{L}^1([a, b], \lambda; Y)$ with the norm $\|(y, g, h)\| := \|y\| + \|g\|_1 + \|h\|_1$ into $C([a, b]; Y)$ with the sup norm.*

**Remark 9.85.** The existence of a solution of a first-order linear differential equation with Bochner integrable coefficients (9.207) is known, see e.g. Daleckiĭ and Kreĭn [35, §3.1]. It is solved via a corresponding integral equation [35, (3.1.2)], i.e. (9.208) above. In [35, (3.1.6)] the solution (of a somewhat more general integral equation) is written in the form of a Peano series similar to that in Corollary 9.52, but in a form adapted to solving inhomogeneous equations. After the proof, Daleckiĭ and Kreĭn's form of the solution will be given and compared with ours.

*Proof.* By Proposition 2.33, $\mu$ and $\nu$ are well-defined additive interval functions satisfying

$$\|\mu(A)\| \le \int_A \|g(s)\| \mathrm{d}s \quad \text{and} \quad \|\nu(A)\| \le \int_A \|h(s)\| \mathrm{d}s \tag{9.209}$$

for all $A \in \mathfrak{I}[a, b]$. It follows that $\mu \in \mathcal{AI}_1([a, b]; \mathbb{B}_Y)$ and $\nu \in \mathcal{AI}_1([a, b]; Y)$. So $\mu$ and $\nu$ satisfy the hypotheses of Corollary 9.82 for $p = 1$. Thus there is a unique function $\phi_3 \in \mathcal{W}_\infty([a, b]; Y)$, in other words a regulated function into $Y$, satisfying (9.204). Then $\phi_3(a) = y$. Also, $\phi_3$ is actually continuous in light of (9.209). To show that $\phi_3$ satisfies (9.208), for each $t \in [a, b]$, we have

$$(Bo) \int_{[a,t]} (g\phi_3 + h)\mathrm{d}\lambda = (Bo) \int_{[a,t]} g\phi_3 \mathrm{d}\lambda + (Bo) \int_{[a,t]} h\mathrm{d}\lambda$$

where the Bochner integrals all exist since $\phi_3$ is bounded and measurable into $Y$ and both $g$ and $h$, thus $g\phi_3$, are Bochner integrable. We have $\nu([a, t)) = (Bo) \int_{[a,t]} h\mathrm{d}\lambda$ by definition of $\nu$ and since $\lambda(\{t\}) = 0$. By (9.204), it will suffice to show that for each $t \in [a, b]$,

$$\oint_{[a,t)} \mathrm{d}\mu \, \phi_3 = (Bo) \int_{[a,t]} g \, \phi_3 \mathrm{d}\lambda. \tag{9.210}$$

Let $(\{A_j\}_{j=1}^k, \{t_j\}_{j=1}^k)$ be any tagged interval partition of $[a, b]$. Then if $\phi_3$ is replaced in (9.210) by the function $\psi := \sum_{j=1}^k 1_{A_j}(\,)\phi_3(t_j)$, the resulting equation holds since we get equal $j$th terms $(Bo) \int_{A_j} g(s)\mathrm{d}\lambda(s) \, \phi_3(t_j)$ on both sides by definition of $\mu$. Now, $\phi_3$, being continuous, can be approximated arbitrarily closely in the sup norm by a sequence of such functions $\psi = \psi_n$, letting the mesh of $\{A_j\}_{j=1}^k$ approach 0. The integrals on each side with $\psi_n$ converge to those for $\phi_3$, on the left side by a Love–Young inequality for $p = 1$ and $q = \infty$, Proposition 3.96, so (9.210) holds. Here $\mathrm{d}\phi_3(t)/\mathrm{d}t$ exists for $\lambda$-almost all $t \in (a, b)$ by Theorem 2.35.

To prove the uniqueness for given $\phi(a)$, let $\phi$ be any continuous solution of (9.207), in other words of (9.208) for the same $y = \phi(a) = \phi_3(a)$. Then for

$\xi := \phi - \phi_3$, we have $\xi(t) = (Bo)\int_{[a,t]} g\xi \, d\lambda$, $a \le t \le b$. Given $0 < \epsilon < 1$ there exists a $\delta > 0$ such that $\int_{[a,a+\delta]} \|g\| \, d\lambda < \epsilon$. Let $s := \sup_{a\le t\le a+\delta} \|\xi(t)\|$. Then $s \le \epsilon s$, and so $s = 0$. Since $\xi$ is continuous it follows that $\xi \equiv 0$ on $[a, a+\delta]$. Iterating this argument as in the proof of Lemma 9.65, it follows that $\xi \equiv 0$ on $[a, b]$, and hence $\phi_3$ is the unique solution of (9.207).

As to analyticity, the maps $I: \; g \mapsto I(g) = \mu$ and $I: \; h \mapsto I(h) = \nu$ are bounded linear operators (with norm 2), thus uniformly entire. Via Theorem 9.76 and composition with the maps $I(\cdot)$, using Theorem 5.38, we get that $\Phi_3$ is uniformly entire. This completes the proof of the theorem. □

To compare the solution as given by Daleckiĭ and Kreĭn [35, (3.1.6)] with ours, for each $t \in J := [a, b]$, $g \in \mathcal{L}_1(J; \mathbb{B}_Y) \equiv \mathcal{L}_1(J, \lambda; \mathbb{B}_Y)$, $h \in \mathcal{L}_1(J; Y) \equiv \mathcal{L}_1(J, \lambda; Y)$, and $k \ge 1$, let

$$U_k(t) := U_k(g, y)(t) := \int_a^t \int_a^{t_k} \cdots \int_a^{t_2} g(t_k) \cdots g(t_1) \, dt_1 \cdots dt_k \, y$$

and

$$V_k(t) := V_k(g, h)(t) := \int_a^t \int_a^{t_k} \cdots \int_a^{t_1} g(t_k) \cdots g(t_1) h(u) \, du \, dt_1 \cdots dt_k,$$

where we write $\int_a^s$ instead of $(Bo)\int_{[a,s]}$. Let $U_0(t) := U_0(g, y)(t) \equiv y$ and $V_0(t) := V_0(g, h)(t) := \int_a^t h(u) \, du$. If $F(t) = \int_a^t v(u) \, du$ where $v$ is Bochner integrable then $F \in \mathcal{W}_1$ with $\|F\|_{[1]} \le 2\|v\|_1$. As in [35, (3.1.5), (3.1.7)] we have for $k \ge 2$ and $a \le t_k \le t$,

$$\int_a^{t_k} \cdots \int_a^{t_1} \|g(t_{k-1})\| \cdots \|g(t_1)\| \|h(u)\| \, du \, dt_1 \cdots dt_{k-1} \le M_k$$

where $M_k := \|h\|_1 \left(\int_a^t \|g(u)\| \, du\right)^{k-1} / (k-1)!$. It follows that

$$\|V_k(g, h)\|_{[1]} \le 2\|g\|_1 M_k \le 2\|g\|_1^k \|h\|_1 / (k-1)!. \tag{9.211}$$

The same bound holds for the $k = 1$ term. Similarly, $\|U_k(g, y)\|_{[1]} \le 2\|g\|_1^k \|y\| / (k-1)!$ for each $k \ge 1$. For $k = 0$ we have $\|U_0(g, y)\|_{[1]} = \|y\|$ and $\|V_0(g, h)\|_{[1]} \le 2\|h\|_1$. It then follows that for each $k$, $V_k(\cdot, \cdot)$ is a $(k+1)$-homogeneous polynomial from $\mathcal{L}_1(J; \mathbb{B}_Y) \times \mathcal{L}_1(J; Y)$ into $\mathcal{W}_1(J; Y)$, and $U_k(\cdot, \cdot)$ is a $(k+1)$-homogeneous polynomial from $\mathcal{L}_1(J; \mathbb{B}_Y) \times Y$ into $\mathcal{W}_1(J; Y)$.

Let $R_1, R_2, R_3$ be any three numbers with $0 < R_j < \infty$. If $\|g\|_1 \le R_1$, $\|h\|_1 \le R_2$ and $\|y\| \le R_3$, then from (9.211), $\|V_k(g, h)\|_{[1]} \le 2R_1^k R_2 / (k-1)!$. Likewise, $\|U_k(g, y)\|_{[1]} \le 2R_1^k R_3 / (k-1)!$, and so

$$\sum_{k=1}^{\infty} \|U_k(g, y)\|_{[1]} + \|V_k(g, h)\|_{[1]} \; \le \; 2(R_2 + R_3) R_1 \exp(R_1) < \infty.$$

Thus the power series $\sum_{k=0}^{\infty} U_k(\cdot,\cdot)$ from $\mathcal{L}_1(J;\mathbb{B}_Y) \times Y$ into $\mathcal{W}_1(J;Y)$ around zero, and the power series $\sum_{k=0}^{\infty} V_k(\cdot,\cdot)$ from $\mathcal{L}_1(J;\mathbb{B}_Y) \times \mathcal{L}_1(J;Y)$ into $\mathcal{W}_1(J;Y)$ around zero, both converge absolutely, and uniformly on any bounded set of $(y,g,h)$. Let $\phi$ be Daleckiĭ and Kreĭn's solution of (9.208) as a function of $(y,g,h) \in Y \times \mathcal{L}_1(J;\mathbb{B}_Y) \times \mathcal{L}_1(J;Y)$, given by [35, (3.1.6)] with their $A$, $f$, $x(\cdot)$, and $x_0$ equal to our $g$, $h$, $\phi(\cdot)$, and $y = \phi(a)$, respectively, and their $g(t)$ equal to our $y + \int_a^t h(s)\,ds$. Then for each $t \in [a,b]$,

$$\phi(t) = \sum_{k=0}^{\infty} U_k(g,y)(t) + \sum_{k=0}^{\infty} V_k(g,h)(t). \tag{9.212}$$

The sum (9.212) gives a Taylor expansion around zero of $\phi$ as a function of $(y,g,h)$. By the uniqueness stated in Theorem 9.84, and the uniqueness of Taylor polynomials, Theorem 5.9, the solution and the respective polynomials equal those in our solution.

## 9.13 Notes

During the long history of its development, product integration became a wide area of analysis mainly related to the theory of differential and integral equations. At least this was the area which motivated the definition of the product integral (9.1) given by Volterra [235, Opere, I, p. 235] in 1887. Earlier results related to product integrals, including equivalence of different forms of Volterra's original product integral, product integration over contours, and differentiability of the product integral with respect to a parameter, are presented in Schlesinger [206], [207], in Rasch [191], and in Volterra and Hostinsky [236, pp. 86–88,223]. Most of these results deal with matrix-valued product integrals. Volterra and Hostinsky's form of the derivative is somewhat obscured by their notation. They also allow the "fixed" endpoints $a, b$ to vary.

In a somewhat different direction is a work [17] of Garrett Birkhoff, in which he extended Volterra's product integral representation of a solution of a matrix-valued differential equation to differential equations describing an evolution in infinite-dimensional families of transformations. Birkhoff defined a product integral of Volterra's form for functions with values in more general spaces and took the limit in (9.1) under refinements of tagged partitions. Birkhoff emphasized a dual relation between product integration and differentiation. Similar ideas on product integration in nonlinear manifolds are discussed in a survey paper of Masani [159]. Earlier, Masani [158, Sect. V] proved among other things that for a function $C$ with values in a Banach algebra, the limit (9.1) exists if and only if $C$ is Riemann integrable.

Schlesinger ([206], [207]) developed product integration as in (9.1) based on approximation by step functions of a matrix-valued function $C$ whose entries are bounded Lebesgue measurable functions. Schmidt [208] proved existence of such a product integral when $C$ is a Bochner integrable Banach space operator-valued function. Another approach to proving existence of the Volterra product integral of a Lebesgue integrable matrix-valued function was indicated by Dollard and Friedman [45, Section 1.8]. Moreover, Dollard and Friedman [45, Theorem 3.5.1] proved that for a Banach space $X$ and the space $\mathbb{B}_X$ of bounded linear operators from $X$ into itself, with operator norm $\|\cdot\|$, such a product integral exists for each function $A\colon\ [a,b] \ni t \mapsto A(t) \in \mathbb{B}_X$ such that for each $f \in X$, $A(\cdot)f$ is the pointwise limit of a sequence of simple functions almost everywhere and $\|A\|$ has finite upper integral on $[a,b]$. Jarník and Kurzweil [111] initiated a study of the product integral obtained by extending the limit in (9.1) as follows. Let $C$ be a matrix-valued function on $[a,b]$. An invertible matrix $Q$ is called the product integral of $C$ if given $\epsilon > 0$ there is a gauge function $\delta(\cdot)$ on $[a,b]$ such that $\|\prod_{i=1}^{n}[I + C(s_i)(t_i - t_{i-1})] - Q\| < \epsilon$ for all $\delta$-fine tagged partitions $(\{t_i\}_{i=0}^{n}, \{s_i\}_{i=1}^{n})$ of $[a,b]$ (see the beginning of Section 2.6). Further extensions of this product integral are due to Schwabik [212], [213].

Another possible definition of product integral is reminiscent of the Riemann–Stieltjes integral. Let $h$ be a function on an interval $[a,b]$ with values in a Banach algebra $\mathbb{B}$ with identity $\mathbb{1}$. Define the product integral with respect to the function $h$ over $[a,b]$ to be the limit

$$\lim_{|\kappa|\downarrow 0}\big[\mathbb{1} + h(t_n) - h(t_{n-1})\big]\cdots\big[\mathbb{1} + h(t_1) - h(t_0)\big] \tag{9.213}$$

if it exists as the mesh $|\kappa|$ of partitions $\kappa = \{t_i\}_{i=0}^{n}$ of $[a,b]$ tends to zero. The product integral (9.213) was treated by MacNerney [152], who discovered reciprocal formulas involving so-called additive and product integrals. Earlier analogous formulas, in which $h$ is an interval function and the limit exists under refinements of interval partitions, were proved by Dobrushin [43]. In particular, MacNerney [152] proved that (9.213) exists if $h$ is continuous and of bounded variation. Hildebrandt [96] removed the continuity requirement. Freedman [70] replaced the bounded variation condition for a continuous function $h$ by the finiteness of the $p$-variation for some $p < 2$. He first proved existence of a unique solution of an appropriate linear integral equation and, second, showed that this solution is the limit (9.213). The product integral defined by (9.213), except that the limit is taken under refinements of partitions $\kappa$, was treated in [54, Part II].

Dollard and Friedman [45, p. 35] differentiate product integrals of the Volterra form along suitable 1-dimensional curves. Thus far, one would envisage differentiability of $\prod_a^b(1 + df)$ in the 1-variation seminorm $\|f\|_{(1)}$, as actually proved for $f$ absolutely continuous, when $\|f\|_{(1)} = \|f'\|_1$. Dollard and Friedman [45, p. 36] briefly state analyticity, also along curves, in a form that suggests it also holds with respect to the seminorm $f \mapsto \|f'\|_1$. Gill and Johansen [81] proved compact differentiability with respect to the supremum norm, although still on $\|\cdot\|_{(1)}$-bounded sets. Freedman [70], by showing existence of the product integral for continuous $f \in \mathcal{W}_p$, opened the way to extending the differentiability.

Gill and Johansen [81], see also Gill [80], made a substantial advance in showing infinite-dimensional differentiability with uniformity and including discontinuous directions $f$. They also gave several applications of the product integral in statistics and probability. It takes a hazard function into a survival function (for real-valued functions), and matrix-valued functions arise for Markov chains in continuous time with non-stationary transitions.

*Notes on Section 9.1.* Dobrushin [43] considered matrix-valued multiplicative and additive interval functions of bounded variation in relation to product integration. He assumed the interval functions to be upper continuous at $\emptyset$.

*Notes on Section 9.2.* Necessary and sufficient conditions for a product integral with respect to a real-valued (point) function $f$ to exist and to be non-zero were proved for the first time to our knowledge in [54, Theorem 4.4 in Part II]. Theorem 9.11 is closely related but is different in regard to atoms of $\mu$ in that a function $f$ could have jumps on both sides of a point.

*Notes on Section 9.4.* Inequalities similar to those in Corollary 9.24 were proved by Freedman [70, Lemma 5.1] using the Love–Young inequality for a Riemann–Stieltjes sum and under additional assumptions. Due to the $(k!)^{1/p}$ in the denominator of the right side of (9.28) these assumptions turn out to be unnecessary. The proof of Theorem 9.23, which yields the inequalities of Corollary 9.24, is based on extending two inequalities of Lyons [151]. The inequality of Lemma 9.25 is part of Theorem 1.1 of Lyons [151], up to the constant 4. The inequality of Lemma 9.26 was proved by Lyons [151, Theorem 1.2] for continuous functions.

*Notes on Section 9.5.* The fact that the finiteness of the $p$-variation with $0 < p < 2$ is sufficient for the existence of the product integral with respect to a Banach-algebra-valued (point) function $f$ was proved in [54, Theorem 4.23 in Part II]. Earlier, Freedman [70] proved existence of a product integral defined as a solution of a linear Riemann–Stieltjes integral equation with respect to a continuous function $f$ having bounded $p$-variation with $1 \leq p < 2$. Theorem 9.28 is a related result for the product integral with respect to an upper continuous additive interval function. The existence of the product integral with respect to a matrix-valued upper continuous additive interval function of bounded variation was proved by Dobrushin [43, Theorem 4].

*Notes on Section 9.6.* The property defining the multiplicative transform was introduced in attempts to prove Duhamel's formula under the conditions of Theorem 9.49. Also, it sometimes appears (e.g. in the proof of Theorem 5 in [43]) in proofs of duality between multiplicative and additive transforms as in Corollary 9.38. For point functions of bounded $p$-variation, $0 < p < 2$, a duality between the product integral operator and a logarithm operator is proved in [54, Part II, Theorems 6.10, 6.12].

*Notes on Section 9.8.* Two-integrand extensions $\int f\,dh\,g$ of the $RS$, $RRS$, and $CY$ integrals for a point function integrator $h$ were considered in Section 3 of [54, Part II]. There $LY$ and $RY$ integrals with two integrands $f$ and $g$ were introduced and used for the same purpose as the Kolmogorov integral with two integrands in this chapter. In that case $f$ and $g$ are not interchangeable even if $\mathbb{B}$ is commutative.

*Notes on Section 9.9.* The Duhamel formula as given by Theorem 9.49 extends the algebraic identity (9.43) to a class of multiplicative transforms using the Kolmogorov integral with two integrands. Other extensions to different types of product integrals were proved by Dollard and Friedman [45, Theorem I.51] and in [54, Part II, Theorem 5.5]. A simpler variant of this formula known under the same name is the following: for any matrices $A, B$, and each $t \geq 0$,

$$\exp(tA) - \exp(tB) = \int_0^t \exp((t-s)A)\, \mathrm{d}s(A-B)\exp(sB).$$

*Notes on Section 9.10.* The fact that the product integral operator (9.92) is a uniformly entire mapping on $\mathcal{AI}_p([a,b];\mathbb{B})$, $1 \leq p < 2$, (Theorem 9.51) is the main result of this chapter. Fréchet differentiability of the product integral operator with respect to a point function at each function of bounded $p$-variation $0 < p < 2$ was proved in [54, Theorem 5.16 in Part II]. Theorem 5.17 in [54, Part II] gives analyticity of this operator when restricted to right- or left-continuous functions of bounded $p$-variation, $0 < p < 2$.

*Notes on Section 9.11.* Using an integral equivalent to the refinement Young–Stieltjes integral, Hildebrandt [96, p. 359] proved that if a matrix-valued function $h$ with bounded variation has a finite number of discontinuities on $[a,b]$ then the integral equation

$$f(t) = \mathbb{I} + \int_a^t \mathrm{d}h\, f, \qquad a \leq t \leq b, \tag{9.214}$$

has a unique solution if and only if the matrices $\mathbb{I} - \Delta^{-}h(x)$ have inverses for all points of discontinuity of $h$. Hildebrandt considers cases where $h$ may have non-zero jumps $(\Delta^{+}h)(u)$ and $(\Delta^{-}h)(u)$ on both sides of the same point and seeks an actual solution of (9.214). The solution is a kind of product integral where the factor at a point $u$ where $h$ has jumps on one side or the other is $[\mathbb{I} + \Delta^{+}h(u)][\mathbb{I} - \Delta^{-}h(u)]^{-1}$.

MacNerney [153] used the left Cauchy and right Cauchy integrals defined as refinement Riemann–Stieltjes, or *RRS*, integrals except that tagged partitions $(\{t_i\}_{i=0}^n, \{s_i\}_{i=1}^n)$ are used such that $s_i = t_{i-1}$ and $s_i = t_i$, respectively. Hönig [100] used the interior *RRS* integral defined as the *RRS* integral except that the tagged partitions $(\{t_i\}_{i=0}^n, \{s_i\}_{i=1}^n)$ defining the integral satisfy the relation $t_{i-1} < s_i < t_i$. The book of Schwabik, Tvrdý, and Vejvoda [214] treats linear integral equations with respect to functions of bounded variation using the Henstock–Kurzweil integral.

In the proof that the solution operators for inhomogeneous integral equations are uniformly entire functions of given $y$, $\mu$, and $\nu$ in Theorem 9.76, we used normed subspaces $M_p$ and $M_p^{\leftarrow}$ of interval functions with norms ostensibly stronger than the two-sided $p$-variation norm $\|\cdot\|_{\overline{[p]}}$. At this writing, we do not know whether $M_p$ and $M_p^{\leftarrow}$ are equal to each other or to $\overline{\mathcal{I}}_p$.

*Notes on Section 9.12.* The main facts in this section on Banach-space-valued functions are corollaries of those in the previous section for Banach-algebra-valued functions. We also consider differential equations with Bochner integrable coefficients and relate these to integral equations in the case $p = 1$. Here a reference, treated in the text, is the book by Daleckiĭ and Kreĭn [35, §3.1]. The available methods for $p = 1$, to the extent that they are based on absolute integrability, seem not to extend to $1 < p < 2$.

# 10

# Nonlinear Differential and Integral Equations

## 10.1 Classical Picard Iteration

Let us recall the classical method of Picard iteration. Let $F$ be a (globally) Lipschitz function from $\mathbb{R}^k$ into $\mathbb{R}^k$,

$$\big|F(x) - F(y)\big| \le K|x-y|$$

for all $x, y \in \mathbb{R}^k$ and some constant $K < \infty$. Suppose we want to solve a nonlinear ordinary differential equation

$$\mathrm{d}f/\mathrm{d}t = F(f(t)), \quad a < t < \infty, \tag{10.1}$$

for $f$ with values in $\mathbb{R}^k$, with the initial conditions $f(a) = c \in \mathbb{R}^k$ and $f(t) \to c$ as $t \downarrow a$. Consider the sequence of functions $f_n$ defined by $f_0 \equiv c$ and

$$f_{n+1}(t) = c + \int_a^t F(f_n(s))\,\mathrm{d}s, \quad a \le t < \infty. \tag{10.2}$$

Then for $n = 0, 1, \ldots$, $f_n$ is a $C^1$ function on $[a, \infty)$ with $f_n(a) = c$. Picard showed that the sequence $\{f_n\}_{n\ge 0}$ converges to a solution. By the Lipschitz condition on $F$, for each $n = 1, 2, \ldots$, and $a \le t < \infty$, we have

$$|(f_{n+1} - f_n)(t)| \le K \int_a^t |(f_n - f_{n-1})(s)|\,\mathrm{d}s. \tag{10.3}$$

Let $G_n(t) := \sup_{a\le s\le t} |(f_n - f_{n-1})(s)|$. Then $G_{n+1}(t) \le \int_a^t KG_n(s)\,\mathrm{d}s$ for each $t \ge a$ and $n = 1, 2, \ldots$. We have $G_1(t) = (t-a)|F(c)|$ for each $t \ge a$. It follows by induction that for $n = 1, 2, \ldots$,

$$G_n(t) \le |F(c)|K^{n-1}(t-a)^n/n!.$$

Since the Taylor series of $\mathrm{e}^{Ky}$ converges absolutely for all $y \in \mathbb{R}$, we get that $f_n(t)$ converges for all $t \ge a$ to some $f(t) := f_\infty(t)$, uniformly on any bounded interval $[a, b]$, and that the function $f$ is a solution of the integral equation

R.M. Dudley and R. Norvaiša, *Concrete Functional Calculus*, Springer Monographs in Mathematics, DOI 10.1007/978-1-4419-6950-7_10,

$$f(t) = c + \int_a^t F(f(s))\,\mathrm{d}s, \quad a \le t < \infty. \tag{10.4}$$

As a uniform limit of continuous functions, $f$ is continuous and so by (10.4) it is $C^1$ and satisfies the differential equation (10.1).

**Theorem 10.1.** *The function $f = f_\infty := \lim_{n\to\infty} f_n$ is the unique solution of the differential equation (10.1) on $(a,b)$ for any $b$ with $a < b \le +\infty$ such that $f(a) = c$ and $\lim_{t\downarrow a} f(t) = c$.*

*Proof.* First, let $\mathcal{L}^\infty_{\mathrm{loc}} := \mathcal{L}^\infty_{\mathrm{loc}}([a,b))$ be the set of all measurable functions from $[a,b)$ into $\mathbb{R}^k$, bounded on each interval $[a,d]$ for $a < d < b$. It will be shown that $f = f_\infty$ is the unique solution of (10.4) in $\mathcal{L}^\infty_{\mathrm{loc}}$. Suppose $g$ is another such solution. For each $t \in [a,b)$, letting $H(t) := \sup_{a\le s\le t} |(f-g)(s)| < \infty$, we have $|(f-g)(t)| \le \int_a^t K|(f-g)(s)|\,\mathrm{d}s$, and so $H(t) \le K(t-a)H(t)$. Thus for $a \le t < \min(b, a+1/K)$ we get $H(t) = 0$. If $b \le a + 1/K$ we are done. Otherwise we can start again at $a + 1/K$ and iterate the argument as many times as needed. It then follows that $f(t) = g(t)$ for all $a \le t < b$, proving that $f_\infty$ is the unique solution of (10.4) in $\mathcal{L}^\infty_{\mathrm{loc}}$.

We have seen that $f = f_\infty$ is a solution of (10.1) as well as (10.4) and that it is $C^1$. Also, $f_\infty$ satisfies the initial condition $f(a) = c$. It is the unique solution of (10.1) that does so: Let $g$ be another such solution. Then being differentiable, it is continuous, and so by (10.1) it is $C^1$. Thus for each $t \ge a$,

$$g(t) - g(a) = \int_a^t g'(s)\,\mathrm{d}s = \int_a^t F(g(s))\,\mathrm{d}s,$$

i.e., $g$ satisfies (10.4). Since it is in $\mathcal{L}^\infty_{\mathrm{loc}}$, it equals $f_\infty$, completing the proof. □

**Example 10.2.** Consider the analytic but not globally Lipschitz function $F(x) = x^2$, $x \in \mathbb{R}$. Solving (10.1) by separation of variables gives $f(t) = 1/((1/c) - t)$ with $c = f(0) \ne 0$. If $c > 0$ we get a solution $f$ of (10.1) and (10.4) for $a = 0 \le t < 1/c$ which goes to $+\infty$ as $t \uparrow 1/c$.

## 10.2 Picard's Method in $p$-Variation Norm, $1 \le p < 2$

A classical non-autonomous nonlinear first-order ordinary differential equation is of the form

$$\mathrm{d}f(t)/\mathrm{d}t = \psi(f(t), t) = (N_\psi f)(t), \quad a < t < \infty,$$

extending (10.1). In a differential form we could write

$$\mathrm{d}f(t) = (N_\psi f)(t)\mathrm{d}t.$$

Here we will extend such equations, replacing $\mathrm{d}t$ by $\mathrm{d}\mu(t)$ where $\mu$ is an interval function in $\mathcal{AI}_p$, and we seek a solution $f \in \mathcal{W}_p$, where $N_\psi$ takes $\mathcal{W}_p$ into $\mathcal{W}_q$ with $p^{-1} + q^{-1} > 1$. We will also add a term $(N_\phi f)(t)\mathrm{d}\nu(t)$ where $\nu$ is of bounded 1-variation, giving a formal equation

$$\mathrm{d}f = (N_\psi f)\cdot\mathrm{d}\mu + (N_\phi f)\cdot\mathrm{d}\nu,$$

to be made precise by the integral equation (10.5), or alternate equations (10.6), (10.7), or (10.8).

Let $X$, $Y$, and $Z$ be Banach spaces with a bounded bilinear map $(x, y) \mapsto x \cdot y\colon X \times Y \to Z$ as usual (1.14), and let $J := [a, b]$ with $-\infty < a < b < \infty$. For $1 \le p < 2$, let $\mu \in \mathcal{AI}_p(J;Y)$ and $\nu \in \mathcal{AI}_1(J;Y)$, that is, let $\mu$ and $\nu$ be $Y$-valued additive and upper continuous interval functions on $J$ having bounded $p$-variation and 1-variation, respectively. Let $\phi, \psi\colon Z \times J \to X$ be such that the Nemytskii operator $N_\psi$ takes $\mathcal{W}_p(J;Z)$ into $\mathcal{W}_q(J;X)$ with $q \ge p$ and $p^{-1} + q^{-1} > 1$ (as before, we identify $\mathcal{W}_\infty(J;X)$ with the class of regulated functions $\mathcal{R}(J;X)$), and the Nemytskii operator $N_\phi$ takes $\mathcal{R}(J;Z)$ into $\mathcal{R}(J;X)$. Explicit sufficient conditions on $\psi$ are given in Theorem 6.58. For $\phi$, conditions will be given in Definition 10.5 and Proposition 10.6($b$). Also let $c \in Z$.

In this section we consider two forward nonlinear integral equations for Kolmogorov integrals:

$$f(t) = c + \oint_{(a,t]} (N_\psi f)_-\cdot\mathrm{d}\mu + \oint_{(a,t]} (N_\phi f)_-\cdot\mathrm{d}\nu, \qquad a \le t \le b, \tag{10.5}$$

$$g(t) = c + \oint_{[a,t)} (N_\psi g)\cdot\mathrm{d}\mu + \oint_{[a,t)} (N_\phi g)\cdot\mathrm{d}\nu, \qquad a \le t \le b, \tag{10.6}$$

and two analogous backward integral equations:

$$f(t) = c + \oint_{[t,b)} (N_\psi f)_+\cdot\mathrm{d}\mu + \oint_{[t,b)} (N_\phi f)_+\cdot\mathrm{d}\nu, \qquad a \le t \le b, \tag{10.7}$$

$$g(t) = c + \oint_{(t,b]} (N_\psi g)\cdot\mathrm{d}\mu + \oint_{(t,b]} (N_\phi g)\cdot\mathrm{d}\nu, \qquad a \le t \le b. \tag{10.8}$$

Recall, as for linear equations in the previous chapter, that use of a closed interval $[a, t]$ in (10.6) or omission of the left limit in (10.5) can lead to contradictions as in (1.8). Analogous considerations hold for the backward equations.

A solution $f$ of (10.5) or $g$ of (10.8), if it exists, will be right-continuous, and solutions of (10.6) and (10.7) will be left-continuous, by additivity and upper continuity at $\emptyset$ of the Kolmogorov integral (Theorem 2.21 and Corollary 2.23, respectively). But solutions $g$ of (10.6) and $f$ of (10.5) may not simply satisfy $g(t) = f(t-)$ for $a < t \le b$, or $f(t) = g(t+)$ for $a \le t < b$, e.g. if $\mu$ or $\nu$ has an atom at $a$ then in general

$$g(a+) = c + (N_\psi g)(a)\cdot\mu(\{a\}) + (N_\phi g)(a)\cdot\nu(\{a\}) \ne c = f(a).$$

If $\mu$ and $\nu$ have no atoms, solutions $f$ and $g$ of (10.5) and (10.6) will be continuous and equal by Lemma 2.49 and Corollary 2.26.

The main result giving the existence and uniqueness of a solution for each of the four integral equations holds for functions $\psi$ and $\phi$ satisfying suitable hypotheses. First recall Definition 6.42 of the class $\mathcal{UH}_\alpha$ and Definition 6.47 of the class $\mathcal{W}_{\alpha,q}$.

**Definition 10.3** (of $\mathcal{HWG}_{\alpha,q}$, $\mathcal{WG}_{\alpha,q}$, and $G_{\alpha,q}$). Let $0 < \alpha \leq 1$, $0 < q < \infty$, and let $J$ be a nondegenerate interval. For Banach spaces $Z$ and $X$, let $\psi \equiv \psi(z,s)\colon Z \times J \to X$, and let $B$ be a nonempty subset of $Z$. We say that $\psi$ is in the class $\mathcal{WG}_{\alpha,q}(B \times J; X)$ if there is a finite constant $C$ such that for each $0 \leq K < \infty$,

$$W_{\alpha,q}(\psi, K; B \times J, X) \leq C(1 + K^\alpha), \tag{10.9}$$

where $W_{\alpha,q}$ is defined to be the minimal $W \geq 0$ such that (6.31) holds for a given $K$. Let $G_{\alpha,q}(\psi) := G_{\alpha,q}(\psi; B \times J, X)$ be the minimal such $C$. Let $\mathcal{HWG}_{\alpha,q}(B \times J; X) := \mathcal{UH}_\alpha(B \times J; X) \cap \mathcal{WG}_{\alpha,q}(B \times J; X)$.

**Remark 10.4.** It follows immediately from the definitions that $\mathcal{WG}_{\alpha,q}(B \times J; X) \subset \mathcal{W}_{\alpha,q}(B \times J; X)$ for any nonempty $B \subset Z$.

For the function $\phi$ we will assume growth conditions of the following kind:

**Definition 10.5** (of $\mathcal{G}_\beta$, $\mathcal{UG}_\beta$, $\mathcal{UCR}$, and $\mathcal{CRG}_\beta$). Let $0 < \beta < \infty$ and let $J$ be a nondegenerate interval. For Banach spaces $Z$ and $X$, let $\phi \equiv \phi(z,s)\colon Z \times J \to X$, and let $B$ be a nonempty subset of $Z$. We say that $\phi$ satisfies the *$s$-uniform $\beta$ growth condition* on $B$, or $\phi \in \mathcal{UG}_\beta(B \times J; X)$, if

$$\|\phi\|_{\mathcal{G}_\beta} := \|\phi\|_{B\times J,\mathcal{G}_\beta} := \sup_{z\in B} \|\phi(z,\cdot)\|_{J,\sup}/(1 + \|z\|^\beta) < \infty.$$

If this holds for $\phi(z,s) \equiv G(z)$ then we write $G \in \mathcal{G}_\beta(B; X)$. We say that $\phi$ is in the class $\mathcal{UCR}(B \times J; X)$ if $(a)$ $B \ni z \mapsto \phi(z,\cdot) \in \ell^\infty(J, X)$ is continuous, and $(b)$ $J \ni s \mapsto \phi(z,s)$ is regulated for each $z \in B$. Let $\mathcal{CRG}_\beta(B \times J; X) := \mathcal{UCR}(B \times J; X) \cap \mathcal{UG}_\beta(B \times J; X)$.

**Proposition 10.6.** $(a)$ *If $B$ is bounded, $\mathcal{UG}_\beta(B \times J; X) = \ell^\infty(B \times J; X)$ for any $\beta$.*

$(b)$ *If $\phi \in \mathcal{UCR}(Z \times J; Y)$ then the Nemytskii operator $N_\phi$ takes $\mathcal{R}(J; Z)$ into $\mathcal{R}(J; X)$.*

*Proof.* Part $(a)$ is clear. For part $(b)$, suppose that $f \in \mathcal{R}(J; Z)$ and $t_n \downarrow t$ or $t_n \uparrow t \in J$ where $t_1 \in J$ and $t_n \neq t$ for all $n$. We need to show that the sequence $\phi(f(t_n), t_n)$ converges to some limit. Let $z_n = f(t_n) \in Z$. Then $z_n$ converge to some limit $z_0$, and $\sup_{t\in J} \|\phi(z_n,t) - \phi(z_0,t)\| \to 0$ as $n \to \infty$. Also $\phi(z_0,\cdot)$ is regulated. Then

$$\begin{aligned} &\|\phi(z_n,t_n)-\phi(z_m,t_m)\| \\ &\quad \le \|\phi(z_n,t_n)-\phi(z_0,t_n)\|+\|\phi(z_0,t_n)-\phi(z_0,t_m)\|+\|\phi(z_0,t_m)-\phi(z_m,t_m)\|, \end{aligned}$$

which approaches 0 as $m,n \to \infty$, so $\{\phi(z_n,t_n)\}_{n\ge 1}$ is a Cauchy sequence and converges. □

Finally, Fréchet $u$-differentiability of $\psi$ and the local classes $\mathcal{HW}^{\mathrm{loc}}_{\alpha,q}, \mathcal{UH}^{\mathrm{loc}}_{\alpha}$ are defined in Definitions 6.8, 6.51, and 6.43, respectively. Now we are ready to formulate the main result of this chapter.

**Theorem 10.7.** *Let $\alpha,\beta \in (0,1]$, $1 \le p < 1+\alpha$, and $J := [a,b]$ with $a < b$. Assuming* (1.14), *let $\psi \in \mathcal{HWG}_{1,p}(Z \times J; X)$ be Fréchet $u$-differentiable on $Z$ with the derivative $\psi_u^{(1)} \in \mathcal{HW}^{\mathrm{loc}}_{\alpha,p/\alpha}(Z \times J; L(Z;X))$, let $\phi \in \mathcal{UG}_\beta(Z \times J; X) \cap \mathcal{UH}^{\mathrm{loc}}_1(Z \times J; X)$ be such that $\phi(z,\cdot) \in \mathcal{R}(J;X)$ for each $z \in Z$, let $\mu \in \mathcal{AI}_p(J;Y)$, $\nu \in \mathcal{AI}_1(J;Y)$, and $c \in Z$. Then each of the four integral equations* (10.5), (10.6), (10.7), *and* (10.8) *has a unique solution in $\mathcal{W}_p(J;Z)$.*

If $\psi(z,s) \equiv F(z)$ and $\phi(z,s) \equiv G(z)$, the theorem just stated implies the following:

**Corollary 10.8.** *Let $\alpha,\beta \in (0,1]$, $1 \le p < 1+\alpha$, and $J := [a,b]$ with $a < b$. Assuming* (1.14), *let $F \in \mathcal{H}_1(Z;X)$ be Fréchet differentiable on $Z$ with derivative $DF \in \mathcal{H}^{\mathrm{loc}}_\alpha(Z;L(Z;X))$, let $G \in \mathcal{G}_\beta(Z;X) \cap \mathcal{H}^{\mathrm{loc}}_1(Z;X)$, let $\mu \in \mathcal{AI}_p(J;Y)$, $\nu \in \mathcal{AI}_1(J;Y)$, and $c \in Z$. Then the conclusion of Theorem* 10.7 *holds with the Nemytskii operators $N_\psi$ and $N_\phi$ replaced by the autonomous Nemytskii operators $N_F$ and $N_G$, respectively.*

The proofs of the above theorem and corollary are given in Section 10.3.

Define the integral transforms $T^{\succ}_{c,u,\mu,\nu}$ and $Q^{\succ}_{c,u,\mu,\nu}$, for any $a \le u \le t \le b$ and $f \in \mathcal{W}_p([u,b];Z)$, letting

$$T^{\succ}_{c,u}f(t) := T^{\succ}_{c,u,\mu,\nu}f(t) := c + \fint_{(u,t]} (N_\psi f)_-\cdot \mathrm{d}\mu + \fint_{(u,t]} (N_\phi f)_-\cdot \mathrm{d}\nu, \quad (10.10)$$

$$Q^{\succ}_{c,u}f(t) := Q^{\succ}_{c,u,\mu,\nu}f(t) := c + \fint_{[u,t)} (N_\psi f)\cdot \mathrm{d}\mu + \fint_{[u,t)} (N_\phi f)\cdot \mathrm{d}\nu, \quad (10.11)$$

provided the Kolmogorov integrals are defined. It will be shown that these transforms are well defined and take $\mathcal{W}_p([u,b];Z)$ into itself under some conditions, as follows:

**Proposition 10.9.** *Let $c \in Z$, $J := [a,b]$ with $a < b$, $1 \le p < 2$, $\nu \in \mathcal{AI}_1(J;Y)$, and $\mu \in \mathcal{AI}_p(J;Y)$. Let $\phi,\psi\colon Z \times J \to X$ be such that the Nemytskii operator $N_\psi$ takes $\mathcal{W}_p(J;Z)$ into $\mathcal{W}_q(J;X)$ with $q \ge p$ and $p^{-1}+q^{-1} > 1$, and assume that $N_\phi$ takes $\mathcal{R}(J;Z)$ into $\mathcal{R}(J;X)$. Let $a \le u < b$.*

(a) *Then the integral transforms $Q^{\succ}_{c,u}$ and $T^{\succ}_{c,u}$ are defined on $\mathcal{W}_p([u,b];Z)$ and take it into itself.*

(*b*) *The hypothesis on* $\psi$ *holds if* $\psi \in \mathcal{HW}_{p/q,q}(Z \times J; X)$ *or more generally if* $\psi \in \mathcal{HW}^{\mathrm{loc}}_{p/q,q}(Z \times J; X)$, *and that on* $\phi$ *holds if* $\phi \in \mathcal{UCR}(Z \times J; X)$.

(*c*) *The hypotheses of Theorem 10.7 imply those of this proposition; specifically they imply conclusion* (*a*).

*Proof.* To prove (*a*) we note that for given $c$ and $u$, the values of $T^{\succ}_{c,u}f$ and $Q^{\succ}_{c,u}f$ on $[u, b]$ depend only on the values of $f$ on $[u, b)$, and apply the Love–Young inequality Proposition 3.96 and Proposition 3.31. Part (*b*) follows directly from Theorem 6.58(*a*) and Proposition 10.6(*b*).

To prove part (*c*), we need to check the conditions on $\psi$ and $\phi$. In Theorem 10.7 it is assumed that $\psi \in \mathcal{HWG}_{1,p}(Z \times J; X)$, which by Definition 10.3 and Remark 10.4 implies that $\psi \in \mathcal{UH}_1(Z \times J; X) \cap \mathcal{W}_{1,p}(Z \times J; X)$. By Proposition 6.54(*b*), it follows that $N_\psi$ takes $\mathcal{W}_p([u, b]; Z)$ into itself (and is a bounded nonlinear operator). The range can also be taken to be $\mathcal{W}_q([u, b]; Z)$ for any $q \geq p$, by Lemma 3.45. Since $2/p > 1$ the hypothesis on $\psi$ for this proposition holds.

For $\phi$, the hypotheses of Theorem 10.7 imply that $\phi \in \mathcal{UH}_1(Z \times J; X)$ and $\phi(z, \cdot) \in \mathcal{R}(J; X)$ for each $z \in Z$. Thus by Definition 10.5, $\phi \in \mathcal{UCR}(Z \times J; X)$ and part (*b*) for $\phi$ applies, proving part (*c*) and the proposition. □

Now, (10.5) is equivalent to $f = T^{\succ}_{c,a,\mu,\nu}f$ on $[a, b]$, and similarly for (10.6). Later the definitions (10.10) and (10.11) will be applied to $v$ in place of $b$ where $u < v \leq b$.

*Picard iterates*

We will show that each of the four integral equations can also be solved by performing an iteration similar to the classical Picard iteration (10.2). To define the iterates, let $f^{\succ}_0 := g^{\succ}_0 := c$, and for each $n \geq 1$ and $t \in J$, let

$$f^{\succ}_n(t) := T^{\succ}_{c,a}f^{\succ}_{n-1}(t) = c + \oint_{(a,t]} (N_\psi f^{\succ}_{n-1})_- \cdot \mathrm{d}\mu + \oint_{(a,t]} (N_\phi f^{\succ}_{n-1})_- \cdot \mathrm{d}\nu, \quad (10.12)$$

$$g^{\succ}_n(t) := Q^{\succ}_{c,a}g^{\succ}_{n-1}(t) = c + \oint_{[a,t)} (N_\psi g^{\succ}_{n-1}) \cdot \mathrm{d}\mu + \oint_{[a,t)} (N_\phi g^{\succ}_{n-1}) \cdot \mathrm{d}\nu. \quad (10.13)$$

Then $\{f^{\succ}_n\}_{n \geq 0}$ and $\{g^{\succ}_n\}_{n \geq 0}$ are two sequences of functions in $\mathcal{W}_p(J; Z)$, which will be called *Picard iterates* associated to the forward integral equations (10.5) and (10.6), respectively, or obtained by the integral transforms $T^{\succ}_{c,u,\mu,\nu}$ and $Q^{\succ}_{c,u,\mu,\nu}$, respectively.

With respect to the two backward integral equations (10.7) and (10.8) we define backward integral transforms and sequences symmetrically as follows. For $f \in \mathcal{W}_p(J; Z)$ and $a \leq t \leq v \leq b$, let

$$T^{\prec}_{c,v}f(t) := T^{\prec}_{c,v,\mu,\nu}f(t) := c + \oint_{[t,v)} (N_\psi f)_+ \cdot \mathrm{d}\mu + \oint_{[t,v)} (N_\phi f)_+ \cdot \mathrm{d}\nu, \quad (10.14)$$

$$Q^{\prec}_{c,v}f(t) := Q^{\prec}_{c,v,\mu,\nu}f(t) := c + \oint_{(t,v]} (N_\psi f) \cdot \mathrm{d}\mu + \oint_{(t,v]} (N_\phi f) \cdot \mathrm{d}\nu. \quad (10.15)$$

**Remark 10.10.** As in Proposition 10.9, under its hypotheses, with now $u < v \le b$, the integral transforms $T^{\prec}_{c,v}$ and $Q^{\prec}_{c,v}$ act from $\mathcal{W}_p([\![u,v]\!];Z)$ into itself.

Let $f_0^{\prec} := g_0^{\prec} := c$, and for each $n \ge 1$ and $t \in J$, let

$$f_n^{\prec}(t) := T^{\prec}_{c,b} f^{\prec}_{n-1}(t) = c + \oint_{[t,b)} (N_\psi f^{\prec}_{n-1})_+ \cdot d\mu + \oint_{[t,b)} (N_\phi f^{\prec}_{n-1})_+ \cdot d\nu, \quad (10.16)$$

$$g_n^{\prec}(t) := Q^{\prec}_{c,b} g^{\prec}_{n-1}(t) = c + \oint_{(t,b]} (N_\psi g^{\prec}_{n-1}) \cdot d\mu + \oint_{(t,b]} (N_\phi g^{\prec}_{n-1}) \cdot d\nu. \quad (10.17)$$

Then $\{f_n^{\prec}\}_{n\ge 0}$ and $\{g_n^{\prec}\}_{n\ge 0}$ are two sequences of functions in $\mathcal{W}_p(J;Z)$, which will be called *Picard iterates* associated to the backward integral equations (10.7) and (10.8), respectively, or obtained by the integral transforms $T^{\prec}_{c,v,\mu,\nu}$ and $Q^{\prec}_{c,v,\mu,\nu}$, respectively.

**Theorem 10.11.** *Under the hypotheses of Theorem* 10.7, *each of the four sequences of Picard iterates* (10.12), (10.13), (10.16) *and* (10.17) *converges in $\mathcal{W}_p(J;Z)$ to the unique solution in $\mathcal{W}_p(J;Z)$ of the corresponding integral equation given by Theorem* 10.7.

If $\psi(z,s) \equiv F(z)$ and $\phi(z,s) \equiv G(z)$, the preceding theorem implies the following:

**Corollary 10.12.** *Under the hypotheses of Corollary* 10.8, *the conclusion of Theorem* 10.11 *holds with the Nemytskii operators $N_\psi$ and $N_\phi$ replaced by the autonomous Nemytskii operators $N_F$ and $N_G$, respectively.*

We will prove Theorem 10.11 and Corollary 10.12 in Section 10.4. Before proving Theorems 10.7 and 10.11 we next present several consequences.

*Examples and corollaries*

First we notice that under the assumption (1.14), the order of integrand and integrator in the four nonlinear integral equations is unimportant. It can be reversed using the formal rule (1.16), and the new equations can be solved as follows. For now, let $\mu$ be an $X$-valued additive and upper continuous interval function on an interval $[a,b]$ having bounded $p$-variation for some $1 \le p < 2$. Also for now, let a function $F\colon Z \to Y$ be $\alpha$-Hölder continuous with $p - 1 < \alpha \le 1$, and let $c \in Z$. Consider two forward nonlinear integral equations for Kolmogorov integrals:

$$f(t) = c + \oint_{(a,t]} d\mu \cdot (F \circ f)_-, \qquad a \le t \le b, \quad (10.18)$$

$$g(t) = c + \oint_{[a,t)} d\mu \cdot (F \circ g), \qquad a \le t \le b. \quad (10.19)$$

The two analogous backward equations can be solved similarly, and are not considered here. Related to the integral equations (10.18) and (10.19), define the associated Picard iterates $\{f_n\}_{n\geq 0}$ and $\{g_n\}_{n\geq 0}$, respectively, by $f_0 := g_0 \equiv c$ and for each $n \geq 1$, let

$$f_n(t) = c + \oint_{(a,t]} \mathrm{d}\mu\cdot(F\circ f_{n-1})_-, \qquad a \leq t \leq b, \tag{10.20}$$

$$g_n(t) = c + \oint_{[a,t)} \mathrm{d}\mu\cdot(F\circ g_{n-1}), \qquad a \leq t \leq b. \tag{10.21}$$

Applying Corollaries 10.8 and 10.12 to the bounded bilinear operator $\widetilde{B}\colon Y \times X \to Z$ defined by $\widetilde{B}(y,x) := B(x,y) := x \cdot y$ we get the following:

**Corollary 10.13.** *Let $0 < \alpha \leq 1$, $1 \leq p < 1+\alpha$, and $J := [a,b]$ with $a < b$. Assuming* (1.14), *let $\mu \in \mathcal{AI}_p(J;X)$, let $F \in \mathcal{H}_1(Z;Y)$ be Fréchet differentiable on $Z$ with derivative $DF \in \mathcal{H}_\alpha^{\rm loc}(Z;L(Z,Y))$, and $c \in Z$. Then (a) each of the two integral equations* (10.18) *and* (10.19) *has a unique solution in $\mathcal{W}_p(J;Z)$, and (b) each of the two sequences of Picard iterates* (10.20) *and* (10.21) *converges in $\mathcal{W}_p(J;Z)$ to the solution of the respective equation from (a).*

Let $Y$ be a Banach space and let $X := \mathbb{B}_Y = L(Y,Y)$ be the Banach algebra of bounded linear operators from $Y$ into itself. Let $Z = Y$ and $B(T,y) = Ty$ for $T \in X$ and $y \in Y$. Suppose that $F \in \mathbb{B}_Y$ is the identity mapping from $Y$ onto itself, denoted by $\mathbb{I}$. Then (10.18) becomes the forward homogeneous linear integral equation (9.203) with $c = y$. In this case using Corollary 2.23 with the reversed order of integrand and integrator, (10.20) can be written recursively in closed form as follows: for $a \leq t \leq b$, and $n \geq 2$,

$$f_n(t) = c + \mu((a,t])c + \oint_{(a,t]} \mu(\mathrm{d}s) \oint_{(a,s)} \mathrm{d}\mu\,(f_{n-2})_- = \cdots$$

$$= \Big\{\mathbb{I} + \mu((a,t]) + \sum_{k=2}^{n} \oint_{(a,t]} \mu(\mathrm{d}s_1) \oint_{(a,s_1)} \mu(\mathrm{d}s_2)\cdots \oint_{(a,s_{k-1})} \mu(\mathrm{d}s_k)\Big\}c.$$

By (9.94), $f_n(t)$ is a partial sum of the Taylor expansion of the product integral with respect to $\mu$ over the interval $A = (a,t]$. Similarly, assuming that $F$ in (10.19) is the identity, the associated Picard iterates $g_n(t)$, $n \geq 2$, $t \in [a,b]$, in (10.21) can be rewritten in the form of partial sums of the Taylor expansion of the product integral with respect to $\mu$ over the intervals $A = [a,t)$, $t \in [a,b]$.

Finally, we show how Corollary 10.12 can be used to solve a nonlinear refinement Riemann–Stieltjes integral equation with respect to a right-continuous point function.

**Corollary 10.14.** *Let $0 < \alpha \leq 1$, $1 \leq p < 1+\alpha$, and $J := [a,b]$ with $a < b$. Assuming* (1.14), *let $F \in \mathcal{H}_1(Z;X)$ be Fréchet differentiable on $Z$ with derivative $DF \in \mathcal{H}_\alpha^{\rm loc}(Z;L(Z,Y))$, let $h \in \mathcal{W}_p(J;Y)$ be right-continuous, and let $c \in Z$. Let $f_0 \equiv c \in Z$, and for $n = 1, 2, \ldots,$*

$$f_n(t) := c + (RRS)\int_a^t (F\circ f_{n-1})_-^{(a)}\cdot dh, \qquad a \le t \le b. \tag{10.22}$$

*Then $f_n \in \mathcal{W}_p(J;Z)$ for all $n$ and the $f_n$ converge in $\mathcal{W}_p(J;Z)$ to some $f$ which is the unique solution in $\mathcal{W}_p(J;Z)$ of*

$$f(t) = c + (RRS)\int_a^t (F\circ f)_-^{(a)}\cdot dh, \qquad a \le t \le b. \tag{10.23}$$

*Proof.* Let $\mu := \mu_{h,J}$ be the $Y$-valued interval function on $J$ defined by (2.2). Then since $h$ is regulated, $\mu$ is upper continuous by Theorem 2.7. Let $\tilde{h}(t) := \mu([a,t])$ for $a \le t \le b$. Since $h$ is right-continuous, by Proposition 2.6, we have $\tilde{h} = R_{\mu,a} = h - h(a)$ for $a \le t \le b$ and $R_{\mu,a}$ defined by (2.3). By Proposition 3.31, $v_p(\mu;J) = v_p(h;J) < \infty$. For an $X$-valued regulated function $g$ on $J$, let $\nu := \mu_{g,J}$ be the $X$-valued interval function on $J$ also defined by (2.2). Then $\nu$ is also upper continuous by Theorem 2.7, $\nu(\emptyset) = 0$, and $g(t-) = \nu([a,t))$ for $a < t \le b$ by Proposition 2.6. Since $\mu(\{a\}) = h(a+) - h(a) = 0$, the Kolmogorov integral $⨍_{(a,b]} g_-\cdot d\mu$ exists if and only if $⨍_J \nu([a,\cdot))\cdot d\mu$ does and if they exist then they are equal by Theorem 2.21 with $A_1 = \{a\}$. Therefore by Proposition 2.89, the Kolmogorov integral $⨍_{(a,b]} g_-\cdot d\mu$ exists if and only if the refinement Riemann–Stieltjes integral $(RRS)\int_a^b g_-^{(a)}\cdot dh$ does, and if they exist then for each $a \le t \le b$,

$$⨍_{(a,t]} g_-\cdot d\mu = (RRS)\int_a^t g_-^{(a)}\cdot dh.$$

Now, each function $f_n$ defined by (10.22) is equal to the function defined by (10.12) with $N_\psi \equiv N_F$ and $N_\phi \equiv 0$. Thus by Corollary 10.12 with $G \equiv 0$ and $\nu \equiv 0$, the $f_n$ converge in $\mathcal{W}_p(J;Z)$ to some function $f$ which is the unique solution in $\mathcal{W}_p(J;Z)$ of (10.5), and hence of (10.23), proving the corollary. □

## 10.3 Existence and Uniqueness

To prepare for the proof of Theorem 10.7 we will characterize relation (10.9) defining the class $\mathcal{HWG}_{\alpha,q}$ given by Definition 10.3. Recall Definition 6.42 of the class $\mathcal{UH}_\alpha$ of $s$-uniformly $\alpha$-Hölder functions $\psi\colon B\times J \to X$ with its seminorm $H_\alpha(\psi) := H_\alpha(\psi; B\times J, X)$, and Definition 6.47 of the seminorm $W_{\alpha,q}(\cdot, K)$, $0 \le K < \infty$.

**Proposition 10.15.** *Let $1 \le p \le q < \infty$, $\alpha := p/q$, and $J := [a,b]$ with $a < b$. For Banach spaces $Z$ and $X$, let $B$ be a nonempty subset of $Z$, and let $\psi\colon Z\times J \to X$ be in $\mathcal{UH}_\alpha(B\times J;X)$. The following statements (a) and (b) are equivalent:*

(a) *$\psi \in \mathcal{HWG}_{\alpha,q}(B \times J; X)$, or equivalently since $\psi \in \mathcal{UH}_\alpha(B \times J; X)$, $\psi \in \mathcal{WG}_{\alpha,q}(B \times J; X)$; in other words, there is a finite constant $C$ such that (10.9) holds for each $0 \le K < \infty$;*
(b) *there is a finite constant $D$ such that for each $g \in \mathcal{W}_p(J; Z)$ with range $\operatorname{ran}(g) \subset B$ and any nondegenerate interval $A \subset J$,*

$$\big\|N_\psi g\big\|_{A,[q]} \le D(1 + \|g\|_{A,[p]}^\alpha).$$

*Moreover, if (a) holds, we can take $D = C + \max\{\|N_\psi(z1_J)\|_{\sup} + \|z\|^\alpha H_\alpha(\psi), 2H_\alpha(\psi)\}$ for any $z \in B$.*

*Proof.* Suppose (a) holds. Then by Lemma 6.53(a) and (10.9), we have

$$\begin{aligned}\|N_\psi g\|_{A,(q)} &\le H_\alpha(\psi)\|g\|_{A,(p)}^\alpha + W_{\alpha,q}(\psi, \|g\|_{A,[p]})\\ &\le C + (C + H_\alpha(\psi))\|g\|_{A,[p]}^\alpha.\end{aligned}$$

Let $z \in B$. By the definition (6.31) of $W_{\alpha,q}(\psi, K)$, $\psi(z, \cdot) \in \mathcal{W}_q(J; X)$, and so $\|N_\psi(z1_J)\|_{\sup} < \infty$. Since $\psi \in \mathcal{UH}_\alpha(B \times J; X)$, it then follows that

$$\|N_\psi g\|_{A,\sup} \le \|N_\psi(z1_J)\|_{\sup} + H_\alpha(\psi)(\|g\|_{A,\sup}^\alpha + \|z\|^\alpha).$$

Thus letting $D := C + \max\{\|N_\psi(z1_J)\|_{\sup} + \|z\|^\alpha H_\alpha(\psi), 2H_\alpha(\psi)\}$, (b) holds.

Now suppose that (b) holds. By Lemma 6.53(c), for $0 \le K < \infty$, we have

$$W_{\alpha,q}(\psi, K; B \times J, X) \le H_\alpha(\psi)K^\alpha + D + DK^\alpha.$$

Thus taking $C := D + H_\alpha(\psi)$, (a) follows, proving the proposition. □

The next three examples give sufficient conditions for statement (a) of the preceding proposition.

**Example 10.16.** Using the notation of Proposition 10.15, let $h \in \mathcal{W}_q(J)$ and let $F\colon Z \to X$ be such that for some constant $M$, $\|F(z)\| \le M(1 + \|z\|^\alpha)$, $z \in Z$ (in the case $Z = X = \mathbb{R}$ this is the $\alpha$ growth condition (7.2)). For each $(z, s) \in Z \times J$, let $\psi(z, s) := F(z)h(s)$. To show that (a) of Proposition 10.15 holds for $\psi$, recall the notation (6.31). Let $0 \le K < \infty$, $B \subset Z$, let $\kappa = \{s_i\}_{i=0}^n$ be a partition of $J$, and let $\mu = \{z_i\}_{i=1}^n \subset B$ be such that $w_{\alpha q}(\mu) \le K$. Then

$$s_q(\psi; \mu, \kappa)^{1/q} \le \|h\|_{(q)} \max_i \|F(z_i)\| \le M\|h\|_{(q)}\big(1 + K^\alpha\big),$$

and so (10.9) holds with $C = M\|h\|_{(q)}$.

**Example 10.17.** Another example is $\psi(z, s) := \sum_{j=1}^m F_j(z)h_j(s)$ with $z \in Z$, $s \in J$, and $1 \le m < \infty$, where $F_j \in \mathcal{H}_\alpha(Z; X)$ and $h_j \in \mathcal{W}_q(J)$. Let $0 \le K < \infty$, $B \subset Z$, let $\kappa = \{s_i\}_{i=0}^n$ be a partition of $J$ and let $\mu = \{z_i\}_{i=1}^n \subset B$ be such that $w_{\alpha q}(\mu) \le K$. For a function $f\colon J \to \mathbb{R}$ and each $i$, let $\Delta_i f := f(s_i) - f(s_{i-1})$. Using the triangle inequality, and then iterating the Minkowski inequality (1.5), it follows that

$$\begin{aligned} s_q(\psi;\mu,\kappa)^{1/q} &\le \Big(\sum_{i=1}^n \Big(\sum_{j=1}^m \|F_j(z_i)\||\Delta_i h_j|\Big)^q\Big)^{1/q} \\ &\le \sum_{j=1}^m \Big(\sum_{i=1}^n \|F_j(z_i)\|^q |\Delta_i h_j|^q\Big)^{1/q} \le \sum_{j=1}^m \|h_j\|_{(q)} \max_i \|F_j(z_i)\| \\ &\le \sum_{j=1}^m \|h_j\|_{(q)}\|F_j(0)\| + K^\alpha \sum_{j=1}^m \|h_j\|_{(q)}\|F_j\|_{(\mathcal{H}_\alpha)}. \end{aligned}$$

Thus (10.9) holds with $C = \sum_{j=1}^m \|h_j\|_{(q)}\|F_j\|_{\mathcal{H}_\alpha}$.

**Example 10.18.** For $j = 1, 2, \ldots$, let $F_j \in \mathcal{H}_\alpha(Z;X)$ and $h_j \in \mathcal{W}_q(J)$ be such that

$$\sum_{j=1}^\infty \|h_j\|_{[q]}\|F_j\|_{\mathcal{H}_\alpha} < \infty.$$

Since $\sum_{j=1}^\infty \|h_j\|_{\sup}\|F_j\|_{\sup} < \infty$, it follows that for each $(z,s) \in Z \times J$, $\psi(z,s) := \sum_{j=1}^\infty F_j(z)h_j(s) \in X$. Let $0 \le K < \infty$, $B \subset Z$, let $\kappa = \{s_i\}_{i=0}^n$ be a partition of $J$ and let $\mu = \{z_i\}_{i=1}^n \subset B$ be such that $w_{\alpha q}(\mu) \le K$. For a function $f\colon J \to \mathbb{R}$ and each $i$, let $\Delta_i f := f(s_i) - f(s_{i-1})$. Then

$$s_q(\psi;\mu,\kappa)^{1/q} \le \Big(\sum_{i=1}^n \Big(\sum_{j=1}^\infty \|F_j(z_i)\||\Delta_i h_j|\Big)^q\Big)^{1/q}.$$

Let $C := \sum_{j=1}^\infty \|h_j\|_{(q)}\|F_j\|_{\mathcal{H}_\alpha} < \infty$. For each $1 \le m < \infty$, as in the preceding example we have

$$\Big(\sum_{i=1}^n \Big(\sum_{j=1}^m \|F_j(z_i)\||\Delta_i h_j|\Big)^q\Big)^{1/q} \le C(1+K^\alpha).$$

Letting $m \to \infty$, it follows that (10.9) holds. Moreover, for any $z, w \in Z$ and $s \in J$, we have

$$\begin{aligned} \|\psi(z,s) - \psi(w,s)\| &\le \sum_j \|F_j(z) - F_j(w)\|\|h_j\|_{\sup} \\ &\le \|z-w\|^\alpha \sum_j \|F_j\|_{(\mathcal{H}_\alpha)}\|h_j\|_{\sup}. \end{aligned}$$

Thus $H_\alpha(\psi; Z \times J; X) < \infty$, and so $\psi \in \mathcal{HWG}_{\alpha,q}(B \times J; Y)$ for each bounded set $B \subset Z$.

We will next give bounds which control $p$-variation norms for the two integral transforms $T^{\succ}_{c,a}$ and $Q^{\succ}_{c,a}$ defined respectively by (10.10) and (10.11). Recall Definitions 10.3 for $G_{\alpha,q}$ and 10.5 for $\mathcal{CRG}_\beta$.

**Lemma 10.19.** *Let $\alpha, \beta \in (0,1]$, $1 \le p < 1+\alpha$, $q := p/\alpha$, and $J := [a,b]$ with $a < b$. Assuming (1.14), let $\psi \in \mathcal{HWG}_{\alpha,q}(Z \times J; X)$, $\phi \in \mathcal{CRG}_\beta(Z \times J; X)$, $\mu \in \mathcal{AI}_p(J;Y)$, and $\nu \in \mathcal{AI}_1(J;Y)$. Suppose that for some $a \le u < v \le b$,*

$$\begin{aligned} \|\mu\|_{(u,v),(p)} &< \rho_1 := \{16K_{p,q}(G_{\alpha,q}(\psi) + \|N_\psi(0)\|_{\sup} + 2H_\alpha(\psi))\}^{-1}, \\ \|\nu\|_{(u,v),(1)} &< \rho_2 := \{16\|\phi\|_{\mathcal{G}_\beta}\}^{-1}, \end{aligned} \tag{10.24}$$

*where $K_{p,q} = \zeta((1+\alpha)/p)$, $\rho_1 := +\infty$ if $\psi \equiv 0$, and $\rho_2 := +\infty$ if $\phi \equiv 0$. Then for any $R \ge 1$ the following two statements hold:*

*(a) $\|T^{\succ}_{c,a}f\|_{[u,v),[p]} \le 2R$ whenever $f \in \mathcal{W}_p([a,v);Z)$, $\|T^{\succ}_{c,a}f(u)\| \le R$ and $\|f\|_{[u,v),[p]} \le 2R$;*
*(b) $\|Q^{\succ}_{c,a}f\|_{(u,v],[p]} \le 2R$ whenever $f \in \mathcal{W}_p([a,v];Z)$, $\|Q^{\succ}_{c,a}f(u+)\| \le R$ and $\|f\|_{(u,v],[p]} \le 2R$.*

*Proof.* If $\psi \equiv 0$ and $\phi \equiv 0$ then $T^{\succ}_{c,a}f = Q^{\succ}_{c,a}f \equiv c$, and the conclusion of the lemma holds for any $a \le u < v \le b$. Suppose that either $\psi \not\equiv 0$ or $\phi \not\equiv 0$.

To prove $(a)$, let $\kappa = \{t_i\}_{i=0}^n$ be a point partition of $[u,v) \subset J$ and let $A := (u,v)$. By additivity of the Kolmogorov integral (Theorem 2.21), the Minkowski inequality (1.5), the inequality $(\sum_i a_i^p)^{1/p} \le \sum_i a_i$ valid for any $a_i \ge 0$, and the Love–Young inequality (3.153), we have

$$\begin{aligned} &s_p(T^{\succ}_{c,a}f;\kappa)^{1/p} \\ &\quad\le \Big(\sum_{i=1}^n \Big\| \oint_{(t_{i-1},t_i]} (N_\psi f)_-\cdot d\mu \Big\|^p\Big)^{1/p} + \sum_{i=1}^n \Big\| \oint_{(t_{i-1},t_i]} (N_\phi f)_-\cdot d\nu \Big\| \\ &\quad\le K_{p,q}\|(N_\psi f)_-\|_{A,[q]} \Big(\sum_{i=1}^n v_p(\mu;(t_{i-1},t_i])\Big)^{1/p} \\ &\qquad + \big\|(N_\phi f)_-\big\|_{A,\sup} \sum_{i=1}^n v_1(\nu;(t_{i-1},t_i]). \end{aligned}$$

Thus by superadditivity of $v_p$ and $v_1$ over an interval partition (3.69),

$$\|T^{\succ}_{c,a}f\|_{[u,v),(p)} \le K_{p,q}\|N_\psi f\|_{A,[q]}\|\mu\|_{A,(p)} + \|N_\phi f\|_{A,\sup}\|\nu\|_{A,(1)}. \tag{10.25}$$

Since $\phi \in \mathcal{UG}_\beta(Z \times J; X)$, we have by Definition 10.5,

$$\|N_\phi f\|_{A,\sup} \le \|\phi\|_{\mathcal{G}_\beta}(1 + \|f\|^\beta_{A,\sup}). \tag{10.26}$$

Since $\psi \in \mathcal{HWG}_{\alpha,q}(Z \times J; Y)$, by Proposition 10.15$(b)$ with $z := 0 \in B := Z$ and $A = (u,v)$, we have

$$\big\|N_\psi f\big\|_{A,[q]} \le D(1 + \|f\|^\alpha_{A,[p]}), \tag{10.27}$$

where $D := G_{\alpha,q}(\psi) + \|N_\psi(0)\|_{\sup} + 2H_\alpha(\psi)$ with $G_{\alpha,q}(\psi)$ given by Definition 10.3. Since $\|T^{\succ}_{c,a}f\|_{[u,v),\sup} \le \|T^{\succ}_{c,a}f(u)\| + \|T^{\succ}_{c,a}f\|_{[u,v),(p)}$, it then follows that

$$
\begin{aligned}
\|T^{\succ}_{c,a}f\|_{[u,v),[p]} &\le \|T^{\succ}_{c,a}f(u)\| + 2\|T^{\succ}_{c,a}f\|_{[u,v),(p)} \\
&\le \|T^{\succ}_{c,a}f(u)\| + 2K_{p,q}D(1+\|f\|^{\alpha}_{A,[p]})\|\mu\|_{A,(p)} + 2\|\phi\|_{\mathcal{G}_\beta}(1+\|f\|^{\beta}_{A,\sup})\|\nu\|_{A,(1)} \\
&\le \|T^{\succ}_{c,a}f(u)\| + 4\Big\{K_{p,q}D\|\mu\|_{A,(p)} + \|\phi\|_{\mathcal{G}_\beta}\|\nu\|_{A,(1)}\Big\}\max\{1, \|f\|_{[u,v),[p]}\}.
\end{aligned} \tag{10.28}
$$

Thus for any $R \ge 1$, assuming that $\|T^{\succ}_{c,a}f(u)\| \le R$ and $\|f\|_{[u,v),[p]} \le 2R$, we have that $\|T^{\succ}_{c,a}f\|_{[u,v),[p]} \le 2R$ by (10.24), proving $(a)$.

To prove $(b)$, in the proof of $(a)$ we can replace $T^{\succ}_{c,a}f$ by $Q^{\succ}_{c,a}f$, $(N_\psi f)_-$ by $N_\psi f$, $(N_\phi f)_-$ by $N_\phi f$, intervals $(t_{i-1}, t_i]$ by intervals $[t_{i-1}, t_i)$, and $[u, v)$ by $(u, v]$. Now in (10.28), since $\|Q^{\succ}_{c,a}f\|_{(u,v],\sup} \le \|Q^{\succ}_{c,a}f(u+)\| + \|Q^{\succ}_{c,a}f\|_{(u,v],(p)}$, we can also replace $\|T^{\succ}_{c,a}f(u)\|$ by $\|Q^{\succ}_{c,a}f(u+)\|$. Thus for any $R \ge 1$, assuming $\|f\|_{(u,v],[p]} \le 2R$ and $\|Q^{\succ}_{c,a}f(u+)\| \le R$, it follows that $\|Q^{\succ}_{c,a}f\|_{(u,v],[p]} \le 2R$ by (10.24), proving $(b)$. The proof of the lemma is now complete. □

For the following recall that the integral transforms $T^{\succ}_{c,u}f$ and $Q^{\succ}_{c,u}f$ are defined respectively by (10.10) and (10.11), for $f \in \mathcal{W}_p(A; Z)$, $1 \le p < 2$, and $A := [u, v)$ or $[u, v]$, $a \le u < v \le b$. For $a \le u \le t \le b$ we have by additivity of Kolmogorov integrals (Theorem 2.21)

$$
T^{\succ}_{c,a}f(t) = T^{\succ}_{c,a}f(u) + T^{\succ}_{0,u}f(t). \tag{10.29}
$$

By additivity and upper continuity of Kolmogorov integrals (Proposition 2.22), we have for $a \le u < t \le b$,

$$
Q^{\succ}_{c,a}f(t) = Q^{\succ}_{c,a}f(u) + Q^{\succ}_{0,u}f(u+) + Q^{\succ}_{0,u+}f(t), \tag{10.30}
$$

where in (10.11) each $\oint_{[u+,t)}$ equals $\oint_{(u,t)}$.

The following gives bounds for the last terms on the right sides of (10.29) and (10.30). Recall the space $\mathcal{HW}^{\mathrm{loc}}_{1+\alpha,q}$ as in Definition 6.61 for $n = 1$.

**Lemma 10.20.** *Let $0 < \alpha \le 1 \le p < 1 + \alpha$, $q := p/\alpha$, $0 < R < \infty$, and $J := [a, b]$ with $a < b$. Assuming (1.14), let $\psi \in \mathcal{HW}^{\mathrm{loc}}_{1+\alpha,q}(Z \times J; X)$, let $\phi \in \mathcal{UH}^{\mathrm{loc}}_1(Z \times J; X)$ be such that $\phi(z, \cdot) \in \mathcal{R}(J; X)$ for each $z \in Z$, let $\mu \in \mathcal{AI}_p(J; Y)$ and $\nu \in \mathcal{AI}_1(J; Y)$. Suppose that an interval $[u, v] \subset J$ is such that*

$$
\begin{aligned}
&\|\mu\|_{(u,v),(p)} \le \theta_1 := \big\{1 + 16K_{p,q}[H_1(\psi) + D(1 + R^{\alpha})]\big\}^{-1}, \\
&\|\nu\|_{(u,v),(1)} \le \theta_2 := \{1 + 16H_1(\phi)\}^{-1},
\end{aligned} \tag{10.31}
$$

*where $H_1(\chi) = H_1(\chi; B_R \times J, X)$ for $\chi = \psi$ or $\phi$ and with $B_R := \{z \in Z\colon \|z\| \le R\}$, $D := H_\alpha(\psi^{(1)}_u; B_R \times J, X) + W_{\alpha,q}(\psi^{(1)}_u, R; B_R \times J, X)$, and $K_{p,q} = \zeta((1+\alpha)/p)$. Let $C := [u, v)$ for part $(a)$ and $[u, v]$ for part $(b)$ below. Let $f, g \in \mathcal{W}_p(C; Z)$ be such that $3\max\{\|f\|_{C,[p]}, \|g\|_{C,[p]}\} \le R$. Then for any $\epsilon > 0$, the following two statements hold:*

(*a*) $\|T^{\succ}_{0,u}f - T^{\succ}_{0,u}g\|_{[u,v),[p]} \le \epsilon/2$ *whenever* $\|f-g\|_{[u,v),[p]} \le 2\epsilon$;
(*b*) $\|Q^{\succ}_{0,u+}f - Q^{\succ}_{0,u+}g\|_{(u,v],[p]} \le \epsilon/2$ *whenever* $\|f-g\|_{(u,v],[p]} \le 2\epsilon$.

*Proof.* To prove $(a)$, let $A := (u,v)$. As in the proof of Lemma 10.19$(a)$, we have

$$\|T^{\succ}_{0,u}f - T^{\succ}_{0,u}g\|_{[u,v),(p)} \\ \le K_{p,q}\|N_\psi f - N_\psi g\|_{A,[q]}\|\mu\|_{A,(p)} + \|N_\phi f - N_\phi g\|_{A,\sup}\|\nu\|_{A,(1)}.$$

By Theorem 6.68 and since $\|\cdot\|_{A,(q)} \le \|\cdot\|_{A,(p)}$ by Lemma 3.45, we have

$$\|N_\psi f - N_\psi g\|_{A,[q]} \le H_1(\psi)\|f-g\|_{A,[q]} + D\big(1+\|g\|^{\alpha}_{A,(p)}\big)\|f-g\|_{A,\sup} \\ \le [H_1(\psi) + D(1+R^{\alpha})]\|f-g\|_{A,[p]}.$$

By the definition of $s$-uniform Lipschitz condition (Definition 6.42), we get

$$\|N_\phi f - N_\phi g\|_{A,\sup} \le H_1(\phi)\|f-g\|_{A,\sup}.$$

Since $(T^{\succ}_{0,u}f - T^{\succ}_{0,u}g)(u) = 0$, we then have

$$\|T^{\succ}_{0,u}f - T^{\succ}_{0,u}g\|_{[u,v),[p]} \\ \le 2\|T^{\succ}_{0,u}f - T^{\succ}_{0,u}g\|_{[u,v),(p)} \tag{10.32} \\ \le \big\{2K_{p,q}[H_1(\psi) + D(1+R^{\alpha})]\|\mu\|_{A,(p)} + 2H_1(\phi)\|\nu\|_{A,(1)}\big\}\|f-g\|_{A,[p]}.$$

For $\epsilon > 0$, assuming $\|f-g\|_{[u,v),[p]} \le 2\epsilon$, by (10.31), it then follows that

$$\|T^{\succ}_{0,u}f - T^{\succ}_{0,u}g\|_{[u,v),[p]} \le \big\{4K_{p,q}\big[H_1(\psi) + D(1+R^{\alpha})\big]\theta_1 + 4H_1(\phi)\theta_2\big\}\epsilon \le \epsilon/2.$$

The proof of statement $(a)$ is complete.

To prove $(b)$, we can replace intervals $(t_{i-1},t_i]$ in the proof of Lemma 10.19$(a)$ by intervals $[t_{i-1},t_i)$, $[u,v)$ by $(u,v]$, and $T^{\succ}_{0,u}$ by $Q^{\succ}_{0,u+}$. In this case to show the analogue of (10.32), we use the fact that

$$\|Q^{\succ}_{0,u+}g - Q^{\succ}_{0,u+}f\|_{(u,v],\sup} \le \|Q^{\succ}_{0,u+}g - Q^{\succ}_{0,u+}f\|_{(u,v],(p)},$$

which follows by Corollary 2.23. The proof of the lemma is complete. □

Finally, before proving Theorem 10.7 we recall the Banach fixed point theorem. Let $(S,d)$ be a metric space. A mapping $T\colon S \to S$ is called a *contraction* if there is a constant $0 \le q < 1$ such that $d(Tx,Ty) \le qd(x,y)$ for all $x,y \in S$. An element $x \in S$ is called a *fixed point* (for $T$) if $Tx = x$.

**Theorem 10.21** (Banach fixed point theorem). *If $(S,d)$ is complete, every contraction has a unique fixed point.*

*Proof.* Let $T^n x := (T \circ T \circ \cdots \circ T)x$ to $n$ iterations. For any $x \in S$, it is easy to check that $\{T^n x\}_{n \geq 1}$ is a Cauchy sequence, with a limit $y$ which is a fixed point, and there is no other fixed point. □

**Proof of Theorem 10.7.** It seems useful to distinguish between conditions we will call "global," namely, when the set $B$ is the entire Banach space $Z$, as in Definition 10.3 or 10.5; and other conditions we call "local," where $B$ is a proper subset, such as a ball in Lemma 10.20, and which may be indicated by the superscript $^{\rm loc}$ in definitions. We begin by considering some global hypotheses in Theorem 10.7. Since $\psi \in \mathcal{HWG}_{1,p}(Z \times J; X)$, by Definition 10.3 with $\alpha = 1$, with $p$ in place of $q$ and $B = Z$, $\psi \in \mathcal{UH}_1(Z \times J; X)$. By Definition 6.42 with $B = Z$, $S = J$ and $Y = X$, it then follows that $H_1(\psi; Z \times J, X) < \infty$. Since $\phi \in \mathcal{UG}_\beta(Z \times J, X)$, by Definition 10.5, $\|\phi\|_{Z \times J, \mathcal{G}_\beta} < \infty$.

Let

$$\begin{aligned} V &:= H(H_1(\psi; Z \times J, X) + \|\phi\|_{Z \times J, \mathcal{G}_\beta}) < \infty, \\ W &:= H(\|N_\psi(0)\|_{\rm sup} + 2\|\phi\|_{Z \times J, \mathcal{G}_\beta}) < \infty, \end{aligned} \tag{10.33}$$

where $H := \sup_{t \in J} \max\{\|\mu(\{t\})\|, \|\nu(\{t\})\|\}$.

We will use Lemmas 10.19 and 10.20, with their local hypotheses, to show that the two integral operators $T^{\succ}_{c,a}$ and $Q^{\succ}_{c,a}$ on suitable classes of functions restricted to sufficiently small intervals are contractions. By Definition 10.5, since

$$\phi \in \mathcal{UH}_1^{\rm loc}(Z \times J; X) \quad \text{and} \quad \phi(z, \cdot) \in \mathcal{R}(J; X) \quad \text{for each} \quad z \in Z, \tag{10.34}$$

$\phi \in \mathcal{UCR}(Z \times J; X)$, and so $\phi \in \mathcal{CRG}_\beta(Z \times J; X)$. This together with the hypothesis $\psi \in \mathcal{HWG}_{1,p}(Z \times J; X)$ shows that $\psi$ and $\phi$ satisfy the hypotheses on them in the second sentence of Lemma 10.19 with $\alpha = 1$ there. By Definition 10.3, $\psi \in \mathcal{WG}_{1,p}(Z \times J; X)$, and so $\psi(0, \cdot) \in \mathcal{W}_p(J; X)$. Thus by Definition 6.61 with $n = 1$,

$$\psi \in \mathcal{HW}^{\rm loc}_{1+\alpha, p/\alpha}(Z \times J; X). \tag{10.35}$$

By (10.34) and (10.35), $\phi$ and $\psi$ satisfy the hypotheses on them in the second sentence of Lemma 10.20.

Let $\rho_1$ and $\rho_2$ be defined by (10.24) with $\alpha = 1$ there. Using Proposition 3.52 twice, one can find a Young interval partition $\{(t_{j-1}, t_j)\}_{j=1}^m$ of $J = [a, b]$ such that $v_p(\mu; (t_{j-1}, t_j)) \leq \rho_1^p$ and $v_1(\nu; (t_{j-1}, t_j)) \leq \rho_2$ for each $1 \leq j \leq m$, and $m \leq 1 + v_p(\mu; J)/\rho_1^p + v_1(\nu; J)/\rho_2$ since the intersections of $m_1$ disjoint open intervals and $m_2$ such intervals yield at most $m_1 + m_2 - 1$ intervals. Thus the conclusion of Lemma 10.19 holds if $t_{j-1} \leq u < v \leq t_j$ for some $j$, which will be used in Lemmas 10.22 and 10.23.

To prove the existence of a unique solution of the integral equation (10.5), let $T^{\succ}_{c,a}$ be the integral operator defined by (10.10). For a function $g \in \mathcal{W}_p(J; Z)$ and $a \leq u < v \leq b$, since $\|g(v)\| \leq \|g\|_{[u,v),\rm sup} + \|\Delta^- g(v)\|$, we have

$$\|g\|_{[u,v],\sup} \le \|g\|_{[u,v),\sup} + \|\Delta^- g(v)\|.$$

Thus by Corollary 3.43($a$), it follows that

$$\|g\|_{[u,v],[p]} \le \|g\|_{[u,v),[p]} + 2\|\Delta^- g(v)\|. \tag{10.36}$$

Applying this to $g = T^{\succ}_{c,a} f$ with $f \in \mathcal{W}_p(J;Z)$, we have

$$\|T^{\succ}_{c,a} f\|_{[u,v],[p]} \le \|T^{\succ}_{c,a} f\|_{[u,v),[p]} + 2\|\Delta^- T^{\succ}_{c,a} f(v)\|.$$

By additivity of the Kolmogorov integral over adjoining disjoint intervals (Theorem 2.21), and then by Corollary 2.23, we have

$$\begin{aligned}
&\|\Delta^- T^{\succ}_{c,a} f(v)\| \\
&\quad = \lim_{u\uparrow v} \Big\| \fint_{(u,v]} (N_\psi f)_- \cdot \mathrm{d}\mu + \fint_{(u,v]} (N_\phi f)_- \cdot \mathrm{d}\nu \Big\| \\
&\quad = \|(N_\psi f)(v-)\cdot\mu(\{v\}) + (N_\phi f)(v-)\cdot\nu(\{v\})\| \\
&\quad \le H\{H_1(\psi)\|f\|_{[u,v),\sup} + \|N_\psi(0)\|_{\sup} + \|\phi\|_{\mathcal{G}_\beta}(1 + \|f\|^\beta_{[u,v),\sup})\} \\
&\quad \le W + V\|f\|_{[u,v),\sup}
\end{aligned}$$

for $V$ and $W$ defined by (10.33). Thus

$$\|T^{\succ}_{c,a} f\|_{[u,v],[p]} \le \|T^{\succ}_{c,a} f\|_{[u,v),[p]} + 2W + 2V\|f\|_{[u,v),\sup}. \tag{10.37}$$

Let $R_1 := \max\{\|c\|, 1\}$ and let $R_{j+1} := 2W + (2+4V)R_j$ for $j = 1, \dots, m$. Let $\mathcal{F}$ be the set of all $f \in \mathcal{W}_p(J;Z)$ such that $\|f\|_{[t_{j-1},t_j),[p]} \le 2R_j$ for $j = 1, \dots, m$ and $\|f(b)\| \le 2R_{m+1}$. By Corollary 3.43($a$) again, for $a \le u < v \le b$, we have

$$\|f\|_{[u,v],[p]} \le \|f\|_{[u,v),[p]} + \|\Delta^- f(v)\| + \|f(v)\|.$$

Then $\|f\|_{[t_{j-1},t_j],[p]} \le 4R_j + 4R_{j+1}$ for each $f \in \mathcal{F}$ and $j = 1, \dots, m$. Applying subadditivity of $\|\cdot\|_{(p)}$ and thus of $\|\cdot\|_{[p]}$ over closed, adjoining intervals (Proposition 3.35($a$)) recursively,

$$\|f\|_{J,[p]} \le \sum_{j=1}^m \|f\|_{[t_{j-1},t_j],[p]} \le 4R_1 + 8\sum_{j=2}^m R_j + 4R_{m+1} =: S < \infty \tag{10.38}$$

for each $f \in \mathcal{F}$.

In the following lemma and its proof, we apply the definitions (10.10) and (10.11) with, in place of $b$, various $B$ with $t_{j-1} \le B \le t_j$.

**Lemma 10.22.** *Let $j \in \{1, \dots, m\}$. Suppose that there is a unique function $h$ in $\mathcal{W}_p([a,t_{j-1}];Z)$ such that $T^{\succ}_{c,a} h = h$, and that $\|h\|_{[t_{r-1},t_r],[p]} \le R_{r+1}$ for $r = 1, \dots, j-1$ if $j \ge 2$. Then there is a unique function $g \in \mathcal{W}_p([a,t_j];Z)$ such that $T^{\succ}_{c,a} g = g$, and we have $\|g\|_{[t_{j-1},t_j],[p]} \le R_{j+1}$.*

*Proof.* Let $R := 3S$. For this value of $R$, let $\theta_1$ and $\theta_2$ be defined by (10.31). For $k = 1, 2$, recalling the definition (10.24) with $\alpha = 1$ of $\rho_k$, let $\sigma_k := \min\{\theta_k, \rho_k\}$ if $m > 1$ and $\sigma_k := \theta_k$ if $m = 1$. Again using Proposition 3.52 twice, one can find a Young interval partition $\{(s_{i-1}, s_i)\}_{i=1}^n$ of $J$ such that $\|\mu\|_{(s_{i-1},s_i),(p)} \leq \sigma_1$ and $\|\nu\|_{(s_{i-1},s_i),(1)} \leq \sigma_2$ for each $1 \leq i \leq n$. We can assume that $\{s_i\}_{i=0}^n$ is a refinement of $\{t_j\}_{j=0}^m$ and $n \leq m + v_p(\mu; J)/\sigma_1^p + v_1(\nu; J)/\sigma_2$.

Since the partition $\{s_i\}_{i=0}^n$ of $J$ is a refinement of $\{t_j\}_{j=0}^m$, $t_{j-1} = s_{i_{j-1}} < \cdots < s_{i_j} = t_j$ for some integers $0 \leq i_{j-1} < i_j \leq n$. Let $u_l := s_{i_{j-1}+l}$ for $l = 0, \ldots, k := i_j - i_{j-1}$, so that $u_0 = t_{j-1}$, and let $h(t) := h(u_0)$ for $t \in (u_0, u_1)$. Let $\mathcal{H}_1 := \mathcal{H}_{1,j}$ be the set of all $f \in \mathcal{W}_p([a, u_1); Z)$ such that $f = h$ on $[a, u_0]$ and $\|f\|_{[u_0,u_1),[p]} \leq 2R_j$. The set $\mathcal{H}_1$ is closed in $\mathcal{W}_p([a, u_1); Z)$ and nonempty since $h \in \mathcal{H}_1$. Thus $\mathcal{H}_1$ is a complete metric space with the metric $d_1(f, g) := \|f - g\|_{[u_0,u_1),[p]}$. By the definition (10.10) of the integral operator $T_{c,a}^{\succ}$, for each $f \in \mathcal{H}_1$, $T_{c,a}^{\succ} f(u_0) = T_{c,a}^{\succ} h(u_0) = h(u_0)$, and so $\|T_{c,a}^{\succ} f(u_0)\| \leq R_j$ by hypothesis if $j \geq 2$ and if $j = 1$, since $h(a) = T_{c,a}^{\succ} h(a) = c$ and by definition of $R_1$. We will apply Lemmas 10.19 and 10.20, both with $u = u_0$ and $v = u_1$. Here (10.24) and (10.31) hold by choice of $\{s_i\}_{i=0}^n$. The hypotheses in the second sentence of each lemma were verified around (10.35). By Lemma 10.19(*a*), with $R$ there equal to $R_j$ here, $T_{c,a}^{\succ}$ takes the set $\mathcal{H}_1$ into itself. Also for each $f \in \mathcal{H}_1$, letting $c_1 := T_{c,a}^{\succ} f(u_0)$, by (10.29) with $u = u_0$, we have $T_{c,a}^{\succ} f = T_{c_1,u_0}^{\succ} f$ on $[u_0, u_1)$.

By definition of the class $\mathcal{H}_1$, for each $f \in \mathcal{H}_1$, we have $\|f\|_{[u_0,u_1),[p]} \leq 2R_j$. We have $2R_j < S = R/3$ by (10.38). Thus we have checked all the hypotheses of Lemma 10.20.

By Lemma 10.20(*a*), for $f_1, f_2 \in \mathcal{H}_1$, taking $\epsilon := (1/2)\|f_1 - f_2\|_{[u_0,u_1),[p]}$, we have

$$\begin{aligned} \|T_{c_1,u_0}^{\succ} f_1 - T_{c_1,u_0}^{\succ} f_2\|_{[u_0,u_1),[p]} &= \|T_{0,u_0}^{\succ} f_1 - T_{0,u_0}^{\succ} f_2\|_{[u_0,u_1),[p]} \\ &\leq (1/4)\|f_1 - f_2\|_{[u_0,u_1),[p]}, \end{aligned} \tag{10.39}$$

and so $T_{c,a}^{\succ}$ is a contraction for $d_1$. Thus by the Banach fixed point theorem (Theorem 10.21), there is a unique function

$$h_1 \in \mathcal{H}_1 \text{ such that } T_{c,a}^{\succ} h_1 = h_1 \text{ on } [a, u_1). \tag{10.40}$$

In particular, $h_1(t_{j-1}) = T_{c,a}^{\succ} h_1(t_{j-1}) = T_{c,a}^{\succ} h(t_{j-1}) = h(t_{j-1})$. Suppose that there is another function $g$ in $\mathcal{W}_p([a, u_1); Z)$ such that $T_{c,a}^{\succ} g = g$ on $[a, u_1)$. We have that $g = h_1 = h$ on $[a, u_0]$ since such a function $h$ is unique by hypothesis. Let $u := \sup\{t \in [u_0, u_1)\colon g(t) = h_1(t)\}$, which is defined since $g(u_0) = h_1(u_0)$. If $u = u_1$ then we are done. Suppose $u < u_1$. We have $u = u_0$, or $u > u_0$ and $g = h_1$ on $[u_0, u)$. In the latter case,

$$\begin{aligned} (g - h_1)(u) &= \Delta^-(g - h_1)(u) \\ &= (N_\psi g - N_\psi h_1)(u-)\cdot\mu(\{u\}) + (N_\phi g - N_\phi h_1)(u-)\cdot\nu(\{u\}) = 0, \end{aligned}$$

by upper continuity of the Kolmogorov integral (Corollary 2.23), and so $g(u) = h_1(u)$. Recall that at the beginning of this proof, $\theta_1$ and $\theta_2$ were defined by

(10.31) for $R := 3S$ and $S$ from (10.38). We will choose $v > u$ close enough to $u$ to satisfy three conditions. First, by Proposition 3.50, the interval function $A \mapsto v_p(\mu; A)$, $A \in \mathfrak{I}(J)$, is upper continuous at $\emptyset$, so for $v \in (u, u_1]$ close enough to $u$ we have $\|\mu\|_{(u,v),(p)} < \theta_1$. Second, likewise, we can and will make $\|\nu\|_{(u,v),(1)} < \theta_2$. Third, by (10.40), we have $h_1 \in \mathcal{H}_1$ and so by definition of $\mathcal{H}_1$, $\|h\|_{[u_0,u_1)} \le 2R_j$. We have $2R_j < S = R/3$ by (10.38). Since $T^{\succ}_{c,a}g = g$ on $[a, u_1)$, by additivity and upper continuity of the Kolmogorov integral (Proposition 2.22), $g$ is right-continuous on $[a, u_1)$. Thus by Proposition 3.42, $\lim_{v\downarrow u} \|g\|_{[u,v],(p)} = 0$. So by the subadditivity of $\|\cdot\|_{(p)}$ over adjoining closed intervals (3.53), we can and do assume that $\|g\|_{[u_0,v],[p]} < S = R/3$. This finishes the choice of $v$.

Let $\epsilon := (1/2)\|g - h_1\|_{[u,v),[p]}$. We have $(g - h_1)(t) = T^{\succ}_{0,u}g(t) - T^{\succ}_{0,u}h_1(t)$ for each $t \in [u, u_1)$. By Lemma 10.20($a$), it follows that

$$2\epsilon = \|g - h_1\|_{[u,v),[p]} = \|T^{\succ}_{0,u}g - T^{\succ}_{0,u}h_1\|_{[u,v),[p]} \le \epsilon/2.$$

Thus $\epsilon = 0$, and hence $g = h_1$ on $[u, v)$, a contradiction, proving $u = u_1$. So $h_1$ is the unique function in $\mathcal{W}_p([a, u_1); Z)$ such that $T^{\succ}_{c,a}h_1 = h_1$.

Now suppose that for some $2 \le l < k$, $h_{l-1}$ is the unique function in $\mathcal{W}_p([a, u_{l-1}); Z)$ such that $T^{\succ}_{c,a}h_{l-1} = h_{l-1}$ on $[a, u_{l-1})$. Also suppose that $h_{l-1}(u_0) = h(u_0)$ and $\|h_{l-1}\|_{[u_0,u_{l-1}),[p]} \le 2R_j$, which hold for $l = 2$ as shown above. Let $h_{l-1}(t) := h_{l-1}(u_{l-1}-)$ for $t \in [u_{l-1}, u_l)$, and let $\mathcal{H}_l$ be the set of all $f \in \mathcal{W}_p([a, u_l); Z)$ such that $f = h_{l-1}$ on $[a, u_{l-1})$ and $\|f\|_{[u_0,u_l),[p]} \le 2R_j$. The set $\mathcal{H}_l$ is closed in $\mathcal{W}_p([a, u_l); Z)$ and nonempty since $h_{l-1} \in \mathcal{H}_l$. Thus $\mathcal{H}_l$ is a complete metric space with the metric $d_l(f, g) := \|f - g\|_{[u_{l-1},u_l),[p]}$. Note that if $f \ne g$ in $\mathcal{H}_l$, since $f \equiv g \equiv h_l$ on $[a, u_{l-1})$, we have $f \ne g$ somewhere in $[u_{l-1}, u_l)$. By the definition (10.10) of the integral operator $T^{\succ}_{c,a}$, for each $f \in \mathcal{H}_l$, $T^{\succ}_{c,a}f(u_0) = T^{\succ}_{c,a}h(u_0) = h(u_0)$, and so $\|T^{\succ}_{c,a}f(u_0)\| \le R_j$. Thus by Lemma 10.19($a$), $T^{\succ}_{c,a}$ takes the set $\mathcal{H}_l$ into itself. By definition of the class $\mathcal{H}_l$, for each $f \in \mathcal{H}_l$, we have $\|f\|_{[u_{l-1},u_l),[p]} \le 2R_j$. We have $2R_j < S = R/3$ by (10.38). Also all the functions in $\mathcal{H}_l$ have the same value $c_l := T^{\succ}_{c,a}h_{l-1}(u_{l-1})$ at $u_{l-1}$, and so by (10.29) with $u = u_{l-1}$, we have $T^{\succ}_{c,a}f = T^{\succ}_{c_l,u_{l-1}}f$ on $[u_{l-1}, u_l)$ for each $f \in \mathcal{H}_l$.

By (10.39) with $c_l$ in place of $c_1$, $[u_{l-1}, u_l)$ in place of $[u_0, u_1)$ and $f_1, f_2 \in \mathcal{H}_l$, $T^{\succ}_{c,a}$ is a contraction for $d_l$. Thus by the Banach fixed point theorem (Theorem 10.21), there is a unique function $h_l \in \mathcal{H}_l$ such that $T^{\succ}_{c,a}h_l = h_l$ on $[a, u_l)$. In particular, $h_l(t_{j-1}) = T^{\succ}_{c,a}h_l(t_{j-1}) = T^{\succ}_{c,a}h(t_{j-1}) = h(t_{j-1})$. It follows as in the case $l = 1$ that $h_l$ is the unique such function in $\mathcal{W}_p([a, u_l); Z)$. By induction, there is a unique function $g \in \mathcal{W}_p([a, t_j); Z)$ such that $T^{\succ}_{c,a}g = g$, and $\|g\|_{[t_{j-1},t_j),[p]} \le R_j$. Let

$$\begin{aligned} g(t_j) &:= g(t_j-) + (N_\psi g)(t_j-)\cdot\mu(\{t_j\}) + (N_\phi g)(t_j-)\cdot\nu(\{t_j\}) \\ &= g(t_j-) + \Delta^-(T^{\succ}_{c,a}g)(t_j) = (T^{\succ}_{c,a}g)(t_j), \end{aligned}$$

where the second equality holds by Corollary 2.23. Then $T^{\succ}_{c,a}g = g$ on $[a, t_j]$ and $\|g\|_{[t_{j-1},t_j],[p]} \leq R_{j+1}$ by (10.37), proving the lemma. □

Now returning to the proof of Theorem 10.7, let $h(a) := c = T^{\succ}_{c,a}h(a)$. Recall that $R_j$ are defined after (10.37). Since the $p$-variation on a singleton is zero, $\|h\|_{\{a\},[p]} = \|h(a)\| \leq R_1$. By Lemma 10.22 with $j = 1$, there is a unique function $g_1 \in \mathcal{W}_p([a, t_1]; Z)$ such that $T^{\succ}_{c,a}g_1 = g_1$, and we have $\|g_1\|_{[p]} \leq R_2$. Suppose that for some $2 \leq j < m$, there is a unique function $g_{j-1}$ in $\mathcal{W}_p([a, t_{j-1}]; Z)$ such that $T^{\succ}_{c,a}g_{j-1} = g_{j-1}$ and $\|g_{j-1}\|_{[t_{r-1},t_r],[p]} \leq R_{r+1}$ for $r = 1, \ldots, j-1$. Then by Lemma 10.22 with $h = g_{j-1}$, there is a unique function $g_j \in \mathcal{W}_p([a, t_j]; Z)$ such that $T^{\succ}_{c,a}g_j = g_j$, and we have $\|g_j\|_{[t_{j-1},t_j],[p]} \leq R_{j+1}$.

Also since $g_j = g_{j-1}$ on $[a, t_{j-1}]$, $\|g_j\|_{[t_{r-1},t_r],[p]} \leq R_{r+1}$ for $r = 1, \ldots, j-1$ as well as $r = j$. Now a unique solution $f = g_m$ of the integral equation (10.5) in $\mathcal{W}_p(J; Z)$ exists by induction.

To prove the existence of a unique solution of the integral equation (10.6) in $\mathcal{W}_p(J; Z)$, let $Q^{\succ}_{c,a}$ be the integral operator defined by (10.11). By additivity of the Kolmogorov integral over adjoining disjoint intervals (Theorem 2.21), and then by Corollary 2.23, for $a \leq u < b$, we have

$$\begin{aligned}
\|\Delta^+ Q^{\succ}_{c,a}f(u)\| &= \lim_{v \downarrow u} \Big\| \oint_{[u,v)} (N_\psi f) \cdot d\mu + \oint_{[u,v)} (N_\phi f) \cdot d\nu \Big\| \\
&= \|(N_\psi f)(u) \cdot \mu(\{u\}) + (N_\phi f)(u) \cdot \nu(\{u\})\| \\
&\leq H\{H_1(\psi)\|f(u)\| + \|N_\psi(0)\|_{\sup} + \|\phi\|_{\mathcal{G}_\beta}(1 + \|f(u)\|^\beta)\} \\
&\leq W + V\|f(u)\|
\end{aligned}$$

for $V$ and $W$ defined by (10.33) and $H$ just after it. Thus assuming $Q^{\succ}_{c,a}f(u) = f(u)$, we have

$$\|Q^{\succ}_{c,a}f(u+)\| \leq \|Q^{\succ}_{c,a}f(u)\| + \|\Delta^+ Q^{\succ}_{c,a}f(u)\| \leq W + (1 + V)\|f(u)\|. \tag{10.41}$$

Let $R_0 := \max\{1, \|c\|\}$ and let $R_j := W + (1 + V)2R_{j-1}$ for $j = 1, \ldots, m$. Let $\mathcal{F}$ be the set of all $f\colon J \to Z$ such that $f(a) = c$ and $\|f\|_{(t_{j-1},t_j],[p]} \leq 2R_j$ for $j = 1, \ldots, m$. By Corollary 3.43($b$), for $a \leq u < v \leq b$, we have

$$\|f\|_{[u,v],[p]} \leq \|f(u)\| + \|\Delta^+ f(u)\| + \|f\|_{(u,v],[p]}.$$

Then $\|f\|_{[t_{j-1},t_j],[p]} \leq 4R_{j-1} + 4R_j$ for each $f \in \mathcal{F}$ and $j = 1, \ldots, m$. Applying subadditivity of $\|\cdot\|_{(p)}$ and thus of $\|\cdot\|_{[p]}$ over closed, adjoining intervals (Proposition 3.35($a$)) recursively, it follows that

$$\|f\|_{J,[p]} \leq \sum_{j=1}^{m} \|f\|_{[t_{j-1},t_j],[p]} \leq 4R_0 + 8\sum_{j=1}^{m-1} R_j + 4R_m =: S < \infty \tag{10.42}$$

for each $f \in \mathcal{F}$.

Here we will use another lemma.

**Lemma 10.23.** *Let $j \in \{1, \ldots, m\}$. Suppose that there is a unique function $h$ in $\mathcal{W}_p([a, t_{j-1}]; Z)$ such that $Q^{\succ}_{c,a}h = h$, and that $\|h\|_{(t_{r-1},t_r],[p]} \leq 2R_r$ for $r = 1, \ldots, j-1$ if $j \geq 2$. Then there is a unique function $g \in \mathcal{W}_p([a, t_j]; Z)$ such that $Q^{\succ}_{c,a}g = g$, and we have $\|g\|_{(t_{j-1},t_j],[p]} \leq 2R_j$.*

*Proof.* Again let $R := 3S$. For this value of $R$, let $\theta_1$ and $\theta_2$ be defined by (10.31). For $k = 1, 2$, recalling the definition (10.24) with $\alpha = 1$ of $\rho_k$, let $\sigma_k := \min\{\theta_k, \rho_k\}$ if $m > 1$ and $\sigma_k := \theta_k$ if $m = 1$. Again using Proposition 3.52 twice, one can find a Young interval partition $\{(s_{i-1}, s_i)\}_{i=1}^n$ of $J$ such that $\|\mu\|_{(s_{i-1},s_i),(p)} \leq \sigma_1$ and $\|\nu\|_{(s_{i-1},s_i),(1)} \leq \sigma_2$ for each $1 \leq i \leq n$. We can assume that $\{s_i\}_{i=0}^n$ is a refinement of $\{t_j\}_{j=0}^m$ and $n \leq m + v_p(\mu; J)/\sigma_1^p + v_1(\nu; J)/\sigma_2$. Then for each $i = 0, \ldots, n-1$ and each $f, g \in \mathcal{W}_p([s_i, s_{i+1}]; Z)$ such that $\max\{\|f\|_{[p]}, \|g\|_{[p]}\} \leq S$, by Lemma 10.20(*b*) with $\epsilon := (1/2)\|f - g\|_{(s_i,s_{i+1}],[p]}$, we have

$$\begin{aligned}\|Q^{\succ}_{z,s_i+}f - Q^{\succ}_{z,s_i+}g\|_{(s_i,s_{i+1}],[p]} &= \|Q^{\succ}_{0,s_i+}f - Q^{\succ}_{0,s_i+}g\|_{(s_i,s_{i+1}],[p]} \\ &\leq (1/4)\|f - g\|_{(s_i,s_{i+1}],[p]} \qquad (10.43)\end{aligned}$$

for any $z \in Z$. Since the partition $\{s_i\}_{i=0}^n$ of $J$ is a refinement of $\{t_j\}_{j=0}^m$, $t_{j-1} = s_{i_{j-1}} < \cdots < s_{i_j} = t_j$ for some integers $0 \leq i_{j-1} < i_j \leq n$. Let $u_l := s_{i_{j-1}+l}$ for $l = 0, \ldots, k := i_j - i_{j-1}$, so that $u_0 = t_{j-1}$, and let $h(t) := h(u_0)$ for $t \in (u_0, u_1]$. Let $\mathcal{H}_1$ be the set of all $f \in \mathcal{W}_p([a, u_1]; Z)$ such that $f = h$ on $[a, u_0]$ and $\|f\|_{(u_0,u_1],[p]} \leq 2R_j$. If $j = 1$, $\|h\|_{(u_0,u_1],[p]} = \|c\|_{(u_0,u_1],[p]} = \|c\| \leq R_0 \leq 2R_1$. The set $\mathcal{H}_1$ is closed in $\mathcal{W}_p([a, u_1]; Z)$ and nonempty since $h \in \mathcal{H}_1$. Thus $\mathcal{H}_1$ is a complete metric space with the metric $d_1(f, g) := \|f - g\|_{(u_0,u_1],[p]}$. Moreover, for each $f \in \mathcal{H}_1$, $\|f\|_{[a,u_1],[p]} \leq R/3$ by (10.42). We have $\|h(u_0)\| \leq 2R_{j-1}$ by hypothesis if $j \geq 2$, and since $h(a) = Q^{\succ}_{c,a}h(a) = c$ if $j = 1$. Thus $\|Q^{\succ}_{c,a}f(u_0+)\| \leq R_j$ for each $f \in \mathcal{H}_1$ by (10.41) and the definition of $R_j$. Also since $Q^{\succ}_{c,a}f = Q^{\succ}_{c,a}h = h$ on $[a, u_0]$, the integral transform $Q^{\succ}_{c,a}$ takes $\mathcal{H}_1$ into itself by Lemma 10.19(*b*), with $R$ there equal to $R_j$ here. We have $c_1 := Q^{\succ}_{c,a}h(u_0+) = Q^{\succ}_{c,a}f(u_0+)$ for each $f \in \mathcal{H}_1$, and so by (10.30) with $u = u_0$, $Q^{\succ}_{c,a}f = Q^{\succ}_{c_1,u_0+}f$ on $(u_0, u_1]$ for each $f \in \mathcal{H}_1$. By (10.43) with $z = c_1$, $Q^{\succ}_{c,a}$ is a contraction for $d_1$, and so by the Banach fixed point theorem (Theorem 10.21), there is a unique function $h_1 \in \mathcal{H}_1$ such that $Q^{\succ}_{c,a}h_1 = h_1$ on $[a, u_1]$. In particular, $\|h_1(t_{j-1})\| = \|h(t_{j-1})\| \leq 2R_{j-1}$. Suppose that there is another function $f$ in $\mathcal{W}_p([a, u_1]; Z)$ such that $Q^{\succ}_{c,a}f = f$ on $[a, u_1]$. We have that $f = h_1 = h$ on $[a, u_0]$ since such a function $h$ is unique by hypothesis. Let $u := \sup\{t \in [u_0, u_1]\colon f(t) = h_1(t)\}$, which is defined since $f(u_0) = h_1(u_0)$. Then $u = u_0$, or $u > u_0$ and $f = h_1$ on $[a, u)$. In the latter case, by the definition (10.11) of $Q^{\succ}_{a,c}$, we also have $f(u) = Q^{\succ}_{c,a}f(u) = Q^{\succ}_{c,a}h_1(u) = h_1(u)$. If $u = u_1$ then we are done. Suppose $u < u_1$. Recall that now $\theta_1$ and $\theta_2$ are defined by (10.31) for $R := 3S$ and $S$ from (10.42). By Proposition 3.50, there exists $v \in (u, u_1]$ such that $\|\mu\|_{(u,v),(p)} < \theta_1$ and $\|\nu\|_{(u,v),(1)} < \theta_2$. Let $\epsilon := (1/2)\|f - h_1\|_{(u,v],[p]}$. We have $(f - h_1)(t) = Q^{\succ}_{0,u+}f(t) - Q^{\succ}_{0,u+}h_1(t)$ for each $t \in (u, u_1]$. By Lemma 10.20(*b*), it follows that

$$2\epsilon = \|f - h_1\|_{(u,v],[p]} = \|Q^{\succ}_{0,u+}f - Q^{\succ}_{0,u+}h_1\|_{(u,v],[p]} \le \epsilon/2.$$

Thus $\epsilon = 0$, and hence $f = h_1$ on $(u, v]$, a contradiction, proving $u = u_1$, and so $h_1$ is the unique function in $\mathcal{W}_p([a, u_1]; Z)$ such that $Q^{\succ}_{c,a}h_1 = h_1$.

Now suppose that for some $2 \le l < k$, $h_{l-1}$ is the unique function in $\mathcal{W}_p([a, u_{l-1}]; Z)$ such that $Q^{\succ}_{c,a}h_{l-1} = h_{l-1}$. Also suppose that $\|h_{l-1}(u_0)\| \le 2R_{j-1}$ and $\|h_{l-1}\|_{(u_0,u_{l-1}],[p]} \le 2R_j$, which hold for $l = 2$ as shown above. Let $h_{l-1}(t) := h_{l-1}(u_{l-1})$ for $t \in (u_{l-1}, u_l]$, and let $\mathcal{H}_l$ be the set of all $f \in \mathcal{W}_p([a, u_l]; Z)$ such that $f = h_{l-1}$ on $[a, u_{l-1}]$ and $\|f\|_{(t_{j-1},u_l],[p]} \le 2R_j$. The set $\mathcal{H}_l$ is closed in $\mathcal{W}_p([a, u_l]; Z)$ and nonempty since $h_{l-1} \in \mathcal{H}_l$. Thus $\mathcal{H}_l$ is a complete metric space with the metric $d_l(f, g) := \|f - g\|_{(u_{l-1},u_l],[p]}$. For each $f \in \mathcal{H}_l$, $\|f\|_{[a,u_l],[p]} \le R/3$ by (10.42). Also for each $f \in \mathcal{H}_l$, $f = h_{l-1}$ on $[a, u_{l-1}]$ implies $\|f(u_0)\| \le 2R_{j-1}$, and so $\|Q^{\succ}_{c,a}f(u_0+)\| \le R_j$ by (10.41) and the definition of $R_j$. Thus the integral transform $Q^{\succ}_{c,a}$ takes $\mathcal{H}_l$ into itself by Lemma 10.19($b$), with $R$ there equal to $R_j$ here, and since $Q^{\succ}_{c,a}f = h_{l-1}$ on $[a, u_{l-1}]$. All the functions in $\mathcal{H}_l$ have the same value $c_l := Q^{\succ}_{c,a}h(u_{l-1}+)$ at $u_{l-1}+$, and so by (10.30) with $u = u_{l-1}$, $Q^{\succ}_{c,a}f = Q^{\succ}_{c_l,u_{l-1}+}f$ on $(u_{l-1}, u_l]$ for each $f \in \mathcal{H}_l$. By (10.43) with $z = c_l$, $Q^{\succ}_{c,a}$ is a contraction for $d_l$, and so by the Banach fixed point theorem (Theorem 10.21), there is a function $h_l \in \mathcal{H}_l$ such that $Q^{\succ}_{c,a}h_l = h_l$ on $[a, u_l]$. It follows as in the case $l = 1$ that $h_l$ is the unique such function in $\mathcal{W}_p([a, u_l]; Z)$. We have $\|h_l\|_{(u_0,u_l],[p]} \le 2R_j$ by definition of $\mathcal{H}_l$. The conclusion of the lemma now follows by induction. □

Now to finish the proof of Theorem 10.7, let $h(a) := c = Q^{\succ}_{c,a}h(a)$. By Lemma 10.23 with $j = 1$, there is a unique function $g_1 \in \mathcal{W}_p([a, t_1]; Z)$ such that $Q^{\succ}_{c,a}g_1 = g_1$ and $\|g_1\|_{(a,t_1],[p]} \le 2R_1$. Suppose that for some $2 \le j < m$, there is a unique function $g_{j-1}$ in $\mathcal{W}_p([a, t_{j-1}]; Z)$ such that $Q^{\succ}_{c,a}g_{j-1} = g_{j-1}$ and $\|g\|_{(t_{r-1},t_r],[p]} \le 2R_r$ for $r = 1, \ldots, j-1$. Then by Lemma 10.23 with $h = g_{j-1}$, there is a unique function $g_j \in \mathcal{W}_p([a, t_j]; Z)$ such that $Q^{\succ}_{c,a}g_j = g_j$, and we have $\|g_j\|_{(t_{j-1},t_j],[p]} \le 2R_j$. Now a unique solution $f = g_m$ of the integral equation (10.6) in $\mathcal{W}_p(J; Z)$ exists by induction.

To prove the existence of a solution of the backward integral equation (10.7) we use a change of variables for the Kolmogorov integral as follows. The function $\theta(t) := a + b - t$ takes $[a, b]$ 1-to-1 onto itself, homeomorphically, interchanging $a$ and $b$, with $\theta \equiv \theta^{-1}$ on $[a, b]$. Let $\widetilde{\mu} := \mu\circ\theta^{-1}$, $\widetilde{\nu} := \nu\circ\theta^{-1}$, $\widetilde{\psi}(z, t) := \psi(z, \theta(t))$, and $\widetilde{\phi}(z, t) := \phi(z, \theta(t))$ for $a \le t \le b$ and $z \in Z$. It is easy to check that $\widetilde{\mu}$, $\widetilde{\nu}$, $\widetilde{\psi}$, and $\widetilde{\phi}$ satisfy the hypotheses of the present theorem, and so by the first part of the proof there exists a solution $\widetilde{f}$ of the integral equation

$$\widetilde{f}(t) = c + \oint_{(a,t]} (N_{\widetilde{\psi}}\widetilde{f})_{-}\cdot d\widetilde{\mu} + \oint_{(a,t]} (N_{\widetilde{\phi}}\widetilde{f})_{-}\cdot d\widetilde{\nu}, \qquad a \le t \le b. \tag{10.44}$$

Let $f(t) := \widetilde{f}(\theta(t))$ for $a \le t \le b$. Then $f(b) = c$ and for $a \le t < b$,

$$f(t) = c + \oint_{(a,a+b-t]} \widetilde{\psi}(\widetilde{f}(s),s)_- \cdot \mathrm{d}\widetilde{\mu}(s) + \oint_{(a,a+b-t]} \widetilde{\phi}(\widetilde{f}(s),s)_- \cdot \mathrm{d}\widetilde{\nu}(s).$$

Now for the first integral on the right side, we have

$$\begin{aligned} &\oint_{(a,a+b-t]} \widetilde{\psi}(\widetilde{f}(s),s)_- \cdot \mathrm{d}\widetilde{\mu}(s) \\ &\quad = \oint_{(a,a+b-t]} \psi(f(a+b-s), a+b-s)_- \cdot \mathrm{d}\widetilde{\mu}(s) \\ &\quad = \oint_{[t,b)} \psi(f(u),u)_+ \cdot \mathrm{d}\mu(u) = \oint_{[t,b)} (N_\psi f)_+ \cdot \mathrm{d}\mu \end{aligned} \tag{10.45}$$

by a change of variables for Kolmogorov integrals, Proposition 2.30. Similarly, for the second integral it follows that for $a \le t < b$,

$$\oint_{(a,a+b-t]} \widetilde{\phi}(\widetilde{f}(s),s)_- \cdot \mathrm{d}\widetilde{\nu}(s) = \oint_{[t,b)} (N_\phi f)_+ \cdot \mathrm{d}\nu. \tag{10.46}$$

Thus $f$ is a solution of the backward integral equation (10.7). To show its uniqueness in $\mathcal{W}_p(J;Z)$, let $g$ be another solution of (10.7), and let $\widetilde{g}(t) := g(\theta(t))$ for $a \le t \le b$. Again using a change of variables for Kolmogorov integrals it follows that $\widetilde{g}$ is a solution of (10.44). Since $\widetilde{g} \neq \widetilde{f}$, this contradicts the uniqueness of the solution of (10.44), shown in the first part of the proof. This contradiction proves uniqueness in $\mathcal{W}_p(J;Z)$ of a solution of (10.7).

The same argument based on the use of a change of variables for Kolmogorov integrals proves that the backward integral equation (10.8) has a unique solution in $\mathcal{W}_p(J;Z)$. The proof of Theorem 10.7 is complete. □

The existence and uniqueness of solutions of the integral equations defined by autonomous Nemytskii operators follow just by verifying the assumptions of Theorem 10.7 as follows.

**Proof of Corollary 10.8.** Let $\psi(z,s) \equiv F(z)$ and $\phi(z,s) \equiv G(z)$. By Example 10.16 with $h \equiv 1$ and $\alpha = 1$ there, statement $(a)$ of Proposition 10.15 with $\alpha = 1$ holds for $\psi$, that is (10.9) holds with $C = 0$ for each $0 \le K < \infty$, and so $\psi \in \mathcal{HWG}_{1,p}(Z \times J; X)$ by Definition 10.3 with $\alpha = 1$ and $q = p$. Returning to $\alpha \in (0,1]$ in the statement of Corollary 10.8, since $W_{\alpha,q}(\psi_u^{(1)}, K) \equiv 0$ with $q = p/\alpha$ for each $0 \le K < \infty$, $\psi_u^{(1)} \in \mathcal{HW}_{\alpha,q}^{\mathrm{loc}}(Z \times J; L(Z;X))$ by Definitions 6.47 and 6.51. Since clearly $\phi \in \mathcal{UG}_\beta(Z \times J; X) \cap \mathcal{UH}_1^{\mathrm{loc}}(Z \times J; X)$ and $\phi(z,\cdot) \in \mathcal{R}(J;X)$ for each $z \in Z$, all the hypotheses on $\psi$ and $\phi$ of Theorem 10.7 hold, and so the conclusion follows. □

## 10.4 Boundedness and Convergence of Picard Iterates

Before proving Theorem 10.11, we will show that the four sequences of Picard iterates defined by (10.12), (10.13), (10.16), and (10.17) are uniformly bounded in $n$ with respect to the $p$-variation norm for $\psi$ and $\phi$ as in Lemma 10.19.

**Proposition 10.24.** *Let $\alpha, \beta \in (0,1]$, $1 \le p < 1+\alpha$, and $J := [a,b]$ with $a < b$. Assuming (1.14), let $\psi \in \mathcal{HWG}_{\alpha,p/\alpha}(Z \times J; X)$, $\phi \in \mathcal{CRG}_\beta(Z \times J; X)$, $\mu \in \mathcal{AI}_p(J;Y)$, and $\nu \in \mathcal{AI}_1(J;Y)$. Then there is an integer $1 \le m \le 1 + v_p(\mu;J)/\rho_1^p + v_1(\nu;J)/\rho_2$, with $\rho_1$ and $\rho_2$ defined by (10.24), such that for all $n = 0,1,2,\dots$,*

$$\max\big\{\|f_n^{\succ}\|_{[p]}, \|f_n^{\prec}\|_{[p]}\big\} \le C2^{m+1}\big(1+2DH\big)^m \tag{10.47}$$

*and*

$$\max\big\{\|g_n^{\succ}\|_{[p]}, \|g_n^{\prec}\|_{[p]}\big\} \le C3^{m+1}\big(1+DH\big)^m, \tag{10.48}$$

*where $C := \max\{1, \|c\|\}$, $D := H_\alpha(\psi) + \|N_\psi(0)\|_{\sup} + 2\|\phi\|_{\mathcal{G}_\beta}$, and $H := \sup_{t\in J}\max\{\|\mu(\{t\})\|, \|\nu(\{t\})\|\}$.*

*Proof.* We can assume that either $\psi \not\equiv 0$ or $\phi \not\equiv 0$. Using Proposition 3.52 twice, one can find a Young interval partition $\{(z_{j-1}, z_j)\}_{j=1}^m$ of $J = [a,b]$ such that $v_p(\mu;(z_{j-1},z_j)) \le \rho_1^p$ and $v_1(\nu;(z_{j-1},z_j)) \le \rho_2$ for each $1 \le j \le m$, and $m \le 1 + v_p(\mu;J)/\rho_1^p + v_1(\nu;J)/\rho_2$ since the intersections of $m_1$ disjoint open intervals and $m_2$ such intervals yield at most $m_1 + m_2 - 1$ intervals.

Let $\{f_n\}_{n\ge 0}$ be either of the two sequences of forward Picard iterates $\{f_n^{\succ}\}$ or $\{g_n^{\succ}\}$. We will show that for $j = 1,\dots,m$,

$$A_{j,n} := \big\|f_n\big\|_{[z_{j-1},z_j],[p]} \le CA^j \qquad \text{for } n = 0,1,2,\dots, \tag{10.49}$$

where $A = 2(1+2DH)$ if $\{f_n\}$ is $\{f_n^{\succ}\}$ and $A = 3(1+DH)$ if $\{f_n\}$ is $\{g_n^{\succ}\}$. Assuming that (10.49) holds, and applying Proposition 3.35(*a*) recursively, it will follow that $\|f_n\|_{[a,b],[p]} \le \sum_{j=1}^m A_{j,n} \le 2CA^m$ for each $n = 0,1,2,\dots$, which is (10.47) and implies (10.48) for the two sequences of forward Picard iterates.

To prove (10.47) and (10.48) for the two sequences of backward Picard iterates we use a time reversal for the Kolmogorov integral as in (10.44) through (10.46). Recall the function $\theta(t) := a+b-t$ taking $[a,b]$ one-to-one onto itself, and that $\widetilde{\mu} \equiv \mu\circ\theta^{-1}$, $\widetilde{\nu} \equiv \nu\circ\theta^{-1}$, $\widetilde{\psi}(z,t) \equiv \psi(z,\theta(t))$, and $\widetilde{\phi}(z,t) \equiv \phi(z,\theta(t))$ for $a \le t \le b$ and $z \in Z$. Let $\{\widetilde{f}_n^{\succ}\}$ be the sequence of Picard iterates obtained by the integral transform $T^{\succ}_{c,a,\widetilde{\mu},\widetilde{\nu}}$ and $\widetilde{\psi}$, $\widetilde{\phi}$. It is easy to check that $\widetilde{\mu}$, $\widetilde{\nu}$, $\widetilde{\psi}$, and $\widetilde{\phi}$ satisfy the hypotheses of the present theorem, and so (10.47) holds for $\{\widetilde{f}_n^{\succ}\}$ in place of $\{f_n^{\succ}\}$. For $n = 1,2\dots$, let $h_n(t) := \widetilde{f}_n^{\succ}(\theta(t))$, $t \in J$, and let $h_0 := c$. Then as in (10.45) and (10.46), it follows that for $n = 1,2,\dots$ and each $t \in J$,

$$h_n(t) = c + \oint_{[t,b)} (N_\psi h_{n-1})_+ \cdot d\mu + \oint_{[t,b)} (N_\phi h_{n-1})_+ \cdot d\nu.$$

Thus $\{h_n\}_{n\ge 0}$ is the sequence $\{f_n^{\prec}\}_{n\ge 0}$ of backward Picard iterates obtained by the integral transform $T^{\prec}_{c,b,\mu,\nu}$ and $\psi$, $\phi$. Since the change of variables $\theta$ does not change the $p$-variation, $\|f_n^{\prec}\|_{[p]} = \|\tilde{f}_n^{\succ}\|_{[p]}$ for each $n$, and so (10.47) holds. The same argument applies to the second sequence of backward Picard iterates, proving (10.48).

To prove (10.49), first let $\{f_n\}$ be $\{f_n^{\succ}\}$ and $A = 2(1+2DH)$. Note that $\|f_0^{\succ}\|_{[p]} = \|c\| \le C$ and $A \ge 1$. Thus it is enough to bound $A_{j,n}$ for each $j \in \{1,\ldots,m\}$ and $n \ge 1$. By (10.36) with $g = f_n^{\succ}$ and $[u,v] = [z_{j-1}, z_j]$, we have

$$A_{j,n} \le \|f_n^{\succ}\|_{[z_{j-1},z_j),[p]} + 2\|\Delta^- f_n^{\succ}(z_j)\|. \tag{10.50}$$

Recalling Definitions 6.47 and 10.3 and Remark 10.4, we have that $\psi \in \mathcal{HW}_{\alpha,p/\alpha}(Z \times J; X)$. By Definition 10.5, we have $\phi \in \mathcal{UCR}(Z \times J; X)$, so the hypotheses of Proposition 10.9($b$) hold. It follows that the integral transform $T^{\succ}_{c,a}$ takes $\mathcal{W}_p([a,b]\!]; Z)$ into itself. Iterating, we see that the Kolmogorov integrals in the definition of each $f_n^{\succ}$ are defined. Thus by Corollary 2.23 applied twice, and then by additivity of the Kolmogorov integral over adjoining disjoint intervals one of which here is $A_2 = \{t\}$ (Theorem 2.21), for each $t \in (a,b]$,

$$\begin{aligned}\|\Delta^- f_n^{\succ}(t)\| &= \|(N_\psi f_{n-1}^{\succ})(t-)\cdot\mu(\{t\}) + (N_\phi f_{n-1}^{\succ})(t-)\cdot\nu(\{t\})\| \\ &\le \big(H_\alpha(\psi)\|f_{n-1}^{\succ}(t-)\|^\alpha + \|N_\psi(0)\|_{\sup} + \|\phi\|_{\mathcal{G}_\beta}(1+\|f_{n-1}^{\succ}(t-)\|^\beta)\big)H,\end{aligned} \tag{10.51}$$

since $\psi$ is $s$-uniformly $\alpha$-Hölder and since $\phi$ satisfies the $s$-uniform $\beta$ growth condition, Definition 10.5. Applying this bound to (10.50), we get that for each $j \in \{1,\ldots,m\}$ and $n \ge 1$,

$$A_{j,n} \le \|f_n^{\succ}\|_{[z_{j-1},z_j),[p]} + 2DH\max\{1, \|f_{n-1}^{\succ}\|_{[z_{j-1},z_j),\sup}\}. \tag{10.52}$$

Using this relation and Lemma 10.19($a$), the claim (10.49) will be proved by induction on $j$. For $j=1$, let $u=a$, $v=z_1$, $R=C$, and $f = f_{n-1}^{\succ}$. Then $\|T^{\succ}_{c,a}f(u)\| = \|f_n^{\succ}(a)\| = \|c\| \le C$ and $\|f_0^{\succ}\|_{[p]} = \|c\| < 2C$. Applying Lemma 10.19($a$) recursively in $n$, it follows that $\|f_n^{\succ}\|_{[a,z_1),[p]} \le 2C$ for $n = 0,1,2,\ldots$. Since $C \ge 1$, using this bound in (10.52), (10.49) holds for $j=1$. Suppose that (10.49) holds for some $1 \le j < m$. It will then be proved for $j+1$. Let $T_j := CA^j$ be the right side of (10.49). Let $u = z_j$, $v = z_{j+1}$, $R = T_j$, and $f = f_{n-1}^{\succ}$. By the induction assumption, $\|T^{\succ}_{c,a}f(u)\| = \|f_n^{\succ}(z_j)\| \le T_j$ for $n = 1,2,\ldots$. Also, $\|f_0^{\succ}(z_j)\| = \|c\| \le T_j$ and $\|f_0^{\succ}\|_{[z_j,z_{j+1}),[p]} = \|c\| \le T_j < 2T_j$. Then again applying Lemma 10.19($a$) recursively in $n$, it follows that $\|f_n^{\succ}\|_{[z_j,z_{j+1}),[p]} \le 2T_j$ for each $n = 0,1,2,\ldots$. Using this bound in (10.52) with $j$ replaced by $j+1$ it follows that (10.49) holds with $j+1$ instead of $j$. Hence by induction, (10.49) holds for all $j = 1,\ldots,m$, when $\{f_n\}$ is the sequence of Picard iterates $\{f_n^{\succ}\}$.

A proof of (10.48) is similar to the preceding proof of (10.47). Recall that now $A = 3(1+DH)$ and let $\{f_n\}$ be the sequence $\{g_n^{\succ}\}$. Now we apply the inequality

$$\begin{aligned} A_{j,n} &= \|g_n^{\succ}\|_{[z_{j-1},z_j],[p]} \\ &\le \|g_n^{\succ}(z_{j-1})\| + \|\Delta^+ g_n^{\succ}(z_{j-1})\| + \|g_n^{\succ}\|_{(z_{j-1},z_j],[p]}, \end{aligned} \tag{10.53}$$

which follows using Corollary 3.43($b$), for each $j \in \{1, \ldots, m\}$ and $n \ge 1$. By additivity and upper continuity of the Kolmogorov integral (Theorem 2.21 and Corollary 2.23), we have for each $t \in [a, b)$,

$$\Delta^+ g_n^{\succ}(t) = (N_\psi g_{n-1}^{\succ})(t)\cdot\mu(\{t\}) + (N_\phi g_{n-1}^{\succ})(t)\cdot\nu(\{t\}). \tag{10.54}$$

Since $\psi$ is $s$-uniformly $\alpha$-Hölder and since $\phi$ satisfies the $s$-uniform $\beta$ growth condition, we thus have for each $t \in [a, b)$,

$$\begin{aligned} &\|\Delta^+ g_n^{\succ}(t)\| \\ &\quad \le \{H_\alpha(\psi)\|g_{n-1}^{\succ}(t)\|^\alpha + \|N_\psi(0)\|_{\sup} + \|\phi\|_{\mathcal{G}_\beta}(1 + \|g_{n-1}^{\succ}(t)\|^\beta)\}H \\ &\quad \le DH \max\{1, \|g_{n-1}^{\succ}(t)\|\}. \end{aligned} \tag{10.55}$$

Again, (10.49) will be proved by induction on $j$. For $j = 1$, since $\|g_k^{\succ}(a)\| = \|c\| \le C$ for each $k \ge 0$, it follows that

$$\|g_n^{\succ}(a+)\| \le \|g_n^{\succ}(a)\| + \|\Delta^+ g_n^{\succ}(a)\| \le C(1 + DH) =: T_0 \tag{10.56}$$

holds for each $n \ge 1$. Then using Lemma 10.19($b$) for $R = T_0$ recursively in $k \ge 0$, it follows that for each $k = 0, 1, 2, \ldots$,

$$\|g_k^{\succ}\|_{(a,z_1],[p]} \le 2T_0. \tag{10.57}$$

Now applying the second inequality in (10.56), and (10.57) with $k = n$, to (10.53) with $j = 1$, we have that $A_{1,n} \le 3T_0 = CA$ for each $n \ge 1$. Since $A_{1,0} = \|c\| \le C$, (10.49) holds for $j = 1$. Suppose that (10.49) holds for some $1 \le j < m$. It then will be shown to hold for $j + 1$ in place of $j$. Letting $R_j := CA^j$, which is the right side of (10.49), by the induction assumption (10.49), we have $\|g_k^{\succ}(z_j)\| \le R_j$ for each $k \ge 0$. Therefore by (10.55), we have that

$$\|g_n^{\succ}(z_j+)\| \le \|g_n^{\succ}(z_j)\| + \|\Delta^+ g_n^{\succ}(z_j)\| \le R_j(1 + DH) =: T_j \tag{10.58}$$

holds for each $n \ge 1$. Using Lemma 10.19($b$) for $R = T_j$ recursively in $k \ge 0$, it follows that for each $k = 0, 1, 2, \ldots$,

$$\|g_k^{\succ}\|_{(z_j,z_{j+1}],[p]} \le 2T_j. \tag{10.59}$$

Now applying the second inequality in (10.58), and (10.59) with $k = n$, to (10.53) with $j + 1$ in place of $j$, it follows that $A_{j+1,n} \le 3T_j = R_j A = CA^{j+1}$ for each $n \ge 1$. Since $A_{j+1,0} = \|c\| \le C$, (10.49) holds for $j + 1$ in place of $j$. By induction, (10.49) holds for each $j = 1, \ldots, m$, when $\{f_n\}$ is the sequence of Picard iterates $\{g_n^{\succ}\}$. The proof of Proposition 10.24 is complete. □

The following is a consequence of Proposition 10.24 when $\psi$ and $\phi$ are autonomous. Recall Definition 10.5 for the class $\mathcal{G}_\beta$.

**Corollary 10.25.** *Let $\alpha, \beta \in (0,1]$, $1 \le p < 1+\alpha$, and $J := [a,b]$ with $a < b$. Assuming (1.14), let $F \in \mathcal{H}_\alpha(Z;X)$, let $G \in \mathcal{G}_\beta(Z;X)$ be continuous, let $\mu \in \mathcal{AI}_p(J;Y)$, and let $\nu \in \mathcal{AI}_1(J;Y)$. Also let*

$$\rho_1 := \{16K_{p,p/\alpha}(\|F(0)\| + 2\|F\|_{(\mathcal{H}_\alpha)})\}^{-1} \quad \text{and} \quad \rho_2 := \{16\|G\|_{\mathcal{G}_\beta}\}^{-1}, \tag{10.60}$$

*where $K_{p,p/\alpha} = \zeta((1+\alpha)/p)$, $\rho_1 := +\infty$ if $F \equiv 0$, and $\rho_2 := +\infty$ if $G \equiv 0$. Then there is an integer $1 \le m \le 1 + v_p(\mu; J)/\rho_1^p + v_1(\nu; J)/\rho_2$ such that (10.47) and (10.48) hold except that here $D = \|F(0)\| + \|F\|_{(\mathcal{H}_\alpha)} + 2\|G\|_{\mathcal{G}_\beta}$.*

*Proof.* We can assume that $F \not\equiv 0$ or $G \not\equiv 0$. Let $\psi(z,s) \equiv F(z)$ and $\phi(z,s) \equiv G(z)$. By Example 10.16 with $h \equiv 1$, $\psi \in \mathcal{HWG}_{\alpha,p/\alpha}(Z \times J; X)$ with $G_{\alpha,q}(\psi) = 0$. Also clearly $\phi \in \mathcal{CRG}_\beta(Z \times J; X)$. Since $F$ is $\alpha$-Hölder continuous, by Lemma 6.22(*a*), for any $a \le u < v \le B$ we have

$$\|N_F f\|_{(u,v),[p/\alpha]} \le \|F(0)\| + 2\|F\|_{(\mathcal{H}_\alpha)}\|f\|^\alpha_{(u,v),[p]},$$

and so (10.27) holds with $D = \|F(0)\| + 2\|F\|_{(\mathcal{H}_\alpha)}$. Since also (10.26) holds with $\|\phi\|_{\mathcal{G}_\beta} = \|G\|_{\mathcal{G}_\beta}$ it follows that Lemma 10.19 holds with $\rho_1$ and $\rho_2$ in (10.24) replaced by (10.60). The conclusion then follows from Proposition 10.24. □

After proving convergence of the four sequences of Picard iterates in the next proposition, we will then prove Theorem 10.11, showing that the limits are the unique solutions of the corresponding four nonlinear integral equations.

Recall again the space $\mathcal{HW}^{\mathrm{loc}}_{1+\alpha,q}$ as in Definition 6.61 for $n = 1$.

**Proposition 10.26.** *Let $0 < \alpha \le 1 \le p < 1+\alpha$, $q := p/\alpha$, and $J := [a,b]$ with $a < b$. Assuming (1.14), let $\psi \in \mathcal{HW}^{\mathrm{loc}}_{1+\alpha,q}(Z \times J; X)$, let $\phi \in \mathcal{UH}^{\mathrm{loc}}_1(Z \times J; X)$ be such that $\phi(z,\cdot) \in \mathcal{R}(J;X)$ for each $z \in Z$, let $\mu \in \mathcal{AI}_p(J;Y)$, and let $\nu \in \mathcal{AI}_1(J;Y)$. For each of the four sequences of Picard iterates, if it is bounded in $\mathcal{W}_p(J;Z)$ then it converges.*

*Proof.* Let $\{f_n\}_{n\ge0}$ be either of the two sequences of forward Picard iterates. Let $R := 3\sup_{n\ge0}\|f_n\|_{[p]} < \infty$ and let $B_R$ be the ball in $Z$ with center at 0 and radius $R$. In particular, the range of each $f_n$ is in the ball $B_R$. If $\psi(z,\cdot) \equiv h_1$ and $\phi(z,\cdot) \equiv h_2$ for each $z$ in $B_R$ and some functions $h_1, h_2\colon J \to X$ then $f_n \equiv f_1$ for each $n > 1$ and so the conclusion holds. Suppose that either $\psi(z,\cdot)$ or $\phi(z,\cdot)$ is non-constant with respect to $z \in B_R$.

Since $\psi \in \mathcal{HW}_{1+\alpha,q}(B_R \times J; X)$, it follows from Definitions 6.61 and 6.47 that $\|\psi^{(1)}_u\|_{B_R\times J,\sup} \le \|\psi^{(1)}_u(\cdot,a)\|_{B_R,\sup} + W_{\alpha,q}(\psi^{(1)}_u, R) < \infty$. Thus by the mean value theorem (5.2), $\psi$ is $s$-uniformly Lipschitz on $B_R$ with $H_1(\psi; B_R \times J, X) \le \|\psi^{(1)}_u\|_{B_R\times J,\sup}$. Also, since $\phi \in \mathcal{UH}^{\mathrm{loc}}_1(Z \times J; X)$, we have

$$C := H_1(\psi; B_R \times J, X) + H_1(\phi; B_R \times J, X) < \infty. \tag{10.61}$$

Let $\theta_1$ and $\theta_2$ be defined by (10.31), with $R$ as defined in this proof. Then $0 < \theta_j < \infty$ for $j = 1, 2$. Using Proposition 3.52, one can find a Young interval partition $\{(t_{j-1}, t_j)\}_{j=1}^m$ of $J$ such that $v_p(\mu; (t_{j-1}, t_j)) \le \theta_1^p$ and $v_1(\nu; (t_{j-1}, t_j)) \le \theta_2$ for each $1 \le j \le m$, and $m \le 1 + v_p(\mu; J)/\theta_1^p + v_1(\nu; J)/\theta_2$.

Let $E := \|f_1\|_{(p)}$, $H := \sup_{t\in(a,b]} \max\{\|\mu(\{t\})\|, \|\nu(\{t\})\|\}$, and $A := 1 + 4CH$. We claim that for each $j = 1, \dots, m$,

$$A_{j,n} := \big\|f_{n+1} - f_n\big\|_{J_j,[p]} \le 2^{-n}(2A)^j E =: B_{j,n} \quad \text{for } n = 0, 1, \dots, \tag{10.62}$$

where $J_j := [t_{j-1}, t_j]$. Assuming that the claim holds, and applying Proposition 3.35($a$) recursively, it follows that

$$\big\|f_{n+1} - f_n\big\|_{J,[p]} \le \sum_{j=1}^m \big\|f_{n+1} - f_n\big\|_{J_j,[p]} \le 2^{-n+1}(2A)^m E$$

for each $n = 0, 1, \dots$. Therefore $\{f_n\}_{n\ge 0}$ is a Cauchy sequence in $\mathcal{W}_p(J; Z)$, and so by Proposition 3.7($d$), it converges to some function in $\mathcal{W}_p(J; Z)$, proving the conclusion for the two sequences of forward Picard iterates. The same conclusion follows for the two sequences of backward Picard iterates by applying time reversal for Kolmogorov integrals as in the proof of Proposition 10.24.

First suppose that $\{f_n\}$ is $\{f_n^{\succ}\}$. For each $n = 0, 1, 2, \dots$ and $u \in J$, let

$$\Delta_{u,n} := \begin{cases} T_{0,u}^{\succ} f_n^{\succ} - T_{0,u}^{\succ} f_{n-1}^{\succ} & \text{if } n \ge 1, \\ T_{0,u}^{\succ} f_0^{\succ} & \text{if } n = 0, \end{cases} \tag{10.63}$$

where $T_{0,u}^{\succ} f$ is defined by (10.10) with $c = 0$.

The claim (10.62) will be proved by induction on $j$. Before starting the induction we show that for each $j \in \{1, \dots, m\}$ and $n \ge 1$,

$$\begin{aligned} A_{j,n} = \|\Delta_{a,n}\|_{J_j,[p]} &\le \big\|\Delta_{a,n}(t_{j-1})\big\| + \big\|\Delta_{t_{j-1},n}\big\|_{[t_{j-1},t_j),[p]} \\ &\quad + 2CH\Big\{\big\|\Delta_{a,n-1}(t_{j-1})\big\| + \big\|\Delta_{t_{j-1},n-1}\big\|_{[t_{j-1},t_j),(p)}\Big\}. \end{aligned} \tag{10.64}$$

For a function $g \in \mathcal{W}_p(J; Z)$ and $a \le u < v \le b$, since $\|g\|_{[u,v),\sup} \le \|g(u)\| + \|g - g(u)\|_{[u,v),\sup}$, by (10.36) we have

$$\|g\|_{[u,v],[p]} \le \|g(u)\| + \|g - g(u)\|_{[u,v),[p]} + 2\|\Delta^- g(v)\|. \tag{10.65}$$

Let $j \in \{1, \dots, m\}$ and $n \ge 1$. Applying (10.65) to $g = \Delta_{a,n} = f_{n+1}^{\succ} - f_n^{\succ}$, $u = t_{j-1}$ and $v = t_j$, since $g - g(t_{j-1}) = \Delta_{t_{j-1},n}$ on $[t_{j-1}, t_j)$ by (10.29), it follows that

$$\|\Delta_{a,n}\|_{J_j,[p]} \le \|\Delta_{a,n}(t_{j-1})\| + \|\Delta_{t_{j-1},n}\|_{[t_{j-1},t_j),[p]} + 2\|\Delta^-(f_{n+1}^{\succ} - f_n^{\succ})(t_j)\|. \tag{10.66}$$

As in (10.51), we have

$$\begin{aligned}
&\|\Delta^-(f_{n+1}^{\succ} - f_n^{\succ})(t_j)\| \\
&\quad = \|(N_\psi f_n^{\succ} - N_\psi f_{n-1}^{\succ})(t_j-)\cdot\mu(\{t_j\}) + (N_\phi f_n^{\succ} - N_\phi f_{n-1}^{\succ})(t_j-)\cdot\nu(\{t_j\})\| \\
&\quad \le H\big[\|N_\psi f_n^{\succ} - N_\psi f_{n-1}^{\succ}\|_{[t_{j-1},t_j),\sup} + \|N_\phi f_n^{\succ} - N_\phi f_{n-1}^{\succ}\|_{[t_{j-1},t_j),\sup}\big].
\end{aligned}$$

For $t_{j-1} \le s < t_j$ we have

$$\|(N_\psi f_n^{\succ} - N_\psi f_{n-1}^{\succ})(s)\| \le H_1(\psi; B_R \times J, X)\|(f_n^{\succ} - f_{n-1}^{\succ})(s)\|$$

and likewise for the term with $\phi$. Thus using notation (10.61), it follows that

$$\|\Delta^-(f_{n+1}^{\succ} - f_n^{\succ})(t_j)\| \le CH\|(f_n^{\succ} - f_{n-1}^{\succ})\|_{[t_{j-1},t_j),\sup}. \tag{10.67}$$

To bound the right side, we have by definition (10.12) of $f_n^{\succ}$, also for $n$ replaced by $n-1$, and for $n \ge 2$ by (10.29) for $f = f_{n-1}^{\succ}$ and $f_{n-2}^{\succ}$, that

$$(f_n^{\succ} - f_{n-1}^{\succ})(s) = \Delta_{t_{j-1},n-1}(s) + \Delta_{a,n-1}(t_{j-1})$$

for each $s \in [t_{j-1}, t_j)$. The same holds for $n = 1$ by the definitions. Since clearly $T_{0,u}^{\succ} f(u) = 0$ for any $f$ and $u$ by definition (10.10) of $T_{c,u}^{\succ}$, we have

$$\|\Delta_{u,k}\|_{[u,t),\sup} \le \|\Delta_{u,k}\|_{[u,t),(p)}$$

for any $a \le u < t \le b$ and $k \ge 1$. Thus

$$\|f_n^{\succ} - f_{n-1}^{\succ}\|_{[t_{j-1},t_j),\sup} \le \|\Delta_{a,n-1}(t_{j-1})\| + \|\Delta_{t_{j-1},n-1}\|_{[t_{j-1},t_j),(p)}.$$

Applying this bound to (10.67) gives

$$\|\Delta^-(f_{n+1}^{\succ} - f_n^{\succ})(t_j)\| \le CH\Big\{\big\|\Delta_{a,n-1}(t_{j-1})\big\| + \big\|\Delta_{t_{j-1},n-1}\big\|_{[t_{j-1},t_j),(p)}\Big\}.$$

By the preceding bound and (10.66), it follows that (10.64) holds.

To begin the proof of (10.62) by induction on $j$, let $j = 1$. We claim that for each $k = 0, 1, 2, \ldots$,

$$\big\|\Delta_{a,k}\big\|_{[a,t_1),[p]} \le 2^{-k+1}E. \tag{10.68}$$

The claim will be proved by induction on $k$. For $k = 0$, we have

$$\|\Delta_{a,0}\|_{[a,t_1),[p]} \le 2\|f_1^{\succ}\|_{[a,t_1),(p)} \le 2E.$$

Suppose (10.68) holds for some $k \ge 0$. To prove it for $k+1$ we use Lemma 10.20($a$) with $g = f_{k+1}^{\succ}$, $f = f_k^{\succ}$, $u = a$, $v = t_1$, and $\epsilon = 2^{-k+1}E$. We have $\|g - f\|_{[u,v),[p]} = \|\Delta_{a,k}\|_{[a,t_1),[p]} \le \epsilon \le 2\epsilon$ by the induction assumption. Thus by Lemma 10.20($a$),

$$\|\Delta_{a,k+1}\|_{[a,t_1),[p]} = \|T_{0,u}^{\succ} g - T_{0,u}^{\succ} f\|_{[u,v),[p]} \le \epsilon/2 = 2^{-k}E.$$

Therefore (10.68) holds with $k+1$ in place of $k$, and hence for all $k = 0, 1, 2, \ldots$, as desired. For $n \geq 1$, applying (10.64) with $j = 1$, then (10.68) with $k = n, n-1$, since $(f_k^{\succ} - f_{k-1}^{\succ})(a) = 0$ for each $k \geq 1$, it follows that

$$\begin{aligned} A_{1,n} &\leq 0 + \|\Delta_{a,n}\|_{[a,t_1),[p]} + 2CH\{0 + \|\Delta_{a,n-1}\|_{[a,t_1),(p)}\} \\ &\leq 2^{-n+1}E + 2CH2^{-n+2}E = 2^{-n+1}EA = B_{1,n}. \end{aligned}$$

For $n = 0$, we have

$$A_{1,0} \leq 2\|f_1^{\succ}\|_{J_1,(p)} \leq 2E \leq B_{1,0}.$$

Thus (10.62) holds for $j = 1$. Now suppose (10.62) holds for some $1 \leq j < m$. It then will be shown to hold for $j+1$ in place of $j$. We claim that for each $k = 0, 1, 2, \ldots$,

$$\left\|\Delta_{t_j,k}\right\|_{[t_j,t_{j+1}),[p]} \leq B_{j,k}. \tag{10.69}$$

Again the claim will be proved by induction on $k$. For $k = 0$, we have

$$\|\Delta_{t_j,0}\|_{[t_j,t_{j+1}),[p]} = \|T_{0,t_j}^{\succ} f_0^{\succ}\|_{[t_j,t_{j+1}),[p]} \leq 2\|f_1^{\succ}\|_{[a,t_{j+1}),(p)} \leq 2E \leq B_{j,0}.$$

Suppose (10.69) holds for some $k \geq 0$. To prove it for $k+1$ we use Lemma 10.20($a$) again, now with $g = f_{k+1}^{\succ}$, $f = f_k^{\succ}$, $u = t_j$, $v = t_{j+1}$, and $\epsilon = B_{j,k}$. By the induction assumption on $j$, that is, by (10.62), we have

$$\|(g-f)(u)\| = \|(f_{k+1}^{\succ} - f_k^{\succ})(t_j)\| \leq A_{j,k} \leq B_{j,k}.$$

By the induction assumption on $k$, that is by (10.69), and by (10.29) for $f = f_k^{\succ}$ and if $k \geq 1$ for $f_{k-1}^{\succ}$, we have

$$\|g-f\|_{[u,v),[p]} \leq \|(g-f)(u)\| + \|\Delta_{t_j,k}\|_{[t_j,t_{j+1}),[p]} \leq 2B_{j,k} = 2\epsilon.$$

For $k = 0$ the same can be seen directly. Thus by Lemma 10.20($a$),

$$\|\Delta_{t_j,k+1}\|_{[t_j,t_{j+1}),[p]} = \|T_{0,u}^{\succ} g - T_{0,u}^{\succ} f\|_{[u,v),[p]} \leq \epsilon/2 = B_{j,k+1}.$$

Therefore (10.69) holds with $k+1$ in place of $k$, and hence for all $k = 0, 1, 2, \ldots$. By (10.64) with $j+1$ in place of $j$, by assumption (10.62) of the induction on $j$, and by (10.69) with $k = n, n-1$, it follows that for each $n \geq 1$,

$$\begin{aligned} A_{j+1,n} &\leq A_{j,n} + \|\Delta_{t_j,n}\|_{[t_j,t_{j+1}),[p]} + 2CH\Big\{A_{j,n-1} + \|\Delta_{t_j,n-1}\|_{[t_j,t_{j+1}),(p)}\Big\} \\ &\leq 2B_{j,n} + 4CHB_{j,n-1} = 2B_{j,n}(1+4CH) = B_{j+1,n}. \end{aligned}$$

Since for $n = 0$,

$$A_{j+1,0} = \left\|f_1^{\succ} - f_0^{\succ}\right\|_{J_{j+1},[p]} \leq 2\|f_1^{\succ}\|_{(a,t_{j+1}],(p)} \leq 2E \leq B_{j+1,0},$$

(10.62) holds for $j+1$ in place of $j$. By induction on $j$, (10.62) holds for each $j = 1, \ldots, m$, proving that claim when $\{f_n\}$ is $\{f_n^{\succ}\}$.

Now, suppose that $\{f_n\}$ is $\{g_n^{\succ}\}$. For each $n = 0, 1, 2, \ldots$ and $u \in [a, b]$, let

$$\Delta_{u,n} := \begin{cases} Q_{0,u}^{\succ} g_n^{\succ} - Q_{0,u}^{\succ} g_{n-1}^{\succ} & \text{if } n \geq 1 \\ Q_{0,u}^{\succ} g_0^{\succ} & \text{if } n = 0, \end{cases} \tag{10.70}$$

where $Q_{0,u}^{\succ} f$ is defined by (10.11) with $c = 0$, and so $\Delta_{a,n} = g_{n+1}^{\succ} - g_n^{\succ}$ for all $n \geq 0$. Let $j \in \{1, \ldots, m\}$ and $n \geq 1$. For each $t \in (t_{j-1}, t_j]$, by (10.30) we have

$$\Delta_{a,n}(t) = \Delta_{a,n}(t_{j-1}) + \Delta_{t_{j-1},n}(t_{j-1}+) + \Delta_{t_{j-1}+,n}(t). \tag{10.71}$$

Therefore

$$\|\Delta_{a,n}\|_{J_j,\sup} \leq \|\Delta_{a,n}(t_{j-1})\| + \|\Delta_{t_{j-1},n}(t_{j-1}+)\| + \|\Delta_{t_{j-1}+,n}\|_{(t_{j-1},t_j],\sup}. \tag{10.72}$$

By the second statement in Corollary 3.43($b$) we have

$$\|\Delta_{a,n}\|_{J_j,(p)} \leq \|\Delta^+(\Delta_{a,n})(t_{j-1})\| + \|\Delta_{a,n}\|_{(t_{j-1},t_j],(p)}.$$

Here $\|\Delta_{a,n}\|_{(t_{j-1},t_j],(p)} = \|\Delta_{t_{j-1}+,n}\|_{(t_{j-1},t_j],(p)}$ because in (10.71) the first two terms on the right do not depend on $t$. Also by (10.71) we have

$$\Delta^+(\Delta_{a,n})(t_{j-1}) = \Delta_{a,n}(t_{j-1}) + \Delta_{t_{j-1},n}(t_{j-1}+) + \Delta_{t_{j-1}+,n}(t_{j-1}+) - \Delta_{a,n}(t_{j-1}),$$

which equals $\Delta_{t_{j-1},n}(t_{j-1}+)$ because $\Delta_{t_{j-1}+,n}(t_{j-1}+) = 0$. It follows that

$$\|\Delta_{a,n}\|_{J_j,(p)} \leq \|\Delta_{t_{j-1},n}(t_{j-1}+)\| + \|\Delta_{t_{j-1}+,n}\|_{(t_{j-1},t_j],(p)}. \tag{10.73}$$

By additivity and upper continuity of the Kolmogorov integral (Theorem 2.21 and Corollary 2.23), as in (10.54), for any $t \in [a, b)$,

$$\Delta_{t,n}(t+) = \big[N_\psi g_n^{\succ} - N_\psi g_{n-1}^{\succ}\big](t) \cdot \mu(\{t\}) + \big[N_\phi g_n^{\succ} - N_\phi g_{n-1}^{\succ}\big](t) \cdot \nu(\{t\}). \tag{10.74}$$

By (10.61), it follows as just before (10.67) that

$$\|\Delta_{t_{j-1},n}(t_{j-1}+)\| \leq CH\|\Delta_{a,n-1}(t_{j-1})\|. \tag{10.75}$$

Adding the two inequalities (10.72) and (10.73), and applying the preceding bound, gives

$$A_{j,n} \leq \|\Delta_{a,n}(t_{j-1})\| + 2CH\|\Delta_{a,n-1}(t_{j-1})\| + \|\Delta_{t_{j-1}+,n}\|_{(t_{j-1},t_j],[p]}. \tag{10.76}$$

To begin the proof of (10.62) by induction on $j$, let $j = 1$. We claim that for each $k = 0, 1, 2, \ldots$,

$$\|\Delta_{a+,k}\|_{(a,t_1],[p]} \leq 2^{-k+1}E. \tag{10.77}$$

The claim will be proved by induction on $k$. For $k = 0$, we have

$$\|\Delta_{a+,0}\|_{(a,t_1],[p]} \leq 2\|g_1^{\succ}\|_{(a,t_1],(p)} \leq 2E,$$

recalling that $E = \|f_1\|_{(p)}$ where now $f_1 = g_1^{\succ}$. Suppose (10.77) holds for some $k \ge 0$. To prove it for $k+1$ we use Lemma 10.20($b$) with $g = g_{k+1}^{\succ}$, $f = g_k^{\succ}$, $u = a$, $v = t_1$, and $\epsilon = 2^{-k+1}E$. If $k \ge 1$, since $g_k^{\succ}(a) = g_{k-1}^{\succ}(a) = c$, and so $\Delta_{a,k}(a+) = 0$ by (10.74) with $t = a$ and $n = k$, we have $\|g - f\|_{(u,v],[p]} = \|\Delta_{a+,k}\|_{(a,t_1],[p]} \le \epsilon \le 2\epsilon$ by the induction assumption. If $k = 0$, we have $\|g - f\|_{(u,v],[p]} = \|g_1^{\succ} - g_1^{\succ}(a)\|_{(a,t_1],[p]} \le 2\|g_1^{\succ}\|_{(p)} \le \epsilon \le 2\epsilon$. Thus by Lemma 10.20($b$),

$$\|\Delta_{a+,k+1}\|_{(a,t_1],[p]} = \|Q_{0,a+}^{\succ}g - Q_{0,a+}^{\succ}f\|_{(u,v],[p]} \le \epsilon/2 = 2^{-k}E.$$

Therefore (10.77) holds with $k+1$ in place of $k$, and hence for all $k = 0, 1, 2, \ldots$. For $n \ge 1$, applying (10.76) with $j = 1$, then (10.77) with $k = n$, since $\Delta_{a,k}(a) = 0$ for each $k \ge 0$, it follows that $A_{1,n} \le 2^{-n+1}E \le B_{1,n}$. For $n = 0$, we have

$$A_{1,0} \le 2\|g_1^{\succ}\|_{J_1,(p)} \le 2E \le B_{1,0}.$$

Thus (10.62) holds for $j = 1$. Now, suppose (10.62) holds for some $1 \le j < m$. It then will be shown to hold for $j+1$ in place of $j$. We claim that for each $k = 0, 1, 2, \ldots$,

$$\big\|\Delta_{t_j+,k}\big\|_{(t_j,t_{j+1}],[p]} \le (1 + 2CH)B_{j,k}. \tag{10.78}$$

Again the claim will be proved by induction on $k$. For $k = 0$, we have

$$\|\Delta_{t_j+,0}\|_{(t_j,t_{j+1}],[p]} \le 2\|g_1^{\succ}\|_{(t_j,t_{j+1}],(p)} \le 2E,$$

so (10.78) holds in this case. Suppose (10.78) holds for some $k \ge 0$. To prove it for $k+1$ we use Lemma 10.20($b$) with $g = g_{k+1}^{\succ}$, $f = g_k^{\succ}$, $u = t_j$, $v = t_{j+1}$, and $\epsilon = (1+2CH)B_{j,k}$. By the induction assumption on $j$, that is, by (10.62), and by (10.71) and (10.75) both with $j$ in place of $j-1$ and $n = k$, we have

$$\|\Delta_{a,k}(t_j+)\| \le B_{j,k} + CHB_{j,k-1} = (1 + 2CH)B_{j,k}.$$

Therefore by (10.30) and by the induction assumption on $k$,

$$\begin{aligned}\|g - f\|_{(u,v],[p]} = \|\Delta_{a,k}\|_{(t_j,t_{j+1}],[p]} &\le \|\Delta_{a,k}(t_j+)\| + \|\Delta_{t_j+,k}\|_{(t_j,t_{j+1}],[p]}\\ &\le 2(1 + 2CH)B_{j,k} = 2\epsilon.\end{aligned}$$

Thus by Lemma 10.20($b$),

$$\|\Delta_{t_j+,k+1}\|_{(t_j,t_{j+1}],[p]} = \|Q_{0,u+}^{\succ}g - Q_{0,u+}^{\succ}f\|_{(u,v],[p]} \le \epsilon/2 = (1 + 2CH)B_{j,k+1}.$$

Therefore (10.78) holds with $k+1$ in place of $k$, and hence for all $k = 0, 1, 2, \ldots$. By (10.76) with $j+1$ in place of $j$, by the induction assumption (10.62) on $j$, and by (10.78) with $k = n$, it follows that for each $n \ge 1$,

$$\begin{aligned}A_{j+1,n} &\le A_{j,n} + 2CHA_{j,n-1} + \|\Delta_{t_j+,n}\|_{(t_j,t_{j+1}],[p]}\\ &\le B_{j,n} + 2CHB_{j,n-1} + (1 + 2CH)B_{j,n} \le 2(1 + 4CH)B_{j,n} = B_{j+1,n}.\end{aligned}$$

For $n = 0$, we have

$$A_{j+1,0} \le 2\|g_1^{\succ}\|_{[a,t_{j+1}],(p)} \le 2E \le B_{j+1,0}.$$

Thus (10.62) holds for $j+1$ in place of $j$, so (10.62) holds for each $j = 1, \ldots, m$ when $\{f_n\}$ is $\{g_n^{\succ}\}$. Thus (10.62) is proved for the two sequences of forward Picard iterates, proving the proposition. □

Now we are ready to prove Theorem 10.11.

**Proof of Theorem 10.11 and Corollary 10.12.** The assumed hypotheses on $\psi$ and $\phi$ imply the respective hypotheses of Proposition 10.24 with $\alpha = 1$: this is clear for $\psi$. For $\phi$, by Definition 10.5, the hypotheses $\phi \in \mathcal{UH}_1^{\rm loc}(Z \times J; X)$ and $\phi(z, \cdot) \in \mathcal{R}(J; X)$ for each $z \in Z$ yield $\phi \in \mathcal{UCR}(Z \times J; X)$, and so $\phi \in \mathcal{CRG}_\beta(Z \times J; X)$. Thus by Proposition 10.24 or Corollary 10.25 in the case of Corollary 10.12, both with $\alpha = 1$, the four sequences of Picard iterates (10.12), (10.13), (10.16), and (10.17) are bounded in $\mathcal{W}_p(J; Z)$.

The assumed hypotheses on $\psi$ and $\phi$ also imply the respective hypotheses of Proposition 10.26: This is clear for $\phi$. For $\psi$, by Definition 10.3, $\psi \in \mathcal{WG}_{1,p}(Z \times J, X)$, and so $\psi(0, \cdot) \in \mathcal{W}_p(J; X)$ by the definition (6.31) of $W_{1,q}(\psi, K)$. Thus $\psi \in \mathcal{HW}_{1+\alpha,q}^{\rm loc}(Z \times J; X)$ with $q = p/\alpha$. Then by Proposition 10.26, the four sequences of Picard iterates converge in $\mathcal{W}_p(J; Z)$. Suppose $f^{\succ}$ is the limit of $\{f_n^{\succ}\}_{n \ge 0}$ in $\mathcal{W}_p(J; Z)$. By Theorem 6.68 with $R := 3 \sup_n \|f_n^{\succ}\|_{[p]} < \infty$, $N_\psi f_n^{\succ} \to N_\psi f^{\succ}$ in $\mathcal{W}_q(J; X)$ as $n \to \infty$. Since $\phi \in \mathcal{UH}_1^{\rm loc}(Z \times J; X)$, it follows that $N_\phi f_n^{\succ} \to N_\phi f^{\succ}$ in $\ell^\infty(J; X)$. Using the fact that the Kolmogorov integral is a continuous bilinear mapping from $\mathcal{W}_q(J; X) \times \mathcal{AI}_p(J; Y)$ into $\mathcal{W}_p(J; Z)$ (Proposition 3.96), it follows that $f^{\succ}$ is a solution of (10.5). Similarly, for each of the other three nonlinear integral equations, if the associated Picard iterates converge in $\mathcal{W}_p(J; Z)$, their limit is a solution. The proof of Theorem 10.11 and Corollary 10.12 is complete. □

## 10.5 Continuity of the Solution Mappings

Let $0 < \alpha \le 1$, $1 \le p < 1 + \alpha$, and $J := [a, b]$. Suppose that $\psi\colon Z \times J \to X$ and $\phi\colon Z \times J \to X$ satisfy the hypotheses of Theorem 10.7. Then for $c \in Z$, $\mu \in \mathcal{AI}_p(J; Y) = \mathcal{AI}_p$, and $\nu \in \mathcal{AI}_1(J; Y) = \mathcal{AI}_1$, the forward nonlinear integral equation (10.5) has a unique solution $f \in \mathcal{W}_p(J; Z) = \mathcal{W}_p$. Define the *solution mapping* $\mathcal{S}_{(a,\cdot]}^{\succ}$ corresponding to the integral equation (10.5) by

$$Z \times \mathcal{AI}_p \times \mathcal{AI}_1 \ni (c, \mu, \nu) \mapsto \mathcal{S}_{(a,\cdot]}^{\succ}(c, \mu, \nu) := f \in \mathcal{W}_p.$$

Likewise, by Theorem 10.7, there exist *solution mappings* $\mathcal{S}^{\succ}_{[a,\cdot)}$, $\mathcal{S}^{\prec}_{(\cdot,b]}$ and $\mathcal{S}^{\prec}_{[\cdot,b)}$ corresponding to the integral equations (10.6), (10.7), and (10.8), respectively. We will prove in this section that the four solution mappings are continuous.

Next is the main result of this section:

**Theorem 10.27.** *Let $\alpha,\beta \in (0,1]$, $1 \le p < 1+\alpha$, and $J := [a,b]$ with $a < b$. Assuming* (1.14), *let $\psi\colon Z \times J \to X$ and $\phi\colon Z \times J \to X$ satisfy the hypotheses of Theorem* 10.7. *Then the four solution mappings $\mathcal{S}^{\succ}_{(a,\cdot]}$, $\mathcal{S}^{\succ}_{[a,\cdot)}$, $\mathcal{S}^{\prec}_{[\cdot,b)}$, and $\mathcal{S}^{\prec}_{(\cdot,b]}$ from $Z \times \mathcal{AI}_p(J;Y) \times \mathcal{AI}_1(J;Y)$ to $\mathcal{W}_p(J;Z)$ corresponding to the integral equations* (10.5), (10.6), (10.7), *and* (10.8), *respectively, are continuous.*

To prepare for the proof we next give bounds for the last terms on the right sides of (10.29) and (10.30). Recall again the space $\mathcal{HW}^{\rm loc}_{1+\alpha,q}$ as in Definition 6.61 for $n = 1$.

**Lemma 10.28.** *Let $0 < \alpha \le 1 \le p < 1+\alpha$, $q := p/\alpha$, $0 < R < \infty$, and $J := [a,b]$ with $a < b$. Assuming* (1.14), *let $\psi \in \mathcal{HW}^{\rm loc}_{1+\alpha,q}(Z \times J;X)$, let $\phi \in \mathcal{UH}^{\rm loc}_1(Z \times J;X)$ be such that $\phi(z,\cdot) \in \mathcal{R}(J;X)$ for each $z \in Z$, let $\mu,\gamma \in \mathcal{AI}_p(J;Y)$, let $\nu,\pi \in \mathcal{AI}_1(J;Y)$, and let $f,g \in \mathcal{W}_p(J;Z)$ be such that $3\max\{\|f\|_{[p]},\|g\|_{[p]}\} \le R$. Let*

$$\tau_1 := 1/\{1+8K_{p,q}[G(\psi)+W_{1,q}(\psi,R)]\} \quad \text{and} \quad \tau_2 := 1/\{1+8G(\phi)\}, \tag{10.79}$$

*where $G(\chi) := \|N_\chi(0)\|_{\sup} + RH_1(\chi)$ and $H_1(\chi) = H_1(\chi;B_R \times J,X)$ for $\chi = \psi$ or $\phi$, $W_{1,q}(\psi,R) = W_{1,q}(\psi,R;B_R \times J,X)$, $B_R := \{z \in Z\colon \|z\| \le R\}$, and $K_{p,q} = \zeta(1/p+1/q)$. Suppose that an interval $[u,v] \subset J$ is such that* (10.31) *holds for $\mu$ and $\nu$. For any $E > 0$ such that $\|\mu-\gamma\|_{J,(p)} \le E\tau_1$ and $\|\nu-\pi\|_{J,(1)} \le E\tau_2$, the following two statements hold:*

(*a*) $\|T^{\succ}_{0,u,\mu,\nu}f - T^{\succ}_{0,u,\gamma,\pi}g\|_{[u,v),[p]} \le E$ *if* $\|f-g\|_{[u,v),[p]} \le 2E$;
(*b*) $\|Q^{\succ}_{0,u+,\mu,\nu}f - Q^{\succ}_{0,u+,\gamma,\pi}g\|_{(u,v],[p]} \le E$ *if* $\|f-g\|_{(u,v],[p]} \le 2E$.

*Proof.* To prove (*a*), for $a \le u \le t \le b$, let

$$R_u(t) := \oint_{(u,t]} (N_\psi g)_- \cdot \mathrm{d}(\mu-\gamma) + \oint_{(u,t]} (N_\phi g)_- \cdot \mathrm{d}(\nu-\pi).$$

Since the Kolmogorov integral is bilinear (Theorem 2.72 and Corollary 2.26), for each $t \in [u,v)$, we have

$$T^{\succ}_{0,u,\mu,\nu}f(t) - T^{\succ}_{0,u,\gamma,\pi}g(t) = T^{\succ}_{0,u,\mu,\nu}f(t) - T^{\succ}_{0,u,\mu,\nu}g(t) + R_u(t). \tag{10.80}$$

In Lemma 10.20, the hypotheses on $J$, $\alpha$, $p$, $q$, $R$, $\psi$, $\phi$, $\mu$, and $\nu$ are the same as in the present lemma. The choice of $u,v \in J$ is the same in both lemmas. The functions $f$ and $g$ here satisfy the hypotheses on them in Lemma 10.20. Since $\|f-g\|_{[u,v),[p]} \le 2E$, by Lemma 10.20(*a*), we have

$$\|T^{\succ}_{0,u,\mu,\nu}f - T^{\succ}_{0,u,\mu,\nu}g\|_{[u,v),[p]} = \|T^{\succ}_{0,u}f - T^{\succ}_{0,u}g\|_{[u,v),[p]} \le E/2, \tag{10.81}$$

and so it is enough to prove that $\|R_u\|_{[u,v),[p]} \le E/2$. Since $\psi \in \mathcal{HW}^{\mathrm{loc}}_{1+\alpha,q}(Z \times J; X)$, by Lemma 6.62 with $n=1$, $\psi \in \mathcal{HW}^{\mathrm{loc}}_{1,q}(Z \times J; X)$. Thus by Lemma 6.53($a$) with $\alpha = 1$ there,

$$\|N_\psi g\|_{J,(q)} \le H_1(\psi)\|g\|_{J,(q)} + W_{1,q}(\psi, \|g\|_{J,[q]}).$$

Since $\psi \in \mathcal{UH}^{\mathrm{loc}}_1(Z \times J; X)$, it follows that

$$\|N_\psi g\|_{J,\sup} \le \|N_\psi(0)\|_{\sup} + H_1(\psi)\|g\|_{J,\sup}.$$

Summing the two bounds, using the fact that $\|g\|_{(q)} \le \|g\|_{(p)}$ (Lemma 3.45), and applying the hypothesis $\|g\|_{J,[p]} \le R$, we get

$$\|N_\psi g\|_{J,[q]} \le G(\psi) + W_{1,q}(\psi, R).$$

Also, recalling some notation given in the paragraph before (2.1), we have $\|(N_\psi g)^{(a)}_-\|_{J,[q]} \le \|N_\psi g\|_{J,[q]}$. Similarly, since $\phi \in \mathcal{UH}^{\mathrm{loc}}_1(Z \times J; X)$, it follows that $\|(N_\phi g)^{(a)}_-\|_{J,\sup} \le G(\phi)$. By additivity of the Kolmogorov integral over disjoint adjoining intervals (Theorem 2.21) and by the Love–Young inequality (3.153), we have

$$\begin{aligned}\|R_u\|_{[u,v),(p)} &\le \Big\| \textstyle\oint (N_\psi g)^{(a)}_- \cdot \mathrm{d}(\mu - \gamma)\Big\|_{J,(p)} + \Big\| \textstyle\oint (N_\phi g)^{(a)}_- \cdot \mathrm{d}(\nu - \pi)\Big\|_{J,(1)} \\ &\le K_{p,q}\big\|(N_\psi g)^{(a)}_-\big\|_{J,[q]}\big\|\mu-\gamma\big\|_{J,(p)} + \big\|(N_\phi g)^{(a)}_-\big\|_{J,\sup}\big\|\nu-\pi\big\|_{J,(1)} \\ &\le K_{p,q}\Big\{G(\psi) + W_{1,q}(\psi, R)\Big\}\|\mu-\gamma\|_{J,(p)} + G(\phi)\|\nu-\pi\|_{J,(1)}.\end{aligned}$$

Using the fact that $R_u(u) = 0$, it then follows that

$$\begin{aligned}\|R_u\|_{[u,v),[p]} &\le 2\|R_u\|_{[u,v),(p)} \\ &\le 2K_{p,q}\Big\{G(\psi) + W_{1,q}(\psi, R)\Big\}\|\mu-\gamma\|_{J,(p)} + 2G(\phi)\|\nu-\pi\|_{J,(1)},\end{aligned}$$

and so $\|R_u\|_{[u,v),[p]} \le E/2$ since $\|\mu-\gamma\|_{J,(p)} \le E\tau_1$ and $\|\nu-\pi\|_{J,(1)} \le E\tau_2$. This together with (10.80) and (10.81) proves ($a$). The proof of statement ($b$) is similar except that we use ($b$) of Lemma 10.20, and therefore is omitted. The proof of the lemma is complete. □

To continue the proof of Theorem 10.27 we have the next fact.

**Proposition 10.29.** *Let* $0 < \alpha \le 1 \le p < 1+\alpha$, $q := p/\alpha$, *and* $J := [a,b]$ *with* $a < b$. *Assuming* (1.14), *let* $\psi \in \mathcal{HW}^{\mathrm{loc}}_{1+\alpha,q}(Z \times J; X)$, *let* $\phi \in \mathcal{UH}^{\mathrm{loc}}_1(Z \times J; X)$ *be such that* $\phi(z,\cdot) \in \mathcal{R}(J;X)$ *for each* $z \in Z$, *let* $\mu, \gamma \in \mathcal{AI}_p(J;Y)$, *let* $\nu, \pi \in \mathcal{AI}_1(J;Y)$, *and let* $c, d \in Z$. *Suppose that* $\{f_n\}_{n\ge0}$ *and* $\{h_n\}_{n\ge0}$ *are two sequences of Picard iterates bounded in* $\mathcal{W}_p(J;Z)$ *obtained, respectively, by* $T^{\succ}_{c,a,\mu,\nu}$ *and by* $T^{\succ}_{d,a,\gamma,\pi}$. *Given* $\epsilon > 0$ *there exist* $\delta_r > 0$, $r = 1,2,3$, *such that if* $\|c-d\| < \delta_1$, $\|\mu-\gamma\|_{(p)} < \delta_2$, *and* $\|\nu-\pi\|_{(1)} < \delta_3$ *then* $\|f_n - h_n\|_{[p]} < \epsilon$ *for each* $n \ge 0$, *and the same holds if* $T^{\succ}$ *is replaced by* $Q^{\succ}$ *for both sequences.*

*Proof.* Let $R := 3\sup_{n\ge 0}\max\{\|f_n\|_{[p]}, \|h_n\|_{[p]}, 1\}$ and let $B_R := \{z \in Z\colon \|z\| \le R\}$. In particular, the range of each $f_n$ and $h_n$ is in the ball $B_R$. If $\psi(z,\cdot) \equiv \chi_1$ and $\phi(z,\cdot) \equiv \chi_2$ for each $z$ in $B_R$ and some functions $\chi_1, \chi_2\colon J \to X$ then for each $n \ge 1$ and $t \in J$,

$$(f_n - h_n)(t) = c - d + \oint_{J_t} \chi_1 \cdot d(\mu - \gamma) + \oint_{J_t} \chi_2 \cdot d(\nu - \pi),$$

where $J_t = (a,t]$ or $[a,t)$, and so the conclusion holds by the Love–Young inequality (3.153). Suppose that either $\psi(z,\cdot)$ or $\phi(z,\cdot)$ is non-constant with respect to $z \in B_R$. Let $\theta_1$ and $\theta_2$ be defined by (10.31). Using Proposition 3.52, one can find a Young interval partition $\{(t_{j-1}, t_j)\}_{j=1}^m$ of $J$ such that

$$v_p(\mu; (t_{j-1}, t_j)) \le \theta_1^p \quad \text{and} \quad v_1(\nu; (t_{j-1}, t_j)) \le \theta_2 \tag{10.82}$$

for each $1 \le j \le m$, and $m \le 1 + v_p(\mu; J)/\theta_1^p + v_1(\nu; J)/\theta_2$. Since $\psi \in \mathcal{H}W^{\rm loc}_{1+\alpha,q}(Z \times J; X)$, by Lemma 6.62, $\psi \in \mathcal{H}W^{\rm loc}_{1,q}(Z \times J; X)$, and so

$$H_1(\psi) := H_1(\psi; B_R \times J, X) < \infty \tag{10.83}$$

and $W_{1,q}(\psi, R) := W_{1,q}(\psi, R; B_R \times J, X) < \infty$. Since $\phi \in \mathcal{UH}_1^{\rm loc}(Z \times J; X)$, $H_1(\phi) := H_1(\phi; B_R \times J, X) < \infty$. Let $A := 3 + 4H[H_1(\psi) + H_1(\phi)] < \infty$, where $H := \sup_{t\in J}\max\{\|\mu(\{t\})\|, \|\nu(\{t\})\|\}$. Given $\epsilon > 0$, let

$$\delta_1 := \epsilon A^{-m-1}/\big\{3 + 4K_{p,q}[H_1(\psi) + D]\|\mu\|_{(p)} + 4H_1(\phi)\|\nu\|_{(1)}\big\}, \tag{10.84}$$

where $K_{p,q} = \zeta(1/p + 1/q)$ and $D := H_\alpha(\psi_u^{(1)}; B_R \times J, X) + W_{\alpha,q}(\psi_u^{(1)}, R; B_R \times J, X)$, let

$$\delta_2 := \epsilon A^{-m-1}/\big\{1 + 8K_{p,q}[G(\psi) + W_{1,q}(\psi, R)]\big\}, \tag{10.85}$$

where $G(\chi) := \|N_\chi(0)\|_{\sup} + RH_1(\chi)$ for $\chi = \psi$ or $\phi$, and let

$$\delta_3 := \epsilon A^{-m-1}/\big\{1 + 8G(\phi)\big\}. \tag{10.86}$$

We claim that if $\|c - d\| < \delta_1$, $\|\mu - \gamma\|_{(p)} < \delta_2$ and $\|\nu - \pi\|_{(1)} < \delta_3$ then for each $j = 1, \dots, m$ and $n = 1, 2, \dots$,

$$A_{j,n} := \|f_n - h_n\|_{J_j,[p]} \le \epsilon A^{j-m-1} =: B_j, \tag{10.87}$$

where $J_j := [t_{j-1}, t_j]$. Assuming that the claim holds and applying subadditivity of $\|\cdot\|_{[p]}$ over closed adjoining intervals (Proposition 3.35(*a*)) recursively, since $A > 2$, it follows that

$$\|f_n - h_n\|_{[p]} \le \sum_{j=1}^m \|f_n - h_n\|_{J_j,[p]} < \epsilon$$

for all $n$, as desired.

To prove (10.87), first suppose that $\{f_n\}$ is the sequence of Picard iterates $\{f_n^{\succ}\}$ obtained by the integral transform $T^{\succ}_{c,a,\mu,\nu}$ and $\{h_n\}$ is the sequence of Picard iterates $\{\xi_n^{\succ}\}$ obtained by the integral transform $T^{\succ}_{d,a,\gamma,\pi}$, where

$$\|c-d\| < \delta_1, \quad \|\mu-\gamma\|_{(p)} < \delta_2, \quad \text{and} \quad \|\nu-\pi\|_{(1)} < \delta_3. \tag{10.88}$$

The claim (10.87) will be proved by induction on $j$. Before starting the induction, we give a preliminary bound on $A_{j,n}$ for each $j\in\{1,\dots,m\}$ and $n\ge 2$. For $a\le u\le t\le b$ and $n=1,2,\dots$, let

$$S^{\psi,\mu,\gamma}_{u,n}(t) := \oint_{(u,t]} (N_\psi f^{\succ}_{n-1})_-\cdot\mathrm{d}\mu - \oint_{(u,t]} (N_\psi \xi^{\succ}_{n-1})_-\cdot\mathrm{d}\gamma,$$

$$S^{\phi,\nu,\pi}_{u,n}(t) := \oint_{(u,t]} (N_\phi f^{\succ}_{n-1})_-\cdot\mathrm{d}\nu - \oint_{(u,t]} (N_\phi \xi^{\succ}_{n-1})_-\cdot\mathrm{d}\pi,$$

and let

$$S_{u,n} := S^{\psi,\mu,\gamma}_{u,n} + S^{\phi,\nu,\pi}_{u,n}. \tag{10.89}$$

Then $S_{u,n}$ is a function on $[u,b]$ such that when $u=a$,

$$S_{a,n} = [f_n^{\succ}-\xi_n^{\succ}] - [c-d]. \tag{10.90}$$

By additivity of the Kolmogorov integral (Theorem 2.21), for each $a\le u\le t\le b$, we have

$$S_{a,n}(t) = S_{a,n}(u) + S_{u,n}(t). \tag{10.91}$$

Let $j\in\{1,\dots,m\}$ and $n\ge 2$. Applying (10.65) to $f=f_n^{\succ}-\xi_n^{\succ}=c-d+S_{a,n}$, $u=t_{j-1}$, and $v=t_j$, since $f-f(t_{j-1})=S_{t_{j-1},n}$ on $[t_{j-1},b]$ by (10.91), it follows that

$$\begin{aligned} A_{j,n} &= \|f_n^{\succ}-\xi_n^{\succ}\|_{J_j,[p]} \qquad (10.92)\\ &\le \|(f_n^{\succ}-\xi_n^{\succ})(t_{j-1})\| + \|S_{t_{j-1},n}\|_{[t_{j-1},t_j),[p]} + 2\|\Delta^- S_{a,n}(t_j)\|. \end{aligned}$$

Clearly $\Delta^- S_{a,n} = \Delta^- S^{\psi,\mu,\gamma}_{a,n} + \Delta^- S^{\phi,\nu,\pi}_{a,n}$. As in (10.51), we have

$$\begin{aligned} \|\Delta^- S^{\psi,\mu,\gamma}_{a,n}(t_j)\| &= \|(N_\psi f^{\succ}_{n-1})(t_j-)\cdot\mu(\{t_j\}) - (N_\psi\xi^{\succ}_{n-1})(t_j-)\cdot\gamma(\{t_j\})\|\\ &\le \|(N_\psi f^{\succ}_{n-1})(t_j-) - (N_\psi\xi^{\succ}_{n-1})(t_j-)\|\,\|\mu(\{t_j\})\| \qquad (10.93)\\ &\quad + \|(N_\psi\xi^{\succ}_{n-1})(t_j-)\|\,\|(\mu-\gamma)(\{t_j\})\|. \end{aligned}$$

For the first term on the right side of (10.93), as in (10.67) and using (10.91), since $\psi$ is $s$-uniformly Lipschitz on $B_R$ by (10.83) and $S_{t_{j-1},n-1}(t_{j-1})=0$, it follows that

$$\begin{aligned} &\|(N_\psi f^{\succ}_{n-1})(t_j-) - (N_\psi\xi^{\succ}_{n-1})(t_j-)\|\\ &\quad\le H_1(\psi)\|f^{\succ}_{n-1}-\xi^{\succ}_{n-1}\|_{[t_{j-1},t_j),\mathrm{sup}}\\ &\quad\le H_1(\psi)\Big\{\|(f^{\succ}_{n-1}-\xi^{\succ}_{n-1})(t_{j-1})\| + \|S_{t_{j-1},n-1}\|_{[t_{j-1},t_j),(p)}\Big\}. \end{aligned}$$

For the second term on the right side of (10.93), again since $\psi$ is $s$-uniformly Lipschitz on $B_R$, we have

$$\|(N_\psi \xi_{n-1}^{\succ})(t_j-)\| \le \|N_\psi(0)\|_{\sup} + H_1(\psi)R = G(\psi).$$

Inserting the preceding two bounds in (10.93) gives

$$\begin{aligned}\|\Delta^- S_{a,n}^{\psi,\mu,\gamma}(t_j)\| &\le G(\psi)\|\mu-\gamma\|_{(p)} \\ &\quad + HH_1(\psi)\Big\{\|(f_{n-1}^{\succ}-\xi_{n-1}^{\succ})(t_{j-1})\|+\|S_{t_{j-1},n-1}\|_{[t_{j-1},t_j),(p)}\Big\}.\end{aligned}$$

Since $\phi$ also is $s$-uniformly Lipschitz on $B_R$, the same bound holds when $\psi$, $\mu$, and $\gamma$ are replaced, respectively, by $\phi$, $\nu$, and $\pi$. Adding the two bounds gives a bound for $\|\Delta^- S_{a,n}(t_j)\|$ in (10.92), and so we have

$$\begin{aligned}A_{j,n} &\le \|(f_n^{\succ}-\xi_n^{\succ})(t_{j-1})\| + \|S_{t_{j-1},n}\|_{[t_{j-1},t_j),[p]} \\ &\quad + 2H(H_1(\psi)+H_1(\phi))\Big\{\|(f_{n-1}^{\succ}-\xi_{n-1}^{\succ})(t_{j-1})\|+\|S_{t_{j-1},n-1}\|_{[t_{j-1},t_j),(p)}\Big\} \\ &\quad + 2G(\psi)\|\mu-\gamma\|_{(p)} + 2G(\phi)\|\nu-\pi\|_{(1)} \end{aligned} \tag{10.94}$$

for each $j \in \{1,\dots,m\}$ and $n \ge 2$.

To begin the induction on $j$, let $j=1$. Recalling (10.90), we claim that for each $k=1,2,\dots,$

$$\big\|S_{a,k}\big\|_{[a,t_1),[p]} \le \epsilon A^{-m-1}. \tag{10.95}$$

For $a \le u \le t \le b$, and for the constant function 1, let

$$I_u^{\psi,\mu,\gamma}(t) := \fint_{(u,t]} \Big[N_\psi(c1)-N_\psi(d1)\Big]_- \cdot d\mu + \fint_{(u,t]} \big[N_\psi(d1)\big]_- \cdot d(\mu-\gamma). \tag{10.96}$$

By Theorem 6.68, since $\max\{\|c\|,\|d\|\} \le R$, we get the bound

$$\|N_\psi(c1)-N_\psi(d1)\|_{J,[q]} \le [H_1(\psi)+D]\|c-d\|.$$

By Lemma 6.53($a$) with $\alpha=1$, $\|N_\psi(d1)\|_{J,(q)} \le W_{1,q}(\psi,R)$, and since $\psi$ is $s$-uniformly Lipschitz on $B_R$,

$$\|N_\psi(d1)\|_{J,\sup} \le \|N_\psi(0)\|_{\sup} + RH_1(\psi) = G(\psi).$$

Thus by the Love–Young inequality (3.153) and by the inequality $\|g_-^{(a)}\|_{J,[q]} \le \|g\|_{J,[q]}$ applied to $g = N_\psi(c1)-N_\psi(d1)$ and $g = N_\psi(d1)$, for any $a \le u < v \le b$, we get

$$\begin{aligned}&\|I_u^{\psi,\mu,\gamma}\|_{[u,v],(p)} \\ &\le K_{p,q}\Big\{\big\|\big[N_\psi(c1)-N_\psi(d1)\big]_-^{(a)}\big\|_{J,[q]}\|\mu\|_{(p)} + \big\|\big[N_\psi(d1)\big]_-^{(a)}\big\|_{J,[q]}\|\mu-\gamma\|_{(p)}\Big\} \\ &\le K_{p,q}\Big\{[H_1(\psi)+D]\|c-d\|\|\mu\|_{(p)} + [G(\psi)+W_{1,q}(\psi,R)]\|\mu-\gamma\|_{(p)}\Big\}.\end{aligned} \tag{10.97}$$

Similarly, letting $I_u^{\phi,\nu,\pi}(t)$ be the right side of (10.96) with $\psi$, $\mu$, $\gamma$ replaced by $\phi$, $\nu$, $\pi$, respectively, by the Love–Young inequality (3.153) with $p = 1$, for any $a \le u < v \le b$, we get

$$\|I_u^{\phi,\nu,\pi}\|_{[u,v],(1)} \le H_1(\phi)\|\nu\|_{(1)}\|c-d\| + G(\phi)\|\nu-\pi\|_{(1)}. \tag{10.98}$$

Recalling (10.89) with $n = 1$, $f_0^{\succ} = c1$, $\xi_0^{\succ} = d1$, and using bilinearity of the Kolmogorov integral (Theorem 2.72 and Corollary 2.26), we have

$$S_{u,1} = S_{u,1}^{\psi,\mu,\gamma} + S_{u,1}^{\phi,\nu,\pi} = I_u^{\psi,\mu,\gamma} + I_u^{\phi,\nu,\pi}.$$

Applying (10.97) and (10.98), and since $\|\cdot\|_{(p)} \le \|\cdot\|_{(1)}$ (by Lemma 3.45) and $S_{u,1}(u) = 0$, it follows that

$$\begin{aligned}
&\|S_{u,1}\|_{[u,v],[p]} \\
&\quad \le 2\|S_{u,1}\|_{[u,v],(p)} \\
&\quad \le 2\Big\{K_{p,q}[H_1(\psi)+D]\|\mu\|_{(p)} + H_1(\phi)\|\nu\|_{(1)}\Big\}\|c-d\| \\
&\qquad + 2K_{p,q}\big[G(\psi)+W_{1,q}(\psi,R)\big]\|\mu-\gamma\|_{(p)} + 2G(\phi)\|\nu-\pi\|_{(1)} \le \epsilon A^{-m-1}
\end{aligned} \tag{10.99}$$

for any $a \le u < v \le b$, by (10.88), (10.84), (10.85), and (10.86). Thus (10.95) holds for $k = 1$. Suppose (10.95) holds for some $k \ge 1$. To prove it for $k+1$ we will use Lemma 10.28($a$) with $f = f_k^{\succ}$, $g = \xi_k^{\succ}$, $u = a$, $v = t_1$, $R$ as in the present proof, and $E = \epsilon A^{-m-1}$. The hypothesis (10.31) holds for $\mu$ and $\nu$ by (10.82). Since $\|c-d\| < \delta_1 \le E$ by (10.84), it follows by (10.90) and the induction assumption (10.95) that

$$\|f-g\|_{[u,v),[p]} = \|f_k^{\succ} - \xi_k^{\succ}\|_{[a,t_1),[p]} \le \|c-d\| + \|S_{a,k}\|_{[a,t_1),[p]} \le 2E.$$

With $\tau_1$ and $\tau_2$ defined by (10.79) and $\delta_2$, $\delta_3$ by (10.85) and (10.86) respectively, we have $\delta_2 = E\tau_1$ and $\delta_3 = E\tau_2$. Using Lemma 10.28($a$), it follows that

$$\|S_{a,k+1}\|_{[a,t_1),[p]} = \Big\|T_{0,a,\mu,\nu}^{\succ} f - T_{0,a,\gamma,\pi}^{\succ} g\Big\|_{[a,t_1),[p]} \le E = \epsilon A^{-m-1}.$$

Therefore (10.95) holds with $k+1$ in place of $k$, and hence for all $k = 1, 2, \ldots$ by induction.

To bound $A_{1,n}$ for $n \ge 2$, recalling (10.88), first apply (10.94) with $j = 1$, then the bound $\|(f_l^{\succ} - \xi_l^{\succ})(a)\| = \|c-d\| < \delta_1 \le \epsilon A^{-m-1}$ for each $l \ge 1$ obtained using (10.84), then (10.95) with $k = n, n-1$, and also (10.85) and (10.86), it follows that

$$A_{1,n} \le \epsilon A^{-m-1}\{3 + 4H[H_1(\psi) + H_1(\phi)]\} = \epsilon A^{-m} = B_1. \tag{10.100}$$

To bound $A_{j,1}$, let $j \in \{1, \ldots, m\}$. By (10.90), (10.99), and (10.88), we have

$$A_{j,1} \le \|c-d\| + \|S_{a,1}\|_{[a,b],[p]} \le 2\epsilon A^{-m-1} < B_j. \tag{10.101}$$

This for $j = 1$ together with (10.100) for $n \geq 2$ proves that (10.87) holds for $j = 1$ and each $n \geq 1$.

Now suppose (10.87) holds for some $1 \leq j < m$. It then will be shown to hold for $j + 1$ in place of $j$. We claim that for each $k = 1, 2, \ldots,$

$$\left\| S_{t_j,k} \right\|_{[t_j,t_{j+1}),[p]} \leq \epsilon A^{j-m-1} = B_j. \tag{10.102}$$

This bound holds for $k = 1$ by (10.99). Suppose (10.102) holds for some $k \geq 1$. To prove it for $k+1$ we will use Lemma 10.28($a$) with $f = f_k^{\succ}$, $g = \xi_k^{\succ}$, $u = t_j$, $v = t_{j+1}$, $R$ as in the present proof, and $E = B_j$. The hypothesis (10.31) holds for $\mu$ and $\nu$ by (10.82). By (10.90) and (10.91), we have

$$\begin{aligned} \|f - g\|_{[u,v),[p]} &= \|f_k^{\succ} - \xi_k^{\succ}\|_{[t_j,t_{j+1}),[p]} \\ &\leq \|(f_k^{\succ} - \xi_k^{\succ})(t_j)\| + \|S_{t_j,k}\|_{[t_j,t_{j+1}),[p]}. \end{aligned}$$

By the induction assumption on $j$, that is, by (10.87), we have

$$\|(f_k^{\succ} - \xi_k^{\succ})(t_j)\| \leq A_{j,k} \leq B_j = E.$$

This and the induction assumption on $k$, which is (10.102), yields that $\|f - g\|_{[u,v),[p]} \leq 2E$. Again since $\delta_2 \leq E\tau_1$ and $\delta_3 \leq E\tau_2$ with $\tau_1$ and $\tau_2$ defined by (10.79), using Lemma 10.28($a$), it follows that

$$\|S_{t_j,k+1}\|_{[t_j,t_{j+1}),[p]} = \left\| T_{0,u,\mu,\nu}^{\succ} f - T_{0,u,\gamma,\pi}^{\succ} g \right\|_{[u,v),[p]} \leq E = B_j.$$

Therefore (10.102) holds with $k+1$ in place of $k$, and hence for all $k = 1, 2, \ldots$ by induction. To bound $A_{j+1,n}$ for $n \geq 2$, first apply (10.94) with $j + 1$ in place of $j$. Next, by the induction assumption, $\|(f_l^{\succ} - \xi_l^{\succ})(t_j)\| \leq A_{j,l} \leq B_j$ for each $l \geq 1$. Then by (10.102) with $k = n, n-1$ and (10.88), it follows that

$$A_{j+1,n} \leq \epsilon A^{j-m-1}\{3 + 4H[H_1(\psi) + H_1(\phi)]\} = \epsilon A^{j-m} = B_{j+1}.$$

Since the same bound holds for $n = 1$ by (10.101), (10.87) holds for $j + 1$ in place of $j$ for each $n \geq 1$. By induction on $j$, (10.87) holds for each $j \in \{1, \ldots, m\}$, proving the claim for the two bounded sequences of Picard iterates $\{f_n^{\succ}\}_{n\geq 0}$ and $\{\xi_n^{\succ}\}_{n\geq 0}$ obtained by the integral transforms $T_{c,a,\mu,\nu}^{\succ}$ and $T_{d,a,\gamma,\pi}^{\succ}$, respectively.

Now suppose that $\{f_n\}$ is the sequence of Picard iterates $\{g_n^{\succ}\}$ obtained by the integral transform $Q_{c,a,\mu,\nu}^{\succ}$ and $\{h_n\}$ is the sequence of Picard iterates $\{\zeta_n^{\succ}\}$ obtained by the integral transform $Q_{d,a,\gamma,\pi}^{\succ}$, where again (10.88) holds. The claim (10.87) will be proved again by induction on $j$. Before starting the induction, we give a preliminary bound on $A_{j,n}$ for each $j \in \{1, \ldots, m\}$ and $n \geq 2$. For $u \in J$ and $n = 1, 2, \ldots$, let

$$S_{u,n} := S_{u,n}^{\psi,\mu,\gamma} + S_{u,n}^{\phi,\nu,\pi} \tag{10.103}$$

be the functions on $[u, b]$ with values for each $t \in [u, b]$,

$$S_{u,n}^{\psi,\mu,\gamma}(t) := \oint_{[u,t)} (N_\psi g_{n-1}^{\succ})\cdot d\mu - \oint_{[u,t)} (N_\psi \zeta_{n-1}^{\succ})\cdot d\gamma,$$

$$S_{u,n}^{\phi,\nu,\pi}(t) := \oint_{[u,t)} (N_\phi g_{n-1}^{\succ})\cdot d\nu - \oint_{[u,t)} (N_\phi \zeta_{n-1}^{\succ})\cdot d\pi.$$

Let $j \in \{1, \ldots, m\}$ and $n \geq 2$. By (10.30), for $a \leq u < t \leq b$, we have

$$S_{a,n}(t) = S_{a,n}(u) + S_{u,n}(u+) + S_{u+,n}(t). \tag{10.104}$$

For a function $f\colon J \to Z$ and any $t \in (u, v] \subset J$, we have $f(t) = f(u) + \Delta^+ f(u) + f(t) - f(u+)$, and so by Corollary 3.43($b$), it follows that

$$\|f\|_{[u,v],[p]} \leq \|f(u)\| + 2\|\Delta^+ f(u)\| + \|f - f(u+)\|_{(u,v],[p]}. \tag{10.105}$$

Since $g_n^{\succ} - \zeta_n^{\succ} = c - d + S_{a,n}$, letting $f := g_n^{\succ} - \zeta_n^{\succ}$ and $[u, v] := [t_{j-1}, t_j] = J_j$, we have $f - f(u+) = S_{t_{j-1}+,n}$ on $(u, b]$, and $\Delta^+ f(u) = S_{t_{j-1},n}(t_{j-1}+)$. Thus by (10.105), we get

$$\begin{aligned} A_{j,n} &= \|g_n^{\succ} - \zeta_n^{\succ}\|_{J_j,[p]} \\ &\leq \|(g_n^{\succ} - \zeta_n^{\succ})(t_{j-1})\| + 2\|S_{t_{j-1},n}(t_{j-1}+)\| + \|S_{t_{j-1}+,n}\|_{(t_{j-1},t_j],[p]}. \end{aligned} \tag{10.106}$$

By additivity and upper continuity of the Kolmogorov integral (Theorem 2.21 and Corollary 2.23), as in (10.54), for any $t \in [a, b)$,

$$\begin{aligned} &\|S_{t,n}^{\psi,\mu,\gamma}(t+)\| \\ &= \|(N_\psi g_{n-1}^{\succ})(t)\cdot\mu(\{t\}) - (N_\psi \zeta_{n-1}^{\succ})(t)\cdot\gamma(\{t\})\| \\ &\leq \|(N_\psi g_{n-1}^{\succ})(t) - (N_\psi \zeta_{n-1}^{\succ})(t)\|\|\mu(\{t\})\| + \|(N_\psi \zeta_{n-1}^{\succ})(t)\|\|(\mu - \gamma)(\{t\})\|. \end{aligned} \tag{10.107}$$

Since $\psi$ is $s$-uniformly Lipschitz on $B_R$, we have the bounds

$$\|(N_\psi g_{n-1}^{\succ})(t) - (N_\psi \zeta_{n-1}^{\succ})(t)\| \leq H_1(\psi)\|(g_{n-1}^{\succ} - \zeta_{n-1}^{\succ})(t)\|$$

and

$$\|(N_\psi \zeta_{n-1}^{\succ})(t)\| \leq \|N_\psi(0)\|_{\sup} + H_1(\psi)R = G(\psi)$$

respectively, since $\|\zeta_{n-1}^{\succ}\|_{\sup} \leq R$, and so applying the preceding two bounds in (10.107) gives

$$\|S_{t,n}^{\psi,\mu,\gamma}(t+)\| \leq HH_1(\psi)\|(g_{n-1}^{\succ} - \zeta_{n-1}^{\succ})(t)\| + G(\psi)\|\mu - \gamma\|_{(p)},$$

recalling $H = \sup_{t\in J} \max\{\|\mu(\{t\})\|, \|\nu(\{t\})\|\}$. Since $\phi$ also is $s$-uniformly Lipschitz on $B_R$, the same bound holds when $\psi$, $\mu$, $\gamma$, and $\|\cdot\|_{(p)}$ are replaced, respectively, by $\phi$, $\nu$, $\pi$, and $\|\cdot\|_{(1)}$. Adding the two bounds gives

$$\begin{aligned} \|S_{t,n}(t+)\| \leq\ & H[H_1(\psi) + H_1(\phi)]\|(g_{n-1}^{\succ} - \zeta_{n-1}^{\succ})(t)\| \\ & + G(\psi)\|\mu - \gamma\|_{(p)} + G(\phi)\|\nu - \pi\|_{(1)}. \end{aligned} \tag{10.108}$$

Taking $t = t_{j-1}$ and applying the preceding bound in (10.106) gives the preliminary bound

$$\begin{aligned} A_{j,n} \le \|(g_n^{\succ} - \zeta_n^{\succ})(t_{j-1})\| &+ \|S_{t_{j-1}+,n}\|_{(t_{j-1},t_j],[p]} \\ &+ 2H[H_1(\psi) + H_1(\phi)]\|(g_{n-1}^{\succ} - \zeta_{n-1}^{\succ})(t_{j-1})\| \\ &+ 2G(\psi)\|\mu - \gamma\|_{(p)} + 2G(\phi)\|\nu - \pi\|_{(1)} \end{aligned} \tag{10.109}$$

for each $j \in \{1, \ldots, m\}$ and $n \ge 2$.

To begin the induction on $j$ for (10.87) in this case, let $j = 1$. We claim that for each $k = 1, 2, \ldots$,

$$\big\|S_{a+,k}\big\|_{(a,t_1],[p]} \le \epsilon A^{-m-1}. \tag{10.110}$$

For $a \le u < t \le b$, let

$$I_u^{\psi,\mu,\gamma}(t) := \oint_{(u,t)} \Big[N_\psi(c1) - N_\psi(d1)\Big] \cdot \mathrm{d}\mu + \oint_{(u,t)} N_\psi(d1) \cdot \mathrm{d}(\mu - \gamma). \tag{10.111}$$

As for (10.96), by the Love–Young inequality, (10.97) holds. Letting $I_u^{\phi,\nu,\pi}$ be the right side of (10.111) with $\psi$, $\mu$, $\gamma$ replaced by $\phi$, $\nu$, $\pi$, it follows similarly that (10.98) holds. Recalling (10.103) with $n = 1$, $g_0^{\succ} = c1$, $\zeta_0^{\succ} = d1$, using bilinearity, additivity, and upper continuity of the Kolmogorov integral (Theorem 2.72, Corollary 2.26, Theorem 2.21, and Corollary 2.23), we have

$$S_{u+,1} = S_{u+,1}^{\psi,\mu,\gamma} + S_{u+,1}^{\phi,\nu,\pi} = I_u^{\psi,\mu,\gamma} + I_u^{\phi,\nu,\pi}.$$

Thus since $S_{u+,1}(u) = 0$, by (10.97) and (10.98), it follows as in (10.99) that

$$\begin{aligned} &\|S_{u+,1}\|_{[u,v],[p]} \\ &\quad \le 2\Big\{K_{p,q}[H_1(\psi) + D]\|\mu\|_{(p)} + H_1(\phi)\|\nu\|_{(1)}\Big\}\|c - d\| \\ &\qquad + 2K_{p,q}\big[G(\psi) + W_{1,q}(\psi, R)\big]\|\mu - \gamma\|_{(p)} + 2G(\phi)\|\nu - \pi\|_{(1)} \le \epsilon A^{-m-1} \end{aligned} \tag{10.112}$$

for any $a \le u < v \le b$. Thus (10.110) holds for $k = 1$. Suppose (10.110) holds for some $k \ge 1$. To prove it for $k + 1$ we will use Lemma 10.28($b$) with $f = g_k^{\succ}$, $g = \zeta_k^{\succ}$, $u = a$, $v = t_1$, $R$ as in the present proof, and $E = \epsilon A^{-m-1}$. The hypothesis (10.31) holds for $\mu$ and $\nu$ by (10.82). Recalling the equality $g_k^{\succ} - \zeta_k^{\succ} = c - d + S_{a,k}$ just after (10.105), and (10.104) with $u = a$, we have

$$\begin{aligned} \|f - g\|_{(a,t_1],[p]} &= \|g_k^{\succ} - \zeta_k^{\succ}\|_{(a,t_1],[p]} \\ &\le \|c - d\| + \|S_{a,k}(a+)\| + \|S_{a+,k}\|_{(a,t_1],[p]}. \end{aligned}$$

Since $(g_{k-1}^{\succ} - \zeta_{k-1}^{\succ})(a) = c - d$, using (10.108) with $t = a$ and $n = k$, it follows that

$$\begin{aligned} &\|c - d\| + \|S_{a,k}(a+)\| \\ &\quad \le \|c - d\|\{1 + H[H_1(\psi) + H_1(\phi)]\} + G(\psi)\|\mu - \gamma\|_{(p)} + G(\phi)\|\nu - \pi\|_{(1)} \\ &\quad \le E \end{aligned}$$

by (10.88), (10.84), (10.85), and (10.86), and using $K_{p,q} > \zeta(2) = \pi^2/6 > 3/2$. This and the induction assumption (10.110) yield that $\|f - g\|_{(a,t_1],[p]} \le 2E$. Since $\delta_2 \le E\tau_1$ and $\delta_3 \le E\tau_2$ with $\tau_1$ and $\tau_2$ defined by (10.79), using Lemma 10.28(*b*), it follows that $\|S_{a+,k+1}\|_{(a,t_1],[p]} \le E$. Therefore (10.110) holds with $k+1$ in place of $k$, and hence for all $k = 1, 2, \ldots$ by induction.

To bound $A_{1,n}$ for $n \ge 2$, applying first (10.109) with $j = 1$, then the bound $\|(g_l^{\succ} - \zeta_l^{\succ})(a)\| = \|c - d\| < \delta_1 \le \epsilon A^{-m-1}$ for each $l \ge 1$ obtained using (10.84), and (10.110) with $k = n$, it follows that

$$A_{1,n} \le \epsilon A^{-m-1}\{3 + 2H[H_1(\psi) + H_1(\phi)]\} < \epsilon A^{-m} = B_1. \tag{10.113}$$

To bound $A_{j,1}$ for $j \in \{1, \ldots, m\}$, note that $(g_1^{\succ} - \zeta_1^{\succ})(t) = [c - d] + S_{a,1}(t)$ and

$$S_{a,1}^{\psi,\mu,\gamma}(t) = \oint_{[a,t)} \Big[N_\psi(c1) - N_\psi(d1)\Big] \cdot \mathrm{d}\mu + \oint_{[a,t)} N_\psi(d1) \cdot \mathrm{d}(\mu - \gamma)$$

for each $t \in [a, b]$. A similar equality holds for $S_{a,1}^{\phi,\nu,\pi}$. By (10.103) with $u = a$ and $n = 1$, the bound $\|S_{a,1}\|_{J,[p]} \le \epsilon A^{-m-1}$ follows as in (10.99). Thus since $S_{a,1}(a) = 0$, as in (10.101) it follows that for each $j \in \{1, \ldots, m\}$,

$$A_{j,1} \le \|c - d\| + \|S_{a,1}\|_{J,[p]} \le 2\epsilon A^{-m-1} < B_1 \le B_j. \tag{10.114}$$

This for $j = 1$ together with (10.113) proves that (10.87) holds for $j = 1$.

Now suppose (10.87) holds for some $1 \le j < m$. It then will be shown to hold for $j + 1$ in place of $j$. We claim that for each $k = 1, 2, \ldots$,

$$\left\|S_{t_j+,k}\right\|_{(t_j,t_{j+1}],[p]} \le \{1.5 + H[H_1(\psi) + H_1(\phi)]\}B_j. \tag{10.115}$$

This bound holds for $k = 1$ by (10.112) with $u = t_j$ and $v = t_{j+1}$ since $\epsilon A^{-m-1} < B_j$. Suppose (10.115) holds for some $k \ge 1$. To prove it for $k + 1$ we will use Lemma 10.28(*b*) with $f = g_k^{\succ}$, $g = \zeta_k^{\succ}$, $u = t_j$, $v = t_{j+1}$, $R$ as in the present proof, and $E$ being the right side of (10.115). The hypothesis (10.31) holds for $\mu$ and $\nu$ by (10.82). To check the other hypotheses of Lemma 10.28(*b*), for a function $h\colon J \to Z$ and any $a \le u < v \le b$, using Corollary 3.43(*b*) as in (10.105), it follows that

$$\|h\|_{(u,v],[p]} \le \|h(u)\| + \|\Delta^+ h(u)\| + \|h - h(u+)\|_{(u,v],[p]}.$$

For $k \ge 2$, taking $h := g_k^{\succ} - \zeta_k^{\succ}$ as in (10.106), then using (10.108) with $t = t_j$ and $n = k$, it follows that

$$\begin{aligned}
&\|f - g\|_{(u,v],[p]} = \|g_k^{\succ} - \zeta_k^{\succ}\|_{(t_j,t_{j+1}],[p]} \\
&\le \|(g_k^{\succ} - \zeta_k^{\succ})(t_j)\| + \|S_{t_j,k}(t_j+)\| + \|S_{t_j+,k}\|_{(t_j,t_{j+1}],[p]} \\
&\le \{1 + H[H_1(\psi) + H_1(\phi)]\} \max_{l \in \{k,k-1\}} \|(g_l^{\succ} - \zeta_l^{\succ})(t_j)\| \\
&\qquad + G(\psi)\|\mu - \gamma\|_{(p)} + G(\phi)\|\nu - \pi\|_{(1)} + \|S_{t_j+,k}\|_{(t_j,t_{j+1}],[p]} \\
&\le \{1.5 + H[H_1(\psi) + H_1(\phi)]\}B_j + \{1.5 + H[H_1(\psi) + H_1(\phi)]\}B_j = 2E,
\end{aligned}$$

where the last inequality holds by (10.88), (10.85), and (10.86) since $A > 1$, and by the induction assumptions on $j$ and on $k$. For $k = 1$, $\|f - g\|_{(u,v],[p]} \le A_{j+1,1} < B_1 \le B_j < 2E$ by (10.114), with $j$ there equal to $j+1$ here, which is valid since $j+1 \le m$. Because $\delta_2 \le E\tau_1$ and $\delta_3 \le E\tau_2$ with $\tau_1$ and $\tau_2$ defined by (10.79), using Lemma 10.28($b$), it follows that

$$\left\|S_{t_j+,k+1}\right\|_{(t_j,t_{j+1}],[p]} \le E = \{1.5 + H[H_1(\psi) + H_1(\phi)]\}B_j.$$

Therefore (10.115) holds with $k+1$ in place of $k$, hence for all $k = 1, 2, \ldots$ by induction.

To bound $A_{j+1,n}$ for $n \ge 2$, we apply first (10.109) with $j+1$ in place of $j$. The induction assumption (10.87) implies $\|(g_l^{\succ} - \zeta_l^{\succ})(t_j)\| \le A_{j,l} \le B_j$ for each $l \ge 1$. Then applying (10.115) with $k = n$ and (10.88), it follows that

$$\begin{aligned} A_{j+1,n} &\le \{1 + 2H[H_1(\psi) + H_1(\phi)]\} \max_{l \in \{n,n-1\}} \|(g_l^{\succ} - \zeta_l^{\succ})(t_j)\| \\ &\quad + \|S_{t_j+,n}\|_{(t_j,t_{j+1}],[p]} + 2G(\psi)\|\mu - \gamma\|_{(p)} + 2G(\phi)\|\nu - \pi\|_{(1)} \\ &\le B_j\{1 + 2H[H_1(\psi) + H_1(\phi)]\} + B_j\{1.5 + H[H_1(\psi) + H_1(\phi)]\} \\ &\quad + B_j/2 \le AB_j = B_{j+1}. \end{aligned}$$

Since the same bound holds for $n = 1$ by (10.114), (10.87) holds for $j + 1$ in place of $j$ for each $n \ge 1$. By induction on $j$, (10.87) holds for each $j \in \{1, \ldots, m\}$, proving the claim for the two bounded sequences of Picard iterates $\{g_n^{\succ}\}_{n\ge 0}$ and $\{\zeta_n^{\succ}\}_{n\ge 0}$ obtained by the integral transforms $Q_{c,a,\mu,\nu}^{\succ}$ and $Q_{d,a,\gamma,\pi}^{\succ}$, respectively. Thus the claim (10.87) is proved for each of the two pairs of sequences of Picard iterates, completing the proof of Proposition 10.29. □

**Proof of Theorem 10.27.** Let $c \in Z$, let $\mu \in \mathcal{AI}_p(J;Y)$, and let $\nu \in \mathcal{AI}_1(J;Y)$. By Theorem 10.7, each of the four integral equations (10.5), (10.6), (10.7), and (10.8) has respectively a solution mapping $\mathcal{S}_{(a,\cdot]}^{\succ}$, $\mathcal{S}_{[a,\cdot)}^{\succ}$, $\mathcal{S}_{[\cdot,b)}^{\prec}$, and $\mathcal{S}_{(\cdot,b]}^{\prec}$ from $Z \times \mathcal{AI}_p(J;Y) \times \mathcal{AI}_1(J;Y)$ to $\mathcal{W}_p(J;Z)$. Let $\mathcal{S} := \mathcal{S}_{(a,\cdot]}^{\succ}$ and let $\epsilon > 0$. We will show that there is a neighborhood $V$ of $(c, \mu, \nu) \in Z \times \mathcal{AI}_p(J;Y) \times \mathcal{AI}_1(J;Y)$ such that $\|\mathcal{S}(c, \mu, \nu) - \mathcal{S}(d, \gamma, \pi)\|_{J,[p]} < \epsilon$ for any $(d, \gamma, \pi) \in V$.

Let $\{f_n\}_{n\ge 0}$ be the sequence of Picard iterates (10.12) obtained by the integral transform $T_{c,a,\mu,\nu}^{\succ}$, converging in $\mathcal{W}_p(J;Z)$ to $\mathcal{S}(c, \mu, \nu)$ by Theorem 10.11. Let $\rho_1 \equiv \rho(p, \alpha, \psi)$ and $\rho_2 \equiv \rho(p, \beta, \phi)$ be defined by (10.24). The assumed hypotheses on $\psi$ and $\phi$ imply the respective hypotheses of Proposition 10.24 with $\alpha = 1$ there as shown in the proof of Theorem 10.11, and so $\rho_1 > 0$ and $\rho_2 > 0$. Let $\upsilon := 1 + (1 + \|\mu\|_{(p)})^p/\rho_1^p + (1 + \|\nu\|_{(1)})/\rho_2$, $C_1 := 1 + \max\{1, \|c\|\}$, and

$$R := C_1 3^{\upsilon+1}(1 + 2D(1 + H))^{\upsilon}, \tag{10.116}$$

where $D := H_1(\psi) + \|N_\psi(0)\|_{\sup} + 2\|\phi\|_{\mathcal{G}_\beta}$ and $H := \sup_{t\in J}\max\{\|\mu(\{t\})\|, \|\nu(\{t\})\|\}$. Let $V_1$ be the set of $(d,\gamma,\pi) \in Z \times \mathcal{AI}_p(J;Y) \times \mathcal{AI}_1(J;Y)$ such that $\|c-d\| \le 1$, $\|\mu-\gamma\|_{(p)} \le 1$, and $\|\nu-\pi\|_{(1)} \le 1$. Given $(d,\gamma,\pi) \in V_1$ let $\{h_n\}_{n\ge 0}$ be the sequence of Picard iterates obtained by the integral transform $T^{\succ}_{d,a,\gamma,\pi}$, converging in $\mathcal{W}_p(J;Z)$ to $\mathcal{S}(d,\gamma,\pi)$ by Theorem 10.11. The right side of (10.47), with $m$, $C$, and $H$ defined for $(c,\mu,\nu)$ or for $(d,\gamma,\pi)$, is less than $R$ defined by (10.116). Thus by Proposition 10.24 with $\alpha = 1$ there, we have $\sup_{n\ge 0}\max\{\|f_n\|_{[p]}, \|h_n\|_{[p]}\} \le R$. The assumed hypotheses on $\psi$ and $\phi$ also imply the respective hypotheses of Proposition 10.29. This is clear for $\phi$. For $\psi$, by Definition 10.3, $\psi \in \mathcal{WG}_{1,p}(Z\times J;X)$, and so $\psi(0,\cdot) \in \mathcal{W}_p(J;X) \subset \mathcal{W}_q(J;X)$ with $q = p/\alpha$ by the definition (6.31) of $W_{1,p}(\psi,K)$, and so $\psi \in \mathcal{HW}^{\rm loc}_{1+\alpha,q}(Z\times J;X)$ as defined in Definition 6.61. Thus by Proposition 10.29, there exist $\delta_1 > 0$, $\delta_2 > 0$, and $\delta_3 > 0$ such that if $\|c-d\| < \delta_1$, $\|\mu-\gamma\|_{(p)} < \delta_2$, and $\|\nu-\pi\|_{(1)} < \delta_3$ then $\|f_n - h_n\|_{[p]} < \epsilon/3$ for all $n$. Let $V$ be the set of $(d,\gamma,\pi) \in Z \times \mathcal{AI}_p(J;Y) \times \mathcal{AI}_1(J;Y)$ such that $\|c-d\| < \min\{1,\delta_1\}$, $\|\mu-\gamma\|_{(p)} < \min\{1,\delta_2\}$, and $\|\nu-\pi\|_{(1)} < \min\{1,\delta_3\}$. Then for any $(d,\gamma,\pi) \in V$, it follows that $\|\mathcal{S}(c,\mu,\nu) - \mathcal{S}(d,\gamma,\pi)\|_{J,[p]} < \epsilon$, proving continuity of $\mathcal{S} = \mathcal{S}^{\succ}_{(a,\cdot]}$. The same proof, except that now (10.48) is used instead of (10.47), applies to the solution mapping $\mathcal{S}^{\succ}_{[a,\cdot)}$ and is therefore omitted.

To prove continuity of the solution mapping $\mathcal{S}^{\prec}_{[\cdot,b)}$, recall that $\theta(t) := a + b - t$, $\widetilde{\psi}(z,t) \equiv \psi(z,\theta(t))$, and $\widetilde{\phi}(z,t) \equiv \phi(z,\theta(t))$ for $a \le t \le b$ and $z \in Z$. As was noted in the proof of Theorem 10.7, $\widetilde{\psi}$ and $\widetilde{\phi}$ satisfy the hypotheses of Theorem 10.7, and so the solution mapping $\tilde{\mathcal{S}}^{\succ}_{(a,\cdot]}$, corresponding to the integral equation (10.5) with $\widetilde{\psi}$ and $\widetilde{\phi}$ in place of $\psi$ and $\phi$, is defined. Let $\epsilon > 0$, let $(c,\mu,\nu)$ be in $Z \times \mathcal{AI}_p(J;Y) \times \mathcal{AI}_1(J;Y)$, and recall that $\widetilde{\mu} \equiv \mu\circ\theta^{-1}$ and $\widetilde{\nu} \equiv \nu\circ\theta^{-1}$. Then $(c,\widetilde{\mu},\widetilde{\nu})$ is in $Z \times \mathcal{AI}_p(J;Y) \times \mathcal{AI}_1(J;Y)$. By the first part of the proof, there exists a neighborhood of zero $V$ such that $\|\tilde{\mathcal{S}}^{\succ}_{(a,\cdot]}(c,\widetilde{\mu},\widetilde{\nu}) - \tilde{\mathcal{S}}^{\succ}_{(a,\cdot]}(d,\widetilde{\gamma},\widetilde{\pi})\|_{[p]} < \epsilon$ for any $(d,\widetilde{\gamma},\widetilde{\pi})$ in $V$. Let $(d,\gamma,\pi)$ be in $V$, let $\widetilde{\gamma} := \gamma\circ\theta^{-1}$, and let $\widetilde{\pi} := \pi\circ\theta^{-1}$. Then $(d,\widetilde{\gamma},\widetilde{\pi})$ is in $V$. Also as in (10.45) and (10.46), it follows that the solutions $f := \mathcal{S}^{\prec}_{[\cdot,b)}(c,\mu,\nu)$ and $h := \mathcal{S}^{\prec}_{[\cdot,b)}(d,\gamma,\pi)$ are obtained by the change of variables with $\theta$ from the solutions $\widetilde{f} := \tilde{\mathcal{S}}^{\succ}_{(a,\cdot]}(c,\widetilde{\mu},\widetilde{\nu})$ and $\widetilde{h} := \tilde{\mathcal{S}}^{\succ}_{(a,\cdot]}(d,\widetilde{\gamma},\widetilde{\pi})$, respectively, and so $\|f-h\|_{[p]} = \|\widetilde{f}-\widetilde{h}\|_{[p]} < \epsilon$, proving continuity of the solution mapping $\mathcal{S}^{\prec}_{[\cdot,b)}$. The same change of variables argument shows continuity of the solution mapping $\mathcal{S}^{\prec}_{(\cdot,b]}$, proving the theorem. □

## 10.6 Notes

*Notes on Section 10.1.* Picard iteration, also known as the method of "successive approximation" of solutions, is apparently named for work in 1893 by É. Picard [186]. Ince [105, p. 63] gives earlier history and references.

*Notes on Section 10.2.* Lyons [151] proved Corollary 10.14 in the case $X = M(d, \mathbb{R})$, the space of $d \times d$ real matrices, $Z = Y = \mathbb{R}^d$, $h$ is continuous, and $F$ is bounded. Banach [6, Théorème II.6, p. 160] proved the fixed point theorem 10.21.

# 11

# Fourier Series

## 11.1 On the Order of Decrease of Fourier Coefficients

Let $T^1 := \{z\colon |z| = 1\} = \{e^{i\theta}\colon 0 \le \theta < 2\pi\}$ be the unit circle in the complex plane $\mathbb{C}$. On $T^1$ let $d\mu = d\theta/2\pi$ be the rotationally invariant probability measure. For $\mathbb{Z} = \{0, \pm 1, \pm 2, \dots\}$, the functions $z \mapsto z^n$ or $\theta \mapsto e^{in\theta}$ for $n \in \mathbb{Z}$ form an orthonormal basis of complex $L^2(T^1, \mu)$ (e.g. [53], Proposition 7.4.2). For $g \in \mathcal{L}^1(T^1, \mu)$, defining the coefficients $c_n := \widehat{g}(n) := \int g(z) z^{-n}\, d\mu$ yields the formal series

$$g \sim \sum_{n\in\mathbb{Z}} c_n z^n. \tag{11.1}$$

If $g \in \mathcal{L}^2(T^1, \mu)$, the series converges unconditionally (as defined in Chapter 1) to $g$ in $\mathcal{L}^2(T^1, \mu)$. For general $g \in \mathcal{L}^1(T^1, \mu)$ and $z \in T^1$, let

$$S_m g(z) := \sum_{|n|\le m} c_n z^n, \qquad m = 0, 1, 2, \dots. \tag{11.2}$$

There is a one-to-one correspondence between functions $f\colon \mathbb{R} \to \mathbb{C}$, periodic of period $2\pi$, and functions $g\colon T^1 \to \mathbb{C}$, given by $g(e^{it}) = f(t)$, $t \in \mathbb{R}$. For such a pair we will write

$$g = \dot{f} \tag{11.3}$$

(the latter notation is as in Edwards [63, p. 15]). We will set $\widehat{f}(n) = c_n = \widehat{g}(n)$. For $f \in \mathcal{L}^1([0, 2\pi]; \mathbb{C})$ (with respect to Lebesgue measure) the formal series (11.1) corresponds to

$$f(\theta) \sim \sum_{n\in\mathbb{Z}} c_n e^{in\theta}, \tag{11.4}$$

which has been called the *exponential Fourier series* of $f$. In terms of trigonometric functions we get another series

$$f(\theta) \sim c_0 + \sum_{n=1}^{\infty} a_n \cos(n\theta) + b_n \sin(n\theta), \tag{11.5}$$

R.M. Dudley and R. Norvaiša, *Concrete Functional Calculus*, Springer Monographs in Mathematics, DOI 10.1007/978-1-4419-6950-7_11,

called the *Fourier series* of $f$ (or of $g$). Here $a_n := c_n + c_{-n}$ and $b_n := \mathrm{i}(c_n - c_{-n})$ for $n \geq 1$. Then

$$c_n = (a_n - \mathrm{i}b_n)/2 \quad \text{and} \quad c_{-n} = (a_n + \mathrm{i}b_n)/2 \quad \text{for } n \geq 1. \tag{11.6}$$

For a complex number $w = u + \mathrm{i}v$ with $u$, $v$ real, the complex conjugate is defined by $\overline{w} := u - \mathrm{i}v$. A function $f$ in $\mathcal{L}^1([0, 2\pi]; \mathbb{C})$ is real-valued (equivalently, $\dot{f}$ in $\mathcal{L}^1(T^1, \mu)$ is) if and only if $c_{-n} = \overline{c_n}$ for all $n$, which holds if and only if $c_0$ and $a_n$, $b_n$ for $n \geq 1$ are all real-valued. In that case $|c_n|^2 = (a_n^2 + b_n^2)/4$ for $n \geq 1$.

The Fourier series converges to $f$ (in whatever sense) if and only if for $g = \dot{f}$, the sequence $S_m g$ converges to $g$ in the same sense as $m \to \infty$. Specifically, the Fourier series of $f$ converges to $f$ at a given $\theta$ if and only if $\lim_{m\to\infty} S_m g(\mathrm{e}^{\mathrm{i}\theta}) = g(\mathrm{e}^{\mathrm{i}\theta})$. Regarding such pointwise convergence, Kolmogorov [119] showed that for $g \in \mathcal{L}^1(T^1, \mu)$ the Fourier series may diverge everywhere. Carleson [29] for $p = 2$ (and thus for $p \geq 2$) showed that if $g \in \mathcal{L}^p(T^1, \mu)$, then the Fourier series converges almost everywhere to $g$. Carleson conjectured, and R. A. Hunt [104] proved, that the same holds for $1 < p < 2$. The Fourier series of $g$ converges to $g$ in $\mathcal{L}^p$ norm for all $g \in \mathcal{L}^p(T^1, \mu)$ if $1 < p < \infty$ (but not for $p = 1$ or $\infty$). This is a consequence of the theorem of Marcel Riesz about boundedness of the Hilbert transform in $L^p$-spaces (see Section 12.10 in Edwards [63]).

For $g \in \mathcal{L}^1(T^1, \mu)$, the partial sums (11.2) may not converge to $g$ in $\mathcal{L}^1$ norm (see e.g. Section II.1.2 in Katznelson [114], or Proposition 1.6.14 in Krantz [126]). For positive integers $N$, let $\sigma_N g$ be the Cesàro sum defined by

$$\sigma_N g(z) := \frac{1}{N+1} \sum_{m=0}^{N} S_m g(z) = \sum_{|n| \leq N} \Big(1 - \frac{|n|}{N+1}\Big) c_n z^n, \qquad z \in T^1.$$

By the Fejér theorem (Theorem III.3.4 in [258, p. 89]),

$$\lim_{N\to\infty} \sigma_N g(\mathrm{e}^{\mathrm{i}\theta}) = \frac{1}{2}\big[g(\mathrm{e}^{\mathrm{i}\theta+}) + g(\mathrm{e}^{\mathrm{i}\theta-})\big] \tag{11.7}$$

at each point $\mathrm{e}^{\mathrm{i}\theta} \in T^1$ such that the right side of (11.7) is defined, and $\sigma_N g$ converges to $g$ uniformly if $g$ is continuous. For a summable function $g$, Lebesgue proved (Theorem III.3.9 in [258, p. 90]) that $\sigma_N g$ converges to $g$ almost everywhere. More precisely, if $z = \mathrm{e}^{\mathrm{i}\theta} \in T^1$ is such that

$$\lim_{h\downarrow 0} \frac{1}{h} \int_0^h \big|g(\mathrm{e}^{\mathrm{i}(\theta+\phi)}) + g(\mathrm{e}^{\mathrm{i}(\theta-\phi)}) - 2g(\mathrm{e}^{\mathrm{i}\theta})\big|\, d\phi = 0$$

then $\sigma_N g(z) \to g(z)$ as $N \to \infty$. In the norm sense for $g \in \mathcal{L}^1(T^1, \mu)$, the Fourier series of $g$ is summable in $\mathcal{L}^1$ to $g$ by means of a family of summability kernels, which includes the Fejér and Poisson kernels (see Section I.2 in Katznelson [114] and Section 1.4 in Krantz [126]). Thus, $\sigma_N g \to g$ as $N \to \infty$ in $\mathcal{L}^1$ norm.

For $1 \le p \le 2$, the Hausdorff–Young inequality (co-named for W. H. Young) states that if $f \in L^p(T^1, \mu)$ then the sequences $\{a_n\}$, $\{b_n\}$, and $\{c_n\}$ are in the set $\ell^q = \{\{d_n\}\colon \sum_n |d_n|^q < \infty\}$, where $1/p + 1/q = 1$. For $2 < p \le \infty$, stronger integrability conditions do not provide any improvement beyond $\ell^2$ on the order of the Fourier coefficients. For $f$ continuous on $T^1$ (= continuous in $\theta$ and periodic of period $2\pi$), Carleman [28] showed that $\{c_n\}$ need not be in $\ell^{2-\epsilon}$ for any $\epsilon > 0$. Orlicz [182] showed that for any $\{d_n\colon n \in \mathbb{Z}\} \notin \ell^2$ one can have $\sum_n |c_n d_n| = +\infty$. Bary [10, pp. 338–340] gives a proof.

If $f$ is real-valued, periodic of period $2\pi$, and on $[0, 2\pi]$ satisfies the Hölder condition

$$|f(\theta') - f(\theta'')| \le K|\theta' - \theta''|^\alpha, \tag{11.8}$$

where $0 < \alpha \le 1$, then $|c_n| = O(|n|^{-\alpha})$ as $n \to \infty$, as will be shown next. Functions Hölder of order $\alpha$ on a bounded interval have bounded $1/\alpha$-variation as shown by (1.7). So more generally, let $f \in \mathcal{W}_p[0, 2\pi]$ for some $1 < p < \infty$ and let $g_n(\theta) = \cos(n\theta)$ or $\sin(n\theta)$ on $[0, 2\pi]$. It is easily seen that $v_q(g_n) \le 2n \cdot 2^q$ for any $1 \le q < \infty$. For any $q \ge 1$ such that $p^{-1} + q^{-1} > 1$, by the Love–Young inequality (3.148),

$$|a_n| = \Big|\frac{1}{\pi}\int_0^{2\pi} f(\theta)\, \mathrm{d}\sin(n\theta)/n\Big| \le 2^{1+1/q}\pi^{-1}K_{p,q}\|f\|_{[p]}n^{-1+1/q} = O(n^{-r})$$

for $r = (1 - 1/q) < 1/p$, and so for any $r < 1/p$. In fact, Marcinkiewicz [156, Theorem 3] proved that this is true for $r = 1/p$. His proof is based on the following observation:

**Lemma 11.1.** *Let $f$ be a real-valued periodic function on $\mathbb{R}$ with period $2\pi$ and of bounded $p$-variation on $[0, 2\pi]$ for some $0 < p < \infty$. Then for $0 < \phi \le \pi$,*

$$\int_0^{2\pi} |f(\theta + \phi) - f(\theta)|^p\, \mathrm{d}\theta \le 2\phi v_p(f; [0, 2\pi]). \tag{11.9}$$

*Proof.* Let $n \ge 2$ be the integer part of $2\pi/\phi$, so $0 \le 2\pi - n\phi < \phi$. Let $(\Delta_\phi f)(\theta) := f(\theta + \phi) - f(\theta)$. By a change of variables, we then have

$$\begin{aligned}
&\int_0^{2\pi-\phi} |\Delta_\phi f(\theta)|^p \mathrm{d}\theta \\
&= \sum_{i=1}^{n-1} \int_0^{\phi} |\Delta_\phi f(\theta + (i-1)\phi)|^p\, \mathrm{d}\theta + \int_0^{2\pi - n\phi} |\Delta_\phi f(\theta + (n-1)\phi)|^p\, \mathrm{d}\theta \\
&= \int_0^{2\pi-n\phi} \sum_{i=1}^{n} |\Delta_\phi f(\theta + (i-1)\phi)|^p\, \mathrm{d}\theta + \int_{2\pi-n\phi}^{\phi} \sum_{i=1}^{n-1} |\Delta_\phi f(\theta + (i-1)\phi)|^p\, \mathrm{d}\theta \\
&\le \phi v_p(f; [0, 2\pi]).
\end{aligned}$$

Also due to periodicity of $f$, $|\Delta_\phi f(\theta)|^p = |f(\theta+\phi-2\pi)-f(\theta)|^p \le v_p(f;[0,2\pi])$ for each $\theta \in [2\pi-\phi, 2\pi]$, and so

$$\int_{2\pi-\phi}^{2\pi} |\Delta_\phi f(\theta)|^p \, d\theta \le \phi v_p(f;[0,2\pi]),$$

proving (11.9). □

Returning to bounds for the Fourier coefficients, we have the following:

**Theorem 11.2** (Marcinkiewicz). *For any $p$ with $1 \le p < \infty$ and $f \in \mathcal{W}_p[0,2\pi]$, its Fourier coefficients are of order $|a_n| \vee |b_n| \le 2|c_n| = O(|n|^{-1/p})$.*

*Proof.* It is easily seen that the identity function $g(\theta) = \theta$, $0 \le \theta \le 2\pi$, is in $\mathcal{W}_p[0,2\pi]$ with $v_p(g) = (2\pi)^p$ and that its Fourier coefficients are $\widehat{g}(n) = \mathrm{i}/n$ for $n \neq 0$, which are $O(|n|^{-1/p})$. Thus we can subtract a constant multiple of $g$ from $f$ and assume that $f(0) = f(2\pi)$. Then $f$ has a unique extension to be periodic of period $2\pi$ on all of $\mathbb{R}$.

Due to periodicity, we then have

$$c_n = \frac{1}{2\pi}\int_0^{2\pi} f(\theta)\mathrm{e}^{-\mathrm{i}n\theta}\, d\theta = \frac{1}{4\pi}\int_0^{2\pi} \Big[f(\theta) - f\Big(\theta + \frac{\pi}{n}\Big)\Big]\mathrm{e}^{-\mathrm{i}n\theta}\, d\theta$$

for each $n \in \mathbb{Z}\setminus\{0\}$. Thus by Hölder's inequality for integrals and (11.9), it follows that

$$\begin{aligned}
|c_n| &\le \frac{1}{4\pi}\int_0^{2\pi} \Big|f(\theta) - f\Big(\theta+\frac{\pi}{n}\Big)\Big|\, d\theta \\
&\le \frac{(2\pi)^{1-(1/p)}}{4\pi}\Big(\int_0^{2\pi}\Big|f(\theta) - f\Big(\theta+\frac{\pi}{n}\Big)\Big|^p d\theta\Big)^{1/p} \\
&\le 2^{-1}|n|^{-1/p}\|f\|_{[0,2\pi],(p)}.
\end{aligned}$$

Using $|c_n|^2 = (a_n^2 + b_n^2)/4$ as mentioned after (11.6) completes the proof. □

**Remark 11.3.** Proposition 3.46 with $h_k \equiv 1$ shows that the previous theorem is sharp in that $|c_n|$ need not be $o(|n|^{-1/p})$. Perhaps surprisingly, the Fourier series of a function $f \in \mathcal{W}_p[0,1]$ or $f \in \mathcal{W}_p[0,2\pi]$ may converge to it uniformly and absolutely while the terms do not approach 0 in $\mathcal{W}_p$.

## 11.2 Uniform Convergence

Let $f$ be a complex-valued periodic function on $\mathbb{R}$ with period $2\pi$, integrable on $[0,2\pi]$, so that it has a well-defined Fourier series (which does not necessarily converge to it anywhere in general). In this section it is proved (Theorem 11.8 below) that the Fourier series of $f$ converges to $f$ uniformly if $f$ is continuous and has bounded $\Phi$-variation under a hypothesis on $\Phi$. In particular, each continuous function of bounded $p$-variation for some $p$ with $0 < p < \infty$ has its Fourier series converging uniformly (Corollary 11.10). It is also shown that the hypothesis on $\Phi$ is best possible (Theorem 11.13).

We begin by recalling a test for convergence of Fourier series. The partial Fourier sums $U_m f$ are defined for each positive integer $m$ and $x \in \mathbb{R}$ by $U_m f(x) := S_m f(\mathrm{e}^{\mathrm{i}x})$, using the definitions (11.2) and (11.3), in other words,

$$U_m f(x) := \sum_{|n|\le m} \widehat{f}(n)\mathrm{e}^{\mathrm{i}nx}.$$

If $U_m f(x) \to f(x)$ uniformly in $x$ as $m \to \infty$ then we say that the Fourier series *converges uniformly*. For each $m \ge 1$ define the *Dirichlet kernel* $D_m$ and the *simplified Dirichlet kernel* $D_m^*$ respectively by

$$D_m(t) := \sum_{|n|\le m} \mathrm{e}^{\mathrm{i}nt} = \begin{cases} \frac{\sin(m+1/2)t}{\sin(t/2)} & \text{if } t \notin 2\pi\mathbb{Z}, \\ 2m+1 & \text{if } t \in 2\pi\mathbb{Z}, \end{cases}$$

and

$$D_m^*(t) := \begin{cases} 2\frac{\sin mt}{t} & \text{if } 0 < |t| \le \pi, \\ 2m & \text{if } t = 0, \end{cases}$$

and extending $D_m^*$ to $\mathbb{R}$ to be periodic of period $2\pi$, as $D_m$ clearly is. Inserting the integral expressions for the Fourier coefficients $\widehat{f}(n)$, by periodicity we have

$$U_m f(x) = \frac{1}{2\pi}\int_0^{2\pi} f(t)D_m(t-x)\,\mathrm{d}t = \frac{1}{2\pi}\int_{-\pi}^{\pi} f(x+t)D_m(t)\,\mathrm{d}t. \tag{11.10}$$

We will see that $D_m$ equals the simpler function $D_m^*$ plus a function which is bounded uniformly in $m$ and $x$. To this aim let $h(t) := 1/\tan(t/2) - 2/t$ if $0 \ne t \in [-\pi, \pi)$ and $h(0) := 0$, where $1/\tan(-\pi/2) := 0$ and $1/\tan(\pi/2) := 0$. Let $h$ be periodic of period $2\pi$. By the Taylor series of the tangent function around 0, we have $|h(t)| = O(|t|)$ as $t \to 0$ and so $h$ is continuous at 0 and thus on $[0, 2\pi]$ and bounded on $\mathbb{R}$. Then we can write for all $t \in \mathbb{R}$,

$$\begin{aligned} D_m(t) &= \frac{\sin(m+1/2)t}{\sin(t/2)} = \frac{\sin mt}{\tan(t/2)} + \cos mt \\ &= D_m^*(t) + h(t)\sin mt + \cos mt, \end{aligned} \tag{11.11}$$

where $h(t)\sin(mt) + \cos(mt)$ is indeed bounded uniformly in $m$ and $t$. We show next that for $f$ regulated,

$$\lim_{m\to\infty} \int_{-\pi}^{\pi} f(x+t)\big[h(t)\sin mt + \cos mt\big]\,\mathrm{d}t = 0 \tag{11.12}$$

uniformly in $x$.

**Lemma 11.4.** *If $f$ and $h$ are regulated and periodic of period $2\pi$, then*

$$\lim_{m\to\infty}\int_{-\pi}^{\pi} f(x+t)h(t)\sin mt\,\mathrm{d}t = \lim_{m\to\infty}\int_{-\pi}^{\pi} f(x+t)h(t)\cos mt\,\mathrm{d}t = 0 \quad (11.13)$$

*uniformly in $x$.*

*Proof.* Let $\eta_x(t) := f(x+t)h(t)$, and let $\Delta_t\rho(s) := \rho(t+s)-\rho(t)$ for $\rho = f$ or $\rho = h$. For a periodic integrable function $\rho$, its integral modulus of continuity $w_1(\rho; u)$ is defined for $u \geq 0$ by

$$w_1(\rho; u) := \sup_{|s|\leq u}\int_{-\pi}^{\pi} \big|\Delta_t\rho(s)\big|\,\mathrm{d}t.$$

By periodicity, for any $x$ we have

$$\begin{aligned}
&2\Big|\int_{-\pi}^{\pi}\eta_x(t)\sin mt\,\mathrm{d}t\Big| \\
&\quad= \Big|\int_{-\pi}^{\pi}\Big[\eta_x(t)-\eta_x\Big(t+\frac{\pi}{m}\Big)\Big]\sin mt\,\mathrm{d}t\Big| \leq \int_{-\pi}^{\pi}\Big|\eta_x(t)-\eta_x\Big(t+\frac{\pi}{m}\Big)\Big|\,\mathrm{d}t \\
&\quad\leq \int_{-\pi}^{\pi}\big|\Delta_{x+t}f(\pi/m)\big|\big|h(t+\pi/m)\big|\,\mathrm{d}t + \int_{-\pi}^{\pi}\big|f(x+t)\big|\big|\Delta_t h(\pi/m)\big|\,\mathrm{d}t \\
&\quad\leq \|h\|_\infty w_1(f;\pi/m) + \|f\|_\infty w_1(h;\pi/m).
\end{aligned}$$

Since $f$ and $h$ are each bounded and continuous except at most on countable sets by Corollary 2.2, the right side of the preceding bound tends to zero as $m \to \infty$ by Lebesgue's dominated convergence theorem. Thus the first limit in (11.13) exists uniformly in $x$ and is zero. The same proof applies to the second limit, proving the lemma. □

Therefore (11.12) holds uniformly in $x$, replacing $h(t)$ by 1 in the cosine term, which together with (11.10) and (11.11) yields the relations

$$\begin{aligned}
U_m f(x) &= \frac{1}{2\pi}\int_{-\pi}^{\pi} f(x+t)D_m^*(t)\,\mathrm{d}t + o(1) \quad \text{as } m\to\infty \qquad (11.14)\\
&= \frac{1}{2\pi}\int_{0}^{\pi}\big[f(x+t)+f(x-t)\big]D_m^*(t)\,\mathrm{d}t + o(1) \quad \text{as } m\to\infty
\end{aligned}$$

uniformly in $x$. The second relation (11.14) holds because $D_m^*(-t) \equiv D_m^*(t)$. Since for $f \equiv 1$, its partial Fourier sums $U_m f$ are identically 1 for each $m \geq 1$, by the preceding relation we have that

$$1 = \frac{1}{\pi}\int_0^{\pi} D_m^*(t)\,\mathrm{d}t + o(1) \quad \text{as } m\to\infty.$$

Multiplying both sides by $g(x)$ for a function $g$, and then combining with (11.14), we can further write

$$U_m f(x)-g(x) = \frac{1}{2\pi}\int_0^\pi \big[f(x+t)+f(x-t)-2g(x)\big]D_m^*(t)\,dt+o(1) \quad \text{as } m\to\infty.$$

We are ready to prove the following test:

**Proposition 11.5.** *For a complex-valued regulated function $f$ periodic of period $2\pi$, its Fourier series converges to $g$ at $x$ (resp. uniformly) if and only if*

$$\lim_{m\to\infty}\int_0^\pi \big[f(x+t)+f(x-t)-2g(x)\big]D_m^*(t)\,dt=0 \tag{11.15}$$

*(respectively, uniformly in $x$). Moreover, for $f\equiv g$, the same equivalences hold if $m\to\infty$ through odd values,*

$$\lim_{n\to\infty}\int_0^\pi \big[f(x+t)+f(x-t)-2f(x)\big]D_{2n+1}^*(t)\,dt=0. \tag{11.16}$$

**Remark 11.6.** The values of $f$ can be changed on a set of measure 0 without changing its Fourier series, in particular, changed on a countable set $\{a_k\}\subset[0,2\pi)$, e.g., if $f$ is regulated, the jumps of $f$, by amounts $c_k\to 0$, while preserving the regulated property. A question is, if the Fourier series converges at a point, to what does it converge. One possibility is, if $f$ is regulated, to set $g(x):=[f(x+)+f(x-)]/2$, which equals $f$ except possibly on the countable set of jumps of $f$, and differs from it there if $f$ is left- or right-continuous. Marcinkiewicz [156, Theorem 5] proved that the Fourier series of $f$ converges pointwise everywhere to $g$ if the $p$-variation of $f$ is bounded for some $p<\infty$. Wiener [241] had proved this fact for $p\le 2$. Recall that the Fourier series of a continuous function can diverge at some points as shown by du Bois-Reymond in 1876; see e.g. [53, Proposition 7.4.3].

*Proof.* The first sentence of the proposition has already been proved. For the second, letting $\Delta_x f(u):=f(x+u)-f(x)$, for each $m\ge 1$ and $x$, the integral in (11.15) with $g\equiv f$ is

$$I_m(x):=\int_0^\pi \big[f(x+t)+f(x-t)-2f(x)\big]D_m^*(t)\,dt=\int_{-\pi}^\pi \Delta_x f(t)D_m^*(t)\,dt.$$

By Lemma 11.4, we have

$$I_{m+1}(x)-I_m(x)=\int_{-\pi}^\pi \Delta_x f(t)\Big[\frac{2(\cos t-1)}{t}\sin mt+\frac{2\sin t}{t}\cos mt\Big]\,dt=o(1) \tag{11.17}$$

as $m\to\infty$ uniformly in $x$. Thus the second sentence of the proposition, showing equivalences with (11.16), follows. □

A more specialized test will be proved assuming in addition that $f$ is continuous. For $x \in \mathbb{R}$ and for a positive integer $n$, let

$$T_n^+(x) := \sum_{j=1}^{n} \frac{1}{2j-1} \left\{ f\left(x + \frac{(2j-1)\pi}{2n+1}\right) - f\left(x + \frac{2j\pi}{2n+1}\right) \right\} \tag{11.18}$$

and

$$T_n^-(x) := \sum_{j=1}^{n} \frac{1}{2j-1} \left\{ f\left(x - \frac{(2j-1)\pi}{2n+1}\right) - f\left(x - \frac{2j\pi}{2n+1}\right) \right\}. \tag{11.19}$$

**Proposition 11.7.** *For a continuous periodic function $f$, its Fourier series converges to $f$ at $x$ (uniformly) if and only if*

$$\lim_{n\to\infty} \int_0^\pi \left[ T_n^+\left(x + \frac{t}{2n+1}\right) + T_n^-\left(x - \frac{t}{2n+1}\right) \right] \sin t \, \mathrm{d}t = 0 \tag{11.20}$$

*for $x$ (respectively, uniformly in $x$).*

*Proof.* For $x \in \mathbb{R}$ and for a positive integer $m$, let

$$I_m^+(x) := \int_0^\pi \Delta_x f(t) D_m^*(t) \, \mathrm{d}t \quad \text{and} \quad I_m^-(x) := \int_0^\pi \Delta_x f(-t) D_m^*(t) \, \mathrm{d}t.$$

Call the integral in (11.16) $I_{2n+1}(x)$. Then $I_{2n+1}(x) = I_{2n+1}^+(x) + I_{2n+1}^-(x)$. We will show that as $n \to \infty$, uniformly in $x$,

$$I_{2n+1}^+(x) = -\frac{2}{\pi} \int_0^\pi T_n^+\left(x + \frac{t}{2n+1}\right) \sin t \, \mathrm{d}t + o(1), \tag{11.21}$$

as well as

$$I_{2n+1}^-(x) = -\frac{2}{\pi} \int_0^\pi T_n^-\left(x - \frac{t}{2n+1}\right) \sin t \, \mathrm{d}t + o(1). \tag{11.22}$$

Then the statement of the proposition will follow by the second part of Proposition 11.5.

Let $w(f; u) := \sup\{|f(t) - f(s)| \colon |t - s| \le u\}$, $u \ge 0$, be the modulus of continuity of $f$, and let $m \ge 3$. Since $\sin x \le x$ for $0 \le x \le \pi$, for the integral $I_m^+$ over the interval $[0, \pi/m]$, we have

$$\left| \int_0^{\pi/m} \Delta_x f(t) D_m^*(t) \, \mathrm{d}t \right| \le 2\pi w(f; \pi/m) = o(1) \quad \text{as } m \to \infty$$

uniformly in $x$. For the integral $I_m^+$ over the remaining interval $[\pi/m, \pi]$, changing variables twice, we have

$$\int_{\pi/m}^{\pi} \Delta_x f(t) D_m^*(t)\,\mathrm{d}t = \sum_{k=0}^{m-2} \int_{(k+1)\pi/m}^{(k+2)\pi/m} \Delta_x f(t) \frac{2\sin mt}{t}\,\mathrm{d}t$$

$$= \sum_{k=0}^{m-2} \int_{\pi/m}^{2\pi/m} \Delta_x f\Big(u + \frac{k\pi}{m}\Big) \frac{(-1)^k 2\sin mu}{u + k\pi/m}\,\mathrm{d}u$$

$$= 2\int_{\pi}^{2\pi} F_m(v,x)\sin v\,\mathrm{d}v,$$

where

$$F_m(v,x) := \sum_{k=0}^{m-2} \Delta_x f\Big(\frac{v+k\pi}{m}\Big)\frac{(-1)^k}{v+k\pi}.$$

We have

$$I_m^+(x) = 2\int_{\pi}^{2\pi} F_m(t,x)\sin t\,\mathrm{d}t + o(1) \qquad \text{as } m \to \infty. \tag{11.23}$$

To prove (11.21), we have

$$F_{2n+1}(t,x) = \sum_{j=1}^{n} \Big\{\Delta_x f\Big(\frac{t+(2j-2)\pi}{2n+1}\Big)\frac{1}{t+(2j-2)\pi}$$
$$-\Delta_x f\Big(\frac{t+(2j-1)\pi}{2n+1}\Big)\frac{1}{t+(2j-1)\pi}\Big\}.$$

For $j = 1,\ldots,n$, the $j$th term $u_j$ of the latter sum can be written as

$$u_j := \Delta_x f\left(\frac{t+(2j-2)\pi}{2n+1}\right)\frac{1}{t+(2j-2)\pi}$$
$$-\Delta_x f\left(\frac{t+(2j-1)\pi}{2n+1}\right)\frac{1}{t+(2j-1)\pi}$$
$$= \left\{f\left(x + \frac{t+(2j-2)\pi}{2n+1}\right) - f\left(x + \frac{t+(2j-1)\pi}{2n+1}\right)\right\}\frac{1}{(2j-1)\pi}$$
$$+\left\{f\left(x + \frac{t+(2j-2)\pi}{2n+1}\right) - f\left(x + \frac{t+(2j-1)\pi}{2n+1}\right)\right\}\times$$
$$\times\left\{\frac{1}{t+(2j-2)\pi} - \frac{1}{(2j-1)\pi}\right\}$$
$$+\ \Delta_x f\left(\frac{t+(2j-1)\pi}{2n+1}\right)\left\{\frac{1}{t+(2j-2)\pi} - \frac{1}{t+(2j-1)\pi}\right\}$$
$$=: \sum_{l=1}^{3} u_{l,j}(t,x).$$

Thus $F_{2n+1}(t,x) = S_{1,n}(t,x) + S_{2,n}(t,x) + S_{3,n}(t,x)$, where $S_{l,n}(t,x) := \sum_{j=1}^{n} u_{l,j}(t,x)$ for $l = 1,2,3$.

We show next that $S_{2,n}$ and $S_{3,n}$ tend to zero as $n \to \infty$ uniformly in $x$ and $t \in [\pi, 2\pi]$. For $\pi \le t \le 2\pi$ and each $j \ge 1$, we have that the maximum

$$\left[\frac{1}{(2j-1)\pi} - \frac{1}{t+(2j-2)\pi}\right] \vee \left[\frac{1}{t+(2j-2)\pi} - \frac{1}{t+(2j-1)\pi}\right]$$

is in the interval $[0, 1/(\pi(2j-1)^2)]$. Letting $j(n)$ be the integer part of $\sqrt{2n+1}$, for $\pi \le t \le 2\pi$ we have

$$\begin{aligned}
&|S_{3,n}(t,x)| \\
&\quad \le \sum_{j=1}^{n} \left| f\left(x + \frac{t+(2j-1)\pi}{2n+1}\right) - f(x)\right| \left|\frac{1}{t+(2j-2)\pi} - \frac{1}{t+(2j-1)\pi}\right| \\
&\quad \le w\left(f; \frac{\pi + 2\pi\sqrt{2n+1}}{2n+1}\right) \sum_{j=1}^{j(n)} \frac{1}{\pi(2j-1)^2} + 2\|f\|_\infty \sum_{j=j(n)+1}^{n} \frac{1}{\pi(2j-1)^2}.
\end{aligned}$$

Since $j(n) \to \infty$ as $n \to \infty$, we get that $|S_{3,n}(t,x)| = o(1)$ as $n \to \infty$ uniformly in $x$ and $t \in [\pi, 2\pi]$. For $S_{2,n}$, we have the bound

$$\begin{aligned}
|S_{2,n}(t,x)| &\le \sum_{j=1}^{n} \left| f\left(x + \frac{t+(2j-2)\pi}{2n+1}\right) - f\left(x + \frac{t+(2j-1)\pi}{2n+1}\right)\right| \times \\
&\qquad \times \left|\frac{1}{t+(2j-2)\pi} - \frac{1}{(2j-1)\pi}\right| \\
&\le w\left(f; \frac{\pi}{2n+1}\right) \sum_{j=1}^{\infty} \frac{1}{\pi(2j-1)^2} = \frac{\pi}{8} w\left(f; \frac{\pi}{2n+1}\right).
\end{aligned}$$

Again the right side, and hence $|S_{2,n}(t,x)|$, is of order $o(1)$ as $n \to \infty$ uniformly in $x$ and $t \in [\pi, 2\pi]$. For $T_n^+$ defined by (11.18), we have $T_n^+(x + (t-\pi)/(2n+1)) = \pi S_{1,n}(t,x)$. Thus

$$\begin{aligned}
&\int_\pi^{2\pi} F_{2n+1}(t,x) \sin t \, \mathrm{d}t \\
&\quad = \frac{1}{\pi} \int_\pi^{2\pi} T_n^+\left(x + \frac{t-\pi}{2n+1}\right) \sin t \, \mathrm{d}t + \int_\pi^{2\pi} \big[S_{2,n}(t,x) + S_{3,n}(t,x)\big] \sin t \, \mathrm{d}t \\
&\quad = -\frac{1}{\pi} \int_0^{\pi} T_n^+\left(x + \frac{t}{2n+1}\right) \sin t \, \mathrm{d}t + o(1) \quad \text{as } n \to \infty
\end{aligned} \tag{11.24}$$

uniformly in $x$ by the dominated convergence theorem. This together with (11.23) proves (11.21).

To prove (11.22), we can apply essentially the same proof to the integral $I_m^-(x)$. Now $F_m$ and $u_j$ are replaced by expressions in which $\Delta_x f(w)$ for each $w$ occurring is replaced by $\Delta_x f(-w)$, and so $f(x+w) - f(x+v) = \Delta_x f(w) - \Delta_x f(v)$ is replaced by $f(x-w) - f(x-v)$ for the $w$ and $v$ in the proof. We then get, instead of (11.23) and (11.24), the following relations with $m = 2n+1$:

$$\begin{aligned} I_m^-(x) &= 2\int_\pi^{2\pi} \Big(\sum_{k=0}^{m-2} \Delta_x f\Big(\frac{-t-k\pi}{m}\Big)\frac{(-1)^k}{t+k\pi}\Big) \sin t \,\mathrm{d}t + o(1) \qquad \text{as } m \to \infty \\ &= -\frac{2}{\pi}\int_0^\pi T_n^-\Big(x - \frac{t}{2n+1}\Big) \sin t \,\mathrm{d}t + o(1) \quad \text{as } n \to \infty \end{aligned}$$

uniformly in $x$. Thus (11.22) holds, proving the proposition. □

Recall that $\mathcal{V}$ is the class of all functions $\Phi$: $[0,\infty) \to [0,\infty)$ which are strictly increasing, continuous, unbounded, and 0 at 0, $\mathcal{CV}$ is the subclass of convex functions in $\mathcal{V}$, and $\mathcal{CV}^*$ is the class of all functions $\Phi \in \mathcal{CV}$ satisfying (3.18). Recalling the definition (3.17) of the complementary function $\Phi^*$, now we are ready to prove the following result of Salem:

**Theorem 11.8.** *Let $\Phi^*$ be the complementary function to $\Phi \in \mathcal{CV}^*$. If*

$$\sum_{j=1}^\infty \Phi^*(1/j) < \infty \tag{11.25}$$

*then every continuous periodic function of bounded $\Phi$-variation has uniformly convergent Fourier series.*

For the proof we use the following fact:

**Lemma 11.9.** *Let $\Psi \in \mathcal{V}$. If for some $a > 0$, $\sum_{k\ge1}\Psi(a/k) < \infty$, then $\sum_{k\ge1}\Psi(b/k) < \infty$ for any $b > 0$, and there is a sequence $c_k \uparrow +\infty$ with $c_k \ge 1$ for all $k$ such that $\sum_{k\ge1}\Psi(c_k/k) < \infty$.*

*Proof.* If $b \le a$ the convergence is clear, so assume $b > a$. Let $N$ be the least integer greater than or equal to the ratio $r := b/a$. Then we have

$$\sum_{k>r}\Psi\Big(\frac{b}{k}\Big) \le \sum_{j=1}^\infty \sum_{Nj\le k<N(j+1)} \Psi\Big(\frac{b}{k}\Big) \le N\sum_{j\ge1}\Psi\Big(\frac{a}{j}\Big) < \infty.$$

This proves the first part of the conclusion and that for each $n \ge 1$, $\sum_{k\ge1}\Psi(n/k) < \infty$. Thus for $n = 1, 2, \ldots$, there is a least integer $k_n$ such that $\sum_{k>k_n}\Psi(n/k) < 1/2^n$. Let $c_k := n$ for $k_n < k \le k_{n+1}$ for each $n$ and $c_k := 1$ for $k \le k_1$. Then

$$\sum_{k>k_1} \Psi\Big(\frac{c_k}{k}\Big) = \sum_{n=1}^{\infty} \sum_{k_n<k\le k_{n+1}} \Psi\Big(\frac{c_k}{k}\Big) \le \sum_{n\ge 1} 2^{-n} < \infty,$$

completing the proof of the lemma. □

**Proof of Theorem 11.8.** By Proposition 11.7 and the dominated convergence theorem, it is enough to prove that $T_n^+(x)$ and $T_n^-(x)$ tend to zero as $n \to \infty$ uniformly in $x \in [0, 2\pi]$. Let $\Phi \in \mathcal{CV}^*$ be such that $\sum_{j\ge 1} \Phi^*(1/j) < \infty$. By Proposition 3.11, $\Phi^* \in \mathcal{CV}^* \subset \mathcal{V}$, and so by Lemma 11.9, there is a sequence $\epsilon_j \downarrow 0$ such that $\sum_{j\ge 1} \Phi^*(1/(j\epsilon_j)) < \infty$. For $j = 1, \ldots, n$, $n \ge 1$, and $x \in [0, 2\pi]$, let

$$u_{j,n}(x) := \frac{1}{\epsilon_j(2j-1)}\Big|f\Big(x + \frac{(2j-1)\pi}{2n+1}\Big) - f\Big(x + \frac{2j\pi}{2n+1}\Big)\Big|. \tag{11.26}$$

Then by the W. H. Young inequality (3.19), for each $n \ge 1$, we have the bound

$$\begin{aligned}\sum_{j=1}^{n} u_{j,n}(x) &\le \sum_{j=1}^{n} \Phi\Big(\Big|f\Big(x + \frac{(2j-1)\pi}{2n+1}\Big) - f\Big(x + \frac{2j\pi}{2n+1}\Big)\Big|\Big) + \sum_{j=1}^{n} \Phi^*\Big(\frac{1}{\epsilon_j(2j-1)}\Big)\\ &\le v_\Phi(f; [0, 3\pi]) + \sum_{j=1}^{\infty} \Phi^*\Big(\frac{1}{j\epsilon_j}\Big) =: A.\end{aligned}$$

Thus for $n > 2$ and $x \in [0, 2\pi]$, it follows by the definition (11.18) of $T_n^+$ that

$$\begin{aligned}|T_n^+(x)| &\le \sum_{j=1}^{k-1} \frac{1}{2j-1}\Big|f\Big(x + \frac{(2j-1)\pi}{2n+1}\Big) - f\Big(x + \frac{2j\pi}{2n+1}\Big)\Big| + \sum_{j=k}^{n} \epsilon_j u_{j,n}(x)\\ &\le w\Big(f; \frac{\pi}{2n+1}\Big)\Big[1 + \frac{1}{3} + \cdots + \frac{1}{2k-3}\Big] + \epsilon_k A \end{aligned}\tag{11.27}$$

for any $1 < k < n$. Since the modulus of continuity $w(f; \pi/(2n+1))$ tends to zero as $n \to \infty$, the right side of (11.27) can be made arbitrarily small by first choosing $k$ large enough so that $\epsilon_k$ is small enough and then $n$ large enough. Therefore $T_n^+(x)$ tends to zero as $n \to \infty$ uniformly in $x \in [0, 2\pi]$. If in (11.26), $\pi$ is replaced by $-\pi$ then the same arguments imply that $T_n^-(x)$ tends to zero as $n \to \infty$ uniformly in $x \in [0, 2\pi]$, proving the theorem. □

The next fact then follows:

**Corollary 11.10.** *Let $0 < p < \infty$. Then every continuous function $f$ of bounded $p$-variation has uniformly convergent Fourier series.*

*Proof.* It suffices to consider $1 < p < \infty$ since then the cases $0 < p \leq 1$ will follow by Lemma 3.45. (Also, for $0 < p < 1$, $f$, being continuous, must be constant by Theorem 2.11 in [54, Part II].) For $\Phi(u) \equiv u^p/p$ we then have $\Phi \in \mathcal{CV}^*$, and its complementary function $\Phi^*(v) \equiv v^q/q$ with $q = p/(p-1) > 1$, as in the example just after (3.19). Thus the hypothesis (11.25) holds and the result follows from Theorem 11.8. □

For a continuous periodic function $f$, let $I_n(f;x)$ be the integral in (11.20), that is, for a positive integer $n$ and $x \in \mathbb{R}$,

$$\begin{aligned} I_n(f;x) & \qquad\qquad\qquad\qquad\qquad\qquad\qquad\qquad\qquad\qquad (11.28) \\ &= \sum_{j=1}^{n} \frac{1}{2j-1} \int_0^\pi \Big\{ f\Big(x + \frac{t+(2j-1)\pi}{2n+1}\Big) - f\Big(x + \frac{t+2j\pi}{2n+1}\Big)\Big\} \sin t\, dt \\ &\quad + \sum_{j=1}^{n} \frac{1}{2j-1} \int_0^\pi \Big\{ f\Big(x - \frac{t+(2j-1)\pi}{2n+1}\Big) - f\Big(x - \frac{t+2j\pi}{2n+1}\Big)\Big\} \sin t\, dt. \end{aligned}$$

By Proposition 11.7, the Fourier series of $f$ converges at $x$ if and only if $I_n(f,x)$ tends to zero as $n \to \infty$. In the rest of this section we show that the hypothesis (11.25) is necessary, that is, if $\Phi \in \mathcal{CV}^*$ is such that $\sum_{k\geq 1} \Phi^*(1/k) = \infty$, then there exists a continuous periodic function $f \in \mathcal{W}_\Phi$ for which $I_n(f;0)$ does not converge to zero.

For $\Phi \in \mathcal{CV}$, let $\ell^\Phi$ be the set of all sequences of real numbers $a = \{a_k\}_{k\geq 1}$ such that $\sum_{k\geq 1} \Phi(|a_k|/c) < \infty$ for some $c > 0$. The space $\ell^\Phi$ equipped with the norm

$$\|a\|_\Phi := \inf\Big\{c > 0 \colon \sum_{k\geq 1} \Phi(|a_k|/c) \leq 1\Big\}$$

is a Banach space, usually called an Orlicz sequence space (see Section 4 in [142]). Luxemburg norms on Orlicz spaces were defined more generally in (1.24). If in addition $\Phi$ satisfies the conditions (3.18), that is, if $\Phi \in \mathcal{CV}^*$, then its complementary function $\Phi^*$ is defined and one can define another norm on $\ell^\Phi$ by

$$|\!|\!| a |\!|\!|_\Phi := \sup\Big\{\sum_{k\geq 1} |a_k|\beta_k \colon \beta_k \geq 0,\ \sum_{k\geq 1} \Phi^*(\beta_k) \leq 1\Big\}.$$

Then $\|\cdot\|_\Phi$ and $|\!|\!|\cdot|\!|\!|_\Phi$ are two equivalent norms on $\ell^\Phi$: for all $a \in \ell^\phi$,

$$\|a\|_\Phi \leq |\!|\!| a |\!|\!|_\Phi \leq 2\|a\|_\Phi. \tag{11.29}$$

The proof of (11.29) is similar to the proof of Proposition 3.12 and we omit it. Let $\phi_*$ be the inverse function to $\Phi^*$. Then for any $k' \geq 1$, taking $\beta_{k'} = \phi_*(1)$ and $\beta_k = 0$ if $k \neq k'$, we get the bound $|a_{k'}|\phi_*(1) \leq |\!|\!| a |\!|\!|_\Phi$, and so

$$\|a\|_{\sup} \leq |\!|\!| a |\!|\!|_\Phi / \phi_*(1). \tag{11.30}$$

By Proposition 3.11, $\Phi^* \in \mathcal{CV}^* \subset \mathcal{V}$, and so by Lemma 11.9, the sufficient condition (11.25) of Theorem 11.8 is equivalent to the condition that the sequence $\{1/k\}_{k\geq 1}$ is in the Orlicz sequence space $\ell^{\Phi^*}$. Also, by Proposition 3.11 again, the norm $\|\cdot\|_{\Phi^*}$ is well defined, and $\Phi^{**} = \Phi$. These facts will be used next to derive certain implications on $\ell^{\Phi}$ from the assumption that $\{1/k\}_{k\geq 1} \notin \ell^{\Phi^*}$.

**Lemma 11.11.** *Let $\Phi^*$ be the complementary function to $\Phi \in \mathcal{CV}^*$. If $\sum_{k\geq 1} \Phi^*(1/k) = +\infty$, then given $\epsilon > 0$, there exists a sequence $a = \{a_k\}$ with all $a_k \geq 0$ such that $a_k = 0$ for all sufficiently large $k$, $\|a\|_{\Phi} \leq \epsilon$, and $\sum_{k\geq 1} k^{-1} a_k > 1$.*

*Proof.* For each integer $n \geq 1$, let $b_k^n := 1/k$ for $k = 1, \ldots, n$, and let $b_k^n := 0$ for $k > n$. Letting $b^n := \{b_k^n\}_{k\geq 1}$ we then have $\|b^n\|_{\Phi^*} \uparrow \infty$ as $n \to \infty$, as follows. Clearly $\|b^n\|_{\Phi^*}$ is nondecreasing in $n$. By (11.29) it suffices to prove that $\|b^n\|_{\Phi^*} \uparrow \infty$. If not, then there is a positive integer $M$ such that $\|b^n\|_{\Phi^*} \leq M$ for all $n$. It then follows that $\sum_{k\geq 1} \Phi^*(1/(Mk)) \leq 1$. Thus by Lemma 11.9, $\sum_{j\geq 1} \Phi^*(1/j) < \infty$, a contradiction, proving the claim. Let $n$ be so large that $\|b^n\|_{\Phi^*} > 2\epsilon^{-1}$. Since $\Phi^{**} = \Phi$, there exists $d = \{d_k\}_{k\geq 1} \in \ell^{\Phi}$ with all $d_k \geq 0$ such that $\sum_{k\geq 1} \Phi(d_k) \leq 1$ and $\sum_{k=1}^{n} k^{-1} d_k > 2\epsilon^{-1}$. We can assume that $d_k = 0$ for all $k > n$. Also $\|d\|_{\Phi} \leq 1$ implies by (11.29) that $\|d\|_{\Phi} \leq 2$. Thus $a := (\epsilon/2)d$ satisfies the conclusion. □

Now assuming $\Phi \in \mathcal{CV}^*$, recall the definition (3.23) of $\|\cdot\|_{J,(\Phi)}$, which on $\widetilde{\mathcal{W}}_{\Phi}$ is a seminorm equivalent to the seminorm $\|\cdot\|_{J,(\Phi)}$ by Proposition 3.12. Let $\|f\|_{(\Phi)} := \|f\|_{[-\pi,\pi],(\Phi)}$ for a function $f$ periodic of period $2\pi$.

**Lemma 11.12.** *For $\Phi \in \mathcal{CV}^*$, let $a = \{a_k\} \in \ell^{\Phi}$ with all $a_k \geq 0$ be such that for some $m > 1$, $a_k = 0$ if $k > m$, let $n \geq m$ and $\epsilon > 0$. Then there exists a function $f$, continuous and periodic of period $2\pi$, with the following three properties:*

(*a*) *$f \geq 0$, $f(t) = 0$ for $t \in [-\pi, \pi/(2n+1)] \cup [2m\pi/(2n+1), \pi]$, and $\sup_t f(t) = \max_k a_k$;*

(*b*) *for $I_n(f;x)$ defined by (11.28),*

$$
\begin{aligned}
I_n(f;0) &= \sum_{j=1}^{m} \frac{1}{2j-1} \int_0^{\pi} \Big\{ f\Big(\frac{t+(2j-1)\pi}{2n+1}\Big) - f\Big(\frac{t+2j\pi}{2n+1}\Big) \Big\} \sin t \, dt \\
&> \sum_{k\geq 1} \frac{a_k}{k} - \epsilon;
\end{aligned}
$$

(*c*) $\|f\|_{(\Phi)} \leq 2\|a\|_{\Phi}$.

*Proof.* Let $u_i := i\pi/(2n+1)$ for $i = 1, \ldots, 2m$, and let $v_i$, $i = 1, \ldots, 2m$, be such that $u_{2j-1} < v_{2j-1} < v_{2j} < u_{2j}$ for $j = 1, \ldots, m$. Define a function $f$ on

$[-\pi,\pi]$ to be zero on $[-\pi,\pi/(2n+1)]\cup[2m\pi/(2n+1),\pi]$ and on $[u_{2j},u_{2j+1}]$, $j=1,\dots,m-1$. Let $f$ be equal to $a_j$ on $[v_{2j-1},v_{2j}]$ for $j=1,\dots,m$, and let $f$ be linear on the remaining intervals and continuous on $[-\pi,\pi]$. For $j=1,\dots,m$, choose $v_{2j-1}$ and $v_{2j}$ such that

$$I_j := \int_0^\pi \Big\{f\Big(\frac{t+(2j-1)\pi}{2n+1}\Big)-f\Big(\frac{t+2j\pi}{2n+1}\Big)\Big\}\sin t\,dt > 2a_j-2^{-j}\epsilon.$$

This is possible because $(t+2j\pi)/(2n+1)\in[u_{2j},u_{2j+1}]$ for $t\in[0,\pi]$. Let $t_{2j-1}:=v_{2j-1}(2n+1)-(2j-1)\pi$ and $t_{2j}:=v_{2j}(2n+1)-(2j-1)\pi$. Then

$$I_j=\int_0^\pi f\Big(\frac{t+(2j-1)\pi}{2n+1}\Big)\sin t\,dt \ge a_j\int_{t_{2j-1}}^{t_{2j}}\sin t\,dt\uparrow 2a_j$$

as $t_{2j-1}\downarrow 0$ and $t_{2j}\uparrow\pi$. Extend $f$ to $\mathbb{R}$ by periodicity. Then $(a)$ clearly holds. For $(b)$, since $f$ is zero on $[-\pi,u_1]\cup[u_{2m},\pi]$, we have

$$I_n(f;0)=\sum_{j=1}^m\frac{1}{2j-1}I_j>\sum_{j=1}^m\frac{2a_j-2^{-j}\epsilon}{2j-1}>\sum_{k\ge1}k^{-1}a_k-\epsilon.$$

To prove $(c)$, let $\kappa=\{t_i\}_{i=0}^n$ be a partition of $[-\pi,\pi]$ and let $\{\beta_i\}_{i=1}^n$ be nonnegative real numbers such that $\sum_{i=1}^n\Phi^*(\beta_i)\le 1$. First suppose that $\kappa$ is a refinement of the partition $\lambda$ of $[-\pi,\pi]$ formed by the points $\{-\pi,0,\pi\}$ and $\{u_i,v_i\colon i=1,\dots,2m\}$, that is, $\lambda\subset\kappa$. For $j=1,\dots,m$, let $A_j:=\{i\colon u_{2j-1}<t_i\le v_{2j-1}\}$ and $B_j:=\{i\colon v_{2j}<t_i\le u_{2j}\}$. The sets $A_j$ and $B_j$ are $2m$ disjoint nonempty sets of values of $i$. Let $\Delta_i f:=f(t_i)-f(t_{i-1})$ for $i=1,\dots,n$. Then $\Delta_i f=0$ if $i\notin\cup_{j=1}^m(A_j\cup B_j)$, and for each $j=1,\dots,m$,

$$\sum_{i\in A_j}|\Delta_i f|=a_j\qquad\text{and}\qquad\sum_{i\in B_j}|\Delta_i f|=a_j.$$

Thus letting $\beta_j':=\max_{i\in A_j}\beta_i$ and $\beta_j'':=\max_{i\in B_j}\beta_i$, we have $\sum_{j=1}^m\Phi^*(\beta_j')\le\sum_{i=1}^n\Phi^*(\beta_i)\le 1$ and $\sum_{j=1}^m\Phi^*(\beta_j'')\le\sum_{i=1}^n\Phi^*(\beta_i)\le 1$, and so

$$\begin{aligned}\sum_{i=1}^n|\Delta_i f|\beta_i&=\sum_{j=1}^m\Big(\sum_{i\in A_j}|\Delta_i f|\beta_i+\sum_{i\in B_j}|\Delta_i f|\beta_i\Big)\\&\le\sum_{j=1}^m a_j\beta_j'+\sum_{j=1}^m a_j\beta_j''\le 2\|a\|_\Phi\end{aligned}\tag{11.31}$$

for each refinement $\kappa$ of $\lambda$. To prove $(c)$ it suffices to prove (11.31) for an arbitrary partition $\kappa$ of $[-\pi,\pi]$, so suppose that $\lambda\setminus\kappa$ is nonempty. Let $\kappa'=\{y_l\colon l=0,\dots,r\}:=\lambda\cup\kappa$, that is, $\kappa'$ is the least common refinement of $\kappa$ and $\lambda$. With each index $i\in\{1,\dots,n\}$ of $\kappa$ we will associate a unique index $\sigma(i)\in\{1,\dots,r\}$ of $\kappa'$ such that

$$[y_{\sigma(i)-1}, y_{\sigma(i)}] \subset [t_{i-1}, t_i] \quad \text{and} \quad |\Delta_i f| \le |f(y_{\sigma(i)}) - f(y_{\sigma(i)-1})|. \tag{11.32}$$

If $f(t_i) = f(t_{i-1})$ then let $\sigma(i)$ be such that $y_{\sigma(i)} = t_i$. Then (11.32) clearly holds. If $f(t_i) > f(t_{i-1})$ then $t_i \in (u_{2j-1}, u_{2j})$ for some $j = 1, \ldots, m$. Let $\sigma(i)$ be such that $y_{\sigma(i)} = t_i$ if $t_i \in (u_{2j-1}, v_{2j-1}]$, or $y_{\sigma(i)} = v_{2j-1}$ if $t_i \in (v_{2j-1}, u_{2j})$. Then $\Delta_i f \le f(y_{\sigma(i)}) - f(y_{\sigma(i)-1})$, and so (11.32) holds. If $f(t_i) < f(t_{i-1})$ then $t_{i-1} \in (u_{2j-1}, u_{2j})$ for some $j = 1, \ldots, m$. Let $\sigma(i)$ be such that $y_{\sigma(i)-1} = t_{i-1}$ if $t_{i-1} \in [v_{2j}, u_{2j})$, or $y_{\sigma(i)-1} = v_{2j}$ if $t_{i-1} \in (u_{2j-1}, v_{2j})$. Then $-\Delta_i f \le f(y_{\sigma(i)-1}) - f(y_{\sigma(i)})$, and (11.32) holds again and so in all cases.

We will show that $\sigma(\cdot)$ is one-to-one. Suppose not. Then for some $\ell$ there are two indices $i' \neq i''$ such that $\ell = \sigma(i') = \sigma(i'')$. By the first relation in (11.32) it follows that the nondegenerate interval $[y_{\ell-1}, y_\ell]$ is a subset of two different intervals $[t_{i'-1}, t_{i'}]$ and $[t_{i''-1}, t_{i''}]$, which yields a contradiction. So $\sigma$ is one-to-one.

For $\ell = 1, \ldots, r$, let $\gamma_\ell := \beta_i$ if $\ell = \sigma(i)$ for some $i \in \{1, \ldots, n\}$, and let $\gamma_\ell := 0$ if $\ell \notin \{\sigma(i)\colon\ i = 1, \ldots, n\}$. Then we have

$$\begin{aligned}\sum_{i=1}^{n} |\Delta_i f|\beta_i &\le \sum_{i=1}^{n} |f(y_{\sigma(i)}) - f(y_{\sigma(i)-1})|\beta_i \\ &= \sum_{\ell=1}^{r} |f(y_\ell) - f(y_{\ell-1})|\gamma_\ell \le 2\|a\|_\Phi\end{aligned}$$

by (11.31) since $\kappa'$ is a refinement of $\lambda$. The proof of (*c*), and hence of the lemma, is complete. □

Now we are ready to prove the result of Baernstein [5]:

**Theorem 11.13.** *Let $\Phi^*$ be the complementary function to $\Phi \in \mathcal{CV}^*$ and let $\sum_{k\ge1} \Phi^*(1/k) = \infty$. Then there exists a continuous function periodic of period $2\pi$ and of bounded $\Phi$-variation with Fourier series divergent at the point* 0.

*Proof.* For each integer $r \ge 1$, by Lemma 11.11 with $\epsilon = 2^{-r-1}$, there exists $a^{(r)} = \{a_k^{(r)}\}_{k\ge1} \in \ell^\Phi$ with all $a_k^{(r)} \ge 0$ and a positive integer $k_r \ge 2$ such that $a_k^{(r)} = 0$ if $k > k_r$, $\|a^{(r)}\|_\Phi \le 2^{-r-1}$, and $\sum_{k\ge1} k^{-1} a_k^{(r)} > 1$. We can assume that $k_r \to \infty$ as $r \uparrow +\infty$. Let $n_1 \ge k_1$ and let $f_1$ be the function given by Lemma 11.12 with $a = a^{(1)}$, $n = n_1$, $m = k_1$, and $\epsilon = 1$. Suppose that we have already chosen functions $f_1, \ldots, f_{r-1}$ and integers $n_1, \ldots, n_{r-1}$ for some $r \ge 2$. Then let $n_r$ be an integer so large that

$$\frac{2k_r + 1}{2n_r + 1} < \frac{1}{2n_{r-1} + 1} \quad \text{and} \quad w\left(\sum_{i=1}^{r-1} f_i; \frac{\pi}{2n_r + 1}\right) < \frac{1}{r\log(4n_{r-1} + 2)}, \tag{11.33}$$

where $w(h; u) := \sup\{|h(t) - h(s)|\colon\ |t - s| \le u\}$, $u \ge 0$. Then $n_r \ge k_r$. Let $f_r$ be the function $f$ of Lemma 11.12 obtained with $a = a^{(r)}$, $n = n_r$, $m = k_r$, and $\epsilon = 1/r$. Continuing this process indefinitely we obtain a sequence of continuous functions $\{f_r\}_{r\geq 1}$ and a sequence of integers $\{n_r\}_{r\geq 1}$ satisfying (11.33). By the first relation in (11.33) and by the property $(a)$ of Lemma 11.12 applied to $f_r$ and $f_{r-1}$, the set $\{t \in [-\pi,\pi]\colon f_r(t) \neq 0\}$ lies strictly to the left of the set $\{t \in [-\pi,\pi]\colon f_{r-1}(t) \neq 0\}$. Thus for each $t \in [-\pi,\pi]$, $f_r(t) \neq 0$ for at most one $r$. Since by (11.30),

$$\sup_t f_r(t) = \max_k a_k^{(r)} \leq \|a^{(r)}\|_{\Phi}/\phi_*(1) \leq 2^{-r-1}(\phi_*(1))^{-1} = o(1)$$

as $r \to \infty$, it then follows that $\sum_{r\geq 1} f_r$ converges uniformly, and so the sum $f := \sum_{r\geq 1} f_r$ is a continuous function. Recall that the space $\widetilde{\mathcal{W}}_{\Phi}[-\pi,\pi]$ is complete (a Banach space) by Theorem 3.7$(d)$ for the norm $\|\cdot\|_{[\Phi]} := \|\cdot\|_{(\Phi)} + \|\cdot\|_{\sup}$. The norm $\|\!|\cdot|\!\|_{[\Phi]} := \|\!|\cdot|\!\|_{(\Phi)} + \|\cdot\|_{\sup}$ is equivalent by Proposition 3.12. Since $\|\!|\cdot|\!\|_{(\Phi)}$ is subadditive, it follows from Lemma 11.12$(c)$ that

$$\|\!|f|\!\|_{(\Phi)} \leq \sum_{i=1}^{\infty} \|\!|f_i|\!\|_{(\Phi)} \leq 2\sum_{i=1}^{\infty} \|a^{(i)}\|_{\Phi} \leq 1.$$

It follows by the completeness that $f \in \widetilde{\mathcal{W}}_{\Phi}[-\pi,\pi]$ and by Proposition 3.12 again that $\|f\|_{(\Phi)} \leq 1$. By Theorem 3.7$(b)$, $f$ has bounded $\Phi$-variation, $f \in \mathcal{W}_{\Phi}[-\pi,\pi]$, in fact $v_{\Phi}(f;[-\pi,\pi]) \leq 1$.

Now for a given integer $r \geq 1$, consider the integral $I_{n_r}(f;0)$ defined by (11.28). For a function $h$, $t \in [0,\pi]$, and integers $j = 1,\dots,n$, $n \geq 1$, let

$$W_{j,n}(h;t) := h\Big(\frac{t+(2j-1)\pi}{2n+1}\Big) - h\Big(\frac{t+2j\pi}{2n+1}\Big). \tag{11.34}$$

Each function $f_i$, $i \geq 1$, is zero on $[-\pi,0]$, by part $(a)$ of Lemma 11.12. Moreover, by the same part $(a)$, specifically $f_i(s) = 0$ for $s \in [2k_i\pi/(2n_i+1),\pi]$, and by the first inequality in (11.33) for $i$ in place of $r$, we have $W_{j,n_r}(f_i;t) = 0$ for each $i > r$, $j = 1,\dots,n_r$, and $t \in [0,\pi]$; in fact, both values of $f_i$ whose difference is taken are 0. Thus

$$I_{n_r}(f;0) \;=\; I_{n_r}(f_r;0) + \sum_{j=1}^{n_r} \frac{1}{2j-1}\int_0^{\pi} W_{j,n_r}\left(\sum_{i=1}^{r-1} f_i;t\right)\sin t\,dt. \tag{11.35}$$

Lemma 11.12$(b)$ with $\epsilon = 1/r$ and the choice of $a_k^{(r)}$ give

$$I_{n_r}(f_r;0) > 1 - r^{-1}. \tag{11.36}$$

To bound the sum in (11.35), let $m_r$ be the largest integer $j \geq 2$ such that $(2j+1)/(2n_r+1) < 1/(2n_{r-1}+1)$, which exists by the first inequality in (11.33) since $k_r \geq 2$. Thus

$$\frac{2m_r+1}{2n_r+1} < \frac{1}{2n_{r-1}+1} \leq \frac{2m_r+3}{2n_r+1},$$

and so

$$1 < \frac{n_r}{m_r} \le 4n_{r-1} + 2.$$

Again by Lemma 11.12($a$), specifically $f_i(s) = 0$ for $s \in [-\pi, \pi/(2n_i+1)]$, and by the definition of $m_r$, $W_{j,n_r}(f_i;t) = 0$ for each $i = 1, \dots, r-1$, $j = 1, \dots, m_r$, and $t \in [0, \pi]$. Therefore by the definition (11.34) of $W_{j,n}$ and the second inequality in (11.33), and since for $j \ge 1$, $(2j-1)^{-1} \le \int_{j-1}^{j} x^{-1}\,dx$, applied for $j \ge m_r + 1 \ge 3$, we have the bound

$$\begin{aligned} &\Big|\sum_{j=1}^{n_r} \frac{1}{2j-1} \int_0^{\pi} W_{j,n_r}\Big(\sum_{i=1}^{r-1} f_i;t\Big) \sin t\,dt\Big| \\ &\quad \le \pi w\Big(\sum_{i=1}^{r-1} f_i; \frac{\pi}{2n_r+1}\Big) \sum_{j=m_r+1}^{n_r} \frac{1}{2j-1} \le \frac{\pi}{r}. \end{aligned} \tag{11.37}$$

In (11.35), applying (11.36) and (11.37), it follows that $I_{n_r}(f;0) > 1 - (1+\pi)r^{-1}$ for each integer $r \ge 1$. Hence $I_{n_r}(f;0) > 1 - o(1)$ as $r \to \infty$. Recalling the definition (11.28) of $I_n(f;0)$, the Fourier series of $f$ diverges at 0 by Proposition 11.7. The proof of Theorem 11.13 is complete. □

To illustrate Theorems 11.8 and 11.13, consider some examples of pairs of functions $\Phi, \Phi^* \in \mathcal{CV}^*$ defined by their derivatives $P, Q$, respectively. For $0 < p < \infty$, let

$$Q(t) := \begin{cases} 0 & \text{if } t = 0, \\ |\log t|^{-p} & \text{if } 0 < t \le e^{-1}, \\ et & \text{if } t > e^{-1}, \end{cases} \quad \text{and } P(s) := \begin{cases} 0 & \text{if } s = 0, \\ \exp\{-s^{-1/p}\} & \text{if } 0 < s \le 1, \\ s/e & \text{if } s > 1. \end{cases}$$

Then $Q$ is the inverse of $P$. Integrating by parts we get

$$\Phi_p^*(v) := \int_0^v Q(t)\,dt \sim v/|\log v|^p \quad \text{as } v \downarrow 0.$$

Also,

$$\Phi_p(u) := \int_0^u P(s)\,ds < \exp\{-u^{-1/p}\}$$

for $0 < u \le 1$. Then $\Phi_p \in \mathcal{CV}^*$ and $\Phi_p^*$ is the complementary function to $\Phi_p$ as shown after (3.22). Now by Theorems 11.8 and 11.13, every continuous function of bounded $\Phi_p$-variation has its Fourier series convergent uniformly if and only if $\sum_{k\ge1} k^{-1}(\log k)^{-p} < \infty$, that is, if and only if $1 < p < \infty$.

## 11.3 Notes

*Notes on Section 11.1.* Much of the early work on Hölder conditions and on $p$-variation was done with Fourier series as one application in view, e.g. Lipschitz

(1864) [143], Wiener (1924) [241], L. C. Young (1936) [244]. The Hausdorff–Young inequality is proved in Edwards [63, Theorem 13.5.1]. Theorem 11.2 is Theorem 3 in Marcinkiewicz [156]. An extension of this fact to several other classes of functions is given by B. I. Golubov [83, §4].

*Notes on Section 11.2.* Salem [202] proved Theorem 11.8. Theorem 11.13 is due to Baernstein [5].

# 12

# Stochastic Processes and $\Phi$-Variation

We refer to the book of Kallenberg [112] for probability terminology and will cite it often, although far from exclusively, for results needed in this chapter.

## 12.1 Processes with Regulated Sample Functions

Let $\mathcal{T} \subset [0, \infty)$ and let $(\Omega, \mathcal{F}, \Pr)$ be a complete probability space. A real-valued *stochastic process* $\{X_t\}_{t \in \mathcal{T}}$, which may also be written $\{X(t)\}_{t \in \mathcal{T}}$ or $\{X(t),\ t \in \mathcal{T}\}$, is a function $(t, \omega) \mapsto X_t(\omega) \equiv X(t)(\omega) \equiv X(t, \omega)$ from $\mathcal{T} \times \Omega$ into $\mathbb{R}$ such that for each $t \in \mathcal{T}$, $X_t$ is a random variable on $\Omega$. If $\mathcal{T} = [0, +\infty)$ we may write $\{X_t\}_{t \geq 0}$, or if $\mathcal{T} = [0, T]$ where $0 < T < \infty$ we may write $\{X_t\}_{0 \leq t \leq T}$. A stochastic process $Y = \{Y_t\}_{t \geq 0}$ on the same probability space is a *modification* of $X$ iff for each $t \geq 0$, $X_t = Y_t$ almost surely (with probability 1). For each $\omega \in \Omega$, the function $X. = X.(\omega)$ is called a *sample function* of $X$. A stochastic process $X = \{X_t\}_{t \geq 0}$ has *regulated sample functions* if $X.(\omega) \in \mathcal{R}[0, \infty)$ for almost all $\omega \in \Omega$. Then we write $X \in \mathcal{R}[0, \infty)$. In probability theory sample functions of stochastic processes are often assumed to be regulated and right-continuous. One sometimes refers to such processes using the French abbreviation "cadlag" processes. Stochastic processes with regulated sample functions then might be called "ladlag" processes.

For any real-valued stochastic process $X = \{X_t\}_{t \geq 0}$ on a probability space $(\Omega, \mathcal{F}, \Pr)$ and finite set $F = \{t_1 < \cdots < t_k\} \subset [0, \infty)$, the mapping $R_F$ : $\omega \mapsto (X_{t_1}(\omega), \ldots, X_{t_k}(\omega))$ gives an image probability measure $\Pr \circ R_F^{-1}$ on the finite-dimensional Euclidean space $\mathbb{R}^F$ or $\mathbb{R}^k$. These measures are called the *finite-dimensional joint distributions* of $X$. Suppose that for each finite set $F \subset [0, \infty)$ a probability measure $P_F$ is given on the Borel sets of $\mathbb{R}^F$, with the natural consistency property that whenever $F \subset G$, the image measure of $P_G$ on $\mathbb{R}^F$ under the natural projection is $P_F$. Then there exists a stochastic process $X$ having $P_F$ as its finite-dimensional joint distributions by a theorem of Kolmogorov, e.g. [99, Section 9.4] or [53, Theorem 12.1.2].

R.M. Dudley and R. Norvaiša, *Concrete Functional Calculus*, Springer Monographs in Mathematics, DOI 10.1007/978-1-4419-6950-7_12,

Let $X \in \mathcal{R}[0,\infty)$ and let $\Omega_X := \{\omega \in \Omega\colon X.(\omega) \in \mathcal{R}[0,\infty)\}$. Due to completeness of the underlying probability space, $\Omega_X \in \mathcal{F}$ and $\Pr(\Omega_X) = 1$. For each $t \in [0,\infty)$, let

$$X_+(t) := X_+(t,\omega) := \begin{cases} X(t+,\omega) & \text{if } \omega \in \Omega_X, \\ X(t,\omega) & \text{otherwise.} \end{cases}$$

Similarly define $X_-(t)$ for each $t \in (0,\infty)$ and let $X_-(0) := X(0)$. For each $t \in [0,\infty)$, $X_+(t)$ is a random variable because it is the limit for $\omega \in \Omega_X$ of the random variables $X(r_n)$ for $r_n$ rational, $r_n \downarrow t$, and equals $X(t)$ otherwise. Likewise, $X(t-)$ is a random variable. Therefore $X_+ = \{X_+(t),\, t \geq 0\}$ and $X_- = \{X_-(t),\, t \geq 0\}$ are stochastic processes on the same probability space as $X$.

Recall that $\Delta^+ X(t) := X_+(t) - X(t)$ and $\Delta^- X(t) := X(t) - X_-(t)$. For each $t \in [0,\infty)$, let

$$\Omega_d(t) := \big\{\omega \in \Omega\colon \text{ either } \Delta^- X(t) \neq 0 \text{ or } \Delta^+ X(t) \neq 0\big\}.$$

A point $t \in [0,\infty)$ is called a point of *fixed discontinuity* of $X$ if $\Pr(\Omega_d(t)) > 0$. If $t$ is not a point of fixed discontinuity of $X$ then

$$\lim_{s \to t} X(s) = X(t) \qquad \text{almost surely.} \tag{12.1}$$

Indeed, for each $t \in (0,\infty)$ and $\omega \in \Omega_X \setminus \Omega_d(t)$, we have

$$\lim_{s \uparrow t} X(s,\omega) = X(t-,\omega) = X(t,\omega) = X(t+,\omega) = \lim_{s \downarrow t} X(s,\omega).$$

Since $\Pr(\Omega_X \setminus \Omega_d(t)) = 1$, (12.1) holds when $t \in (0,\infty)$. The same argument yields (12.1) when $t = 0$. If the stochastic process $X$ has no points of fixed discontinuity then the three processes $X_-$, $X_+$, and $X$ are modifications of each other.

A point $t \in [0,\infty)$ is a point of *stochastic continuity* of $X$ if $X(s) \to X(t)$ in probability as $s \to t$, otherwise a point of *stochastic discontinuity.*

**Theorem 12.1.** *Let $X \in \mathcal{R}[0,\infty)$. Then the set of points of fixed discontinuity of $X$ is at most countable and coincides with the set of points of stochastic discontinuity of $X$.*

*Proof.* Since the limit of a sequence convergent in probability is unique almost surely, one can show that $X$ is not stochastically continuous at some $t \in [0,\infty)$ if $t$ is a point of fixed discontinuity. Also, $X$ is stochastically continuous at $t$ whenever $t$ is not a point of fixed discontinuity. This yields the second part of the claim. The first part follows from Theorem 11.1, Ch. VII, of Doob [46]. $\square$

By the preceding theorem, the set of fixed discontinuities of any stochastic process with regulated sample functions is at most countable. However, almost every sample function of such a process may have a non-fixed discontinuity, e.g. $X(t,\omega) = 1_{t\geq\omega}$, where $\omega$ has a uniform distribution in $[0,1]$.

Let $(\Omega, \mathcal{F}, \Pr)$ be a probability space. Then a *filtration* $\mathbb{F} = \{\mathcal{F}_t\colon\ t \geq 0\}$ is a family of sub-$\sigma$-algebras of $\mathcal{F}$ such that $\mathcal{F}_s \subset \mathcal{F}_t$ for $0 \leq s \leq t$. A real-valued stochastic process $X = \{X_t\}_{t\geq 0}$ is *adapted* to $\mathbb{F}$ iff $X_t$ is $\mathcal{F}_t$ measurable for each $t$. Given $X$, the filtration $\mathbb{F}(X) = \{\mathcal{F}_t\}_{t\geq 0}$ *generated by* $X$ is defined by letting $\mathcal{F}_t$ be the smallest $\sigma$-algebra for which $X_s$ are measurable for $0 \leq s \leq t$. Then clearly $X$ is adapted to $\mathbb{F}(X)$ and if $\{\mathcal{G}_t\}_{t\geq 0}$ is any other filtration to which $X$ is adapted then $\mathcal{F}_t \subset \mathcal{G}_t$ for all $t \geq 0$.

Suppose that almost all sample functions of a stochastic process $X$ have bounded $p$-variation on $[0,t]$ for some $0 < p < \infty$ and $0 < t < \infty$. Since the set of partitions PP $[0,t]$ is uncountable, the function

$$\omega \mapsto v_p(X(\cdot,\omega);[0,t]) \tag{12.2}$$

need not be measurable. For example, let Pr be Lebesgue measure on $\Omega := [0,1]$. For a non-Lebesgue measurable set $A \subset [0,1]$, let $X$ be a stochastic process defined by $X(t,\omega) = 1_{t=\omega\in A}$, $t \in [0,1]$. Then $v_p(X;[0,1]) = 0$ if $\omega \notin A$, and $= 2$ if $\omega \in A$. We will show in Theorem 12.3 that in many cases there is a stochastic process adapted to the filtration $\mathbb{F}(X)$ generated by $X$ and whose sample functions almost surely agree with (12.2).

Recall that $\mathcal{V}$ is the class of all functions $\Phi\colon\ [0,\infty) \to [0,\infty)$ which are strictly increasing, continuous, unbounded, and 0 at 0.

**Definition 12.2.** Let $X = \{X_t\}_{t\geq 0}$ be a stochastic process, and let $\Phi \in \mathcal{V}$. We say that $X$ has *locally bounded* $\Phi$*-variation* (or locally bounded $p$-variation if $\Phi(u) \equiv u^p$) and write $X \in \mathcal{W}_\Phi^{\mathrm{loc}}$ if

(*a*) for almost all $\omega \in \Omega$,

$$v_\Phi(X(\cdot,\omega);[0,T]) < +\infty \qquad \forall T > 0; \tag{12.3}$$

(*b*) there exists an $\mathbb{F}(X)$-adapted stochastic process $v_\Phi(X) = \{v_\Phi(X;t)\}_{t\geq 0}$ such that for a Pr-null set $\Omega_0$, if $\omega \notin \Omega_0$ then

$$v_\Phi(X(\cdot,\omega);[0,t]) = v_\Phi(X;t)(\omega) \qquad \forall t > 0. \tag{12.4}$$

We call $v_\Phi(X)$ the $\Phi$*-variation stochastic process* (or $p$-variation stochastic process if $\Phi(u) \equiv u^p$) of $X$.

A stochastic process $X = \{X_t\}_{t\geq 0}$ is called *separable* if there exist a countable set $S \subset [0,\infty)$ and a Pr-null set $\Omega_0$ such that for every closed subset $A$ of $\mathbb{R}$ and open subset $I$ of $[0,\infty)$, we have

$$\big\{\omega\colon\ X(t,\omega) \in A, \quad t \in S \cap I\big\} \setminus \big\{\omega\colon\ X(t,\omega) \in A, \quad t \in I\big\} \subset \Omega_0. \tag{12.5}$$

The set $S$ in the definition of separability is called a *separating* set.

The following shows that $(b)$ follows from $(a)$ in Definition 12.2 under some conditions.

**Theorem 12.3.** *Let $X = \{X_t\}_{t\ge 0}$ be either a separable stochastic process continuous in probability, or a cadlag stochastic process. If for some $\Phi \in \mathcal{V}$, (12.3) holds for almost all $\omega \in \Omega$, then $X$ has locally bounded $\Phi$-variation.*

*Proof.* Let $S = \{t_k\}_{k=0}^{\infty}$ be a countable dense subset of $[0,\infty)$ with $t_0 = 0$. Suppose first that $X$ is a separable stochastic process continuous in probability. Then by Theorem 2.2 of Doob [46, Section II.2], $S$ is a separating set. The following consequence of separability is what we need. Let $I$ be an open subset of $[0,\infty)$ and let $\Omega_0$ be the Pr-null set in (12.5). Then for $\omega \notin \Omega_0$,

$$\sup_{t\in S\cap I} X(t,\omega) = \sup_{t\in I} X(t,\omega) \quad \text{and} \quad \inf_{t\in S\cap I} X(t,\omega) = \inf_{t\in I} X(t,\omega). \tag{12.6}$$

Second, if $X$ is a cadlag stochastic process then (12.5) holds with $\Omega_0 = \emptyset$, which also implies (12.6). To construct the $\Phi$-variation stochastic process, for each $n \ge 2$, let $S_n := \{t_0^n, \dots, t_n^n\}$ be the elements $\{t_0, \dots, t_n\}$ arranged in increasing order, and let $[0,t] \Cap S_n := \{t_0^n, \dots, t_k^n, t\}$, where $k$ is the largest $j = 1, \dots, n$ such that $t_j^n < t$. We can and do assume that (12.3) holds for each $\omega \notin \Omega_0$. For each $t \ge 0$, let

$$v_\Phi(X;t)(\omega) := \begin{cases} \sup_n v_\Phi(X(\cdot,\omega); [0,t] \Cap S_n) & \text{if } \omega \notin \Omega_0, \\ 0 & \text{if } \omega \in \Omega_0, \end{cases}$$

where the $\Phi$-variation over a finite set $E$ is defined by restricting partitions to be subsets of $E$. It is easily seen that for each $t \ge 0$, $v_\Phi(X;t)(\cdot)$ is measurable with respect to the $\sigma$-algebra generated by $\{X(s)\colon s \le t\}$. Thus $v_\Phi(X) = \{v_\Phi(X;t)\}_{t\ge 0}$ is adapted to the filtration $\mathbb{F}(X)$ generated by $X$. To prove (12.4) notice that

$$v_\Phi(X(\cdot,\omega);[0,t]) \ge v_\Phi(X;t)(\omega) \tag{12.7}$$

for each $\omega \in \Omega$ and each $t \ge 0$. To prove the converse inequality let $\omega \notin \Omega_0$ and let $t > 0$. Given $\epsilon > 0$ let $\kappa_\omega \in \mathrm{PP}\,[0,t]$ be such that

$$s_\Phi(X(\cdot,\omega);\kappa_\omega) \ge v_\Phi(X(\cdot,\omega);[0,t]) - \epsilon.$$

Taking non-intersecting open neighborhoods around each point of $\kappa_\omega \setminus \{0,t\}$ and using (12.6), one can find an integer $n = n_\omega$ and a subset $\lambda_\omega$ of $[0,t] \Cap S_n$ including $\{0,t\}$ such that

$$s_\Phi(X(\cdot,\omega);\lambda_\omega) \ge s_\Phi(X(\cdot,\omega);\kappa_\omega) - \epsilon.$$

Since $\epsilon > 0$ is arbitrary, the inequality converse to (12.7) holds. Thus $v_\Phi(X) = \{v_\Phi(X;t)\}_{t\ge 0}$ is the $\Phi$-variation stochastic process by Definition 12.2. ☐

Notice that the ladlag (regulated) property of $X$ is not sufficient for the theorem by the example given before Definition 12.2.

## 12.2 Brownian Motion

A *Brownian motion* or *Wiener* process is a sample-continuous real-valued Gaussian process $B = \{B_t\}_{t\ge 0}$ with mean 0 and covariance $EB_tB_u = \min\{t,u\}$ for $t,u \ge 0$. Recall that a stochastic process $\{X_t\}_{t\ge 0}$ is said to have *stationary increments* if for any $0 \le t_0 < t_1 < \cdots < t_n$, the joint distribution of $\{X_{t_j+h} - X_{t_{j-1}+h}\}_{1\le j\le n}$ for $h \ge 0$ does not depend on $h$. It is easily seen that Brownian motion has stationary increments.

The local law of the iterated logarithm for 1-dimensional Brownian motion is well known and not hard to prove [53, Theorem 12.5.2], noting that $\{B(t)\}_{t>0}$ and $\{tB(1/t)\}_{t>0}$ are equal in distribution: with probability 1,

$$\limsup_{t\downarrow 0} \frac{|B(t)|}{\sqrt{2t\log\log(1/t)}} = 1. \tag{12.8}$$

The same holds if $|B(t)|$ is replaced by $|B(u+t) - B(u)|$ for any fixed $u \ge 0$, by the stationary increments property.

Next is P. Lévy's global Hölder condition for Brownian motion:

**Proposition 12.4.** *With probability* 1,

$$\limsup_{s\downarrow 0} \sup_{0\le t\le 1} \frac{|B(t+s) - B(t)|}{\sqrt{2s\log(1/s)}} = 1. \tag{12.9}$$

*For almost every $\omega$ there is a $K(\omega) < \infty$ such that for $0 \le t \le t+s \le 1$,*

$$|B(t+s) - B(t)|(\omega) \ \le\ K(\omega)\sqrt{s\log(3/s)}. \tag{12.10}$$

*Proof.* Equation (12.9) is proved e.g. in Section 1.9 of Itô and McKean [109]. Then, for almost all $\omega$ there is a $\delta(\omega) > 0$ such that (12.10) holds with $K(\omega)$ replaced by 2 if $0 < s \le \delta(\omega)$. Let $M(\omega) := \max_{0\le t\le 1}|B(t,\omega)| < +\infty$. Since $s \mapsto s\log(3/s)$ is increasing for $0 < s \le 1$, (12.10) then holds for

$$K(\omega) \ :=\ \max(2, 2M(\omega)/\sqrt{\delta(\omega)\log(3/\delta(\omega))}),$$

completing the proof. □

Brownian paths have bounded $p$-variation only for $p > 2$, as P. Lévy proved.

**Proposition 12.5** (P. Lévy). *For a Brownian motion $B$ and $0 < T < \infty$, we have almost surely*

$$v_p(B;[0,T]) \begin{cases} < +\infty \ \textit{if}\ p > 2, \\ = +\infty \ \textit{if}\ p \le 2. \end{cases} \tag{12.11}$$

*Proof.* The fact that $v_p(B; [0,T]) < \infty$ a.s. (almost surely) for $p > 2$ follows from Lévy's Hölder condition, Proposition 12.4, for $T = 1$. For general $T > 0$, $B(Tu) = \sqrt{T}B(u)$ in distribution as processes for $0 \le u \le 1$.

It will be shown that $(v_2(B(\cdot,\omega); [0,1])) = +\infty$ almost surely. Let $J := (1/4, 3/4)$ and $0 < M < \infty$. By sample continuity and the local law of the iterated logarithm (12.8), for any $t \in (1/4, 3/4)$, almost surely there exist rational $q, r$ with $1/4 \le q < t < r \le 3/4$, $r - q < 1/4$, and

$$(B(r) - B(q))^2 \ > \ M(r-q). \tag{12.12}$$

Let $A$ be the set of $(t,\omega) \in J \times \Omega$ for which such $r$ and $q$ exist. Then $A$ is evidently jointly measurable. For each $t \in J$, $\Pr(\{\omega: \ (t,\omega) \in A\}) = 1$, so by the Fubini theorem, with $\lambda$ equal to Lebesgue measure, $(\lambda \times \Pr)(A) = 1/2$. Let $\Omega_1 := \{\omega: \ (t,\omega) \in A \ \text{ for } \ \lambda\text{-almost all } \ t \in J\}$. Then $\Pr(\Omega_1) = 1$. For each $\omega \in \Omega_1$, $\{t: \ (t,\omega) \in A\}$ is a countable union of open intervals $(q_n, r_n)$ satisfying (12.12). By a Vitali lemma, e.g. [53, Lemma 7.2.2], there is a finite, disjoint set $(q_{n_j}, r_{n_j})$ of those intervals such that $\sum_j r_{n_j} - q_{n_j} > 1/(3 \cdot 2)$ and thus by (12.12),

$$v_2(B(\cdot,\omega); [0,1]) \ \ge \ \sum_j (B(r_{n_j}) - B(q_{n_j}))^2 \ > \ M/6.$$

So $v_2(B; [0,1]) \ > \ M/6$ almost surely and since $M \ < \ +\infty$ was arbitrary, $v_2(B; [0,1]) = +\infty$ almost surely. It follows by Lemma 3.45 that $v_p(B; [0,1]) = +\infty$ a.s. for $p \le 2$, and the proof is complete. □

Rather than the global modulus of continuity (Hölder condition) as in Proposition 12.4, S. J. Taylor showed that the local modulus of continuity (log log law), as in (12.8) and the proof just given, determines the precise $\Phi$-variation behavior of Brownian motion. His theorem will follow as a special case of facts about fractional Brownian motion, in Corollary 12.24.

## 12.3 Martingales

A complete probability space $(\Omega, \mathcal{F}, \Pr)$ and a filtration $\mathbb{F} = \{\mathcal{F}_t\colon \ t \ge 0\}$ will be said to satisfy the usual hypotheses if $\mathbb{F}$ is $(i)$ complete, meaning that all sets of probability zero are in $\mathcal{F}_0$, and $(ii)$ right-continuous, meaning that $\mathcal{F}_t = \cap_{s>t}\mathcal{F}_s$ for each $0 \le t < \infty$. We assume throughout this section that the usual hypotheses hold.

Let $X = \{X_t\}_{t \ge 0}$ be a real-valued stochastic process defined on $\Omega$. The pair $(X, \mathbb{F})$ is a *martingale* iff $X$ is adapted to $\mathbb{F}$, the expectation $E|X(t)|$ is finite for all $t \ge 0$, and the conditional expectation $E(X(t)\big|\mathcal{F}_s)$ equals $X(s)$ almost surely for $0 \le s \le t$. (Conditional expectations are defined e.g. in [53, Section 10.1] and [112, Theorem 6.1].)

It is known that under the usual hypotheses, a martingale $(X, \mathbb{F})$ has a modification $M = \{M_t\}_{t\geq 0}$ which is a cadlag stochastic process and such that $(M, \mathbb{F})$ is still a martingale ([112, Theorem 7.27]).

For a filtration $\mathbb{F} = \{\mathcal{F}_t \colon t \geq 0\}$, a random variable $\tau$ on $\Omega$ with values in $[0, +\infty]$ is called a *stopping time* iff $\{\tau \leq t\} \in \mathcal{F}_t$ for every $t \in [0, \infty)$. Suppose that on $(\Omega, \mathcal{F}, \Pr)$, $\mathbb{F}$ is a filtration and $M$ is a cadlag stochastic process. The pair $(M, \mathbb{F})$ is a *local martingale* iff $M$ is adapted to $\mathbb{F}$ and there is a nondecreasing sequence of stopping times $\{\tau_n\}_{n\geq 1}$ such that $\lim_n \tau_n = +\infty$ almost surely and $M_n := \{M_{t\wedge\tau_n}\}_{t\geq 0}$ is a uniformly integrable martingale for each $n$. (Uniform integrability is defined e.g. in [53, p. 355] and [112, p. 67].)

For a cadlag stochastic process $X = \{X_t\}_{t\geq 0}$ adapted to $\mathbb{F}$, the pair $(X, \mathbb{F})$ is a *semimartingale* iff almost surely $X(t) = M(t) + A(t)$ for each $t \geq 0$, where $(M, \mathbb{F}) = \{M_t, \mathcal{F}_t\}_{t\geq 0}$ is a local martingale and $A = \{A(t)\}_{t\geq 0}$ is a stochastic process having locally bounded variation (as in Definition 12.2 for $p = 1$).

If $B = \{B(t)\}_{t\geq 0}$ is a stochastic process and $\mathbb{F} = \{\mathcal{F}_t \colon t \geq 0\}$ is a filtration, we will say that $(B, \mathbb{F})$ is a *Brownian motion* if $B$ is a Brownian motion, it is adapted to $\mathbb{F}$, and for $0 \leq t \leq u$, the increments $B(u) - B(t)$ are independent of $\mathcal{F}_t$. We say that a stochastic process $X$ is *equivalent to a time change of Brownian motion* if there exist a Brownian motion $(B, \mathbb{F})$ and a stochastic process $\{\tau_t\}_{t\geq 0}$ such that each $\tau_t$ is an $\mathbb{F}$ stopping time, $0 \leq \tau_s \leq \tau_t$ whenever $0 \leq s \leq t$, and the stochastic process $\{B(\tau_t)\}_{t\geq 0}$ has the same finite-dimensional joint distributions as $X$. Here $X$ and $B$ may be defined on different probability spaces.

**Theorem 12.6** (Monroe [174]). *A real-valued stochastic process is equivalent to a time change of Brownian motion if and only if it is a semimartingale.*

This fact can be used to bound the $p$-variation of a semimartingale as follows. Given $p > 2$, it follows from Proposition 12.5 that $v_p(B; [0, T]) < \infty$ almost surely for any $T < \infty$ and thus for any finite-valued random variable $\tau$, $v_p(B; [0, \tau]) < \infty$ almost surely. Let $X$ be a semimartingale, and let $t > 0$. By Theorem 12.6, for a suitable finite-valued stopping time $\tau_t$, $v_p(X; [0, t])$ has the same distribution as $v_p(B\circ\tau; [0, t])$. We have $v_p(B\circ\tau; [0, t]) \leq v_p(B; [0, \tau_t])$ almost surely, since $s \mapsto \tau_s$ is nondecreasing and restricting partitions to consist of points of its range gives a supremum over a smaller set of $p$-variation sums. So $v_p(X; [0, t]) < \infty$ almost surely. Thus we have proved the following:

**Corollary 12.7.** *A semimartingale is of bounded $p$-variation on any bounded interval for each $p > 2$.*

If $M$ is a martingale such that $EM(t) = 0$, $0 \leq t \leq T$, $EM(T)^2 < \infty$, and $B(u)$, $u \geq 0$, is a Brownian motion with stopping times $\tau_t$, $0 \leq t \leq T$, as in the definition before Theorem 12.6, such that $B(\tau_t)$ and $M(t)$ for $0 \leq t \leq T$ each have the same finite-dimensional joint distributions, then $E\tau_T = EM(T)^2$ (Monroe, [173, Theorems 5 and 11]). Thus for any sequence of martingales $M_i = \{M_i(t),\ 0 \leq t \leq T\}$, if $\sup_i EM_i(T)^2 < +\infty$ then for each $p > 2$,

$v_p(M_i; [0,T])$ is bounded in probability uniformly in $i$. It is well known that if $(B, \mathbb{F})$ is a Brownian motion and $\tau$ is a stopping time for $\mathbb{F}$ with $E\tau < \infty$ then $EB(\tau)^2 = E\tau$, e.g. [53, Theorem 12.4.1].

It was also well known that sample functions of martingales which are continuous and nonconstant must have unbounded 1-variation (Lemma 3.2.1 of Fisk [67]). In fact, we will see that they must have unbounded $p$-variation for every $p < 2$:

**Theorem 12.8.** *If for some $p < 2$ a sample continuous local martingale has locally bounded $p$-variation then almost surely its sample functions are constants.*

A proof of this statement will be given, based on a stopping time technique, as follows. We will say that a stochastic process $X$ is sample uniformly continuous if for each $0 < T < \infty$ there exists a null set $\Omega_0$ such that for any $\epsilon > 0$ one can choose $\delta > 0$ such that $|X(t,\omega) - X(s,\omega)| < \epsilon$ whenever $|t - s| < \delta$, $t, s \in [0,T]$, and $\omega \notin \Omega_0$.

For a stopping time $\tau$, let $X^\tau := \{X(t \wedge \tau),\ t \geq 0\}$.

**Lemma 12.9.** *Let $X$ be a sample continuous stochastic process adapted to a filtration $\mathbb{F}$. For each $0 < T < \infty$ there exists a nondecreasing sequence $\{\tau_n\}_{n\geq 1}$ of $\mathbb{F}$ stopping times $\tau_n \leq T$ such that $\Pr(\cup_{n=1}^\infty\{\tau_n = T\}) = 1$, and each $X^{\tau_n}$ is sample uniformly continuous and bounded by $n$.*

*Proof.* Let $0 < T < \infty$. Since almost every sample function of $X$ is uniformly continuous on $[0,T]$, there exists a double sequence $\{\delta_{kn}\}_{k,n\geq 1}$ of positive numbers such that for each $k$, $\delta_{k1} > \delta_{k2} > \cdots$, and for each $k, n$,

$$\Pr\left(\{\sup\{|X(t) - X(s)|\colon\ |t - s| \leq \delta_{kn},\ s,t \in [0,T]\} \geq 1/k\}\right) \leq 2^{-n-k}. \tag{12.13}$$

Let $\tau'_{kn}$ be the least $r \in [0,T]$ such that

$$\sup\left\{|X(t) - X(s)|\colon\ |t - s| \leq \delta_{kn},\ s,t \in [0,r]\right\} \geq 1/k.$$

If there is no such $r$, let $\tau'_{kn} := T$. Since $X$ is sample continuous each $\tau'_{kn}$ is an $\mathbb{F}$ stopping time. Let $\tau'_n := \inf_k \tau'_{kn}$. Then for each $n$ and any $t \in [0,\infty)$, $\{\tau'_n < t\} = \cup_k\{\tau'_{kn} < t\} \in \mathcal{F}_t$. Since $\mathbb{F}$ is right-continuous, each $\tau'_n$ is an $\mathbb{F}$ stopping time. Then each stopped process $X^{\tau'_n}$ is sample uniformly continuous.

For $n \geq 1$, since $\delta_{kn} > \delta_{k,n+1}$ for each $k$, it follows that $\tau'_{kn} \leq \tau'_{k,n+1}$ for each $k$, and so $\tau'_n \leq \tau'_{n+1}$. Also, by (12.13), $\Pr(\{X^{\tau'_n} \neq X \text{ on } [0,T]\}) \leq 2^{-n}$. It then follows from the Borel–Cantelli lemma that $\Pr(\cup_n\{\tau'_n = T\}) = 1$.

Let $\tau''_n$ be the least $t \in [0,T]$ such that $|X(t)| \geq n$. If no such $t$ exists, let $\tau''_n := T$. Then each $\tau''_n$ is an $\mathbb{F}$ stopping time and $X^{\tau''_n}$ is bounded by $n$. Clearly the sequence of $\mathbb{F}$ stopping times $\tau_n := \tau'_n \wedge \tau''_n$, $n \geq 1$, satisfies the conclusion of the lemma. □

Now we are ready to prove Theorem 12.8.

**Proof of Theorem 12.8.** Let a sample continuous local martingale $(X, \mathbb{F}) = \{X_t, \mathcal{F}_t\}_{t\geq 0}$ have locally bounded $p$-variation for some $p < 2$, and let $0 < T < \infty$. It is enough to prove that almost surely, $X(t) = X(0)$ for each $0 \leq t \leq T$. Since $X$ is sample continuous, the $p$-variation stochastic process $v_p(X)$ is also sample continuous by Propositions 3.35($a$) and 3.42. By Lemma 12.9 applied to $X$ and $v_p(X)$, taking the minimum of stopping times for each process, and because $X$ is a local martingale, there exists a nondecreasing sequence $\{\tau_n\}_{n\geq 1}$ of $\mathbb{F}$ stopping times such that $\Pr(\bigcup_n \{\tau_n = T\}) = 1$, each $X^{\tau_n}$ is a sample uniformly continuous martingale by optional sampling or "localization" [112, Lemma 17.1], bounded by $n$, and each $v_p(X)^{\tau_n}$ is bounded by $n$. Notice that for each $n$, $v_p(X)^{\tau_n}$ is the $p$–variation process of $X^{\tau_n}$. Let $\lambda = \{\lambda_m\}_{m\geq 1}$ be a nested sequence of partitions $\lambda_m = \{t_i^m\}_{i=0,\dots,k(m)}$ of $[0, T]$ such that $\max_i(t_i^m - t_{i-1}^m) \to 0$ as $m \to \infty$. Then $\cup_m \lambda_m$ is dense in $[0, T]$. For each $n, m \geq 1$, let

$$\epsilon_{n,m} := \Big\| \max_i |X^{\tau_n}(t_i^m) - X^{\tau_n}(t_{i-1}^m)| \Big\|_\infty.$$

Let $t \in \cup_m \lambda_m$. Then there exists an integer $m_t$ such that $t \in \lambda_m$ for each $m \geq m_t$. For any $0 \leq s_1 < t_1 \leq s_2 < t_2 \leq T$, let $\Delta_i X := X^{\tau_n}(t_i) - X^{\tau_n}(s_i)$ for $i = 1, 2$. Since $X^{\tau_n}$ is uniformly bounded it follows that

$$E\big[\Delta_1 X \Delta_2 X\big] = E\big[\Delta_1 X \, E\big[\Delta_2 X \big| \mathcal{F}_{s_2}\big]\big] = 0$$

($X^{\tau_n}$ has orthogonal increments), using the martingale property and properties of conditional expectations, e.g. [53, Theorem 10.1.9] or [112, Theorem 6.1]. Thus for each $m \geq m_t$, we have

$$\begin{aligned} E\big[X^{\tau_n}(t) - X^{\tau_n}(0)\big]^2 &= E\Big(\sum_{i=1}^{k(m)} [X^{\tau_n}(t_i^m) - X^{\tau_n}(t_{i-1}^m)]^2\Big) \\ &\leq E\big(\epsilon_{n,m}^{2-p} v_p(X)^{\tau_n}(t)\big) \leq nE\big(\epsilon_{n,m}^{2-p}\big). \end{aligned}$$

Since for each $n \geq 1$, $\epsilon_{n,m} \leq 2n$ and $\epsilon_{n,m} \to 0$ almost surely as $m \to \infty$, by the bounded convergence theorem, the left side of the preceding inequality is zero for each $n \geq 1$. Therefore, almost surely, for each $n \geq 1$, $X^{\tau_n}(t) = X^{\tau_n}(0)$ for each $t \in \cup_m \lambda_m$. Since $X$ is sample continuous and for almost every $\omega \in \Omega$, there is an $n \geq 1$ such that $X^{\tau_n}(t, \omega) = X(t, \omega)$ for $0 \leq t \leq T$, it follows that almost surely, $X(t) = X(0)$ for each $0 \leq t \leq T$. The proof of Theorem 12.8 is complete. □

## 12.4 Gaussian Stochastic Processes

A *Gaussian stochastic process* is a process all of whose finite-dimensional joint distributions are normal measures. To recall the definition of normal probability measure on finite-dimensional Euclidean spaces, first on $\mathbb{R}$, for any

$\mu \in \mathbb{R}$ and $\sigma > 0$, the normal or Gaussian probability measure $N(\mu, \sigma^2)$ is defined as having the density $(\sigma\sqrt{2\pi})^{-1} \exp(-(x-\mu)^2/(2\sigma^2))$ with respect to Lebesgue measure. If $\sigma = 0$ then $N(\mu, \sigma^2)$ is defined as a point mass $\delta_\mu$ at $\mu$. A Borel probability measure $P$ on $\mathbb{R}^k$ is called *normal* or *Gaussian* iff when $X = (X_1, \ldots, X_k)$ has distribution $P$, for any $t \in \mathbb{R}^k$, the inner product $(t, X)$ has a 1-dimensional normal distribution $N(\mu(t), \sigma^2(t))$ where $\sigma^2(t) \geq 0$. Then $X$ has a well-defined mean vector $\mu = EX$ and a covariance matrix $C$ with $C_{rs} = E[(X_r - \mu_r)(X_s - \mu_s)]$ for $r, s = 1, \ldots, k$. Here $C$ is nonnegative definite and symmetric. Conversely, for any such matrix $C$ and any $\mu \in \mathbb{R}^k$ there is a unique normal measure, called $N(\mu, C)$, with mean vector $\mu$ and covariance matrix $C$, e.g. by [53, Theorems 9.5.7 and 9.5.13], where Theorem 9.5.7 for $\mu = 0$ extends by translation to other $\mu$.

Let $X = \{X_t\}_{t \geq 0}$ be a separable Gaussian stochastic process with mean 0, and let $f_X(t) := X(t, \cdot) \in L^2(\Omega, \Pr)$ for $t \geq 0$. Recall that $\mathcal{CV}$ is the subclass of convex functions in the class $\mathcal{V}$ of all functions $\Psi\colon [0, \infty) \to [0, \infty)$ which are strictly increasing, continuous, unbounded, and 0 at 0. For $\Psi \in \mathcal{CV}$ and $0 < T < \infty$, recall the $\Psi$-variation $v_\Psi(f) = v_\Psi(f; J)$ as defined in (3.1), which will be applied for $f = f_X$ and $J = [0, T]$.

By Lemma 3.79(a), for $1 < p < \infty$ there is a $\delta$ with $0 < \delta \leq \mathrm{e}^{-\mathrm{e}}$ such that

$$\Psi_{p,2}(u) := [u(\log\log(1/u))^{1/2}]^p \tag{12.14}$$

is convex for $0 < u \leq \delta$. By (3.125) we can set $\Psi_{p,2}(0) := 0$ and extend $\Psi_{p,2}$ to be linear on $[\delta, +\infty)$ in such a way that $\Psi_{p,2}$ is convex on $[0, \infty)$ and so in $\mathcal{CV}$. In Theorem 12.13 we will show that for $1 < p < \infty$, $X$ has almost all sample functions of bounded $p$-variation on $[0, T]$, assuming boundedness of $v_{\Psi_{p,2}}(f_X; [0, T])$.

**Proposition 12.10.** *Let $X$ be a separable Gaussian stochastic process with mean 0, and let $1 \leq p < \infty$. If $X$ has locally bounded $p$-variation then $v_p(f_X; [0, T]) < \infty$ for each $0 < T < \infty$. The converse implication holds for $p = 1$.*

*Proof.* First let $p = 1$. It will be shown that $X$ has almost all sample functions of bounded 1-variation on $[0, T]$ if and only if

$$v_1(f_X; [0, T]) < \infty. \tag{12.15}$$

Indeed, suppose (12.15) holds. By separability of $X$, there exists a nested sequence $\{\kappa_n\}_{n \geq 1}$ of partitions of $[0, T]$ formed by points of a separating set $S$ such that $v_1(X; [0, T]) = \sup_{n \geq 1} s_1(X; \kappa_n)$ almost surely and $\bigcup_n \kappa_n = S$. Note that for any function $f$ and partitions $\kappa_n \subset \kappa_{n+1}$, $s_1(f; \kappa_n) \leq s_1(f; \kappa_{n+1})$. By the monotone convergence theorem,

$$E \sup_{n \geq 1} s_1(X; \kappa_n) = \sup_{n \geq 1} E s_1(X; \kappa_n) = K_1 v_1(f_X; [0, T]) < \infty, \tag{12.16}$$

where $K_1 := E|Z|$ for a standard normal random variable $Z$, i.e. one having distribution $N(0,1)$, and so $v_1(X;[0,T]) < \infty$ almost surely. Conversely, let $V := \mathbb{R}^S$ be the set of all real-valued functions on a separating set $S \subset [0,T]$ with product topology. Then $V$ is a measurable vector space and the distribution of $X$ restricted to $S$ is a centered Gaussian measure $P$ on $V$, both as defined on p. 25 of [52]. Now $\|\cdot\| := \|\cdot\|_{(1),[0,T]}$ is a measurable function from $V$ to $[0,\infty]$ and by assumption, we have $P(\|\cdot\| < \infty) = 1$. Therefore $\int \|x\|\, dP(x) < \infty$, in fact $\int \exp\{\alpha\|x\|^2\}\, dP(x) < \infty$ for some $\alpha > 0$, by a theorem of Fernique [52, Lemma 2.2.5]. Thus (12.15) holds by (12.16).

Now let $1 < p < \infty$ and suppose $X$ has almost all sample functions of bounded $p$-variation. Then by the same theorem of Fernique applied to the seminorm $\|\cdot\| = \|\cdot\|_{(p),[0,T]}$, we have

$$\infty > E\|X\|^p \geq \sup\big\{Es_p(X;\kappa)\colon\ \kappa \in PP[0,T]\big\} = K_p v_p(f_X;[0,T]),$$

where $K_p := E|Z|^p$ for a standard normal random variable $Z$. Thus the proposition holds for $1 < p < \infty$. □

The converse in Proposition 12.10 does not hold for $p > 1$, however, as (12.11) with $p = 2$ shows for a Brownian motion.

By Lemma 3.79(a), for $1 < p < \infty$ there is a $\delta$ with $0 < \delta \leq e^{-e}$ such that $\Psi_{p,1}(u) := [u(\log(1/u))^{1/2}]^p$ is convex for $0 < u \leq \delta$. By (3.125) we can set $\Psi_{p,1}(0) := 0$ and extend $\Psi_{p,1}$ to be linear on $[\delta,+\infty)$ in such a way that $\Psi_{p,1}$ is convex on $[0,\infty)$ and so in $\mathcal{CV}$. Letting $\Psi_p(u) \equiv u^p$, it is clear that $\Psi_p \leq \Psi_{p,2} \leq \Psi_{p,1}$ on $[0,\delta]$ for some $\delta > 0$. The following shows that $v_{\Psi_{p,1}}(f_X) < \infty$ is a sufficient condition for a Gaussian process $X$ to have locally bounded $p$-variation.

**Lemma 12.11.** *Let $X = \{X_t\}_{t\geq 0}$ be a separable Gaussian stochastic process with mean zero such that $v_p(f_X;[0,T]) < \infty$ for some $1 < p < \infty$ and $0 < T < \infty$. Then for almost all $\omega \in \Omega$ there is a $C(\omega) < +\infty$ such that for $s,t \in [0,T]$,*

$$|X(t,\omega) - X(s,\omega)| \leq C(\omega)\big[\Psi_{p,1}(d(f_X(t), f_X(s)))\big]^{1/p}, \tag{12.17}$$

*where $d$ is the usual metric on the Hilbert space $H = L^2(\Omega,\Pr)$.*

*Proof.* Let $0 < T < \infty$ and $1 < p < \infty$ be such that $v_p(f_X;[0,T]) < \infty$. Let $R := \{f_X(t)\}_{0\leq t\leq T} \subset L^2(\Omega,\Pr)$. For each $\epsilon > 0$, let $N(R,\epsilon)$ be the minimal number of balls of radius at most $\epsilon$ with respect to $d$ that cover $R$. Then for each $0 < t \leq T$ let

$$F(t) := \sup\Big\{\sum_{i=1}^n E|X(t_i) - X(t_{i-1})|^p\colon\ \{t_i\}_{i=0}^n \in \mathrm{PP}\,[0,t]\Big\} = K_p v_p(f_X;[0,t]).$$

Thus for each $0 \le s < t \le T$,

$$K_p^{1/p} d(f_X(s), f_X(t)) = \left(E|X(t) - X(s)|^p\right)^{1/p} \le \left[F(t) - F(s)\right]^{1/p},$$

and so there is a finite constant $A$ such that $N(R, \epsilon) \le A\epsilon^{-p}$. By Corollary 2.3 of [47], for almost all $\omega$ there exists a finite $C(\omega)$ such that (12.17) holds. □

For $1 \le p < \infty$, let $\psi_{p,2}(v) := v^{1/p}/\sqrt{\log\log(1/v)}$ if $0 < v \le \mathrm{e}^{-\mathrm{e}}$, $\psi_{p,2}(v) := v^{1/p}$ if $v > \mathrm{e}^{-\mathrm{e}}$ and $\psi_{p,2}(0) := 0$. For $p > 1$, by Lemma 3.79, there is a $\delta' = \delta'(p) > 0$ such that $\psi_{p,2}$ is concave on $[0, \delta']$.

**Lemma 12.12.** *Let $1 < p < \infty$, and for some $0 < T < \infty$ let $X = \{X_t\}_{0 \le t \le T}$ be a separable Gaussian stochastic process with mean zero. Suppose that there are a nondecreasing function $F$ on $[0, T]$ with $F(0) = 0$, a finite constant $\gamma$, and a $\delta$ with $0 < \delta < \mathrm{e}^{-\mathrm{e}}$ such that $\psi_{p,2}$ is concave on $[0, \delta]$ and*

$$\sigma_X(s, t) := [E((f_X(s) - f_X(t))^2)]^{1/2} \le \gamma\psi_{p,2}(F(t) - F(s)) \tag{12.18}$$

*whenever $s, t \in [0, T]$ and $F(t) - F(s) \le \delta$. Then there is a finite constant $K$ such that for any $k > 0$, and any $a, b \in \mathbb{R}$ with $0 < b - a < \delta$ and $(a, b) \cap [0, F(T)] \ne \emptyset$,*

$$\Pr\left(\left\{\sup_{s,t \in I(a,b)} |X(t) - X(s)| > 2(K + \gamma\sqrt{2k})(b - a)^{1/p}\right\}\right) \le \left[\log\frac{1}{b - a}\right]^{-k},$$

*where $I(a, b) := \{t \in [0, T]\colon\ a < F(t) < b\}$.*

*Proof.* Let $k > 0$ and let $a, b \in \mathbb{R}$ be such that $0 < b - a < \delta$ and $(a, b) \cap [0, F(T)] \ne \emptyset$. Let $R := (a, b) \cap \{F(t)\colon\ 0 \le t \le T\}$. If $R = \emptyset$ the conclusion holds vacuously, so suppose $R \ne \emptyset$. Let $V := (a, b) \setminus R$. We will show that there exists a separable, Gaussian stochastic process with mean zero $Y = \{Y_u\}_{a<u<b}$ such that $X(t, \omega) = Y(F(t), \omega)$ for all $t \in I(a, b)$ for almost all $\omega \in \Omega$ and for each $u, v \in (a, b)$,

$$\sigma_Y(u, v) \le \gamma\psi_{p,2}(|u - v|) \le \gamma|u - v|^{1/p}. \tag{12.19}$$

First we will show that $V = \cup_{i=1}^{\infty} V_i$, where all $V_i$ are disjoint open or half-open intervals. Indeed, if $F(t) = c \in (a, b)$ for some $t \in [0, T]$, $J = (a, c)$ or $J = (c, b)$, and $J \subset V$, then $J$ is clearly a maximal interval included in $V$. If $F(t) < F(t+)$ for some $t \in [0, T)$ and $F(t+) \in (a, b)$, then either the open interval $(a \vee F(t), F(t+))$ or the half-open interval $(a \vee F(t), F(t+)]$ is a maximal interval included in $V$, depending on whether $F(t+) \in R$ or $F(t+) \in V$. Similarly, if $F(t-) < F(t)$ for some $t \in (0, T]$ and $F(t-) \in (a, b)$, then either $(F(t-), F(t) \wedge b)$ or $[F(t-), F(t) \wedge b)$ is a maximal interval included in $V$. To define $Y$, let $Y(u) := X(t)$ if $u \in R$ and $u = F(t)$ for some $t$. Then the set $\{t\colon F(t) = u\}$, if not empty, is a singleton or a nondegenerate interval, $I_u$, in which case $X$ is almost surely constant on $I_u$ by (12.18) and separability. Since there are at most countably many such intervals the almost sure constancy holds for all of them.

For each $t$ such that $F(t) < F(t+) \in (a,b)$ there is an $\epsilon > 0$ such that $F(s) < F(t+) + \delta$ for $t < s < t + \epsilon$, and so $X(t+)$, defined as the limit in $L^2(\Omega, \Pr)$ of $X(s)$ as $s \downarrow t$, exists by (12.18). Similarly, if $F(t) > F(t-) \in (a,b)$, we can define $X(t-)$. Now, if $u \in (a,b)$ is a right endpoint of $V_i$ for some $i$, then either $u = F(t+)$ for some $t$ and we let $Y(u) := X(t+)$, or $u = F(t)$ and $Y(u)$ is already defined. Similarly, if $u \in (a,b)$ and $u = F(t-)$ is a left endpoint of $V_i$, then let $Y(u) := X(t-)$. On the interior of $V_i$ the paths of $Y$ are defined either by linearity if both endpoints of $V_i$ are in $(a,b)$ or equal to the value of $Y$ at the one endpoint in $(a,b)$.

The stochastic process $Y = \{Y_u\}_{a<u<b}$ is Gaussian with mean zero, and $X(\cdot,\omega) = Y(F(\cdot),\omega)$ on $I(a,b)$ for almost all $\omega \in \Omega$. Moreover, $Y$ is separable with separating set $(F(S) \cap (a,b)) \cup S_1$, where $S$ is a separating set for $X$ and $S_1$ is any countable dense subset of $V$.

To show (12.19) for $u,v \in (a,b)$ note that it holds by (12.18) if $u,v \in R$. If $u,v \in V_i$ and $a$ or $b$ is an endpoint of $V_i$ there is no problem. Otherwise if $u,v \in V_i = (u_1,u_2]$ for some $u_1,u_2 \in (a,b)$ and $\lambda := (v-u)/(u_2-u_1) > 0$, then since $\psi_{p,2}$ is concave on $[0,\delta]$,

$$\sigma_Y(u,v) = \lambda\sigma_Y(u_1,u_2) \le \gamma\lambda\psi_{p,2}(u_2-u_1) \le \gamma\psi_{p,2}(v-u).$$

Now let $v \in V_i = (u_1,u_2]$ with $Y(u_2) = X(t+)$, $Y(u_1) = X(t)$, and $u = F(s) \in (a,u_1)$ for some $s,t \in I(a,b)$. Letting $\lambda := (v-u_1)/(u_2-u_1)$, we have

$$Y(v) - Y(u) = (1-\lambda)[X(t) - X(s)] + \lambda[X(t+) - X(s)].$$

Thus by concavity of $\psi_{p,2}$ again it follows that

$$\begin{aligned}\sigma_Y(u,v) &\le (1-\lambda)\sigma_X(s,t) + \lambda\limsup_{r\downarrow t}\sigma_X(s,r)\\ &\le \gamma\Big[(1-\lambda)\psi_{p,2}(u_1-u) + \lambda\psi_{p,2}(u_2-u)\Big] \le \gamma\psi_{p,2}(v-u).\end{aligned}$$

Similarly, one can also check that (12.19) holds for any $u,v \in (a,b)$. This finishes the proof of existence of $Y$ with the stated properties.

Let $c := (a+b)/2$. Let $Z(u) := Y(u) - Y(c)$ for $u \in (a,b)$ and let $\|Z\|_{\sup} := \sup_{u\in(a,b)} |Z(u)|$. Endowed with the pseudometric $\sigma_Y$, $(a,b)$ is a totally bounded pseudometric space. Denote by $N(r) = N((a,b),\sigma_Y,r)$ the smallest number of open balls of radius $r > 0$ in the pseudometric $\sigma_Y$ which form a covering of $(a,b)$. By (12.19), $N(r) \le 1 + (b-a)\gamma^p/r^p$ for $0 < r \le \bar\sigma_Y(a,b) := \sup\{\sigma_Y(u,v)\colon u,v \in (a,b)\}$. Clearly $N(r) = 1$ if $r > \bar\sigma_Y(a,b)$. Then by the change of variables $s = (r/\gamma)^p(b-a)^{-1}$, where $N(r) = 1$ for $s > 1$, we get

$$\int_0^\infty \sqrt{\log N(r)}\,\mathrm{d}r \leq p^{-1}\gamma(b-a)^{1/p}\int_0^1 s^{(1/p)-1}\sqrt{\log(1+(1/s))}\,\mathrm{d}s < \infty.$$

It follows that the sample functions of the process $\{Z(u): \ a < u < b\}$ can be taken to be in the separable Banach space $E$ of all real-valued uniformly continuous functions on $(a,b)$ for $\sigma_Y$, e.g. [135, Theorems 6.1 and 6.2] or [52, Theorem 2.6.1], with supremum norm. Because $Z(c)=0$, we have

$$\|Z\|_{\sup} = \sup_{u\in(a,b)}|Z(u)| \leq \sup_{u\in(a,b)} Z(u) + \sup_{u\in(a,b)} -Z(u).$$

By the metric entropy bound for Gaussian processes in expectation (cf. e.g. [135, Theorem 6.1]), we have

$$E\|Z\|_{\sup} \leq 84\int_0^\infty \sqrt{\log N(r)}\,\mathrm{d}r \leq K(b-a)^{1/p}$$

for some finite constant $K$ depending only on $p$ and $\gamma$. By (12.19), we have

$$\sup_{u\in(a,b)} (EZ(u)^2)^{1/2} \leq \gamma\psi_{p,2}(b-a).$$

Then by the Gaussian concentration inequality of Ledoux [135, (4.4)], it follows that

$$\begin{aligned}
&\Pr\big(\{\sup_{s,t\in I(a,b)}|X(t)-X(s)| > 2(K+\gamma\sqrt{2k})(b-a)^{1/p}\}\big)\\
&\leq \Pr\big(\{\|Z\|_{\sup} > (K+\gamma\sqrt{2k})(b-a)^{1/p}\}\big)\\
&\leq \Pr\big(\{\|Z\|_{\sup} > E\|Z\|_{\sup} + \gamma\sqrt{2k}(b-a)^{1/p}\}\big)\\
&\leq \exp\Big\{-\frac{2k\gamma^2(b-a)^{2/p}}{2\gamma^2\psi_{p,2}^2(b-a)}\Big\} = \Big[\log\frac{1}{b-a}\Big]^{-k},
\end{aligned}$$

proving the lemma. □

Now we are ready to prove the main result of this section, which gives a sufficient condition for boundedness of the $p$-variation of almost all sample functions of a Gaussian stochastic process. Proposition 12.14 will show that in some cases the condition is necessary.

**Theorem 12.13.** *Let $X$ be a separable Gaussian stochastic process with mean zero, and let $1 < p < \infty$. If $v_{\Psi_{p,2}}(f_X;[0,T]) < \infty$ for each $0 < T < \infty$, then $X$ has locally bounded $p$-variation.*

*Proof.* Let $0 < T < \infty$ and for each $0 \leq t \leq T$, let $F(t) := v_{\Psi_{p,2}}(f_X;[0,t])$. Recalling $\psi_{p,2}$ as defined before Lemma 12.12, note that we have

$$\Psi_{p,2}^{-1}(v) \sim \psi_{p,2}(v) \tag{12.20}$$

as $v \downarrow 0$. This and Lemma 3.79 yield that there is a $\delta > 0$ such that $\psi_{p,2}$ is concave on $[0, \delta]$ and for $0 \le s < t \le T$,

$$\sigma_X(s,t) \le \mathrm{e}\psi_{p,2}\big(F(t) - F(s)\big) \le \mathrm{e}\big[F(t) - F(s)\big]^{1/p} \tag{12.21}$$

if $F(t) - F(s) \le \delta$. Also we assume that $\delta$ is so small that $\delta \le (\delta_1/\mathrm{e})^p$ where $\delta_1$ is such that $\Psi_{p,1}(u) = [u(\log(1/u))^{1/2}]^p$ for $0 < u \le \delta_1$ as defined before Lemma 12.11.

Let $m_0 \ge 4$ be the smallest integer such that $2\mathrm{e}^{-m_0} < \delta$, and let $K$ be the constant obtained from Lemma 12.12 with $\gamma = \mathrm{e}$. For each $m \ge m_0$ and each $j = 0, 1, \ldots, j_m := \lfloor \mathrm{e}^m F(T) \rfloor$, let $S_{m,j} := I((j-1)\mathrm{e}^{-m}, (j+1)\mathrm{e}^{-m})$, where $I(a,b) = \{t \in [0,T]\colon\ a < F(t) < b\}$, and let

$$V_{m,j} := \big\{\omega \in \Omega\colon \sup_{t,s \in S_{m,j}} |X(t,\omega) - X(s,\omega)| > Me^{-m/p}\big\},$$

where $M := 2^{1+1/p}(K + \mathrm{e}\sqrt{2k})$ and $k$ is such that $k > 2 + p/2$. The intervals $S_{m,j}$, $j = 0, \ldots, j_m$, overlap and cover $[0,T]$. We shall bound the number $Z_m(\omega) := \mathrm{card}\big\{j = 0, \ldots, j_m\colon\ \omega \in V_{m,j}\big\}$. By Lemma 12.12 with $a = (j-1)\mathrm{e}^{-m}$ and $b = (j+1)\mathrm{e}^{-m}$, we have $\Pr\big(V_{m,j}\big) \le 2^k m^{-k}$ for all $j = 0, \ldots, j_m$ and all $m \ge m_0$. Thus for all such $m$, it follows that

$$EZ_m \le \sum_{j=0}^{j_m} \Pr\big(\{\sup_{t,s \in S_{m,j}} |X(t) - X(s)| > M\mathrm{e}^{-m/p}\}\big) \le (1 + F(T))\mathrm{e}^m 2^k m^{-k}.$$

We choose $\nu$ such that $2 + p/2 < \nu + 1 < k$. Then we have

$$\sum_m \Pr\big(\{Z_m > e^m m^{-\nu}\}\big) \le \sum_m EZ_m e^{-m} m^{\nu} \le (1 + F(T))2^k \sum_m m^{-(k-\nu)} < \infty.$$

By the Borel–Cantelli lemma, for almost all $\omega \in \Omega$, there exists an integer $m_1(\omega) \ge m_0$ such that $Z_m(\omega) \le \mathrm{e}^m m^{-\nu}$ for all $m \ge m_1(\omega)$.

Since $F$ is regulated on $[0,T]$, by Theorem 2.1, there exists a Young interval partition $\{(z_{j-1}, z_j)\}_{j=1}^m$ of $[0,T]$ such that the oscillation of $F$ on each $(z_{j-1}, z_j)$ is $< \mathrm{e}^{-m_0}$. By Proposition 3.36, it suffices to prove the bounded $p$-variation of paths of $X$ on each interval $(z_{j-1}, z_j)$, or, on any open interval $(u,v) \subset [0,T]$ on which the oscillation of $F$, namely $F(v-) - F(u+)$, is less than $\mathrm{e}^{-m_0}$. Then, $X'(u) := X(u+)$ and $X'(v) := X(v-)$ exist as limits in quadratic mean. Defining $X'(t) := X(t)$ for $u < t < v$ we get a process $X'$ on the closed interval $[u,v]$. It suffices to prove the almost sure finite $p$-variation of paths of $X'$ on $[u,v]$. Thus, without loss of generality, it suffices to prove this for $X$ on $[u,v]$ assuming that $F(v) - F(u) < \mathrm{e}^{-m_0}$.

Let $\kappa = \{t_i\}_{i=0}^n$ be a partition of $[u,v]$, and for $\omega \in \Omega$, let

$$E(\omega) := \big\{i\colon\ |X(t_i,\omega) - X(t_{i-1},\omega)|^p > \mathrm{e}M^p[F(t_i) - F(t_{i-1})]\big\}.$$

For $m \geq m_0$, let

$$\Lambda_m := \{i\colon \mathrm{e}^{-m-1} \leq F(t_i) - F(t_{i-1}) < \mathrm{e}^{-m}\}.$$

Then

$$\sum_{i \notin E(\omega)} \big|X(t_i,\omega) - X(t_{i-1},\omega)\big|^p < \mathrm{e}^{1-m_0} M^p. \tag{12.22}$$

Since $v_p(f_X) \leq v_{\Psi_{p,2}}(f_X) < \infty$, by Lemma 12.11, for almost all $\omega \in \Omega$, there exists a $C = C(\omega) < \infty$ such that (12.17) holds. Moreover, $\mathrm{card}\{E(\omega) \cap \Lambda_m\} \leq 5Z_m(\omega)$. Since for each $i$, $F(t_i) - F(t_{i-1}) < \mathrm{e}^{-m_0}$, there is an $m \geq m_0$ such that $i \in \Lambda_m$. For $\delta_1$ mentioned near the beginning of the proof, since $\delta \leq (\delta_1/\mathrm{e})^p$, by (12.21) we have $u_i := \sigma_X(t_{i-1}, t_i) \leq \delta_1$, and in (12.17) we have $\Psi_{p,1}(u_i)$ for $0 < u_i \leq \delta_1$ and the formula for $\Psi_{p,1}(u)$ for small $u > 0$ applies. Moreover, for any $m \geq m_0$ and $i \in \Lambda_m$ we have $\sigma_X(t_{i-1}, t_i) \leq \delta_1 \mathrm{e}^{-(m-m_0)/p}$. Recalling that $d(f_X(t), f_X(s)) = \sigma_X(s,t)$, we have

$$\begin{aligned}
&\sum_{i \in E(\omega)} \big|X(t_i,\omega) - X(t_{i-1},\omega)\big|^p \\
&\quad = \sum_{m=m_0}^{\infty} \sum_{i \in E(\omega) \cap \Lambda_m} \big|X(t_i,\omega) - X(t_{i-1},\omega)\big|^p \\
&\quad \leq C(\omega)^p \sum_{m=m_0}^{\infty} \sum_{i \in E(\omega) \cap \Lambda_m} \Big[\sigma_X(t_{i-1},t_i)\sqrt{\log(1/\sigma_X(t_{i-1},t_i))}\Big]^p \\
&\quad \leq 5\delta_1^p C(\omega)^p \sum_{m=m_0}^{\infty} Z_m(\omega) \mathrm{e}^{-m+m_0} \Big[\log(1/\delta_1) + (m-m_0)/p\Big]^{p/2}.
\end{aligned}$$

The last series converges because $Z_m(\omega) \leq \mathrm{e}^m m^{-\nu}$ for all $m \geq m_1(\omega) \geq m_0$, and because $\sum_{m \geq m_1(\omega)} m^{p/2-\nu} < \infty$ by the choice of $\nu$. This together with (12.22) gives a bound on $s_p(X(\cdot,\omega);\kappa)$ for almost all $\omega \in \Omega$, proving the theorem. □

The next fact shows that boundedness of $v_{\Psi_{p,2}}(f_X;[0,1])$ in the preceding theorem is necessary in some cases.

**Proposition 12.14.** *Let $1 < p < \infty$, let $M > 1$ be an even integer such that $M^{1-(1/p)} > 4$, let $\{\xi_k\}_{k \geq 1}$ be independent identically distributed $N(0,1)$ random variables on $(\Omega, \mathcal{F}, \Pr)$, let $h = \{h_k\}_{k \geq 1}$ be a nondecreasing sequence of positive numbers such that (3.66) holds, and for $0 \leq t \leq 1$ and $\omega \in \Omega$ let*

$$X(t,\omega) := \sum_{k=1}^{\infty} \frac{\xi_k(\omega)}{\sqrt{\log(k+1)}} h_k M^{-k/p} \sin(2\pi M^k t). \tag{12.23}$$

*Then $X = \{X_t\}_{0 \leq t \leq 1}$ is a Gaussian stochastic process with mean zero having almost all sample functions continuous, and the following three statements are equivalent:*

(*a*) $v_p(X(\cdot,\omega);[0,1]) < \infty$ *for almost all* $\omega \in \Omega$;
(*b*) $\|h\|_{\sup} = \sup_{k\geq 1} h_k < \infty$;
(*c*) $v_{\Psi_{p,2}}(f_X;[0,1]) < \infty$.

*Proof.* By the Borel–Cantelli lemma, $\Pr(\cup_{n\geq 1}\cap_{k\geq n}\{|\xi_k| \leq 2\sqrt{\log(k+1)}\}) = 1$, and so $\sup_{k\geq 1}|\xi_k(\omega)|/\sqrt{\log(k+1)} < \infty$ for almost all $\omega \in \Omega$. Thus by (3.66), for almost all $\omega \in \Omega$, the series (12.23) converges uniformly for $t \in [0,1]$, and so $X$ is a well-defined Gaussian stochastic process with mean zero having almost all sample functions continuous.

(*a*) $\Rightarrow$ (*b*). Suppose that $\|h\|_{\sup} = \infty$. By (*a*) and Proposition 3.46, for almost all $\omega \in \Omega$, $\sup_{k\geq 1} h_k|\xi_k(\omega)|/\sqrt{\log(k+1)} < \infty$. Since $\{h_k\}_{k\geq 1}$ is nondecreasing, $|\xi_k|/\sqrt{\log(k+1)} \to 0$ as $k \to \infty$ with probability 1. On the other hand, by the Borel–Cantelli lemma, $\Pr(\cap_{n\geq 1}\cup_{k\geq n}\{|\xi_k| > \sqrt{\log(k+1)}\}) = 1$, a contradiction, proving (*b*).

(*b*) $\Rightarrow$ (*c*). It will be shown that there exist a $\delta > 0$ and a finite constant $C$ such that

$$\sigma_X(s,t) \leq C\psi_{p,2}(|t-s|) \qquad \text{if } s,t \in [0,1] \text{ and } |s-t| < \delta, \tag{12.24}$$

where $\sigma_X$ is defined in (12.18) and $\psi_{p,2}$ is defined before Lemma 12.12. To show that (12.24) suffices, by definition (12.14) of $\Psi_{p,2}$, if $K > 1$ and $0 < u \leq \delta'/K$ for some $0 < \delta' \leq \mathrm{e}^{-\mathrm{e}}$, we have

$$\Psi_{p,2}(Ku) = K^p \frac{(\log\log[1/(Ku)])^{p/2}}{(\log\log(1/u))^{p/2}}\Psi_{p,2}(u) \leq K^p\Psi_{p,2}(u)$$

since $\log\log$ is a positive and nondecreasing function on $[\mathrm{e}^{\mathrm{e}},\infty)$. We can assume that $\delta$ in (12.24) is such that $\psi_{u,2}(u) \leq 2\Psi_{p,2}^{-1}(u)$ and $\Psi_{p,2}^{-1}(u) \leq \delta'/K$ for $0 < u < \delta$ and $K = \max\{1, 2C\}$ by (12.20) and continuity of $\Psi_{p,2}^{-1}$ at 0. Therefore if $s,t \in [0,1]$ and $|s-t| < \delta$, then

$$\Psi_{p,2}(\sigma_X(s,t)) \leq \Psi_{p,2}(2C\Psi_{p,2}^{-1}(|s-t|)) \leq \max\{1,(2C)^p\}|s-t|,$$

so indeed (12.24) suffices. To prove (12.24), for $s,t \in [0,1]$, we have

$$\sigma_X^2(s,t) = \sum_{k=1}^{\infty} \frac{h_k^2 M^{-2k/p}}{\log(k+1)}\big[\sin(2\pi M^k t) - \sin(2\pi M^k s)\big]^2.$$

If $|t-s| < 1/M$ let $k_0 = k_0(|t-s|) \geq 1$ be the maximal integer such that $M^{k_0} \leq 1/|t-s|$ and let $k_0 = 1$ otherwise. Then as in the proof of Lemma 3.47, it follows that $\sigma_X^2(s,t) \leq 4\|h\|_{\sup}^2[\pi^2(t-s)^2 S(k_0) + T(k_0)]$, where letting $q := 1-(1/p)$,

$$S(k_0) := \sum_{k=1}^{k_0} \frac{M^{2kq}}{\log(k+1)} \quad \text{and} \quad T(k_0) := \sum_{k>k_0} \frac{M^{-2k/p}}{\log(k+1)}. \tag{12.25}$$

For $x \ge 1$, let $L(x) := 1/\log(2 \vee x)$. Then changing variables in the integral, we get the bound

$$S(k_0) \le \int_1^{k_0+1} L(x) M^{2qx}\,\mathrm{d}x = \frac{1}{\log M} \int_M^{M^{k_0+1}} L\Big(\frac{\log y}{\log M}\Big) y^{2q-1} \mathrm{d}y =: \tilde{S}(k_0).$$

It is easily seen that $L((\log y)/\log M) \sim 1/(\log\log y)$ as $y \to +\infty$. Also, $L(x) \le 1/\log 2$ for all $x$. So since $p > 1$, as $k_0 \to \infty$, we have (cf. Karamata's theorem, e.g. [15, Proposition 1.5.8])

$$\tilde{S}(k_0) \sim \frac{M^{2q(k_0+1)}}{2q \log M} L(k_0+1) \le \frac{M^{2q}}{q \log M} \frac{|t-s|^{-2q}}{\log\log(1/|t-s|)},$$

where the last inequality holds by the definitions of $L$ and $k_0$, provided $M \le k_0 + 1$. For the second sum in (12.25), we have

$$T(k_0) \le M^{2/p} \int_{k_0+1}^{\infty} \frac{L(x)\,\mathrm{d}x}{M^{2x/p}} = \frac{M^{2/p}}{\log M} \int_{M^{k_0+1}}^{\infty} L\Big(\frac{\log y}{\log M}\Big) \frac{\mathrm{d}y}{y^{1+2/p}} =: \tilde{T}(k_0).$$

By monotonicity of the logarithm and by the definition of $k_0$, it follows that

$$\tilde{T}(k_0) \le \frac{pM^{2/p}}{2\log M} M^{-(k_0+1)(2/p)} L(k_0+1) \le \frac{pM^{2/p}}{\log M} \frac{|t-s|^{2/p}}{\log\log(1/|t-s|)}$$

provided $M \le k_0 + 1$. Because $k_0 \to \infty$ as $|t-s| \downarrow 0$, there exist $\delta > 0$ and a finite constant $C$ depending on $M$, $p$, and $\|h\|_{\sup}$ such that (12.24) holds, proving (*c*). Since (*a*) follows from (*c*) by Theorem 12.13, the proof of the proposition is complete. □

For a function $f$ on $[a,b]$, define the *index of $p$-variation* by

$$\upsilon(f;[a,b]) := \begin{cases} \inf\{p > 0\colon\ v_p(f) < \infty\} & \text{if the set is nonempty,} \\ +\infty & \text{otherwise.} \end{cases}$$

Let $X$ be a separable mean-zero Gaussian stochastic process. If for some $1 < p < \infty$, $v_p(f_X) < \infty$ and if $p' > p$, then $v_{\Psi_{p',2}}(f_X) < \infty$. Thus by Theorem 12.13, $X$ has locally bounded $p'$-variation. On the other hand, if for some $1 < p < \infty$, $X$ has locally bounded $p$-variation then $v_p(f_X) < \infty$ by Proposition 12.10. These considerations yield the following fact.

**Corollary 12.15.** *Let $X$ be a separable mean-zero Gaussian stochastic process and let $0 < T < \infty$. Then with probability* 1

$$\upsilon(X;[0,T]) = \inf\{p > 0\colon\ v_p(f_X;[0,T]) < \infty\}.$$

*Fractional Brownian motion*

A fractional Brownian motion with Hurst index $H \in (0,1)$ is a mean-zero Gaussian stochastic process $B_H = \{B_H(t),\ t \geq 0\}$ with the covariance function

$$E\big[B_H(t)B_H(s)\big] = \frac{1}{2}\Big[t^{2H} + s^{2H} - |t-s|^{2H}\Big] \qquad \text{for } t,s \geq 0. \tag{12.26}$$

**Proposition 12.16.** *For each $H$ with $0 < H < 1$, (12.26) does define a (nonnegative definite) covariance, and fractional Brownian motion exists. Moreover, it can be taken to be sample-continuous.*

*Proof.* Given $H$, there is an isometric embedding $f_H$ of the half-line $[0,\infty)$ with the metric $d_H(s,t) := |s-t|^H$ into a Hilbert space $\mathcal{H}$, as shown by Schoenberg [209, Theorem 1], or more generally [210, Corollary 1]. We can assume that $f_H(0) = 0 \in \mathcal{H}$. Then for each $s,t \geq 0$ we have for the norm $\|\cdot\|$ on $\mathcal{H}$ that $\|f_H(t)\|^2 = t^{2H}$ and

$$|t-s|^{2H} \;=\; \|f_H(t) - f_H(s)\|^2 \;=\; t^{2H} + s^{2H} - 2(f_H(s), f_H(t))$$

where $(\cdot,\cdot)$ is the inner product in $\mathcal{H}$. Then the expression in (12.26) equals $(f_H(s), f_H(t))$ and so it gives a covariance. Fractional Brownian motions thus exist by the general theorem on existence of Gaussian processes (with mean 0), e.g. [53, Theorem 12.1.3]. They can be taken to be sample-continuous by the metric entropy sufficient condition, e.g. [52, Theorem 2.6.1]. □

**Remark 12.17.** For $H = 1$ the proposition still holds but the resulting process is a trivial one with $B_1(t) \equiv tZ$ for one standard normal random variable $Z$. For $H > 1$, (12.26) does not define a covariance, because $[E(B_H(t) - B_H(s))^2]^{1/2}$ would equal $|s-t|^H$, which is not a metric on $[0,\infty)$.

Since the right side of (12.26) is equal to $t \wedge s$ for $H = 1/2$, $B_H$ is a Brownian motion in this case. It follows from (12.26) that for each $t,u \geq 0$,

$$\sigma_H(u) := \sigma_{B_H}(t,t+u) = \Big[E\big[B_H(t+u) - B_H(t)\big]^2\Big]^{1/2} = u^H.$$

For the $p$-variation of fractional Brownian motions we have the following. Let $0 < H < 1$ and let $1 < p < \infty$. Then for any $0 < T < \infty$,

$$v_p(f_{B_H};[0,T]) = \begin{cases} +\infty & \text{if } pH < 1, \\ T^{pH} & \text{if } pH \geq 1. \end{cases}$$

Thus if a fractional Brownian motion $B_H$ is of locally bounded $p$-variation then $p \geq 1/H$ by Proposition 12.10. More precisely, by Corollary 12.15, the $p$-variation index of a fractional Brownian motion $B_H$ is given by

$$\upsilon(B_H) = 1/H \quad \text{with probability 1.}$$

Corollary 12.23 below will show that $v_p(B_H) = +\infty$ almost surely for $p = 1/H$. It will also be proved in Corollary 12.23 that a fractional Brownian motion has locally bounded $\Phi$-variation for a suitable $\Phi \in \mathcal{CV}$ which is best possible.

Let $H \in (0,1)$. By Lemma 3.79(a) there is a $\delta$ with $0 < \delta \le \mathrm{e}^{-\mathrm{e}}$ such that $\Phi_H(u) := \left[u/\sqrt{2\log\log(1/u)}\right]^{1/H}$ is convex for $0 < u \le \delta$. By (3.125) we can set $\Phi_H(0) := 0$ and extend $\Phi_H$ to be linear on $[\delta, +\infty)$ in such a way that $\Phi_H$ is convex on $[0,\infty)$ and so in $\mathcal{CV}$. Moreover, the function $\Phi_H$ is regularly varying near zero of order $1/H$, that is, for each $\lambda > 0$,

$$\lim_{u\downarrow 0} \Phi_H(\lambda u)/\Phi_H(u) = \lambda^{1/H}. \tag{12.27}$$

Let $\phi_H(v) := v^H\sqrt{2\log\log(1/v)}$ for $0 < v \le \mathrm{e}^{-\mathrm{e}}$ and $\phi_H(0) := 0$. Then we have for the inverse function $\Phi_H^{-1}(v) \sim \phi_H(v)$ as $v \downarrow 0$. It will be proved that a fractional Brownian motion with Hurst index $H$ has locally bounded $\Phi_H$-variation and for any $\Phi \in \mathcal{V}$ such that $\Phi_H(u) = o(\Phi(u))$ as $u \downarrow 0$, the $\Phi$-variation of a sample function of such a process is infinite almost surely. We begin with preliminary results.

We will use the following probability bound, which will be shown to be a simple consequence of the Landau–Shepp–Marcus–Fernique theorem.

**Lemma 12.18.** *Let $S$ be a compact set in a Banach space such that $cS \subset S$ for each $c \in (0,1]$. Let $X = \{X_t\}_{t\in S}$ be a mean-zero sample-continuous Gaussian stochastic process such that for some $0 < H < 1$ and each $c \in (0,1]$, $\{X(ct)\}_{t\in S}$ is equal in distribution to $c^H X$. Let $S(1)$ be a closed subset of $S$ and define $S(\delta) := \delta S(1)$ for $0 < \delta < 1$. For each $\delta \in (0,1]$, let $\sigma_\delta^2 := \sup\{EX(t)^2 \colon t \in S(\delta)\}$. Then $\sigma_\delta^2 = \delta^{2H}\sigma_1^2$ for $0 < \delta < 1$. For each $\theta \in (0,1)$ there is a finite $M = M(\theta)$ such that for each $\delta \in (0,1]$ and $x > 0$,*

$$\Pr\left(\left\{\sup_{t\in S(\delta)} |X(t)| > x\right\}\right) \le M \exp\left\{-\frac{\theta x^2}{2\sigma_\delta^2}\right\}. \tag{12.28}$$

*Proof.* The first part of the conclusion follows easily from the assumptions. To prove the second part, for each $\delta > 0$ let $C_\delta$ be the separable Banach space of all continuous real-valued functions on $S(\delta)$ with supremum norm $\|\cdot\|_{\sup}$. Then $X_\delta$, defined as $X$ restricted to $S(\delta)$, is a random element of $C_\delta$ with a law $P$, defined on the Borel sets of $C_\delta$, which is centered and Gaussian. Let $D_\delta$ be a countable dense subset of $S(\delta)$ and consider the sequence $\{e_x \colon x \in D_\delta\}$ of continuous linear forms $e_x \colon f \mapsto f(x)$ on $C_\delta$. Then $\sup\{\int e_x^2\,\mathrm{d}P \colon x \in D_\delta\} = \sigma_\delta^2$. Let $A_\delta := \sup\{|X(t)| \colon t \in S(\delta)\}$. By the Landau–Shepp–Marcus–Fernique theorem (e.g. [52, Theorem 2.2.8]), we have

$$M(\alpha,\delta) := E\mathrm{e}^{\alpha A_\delta^2} = \int_{C_\delta} \mathrm{e}^{\alpha\|f\|_{\sup}^2}\,P(\mathrm{d}f) < \infty \quad \text{if and only if} \quad \alpha < \frac{1}{2\sigma_\delta^2}. \tag{12.29}$$

For $\theta \in (0,1)$, let $\alpha_\delta := \theta/(2\sigma_\delta^2)$. For any $c > 0$, $A_{c\delta}$ is equal in distribution to $c^H A_\delta$ and $\alpha_{c\delta} = c^{-2H}\alpha_\delta$, and so $\alpha_{c\delta}A_{c\delta}^2 = \alpha_\delta A_\delta^2$ in distribution. Thus $M := M(\theta) := M(\alpha_\delta, \delta)$ depends on $\theta$ but not on $\delta$. Now (12.28) follows from (12.29), proving the lemma. □

Next is the local law of the iterated logarithm for fractional Brownian motion.

**Proposition 12.19.** *For $0 < H < 1$, with probability 1,*

$$\limsup_{t\downarrow 0} \frac{|B_H(t)|}{t^H\sqrt{2\log\log(1/t)}} = 1. \tag{12.30}$$

To prove the proposition we use the following:

**Lemma 12.20.** *Let $\{\xi_k\}_{k\geq 1}$ be a sequence of jointly normal random variables with mean zero and variance 1, such that for some $0 < \theta < 1/2$,*

$$\limsup_{n\to\infty} \max\left\{E\left[\xi_k\xi_m\right]: \ k,m \in (n,2n], \ k \neq m\right\} < \theta. \tag{12.31}$$

*Then with probability 1,*

$$\limsup_{k\to\infty} \frac{\xi_k}{\sqrt{2\log k}} \geq 1 - 2\theta. \tag{12.32}$$

*Proof.* By the Borel–Cantelli lemma, it is enough to prove that

$$\sum_{n\geq 1} \Pr\left(\left\{\max_{2^n<k\leq 2^{n+1}} \xi_k/\sqrt{2\log k} < 1 - 2\theta\right\}\right) < \infty. \tag{12.33}$$

Let $\{\eta, \eta_k\}_{k\geq 1}$ be a sequence of independent normal random variables with mean zero, $E\eta^2 = \theta$ and $E\eta_k^2 = 1 - \theta$ for each $k$. Let $N$ be a positive integer such that for each $n \geq N$,

$$\max\left\{E\left[\xi_k\xi_m\right]: \ 2^n < k,m \leq 2^{n+1}, \ k \neq m\right\} < \theta,$$

and let $n \geq N$. Then $E\left[\xi_k\xi_m\right] \leq \theta = E\left[(\eta+\eta_k)(\eta+\eta_m)\right]$ for $2^n < k, m \leq 2^{n+1}$ with $k \neq m$, and $E\xi_k^2 = 1 = E[\eta + \eta_k]^2$. By the Slepian lemma (e.g. Corollary 3.12 in Ledoux and Talagrand [136]), we have

$$\begin{aligned}
&\Pr\left(\left\{\max_{2^n<k\leq 2^{n+1}} \xi_k/\sqrt{2\log k} \leq 1 - 2\theta\right\}\right) \\
&\quad \leq \Pr\left(\left\{\max_{2^n<k\leq 2^{n+1}} \xi_k \leq (1-2\theta)\sqrt{2\log 2^{n+1}}\right\}\right) \\
&\quad \leq \Pr\left(\left\{\eta + \max_{2^n<k\leq 2^{n+1}} \eta_k \leq (1-2\theta)\sqrt{2\log 2^{n+1}}\right\}\right) \\
&\quad \leq \Pr\left(\left\{\eta \leq -\theta\sqrt{2\log 2^{n+1}}\right\}\right) + \left(\Pr\left(\left\{\eta_1 \leq (1-\theta)\sqrt{2\log 2^{n+1}}\right\}\right)\right)^{2^n}.
\end{aligned}$$

Because $\eta/\sqrt{\theta}$ has law $N(0,1)$, it follows that

$$\Pr\left(\{\eta \le -\theta\sqrt{2\log 2^{n+1}}\}\right) \le 2^{-\theta(n+1)}.$$

Similarly since $\eta_1/\sqrt{1-\theta}$ has law $N(0,1)$ and $x < \mathrm{e}^{-(1-x)}$ for $0 < x < 1$, there is a finite constant $C = C(\theta)$ such that for all large enough $n$,

$$\Big(\Pr\left(\{\eta_1 \le (1-\theta)\sqrt{2\log 2^{n+1}}\}\right)\Big)^{2^n} < \exp\big\{-C2^{\theta n}/\sqrt{n+1}\big\}.$$

Thus (12.33) holds, proving the lemma. □

**Proof of Proposition 12.19.** For each $r \in (0,1)$, since $\log(k(\log(1/r)) \sim \log k$ as $k \to \infty$, we have

$$\limsup_{t\downarrow 0} \frac{B_H(t)}{t^H\sqrt{2\log\log(1/t)}} \ge \limsup_{k\to\infty} \frac{B_H(r^k)}{r^{kH}\sqrt{2\log k}}. \tag{12.34}$$

For $k = 1, 2, \ldots$, let $\xi_k := B_H(r^k)/r^{kH}$. Then $\{\xi_k\}_{k\ge 1}$ is a sequence of jointly normal random variables with mean zero and variance 1, and for $1 \le k < m < \infty$,

$$E\big[\xi_k\xi_m\big] = \frac{1}{2}\Big[r^{(m-k)H} + r^{(k-m)H} - \big(r^{(k-m)/2} - r^{(m-k)/2}\big)^{2H}\Big].$$

For any $x$ with $0 < x < 1$, using a derivative if $1 < 2H < 2$ and concavity if $0 < 2H \le 1$, it follows that

$$x^{2H} + x^{-2H} - \big(x^{-1} - x\big)^{2H} \le \begin{cases} x^{2H} + 2Hx^{2-2H} & \text{if } 1 < 2H < 2, \\ 2x^{2H} & \text{if } 0 < 2H \le 1. \end{cases}$$

Thus for any $\epsilon \in (0, 1/2)$, there is an $r \in (0,1)$ small enough such that $E[\xi_k\xi_m] < \epsilon$ for each $k \ne m$, and so by Lemma 12.20, we have

$$\limsup_{k\to\infty} \frac{\xi_k}{\sqrt{2\log k}} > 1 - 2\epsilon$$

with probability 1. Since $\epsilon \in (0, 1/2)$ is arbitrary, this together with (12.34) yields (12.30) with "$\ge$" instead of "$=$".

To prove the reverse inequality first note that given $\epsilon > 0$ and $r \in (0,1)$, for each positive integer $k$ large enough so that $r^k < \mathrm{e}^{-\mathrm{e}}$,

$$\begin{aligned} &\Pr\left(\{|B_H(r^k)| \ge (1+\epsilon)\phi_H(r^k)\}\right) \\ &\quad = 2N(0,1)([(1+\epsilon)\sqrt{2\log(k\log(1/r))}, +\infty)) \le 2\big[k\log(1/r)\big]^{-(1+\epsilon)^2}, \end{aligned}$$

and so by the Borel–Cantelli lemma, with probability 1 we have

$$\limsup_{k\to\infty} |B_H(r^k)|/\phi_H(r^k) \le 1+\epsilon. \tag{12.35}$$

Next, given $\epsilon \in (0,1/2)$, let $r \in (0,1)$ be such that $2(1-r)^H \le \epsilon$ and $r^H \ge (1+\epsilon)/(1+2\epsilon)$. Then we will show that with probability 1,

$$\limsup_{k\to\infty} \sup\left\{\left|\frac{B_H(t)}{\phi_H(t)} - \frac{B_H(r^k)}{\phi_H(r^k)}\right|:\ r^{k+1} \le t \le r^k\right\} < 3\epsilon. \tag{12.36}$$

Given $k = 1,2,\ldots$ let $X_k(s) := B_H((1-s)r^k) - B_H(r^k)$ for $s \in [0,1]$. Then $X_k$ is a mean-zero Gaussian stochastic process with covariance equal to that of $\{B_H(sr^k),\ s\in[0,1]\}$. Since

$$\sup\{|B_H(t) - B_H(r^k)|:\ r^{k+1} \le t \le r^k\} = \sup\{|X_k(s)|:\ s \in [0,1-r]\}$$

it then follows that

$$\begin{aligned} &\Pr\big(\{\sup\{|B_H(t)-B_H(r^k)|:\ r^{k+1}\le t\le r^k\} > \epsilon\phi_H(r^k)\}\big) \\ &\quad = \Pr\big(\{\sup\{|B_H(s)|:\ s\in[0,r^k(1-r)]\} > \epsilon\phi_H(r^k)\}\big) =: A. \end{aligned}$$

Applying Lemma 12.18 with $X = B_H$, $S = [0,1]$, $S(\delta) = [0,\delta]$, $0 < \delta < 1$, $\theta = 1/2$, and $M = M(1/2)$, we have for $r^k < \mathrm{e}^{-\mathrm{e}}$

$$\begin{aligned} A &\le M\exp\left\{-\frac{\epsilon^2\phi_H(r^k)^2}{4(1-r)^{2H}r^{2kH}}\right\} \\ &= M\big(k\log(1/r)\big)^{-\epsilon^2/(2(1-r)^{2H})} \le M\big(k\log(1/r)\big)^{-2} \end{aligned}$$

by the choice of $r$. Thus by the Borel–Cantelli lemma, with probability 1, we have

$$\limsup_{k\to\infty} \sup\{|B_H(t)-B_H(r^k)|/\phi_H(r^k):\ r^{k+1}\le t\le r^k\} \le \epsilon. \tag{12.37}$$

Since $\phi_H$ is increasing on $[0,\delta]$ for some $\delta > 0$, it follows by the choice of $r$ again that

$$\lim_{k\to\infty} \sup\left\{\left|\frac{\phi_H(r^k)}{\phi_H(t)} - 1\right|:\ r^{k+1}\le t\le r^k\right\} = r^{-H} - 1 \le \frac{\epsilon}{1+\epsilon}.$$

This together with (12.35) and (12.37) yields (12.36). Finally, by (12.35) and (12.36), letting $\epsilon \downarrow 0$, (12.30) holds with probability 1 and with "$\le$" instead of "$=$", proving the proposition. □

**Lemma 12.21.** *Let $H \in (0,1)$. For each $t > 0$, with probability 1,*

$$\lim_{\delta\downarrow 0}\left[\sup_{\substack{0\le u,v\\ 0<u+v\le\delta}} \frac{|B_H(t+u) - B_H(t-v)|}{\phi_H(u+v)}\right] = 1. \tag{12.38}$$

*Proof.* Let $t > 0$ and $\tau := \min\{t, \mathrm{e}^{-\mathrm{e}}\}$. For $0 < \delta \le 1$, let $S(\delta) := \{(u, v) \in [0, \infty) \times [0, \infty) \colon 0 < u + v \le \delta\tau\}$ and

$$D(\delta) := \sup\left\{\frac{|B_H(t+u) - B_H(t-v)|}{\phi_H(u+v)} \colon (u, v) \in S(\delta)\right\}.$$

Clearly $D(\delta)$ is nondecreasing in $\delta$, and $\lim_{\delta\downarrow 0} D(\delta) \ge 1$ with probability 1 by the local law of the iterated logarithm (12.30). To bound $D(\delta)$ let $\epsilon \in (0, 1/2)$. For each $n = 8, 9, \ldots$, let $\delta_n := \exp\{-n^{1-\epsilon}\}$, $\phi_n := (\delta_n\tau)^H\sqrt{2\log\log(1/\delta_n)}$, $S_n := S(\delta_n)$, and

$$E_n := \bigl\{\omega \in \Omega \colon \sup_{(u,v)\in S_n} |B_H(t+u) - B_H(t-v)| \ge (1+2\epsilon)\phi_n\bigr\}.$$

To bound $\Pr(E_n)$ we apply Lemma 12.18. For $(u, v) \in S := [0, \tau] \times [0, \tau]$, let $X(u, v) := B_H(t+u) - B_H(t-v)$, $Y(u) := B_H(t+u) - B_H(t)$, and $Z(v) := B_H(t) - B_H(t-v)$. Then $X(u, v) \equiv Y(u) + Z(v)$, and $X$ is a Gaussian process satisfying the hypotheses of Lemma 12.18 with $S$ and $S(1)$ as defined in this proof. Then $\sigma_\delta^2 = (\tau\delta)^{2H}$ for $0 < \delta \le 1$. By (12.28) with $\theta = 1/(1+2\epsilon)$, $M = M(\theta)$, and $x = (1+2\epsilon)y(\tau\delta)^H$, the inequality

$$\Pr(\{\sup_{(u,v)\in S(\delta)} |X(u, v)| > (1+2\epsilon)y(\tau\delta)^H\}) \le M\exp\bigl\{-(1+2\epsilon)y^2/2\bigr\} \quad (12.39)$$

holds for each $y > 0$ and $0 < \delta \le 1$.

For each $n \ge 8$, setting $\delta = \delta_n$ and $y := \sqrt{2\log\log(1/\delta_n)}$, by (12.39), we have

$$\Pr(E_n) \le M\exp\bigl\{-(1+2\epsilon)\log\log(1/\delta_n)\bigr\} = Mn^{-(1-\epsilon)(1+2\epsilon)}.$$

By the Borel–Cantelli lemma, with probability 1, there exists $n_0 = n_0(\omega)$ such that for each $n \ge n_0$,

$$\sup_{(u,v)\in S_n} |B_H(t+u) - B_H(t-v)| \le (1+2\epsilon)\phi_n.$$

Now for $\delta \le \delta_n$ and $n \ge n_0$, we have

$$\begin{aligned} D(\delta) &\le \sup_{m\ge n}\sup\left\{\frac{|B_H(t+u) - B_H(t-v)|}{\phi_H(u+v)} \colon (u, v) \in S_m \setminus S_{m+1}\right\} \\ &\le \sup_{m\ge n}\sup_{(u,v)\in S_m} \frac{|B_H(t+u) - B_H(t-v)|}{(\tau\delta_{m+1})^H\sqrt{2\log\log(1/(\tau\delta_m))}} \\ &\le (1+2\epsilon)\sup_{m\ge n}\Bigl(\frac{\delta_m}{\delta_{m+1}}\Bigr)^H\sqrt{\frac{\log\log(1/\delta_m)}{\log\log(1/(\tau\delta_m))}}. \end{aligned}$$

Since $\delta_m/\delta_{m+1} \to 1$ as $m \to \infty$, the function $x \mapsto \log\log x$ is slowly varying as $x \to \infty$ and $\epsilon > 0$ is arbitrary, (12.38) holds with probability 1, proving the lemma. $\square$

The stationary increments property of a stochastic process was recalled before (12.8). Fractional Brownian motions clearly have the property where now, since for $H \neq 1/2$ $B_H$ does not have independent increments, we need the full definition as given for joint distributions, not only for one-dimensional distributions.

**Theorem 12.22.** *Let $H \in (0,1)$ and $0 < T < \infty$. For the fractional Brownian motion $B_H$, almost surely*

$$\lim_{\delta \downarrow 0} \sup \big\{ s_{\Phi_H}(B_H; \kappa) \colon \kappa \in \mathrm{PP}\,[0,T],\ |\kappa| \leq \delta \big\} = T. \tag{12.40}$$

*Proof.* The fact that the left side of (12.40) is at least $T$ will be obtained using the local law of the iterated logarithm (12.30) and the Vitali covering lemma. Let $0 < \epsilon < 1$. For each $\delta > 0$, let $E_\delta$ be the set of all $(t,\omega) \in [0,T] \times \Omega$ such that

$$\Phi_H(|B_H(t+s,\omega) - B_H(t,\omega)|) > (1-\epsilon)s \tag{12.41}$$

for some (rational) $s \in (0,\delta)$. By continuity of $\Phi_H$ and sample continuity of $B_H$, the set $E_\delta$ is jointly measurable, with Lebesgue measure on $[0,T]$. Since $\Phi_H$ is regularly varying of order $1/H$ (12.27) and is asymptotic near zero to $\phi_H^{-1}$, for each $c > 0$, we have

$$\lim_{v \downarrow 0} \Phi_H(c\phi_H(v))/v = \lim_{u \downarrow 0} \Phi_H(cu)/\phi_H^{-1}(u) = c^{1/H}. \tag{12.42}$$

Taking $c = (1-\epsilon/2)^H$, there is a $\xi(\epsilon) > 0$ such that

$$\Phi_H((1-\epsilon/2)^H \phi_H(v)) > (1-\epsilon)v$$

for each $0 < v < \xi(\epsilon)$. Since $B_H$ has stationary increments, by Proposition 12.19, for each $t \in [0,T]$ and $\delta > 0$, we have $\Pr\big(\{\omega \in \Omega \colon (t,\omega) \in E_\delta\}\big) = 1$. The Fubini theorem then yields

$$\int_\Omega \int_{[0,T]} \Big[1 - 1_{E_\delta}(t,\omega)\Big]\, \mathrm{d}t\, \Pr(\mathrm{d}\omega) = \int_{[0,T]} \int_\Omega \Big[1 - 1_{E_\delta}(t,\omega)\Big] \Pr(\mathrm{d}\omega)\, \mathrm{d}t = 0.$$

For Lebesgue measure $\lambda$ and any Lebesgue measurable set $A$, let $|A| := \lambda(A)$. Thus for each $\delta > 0$, $\Pr\big(\{\omega \in \Omega \colon \big|\{t \in [0,T] \colon (t,\omega) \in E_\delta\}\big| = T\}\big) = 1$. Let $E := \bigcap_{0<\delta\leq 1} E_\delta = \bigcap_{k=1}^\infty E_{1/k}$ and

$$\Omega_0 := \{\omega \in \Omega \colon \big|\{t \in [0,T] \colon (t,\omega) \in E\}\big| = T\}.$$

By continuity of Lebesgue measure, $\Pr(\Omega_0) = 1$. Let $\omega \in \Omega_0$ and $\delta > 0$. Then the set $\mathcal{F}_\delta(\omega)$ of all intervals $[t, t+s]$, $t \in [0,T]$, $s \in (0,\delta]$, such that (12.41) holds is a Vitali cover of $E(\omega) := \{t \in [0,T] \colon (t,\omega) \in E\}$, that is, for each $\eta > 0$ one can extract a subcover of $E(\omega)$ by intervals in $\mathcal{F}_\delta(\omega)$ of length less than $\eta$. By the Vitali lemma (e.g. Corollary II.17.2 in [40]), one can pick a finite subcollection of intervals from $\mathcal{F}_\delta(\omega)$ which are disjoint and have total length at least $T-\epsilon$. Let $\kappa = \{x_i\}_{i=0}^n$ be a partition of $[0,T]$ with mesh $|\kappa| < \delta$ such that for each of the disjoint intervals $[t_j, t_j + s_j]$ from $\mathcal{F}_\delta(\omega)$ with total length at least $T-\epsilon$, there is some $i$ with $x_{i-1} = t_j$ and $x_i = t_j + s_j$. Then the bound

$$
\begin{aligned}
s_{\Phi_H}(B_H(\cdot,\omega);\kappa) &\geq \sum_j \Phi_H(|B_H(t_j+s_j,\omega) - B_H(t_j,\omega)|) \\
&> (1-\epsilon)\sum_j s_j > (1-\epsilon)(T-\epsilon)
\end{aligned}
$$

holds for each $\omega \in \Omega_0$. Thus almost surely

$$\sup\big\{s_{\Phi_H}(B_H;\kappa)\colon\ \kappa \in \mathrm{PP}\,[0,T],\ |\kappa| \leq \delta\big\} > (1-\epsilon)(T-\epsilon).$$

Since $\epsilon > 0$ and $\delta > 0$ are arbitrary, (12.40) holds with "$\geq$" instead of "$=$".

To prove the reverse inequality let $\epsilon > 0$. For any partition $\kappa = \{t_i\}_{i=0}^n$ of $[0,T]$, let $\Delta_i := t_i - t_{i-1}$ and $\Delta_i B_H := B_H(t_i) - B_H(t_{i-1})$ for each $i = 1,\dots,n$. Also let $I_1$, $I_2$, and $I_3$ be the sets of $i \in \{1,\dots,n\}$ for which $|\Delta_i B_H|$ is, respectively, in $[0,(1+\epsilon)\phi_H(\Delta_i))$, $[(1+\epsilon)\phi_H(\Delta_i), A\phi_H(\Delta_i))$, and $[A\phi_H(\Delta_i),\infty)$ with

$$A := \sqrt{8\gamma}\mathrm{e}^{2H} \quad \text{and} \quad \gamma := 4 + 1/(2H), \tag{12.43}$$

and so the $\Phi_H$-variation sum $s_{\Phi_H}(B_H;\kappa)$ splits into three sums, to be bounded as follows.

By (12.42), for any $c > 0$ there is an $\eta(c,\epsilon) > 0$ such that

$$\Phi_H(c\phi_H(v)) \leq [c^{1/H} + \epsilon]v \tag{12.44}$$

for each $0 < v < \eta(c,\epsilon)$. Letting $\delta_1 := \eta(1+\epsilon,\epsilon)$, for a partition $\kappa$ of $[0,T]$ with mesh $|\kappa| < \delta_1$, we then have

$$\sum_{i\in I_1} \Phi_H(|\Delta_i B_H|) \leq \sum_{i\in I_1} \Phi_H((1+\epsilon)\phi_H(\Delta_i)) \leq \big[(1+\epsilon)^{1/H} + \epsilon\big]T. \tag{12.45}$$

We will show that the sum of $\Delta_i$ for $i \in I_1$ is close to $T$ as the mesh of $\kappa$ becomes small enough, while the contribution of $I_2 \cup I_3$ is negligible. To this aim, for a given $\delta > 0$, let $U_\delta$ be the set of all $(t,\omega) \in [0,T] \times \Omega$ such that

$$|B_H(t+u,\omega) - B_H(t-v,\omega)| < (1+\epsilon)\phi_H(u+v)$$

for all $u, v \geq 0$ with $u + v \leq \delta$ and $v \leq t$. By Lemma 12.21, for each $t \in (0,T)$, we have

$$\Pr\big(\big\{\omega \in \Omega\colon\ \lim_{\delta\downarrow 0} 1_{U_\delta}(t,\omega) = 1\big\}\big) = 1.$$

By Fatou's lemma and the Fubini theorem, it follows that

$$
\begin{aligned}
\int_\Omega \Big[\liminf_{\delta\downarrow 0} \int_{[0,T]} 1_{U_\delta}(t,\omega)\,dt\Big]\Pr(d\omega) &\geq \int_\Omega \int_{[0,T]} \Big[\liminf_{\delta\downarrow 0} 1_{U_\delta}(t,\omega)\Big]\,dt\Pr(d\omega) \\
&= \int_{[0,T]} \int_\Omega \Big[\lim_{\delta\downarrow 0} 1_{U_\delta}(t,\omega)\Big]\Pr(d\omega)\,dt = T,
\end{aligned}
$$

where we can take $\delta \downarrow 0$ through the sequence $\{1/k\}_{k\ge 1}$. Hence

$$\Pr\left(\left\{\omega \in \Omega\colon \lim_{\delta\downarrow 0}\int_{[0,T]} 1_{U_\delta}(t,\omega)\,\mathrm{d}t = T\right\}\right) = 1.$$

Let $\Omega_1 \subset \Omega$ be a set with probability 1 such that for each $\omega \in \Omega_1$ there exists a $\delta_2(\omega) > 0$ with the property

$$\big|\{t \in [0,T]\colon (t,\omega) \notin U_\delta\}\big| = \int_{[0,T]} \big[1 - 1_{U_\delta}(t,\omega)\big]\,\mathrm{d}t \le \epsilon$$

for all $\delta \le \delta_2(\omega)$. We can and do choose $\delta_2(\omega) \le \eta(A,\epsilon)$. If the partition $\kappa$ has mesh $|\kappa| \le \delta_2(\omega)$ and an interval $[t_{i-1}, t_i]$ for it contains a point of $U_{\delta_2(\omega)}$, then $i \in I_1$. Also, the total length of such intervals is at least $T - \epsilon$. In particular, $\sum_{i\in I_2} \Delta_i \le \epsilon$. By (12.44) with $c = A$, if $|\kappa| \le \delta_2(\omega)$ then

$$\begin{aligned}\sum_{i\in I_2} \Phi_H(|\Delta_i B_H|) &\le \sum_{i\in I_2} \Phi_H(A\phi_H(\Delta_i)) \\ &\le (A^{1/H}+\epsilon)\sum_{i\in I_2}\Delta_i \le (A^{1/H}+\epsilon)\epsilon.\end{aligned} \tag{12.46}$$

It remains to bound the sum of terms $\Phi_H(|\Delta_i B_H|)$ with $i \in I_3$.

For each integer $m \ge 3$ and each $j = 0, 1, \ldots, j_m := \lfloor 2T\mathrm{e}^m\rfloor - 1$, where $\lfloor x\rfloor$ is the largest integer $\le x$, let $S_{m,j} := [(j/2)\mathrm{e}^{-m}, (1+j/2)\mathrm{e}^{-m}]$ and

$$V_{m,j} := \left\{\omega \in \Omega\colon \sup_{t,s\in S_{m,j}} |B_H(t,\omega) - B_H(s,\omega)| \ge A\phi_H(\mathrm{e}^{-m-2})\right\}.$$

The intervals $S_{m,j}$, $j = 0, \ldots, j_m$, overlap and cover $[0,T]$. Moreover, we have

$$\begin{aligned}&\operatorname{card}\big\{i \in I_3(\omega)\colon \mathrm{e}^{-m-1} < 2\Delta_i \le \mathrm{e}^{-m}\big\} \\ &\quad\le 5\operatorname{card}\big\{j = 0,\ldots,j_m\colon \omega \in V_{m,j}\big\} =: Z_m(\omega)\end{aligned} \tag{12.47}$$

for each $\omega \in \Omega$. To bound $Z_m$ we apply Lemma 12.18. For a given $u \ge 0$, let $X(t) := B_H(t+u) - B_H(u)$ for $t \in [0,1]$. Then $X$ is a Gaussian process with the covariance of $B_H$ and satisfying the hypotheses of Lemma 12.18 with $S = S(1) = [0,1]$ and $\sigma_\delta^2 = \delta^{2H}$ for $0 < \delta \le 1$. By (12.28) with $\theta = 1/2$, so $M = M(1/2)$, and $x = 2y\delta^H$, the inequality

$$\Pr\big(\{\sup_{t\in[0,\delta]} |X(t)| > 2y\delta^H\}\big) \le M\exp\big\{-y^2\big\}$$

holds for each $y > 0$ and $\delta \in (0,1]$. For each $m \ge 3$ and $j = 0, \ldots, j_m$, setting $u := (j/2)\mathrm{e}^{-m}$, $\delta := \mathrm{e}^{-m}$ and $y := \sqrt{\gamma\log(m+2)}$, recalling (12.43), it then follows that

$$\begin{aligned}\Pr(V_{m,j}) &= \Pr\big(\{\sup_{t,s\in S_{m,j}} |B_H(t)-B_H(s)| \ge A\phi_H(\mathrm{e}^{-m-2})\}\big) \\ &\le \Pr(\{\sup_{t\in[0,\delta]} |X(t)| \ge 2y\delta^H\}) \le M\exp\{-y^2\} = M(m+2)^{-\gamma}.\end{aligned} \tag{12.48}$$

Thus $EZ_m \le (2{\cdot}5)TM\mathrm{e}^m m^{-\gamma}$, and so

$$\sum_{m\ge 3}\Pr(\{Z_m > \mathrm{e}^m m^{-\gamma+2}\}) \le 10TM\sum_{m\ge 3} m^{-2} < \infty.$$

By the Borel–Cantelli lemma, there is a set $\Omega_2 \subset \Omega$ with probability 1 such that for each $\omega \in \Omega_2$ there exists an integer $m_1(\omega) \ge 3$ such that $Z_m(\omega) \le \mathrm{e}^m m^{-\gamma+2}$ for all $m \ge m_1(\omega)$. Let $m_1'$ be large enough so that $\delta' := \exp(-m_1') \le \delta$ in the definition of $\Psi_{p,1}$ before Lemma 12.11. We can assume that $m_1(\omega) \ge m_1'$ for all $\omega \in \Omega_2$. By Lemma 12.11, there is another set $\Omega_3 \subset \Omega$ with probability 1 such that for each $\omega \in \Omega_3$ there exists a finite constant $K(\omega)$ such that for $0 \le s < t \le \delta'$,

$$|B_H(t,\omega)-B_H(s,\omega)| \le K(\omega)(t-s)^H\sqrt{\log(1/(t-s))}. \tag{12.49}$$

Let $\omega \in \Omega_2\cap\Omega_3$. Let $m_2(\omega) \ge m_1(\omega)$ be such that

$$C(\omega) := 2^{1/(2H)-1}K(\omega)^{1/H}\sum_{m\ge m_2(\omega)} m^{-2} < \epsilon,$$

and let $\delta_3(\omega) := \mathrm{e}^{-m_2(\omega)}/2$. Let $\kappa = \{t_i\}_{i=0}^n$ be a partition of $[0,T]$ with mesh $|\kappa| < \delta_3(\omega)$, and for each $m \ge m_2(\omega)$, let $\Lambda_m := \{i = 1,\ldots,n\colon \mathrm{e}^{-m-1} < 2\Delta_i \le \mathrm{e}^{-m}\}$. Then for each $i = 1,\ldots,n$, $2\Delta_i \le \mathrm{e}^{-m_2(\omega)}$, so $i \in \Lambda_m$ for some $m \ge m_2(\omega)$. For each $m$, if $i \in \Lambda_m$ then $[t_{i-1},t_i] \subset S_{m,j}$ for some $j = 0,\ldots,j_m$. Thus for such a $\kappa$, by (12.47) and (12.49), we have

$$\begin{aligned}&\sum_{i\in I_3}\Phi_H(|\Delta_i B_H|) \\ &\quad= \sum_{m\ge m_2(\omega)}\sum_{i\in I_3\cap\Lambda_m}\Phi_H(|\Delta_i B_H|) \\ &\quad\le \sum_{m\ge m_2(\omega)} Z_m(\omega)\Phi_H\Big(K(\omega)(\mathrm{e}^{-m}/2)^H\sqrt{\log(2\mathrm{e}^{m+1})}\Big) \\ &\quad\le \sum_{m\ge m_2(\omega)} \mathrm{e}^m m^{-\gamma+2}K(\omega)^{1/H}(\mathrm{e}^{-m}/2)(m+2)^{1/(2H)} \le C(\omega) < \epsilon\end{aligned} \tag{12.50}$$

since $\Phi_H(u) < u^{1/H}$ for $0 \le u \le \mathrm{e}^{-\mathrm{e}}$ and $\gamma = 4 + 1/(2H)$.

Finally, let $\omega \in \Omega_1\cap\Omega_2\cap\Omega_3$. Taking $\delta \le \min\{\delta_1,\delta_2(\omega),\delta_3(\omega)\}$, by (12.50), (12.46), and (12.45), for each partition $\kappa$ of $[0,T]$ with mesh $|\kappa| < \delta(\omega)$ we have the bound

$$s_{\Phi_H}(B_H;\kappa) \le \big[(1+\epsilon)^{1/H}+\epsilon\big]T + (A^{1/H}+\epsilon)\epsilon + \epsilon.$$

Since $\epsilon > 0$ is arbitrary, (12.40) holds, proving the theorem. □

Now we are ready to show that a fractional Brownian motion $B_H$ has locally bounded $\Phi_H$-variation and $\Phi_H$ is best possible in a sense to be given.

**Corollary 12.23.** *Let $H \in (0,1)$. A fractional Brownian motion $B_H$ has locally bounded $\Phi_H$-variation and for each $0 < T < \infty$,*

$$\Pr(\{T \leq v_{\Phi_H}(B_H;[0,T]) < \infty\}) = 1.$$

*If $\Phi \in \mathcal{V}$ and $\Phi_H(u) = o(\Phi(u))$ as $u \downarrow 0$, then $v_\Phi(B_H;[0,T]) = +\infty$ almost surely. In particular, $v_p(B_H;[0,T]) = +\infty$ almost surely for $p = 1/H$.*

*Proof.* Let $0 < T < \infty$. The almost sure bound $v_{\Phi_H}(B;[0,T]) \geq T$ follows from Theorem 12.22. Moreover, by the same theorem, for almost every $\omega \in \Omega$, there is a $\delta(\omega) > 0$ such that

$$\sup\left\{s_{\Phi_H}(B_H(\cdot,\omega);\kappa)\colon\ \kappa \in \mathrm{PP}\,[0,T],\ |\kappa| \leq \delta(\omega)\right\} \leq T+1.$$

If $\kappa$ is a partition of $[0,T]$ then the number of intervals of the partition of length at least $\delta(\omega)$ cannot be more than $T/\delta(\omega)$. Thus

$$v_{\Phi_H}(B_H(\cdot,\omega);[0,T]) \leq 1 + T + T\delta(\omega)^{-1}\Phi_H(\mathrm{Osc}(B(\cdot,\omega);[0,T]))$$

and the right side is finite for almost all $\omega$. The first part of the corollary now follows from Theorem 12.3 since almost all sample functions of $B_H$ are continuous by Proposition 12.16.

Let $\Phi \in \mathcal{V}$ be such that $\Phi_H(u) = o(\Phi(u))$ as $u \downarrow 0$. Then for any partition $\kappa$ of $[0,T]$, we have

$$s_\Phi(B_H;\kappa) \geq s_{\Phi_H}(B_H;\kappa)\min_i\left\{\Phi(|\Delta_i B_H|)/\Phi_H(|\Delta_i B_H|)\right\},$$

where $\Delta_i B_H$ are non-zero increments of $B_H$ over intervals of $\kappa$. Letting the mesh of $\kappa$ be arbitrarily small, the second part of the corollary now follows from Theorem 12.22. □

Letting $H = 1/2$, the preceding theorem and corollary yield the following result for Brownian motion:

**Corollary 12.24** (S. J. Taylor [231]). *For Brownian motion $B$ and for $0 < T < \infty$, almost surely*

$$\lim_{\delta\downarrow 0}\sup\left\{s_{\Phi_{1/2}}(B;\kappa)\colon\ \kappa \in \mathrm{PP}\,[0,T],\ |\kappa| \leq \delta\right\} = T.$$

*If $\Phi \in \mathcal{V}$ and $\Phi_{1/2}(u) = o(\Phi(u))$ as $u \downarrow 0$, then $v_\Phi(B;[0,T]) = +\infty$ almost surely.*

## 12.5 Markov Processes

This section contains one main result, Theorem 12.27, showing that Markov processes with certain power-type bounds on their transition probabilities can be chosen to have trajectories of bounded $p$-variation.

A *measurable space* $(S, \mathcal{S})$ is a set $S$ together with a $\sigma$-algebra $\mathcal{S}$ of subsets of $S$. It will be called a *Borel space* if it is measurably isomorphic to a Borel subset of a Polish (complete separable metric) space with Borel $\sigma$-algebra. (The terminology follows [112, p. 7].) In this section we consider measurable functions with values in $S$. In particular, given a set $\mathcal{T}$ and a probability space $(\Omega, \mathcal{F}, \Pr)$, a stochastic process $\{X_t\}_{t\in\mathcal{T}}$ is a function $(t, \omega) \mapsto X_t(\omega)$, $t \in \mathcal{T}$, $\omega \in \Omega$, with values in $S$, such that for each $t \in \mathcal{T}$, $X_t(\cdot)$ is measurable.

For any measurable space $(\Omega, \mathcal{F})$ and measurable function $\xi$ from $\Omega$ to another measurable space, let $\sigma(\xi)$ denote the smallest $\sigma$-algebra for which $\xi$ is measurable. Thus $\sigma(\xi) \subset \mathcal{F}$. For a probability space $(\Omega, \mathcal{F}, \Pr)$, a stochastic process $\{X_t\}_{t\in\mathcal{T}}$ defined on $\Omega$, and a subset $V \subset \mathcal{T}$, let $\sigma(\{X(s)\colon\ s \in V\}) \equiv \sigma(\{X(s)\}_{s\in V}) \subset \mathcal{F}$ be the smallest $\sigma$-algebra for which all $X(s)$ for $s \in V$ are measurable.

For any sub-$\sigma$-algebra $\mathcal{B} \subset \mathcal{F}$, let $\Pr_{\mathcal{B}}$ denote the restriction of $\Pr$ to $\mathcal{B}$. For any $W \in \mathcal{F}$, the *conditional probability* $\Pr(W|\mathcal{B})$ is a $\mathcal{B}$-measurable real-valued random variable satisfying, for each $C \in \mathcal{B}$,

$$\Pr(W \cap C) = \int_C \Pr(W|\mathcal{B})\, \mathrm{d}\Pr. \tag{12.51}$$

Then $\Pr(W|\mathcal{B})$ is unique up to equality $\Pr_{\mathcal{B}}$-almost surely.

Let $\xi$ and $\eta$ be two random variables on $(\Omega, \mathcal{F}, \Pr)$ with values in measurable spaces $(S, \mathcal{S})$ and $(M, \mathcal{M})$, respectively. For each $A \in \mathcal{M}$, we can then define

$$\Pr(\{\eta \in A\}|\sigma(\xi)) \equiv \Pr(\eta^{-1}(A)|\sigma(\xi)) \tag{12.52}$$

since $\{\eta \in A\}$ is just another notation for the set $\eta^{-1}(A)$. Let $P_\xi := \Pr\circ\xi^{-1}$, in other words $P_\xi(B) := \Pr(\xi^{-1}(B))$ for each $B \in \mathcal{S}$. The conditional probability $\Pr(\eta^{-1}(A)|\sigma(\xi))$ can be represented as a function $f_A(\xi) = f_A \circ \xi$, where $f_A$ is an $\mathcal{S}$-measurable function from $S$ into $[0, 1]$ (see e.g. [112, Lemma 1.13]), since $[0, 1]$ with its Borel $\sigma$-algebra is a Borel space. Then $f_A$ is unique up to equality $P_\xi$-a.s. We also write $\Pr(\eta^{-1}(A)|\{\xi = x\}) := f_A(x)$ for $x \in S$. By a change of variables (image measure theorem, e.g. [53, Theorem 4.1.11]) in (12.51) with $C = \xi^{-1}(B)$ and $W = \eta^{-1}(A)$, we have

$$\Pr(\eta^{-1}(A) \cap \xi^{-1}(B)) = \int_B \Pr(\eta^{-1}(A)|\{\xi = x\})\, P_\xi(\mathrm{d}x) \tag{12.53}$$

for each $B \in \mathcal{S}$.

For any two measurable spaces $(S, \mathcal{S})$ and $(M, \mathcal{M})$, a *probability kernel* on $(S, \mathcal{S}|M, \mathcal{M})$ will mean a function $Q$ from $S \times \mathcal{M}$ into $[0, 1]$ such that $Q(x, \cdot)$ is a probability measure on $(M, \mathcal{M})$ for all $x \in S$ and $Q(\cdot, A)$ is an

$\mathcal{S}$-measurable function on $S$ for all $A \in \mathcal{M}$. A probability kernel on $(S, \mathcal{S})$ will mean one on $(S, \mathcal{S}|S, \mathcal{S})$. A *regular conditional distribution* of $\eta$ given $\xi$ is a probability kernel $Q_{\xi,\eta}$ on $(S, \mathcal{S}|M, \mathcal{M})$ such that for each $A \in \mathcal{M}$, $Q_{\xi,\eta}(\cdot, A)$ is a conditional probability $\Pr(\eta^{-1}(A)|\{\xi = \cdot\})$. If $(M, \mathcal{M})$ is a Borel space, then a regular conditional distribution $Q_{\xi,\eta}$ exists, and if $Q'_{\xi,\eta}$ is another regular conditional distribution, then for some $B \in \mathcal{S}$ with $P_\xi(B) = 0$, if $x \notin B$, then $Q'_{\xi,\eta}(x, A) = Q_{\xi,\eta}(x, A)$ for all $A \in \mathcal{M}$ (e.g. Theorems 10.2.2 and 13.1.1 in [53]).

Thus $Q_{\xi,\eta}$ is a probability kernel such that for each $A \in \mathcal{M}$ and $B \in \mathcal{S}$,

$$\Pr(\eta^{-1}(A) \cap \xi^{-1}(B)) = \int_B Q_{\xi,\eta}(x, A)\, P_\xi(dx). \tag{12.54}$$

Most often, stochastic processes $X(t)$ are considered starting at time $t = 0$, but we will also consider processes starting at times $s_0 \geq 0$ with a view to the behavior beginning at time $s_0$ of a process conditional on having a certain value at time $s_0$. In what follows, $[s_0, T]\!]$ will mean $[s_0, \infty)$ if $T = +\infty$ but for $s_0 < T < \infty$ it may mean, as usual, either $[s_0, T)$ or $[s_0, T]$. Let $(S, \mathcal{S})$ be a Borel space and $s_0 < T \leq +\infty$. Then $(s, t) \mapsto P_{s,t}(\cdot, \cdot)$ for $s_0 \leq s < t \in (s_0, T]\!]$ will be called a *Markov (transition) kernel on* $(S, \mathcal{S})$ if for any $s_0 \leq s < t \in (s_0, T]\!]$, $P_{s,t}$ is a probability kernel on $(S, \mathcal{S})$ and for any $s_0 \leq s < t < u \in (s_0, T]\!]$, $x \in S$, and $A \in \mathcal{S}$, the following *Chapman–Kolmogorov equation* holds:

$$P_{s,u}(x, A) \;=\; \int P_{t,u}(y, A) P_{s,t}(x, \mathrm{d}y). \tag{12.55}$$

A stochastic process $\{X_t\}_{t \in [s_0, T]\!]}$ for $T > s_0$ with values in $S$ will be called a *Markov process* if there exists a Markov kernel $(s, t) \mapsto P_{s,t}(\cdot, \cdot)$ on $(S, \mathcal{S})$ for $s_0 \leq s < t \in (s_0, T]\!]$ such that $P_{s,t}$ is a regular conditional distribution of $X(t)$ given $X(s)$ for $s_0 \leq s < t \in (s_0, T]\!]$. (Actually, if $P_{s,t}(x, A)$ is just a conditional probability that $X(t) \in A$ given $X(s) = x$, the regular conditional probability properties follow from those assumed for Markov kernels.) Given $P_{s,t}$ for $s < t$ one can define $P_{t,t}(x, A) \;:=\; \delta_x(A) \equiv 1_A(x)$. This is clearly a probability kernel, and (12.55) still holds if $t = s$ or $t = u$. If $s_0$ is not mentioned explicitly, then it is meant that $s_0 = 0$, which is usually the case in the main definition of a Markov kernel or process. Then in proofs or auxiliary statements, other values of $s_0$ may be used.

For any nonempty set $J$ recall that $S^J$ denotes the set of all functions from $J$ into $S$. For each $t \in J$, we have the function $\xi_t$ on $S^J$ defined by $\xi_t(f) \;:=\; f(t)$ ($\xi_t$ may be called the $t$ *coordinate* function on $S^J$.) On $S^J$ let $\mathcal{S}^J$ be the smallest $\sigma$-algebra for which all $\xi_t$ for $t \in J$ are measurable. The following is known:

**Theorem 12.25** (Kolmogorov–Doob). *Let $(S, \mathcal{S})$ be a Borel space, let $0 \leq s_0 < T \leq +\infty$, and suppose $P_{s,t}(\cdot, \cdot)$ is a Markov kernel on $(S, \mathcal{S})$ for $s_0 \leq s < t \in (s_0, T]\!]$ and a probability measure $\nu$ on $(S, \mathcal{S})$. Then there are a probability* *space* $(\Omega, \mathcal{A}, \Pr)$ *and a Markov process* $X(t) := X(t, \omega)$ *for* $t \in J := [s_0, T]\!]$ *and* $\omega \in \Omega$ *having the kernel* $P_{s,t}$ *and for which* $X(s_0)$ *has distribution* $\nu$*. Specifically, we can take* $\Omega = S^J$, $\mathcal{A} = \mathcal{S}^J$, *and* $X(t) = \xi_t$, *and then the probability measure* $\mathbb{P}_{s_0;\nu} := \Pr$ *is unique.*

*Proof.* A proof is given by Kallenberg [112, Theorem 8.4] and will also be sketched here. It suffices to consider the case $\Omega = S^J$, $\mathcal{A} = \mathcal{S}^J$, and $X(t) := \xi_t$. Let $n = 1, 2, \ldots$, and let $s_0 < s_1 < \cdots < s_n \in (s_0, T]\!]$. If $B$ is a measurable set in $S \times S$, let

$$\mathbb{P}_{s_0;\nu}\left((\xi_{s_0}, \xi_{s_1}) \in B\right) := \int\!\!\int 1_B(u, v) P_{s_0,s_1}(u, \mathrm{d}v)\mathrm{d}\nu(u).$$

Beginning with measurable rectangles $B = C \times D$, continuing via finite (disjoint) unions of them, and using monotone classes (e.g. [53, §4.4, Theorem 10.2.1]), one gets that $\mathbb{P}_{s_0;\nu}$ is a countably additive probability measure on such sets. Then, if $A$ is a measurable set in $S^{\{s_0,\ldots,s_n\}}$, let

$$\begin{aligned}&\mathbb{P}_{s_0;\nu}\left((\xi_{s_0}, \ldots, \xi_{s_n}) \in A\right)\\ &\quad := \int\!\!\int \cdots \int 1_A(u_0, \ldots, u_n) P_{s_{n-1},s_n}(u_{n-1}, \mathrm{d}u_n) \cdots P_{s_0,s_1}(u_0, \mathrm{d}u_1)\mathrm{d}\nu(u_0).\end{aligned}$$

By the argument for $n = 1$ and induction, we get a countably additive probability measure on $\mathcal{S}^{\{s_0,\ldots,s_n\}}$. The resulting joint distributions of $\xi_{s_0}, \ldots, \xi_{s_n}$ are mutually consistent by the Chapman–Kolmogorov equation (12.55). So, by Kolmogorov's general existence theorem for stochastic processes, e.g. [53, Theorem 12.1.2], and since $(S, \mathcal{S})$ is a Borel space, $\mathbb{P}_{s_0;\nu}$ extends to a probability measure on $(S^J, \mathcal{S}^J)$. The extension is unique because the union of the $\sigma$-algebras $\mathcal{S}^F$ for finite sets $F \subset J$ is an algebra generating $\mathcal{S}^J$, by a monotone class theorem (e.g. [53, Theorem 4.4.2]). By construction, $\{\xi_t\}_{t\in J}$ is then a Markov process with kernel $P_{s,t}$ and $\xi_{s_0}$ has distribution $\nu$. □

If $\nu$ is a point mass $\delta_x$ at a point $x \in S$, then let $\mathbb{P}_{s;x} := \mathbb{P}_{s;\delta_x}$.

**Remark 12.26.** The $\sigma$-algebra $\mathcal{S}^J$ is a relatively small one, since it is the union of $\mathcal{S}^V$ over all countable subsets $V$ of $J$ (this union is easily seen to be a $\sigma$-algebra with respect to which all $\xi_t$ for $t \in J$ are measurable). In other words, each individual set in $\mathcal{S}^J$ depends on only countably many coordinates $\xi_t$. It follows that a real-valued function on $S^J$, measurable for $\mathcal{S}^J$, can depend on only countably many coordinates. On the other hand, if we take a $\sigma$-algebra larger than $\mathcal{S}^J$, specifically, larger than its completion for a given $\mathbb{P}_{s_0;\nu}$, then $\mathbb{P}_{s_0;\nu}$ will not extend uniquely to such a $\sigma$-algebra.

A frequent alternative definition of Markov process is as follows. Given a probability space $(\Omega, \mathcal{F}, \Pr)$, a stochastic process $X = \{X_t\}_{t\in[s_0,T]\!]}$ has the *Markov property* if for each $t \in (s_0, T]\!]$, the $\sigma$-algebras $\sigma(X(s)\colon s \geq t)$ and

$\mathcal{F}_t^X := \sigma(X(s)\colon s_0 \le s \le t)$ are conditionally independent given $\sigma(X(t))$. It is easily seen that any Markov process as defined above has the Markov property. Conversely, for a given process $X$ with the Markov property and values in a Borel space, we can define probability kernels $P_{s,t}$ as regular conditional distributions of $X(t)$ given $X(s)$. Then the Chapman–Kolmogorov equation (12.55) will hold for almost all $x$ with respect to the distribution of $X(s)$ [112, Corollary 7.3]. In order that the Markov kernel be defined uniquely rather than only in an almost sure sense, following Kallenberg [112, p. 140] and others, we assume the definition of Markov process as given above after (12.55).

Let $(S, d)$ be a complete separable metric space with Borel $\sigma$-algebra $\mathcal{S}$ and $0 < T < \infty$. Let $\mathcal{T}$ be a linearly ordered set, specifically, $[0, T]$ or a subset of it. Let $f$ be a function from $\mathcal{T}$ into $S$. Let $\kappa$ be a point partition of $\mathcal{T}$. The $p$-variation sum $s_p(f;\kappa)$ is defined as in and before (1.19) except with $d(f(t_{i-1}), f(t_i))^p$ in place of $\|f(t_i) - f(t_{i-1})\|^p$. Then the $p$-variation $v_p(f) := v_p(f;\mathcal{T})$ is defined as in (1.19), namely as the supremum of $s_p$ over all point partitions of $\mathcal{T}$.

Let $P_{s,t}(\cdot,\cdot)$, $0 \le s < t \le T$, be a Markov kernel on $(S, \mathcal{S})$. For $h \in [0, T)$, $0 \le s_0 < T - h$, and $r > 0$, let $\alpha_{s_0,T}(h, r)$ be defined by $\alpha_{s_0,T}(0, r) := 0$ and for $h > 0$,

$$\alpha_{s_0,T}(h, r) := \sup\bigl\{P_{s,t}(x, \{y\colon d(x, y) \ge r\})\colon x \in S,\ s_0 \le s \le t \le (s+h)\wedge T\bigr\}.$$

If $s_0 = 0$ then let $\alpha_T(h, r) := \alpha_{0,T}(h, r)$.

For $\beta \ge 1$ and $\gamma > 0$, the kernel $P_{\cdot,\cdot}(\cdot,\cdot)$ will be said to belong to the class $\mathcal{M}(\beta,\gamma)$ if there exist constants $r_0 > 0$ and $K > 0$ such that

$$\alpha_T(h, r) \le K\frac{h^\beta}{r^\gamma} \tag{12.56}$$

for all $h \in [0, T]$ and $r \in (0, r_0]$.

Here is the main theorem of this section. As Manstavičius [155] pointed out, the conclusion is sharp: it can fail when $p = \gamma/\beta$, for symmetric stable Lévy processes as will be shown in Remark 12.39.

**Theorem 12.27** (M. Manstavičius). *Let $(S, d)$ be a complete separable metric space with Borel $\sigma$-algebra $\mathcal{S}$ and let $\nu$ be any probability measure on $(S, \mathcal{S})$. Let $\beta \ge 1$ and $\gamma > 0$. For any $p > \gamma/\beta$ and $0 < T < \infty$, and any Markov kernel $\{P_{s,t}(\cdot,\cdot)\}_{0\le s<t\le T}$ on $(S, \mathcal{S})$ in the class $\mathcal{M}(\beta,\gamma)$, there exists a Markov process $\{Y_t\}_{0\le t\le T}$ with this kernel, such that $Y(0) \equiv Y_0$ has distribution $\nu$, and such that the trajectories $t \mapsto Y(t, \omega)$, $0 \le t \le T$, have bounded $p$-variation on $[0, T]$ and are right-continuous on $[0, T)$ almost surely.*

To begin the proof, let $\{X_t\}_{0\le t\le T}$ be a Markov process with the given kernel and initial distribution by Theorem 12.25, namely, let $\Omega := S^{[0,T]}$, $\mathcal{A} := \mathcal{S}^{[0,T]}$, $X(t) = \xi_t$ for $0 \le t \le T$, and $\Pr = \mathbb{P}_{0;\nu}$. Let $\mathcal{T}$ be a countable dense subset of $[0, T]$ containing 0 and $T$. Let $Y(t) := X(t)$ for $t \in \mathcal{T}$. Then $Y(0)$ has distribution $\nu$. The plan of the proof is to bound the $p$-variation of

$\{Y_t\}_{t\in\mathcal{T}}$ over finite subsets $F$ of $\mathcal{T}$, uniformly in $F$ and thus over $\mathcal{T}$, then to extend to $[0,T]$.

By rescaling time we can assume that $T=1$. From here on let $\alpha(\cdot,\cdot) := \alpha_1(\cdot,\cdot) := \alpha_T(\cdot,\cdot)$ for $T=1$. For a countable set $F$ and subinterval $J$ of $[0,1]$, let $J_F := J\cap F$.

For a function $g$ from a set $V$ into a metric space $X$ with metric $d$, the oscillation of $g$ on $V$ is defined by $\mathrm{Osc}(g;V) := \sup\{d(g(u),g(v)) : u,v\in V\}$. The proof continues with the following inequality of Ottaviani type.

**Lemma 12.28.** *Let the Markov process $X(v,\cdot)=\xi_v$, $v\in[0,1]$, and $\alpha(\cdot,\cdot)$ be as defined above. Let $0\le s<t\le 1$ and suppose that $u>0$ is such that $\alpha(t-s,u/4)<1$. Let $F$ be a countable subset of $[0,1]$. Then almost surely*

$$\Pr\big(\{\mathrm{Osc}(X;[s,t]_F)>u\}\big|\sigma(X(s))\big) \le \frac{\Pr(\{d(X(t),X(s))>u/4\}\big|\sigma(X(s)))}{1-\alpha(t-s,u/4)}. \tag{12.57}$$

*Proof.* Let $G := [s,t]_F\cup\{s,t\}$, which is countable. The oscillation, $X(s)$, and $X(t)$ are all measurable with respect to $\mathcal{S}^G$. For the given $s$, and each $x\in S$, we have the probability measure $\mathbb{P}_{s;x}$ defining a Markov process on $S^{[s,1]}$, which is uniquely determined on $\mathcal{S}^{[s,1]}$ and thus on $\mathcal{S}^G$.

It will be shown that for each $x\in S$,

$$\mathbb{P}_{s;x}\big(\{\mathrm{Osc}(X;[s,t]_F)>u\}\big) \le \frac{\mathbb{P}_{s;x}(\{d(X(t),X(s))>u/4\})}{1-\alpha(t-s,u/4)}. \tag{12.58}$$

For any event $A\in\mathcal{S}^{[0,1]}$, the conditional probability $\Pr\big(A|\sigma(X(s))\big)$ equals $g_A(X(s))$ almost surely for some $\mathcal{S}$-measurable function $g_A\equiv\Pr(A|X(s)=\cdot)$, as noted above after (12.52), and $\Pr=\mathbb{P}_{0;\nu}$. If, moreover, $A\in\mathcal{S}^G$, then

$$\Pr(A|X(s)=x)=\mathbb{P}_{s;x}(A) \tag{12.59}$$

almost surely for $P_{X(s)}=\Pr\circ X(s)^{-1}$. To see this, first suppose that $A\in\mathcal{S}^H$ for a finite set $H\subset G$. Then by the constructions of $\mathbb{P}_{0;\nu}$ and $\mathbb{P}_{s;x}$ in the proof of Theorem 12.25, we have

$$\Pr(A\cap\{X(s)\in B\}) = \int_B \mathbb{P}_{s;x}(A)\,P_{X(s)}(dx)$$

for each $B\in\mathcal{S}$. Thus (12.59) holds if $A\in\mathcal{S}^H$ for $P_{X(s)}$-almost all $x$ by a.s. uniqueness of conditional probabilities as in (12.51), (12.52), and (12.53). We can let the finite sets $H$ increase up to the countable set $G$ via a monotone class theorem. So, (12.58) will imply (12.57).

To prove (12.58), for any fixed $x$, let $\Pr=\mathbb{P}_{s;x}$ in the rest of the proof. Note that the kernel for the Markov process defined by each $\mathbb{P}_{s;x}$ is the subset $\{P_{v,w} : s\le v<w\le 1\}$ of the original one, where clearly $\alpha_{s,1}(h,r)\le\alpha_1(h,r)$

for all $h > 0$ and $r > 0$ and the bound on the right side of (12.56) also applies to $\alpha_{s,1}$.

First let $F$ be finite. We can assume that $s \in F$ and $t \in F$ since this can only enlarge the oscillation. Let $[s,t]_F = \{t_j\}_{j=0}^m$ where $s = t_0 < t_1 < \cdots < t_m = t$. For $k = 2, \ldots, m$, let $B_k$ be the set of all $(x_0, x_1, \ldots, x_k) \in S^{k+1}$ such that $d(x_j, x_0) \le u/2$ for $j = 1, \ldots, k-1$ and $d(x_k, x_0) > u/2$, and let $B_1$ be the set of all $(x_0, x_1) \in S^2$ such that $d(x_1, x_0) > u/2$. Letting $U_k := U_k(\omega) := (X(t_0,\omega), \ldots, X(t_k,\omega))$, $\omega \in \Omega$, we have

$$\left\{\omega\colon \max_{v \in [s,t]_F} d(X(v,\omega), X(s,\omega)) > u/2\right\} = \bigcup_{k=1}^m \{\omega\colon U_k(\omega) \in B_k\}, \quad (12.60)$$

a union of disjoint sets. For $k = 1, \ldots, m$, let $V_k := (X(t_k), X(t))$, and let $C$ be the set of all $(x,y) \in S^2$ such that $d(x,y) < u/4$. Then

$$\{\omega\colon d(X(t,\omega), X(s,\omega)) > u/4\} \supset \bigcup_{k=1}^m \{U_k \in B_k\} \cap \{V_k \in C\}. \quad (12.61)$$

Due to the Markov property of $X$, for each $k \in \{1, \ldots, m\}$ the events $\{U_k \in B_k\}$ and $\{V_k \in C\}$ are conditionally independent given $\sigma(X(t_k))$, and so using the definitions (12.51), (12.52) of conditional probabilities and a change of variables, we have

$$\begin{aligned}\Pr\big(\{U_k \in B_k\} \cap \{V_k \in C\}\big) &= \int \Pr\big(\{U_k \in B_k\} \cap \{V_k \in C\} \big| \sigma(X(t_k))\big)\,\mathrm{d}\Pr \\ &= \int \Pr\big(\{U_k \in B_k\} \big| \{X(t_k) = x\}\big) \times \qquad (12.62) \\ &\qquad \times \Pr\big(\{V_k \in C\} \big| \{X(t_k) = x\}\big)\, P_{X(t_k)}(\mathrm{d}x).\end{aligned}$$

By the definitions of $V_k$ and $C$, it follows that for each $k \in \{1, \ldots, m\}$, and almost all $x$ for $P_{X(t_k)}$,

$$\begin{aligned}\Pr\big(\{V_k \in C\} | \{X(t_k) = x\}\big) &= 1 - P_{t_k,t}(x, \{y\colon d(y,x) \ge u/4\}) \\ &\ge 1 - \alpha(t-s, u/4).\end{aligned} \quad (12.63)$$

By (12.62), (12.63), and (12.51) with $C$ there $= \Omega$ and $W = \{U_k \in B_k\}$ in (12.51), we then have

$$\Pr\big(\{U_k \in B_k\} \cap \{V_k \in C\}\big) \ge [1 - \alpha(t-s, u/4)] \Pr\big(\{U_k \in B_k\}\big).$$

Since the sets $\{U_k \in B_k\}$ are disjoint, by (12.60) and (12.61), it then follows that

$$\begin{aligned}&\Pr\big(\{d(X(t), X(s)) > u/4\}\big) \\ &\quad \ge \sum_{k=1}^m \Pr\big(\{U_k \in B_k\} \cap \{V_k \in C\}\big) \\ &\quad \ge \Big[1 - \alpha\Big(t - s, \frac{u}{4}\Big)\Big] \Pr\big(\{\max_{v \in [s,t]_F} d(X(v), X(s)) > u/2\}\big).\end{aligned}$$

Letting the finite set $F$ increase up to a countable set, this yields (12.58) and so (12.57), proving the lemma. $\square$

Now continuing with the proof of Theorem 12.27, recall that $\mathcal{T}$ is a countable dense subset of $[0,1]$ containing 0 and 1, and $Y(t) = X(t) = \xi_t$ for $t \in \mathcal{T}$. Let $F$ be any finite subset of $\mathcal{T}$ with $1 \in F$. For $t \geq 0$, let $\mathcal{F}_t^{Y,F} := \sigma(Y(s)\colon\ 0 \leq s \leq t,\ s \in F)$. Thus a random variable $\tau \geq 0$ is a stopping time for the process $Y = \{Y_t\}_{t \in F}$ if and only if $\{\tau \leq t\} \in \mathcal{F}_t^{Y,F}$ for each $t \geq 0$. If $\tau$ is any stopping time for $\{Y_t\}_{t\in F}$ then the $\sigma$-algebra $\mathcal{F}_\tau := \mathcal{F}_\tau^{Y,F}$ is defined as the collection of all measurable sets $A$ such that for each $t \geq 0$, $A \cap \{\tau \leq t\} \in \mathcal{F}_t^{Y,F}$.

**Lemma 12.29.** *Let $Y$, $F$, and $\mathcal{F}_t = \mathcal{F}_t^{Y,F}$, $t \geq 0$, be as above, let $\tau$ be a stopping time for the filtration $\{\mathcal{F}_t\colon t \geq 0\}$ such that $\tau$ has values in the finite set $F \cup \{2\}$, and let $A(s) \in \sigma(Y(t)\colon\ t \geq s,\ t \in F)$ for each $s \in F$ and $A(2) := \emptyset$. Let $A(\tau) := \bigcup_{t \in F\cup\{2\}} A(t) \cap \{\tau = t\}$. Then almost surely on $\{\tau \leq 1\}$,*

$$\Pr(A(\tau)|\mathcal{F}_\tau) = \Pr(A(\tau)|\sigma(\tau, Y(\tau))). \tag{12.64}$$

*Proof.* Note that in the definition of $A(\tau)$, $F\cup\{2\}$ can be replaced equivalently by $F$. Let $s \in F$. We have $\{\tau = s\} = \{\tau \leq s\} \cap C_s \in \mathcal{F}_s$ where $C_s = \Omega$ if $s = \min\{t :\ t \in F\}$. Otherwise, if $u = \max\{v \in F :\ v < s\}$, we can let $C_s = \Omega \setminus \{\tau \leq u\} \in \mathcal{F}_u \subset \mathcal{F}_s$. Since $Y$ is a Markov process and $A(s) \in \sigma(Y(t)\colon\ t \geq s,\ t \in F)$, by the Markov property, we have almost surely

$$\Pr(A(s)|\mathcal{F}_s) = \Pr(A(s)|\sigma(Y(s))) = \mathbb{P}_{s;Y(s)}(A(s)), \tag{12.65}$$

where the last equality holds by (12.59). Let $B \in \mathcal{F}_\tau$. Then $B \cap \{\tau = s\} \in \mathcal{F}_s$, since $B \cap \{\tau \leq u\} \in \mathcal{F}_u \subset \mathcal{F}_s$ for $u < s$, and so

$$\begin{aligned}
&\Pr(A(\tau) \cap B \cap \{\tau = s\}) \\
&\quad = \int_{B\cap\{\tau=s\}} \Pr(A(s)|\mathcal{F}_s)\, d\Pr \\
&\quad = \int_{B\cap\{\tau=s\}} \Pr(A(s)|\sigma(Y(s)))\, d\Pr = \int_{B\cap\{\tau=s\}} \mathbb{P}_{\tau;Y(\tau)}(A(\tau))\, d\Pr
\end{aligned}$$

by (12.65). Summing over the finitely many values of $\tau$ it follows that

$$\Pr(A(\tau) \cap B) = \int_{B\cap\{\tau\leq 1\}} \mathbb{P}_{\tau;Y(\tau)}(A(\tau))\, d\Pr$$

for any $B \in \mathcal{F}_\tau$ since $A(2) = \emptyset$. Now on $\{\tau \leq 1\}$, $\mathbb{P}_{\tau;Y(\tau)}(A(\tau)) \equiv \sum_{s\in F} 1_{\tau=s}\mathbb{P}_{s;Y(s)}(A(s))$ and for each $s \in F$, $\mathbb{P}_{s;Y(s)}(A(s))$ is a measurable function of $Y(s)$. Thus $\mathbb{P}_{\tau;Y(\tau)}(A(\tau))$ is measurable for $\sigma(\tau, Y(\tau))$ and so for $\mathcal{F}_\tau$ on $\{\tau \leq 1\}$. Then $\Pr(A(\tau)|\mathcal{F}_\tau) = \mathbb{P}_{\tau;Y(\tau)}(A(\tau))$. Similarly we have on $\{\tau \leq 1\}$ that $\Pr(A(\tau)|\sigma(\tau, Y(\tau))) = \mathbb{P}_{\tau;Y(\tau)}(A(\tau))$, proving (12.64). $\square$

For integers $k = 0, \pm 1, \pm 2, \ldots$, let $M_k := 2^{-k-1}$ and let $\{\tau_{i,k}\}_{i=0,1,2,\ldots}$ be the sequence of random variables defined by $\tau_{0,k} := 0$ and

$$\tau_{i,k} := \begin{cases} \min\left\{t \in [\tau_{i-1,k}, 1]_F \colon \operatorname{Osc}(Y; [\tau_{i-1,k}, t]_F) > M_k\right\} & \text{if such a } t \text{ exists,} \\ 2 & \text{otherwise.} \end{cases}$$

Here $\tau_{i,k}^F := \tau_{i,k}$ depends on $F$. For each integer $k$, let $i_0 = i_0(k)$ be the smallest integer $i \geq 0$ such that $\tau_{i,k} \geq 1$. We have $0 = \tau_{0,k} < \tau_{1,k} < \cdots < \tau_{i_0-1,k} < 1 \leq \tau_{i_0,k}$ almost surely. Moreover, $\{\tau_{i,k} \leq t\} \in \mathcal{F}_t^{Y,F}$ for each $t \geq 0$, and so $\tau_{i,k}$ is a stopping time for each $i, k$.

For each integer $k$ and $i = 1, 2, \ldots$, let $\zeta_{i,k} := \zeta_{i,k}^F := \tau_{i,k}^F - \tau_{i-1,k}^F$. For random variables $\xi$ and $\eta$ on $(\Omega, \mathcal{F}, \Pr)$ and $A \in \mathcal{F}$, we say that $\xi \leq \eta$ almost surely on $A$ if $\Pr(\{\xi \leq \eta\} \cap A) = \Pr(A)$. If $\Pr(A) > 0$, this is equivalent to $\Pr(\{\xi \leq \eta\} | A) = 1$.

**Lemma 12.30.** *For an integer $k$, let $h_0 \in (0,1]$ be such that $\alpha(h_0, M_{k+2}) < 1$. Then for any finite $F \subset T$, each $i = 1, 2, \ldots$, and $h \in [0, h_0]$, almost surely on $\{\tau_{i-1,k} < 1\}$,*

$$\Pr\left(\{\zeta_{i,k}^F \leq h\} \middle| \mathcal{F}_{\tau_{i-1,k}}^{Y,F}\right) \leq \frac{\alpha(h, M_{k+2})}{1 - \alpha(h, M_{k+2})}.$$

*Proof.* Let $i$ be such that $\Pr(\{\tau_{i-1,k} < 1\}) > 0$. If $\tau_{i,k} = 2$ then the conclusion holds. Otherwise, $\tau_{i,k} \leq 1$. Let $h \in (0, h_0]$. If $\zeta_{i,k} \leq h$ and $\tau_{i-1,k} < 1$, then $\tau_{i,k} \leq 1$. By definition of the stopping times, we have

$$\begin{aligned} &\left\{\zeta_{i,k} \leq h,\ \tau_{i-1,k} < 1\right\} \\ &\qquad = \left\{\operatorname{Osc}(Y; [\tau_{i-1,k}, (\tau_{i-1,k} + h) \wedge 1]_F) > M_k,\ \tau_{i-1,k} < 1\right\}. \end{aligned} \tag{12.66}$$

Now $\tau_{i-1,k}$ has possible values less than 1 in a finite set $F_i = F_{i,k} \subset [0,1)_F$. Let $\sigma_{i,k} := \sigma(\tau_{i-1,k}, Y(\tau_{i-1,k}))$. Since $\{Y_t\}_{t \in F}$ is a Markov process, for each $s \in F_i$, the two events $\{\operatorname{Osc}(Y; [s, (s+h) \wedge 1]_F) > M_k\}$ and $\{d(Y((s+h) \wedge 1), Y(s)) > M_{k+2}\}$ are each conditionally independent of $\{\tau_{i-1,k} = s\}$ given $\sigma(Y(s))$. To apply Lemma 12.29, note that $\{\zeta_{i,k} \leq h\}$ is an event $A(\tau)$ for $\tau := \tau_{i-1,k}$. Thus by Lemmas 12.29 and 12.28, and since $M_{k+2} \equiv M_k/4$, we have

$$\begin{aligned} &\Pr\left(\{\zeta_{i,k} \leq h\} \middle| \mathcal{F}_{\tau_{i-1,k}}^{Y,F}\right) \\ &= \Pr\left(\{\zeta_{i,k} \leq h\} \middle| \sigma_{i,k}\right) \\ &\leq \Pr\left(\left\{\operatorname{Osc}(Y; [\tau_{i-1,k}, (\tau_{i-1,k} + h) \wedge 1]_F) > M_k\right\} \middle| \sigma_{i,k}\right) \\ &= \sum_{s \in F_i} \Pr(\tau_{i-1,k} = s | \sigma(Y(s))) \Pr\left(\{\operatorname{Osc}(Y; [s, (s+h) \wedge 1]_F) > M_k\} \middle| \sigma(Y(s))\right) \\ &\leq \sum_{s \in F_i} \Pr(\tau_{i-1,k} = s | \sigma(Y(s))) \frac{\Pr(\{d(Y((s+h) \wedge 1), Y(s)) > M_{k+2}\} | \sigma(Y(s)))}{1 - \alpha(h, M_{k+2})} \\ &= \Pr(\{d(Y((\tau_{i-1,k} + h) \wedge 1), Y(\tau_{i-1,k})) > M_{k+2}\} | \sigma_{i,k}) / [1 - \alpha(h, M_{k+2})] \end{aligned}$$

almost surely on $\{\tau_{i-1,k} < 1\}$ by (12.66). For a fixed $h > 0$, $(\tau_{i-1,k}+h)\wedge 1$ has possible values in a finite set $G_{i,h}$. For each $u \in F_i$ and $v := (u+h)\wedge 1 \in G_{i,h}$, we have $0 < v-u \le h$. Applying the definition of $\alpha$ to the Markov kernel $P_{u,v}$ for all such $(u,v)$ and noting that the supremum over $x$ includes all possible values of $Y(\tau_{i-1,k})$, we see that the numerator in the last displayed fraction is at most $\alpha(h, M_{k+2})$. The conclusion of the lemma follows. □

We also need to deal with the incomplete gamma function defined for $a > 0$ and $x \ge 0$ by

$$\gamma(a,x) := \int_0^x u^{a-1}\mathrm{e}^{-u}\,du. \tag{12.67}$$

**Lemma 12.31.** *If* $0 < x < 3(3+a)(2+a)^{-1}$ *then*

$$\gamma(a,x) < \frac{x^a}{a}\Big\{1 - \frac{a}{(a+1)}x + \frac{a}{2(a+2)}x^2\Big\}.$$

*Proof.* By equations (6.5.2), (6.5.4), and (6.5.29) from [38, pp. 260 and 262], the incomplete gamma function has the following series representation: for any $a > 0$ and $x \ge 0$,

$$\gamma(a,x) = \sum_{k=0}^{\infty} \frac{(-1)^k x^{k+a}}{k!(k+a)}.$$

For $k \ge 2$ and $x$ as specified, the ratio of absolute values of successive terms is

$$\frac{x(k+a)}{(k+1)(k+a+1)} < 1,$$

as can be seen since $(u+a)/[(u+1)(u+a+1)]$ is decreasing for $u \ge 2$. Since the series is alternating and the term with index $k = 3$ is negative, one can discard all terms with $k \ge 3$ and get the stated bound, completing the proof of the lemma. □

For all $h \in [0,1]$ and $r > 0$, since $\alpha(h,r)$ is nondecreasing as $r$ decreases, we have for $r_0$ as in the definition of the class $\mathcal{M}(\beta,\gamma)$,

$$\alpha(h,r) \le K\frac{h^\beta}{(r\wedge r_0)^\gamma}. \tag{12.68}$$

**Lemma 12.32.** *Suppose that for an integer* $k$, *we have*

$$T_k := \min\Big\{\Big(\frac{(r_0\wedge M_{k+2})^\gamma}{2K}\Big)^{1/\beta}, 1\Big\} < 1. \tag{12.69}$$

*Then for each finite* $F \subset \mathcal{T}$ *and* $j = 1, 2, \ldots$,

$$\Pr(\{\tau_{j,k}^F \le 1\}) \le \mathrm{e}\left(\beta\gamma(\beta, T_k)/T_k^\beta\right)^j.$$

*Proof.* By condition (12.68) and by the definition of $T_k$, for each $k$ such that (12.69) holds, we have

$$\alpha(T_k, M_{k+2}) \le KT_k^\beta (M_{k+2} \wedge r_0)^{-\gamma} = 1/2.$$

Let $i$ be a positive integer. Recall that $\zeta_{i,k} = \tau_{i,k} - \tau_{i-1,k}$. Using Theorem 10.2.5 in [53], integration by parts, Lemma 12.30, and (12.68), we have

$$\begin{aligned}
&E(\mathrm{e}^{-\zeta_{i,k}}|\mathcal{F}^{Y,F}_{\tau_{i-1,k}}) \\
&\quad = \int_0^2 \mathrm{e}^{-x}\, \mathrm{d}\Pr\left(\{\zeta_{i,k} \le x|\mathcal{F}^{Y,F}_{\tau_{i-1,k}}\}\right) \\
&\quad \le \mathrm{e}^{-2} + \int_{T_k}^2 \mathrm{e}^{-x}\, \mathrm{d}x + \int_0^{T_k} \Pr\left(\{\zeta_{i,k} \le x|\mathcal{F}^{Y,F}_{\tau_{i-1,k}}\}\right)\mathrm{e}^{-x}\, \mathrm{d}x \\
&\quad \le \mathrm{e}^{-T_k} + 2\int_0^{T_k} \alpha(x, M_{k+2})\mathrm{e}^{-x}\, \mathrm{d}x \le \mathrm{e}^{-T_k} + \frac{1}{T_k^\beta}\int_0^{T_k} x^\beta \mathrm{e}^{-x}\, \mathrm{d}x
\end{aligned}$$

almost surely on $\{\tau_{i-1,k} < 1\}$. Recalling the incomplete gamma function (12.67), integration by parts once again yields

$$0 \le \gamma(\beta+1, T_k) = \int_0^{T_k} x^\beta \mathrm{e}^{-x}\, \mathrm{d}x = -T_k^\beta \mathrm{e}^{-T_k} + \beta\gamma(\beta, T_k).$$

Thus almost surely on $\{\tau_{i-1,k} < 1\}$, we have

$$E(\mathrm{e}^{-\zeta_{i,k}}|\mathcal{F}^{Y,F}_{\tau_{i-1,k}}) \le \beta\gamma(\beta, T_k)/T_k^\beta. \tag{12.70}$$

Let $j \ge 1$ be an integer. Using properties of conditional expectations, Markov's inequality, and the bound (12.70) applied $j$ times, it follows that

$$\begin{aligned}
&\Pr(\{\tau_{j,k} \le 1\}) = \Pr(\{\tau_{j,k} \le 1,\ \tau_{j-1,k} < 1\}) \\
&\quad = E\Big(1_{\{\tau_{j-1,k}<1\}} \Pr(\{\tau_{j,k} \le 1|\mathcal{F}^{Y,F}_{\tau_{j-1,k}}\})\Big) \\
&\quad = E\Big(1_{\{\tau_{j-1,k}<1\}} \Pr(\{\mathrm{e}^{-\tau_{j,k}} \ge \mathrm{e}^{-1}|\mathcal{F}^{Y,F}_{\tau_{j-1,k}}\})\Big) \\
&\quad \le \mathrm{e}E\Big(1_{\{\tau_{j-1,k}<1\}} E(\mathrm{e}^{-\tau_{j,k}}|\mathcal{F}^{Y,F}_{\tau_{j-1,k}})\Big) \\
&\quad = \mathrm{e}E\left(\prod_{i=1}^{j-1}\left(1_{\{\tau_{i,k}<1\}}\mathrm{e}^{-\zeta_{i,k}}\right) E(\mathrm{e}^{-\zeta_{j,k}}|\mathcal{F}^{Y,F}_{\tau_{j-1,k}})\right) \\
&\quad \le \mathrm{e}\frac{\beta\gamma(\beta, T_k)}{T_k^\beta} E\left(\prod_{i=1}^{j-2}\left(1_{\{\tau_{i,k}<1\}}\mathrm{e}^{-\zeta_{i,k}}\right) E(\mathrm{e}^{-\zeta_{j-1,k}}|\mathcal{F}^{Y,F}_{\tau_{j-2,k}})\right) \\
&\quad \le \cdots \le \mathrm{e}\left(\frac{\beta\gamma(\beta, T_k)}{T_k^\beta}\right)^j,
\end{aligned}$$

proving the lemma. □

Now to complete the proof of Theorem 12.27, we can and do assume that $0 < r_0 \le 1$ and $K \ge 1$. For $u > 0$ and $\omega \in \Omega$, let $\nu_u(\omega)$ be the supremum of integers $m$ such that for some $\{t_i\}_{i=1}^{2m} \subset \mathcal{T}$ such that $0 \le t_1 < t_2 \le t_3 < \cdots < t_{2m} \le 1$, we have $\Delta_i Y(\cdot,\omega) := d(Y(t_{2i},\omega), Y(t_{2i-1},\omega)) > u$ for each $1 \le i \le m$. For an integer $k$, a partition $\kappa = \{t_i\}_{i=0}^{2n} \subset \mathcal{T}$ of $[0,1]$, and $\omega \in \Omega$, let

$$I_k = I_k(\kappa,\omega) := \big\{i \in \{1,\dots,n\}\colon\, 2^{-k-1} \le \Delta_i Y(\cdot,\omega) < 2^{-k}\big\}$$

and $N_k(\omega) := \sup\{\operatorname{card} I_k(\kappa,\omega)\colon\, \kappa \in PP[0,1],\ \kappa \subset \mathcal{T}\}$. Let $k_1$ be the largest integer less than or equal to $-(\log_2 r_0 + 3) \ge -3$. Thus $k_1 \ge -3$ and $2^{-k-1} \ge 4r_0 > r_0$ for each $k \le k_1$. Then we have

$$\begin{aligned} v_p\big(Y(\cdot,\omega);\mathcal{T}\big) &\le \sum_{k>k_1} \sup_{\kappa} \sum_{i\in I_k} \Delta_i Y(\cdot,\omega)^p + \sup_{\kappa} \sum_{k\le k_1} \sum_{i\in I_k} \Delta_i Y(\cdot,\omega)^p \\ &\le \sum_{k>k_1} 2^{-kp} N_k(\omega) + \operatorname{Osc}(Y(\cdot,\omega);\mathcal{T})^p \nu_{r_0}(\omega). \end{aligned} \tag{12.71}$$

It will next be proved that

$$\sum_{k>k_1} 2^{-kp} E N_k < \infty. \tag{12.72}$$

Let $k > k_1$. Then we have

$$EN_k = \int_0^\infty \Pr(\{N_k \ge x\})\, dx \le \sum_{j=0}^\infty \Pr(\{N_k \ge j\}). \tag{12.73}$$

By definition of $k_1$, we have $2^{-k-3} < r_0 \le 1$, and so

$$T_k = \min\{(2^{-(k+3)\gamma}/(2K))^{1/\beta}, 1\} < 1.$$

By Lemma 12.32, for each $j$

$$\Pr(\{N_k \ge j\}) \le \sup_{F\subset\mathcal{T}} \Pr(\{\tau_{j,k}^F \le 1\}) \le \mathrm{e}\left\{\frac{\beta\gamma(\beta,T_k)}{T_k^\beta}\right\}^j, \tag{12.74}$$

where the supremum is over finite sets $F \subset \mathcal{T}$ or specifically over a sequence $F = F_n \uparrow \mathcal{T}$. By Lemma 12.31, since $T_k < 1 < 2(\beta+2)/(\beta+1)$, we have $\beta\gamma(\beta,T_k) < T_k^\beta$ and further

$$\frac{1}{T_k^\beta - \beta\gamma(\beta,T_k)} \le \frac{1}{\frac{\beta}{\beta+1}T_k^{\beta+1} - \frac{\beta}{2(\beta+2)}T_k^{\beta+2}} \le \frac{2(\beta+1)}{\beta T_k^{\beta+1}} \le \frac{4}{T_k^{\beta+1}},$$

recalling that $\beta \ge 1$ is assumed. Thus by (12.73) and (12.74),

$$EN_k \le \mathrm{e}\sum_{j=0}^\infty \left\{\frac{\beta\gamma(\beta,T_k)}{T_k^\beta}\right\}^j = \frac{\mathrm{e}T_k^\beta}{T_k^\beta - \beta\gamma(\beta,T_k)} \le \frac{4\mathrm{e}}{T_k}.$$

Still for $k > k_1$, for some constant $C$,

$$T_k^{-1} = (2K)^{1/\beta} 2^{(k+3)\gamma/\beta} \le C 2^{k\gamma/\beta}.$$

Summing a geometric series, since $p > \gamma/\beta$, it follows that

$$\sum_{k>k_1} 2^{-kp} EN_k \le 4\mathrm{e} \sum_{k>k_1} 2^{-kp} T_k^{-1} \le 4\mathrm{e}C \sum_{k>k_1} 2^{-k(p-\gamma/\beta)}$$

$$= 4\mathrm{e}C \frac{2^{-(k_1+1)(p-\gamma/\beta)}}{1 - 2^{-(p-\gamma/\beta)}} < \infty,$$

proving (12.72).

Now, for any $k > k_1$, we have $\beta\gamma(\beta, T_k) < T_k^\beta$, and so the right side of (12.74) tends to 0 as $j \to \infty$. For $u = 2^{-k-1}$ it follows that $\Pr(\{\nu_u < \infty\}) = 1$. For a fixed $k$, e.g. $k = k_1 + 1$, we get for almost all $\omega$ that for any fixed $x_0 \in S$, $\sup_{t\in\mathcal{T}} d(Y_t(\omega), x_0) < \infty$. Also, since $2^{-k-1}$ can be arbitrarily small, it follows that $\Pr(\{\nu_u < \infty\}) = 1$ for every $u > 0$, in particular for $u = r_0$. It then follows from (12.71) and (12.72) that $v_p(Y(\cdot,\omega); \mathcal{T}) < \infty$ almost surely.

Just as a function of bounded $p$-variation (or $\Phi$-variation) from $[0,1]$ into a Banach space must be regulated (Proposition 3.33), it follows that for some $A_0$ with $\Pr(A_0) = 0$, for all $\omega \notin A_0$, and every $t \in [0,1)$, the right limit $Y(t+,\omega) := \lim_{s\downarrow t, s\in\mathcal{T}} Y(s,\omega)$ exists in $S$ (here we use completeness of $S$). From the assumptions and definition of $\alpha$, for each $t \in [0,1)$, $Y(s) = X(s) \to X(t)$ in probability as $s \downarrow t$ with $s \in \mathcal{T}$. Thus $X(t) = Y(t+)$ a.s. In particular, $Y(t) = Y(t+)$ a.s. for all $t \in \mathcal{T} \cap [0,1)$. Since $\mathcal{T}$ is countable, for some $A_1$ with $\Pr(A_1) = 0$, we have $Y(t,\omega) = Y(t+,\omega)$ for all $t \in \mathcal{T} \cap [0,1)$ and $\omega \notin A_1$. Define $Y(t,\omega) = Y(t+,\omega)$ for all $t \in [0,1)$ and $\omega \notin A_0 \cup A_1$, which leaves $Y(t,\omega)$ unchanged for $t \in \mathcal{T}$. For a fixed $x_0 \in S$ let $Y(t,\omega) = x_0$ for $\omega \in A_0 \cup A_1$ and all $t \in [0,1]$. Then $Y(t,\omega)$, $0 \le t \le 1$, is a Markov process with $\Pr(\{\omega\colon Y(t,\omega) = X_t(\omega)\}) = 1$ for each $t$. So, $Y$ has the same kernel $P_{\cdot,\cdot}(\cdot,\cdot)$ as for $X$. Clearly the $Y$ process has right-continuous trajectories on $[0,1)$. Because of that, and because $1 \in \mathcal{T}$, $Y(\cdot,\omega)$ has the same $p$-variation on $[0,1]$ as on $\mathcal{T}$, which is finite. The proof of Theorem 12.27 is complete. □

## 12.6 Lévy Processes

### *Infinitely divisible laws and the Lévy–Khinchin formula*

Recall that for two laws (probability measures) $P$ and $Q$, defined in this section on the Borel sets of the real line, the convolution $P * Q$ is defined by $(P*Q)(A) = \int_{-\infty}^{\infty} P(A-x)\,\mathrm{d}Q(x)$. Convolution is a commutative, associative operation. For $n = 1, 2, \ldots$, the $n$th convolution power of a law $P$ is defined by $P^{*n} := (P * P * \cdots * P)$ to $n$ factors, with $P^{*0} := \delta_0$.

A law $P$ is said to be *infinitely divisible* iff for all $n$ there is a law we call $P_{1/n}$ such that $P_{1/n}^{*n} = P$. Recall that the characteristic function of a law $P$ is defined by $f_P(u) := \int_{-\infty}^{+\infty} \mathrm{e}^{iux}\, \mathrm{d}P(x)$ for $u \in \mathbb{R}$. Then we recall the following fact, called a Lévy–Khinchin formula (e.g. [23, Chapter 9], [64, §XVII.2]):

**Theorem 12.33.** *A law $P$ on $\mathbb{R}$ is infinitely divisible if and only if its characteristic function can be written in the form $f_P(u) \equiv \mathrm{e}^{\eta(u)}$ where*

$$\eta(u) := \eta_{a,\sigma,L}(u) := \mathrm{i}au - \frac{\sigma^2}{2}u^2 + \int_{\mathbb{R}\setminus\{0\}} \left(\mathrm{e}^{\mathrm{i}ux} - 1 - \mathrm{i}uh(x)\right) L(\mathrm{d}x) \quad (12.75)$$

*for $u \in \mathbb{R}$, where $h(x) := x/(1+x^2)$, $a$ and $\sigma \geq 0$ are real numbers, and $L$ is a measure on $\mathbb{R} \setminus \{0\}$ such that $\int_{\mathbb{R}\setminus\{0\}} (1 \wedge x^2)\, L(\mathrm{d}x) < \infty$. Here the numbers $a$ and $\sigma$ and the measure $L$ are uniquely determined.*

A measure $\mu$ on $\mathbb{R}$ or $\mathbb{R} \setminus \{0\}$ is called *symmetric* if it is preserved by $x \mapsto -x$. We then have the following:

**Corollary 12.34.** *An infinitely divisible law $P$ is symmetric if and only if in the representation (12.75), $a = 0$ and $L$ is symmetric. Then the integrand can be replaced by its real part* $\cos(ux) - 1$.

*Proof.* If $P$ is symmetric, it is easily seen that its characteristic function $f_P$ is real-valued and even, $f_P(-u) \equiv f_P(u)$. Since in (12.75), $\eta(0) = 0$, $\eta$, being continuous by dominated convergence, must also be real-valued and even. From (12.75) with $-u$ in place of $u$ and uniqueness of $a$ and $L$ we see that $a = 0$ and $L$ is symmetric. Then since the imaginary part of the integrand is an odd function of $x$, its integral vanishes and $\eta$ has the stated form.

Conversely, the given form implies that $\eta$ and so $f_P$ are real-valued and even, which by uniqueness of characteristic functions implies that $P$ is symmetric. □

The function $h$ in Theorem 12.33 is far from uniquely determined. Its essential properties are that it is bounded and Borel measurable on $\mathbb{R}$ and is $x + O(x^2)$ as $x \to 0$. Different choices of $h$ satisfying these conditions can give different values of $a$. The choice of $h$ given in the theorem as an odd function is convenient, as in the proof of Corollary 12.34. Under the conditions of Theorem 12.33, $\eta$ is called a *characteristic exponent* and $L$ a *Lévy measure.*

A family $\{P_t\}_{t\geq 0}$ of laws on $\mathbb{R}$ will be called a *convolution semigroup* if $P_s * P_t = P_{s+t}$ for all $s, t \geq 0$. A convolution semigroup $\{P_t\}$ will be called *continuous* iff as $t \downarrow 0$, $P_t \to \delta_0$ in the sense of weak convergence, in other words, $\int_{-\infty}^{+\infty} f(x)\, \mathrm{d}P_t(x) \to f(0)$ as $t \downarrow 0$ for every bounded continuous real-valued function $f$ on $\mathbb{R}$. The following fact will then be easy to prove, and the one-to-one correspondence will show that the notation $P_{1/n}$ is unambiguous for infinitely divisible laws.

**Theorem 12.35.** *There is a one-to-one correspondence between infinitely divisible laws $P$ on $\mathbb{R}$ and continuous convolution semigroups $\{P_t\}_{t\geq 0}$ such that $P = P_1$.*

*Proof.* Clearly, each law, specifically $P_1$, in a convolution semigroup is infinitely divisible. Conversely, let $P$ be infinitely divisible. It is immediate that if a Lévy measure $L$ or characteristic exponent $\eta$ is multiplied by a constant $c \geq 0$, then it remains a Lévy measure or characteristic exponent respectively. Let $\eta$ be the characteristic exponent of $P$. Then for each $t \geq 0$, there exists a law $P_t$ with characteristic function $e^{t\eta(\cdot)}$, and by properties of characteristic functions (e.g. [53, Theorems 9.4.3, 9.5.1, 9.8.2]), $P_t$ for $t \geq 0$ form a continuous convolution semigroup, with $P = P_1$. Let $\{Q_t\}_{t\geq 0}$ be another continuous convolution semigroup with $Q_1 = P$. For each $n = 2, 3, \ldots$, the characteristic exponent of $Q_{1/n}$ is $\eta/n$ by the uniqueness of $a, \sigma$, and $L$ in Theorem 12.33. It follows that $Q_t = P_t$ for all rational $t \geq 0$ and thus by continuity of both semigroups for all $t \geq 0$, finishing the proof. □

Let $X = \{X_t\}_{t\geq 0}$ be a real-valued stochastic process with independent increments. Then the finite-dimensional joint distributions of the process, in other words the distributions of $\{X(t)\}_{t\in F}$ for each finite set $F$, are uniquely determined by those of the increments $X(t) - X(s)$ for $0 \leq s < t$. Thus the process has stationary increments (the definition is recalled before (12.8)) if and only if the distribution of $X(t+s) - X(t)$ with $t, s \geq 0$ does not depend on $t$. If in addition $X(0) \equiv 0$, then the finite-dimensional joint distributions of the process are uniquely determined by the distributions $P_t$ of $X(t)$ for each $t \geq 0$, which clearly form a convolution semigroup. If the process is continuous in probability at 0, i.e. $X(t) \to 0$ in probability as $t \downarrow 0$, then the convolution semigroup is continuous at 0.

We then have the following fact:

**Theorem 12.36.** *Let $\{P_t\}_{t\geq 0}$ be any continuous convolution semigroup of laws on $\mathbb{R}$. Then there exists a stochastic process $\{X_t\}_{t\geq 0}$ with stationary, independent increments, with $X(0) \equiv 0$, where for any $0 \leq s < t$, the distribution of $X(t) - X(s)$ is $P_{t-s}$. Moreover, the process can be chosen so that almost all its sample functions are right continuous with left limits, and for each fixed $t$, the probability that $X(\cdot)$ is continuous at $t$ is 1.*

*Proof.* The existence of a process with the given properties except those in the last sentence follows from the Kolmogorov consistency theorem, e.g. [53, Theorem 12.1.2]. The last sentence follows e.g. from Kallenberg [112, Theorem 13.1]. □

A process $X$ with the properties given in Theorem 12.36 is called a *Lévy process* (cf. Itô [107, Section 1.3]). For a Lévy process $X$, the characteristic function of $X(t)$ for each $t$ is given by $E\exp\{\mathrm{i}uX(t)\} = \exp\big\{t\eta(u)\big\}$ for each $t \geq 0$ and $u \in \mathbb{R}$, where $\eta$ is the characteristic exponent of the distribution of $X(1)$.

A Lévy process $X$ has the Itô representation (cf. Itô [107, Section 1.7]), for each $t \geq 0$,

$$\begin{aligned} X(t) &= at + \sigma B(t) \\ &\quad + \lim_{\epsilon\downarrow 0}\Big[\sum\Big\{\Delta^- X(s)\colon\ s \leq t,\ |\Delta^- X(s)| > \epsilon\Big\} - t\int_{|x|>\epsilon} h(x)\,L(\mathrm{d}x)\Big], \end{aligned} \tag{12.76}$$

where $B$ is a Brownian motion independent of $X - \sigma B$.

It is well known that sample functions of a Lévy process $X$ with the characteristic exponent (12.75) are of bounded variation if and only if $\sigma = 0$ and

$$\int_{\mathbb{R}\setminus\{0\}} (1 \wedge |x|)\,L(\mathrm{d}x) < \infty \tag{12.77}$$

(cf. e.g. Gihman and Skorohod [78, Theorem 3, p. 279]). The following result of Bretagnolle [24, Théorème III b] is less well known. It sharpened an earlier result of Blumenthal and Getoor [19] and Monroe [172].

**Theorem 12.37** (J. Bretagnolle). *Let* $1 < p < 2$ *and let* $X = \{X_t\}_{t\geq 0}$ *be a mean-zero Lévy process with the characteristic exponent* (12.75) *such that* $\sigma = 0$. *Then* $v_p(X;[0,1]) < \infty$ *with probability* 1 *if and only if*

$$\int_{\mathbb{R}\setminus\{0\}} (1 \wedge |x|^p)\,L(\mathrm{d}x) < \infty. \tag{12.78}$$

*If* (12.78) *fails then* $v_p(X;[0,1]) = +\infty$ *almost surely.*

The theorem is used in Remark 12.39 to show that the main result in the preceding section (Theorem 12.27) is sharp.

**Remark 12.38.** Suppose a Lévy process $X$ is symmetric in the sense that the process $-X$ is the same in distribution, or equivalently that in the corresponding convolution semigroup $\{P_t\}_{t\geq 0}$ given by Theorem 12.36 and the discussion before it, each $P_t$, or equivalently $P_1$, is symmetric. Then by Corollary 12.34, in (12.75) and (12.76), $a = 0$ and $L$ is symmetric. Thus the integral in (12.76) is 0 for all $\epsilon > 0$. If in addition as in the preceding Theorem 12.37, $\sigma = 0$, then the Itô representation (12.76) reduces to

$$X(t) \;=\; \lim_{\epsilon\downarrow 0}\sum\Big\{\Delta^- X(s)\colon\ s \leq t,\ |\Delta^- X(s)| > \epsilon\Big\}$$

for each $t \geq 0$. Thus the process is determined entirely by its jumps, although if the integral in (12.77) diverges, these will not be absolutely summable. On a finite, nondegenerate interval, the jumps will sum up in a nontrivial way.

*$\alpha$-stable Lévy motion*

A Lévy process $X$ is an *$\alpha$-stable Lévy motion* with $\alpha \in (0,2)$ if in its characteristic exponent (12.75), $\sigma = 0$ and the Lévy measure $L$ has the form $L = L_\alpha = L_{\alpha,r,q}$, where

$$L_{\alpha,r,q}(\mathrm{d}x) := \begin{cases} rx^{-1-\alpha}\mathrm{d}x & \text{if } x > 0, \\ q(-x)^{-1-\alpha}\mathrm{d}x & \text{if } x < 0, \end{cases}$$

for $r, q \geq 0$ with $r + q > 0$. It is easy to see that (12.78) with $L = L_\alpha$ holds if and only if $p > \alpha$. Therefore $v_p(X; [0,1]) < \infty$ with probability 1 for each $p > \alpha$.

An $\alpha$-stable Lévy motion $X$ is called *symmetric* if in (12.75) also $a = 0$ and $r = q$. In that case, taking $X_0 \equiv 0$, the distribution of $X_1$ is a symmetric $\alpha$-stable law, i.e. one whose characteristic function $\mathrm{e}^\eta$ for $\eta$ given by (12.75) has $a = \sigma = 0$ and Lévy measure $L = L_{\alpha,r,r}$ for some $r > 0$. Symmetry of the law of $X_1$ implies $a = 0$ and $q = r$ by Corollary 12.34. It is well known, e.g. [204, p. 86, Theorem 14.15], that in this case for $\eta$ in (12.75), $\eta(u) \equiv c|u|^\alpha$ for some $c > 0$. Moreover, for $0 < s < t$, the distribution of $X_t - X_s$, also conditional on any value of $X_s$, equals that of $(t-s)^{1/\alpha}X_1$ by equality of characteristic functions (e.g. [53, Theorem 9.5.1]).

**Remark 12.39.** It will be shown that for a symmetric $\alpha$-stable process with $0 < \alpha < 2$, Theorem 12.27 of Manstavičius is sharp in that $p > \gamma/\beta$ cannot be replaced by $p \geq \gamma/\beta$. Given $\alpha$, there is a $b$ with $0 < b < \infty$ such that as $x \to +\infty$, $x^\alpha \Pr(X_1 > x) \to b$, e.g. [203, p. 16, Property 1.2.15]. Thus for some $x_0$ with $0 < x_0 < \infty$ we have $\Pr(X_1 > x) \leq 2bx^{-\alpha}$ for all $x \geq x_0$. Clearly $\Pr(X_1 > x) \leq (x_0/x)^\alpha$ for $0 < x \leq x_0$. Since $X_1$ is symmetric, it follows that for some $K < \infty$, $\Pr(|X_1| > x) \leq K/x^\alpha$ for all $x > 0$. It follows for any $h > 0$, $a > 0$, and $t \geq 0$ that

$$\Pr(|X_{t+h} - X_t| > a) \;=\; \Pr(|X_1| > ah^{-1/\alpha}) \;\leq\; Kh/a^\alpha,$$

also conditional on any value of $X_t$. Thus, the condition (12.56) holds with $\beta = 1$ and $\gamma = \alpha$ and so $\gamma/\beta = \alpha$. Since the process for given $\alpha$ does not have bounded $p$-variation for $p = \alpha$, this shows that the assumption $p > \gamma/\beta$ in Theorem 12.27 is sharp. Moreover, for this, according to Manstavičius [155], one does not need the full strength of Bretagnolle's theorem as quoted above; the earlier results of Blumenthal and Getoor [19] suffice.

## 12.7 Empirical Processes

Let $P$ be a probability distribution on $\mathbb{R}$ with distribution function $F$: $F(x) := P((-\infty, x])$ for all $x$. Then $F$ is nondecreasing and right-continuous. For $0 <$

$y < 1$ let $F^{\leftarrow}(y) := \inf\{x : F(x) \geq y\}$. Then the *quantile function* $F^{\leftarrow}$ is nondecreasing and left-continuous from $(0, 1)$ into $\mathbb{R}$.

Let $X_1, X_2, \ldots, X_n$ be independent with law $P$ and let $F_n$ be the empirical distribution function $F_n(t) := n^{-1}\sum_{j=1}^{n} 1_{\{X_j \leq t\}}$ for $t \in \mathbb{R}$. Let $U$ be the $U[0, 1]$ distribution function $U(t) := \max(0, \min(1, t))$ for $t \in \mathbb{R}$, and $U_n$ an empirical distribution function for it. Clearly $U \circ F \equiv F$, and we can take $F_n \equiv U_n \circ F$ [53, Lemma 11.4.3]. Then letting $\beta_n := n^{1/2}(U_n - U)$ on $[0, 1]$,

$$\alpha_n := n^{1/2}(F_n - F) = \beta_n \circ F \tag{12.79}$$

is the classical empirical process.

For any distribution function $F$ and function $h$ on $[0, 1]$ it follows from the definition of $p$-variation that

$$\|h \circ F\|_{[p]} \leq \|h\|_{[p]}. \tag{12.80}$$

Here, equality holds if $F$ is continuous and $h$ is right-continuous at 0 and left-continuous at 1. Almost surely $U_n$ is continuous at 0 and at 1. Then for $F$ continuous, by (12.79), the $p$-variation of $\alpha_n$ on $\mathbb{R}$ equals that of $\beta_n$. In studying the $p$-variation norm of $\alpha_n$ for $F$ continuous we can thus assume $F \equiv U$ and $\alpha_n = \beta_n$. Another simplification comes from the fact that for a function $f\colon [0, 1] \to \mathbb{R}$ which is 0 at the endpoints of $[0, 1]$, as $\beta_n$ is, we have for $1 \leq p < \infty$,

$$\|f\|_{\sup} \leq \|f\|_{(p)} \quad \text{and so} \quad \|f\|_{[p]} \leq 2\|f\|_{(p)}. \tag{12.81}$$

Qian [189, Theorems 3.2 and 4.2] proved the following:

**Theorem 12.40** (J. Qian). *Let $1 \leq p < 2$. Then there are constants $M_p, \lambda_p < \infty$, depending only on $p$, such that*

$$Ev_p(\alpha_n) \leq M_p n^{1-p/2} \quad \textit{and} \quad \limsup_{n\to\infty} v_p(\alpha_n)/n^{1-p/2} \leq \lambda_p$$

*almost surely.*

Conversely, if $F$ is continuous, the $X_i$ are almost surely distinct, so we have $v_p(\alpha_n) \geq n^{1-p/2}$, and the above theorem is sharp up to the constants $M_p$ and $\lambda_p$.

For $p = 2$ we have the following:

**Theorem 12.41.** *Let $F$ be a distribution function on $\mathbb{R}$.*

(*a*) $v_2(\alpha_n) = O_{\Pr}(\log\log n)$ *as* $n \to \infty$.
(*b*) *If $F$ is continuous and $0 < c < 1/12$, then* $\Pr(\{v_2(\alpha_n) > c\log\log n\}) \to 1$ *as* $n \to \infty$.

*Proof.* (*a*): By (12.80) with $h = \beta_n$ we can assume that $F$ is continuous. For a given $n$ let $f := \alpha_n$. For a partition $\kappa = \{t_i\}_{i=0}^m$ of $\mathbb{R}$ let $\Delta_i := f(t_i) - f(t_{i-1})$, $i = 1, \ldots, m$. We can assume that $m \geq 2$ and

$$\inf\{t \in \mathbb{R}\colon\ F(t) > 0\} < t_1 \leq t_{m-1} < \sup\{t \in \mathbb{R}\colon\ F(t) < 1\}.$$

To maximize the 2-variation sums $s_2(f; \kappa) = \sum_{i=1}^m \Delta_i^2$ while using only elements of $\kappa$ we can assume that all $\Delta_i \neq 0$ and that the $\Delta_i$ are alternating in sign, since $(A+B)^2 > A^2 + B^2$ for $AB > 0$, so any adjoining differences of the same sign should be combined. Moreover, if at any stage of the following proof, points are adjoined to or deleted from $\kappa$, giving a new partition $\kappa'$ such that $s_2(f; \kappa') \geq s_2(f; \kappa)$, and if $\kappa'$ gives any $\Delta_i = 0$, or any adjoining $\Delta_i$ of the same sign, points can be deleted from $\kappa'$ until we get a partition $\kappa''$ with no $\Delta_i = 0$, all $\Delta_i$ alternating in sign, and $s_2(f; \kappa'') \geq s_2(f; \kappa') \geq s_2(f; \kappa)$.

Claim 1: while only increasing $m$ by at most 3, we can always assume that $\Delta_1 < 0$.

To prove the claim, if $F(t_0) = 0$ then for $X_{(1)}$, the smallest $X_j$, we have $X_{(1)} > t_0$ and $F(X_{(1)}) > 0$ (with probability 1) and if $\Delta_1 \geq 0$, then $X_{(1)} \leq t_1$ and we can enlarge $s_2(f; \kappa)$ by inserting a new point just to the left of $X_{(1)}$, giving a new $\Delta_1' < 0$ as desired and $\Delta_2' > \Delta_1$.

If $F(t_0) > 0$ and $X_{(1)} \geq t_0$ then we can enlarge $s_2(f; \kappa)$ by adjoining two new points $t_{-2} < t_{-1} < t_0$ with $F(t_{-2}) < F(t_{-1})$ and so again $\Delta_1' < 0$. So we can assume $X_{(1)} < t_0$ but then we adjoin three points, $t_{-1} = X_{(1)}$, $t_{-2}$ just to the left of $X_{(1)}$, and $t_{-3} < t_{-2}$ where $F(t_{-3}) < F(t_{-2})$. In this last case we again get a $\Delta_1' < 0$, proving Claim 1.

Claim 2: In $s_2(f; \kappa)$ we can assume, adjoining at most one point to the partition $\kappa$, that

(1) $\Delta_i > 2/(3n^{1/2})$ whenever $\Delta_i > 0$, and
(2) if $\Delta_i < 0$, then either $i = 1$ or $m$ or $|\Delta_i| \geq 1/(3n^{1/2})$.

For (1) let $\Delta_i > 0$, so $i \geq 2$ by Claim 1. There is at least one $X_j$ (in the sample on which $F_n$ is based) with $t_{i-1} < X_j \leq t_i$. For the largest such $X_j$, we can take $t_i = X_j$ since this can only enlarge both $\Delta_i$ and $|\Delta_{i+1}|$ (or a new $|\Delta_{m+1}|$ if $i = m$). Likewise, we can let $t_{i-1}$ increase toward (but not quite equal) some $X_k \leq X_j$. If $X_k = X_j$ then (1) holds. Otherwise $X_k < X_j$ and if $\Delta_i \leq 2/(3n^{1/2})$ we can move $t_{i-1}$ to a point just less than $X_j$. This will increase both $|\Delta_{i-1}|$ and $\Delta_i$ and make $\Delta_i > 2/(3n^{1/2})$, so (1) is proved.

For (2) let $\Delta_i < 0$ and $i \geq 2$. By Claim 1, we can assume $\Delta_1 < 0$ and therefore $\Delta_2 > 0$. Thus in fact $i \geq 3$ and $X_{(1)} \leq t_2$. The sum $s_2(f; \kappa)$ is not decreased if we let $t_{i-1}$ decrease until $t_{i-1} = X_j$ for some $j$. (If the largest $X_j < t_{i-1}$ also satisfies $X_j \leq t_{i-2}$ then $\Delta_{i-1} < 0$, contradicting the alternation in sign.) If $i < m$, we can also let $t_i$ increase nearly up to another observation $X_k$. Suppose (2) fails for $\Delta_i$ and there is another $r$ with $X_j < X_r < X_k$. Then by inserting new division points at $X_r$ and just below it, say at $y$, we can enlarge the 2-variation sum, as follows. Let $A := f(y) - f(t_{i-1})$, $B := f(X_r) - f(y)$, and $C := f(t_i) - f(X_r)$. Then by assumption $\Delta_i^2 = (A+B+C)^2 \le 1/(9n)$, whereas $A^2+B^2+C^2 \ge B^2$ which is as close as we want to $1/n$ and so is larger. If the partition $\kappa'$ resulting from the proof so far has any $\Delta_i = 0$ or any $\Delta_i\Delta_{i+1} > 0$, points can be deleted from $\kappa'$ until we have a partition $\kappa''$ where $\Delta_i$ alternate in sign and $s_2(f;\kappa'') \ge s_2(f;\kappa') \ge s_2(f;\kappa)$. Then (1) is preserved. Iterating, we can assume that there is no such $r$. By (1), we have $A := \Delta_{i-1} > 2/(3n^{1/2})$ and $B := \Delta_{i+1} > 2/(3n^{1/2})$, while $0 < C := -\Delta_i < 1/(3n^{1/2})$. Since $C < (1/2)\min\{A,B\}$ and $A+B \le 2\max\{A,B\}$ it follows that $AB > C(A+B)$ and so

$$(f(t_{i+1}) - f(t_{i-2}))^2 = (A+B-C)^2 > A^2+B^2+C^2.$$

Therefore the 2-variation sum is increased by deleting $t_{i-1}$ and $t_i$. Since $A+B-C > 2/(3n^{1/2}) > 0$, again (1) is preserved, and, iterating, (2) is proved.

For $\Phi := \Phi_{1/2}$ defined before (12.27) we have $\Phi(u) = u^2/(2\log\log(1/u))$ if $0 < u \le \delta$ for some $0 < \delta < \mathrm{e}^{-\mathrm{e}}$ and $\Phi$ is linear on $[\delta,\infty)$. Since $|\Delta_i| \ge 1/(3n^{1/2})$ for $i = 1,\dots,m$ except possibly for $i = 1, m$, there is a finite constant $K$ such that

$$s_2(f;\kappa) \le K\big[v_\Phi(f)\log\log n + \|f\|_{\sup} v_\Phi(f) + 1/n\big].$$

Dudley [50, §5] proved that $v_\Phi(\alpha_n)$ is bounded in probability. It is well known that $\|\alpha_n\|_{\sup}$ is bounded in probability (in connection with Kolmogorov–Smirnov statistics and Dvoretzky–Kiefer–Wolfowitz inequalities, e.g. Massart [160]). Conclusion $(a)$ then follows.

Qian [189, Theorem 3.1] proved (b). □

We do not know at this writing whether the bound in Theorem 12.41(a) holds almost surely.

Let $B^\circ = \{B_t^\circ\}_{0\le t\le 1}$ be the Brownian bridge process, a Gaussian stochastic process with mean 0 and $EB^\circ(t)B^\circ(s) = t(1-s)$ for $0 \le t \le s \le 1$, and such that $t \mapsto B^\circ(t,\omega)$ is continuous for almost all $\omega$. We can write $B^\circ(t) = B(t) - tB(1)$ for $0 \le t \le 1$ where $B(\cdot)$ is a Brownian motion since this gives the correct mean 0 and covariances. Thus, as for Brownian motion (12.11), almost surely

$$v_p(B^\circ(\cdot,\omega);[0,1]) \begin{cases} < +\infty \text{ if } p > 2, \\ = +\infty \text{ if } p \le 2. \end{cases} \tag{12.82}$$

For $1 \le p < \infty$ and $y \ge 0$, let

$$\phi_p(y) \ := \ (\mathrm{e}y/p)^p 1_{0\le y<p} + \mathrm{e}^y 1_{y\ge p}. \tag{12.83}$$

The left and right limits and derivatives of $\phi_p$ at $y = p$ all equal $\mathrm{e}^p$, and so $\phi_p$ is a $C^1$ function. Also, $\phi_p$ is strictly increasing and convex from $[0,\infty)$ onto itself, that is, $\phi_p \in \mathcal{CV}$. Let $\|\cdot\|_{\phi_p}$ be the corresponding Luxemburg norm on the Orlicz space $L^{\phi_p} = L^{\phi_p}(\Omega,\mathcal{F},\Pr;\mathbb{R})$ as defined in (1.24). The following fact, to be proved, is from Huang and Dudley [103].

**Theorem 12.42.** *For $2 < p < \infty$ there is a finite constant $A(p)$ such that if $F$ is any probability distribution function on $\mathbb{R}$, then on some probability space there exist independent random variables $X_1, X_2, \ldots$ with distribution function $F$ and Brownian bridges $B_n^\circ$ such that for all $n$, if $F_n$ is the empirical d.f. based on $X_1, \ldots, X_n$, then*

$$\big\| \big\|n^{1/2}(F_n - F) - B_n^\circ \circ F\big\|_{[p]}\big\|_{\phi_p} \le A(p) n^{\frac{1}{p}-\frac{1}{2}}. \tag{12.84}$$

The bound (12.84) in $L^{\phi_p}$-norm is sharp since the bound in $L^1$-norm it implies is sharp according to the following corollary (the first result in this section to be proved; see the Notes):

**Corollary 12.43** (Y.-C. Huang). *For $2 < p < \infty$ and $B_n^\circ$, $F_n$, and $F$ as in Theorem 12.42, there is a finite constant $C(p)$ such that*

$$E\big\|n^{1/2}(F_n - F) - B_n^\circ \circ F\big\|_{[p]} \le C(p) n^{\frac{1}{p}-\frac{1}{2}}. \tag{12.85}$$

*If $F$ is continuous, then for $V = B_n^\circ \circ F$ or any other sample-continuous stochastic process $V$,*

$$\big\|n^{1/2}(F_n - F) - V\big\|_{(p)} \ge n^{\frac{1}{p}-\frac{1}{2}} \qquad \textit{almost surely.} \tag{12.86}$$

*Proof.* (12.85) follows from (12.84) once we show that for any nonnegative random variable $Y$ and $1 < p < \infty$,

$$EY \le (1 + p/\mathrm{e})\|Y\|_{\phi_p}. \tag{12.87}$$

By homogeneity we can assume that $\|Y\|_{\phi_p} = 1$. Thus by Fatou's lemma, it follows that $E\phi_p(Y) \le 1$. Then by Hölder's inequality and the definition (12.83) of $\phi_p$,

$$EY1_{\{Y<p\}} \le \big(EY^p 1_{\{Y<p\}}\big)^{1/p} \le p/\mathrm{e}.$$

This together with the bound $EY1_{\{Y\ge p\}} \le 1$ yields (12.87). Thus (12.85) holds by (12.84).

For the lower bound (12.86), since $F$ is continuous, the random variables $X_1, X_2, \ldots$ are almost surely distinct, so the function $n^{1/2}(F_n - F)$ has $n$ distinct jumps of height $n^{-1/2}$, and (12.86) follows. □

Now we will prove Theorem 12.42. First, by (12.79) and (12.80), we can and do assume that $F$ is the $U[0,1]$ distribution function $U$. Next, for any $f\colon [0,1] \to \mathbb{R}$ and integer $r \ge 0$, we define a piecewise linear interpolation $[f]_r$ to be the function equal to $f$ at $k/2^{r+1}$ for $k = 0, 1, \ldots, 2^{r+1}$ and linear in between. If $f(0) = f(1) = 0$, as holds for $f = \beta_n$ or $B_n^\circ$, then $[f]_r$ can be written as a sum as follows. For $j = 0, 1, \ldots$, and $k = 1, \ldots, 2^j$, let $T_{j,k}$ be the "triangle function" such that $T_{j,k} = 0$ outside the interval $((k-1)/2^j, k/2^j)$,

$T_{j,k} = 1$ at the midpoint $(2k-1)/2^{j+1}$, and $T_{j,k}$ is linear in between. For each $j$ and $k$, let

$$f_{j,k} := W_{j,k}(f) := f((2k-1)/2^{j+1}) - \frac{1}{2}[f((k-1)/2^j) + f(k/2^j)]. \quad (12.88)$$

Then for $r = 0, 1, \ldots,$ we have

$$[f]_r = \sum_{j=0}^{r} \sum_{k=1}^{2^j} f_{j,k} T_{j,k}. \quad (12.89)$$

Using the piecewise linear interpolations $[\beta_n]_r$ and $[B_n^\circ]_r$ of $\beta_n$ and $B_n^\circ$, we bound $\|\beta_n - B_n^\circ\|_{(p)}$ above by

$$\|\beta_n - [\beta_n]_r\|_{(p)} + \|[\beta_n]_r - [B_n^\circ]_r\|_{(p)} + \|[B_n^\circ]_r - B_n^\circ\|_{(p)}. \quad (12.90)$$

The goal is to find a good bound for the Luxemburg norm $\|\cdot\|_{\phi_p}$ of each term in (12.90). We will fix an integer $r$ at first; then an appropriate $r$ will be chosen to have a good total bound.

We can bound the function $\phi_p$ defined by (12.83) as follows:

**Lemma 12.44.** *For each $p$ with $1 < p < \infty$, there is a finite constant $C_p$ such that for all $y \geq 0$,*

$$\phi_p(y) \leq \exp(C_p y) - 1. \quad (12.91)$$

*Proof.* For $y \geq p$, since $1 \leq e^y$, (12.91) holds if $e^y \leq \exp(C_p y)/2$, which in turn holds if $C_p \geq 1 + (\log 2)/p$. For $0 \leq y < p$, since $e^x \geq 1 + x$ for any $x$, (12.91) holds if $\kappa_p y^p \leq C_p y$ with $\kappa_p := (e/p)^p$. Since $y \mapsto y^p$ is convex, the latter bound holds if it holds for $y = p$. So we can set $C_p := \max\{\kappa_p p^{p-1}, 1 + (\log 2)/p\}$. □

It will be useful to consider another Luxemburg norm closely related to $\|\cdot\|_{\phi_p}$. For $1 < p < \infty$ and $y \geq 0$, let $\psi_p(y) := \phi_p(y^{1/p})$. Then $\psi_p$ is also $C^1$ for $y > 0$. We have $\psi_p(y) = \kappa_p y$ for $0 \leq y \leq y_p$, where $\kappa_p = (e/p)^p$ and $y_p = p^p$. For $y > y_p$, $\psi_p(y) = \exp\{y^{1/p}\}$ and one can check that $\psi_p''(y) > 0$. Thus $\psi_p$ is convex and so is in the class $\mathcal{CV}$.

It is easily seen that for a random variable $Y$, $Y \in L^{\phi_p}$ if and only if $|Y|^p \in L^{\psi_p}$, with

$$\|Y\|_{\phi_p}^p = \| |Y|^p \|_{\psi_p}. \quad (12.92)$$

In particular, for any stochastic process $X$ such that $v_p(X)$ is measurable,

$$\big\| \|X\|_{(p)} \big\|_{\phi_p} = \big\| v_p(X) \big\|_{\psi_p}^{1/p}. \quad (12.93)$$

**Lemma 12.45.** *For each $p$ with $1 < p < \infty$, there is a finite constant $K_p$ such that whenever $Y$ has a binomial $b(n, q)$ distribution with $nq \geq 1$, we have $\|Y^p\|_{\psi_p} \leq K_p (nq)^p$.*

*Proof.* By (12.92), it is enough to show that $E\phi_p(Y/(K_p^{1/p}nq)) \le 1$. Let $K_p := (3C_p)^p$ with $C_p$ from Lemma 12.44. Then by (12.91), it is enough to show that $E\exp(Y/(3nq)) \le 2$. We have $Ee^{uY} = (qe^u + 1 - q)^n$ and $1 + x \le e^x \le 1 + 2x$ for $0 \le x \le 1$. Thus $E\exp(Y/(3nq)) \le (1 + 2/(3n))^n \le e^{2/3} < 2$, as desired. □

**Lemma 12.46.** *Let $\{B_t^\circ\}_{0\le t\le 1}$ be a Brownian bridge. For $0 \le u < v \le 1$ and $t \in [0,1]$, let*

$$M(t;[u,v]) := (v-u)^{-1/2}\big[B^\circ(tv+(1-t)u) - tB^\circ(v) - (1-t)B^\circ(u)\big]. \quad (12.94)$$

*Then $\{M(t;[u,v])\}_{0\le t\le 1}$ is a Brownian bridge. Also, the variable $W_{j,k}(B^\circ)$ has a $N(0,2^{-j-2})$ distribution for all $j$ and $k$ for which (12.88) is defined. The variables $W_{j,k}(B^\circ)$ are all jointly independent.*

*Proof.* Let $Z$ be a standard normal random variable independent of the process $B^\circ$. Then $B(t) := B^\circ(t) + Zt$, $0 \le t \le 1$, gives a Brownian motion on $[0,1]$. Moreover, $W_{j,k}(B^\circ) = W_{j,k}(B)$ by definition (12.88). It is then easy to check that this has distribution $N(0,2^{-j-2})$. For $t \in [0,1]$, letting

$$W(t) := (v-u)^{-1/2}\big[B(tv + (1-t)u) - B(u)\big],$$

it then follows that $M(t;[u,v]) = W(t) - tW(1)$. Again, it is easy to check by covariances that $\{W_t\}_{0\le t\le 1}$ is a Brownian motion, and so the first part of the conclusion follows. The independence of the $W_{j,k}(B^\circ)$ follows from zero covariances of $W_{j,k}(B)$. □

**Lemma 12.47.** *For any $p \in (2,\infty)$ and Brownian bridge $B^\circ$, there exists a finite constant $C_B(p)$ such that for $r = 1,2,\ldots,$*

$$\big\| \|B^\circ - [B^\circ]_{r-1}\|_{(p)}\big\|_{\phi_p} \le C_B(p)(2^r)^{-(p-2)/(2p)}. \quad (12.95)$$

*Proof.* Since $B^\circ - [B^\circ]_{r-1}$ is equal to 0 at $k/2^r$ for $k = 0,1,\ldots,2^r$, by Proposition 3.37, we have

$$v_p(B^\circ - [B^\circ]_{r-1};[0,1]) \le 2^{p-1}\sum_{k=1}^{2^r} v_p(B^\circ - [B^\circ]_{r-1};J_{r,k}), \quad (12.96)$$

where $J_{r,k} := [(k-1)/2^r, k/2^r]$. Using the notation (12.94) and $M_{r,k} := M(\cdot;J_{r,k})$, for $k = 1,\ldots,2^r$ and $t \in J_{r,k}$, we have $(B^\circ - [B^\circ]_{r-1})(t) = (2^r)^{-1/2}M_{r,k}(s_t)$ with $s_t := t2^r - k + 1 \in [0,1]$. To see this in more detail, note that $t \mapsto s_t$ is the unique linear, increasing map from $J_{r,k}$ onto $[0,1]$. For any $u < v$, $s \mapsto u + s(v-u) = (1-s)u + sv$ is the unique increasing linear map from $[0,1]$ onto $[u,v]$. Thus for $[u,v] = J_{r,k}$, these maps must be inverses of each other. So for $t \in J_{r,k}$, we have $2^{-r/2}M_{r,k}(s_t) = B^\circ(t) - s_t B^\circ(v) - (1-s_t)B^\circ(u)$. Moreover, for $t \in J_{r,k}$, $t \mapsto s_t B^\circ(v) + (1-s_t)B^\circ(u)$ is a linear random function equal to $B^\circ$ at the endpoints; thus it equals $[B^\circ]_{r-1}$ on $J_{r,k}$ as stated. So,

$$v_p(B^\circ - [B^\circ]_{r-1}; J_{r,k}) = (2^r)^{-(p-2)/2}\frac{1}{2^r}v_p(M_{r,k}; [0,1]).$$

By Lemma 12.46, $M_{r,k}$ is a Brownian bridge and $v_p(B^\circ; [0,1]) < \infty$ a.s. by (12.82). Using (12.93), the Landau–Shepp–Marcus–Fernique theorem (e.g. Lemma 2.2.5 in [52]) and Lemma 12.44, we have that $\|v_p(B^\circ; [0,1])\|_{\psi_p} = \| \|B^\circ\|_{(p)}\|_{\phi_p}^p < \infty$. Thus by (12.96), it follows that

$$\left\|v_p(B^\circ - [B^\circ]_{r-1}; [0,1])\right\|_{\psi_p} \le 2^{p-1}(2^r)^{-(p-2)/2}\| \|B^\circ\|_{(p)}\|_{\phi_p}^p.$$

Applying (12.93) to the left side of the preceding inequality, (12.95) follows with $C_B(p) = 2^{1-1/p}\| \|B^\circ\|_{(p)}\|_{\phi_p}$. □

Next we bound the Luxemburg norm $\|\cdot\|_{\phi_p}$ of the first term in (12.90).

**Lemma 12.48.** *For any $p \in (2, \infty)$, there exists a finite constant $C_\beta(p)$ such that for $r = 1, 2, \ldots$ and $n \ge 2^r$,*

$$\left\| \|\beta_n - [\beta_n]_{r-1}\|_{(p)}\right\|_{\phi_p} \le C_\beta(p) n^{1/2}(2^r)^{-(1-1/p)}. \tag{12.97}$$

*Proof.* Let $n, r$ be a pair of positive integers with $n/2^r \ge 1$. As in the preceding proof, since $\beta_n - [\beta_n]_{r-1}$ is equal to 0 at $k/2^r$ for $k = 0, 1, \ldots, 2^r$, by Proposition 3.37 again, we have

$$v_p(\beta_n - [\beta_n]_{r-1}; [0,1]) \le 2^{p-1}\sum_{k=1}^{2^r} v_p(\beta_n - [\beta_n]_{r-1}; J_{r,k}),$$

where $J_{r,k} = [(k-1)/2^r, k/2^r]$. Note that the map $f \mapsto [f]_{r-1}$ is linear and that $[U]_{r-1}$ equals the $U[0,1]$ distribution function $U$. Thus $\beta_n - [\beta_n]_{r-1} \equiv \sqrt{n}(U_n - [U_n]_{r-1})$. For $k = 1, \ldots, 2^r$, let $Y_{r,k}$ be the number of values of $j = 1, \ldots, n$ such that $X_j \in ((k-1)/2^r, k/2^r]$. Then $Y_{r,k}$ almost surely equals the number of $i = 1, \ldots, n$ with $X_i$ in the closure $J_{r,k}$. By convexity of $y \mapsto y^p$, $y \ge 0$, and then, since $v_p(H) = (\sup H - \inf H)^p$ for any monotone function $H$, we have

$$\begin{aligned} v_p(\beta_n - [\beta_n]_{r-1}; J_{r,k}) &\le 2^{p-1}n^{-p/2}\left[v_p(nU_n; J_{r,k}) + v_p([nU_n]_{r-1}; J_{r,k})\right] \\ &\le 2^p n^{-p/2} Y_{r,k}^p. \end{aligned}$$

Since $Y_{r,k}$ has a binomial $b(n, 1/2^r)$ distribution for each $k$, by Lemma 12.45, it follows that

$$\|v_p(\beta_n - [\beta_n]_{r-1}; [0,1])\|_{\psi_p} \le 4^p n^{-p/2} \sum_{k=1}^{2^r} \|Y_{r,k}^p\|_{\psi_p} \le 4^p K_p n^{p/2} 2^{-r(p-1)}.$$

Then (12.97) with $C_\beta(p) = 4K_p^{1/p}$ follows from (12.93). □

To bound the middle term in (12.90), the Komlós, Major, and Tusnády construction will be used to link empirical processes and Brownian bridges. For $N = 0, 1, \ldots$ and $k = 0, 1, \ldots, N$, let $P_{N,1/2}(k) = \binom{N}{k} 2^{-N}$, the binomial $b(N, 1/2)$ distribution. Let $F_{(N)}$ be its distribution function. For its quantile function let $H(t|N) := F_{(N)}^{\leftarrow}(t)$ from $(0,1)$ onto $\{0, 1, \ldots, N\}$. Let $\Phi$ be the standard normal distribution function, $\Phi(x) := \int_{-\infty}^x \exp(-u^2/2) du$. Given a positive integer $n$ and a Brownian bridge $B_n^\circ$, random variables $Y_{j,k}^*$, $j = 0, 1, \ldots$, $k = 1, \ldots, 2^j$ will be constructed iteratively as follows. Let $Y_{0,1}^* := n$. Next, let $Y_{1,1}^* := H(\Phi(2W_{0,1}(B_n^\circ))|Y_{0,1}^*)$ and $Y_{1,2}^* := Y_{0,1}^* - Y_{1,1}^*$. Then, given $Y_{j-1,k}^*$, for $j \ge 2$, for each $k = 1, \ldots, 2^{j-1}$, let

$$Y_{j,2k-1}^* := H\big(\Phi\big(2^{(j+1)/2} W_{j-1,k}(B_n^\circ)\big) \big| Y_{j-1,k}^*\big) \tag{12.98}$$

and $Y_{j,2k}^* := Y_{j-1,k}^* - Y_{j,2k-1}^*$. By Lemma 12.46, each $2^{(j+1)/2} W_{j-1,k}(B_n^\circ)$ has law $N(0,1)$, and so $\Phi$ of it has law $U[0,1]$. In fact, if $Y$ is any random variable having a continuous distribution function $F$, then $F(Y)$ has law $U[0,1]$, since for $0 < t < 1$,

$$P(F(Y) \ge t) = P(Y \ge F^{\leftarrow}(t)) = 1 - F(F^{\leftarrow}(t)) = 1 - t$$

by continuity. For $\Phi$, simply $\Phi^{\leftarrow} = \Phi^{-1}$. Therefore each $Y_{j,2k-1}^*$ has the binomial law $b(Y_{j-1,k}^*, 1/2)$, given $Y_{j-1,k}^*$, e.g. by [53, Proposition 9.1.2]. By interpreting $Y_{j,k}^*$ as the number of points in the interval $I_{j,k} := ((k-1)/2^j, k/2^j]$ for each $j$ and $k$, and letting $j \to \infty$, the $Y_{j,k}^*$ define $n$ points in the interval $(0,1)$, so that intervals $I_{j,k}$ containing the points have $Y_{j,k}^* = 1$ for $j$ large enough. Let these $n$ points be the ordered sample $X_{(1)}^*, X_{(2)}^*, \ldots, X_{(n)}^*$ and define $X_1^*, X_2^*, \ldots, X_n^*$ by a random permutation $\pi$ of $\{1, \ldots, n\}$, independent of $X_{(\cdot)}^*$: $X_i^* := X_{(\pi(i))}^*$, $i = 1, 2, \ldots, n$. Let $X_1, \ldots, X_n$ be a sample of independent uniform $(0,1)$ random variables and $Y_{j,k}$ the number of $X_i$ in $I_{j,k}$. To show that $Y_{j,k}$ and the $Y_{j,k}^*$ have the same joint distribution, note that the $Y_{j,k}^*$ are obtained by starting with $n$ points in the interval $(0,1]$ and successively bisecting intervals, where if an interval $I_{j,k}$ has $Y_{j,k}^*$ points, the number of points in the left half $I_{j+1,2k-1}$ is binomial $b(Y_{j,k}^*, 1/2)$ given $Y_{j,k}^*$. Because the variables $W_{j,k}(B_n^\circ)$ are all jointly independent (by Lemma 12.46), the array of random variables $\{Y_{j,k}^*\}$ has the same joint distribution as $\{Y_{j,k}\}$, namely the distribution of the increments of $U_n$ over the intervals $I_{j,k}$. So $X_1^*, X_2^*, \ldots, X_n^*$ are indeed i.i.d. with uniform $(0,1)$ distribution. By virtue of this fact, we will drop the superscript $*$ from now on and treat $Y_{j,k}^*$ and $Y_{j,k}$, and thus $X_i^*$ and $X_i$, as the same.

In sum, the so-called *KMT construction* defines a joint distribution of a Brownian bridge $B_n^\circ$ and empirical process $\beta_n = n^{1/2}(U_n - U)$ as follows. Begin with $B_n^\circ$, which has continuous sample functions, and for it, extend (12.89) to

$$B_n^\circ = \sum_{j=0}^{\infty} \sum_{k=1}^{2^j} W_{j,k}(B_n^\circ) T_{j,k},$$

which by piecewise linear interpolation clearly converges uniformly on $[0,1]$. Next let $U_n(0) := 0$ and $U_n(1) := 1$. Apply (12.98) repeatedly to get

$$\begin{aligned} U_n(1/2) &:= Y_{1,1}^*/n, \quad U_n(1/4) := Y_{2,1}^*/n, \\ U_n(3/4) &:= U_n(1/2) + Y_{2,3}^*/n, \end{aligned} \tag{12.99}$$

and so on. Then $U_n(k/2^j)$ for $j = 0, 1, \ldots$, $k = 0, 1, \ldots, 2^j$, have their correct joint distribution, and $U_n(t)$ for $0 \le t \le 1$ is then defined by monotonicity and right-continuity.

We note that Theorem 8.1 of Major [154] implies that if $Z$ has law $N(0,1)$, then $Y = H(\Phi(Z)|m)$ minimizes $E(|m^{-1/2}(2Y - m) - Z|^p)$ for $1 \le p < \infty$ among all $Y$ with law $b(m, 1/2)$. This motivates the choice (12.98).

To prove Theorem 12.42, we need to define a distribution of $\{(\beta_n, B_n^\circ)\}_{n=1}^\infty$, although the joint distribution for different $n$ has no effect on (12.84). One way is as follows. Let $X_1, X_2, \ldots$ be independent and identically distributed random variables with uniform law $U[0,1]$ and define $\beta_n$ from $X_1, \ldots, X_n$ as usual. For each $n$, let $((X_1, X_2, \ldots, X_n), B_n^\circ)$ have the KMT joint distribution as just defined. Let $\xi := \{X_i\}_{i=1}^\infty$. Then for each $n$, since $C[0,1]$ is a complete separable metric space, $B_n^\circ$ has a conditional distribution $\mu_n(\cdot|\xi)$ given $\xi$, where for each $\xi$, $\mu_n(\cdot|\xi)$ is a probability measure on the Borel $\sigma$-algebra of $C[0,1]$ [53, Theorem 10.2.2]. It depends on $\xi$ only through $\{X_i\}_{i=1}^n$. Moreover, for each Borel set $A \subset C[0,1]$, $\mu_n(A|\xi)$ is a measurable function of $\{X_i\}_{i=1}^n$. Let $C_n$ be a copy of $C[0,1]$ on which $\mu_n$ is defined. On the countable Cartesian product $C[0,1]^\infty := \Pi_{n=1}^\infty C_n$, we can form the product probability measure $\mu(\cdot|\xi) := \Pi_{n=1}^\infty \mu_n(\cdot|\xi)$ for any fixed $\xi$, e.g. [53, Theorem 8.2.2]. For any Borel sets $A_j \subset C_j$, $j = 1, \ldots, n$, $n = 1, 2, \ldots$, $\Pi_{j=1}^n \mu_j(A_j|\xi)$ is clearly a measurable function of $\xi$. It follows by monotone convergence as in the proofs of [53, Theorems 4.4.4, 8.2.2] that $\xi \mapsto \mu(E|\xi)$ is measurable for any measurable set $E \subset C[0,1]^\infty$ for the product $\sigma$-algebra. Let $\mathcal{U}$ be the uniform probability measure on the Borel sets of $[0,1]$ whose distribution function is $U$. We have also the product measure $\mathcal{U}^\infty = \Pi_{i=1}^\infty \mathcal{U}_i$ of copies $\mathcal{U}_i$ of $\mathcal{U}$ on the infinite-dimensional cube $I^\infty := \Pi_{i=1}^\infty [0,1]_i$ of possible values of $\xi$. Thus, for any measurable $A \subset I^\infty$ we can define a joint distribution for $\{B_n^\circ\}_{n=1}^\infty$ and $\xi = \{X_i\}_{i=1}^\infty$ for which

$$\Pr(\xi \in A, \{B_n^\circ\}_{n=1}^\infty \in E) = \int_A \mu(E|\xi) d\mathcal{U}^\infty(\xi)$$

for any measurable sets $A$ and $E$ as described. The joint distribution then extends from rectangles $A \times E$ to general product measurable subsets of $I^\infty \times C[0,1]^\infty$ by the usual extension for measures on a product space. The $B_n^\circ$ are conditionally independent given $\{X_i\}_{i=1}^\infty$.

Recall that for each $n$, $(\beta_n, B_n^\circ)$ have the joint distribution given by the KMT construction. For the middle term in (12.90), by (12.89), we obtain the representation

$$[\beta_n]_r - [B_n^\circ]_r = \sum_{j=0}^{r} \sum_{k=1}^{2^j} [W_{j,k}(\beta_n - B_n^\circ)] T_{j,k} \tag{12.100}$$

for $r = 1, 2 \ldots$. For $1 \leq p < \infty$ and each $j$, by Proposition 3.37, since $v_p(T_{j,k}) = 2$, we have

$$\left\| v_p\Big( \sum_{k=1}^{2^j} [W_{j,k}(\beta_n - B_n^\circ)] T_{j,k} \Big) \right\|_{\psi_p} \leq 2^p \sum_{k=1}^{2^j} \left\| \left| W_{j,k}(\beta_n - B_n^\circ) \right|^p \right\|_{\psi_p}. \tag{12.101}$$

Let $Z$ be a standard normal random variable. Let $\gamma(Z, N) := F_{(N)}^{\leftarrow}(\Phi(Z)) - N/2$. To bound the right side of (12.100), we use a lemma of Tusnády whose first published proof was given as far as we know by Bretagnolle and Massart [25], Lemma 4: (almost surely)

$$\left| \gamma(Z, N) - \frac{\sqrt{N}}{2} Z \right| \leq 1 + Z^2/8. \tag{12.102}$$

Letting $N := Y_{j-1,k}^*$ and $Z := Z_{j,k} := 2^{(j+1)/2} W_{j-1,k}(B_n^\circ)$, by (12.88), (12.99), and (12.98), it follows that

$$W_{j-1,k}(\beta_n) \equiv \sqrt{n} W_{j-1,k}(U_n) \equiv \frac{1}{\sqrt{n}} \left[ Y_{j,2k-1}^* - \frac{Y_{j-1,k}^*}{2} \right] \equiv \gamma(Z, N)/\sqrt{n}.$$

Since $EN = n/2^{j-1}$, by (12.102), and then using the inequality $2ab \leq a^2 + b^2$, we have the bound

$$\begin{aligned} \sqrt{n} \left| W_{j-1,k}(\beta_n - B_n^\circ) \right| &= \left| \gamma(Z, N) - \frac{\sqrt{EN}}{2} Z \right| \\ &\leq 1 + \frac{3}{8} Z^2 + \frac{1}{4} |\sqrt{N} - \sqrt{EN}|^2. \end{aligned}$$

For a standard normal variable $Z$ such as $Z_{j,k}$, $\| \, |Z|^{2p} \|_{\psi_p}$ is a finite constant depending on $p$. Therefore using the inequality $(a+b+c)^p \leq 3^{p-1}(a^p + b^p + c^p)$, we get finite constants $A_p$, $B_p$ such that

$$n^{p/2} \left\| \left| W_{j-1,k}(\beta_n - B_n^\circ) \right|^p \right\|_{\psi_p} \leq A_p + B_p \left\| \left| \sqrt{N} - \sqrt{EN} \right|^{2p} \right\|_{\psi_p}. \tag{12.103}$$

Since for $1 < p < \infty$ and $y \geq 0$, $\psi_p(y) = \phi_p(y^{1/p})$, by Lemma 12.44, there is a finite constant $C_p$ such that for each $t > 0$,

$$E\psi_p\Big(\big|\sqrt{N}-\sqrt{EN}\big|^{2p}/t\Big) \le E\exp\Big\{C_p\big(\sqrt{N}-\sqrt{EN}\big)^2/t^{1/p}\Big\}-1.$$

For each $s>0$, let $\eta_N(s) := E\exp\{s(\sqrt{N}-\sqrt{EN})^2\}$. If $\eta_N(C_p/t^{1/p}) \le 2$, then we can bound the Luxemburg norm by

$$\| \, |\sqrt{N}-\sqrt{EN}|^{2p}\|_{\psi_p} \le t. \tag{12.104}$$

The following will be used to bound such $t$:

**Lemma 12.49.** *For any binomial $b(n,q)$ random variable $N$ and $s \in (0,1/2)$, $\eta_N(s) \le \mathrm{e}^s/\sqrt{1-2s}$.*

*Proof.* Since $E\sqrt{N} \le \sqrt{EN}$, we have

$$(\sqrt{N}-\sqrt{EN})^2 \le (\sqrt{N}-E\sqrt{N})^2 + EN - (E\sqrt{N})^2. \tag{12.105}$$

By Hölder's inequality $(Efg)^3 \le (Ef^{3/2})^2Eg^3$ for $f = N^{1/3} \in \mathcal{L}^{3/2}$ and $g = N^{2/3} \in \mathcal{L}^3$. It follows that $-(E\sqrt{N})^2 \le -(EN)^3/E(N^2)$. Next, since $[E(N^2)-(EN)^2]^2 \ge 0$, we have the bound

$$EN-(E\sqrt{N})^2 \le EN-(EN)^3/E(N^2) \le \mathrm{Var}(N)/EN = 1-q \le 1.$$

Thus if $N'$ and $N$ are independent and identically distributed random variables, then $E\sqrt{N'} = E\sqrt{N}$ and

$$\eta_N(s) \le \mathrm{e}^s E\exp\Big\{s\big(\sqrt{N}-E\sqrt{N}\big)^2\Big\} \le \mathrm{e}^s E\exp\Big\{s\big(\sqrt{N}-\sqrt{N'}\big)^2\Big\}$$

by (12.105) and Jensen's inequality, taking expectations with respect to $\sqrt{N'}$ for each fixed $N$. Let $U := N+N'$ and $V := N-N'$. Then

$$\big(\sqrt{N}-\sqrt{N'}\big)^2 \equiv U\big(1-\sqrt{1-(V/U)^2}\big) \le V^2/U$$

as is easily checked. Now write $V = N-N' = \sum_{i=1}^n \xi_i$, where $\xi_i = \rho_i-\rho_i'$ and $\rho_i, \rho_i'$, $i=1,\dots,n$ are independent and identically distributed Bernoulli $(q)$ random variables (i.e., $\Pr(\rho_i=1) = 1-\Pr(\rho_i=0) = q$). Then $\rho_i \equiv \rho_i^2$ and $\rho_i' \equiv \rho_i'^2$, so $U \ge \sum_{i=1}^n \xi_i^2$. Conditional on $\xi_i^2$ for $i=1,\dots,n$, $V$ is equal in distribution to $\sum_{i=1}^n \epsilon_i\xi_i$, where $\epsilon_i$ are Rademacher random variables, equal to $\pm1$ with probability $1/2$ each, independent of each other and the $\xi_i$. It then follows that

$$\eta_N(s) \le \mathrm{e}^s \sup\left\{E\exp\left(s\Big[\sum_{i=1}^n \alpha_i\epsilon_i\Big]^2\right) : \sum_{i=1}^n \alpha_i^2 \le 1\right\}. \tag{12.106}$$

To see this, write the expectation as that of the conditional expectation given the $\xi_i$. If not all $\xi_i$ are 0, let $\alpha_i := \xi_i/(\sum_{j=1}^n \xi_j^2)^{1/2}$. If all $\xi_i$ are 0, let $\alpha_i = 0$ for all $i$.

We next need the following bound, which is asymptotically sharp, letting $\alpha_i = 1/\sqrt{n}$, $n \to \infty$.

**Lemma 12.50.** *For $0 < s < 1/2$, a $N(0,1)$ random variable $Z$, any $\alpha_i$ with $\sum_{i=1}^n \alpha_i^2 \le 1$, and independent Rademacher $\epsilon_i$,*

$$E\exp\left(s\left[\sum_{i=1}^n \alpha_i\epsilon_i\right]^2\right) \le E\exp\left(sZ^2\right).$$

*Proof.* Let $Y := \sum_{i=1}^n \alpha_i\epsilon_i$. It suffices to show that $E(Y^{2k}) \le E(Z^{2k})$ for $k = 0, 1, 2, \ldots$. We have from $Ee^{tZ} = \exp(t^2/2)$ that $E(Z^{2k}) = (2k)!/(2^k k!)$. For any real $u$, $Ee^{uY} = \Pi_{i=1}^n E\exp(u\epsilon_i\alpha_i) = \Pi_{i=1}^n \cosh(u\alpha_i)$, where for each real $v$, $\cosh(v) = \sum_{j=0}^\infty v^{2j}/(2j)!$. For given nonnegative integers $k$ and $n$ let $(\sum\Pi)_{n,k}$ denote the sum, over all $n$-tuples $\{j_i\}_{i=1}^n$ of nonnegative integers $j_i$ such that $\sum_{i=1}^n j_i = k$, of products from $i = 1$ to $n$ of given functions of $j_i$. Thus $E(Y^{2k}) = (2k)!(\sum\Pi)_{n,k}\alpha_i^{2j_i}/(2j_i)!$, and $1 \ge (\alpha_1^2 + \cdots + \alpha_n^2)^k = k!(\sum\Pi)_{n,k}\alpha_i^{2j_i}/j_i!$. So it will be enough to prove that for any integers $j_i \ge 0$ with $\sum_{i=1}^n j_i = k$, $\Pi_{i=1}^n 1/(2j_i)! \le 2^{-k}\Pi_{i=1}^n 1/j_i!$, which is clear since $2^j j! \le (2j)!$ for each $j = j_i$ (with equality for $j = 0$ or 1, which helps to explain the asymptotic sharpness for large $n$ and fixed $k$). □

By Lemma 12.50 and (12.106), since $E\exp(sZ^2) = 1/\sqrt{1-2s}$ for $s < 1/2$, Lemma 12.49 follows. □

Lemma 12.49 gives that $\eta_N(0.28) < 2$. Thus one can replace $t$ in (12.104) by $(3.6C_p)^p$, and so by (12.103), there is a finite constant $D_p$ such that

$$n^{p/2}\left\|\left|W_{j,k}(\beta_n - B_n^\circ)\right|^p\right\|_{\psi_p} \le D_p$$

for each $j = 0, 1, 2, \ldots$, $k = 1, \ldots, 2^j$, and $n = 1, 2, \ldots$. Then by (12.100), (12.93), and (12.101), it follows that

$$\begin{aligned}\left\| \|[\beta_n]_r - [B_n^\circ]_r\|_{(p)}\right\|_{\phi_p} &\le \sum_{j=0}^r \left\|v_p\Big(\sum_{k=1}^{2^j}[W_{j,k}(\beta_n - B_n^\circ)]T_{j,k}\Big)\right\|_{\psi_p}^{1/p} \\ &\le 2D_p^{1/p}n^{-1/2}(2^r)^{1/p}/(1-2^{-1/p})\end{aligned}$$

for each $r = 1, 2, \ldots$. We choose $r = r(n) = \lfloor \log_2(n)\rfloor$, where $\lfloor x\rfloor$ is the largest integer less than or equal to $x$. Then $n \ge 2^r$, $2^{-r} < 2/n$, and the desired bound (12.84) follows by (12.90), (12.95), (12.97), and the preceding bound. The proof of Theorem 12.42 is complete. □

## 12.8 Differentiability of Operators on Processes

Some operators are Fréchet differentiable with respect to $p$-variation norms only for $p < 2$. The product integral need not exist for interval functions in $\mathcal{AI}_2(J)$ but not in $\mathcal{AI}_2^*(J)$ (Proposition 9.19) or in $\mathcal{AI}_p(J)$ for all $p > 2$ (Proposition 9.22).

In the Love–Young inequality (3.129) and the inequality for convolutions (3.167), at least one of the two functions must be in $\mathcal{W}_p$ for $p < 2$. On the other hand, paths of Brownian motion and certain other processes have bounded $p$-variation only for $p > 2$, and the empirical process converges in $p$-variation norm to the Brownian bridge only for $p > 2$. The question arises how the differentiability can still be useful. We will examine this question for the empirical process.

The degree of smoothness of an operator $T$ on distribution functions (remainder bounds of sufficiently small order) and properties of the derivative operator $DT$ may compensate for the relatively slow convergence of $F_n$ to $F$ in $\|\cdot\|_{(p)}$ for $p < 2$. We have for example the following.

**Proposition 12.51.** *Let $1 < p < 2$, let $F$ be a distribution function on $\mathbb{R}$, and let $T$ be an operator defined on an open neighborhood $U$ of $F$ in $\mathcal{W}_p(\mathbb{R})$, Fréchet differentiable at $F$, with derivative $DT$ at $F$, and such that for some $\gamma > 1$, as $g \to F$ in $\mathcal{W}_p$,*

$$\|T(g) - T(F) - (DT)(g - F)\| = O(\|g - F\|_{[p]}^{\gamma}). \tag{12.107}$$

*If $p > (2\gamma)/(2\gamma - 1)$ then for the empirical distribution functions $F_n$ we have*

$$\begin{aligned}\sqrt{n}\big(T(F_n) - T(F)\big) &= (DT)\big(\sqrt{n}(F_n - F)\big) + O_{\Pr}\big(n^{[p+2\gamma(1-p)]/(2p)}\big)\\ &= (DT)\big(\sqrt{n}(F_n - F)\big) + o_{\Pr}(1),\end{aligned}$$

*where $T(F_n)$ is defined with probability converging to 1 as $n \to \infty$.*

*Proof.* We have $F_n \in U$ with probability converging to 1 as $n \to \infty$ by Theorem 12.40, which also then implies the first conclusion. The second holds since the hypothesis on $p$ is equivalent to $p+2\gamma(1-p) < 0$, and the proposition is proved. □

**Remark 12.52.** Since $1 < 2\gamma/(2\gamma - 1) < 2$, we can choose $p < 2$. If $\gamma = 2$, e.g. if $T$ is twice differentiable at $F$, then for $p$ close enough to 2, the remainder becomes $O_{\Pr}(n^{\epsilon-1/2})$ for any $\epsilon > 0$.

Let $Y$ be the range Banach space of $T$, so that $DT$ is a bounded linear operator from all of $\mathcal{W}_p(\mathbb{R})$ into $Y$. Even though $\sqrt{n}(F_n - F)$ does not converge in distribution to a Brownian bridge evaluated at $F$ in the sense that it does for $p > 2$, it may well be that $(DT)(\sqrt{n}(F_n - F))$ converges in distribution in $Y$ to $(DT)(B^\circ \circ F)$ for a Brownian bridge $B^\circ$, for example if $Y$ is finite-dimensional, and then so does $\sqrt{n}(T(F_n) - T(F))$.

## 12.9 Integration

One might be tempted to think that $\int_0^1 B\,\mathrm{d}B = B(1)^2/2$ for almost all paths of Brownian motion $B$, but such integrals do not exist in any of the pathwise senses we have treated up to this point:

**Proposition 12.53.** *For almost all paths $B = \{B_t\}_{t\geq 0}$ of a Brownian motion, $(A)\int_0^1 B\,\mathrm{d}B$ is not defined for $A = RS$, $RRS$, $LS$, $RYS$, $CY$, or $HK$.*

*Proof.* It suffices to consider $CY$ and $HK$ integrals by the implications shown in Figure 2.1 in Section 2.9. If the $CY$ integral (defined in Section 2.5) existed it would equal the $RRS$ integral by sample continuity and the definition of the integral, so the refinement Young–Stieltjes integral would exist. Brownian paths almost surely have unbounded 2-variation on every interval $[a, b]$, $a < b$, by (12.11). By continuity, there are arbitrarily large sums

$$\sum_{i=1}^{n} \big[B(t_i) - B(t_{i-1})\big]\big[B(v_i) - B(u_i)\big]$$

with $a = t_0 < \cdots < t_{i-1} < u_i < v_i < t_i < \cdots < t_n = b$ for each $i = 1, \ldots, n$. Thus the sums $\sum_{i=1}^n [B(t_i) - B(t_{i-1})]B(w_i)$ with $t_{i-1} < w_i < t_i$ for each $i$ are unbounded. Since these are Young sums, $(RYS)\int_0^1 B\,\mathrm{d}B$ does not exist. Nor does $(HK)\int_0^1 B\,\mathrm{d}B$ by Proposition 2.70 since the corresponding sums are unbounded on any subinterval of $[0, 1]$. □

*Stochastic integrals*

Integrals have been defined (for more than half a century) by probabilistic methods.

We recall first the Itô integral. Let $B = \{B_t\}_{t\geq 0}$ be a Brownian motion on a probability space $(\Omega, \mathcal{F}, \Pr)$. For $t \geq 0$, let $\mathcal{F}_t$ be the $\sigma$-algebra generated by the random variables $\{B(u)\colon\ 0 \leq u \leq t\}$. Then for $0 \leq s < t < \infty$ we have $\{\emptyset, \Omega\} := \mathcal{F}_0 \subseteq \mathcal{F}_s \subseteq \mathcal{F}_t \subseteq \mathcal{F}$. For $0 < T < \infty$ consider the class $\mathcal{H}[0, T]$ of stochastic processes $H = \{H_t\}_{0\leq t\leq T}$ satisfying the following properties:

(1) $(t, \omega) \mapsto H(t, \omega)$ is $\mathcal{B} \times \mathcal{F}$-measurable, where $\mathcal{B}$ is the Borel $\sigma$-algebra on $[0, T]$;
(2) $H$ is $\mathcal{F}_t$-adapted, i.e. $H(t, \cdot)$ is $\mathcal{F}_t$-measurable for all $0 \leq t \leq T$;
(3) $E\big[\int_0^T H(t)^2\,\mathrm{d}t\big] < \infty$.

We now define the Itô integral $(I)\int_0^T H\,\mathrm{d}B$ for each $H \in \mathcal{H}([0, T])$. First suppose that $H$ is elementary, that is,

$$H(t, \omega) = h_0 1_{\{0\}}(t) + \sum_{i=1}^{n} h_i(\omega) 1_{(t_{i-1}, t_i]}(t)$$

for $0 \le t \le T$ and some $0 = t_0 < t_1 < \cdots < t_n = T$, where $h_0$ is a constant, each $h_i$, $i = 1, \ldots, n$, is $\mathcal{F}_{t_{i-1}}$-measurable, and $Eh_i^2 < \infty$. For such $H$ define the Itô integral by

$$(I)\int_0^T H\,\mathrm{d}B := \sum_{i=1}^n h_i\big[B(t_i) - B(t_{i-1})\big]. \tag{12.108}$$

It is easy to check that

$$E\Big[(I)\int_0^T H\,\mathrm{d}B\Big]^2 = E\Big[\int_0^T H(t)^2\,\mathrm{d}t\Big] \tag{12.109}$$

for each elementary $H \in \mathcal{H}[0,T]$. The set of all elementary stochastic processes is dense in $\mathcal{H}[0,T]$ in the norm of $L^2([0,T],\lambda)$ (see e.g. Section 2.2 in McKean [164]): for an arbitrary $H \in \mathcal{H}[0,T]$, there exists a sequence $\{H_m\}_{m\ge1}$ of elementary processes $H_m$ such that

$$\lim_{m\to\infty} E\Big[\int_0^T \big(H(t) - H_m(t)\big)^2\,\mathrm{d}t\Big] = 0.$$

Then the Itô integral for $H$ is defined as the limit

$$(I)\int_0^T H\,\mathrm{d}B := \lim_{m\to\infty}(I)\int_0^T H_m\,\mathrm{d}B \quad \text{in } L^2(\Omega,\Pr),$$

which by (12.109) does not depend on the actual sequence $\{H_m\}_{m\ge1}$ chosen. The identity (12.109) extends to all processes in $\mathcal{H}[0,T]$ and then is called the Itô isometry.

The following equality for a smooth function of a Brownian motion is an important tool in stochastic calculus: for a $C^2$ function $F$ the composition $F'{\circ}B$ is Itô integrable with respect to $B$ and for any $0 \le s < t < \infty$,

$$(F{\circ}B)(t) - (F{\circ}B)(s) = (I)\int_s^t (F'{\circ}B)\,\mathrm{d}B + \frac{1}{2}\int_s^t (F''{\circ}B)\,\mathrm{d}t. \tag{12.110}$$

This identity is called the *Itô formula* for a Brownian motion, and extensions of it hold for a large class of stochastic processes (Theorem 6 in [106]). The Itô formula is sometimes written in a differential form as

$$\mathrm{d}F(B) = F'(B)\mathrm{d}B + F''(t)\mathrm{d}t/2,$$

whose precise meaning is just (12.110). For example, the integral $\int_0^1 B\,\mathrm{d}B$, shown not to exist in several senses in Proposition 12.53, exists in the Itô sense: for $F(u) \equiv u^2/2$, the formula (12.110) gives

$$(I)\int_0^1 B\,\mathrm{d}B = (B(1)^2 - 1)/2, \tag{12.111}$$

differing by $-1/2$ from the value if $B$ is replaced by a $C^1$ function of $t$.

Let $f$ be a real-valued continuous function of bounded variation on $[a,b]$ and let $F$ be a $C^1$ function. By Theorem 2.87, the composition $F' \circ f$ is Riemann–Stieltjes integrable with respect to $f$ and

$$(F \circ f)(b) - (F \circ f)(a) = (RS) \int_a^b (F' \circ f)\, \mathrm{d}f. \tag{12.112}$$

The Itô formula (12.110) differs from the "chain rule" equality (12.112) by a correction term.

Let $f, h\colon [a,b] \to \mathbb{R}$. For a partition $\kappa = \{x_i\}_{i=0}^n$ of $[a,b]$, let

$$S_{MS}(f, \mathrm{d}h; \kappa) := \frac{1}{2} \sum_{i=1}^{n} \big[f(x_{i-1}) + f(x_i)\big] \big[h(x_i) - h(x_{i-1})\big]$$

and $|\kappa| := \max_i (x_i - x_{i-1})$. Define the (mesh) mean Stieltjes integral by

$$(MS) \int_a^b f\, \mathrm{d}h := \lim_{|\kappa| \to 0} S_{MS}(f, \mathrm{d}h; \kappa) \tag{12.113}$$

provided the limit exists. Hildebrandt [97, Section 19.1] treats this integral and gives some of its history. Let $f$ be any real-valued function on $[a,b]$. Then for any partition $\kappa$ of $[a,b]$, we have

$$S_{MS}(f, \mathrm{d}f; \kappa) = [f^2(b) - f^2(a)]/2,$$

and so the integral $(MS)\int_a^b f\, \mathrm{d}f$ exists and equals the right side. In particular, for almost all sample functions of a Brownian motion $B$ and each $t > 0$, we have $(MS)\int_0^t B\, \mathrm{d}B = B(t)^2/2$, in contrast to (12.111).

Let $X$ and $Y$ be two real-valued stochastic processes defined for $0 \le t < \infty$. Then for each $0 < T < \infty$, the limit

$$[X, Y](T) := \lim_{|\{t_i\}| \to 0} \sum_{i=1}^{n} [X(t_i) - X(t_{i-1})][Y(t_i) - Y(t_{i-1})], \tag{12.114}$$

if it exists in probability, is called the *quadratic covariation* between $X$ and $Y$ on $[0,T]$. It exists under some conditions, see e.g. Proposition 17.17 in O. Kallenberg [112]. In particular $[X,X](T)$ is called the *quadratic variation* of $X$ on $[0,T]$ and satisfies $[X,X](T) \le v_2(X; [0,T])$ a.s. For example, it is well known that for Brownian motion $B$, $[B,B](T) = T$ a.s. Recall however that $v_2(B; [0,T]) = +\infty$ a.s. by (12.11).

### *Nonexistence of pathwise integrals for some marginal Brownian motions*

In Proposition 12.53 we saw that for a Brownian motion $B$, $(A)\int_0^1 B\, \mathrm{d}B$ does not exist for several integrals of Riemann–Stieltjes type. The Itô stochastic integral $(I)\int_0^t B\, \mathrm{d}B$, given by (12.111), exists.

If $X$ and $Y$ are two independent Brownian motions, then the Itô integrals $(I)\int_0^t X\,\mathrm{d}Y$ and $(I)\int_0^t Y\,\mathrm{d}X$ exist. (It is easily seen that the quadratic covariation $[X,Y](T)$, defined by (12.114), is 0 for all $T>0$.) If $X$ and $Y$ are Brownian motions with an arbitrary joint distribution, however, it will be shown that the sums "approximating" $(I)\int_0^t Y\,\mathrm{d}X$ may be unbounded in probability. Nonexistence of such an integral can occur because $X$ is anticipating with respect to $Y$. Thus, it appears that not only properties of the individual sample functions $X$ and $Y$, but probabilistic properties are needed for existence of such integrals.

For a partition $\kappa=\{t_r\}_{r=0}^n$ of $[a,b]$ and two real functions $f$, $g$ on $[a,b]$, let

$$S_{LC}(f,\mathrm{d}g;\kappa) := \sum_{i=1}^n f(t_{i-1})\big[g(t_i)-g(t_{i-1})\big].$$

Here $LC$ denotes "left Cauchy." Also, the sums $S_{LC}$ are of the form of approximating sums for Itô integrals as in (12.108), where $h_i$ is $\mathcal{F}_{t_{i-1}}$ measurable.

Let

$$\begin{aligned} A(\kappa) := A(f,g;\kappa) &:= \frac{1}{2}\sum_{r=1}^n \big[f(t_{r-1})g(t_r)-f(t_r)g(t_{r-1})\big] \\ &= \frac{1}{2}\Big\{S_{LC}(f,\mathrm{d}g;\kappa)-S_{LC}(g,\mathrm{d}f;\kappa)\Big\}. \end{aligned}$$

If $t\mapsto(f,g)(t)$, $a\le t\le b$, defines a simple, piecewise $C^1$, counterclockwise closed curve in $\mathbb{R}^2$ (so that it has winding number 1 around each point in the bounded part $U$ of its complement), then by a corollary of Green's theorem (e.g. [157, Section 7.1, Theorem 2]) the area $A(U)$ of $U$ is given by

$$A(U)=\frac{1}{2}\int_a^b f\,\mathrm{d}g-g\,\mathrm{d}f.$$

A simple example is $(f,g)(\theta):=(\cos\theta,\sin\theta)$, $0\le\theta\le 2\pi$. Extending this, if $f$ and $g$ are continuous on $[a,b]$ and $(f,g)(b)=(f,g)(a)$, then the limit of $A(\kappa)$ as the mesh $|\kappa|\downarrow 0$, if it exists, has been defined as the area enclosed by the closed curve $(f,g)$, e.g. W. H. Young [255]. Here, $U$ will be a union of disjoint connected open sets $U_n$. The winding number of the curve around each point $x\in U_n$ will be an integer $z_n\in\mathbb{Z}$. One can see in some elementary cases that

$$\frac{1}{2}\int_a^b f\,\mathrm{d}g-g\,\mathrm{d}f=\sum_n z_nA(U_n). \tag{12.115}$$

For more complicated nonsmooth curves, we can have $\sup_n|z_n|=+\infty$, and the sum in (12.115) may not converge. We will see in part (c) of the next proposition that for $f$ and $g$ Brownian bridges, the integral in (12.115) need not exist (even as a stochastic integral).

**Proposition 12.54.** *For $n = 1, 2, \ldots$ let $\kappa_n$ be the partition $\{r/n\}_{r=0}^n$ of $[0,1]$. On some probability space $(\Omega, \mathcal{F}, \Pr)$, there exists a Gaussian stochastic process $\{(X_t, Y_t)\}_{0 \le t \le 1}$ with values in $\mathbb{R}^2$ such that $X$ and $Y$ are each Brownian motions, and $S_n$ is unbounded in probability as $n \to \infty$, where for each $n$,*

*(a)* $S_n = S_{LC}(Y, \mathrm{d}X; \kappa_n)$;
*(b)* $S_n = S_{MS}(Y, \mathrm{d}X; \kappa_n)$;
*(c)* $S_n = A(Y_0, X_0; \kappa_n)$ *where* $X_0 = \{X(t) - X(1)t\}_{0 \le t \le 1}$ *and* $Y_0 = \{Y(t) - Y(1)t\}_{0 \le t \le 1}$ *are Brownian bridges.*

*Proof.* Let $L$ be the isonormal Gaussian process on a probability space $(\Omega, \mathcal{F}, \Pr)$, indexed by the Hilbert space $L^2[0,1] = L^2([0,1], \lambda)$ with Lebesgue measure $\lambda$, that is, $\{L(f),\ f \in L^2[0,1]\}$ is a Gaussian process with mean 0 and covariance $E[L(f)L(g)] = (f, g)$, the inner product in the Hilbert space $L^2[0,1]$. In particular, the stochastic process $B = \{B_t := L(1_{[0,t]})\}_{0 \le t \le 1}$ is a Brownian motion, since it has mean zero and the correct covariance $\min\{s, t\}$. Let $c_0(x) \equiv 1$, and for $k = 1, 2, \ldots$, let $c_k(x) := \sqrt{2}\cos(2\pi kx)$ and $s_k(x) := \sqrt{2}\sin(2\pi kx)$ for $x \in [0,1]$. Then $\{c_0, c_k, s_k\}_{k \ge 1}$ form an orthonormal basis of $L^2[0,1]$ and we have the expansion

$$B(t) = \xi_0 t + \frac{1}{2\pi} \sum_{k=1}^{\infty} \frac{1}{k} \Big\{ \xi_k s_k(t) + \eta_k (\sqrt{2} - c_k(t)) \Big\}, \tag{12.116}$$

where $\xi_k$, $\eta_k$ are i.i.d. $N(0,1)$ random variables on the probability space $(\Omega, \mathcal{F}, \Pr)$. The series converges a.s. uniformly on $[0,1]$, e.g. [52, Theorem 2.5.5]. Recall the notion of unconditional convergence of series in a Banach space from Section 1.4. For any given $t$, the series (12.116) converges unconditionally in $H = L^2(\Omega, \mathcal{F}, \Pr)$. For $0 \le t \le 1$, let

$$X_0(t) := \frac{1}{2\pi} \sum_{k=1}^{\infty} \frac{1}{k} \Big\{ \xi_k s_k(t) + \eta_k [\sqrt{2} - c_k(t)] \Big\}$$

and

$$Y_0(t) := \frac{1}{2\pi} \sum_{k=1}^{\infty} \frac{1}{k} \Big\{ \eta_k s_k(t) + \xi_k [c_k(t) - \sqrt{2}] \Big\}.$$

Then by (12.116), $X(t) := B(t) = X_0(t) + \xi_0 t$ is a Brownian motion. The processes $X_0$ and $Y_0$ are the same in distribution since $\{\eta_k, -\xi_k\}_{k \ge 1}$ are i.i.d. $N(0,1)$. Thus $Y(t) := Y_0(t) + \eta_0 t$, where $\eta_0$ is such that $\{\xi_k, \eta_k\}_{k \ge 0}$ are i.i.d. $N(0,1)$, gives another Brownian motion. Clearly $(X, Y)$ is Gaussian. So the first statement in the proposition is proved. Here $X$ and $Y$ have an unusual joint distribution.

It is enough to prove statement *(a)* for the Gaussian process $(X_0, Y_0)$ instead of $(X, Y)$. Indeed, for a function $\psi$ defined on $[0,1]$ and $t_r := r/n$, let $(\Delta\psi)(t_r) := \psi(t_r) - \psi(t_{r-1})$ for $r = 1, \ldots, n$. For each $n$ we then have $S_{LC}(Y, \mathrm{d}X; \kappa_n) = S_{LC}(Y_0, \mathrm{d}X_0; \kappa_n) + R_n$, where

$$R_n := Y(1) \sum_{r=1}^{n} t_{r-1}(\Delta X)(t_r) + X(1) \sum_{r=1}^{n} Y(t_{r-1}) \frac{1}{n} - X(1)Y(1) \frac{n-1}{2n}.$$

Here $(\Delta X)(t_r)$ for $r = 1, \ldots, n$ are i.i.d. $N(0, 1/n)$ random variables. Thus the first sum in $R_n$ has mean 0 and variance $< 1$ and so it is bounded in probability as $n$ varies. Applying summation by parts to the second sum in $R_n$, we get a representation $Y(1) - \sum_{r=1}^{n} t_r(\Delta Y)(t_r)$, which is again bounded in probability. Therefore for any $\epsilon > 0$, there exists a finite $C$ such that $\sup_n \Pr(\{|R_n| > C\}) < \epsilon$. So for each finite $M$ and each $n$, we have

$$\Pr(\{|S_{LC}(Y, \mathrm{d}X; \kappa_n)| \le M\}) \le \Pr(\{|S_{LC}(Y_0, \mathrm{d}X_0; \kappa_n)| \le M + C\} + \epsilon,$$

proving the claim.

Let $\eta := (\sqrt{2}/(2\pi)) \sum_{k=1}^{\infty} \eta_k/k$ and $\xi := (\sqrt{2}/(2\pi)) \sum_{k=1}^{\infty} \xi_k/k$. For each $k = 1, \ldots, n$, let $N(k) := \{k, k+n, k+2n, \ldots\}$ be the set of all positive integers $j$ such that $j - k$ is divisible by $n$. Then for all $j \in N(k)$ and all $r = 0, 1, \ldots, n$, $c_j(t_r) = c_k(t_r)$ and $s_j(t_r) = s_k(t_r)$. It follows that

$$X_0(t_r) - \eta \ = \ \frac{1}{2\pi} \sum_{k=1}^{n} G_k s_k(t_r) - H_k c_k(t_r) \tag{12.117}$$

where $G_k := \sum_{j \in N(k)} \xi_j/j$ and $H_k := \sum_{j \in N(k)} \eta_j/j$, each converging almost surely, and unconditionally in $H$, as follows from the unconditional convergence mentioned after (12.116). Each of $G_k$ and $H_k$ has distribution $N(0, \sigma_k^2)$ where

$$\sigma_k^2 \ := \ \sum_{j \in N(k)} 1/j^2. \tag{12.118}$$

Likewise

$$Y_0(t_r) + \xi \ = \ \frac{1}{2\pi} \sum_{k=1}^{n} H_k s_k(t_r) + G_k c_k(t_r). \tag{12.119}$$

Then we have

$$\begin{aligned} S(\kappa_n) &:= S_{LC}(Y_0, \mathrm{d}X_0; \kappa_n) \\ &= S_{LC}((Y_0 + \xi), \mathrm{d}(X_0 - \eta); \kappa_n) = S_1(n) + S_2(n), \end{aligned} \tag{12.120}$$

where

$$S_1(n) := \frac{1}{4\pi^2} \sum_{k,l=1}^{n} G_k G_l S_{LC}(c_k, \mathrm{d}s_l; \kappa_n) - H_k H_l S_{LC}(s_k, \mathrm{d}c_l; \kappa_n), \tag{12.121}$$

$$S_2(n) := \frac{1}{4\pi^2} \sum_{k,l=1}^{n} H_k G_l S_{LC}(s_k, \mathrm{d}s_l; \kappa_n) - G_k H_l S_{LC}(c_k, \mathrm{d}c_l; \kappa_n). \tag{12.122}$$

We have $s_n(t_r) = 0$ and $c_n(t_r) = 1$ for all $r$, so the difference between any two values of $c_n$ is 0. Thus in both sums $S_1(n)$ and $S_2(n)$ we can take $\sum_{k,l=1}^{n-1}$. In proving the unboundedness in probability we can and will assume that $n$ is odd. Let $n$ be fixed. The following relations, no doubt well known, about the discrete Fourier transform, are given here for completeness. For $k, l = 1, \ldots, n-1$ we have

$$T_{k,l,1} := \sum_{r=1}^{n} \exp(2\pi i k t_{r-1}) \exp(2\pi i l t_{r-1}) \;=\; n 1_{\{k+l=n\}}, \tag{12.123}$$

$$T_{k,l,2} := \sum_{r=1}^{n} \exp(2\pi i k t_{r-1}) \exp(2\pi i l t_{r}) \;=\; \exp(2\pi i l/n) T_{k,l,1}. \tag{12.124}$$

Taking real and imaginary parts it follows that if $k + l \neq n$,

$$\begin{aligned} 0 &= \sum_{r=1}^{n} \cos(2\pi k t_{r-1}) \cos(2\pi l t_{r-1}) - \sin(2\pi k t_{r-1}) \sin(2\pi l t_{r-1}) \\ &= \sum_{r=1}^{n} \cos(2\pi k t_{r-1}) \sin(2\pi l t_{r-1}) + \sin(2\pi k t_{r-1}) \cos(2\pi l t_{r-1}) \end{aligned} \tag{12.125}$$

and

$$\begin{aligned} 0 &= \sum_{r=1}^{n} \cos(2\pi k t_{r-1}) \cos(2\pi l t_{r}) - \sin(2\pi k t_{r-1}) \sin(2\pi l t_{r}) \\ &= \sum_{r=1}^{n} \cos(2\pi k t_{r-1}) \sin(2\pi l t_{r}) + \sin(2\pi k t_{r-1}) \cos(2\pi l t_{r}). \end{aligned} \tag{12.126}$$

On the other hand, likewise

$$T_{k,l,3} := \sum_{r=1}^{n} \exp(2\pi i k t_{r-1}) \exp(-2\pi i l t_{r-1}) \;=\; n 1_{\{k=l\}}, \tag{12.127}$$

$$T_{k,l,4} := \sum_{r=1}^{n} \exp(2\pi i k t_{r-1}) \exp(-2\pi i l t_{r}) \;=\; \exp(-2\pi i l/n) T_{k,l,3}. \tag{12.128}$$

Taking real and imaginary parts again gives for $k \neq l$,

$$\begin{aligned} 0 &= \sum_{r=1}^{n} \cos(2\pi k t_{r-1}) \cos(2\pi l t_{r-1}) + \sin(2\pi k t_{r-1}) \sin(2\pi l t_{r-1}) \\ &= \sum_{r=1}^{n} - \cos(2\pi k t_{r-1}) \sin(2\pi l t_{r-1}) + \sin(2\pi k t_{r-1}) \cos(2\pi l t_{r-1}) \end{aligned} \tag{12.129}$$

and

$$
\begin{aligned}
0 &= \sum_{r=1}^{n} \cos(2\pi k t_{r-1})\cos(2\pi l t_r) + \sin(2\pi k t_{r-1})\sin(2\pi l t_r) \\
&= \sum_{r=1}^{n} -\cos(2\pi k t_{r-1})\sin(2\pi l t_r) + \sin(2\pi k t_{r-1})\cos(2\pi l t_r).
\end{aligned} \tag{12.130}
$$

If $k \neq l$ and $k + l \neq n$, then combining (12.125) and (12.129) gives

$$
\begin{aligned}
0 &= \sum_{r=1}^{n} \cos(2\pi k t_{r-1})\cos(2\pi l t_{r-1}) = \sum_{r=1}^{n} \sin(2\pi k t_{r-1})\sin(2\pi l t_{r-1}) \\
&= \sum_{r=1}^{n} \cos(2\pi k t_{r-1})\sin(2\pi l t_{r-1}) = \sum_{r=1}^{n} \sin(2\pi k t_{r-1})\cos(2\pi l t_{r-1}),
\end{aligned} \tag{12.131}
$$

and likewise combining (12.126) and (12.130) gives

$$
\begin{aligned}
0 &= \sum_{r=1}^{n} \cos(2\pi k t_{r-1})\cos(2\pi l t_r) = \sum_{r=1}^{n} \sin(2\pi k t_{r-1})\sin(2\pi l t_r) \\
&= \sum_{r=1}^{n} \cos(2\pi k t_{r-1})\sin(2\pi l t_r) = \sum_{r=1}^{n} \sin(2\pi k t_{r-1})\cos(2\pi l t_r).
\end{aligned} \tag{12.132}
$$

Subtracting each of the four sums in (12.131) from the corresponding sum in (12.132) gives also for $k \neq l$ and $k + l \neq n$,

$$0 = S_{LC}(c_k, \mathrm{d}c_l; \kappa_n) = S_{LC}(s_k, \mathrm{d}s_l; \kappa_n) = S_{LC}(c_k, \mathrm{d}s_l; \kappa_n) = S_{LC}(s_k, \mathrm{d}c_l; \kappa_n).$$

Thus when neither $k = n - l$ nor $k = l$, all the terms in $S_1(n)$ and $S_2(n)$ are 0. So $S_1(n) = S_{11}(n) + S_{12}(n)$ and $S_2(n) = S_{21}(n) + S_{22}(n)$ where

$$
\begin{aligned}
S_{11}(n) &:= \frac{1}{4\pi^2}\sum_{k=1}^{n-1} G_k^2 S_{LC}(c_k, \mathrm{d}s_k; \kappa_n) - H_k^2 S_{LC}(s_k, \mathrm{d}c_k; \kappa_n), \\
S_{12}(n) &:= \frac{1}{4\pi^2}\sum_{k=1}^{n-1} G_k G_{n-k} S_{LC}(c_k, \mathrm{d}s_{n-k}; \kappa_n) - H_k H_{n-k} S_{LC}(s_k, \mathrm{d}c_{n-k}; \kappa_n), \\
S_{21}(n) &:= \frac{1}{4\pi^2}\sum_{k=1}^{n-1} G_k H_k \left[S_{LC}(s_k, \mathrm{d}s_k; \kappa_n) - S_{LC}(c_k, \mathrm{d}c_k; \kappa_n)\right], \\
S_{22}(n) &:= \frac{1}{4\pi^2}\sum_{k=1}^{n-1} H_k G_{n-k} S_{LC}(s_k, \mathrm{d}s_{n-k}; \kappa_n) - G_k H_{n-k} S_{LC}(c_k, \mathrm{d}c_{n-k}; \kappa_n) \\
&= \frac{1}{4\pi^2}\sum_{k=1}^{n-1} H_k G_{n-k}[S_{LC}(s_k, \mathrm{d}s_{n-k}; \kappa_n) - S_{LC}(c_{n-k}, \mathrm{d}c_k; \kappa_n)].
\end{aligned}
$$

(Because $n$ is odd, the cases $k = l$ and $k + l = n$ do not overlap.) Let $m := \lfloor n/2 \rfloor = (n-1)/2$ since $n$ is odd. Then to write $S_{12}(n)$ as a sum of independent terms, we have $S_{12}(n) = S_{121}(n) + S_{122}(n)$ where using

$$c_{n-k} \equiv c_k \text{ and } s_{n-k} \equiv -s_k, \tag{12.133}$$

$$S_{121}(n) := \frac{1}{4\pi^2}\sum_{k=1}^{m} G_k G_{n-k}[S_{LC}(c_k, \mathrm{d}s_{n-k}; \kappa_n) + S_{LC}(c_{n-k}, \mathrm{d}s_k; \kappa_n)] \equiv 0,$$

$$S_{122}(n) := \frac{1}{4\pi^2}\sum_{k=1}^{m} -H_k H_{n-k}[S_{LC}(s_k, \mathrm{d}c_{n-k}; \kappa_n) + S_{LC}(s_{n-k}, \mathrm{d}c_k; \kappa_n)] \equiv 0.$$

Thus $S_{12}(n) \equiv 0$. Clearly $ES_{ij}(n) = 0$ except for $i = j = 1$. So

$$E(S(\kappa_n)) = \frac{1}{4\pi^2}\sum_{k=1}^{n-1} \sigma_k^2 \left(S_{LC}(c_k, \mathrm{d}s_k; \kappa_n) - S_{LC}(s_k, \mathrm{d}c_k; \kappa_n)\right). \tag{12.134}$$

We have clearly $E[S_{ij}(n)S_{uv}(n)] = 0$ except when $u = i$ and $v = j$. It follows that

$$\mathrm{Var}(S(\kappa_n)) = \frac{1}{16\pi^4}\sum_{k=1}^{n-1} \sigma_k^4 (A_k + C_k + D_k) \tag{12.135}$$

where again using (12.133),

$$\begin{aligned} A_k &:= 2[S_{LC}(c_k, \mathrm{d}s_k; \kappa_n)^2 + S_{LC}(s_k, \mathrm{d}c_k; \kappa_n)^2], \\ C_k &:= [S_{LC}(s_k, \mathrm{d}s_k; \kappa_n) - S_{LC}(c_k, \mathrm{d}c_k; \kappa_n)]^2, \\ D_k &:= [S_{LC}(s_k, \mathrm{d}s_k; \kappa_n) + S_{LC}(c_k, \mathrm{d}c_k; \kappa_n)]^2. \end{aligned} \tag{12.136}$$

When $k = l$, we have $k + l \neq n$ since $n$ is odd, so (12.125) and (12.126) hold. By (12.127) in this case, in (12.129) the second line still equals 0, but now the right side of the first line equals $n$ in an obvious identity. Combining gives for each $k = 1, \ldots, n$,

$$\begin{aligned} \sum_{r=1}^{n} s_k(t_{r-1})c_k(t_{r-1}) &= 0, \\ \sum_{r=1}^{n} s_k^2(t_{r-1}) &= \sum_{r=1}^{n} c_k^2(t_{r-1}) = n. \end{aligned} \tag{12.137}$$

By (12.128) for $k = l$, the second line of (12.130) equals $-n\sin(2\pi k/n)$ and the right side of its first line equals $n\cos(2\pi k/n)$. Combining these equations with (12.126) gives

$$\sum_{r=1}^{n} c_k(t_{r-1})s_k(t_r) = n\sin(2\pi k/n) = -\sum_{r=1}^{n} s_k(t_{r-1})c_k(t_r) \tag{12.138}$$

and

$$\sum_{r=1}^{n} s_k(t_{r-1})s_k(t_r) = \sum_{r=1}^{n} c_k(t_{r-1})c_k(t_r) = n\cos(2\pi k/n). \tag{12.139}$$

Combining in the last four displays gives

$$S_{LC}(s_k, \mathrm{d}s_k; \kappa_n) = S_{LC}(c_k, \mathrm{d}c_k; \kappa_n) = n[\cos(2\pi k/n) - 1] < 0, \quad (12.140)$$
$$S_{LC}(c_k, \mathrm{d}s_k; \kappa_n) = n\sin(2\pi k/n) = -S_{LC}(s_k, \mathrm{d}c_k; \kappa_n). \quad (12.141)$$

We then get from (12.134) that

$$E(S(\kappa_n)) = \frac{1}{4\pi^2}\sum_{k=1}^{n-1} 2n\sigma_k^2 \sin(2\pi k/n) = \frac{n}{2\pi^2}\sum_{k=1}^{m}(\sigma_k^2 - \sigma_{n-k}^2)\sin(2\pi k/n).$$

For $1 \le k \le m < n/2$ we have $\sin(2\pi k/n) > 0$, and from the definition (12.118) of $\sigma_k$ we have that each term in the sum defining $\sigma_{n-k}^2$ is smaller than the corresponding term in the sum for $\sigma_k^2$. It follows that $\sigma_k^2 - \sigma_{n-k}^2 > k^{-2} - (n-k)^{-2} > 0$. Moreover, a straightforward calculation shows that for $k \le n/4$ we have $\sigma_k^2 - \sigma_{n-k}^2 \ge 8/(9k^2)$. We also have $\sin(x) \ge 2x/\pi$ for $0 \le x \le \pi/2$ by concavity. We thus have

$$\begin{aligned} ES(\kappa_n) &\ge \frac{4n}{9\pi^2}\sum_{k=1}^{\lfloor n/4\rfloor}\frac{1}{k^2}\sin\left(\frac{2\pi k}{n}\right) \\ &\ge \frac{16}{9\pi^2}\sum_{k=1}^{\lfloor n/4\rfloor}\frac{1}{k} \\ &\sim \frac{16}{9\pi^2}\log\left(\frac{n}{4}\right) \to \infty \end{aligned}$$

as $n \to \infty$, $n$ odd. It will next be shown that $\mathrm{Var}(S(\kappa_n))$ is bounded for $n$ odd. In (12.135) and (12.136) we have by (12.140) and (12.141) that $A_k = 4n^2\sin^2(2\pi k/n)$ and $C_k + D_k = 4n^2(1-\cos(2\pi k/n))^2$. For any real $x$, it is well known and easily checked by derivatives that $|\sin x| \le |x|$ and $1-\cos x \le x^2/2$. It follows that

$$\begin{aligned} \mathrm{Var}(S(\kappa_n)) &\le \frac{1}{16\pi^4}\sum_{k=1}^{n-1}\sigma_k^4(16\pi^2k^2 + 16\pi^4k^4/n^2) \\ &\le \frac{1}{\pi^2}\sum_{k=1}^{n-1}k^2\sigma_k^4(1+\pi^2). \end{aligned}$$

From (12.118) we have since $k \le n$ that

$$\sigma_k^2 \le k^{-2}\sum_{j=1}^{\infty} j^{-2} = \pi^2/(6k^2);$$

thus $\sum_{k=1}^{n-1} k^2\sigma_k^4 \le (\pi^2/6)^3$, so indeed the variance of $S(\kappa_n)$ remains bounded, and since its expectation is unbounded, $S(\kappa_n)$ is unbounded in probability for $n$ odd, proving (a).

For $(b)$, for each $n$, we have

$$S_{MS}(Y, \mathrm{d}X; \kappa_n) = S_{LC}(Y, \mathrm{d}X; \kappa_n) + \frac{1}{2}\sum_{r=1}^{n} \Delta Y(t_r)\Delta X(t_r).$$

The quadratic variation of a Brownian motion exists, as noted after (12.114) (for the given sequence of partitions, even in an almost sure sense: [47, Theorem 4.5]), so by the Cauchy–Schwarz inequality, the sum on the right side is bounded in probability, that is, given $\epsilon > 0$, there is a finite $C$ such that

$$\sup_n \Pr\left(\left\{\frac{1}{2}\left|\sum_{r=1}^{n} \Delta Y(t_r)\Delta X(t_r)\right| > C\right\}\right) < \epsilon.$$

So for any finite $M$ and each $n$, we have

$$\Pr\left(\{|S_{MS}(Y, \mathrm{d}X; \kappa_n)| \le M\}\right) \le \Pr\left(\{|S_{LC}(Y, \mathrm{d}X; \kappa_n)| \le M + C\}\right) + \epsilon.$$

The right side is less than $2\epsilon$ for all sufficiently large $n$ odd by the proof of $(a)$, proving $(b)$.

For $(c)$, by (12.120), we have $A(\kappa_n) := A(Y_0, X_0; \kappa_n) = A_1(n) + A_2(n)$, where

$$A_1(n) := \frac{1}{8\pi^2}\sum_{k,l=1}^{n} \left[G_kG_l + H_kH_l\right]\left[S_{LC}(c_k, \mathrm{d}s_l; \kappa_n) - S_{LC}(s_k, \mathrm{d}c_l; \kappa_n)\right]$$

and

$$A_2(n) := \frac{1}{8\pi^2}\sum_{k,l=1}^{n} \left[H_kG_l - G_kH_l\right]\left[S_{LC}(s_k, \mathrm{d}s_l; \kappa_n) + S_{LC}(c_k, \mathrm{d}c_l; \kappa_n)\right].$$

Let $n$ be odd, as in the proof of (a). Since $s_n(t_r) = 0$ and $c_n(t_r) = 1$ for all $r$ in both sums $A_1(n)$ and $A_2(n)$ we can take $\sum_{k,l=1}^{n-1}$. When neither $k = n - l$ nor $k = l$, all the terms in $A_1(n)$ and $A_2(n)$ are 0. Thus $A_1(n) = A_{11}(n) + A_{12}(n)$ and $A_2(n) = A_{21}(n) + A_{22}(n)$ where

$$A_{11}(n) := \frac{1}{8\pi^2}\sum_{k=1}^{n-1} \left[G_k^2 + H_k^2\right]\left[S_{LC}(c_k, \mathrm{d}s_k; \kappa_n) - S_{LC}(s_k, \mathrm{d}c_k; \kappa_n)\right],$$

$$A_{12}(n) := \frac{1}{8\pi^2}\sum_{k=1}^{n-1}\left[G_kG_{n-k} + H_kH_{n-k}\right]\left[S_{LC}(c_k, \mathrm{d}s_{n-k}; \kappa_n) - S_{LC}(s_k, \mathrm{d}c_{n-k}; \kappa_n)\right],$$

$$A_{21}(n) := \frac{1}{8\pi^2}\sum_{k=1}^{n-1}\left[H_kG_k - G_kH_k\right]\left[S_{LC}(s_k, \mathrm{d}s_k; \kappa_n) + S_{LC}(c_k, \mathrm{d}c_k; \kappa_n)\right] = 0,$$

$$A_{22}(n) := \frac{1}{8\pi^2}\sum_{k=1}^{n-1}\big[H_kG_{n-k}-G_kH_{n-k}\big]\big[S_{LC}(s_k,\mathrm{d}s_{n-k};\kappa_n)+S_{LC}(c_k,\mathrm{d}c_{n-k};\kappa_n)\big].$$

By (12.133) and (12.141) it follows that $A_{12}(n)=0$. By (12.133) and (12.140) it follows that $A_{22}(n)=0$. Thus $A(\kappa_n)\equiv A_{11}(n)$, and $EA(\kappa_n)=ES(\kappa_n)$ in the proof of (a), which as shown there is unbounded for $n$ odd. The variance of $A_{11}(n)$ is bounded similarly as but more easily than in the proof of (a). This completes the proof of (c) and the proposition. □

## 12.10 Notes

*Notes on Section 12.2.* The book Kallenberg [112], which we often cite throughout the chapter, is the second edition of a book that first appeared in 1997. The second edition has, perhaps most notably, newly inserted chapters 2: "Measure Theory: Key Results," 11: "Special notions of symmetry and invariance," 20: "Ergodic properties of Markov processes," and 27: "Large deviations."

For $p>2$ and for almost all sample functions of Brownian motion $B=B(\cdot,\omega)$, the boundedness of the $p$-variation of $B$, which is the first part of (12.11), follows from the $\alpha$-Hölder property for $B$ with $\alpha=1/p$. The latter fact was established by N. Wiener in [240, §7]. The second part of (12.11), that is, the almost sure unboundedness of the 2-variation of $B$, was proved by P. Lévy in [140, Theorem 9, p. 516].

*Notes on Section 12.4.* Theorem 12.13 and Proposition 12.14 are due to Jain and Monrad [110]. The local law of the iterated logarithm for a fractional Brownian motion follows from a functional version of the law of the iterated logarithm for a class of Gaussian processes proved by Oodaira [181] (see Lai [132] for some corrections). The direct proof given here (Proposition 12.19) is based on ideas of Nisio [180] used to prove related results. A probability bound similar to the one in Lemma 12.18 can also be obtained using the Gaussian concentration inequality (4.4) and Ledoux [135, Theorem 6.1].

*Notes on Section 12.5.* Regarding Theorem 12.25 on existence of Markov processes, J. L. Doob's book [46], Chapter 2, §6, between (6.18) and (6.19), states that for given, not necessarily stationary, transition probabilities and an initial distribution, first consistent finite-dimensional joint distributions can be defined by iterated integrals, and then a Markov process with the given transition probabilities exists by the general Kolmogorov existence theorem.

The rest of the section is based on a paper by M. Manstavičius [155]. The lemmas and inequalities are virtually the same as in his paper, but the theoretical framework has been arranged differently here. Most of the accessible theory of Markov processes has restrictions that were inconvenient for our purposes.

Many authors deal with processes taking values in a locally compact metric space, with conditions having to do with compact sets, and/or the time-homogeneous case, where $P_{s,t}$ depends only on the difference $t - s$; see e.g. Kallenberg [112, Chapter 19]. Some more elementary parts of the theory are developed without time homogeneity [112, Chapter 8].

One of the main tools of Markov process theory is the strong Markov property, saying that a suitable Markov process gets a fresh start at Markov times such as stopping times. This is easiest to formulate for time-homogeneous processes.

Dynkin [59, §2.1] defines general (not necessarily time-homogeneous) Markov processes in general state spaces, where not only the transition probabilities $P_{s,t}$ but the full distribution $\mathbb{P}_{s;x}$ of a process starting at $x$ at time $s$ are included in the definition. Also, beyond a filtration $\mathcal{F}_t$ of events occurring between times 0 and $t$, Dynkin's definition includes $\sigma$-algebras $\mathcal{F}_t^s$ of events occurring between times $s$ and $t$ for $0 \leq s \leq t$. Dynkin [59, §5.2] gives a definition of strong ("strict") Markov property and, in §5.4, he gives sufficient conditions for this in a class of topological measurable spaces including metric spaces, for right-continuous processes. Instead of trying to prove the theorem just as stated by Manstavičius [155, Theorem 1.3] through Dynkin's formulations, we decided to give a more direct, self-contained formulation and proof. Thus, we do not need the strong Markov property since we consider stopping times with finitely many values, where the ordinary Markov property suffices.

The classical theorem of Kinney [116], saying that if for each $r > 0$, $\alpha(h, r) \downarrow 0$ as $h \downarrow 0$, trajectories can be taken to be cadlag (right-continuous with left limits), was formulated for real-valued functions. Kallenberg [112, Theorem 17.15] gives a version in locally compact metric spaces. Gihman and Skorohod [79, Vol. I, Chapter III, §4] treat separable processes with values in a complete metric space. Kinney's result is extended, first with regulated trajectories in Theorem 4, then with right-continuity in Theorem 5. These results are derived from a fact about general stochastic processes where the bound $\alpha(h, r)$ is assumed on suitable conditional probabilities in place of Markov kernels (Theorem 2). Gihman and Skorohod also [79, Vol. II, Chapter I, §6, Theorem 2] extend Kinney's theorem to "Markov families" $\mathbb{P}_{s,x}, \mathcal{F}_t^s$, i.e. to Markov processes with non-stationary transition probabilities indexed by different starting times and points, as defined by Dynkin. The processes are assumed to be separable, with values in a complete metric space. Existence of separable versions is proved assuming that the state space is a separable locally compact metric space. The local compactness is used by way of a compactification such as a 1-point compactification, e.g. [79, Vol. I, Chapter III, Theorem 2]. Indeed, the separable version may take values in the compactification, not necessarily in the original sample space. Thus, for processes with values in metric spaces that are not locally compact, separability seems to be a nontrivial assumption, and results assuming it may not be as useful as they first appear.

In our proof, instead of calling on some version of Kinney's theorem, we just obtain right-continuity with left limits of the trajectories as a byproduct near the end of the proof. The right continuity is virtually necessary for the strong Markov property, and so it is generally adopted in the literature. For $p$-variation alone, as seen in our proof, one could if desired take left continuity instead.

*Notes on Section 12.6.* An exact result on $\Phi$-variation is due to Fristedt and Taylor [72, Theorem 2]. Namely for a suitable $\alpha$-stable Lévy motion $X_\alpha$ with $\alpha \in (0,2)$ and any $\Phi \in \mathcal{V}$, with probability 1,

$$\lim_{|\kappa|\to 0,\ \kappa\in \mathrm{PP}\,[0,1]} s_\Phi(X_\alpha;\kappa) = \sum_{(0,1]} \Phi(|\Delta^- X_\alpha|),$$

where the right side is finite with probability 1 if and only if

$$\int_{\mathbb{R}\setminus\{0\}} (1\wedge \Phi(|x|))\,\frac{\mathrm{d}x}{|x|^{1+\alpha}} < \infty.$$

Xu [242, Section 3] established necessary and/or sufficient conditions for the boundedness of the $p$-variation of a symmetric $\alpha$-stable process with possibly dependent increments. More information about $p$-variation of stable processes can be found in Fristedt [71].

*Notes on Section 12.7.* For a sequence of independent real random variables $X_1, X_2, \ldots$ with a distribution function $F$, let $Y_i(t) := 1_{\{X_i\le t\}} - F(t)$ for $t\in\mathbb{R}$ and $i = 1,2,\ldots$. Theorem 12.41($a$) then states that the sequence

$$\Big\{(n\log\log n)^{-1/2}\Big\|\sum_{i=1}^n Y_i\Big\|_{(2)}\Big\}_{n\ge 3}$$

is bounded in probability. The question whether the bound in this theorem holds almost surely is a question whether the bounded law of the iterated logarithm holds for the random elements $Y_1, Y_2, \ldots$ with respect to the 2-variation seminorm.

In historical order, first Y.-C. Huang [101], [102] proved Corollary 12.43, then Dudley [51] proved Theorem 12.41(a), then Qian [189] proved Theorem 12.40 and Theorem 12.41(b), then Huang and Dudley [103] proved Theorem 12.42. The bound $E(Y^{2k}) \le E(Z^{2k})$, used in the proof of Lemma 12.50, was proved by Whittle [239, Theorem 1] by a different method.

Komlós, Major, and Tusnády [121] stated that on some probability space there exist independent random variables $X_1, X_2, \ldots$ having $U[0,1]$ distribution and Brownian bridges $B_n^\circ$ such that

$$\Pr\Big(\Big\{\sup_{0\le t\le 1}|\sqrt{n}(\beta_n(t)-B_n^\circ(t))| > x + c\log n\Big\}\Big) < K\mathrm{e}^{-\lambda x}$$

for all $n$ and positive real numbers $x$, where $c$, $K$, and $\lambda$ are positive absolute constants. Komlós, Major, and Tusnády formulated the KMT construction giving a joint distribution of $\beta_n$ and $B_n^\circ$ also used to prove Theorem 12.42. Bretagnolle and Massart [25] gave a proof of the stated inequality for $n \geq 2$ with constants $c = 12$, $K = 2$, and $\lambda = 1/6$.

*Notes on Section 12.9.* The construction of the Itô integral can be extended in several ways (see e.g. Dellacherie and Meyer [39, Chapter VIII]). Condition 3 can be relaxed to $\int_0^T H(t)^2\,dt < \infty$ almost surely (see e.g. Section 2.2 in McKean [164]). Then $(I)\int_0^T H\,dB$ need no longer be square-integrable but can be an arbitrary $\mathcal{F}_T$-measurable random variable (see [48]). Researchers in stochastic analysis some decades ago defined $\int H\,dX$ rather generally, specifically when $X$ is a semimartingale (for example, a right-continuous martingale) and $H$ is a locally bounded predictable process (for example, an adapted process with regulated, left-continuous paths); see Doléans-Dade and Meyer [44], Dellacherie and Meyer [39]. We will just mention that there are also ways to define stochastic integrals with anticipating integrands: Skorohod [222].

Let a closed curve be defined by continuous functions $x(t)$, $y(t)$, $t \in [a, b]$, such that $x(a) = x(b)$ and $y(a) = y(b)$. W. H. Young [255, p. 346] defined the area to be the limit

$$A = \frac{1}{2} \lim_{|\kappa| \to 0} \sum_{i=1}^{n} \Big[ y(t_i)x(t_{i-1}) - y(t_{i-1})x(t_i) \Big],$$

if it exists, as the mesh of partitions $\kappa = \{t_i\}_{i=0}^n$ of $[a, b]$ tends to zero.

Paul Lévy [141] wrote that if $(X(t), Y(t))$ is 2-dimensional Brownian motion, then the area $S$ included by its graph $C$ for $0 \leq t \leq T$ and the chord $D$ which is the straight line segment from $(0, 0)$ to $(X(T), Y(T))$ "may be *formally* defined by the formula

$$S = \frac{1}{2} \int_0^{2\pi} X(t)\mathrm{d}Y(t) - Y(t)\mathrm{d}X(t)."$$

Lévy noted that a formal calculation with Fourier series gave an almost surely convergent series of random variables, and he went on to consider the area further. The "Lévy area" seems to have been named in honor of this work.

The proof of Proposition 12.54 is based on the paper by Lyons [150].

# Appendix

## Nonatomic Measure Spaces

In this appendix we collect some facts about finite measure spaces without (and with) atoms, which are applied in Chapter 7.

Let $(\Omega, \mathcal{S}, \mu)$ be a $\sigma$-finite measure space. Then an *atom* of $\mu$ is a set $A \in \mathcal{S}$ with $\mu(A) > 0$ such that for all $C \in \mathcal{S}$ with $C \subset A$, either $\mu(C) = 0$ or $\mu(C) = \mu(A)$. By $\sigma$-finiteness, we have $\mu(A) < +\infty$. $(\Omega, \mathcal{S}, \mu)$ or $\mu$ is called *nonatomic* if it has no atoms.

**Proposition A.1.** *Let $(\Omega, \mathcal{S}, \mu)$ be a nonatomic finite measure space with $\mu(\Omega) > 0$. Then for any $c$ with $0 < c < \mu(\Omega)$, there is an $A \in \mathcal{S}$ with $\mu(A) = c$.*

*Proof.* We can assume that $\mu(\Omega) = 1$. It will first be shown that for some $C \in \mathcal{S}$, $1/3 \leq \mu(C) \leq 2/3$. Suppose not. Let $p := \sup\{\mu(B)\colon \mu(B) < 1/3\} \leq 1/3$. Then $p > 0$ since $\Omega$ is not an atom and if $2/3 < \mu(D) < 1$, let $B := \Omega \setminus D$. Take $B_n \in \mathcal{S}$ with $\mu(B_n) \uparrow p$. Let $E_n := \bigcup_{j=1}^{n} B_j$. It will be shown by induction that $\mu(E_n) < 1/3$ for all $n$. This is true for $n = 1$. Assuming it holds for a given $n$, we have $\mu(E_{n+1}) = \mu(E_n \cup B_{n+1}) < (1/3) + (1/3) = 2/3$. Thus by the assumption that $\mu$ takes no values in $[1/3, 2/3]$, $\mu(E_{n+1}) < 1/3$, completing the induction. Let $E := \bigcup_{n=1}^{\infty} B_n$. Then $\mu(E) = p$, so $p < 1/3$.

Now, $\Omega \setminus E$ is not an atom, so take $F \subset \Omega \setminus E$ with $0 < \mu(F) < 1 - p$. If $\mu(F) \leq (2/3) - p$, then $p < \mu(E \cup F) \leq 2/3$, contradicting the assumption or the definition of $p$. But if $\mu(F) > (2/3) - p$, replacing $F$ by $\Omega \setminus (E \cup F)$ leads to the same contradiction. So, it is shown that for some $C \in \mathcal{S}$, $1/3 \leq \mu(C) \leq 2/3$.

Next, an *inclusion-chain* will be a finite sequence $\{C_j\}_{j=0}^{n} \subset \mathcal{S}$ with $\emptyset = C_0 \subset C_1 \subset \cdots \subset C_n = \Omega$. One inclusion-chain $\{C_j\}_{j=0}^{n}$ will be a *refinement* of another $\{D_i\}_{i=0}^{k}$ iff for each $i$, $D_i = C_j$ for some $j$. Let $\mathcal{C}_0 := \{\emptyset, \Omega\}$. Given an inclusion-chain $\mathcal{C}_n = \{C_j^n\}_{j=0}^{2^n}$, define a refinement $\mathcal{C}_{n+1}$ of it by adjoining, for each $j = 1, \ldots, 2^n$, a set $C_{j-1} \cup A_j$ where $A_j \in \mathcal{S}$, $A_j \subset C_j^n \setminus C_{j-1}^n$, and $\mu(A_j)/\mu(C_j^n \setminus C_{j-1}^n) \in [1/3, 2/3]$. Thus $\mathcal{C}_n$ are defined recursively for all $n$. Clearly, for each $n$ and $j = 1, \ldots, 2^n$, if $\mathcal{C}_n = \{C_j^n\}_{j=0}^{2^n}$, then $\mu(C_j^n \setminus C_{j-1}^n) \leq$

R.M. Dudley and R. Norvaiša, *Concrete Functional Calculus*, Springer Monographs in Mathematics, DOI 10.1007/978-1-4419-6950-7_1,

$(2/3)^n \to 0$ as $n \to \infty$. Thus the values $\{\mu(C_j^n)\}_{j=0}^{2^n}$ become dense in $[0,1]$ as $n \to \infty$. For any $t \in [0,1]$, let $B_t := \bigcup_n \bigcup_{j=0}^{2^n} \{C_j^n : \mu(C_j^n) \le t\}$. Then $B_t \in \mathcal{S}$ and $\mu(B_t) = t$, completing the proof. □

Proposition A.1 and induction on $n$ give directly the following:

**Corollary A.2.** *Let $(\Omega, \mathcal{S}, \mu)$ be a finite nonatomic measure space. Let $r_i$ for $i = 1, \ldots, n$ be numbers with $r_i > 0$ and $\sum_{i=1}^n r_i = \mu(\Omega)$. Then $\Omega$ can be decomposed as a union of disjoint sets $A_i \in \mathcal{S}$ with $\mu(A_i) = r_i$ for $i = 1, \ldots, n$.*

Here are sufficient conditions under which atoms equal singletons up to sets of measure 0, as holds in familiar cases. A measurable space $(\Omega, \mathcal{S})$ is called *separated* iff for every $x \ne y$ in $\Omega$, there is some $C \in \mathcal{S}$ containing just one of $x$ and $y$.

**Proposition A.3.** *Let $(\Omega, \mathcal{S}, \mu)$ be a $\sigma$-finite measure space such that $(\Omega, \mathcal{S})$ is separated and $\mathcal{S}$ is countably generated, i.e. for some countably many $A_j \in \mathcal{S}$, $\mathcal{S}$ is the smallest $\sigma$-algebra containing all $A_j$. Then for any atom $A$ of $\mu$ there is an $x \in A$ such that $\mu(A \setminus \{x\}) = 0$. So, the singleton $\{x\}$ is also an atom. This holds in particular if there is a metric on $\Omega$ for which it is separable and $\mathcal{S}$ is its Borel $\sigma$-algebra.*

*Proof.* Let $A$ be an atom of $\mu$. By $\sigma$-finiteness we have $0 < \mu(A) < +\infty$. For each $j$, let $B_j := A_j$ if $\mu(A \cap A_j) = \mu(A)$ or $B_j := A_j^c$ if $\mu(A \cap A_j) = 0$. Let $C := A \cap \bigcap_{j=1}^\infty B_j$. Then $\mu(C) = \mu(A)$ and $C$ is also an atom. If there exist $x \ne y$ in $C$ then by the separated assumption there must exist $A_j$ such that $A_j$ contains one of $x$ and $y$ but not both, contradicting the definition of $C$. Thus $C$ is a singleton $\{x\}$ and $\mu(A \setminus \{x\}) = 0$ as stated.

The Borel $\sigma$-algebra of a separable metric space is generated by the countably many open balls with centers in a countable dense set and rational radii, and it clearly separates points. This completes the proof. □

Let $\mathcal{D}_\lambda = \mathcal{D}_\lambda(\Omega, \mathcal{S}, \mu)$ be the set of all functions $G \in \mathcal{L}^0(\Omega, \mathcal{S}, \mu)$ such that $\mu \circ G^{-1}$ has a bounded density $\xi_G$ with respect to Lebesgue measure $\lambda$ on $\mathbb{R}$. Such functions were characterized in Theorem 7.24. If the set $\mathcal{D}_\lambda(\Omega, \mathcal{S}, \mu)$ is nonempty then $\mu$ must be nonatomic as the following shows:

**Proposition A.4.** *Let $(\Omega, \mathcal{S}, \mu)$ be a finite measure space and $G \in \mathcal{L}^0(\Omega, \mathcal{S}, \mu)$. If $\mu$ has an atom then $\mu \circ G^{-1}$ has an atom $\{y\}$ for some $y$.*

*Proof.* Suppose that $A \in \mathcal{S}$ is an atom of $\mu$ with $c := \mu(A) > 0$. Restricting $\mu$ to $A$, we can assume that $\Omega = A$. Let $F(x) := \mu(G^{-1}((-\infty, x])) \in [0, c]$ for each $x \in \mathbb{R}$. Then $F(x) \downarrow 0$ as $x \to -\infty$ and $F(x) \uparrow c$ as $x \to +\infty$. If $F(x) \in (0, c)$ for some $x \in \mathbb{R}$, then the set $G^{-1}((-\infty, x]) \in \mathcal{S}$ gives a contradiction to the fact that $A$ is an atom. Thus $F$ has only one point of increase, say $y \in \mathbb{R}$, with $F(x) = 0$ for all $x < y$ and $F(x) = c$ for all $x \geq y$, and so $\{y\}$ is an atom of $\mu \circ G^{-1}$. This completes the proof. □

On the other hand if $(\Omega, \mathcal{S}, \mu)$ is nonatomic, then the set $\mathcal{D}_\lambda(\Omega, \mathcal{S}, \mu)$ is rich enough in the sense stated by the following two propositions.

**Proposition A.5.** *For any nonatomic finite measure space $(\Omega, \mathcal{S}, \mu)$ with $\mu(\Omega) > 0$ and $-\infty < a < b < \infty$, there is a $G \in \mathcal{D}_\lambda$ such that the density $\xi_G$ equals $c1_{[a,b]}$, where $c = \mu(\Omega)/(b-a)$.*

*Proof.* If $G$ exists such that the density $\xi_G$ equals $\mu(\Omega)1_{[0,1]}$, then $a + (b-a)G$ satisfies the conclusion. So we can assume $a = 0$ and $b = 1$. Also, multiplying $\mu$ by a constant, we can assume $\mu(\Omega) = 1$.

By Proposition A.1, take $A_{11} \in \mathcal{S}$ with $\mu(A_{11}) = 1/2$. Let $A_{12} := \Omega \setminus A_{11}$, and $G_1 := (1/2)1_{A_{12}}$. At the $j$th stage we will have a decomposition of $\Omega$ into $2^j$ disjoint sets $A_{jk} \in \mathcal{S}$, $k = 1, \ldots, 2^j$, with $\mu(A_{jk}) = 1/2^j$ and $A_{j,k} = A_{j+1,2k-1} \cup A_{j+1,2k}$ for each $j \geq 1$ and $k = 1, \ldots, 2^j$. Let $G_j := (k-1)/2^j$ on $A_{jk}$ for $k = 1, \ldots, 2^j$. Then $\mu \circ G_j^{-1}$ is uniformly distributed over the points $(k-1)/2^j$ for $k = 1, \ldots, 2^j$. As $j \to \infty$, $G_j$ converges uniformly to a $\mu$-measurable function $G$. So the probability laws $\mu \circ G_j^{-1}$ converge to $\mu \circ G^{-1}$ with the density $\xi_G = 1_{[0,1]}$, completing the proof of the proposition. □

**Proposition A.6.** *For any nonatomic finite measure space $(\Omega, \mathcal{S}, \mu)$ and $1 \leq s < \infty$, bounded elements of $\mathcal{D}_\lambda$ are dense in $L^s(\Omega, \mathcal{S}, \mu)$.*

*Proof.* We know that $\mu$-simple functions $h := \sum_{i=1}^k c_i 1_{A_i}$ are dense. By Proposition A.5, for any $\delta > 0$ and $i = 1, \ldots, k$, there is a $\mu$-measurable function $f_i$ on $A_i$ such that for the restriction $\mu_i$ of $\mu$ to $A_i$, $\mu_i \circ f_i^{-1}$ has density $\mu(A_i)1_{[c_i, c_i+\delta]}/\delta$ with respect to Lebesgue measure. Let $G := \sum_{i=1}^k f_i 1_{A_i}$. Then $\|G - h\|_s \leq \delta \mu(\Omega)^{1/s}$ and $\mu \circ G^{-1}$ has density $\xi_G \leq \mu(\Omega)/\delta < \infty$. The proof of the proposition is complete. □

The following is used to prove the necessity of a suitable measurability assumption on Nemytskii operators acting on $L^p$ spaces (Theorem 7.13(c)).

**Proposition A.7.** *Let $(\Omega, \mathcal{S}, \mu)$ be any nonatomic probability space and let $Q$ be any probability measure on the Borel sets of $\mathbb{R}$. Then there is a $G \in \mathcal{L}^0(\Omega, \mathcal{S}, \mu)$ such that $\mu \circ G^{-1} = Q$.*

*Proof.* By Proposition A.5, take $H \in \mathcal{L}^0(\Omega, \mathcal{S}, \mu)$ such that $\mu \circ H^{-1} = U[0,1]$ (Lebesgue measure restricted to $[0,1]$). Let $F(x) := Q((-\infty, x])$ for all $x \in \mathbb{R}$ and $F^{\leftarrow}(y) := \inf\{x\colon F(x) \geq y\}$ for $0 < y < 1$. Then $U[0,1] \circ (F^{\leftarrow})^{-1} = Q$ [53, Proposition 9.1.2]. Since $F^{\leftarrow}$ is Borel measurable, $G := F^{\leftarrow} \circ H \in \mathcal{L}^0(\Omega, \mathcal{S}, \mu)$ and $\mu \circ G^{-1} = Q$. □

The following is used to show that continuity of a Nemytskii operator implies its boundedness and that several smoothness properties of Nemytskii operators valid on a nonempty open subset of $L^s$ can be extended to all of $L^s$.

**Proposition A.8.** *Let $(\Omega, \mathcal{S}, \mu)$ be a nonatomic $\sigma$-finite measure space. Let $g \in \mathcal{L}^1(\Omega, \mathcal{S}, \mu)$, $g \geq 0$, and $\int g \, d\mu > 0$. For each $E \in \mathcal{S}$, let $\nu(E) := \int_E g \, d\mu$. Then $(\Omega, \mathcal{S}, \nu)$ is a nonatomic finite measure space.*

*Proof.* Clearly $\nu$ is a finite measure. Suppose that $A$ is an atom of $\nu$. Then $\nu(A) > 0$ and $\mu(A) > 0$. Let $B := \{\omega \in A\colon\ g(\omega) \neq 0\}$. Then $\mu(B) > 0$. Take $C \in \mathcal{S}$ such that $C \subset B$ and $0 < \mu(C) < \mu(B)$. Then both $\nu(C) > 0$ and $\nu(A \setminus C) = \nu(B \setminus C) > 0$, a contradiction, proving that $\nu$ is nonatomic. □

**Proposition A.9.** *Let $(\Omega, \mathcal{S}, \mu)$ be a nonatomic finite measure space with $\mu(\Omega) > 0$, and let $1 \leq s < \infty$. For each $c > 0$ and $h \in L^s(\Omega, \mathcal{S}, \mu)$ such that $\|h\|_s > c$ there exists a partition $\{A_i\}_{i=0}^n$ of $\Omega$ into measurable sets such that $\int_{A_i} |h|^s \, d\mu = \|h\|_s^s/(n+1) \in (c^s/2, c^s]$ for $i = 0, \dots, n$.*

*Proof.* Let $g := |h|^s$ and apply Proposition A.8. Then $(\Omega, \mathcal{S}, \nu)$ is nonatomic. Let $n$ be the unique positive integer such that $nc^s < \|h\|_s^s \leq (n+1)c^s$. Then the desired partition $\{A_i\}_{i=0}^n$ of $\Omega$ exists by Corollary A.2. □

*Notes.* The statement of Proposition A.1 is [53, §3.5 Problem 11] and [87, §41 Problem 2].

Regarding Proposition A.3, it is known, e.g. Cohn [34, Corollary 8.6.4], that if a measurable space $(\Omega, \mathcal{S})$ is separated and $\mathcal{S}$ is countably generated then there exists a metric $d$ on $\Omega$ for which it is separable and $\mathcal{S}$ is the Borel $\sigma$-algebra. Thus the "particular" case is not really special.

# References


1. L. V. Ahlfors. *Complex Analysis.* 2d ed., McGraw-Hill, New York, 1966.
2. A. Alexiewicz and W. Orlicz. Analytic operations in real Banach spaces. *Studia Math.* **14** (1954), 57–78.
3. J. Appell and P. P. Zabrejko. *Nonlinear Superposition Operators.* Cambridge University Press, 1990.
4. J. Appell and P. P. Zabrejko. On the degeneration of the class of differentiable superposition operators in function spaces. *Analysis* **7** (1987), 305–312.
5. A. Baernstein. On the Fourier series of functions of bounded $\Phi$-variation. *Studia Math.* **42** (1972), 243–248.
6. S. Banach. Sur les opérations dans les ensembles abstraits et leur application aux équations intégrales. *Fund. Math.* **3** (1922), 133–181.
7. S. Banach. *Teorja Operacyj,* tom I. *Operacje Liniowe* (in Polish). Kasa im. Mianowskiego, Warszawa, 1931. Web address of a photocopy: `http://banach.univ.gda.pl/teoria-operacji-pol.html`
8. S. Banach. *Théorie des opérations linéaires.* Monografie Matematyczne 1. PWN, Warszawa, 1932.
9. S. Banach. Über homogene Polynome in $(L^2)$. *Studia Math.* **7** (1938), 36–44.
10. N. K. Bari (Bary). *A Treatise on Trigonometric Series.* 2 vols., Fizmatgiz, Moscow, 1961; transl. by M. F. Mullins, Macmillan, New York, 1964.
11. R. G. Bartle. A general bilinear vector integral. *Studia Math.* **15** (1956), 337–352.
12. A. F. Beardon. *Complex Analysis.* Wiley, Chichester, 1979.
13. V. A. Belfi and R. S. Doran. Norm and spectral characterizations in Banach algebras. *L'Enseignement Mathématique* (Ser. 2) **26**, fasc. 1-2 (1981), 103–130.
14. A. Beurling. Analyse harmonique de pseudomesures. Intégration par rapport aux pseudomesures. *C. R. Acad. Sci. Paris* **258** (1964), 1984–1987.
15. N. H. Bingham, C. M. Goldie and J. L. Teugels. *Regular Variation. Encyclopedia of Math. and its Appl.*, Cambridge University Press, 1987.
16. Garrett Birkhoff. Integration of functions with values in a Banach space. *Trans. Amer. Math. Soc.* **38** (1935), 357–378.
17. Garrett Birkhoff. On product integration. *J. Math. Phys.* (MIT, Cambridge, Mass.), **16** (1937), 104–132.
18. Z. Birnbaum and W. Orlicz. Über die Verallgemeinerung des Begriffes der zueinander konjugierten Potenzen. *Studia Math.* **3** (1931), 1–67.

19. R. M. Blumenthal and R. K. Getoor. Sample functions of stochastic processes with stationary independent increments. *J. Math. and Mech.* **10** (1961), 493–516.
20. J. Bochnak and J. Siciak. Polynomials and multilinear mappings in topological vector spaces. *Studia Math.* **39** (1971), 59–76.
21. S. Bochner. Integration von Funktionen, deren Werte die Elemente eines Vektorraumes sind. *Fund. Math.* **20** (1933), 262-276.
22. F. F. Bonsall and J. Duncan. *Complete Normed Algebras.* Ergebnisse der Mathematik und ihrer Grenzgebiete. Berlin, Springer-Verlag, 1973.
23. L. Breiman. *Probability.* Addison-Wesley, Reading, Mass., 1968.
24. J. Bretagnolle. $p$-variation de fonctions aléatoires. In: *Séminaire de Probabilités VI, Lect. Notes in Math.* (Springer-Verlag) **258** (1972), 51–71.
25. J. Bretagnolle and P. Massart. Hungarian constructions from the nonasymptotic viewpoint. *Ann. Probab.* **17** (1989), 239–256.
26. M. Brokate and F. Colonius (1990). Linearizing equations with state-dependent delays. *Appl. Math. Optimiz.* **21**, 45–52.
27. J. C. Burkill. Differential properties of Young-Stieltjes integrals. *J. London Math. Soc.* **23** (1948), 22–28.
28. T. Carleman. Über die Fourierkoeffizienten einer stetigen Funktion. *Acta Math.* (Sweden) **41** (1918), 377–384.
29. L. Carleson. On convergence and growth of partial sums of Fourier series. *Acta Math.* (Sweden) **116** (1966), 135–157.
30. H. Cartan. *Calcul differéntiel.* Hermann, Paris, 1967.
31. A. Cena. *Geometric structures on the non-parametric statistical manifold.* Doctoral thesis, Univ. degli studi di Milano, 2002-2003.
32. S. B. Chae. *Holomorphy and Calculus in Normed Spaces.* Dekker, New York, 1985.
33. J. Ciemnoczołowski and W. Orlicz. Composing functions of bounded $\phi$-variation. *Proc. Amer. Math. Soc.* **96** (1986), 431–436.
34. D. L. Cohn. *Measure Theory.* Birkhäuser, Boston, 1980.
35. Ju. L. Daleckiĭ and M. G. Kreĭn. *Stability of Solutions of Differential Equations in Banach Space.* Amer. Math. Soc. *Transls. of Math. Monographs* **43**, 1974. Russian orig. Nauka, Moscow, 1970.
36. H. G. Dales. *Banach Algebras and Automatic Continuity.* Clarendon Press, Oxford, 2000.
37. P. J. Daniell. A general form of integral. *Ann. Math.* (Ser. 2) **19** (1918), 279–294.
38. P. J. Davis. Gamma Function and Related Functions. In: *Handbook of Mathematical Functions* (M. Abramowitz and I. A. Stegun, eds.), chap. 6, pp. 253–293. Dover, New York, 1970.
39. C. Dellacherie and P.-A. Meyer. *Probabilities and potential B.* Transl. by J. P. Wilson. North-Holland, Amsterdam, 1982.
40. E. DiBenedetto. *Real Analysis.* Birkhäuser, Boston, 2002.
41. J. Diestel and J. J. Uhl. *Vector Measures. Mathematical Surveys* **15**, AMS, Providence, Rhode Island, 1977.
42. J. Dieudonné. *Foundations of Modern Analysis.* Academic Press, New York, 1960.
43. R. L. Dobrushin. Generalization of Kolmogorov's equations for Markov processes with a finite number of possible states (in Russian). *Mat. Sb.* **33** (1953), 567–596.

44. C. Doléans-Dade and P. A. Meyer. Intégrales stochastiques par rapport aux martingales locales. In: *Séminaire de Probabilités IV*, Univ. Strasbourg, *Lecture Notes in Math.* (Springer-Verlag) **124** (1970), 77–107.
45. J. D. Dollard and C. N. Friedman. *Product Integration with Applications to Differential Equations. Encyclopedia of Math. and its Appl.*, Vol. 10, Addison-Wesley, Reading, Massachusetts, 1979.
46. J. L. Doob. *Stochastic Processes.* Wiley, New York, 1953.
47. R. M. Dudley. Sample functions of the Gaussian process. *Ann. Probab.* **1** (1973), 66–103.
48. R. M. Dudley. Wiener functionals as Itô integrals. *Ann. Probab.* **5** (1977), 140–141.
49. R. M. Dudley. Fréchet differentiability, $p$-variation and uniform Donsker classes. *Ann. Probab.* **20** (1992), 1968–1982.
50. R. M. Dudley. The order of the remainder in derivatives of composition and inverse operators for $p$-variation norms. *Ann. Statist.* **22** (1994), 1–20.
51. R. M. Dudley. Empirical processes and $p$-variation. In: *Festschrift for Lucien Le Cam* (D. Pollard, E. Torgersen, and G. L. Yang, eds.), 219–233. Springer-Verlag, New York, 1997.
52. R. M. Dudley. *Uniform Central Limit Theorems.* Cambridge University Press, 1999.
53. R. M. Dudley. *Real Analysis and Probability.* 2d ed., Cambridge University Press, 2002.
54. R. M. Dudley and R. Norvaiša. *Differentiability of Six Operators on Nonsmooth Functions and p-Variation. Lecture Notes in Math.* (Springer-Verlag) **1703** (1999).
55. R. M. Dudley and R. Norvaiša. *An Introduction to p-variation and Young Integrals.* MaPhySto Lecture Notes no. 1, 1998, Aarhus, Denmark.
56. N. Dunford. Integration of abstract functions (abstract). *Bull. Amer. Math. Soc.* **42** (1936), 178.
57. N. Dunford and J. T. Schwartz. *Linear Operators, Part I: General Theory.* Wiley Classics Library Ed., New York, 1988. 1st ed. 1958.
58. J. Durdil. On the differentiability of Urysohn and Nemyckij operators. *Comment. Math. Univ. Carolinae* **8** (1967), 515–553.
59. E. B. Dynkin. *Theory of Markov Processes.* English transl. by D. E. Brown, ed. T. Kövary. Prentice-Hall, Englewood Cliffs, NJ, 1961. Russian orig. *Osnovaniy Teorii Markovskikh Protsessov.* Fizmatgiz, Moscow, 1959. German transl. by Joseph Wloka, *Die Grundlagen der Theorie der Markoffschen Prozesse.* Springer-Verlag, Berlin, 1961. French transl. by Ch. Sarthou, *Théorie des processus markoviens.* Dunod, Paris, 1963.
60. A. M. D'yačkov. Conditions for the existence of Stieltjes integral of functions of bounded generalized variation. *Anal. Math.* (Budapest) **14** (1988), 295–313.
61. D. G. Ebin and J. Marsden. Groups of diffeomorphisms and the motion of an incompressible fluid. *Ann. Math.* **92** (1970), 102–163.
62. C. H. Edwards, Jr. *Advanced Calculus of Several Variables.* Dover, New York, 1973.
63. R. E. Edwards. *Fourier Series: A Modern Introduction.* 2d ed., Springer-Verlag, New York, 1982.
64. W. Feller. *An Introduction to Probability Theory and Its Applications*, vol. 2, 2d ed. Wiley, New York, 1971.

65. L. T. Fernholz. *von Mises Calculus for Statistical Functionals. Lect. Notes in Statist.* (Springer-Verlag) **19** (1983).

66. *B. de Finetti and M. Jacob. Sull' integrale di Stieltjes-Riemann. *Giorn. Istituto Ital. Attuari* **6** (1935), 303–319.

67. D. L. Fisk. Quasi-martingales. *Trans. Amer. Math. Soc.* **120** (1965), 369–389.

68. M. Fréchet. La notion de différentielle dans l'analyse générale. *C. R. Acad. Sc. Paris* **180** (1925), 806–809.

69. M. Fréchet. La notion de différentielle dans l'analyse générale. *Ann. Ecole Norm. Sup.* **42** (1925), 293–323.

70. M. A. Freedman. Operators of $p$-variation and the evolution representation problem. *Trans. Amer. Math. Soc.* **279** (1983), 95–112.

71. B. Fristedt. Sample functions of stochastic processes with stationary, independent increments. In: *Advances in Probability and Related Topics*, **3** 241–396. Dekker, New York, 1974.

72. B. Fristedt and S. J. Taylor. Strong variation for the sample functions of a stable process. *Duke Math. J.* **40** (1973), 259–278.

73. B. M. Garay. Hyperbolic structures in ODE's and their discretization. In *Nonlinear Analysis and Boundary Value Problems for ODEs* (Proc. CISM, Udine, 1995; F. Zanolin, ed.). *CISM Courses and Lectures* **371** (1996), 149–174. Springer-Verlag, Vienna.

74. F. W. Gehring. A study of $\alpha$-variation. I. *Trans. Amer. Math. Soc.* **76** (1954), 420–443.

75. I. M. Gelfand. On normed rings. *Dokl. Akad. Nauk SSSR* **23** (1939), 430–432. In: *Izrail M. Gelfand: Collected Papers* **1** (S. G. Gindikin et al., eds.), 169–171. Springer-Verlag, Berlin, 1987.

76. I. M. Gelfand. Normierte Ringe. *Rec. Math.* (*Matem. Sbornik*) N. S. **9** (1941), 3–24. In: *Izrail M. Gelfand: Collected Papers* **1** (S. G. Gindikin et al., eds.), 181–201. Springer-Verlag, Berlin, 1987.

77. I. M. Gelfand, D. Raikov, and G. Shilov. *Commutative Normed Rings.* English ed. Chelsea, New York, 1964.

78. I. I. Gihman [Gikhman] and A. V. Skorohod [Skorokhod]. *Introduction to the Theory of Random Processes.* W. B. Saunders, Philadelphia, 1969.

79. I. I. Gihman and A. V. Skorohod. *The Theory of Stochastic Processes I, II, III.* Fizmatgiz, Moscow, 1971, 1973, 1975; English transl. by S. Kotz, Springer-Verlag, Berlin, 1974, 1975, 1979.

80. R. D. Gill. Lectures on Survival Analysis. In: *Ecole d'été de Probabilités de Saint-Flour* (P. Bernard, ed.), *Lect. Notes in Math.* (Springer-Verlag) **1581** (1994), 115–241.

81. R. D. Gill and S. Johansen. A survey of product-integration with a view toward application in survival analysis. *Ann. Statist.* **18** (1990), 1501–1555.

82. E. Ch. Gochman. *The Stieltjes integral and its applications* (in Russian). Moscow, GIFML, 1958.

83. B. I. Golubov. Functions of bounded $p$-variation (in Russian). *Izv. Akad. Nauk SSSR Ser. Mat.* **32** (1968), 837–858. English transl. in: *Math. USSR Izv.* **2** (1968), 799–819.

84. R. A. Gordon. *The Integrals of Lebesgue, Denjoy, Perron, and Henstock.* Amer. Math. Soc., Providence, 1994.

85. A. Gray. Differentiation of composites with respect to a parameter. *J. Austral. Math. Soc.* (Ser. A) **19** (1975), 121–128.

86. R. E. Greene and S. G. Krantz. *Function Theory of One Complex Variable.* 2d ed., *Graduate Studies in Mathematics* **40**, AMS, Providence, Rhode Island, 2002.
87. P. Halmos. *Measure Theory.* Van Nostrand, Princeton, 1950.
88. F. Hartung and J. Turi. On differentiability of solutions with respect to parameters in state-dependent delay equations. *J. Differential Equations* **135** (1997), 192–237.
89. F. Hausdorff. Zur Theorie der linearen metrischen Räume. *J. Reine Angew. Math.* **167** (1932), 294–311.
90. R. Henstock. Definitions of Riemann type of the variational integrals. *Proc. London Math. Soc.* (3) **11** (1961), 402–418.
91. R. Henstock. *Theory of Integration.* Butterworths, London, 1963.
92. R. Henstock. *The General Theory of Integration.* Clarendon Press, Oxford, 1991.
93. E. Hewitt. Integration by parts for Stieltjes integrals. *Amer. Math. Monthly* **67** (1960), 419–423.
94. T. H. Hildebrandt. Definitions of Stieltjes integrals of the Riemann type. *Amer. Math. Monthly* **45** (1938), 265–278.
95. T. H. Hildebrandt. Integration in abstract spaces. *Bull. Amer. Math. Soc.* **59** (1953), 111–139.
96. T. H. Hildebrandt. On systems of linear differentio-Stieltjes-integral equations. *Illinois J. Math.* **3** (1959), 352–373.
97. T. H. Hildebrandt. *Introduction to the Theory of Integration.* Academic Press, New York, 1963.
98. W. Hoeffding. Probability inequalities for sums of bounded random variables. *J. Amer. Statist. Assoc.* **58**, 13–30.
99. J. Hoffmann-Jørgensen. *Probability with a View toward Statistics*, Vols. I, II. Chapman & Hall, New York, 1994.
100. C. S. Hönig. Volterra Stieltjes-Integral Equations. In: *Functional Differential Equations and Bifurcation* (A. F. Izé, ed.), *Lect. Notes in Math.* (Springer-Verlag) **799** (1980), 173–216.
101. Yen-Chin Huang. *Empirical distribution function statistics, speed of convergence, and p-variation.* Ph.D. thesis, Massachusetts Institute of Technology, 1994.
102. Yen-Chin Huang. Speed of convergence of classical empirical processes in $p$-variation norm. Preprint, 1998.
103. Yen-Chin Huang and R. M. Dudley. Speed of convergence of classical empirical processes in $p$-variation norm. *Ann. Probab.* **29** (2001), 1625–1636.
104. R. A. Hunt. On the convergence of Fourier series. In: *Orthogonal Expansions and Their Continuous Analogues* (D. T. Haimo, ed.), 235–255. Southern Illinois Univ. Press [Carbondale], 1968.
105. E. L. Ince. *Ordinary Differential Equations.* Longmans, London, 1927. Repr. Dover, New York, 1956.
106. K. Itô. On a formula concerning stochastic differentials. *Nagoya Math. J.* **3** (1951), 55–65. Reprinted in: Kiyosi Itô. Selected papers (D. W. Stroock, S. R. S. Varadhan eds.), 169–179. Springer-Verlag, New York, 1987.
107. K. Itô. Stochastic processes. *Lecture Notes Series* **16**. Aarhus Universitet, 1969.
108. K. Itô. Stochastic differentials. *Applied Mathematics and Optimization* **1** (1974), 374–381.

109. K. Itô and H. P. McKean, Jr. *Diffusion Processes and their Sample Paths.* Academic Press, New York, 1965.
110. N. C. Jain and D. Monrad. Gaussian measures in $B_p$. *Ann. Probab.* **11** (1983), 46–57.
111. J. Jarník and J. Kurzweil. A general form of the product integral and linear ordinary differential equations. *Czechoslovak Math. J.* **37** (1987), 642–659.
112. O. Kallenberg. *Foundations of Modern Probability.* 2d ed., Springer-Verlag, New York, 2002.
113. H. S. Kaltenborn. Existence conditions and a substitution theorem for Stieltjes mean integrals. *Tohoku Math. J.* **44** (1938), 1–11.
114. Y. Katznelson. *An Introduction to Harmonic Analysis.* Wiley, New York, 1968.
115. T. Kawada and N. Kôno. On the variation of Gaussian processes. In: *Proc. Second Japan-USSR Symp. on Probab. Theory* (G. Maruyama and Yu. V. Prokhorov, eds.), *Lect. Notes in Math.* (Springer-Verlag) **330** (1973), 176–192.
116. J. R. Kinney. Continuity properties of sample functions of Markov processes. *Trans. Amer. Math. Soc.* **74** (1953), 280–302.
117. P. Kirwan. On an error by Angus E. Taylor. *Irish Math. Soc. Bull.* **44** (2000), 66–75.
118. I. Kluvánek. *Integration Structures. Proceedings of the Centre for Mathematical Analysis*, Australian National University, Vol. 18, 1988.
119. A. N. Kolmogorov. Une série de Fourier-Lebesgue divergente partout. *C. R. Acad. Sci. Paris* **183** (1926), 1327–1328.
120. A. N. Kolmogorov. Untersuchungen über den Integralbegriff. *Math. Ann.* **103** (1930), pp. 654–696. Russian transl. by P. Osvald in: *A. N. Kolmogorov. Selected Works. Mathematics and Mechanics.* Nauka, Moscow, 1985, pp. 96–134. English transl. by V. M. Volosov in: *Selected Works of A. N. Kolmogorov. Vol. I. Mathematics and mechanics.* (V. M. Tikhomirov, ed.) Kluwer, Dordrecht, 1993, pp. 100–143.
121. J. Komlós, P. Major, and G. Tusnády. An approximation of partial sums of independent RV'-s and the sample DF. I. *Z. Wahrscheinlichkeitstheorie verw. Geb.* **32** (1975), 111–131.
122. *Gyula König. On the theory of definite integrals (in Hungarian). *Math. és Term. tud. Értesitö* **15** (1897), 380–394.
123. J. Kopeć and J. Musielak. On the estimation of the norm of the $n$-linear symmetric operation. *Studia Math.* **15** (1956), 29–30.
124. G. L. Krabbe. Integration with respect to operator-valued functions. *Bull. Amer. Math. Soc.* **67** (1961), 214–218.
125. G. L. Krabbe. Integration with respect to operator-valued functions. *Acta Sci. Math.* (Szeged) **22** (1961), 301–319.
126. S. G. Krantz. *A Panorama of Harmonic Analysis.* Carus Math. Monographs, No. 27. Math. Assoc. America, USA, 1999.
127. M. A. Krasnosel'skiĭ. *Topological Methods in the Theory of Nonlinear Integral Equations* (in Russian). Gostekhizdat, Moscow, 1956. English transl. Macmillan, New York, 1964.
128. M. A. Krasnosel'skiĭ and Ya. B. Rutickiĭ. *Convex Functions and Orlicz Spaces* (in Russian). Fizmatgiz, Moscow, 1958. English transl. by L. F. Boron. Noordhoff, Groningen, 1961.
129. J. Kurzweil. Generalized ordinary differential equations and continuous dependence on a parameter. *Czechoslovak Math. J.* **7** (1957), 418–446.

130. J. Kurzweil. On integration by parts. *Czechoslovak Math. J.* **8** (1958), 356–359.
131. J. Kurzweil. Addition to my paper "Generalized ordinary differential equations and continuous dependence on a parameter." *Czechoslovak Math. J.* **9** (1959), 564–573.
132. T. L. Lai. Reproducing kernel Hilbert spaces and the law of the iterated logarithm for Gaussian processes. *Z. Wahrscheinlichkeitstheorie verw. Geb.* **29** (1974), 7–19.
133. R. Larsen. *Banach Algebras: An Introduction.* Dekker, New York, 1973.
134. H. Lebesgue. Sur l'intégrale de Stieltjes et sur les opérations fonctionnelles linéaires. *C. R. Acad. Sci. Paris* **150** (1910), 86–88.
135. M. Ledoux. Isoperimetry and Gaussian analysis. In: *Lectures on Probability Theory and Statistics*, Saint-Flour, 1994, *Lect. Notes in Math.* (Springer-Verlag) **1648** (1996), 165–294.
136. M. Ledoux and M. Talagrand. *Probability in Banach Spaces.* Springer-Verlag, Berlin, 1991.
137. E. L. Lehmann. *Testing Statistical Hypotheses.* 2d ed., Wiley, New York, 1986, repr. Springer-Verlag, New York, 1997.
138. R. Leśniewicz. On Hardy-Orlicz spaces, I. *Comment. Math. Prace Mat.* **15** (1971), 3–56.
139. R. Leśniewicz and W. Orlicz. On generalized variations (II). *Studia Math.* **45** (1973), 71–109.
140. P. Lévy. Le mouvement brownien plan. *Amer. J. Math.* **62** (1940), 487–550.
141. P. Lévy. Wiener's random function, and other Laplacian random functions. In: *Proc. 2d Berkeley Symp. Math. Statist. Probability* (J. Neyman, ed.), 171–187. Univ. Calif. Press, Berkeley and Los Angeles, 1951.
142. J. Lindenstrauss and L. Tzafriri. *Classical Banach Spaces I. Sequence Spaces.* Springer-Verlag, Berlin, 1977.
143. R. Lipschitz. De explicatione per series trigonometricas instituenda functiorum unius variabilis arbitrariarum ... *J. reine angew. Math.* **63** (1864), 296–308. French. transl. by P. Montel in: *Acta Math.* (Sweden) **36** (1912-13), 281–295.
144. E. R. Love. The refinement-Ross-Riemann-Stieltjes ($R^3S$) integral. In: *Analysis, Geometry and Groups: A Riemann Legacy Volume* (H. M. Srivastava and Th. M. Rassias, eds.), 289–312. Hadronic Press, Palm Harbor, FL, 1993.
145. E. R. Love. Integration by parts and other theorems for $R^3S$-integrals. *Real Anal. Exchange* **24** (1998/9), 315–336.
146. E. R. Love and L. C. Young. Sur une classe de fonctionelles linéaires. *Fund. Math.* **28** (1937), 243–257.
147. E. R. Love and L. C. Young. On fractional integration by parts. *Proc. London Math. Soc.* (Ser. 2) **44** (1938), 1–35.
148. W. A. J. Luxemburg. Banach function spaces. Ph.D. Dissertation, Delft, 1955.
149. W. A. J. Luxemburg and A. C. Zaanen. Notes on Banach function spaces I. *Indag. Math.* **7** = *Akad. Wetenschappen Amsterdam Proc. Ser. A Math.* **66** (1963), 135–147.
150. T. Lyons. On the non-existence of path integrals. *Proc. Roy. Soc. Lond.* (Ser. A) **432** (1991), 281–290.
151. T. Lyons. Differential equations driven by rough signals (I): an extension of an inequality of L. C. Young. *Math. Res. Letters* **1** (1994), 451–464.
152. J. S. MacNerney. Stieltjes integrals in linear spaces. *Ann. Math.* **61** (1955), 354–367.

153. J. S. MacNerney. Integral equations and semigroups. *Illinois J. Math.* **7** (1963), 148–173.
154. P. Major. On the invariance principle for sums of independent identically distributed random variables. *J. Multivariate Anal.* **8** (1978), 487–517.
155. M. Manstavičius. $p$-variation of strong Markov processes. *Ann. Probab.* **32** (2004), 2053–2066.
156. J. Marcinkiewicz. On a class of functions and their Fourier series. *C. R. Soc. Sci. Varsovie* **26** (1934), 71–77. Reprinted in: *J. Marcinkiewicz. Collected Papers* (A. Zygmund, ed.), 36–41. Państwowe Wydawnictwo Naukowe, Warsaw, 1964.
157. J. E. Marsden and A. J. Tromba. *Vector Calculus.* 2d ed., W. H. Freeman, San Francisco, 1981.
158. P. R. Masani. Multiplicative Riemann integration in normed rings. *Trans. Amer. Math. Soc.* **61** (1947), 147–192.
159. P. R. Masani. The place of multiplicative integration in modern analysis. Appendix II in [45].
160. P. Massart. The tight constant in the Dvoretzky–Kiefer–Wolfowitz inequality. *Ann. Probab.* **18** (1990), 1269–1283.
161. J. Matkowski and J. Miś. On a characterization of Lipschitzian operators of substitution in the space $BV[a, b]$. *Math. Nachr.* **117** (1984), 155–159.
162. S. Mazur. Sur les anneaux linéaires. *C. R. Acad. Sci. Paris* **207** (1938), 1025–1027.
163. S. Mazur and W. Orlicz. Grundlegende Eigenschaften der polynomischen Operationen, I, II. *Studia Math.* **5** (1935), 50–68, 179–189.
164. H. P. McKean, Jr. *Stochastic Integrals.* Academic Press, New York, 1969.
165. R. M. McLeod. *The Generalized Riemann Integral.* Carus Math. Monographs, No. 20. Math. Assoc. America, 1980.
166. E. J. McShane. *A Riemann-type Integral that Includes Lebesgues-Stieltjes, Bochner and Stochastic Integrals.* Memoirs AMS, **88** (1969). Providence, Rhode Island.
167. E. J. McShane. *Unified Integration.* Academic Press, Orlando, 1983.
168. F. A. Medvedev. *Development of the Concept of Integral* (in Russian). Nauka, Moscow, 1974.
169. P.-A. Meyer. *Probability and Potentials.* Blaisdell, Waltham, MA, 1966.
170. P. Meyer-Nieberg. *Banach Lattices.* Springer-Verlag, Berlin, 1991.
171. J. Milnor. Remarks on infinite-dimensional Lie groups. In: *Relativité, groupes et topologie* II, Les Houches 1983 (B. S. DeWitt and R. Stora, eds.), 1007–1057. North-Holland, Amsterdam, 1984.
172. I. Monroe. On the $\gamma$-variation of processes with stationary independent increments. *Ann. Math. Statist.* **43** (1972), 1213–1220.
173. I. Monroe. On embedding right continuous martingales in Brownian motion. *Ann. Math. Statist.* **43** (1972), 1293–1311.
174. I. Monroe. Processes that can be embedded in Brownian motion. *Ann. Probab.* **6** (1978), 42–56.
175. E. H. Moore. Definition of limit in general integral analysis. *Proc. Nat. Acad. Sci. USA* **1** (1915), 628.
176. J. Musielak and W. Orlicz. On generalized variations (I). *Studia Math.* **18** (1959), 11–41.

177. M. A. Naimark. *Normed Algebras.* Transl. from the 2d Russian ed. by L. F. Boron. Wolters-Noordhoff, Groningen, 1972. *Normirovannye Kol'tsa.* 2d ed. revised, Nauka, Moscow, 1968.
178. M. H. A. Newman. *Elements of the Topology of Plane Sets of Points.* 2d ed., Cambridge University Press, 1951.
179. T. Nguyen. A lower bound on the radius of analyticity of a power series in a real Banach space. *Studia Math.* **191** (2009), 171–179,
180. M. Nisio. On the extreme values of Gaussian processes. *Osaka J. Math.* **4** (1967), 313–326.
181. H. Oodaira. On Strassen's version of the law of the iterated logarithm for Gaussian processes. *Z. Wahrscheinlichkeitstheorie verw. Geb.* **21** (1972), 289–299.
182. W. Orlicz. Beiträge zur Theorie der Orthogonalentwicklungen (III). *Bull. Acad. Pol. Sci.* **1932**, 229–238.
183. G. Peano. Intégration par séries des équations différentielles linéaires. *Math. Ann.* **32** (1888), 450–456.
184. Lee Peng-Yee. *Lanzhou Lectures on Henstock Integration.* World Scientific, Singapore, 1989.
185. W. F. Pfeffer. *The Riemann Approach to Integration: Local Geometric Theory.* Cambridge University Press, 1993.
186. É. Picard. Sur l'application des méthodes d'approximation successives a l'étude de certaines équations différentielles ordinaires. *J. Math. Pures et Appliquées* (Ser. 4) **9** (1893), 217–271.
187. S. Pollard. The Stieltjes' integral and its generalisations. *Quart. J. Pure Appl. Math.* **49** (1923), 73–138.
188. A. V. Ponosov. On the Nemytskij conjecture (in Russian). *Doklady Akad. Nauk SSSR* **289** (1986), 1308–1311. English transl. in: *Soviet Math. Doklady* **34** (1987), 231–233.
189. Jinghua Qian. The $p$-variation of partial sum processes and the empirical process. *Ann. Probab.* **26** (1998), 1370–1383.
190. J. Radon. Theorie und Anwendungen der absolut additiven Mengenfunktionen. *Sitzungsber. K. Akad. Wiss. Wien,* Math.-Naturwiss. Kl., Abt. IIa, **122** (1913), 1295–1438.
191. G. Rasch. Zur Theorie und Anwendung des Produktintegrals. *J. reine angew. Math.* **171** (1934), 65–119.
192. J. A. Reeds III. *On the definition of von Mises functionals.* Ph.D. thesis, Statistics, Harvard University (1976).
193. F. Riesz. Sur les opérations fonctionnelles linéaires. *C. R. Acad. Sci. Paris* **149** (1909), 974–977.
194. P. Rosenthal. Review of Shapiro (1993) and Singh and Manhas (1993). *Bull. Amer. Math. Soc.* **32** (1995), 150–153.
195. K. A. Ross. *Elementary Analysis: The Theory of Calculus.* Springer-Verlag, New York, 1980.
196. W. T. Ross and H. S. Shapiro. *Generalized Analytic Continuation. Amer. Math. Soc. University Lect. Ser.* **25**, 2002.
197. A. I. Rubinšteĭn. On $\omega$-lacunary series and functions of the class $H^\omega$ (in Russian). *Mat. Sb.* (N. S.) **65** (107) (1964), 239–271.
198. W. Rudin. *Real and Complex Analysis.* 2d ed., McGraw-Hill, New York, 1974.
199. T. Runst and W. Sickel. *Sobolev Spaces of Fractional Order, Nemytskij Operators, and Nonlinear Partial Differential Equations.* de Gruyter, Berlin, 1996.

200. A. F. Ruston. *Fredholm Theory in Banach Spaces.* Cambridge University Press, 1986.
201. S. Saks. *Theory of the Integral.* 2d ed. English transl. by L. C. Young. Hafner, New York. Repr. corrected Dover, New York, 1964.
202. R. Salem. Sur un test général pour la convergence uniforme des séries de Fourier. *C. R. Acad. Sci. Paris* **207** (1938), 662–665. Repr. as: La convergence uniforme des séries de Fourier. Chap. VI. *Essais sur les séries trigonométriques. Ensembles et fonctions.* Exposés publiés par A. Denjoy. *Actual. Sci. et Industr.*, **862**, Hermann, Paris, 1940, and in: *Raphaël Salem Œuvres Mathématiques*, pp. 156–161. Hermann, Paris, 1967.
203. G. Samorodnitsky and M. S. Taqqu. *Stable Non-Gaussian Random Processes. Stochastic Models with Infinite Variance.* Chapman & Hall, New York, 1994.
204. K. Sato. *Lévy Processes and Infinitely Divisible Distributions.* Cambridge University Press, 1999.
205. R. Schatten. *A Theory of Cross-Spaces.* Annals of Mathematics Studies, **26**. Princeton University Press, 1950.
206. L. Schlesinger. Neue Grundlagen für einen Infinitesimalkalkul der Matrizen. *Math. Z.* **33** (1931), 33–61.
207. L. Schlesinger. Weitere Beiträge zum Infinitesimalkalkul der Matrizen. *Math. Z.* **35** (1932), 485–501.
208. G. Schmidt. On multiplicative Lebesgue integration and families of evolution operators. *Math. Scand.* **29** (1971), 113–133.
209. I. J. Schoenberg. On certain metric spaces arising from Euclidean spaces by a change of metric and their imbedding in Hilbert space. *Ann. Math* (Ser. 2) **38** (1937), 787–793.
210. I. J. Schoenberg. Metric spaces and positive definite functions. *Trans. Amer. Math. Soc.* **44** (1938), 522–536. Reprinted in: *I. J. Schoenberg. Selected Papers* (Carl de Boor, ed.), Vol. 1, 100–114. Birkhäuser, Boston, 1988.
211. Š. Schwabik. On the relation between Young's and Kurzweil's concept of Stieltjes integral. *Časopis Pěst. Mat.* **98** (1973), 237–251.
212. Š. Schwabik. The Perron product integral and generalized linear differential equations. *Časopis Pěst. Mat.* **115** (1990), 368–404.
213. Š. Schwabik. Bochner product integration. *Math. Bohem.* **119** (1994), 305–335.
214. Š. Schwabik, M. Tvrdý and O. Vejvoda. *Differential and Integral Equations.* Reidel, Dordrecht, 1979.
215. H.-U. Schwarz. *Banach Lattices and Operators.* Teubner, Leipzig, 1984.
216. I. E. Segal and R. A. Kunze. *Integrals and Operators.* McGraw-Hill, New York.
217. J. H. Shapiro. *Composition Operators and Classical Function Theory.* Springer-Verlag, New York, 1993.
218. I. V. Shragin. Conditions for the measurability of superpositions (in Russian). *Doklady Akad. Nauk SSSR* **197** (1971), 295–298. English transl. in: *Soviet Math. Doklady* **12** (1971), 465–470. Erratum: *Soviet Math. Doklady* **12** No. 4 (1971), vi.
219. I. V. Shragin. Superpositional measurability (in Russian). *Izvestija Vyssh. Uchebn. Zaved. Mat.* **1** (1975), 82–89.
220. I. V. Shragin. On the Carathéodory conditions (in Russian). *Uspekhi Mat. Nauk* **34** 3 (1979), 219–220. English transl. in: *Russian Math. Surveys* **34**, 3 (1979), 220–221.
221. R. K. Singh and J. S. Manhas. *Composition Operators on Function Spaces.* North-Holland, Amsterdam, 1993.

222. A. V. Skorohod. On a generalization of the stochastic integral. *Theory Probab. Appl.* **20** (1975), 219–233.
223. H. L. Smith. On the existence of the Stieltjes integral. *Trans. Amer. Math. Soc.* **27** (1925), 491–515.
224. E. M. Stein and R. Shakarchi. *Complex Analysis.* Princeton University Press, Princeton, 2003.
225. J. O. Stevenson. Holomorphy of composition. *Bull. Amer. Math. Soc.* **80** (1974), 300–303.
226. J. O. Stevenson. Holomorphy of composition. In: *Infinite Dimensional Holomorphy and Applications* (M. C. Matos, ed.), 397–424. North-Holland, Amsterdam, 1977.
227. *T. J. Stieltjes. Recherches sur les fractions continues. *Ann. Fac. Sci. Toulouse* Sér. 1 **8** (1894), 1–122; **9** (1895), 1–47.
228. B. Szénássy. *History of Mathematics in Hungary until the 20th Century.* English transl. by J. Pokoly, ed. by J. Bognár. Springer-Verlag, New York, 1992.
229. A. E. Taylor. Historical notes on amalyticity as a concept in functional analysis. In: *Problems in Analysis. A Symposium in Honor of Solomon Bochner* (R. C. Gunning, ed.), 325–343. Princeton University Press, Princeton, 1970. Reprinted in [32].
230. A. E. Taylor. A study of Maurice Fréchet: III. Fréchet as analyst, 1909–1930. *Arch. Hist. Exact Sci.* **37** (1987), 25–76.
231. S. J. Taylor. Exact asymptotic estimates of Brownian path variation. *Duke Math. J.* **39** (1972), 219–241.
232. M. M. Vaĭnberg. On the continuity of some special operators (in Russian). *Doklady Akad. Nauk SSSR* **73** (1950), 253–255.
233. M. M. Vaĭnberg. Some questions of the differential calculus in linear spaces (in Russian). *Uspehi Matem. Nauk (N.S.)* **7** (1952), no. 4(50), 55–102.
234. Yu. R. Vaĭnerman. Sufficient conditions for existence of the Stieltjes integral (in Russian). *Vestnik Leningrad Univ.* **1972** no. 13, 15–20.
235. V. Volterra. Sui fondamenti della teoria delle equazioni differenziali lineari. *Mem. Soc. Ital. Sci.* (XL), ser. III, vol. 6, no. 8, 1887. In: *Opere Matematiche* I, 209–289. Accad. Nazionale dei Lincei, Rome, 1954.
236. V. Volterra and B. Hostinsky. *Opérations Infinitésimales Linéaires. Applications aux équations différentielles et fonctionnelles.* Gauthier-Villars, Paris, 1938.
237. Wang Sheng-Wang. Differentiability of the Nemyckii operator. *Doklady Akad. Nauk SSSR* **150** (1963), 1198–1201 (in Russian); *Sov. Math. Doklady* **4** (1963), 834–837.
238. A. J. Ward. The Perron–Stieltjes integral. *Math. Z.* **41** (1936), 578–604.
239. P. Whittle. Bounds for the moments of linear and quadratic forms in independent variables. *Theory Probab. Appl.* **5** (1960), 302–305.
240. N. Wiener. Differential-Space. *J. Math. Phys.* (MIT, Cambridge, Mass.) **2** (1923), 131–174. In: *Norbert Wiener: Collected Works* **1**, 455–498. MIT Press, Cambridge, 1976.
241. N. Wiener. The quadratic variation of a function and its Fourier coefficients. *J. Math. Phys.* (MIT, Cambridge, Mass.) **3** (1924), 72–94. In: *Norbert Wiener: Collected Works* **2**, 36–58. MIT Press, Cambridge, 1979.
242. Quanhua Xu. Espaces d'interpolation réels entre les espaces $V_p$: propriétés géométriques et applications probabilistes. In: *Séminaire d'Analyse Fonctionnelle* 1985/1986/1987, *Publ. Math. Univ. Paris VII* **28** 1988, 77–123.

243. L. C. Young. *The Theory of Integration.* Cambridge University Press, 1927. Repr. Stechert-Hafner, New York, 1964.
244. L. C. Young. An inequality of the Hölder type, connected with Stieltjes integration. *Acta Math.* (Sweden) **67** (1936), 251–282.
245. L. C. Young. Inequalities connected with bounded $p$-th power variation in the Wiener sense and with integrated Lipschitz conditions. *Proc. London Math. Soc.* (Ser. 2) **43** (1937), 449–467.
246. L. C. Young. Sur une généralisation de la notion de variation de puissance $p^{\text{ième}}$ bornée au sens de M. Wiener, et sur la convergence des séries de Fourier. *C. R. Acad. Sc. Paris* **204** (1937), 470–472.
247. L. C. Young. General inequalities for Stieltjes integrals and the convergence of Fourier series. *Math. Ann.* **115** (1938), 581–612.
248. L. C. Young. Some new stochastic integrals and Stieltjes integrals. Part I. Analogues of Hardy-Littlewood classes. In: *Advanccs in Probability and Related Topics*, Vol. II (P. Ney, ed.), 161–240. Dekker, New York, 1970.
249. L. C. Young. Some new stochastic integrals and Stieltjes integrals. Part II. In: *Advances in Probability and Related Topics*, Vol. III (P. Ney, ed.), 101–178. Dekker, New York, 1974.
250. L. C. Young. A converse to some inequalities and approximations in the theory of Stieltjes and stochastic integrals, and for $n$th derivatives. *Studia Math.* **55** (1976), 215–223.
251. R. C. Young. On Riemann integration with respect to a continuous increment. *Math. Ann.* **29** (1928), 217–233.
252. W. H. Young. *The Fundamental Theorems of the Differential Calculus.* Cambridge University Press, 1910. Repr. Hafner, New York, 1971.
253. W. H. Young. On classes of summable functions and their Fourier series. *Proc. Roy. Soc. London* (Ser. A) **87** (1912), 225–229.
254. W. H. Young. On integration with respect to a function of bounded variation. *Proc. London Math. Soc.* (Ser. 2) **13** (1914), 109–150.
255. W. H. Young. On a formula for an area. *Proc. London Math. Soc.* (Ser. 2) **18** (1919), 339–345.
256. A. C. Zaanen. *Riesz Spaces II.* North-Holland, Amsterdam, 1983.
257. E. Zeidler. *Nonlinear Functional Analysis and its Applications.* I: *Fixed-Point Theorems.* Springer-Verlag, New-York, 1986.
258. A. Zygmund. *Trigonometric series.* 2d ed., Cambridge University Press, 1959, repr. with corrections and some additions 1968, repr. 1988.

An asterisk (*) indicates works we have seen cited in secondary sources but have not seen in the original.

# Subject Index

**Page numbers in italics indicate pages with the main or most general definition of each term**

# Author Index

# Index of Notation

2761065R00366

Made in the USA
San Bernardino, CA
01 June 2013